Air (dry, at 20° C and 1 atm)
- Density — 1.29 kg/meter^3
- Specific heat at constant pressure — 1.00×10^3 joules/kg C°
  0.240 cal/gm C°
- Ratio of specific heats (γ) — 1.40
- Speed of sound — 331 meter/sec
  1090 ft/sec

Water (20° C and 1 atm)
- Density — 1.00×10^3 kg/meter^3
  1.00 gm/cm^3
- Speed of sound — 1460 meter/sec
  4790 ft/sec
- Index of refraction ($\lambda = 5890A$) — 1.33
- Specific heat at constant pressure — 4180 joules/kg C°
  1.00 cal/gm C°
- Heat of fusion (0° C) — 3.33×10^5 joules/kg
  79.7 cal/gm
- Heat of vaporization (100° C) — 2.26×10^6 joules/kg
  539 cal/gm

The Earth
- Mass — 5.98×10^{24} kg
- Mean radius — 6.37×10^6 meters
  3960 miles
- Mean earth-sun distance — 1.49×10^8 km
  9.29×10^7 miles
- Mean earth-moon distance — 3.80×10^5 km
  2.39×10^5 miles
- Standard gravity — 9.81 meters/sec^2
  32.2 ft/sec^2
- Standard atmosphere — 1.01×10^5 nt/meter^2
  14.7 lb/in.^{2}
  760 mm-Hg

Fundamentals of
PHYSICS

with the assistance of

W. FARRELL EDWARDS

UTAH STATE UNIVERSITY

JOHN MERRILL

BRIGHAM YOUNG UNIVERSITY

Fundamentals of
PHYSICS

Revised Printing

DAVID HALLIDAY

University of Pittsburgh

ROBERT RESNICK

Rensselaer Polytechnic Institute

John Wiley & Sons, Inc.

NEW YORK · LONDON · SYDNEY · TORONTO

Library of Congress Catalogue Card Number: 70   102867

Halliday, David.
  Fundamentals of physics.

  1. Physics. I. Resnick, Robert, 1923–   joint
author. II. Title.
QC21.2.H35   1974   530   74-1023

ISBN   0-471-34431-1

Printed in the United States of America

10 9 8 7 6 5 4

Preface to the Revised Printing

We have prepared this revised printing of *Fundamentals of Physics* in response to instructors who asked for new problems in the book. Because of the success of the original set of 1220 problems, we have retained essentially all of them, but we have added a set of 450 new ones—an increase of about 37 percent. The new problems, distributed evenly over all of the chapters, form a set that has about the same spread of level as the original problems.

We took advantage of this opportunity to improve a significant number of the original problems in ways learned from teaching experience with them; for instance, by simplifying and clarifying the wording, adding hints or explanatory remarks, introducing helpful data and references, and correcting misprints. In this new printing, scores of similar small improvements also were made in the text proper. Throughout the problem sets we have shifted somewhat more toward the metric (SI) system than before.

To assist students and instructors in organizing and evaluating this large number of problems, we have done several things. First, we have grouped problems within each chapter by section number, the first section needed to be covered in order to be able to work out the problem. Then, within each set of section problems, we have arranged the problems in the approximate order of increasing difficulty. Naturally, neither the assignment by section nor by difficulty is absolute, given different ways of solving some problems and different pedagogic values and tastes. A problem labeled **17(4),** for example, means the seventeenth problem in the chapter, requiring completion of the material in the fourth section of the chapter for solution. Finally, we have coded the illustrations to the problems and put the answers to the odd-numbered problems right at the end of the problem rather than in the back of the book.

Although this book is much more than a new corrected printing of the original edition, as we have explained above, it is not a new edition, strictly speaking. We call it a revised printing, and we regard it as a significantly more useful teaching instrument than the earlier printings. Instead of preparing a new edition of this book, we are currently working on a new introductory physics text.

We gratefully acknowledge the assistance of Farrell Edwards and John Merrill in the preparation of this revised printing of *Fundamentals of Physics*.

January, 1974

David Halliday
University of Pittsburgh
Pittsburgh, Pennsylvania, 15260

Robert Resnick
Rensselaer Polytechnic Institute
Troy, New York, 12181

Preface

During the past few years, we have observed gradual but steady changes in the character of the calculus-based introductory physics course. In a large number of schools, less time is available than was previously the case. In others, it appears that not so much attention may be given to a careful and detailed explanation of the subject matter. Also, changes in course requirements at the upper levels of science and engineering curricula have indicated the need for a shorter book than *Physics.**

For these reasons and others, we have been persuaded to prepare an alternate version of the earlier book. The modifications have taken several forms. One has been to remove a large amount of the supplementary material and some of the appendix material that appear in *Physics*. Another has been to condense several chapters and to combine some of them. This necessarily led to careful rewriting and we took advantage of this opportunity to revise some of the material of Part II as well. Other alterations were made in the questions and problems. Many new problems have been added and some, such as those applying to deleted material, have been removed. A number of the new problems fall into what we describe as a "confidence-building" category.

The effect of this surgery has been to decrease appreciably the size of the book and to reduce somewhat its level of sophistication without sacrificing a broad coverage of the fundamentals. *Fundamentals of Physics* emerges then as both a shorter and an easier alternative to *Physics*. This book consequently will be relevant to those courses in which time and prior preparation of the student do not permit the use of the more thorough treatment and somewhat more rigorous pace set by *Physics*.

Authors are perhaps the least qualified persons to engage in an abridgment of their works. We were very fortunate to have had the active assistance of Farrell Edwards and John Merrill, both of Utah State University, in this task. Professors Edwards and Merrill have taught a one-year physics course at their school for some time and are experienced in ways to achieve the objectives we sought. They

---

* David Halliday and Robert Resnick, *Physics*, Wiley, New York, 1966.

discussed with us in detail where material ought to be removed or condensed. They joined us in doing much of the required rewriting and contributed a number of new problems. And, equally important, they did much of the work involved in seeing the book safely through production. To them go our heartfelt thanks. Completing this book in a reasonable length of time would not have been possible without their contributions.

We are grateful to Wiley for its outstanding cooperation and especially to Donald Deneck, Physics Editor, whose management of this task was superb. Richard Martin, of Alfred University, was also helpful in providing additional material for the problem sets of most of the chapters.

We believe that *Fundamentals of Physics* is relevant and appropriate to a new spectrum of students and courses, and hope it will contribute to the improvement of physics education.

January, 1970

David Halliday
University of Pittsburgh
Pittsburgh, Pennsylvania, 15260

Robert Resnick
Rensselaer Polytechnic Institute
Troy, New York, 12181

Contents

1

Measurement 1

1–1 Physical Quantities, Standards, and Units **1**
1–2 Reference Frames **2**
1–3 Standard of Length **3**
1–4 Standard of Time **4**
1–5 Systems of Units **7**

2

Vectors 11

2–1 Vectors and Scalars **11**
2–2 Addition of Vectors, Geometrical Method **12**
2–3 Resolution and Addition of Vectors, Analytic Method **13**
2–4 Multiplication of Vectors **18**

3

Motion in One Dimension 25

3–1 Mechanics **25**
3–2 Particle Kinematics **25**
3–3 Average Velocity **26**
3–4 Instantaneous Velocity **27**
3–5 One-Dimensional Motion—Variable Velocity **28**
3–6 Acceleration **31**
3–7 One-Dimensional Motion—Variable Acceleration **32**
3–8 One-Dimensional Motion—Constant Acceleration **32**
3–9 Consistency of Units and Dimensions **35**
3–10 Freely Falling Bodies **36**

4

Motion in a Plane   43

4-1   Displacement, Velocity, and Acceleration   **43**
4-2   Motion in a Plane with Constant Acceleration   **44**
4-3   Projectile Motion   **45**
4-4   Uniform Circular Motion   **48**
4-5   Relative Velocity and Acceleration   **51**

5

Particle Dynamics   59

5-1   Introduction   **59**
5-2   Classical Mechanics   **59**
5-3   Newton's First Law   **61**
5-4   Force   **62**
5-5   Mass; Newton's Second Law   **63**
5-6   Newton's Third Law   **65**
5-7   Systems of Mechanical Units   **68**
5-8   The Force Laws   **69**
5-9   Weight and Mass   **70**
5-10 A Static Procedure for Measuring Forces   **72**
5-11 Some Applications of Newton's Laws of Motion   **72**
5-12 Frictional Forces   **78**
5-13 The Dynamics of Uniform Circular Motion   **82**

6

Work and Energy   95

6-1   Introduction   **95**
6-2   Work Done by a Constant Force   **96**
6-3   Work Done by a Variable Force—One Dimensional Case   **99**
6-4   Work Done by a Variable Force—Two-Dimensional Case   **101**
6-5   Kinetic Energy and the Work-Energy Theorem   **102**
6-6   Significance of the Work-Energy Theorem   **105**
6-7   Power   **105**

7

The Conservation of Energy   109

7-1   Introduction   **109**
7-2   Conservative Forces   **109**

7–3   Potential Energy   **113**

7–4   One-Dimensional Conservative Systems   **116**

7–5   Total Energy and the Potential Energy Curve   **120**

7–6   Two- and Three-Dimensional Conservative Systems   **121**

7–7   Nonconservative Forces   **123**

7–8   The Conservation of Energy   **125**

7–9   Mass and Energy   **126**

8
Conservation of Linear Momentum   135

8–1   Center of Mass   **135**

8–2   Motion of the Center of Mass   **139**

8–3   Linear Momentum of a Particle   **141**

8–4   Linear Momentum of a System of Particles   **142**

8–5   Conservation of Linear Momentum   **143**

9
Collisions   153

9–1   What is a Collision?   **153**

9–2   Impulse and Momentum   **155**

9–3   Conservation of Momentum during Collisions   **155**

9–4   Collisions in One Dimension   **157**

9–5   Collisions in Two and Three Dimensions   **161**

9–6   Cross Section   **164**

9–7   Reactions and Decay Processes   **165**

10
Rotational Kinematics   173

10–1   Rotational Motion   **173**

10–2   Rotational Kinematics—The Variables   **174**

10–3   Rotation with Constant Angular Acceleration   **176**

10–4   Relation between Linear and Angular Kinematics for a Particle in Circular Motion   **177**

11
Rotational Dynamics and the Conservation of Angular Momentum   183

11–1   Introduction   **183**

11–2   Torque Acting on a Particle   **183**

11–3   Angular Momentum of a Particle   **186**

11–4   Systems of Particles   **189**

11–5   Kinetic Energy of Rotation and Rotational Inertia   **190**

11–6   Rotational Dynamics of a Rigid Body   **193**

11–7   Conservation of Angular Momentum   **199**

11–8   Rotational Dynamics—A Review   **204**

12
Equilibrium of Rigid Bodies   209

12–1   The Equilibrium of a Rigid Body   **209**

12–2   Center of Gravity   **211**

12–3   Examples of Equilibrium   **213**

13
Oscillations   223

13–1   Oscillations   **223**

13–2   The Simple Harmonic Oscillator   **225**

13–3   Simple Harmonic Motion   **228**

13–4   Energy Considerations in Simple Harmonic Motion   **232**

13–5   Applications of Simple Harmonic Motion   **236**

13–6   Relation between Simple Harmonic Motion and Uniform Circular Motion   **238**

13–7   Combinations of Harmonic Motions   **241**

14
Gravitation   247

14–1   The Law of Universal Gravitation   **247**

14–2   The Constant of Universal Gravitation, G **250**

14–3   Inertial and Gravitational Mass and the Principle of Equivalence   **253**

14–4   Gravitational Effect of a Spherical Distribution of Mass   **255**

14–5   Gravitational Acceleration, g **258**

14–6   The Gravitational Field   **261**

14–7   The Motions of Planets and Satellites   **262**

14–8   Gravitational Potential Energy   **265**

14–9   Potential Energy for Many-Particle Systems   **268**

14–10 Energy Considerations in the Motions of Planets and Satellites   **269**

15
Fluid Mechanics   277

15–1   Fluids   **277**

15–2   Pressure and Density   **277**

15–3   The Variation of Pressure in a Fluid at Rest   **278**

15–4   Pascal's Principle and Archimedes' Principle   **281**

15–5   Measurement of Pressure   **283**

15–6   Fluid Dynamics   **284**

15–7   Streamlines and the Equation of Continuity   **286**

15–8   Bernoulli's Equation   **287**

15–9   Applications of Bernoulli's Equation and the Equation of Continuity   **289**

16

Waves in Elastic Media   299

16–1   Mechanical Waves   **299**

16–2   Types of Waves   **300**

16–3   Traveling Waves   **302**

16–4   Wave Speed in a Stretched String   **305**

16–5   Power and Intensity in Wave Motion   **308**

16–6   The Superposition Principle   **309**

16–7   Interference of Waves   **310**

16–8   Standing Waves   **313**

16–9   Resonance   **316**

17

Sound Waves   323

17–1   Audible, Ultrasonic, and Infrasonic Waves   **323**

17–2   Propagation and Speed of Longitudinal Waves   **324**

17–3   Traveling Longitudinal Waves   **327**

17–4   Vibrating Systems and Sources of Sound   **329**

17–5   Beats   **332**

17–6   The Doppler Effect   **334**

18

Temperature   343

18–1   Macroscopic and Microscopic Descriptions   **343**

18–2   Thermal Equilibrium—The Zeroth Law of Thermodynamics   **344**

18–3   Measuring Temperature   **345**

18–4   Ideal Gas Temperature Scale   **347**

18–5   The Celsius and Fahrenheit Scales   **348**

18–6   The International Practical Temperature Scale   **349**

18–7   Temperature Expansion   **350**

19

Heat and the First Law of Thermodynamics   357

19–1   Heat, a Form of Energy   357
19–2   Quantity of Heat and Specific Heat   358
19–3   Heat Conduction   360
19–4   The Mechanical Equivalent of Heat   362
19–5   Heat and Work   363
19–6   The First Law of Thermodynamics   365
19–7   Some Applications of the First Law of Thermodynamics   366

20

Kinetic Theory of Gases   375

20–1   Introduction   375
20–2   Ideal Gas—A Macroscopic Description   376
20–3   Ideal Gas—A Microscopic Description   378
20–4   Kinetic Calculation of the Pressure   379
20–5   Kinetic Interpretation of Temperature   382
20–6   Specific Heats of an Ideal Gas   383
20–7   Equipartition of Energy   386
20–8   Mean Free Path   391
20–9   Distribution of Molecular Speeds   393

21

Entropy and the Second Law of Thermodynamics   401

21–1   Introduction   401
21–2   Reversible and Irreversible Processes   401
21–3   The Carnot Cycle   403
21–4   The Second Law of Thermodynamics   407
21–5   The Efficiency of Engines   409
21–6   Entropy—Reversible Processes   411
21–7   Entropy—Irreversible Processes   413
21–8   Entropy and the Second Law   415

PHY 408

22

Charge and Matter   421

22–1   Electromagnetism   421
22–2   Electric Charge   422
22–3   Conductors and Insulators   423
22–4   Coulomb's Law   423

22–5   Charge is Quantized   **427**

22–6   Charge and Matter   **427**

22–7   Charge is Conserved   **429**

23
The Electric Field   433

23–1   The Electric Field   **433**

23–2   The Electric Field **E**   **434**

23–3   Lines of Force   **435**

23–4   Calculation of **E**   **437**

23–5   A Point Charge in an Electric Field   **440**

23–6   A Dipole in an Electric Field   **442**

24
Gauss's Law   449

24–1   Flux of the Electric Field   **449**

24–2   Gauss's Law   **452**

24–3   Gauss's Law and Coulomb's Law   **452**

24–4   An Insulated Conductor   **453**

24–5   Experimental Proof of Gauss's and Coulomb's Laws   **454**

24–6   Gauss's Law—Some Applications   **455**

25
Electric Potential   465

25–1   Electric Potential   **465**

25–2   Potential and the Electric Field   **468**

25–3   Potential Due to a Point Charge   **470**

25–4   A Group of Point Charges   **472**

25–5   Potential Due to a Dipole   **474**

25–6   Electric Potential Energy   **475**

25–7   Calculation of **E** From V **478**

25–8   An Insulated Conductor   **480**

25–9   The Electrostatic Generator   **481**

26
Capacitors and Dielectrics   489

26–1   Capacitance   **489**

26–2   Calculating Capacitance   **492**

26–3 Parallel-Plate Capacitor with Dielectric **494**

26–4 Dielectrics—An Atomic View **496**

26–5 Dielectrics and Gauss's Law **498**

26–6 Energy Storage in an Electric Field **499**

27

Current and Resistance 507

27–1 Current and Current Density **507**

27–2 Resistance, Resistivity, and Conductivity **510**

27–3 Ohm's Law **512**

27–4 Resistivity—An Atomic View **514**

27–5 Energy Transfers in an Electric Circuit **516**

28

Electromotive Force and Circuits 521

28–1 Electromotive Force **521**

28–2 Calculating the Current **523**

28–3 Other Single-Loop Circuits **524**

28–4 Potential Differences **525**

28–5 Multiloop Circuits **528**

28–6 *RC* Circuits **530**

29

The Magnetic Field 537

29–1 The Magnetic Field **537**

29–2 The Definition of **B** **538**

29–3 Magnetic Force on a Current **541**

29–4 Torque on a Current Loop **542**

29–5 The Hall Effect **545**

29–6 Circulating Charges **546**

29–7 The Cyclotron **548**

29–8 Thomson's Experiment **550**

30

Ampère's Law 557

30–1 Ampère's Law **557**

30–2 **B** Near a Long Wire **561**

30–3 Lines of **B** **562**

30–4   Two Parallel Conductors   **563**
30–5   **B** for a Solenoid   **565**
30–6   The Biot-Savart Law   **568**

31
Faraday's Law   577

31–1   Faraday's Experiments   **577**
31–2   Faraday's Law of Induction   **578**
31–3   Lenz's Law   **579**
31–4   Induction—A Quantitative Study   **581**
31–5   Time-Varying Magnetic Fields   **584**
31–6   The Betatron   **587**

32
Inductance   597

32–1   Inductance   **597**
32–2   Calculation of Inductance   **598**
32–3   An *LR* Circuit   **600**
32–4   Energy and the Magnetic Field   **603**
32–5   Energy Density and the Magnetic Field   **605**

33
Magnetic Properties of Matter   611

33–1   Poles and Dipoles   **611**
33–2   Gauss's Law for Magnetism   **614**
33–3   Paramagnetism   **615**
33–4   Diamagnetism   **617**
33–5   Ferromagnetism   **619**

34
Electromagnetic Oscillations   625

34–1   *LC* Oscillations   **625**
34–2   Analogy to Simple Harmonic Motion   **628**
34–3   Electromagnetic Oscillations—Quantitative   **629**
34–4   Induced Magnetic Fields   **632**
34–5   Displacement Current   **634**
34–6   Maxwell's Equations   **635**

35

Electromagnetic Waves   639

35-1   Introduction   639
35-2   Radiation Sources   640
35-3   Traveling Waves and Maxwell's Equations   641
35-4   Energy and the Poynting Vector   646
35-5   Momentum   648
35-6   Polarization   649
35-7   The Electromagnetic Spectrum   653
35-8   The Speed of Light   654
35-9   Moving Sources and Observers   657
35-10 Doppler Effect   660

36

Geometrical Optics   669

36-1   Geometrical Optics   669
36-2   Reflection and Refraction—Plane Waves and Plane Surfaces   669
36-3   Huygens' Principle   672
36-4   The Law of Refraction   673
36-5   Total Internal Reflection   675
36-6   Brewster's Law   676
36-7   Spherical Waves—Plane Mirror   678
36-8   Spherical Waves—Spherical Mirror   681
36-9   Spherical Waves—Spherical Refracting Surface   686
36-10 Thin Lenses   689

37

Interference   703

37-1   Wave Optics   703
37-2   Young's Experiment   705
37-3   Coherence   708
37-4   Intensity of Interfering Waves   710
37-5   Interference from Thin Films   714
37-6   Michelson's Interferometer   718

38

Diffraction, Gratings, and Spectra   725

38-1   Diffraction   725
38-2   Single Slit   728

38–3   Diffraction from a Single Slit—Qualitative   **730**

38–4   Diffraction from a Single Slit—Quantitative   **732**

38–5   Diffraction from a Circular Aperture   **735**

38–6   Diffraction from a Double Slit   **738**

38–7   Multiple Slits   **741**

38–8   Diffraction Gratings   **744**

38–9   Resolving Power of a Grating   **746**

38–10 X-ray Diffraction   **748**

39
Light and Quantum Physics   757

39–1   Sources of Light   **757**

39–2   Cavity Radiators   **758**

39–3   Planck's Radiation Formula   **760**

39–4   Photoelectric Effect   **763**

39–5   Einstein's Photon Theory   **765**

39–6   The Compton Effect   **766**

39–7   Line Spectra   **770**

39–8   Atomic Models—The Bohr Hydrogen Atom   **771**

39–9   The Correspondence Principle   **776**

40
Waves and Particles   781

40–1   Matter Waves   **781**

40–2   Atomic Structure and Standing Waves   **784**

40–3   Wave Mechanics   **784**

40–4   Meaning of Ψ **787**

40–5   The Uncertainty Principle   **789**

Appendices

A   Physical Standards and Constants   **795**

B   Some Terrestrial Data   **797**

C   The Solar System   **798**

D   Periodic Table of the Elements   **799**

E   Conversion Factors   **800**

F   Mathematical Symbols and the Greek Alphabet   **807**

G   Mathematical Formulas   **808**

H   Values of Trigonometric Functions   **811**

I   Nobel Prize Winners in Physics   **813**

Index   817

Fundamentals of
PHYSICS

CHAPTER 1

Measurement

1-1 Physical Quantities, Standards, and Units

The building blocks of physics are the physical quantities in terms of which the laws of physics are expressed. Among these are force, time, velocity, density, temperature, charge, magnetic susceptibility, and numerous others. Many of these terms, such as force and temperature, are part of our everyday vocabulary. When these terms are so used, their meanings may be vague or may differ from their scientific meanings.

For the purposes of physics the basic quantities must be defined clearly and precisely. One view is that the definition of a physical quantity has been given when the procedures for measuring that quantity are specified. This is called the *operational* point of view because the definition is, at root, a set of laboratory operations leading to a number with a unit. The operations may include mathematical calculations.

Physical quantities are often divided into *fundamental quantities* and *derived quantities*. Such a division is arbitrary in that a given quantity can be regarded as fundamental in one set of operations and as derived in another. Derived quantities are those whose defining operations are based on measurements of other physical quantities. Examples of quantities usually viewed as derived are velocity, acceleration, and volume. Fundamental quantities are not defined in terms of other physical quantities. The number of quantities regarded as fundamental is the minimum number needed to give a consistent and unambiguous description of all the quantities of physics. Examples of quantities usually viewed as fundamental are length and time. Their operational definitions involve two steps: first, the choice of a *standard*, and second, the establishment of procedures for comparing the standard in such a way that a number and a unit are determined as the measure of that quantity.

An ideal standard has two principal characteristics: it is accessible and it is invariable. These two requirements are often incompatible and a compromise has to be made between them. At first greater emphasis was placed on accessibility, but the growing requirements of science and technology introduced the need for greater invariability. The familiar yard, foot, and inch, for example, come from the practice of using the human arm, foot, and upper thumb directly as standards. Today, of course, such rough measures of length are rarely satisfactory; much less variable standards must be used even at the expense of accessibility.

Suppose that we have chosen our standard of length to be a bar whose length we define as one meter. If by direct comparison of this bar with a second bar we conclude that the second bar is three times as long as the standard, we say that the second bar has a length of three meters. Such direct comparisons with a primary standard can seldom be made, however. Indirect approaches, using more involved procedures, are usually necessary. Certain assumptions are then necessary to relate the results of such an indirect measurement to the direct operation. For example, astronomical distances, such as the distances of stars from the earth, cannot be measured in a direct way. Similarly, very small distances, such as those within atoms and molecules, must also be measured by indirect methods.

1–2   Reference Frames

The same physical quantity may have different values if it is measured by observers who are moving with respect to each other. The velocity of a train has one value if measured by an observer on the ground, a different value if measured from a speeding car, and the value zero if measured by an observer sitting in the train itself. None of these values has any fundamental advantage over any other; each is equally "correct" from the point of view of the observer making the measurement. This should become clear to you as you proceed through the text.

In general, the measured value of a physical quantity depends on the reference frame of the observer who is making the measurement. This is easy enough to see if the physical quantity is a velocity, as above. It is also true, however, if the physical quantity is, say, a displacement of a particle, a time interval between two events, an electric field, or a magnetic field; a full appreciation of these four special examples must await the study of the theory of relativity.

In earlier days physicists believed that one particular reference frame, a so-called absolute frame, existed and had a fundamental advantage over all other frames. For an observer at rest in such a frame physical quantities would have their "true" or "absolute" values. This viewpoint has now been abandoned because, over many decades, experimental efforts to find this absolute reference frame have failed completely.

Consider reference frames moving with uniform velocity with respect to each other and with respect to the fixed stars. Such (unaccelerated, nonrotating) reference frames are called *inertial reference frames.* Experiment shows that all inertial reference frames are equivalent for the explanation of physical phenomena. Observers in different frames may obtain different numerical values for measured physical quantities, but the relationships between the measured quantities, that is, the laws of physics, will be the same for all observers.

For example, suppose that observers in different inertial frames measure the momenta of the particles involved in an atomic collision. They will obtain different numerical values both for the momenta of the individual particles and for the total momentum of the system of particles. Each observer, however, will note that the total momentum of the system of particles, whatever value he measured it to be, is the same after the collision as before. In other words each observer will note that the collision obeys the law of conservation of momentum; we shall discuss this law in detail in Chapter 8.

So we see that physical laws are the same in all reference frames, and that the measured values of the physical quantities may or may not be. In working problems you should always understand clearly what reference frames you are using.

1–3 Standard of Length*

The first truly international standard of length was a bar of platinum-iridium alloy called the *standard meter*, kept at the International Bureau of Weights and Measures near Paris, France. The distance between two fine lines engraved on gold plugs near the ends of the bar (when the bar was at 0.00° C and supported mechanically in a prescribed way) was defined to be *one meter*. Historically, the meter was intended to be a convenient fraction (one ten-millionth) of a distance from pole to equator along the meridian line through Paris. However, accurate measurements taken after the standard meter bar was constructed show that it differs slightly (about 0.023%) from its intended value.

Because the standard meter was not very accessible, accurate master copies of it were made and sent to standardizing laboratories throughout the civilized world. These secondary standards were used to calibrate other, still more accessible, measuring rods. Thus until recently every ruler, micrometer, or vernier caliper derived its legal authority from the standard meter through a complicated chain of comparisons using microscopes and dividing engines. This statement was also true for the *yard* used in English-speaking countries. Since 1959, however, the definition of the yard, by international agreement, has been

$$1 \text{ yard} = 0.9144 \text{ meter, exactly,}$$

* See "The Metre" by H. Barrell, in *Contemporary Physics*, Vol. 3, p. 415, 1962, for an excellent discussion of the standard of length.

Table 1–1 SOME MEASURED LENGTHS

	Meters
Distance to the most-distant quasar yet detected* (1964)	6×10^{25}
Distance to the nearest nebula (Great Nebula in Andromeda)	2×10^{22}
Radius of our galaxy	6×10^{19}
Distance to the nearest star (Alpha Centauri)	4.3×10^{16}
Mean orbit radius for the most distant planet (Pluto)	5.9×10^{12}
Radius of the sun	6.9×10^{8}
Radius of the earth	6.4×10^{6}
Highest free balloon ascension (1959)	4.6×10^{4}
Height of a man	1.8×10^{0}
Thickness of a page in this book	1×10^{-4}
Size of a poliomyelitis virus	1.2×10^{-8}
Radius of a hydrogen atom	5.0×10^{-11}
Effective radius of a proton	1.2×10^{-15}

* *quasar* = quasi-stellar radio source.

which is equivalent to

$$1 \text{ in.} = 2.54 \text{ cm, exactly.}$$

There are several objections to the meter bar as the primary standard of length: It is potentially destructible, by fire or war for example; it is not accurately reproducible; it is not very accessible. Most important, the accuracy with which the necessary intercomparisons of length can be made by the technique of comparing fine scratches, using a microscope, is no longer great enough to meet modern requirements of science and technology. The maximum accuracy obtainable with the standard meter as a reference is about 1 part in 10^7; an error of this amount in the borehole of a guidance gyroscope could cause a space shot aimed at the moon to miss by a thousand miles.

The suggestion that the length of a light wave be used as a length standard was first made in 1864 by Hippolyte Louis Fizeau (1819–1896). The later development of the *interferometer* (see Chapter 37) provided scientists with a precision optical device in which light waves can be used as a length comparison probe. Light waves are about 5×10^{-5} cm long and length measurements of bars even many centimeters long can be made to a very small fraction of a wavelength. An accuracy of 1 part in 10^9 in the intercomparison of lengths using light waves is inherently possible. As the need for this increased accuracy in length comparisons arose, efforts were made to determine the best light source.

In 1961 an atomic standard of length was adopted by international agreement. The wavelength in vacuum of a particular orange radiation (identified by the spectroscopic notation $2p_{10} - 5d_5$) emitted by atoms of a particular isotope of krypton (Kr^{86}) in an electrical discharge was chosen. Specifically, one meter is now defined to be 1,650,763.73 wavelengths of this light. This number of wavelengths was arrived at by carefully measuring the length of the standard meter bar in terms of these light waves. This comparison was done so that the new standard, based on the wave length of light, would be as consistent as possible with the old standard, based on the meter bar. Figure 1–1 shows a krypton-86 light source, used as the basis of the length standard.

The choice of an atomic standard offers advantages other than increased precision in length measurements. The atoms that generate light are available everywhere and all atoms of a given species are identical and emit light of the same wavelength. Hence such an atomic standard is both accessible and invariable. The particular wavelength chosen is uniquely characteristic of krypton-86 and is very sharply defined. The isotope can be obtained with great purity relatively easily and cheaply.

1–4 Standard of Time

The measurement of time has two different aspects. For civil and for some scientific purposes, we want to know the time of day so that we can order events in sequence. In most scientific work, we want to know how long an event lasts. Thus any time standard must be able to answer both the question "What time is it?" and the question "How long does it last?"* Table 1–2 shows the wide range of time intervals that can be measured.

Any phenomenon that repeats itself can be used as a measure of time; the measurement consists of counting the repetitions. An oscillating pendulum, coiled spring, or

* See "Accurate Measurement of Time" by Louis Essen, in *Physics Today*, July 1960, for an excellent discussion of the standard of time.

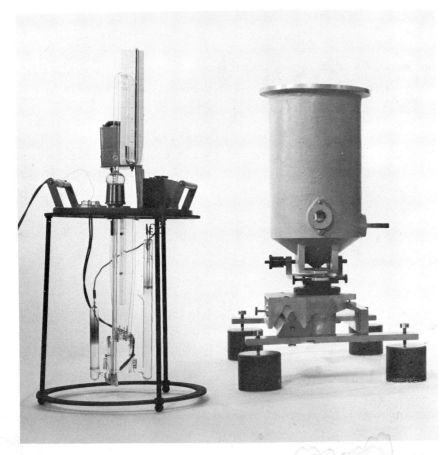

FIGURE 1–1 A Kr⁸⁶ light source shown removed from the container in which it is housed. In operation the lamp is cooled with liquid nitrogen. (Courtesy the National Physical Laboratories, Teddington, England. Crown copyright reserved.)

Table 1–2 SOME MEASURED TIME INTERVALS

	Seconds
Age of the earth	1.3×10^{17}
Age of the pyramid of Cheops	1.5×10^{11}
Human life expectancy (USA)	2×10^{9}
Time of earth's orbit around the sun (1 year)	3.1×10^{7}
Time of earth's rotation about its axis (1 day)	8.6×10^{4}
Period of the Echo II satellite	5.1×10^{3}
Half-life of the free neutron	7.0×10^{2}
Time between normal heartbeats	8.0×10^{-1}
Period of concert-A tuning fork	2.3×10^{-3}
Half-life of the muon	2.2×10^{-6}
Period of oscillation of 3-cm microwaves	1.0×10^{-10}
Typical period of rotation of a molecule	1×10^{-12}
Half-life of the neutral pion	2.2×10^{-16}
Period of oscillation of a 1-MeV gamma ray (calculated)	4×10^{-21}
Time for a fast elementary particle to pass through a medium-sized nucleus (calculated)	2×10^{-23}

quartz crystal can be used, for example. Of the many repetitive phenomena occurring in nature, the rotation of the earth on its axis, which determines the length of the day, has been used as a time standard from earliest times. It is still the basis of our civil and legal time standard, one (mean solar) second being defined to be 1/86,400 of a (mean solar) day. Time defined in terms of the rotation of the earth is called *universal time (UT)*.

Universal time must be determined by astronomical observations. Since these observations must be extended over several weeks, a good secondary terrestrial clock, calibrated by the astronomical observations, is needed. Quartz crystal clocks, based on the electrically sustained natural periodic vibrations of a quartz wafer, serve well as secondary time standards. The best of these have kept time for a year with a maximum error of 0.02 sec.

One of the most common uses of a time standard is the determination of frequencies. In the radio range, frequency comparisions to a quartz clock can be made electronically to a precision of at least 1 part in 10^{10} and, indeed, many situations require such precision. However, this precision is about one hundred times greater than that with which a quartz clock itself can be calibrated by astronomical observations. To meet the need for a better time standard, atomic clocks have been developed in several countries, using periodic atomic vibrations as a standard.

A particular type of atomic clock, based on a characteristic frequency associated with the cesium atom, has been in continuous operation at the National Physical Laboratory in England since 1955. Figure 1–2 shows a similar clock at the U. S. Bureau of Standards.

In 1967, the second based on the cesium clock was adopted as an international

FIGURE 1–2 This cesium atomic clock at the Boulder Laboratories of the National Bureau of Standards measures frequency and time intervals to an accuracy equivalent to the loss of less than 1 sec in 3000 years.

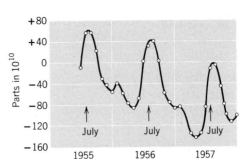

FIGURE 1–3 Variation in the rate of rotation of the earth as revealed by comparison with a cesium clock. (Adapted from L. Essen, *Physics Today*, July 1960.)

standard by the Thirteenth General Conference of Weights and Measures meeting in Paris. The action increases the accuracy of time measurements to 1 part in 10^{11}, an improvement over the accuracy associated with astronomical methods of about 200. If two cesium clocks are operated at this precision, and if there are no other sources of error, the clocks will differ by no more than one second after running 3000 years.

Figure 1–3 shows, by comparison with the cesium clock, variations in the rate of rotation of the earth over nearly a three-year period. Note that the earth's rotation rate is high in summer and low in winter (northern hemisphere) and exhibits a steady decrease from year to year. In this connection it is legitimate to ask how we can be sure that the rotating earth and not the cesium clock is "at fault." There are two answers. (1) The relative simplicity of the atom compared to the earth leads us to ascribe any differences between these two timekeepers to physical phenoma on the earth. Tidal friction between the water and the land, for example, causes a slowing down of the earth's rotation. Also the seasonal motion of the winds introduces a regular seasonal variation in the rotation. Other variations may be associated with the melting of polar icecaps and shifts of other earth masses. (2) The solar system contains other timekeepers, such as the orbiting planets and the orbiting moons of the planets; the rotation of the earth shows variations with respect to these, too, which are similar to, but less accurately observable than, the variation exhibited in Fig. 1–3.

The time standard can be made available at remote locations by radio transmission. Station WWV located in Fort Collins, Colorado and station WWVH in Hawaii, both operated by the National Bureau of Standards broadcast on frequencies of 2.5, 5, 10, 15, 20 and 25 $\times$ 10^6 Hz, stabilized to 1 part in 10^{11} by comparison to a cesium clock. One Hertz (abbr. Hz) is one cycle/sec. At 5-min intervals, both stations alternately broadcast an accurate 440 Hz tone (concert A) and a 600 Hz tone. Ten times per hour it broadcasts time signals, using a binary digit coding system; the signals are based on the earth's rotation, that is, they refer to universal time. Corrections are made for the wandering of the earth's axis and the annual variation in the earth's rotational speed.

1–5 Systems of Units

As already pointed out, there is a certain amount of arbitrariness in the choice of the fundamental quantities.* For example, length, time, and mass can be chosen as fundamental quantities; all other mechanical quantities, such as force, torque, density, etc., can be expressed in terms of these fundamental quantities. However,

---

* See "Dimensions, Units, and Standards" by A. G. McNish, in *Physics Today*, April 1957.

Table 1–3 PREFIXES USED FOR MULTIPLES AND
SUBMULTIPLES OF METRIC QUANTITIES

10^{-1}	deci-	deca-	10^1
10^{-2}	centi-	hecto-	10^2
10^{-3}	milli-	kilo-	10^3
10^{-6}	micro-	mega-	10^6
10^{-9}	nano-	giga-	10^9
10^{-12}	pico-	tera-	10^{12}

we might equally well choose force instead of mass as a fundamental quantity. However, having picked the fundamental quantities and determined units for them, we thereby automatically determine the units of the derived quantities.

Three different systems of units are most commonly used in science and engineering. They are the *meter-kilogram-second* or mks system, the Gaussian system, in which the fundamental mechanical units are the *c*entimeter, the *g*ram, and the *s*econd (a cgs system), and the British engineering system (a *foot-pound-second* or fps system). The gram and kilogram are mass units, and the pound is a force unit; these will be defined and discussed in Chapter 5.

We shall use the mks system principally throughout the text, except in mechanics where the fps system will also be used. The metric system is used universally in scientific work and provides the common units of commerce in most countries of the world.

Some prefixes used to identify multiples and submultiples of metric quantities are shown in Table 1–3. Thus 1 millimeter $= 10^{-3}$ meter, 1 nanosecond $= 10^{-9}$ sec, 1 megavolt $= 10^6$ volt, etc.

QUESTIONS

1. Do you think that a definition of a physical quantity for which no method of measurement is known or given has meaning?

2. According to operational philosophy, if we cannot prescribe a feasible operation for determining a physical quantity, the quantity is undetectable by physical means and should be given up as having no physical reality. Not all scientists accept this view. What are the merits and drawbacks of this point of view in your opinion?

3. What characteristics, other than accessibility and invariability, would you consider desirable for a physical standard?

4. If someone told you that every dimension of every object had shrunk to half its former value overnight, how could you refute his statement?

5. How would you criticize the following statement: "Once you have picked a physical standard, by the very meaning of standard it is invariable?"

6. What does an observer on the earth mean by "up" and "down"? Do all such observers use the same reference frame? How could one make the meaning clearly understood to *any* observer?

7. Why was it necessary to specify the temperature at which comparisons with the standard meter bar were to be made? Can length be called a fundamental quantity if another physical quantity, such as temperature, must be specified in choosing a standard?

8. Can length be measured along a curved line? If so, how?

9. Can you suggest a way to measure (*a*) the radius of the earth; (*b*) the distance between the sun and the earth; (*c*) the radius of the sun?

10. Can you suggest a way to measure (*a*) the thickness of a sheet of paper; (*b*) the thickness of a soap bubble film; (*c*) the diameter of an atom?

11. What criteria should a good clock satisfy?

12. Name several repetitive phenomena occurring in nature which could serve as reasonable time standards.

13. The time it takes the moon to return to a given position as seen against the background of the fixed stars is called a siderial month. The time interval between identical phases of the moon is called a lunar month. The lunar month is longer than a siderial month. Why?

14. When man colonizes other planets, what drawbacks would our older non-atomic standards of length and time have? What drawbacks would atomic standards have?

15. Can you think of a way to define a length standard in terms of a time standard or vice versa? (Think about a pendulum clock.) If so, can length and time both be considered as fundamental quantities?

PROBLEMS *

1(3). What is your height in meters?

Answer: 6.00 ft = 1.83 meters, etc.

2(3). Calculate the number of kilometers in 20 miles using only the following conversion factors: 1 mile = 5280 ft, 1 ft = 12 in., 1 in. = 2.54 cm, 1 meter = 100 cm, and 1 km = 1000 meters.

3(3). A rocket attained a height of 300 km. What is this distance in miles?

Answer: 186 miles.

4(3). In track meets both 100 yards and 100 meters are used as distances for dashes. Which is longer? By how many meters is it longer? By how many feet?

5(3). Astronomical distances are so large compared to terrestrial ones that much larger units of length are used for easy comprehension of the relative distances of astronomical objects. An *astronomical unit* (AU) is equal to the average distance from the earth to the sun, about 92.9×10^6 miles. A *parsec* is the distance at which one astronomical unit would subtend an angle of 1 sec of arc. A *light-year* is the distance that light, traveling through a vacuum with a speed of 186,000 miles/sec, would cover in one year. (*a*) Express the distance from earth to sun in parsecs and in light years. (*b*) Express a light year and a parsec in miles.

Answer: (*a*) 4.85×10^{-6} parsec; 1.58×10^{-5} light-years. (*b*) 5.87×10^{12} miles; 1.91×10^{13} miles.

6(3). Master machinists would like to have master gauges (1 in. long, for example) good to 0.0000001 in. Show that the platinum-iridium meter is not measurable to this accuracy but that the krypton-86 meter is. Use data given in this chapter.

7(3). Give the relation between (*a*) a square inch and a square centimeter; (*b*) a square mile and a square kilometer; (*c*) a cubic meter and a cubic centimeter; (*d*) a square foot and a square yard.

Answer: (*a*) 1 in.² = 6.45 cm². (*b*) 1 mi² = 2.60 km². (*c*) 1 meter³ = 10^6 cm³. (*d*) 1 ft² = 0.111 yd².

8(3). Assume that the average distance of the sun from the earth is 400 times the average distance of the moon from the earth. Now consider a total eclipse of the sun and state conclusions that can be drawn about (*a*) the relation between the sun's diameter and the moon's diameter; (*b*) the relative volumes of the sun and the moon. (*c*) Find the angle intercepted at the eye by a dime that just eclipses the full moon and from this experimental result and the given distance between the moon and the earth (= 3.80×10^5 km) estimate the diameter of the moon.

9(4). (*a*) A unit of time sometimes used in microscopic physics is the *shake*. One shake equals 10^{-8} sec. Are there more shakes in a second than there are seconds in a year? (*b*) Mankind has ex-

---

* The number in parentheses identifies the section in this chapter to which the problem applies. For example, Problem 2(3) refers to Section 1–3.

isted for about 10^6 years, whereas the universe is about 10^{10} years old. If the age of the universe is taken to be one day, for how many seconds has mankind existed?
Answer: (*a*) Yes. (*b*) 8.6 sec.

10(4). From Fig. 1–3, calculate by what length of time the earth's rotation period in midsummer differs from that in the following spring.

11(4). Five clocks are being tested in a laboratory. Exactly at noon, as determined by the WWV time signal, on the successive days of a week the clocks read as follows:

Clock	Sun.	Mon.	Tues.	Wed.	Thurs.	Fri.	Sat.
A	12:36:40	12:36:56	12:37:12	12:37:27	12:37:44	12:37:59	12:38:14
B	11:59:59	12:00:02	11:59:57	12:00:07	12:00:02	11:59:56	12:00:03
C	15:50:45	15:51:43	15:52:41	15:53:39	15:54:37	15:55:35	15:56:33
D	12:03:59	12:02:52	12:01:45	12:00:38	11:59:31	11:58:24	11:57:17
E	12:03:59	12:02:49	12:01:54	12:01:52	12:01:32	12:01:22	12:01:12

How would you arrange these five clocks in the order of their relative value as good timekeepers? Justify your choice.
Answer: C, D, A, B, E (best to worst). The important criterion is the constancy of the daily variation, not its magnitude.

12(4). Assuming that the length of the day uniformly increases by 0.001 sec in a century, calculate the cumulative effect on the measure of time over twenty centuries. Such a slowing down of the earth's rotation is indicated by observations of the occurrences of solar eclipses during this period.

13(4). Express the speed of light, 3×10^8 meters/sec, in (*a*) feet/nanosecond and (*b*) in milli-meters/picosecond.
Answer: (*a*) 0.98 ft/nanosecond. (*b*) 0.33 millimeters/picosecond.

14(4). An astronomical unit is the average distance of the earth from the sun, approximately 149,000,000 km. The speed of light is about 3.0×10^8 meters/sec. Express the speed of light in terms of astronomical units per minute.

15(4). A certain spaceship has a speed of 18,600 miles/hr. What is its speed in light-years per century? A light-year is the distance light travels in one year with a speed of 186,000 miles/sec.
Answer: 2.8×10^{-3} light-years/century.

16(4). (*a*) The radius of the proton is about 10^{-15} meter; the radius of the observable universe is about 10^{28} cm. Identify a physically meaningful distance which is approximately halfway between these two extremes on a logarithmic scale. (*b*) The mean life of a neutral pion (an elementary particle) is about 2×10^{-16} sec. The age of the universe is about 4×10^9 years. Identify a physically meaningful time interval that is approximately halfway between these two extremes on a logarithmic scale.

CHAPTER 2

Vectors

THOMAS p. 387 - 401

2-1 Vectors and Scalars

A change of position of a particle is called a *displacement*. If a particle moves from position A to position B (Fig. 2–1a), we can represent its displacement by drawing a line from A to B; the direction of displacement can be shown by putting an arrowhead at B indicating that the displacement was from A to B. The path of the particle need not necessarily be a straight line from A to B; the arrow represents only the net effect of the motion, not the actual motion.

In Fig. 2–1b, for example, we plot a path followed by a particle from A to B. The path is not the same as the displacement AB. If we were to take snapshots of the particle when it was at A and, later, when it was at some intermediate position P, we could obtain the displacement vector AP, representing the net effect of the motion during this interval, even though we would not know the actual path taken between

FIGURE 2–1 Displacement vectors. (*a*) Vectors AB and $A'B'$ are identical since they have the same length and point in the same direction. (*b*) The actual *path* of the particle in moving from A to B may be the curve shown; the *displacement* remains the vector AB. At some intermediate point P the displacement from A is the vector AP. (*c*) After displacement AB the particle undergoes another displacement BC. The net effect of the two displacements is represented by the vector AC.

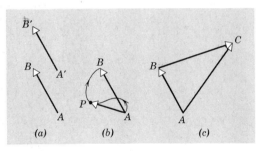

(a) (b) (c)

these points. Furthermore, a displacement such as $A'B'$ (Fig. 2–1a), which is parallel to AB, similarly directed, and equal in length to AB, represents the same change in position as AB. We make no distinction between these two displacements. A displacement is therefore characterized by a length and a direction.

In a similar way, we can represent a subsequent displacement from B to C (Fig. 2–1c). The net effect of the two displacements will be the same as a displacement from A to C. We speak then of AC as the *sum* or *resultant* of the displacements AB and BC. Notice that this sum is not an algebraic sum and that a number alone cannot uniquely specify it.

Quantities that behave like displacements are called *vectors*. Vectors, then, are quantities that have both magnitude and direction and combine according to certain rules of addition. These rules are stated below. The displacement vector can be considered as the prototype. Some other physical quantities which are vectors are force, velocity, acceleration, momentum, electric field, and magnetic field. Many of the laws of physics can be expressed in compact form using vectors; derivations involving these laws are often greatly simplified if this is done.

Quantities that can be completely specified by a number and unit and that therefore have magnitude only are called *scalars*. Some physical quantities which are scalars are mass, length, time, density, energy, and temperature. Scalars can be manipulated by the rules of ordinary algebra.

2–2 Addition of Vectors, Geometrical Method

To represent a vector on a diagram we draw an arrow. We choose the length of the arrow proportional to the magnitude of the vector (that is, we choose a scale), and we choose the direction of the arrow to be the direction of the vector, with the arrowhead giving the sense of the direction. A vector such as this is represented conveniently in printing by a boldface symbol such as **d**. In handwriting it is convenient to put an arrow above the symbol to denote a vector quantity, such as $\vec{d}$.

Often we shall be interested only in the magnitude of the vector and not in its direction. The magnitude of **d** may be written as $|\mathbf{d}|$, called the absolute value of **d**; more frequently we represent the magnitude alone by the italic letter symbol, such as d. The boldface symbol is meant to signify both properties of the vector, magnitude and direction.

Consider now Fig. 2–2 in which we have redrawn and relabeled the vectors of Fig. 2–1c. The relation among these displacements (vectors) can be written as

$$\mathbf{a} + \mathbf{b} = \mathbf{r}. \tag{2–1}$$

The rules to be followed in performing this (vector) addition geometrically are these: On a diagram drawn to scale lay out the displacement vector **a;** then draw **b** with its tail at the head of **a**, and draw a line from the tail of **a** to the head of **b** to construct the vector sum **r**. This is a displacement equivalent in length and direction to the successive displacements **a** and **b**. This procedure can be generalized to obtain the sum of any number of successive displacements.

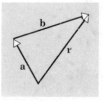

FIGURE 2–2 . The vector sum **a** + **b** = **r**. Compare with Fig. 2–1c.

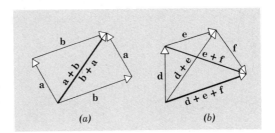

FIGURE 2–3 (a) The commutative law for vector sums, which states that $\mathbf{a} + \mathbf{b} = \mathbf{b} + \mathbf{a}$. (b) The associative law, which states that $\mathbf{d} + (\mathbf{e} + \mathbf{f}) = (\mathbf{d} + \mathbf{e}) + \mathbf{f}$.

Since vectors are new quantities we must expect new rules for their manipulation. The symbol "+" in Eq. 2–1 simply has a different meaning than it has in arithmetic or ordinary algebra. It tells us to carry out a different set of operations.

Using Fig. 2-3 we can prove two important properties of vector addition:

$$\mathbf{a} + \mathbf{b} = \mathbf{b} + \mathbf{a}, \qquad \text{(commutative law)} \qquad (2\text{--}2)$$

and

$$\mathbf{d} + (\mathbf{e} + \mathbf{f}) = (\mathbf{d} + \mathbf{e}) + \mathbf{f}. \qquad \text{(associative law)} \qquad (2\text{--}3)$$

These laws assert that it makes no difference in what order or in what grouping we add vectors; the sum is the same. In this respect, vector addition and scalar addition follow the same rules.

The operation of subtraction can be included in our vector algebra by defining the negative of a vector to be another vector of equal magnitude but opposite direction. Then

$$\mathbf{a} - \mathbf{b} = \mathbf{a} + (-\mathbf{b}) \qquad (2\text{--}4)$$

as shown in Fig. 2-4.

Remember that, although we have used displacements to illustrate these operations, the rules apply to all vector quantities.

2–3 Resolution and Addition of Vectors, Analytic Method

The geometrical method of adding vectors is not very useful for vectors in three dimensions; often it is even inconvenient for the two-dimensional case. Another way of adding vectors is the analytic method, involving the resolution of a vector into components with respect to a particular coordinate system.

Figure 2–5a shows a vector $\mathbf{a}$ whose tail has been placed at the origin of a rectangular coordinate system. If we drop perpendicular lines from the head of $\mathbf{a}$ to the axes the quantities a_x and a_y so formed are called the *components* of the vector $\mathbf{a}$. The process

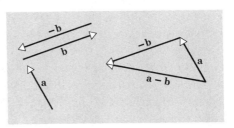

FIGURE 2–4 The vector difference $\mathbf{a} - \mathbf{b} = \mathbf{a} + (-\mathbf{b})$.

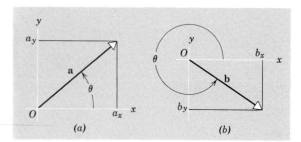

FIGURE 2–5 Two examples of the resolution of a vector into its scalar components in a particular coordinate system.

is called *resolving a vector into its components*. Figure 2–5 shows a two-dimensional case for convenience; the extension of our conclusions to three dimensions will be clear.

A vector may have many sets of components. For example, if we rotate the x-axis and y-axis in Fig. 2–5a by 10° counterclockwise, the components of **a** would be different. Furthermore, we may use a nonrectangular coordinate system, that is, the angle between the two axes need not be 90°. Thus the components of a vector are only uniquely specified if we specify the particular coordinate system being used. The vector need not be drawn with its tail at the origin of the coordinate system to find its components—although we have done so for convenience; the vector may be moved anywhere in the coordinate space and, as long as its angles with the coordinate directions are maintained, its components will be unchanged.

The components a_x and a_y in Fig. 2–5a are found readily from

$$a_x = a \cos \theta \qquad \text{and} \qquad a_y = a \sin \theta, \qquad (2\text{–}5)$$

where θ is the angle that the vector **a** makes with the positive x-axis, measured counterclockwise from this axis. Note that, depending on the angle θ, a_x and a_y can be positive or negative. For example, in Fig. 2–5b, b_y is negative and b_x is positive. The components of a vector behave like scalar quantities because, in any particular coordinate system of a given reference frame, only a number, with an algebraic sign, is needed to specify them.

Once a vector is resolved into its components, the components themselves can be used to specify the vector. Instead of the two numbers a (magnitude of the vector) and θ (direction of the vector relative to the x-axis), we now have the two numbers a_x and a_y. We can pass back and forth between the description of a vector in terms of its components a_x, a_y and the equivalent description in terms of magnitude and direction a and θ. To obtain a and θ from a_x and a_y, we note from Fig. 2–5a that

$$a = \sqrt{a_x^2 + a_y^2} \qquad (2\text{–}6a)$$

and

$$\tan \theta = a_y/a_x. \qquad (2\text{–}6b)$$

The quadrant in which the vector lies is determined from the signs of a_x and a_y.

When resolving a vector into components it is sometimes useful to introduce a vector of unit length in a given direction. Thus vector **a** in Fig. 2–6a may be written, for example, as

$$\mathbf{a} = \mathbf{u}_a a, \qquad (2\text{–}7)$$

where $\mathbf{u}_a$ is a *unit vector* in the direction of **a**. Often it is convenient to draw unit vectors along the particular coordinate axes chosen. In the rectangular coordinate system the special symbols **i**, **j**, and **k** are usually used for unit vectors in the positive x-, y-,

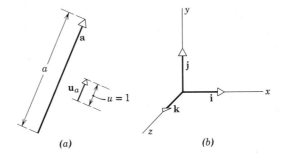

FIGURE 2–6   (a) The vector **a** may be written as $\mathbf{u}_a a$ in which $\mathbf{u}_a$ is a unit vector in the direction of **a**. (b) The unit vectors **i, j,** and **k,** used to specify the positive x-, y-, and z-directions respectively.

and z-directions, respectively; see Fig. 2–6b. Note that **i, j,** and **k** need not be located at the origin. Like all vectors, they can be translated anywhere in the coordinate space as long as their directions with respect to the coordinate axes are not changed.

The vectors **a** and **b** of Fig. 2–5 may be written in terms of their components and the unit vectors as

$$\mathbf{a} = \mathbf{i}a_x + \mathbf{j}a_y \tag{2–8a}$$

and

$$\mathbf{b} = \mathbf{i}b_x + \mathbf{j}b_y; \tag{2–8b}$$

see Fig. 2–7. The vector relation Eq. 2–8a is equivalent to the scalar relations Eqs. 2–6; each relates the vector (**a,** or a and θ) to its components (a_x and a_y). Sometimes we will call quantities such as $\mathbf{i}a_x$ and $\mathbf{j}a_y$ in Eq. 2–8a the *vector components* of **a;** they are drawn as vectors in Fig. 2–7a. The word *component* alone will continue to refer to the scalar quantities a_x and a_y.

We now consider the addition of vectors by the analytic method. Let **r** be the sum of the two vectors **a** and **b** lying in the x-y plane, so that

$$\mathbf{r} = \mathbf{a} + \mathbf{b}. \tag{2–9}$$

In a given coordinate system, two vectors such as **r** and **a** + **b** can only be equal if their corresponding components are equal, or

$$r_x = a_x + b_x \tag{2–10a}$$

and

$$r_y = a_y + b_y. \tag{2–10b}$$

These two algebraic equations, taken together, are equivalent to the single vector relation Eq. 2–9. From Eqs. 2–6 we may find r and the angle θ that **r** makes with the x-axis; that is,

$$r = \sqrt{r_x{}^2 + r_y{}^2}$$

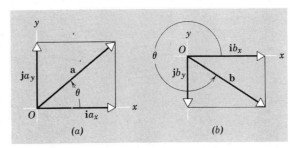

FIGURE 2–7   Two examples of the resolution of a vector into its vector components in a particular coordinate system; compare with Fig. 2–5.

and

$$\tan \theta = r_y/r_x.$$

Thus we have the following analytic rule for adding vectors: Resolve each vector into its components in a given coordinate system; the algebraic sum of the individual components along a particular axis is the component of the sum vector along that same axis; the sum vector can be reconstructed once its components are known. This method for adding vectors may be generalized to many vectors and to three dimensions (see Problems 10 and 17).

The advantage of the method of breaking up vectors into components, rather than adding directly with the use of suitable trigonometric relations, is that we always deal with right triangles and thus simplify the calculations.

When you are adding vectors by the analytic method, your choice of coordinate axes may simplify or complicate the process. Sometimes the components of the vectors with respect to a particular set of axes are known to begin with, so that the choice of axes is obvious. Other times a judicious choice of axes can greatly simplify the job of resolution of the vectors into components. For example, the axes can be oriented so that at least one of the vectors lies parallel to an axis.

Example 1. An airplane traveled 130 miles on a straight course making an angle of 22.5° east of due north. How far north and how far east did the plane travel from its starting point?

We choose the positive x-direction to be east and the positive y-direction to be north. Next (Fig. 2–8) we draw a displacement vector from the origin (starting point), making an angle of 22.5° with the y-axis (north) inclined along the positive x-direction (east). The length of the vector is chosen to represent a magnitude of 130 miles. If we call this vector **d,** then d_x gives the distance traveled east of the starting point and d_y gives the distance traveled north of the starting point. We have

$$\theta = 90.0° - 22.5° = 67.5°,$$

so that (see Eqs. 2–5)

$$d_x = d \cos \theta = (130 \text{ miles}) \cos 67.5° = 50 \text{ miles},$$

and

$$d_y = d \sin \theta = (130 \text{ miles}) \sin 67.5° = 120 \text{ miles}.$$

Example 2. An automobile travels due east on a level road for 30 miles. It then turns due north at an intersection and travels 40 miles before stopping. Find the resultant displacement of the car.

We choose a reference fixed with respect to the earth, with the positive x-direction of our coordinate system pointing east and the positive y-direction pointing north. The two successive displacements, **a** and **b,** are then drawn as shown in Fig. 2–9. The resultant displacement **r** is obtained from **r = a + b.** Since **b** has no x-component and **a** has no y-component, we obtain (see Eqs. 2–10)

$$r_x = a_x + b_x = 30 \text{ miles} + 0 = 30 \text{ miles},$$

$$r_y = a_y + b_y = 0 + 40 \text{ miles} = 40 \text{ miles}.$$

The magnitude and direction of **r** are then (see Eqs. 2–6)

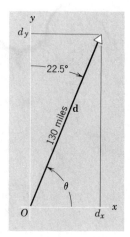

FIGURE 2–8 Example 1.

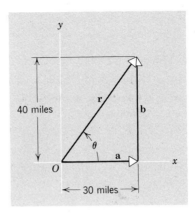

FIGURE 2–9   Example 2.

$$r = \sqrt{r_x{}^2 + r_y{}^2} = \sqrt{(30 \text{ miles})^2 + (40 \text{ miles})^2} = 50 \text{ miles},$$

$$\tan \theta = r_y/r_x = \frac{40 \text{ miles}}{30 \text{ miles}} = 1.33 \qquad \theta = \tan^{-1}(1.33) = 53°.$$

The resultant vector displacement **r** has a magnitude of 50 miles and makes an angle of 53° north of east.

Example 3. Three coplanar vectors are expressed, with respect to a certain rectangular coordinate system, as

$$\mathbf{a} = 4\mathbf{i} - \mathbf{j},$$
$$\mathbf{b} = -3\mathbf{i} + 2\mathbf{j},$$

and

$$\mathbf{c} = -3\mathbf{j},$$

in which the components are given in arbitrary units. Find the vector **r** which is the sum of these vectors.

From Eqs. 2–10 we have

$$r_x = a_x + b_x + c_x = 4 - 3 + 0 = 1,$$

and

$$r_y = a_y + b_y + c_y = -1 + 2 - 3 = -2.$$

Thus

$$\mathbf{r} = \mathbf{i}r_x + \mathbf{j}r_y$$
$$= \mathbf{i} - 2\mathbf{j}.$$

Figure 2–10 shows the four vectors. From Eqs. 2–6 we can calculate that the magnitude of **r** is $\sqrt{5}$ and that the angle that **r** makes with the positive *x*-axis, measured counterclockwise from

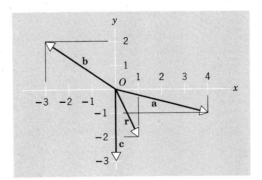

FIGURE 2–10   Three vectors **a**, **b**, and **c**, and their vector sum **r**.

that axis, is

$$\tan^{-1}(-2/1) = 297°.$$

2–4 Multiplication of Vectors*

We have assumed in the previous discussion that the vectors being added together are of like kind; that is, displacement vectors are added to displacement vectors, or velocity vectors are added to velocity vectors. Just as it would be meaningless to add together scalar quantities of different kinds, such as mass and temperature, so it would be meaningless to add together vector quantities of different kinds.

However, like scalars, vectors of different kinds can be multiplied by one another to generate quantities of new physical dimensions. Because vectors have direction as well as magnitude, vector multiplication cannot follow exactly the same rules as the algebraic rules of scalar multiplication. We must establish new rules of multiplication for vectors.

We find it useful to define three kinds of multiplication operations for vectors: (1) multiplication of a vector by a scalar, (2) multiplication of two vectors in such a way as to yield a scalar, and (3) multiplication of two vectors in such a way as to yield another vector. There are still other possibilities, but we shall not consider them here.

The multiplication of a vector by a scalar has a simple meaning: The product of a scalar k and a vector **a**, written k**a**, is defined to be a new vector whose magnitude is k times the magnitude of **a**. The new vector has the same direction as **a** if k is positive and the opposite direction if k is negative. To divide a vector by a scalar we simply multiply the vector by the reciprocal of the scalar.

When we multiply a vector quantity by another vector quantity, we must distinguish between the *scalar* (or *dot*) *product* and the *vector* (or *cross*) *product*. The *scalar product* of two vectors **a** and **b**, written as **a · b**, is defined to be

$$\mathbf{a \cdot b} = ab \cos \phi, \tag{2–11}$$

where a is the magnitude of vector **a**, b is the magnitude of vector **b**, and $\cos \phi$ is the cosine of the (smallest) angle ϕ between the two vectors (see Fig. 2–11).

Since a and b are scalars and $\cos \phi$ is a pure number, the scalar product of two

---

* The material of this section will be used later in the text. The scalar product is used first in Chapter 6 and the vector product in Chapter 11. The instructor can postpone this section accordingly if he wishes. Its presentation here gives a unified treatment of vector algebra and serves as a convenient reference for later work.

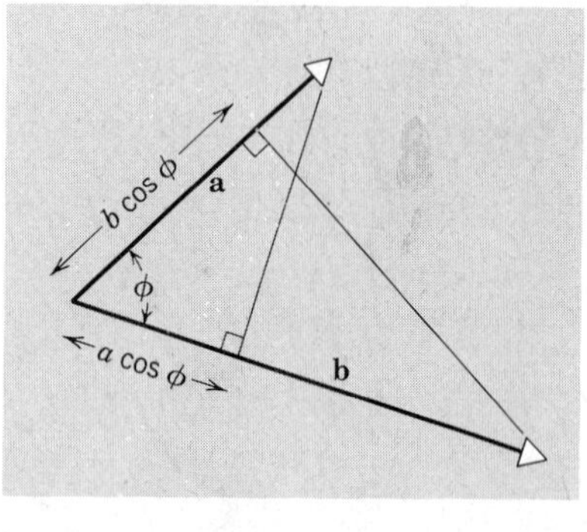

FIGURE 2–11 The scalar product **a · b** ($= ab \cos \phi$) is the product of the magnitude of either vector (a, say) by the component of the other vector in the direction of the first vector ($b \cos \phi$, say).

vectors is a scalar. The scalar product of two vectors can be regarded as the product of the magnitude of one vector and the component of the other vector in the direction of the first. Because of the notation, $\mathbf{a} \cdot \mathbf{b}$ is also called the dot product of $\mathbf{a}$ and $\mathbf{b}$ and is spoken as "$\mathbf{a}$ dot $\mathbf{b}$."

We could have defined $\mathbf{a} \cdot \mathbf{b}$ to be any operation we want, for example, to be $a^{1/3}b^{1/3}$ tan $(\phi/2)$, but this would turn out to be of no use to us in physics. With our definition of the scalar product, a number of important physical quantities can be described as the scalar product of two vectors. Some of them are mechanical work, gravitational potential energy, electrical potential, electric power, and electromagnetic energy density.

The *vector product* of two vectors $\mathbf{a}$ and $\mathbf{b}$ is written as $\mathbf{a} \times \mathbf{b}$ and is another vector $\mathbf{c}$, where $\mathbf{c} = \mathbf{a} \times \mathbf{b}$. The *magnitude* of $\mathbf{c}$ is defined by

$$c = ab \sin \phi, \tag{2–12}$$

where ϕ is the angle between $\mathbf{a}$ and $\mathbf{b}$. $\vec{c} = ab|\sin \phi| \vec{e_c}$

The *direction* of $\mathbf{c}$, the vector product of $\mathbf{a}$ and $\mathbf{b}$, is defined to be perpendicular to the plane formed by $\mathbf{a}$ and $\mathbf{b}$. To specify the sense of the vector $\mathbf{c}$ refer to Fig. 2–12. Imagine rotating a right-handed screw whose axis is perpendicular to the plane formed by $\mathbf{a}$ and $\mathbf{b}$ so as to turn it from $\mathbf{a}$ to $\mathbf{b}$ through the angle ϕ between them. Then the direction of advance of the screw gives the direction of the vector product $\mathbf{a} \times \mathbf{b}$ (Fig. 2–12a). Another convenient way to obtain the direction is to imagine an axis perpendicular to the plane of $\mathbf{a}$ and $\mathbf{b}$ through their origin. Now wrap the fingers of the right hand around this axis and push the vector $\mathbf{a}$ into the vector $\mathbf{b}$ through the

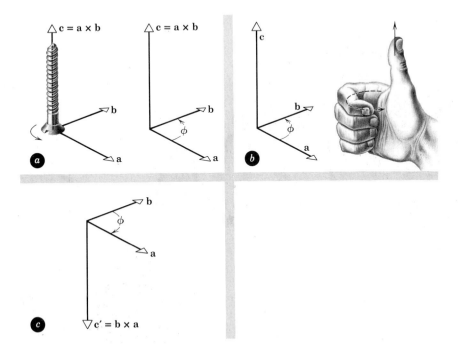

FIGURE 2–12 The vector product. (*a*) In $\mathbf{c} = \mathbf{a} \times \mathbf{b}$, the direction of $\mathbf{c}$ is that in which a right-handed screw advances when turned from $\mathbf{a}$ to $\mathbf{b}$ through the smaller angle. (*b*) The direction of $\mathbf{c}$ can also be obtained from the "right-hand rule": If the right hand is held so that the curled fingers follow the rotation of $\mathbf{a}$ into $\mathbf{b}$, the extended right thumb will point in the direction of $\mathbf{c}$. (*c*) The vector product changes sign when the order of the factors is reversed: $\mathbf{a} \times \mathbf{b} = -(\mathbf{b} \times \mathbf{a})$.

smaller angle between them with the fingertips, keeping the thumb erect; the direction of the erect thumb then gives the directioɴ of the vector product **a** ✕ **b** (Fig. 2–12*b*). Because of the notation, **a** ✕ **b** is also called the cross product of **a** and **b** and is spoken as "**a** cross **b**."

Notice that **b** ✕ **a** is not the same vector as **a** ✕ **b,** so that the order of factors in a vector product is important. This is not true for scalars because the order of factors in algebra or arithmetic does not affect the resulting product. Actually, **a** ✕ **b** = −(**b** ✕ **a**) (Fig. 2–12*c*). This can be deduced from the fact that the magnitude $ab \sin \phi$ equals the magnitude $ba \sin \phi$, but the direction of **a** ✕ **b** is opposite to that of **b** ✕ **a;** this is so because the right-handed screw advances in one direction when rotated from **a** to **b** through ϕ but advances in the opposite direction when rotated from **b** to **a,** through ϕ. The student can obtain the same result by applying the right-hand rule.

If ϕ is 90°, **a, b,** and **c** (= **a** ✕ **b**) are all at right angles to one another and give the directions of a three-dimensional right-handed coordinate system.

The reason for defining the vector product in this way is that it proves to be useful in physics. We often encounter physical quantities that are vectors whose product, defined as above, is a vector quantity having important physical meaning. Some examples of physical quantities that are vector products are torque, angular momentum, the force on a moving charge in a magnetic field, and the flow of electromagnetic energy. When such quantities are discussed later, your attention will be drawn to their connection with the vector product.

The scalar product is the simplest product of two vectors. The order of multiplication does not affect the product. The vector product is the next simplest case. Here the order of multiplication does affect the product, but only by a factor of minus one, which implies a direction reversal. Other products of vectors are useful but more involved. For example, a tensor can be generated by multiplying each of the three components of one vector by the three components of another vector. Hence a tensor (of the second rank) has nine numbers associated with it, a vector three, and a scalar only one. Some physical quantities that can be represented by tensors are mechanical and electrical stress, moments and products of inertia, and strain. Still more complex physical quantities are possible. In this book, however, we are concerned only with scalars and vectors.

QUESTIONS

1. Can two vectors of different magnitude be combined to give a zero resultant? Can three vectors?

2. Can a vector be zero if one of its components is not zero?

3. Does it make any sense to call a quantity a vector when its magnitude is zero?

4. Name several scalar quantities. Is the value of a scalar quantity dependent on the references frame chosen?

5. We can order events in time. For example, event *b* may precede event *c* but follow event *a*, giving us a time order of events *a*, *b*, *c*. Hence there is a sense of time, distinguishing past, present, and future. Is time a vector therefore? If not, why not?

6. Do the commutative and associative laws apply to vector subtraction?

7. Can a scalar product be a negative quantity?

8. On the basis of your present understanding, which of the following are vectors, which are scalars, and which are neither: position, velocity, mass, speed, light, temperature, time, force, energy, and area? Later chapters will help to clarify the situation.

PROBLEMS

1(2). A person walks in the following pattern: 3.1 miles north, then 2.4 miles west, and finally 5.2 miles south. (*a*) Construct the vector diagram that represents his motion. (*b*) How far and in what direction would a bird fly in a straight line to arrive at the same final point?
Answer: (*b*) 3.2 miles, at an angle of 41° south of west.

2(2). A golfer takes three strokes to get his ball into the hole once he is on the green. The first stroke displaces the ball 12 ft north, the second stroke 6.0 ft southeast, and the third stroke 3.0 ft southwest. What displacement was needed to get the ball into the hole on the first stroke?

3(2). A car is driven east for a distance of 50 miles, then north for 30 miles, and then in a direction 30° east of north for 25 miles. Draw the vector diagram and determine the displacement of the car from its starting point.
Answer: 81 miles, 40° north of east.

4(2). Consider two displacements, one of magnitude 3 meters and another of magnitude 4 meters. Show how the displacement vectors may be combined to get a resultant displacement of magnitude (*a*) 7 meters, (*b*) 1 meter, and (*c*) 5 meters.

5(2). Vector **A** has a magnitude of 5.0 units and is directed east. Vector **B** is directed 45° west of north and has a magnitude of 4.0 units. Construct vector diagrams for calculating (**A** + **B**) and (**B** − **A**). Estimate the magnitudes and directions of (**A** + **B**) and (**B** − **A**) from your diagrams.
Answer: (**A** + **B**) points 37° east of north and has a magnitude of 3.6 units. (**B** − **A**) points 70° west of north and has a magnitude of 8.3 units.

6(2). Two vectors **a** and **b** are added. Show that the magnitude of the resultant cannot be greater than $a + b$ or smaller than $|a - b|$, where the vertical bars signify absolute value.

7(2). What are the properties of the two vectors **a** and **b** such that

(*a*) $\qquad\qquad\qquad$ **a** + **b** = **c** $\quad$ and $\quad$ $a + b = c$,

(*b*) $\qquad\qquad\qquad$ **a** + **b** = **a** − **b**,

(*c*) $\qquad\qquad\qquad$ **a** + **b** = **c** $\quad$ and $\quad$ $a^2 + b^2 = c^2$.

Answer: (*a*) **a** ∥ **b**. (*b*) **b** = 0. (*c*) **a** ⊥ **b**.

8(2). Three vectors **A**, **B**, and **C** each having a magnitude of 50 units lie in the *x-y* plane and make angles of 30°, 195°, and 315° with the positive *x*-axis respectively. Find graphically the magnitudes and directions of the vectors (*a*) **A** + **B** + **C**, (*b*) **A** − **B** + **C**, (*c*) a vector **D** such that (**A** + **B**) − (**C** + **D**) = 0.

9(3). The *x*-component of a certain vector is −25 units and the *y*-component is +40 units. (*a*) What is the magnitude of the vector? (*b*) What is the angle between the direction of the vector and the positive *x*-axis?
Answer: (*a*) 47 units. (*b*) 122°.

10(3). What are the magnitude and direction of a vector if its *x*- and *y*-components are −45 units and −30 units respectively?

11(3). A person desires to reach a point that is 3.4 miles from his present location and in a direction that is 35° north of east. However, he must travel along streets that go either north-south or east-west. What is the minimum distance he could travel to reach his destination?
Answer: 4.7 miles.

12(3). What are the components of a vector **A** in the *x-y* plane if its direction is 250° counterclockwise from the positive *x*-axis and its magnitude is 7.3 units?

13(3). A particle undergoes three successive displacements in a plane, as follows: 4.0 meters southwest, 5.0 meters east, 6.0 meters in a direction 60° north of east. Choose the *y*-axis pointing north and the *x*-axis pointing east and find (*a*) the components of each displacement, (*b*) the components of the resultant displacement, (*c*) the magnitude and direction of the resultant displacement, and (*d*) the displacement that would be required to bring the particle back to the starting point. *Answer:* (*a*) $a_x = -2.8$ meters, $a_y = -2.8$ meters;

$\qquad b_x = +5.0$ meters, $b_y = 0$;

$\qquad c_x = +3.0$ meters, $c_y = +5.2$ meters.

(*b*) $d_x = +5.2$ meters, $d_y = +2.4$ meters.

(*c*) 5.7 meters, 25° north of east.

(*d*) 5.7 meters, 25° south of west.

14(3). Calculate the components of a vector **A** in the *x*-*y* plane whose magnitude is 6.5 units and whose direction is 125° counterclockwise from the positive *x*-axis.

15(3). Find the vector components of the sum **r** of the vector displacements **c** and **d** whose components in miles along three perpendicular directions are

$$c_x = 5.0, \ c_y = 0, \ c_z = -2.0; \ d_x = -3.0, \ d_y = 4.0, \ d_z = 6.0.$$

Answer: $r_x = +2.0$ miles; $r_y = r_z = +4.0$ miles.

16(3). Two vectors **a** and **b** have equal magnitudes, say 10 units. They are oriented as shown in Fig. 2–13 and their vector sum is **r**. Find (*a*) the *x*- and *y*-components of **r**, (*b*) the magnitude of **r**, and (*c*) the angle **r** makes with the *x*-axis.

17(3). Two vectors **A** and **B** have components, in arbitrary units $A_x = 3.2$, $A_y = 1.6$; $B_x = 0.50$, $B_y = 4.5$. (*a*) Find the angle between **A** and **B**. (*b*) Find the components of a vector **C** which is perpendicular to **A**, is in the *x*-*y* plane, and has a magnitude of 5.0 units. *Answer:* (*a*) 57°. (*b*) $C_x = \pm 2.2$ units; $C_y = \mp 4.5$ units

18(3). A room has the dimensions 10 ft × 12 ft × 14 ft. A fly starting at one corner ends up at a diametrically opposite corner. (*a*) What is the magnitude of its displacement? (*b*) Could the length of its path be less than this distance? Greater than this distance? Equal to this distance? (*c*) Choose a suitable coordinate system and find the components of the displacement vector in this frame.

19(3). In Problem 18(3), if the fly does not fly but crawls, what is the length of the shortest path it can take? *Answer:* 26 ft.

20(3). Two vectors of lengths *a* and *b* make an angle *θ* with each other when placed tail to tail. Prove, by taking components along two perpendicular axes, that the length of the resultant vector is

$$r = \sqrt{a^2 + b^2 + 2ab \cos \theta}.$$

21(3). Determine the components of each of the vectors of Problem 8(2) and analytically calculate the same quantities.

22(3). (*a*) A man leaves his front door, walks 1000 ft east, 2000 ft north, and then takes a penny from his pocket and drops it from a cliff 500 ft high. Set up a coordinate system and write down an expression, using unit vectors, for the displacement of the penny. (*b*) The man then returns to his

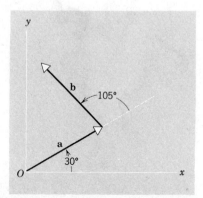

FIGURE 2-13 Problem 16(3).

front door, following a different path on the return trip. What is his resultant displacement for the round trip?

23(3). (a) What is the sum in unit vector notation of the two vectors $a = 4i + 3j$ and $b = -3i + 7j$? (b) What are the magnitude and direction of $a + b$?
Answer: (a) $i + 10j$. (b) ~10 units, at ~84° with the x-axis.

24(3). Calculate the components, magnitudes and directions of (a) $A + B$, and (b) $B - A$, if $A = 3i + 4j$ and $B = 5i - 2j$.

25(3). Two vectors are given by $A = 4i - 3j + k$ and $B = -i + j + 4k$. Find (a) $A + B$, (b) $A - B$, and (c) a vector C such that $A - B + C = 0$.
Answer: (a) $3i - 2j + 5k$. (b) $5i - 4j - 3k$. (c) Negative of (b).

26(3). For the vectors A and B of the previous problem, find (a) $3A - 4B$, and (b) $A/4 + 4B$.

27(3). Given two vectors $a = 4i - 3j$ and $b = 6i + 8j$, find the magnitude and direction of a, of b, of $a + b$, of $b - a$, and of $a - b$.
Answer: The magnitudes are 5, 10, 11, 11, and 11. The angles with the positive x-axis are 320°, 53°, 27°, 80°, and 260°.

28(3). A particle moves in the x-y plane as shown in the table:

t (sec)	0.0	0.1	0.2	0.3	0.4
x (meter)	+1.0	+0.707	0	−0.707	−1.0
y (meter)	0.0	+0.707	+1.0	+0.707	0.0

What is the vector displacement from $t = 0$ to $t = 0.4$ sec? Express in unit vector notation.

29(3). Write in unit vector notation the vector in the x-y plane whose magnitude is 9.1 units and whose direction is 49° counterclockwise from the x-axis.
Answer: $6.0i + 6.9j$.

30(3). (a) Determine the components and magnitude of $R = A - B + C$ if $A = 5i + 4j - 6k$, $B = -2i + 2j + 3k$, and $C = 4i + 3j + 2k$. (b) Calculate the angle between R and the positive z-axis.

31(4). A vector d has a magnitude 2.5 meters and points north. What are the magnitudes and directions of the vectors (a) $-d$, (b) $d/2.0$, (c) $-2.5d$, and (d) $4.0d$?
Answer: (a) 2.5 meters, south. (b) 1.3 meters, north. (c) 6.3 meters, south. (d) 10 meters, north.

32(4). In the coordinate system of Fig. 2–6b show that

$$i \cdot i = j \cdot j = k \cdot k = 1$$

and

$$i \cdot j = j \cdot k = k \cdot i = 0.$$

33(4). *Scalar product in unit vector notation.* Let two vectors be represented in terms of their coordinates as

$$a = ia_x + ja_y + ka_z$$

and

$$b = ib_x + jb_y + kb_z$$

Show analytically that

$$a \cdot b = a_x b_x + a_y b_y + a_z b_z.$$

[Hint: See Problem 32(4).]

34(4). Use the definition of scalar product $a \cdot b = ab \cos \phi$ and the fact that $a \cdot b = a_x b_x + a_y b_y + a_z b_z$ [see Problem 33(4)] to obtain the angle between the two vectors given by $a = 3i + 3j - 3k$ and $b = 2i + j + 3k$.

35(4). A certain vector A in the x-y plane is 250° counterclockwise from the positive x-axis and has magnitude 7.4 units. Vector B has magnitude 5.0 units and is directed parallel to the z-axis. Calculate (a) $A \cdot B$ and (b) $A \times B$.
Answer: (a) $A \cdot B = 0$. (b) $|A \times B| = 37$; its angle with the positive x-axis is 160°; $A \times B$ lies in the x-y plane.

36(4). Two vectors, R and S, lie in the x-y plane. Their magnitudes are 4.5 and 7.3 units, respectively while their directions are 320° and 85° measured counterclockwise from the positive x-axis. What are the values of (a) $\mathbf{R} \cdot \mathbf{S}$ and (b) $\mathbf{R} \times \mathbf{S}$?

37(4). A vector **a** of magnitude ten units and another vector **b** of magnitude six units point in directions differing by 60°. Find (a) scalar product of the two vectors and (b) the vector product of the two vectors.

Answer: (a) 30. (b) A vector of magnitude 52, pointing as given by the right-hand rule.

38(4). Show for any vector **a** that $\mathbf{a} \cdot \mathbf{a} = a^2$ and that $\mathbf{a} \times \mathbf{a} = 0$.

39(4). Given vector **a** in the +x-direction, vector **b** in the +y-direction, and the scalar quantity d: (a) What is the direction of $\mathbf{a} \times \mathbf{b}$? (b) What is the direction of $\mathbf{b} \times \mathbf{a}$? (c) What is the direction of $\mathbf{b}/d$? (d) What is the magnitude of $\mathbf{a} \cdot \mathbf{b}$?

Answer: (a) The positive z-direction. (b) The negative z-direction. (c) The positive y-direction for $d > 0$ and the negative y-direction for $d < 0$. (d) Zero.

40(4). Two vectors are given by $\mathbf{A} = 3.0\mathbf{i} + 5.0\mathbf{j}$ and $\mathbf{B} = -2.0\mathbf{i} + 4.0\mathbf{j}$, find (a) $\mathbf{A} \times \mathbf{B}$, (b) $\mathbf{A} \cdot \mathbf{B}$, and (c) $(\mathbf{A} + \mathbf{B}) \cdot \mathbf{B}$.

41(4). In the right-handed coordinate system of Fig. 2–6b show that

$$\mathbf{i} \times \mathbf{i} = \mathbf{j} \times \mathbf{j} = \mathbf{k} \times \mathbf{k} = 0$$

$$\mathbf{i} \times \mathbf{j} = \mathbf{k}; \quad \mathbf{k} \times \mathbf{i} = \mathbf{j}; \quad \mathbf{j} \times \mathbf{k} = \mathbf{i}.$$

42(4). Find (a) "north" cross "west," (b) "down" dot "south," (c) "east" cross "up," (d) "west" dot "west," and (e) "south" cross "south." Let each vector have unit magnitude.

43(4). *Vector product in unit vector notation.* Show analytically that $\mathbf{a} \times \mathbf{b} = \mathbf{i}(a_y b_z - a_z b_y) + \mathbf{j}(a_z b_x - a_x b_z) + \mathbf{k}(a_x b_y - a_y b_z)$. [Hint: See Problem 41(4).]

44(4). (a) We have seen that the commutative law does *not* apply to vector products, that is, $\mathbf{a} \times \mathbf{b}$ does not equal $\mathbf{b} \times \mathbf{a}$. Show that the commutative law *does* apply to scalar products, that is, $\mathbf{a} \cdot \mathbf{b} = \mathbf{b} \cdot \mathbf{a}$. (b) Show that the distributive law applies to both scalar products and vector products, that is, show that

$$\mathbf{a} \cdot (\mathbf{b} + \mathbf{c}) = \mathbf{a} \cdot \mathbf{b} + \mathbf{a} \cdot \mathbf{c} \text{ and that } \mathbf{a} \times (\mathbf{b} + \mathbf{c}) = \mathbf{a} \times \mathbf{b} + \mathbf{a} \times \mathbf{c}.$$

(c) Does the associative law apply to vector products, i.e., does $\mathbf{a} \times (\mathbf{b} \times \mathbf{c})$ equal $(\mathbf{a} \times \mathbf{b}) \times \mathbf{c}$? (d) Does it make any sense to talk about an associative law for scalar products?

45(4). Three vectors are given by $\mathbf{A} = 3\mathbf{i} + 3\mathbf{j} - 2\mathbf{k}$, $\mathbf{B} = -\mathbf{i} - 4\mathbf{j} + 2\mathbf{k}$ and $\mathbf{C} = 2\mathbf{i} + 2\mathbf{j} + \mathbf{k}$, find (a) $\mathbf{A} \cdot (\mathbf{B} \times \mathbf{C})$, (b) $\mathbf{A} \cdot (\mathbf{B} + \mathbf{C})$ and (c) $\mathbf{A} \times (\mathbf{B} + \mathbf{C})$.

Answer: (a) −21. (b) −9. (c) $5\mathbf{i} - 11\mathbf{j} - 9\mathbf{k}$.

46(4). (a) Show that $\mathbf{A} \cdot (\mathbf{B} \times \mathbf{A})$ is zero for all vectors **A** and **B**. (b) What is the value of $\mathbf{A} \times (\mathbf{B} \times \mathbf{A})$ if there is an angle φ between the directions of **A** and **B**?

47(4). For the two vectors in Problem 16(3), find (a) $\mathbf{a} \cdot \mathbf{b}$, and (b) $\mathbf{a} \times \mathbf{b}$.

Answer: (a) 26. (b) $97\mathbf{k}$.

CHAPTER 3

Motion in One Dimension

3-1 Mechanics

Mechanics, the oldest of the physical sciences, is the study of the motion of objects. The calculation of the path of an artillery shell or of a space probe sent from Earth to Mars are among its problems. So too is the analysis of tracks formed in bubble chambers, representing the decay and interactions of elementary particles.

When we describe motion we are dealing with that part of mechanics called *kinematics*. When we relate motion to the forces associated with it and to the properties of the moving objects, we are dealing with *dynamics*. In this chapter we shall define some kinematical quantities and study them in detail for the special case of motion in one dimension. In Chapter 4 we discuss some cases of two- and three-dimensional motion. Chapter 5 deals with the more general case of dynamics.

3-2 Particle Kinematics

The actual motion of most real objects may be quite complicated. In general, an object will rotate or vibrate internally while it is moving as a whole in some trajectory. Sometimes these internal motions may be neglected when it is wished to calculate only the gross motion. Usually, in such cases the dimensions of the body are much smaller than the dimensions of the trajectory so that the body may be considered to be a mathematical point. The body is then referred to as a *particle*. For example, the earth may be considered as a particle if we wish to calculate with reasonable accuracy only the motion of the earth about the sun. It is certainly not a particle if we wish to consider the ocean tides, atmospheric disturbances, or earthquakes. Similarly, a gas molecule may be considered to be a particle if we are interested in the relationships between the pressure, density, and temperature of the gas but may not

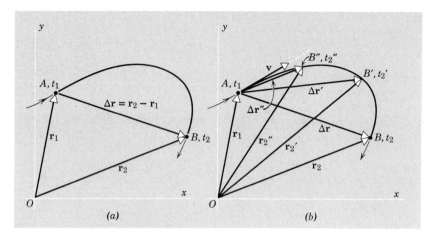

FIGURE 3–1   (*a*) A particle moves from *A* to *B* in time Δt (= $t_2 - t_1$) undergoing a displacement $\Delta \mathbf{r}$ (= $\mathbf{r}_2 - \mathbf{r}_1$). *The average velocity* $\bar{\mathbf{v}}$ between *A* and *B* is in the direction of $\Delta \mathbf{r}$. (*b*) As *B* moves closer to *A* the average velocity approaches the *instantaneous velocity* $\mathbf{v}$ at *A*; $\mathbf{v}$ is tangent to the path at *A*.

be considered a particle when considering phenomena for which the internal motions of the molecule are important.

Even if a body is too large to be considered a particle for a particular problem, it can always be thought of as made up of a large number of particles, and the results of particle motion may be useful in analyzing the problem. As a simplification, therefore, we confine our present treatment to the motion of a particle.

The motion of particles, for which there are no rotations or vibrations, is referred to as *translational* motion.

3–3   Average Velocity

The velocity of a particle is the rate at which its position changes with time. The position of a particle in a particular reference frame is given by a *position vector* drawn from the origin of that frame to the particle. At time t_1, let a particle be at point *A* in Fig. 3–1*a*, its position in the *x-y* plane being described by position vector $\mathbf{r}_1$. For simplicity we treat motion in two dimensions only; the extension to three dimensions will not be difficult.

At a later time t_2 let the particle be at point *B*, described by position vector $\mathbf{r}_2$. The *displacement vector* describing the *change* in position of the particle as it moves from *A* to *B* is $\Delta \mathbf{r}$ (= $\mathbf{r}_2 - \mathbf{r}_1$) and the elasped time for the motion between these points is Δt (= $t_2 - t_1$). The *average velocity* for the particle during this interval is defined by

$$\bar{\mathbf{v}} = \frac{\Delta \mathbf{r}}{\Delta t} = \frac{\text{displacement (a vector)}}{\text{elapsed time (a scalar)}} \tag{3-1}$$

A bar above a symbol indicates an average value for the quantity in question.

The quantity $\bar{\mathbf{v}}$ is a vector, for it is obtained by dividing the vector $\Delta \mathbf{r}$ by the scalar Δt. Velocity, therefore, involves both direction and magnitude. Its direction is the direction of $\Delta \mathbf{r}$ and its magnitude is $|\Delta \mathbf{r}/\Delta t|$. The magnitude is expressed in distance units divided by time units, as, for example, meters per second or miles per hour.

Notice that the *average* velocity defined by Eq. 3–1 involves only the total displacement and the total elapsed time and does not tell us anything about the details of the motion between *A* and *B*. For example, *A* and *B* could be the same point (for a closed path trajectory) in which case the average velocity for this particular time interval Δt is zero.

If we were to measure the time of arrival of the particle at each of many points along the actual path between A and B in Fig. 3–1a, we could describe the motion in more detail. If the average velocity turned out to be the same (in magnitude and direction) between any two points along the path, we would conclude that the particle moved with constant velocity, that is, along a straight line (constant direction) at a uniform rate (constant magnitude).

3–4 Instantaneous Velocity

Suppose that a particle is moving in such a way that its average velocity, measured for a number of different time intervals, does *not* turn out to be constant. This particle is said to move with variable velocity. Then we must seek to determine a velocity of the particle at any given instant of time, called the *instantaneous velocity*.

Velocity can vary by a change in magnitude, by a change in direction, or both. For the motion portrayed in Fig. 3–1a, the average velocity during the time interval $t_2 - t_1$ may differ both in magnitude and direction from the average velocity obtained during another time interval $t_2' - t_1$. In Fig. 3–1b we illustrate this by choosing the point B to be successively closer to point A. Points B' and B'' show two intermediate positions of the particle corresponding to the times t_2' and t_2'' and described by position vectors $\mathbf{r}_2'$ and $\mathbf{r}_2''$, respectively. The vector displacements $\Delta\mathbf{r}$, $\Delta\mathbf{r}'$, and $\Delta\mathbf{r}''$ differ in direction and become successively smaller. Likewise, the corresponding time intervals $\Delta t \,(= t_2 - t_1)$, $\Delta t' \,(= t_2' - t_1)$, and $\Delta t'' \,(= t_2'' - t_1)$ become successively smaller.

As we continue this process, letting B approach A, we find that the ratio of displacement to elapsed time approaches a definite limiting value. Although the displacement in this process becomes extremely small, the time interval by which we divide it becomes small also and the ratio is not necessarily a small quantity. Similarly, while growing smaller, the displacement vector approaches a limiting direction, that of the tangent to the path of the particle at A. This limiting value of $\Delta\mathbf{r}/\Delta t$ is called the *instantaneous* velocity at the point A, or the velocity of the particle at the instant t_1.

If $\Delta\mathbf{r}$ is the displacement in a small interval of time Δt, following the time t, the velocity at the time t is the limiting value approached by $\Delta\mathbf{r}/\Delta t$ as both $\Delta\mathbf{r}$ and Δt approach zero. That is, if we let $\mathbf{v}$ represent the instantaneous velocity,

$$\mathbf{v} = \lim_{\Delta t \to 0} \frac{\Delta\mathbf{r}}{\Delta t}.$$

The direction of $\mathbf{v}$ is the limiting direction that $\Delta\mathbf{r}$ takes as B approaches A or as Δt approaches zero. As we have seen, this limiting direction is that of the tangent to the path of the particle at point A.

In the notation of the calculus, the limiting value of $\Delta\mathbf{r}/\Delta t$ as Δt approaches zero is written $d\mathbf{r}/dt$ and is called the *derivative* of $\mathbf{r}$ with respect to t. We have then

$$\mathbf{v} = \lim_{\Delta t \to 0} \frac{\Delta\mathbf{r}}{\Delta t} = \frac{d\mathbf{r}}{dt}. \tag{3–2}$$

The magnitude v of the instaneous velocity is called the *speed* and is simply the absolute value of $\mathbf{v}$. That is,

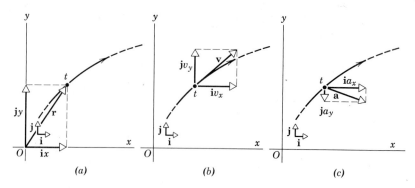

FIGURE 3–2   A particle at time t has (a) a position described by **r**, (b) an instantaneous velocity **v**, and (c) an instantaneous acceleration **a**. The vector components $\mathbf{i}x$ and $\mathbf{j}y$ of Eq. 3–4, $\mathbf{i}v_x$ and $\mathbf{j}v_y$ of Eq. 3–5, and $\mathbf{i}a_x$ and $\mathbf{j}a_y$ of Eq. 3–10 are also shown, as are the unit vectors **i** and **j**.

$$v = |\mathbf{v}| = |d\mathbf{r}/dt|. \tag{3-3}$$

Speed, being the magnitude of a vector, is intrinsically positive.

3–5 One-Dimensional Motion—Variable Velocity

Figure 3–2 shows a particle moving along a path in the x-y plane. At time t its position with respect to the origin is described by position vector **r** (see Fig. 3–2a) and it has a velocity **v** (see Fig. 3–2b) tangent to its path as shown. We can write (see Eq. 2–8)

$$\mathbf{r} = \mathbf{i}x + \mathbf{j}y, \tag{3-4}$$

where **i** and **j** are unit vectors in the positive x- and y-directions, respectively, and x and y are the (scalar) components of vector **r**. Because **i** and **j** are constant vectors, we have, on combining Eqs. 3–2 and 3–4,

$$\mathbf{v} = \frac{d\mathbf{r}}{dt} = \mathbf{i}\frac{dx}{dt} + \mathbf{j}\frac{dy}{dt},$$

which we can express as

$$\mathbf{v} = \mathbf{i}v_x + \mathbf{j}v_y \qquad \text{(two-dimensional motion)}, \tag{3-5}$$

where $v_x\,(= dx/dt)$ and $v_y\,(= dy/dt)$ are the (scalar) components of the vector **v**.

We now consider motion in one dimension only, chosen for convenience to be the x-axis. We must then have $v_y = 0$ so that Eq. 3–5 reduces to

$$\mathbf{v} = \mathbf{i}v_x \qquad \text{(one-dimensional motion)}. \tag{3-6}$$

Since **i** points in the positive x-direction, v_x will be positive (and equal to $+v$) when **v** points in that direction, and negative (and equal to $-v$) when it points in the other direction. Since, in one-dimensional motion, there are only two choices as to the direction of **v**, the full power of the vector method is not needed; we can work with the velocity component v_x alone.

Example 1. *The limiting process.* As an illustration of the limiting process in one dimension, consider the following table of data taken for motion along the x-axis. The first four columns are experimental data. The symbols refer to Fig. 3–3 in which the particle is moving from left to right, that is, in the positive x-direction. The particle was at position x_1 (100 cm from the origin) at time t_1 (1.00 sec). It was at position x_2 at time t_2. As we consider different values for x_2, and different corresponding times t_2, we find the following values:

x_1, cm	t_1, sec	x_2, cm	t_2, sec	$x_2 - x_1$ $= \Delta x$, cm	$t_2 - t_1$ $= \Delta t$, sec	$\Delta x/\Delta t$, cm/sec
100.0	1.00	200.0	11.00	100.0	10.00	10.0
100.0	1.00	180.0	9.60	80.0	8.60	9.3
100.0	1.00	160.0	7.90	60.0	6.90	8.7
100.0	1.00	140.0	5.90	40.0	4.90	8.2
100.0	1.00	120.0	3.56	20.0	2.56	7.8
100.0	1.00	110.0	2.33	10.0	1.33	7.5
100.0	1.00	105.0	1.69	5.0	0.69	7.3
100.0	1.00	103.0	1.42	3.0	0.42	7.1
100.0	1.00	101.0	1.14	1.0	0.14	7.1

Equation 3–2, which holds for the general case of motion in three dimensions, is

$$\mathbf{v} = \lim_{\Delta t \to 0} \frac{\Delta \mathbf{r}}{\Delta t} = \frac{d\mathbf{r}}{dt}.$$

For one-dimensional motion along the x-axis we have a similar relation, scalar in character, in which each vector quantity is replaced by its corresponding component or

$$v_x = \lim_{\Delta t \to 0} \frac{\Delta x}{\Delta t} = \frac{dx}{dt}. \tag{3–7}$$

It is clear from the table that as we select values of x_2 closer to x_1, Δt approaches zero and the ratio $\Delta x/\Delta t$ approaches the apparent limiting value $+7.1$ cm/sec. At time t_1, therefore, $v_x = +7.1$ cm/sec, as closely as we are able to determine from the data. Since v_x is positive, the velocity $\mathbf{v}$ ($= \mathbf{i}v_x$; see Eq. 3–6) points to the right in Fig. 3–3. This is tangent to the path in the direction of motion, as it must be.

Example 2. Figure 3–4a shows six successive "snapshots" of a particle moving along the x-axis with variable velocity. At $t = 0$ it is at position $x = +1.00$ ft to the right of the origin; at $t = 2.5$ sec it has come to rest at $x = +5.00$ ft; at $t = 4.0$ sec it has returned to $x = +1.40$ ft. Figure 3–4b is a plot of position x versus time t for this motion. The *average velocity* for the entire 4.0-sec interval is the net displacement or change of position ($+0.40$ ft) divided by the elapsed time (4.0 sec) or $\bar{v}_x = +0.10$ ft/sec. (We call $\bar{v}_x$ average velocity and v_x velocity, for one-dimensional motion, even though velocity is a vector and not a scalar. This conforms to common usage and should cause no misundertsandings. These quantities are not speeds because they may be negative, whereas speed is intrinsically positive.) The average velocity vector $\bar{\mathbf{v}}$ points in the positive x-direction (that is, to the right in Fig. 3–4a) because the net displacement points in this direction. The quantity $\bar{v}_x$ can be obtained directly from the slope of the dashed line af in Fig. 3–4b, where by slope we mean the ratio of the net displacement gf to the elapsed time ga. (The slope is *not* the tangent of the angle fag measured on the graph with a protractor. This angle is arbitrary because it depends on the scales we choose for x and t.)

The velocity v_x at any instant is found from the slope of the curve of Fig. 3–4b at that instant. Equation 3–7 is in fact the relation by which the slope of the curve is defined in the calculus. In our example the slope at b, which is the value of v_x at b, is $+1.7$ ft/sec; the slope at d is zero and the

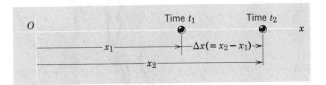

FIGURE 3–3 A particle is moving to the right along the x-axis.

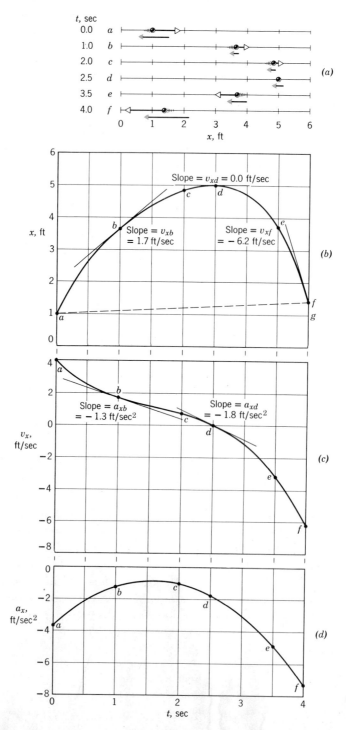

FIGURE 3-4   (*a*) Six consecutive "snapshots" of a particle moving along the *x*-axis. The vector joined to the particle is its instantaneous velocity; that below the particle is its instantaneous acceleration. (*b*) A plot of *x* versus *t* for the motion of the particle. (*c*) A plot of v_x versus *t*. (*d*) A plot of a_x versus *t*.

slope at *f* is -6.2 ft/sec. When we determine the slope dx/dt at each instant *t*, we can make a plot of v_x versus *t*, as in Fig. 3–4*c*. Note that for the interval $0 < t < 2.5$ sec, v_x is positive so that the velocity vector **v** points to the right in Fig. 3–4*a*; for the interval 2.5 sec $< t <$ 4.0 sec v_x is negative so that **v** points to the left in Fig. 3–4*a*.

3–6   Acceleration

Often the velocity of a moving body changes either in magnitude, in direction, or both as the motion proceeds. The body is then said to have an acceleration. *The acceleration of a particle is the rate of change of its velocity with time.* Suppose that at the instant t_1 a particle, as in Fig. 3–5, is at point A and is moving in the x-y plane with the instantaneous velocity $\mathbf{v}_1$, and at a later instant t_2 it is at point B and moving with the instantaneous velocity $\mathbf{v}_2$. The *average acceleration* $\bar{\mathbf{a}}$ during the motion from A to B is defined to be the change of velocity divided by the time interval, or

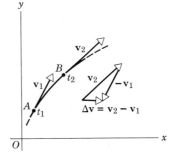

FIGURE 3–5   A particle has velocity $\mathbf{v}_1$ at point A and moves to point B, where its velocity is $\mathbf{v}_2$. The triangle shows the (vector) change in velocity $\Delta\mathbf{v}$ ($= \mathbf{v}_2 - \mathbf{v}_1$) experienced by the particle as it moves from A to B.

$$\bar{\mathbf{a}} = \frac{\mathbf{v}_2 - \mathbf{v}_1}{t_2 - t_1} = \frac{\Delta\mathbf{v}}{\Delta t}. \qquad (3\text{–}8)$$

The quantity $\bar{\mathbf{a}}$ is a vector, for it is obtained by dividing a vector $\Delta\mathbf{v}$ by a scalar Δt. Acceleration is therefore characterized by magnitude and direction. Its direction is the direction of $\Delta\mathbf{v}$ and its magnitude is $|\Delta\mathbf{v}/\Delta t|$. The magnitude of the acceleration is expressed in velocity units divided by time units, as for example meter/sec per sec (written meters/sec² and read "meters per second squared"), cm/sec², and ft/sec².

We call $\bar{\mathbf{a}}$ of Eq. 3–8 the *average* acceleration because nothing has been said about the time variation of velocity *during* the interval Δt. We know only the net change in velocity and the total elapsed time. If the change in velocity (a vector) divided by the corresponding elapsed time, $\Delta\mathbf{v}/\Delta t$, were to remain constant, regardless of the time intervals over which we measured the acceleration, we would have *constant* acceleration. Constant acceleration, therefore, implies that the *change* in velocity is uniform with time in direction and magnitude. If there is no change in velocity, that is, if the velocity were to remain constant both in magnitude and direction, then $\Delta\mathbf{v}$ would be zero for all time intervals and the acceleration would be zero.

If a particle is moving in such a way that its average acceleration, measured for a number of different time intervals, does *not* turn out to be constant, the particle is said to have a variable acceleration. The acceleration can vary in magnitude, or in direction, or both. In such cases we seek to determine the acceleration of the particle at any given time, called the instantaneous acceleration.

The *instantaneous acceleration* is defined by

$$\mathbf{a} = \lim_{\Delta t \to 0} \frac{\Delta\mathbf{v}}{\Delta t} = \frac{d\mathbf{v}}{dt}. \qquad (3\text{–}9)$$

That is, the acceleration of a particle at time t is the limiting value of $\Delta\mathbf{v}/\Delta t$ at time t as both $\Delta\mathbf{v}$ and Δt approach zero. The direction of the instantaneous acceleration $\mathbf{a}$ is the limiting direction of the vector change in velocity $\Delta\mathbf{v}$. The magnitude a of the instantaneous acceleration is simply $a = |\mathbf{a}| = |d\mathbf{v}/dt|$. When the acceleration is constant the instantaneous acceleration equals the average acceleration. Note that the relation of $\mathbf{a}$ to $\mathbf{v}$, in Eq. 3–9, is the same as that of $\mathbf{v}$ to $\mathbf{r}$, in Eq. 3–2.

Two special cases illustrate that acceleration can arise from a change in either the magnitude or the direction of the velocity. In one case we have motion along a straight line with uniformly changing speed (as in Section 3–8). Here the velocity does not change in direction but its magnitude changes uniformly with time. This is a case of constant acceleration. In the second case we have motion in a circle at constant speed (Section 4–4). Here the velocity vector changes continuously in direction but the direction of the acceleration vector is not constant. Later we will encounter other important cases of accelerated motion.

3–7   One-Dimensional Motion—Variable Acceleration

From Eqs. 3–5 and 3–9 we can write, for motion in two dimensions as in Fig. 3–2,

$$\mathbf{a} = \frac{d\mathbf{v}}{dt} = \mathbf{i}\,\frac{dv_x}{dt} + \mathbf{j}\,\frac{dv_y}{dt}$$

or

$$\mathbf{a} = \mathbf{i}a_x + \mathbf{j}a_y, \tag{3–10}$$

where $a_x\ (= dv_x/dt)$ and $a_y\ (= dv_y/dt)$ are the scalar components of the acceleration vector $\mathbf{a}$ (see Fig. 3–2c).

We again restrict ourselves to motion in one dimension only, chosen for convenience to be the x-axis. Since v_y for such motion does not change with time (and is, in fact, zero), a_y, which is dv_y/dt, must also be zero so that

$$\mathbf{a} = \mathbf{i}a_x. \tag{3–11}$$

Since $\mathbf{i}$ points in the positive x-direction, a_x will be positive when $\mathbf{a}$ points in this direction and negative when it points in the other direction.

Example 3. The motion of Fig. 3–4a is one of variable acceleration along the x-axis. To find the acceleration* a_x at each instant we must determine dv_x/dt at each instant. This is simply the slope of the curve of v_x versus t at that instant. The slope of Fig. 3–4c at point b is −1.3 ft/sec² and that at point d is −1.8 ft/sec², as shown in the figure. The result of calculating the slope for all points is shown in Fig. 3–4d. Notice that a_x is negative at all instants, which means that the acceleration vector $\mathbf{a}$ points in the negative x-direction. This means that v_x is always decreasing with time, as is clearly seen from Fig. 3–4c. The motion is one in which the acceleration vector has a constant direction but varies in magnitude (see Fig. 3–4a).

3–8   One-Dimensional Motion—Constant Acceleration

Let us now further restrict our considerations to motion which not only occurs in one dimension (the x-axis) but for which $a_x =$ a constant. For such *constant acceleration* the average acceleration for any time interval is equal to the (constant) instantaneous acceleration a_x. Let $t_1 = 0$ and let t_2 be any arbitrary time t. Let v_{x0} be the value of v_x at $t = 0$ and let v_x be its value at the arbitrary time t. With this notation we find a_x (see Eq. 3–8) from

$$a_x = \frac{\Delta v}{\Delta t} = \frac{v_x - v_{x0}}{t - 0}$$

or

$$v_x = v_{x0} + a_x t. \tag{3–12}$$

This equation states that the velocity v_x at time t is the sum of its value v_{x0} at time $t = 0$ plus the change in velocity during time t, which is $a_x t$.

---

* As for velocity, we commonly call a_x for one-dimensional motion the acceleration even though acceleration is a vector and a_x is correctly an acceleration component. For one-dimensional motion there is only one component if the axis is chosen along the line of the motion.

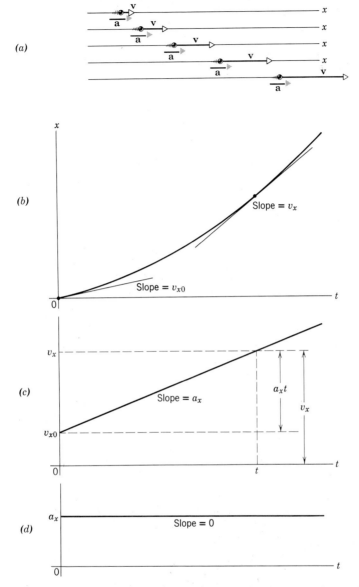

FIGURE 3–6 (a) Five successive "snapshots" of rectilinear motion with constant acceleration. The arrows on the spheres represent **v**; those below represent **a**. (b) The displacement increases quadratically according to $x = v_{x0}t + \frac{1}{2}a_x t^2$. Its slope increases uniformly and at each instant has the value v_x, the velocity. (c) The velocity v_x increases uniformly according to $v_x = v_{x0} + a_x t$. Its slope is constant and at each instant has the value a_x, the acceleration. (d) The acceleration a_x has a constant value; its slope is zero. Figure 3–4 shows similar plots for one-dimensional motion in which the acceleration is *not* constant.

Figure 3–6c shows a graph of v_x versus t for constant acceleration; it is a graph of Eq. 3–12. Notice that the slope of the velocity curve is constant, as it must be because the acceleration $a_x \ (= dv_x/dt)$ has been assumed to be constant, as Fig. 3–6d shows.

When the velocity v_x changes uniformly with time, its average value over any time interval equals one-half the sum of the values of v_x at the beginning and at the end of the interval. That is, the average velocity $\bar{v}_x$ between $t = 0$ and $t = t$ is

$$\bar{v}_x = \tfrac{1}{2}(v_{x0} + v_x). \tag{3–13}$$

This relation would not be true if the acceleration were not constant, for then the curve of v_x versus t would not be a straight line.

If the position of the particle at $t = 0$ is x_0, the position x at $t = t$ can be found from

$$x = x_0 + \bar{v}_x t$$

which can be combined with Eq. 3–13 to yield

$$x = x_0 + \tfrac{1}{2}(v_{x0} + v_x)t. \tag{3–14}$$

The displacement due to the motion in time t is $x - x_0$. Often the origin is chosen so that $x_0 = 0$.

Notice that aside from initial conditions of the motion, that is, the values of x and v_x at $t = 0$ (taken here as $x = x_0$ and $v_x = v_{x0}$), there are four parameters of the motion. These are x, the displacement; v_x, the velocity; a_x, the acceleration; and t, the elapsed time. If we know only that the acceleration is constant, but not necessarily its value, from any two of these parameters we can obtain the other two. For example, if a_x and t are known, Eq. 3–12 gives v_x, and having obtained v_x, we find x from Eq. 3–14.

In most problems in uniformly accelerated motion, two parameters are known and a third is sought. It is convenient, therefore, to obtain relations between any three of the four parameters. Equation 3–12 contains v_x, a_x, and t, but *not x*; Eq. 3–14 contains x, v_x, and t but *not a_x*. To complete our system of equations we need two more relations, one containing x, a_x, and t but *not v_x* and another containing x, v_x, and a_x but *not t*. These are easily obtained by combining Eqs. 3–12 and 3–14.

Thus, if we substitute into Eq. 3–14 the value of v_x from Eq. 3–12, we thereby eliminate v_x and obtain

$$x = x_0 + v_{x0}t + \tfrac{1}{2}a_x t^2. \tag{3–15}$$

When Eq. 3–12 is solved for t and this value for t is substituted into Eq. 3–14, we obtain

$$v_x{}^2 = v_{x0}{}^2 + 2a_x(x - x_0). \tag{3–16}$$

Equations 3–12, 3–14, 3–15, and 3–16 (see Table 3–1) are the complete set of equations for motion along a straight line with constant acceleration.

A special case of motion with constant acceleration is one in which the acceleration is zero, that is, $a_x = 0$. In this case, the four equations in Table 3–1 reduce to the expected results $v_x = v_{x0}$ (the velocity does not change) and $x = x_0 + v_{x0}t$ (the displacement changes linearly with time).

Table 3–1   KINEMATIC EQUATIONS FOR STRAIGHT LINE MOTION WITH CONSTANT ACCELERATION   (THE POSITION x_0 AND THE VELOCITY v_{x0} AT THE INITIAL INSTANT $t = 0$ ARE THE GIVEN INITIAL CONDITIONS)

Equation Number	Equation	Contains			
		x	v_x	a_x	t
3–12	$v_x = v_{x0} + a_x t$	×	√	√	√
3–14	$x = x_0 + \tfrac{1}{2}(v_{x0} + v_x)t$	√	√	×	√
3–15	$x = x_0 + v_{x0}t + \tfrac{1}{2}a_x t^2$	√	×	√	√
3–16	$v_x{}^2 = v_{x0}{}^2 + 2a_x(x - x_0)$	√	√	√	×

Example 4. The curve of Fig. 3–6*b* is a displacement-time graph for motion with constant acceleration; that is, it is a graph of Eq. 3–15 in which $x_0 = 0$. The slope of the tangent to the curve at time t equals the velocity v_x at that time. Notice that the slope increases continuously with time from v_{x0} at $t = 0$. The *rate of increase* of this slope should give the acceleration a_x, which is constant in this case. The curve of Fig. 3–6*b* is a parabola since Eq. 3–15 is the equation for a parabola having slope v_{x0} at $t = 0$. We obtain, on successive differentiation of Eq. 3–15,

$$x = x_0 + v_{x0}t + \tfrac{1}{2}a_x t^2$$

$$dx/dt = v_{x0} + a_x t \quad \text{or} \quad v_x = v_{x0} + a_x t,$$

which gives the velocity v_x at time t (compare Eq. 3–12), and

$$dv_x/dt = a_x,$$

the constant acceleration. The displacement-time graph for uniformly accelerated rectilinear motion will therefore always be parabolic.

3–9 Consistency of Units and Dimensions

You should not feel compelled to memorize relations such as those of Table 3–1. The important thing is to be able to follow the line of reasoning used to obtain the results. These relations will be recalled automatically after you have used them repeatedly to solve problems, partly as a result of the familiarity acquired but chiefly as a result of the better understanding obtained through application.

We can use any convenient units of time and distance in these equations. If we choose to express time in seconds and distance in feet, for self-consistency we must express velocity in ft/sec and acceleration in ft/sec². If we are given data in which the units of one quantity, as velocity, are not consistent with the units of another quantity, as acceleration, then before using the data in our equations we should transform both quantities to units that are consistent with one another. Having chosen the units of our fundamental quantities, we automatically determine the units of our derived quantities consistent with them. In carrying out any calculation, always remember to attach the proper units to the final result, for the result is meaningless without this label.

Example 5. Suppose we wish to find the speed of a particle which has a uniform acceleration of 5.00 cm/sec² for an interval of $\tfrac{1}{2}$ hr if the particle has a speed of 10.0 ft/sec at the beginning of this interval. We decide to choose the foot as our length unit and the second as our time unit. Then

$$a_x = 5.00 \text{ cm/sec}^2 = 5.00 \text{ cm/sec}^2 \times \left(\frac{1 \text{ in.}}{2.54 \text{ cm}}\right) \times \left(\frac{1 \text{ ft}}{12 \text{ in.}}\right) = \frac{5.00}{30.5} \text{ ft/sec}^2 = 0.164 \text{ ft/sec}^2.$$

The time interval

$$\Delta t = t - t_0 = \frac{1}{2} \text{ hr} \times \left(\frac{60 \text{ min}}{1 \text{ hr}}\right) \times \left(\frac{60 \text{ sec}}{1 \text{ min}}\right) = 1800 \text{ sec}.$$

Note that the conversion factors in large parentheses are equal to unity. Taking the initial time $t_0 = 0$, as in Eq. 3–12, we then have

$$v_x = v_{x0} + a_x t = 10.0 \text{ ft/sec} + (0.164 \text{ ft/sec}^2)(1800 \text{ sec})$$

$$= 305 \text{ ft/sec}.$$

One way to spot an erroneous equation is to check the *dimensions* of all its terms. The dimensions of any physical quantity can always be expressed as some combination of the fundamental quantities, such as mass, length, and time, from which they are derived. The dimensions of velocity are length (L) divided by time (T); the dimensions of acceleration are length divided by time squared, etc. *In any legitimate physical equation the dimensions of all the terms must be the same.* This means, for example, that we cannot equate a term whose total dimension is a velocity to one whose total dimension is an acceleration. The dimensional labels attached to various quantities may be treated just like algebraic quantities and may be combined, canceled, etc., just as if they were factors in the equation. For example, to check Eq. 3–15, $x = x_0 + v_{x0}t + \frac{1}{2}a_xt^2$, dimensionally, we note that x and x_0 have the dimension of a length. Therefore the two remaining terms must also have the dimension of a length. The dimension of the term $v_{x0}t$ is

$$\frac{\text{length}}{\text{time}} \times \text{time} = \text{length} \qquad \text{or} \qquad \frac{L}{T} \times T = L,$$

and that of $\frac{1}{2}a_xt^2$ is

$$\frac{\text{length}}{\text{time}^2} \times \text{time}^2 = \text{length} \qquad \text{or} \qquad \frac{L}{T^2} \times T^2 = L.$$

The equation is therefore *dimensionally correct.* You should check the dimensions of all the equations you use.

Example 6. The nucleus of a helium atom (alpha-particle) travels along the inside of a straight hollow tube 2.0 meters long which forms part of a particle accelerator. (*a*) If one assumes uniform acceleration, how long is the particle in the tube if it enters at a speed of 1.0×10^4 meters/sec and leaves at 5.0×10^6 meters/sec? (*b*) What is its acceleration during this interval?

(*a*) We choose an x-axis parallel to the tube, its positive direction being that in which the particle is moving and its origin at the tube entrance. We are given x and v_x and we seek t. The acceleration a_x is not involved. Hence we use Eq. 3–14, $x = x_0 + \frac{1}{2}(v_{x0} + v_x)t$ with $x_0 = 0$ or

$$t = \frac{2x}{v_{x0} + v_x},$$

$$t = \frac{(2)(2.0 \text{ meters})}{(500 + 1) \times 10^4 \text{ meters/sec}} = 8.0 \times 10^{-7} \text{ sec},$$

or 0.80 microseconds.

(*b*) The acceleration follows from Eq. 3–12, $v_x = v_{x0} + a_xt$, or

$$a_x = \frac{v_x - v_{x0}}{t} = \frac{(500 - 1) \times 10^4 \text{ meters/sec}}{8.0 \times 10^{-7} \text{ sec}} = 6.3 \times 10^{12} \text{ meters/sec}^2,$$

Although this acceleration is enormous by ordinary standards, it occurs over an extremely short time. The acceleration **a** is in the positive x-direction, that is, in the direction in which the particle is moving, because a_x is positive.

3–10 Freely Falling Bodies

The most common example of motion with (nearly) constant acceleration is that of a body falling toward the earth. In the absence of air resistance it is found that all bodies, regardless of their size, weight, or composition, fall with the same acceleration at the same point of the earth's surface, and if the distance covered is not too great, the acceleration remains constant throughout the fall. This ideal motion, in which air resistance and the small change in acceleration with altitude are neglected, is called "free fall."

The acceleration of a freely falling body is called the acceleration due to gravity and is denoted by the symbol **g.** Near the earth's surface its magnitude is approximately 32 ft/sec², 9.8 meters/sec², or 980 cm/sec², and it is directed down toward the center of the earth. The variation of the exact value with latitude and altitude will be discussed later (Chapter 14).

We shall select a reference frame rigidly attached to the earth. The *y*-axis will be taken as positive vertically upward. Then the acceleration due to gravity **g** will be a vector pointing vertically down (toward the center of the earth) in the negative *y*-direction. (This choice is arbitrary. In other problems it may be convenient to choose down as positive.) Our equations for constant acceleration are applicable here. We simply replace *x* by *y* and set $y_0 = 0$ in Eqs. 3–12, 3–14, 3–15, and 3–16, obtaining

$$v_y = v_{y0} + a_y t,$$

$$y = \tfrac{1}{2}(v_{y0} + v_y)t,$$

$$y = v_{y0}t + \tfrac{1}{2}a_y t^2, \tag{3-17}$$

$$v_y{}^2 = v_{y0}{}^2 + 2a_y y,$$

and, for problems in free fall, we set $a_y = -g$. Notice that we have chosen the initial position as the origin, that is we have chosen $y_0 = 0$ at $t = 0$. Note also that g is the magnitude of the acceleration due to gravity.

Example 7. A body is dropped from rest and falls freely. Determine the position and speed of the body after 1.0, 2.0, 3.0, and 4.0 sec have elapsed.

We choose the starting point as the origin. We know the initial speed and the acceleration and we are given the time. To find the position we use

$$y = v_{y0}t - \tfrac{1}{2}gt^2.$$

Then, $v_{y0} = 0$ and $g = 32$ ft/sec², and with $t = 1.0$ sec we obtain

$$y = 0 - \tfrac{1}{2}(32 \text{ ft/sec}^2)(1.0 \text{ sec})^2 = -16 \text{ ft.}$$

To find the speed with $t = 1.0$ sec, we use

$$v_y = v_{y0} - gt$$

and obtain $$v_y = 0 - (32 \text{ ft/sec}^2)(1.0 \text{ sec}) = -32 \text{ ft/sec.}$$

After 1.0 sec of falling from rest, the body is 16 ft below its starting point and has a velocity directed downward whose magnitude is 32 ft/sec; the negative signs for y and v_y show that the associated vectors each point in the negative *y*-direction, that is, downward.

You should now show that the values of y, v_y, and a_y obtained at times $t = 2.0$, 3.0, and 4.0 sec are those shown in Fig. 3–7.

Example 8. A ball is thrown vertically upward from the ground with a speed of 80 ft/sec.

(*a*) How long does it take to reach its highest point? At its highest point, $v_y = 0$, and we have $v_{y0} = +80$ ft/sec. To obtain the time *t* we use $v_y = v_{y0} - gt$, or

$$t = \frac{v_{y0} - v_y}{g}$$

$$= \frac{(80 - 0) \text{ ft/sec}}{32 \text{ ft/sec}^2} = 2.5 \text{ sec.}$$

(*b*) How high does the ball rise? Using only the original data, we choose the relation $v_y{}^2 = v_{y0}{}^2 - 2gy$, or

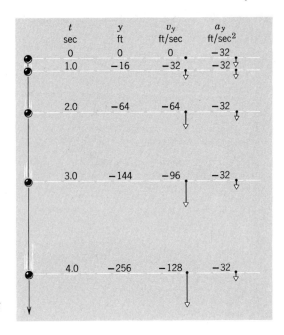

t sec	y ft	v_y ft/sec	a_y ft/sec^2
0	0	0	-32
1.0	-16	-32	-32
2.0	-64	-64	-32
3.0	-144	-96	-32
4.0	-256	-128	-32

FIGURE 3–7 A body in free fall; showing y, v_y, and a_y at particular times t.

$$y = \frac{v_{y0}{}^2 - v_y{}^2}{2g},$$

$$= \frac{(80 \text{ ft/sec})^2 - 0}{2 \times 32 \text{ ft/sec}^2} = +100 \text{ ft}.$$

(*c*) At what times will the ball be 96 ft above the ground? Using $y = v_{y0}t - \frac{1}{2}gt^2$, we have

$$\tfrac{1}{2}gt^2 - v_{y0}t + y = 0,$$

$$\tfrac{1}{2}(32 \text{ ft/sec}^2)t^2 - (80 \text{ ft/sec})t + 96 \text{ ft} = 0,$$

or

$$t^2 - 5.0t + 6.0 = 0,$$

which yields $t = 2.0$ sec and $t = 3.0$ sec. At $t = 2.0$ sec, the ball is moving upward with a speed of 16 ft/sec, for

$$v_y = v_{y0} - gt = 80 \text{ ft/sec} - (32 \text{ ft/sec}^2)(2.0 \text{ sec}) = +16 \text{ ft/sec}.$$

At $t = 3.0$ sec, the ball is moving downward with the same speed, for

$$v_y = v_{y0} - gt = 80 \text{ ft/sec} - (32 \text{ ft/sec}^2)(3.0 \text{ sec}) = -16 \text{ ft/sec}.$$

Notice that in this 1.0-sec interval the velocity changed by -32 ft/sec, corresponding to an acceleration of -32 ft/sec^2.

You should be able to convince yourself that in the absence of air resistance the ball will take as long to rise as to fall the same distance, and that it will have the same speed going down at each point as it had going up.

QUESTIONS

1. Can you think of physical phenomena involving the earth in which the earth cannot be treated as a particle?

2. Each second a rabbit moves half the remaining distance from his nose to a head of lettuce.

Does he ever get to the lettuce? What is the limiting value of his average velocity? Draw graphs showing his velocity and position as time increases.

3. Average speed can mean the magnitude of the average velocity vector. Another meaning given to it is that average speed is the total length of path traveled divided by the elapsed time. Are these meanings different? If so, give an example.

4. When the velocity is constant, does the average velocity over any time interval differ from the instantaneous velocity at any instant?

5. Is the average velocity of a particle moving along the x-axis $\frac{1}{2}(v_{x0} + v_x)$ when the acceleration is not uniform? Prove your answer with the use of graphs.

6. Does the speedometer on an automobile register speed as we defined it?

7. (*a*) Can a body have zero velocity and still be accelerating? (*b*) Can a body have a constant speed and still have a varying velocity? (*c*) Can a body have a constant velocity and still have a varying speed?

8. Can an object have an eastward velocity while experiencing a westward acceleration?

9. Can the direction of the velocity of a body change when its acceleration is constant?

10. If a particle is released from rest ($v_{y0} = 0$) at $y_0 = 0$ at the time $t = 0$, Eq. 3–17 for constant acceleration says that it is at position y at two different times, namely, $+ \sqrt{2y/a_y}$ and $- \sqrt{2y/a_y}$. What is the meaning of the negative root of this quadratic equation?

11. What happens to our kinematic equations under the operation of time reversal, that is, replacing t by $-t$? Explain.

12. Consider a ball thrown vertically up. Taking air resistance into account, would you expect the time during which the ball rises to be longer or shorter than the time during which it falls?

13. Can there be motion in two dimensions with an acceleration in only one dimension?

14. A man standing on the edge of a cliff at some height above the ground below throws one ball straight up with initial speed u and then throws another ball straight down with the same initial speed. Which ball, if either, has the larger speed when it hits the ground? Neglect air resistance.

15. From what you know about angular measure, what dimensions would you assign to an angle? Can a quantity have units without having dimensions?

PROBLEMS

1(3). Compare your average speed in the following two cases. (*a*) You walk 240 ft at a speed of 4.0 ft/sec and then run 240 ft at a speed of 10 ft/sec along a straight track. (*b*) You walk for 1.0 min at a speed of 4.0 ft/sec and then run for 1.0 min at 10 ft/sec along a straight track.
Answer: (*a*) 5.7 ft/sec. (*b*) 7.0 ft/sec.

2(3). A train moving at an essentially constant speed of 60 miles/hr moves east for 40 min, then in a direction 45° east of north for 20 min, and finally west for 50 min. What is the average velocity of the train during this run?

3(3). An automobile travels on a straight road for 40 miles at 30 miles/hour. It then continues in the same direction for another 40 miles at 60 miles/hr. What is the average velocity of the car during this 80-mile trip?
Answer: 40 miles/hr.

4(3). The position of an object moving in a straight line is given by $x = 3.0t - 4.0t^2 + t^3$, where x is in meters and t in seconds. (*a*) What is the position of the object at $t = 1, 2, 3$, and 4 sec? (*b*) What is the total distance traveled between $t = 0$ and $t = 4$ sec? (*c*) What is the average velocity for the time interval from $t = 2$ to $t = 4$ sec?

5(3). A particle moves with constant speed around the circumference of a circle with radius r (5 meters). See Fig. 3–8. It completes one revolution each 20 seconds. (*a*) What is the average velocity of the particle during the first 5 sec of its motion? (*b*) What is the average velocity of the particle during the first 25 sec of its motion? Use unit vector notation.
Answer: (*a*) $-\mathbf{i} + \mathbf{j}$, meters/sec. (*b*) $-0.2\mathbf{i} + 0.2\mathbf{j}$, meters/sec.

6(4). Construct a graph of x versus t for the object of Problem 4(3) during the time interval from $t = 0$ to $t = 4$ sec. From your graph estimate (*a*) the initial velocity of the object and (*b*) the instantaneous velocities at $t = 1, 2, 3,$ and 4 sec.

7(4). Two trains, each having a speed of 30 miles/hr, are headed at each other on the same straight track. A bird that can fly 60 miles/hr, flies off one train when they are 60 miles apart and heads directly for the other train. On reaching the other train it flies directly back to the first train, and so forth. (*a*) How many trips can the bird make from one train to the other before they crash? (*b*) What is the total distance the bird travels?

Answer: (*a*) An infinite number. (*b*) 60 miles.

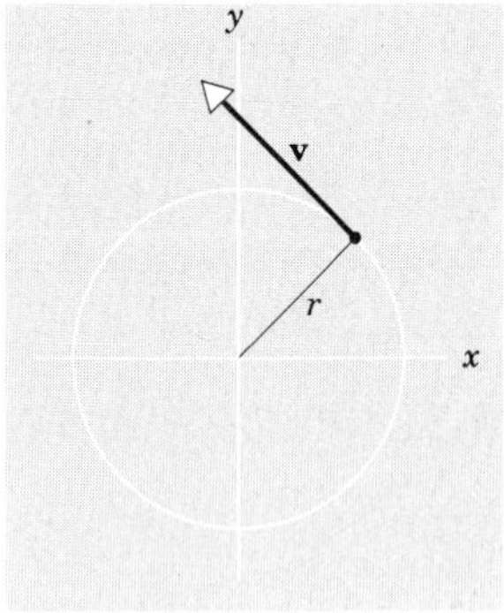

8(6). A particle had a velocity of 18 meters/sec and 2.4 sec later its velocity was 30 meters/sec in the opposite direction. What was the average acceleration of the particle during this 2.4-sec interval?

9(6). A particle moving along the positive x-axis has the following positions at various times:

x(meters)	0.080	0.050	0.040	0.050	0.080	0.13	0.68
t(sec)	0.0	1.0	2.0	3.0	4.0	5.0	10

(*a*) Plot displacement (not position) versus time. (*b*) Find the average velocity of the particle in the intervals 0.0 to 1.0 sec, 0.0 to 2.0 sec, 0.0 to 3.0 sec, 0.0 to 4.0 sec. (*c*) Find the slope of the curve drawn in part *a* at the points $t = 0.0, 1.0, 2.0, 3.0, 4.0,$ and 5.0 sec. (*d*) Plot the slope (units?) versus time. (*e*) From the curve of part *d* determine the acceleration of the particle at times $t = 2.0, 3.0$ and 4.0 sec.

Answer: (*b*) $-0.030, -0.020, -0.010, 0.0$ meters/sec. (*c*) $-0.040, -0.020, 0.0, +0.20, +0.040, +0.060$, meters/sec. (*e*) $+0.020$ meters/sec^2.

10(7). The graph of x versus t (see Fig. 3–9*a*) is for a particle in straight line motion. (*a*) State for each interval whether the velocity v_x is $+, -,$ or 0, and whether the acceleration a_x is $+, -,$ or 0. The intervals are *OA, AB, BC,* and *CD*. (*b*) From the curve is there any interval over which the acceleration is obviously not constant? (Ignore the behavior at the end points of the intervals.)

11(7). Answer the previous questions for the motion described by the graph of Fig. 3–9*b*.

Answer: (*a*)

	v_x	a_x
OA	+	−
AB	0	0
BC	+	+
CD	+	0

(*b*) No.

12(7). A particle moves along the x-axis with a displacement versus time as shown in Fig. 3–10. Sketch roughly curves of velocity versus time and acceleration versus time for this motion.

13(7). A particle moves along the x-axis according to the equation $x = 50t + 10t^2$, meters. Calculate (*a*) the average velocity of the particle during the first 3 sec of its motion, (*b*) the instantaneous velocity of the particle at $t = 3.0$ sec, and (*c*) the instantaneous acceleration of the particle at $t = 3.0$ sec.

Answer: (*a*) 80 meters/sec. (*b*) 110 meters/sec. (*c*) 20 meters/sec^2.

FIGURE 3-8   Problem 5(3).

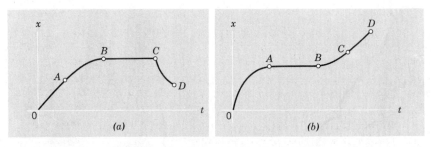

FIGURE 3-9   Problems 10(7) and 11(7).

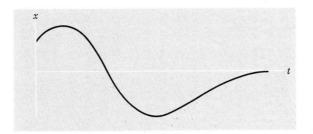

FIGURE 3-10 Problem 12(7).

14(8). A rocket-driven sled running on a straight level track is used to investigate the physiological effects of large accelerations on humans. One such sled can attain a speed of 1000 miles/hr in 1.8 sec starting from rest. (*a*) Assume the acceleration is constant and compare it to *g*. (*b*) What is the distance traveled in this time?

15(8). A certain drag racer can accelerate from 0 to 60 miles/hr in 5.0 sec. (*a*) What is its average acceleration, in *g*s, during this time? (*b* How far will it travel during the 5.0 sec, assuming its acceleration to be constant? (*c*) How much time would be required to go a distance of 0.25 mile if the acceleration could be maintained at the same value?

Answer: (*a*) 0.55 *g*. (*b*) 0.042 mile. (*c*) 12 sec.

16(8). A rocketship in free space moves with constant acceleration equal to 32 ft/sec². (*a*) If it starts from rest, how long will it take to acquire a speed one-tenth that of light? (*b*) How far will it travel in so doing?

17(8). An arrow while being shot from a bow was accelerated over a distance of 2.0 ft. If its speed at the moment it left the bow was 200 ft/sec what was the average acceleration imparted by the bow?

Answer: 1.0×10^4 ft/sec².

18(8). The speed of an automobile traveling east is uniformly reduced from 45 miles/hr to 30 miles/hr in a distance of 264 ft. (*a*) What is the magnitude and direction of the constant acceleration? (*b*) How much time has elapsed during this deceleration? (*c*) If the car continues to decelerate at the same rate how much time would elapse in bringing it to rest from 45 miles/hr? (*d*) What distance is required to bring the car to rest from 45 miles/hr?

19(8). A train started from rest and moved with constant acceleration. At one time it was traveling 30 ft/sec and 160 ft farther on it was traveling 50 ft/sec. Calculate (*a*) the acceleration, (*b*) the time required to travel the 160 ft mentioned. (*c*) the time required to attain the speed of 30 ft/sec, (*d*) the distance moved from rest to the time the train had a speed of 30 ft/sec.

Answer: (*a*) 5.0 ft/sec². (*b*) 4.0 sec. (*c*) 6.0 sec. (*d*) 90 ft.

20(8). A car moving with constant acceleration covers the distance between two points 180 ft apart in 6.0 sec. Its speed as it passes the second point is 45 ft/sec. (*a*) What is its speed at the first point? (*b*) What is its acceleration? (*c*) At what prior distance from the first point was the car at rest?

21(8). An electron with initial velocity $v_{x0} = 1.0 \times 10^4$ meters/sec enters a region where it is electrically accelerated (Fig. 3–11). It emerges with a velocity $v_x = 4.0 \times 10^6$ meters/sec. What was its acceleration, assumed constant? (Such a process occurs in the electron gun in a cathode-ray tube, used in television receivers and oscilloscopes.)

Answer: 8.0×10^{14} meters/sec².

22(8). A muon (an elementary particle) is shot with constant speed 5.00×10^6 meters/sec into a region where an electric field produces an acceleration of the muon of 1.25×10^{14} meters/sec² directed opposite to the initial velocity. How far does the muon travel before coming to rest?

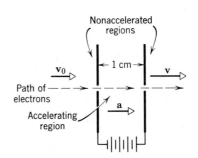

FIGURE 3-11 Problem 21(8).

23(8). A fast train running at 60 miles/hr rounds a curve onto a straightaway. The engineer observes, 200 ft ahead, a slower train running in the same direction on the same track at 30 miles/hr. The engineer instantly applies his brakes. What must be the resulting constant acceleration if a collision is to be avoided?

Answer: $a > 4.8$ ft/sec².

24(8). (*a*) If the maximum acceleration which would be tolerable for passengers in a subway train is 3.0 mph/sec and subway stations are located one-half mile apart what is the maximum speed a subway train could attain in this distance? (*b*) What would be the time between stations? (*c*) If the subway train stops for a 20-sec interval at each station, what is the maximum average speed of a subway train?

25(8). At the instant the traffic light turns green, an automobile starts with a constant acceleration a_x of 6.0 ft/sec². At the same instant a truck, traveling with a constant speed of 30 ft/sec, overtakes and passes the automobile. (*a*) How far beyond the starting point will the automobile overtake the truck? (*b*) How fast will the car be traveling at that instant? (It is instructive to plot a qualitative graph of *x* versus *t* for each vehicle.)

Answer: (*a*) 300 ft. (*b*) 60 ft/sec.

26(8). A particle is moving in the direction of the positive *x*-axis with a constant speed of 3.0 meter/sec. As it passes the origin a second particle, initially at rest at $x = 10$ meters, begins accelerating toward the origin with a constant acceleration whose magnitude is 2.0 meters/sec². (*a*) Where and when do the two particles meet? (*b*) What are the velocities of the two particles just before they collide?

27(8). A driver's handbook states that an automobile with good brakes and going 50 miles/hr can stop in a distance of 186 ft. The corresponding distance for 30 miles/hr is 80 ft. Assume that the driver reaction time, during which the acceleration is zero, and the acceleration after he applies the brakes are both the same for the two speeds. Calculate (*a*) the driver reaction time and (*b*) the acceleration.

Answer: (*a*) 0.75 sec. (*b*) −20 ft/sec².

28(9). The position of a particle moving along the *x*-axis depends on the time according to the equation

$$x = at^2 - bt^3,$$

where *x* is in feet and *t* in seconds. (*a*) What dimensions and units must *a* and *b* have? For the following, let their numerical values be 3.0 and 1.0, respectively. (*b*) At what time does the particle reach its maximum positive *x*-position? (*c*) What total length of path does the particle cover in the first 4.0 sec? (*d*) What is its displacement during the first 4.0 sec? (*e*) What is the particle's speed at the end of each of the first four seconds? (*f*) What is the particle's acceleration at the end of each of the first four seconds?

29(10). (*a*) With what speed must a ball be thrown vertically up in order to rise to a height of 50 ft? (*b*) How long will it be in the air?

Answer: (*a*) 57 ft/sec. (*b*) 3.6 sec.

30(10). A ball is thrown down vertically with an initial speed of 20 ft/sec from a height of 60 ft. (*a*) What will be its speed just before it strikes the ground? (*b*) How long will it take for the ball to reach the ground? (*c*) What would be the answers to (*a*) and (*b*) if the ball were thrown directly up from the same height and with the same initial speed?

31(10). A tennis ball is dropped onto the floor from a height of 4.0 ft. It rebounds to a height of 3.0 ft. If the ball was in contact with the floor for 0.010 sec, what was its average acceleration during contact?

Answer: 3000 ft/sec², up.

32(10). A ball of clay falls to the ground from a height *h* (49 ft). It is in contact with the ground for a time *t* (0.20 sec) before coming to rest. What is the average acceleration of the clay during the time it is in contact with the ground?

33(10). A lead ball is dropped into a lake from a diving board 16 ft above the water. It hits the water with a certain velocity and then sinks to the bottom with this same constant velocity. It reaches the bottom 5.0 sec after it is dropped. (*a*) How deep is the lake? (*b*) What is the average velocity of the ball? (*c*) Suppose all the water is drained from the lake. The ball is thrown from the diving board so that it again reaches the bottom in 5.0 sec. What is the initial velocity of the ball? *Answer:* (*a*) 130 ft. (*b*) 29 ft/sec. (*c*) 51 ft/sec, up.

34(10). A rocket is fired vertically and ascends with a constant vertical acceleration of 64 ft/sec² for 1.0 min. Its fuel is then all used and it continues as a free particle. (*a*) What is the maximum altitude reached? (*b*) What is the total time elapsed from take-off until the rocket strikes the earth?

35(10). A balloon is ascending at the rate of 12 meters/sec at a height 80 meters above the ground when a package is dropped. How long does it take the package to reach the ground? *Answer:* 5.5 sec.

36(10). An object falls from a bridge that is 144 ft above the water. It falls directly into a boat moving with constant velocity that was 40 ft from the point of impact when the object was released. What was the speed of the boat?

37(10). A stone is dropped into the water from a bridge 144 ft above the water. Another stone is thrown vertically down 1.0 sec after the first is dropped. Both stones strike the water at the same time. (*a*) What was the initial speed of the second stone? (*b*) Plot speed versus time on a graph for each stone, taking zero time as the instant the first stone was released. *Answer:* (*a*) 40 ft/sec.

38(10). An open elevator is ascending with a constant speed *v* (32 ft/sec). A ball is thrown straight up by a boy on the elevator when it is a height *h* (100 ft) above the ground. The initial speed of the ball with respect to the elevator is 64 ft/sec. (*a*) What is the maximum height attained by the ball? (*b*) How long does it take for the ball to return to the elevator?

39(10). An arrow is shot straight up in the air with an initial speed of 250 ft/sec. If on striking the ground it imbeds itself 6.0 in. into the ground, find (*a*) the acceleration (assumed constant) required to stop the arrow and (*b*) the time required for it to come to rest. Neglect air resistance during the arrow's flight. *Answer:* (*a*) -6.3×10^4 ft/sec². (*b*) 4.0×10^{-3} sec.

40(10). A parachutist after bailing out falls 50 meters without friction. When the parachute opens, he decelerates downward 2.0 meters/sec². He reaches the ground with a speed of 3.0 meters/sec. (*a*) How long is the parachutist in the air? (*b*) At what height did he bail out?

41(10). A shell is fired directly up from a gun; a rocket, propelled by burning fuel, takes off vertically from a launching area. Plot qualitatively (numbers nor required) possible graphs of a_y versus *t*, of v_y versus *t*, and of *y* versus *t* for each. Take $t = 0$ at the instant the shell leaves the gun barrel or the rocket leaves the ground. Continue the plots until the rocket and the shell fall back to earth; neglect air resistance; assume that up is positive and down is negative.

42(10). If a body travels half its total path in the last second of its fall from rest, find (*a*) the time and (*b*) height of its fall. (*c*) Explain the physically unacceptable solution of the quadratic time equation.

43(10). A steel ball bearing is dropped from the roof of a building (the initial velocity of the ball is zero). An observer standing in front of a window 4.0 ft high notes that the ball takes ⅛ sec to fall from the top to the bottom of the window. The ball bearing continues to fall, makes a completely elastic collision with a horizontal sidewalk, and reappears at the bottom of the window 2.0 sec after passing it on the way down. How tall is the building? (The ball will have the same speed at a point going up as it had going down after a completely elastic collision.) *Answer:* 68 ft.

44(10). Water drips from the nozzle of a shower onto the floor 81 in. below. The drops fall at regular intervals of time, the first drop striking the floor at the instant the fourth drop begins to fall. Find the location of the individual drops when a drop strikes the floor.

45(10). An elevator ascends with an upward acceleration of 4.0 ft/sec². At the instant its upward speed is 8.0 ft/sec, a loose bolt drops from the ceiling of the elevator 9.0 ft from the floor. Calculate (*a*) the time of flight of the bolt from ceiling to floor and (*b*) the distance it has fallen relative to the elevator shaft. *Answer:* (*a*) 0.71 sec. (*b*) 2.3 ft.

46(10). A dog sees a flowerpot sail up and then back past a window 5.0 ft high. If the total time the pot is in sight is 1.0 sec, find the height above the window that the pot rises.

CHAPTER 4

Motion in a Plane

4–1 Displacement, Velocity, and Acceleration

In this chapter we return to a consideration of motion in two dimensions taken, for convenience, to be the x-y plane. Figure 4–1 shows a particle at time t moving along a curved path in this plane. Its position, or displacement from the origin, is measured by the vector **r**; its velocity is indicated by the vector **v** which, as we have seen in Section 3–4, must be tangent to the path of the particle. The acceleration is indicated by the vector **a**; the direction of **a**, as we shall see more explicitly later, does not bear any unique relationship to the path of the particle but depends rather on the rate at which the velocity **v** changes with time as the particle moves along its path.

The vectors **r**, **v**, and **a** are interrelated (see Eqs. 3–4, 3–5, and 3–10) and can be expressed in terms of their components, using unit vector notation, as

$$\mathbf{r} = \mathbf{i}x + \mathbf{j}y, \tag{4-1}$$

$$\mathbf{v} = \frac{d\mathbf{r}}{dt} = \mathbf{i}v_x + \mathbf{j}v_y, \tag{4-2}$$

and

$$\mathbf{a} = \frac{d\mathbf{v}}{dt} = \mathbf{i}a_x + \mathbf{j}a_y. \tag{4-3}$$

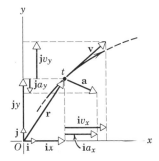

FIGURE 4–1   A particle moves along a curved path in the x-y plane. Its position **r**, velocity **v**, and acceleration **a** are shown at time t, along with the vector components of those vectors. Note that $x, y, v_x, v_y,$ and a_x are positive but that a_y is negative. Compare to Fig. 3–2.

43

You can easily extend these equations to three dimensions by adding to them the terms $\mathbf{k}z$, $\mathbf{k}v_z$, and $\mathbf{k}a_z$, respectively, in which $\mathbf{k}$ is a unit vector in the z-direction.

In Chapter 3 we considered the special case in which the particle moved in one direction only, say along the x-axis, where the vectors $\mathbf{r}$, $\mathbf{v}$, and $\mathbf{a}$ were directed along this axis, either in the positive x-direction or the negative x-direction. The components y, v_y, and a_y were zero and we described the motion in terms of equations relating the scalar quantities x, v_x, and a_x. Or, when the particle moved along the y-axis, the components x, v_x, and a_x were zero and the motion was described in terms of equations relating the scalar quantities y, v_y, and a_y. In this chapter we consider motion in the x-y plane so that, in general, both sets of components have nonzero values.

4–2   Motion in a Plane with Constant Acceleration

Let us consider first the special case of motion in a plane with constant acceleration. Here, as the particle moves, the acceleration $\mathbf{a}$ does not vary either in magnitude or in direction. Hence the components of $\mathbf{a}$ in any particular reference frame also will not vary, that is, $a_x = $ constant and $a_y = $ constant. We then have a situation which can be described as the sum of two component motions occurring simultaneously with constant acceleration along each of two perpendicular directions. The particle will move, in general, along a curved path in the plane. This may be so even if one component of the acceleration, say a_x, is zero, for then the corresponding component of the velocity, say v_x, may have a constant, nonzero value. An example of this latter situation is the motion of an artillery shell which follows a curved path in a vertical plane and, neglecting the effects of air resistance, is subject to a constant acceleration $\mathbf{g}$ directed down along the y-axis only.

We can obtain the general equations for plane motion with constant $\mathbf{a}$ simply by setting

$$a_x = \text{constant} \quad \text{and} \quad a_y = \text{constant}.$$

The equations for constant acceleration, summarized in Table 3–1, then apply to both the x- and y-components of the position vector $\mathbf{r}$, the velocity vector $\mathbf{v}$, and the acceleration vector $\mathbf{a}$ (see Table 4-1).

The two sets of equations in Table 4–1 are related in that the time parameter t is the same for each, since t represents the time at which the particle, moving in a curved path in the x-y plane, occupies a position described by the position components x and y.

The equations of motion in Table 4–1 may also be expressed in vector form. For example, substituting Eqs. 4–4a, 4a′ into Eq. 4–2 yields

$$\mathbf{v} = \mathbf{i}v_x + \mathbf{j}v_y$$

$$= \mathbf{i}(v_{x0} + a_xt) + \mathbf{j}(v_{y0} + a_yt)$$

$$= (\mathbf{i}v_{x0} + \mathbf{j}v_{y0}) + (\mathbf{i}a_x + \mathbf{j}a_y)t.$$

Table 4–1   MOTION WITH CONSTANT ACCELERATION IN THE x-y PLANE

Equation No.	x-Motion Equations	Equation No.	y-Motion Equations
4–4a	$v_x = v_{x0} + a_xt$	4–4a′	$v_y = v_{y0} + a_yt$
4–4b	$x = x_0 + \frac{1}{2}(v_{x0} + v_x)t$	4–4b′	$y = y_0 + \frac{1}{2}(v_{y0} + v_y)t$
4–4c	$x = x_0 + v_{x0}t + \frac{1}{2}a_xt^2$	4–4c′	$y = y_0 + v_{y0}t + \frac{1}{2}a_yt^2$
4–4d	$v_x^2 = v_{x0}^2 + 2a_x(x - x_0)$	4–4d′	$v_y^2 = v_{y0}^2 + 2a_y(y - y_0)$

The first quantity in parentheses is the initial velocity vector $\mathbf{v}_0$ (see Eq. 4–2) and the second is the (constant) acceleration vector $\mathbf{a}$ (see Eq. 4–3). Thus the vector relation

$$\mathbf{v} = \mathbf{v}_0 + \mathbf{a}t \tag{4–5a}$$

is equivalent to the scalar relations Eqs. 4–4a, a′ in Table 4–1. It shows simply and compactly that the velocity $\mathbf{v}$ at time t is the sum of the initial velocity $\mathbf{v}_0$ which the particle would have in the absence of acceleration plus the (vector) change in velocity, $\mathbf{a}t$, acquired during the time t under the constant acceleration $\mathbf{a}$. Similarly, the scalar equations 4–4c, c′ are equivalent to the single vector equation

$$\mathbf{r} = \mathbf{r}_0 + \mathbf{v}_0 t + \tfrac{1}{2}\mathbf{a}t^2, \tag{4–5b}$$

which is also easily interpreted.

4–3 Projectile Motion

An example of curved motion with constant acceleration is projectile motion. This is the two-dimensional motion of a particle thrown obliquely into the air. The ideal motion of a baseball, a golf ball, or a bullet is an example of projectile motion.* We assume that the effect the air itself would have on their motions can be neglected.

The motion of a projectile is one of constant acceleration $\mathbf{g}$, directed downward, and thus should be described by the equations in Table 4–1. There is no horizontal component of acceleration. If we choose a reference frame with the positive y-axis vertically upward, we may put $a_y = -g$ and $a_x = 0$ in these equations.

Let us further choose the origin of our reference frame to be the point at which the projectile begins its flight (see Fig. 4–2). Hence the origin will be the point at which the ball leaves the thrower's hand or the fuel in the rocket burns out, for example. In Table 4–1 this choice of origin implies that $x_0 = y_0 = 0$. The velocity at $t = 0$, the instant the projectile begins its flight, is $\mathbf{v}_0$, which makes an angle θ_0 with the positive x-direction. The x- and y-components of $\mathbf{v}_0$ (see Fig. 4–2) are then

$$v_{x0} = v_0 \cos \theta_0 \qquad \text{and} \qquad v_{y0} = v_0 \sin \theta_0.$$

Because there is no horizontal component of acceleration, the horizontal component of the velocity will be constant. In Eq. 4–4a we set $a_x = 0$ and $v_{x0} = v_0 \cos \theta_0$, so that

$$v_x = v_0 \cos \theta_0. \tag{4–6a}$$

The horizontal velocity component retains its initial value throughout the flight.

The vertical component of the velocity will change with time in accordance with vertical motion with constant downward acceleration. In Eq. 4–4a′ we set

$$a_y = -g \qquad \text{and} \qquad v_{y0} = v_0 \sin \theta_0,$$

so that

$$v_y = v_0 \sin \theta_0 - gt. \tag{4–6a′}$$

The vertical velocity component is that of free fall. Indeed, if we view the motion of Fig. 4–2 from a reference frame that moves to the right with a speed v_{x0}, the motion will be that of an object thrown vertically upward with an initial speed $v_0 \sin \theta_0$.

---

* See Galileo Galilei, *Dialogues Concerning Two New Sciences*, the "Fourth Day," for a fascinating discussion of Galileo's research on projectiles.

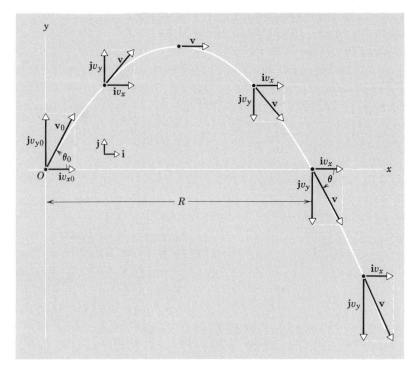

FIGURE 4–2 The trajectory of a projectile, showing the initial velocity **v₀** and its vector components and also the velocity **v** and its vector components at five later times. Note that $v_x = v_{x0}$ throughout the flight. The distance R is the horizontal range.

The magnitude of the resultant velocity vector at any instant is

$$v = \sqrt{v_x{}^2 + v_y{}^2}. \qquad (4\text{–}7a)$$

The angle θ that the velocity vector makes with the horizontal at that instant is given by

$$\tan \theta = \frac{v_y}{v_x}. \qquad (4\text{–}7b)$$

The velocity vector is tangent to the path of the particle at every point, as shown in Fig. 4–2.

The x-coordinate of the particle's position at any time, obtained from Eq. 4–4c with $x_0 = 0$, $a_x = 0$, and $v_{x0} = v_0 \cos \theta_0$, is

$$x = (v_0 \cos \theta_0)t. \qquad (4\text{–}6c)$$

The y-coordinate, obtained from Eq. 4–4c′ with $y_0 = 0$, $a_y = -g$, and $v_{y0} \sin \theta_0$, is

$$y = (v_0 \sin \theta_0)t - \tfrac{1}{2}gt^2. \qquad (4\text{–}6c')$$

Equations 4–6c, c′ give us x and y as functions of the common parameter t, the time in flight. By combining and eliminating t from them, we obtain

$$y = (\tan \theta_0)x - \frac{g}{2(v_0 \cos \theta_0)^2}\, x^2, \qquad (4\text{–}8)$$

which relates y to x and is the equation of the trajectory of the projectile. Since v_0, θ_0, and g are constants, this equation has the form

$$y = bx - cx^2,$$

the equation of a parabola. Hence the trajectory of a projectile is parabolic.

Example 1. A bomber is flying at a constant horizontal velocity of 820 miles/hr at an elevation of 52,000 ft toward a point directly above its target. At what angle of sight ϕ should a bomb be released to strike the target (Fig. 4–3)? Neglect air resistance.

We choose a reference frame fixed with respect to the earth, its origin O being the bomb release point. The motion of the bomb at the instant of release is the same as that of the bomber. Hence the initial projectile velocity $\mathbf{v}_0$ is horizontal and its magnitude is 820 miles/hr or 1200 ft/sec. The angle of projection θ_0 is zero.

The time of fall is obtained from Eq. 4–6c'. With $\theta_0 = 0$ and $y = -52,000$ ft, this gives

$$t = \sqrt{-\frac{2y}{g}} = \sqrt{-\frac{2(-52,000)\text{ft}}{32 \text{ ft/sec}^2}} = 57 \text{ sec}.$$

Note that the time of fall of the bomb does not depend on the speed of the plane for a horizontal projection.

The horizontal distance traveled by the bomb in this time is given by Eq. 4–6c, $x = (v_0 \cos \theta_0)t$, or $x = (1200 \text{ ft/sec})(57 \text{ sec}) = 68,000$ ft so that the angle of sight (Fig. 4–3) should be

$$\phi = \tan^{-1} \frac{x}{|y|} = \tan^{-1} \frac{68,000}{52,000} = 53°.$$

Does the motion of the bomb appear to be parabolic when viewed from a reference frame fixed with respect to the bomber?

Example 2. A soccer player kicks a ball at an angle of 37° from the horizontal with an initial speed of 50 ft/sec. (A right triangle, one of whose angles is 37°, has sides in the ratio 3:4:5, or 6:8:10.) Assuming that the ball moves in a vertical plane:

(*a*) Find the time t_1 at which the ball reaches the highest point of its trajectory. At the highest point, the vertical component of velocity v_y is zero. Solving Eq. 4–6a' for t, we obtain

$$t = \frac{v_0 \sin \theta_0 - v_y}{g}.$$

With

$$v_y = 0, \qquad v_0 = 50 \text{ ft/sec}, \qquad \theta_0 = 37°, \qquad g = 32 \text{ ft/sec}^2,$$

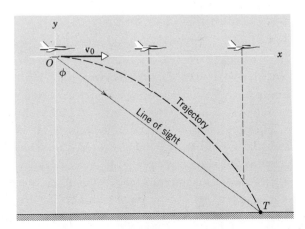

FIGURE 4–3   Example 1. A bomb is released from an airplane with horizontal velocity $\mathbf{v}_0$.

we have

$$t_1 = \frac{[50(\frac{6}{10}) - 0] \text{ ft/sec}}{32 \text{ ft/sec}^2} = \frac{15}{16} \text{ sec.}$$

(*b*) How high does the ball go? The maximum height is reached at $t = \frac{15}{16}$ sec. By using Eq. 4–6*c*′,

$$y = (v_0 \sin \theta_0)t - \tfrac{1}{2}gt^2,$$

we have

$$y_{\text{max}} = (50 \text{ ft/sec})(\tfrac{6}{10})(\tfrac{15}{16} \text{ sec}) - \tfrac{1}{2}(32 \text{ ft/sec}^2)(\tfrac{15}{16})^2 \text{ sec}^2 = 14 \text{ ft.}$$

(*c*) What is the horizontal range of the ball and how long is it in the air?

The horizontal distance from the starting point at which the ball returns to its original elevation (ground level) is the *range* R. We set $y = 0$ in Eq. 4–6*c*′ and find the time t_2 required to transverse this range. We obtain

$$t_2 = \frac{2v_0 \sin \theta_0}{g} = \frac{2(50 \text{ ft/sec})(\tfrac{6}{10})}{32 \text{ ft/sec}^2} = \frac{15}{8} \text{ sec.}$$

Notice that $t_2 = 2t_1$. This corresponds to the fact that the same time is required for the ball to go up (reach its maximum height from ground) as is required for the ball to come down (reach the ground from its maximum height).

The range R can be obtained by inserting this value t_2 for t in Eq. 4–6*c*. We obtain, from $x = (v_0 \cos \theta_0)t$,

$$R = (v_0 \cos \theta_0)t_2 = (50 \text{ ft/sec})(\tfrac{8}{10})(\tfrac{15}{8} \text{ sec}) = 75 \text{ ft.}$$

(*d*) What is the velocity of the ball as it strikes the ground? From Eq. 4–6*a* we obtain

$$v_x = v_0 \cos \theta_0 = (50 \text{ ft/sec})(\tfrac{8}{10}) = 40 \text{ ft/sec.}$$

From Eq. 4–6*a*′ we obtain for $t = t_2 = \frac{15}{8}$ sec,

$$v_y = v_0 \sin \theta_0 - gt = (50 \text{ ft/sec})(\tfrac{6}{10}) - (32 \text{ ft/sec}^2)(\tfrac{15}{8} \text{ sec}) = -30 \text{ ft/sec.}$$

Hence, from Eqs. 4–7,

$$v = \sqrt{v_x{}^2 + v_y{}^2} = \sqrt{(40 \text{ ft/sec})^2 + (-30 \text{ ft/sec})^2} = 50 \text{ ft/sec,}$$

and

$$\tan \theta = v_y/v_x = -\tfrac{30}{40},$$

so that $\theta = -37°$, or 37° clockwise from the *x*-axis. Notice that $\theta = -\theta_0$, as we expect from symmetry (Fig. 4–2).

4–4 Uniform Circular Motion

In Section 3–6 we saw that acceleration arises from a change in velocity. In the simple case of free fall the velocity changed in magnitude only, but not in direction. In the present case, called uniform circular motion, the particle moves in a circle with constant speed; the velocity vector changes continuously in direction but not in magnitude. We seek now to obtain the acceleration in uniform circular motion.

The situation is shown in Fig. 4–4*a*. Let P be the position of the particle at the time t and P' its position at the time $t + \Delta t$. The velocity at P is **v**, a vector tangent to the curve at P. The velocity at P' is **v′**, a vector tangent to the curve at P'. Vectors **v** and **v′** are equal in magnitude, the speed being constant, but their directions are different. The length of path traversed during Δt is the arc length PP', which is equal to $v \Delta t$, v being the constant speed.

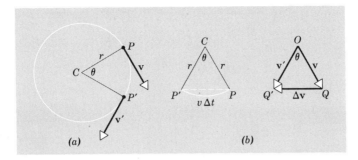

FIGURE 4–4 Uniform circular motion. The particle travels around a circle at constant speed. Its velocity at two points P and P' is shown. Its change in velocity in going from P to P' is $\Delta \mathbf{v}$.

Now redraw the vectors $\mathbf{v}$ and $\mathbf{v}'$, as in Fig. 4–4b, so that they originate at a common point. We are free to do this as long as the magnitude and direction of each vector are the same as in Fig. 4–4a. This diagram (Fig. 4–4b) enables us to see clearly the change in velocity as the particle moved from P to P'. This change, $\mathbf{v}' - \mathbf{v} = \Delta \mathbf{v}$, is the vector which must be added to $\mathbf{v}$ to get $\mathbf{v}'$. Notice that it points inward, approximately toward the center of the circle.

Now the triangle OQQ' formed by $\mathbf{v}$, $\mathbf{v}'$, and $\Delta \mathbf{v}$ is similar to the triangle CPP' formed by the chord PP' and the radii CP and CP'. This is so because both are isosceles triangles having the same vertex angle; the angle θ between $\mathbf{v}$ and $\mathbf{v}'$ is the same as the angle PCP' because $\mathbf{v}$ is perpendicular to CP and $\mathbf{v}'$ is perpendicular to CP'. We can therefore write

$$\frac{\Delta v}{v} = \frac{v\,\Delta t}{r}, \qquad \text{approximately,}$$

the chord PP' being taken equal to the arc length PP'. This relation becomes more nearly exact as Δt is diminished, since the chord and the arc then approach each other. Notice also that $\Delta \mathbf{v}$ approaches closer and closer to a direction perpendicular to $\mathbf{v}$ and $\mathbf{v}'$ as Δt is diminished and therefore approaches closer and closer to a direction pointing to the exact center of the circle. It follows from this relation that

$$\frac{\Delta v}{\Delta t} = \frac{v^2}{r}, \qquad \text{approximately,}$$

and in the limit when $\Delta t \to 0$ this expression becomes exact. We therefore obtain

$$a = \lim_{\Delta t \to 0} \frac{\Delta v}{\Delta t} = \frac{v^2}{r} \qquad (4\text{–}9)$$

as the magnitude of the acceleration. The direction of $\mathbf{a}$ is instantaneously along a radius inward toward the center of the circle.

Figure 4–5 shows the instantaneous relation between $\mathbf{v}$ and $\mathbf{a}$ at various points of the motion. The magnitude of $\mathbf{v}$ is constant, but its direction changes continuously. This gives rise to an acceleration $\mathbf{a}$ which is also constant in magnitude (but not zero) but continuously changing in direction. The velocity $\mathbf{v}$ is always tangent to the circle in the direction of motion; the acceleration $\mathbf{a}$ is always directed radially inward. Because of this, $\mathbf{a}$ is called a *radial,* or *centripetal,* acceleration. *Centripetal* means "seeking a center."

Both in free fall and in projectile motion $\mathbf{a}$ is constant in direction and magnitude

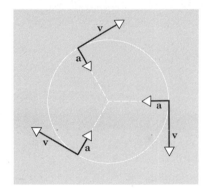

FIGURE 4–5 In uniform circular motion the acceleration **a** is always directed toward the center of the circle and hence is perpendicular to **v**.

and we can use the equations developed for constant acceleration (see Table 4–1). We cannot use these equations for uniform circular motion because in the latter case **a** varies in direction and is therefore not constant.

The units of acceleration must be the same no matter how it arises. The dimensions of centripetal acceleration are

$$\frac{v^2}{r} = \left(\frac{\text{length}}{\text{time}}\right)^2 \bigg/ \text{length} = \frac{\text{length}}{\text{time}^2} \quad \text{or} \quad \frac{L}{T^2},$$

which are the dimensions of acceleration previously encountered. The units therefore may be ft/sec², meters/sec², among others.

The acceleration resulting from a change in direction of a velocity is just as real and just as much an acceleration in every sense as that arising from a change in magnitude of a velocity. By definition, acceleration is the time rate of change of velocity, and velocity, being a vector, can change in direction as well as magnitude. If a physical quantity is a vector, its directional aspects cannot be ignored, for their effects will prove to be every bit as important and real as those produced by changes in magnitude.

It is worth emphasizing at this point that there need not be any motion in the direction of an acceleration and that there is no fixed relation in general between the directions of **a** and **v**. In Fig. 4–6 we give examples in which the angle between **v** and **a** varies from 0 to 180°. Only in one case, $\theta = 0°$, is the motion in the direction of **a**.

▨ **Example 3.** The moon revolves about the earth, making a complete revolution in 27.3 days. Assume that the orbit is circular and has a radius of 239,000 miles. What is the magnitude of the acceleration of the moon toward the earth?

We have $r = 239{,}000$ miles $= 3.85 \times 10^8$ meters. The time for one complete revolution, called the period, is $T = 27.3$ days $= 2.36 \times 10^6$ sec. The speed of the moon (assumed constant) is

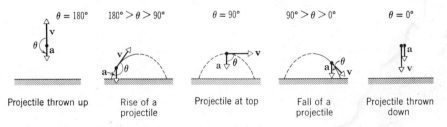

$\theta = 180°$	$180° > \theta > 90°$	$\theta = 90°$	$90° > \theta > 0°$	$\theta = 0°$
Projectile thrown up	Rise of a projectile	Projectile at top	Fall of a projectile	Projectile thrown down

FIGURE 4–6   Showing the relation between **v** and **a** for various motions.

therefore

$$v = 2\pi r/T = 1020 \text{ meters/sec.}$$

The centripetal acceleration is

$$a = \frac{v^2}{r} = \frac{(1020 \text{ meters/sec})^2}{3.85 \times 10^8 \text{ meters}} = 0.00273 \text{ meter/sec}^2, \quad \text{or only } 2.78 \times 10^{-4}g.$$

Example 4. Calculate the speed of an artificial earth satellite, assuming that it is traveling at an altitude h of 140 miles above the surface of the earth where $g = 30$ ft/sec^2. The radius of the earth R is 3960 miles.

Like any free object near the earth's surface the satellite has an acceleration g toward the earth's center. It is this acceleration that causes it to follow the circular path. Hence the centripetal acceleration is g, and from Eq. 4–9, $a = v^2/r$, we have

$$g = v^2/(R + h),$$

or

$$v = \sqrt{(R + h)g} = \sqrt{(3960 \text{ miles} + 140 \text{ miles})(5280 \text{ ft/mile})(30 \text{ ft/sec}^2)}$$

$$= 2.55 \times 10^4 \text{ ft/sec} = 17,400 \text{ miles/hr.}$$

4–5 Relative Velocity and Acceleration

In earlier sections we considered the addition of velocities in a particular reference frame. Let us now consider the relation between the velocity of an object as determined by one observer S (= reference frame S) and the velocity of the same object as determined by another observer S' (= reference frame S') who is moving with respect to the first.

Consider observer S fixed to the earth, so that his reference frame is the earth. The other observer S' is moving relative to the earth—for example, a passenger sitting on a moving train—so that his reference frame is the train. They each follow the motion of the same object, say a ball moving on a flatcar. Each observer will record a displacement, a velocity, and an acceleration for this object measured relative to his reference frame. How will these measurements compare? We will consider the case in which the second frame is in motion with respect to the first with a constant velocity **u**.

In Fig. 4–7 the reference frame S represented by the x- and y-axes can be thought of as fixed to the earth. The shaded region indicates the other reference frame S', represented by x'- and y'-axes, which moves along the x-axis with a constant velocity **u**, as measured in the S-system; it can be thought of as drawn on the floor of the flatcar.

Initially, the ball is at a position called A in the S-frame and called A' in the S'-frame. At a time t later the flatcar and its S' reference frame have moved a distance ut to the right and the particle has moved to B. The displacement of the particle from its initial position in the S-frame is the vector **r** from A to B. The displacement of the particle from its initial position in the S'-frame is the vector **r'** from A' to B. These are different vectors because the reference point A' of the moving frame has been displaced a distance ut along the x-axis during the motion. From the figure we see that **r** is the vector sum of **r'** and **u**t:

$$\mathbf{r} = \mathbf{r'} + \mathbf{u}t. \tag{4–10}$$

Differentiating Eq. 4–10 leads to

$$\frac{d\mathbf{r}}{dt} = \frac{d\mathbf{r'}}{dt} + \mathbf{u}.$$

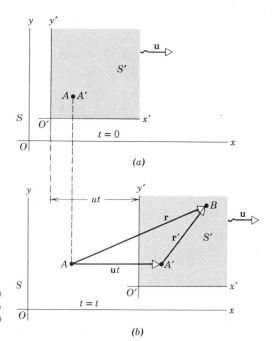

FIGURE 4–7 Two reference frames, S ($= x, y$) and S' ($= x', y'$); S' moves to the right, relative to S, with speed u. (*a*) The situation at time $t = 0$ (*b*) The situation at a later time.

But $d\mathbf{r}/dt = \mathbf{v}$, the instantaneous velocity of the particle measured in the S-frame, and $d\mathbf{r}'/dt = \mathbf{v}'$, the instantaneous velocity of the same particle measured in the S' frame, so that

$$\mathbf{v} = \mathbf{v}' + \mathbf{u}. \qquad (4\text{-}11)$$

Hence the velocity of the particle relative to the S-frame, $\mathbf{v}$, is the vector sum of the velocity of the particle relative to the S'-frame, $\mathbf{v}'$, and the velocity $\mathbf{u}$ of the S'-frame relative to the S-frame.

■ **Example 5.** (*a*) The compass of an airplane indicates that it is heading due east. Ground information indicates a wind blowing due north. Show on a diagram the velocity of the plane with respect to the ground.

The object is the airplane. The earth is one reference frame (S) and the air is the other reference frame (S') moving with respect to the first. Then

$\mathbf{u}$ is the velocity of the air with respect to the ground.
$\mathbf{v}'$ is the velocity of the plane with respect to the air.
$\mathbf{v}$ is the velocity of the plane with respect to the ground.

In this case $\mathbf{u}$ points north and $\mathbf{v}'$ points east. Then the relation $\mathbf{v} = \mathbf{v}' + \mathbf{u}$ determines the velocity of the plane with respect to the ground, as shown in Fig. 4–8*a*.

The angle α is the angle N of E of the plane's course with respect to the ground and is given by

$$\tan \alpha = u/v'.$$

The airplane's speed with respect to the ground is given by

$$v = \sqrt{(v')^2 + u^2}.$$

For example, if the air-speed indicator shows that the plane is moving relative to the air at a speed of 200 miles/hr, and if the speed of the wind with respect to the ground is 40.0 miles/hr, then

$$v = \sqrt{(200)^2 + (40.0)^2} \text{ miles/hr} = 204 \text{ miles/hr}$$

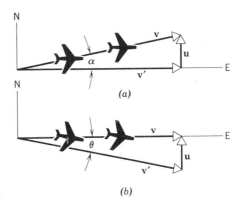

FIGURE 4–8   Example 5.

(a)

(b)

is the ground speed of the plane and

$$\alpha = \tan^{-1} \frac{40.0}{200} = 11°20'$$

gives the course of the plane N of E.

(b) Now draw the vector diagram showing the direction the pilot must steer the plane through the air for the plane to travel due east with respect to the ground.

He would naturally head partly into the wind. His speed relative to the earth will therefore be less than before. The vector diagram is shown in Fig. 4–8b. You should calculate θ and v, using the previous data for **u** and **v′**.

We have seen that different velocities are assigned to a particle by different observers when the observers are in relative motion. These velocities always differ by the relative velocity of the two observers, which here is a constant velocity. It follows that when the particle velocity changes, the change will be the same for both observers. Hence they each measure the same acceleration for the particle. The acceleration of a particle is the same in all reference frames moving relative to one another with constant velocity; that is, **a** = **a′**. This result follows in a formal way if we differentiate Eq. 4–11. Thus $d\mathbf{v}/dt = d\mathbf{v'}/dt + d\mathbf{u}/dt$; but $d\mathbf{u}/dt = 0$ when **u** is constant, so that **a** = **a′**.

QUESTIONS

1. In projectile motion when air resistance is negligible, is it ever necessary to consider three-dimensional motion rather than two-dimensional?

2. In broad jumping does it matter how high you jump? What factors determine the span of the jump?

3. Why doesn't the electron in the beam from an electron gun fall as much because of gravity as a water molecule in the stream from a hose? Assume horizontal motion initially in each case.

4. An aviator, pulling out of a dive, follows the arc of a circle. He was said to have "experienced 3g's" in pulling out of the dive. Explain what this statement means.

5. Describe qualitatively the acceleration acting on a bead which moves inward with constant speed along a spiral.

6. A boy sitting in a railroad car moving at constant velocity throws a ball straight up into the air. Will the ball fall behind him? In front of him? Into his hand? What happens if the car accelerates forward or goes around a curve while the ball is in the air?

7. A man on the observation platform of a train moving with constant velocity drops a coin while leaning over the rail. Describe the path of the coin as seen by (*a*) the man on the train, (*b*) a person standing on the ground near the track, and (*c*) a person in a second train moving in the opposite direction to the first train on a parallel track.

8. A bus with a vertical windshield moves along in a rainstorm at speed v_b. The raindrops fall vertically with a terminal speed v_r. At what angle do the raindrops strike the windshield?

9. Drops are falling vertically in a steady rain. In order to go through the rain from one place to another in such a way as to encounter the least number of raindrops, should you move with the greatest possible speed, the least possible speed, or some intermediate speed?

10. An elevator is descending at a constant speed. A passenger takes a coin from his pocket and drops it to the floor. What accelerations would (*a*) the passenger and (*b*) a person at rest with respect to the elevator shaft observe for the falling coin?

PROBLEMS

1(1). A plane flies 300 miles east from city A to city B in 45 min. and then 600 miles south from city B to city C in 1.5 hr. (*a*) What are the magnitude and direction of the displacement vector which represents the total trip? What are (*b*) the average velocity vector, and (*c*) the average speed for the trip?
Answer: (*a*) 670 miles, 63° south of east. (*b*) 300 miles/hr, 63° south of east. (*c*) 400 miles/hr.

2(1). At time t_0 the velocity of an object is given, in ft/sec, by $\mathbf{v}_0 = 125\mathbf{i} + 25\mathbf{j}$. At 3.0 sec later the velocity is $\mathbf{v} = 100\mathbf{i} - 75\mathbf{j}$. What was the average acceleration of the object during this time interval?

3(1). If the coordinates of a particle moving in a plane are given by $x = 3.0t - 4.0t^2$ and $y = -6.0t^2 + t^3$ where x and y are in meters and t in seconds, find (*a*) the coordinates and displacement vector at $t = 3.0$ sec, (*b*) the average velocity during the first 3.0 sec, (*c*) the instantaneous velocity at $t = 3.0$ sec, (*d*) the average acceleration during the first three seconds and (*e*) the instantaneous acceleration at $t = 3.0$ sec.
Answer: (*a*) $x = -27$ meters, $y = -27$ meters, $\mathbf{r} = 38$ meters at 230°. (*b*) 13 meters/sec at 230°. (*c*) 23 meters/sec at 200°. (*d*) 8.5 meters/sec² at 200°. (*e*) 10 meters/sec² at 140°.

4(1). Prove that for a vector **a** defined by

$$\mathbf{a} = \mathbf{i}a_x + \mathbf{j}a_y + \mathbf{k}a_z$$

the scaler components are given by

$$a_x = \mathbf{i} \cdot \mathbf{a}, \qquad a_y = \mathbf{j} \cdot \mathbf{a}, \qquad \text{and} \qquad a_z = \mathbf{k} \cdot \mathbf{a}.$$

5(2). A particle moves so that its position as a function of time is

$$\mathbf{r}(t) = \mathbf{i} + 4t^2\mathbf{j} + t\mathbf{k}.$$

(*a*) Write expressions for its velocity and acceleration as functions of time. (*b*) What is the shape of the particle's trajectory?
Answer: (*a*) $\mathbf{v} = 8t\mathbf{j} + \mathbf{k}$; $\mathbf{a} = 8\mathbf{j}$. (*b*) A parabola.

6(2). A particle leaves the origin with an initial velocity **v** (3.0 meters/sec) in the direction of the positive x-axis. It experiences a constant acceleration $\mathbf{a} = -1.0\mathbf{i} - 0.5\mathbf{j}$, in meters/sec². (*a*) What is the velocity of the particle when it reaches its maximum x-coordinate? (*b*) Where is the particle at this time?

7(2). A particle *A* moves along the line $y = d$ (30 meters) with a constant velocity **v** ($v = 3.0$ meters/sec) directed parallel to the positive x-axis (Fig. 4–9). A second particle *B* starts at the origin with zero speed and constant acceleration **a** ($a = 0.40$ meters/sec²) at the same i stant that particle *A* passes the y-axis. What angle θ between **a** and the positive y-axis would result in a collision between these two particles?
Answer: 60°.

8(3). A ball rolls off the edge of a horizontal table top 4.0 ft high. If it strikes the floor at a point 5.0 ft horizontally away from the edge of the table, what was its speed at the instant it left the table?

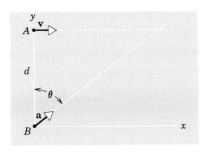

FIGURE 4-9 Problem 7(2).

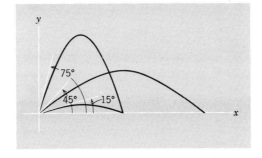

FIGURE 4-10 Problems 12(3) and 26(3).

9(3). A rifle is aimed horizontally at a target 100 ft away. The bullet hits the target 3.0 in. below the aiming point. What is the muzzle velocity of the rifle?
Answer: 800 ft/sec.

10(3). A ball is thrown horizontally from a height of 50 ft and hits the ground with a speed that is three times its initial speed. What was the initial speed?

11(3). A rifle with a muzzle velocity of 1500 ft/sec shoots a bullet at a target 150 ft away. How high above the target must the rifle be aimed so that the bullet will hit the target?
Answer: 1.9 in.

12(3). Show that the horizontal range of a projectile having an initial speed v_0 and angle of projection θ_0 is $R = (v_0^2/g) \sin 2\theta_0$. Then show that a projection angle of 45° gives the maximum horizontal range (Fig. 4-10).

13(3). Show that the maximum height reached by a projectile is $y_{max} = (v_0 \sin \theta_0)^2/2g$.

14(3). Find the angle of projection at which the horizontal range and the maximum height of a projectile are equal.

15(3). A certain airplane has a speed of 180 miles/hr and is diving at an angle of 30° below the horizontal when an object is dropped. The horizontal distance between the release point and the point where the object first strikes the ground is 2300 ft. (*a*) How high was the plane when the object was released? (*b*) How long was the object in the air?
Answer: (*a*) 2900 ft. (*b*) 10 sec.

16(3). A projectile is fired horizontally from a gun located 144 ft above a horizontal plane with a muzzle speed of 800 ft/sec. (*a*) How long does the projectile remain in the air? (*b*) At what horizontal distance does it strike the ground? (*c*) What is the magnitude of the vertical component of its velocity as it strikes the ground?

17(3). A ball is thrown from the ground into the air. At a height of 30 ft the velocity is observed to be $\mathbf{v} = 25\mathbf{i} + 20\mathbf{j}$ in ft/sec. (*a*) To what maximum height will the ball rise? (*b*) What will be the total horizontal distance traveled by the ball? (*c*) What is the velocity of the ball (magnitude and direction) the instant before it hits the ground?
Answer: (*a*) 36 ft. (*b*) 75 ft. (*c*) 54 ft/sec, 62° below the horizontal.

18(3). Electrons, like all forms of matter, fall under the influence of gravity. If an electron is projected horizontally with a speed of 3.0×10^7 meters/sec ($\frac{1}{10}$ the speed of light), how far will it fall in traversing 1.0 meter?

19(3). What is the maximum vertical height to which a baseball player can throw a ball if he can throw it a maximum distance of 200 ft? Hint: see Probs. 12(3) and 13(3).
Answer: 100 ft.

20(3). A plane, diving at an angle of 53° with the vertical, releases a projectile at an altitude of 2400 ft. The projectile hits the ground 5.0 sec after being released. (*a*) What is the speed of the plane? (*b*) How far did the projectile travel horizontally during its flight? (*c*) What were the horizontal and vertical components of its velocity just before striking the ground?

21(3). A football is kicked off with an initial speed of 64 ft/sec at a projection angle of 45°. A receiver on the goal line 60 yd away in the direction of the kick starts running to meet the ball at

that instant. What must his speed be if he is to catch the ball before it hits the ground?
Answer: 19 ft/sec.

22(3). In a detective story a body is found 15 ft from the base of a building and beneath an open window 80 ft above. Would you guess the death to be accidental or not? Why?

23(3). In a cathode-ray tube a beam of electrons is projected horizontally with a speed of 1.0×10^9 cm/sec into the region between a pair of horizontal plates 2.0 cm long. An electric field between the plates exerts a constant downward acceleration on the electrons of magnitude 1.0×10^{17} cm/sec². Find (*a*) the vertical displacement of the beam in passing through the plates and (*b*) the velocity of the beam (direction and magnitude) as it emerges from the plates.
Answer: (*a*) 2.0 mm. (*b*) $v_{\text{hor}} \cong 1.0 \times 10^9$ cm/sec; $v_{\text{vert}} = -0.20 \times 10^9$ cm/sec.

24(3). Consider a projectile at the top of its trajectory. (*a*) What is its speed in terms of v_0 and θ_0? (*b*) What is its acceleration? (*c*) How is the direction of its acceleration related to that of its velocity? (*d*) Over a short distance a circular arc is a good approximation to a parabola. What then is the radius of the circular arc approximating the projectile's motion near the top of its path?

25(3). A batter hits a pitched ball at a height 4.0 ft above ground so that its angle of projection is 45° and its horizontal range is 350 ft. The ball is fair down the left field line where a 24-ft-high fence is located 320 ft from home plate. Will the ball clear the fence?
Answer: Yes.

26(3). In Galileo's *Two New Sciences* the author states that "for elevations (angles of projection) which exceed or fall short of 45° by equal amounts, the ranges are equal. . . ." Prove this statement. See Fig. 4–10.

27(3). A ball rolls off the top of a stairway with a horizontal velocity of magnitude 5.0 ft/sec. The steps are 8.0 in. high and 8.0 in. wide. Which step will the ball hit first?
Answer: The third.

28(3). A radar observer on the ground is "watching" an approaching projectile. At a certain instant he has the following information: (*a*) the projectile has reached maximum altitude and is moving horizontally with a speed v; (*b*) the straight-line distance to the projectile is l; (*c*) the line of sight to the projectile is an angle θ above the horizontal. Find the distance D between the observer and the point of impact of the projectile. D is to be expressed in terms of the observed quantities v, l, and θ and the known value of g. Assume a flat earth; assume also that the observer lies in the plane of the projectile's trajectory.

29(3). Projectiles are hurled at a horizontal distance R from the edge of a cliff of height h in such a way as to land a horizontal distance x from the bottom of the cliff. If you want x to be as small as possible, how would you adjust θ_0 and v_0, assuming that v_0 can be varied from zero to some maximum finite value and that θ_0 can be varied continuously? Only one collision with the ground is allowed (see Fig. 4–11).
Answer: Select θ_0 to satisfy $v^2_{\text{max}} \sin 2\theta_0 = gR$ in which v_{max} is the maximum value permitted for v_0.

30(4). An earth satellite moves in a circular orbit 400 miles above the earth's surface. The time for one revolution is 98 min. What is the acceleration of gravity at the orbit?

31(4). What is the centripetal acceleration of an object on the earth's equator due to the rotation of the earth?
Answer: 0.034 meter/sec².

32(4). Find the magnitude of the centripetal acceleration of a particle on the tip of a fan blade, 0.30 meter in diameter, rotating at 1200 rev/min.

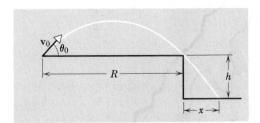

FIGURE 4-11   Problem 29(3).

33(4). A space station is constructed in the form of a rotating wheel 100 meters in diameter. The acceleration produced at the wheel's rim is to be equal to the gravitational acceleration on the moon, namely $g/6$. (*a*) With what angular speed must the wheel rotate? (*b*) What period of rotation would be required?

Answer: (*a*) 0.18 radian/sec. (*b*) 35 sec.

34(4). Certain neutron stars (extremely dense stars) are believed to be rotating at about 1 rev/sec. If such a star has a radius of 20 km, what is the centripetal acceleration of an object on the equator of the star?

35(4). A magnetic field will deflect a charged particle perpendicular to its direction of motion. An electron experiences a radial acceleration of 3.0×10^{14} meters/sec^2 in one such field. What is its speed if the radius of its curved path is 0.15 meter?

Answer: 6.7×10^6 meters/sec.

36(4). In Bohr's model of the hydrogen atom an electron revolves around a proton in a circular orbit of radius 5.28×10^{-11} meter with a speed of 2.18×10^6 meters/sec. What is the acceleration of the electron in the hydrogen atom?

37(4). By what factor would the speed of the earth's rotation have to increase for a body on the equator to require a centripetal acceleration of g to keep it on the earth?

Answer: 18.

38(4). A boy whirls a stone in a horizontal circle 6.0 ft above the ground by means of a string 4.0 ft long. The string breaks, and the stone flies off horizontally, striking the ground 30 ft away. What was the centripetal acceleration while in circular motion?

39(4). A carnival Ferris wheel has a 15-meter radius and completes five turns about its horizontal axis every minute. (*a*) What is the acceleration of a passenger at his highest point? (*b*) What is his acceleration at the lowest point?

Answer: (*a*) 4.1 meters/sec, down. (*b*) 4.1 meters/sec, up.

40(4). A particle P travels with constant speed on a circle of radius 3.0 meters and completes 1 rev in 20 sec (Fig. 4–12). The particle passes through O at $t = 0$. Starting from the origin O, find (*a*) the magnitude and direction of the position vectors 5.0 sec, 7.5 sec, and 10 sec later; (*b*) the magnitude and direction of the displacement in the 5.0-sec interval from the fifth to the tenth second; (*c*) the average velocity vector in this interval; (*d*) the instantaneous velocity vector at the beginning and at the end of this interval; (*e*) the average acceleration vector in this interval; and (*f*) the instantaneous acceleration vector at the beginning and at the end of this interval.

41(4). (*a*) Use the data of Appendix C to calculate the ratio of the centripetal accelerations of the Earth and Saturn due to their revolution about the sun. Assume that both planets move in circular orbits with constant speed. (*b*) What is the ratio of the distances of these two planets from the sun? (*c*) Compare your answers in parts (*a*) and (*b*) and suggest a simple relationship between centripetal acceleration and distance from the sun. Check your hypothesis by calculating the same ratios for another pair of planets.

Answer: (*a*) 92. (*b*) 9.6. (*c*) $92 = (9.6)^2$.

42(4). (*a*) Write an expression for the position vector **r** for a particle describing uniform circular motion, using rectangular coordinates and the unit vectors **i** and **j**. (*b*) From (*a*) derive vector expressions for the velocity **v** and the acceleration **a**. (*c*) Prove that the acceleration is directed toward the center of the circular motion.

43(5). Snow is falling vertically at a constant speed of 8.0 meters/sec. At what angle from the vertical do the snowflakes appear to be falling as viewed by the driver of a car traveling on a straight road with a speed of 50 km/hr?

Answer: 60°.

44(5). A train travels due south at 88.2 ft/sec (relative to ground) in a rain that is blown toward the south by the wind. The path of each raindrop makes the angle 21.6° with the vertical, as measured by an observer stationary on the earth.

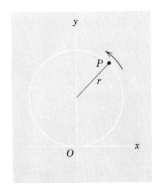

FIGURE 4–12   Problem 40(4).

An observer seated in the train, however, sees perfectly vertical tracks of rain on the windowpane. Determine the speed of each raindrop relative to the earth.

45(5). An airplane has a speed of 135 miles/hr in still air. It is flying straight north so that it is at all times directly above a north-south highway. A ground observer tells the pilot by radio that a 70-miles/hr wind is blowing, but neglects to tell him the wind direction. The pilot observes that in spite of the wind he can travel 135 miles along the highway in one hour. In other words, his ground speed is the same as if there were no wind. (*a*) What is the direction of the wind? (*b*) What is the heading of the plane, that is, the angle between its axis and the highway?
Answer: (*a*) From 75° east of south. (*b*) 30° east of north. Substituting west for east above yields a second solution.

46(5). A man wants to cross a river 500 meters wide. His rowing speed (relative to the water) is 3000 meters/hr. The river flows at a speed of 2000 meters/hr. If the man's walking speed on shore is 5000 meters/hr, (*a*) find the path (combined rowing and walking) he should take to get to the point directly opposite his starting point in the shortest time. (*b*) How long does it take?

47(5). A man can row a boat 4.0 miles/hr in still water. (*a*) If he is crossing a river where the current is 2.0 miles/hr, in what direction will his boat be headed if he wants to reach a point directly opposite from his starting point? (*b*) If the river is 4.0 miles wide, how long will it take him to cross the river? (*c*) How long will it take him to row 2.0 miles *down* the river and then back to his starting point? (*d*) How long will it take him to row 2.0 miles *up* the river and then back to his starting point? (*e*) In what direction should he head the boat if he wants to cross in the shortest possible time?
Answer: (*a*) 30° upstream. (*b*) 69 min. (*c*) 80 min. (*d*) 80 min. (*e*) Perpendicular to the current; the least time is 60 min.

48(5). A man launches a boat at a bridge and rows upstream a distance of one mile where he drops a bottle in the water. He then continues to row upstream for an additional 20 min. At that point he turns around and rows downstream, arriving at the bridge at the same time as the bottle. What is the speed of the water in the river? Assume only that the man rows at the same speed relative to the water at all times.

49(5). Find the speeds of two objects if, (*a*) when they move directly toward each other, they get 4.0 meters closer each second, and (*b*) when they move in the same direction with the original speeds, they get 4.0 meters closer each 10 sec.
Answer: (*a*) 2.2 meters/sec. (*b*) 1.8 meters/sec.

50(5). A person walks up a stalled escalator in 90 sec. When standing on the same escalator, now moving, he is carried up in 60 sec. How much time would it take him to walk up the moving escalator?

CHAPTER 5

Particle Dynamics

5-1 Introduction

In Chapters 3 and 4, we studied the motion of a particle, with emphasis on motion along a straight line or in a plane. We did not ask what "caused" the motion; we simply described it in terms of the vectors **r, v,** and **a.** Our discussion was thus largely geometrical. In this chapter we discuss the causes of motion, an aspect of mechanics called dynamics. As before, bodies will be treated as though they were single particles. Later in the book we shall treat groups of particles and extended bodies as well.

The concept of force and Newton's laws of motion will first be discussed. Then the particle dynamics for bodies subject to a force that is constant in both magnitude and direction will be presented. Gravitational and elastic forces, exerted respectively by the earth or by taut cords, will be introduced as illustrations.

Forces that are not constant will be considered in later sections. The dynamics of uniform circular motion in which the force, although constant in magnitude, changes in direction with time is also treated. In Chapter 9 transient forces, such as in a collision, where the direction is constant but the magnitude changes with time shall be considered. Later, in Chapter 13, problems in which the force changes in both magnitude and direction are treated.

5-2 Classical Mechanics

The motion of a given particle is determined by the nature and the arrangement of the other bodies that form its *environment*. In general, only nearby objects need to be included in the environment, the effects of more distant objects usually being negligible. Table 5-1 shows some "particles" and possible environments for them.

Table 5–1

	System	The "Particle"	The Environment
1.		A block	The spring; the rough surface
2.		A golf ball	The earth
3.		An artificial satellite	The earth
4.		An electron	A large uniformly charged sphere
5.		A bar magnet	A second bar magnet

In what follows, we limit ourselves to the very important special case of gross objects moving at speeds that are small compared to c, the speed of light; this is the realm of *classical mechanics*. Specifically, we shall not inquire here into such questions as the motion of an electron in a uranium atom or the collision of two protons whose speeds are, say $0.90c$. The first inquiry would involve us with the quantum theory and the second with the theory of relativity. We leave consideration of these theories, of which classical mechanics is a special case, until later.

The central problem of classical particle mechanics is this; (1) We are given a particle whose characteristics (mass, charge, magnetic dipole moment, etc.) we know. (2) We place this particle, with a known initial velocity, in an environment of which we have a complete description. (3) Problem: what is the subsequent motion of the particle?

This problem was solved, a least for a large variety of environments, by Isaac Newton (1642–1727) when he put forward his laws of motion and formulated his law of universal gravitation. The program for solving this problem, in terms of our present understanding of classical mechanics,* is: (1) We introduce the concept of *force* **F** and define it in terms of the acceleration **a** experienced by a particular standard body. (2) We develop a procedure for assigning a *mass m* to a body so that we may understand the fact that different particles of the same kind experience different accelerations in the same environment. (3) Finally, we try to find ways of calculating the forces that act on particles from the properties of the particle and of its environment; that is, we look for *force laws*. Force, which is at root a technique for relating the

---

* See "Presentation of Newtonian Mechanics" by Norman Austern, in *American Journal of Physics,* September 1961.

environment to the motion of the particle, appears both in the laws of motion (which tell us what acceleration a given body will experience under the action of a given force) and in the force laws (which tell us how to calculate the force that will act on a given body in a given environment). The laws of motion and the force laws, taken together, constitute the laws of mechanics.

The program of mechanics cannot be tested piecemeal. We must view it as a unit and we shall judge it to be successful if we can say "yes" to these two questions. (1) Does the program yield results that agree with experiment? (2) Are the force laws simple in form? It is the crowning glory of Newtonian mechanics that for a fantastic variety of phenomena we can indeed answer each of these questions in the affirmative. The exceptions can usually be handled using the two extensions of Newtonian mechanics, namely, quantum mechanics and Einstein's special theory of relativity.

In this section we have used the terms force and mass rather unprecisely, having identified force with the influence of the environment, and mass with the resistance of a body to being accelerated when a force acts on it, a property often called inertia. In later sections we shall refine these primitive ideas about force and mass.

5–3   Newton's First Law

For centuries the problem of motion and its causes was a central theme of natural philosophy. It was not until the time of Galileo and Newton, however, that dramatic progress was made. Isaac Newton, born in England in the year of Galileo's death, was the principal architect of classical mechanics. He carried to full fruition the ideas of Galileo and others who preceded him. His three laws of motion were first presented (in 1686) in his *Principia Mathematica Philosophiae Naturalis*.

Before Galileo's time most philosophers thought that some influence or "force" was needed to keep a body moving. To them, a body was in its "natural state" when it was at rest. For a body to move in a straight line at constant speed, for example, they believed that some external agent had to continually propel it; otherwise it would "naturally" stop moving.

If we wanted to test these ideas experimentally, we would first have to find a way to free a body from all influences of its environment or from all forces. This is hard to do, but in certain cases we can make the forces very small. If we study the motions as we make the forces smaller and smaller, we shall have some idea of what the motion would be like if the external forces were truly zero.

Let us place a test body, say a block, on a rigid horizontal plane. If we let the block slide along this plane, we notice that it gradually slows down and stops. This observation was used, in fact, to support the idea that motion stopped when the external force, in this case the hand initially pushing the block, was removed. Galileo argued against this idea, however, reasoning somewhat as follows: Repeat the experiment using a smoother block and a smoother plane and providing a lubricant. You will notice that the velocity decreases more slowly than before. Now use still smoother blocks and surfaces and better lubricants. As you might expect, you will find that the block decreases in velocity at a slower and slower rate and travels farther each time before coming to rest. We can now extrapolate and say that if all friction could be eliminated, the body would continue indefinitely in a straight line with constant speed. This was Galileo's conclusion. Galileo asserted that some external force was necessary to *change* the velocity of a body but that no external force was necessary to *maintain* the velocity of a body. Our hand, for example, exerts a force on the block when it sets it in motion. The rough plane exerts a force on it when it slows it down. Both of these forces produce a change in the velocity, that is, they produce an acceleration.

This principle of Galileo was adopted by Newton as the first of his three laws of motion. Newton stated his first law in these words: "Every body persists in its state of rest or of uniform motion in a straight line unless it is compelled to change that state by forces impressed on it."

Newton's first law is really a statement about reference frames. For, in general, the acceleration of a body depends on the reference frame relative to which it is measured. The first law tells us that, if there are no nearby objects (and by this we mean that there are no forces because every force must be associated with an object in the environment) then it is possible to find a family of reference frames in which a particle has no acceleration. The fact that bodies stay at rest or retain their uniform linear motion in the absence of applied forces is often described by assigning a property to matter called inertia. Newton's first law is often called the law of inertia and the reference frames to which it applies are therefore called inertial frames. Such frames are either fixed with respect to the distant stars or moving at uniform velocity with respect to them.

Notice that there is no distinction in the first law between a body at rest and one moving with a constant velocity. Both motions are "natural" in the absence of forces. That this is so becomes clear when a body at rest in one inertial frame is viewed from a second inertial frame, that is, a frame moving with constant velocity with respect to the first. An observer in the first frame finds the body to be at rest; an observer in the second frame finds the same body to be moving with uniform velocity. Both observers find the body to have no acceleration, that is, no change in velocity, and both may conclude from the first law that no force acts on the body.

Notice, too, that by implication there is no distinction in the first law between the absence of all forces and the presence of forces whose resultant is zero. For example, if the push of our hand on a block exactly counteracts the force of friction on it, the block will move with uniform velocity. Hence another way of stating the first law is: If no net force acts on a body its acceleration **a** is zero.

If there is an interaction between the body and objects present in the environment, the effect may be to change the "natural" state of the body's motion. To investigate this we must now examine carefully the concept of force.

5–4   Force

Let us refine our concept of force by defining it operationally. In our everyday language force is associated with a push or a pull, perhaps exerted by our muscles. In physics, however, we need a more precise definition. We define force here in terms of the acceleration that a given standard body experiences when placed in a suitable environment.

As a standard body we find it convenient to use (or rather to imagine that we use!) a particular platinum cylinder carefully preserved at the International Bureau of Weights and Measures near Paris, and called the *standard kilogram*. For use here we state that this has been assigned, by definition, a mass m_0 of exactly 1 kg. Later we will describe how masses are assigned to other bodies.

As for an environment we place the standard body on a horizontal table having negligible friction and we attach a spring to one end. We hold the other end of the spring in our hand, as in Fig. 5–1a. Now we pull the spring horizontally to the right

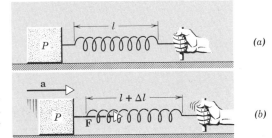

FIGURE 5–1 (*a*) A "particle" *P* (the standard kilogram) at rest on a horizontal frictionless surface. (*b*) The body is accelerated by pulling the spring to the right.

so that by trial and error the standard body experiences a measured uniform acceleration of 1.00 meter/sec^2. We then declare, as a matter of definition, that the spring (which is the significant body in the environment) is exerting a constant force, whose magnitude we will call "1.00 newton," on the standard body. We note that, in imparting this force, the spring is kept stretched an amount Δl beyond its normal unextended length, as Fig. 5–1*b* shows.

We can repeat the experiment, either stretching the spring more or using a stiffer spring, so that we measure an acceleration of 2.00 meters/sec^2 for the standard body. We now declare that the spring is exerting a force of 2.00 newtons on the standard body. In general, if we observe this particular standard body to have an acceleration *a* in a particular environment, we then say that the environment is exerting a force *F* on the standard body, where *F* (in newtons) is numerically equal to *a* (in meters/sec^2).

Now let us see whether force, as we have defined it, is a vector quantity. In Fig. 5–1*b* we assigned a magnitude to the force *F*, and it is a simple matter to assign a direction to it as well, namely, the direction of the acceleration that the force produces. However, to be a vector it is not enough for a quantity to have magnitude and direction; it must also obey the laws of vector addition described in Chapter 2. We can learn only from experiment whether forces, as we defined them, do indeed obey these laws.

We could arrange to exert a 4.00-newton force along the *x*-axis and a 3.00-newton force along the *y*-axis. Let us apply these forces simultaneously to the standard body placed, as before, on a horizontal, frictionless surface. What will be the acceleration of the standard body? We would find by experiment that it was 5.00 meters/sec^2, directed along a line that makes an angle of 37° with the *x*-axis. In other words, we would say that the standard body was experiencing a force of 5.00 newtons in this same direction. This same result can be obtained by adding the 4.00-newton and 3.00-newton forces vectorially according to the parallelogram method. Experiments of this kind show conclusively that forces are vectors; they have magnitude; they have direction; they add according to the parallelogram law.

The result of experiments of this general type is often stated as follows: When several forces act on a body, each produces its own acceleration independently. The resulting acceleration is the vector sum of the several independent accelerations.

5–5 Mass; Newton's Second Law

In Section 5–4 we considered only the accelerations given to one particular object, the standard kilogram. We were able thereby to define forces quantitatively. What effect would these forces have on other objects? Since our standard body was chosen arbitrarily in the first place, we know that for any given object the acceleration will be directly proportional to the force applied. The significant question remaining then is: What effect will the same force have on different objects? Everyday experience gives us a qualitative answer. The same force will produce different accelerations on different bodies. A baseball will be accelerated more by a given force than will an automobile. In order to obtain a quantitative answer to this question we need a method to measure mass, the property of a body which determines its resistance to a change in its motion.

Let us attach a spring to our standard body (the standard kilogram, to which we have arbitrarily assigned a mass $m_0 = 1.00$ kg, exactly) and arrange to give it an acceleration a_0 of 2.00 meters/sec^2 using the method of Fig. 5–1b. Let us measure carefully the extension Δl of the spring associated with the force that the spring is exerting on the block.

Now we remove the standard kilogram and substitute an arbitrary body, whose mass we label m_1. We apply the same force (the one that accelerated the standard kilogram 2.00 meters/sec^2) to the arbitrary body (by stretching the spring by the same amount) and we measure an acceleration a_1 of 0.50 meter/sec^2.

We define the ratio of the masses of the two bodies to be the inverse ratio of the acceleration given to these bodies by the same force, or

$$m_1/m_0 = a_0/a_1 \qquad \text{(same force } \mathbf{F} \text{ acting)}.$$

In this example we have, numerically,

$$m_1 = m_0(a_0/a_1) = 1.00 \text{ kg } [(2.00 \text{ meters/sec}^2)/(0.50 \text{ meters/sec}^2)] = 4.00 \text{ kg}.$$

The second body, which has only one-fourth the acceleration of the first body when the same force acts on it, has, by definition, four times the mass of the first body. Hence mass may be regarded as a quantitative measure of inertia.

If we repeat the preceding experiment with a different common force acting, we find the ratio of the accelerations, a_0'/a_1', to be the same as in the previous experiment, or

$$m_1/m_0 = a_0/a_1 = a_0'/a_1'.$$

The ratio of the masses of two bodies is thus independent of the common force used.

Furthermore, experiment shows that we can consistently assign masses to any body by this procedure. For example, let us compare a second arbitrary body with the standard body, and thus determine its mass, say m_2. We can now compare the two arbitrary bodies, m_2 and m_1, directly, obtaining accelerations a_2'' and a_1'' when the same force is applied. The mass ratio, defined as usual from

$$m_2/m_1 = a_1''/a_2'', \qquad \text{(same force acting)}$$

turns out to have the same value that we obtain by using the masses m_2 and m_1 determined previously by direct comparison with the standard.

We can show, in still another experiment of this type, that if objects of mass m_1 and m_2 are fastened together they behave mechanically as a single object of mass $(m_1 + m_2)$. In other words, masses add like (and are) scalar quantities.

Table 5–2 shows the range of values over which masses can be determined, using various techniques.

Table 5–2 SOME MEASURED MASSES

Object	Mass (kg)
Our galaxy	2.2×10^{41}
The sun	2.0×10^{30}
The earth	6.0×10^{24}
The moon	7.4×10^{22}
Mass of all the water in the oceans	1.4×10^{21}
An ocean liner	7.2×10^{7}
An elephant	4.5×10^{3}
A man	7.3×10^{1}
A grape	3.0×10^{-3}
A tobacco mosaic virus	6.7×10^{-10}
A speck of dust	2.3×10^{-13}
A penicillin molecule	5.0×10^{-17}
A uranium atom	4.0×10^{-25}
A proton	1.7×10^{-27}
An electron	9.1×10^{-31}

We can now summarize all the experiments and definitions described above in one equation of classical mechanics,

$$\mathbf{F} = m\mathbf{a}. \tag{5–1}$$

In this equation $\mathbf{F}$ is the (vector) sum of all the forces acting on the body, m is the mass of the body, and $\mathbf{a}$ is its (vector) acceleration. Equation 5–1 may be taken as a statement of Newton's second law. If we write it in the form $\mathbf{a} = \mathbf{F}/m$, we can see easily that the acceleration of the body is directly proportional to the resultant force acting on it and parallel in direction to this force and that the acceleration, for a given force, is inversely proportional to the mass of the body.

Notice that the first law of motion is contained in the second law as a special case, for if $\mathbf{F} = 0$, then $\mathbf{a} = 0$. In other words, if the resultant force on a body is zero, the acceleration of the body is zero. Therefore in the absence of applied forces a body will move with constant velocity or be at rest (zero velocity), which is what the first law of motion says. Therefore, of Newton's three laws of motion only two are independent, the second and the third (Section 5–6). The division of translational particle dynamics that includes only systems for which the resultant force $\mathbf{F}$ is zero is called *statics*.

Equation 5–1 is a vector equation. We can write this single vector equation as three scalar equations,

$$F_x = ma_x, \qquad F_y = ma_y, \qquad \text{and} \quad F_z = ma_z, \tag{5–2}$$

relating the x, y, and z components of the resultant force (F_x, F_y, and F_z) to the x, y, and z components of acceleration (a_x, a_y, and a_z) for the mass m.

5–6 Newton's Third Law

Forces acting on a body originate in other bodies that make up its environment. Any single force is only one aspect of a mutual interaction between two bodies. We find by experiment that when one body exerts a force on a second body, the second body always exerts a force on the first. Furthermore, we find that these forces are equal in magnitude but opposite in direction. A single isolated force is therefore an impossibility.

If one of the two forces involved in the interaction between two bodies is called an "action" force, the other is called the "reaction" force. Either force may be considered the "action" and the other the "reaction." Cause and effect is not implied here, but a mutual simultaneous interaction is implied.

This property of forces was first stated by Newton in his third law of motion: *"To every action there is always opposed an equal reaction; or, the mutual actions of two bodies upon each other are always equal, and directed to contrary parts."*

In other words, if body A exerts a force on body B, body B exerts an equal but oppositely directed force on body A; and furthermore the forces lie along the line joining the bodies. Notice that the action and reaction forces, which always occur in pairs, act on different bodies. If they were to act on the same body, we could never have accelerated motion because the resultant force on every body would always be zero.

Imagine a boy kicking open a door. The force exerted by the boy B on the door D accelerates the door (it flies open); at the same time, the door D exerts an equal but opposite force on the boy B, which decelerates the boy (his foot loses forward velocity). The boy will be painfully aware of the "reaction" force to his "action."

The following examples illustrate the application of the third law and clarify its meaning.

Example 1. Consider a man pulling horizontally on a rope attached to a block on a horizontal table as in Fig. 5–2. The man pulls on the rope with a force $\mathbf{F}_{MR}$. The rope exerts a reaction force $\mathbf{F}_{RM}$ on the man. According to Newton's third law, $\mathbf{F}_{MR} = -\mathbf{F}_{RM}$. Also, the rope exerts a force $\mathbf{F}_{RB}$ on the block, and the block exerts a reaction force $\mathbf{F}_{BR}$ on the rope. Again according to the third law, $\mathbf{F}_{RB} = -\mathbf{F}_{BR}$.

Suppose that the rope has a mass m_R. Then, in order to start the block and rope moving from rest, we must have an acceleration, say $\mathbf{a}$. The only forces acting on the rope are $\mathbf{F}_{MR}$ and $\mathbf{F}_{BR}$, so that the resultant force on it is $\mathbf{F}_{MR} + \mathbf{F}_{BR}$, and this must be different from zero if the rope is to accelerate. In fact, from the second law we have

$$\mathbf{F}_{MR} + \mathbf{F}_{BR} = m_R\mathbf{a}$$

FIGURE 5–2 Example 1. A man pulls on a rope attached to a block. (*a*) The forces exerted on the rope by the block and by the man are equal and opposite. Thus the resultant horizontal force on the rope is zero, as is shown in the free-body diagram. The rope does not accelerate. (*b*) The force exerted on the rope by the man exceeds that exerted by the block. The net horizontal force has magnitude $F_{MR} - F_{BR}$ and points to the right. Thus the rope is accelerated to the right. The block is also acted upon by a frictional force not shown here.

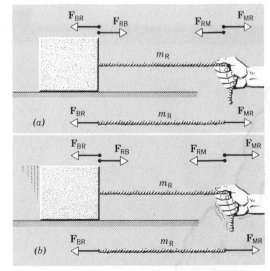

Since the forces and the acceleration are along the same line, we can drop the vector notation and write the relation between the magnitudes of the vectors, namely

$$F_{MR} - F_{BR} = m_R a.$$

We see therefore that in general $\mathbf{F}_{MR}$ does not have the same magnitude as $\mathbf{F}_{BR}$ (Fig. 5–2b). These two forces act on the same body and are not action and reaction pairs.

According to Newton's third law the magnitude of $\mathbf{F}_{MR}$ always equals the magnitude of $\mathbf{F}_{RM}$, and the magnitude of $\mathbf{F}_{RB}$ always equals the magnitude of $\mathbf{F}_{BR}$. However, only if the acceleration **a** of the system is zero will we have the pair of forces $\mathbf{F}_{MR}$ and $\mathbf{F}_{RM}$ equal in magnitude to the pair of forces $\mathbf{F}_{RB}$ and $\mathbf{F}_{BR}$ (Fig. 5-2a). In this special case only, we could imagine that the rope merely transmits the force exerted by the man to the block without change. This same result holds in principle if $m_R = 0$. In practice, we never find a massless rope. However, we can often neglect the mass of a rope, in which case the rope is assumed to transmit a force unchanged. The force exerted at any point in the rope is called the *tension* at that point. We may measure the tension at any point in the rope by cutting a suitable length from it and inserting a spring scale; the tension is the reading of the scale. The tension is the same at all points in the rope only if the rope is unaccelerated or assumed to be massless.

Example 2. Consider a spring attached to the ceiling and at the other end holding a block at rest (Fig. 5–3a). Since no body is accelerating, all the forces on any body will add vectorially to zero. For example, the forces on the suspended block are **T,** the tension in the stretched spring, pulling vertically up on the mass, and **W,** the pull of the earth acting vertically down on the body, called its weight. These are drawn in Fig. 5–3b, where we show only the block for clarity. There are no other forces on the block.

In Newton's second law, **F** represents the sum of all the forces acting on a body, so that for the block

$$\mathbf{F} = \mathbf{T} + \mathbf{W}.$$

The block is at rest so that its acceleration is zero, or **a** $= 0$. Hence, from the relation **F** $= m\mathbf{a}$, we obtain **T** $+$ **W** $= 0$, or

$$\mathbf{T} = -\mathbf{W}.$$

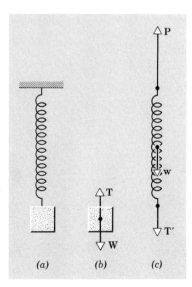

FIGURE 5–3 Example 2. (*a*) A block is suspended by a spring. (*b*) A free-body diagram showing all the vertical forces exerted on the block. (*c*) A similar diagram for the vertical forces on the spring.

The forces act along the same line, so that their magnitudes are equal, or

$$T = W.$$

Therefore the tension in the spring is an exact measure of the weight of the block. We shall use this result later in presenting a static procedure for measuring forces.

It is instructive to examine the forces exerted on the spring; they are shown in Fig. 5–3c. **T'** is the pull of the block on the spring and is the reaction force of the action force **T**. **T'** therefore has the same magnitude as **T,** which is W. **P** is the upward pull of the ceiling on the spring, and **w** is the weight of the spring, that is, the pull of the earth on it. Since the spring is at rest and all forces act along the same line, we have

$$\mathbf{P} + \mathbf{T'} + \mathbf{w} = 0,$$

or

$$P = W + w.$$

The ceiling therefore pulls up on the spring with a force whose magnitude is the sum of the weights of the block and spring.

From the third law of motion, the force exerted by the spring on the ceiling, **P',** must be equal in magnitude to **P,** which is the reaction force to the action force **P'**. **P'** therefore has a magnitude $W + w$.

In general, the spring exerts different forces on the bodies attached at its different ends, for $P' \neq T$. In the special case in which the weight of the spring is negligible, $w = 0$ and $P' = W = T$. Therefore a *weightless* spring (or cord) may be considered to transmit a force from one end to the other without change.

It is instructive to classify all the forces in this problem according to action and reaction pairs. The reaction to **W,** a force exerted by the earth on the block, must be a force exerted by the block on the earth. Similarly, the reaction to **w** is a force exerted by the spring on the earth. Because the earth is so massive, we do not expect these forces to impart a noticeable acceleration to the earth. Since the earth is not shown in our diagrams, these forces have not been shown. The forces **T** and **T'** are action-reaction pairs, as are **P** and **P'**. Notice that although **T** $= -$**W** in our problem, these forces are not an action-reaction pair because they act on the same body. ■

5–7 Systems of Mechanical Units

Unit force is defined as a force that causes a unit of acceleration when applied to a unit mass. The mks (meter, kilogram, second) unit of mass is the *kilogram*. The cgs (centimeter, gram, second) unit of mass is the *gram,* defined as one-thousandth of the kilogram mass.

In the mks system unit force is the force that will accelerate a one-kilogram mass at one meter/sec^2; we have seen that this unit is called the *newton*. In the cgs system unit force is the force that will accelerate a one-gram mass at one cm/sec^2; this unit is called the *dyne*. Since 1 kg $= 10^3$ gm and 1 meter/sec^2 $= 10^2$ cm/sec^2, it follows that 1 nt $= 10^5$ dynes.

In each of our systems of units we have chosen mass, length, and time as our fundamental quantities. Standards were adopted for these fundamental quantities and units defined in terms of these standards. Force appears as a derived quantity, determined from the relation **F** $= m$**a.**

In the British engineering system of units, however, *force,* length, and time are chosen as the fundamental quantities and mass is a derived quantity. In this system, mass is determined from the relation $m = F/a$. The standard and unit of force in this system is the *pound*. Actually, the pound of force was originally defined to be the pull of the earth on a certain standard body at a certain place on the earth. We can get this force in an operational way by hanging the standard body from a spring at the particular point where the earth's pull on it is defined to be 1 lb of force. If the body is at rest, the earth's pull on the body, its weight W, is balanced by the tension in the spring. Therefore $T = W = 1$ lb, in this instance. We can now use this spring (or any other one thus calibrated) to exert a force of 1 lb on any other body; to do this we simply attach the spring to another body and stretch it the same amount as the pound force had stretched it. The standard body can be compared to the kilogram and it is found to have the mass 0.45359237 kg. The acceleration due to gravity at the certain place on the earth is found to be 32.1740 ft/sec². The pound of force can therefore be defined from $F = ma$ as the force that accelerates a mass of 0.45359237 kg at the rate of 32.1740 ft/sec².

This procedure enables us to compare the pound-force with the newton. Using the fact that 32.1740 ft/sec² equals 9.8066 meters/sec², we find that

$$1 \text{ lb} = (0.45359237 \text{ kg})(32.1740 \text{ ft/sec}^2)$$
$$= (0.45359237 \text{ kg})(9.8066 \text{ meters/sec}^2)$$
$$\cong 4.45 \text{ nt}.$$

The unit of mass in the British engineering system can now be derived. It is defined as the mass of a body whose acceleration is 1 ft/sec² when the force on it is 1 lb; this mass is called the *slug*. Thus, in this system

$$F \text{ [lb]} = m \text{ [slugs]} \times a \text{ [ft/sec}^2\text{]}.$$

In this book only forces will be measured in pounds. The corresponding unit of mass is the slug. The units of force, mass, and acceleration in the three systems are summarized in Table 5–3.

The dimensions of force are the same as those of mass times acceleration. In a system in which mass, length, and time are the fundamental quantities, the dimensions of force are, therefore, mass × length/time² or MLT^{-2}.

5–8 The Force Laws

The three laws of motion that we have described are only part of the program of mechanics that we outlined in Section 5–2. It remains to investigate the force laws, which are the procedures by which we calculate the force acting on a given body in terms of the properties of the body and its environment. Newton's second law

$$\mathbf{F} = m\mathbf{a} \tag{5–3}$$

is essentially not a law of nature but a definition of force. We need to identify various functions of the type:

$$\mathbf{F} = \text{a function of the properties of the particle and of the environment} \tag{5–4}$$

Table 5–3 Units in $F = ma$

Systems of Units	Force	Mass	Acceleration
Mks	newton (nt)	kilogram (kg)	meter/sec²
Cgs (Gaussian)	dyne	gram (gm)	cm/sec²
Engineering	pound (lb)	slug	ft/sec²

so that we can, in effect, eliminate **F** between Eqs. 5–3 and 5–4, thus obtaining an equation that will let us calculate the acceleration of a particle in terms of the properties of the particle and its environment. Force is a concept that connects the acceleration of the particle on the one hand with the properties of the particle and its environment on the other. We indicated earlier that one criterion for declaring the program of mechanics to be successful would be the discovery that *simple* laws of the type of Eq. 5–4 do indeed exist. This turns out to be the case, and this fact constitutes the essential reason that we "believe" the laws of classical mechanics. If the force laws had turned out to be very complicated, we would not be left with the feeling that we had gained much insight into the workings of nature.

The number of possible environments for an accelerated particle is so great that a detailed discussion of all the force laws is not feasible in this chapter. We shall, however, indicate in Table 5–4 the force laws that apply to the five particle-plus-environment situations of Table 5–1. At appropriate places throughout the text we discuss these and other force laws in detail; several of the laws in Table 5–4 are approximations or special cases.

5–9 Weight and Mass

The *weight* of a body is the gravitational force exerted on it by the earth. Weight, being a force, is a vector quantity. The direction of this vector is the direction of the gravitational force, that is, toward the center of the earth. The magnitude of the weight is expressed in force units, such as pounds or newtons.

When a body of mass m is allowed to fall freely, its acceleration is that of gravity **g** and the force acting on it is its weight **W**. Newton's second law, $F = m\mathbf{a}$, when applied to a freely falling body, gives us $\mathbf{W} = m\mathbf{g}$. Both **W** and **g** are vectors directed toward

Table 5–4   The Force Laws for the Systems of Table 5–1

System	Force Law
1. A block propelled by a stretched spring over a rough horizontal surface	(a) Spring force: $F = -kx$, where x is the extension of the spring and k is a constant that describes the spring; **F** points to the right; see Chapter 13
	(b) Friction force: $F = \mu mg$, where μ is the coefficient *of friction* and mg is the weight of the block; **F** points to the left; see Section 5–12
2. A golf ball in flight	$F = mg$; **F** points down (see Section 5–9)
3. An artificial satellite	$F = GmM/r^2$, where G is the *gravitational constant*, M the mass of the earth, and r the orbit radius; **F** points toward the center of the earth; see Chapter 14. This is *Newton's law of universal gravitation*
4. An electron near a charged sphere	$F = (1/4\pi\epsilon_0)eQ/r^2$, where ϵ_0 is a constant, e is the electron charge, Q is the charge on the sphere, and r is the distance from the electron to the center of the sphere; **F** points to the right; see Chapter 22. This is *Coulomb's law of electrostatics*
5. Two bar magnets	$F = (3\mu_0/2\pi)\mu^2/r^4$, where μ_0 is a constant, μ is the *magnetic dipole* moment of each magnet, and r is the center-to-center separation of the magnets; we assume that $r \gg l$, where l is the length of each magnet; **F** points to the right

the center of the earth. We can therefore write

$$W = mg, \qquad (5\text{–}5)$$

where W and g are the magnitudes of the weight and acceleration vectors. To keep an object from falling we have to exert on it an upward force equal in magnitude to W, so as to make the net force zero. In Fig. 5–3a the tension in the spring supplies this force.

We stated previously that g is found experimentally to have the same value for all objects at the same place. From this it follows that the ratio of the weights of two objects must be equal to the ratio of their masses. Therefore a chemical balance, which actually is an instrument for comparing two downward forces, can be used in practice to compare masses. If a sample of salt in one pan of a balance is pulling down on that pan with the same force as is a standard one gram-mass on the other pan, we infer that the mass of salt is equal to one gram. We are likely to say that the salt "weighs" one gram, although a gram is a unit of mass, not weight. However, it is always important to distinguish carefully between weight and mass.

We have seen that the weight of a body, the downward pull of the earth on that body, is a vector quantity. The mass of a body is a scalar quantity. The quantitative relation between weight and mass is given by $\mathbf{W} = m\mathbf{g}$. Because $\mathbf{g}$ varies from point to point on the earth, $\mathbf{W}$, the weight of a body of mass m, is different in different localities. Thus, the weight of a one kg-mass in a locality where g is 9.80 meters/sec² is 9.80 nt; in a locality where g is 9.78 meters/sec², the same one kg-mass weighs 9.78 nt. If these weights were determined by measuring the amount of stretch required in a spring to balance them, the difference in weight of the same one kg-mass at the two different localities would be evident in the slightly different stretch of the spring at these two localities. Hence, unlike the mass of a body, which is an intrinsic property of the body, the weight of a body depends on its location relative to the center of the earth. Spring scales read differently, balances the same, at different parts of the earth.

We shall generalize the concept of weight in Chapter 14 in which we discuss universal gravitation. There we shall see that the weight of a body is zero in regions of space where the gravitational effects are nil, although the inertial effects, and hence the mass of the body, remain unchanged from those on earth. In a space ship free from the influence of gravity it is a simple matter to "lift" a large block of lead ($\mathbf{W}=0$), but the astronaut would still stub his toe if he were to kick the block ($m \neq 0$).

It takes the same force to accelerate a body in gravity-free space as it does to accelerate it along a horizontal frictionless surface on earth, for its mass is the same in each place.

Often, instead of being given the mass, we are given the weight of a body on which forces are exerted. The acceleration $\mathbf{a}$ produced by the force $\mathbf{F}$ acting on a body whose weight has a magnitude W can be obtained by combining Eq. 5–3 and Eq. 5–5. Thus from $\mathbf{F} = m\mathbf{a}$ and $W = mg$ we obtain

$$m = W/g, \qquad \text{so that} \qquad \mathbf{F} = (W/g)\mathbf{a}. \qquad (5\text{–}6)$$

The quantity W/g plays the role of m in the equation $F = ma$ and is, in fact, the mass of a body whose weight has the magnitude W. For example, a man whose weight is 160 lb at a point where $g = 32.0$ ft/sec² has a mass $m = W/g = (160 \text{ lb})/(32.0 \text{ ft/sec}^2) = 5.00$ slugs. Notice that his weight at another point where $g = 32.2$ ft/sec² is $W = mg = (5.00 \text{ slugs})(32.2 \text{ ft/sec}^2) = 161$ lb.

5–10 A Static Procedure for Measuring Forces

In Section 5–4 we defined force by measuring the acceleration imparted to a standard body by pulling on it with a stretched spring. That may be called a dynamic method for measuring force. Although convenient for the purposes of definition, it is not a particularly practical procedure for the measurement of forces. Another method for measuring forces is based on measuring the change in shape or size of a body (a spring, say) on which the force is applied when the body is unaccelerated. This may be called the static method of measuring forces.

The idea of the static method is to use the fact that when a body, under the action of several forces, has zero acceleration, the vector sum of all the forces acting on the body must be zero. This is, of course, just the content of the first law of motion.

The instrument most commonly used to measure forces in this way is the spring balance. It consists of a coiled spring having a pointer at one end that moves over a scale. A force exerted on the balance changes the length of the spring. If a body weighing 1.00 lb is hung from the spring, the spring stretches until the pull of the spring on the body is equal in magnitude but opposite in direction to its weight. A mark can be made on the scale next to the pointer and labeled "1.00-lb force." Similarly, 2.00-lb, 3.00-lb, etc., weights may be hung from the spring and corresponding marks can be made on the scale next to the pointer in each case. In this way the spring is calibrated. We assume that the force exerted on the spring is always the same when the pointer stands at the same position. The calibrated balance can now be used to measure any suitable unknown force, not merely the pull of the earth on some body.

The third law is tacitly used in our static procedure because we assume that the force exerted by the spring on the body is the same in magnitude as the force exerted by the body on the spring. This latter force is the force we wish to measure.

5–11 Some Applications of Newton's Laws of Motion

It will be helpful to write down some procedures for solving problems in classical mechanics and to illustrate them by several examples. Newton's second law states that the vector sum of all the forces acting on a body is equal to its mass times its acceleration. The first step in problem solving is therefore: (1) Identify the body to whose motion the problem refers. As obvious as this seems, lack of clarity on the point as to what has been or should be picked as "the body" is a major source of mistakes. (2) Having selected "the body," we next turn our attention to the objects in "the enviornment" because these objects (inclined planes, springs, cords, the earth, etc.) exert forces on the body. We must be clear as to the nature of these forces. (3) The next step is to select a suitable (inertial) reference frame. We should position the origin and orient the coordinate axes so as to simplify the task of our next step as much as possible. (4) We now make a separate diagram of the body alone, showing the reference frame and all of the forces acting on the body. This is called a *free-body* diagram. (5) Finally we apply Newton's second law, in the form of Eq. 5–2, to each component of force and acceleration.

The following examples illustrate the method of analysis used in applying Newton's laws of motion. Each body is treated as if it were a particle of definite mass, so that the forces acting on it may be assumed to act at a point. Strings and pulleys are considered to have negligible mass. Although some of the situations picked for analysis may seem simple and artificial, they are the prototypes for many interesting real situations; but, more important, the method of analysis—which is the chief thing to understand—is applicable to all the modern and sophisticated situations of classical mechanics, even sending a spaceship to Mars.

Example 3. Figure 5–4*a* shows a weight W hung by strings. Consider the knot at the junction of the three strings to be "the body." The body remains at rest under the action of the three forces shown in Fig. 5–4*b*. Suppose we are given the magnitude of one of these forces. How can we find the magnitude of the other forces?

$\mathbf{F}_A$, $\mathbf{F}_B$, and $\mathbf{F}_C$ are *all* the forces acting *on* the body. Since the body is unaccelerated (actually at rest), $\mathbf{F}_A + \mathbf{F}_B + \mathbf{F}_C = 0$. Choosing the *x*- and *y*-axes as shown, we can write this vector equation as three scalar equations:

$$F_{Ax} + F_{Bx} = 0,$$

$$F_{Ay} + F_{By} + F_{Cy} = 0,$$

using Eq. 5–2. The third scalar equation for the *z*-axis is simply

$$F_{Az} = F_{Bz} = F_{Cz} = 0.$$

That is, the vectors all lie in the *x-y* plane so that they have no *z*-components.

From the figure we see that

$$F_{Ax} = -F_A \cos 30° = -0.866F_A,$$

$$F_{Ay} = F_A \sin 30° = 0.500F_A,$$

and

$$F_{Bx} = F_B \cos 45° = 0.707F_B,$$

$$F_{By} = F_B \sin 45° = 0.707F_B.$$

Also,

$$F_{Cy} = -F_C = -W,$$

because the string C merely serves to transmit the force on one end to the junction at its other end. Substituting these results into our original equations, we obtain

$$-0.866F_A + 0.707F_B = 0,$$

$$0.500F_A + 0.707F_B - W = 0.$$

If we are given the magnitude of any one of these three forces, we can solve these equations for the other two. For example, if $W = 100$ lb, we obtain $F_A = 73.3$ lb and $F_B = 89.6$ lb.

Example 4. We wish to analyze the motion of a block on a smooth incline.

(*a*) *Static case.* Figure 5–5*a* shows a block of mass m kept at rest on a smooth plane, inclined at an angle θ with the horizontal, by means of a string attached to the vertical wall. The forces

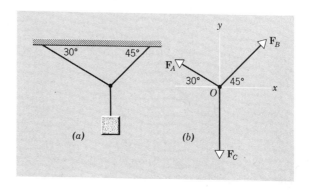

FIGURE 5–4 Example 3. (*a*) A mass is suspended by strings. (*b*) A free-body diagram showing all the forces acting on the knot. The strings are assumed to be weightless.

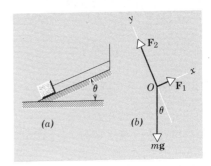

FIGURE 5-5   Example 4. (*a*) A block is held on a smooth inclined plane by a string. (*b*) A free-body diagram showing all the forces acting on the block.

acting on the block are shown in Fig. 5–5*b*. $\mathbf{F}_1$ is the force exerted on the block by the string; $m\mathbf{g}$ is the force exerted on the block by the earth, that is, its weight; and $\mathbf{F}_2$ is the force exerted on the block by the inclined surface. $\mathbf{F}_2$, called the normal force, is normal to the surface of contact because there is no frictional force between the surfaces. If there were a frictional force, $\mathbf{F}_2$ would have a component parallel to the incline. Because we wish to analyze the motion of the block, we choose ALL the forces acting ON the block. You should note that the block will exert forces on other bodies in its environment (the string, the earth, the surface of the incline) in accordance with the action-reaction principle; these forces, however, are not needed to determine the motion of the block because they do not act on the block.

Suppose θ and m are given. How do we find F_1 and F_2? Since the block is unaccelerated, we obtain

$$\mathbf{F}_1 + \mathbf{F}_2 + m\mathbf{g} = 0.$$

It is convenient to choose the *x*-axis of our reference frame to be along the incline and the *y*-axis to be normal to the incline (Fig. 5–5*b*) With this choice of coordinates, only one force, $m\mathbf{g}$, must be resolved into components in solving the problem. The two scalar equations obtained by resolving $m\mathbf{g}$ along the *x*- and *y*-axes are

$$F_1 - mg \sin \theta = 0, \quad \text{and} \quad F_2 - mg \cos \theta = 0,$$

from which F_1 and F_2 can be obtained if θ and m are given.

(*b*) *Dynamic case.* Now suppose that the string is cut. Then the force $\mathbf{F}_1$, the pull of the string on the block, will be removed. The resultant force on the block will no longer be zero, and the block will accelerate. What is its acceleration?

From Eq. 5–2 we have $F_x = ma_x$ and $F_y = ma_y$. Using these relations we obtain

$$F_2 - mg \cos \theta = ma_y = 0,$$

and

$$-mg \sin \theta = ma_x,$$

which yield

$$a_y = 0, \quad a_x = -g \sin \theta.$$

The acceleration is directed down the incline with a magnitude of $g \sin \theta$.

Example 5. Figure 5–6*a* shows a block of mass m_1 on a smooth horizontal surface pulled by a string which is attached to a block of mass m_2 hanging over a pulley. We asume that the pulley has no mass and is frictionless and that it merely serves to change the direction of the tension in the string at that point. Find the acceleration of the system and the tension in the string.

Suppose we choose the block of mass m_1 as the body whose motion we investigate. The forces on this block, taken to be a particle, are shown in Fig. 5–6*b*. **T**, the tension in the string, pulls on the block to the right; $m_1\mathbf{g}$ is the downward pull of the earth on the block and **N** is the vertical force exerted on the block by the smooth table. The block will accelerate in the *x*-direction only, so that $a_{1y} = 0$. We therefore, can write

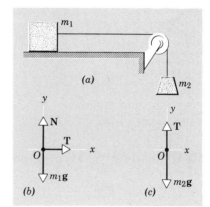

FIGURE 5-6   Example 5. (*a*) Two masses are connected by a string; m_1 lies on a smooth table, m_2 hangs freely. (*b*) A free-body diagram showing all the forces acting on m_1. (*c*) A similar diagram for m_2.

$$N - m_1 g = 0 = m_1 a_{1y},$$

and
$$T = m_1 a_{1x}. \tag{5-7}$$

From these equations we conclude that $N = m_1 g$. We do not know T, so we cannot solve for a_{1x}.

To determine T we must consider the motion of the block m_2. The forces acting on m_2 are shown in Fig. 5-6*c*.

If we choose a downward acceleration of m_2 as being positive, the equation of motion for the suspended block is

$$m_2 g - T = m_2 a_{2y}. \tag{5-8}$$

Because the string has a fixed length, it is clear that

$$a_{2y} = a_{1x},$$

so that we can represent the acceleration of the system as simply a. We then obtain from Eqs. 5-7 and 5-8

$$m_2 g - T = m_2 a, \tag{5-9}$$

and
$$T = m_1 a.$$

These yield
$$m_2 g = (m_1 + m_2)a, \tag{5-10}$$

or
$$a = \frac{m_2}{m_1 + m_2} g,$$

and
$$T = \frac{m_1 m_2}{m_1 + m_2} g, \tag{5-11}$$

which gives us the acceleration of the system a and the tension in the string T.

Notice that the tension in the string is always less than $m_2 g$. This is clear from Eq. 5-11, which can be written

$$T = m_2 g \frac{m_1}{m_1 + m_2}.$$

Notice also that a is always less than g, the acceleration due to gravity. Only when m_1 equals zero, which means that there is no block at all on the table, do we obtain $a = g$ (from Eq. 5-10). In this case $T = 0$ (from Eq. 5-9).

We can interpret Eq. 5-10 in a simple way. The net unbalanced force on the system of mass $m_1 + m_2$ is represented by $m_2 g$. Hence, from $F = ma$, we obtain Eq. 5-10 directly.

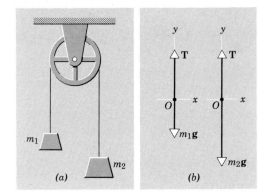

FIGURE 5–7 Example 6. (*a*) Two unequal masses are suspended by a string from a pulley (Atwood's machine). (*b*) Free-body diagrams for m_1 and m_2.

To make the example specific, suppose $m_1 = 2.0$ kg and $m_2 = 1.0$ kg. Then

$$a = \frac{m_2}{m_1 + m_2} g = \tfrac{1}{3}g = 3.3 \text{ meters/sec}^2,$$

and

$$T = \frac{m_1 m_2}{m_1 + m_2} g = (\tfrac{2}{3})(9.8) \text{ kg-m/sec}^2 = 6.5 \text{ nt}.$$

Example 6. Consider two unequal masses connected by a string which passes over a frictionless and massless pulley, as shown in Fig. 5–7*a*. Let m_2 be greater than m_1. Find the tension in the string and the acceleration of the masses.

We consider an upward acceleration positive. If the acceleration of m_1 is a, the acceleration of m_2 must be $-a$. The forces acting on m_1 and on m_2 are shown in Fig. 5–7*b* in which T represents the tension in the string.

The equation of motion for m_1 is

$$T - m_1 g = m_1 a$$

and for m_2 is

$$T - m_2 g = -m_2 a.$$

Combining these equations, we obtain

$$a = \frac{m_2 - m_1}{m_2 + m_1} g, \tag{5–12}$$

and

$$T = \frac{2m_1 m_2}{m_1 + m_2} g.$$

For example, if $m_2 = 2.0$ slugs and $m_1 = 1.0$ slug,

$$a = (32/3.0) \text{ ft/sec}^2 = g/3,$$

$$T = (\tfrac{4}{3})(32) \text{ slug-ft/sec}^2 = 43 \text{ lb}.$$

Notice that the magnitude of T is always intermediate between the weight of the mass m_1 (32 lb in our example) and the weight of the mass m_2 (64 lb in our example). This is to be expected, since T must exceed $m_1 g$ to give m_1 an upward acceleration, and $m_2 g$ must exceed T to give m_2 a downward acceleration. In the special case when $m_1 = m_2$, we obtain $a = 0$ and $T = m_1 g = m_2 g$, which is the static result to be expected.

Example 7. Consider an elevator moving vertically with an acceleration **a**. We wish to find the force exerted by a passenger on the floor of the elevator.

Acceleration will be taken positive upward and negative downward. Thus positive acceleration in this case means that the elevator is either moving upward with increasing speed or is moving downward with decreasing speed. Negative acceleration means that the elevator is moving upward with decreasing speed or downward with increasing speed.

From Newton's third law the force exerted by the passenger on the floor will always be equal in magnitude but opposite in direction to the force exerted by the floor on the passenger. We can therefore calculate either the action force or the reaction force. When the forces acting on the passenger are used, we solve for the latter force. When the forces acting on the floor are used, we solve for the former force.

The situation is shown in Fig. 5–8: The passenger's true weight is **W** and the force exerted on him by the floor, called **P,** is his *apparent* weight in the accelerating elevator. The resultant force acting on him is **P** + **W.** Forces will be taken as positive when directed upward. From the second law of motion we have

$$F = ma,$$

or

$$P - W = ma, \qquad\qquad (5\text{-}13)$$

where m is the mass of the passenger and a is his (and the elevator's) acceleration.

Suppose, for example, that the passenger weighs 160 lb and the acceleration is 2.0 ft/sec² upward. We have

$$m = \frac{W}{g} = \frac{160 \text{ lb}}{32 \text{ ft/sec}^2} = 5.0 \text{ slugs},$$

and from Eq. 5–13,

$$P - 160 \text{ lb} = (5.0 \text{ slugs})(2.0 \text{ ft/sec}^2)$$

or

$$P = \text{apparent weight} = 170 \text{ lb}.$$

If we were to measure this force directly by having the passenger stand on a spring scale fixed to the elevator floor (or suspended from the ceiling), we would find the scale reading to be 170 lb for a man whose weight is 160 lb. The passenger feels himself pressing down on the floor with greater force (the floor is pressing upward on him with greater force) than when he and the elevator are at rest. Everyone experiences this feeling when an elevator starts upward from rest.

If the acceleration were taken as 2.0 ft/sec² downward, then $a = -2.0$ ft/sec² and $P = 150$ lb

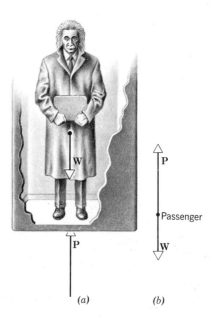

FIGURE 5–8 Example 7. (*a*) A passenger stands on the floor of an elevator. (*b*) A free-body diagram for the passenger.

for the passenger. The passenger who weighs 160 lb feels himself pressing down on the floor with less force than when he and the elevator are at rest.

If the elevator cable were to break and the elevator were to fall freely with an acceleration $a = -g$, then P would equal $W + (W/g)(-g) = 0$. Then the passenger and floor would exert no forces on each other. The passenger's apparent weight, as indicated by the spring scale on the floor, would be zero.

5–12 Frictional Forces*

We shall now consider the important forces that are classified under friction. Because a most simple and relatively general law describes many frictional forces, we will be able to solve many problems without necessarily understanding the origin of frictional forces themselves. Before considering applications of the frictional force law, however, we shall first consider briefly some ideas concerning the origin and nature of these forces.

If we project a block of mass m with initial velocity $\mathbf{v}_0$ along a long horizontal table, it eventually comes to rest. This means that, while it is moving, it experiences an average acceleration $\bar{\mathbf{a}}$ that points in the direction opposite to its motion. If (in an inertial frame) we see that a body is being accelerated, we always associate a force, defined from Newton's second law, with the motion. In this case we declare that the table exerts a *force of friction*, whose average value is $m\bar{\mathbf{a}}$, on the sliding block.

Actually, whenever the surface of one body slides over that of another, each body exerts a frictional force on the other, parallel to the surfaces. The frictional force on each body is in a direction opposite to its motion relative to the other body. Even when there is no relative motion, frictional forces may exist between surfaces.

Although we have ignored its effects up to now, friction is very important in our daily lives. Left to act alone it brings every rotating shaft to a halt. In an automobile, about 20% of the engine power is used to counteract frictional forces. On the other hand, without friction we could not walk as we now do; we could not hold a pencil in our hand and if we could it would not write; wheeled transport as we know it would not be possible.

We want to know how to express frictional forces in terms of the properties of the body and its environment; that is, we want to know the force law for frictional forces. In what follows we consider the sliding (not rolling) of one dry (unlubricated) surface over another. As we shall see later, friction, viewed at the microscopic level, is a very complicated phenomenon† and the force laws for dry, sliding friction are empirical in character and approximate in their predictions. They do not have the elegant simplicity and accuracy that we find for the gravitational force law (Chapter 14) or for the electrostatic force law (Chapter 22). It is remarkable, however, considering the enormous diversity of surfaces one encounters, that many aspects of frictional behavior can be understood qualitatively on the basis of a few simple mechanisms.

Consider a block at rest on a horizontal table as in Fig. 5–9. Attach a spring to it to measure the force required to set the block in motion. We find that the block will not move even though we apply a small force. We say that our applied force is balanced by an opposite frictional force exerted on the block by the table, acting along the surface of contact. As we increase the applied force we find some definite force at

---

* See "The Friction of Solids" by E. H. Freitag, in *Contemporary Physics*, Vol. 2, 1961, p. 198, for a good general reference; see also the article "Friction" in the *Encyclopaedia Britannica*.

† See, for example, "Stick and Slip" by Ernest Rabinowicz, in *Scientific American*, May 1956.

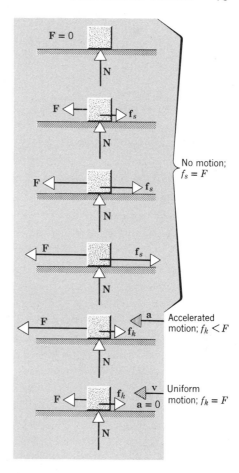

FIGURE 5–9 A block being put into motion as applied force **F** overcomes frictional forces. In the first four drawings the applied force is gradually increased from zero to magnitude $\mu_s N$. No motion occurs until this point because the frictional force always just balances the applied force. The instant F becomes greater than $\mu_s N$, the block goes into motion, as is shown in the fifth drawing. In general, $\mu_k N < \mu_s N$; this leaves an unbalanced force to the left and the block accelerates. In the last drawing F has been reduced to equal $\mu_k N$. The net force is zero, and the block continues with constant velocity.

which the block just begins to move. Once motion has started, this same force produces accelerated motion. By reducing the force once motion has started, we find that it is possible to keep the block in uniform motion without acceleration; this force may be small, but it is never zero.

The frictional forces acting between surfaces at rest with respect to each other are called forces of *static friction*. The maximum force of static friction will be the same as the smallest force necessary to start motion. Once motion is started, the frictional forces acting between the surfaces usually decrease so that a smaller force is necessary to maintain uniform motion. The force acting between surfaces in relative motion are called forces of *kinetic friction*.

The maximum force of static friction between any pair of dry unlubricated surfaces follows these two empirical laws. (1) It is approximately independent of the area of contact, over wide limits and (2) it is proportional to the normal force. The normal force, sometimes called the loading force, is the one which either body exerts on the other at right angles to their mutual interface. It arises from the elastic deformation of the bodies in contact, such bodies never really being entirely rigid. For a block resting on a horizontal table or sliding along it, the normal force is equal in magnitude to the weight of the block. Because the block has no vertical acceleration, the table must be exerting a force on the block that is directed upward and is equal in magnitude to the downward pull of the earth on the block, that is, equal to the block's weight.

The ratio of the magnitude of the maximum force of static friction to the magnitude of the normal force is called the *coefficient of static friction* for the surfaces involved. If f_s represents the magnitude of the force of static friction, we can write

$$f_s \leq \mu_s N, \tag{5-14}$$

where μ_s is the coefficient of static friction and N is the magnitude of the normal force. The equality sign holds only when f_s has its maximum value.

The force of kinetic friction f_k between dry, unlubricated surfaces follows the same two laws as those of static friction. (1) It is approximately independent of the area of contact over wide limits and (2) it is proportional to the normal force. The force of kinetic friction is also reasonably independent of the relative speed with which the surfaces move over each other.

The ratio of the magnitude of the force of kinetic friction to the magnitude of this normal force is called the *coefficient of kinetic friction*. If f_k represents the magnitude of the force of kinetic friction,

$$f_k = \mu_k N, \tag{5-15}$$

where μ_k is the coefficient of kinetic friction.

Both μ_s and μ_k are dimensionless constants, each being the ratio of (the magnitudes of) two forces. Usually, for a given pair of surfaces $\mu_s > \mu_k$. The actual values of μ_s and μ_k depend on the nature of both the surfaces in contact. Both μ_s and μ_k can exceed unity, although commonly they are less than one. Notice that Eqs. 5–14 and 5–15 are relations between the magnitudes only of the normal and frictional forces. These forces are always directed perpendicularly to one another.

On the atomic scale even the most finely polished surface is far from plane. When two bodies are placed in contact, the actual microscopic area of contact is much less than the apparent macroscopic area of contact; in a particular case these areas can be easily in the ratio of 1 to 10^4.

The actual (microscopic) area of contact is proportional to the normal force, because the contact points deform plastically under the great stresses that develop at these points. The actual contact area remains the same even when the apparent contact area is reduced because increased normal force per unit actual area produces further plastic deformation. Many contact points actually become "cold-welded" together. This phenomenon, *surface adhesion,* occurs because at the contact point the molecules on opposite sides of the surface are so close together that they exert strong intermolecular forces on each other.

When one body is pulled across another, the frictional resistance is associated with the rupturing of these thousands of tiny welds, which continually reform as new chance contacts are made (see Fig. 5–10).

The coefficient of friction depends on many variables, such as the nature of the materials, surface finish, surface films, temperature, and extent of contamination. For example, if two carefully cleaned metal surfaces are placed in a highly evacuated chamber so that surface oxide films do not form, the coefficient of friction rises to enormous values and the surfaces actually become firmly "welded" together. The admission of a small amount of air to the chamber so that oxide films may form on the opposing surfaces reduces the coefficient of friction to its "normal" value.

The frictional force that opposes one body rolling over another is much less than that for a sliding motion and this, indeed, is the advantage of the wheel over the

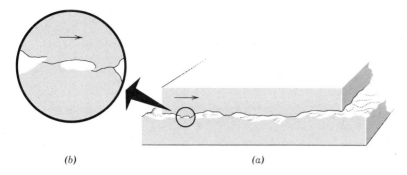

FIGURE 5–10 Sliding friction. (*a*) The upper body is sliding to the right over the lower body in this enlarged diagram. (*b*) A further enlarged view showing two spots where surface adhesion has occurred. Force is required to break these welds apart and maintain the motion.

sledge. This reduced friction is due in large part to the fact that, in rolling, the microscopic contact welds are "peeled" apart rather than "sheared" apart as in sliding friction. This may reduce the frictional force by as much as 1000-fold.

Examples of the application of the empirical force law for friction follow. The coefficients of friction given are assumed to be constant. Actually μ_k can be regarded as a good average value that is not greatly different from the value at any particular speed in the range.

Example 8. Consider an automobile moving along a straight horizontal road with a speed v_0. If the coefficient of static friction between the tires and the road is μ_s, what is the shortest distance in which the automobile can be stopped?

The forces acting on the automobile, considered to be a particle, are shown in Fig. 5–11. The car is assumed to be moving in the positive x-direction. If we assume that f_s is a constant force, we have uniformly decelerated motion.

From the relation (see Eq. 3–16)

$$v^2 = v_0{}^2 + 2ax,$$

with the final speed $v = 0$, we obtain

$$x = -v_0{}^2/2a,$$

where the minus sign means that **a** points in the negative x-direction.

To determine a, apply the second law of motion to the x-component of the motion:

$$-f_s = ma = (W/g)a \qquad \text{or} \qquad a = -g(f_s/W).$$

From the y-component we obtain

$$N - W = 0 \qquad \text{or} \qquad N = W,$$

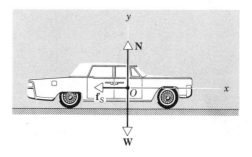

FIGURE 5–11 Example 8. The forces on a decelerating automobile.

so that $$\mu_s = f_s/N = f_s/W$$

and $$a = -\mu_s g.$$

Then the distance of stopping is

$$x = -v_0{}^2/2a = v_0{}^2/2g\mu_s \tag{5-16}$$

The greater the initial speed, the longer the distance required to come to a stop; in fact, this distance varies as the square of the initial speed. Also, the greater the coefficient of static friction between the surfaces, the less the distance required to come to a stop.

We have used the coefficient of static friction in this problem, rather than the coefficient of sliding friction, because we assume there is no sliding between the tires and the road. We have neglected rolling friction. Furthermore, we have assumed that the maximum force of static friction ($f_s = \mu_s N$) operates because the problem seeks the shortest distance for stopping. With a smaller static frictional force the distance for stopping would obviously be greater. The correct braking technique required here is to keep the car just on the verge of skidding. If the surface is smooth and the brakes are applied fully, sliding may occur. In this case μ_k replaces μ_s, and the distance required to stop is seen to increase from Eq. 5–16.

As a specific example, if $v_0 = 60$ miles/hr $= 88$ ft/sec, and $\mu_s = 0.60$ (a typical value) we obtain

$$x = \frac{v_0{}^2}{2\mu_s g} = \frac{(88 \text{ ft/sec})^2}{2(0.60)(32 \text{ ft/sec}^2)} = 200 \text{ ft.}$$

Notice that the mass of the car does not appear in Eq. 5–16. How can you explain the practice of "weighing down" a car in order to increase safety in driving on icy roads?

The student should now investigate how, in principle, forces of friction would modify the results of the examples of Section 5–11.

5–13   The Dynamics of Uniform Circular Motion

In Section 4–4 we pointed out that if a body is moving at uniform speed v in a circle of radius r, it experiences a centripetal acceleration **a** whose magnitude is v^2/r. The direction of **a** is always radially inward toward the center of rotation. Thus **a** is a variable vector because, even though its magnitude remains constant, its direction changes continuously as the motion progresses.

Recall that there need not be any motion in the direction of an acceleration. In general, there is no fixed relation between the directions of the acceleration **a** and the velocity **v** of a particle, as Fig. 4–6 shows. For a particle in uniform circular motion the acceleration **a** and velocity **v** are always at right angles to each other.

Every accelerated body must have a force **F** acting on it, defined by Newton's second law (**F** $= m$**a**). Thus (assuming that we are in an inertial frame), if we see a body undergoing uniform circular motion, we can be certain that a net force **F**, given in magnitude by

$$F = ma = mv^2/r$$

must be acting on the body; the body is not in equilibrium. The direction of **F** at any instant must be the direction of **a** at that instant, namely, radially inward. We must always be able to account for this force by pointing to a particular object in the environment that is exerting the force on the accelerating, circulating body.

If the body in uniform circular motion is a disk on the end of a string moving in a circle on a frictionless horizontal table as in Fig. 5–12, the force **F** on the disk is provided by the tension **T** in the string. This force **T** is the net force acting on the disk.

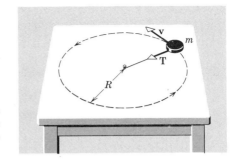

FIGURE 5–12 A disk m moves with constant speed in a circular path on a horizontal frictionless surface. The only horizontal force acting on m is the centripetal force **T** with which the string pulls on the body.

It accelerates the disk by constantly changing the direction of its velocity so that the disk moves in a circle. **T** is always directed toward the pin at the center and its magnitude is mv^2/R. If the string were to be cut where it joins the disk, there would be no net force exerted on the disk. The disk would then move with constant speed in a straight line along the direction of the tangent to the circle at the point at which the string was cut. Hence, to keep the disk moving in a circle, a force must be supplied to it pulling it inward toward the center.

Forces responsible for uniform circular motion are called *centripetal* forces because they are directed "toward the center" of the circular motion. To label a force as "centripetal," however, simply means that it always points radially inward; the name tells us nothing about the nature of the force or about the body that is exerting it. Thus, for the revolving disk of Fig. 5–12, the centripetal force is an elastic force provided by the string; for the moon revolving around the earth (in an approximately circular orbit) the centripetal force is the gravitational pull of the earth on the moon; for an electron circulating about an atomic nucleus the centripetal force is electrostatic. A centripetal force is not a new kind of force but simply a way of decribing the behavior with time of forces that are attributable to specific bodies in the environment. Thus a force can be centripetal and elastic, centripetal and gravitational, or centripetal and electrostatic, among other possibilities.

Let us consider some examples of forces that act centripetally.

■ **Example 9.** *The conical pendulum.* Figure 5–13a represents a small body of mass m revolving in a horizontal circle with constant speed v at the end of a string of length L. As the body swings around, the string sweeps over the surface of a cone. This device is called a *conical pendulum*. Find the time required for one complete revolution of the body.

If the string makes an angle θ with the vertical, the radius of the circular path is $R = L \sin \theta$. The forces acting on the body of mass m are **W**, its weight, and **T**, the pull of the string, as shown in Fig. 5–13b. It is clear that $\mathbf{T} + \mathbf{W} \neq 0$. Hence, the resultant force acting on the body is nonzero, which is as it should be because a force is required to keep the body moving in a circle with constant speed.

We can resolve **T** at any instant into a radial and a vertical component.

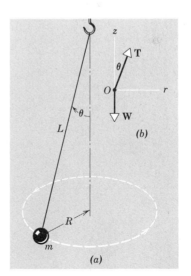

FIGURE 5–13 Example 9. (*a*) A mass m suspended from a string of length L swings so as to describe a circle. The string describes a right circular cone of semiangle θ. (*b*) A free-body force diagram for m.

$$T_r = T \sin \theta \quad \text{and} \quad T_z = T \cos \theta.$$

Since the body has no vertical acceleration,

$$T_z - W = 0.$$

But

$$T_z = T \cos \theta \quad \text{and} \quad W = mg,$$

so that

$$T \cos \theta = mg.$$

The radial acceleration is v^2/R. This acceleration is supplied by T_r, the radial component of **T**, which is the centripetal force acting on m. Hence

$$T_r = T \sin \theta = mv^2/R.$$

Dividing this equation by the preceding one, we obtain

$$\tan \theta = v^2/Rg, \quad \text{or} \quad v^2 = Rg \tan \theta,$$

which gives the constant speed of the bob. If we let τ represent the time for one complete revolution of the body, then

$$v = \frac{2\pi R}{\tau} = \sqrt{Rg \tan \theta}$$

or

$$\tau = \frac{2\pi R}{v} = \frac{2\pi R}{\sqrt{Rg \tan \theta}} = 2\pi \sqrt{R/(g \tan \theta)}.$$

But $R = L \sin \theta$, so that

$$\tau = 2\pi \sqrt{(L \cos \theta)/g}.$$

This equation gives the relation between τ, L, and θ. Notice that τ, called the *period* of motion, does not depend on m.

If $L = 3.0$ ft and $\theta = 30°$, what is the period of the motion? We have

$$\tau = 2\pi \sqrt{\frac{(3.0 \text{ ft})(0.866)}{32 \text{ ft/sec}^2}} = 1.8 \text{ sec.}$$

Example 10. *The rotor.* In many amusement parks we find a device called the *rotor*. The rotor is a hollow cylindrical room which can be set rotating about the central vertical axis of the cylinder. A person enters the rotor, closes the door, and stands up against the wall. The rotor gradually increases its rotational speed from rest until, at a predetermined speed, the floor below the person is opened downward, revealing a deep pit. The passenger does not fall but remains "pinned up" against the wall of the rotor. Find the coefficient of friction necessary to prevent falling.

The forces acting on the passenger are shown in Fig. 5–14. **W** is the passenger's weight, **f**$_s$ is the force of static friction between passenger and rotor wall, and **P** is the centripetal force exerted by the wall on the passenger necessary to keep him moving in a circle. Let the radius of the rotor be R and the final speed of the passenger be v. Since the passenger does not move vertically, but experiences a radial acceleration v^2/R at any instant, we have

$$f_s - W = 0$$

and

$$P(= ma) = (W/g)(v^2/R)$$

If μ_s is the coefficient of static friction between passenger and wall necessary to prevent slipping, then $f_s = \mu_s P$ and

$$f_s = W = \mu_s P$$

or

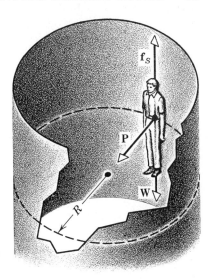

FIGURE 5-14 The forces on a person in a "rotor" of radius R.

$$\mu_s = \frac{W}{P} = \frac{gR}{v^2}.$$

This equation gives the minimum coefficient of friction necessary to prevent slipping for a rotor of radius R when a particle on its wall has a speed v. Notice that the result does not depend on the passenger's weight.

As a practical matter the coefficient of friction between the textile material of clothing and a typical rotor wall (canvas) is about 0.40. For a typical rotor the radius is 7.0 ft, so that v must be about 24 ft/sec or 16 miles/hr or more.

Example 11. Let the block in Fig. 5-15a represent an automobile or railway car moving at constant speed v on a level road-bed around a curve having a radius of curvature R. In addition to the two vertical forces, namely the force of gravity **W** and a normal force **N**, a horizontal centripetal force **P** acts on the car. In the case of the automobile this centripetal force is supplied by a sidewise frictional force exerted by the road on the tires; in the case of the railway car the centripetal force is supplied by the rails exerting a sidewise force on the inner rims of the car's wheels.

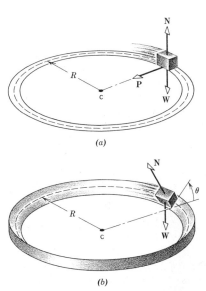

FIGURE 5-15

Neither of these sidewise forces can be safely relied upon to be large enough at all times and both cause unnecessary wear. Hence, the roadbed is banked on curves, as shown in Fig. 5–15*b*. In this case, the normal force **N** has not only a vertical component, as before, but also a horizontal component which supplies the centripetal force necessary for uniform circular motion; no additional sidewise forces are needed, therefore, with a properly banked roadbed.

The correct angle θ of banking can be obtained as follows. There is no vertical acceleration, so that

$$N \cos \theta = W.$$

The centripetal force is $N \sin \theta$, so that $N \sin \theta = mv^2/R$. Dividing the latter equation by the former and setting $W = mg$, we obtain

$$\tan \theta = v^2/Rg$$

Notice that the proper angle of banking depends upon the speed of the car and the curvature of the road. For a given curvature, the road is banked at an angle corresponding to an expected average speed. Often curves are marked by signs giving the proper speed for which the road was banked.

Check the banking formula for the limiting cases $v = 0$; $R \rightarrow \infty$; v large; and R small. You should note that great similarity between Fig. 5–13 of Example 9 and Fig. 5–15*b* of this example.

QUESTIONS

1. What is your mass in slugs? Your weight in newtons?

2. Why do you fall forward when a moving train decelerates to a stop and fall backward when a train accelerates from rest? What would happen if the train rounded a curve at constant speed?

3. A horse is urged to pull a wagon. The horse refuses to try, citing Newton's third law as his defense: "'The pull of the horse on the wagon is equal but opposite to the pull of the wagon on the horse.' If I can never exert a greater force on the wagon than it exerts on me, how can I ever start the wagon moving?" asks the horse. How would you reply?

4. A block of mass m is supported by a cord C from the ceiling, and another cord D is attached to the bottom of the block (Fig. 5–16). Explain this: If you give a sudden jerk to D it will break, but if you pull on D steadily, C will break.

5. Two 10-lb weights are attached to a spring scale as shown in Fig. 5–17. Does the scale read 0 lb, 10 lb, 20 lb, or give some other reading?

6. Criticize the statement, often made, that the mass of a body is a measure of the "quantity of matter" in it.

7. Using force, length, and time as fundamental quantities, what are the dimensions of mass?

8. Is the definition of mass that we have given limited to objects initially at rest?

9. Is the current standard of mass accessible, invariable, reproducible, indestructible? Does it have simplicity for comparison purposes? Would an atomic standard be better in any respect?

10. Suppose the carbon atom was chosen as the standard of mass. What information would be needed to express the mass of the standard kilogram in terms of this atomic standard? How could this information be obtained?

11. In a tug of war, three men pull on a rope to the left at A and three men pull to the right at B with forces of equal magnitude. Now a weight of 5.0 lb is hung vertically from the center of the rope.

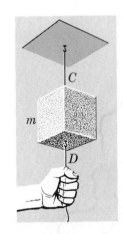

FIGURE 5–16

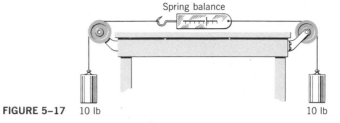

FIGURE 5-17   10 lb   10 lb

(*a*) Can the men get the rope *AB* to be horizontal? (*b*) If not, explain. If so, determine the magnitude of the forces required at *A* and *B* to do this.

12. Both the following statements are true; explain them. Two teams having a tug of war must always pull equally hard on one another. The team that pushes harder against the ground wins.

13. Under what circumstances would your weight be zero? Does your answer depend on the choice of a reference system?

14. Two objects of equal mass rest on opposite pans of a trip scale. Does the scale remain balanced when it is accelerated up or down in an elevator?

15. A massless rope is strung over a frictionless pulley. A monkey holds onto one end of the rope and a mirror, having the same weight as the monkey, is attached to the other end of the rope at the monkey's level. Can the monkey get away from his image seen in the mirror (*a*) by climbing up the rope, (*b*) by climbing down the rope, (*c*) by releasing the rope?

16. A student standing on the large platform of a spring scale notes his weight. He then takes a step on this platform and notices that the scale reads less than his weight at the beginning of the step and more than his weight at the end of the step. Explain.

17. In 1920 a prominent newspaper editorialized as follows about the pioneering rocket experiments of Robert H. Goddard, dismissing the notion that a rocket could operate in a vacuum: "That Professor Goddard, with his 'chair' in Clark College and the countenancing of the Smithsonian Institution, does not know the relation of action to reaction, and of the need to have something better than a vacuum against which to react—to say that would be absurd. Of course, he only seems to lack the knowledge ladled out daily in high schools." What is wrong with this argument?

18. There is a limit beyond which further polishing of a surface *increases* rather than decreases frictional resistance. Can you explain this?

19. Is it reasonable to expect a coefficient of friction to exceed unity?

20. How could a person who is at rest on completely frictionless ice covering a pond reach shore? Could he do this by walking, rolling, swinging his arms, or kicking his feet? How could a person be placed in such a position in the first place?

21. Explain how the range of your car's headlights limits the safe driving speed at night.

22. Your car skids across the center line on an icy highway. Should you turn the front wheels in the direction of skid or in the opposite direction (*a*) when you want to avoid a collision with an oncoming car, (*b*) when no other car is near but you want to regain control of the steering?

23. If you want to stop the car in the shortest distance on an icy road, should you (*a*) push hard on the brakes to lock the wheels, (*b*) push hard enough to prevent slipping, or (*c*) "pump" the brakes?

24. A cube of weight *W* rests on a rough inclined plane which makes an angle θ with the horizontal. Compare the minimum force necessary to start the cube moving down the plane with that necessary to start the cube moving up the plane. How do these compare with the minimum *horizontal* force (transverse to the slope) that will cause the cube to move down the plane?

25. Why are the train roadbeds and highways banked on curves?

26. How does the earth's rotation affect the apparent weight of a body at the equator?

27. A car is riding on a country road that resembles a roller coaster track. If the car travels with uniform speed, compare the force it exerts on a horizontal section of the road to the force it exerts on the road at the top of a hill and at the bottom of a hill. Explain.

28. Suppose you need to measure whether a table top in a train is truly horizontal. If you use a spirit level can you determine this when the train is moving down or up a grade? When the train is moving along a curve? (Hint: there are two horizontal components.)

29. In the conical pendulum of Example 9 what happens to the period τ and the speed v when $\theta = 90°$? Why is this angle not achievable physically? Discuss the case for $\theta = 0°$.

30. A coin is put on a phonograph turntable. The motor is started, but before the final speed of rotation is reached, the coin flies off. Explain.

31. A passenger in the front seat of a car finds himself sliding toward the door as the driver makes a sudden turn. Describe the forces on the passenger and on the car at this instant.

32. While standing in an elevator, a person hangs unequal masses m and M over a pulley that is free to turn. He observes the curious fact that the masses remain balanced, that is, there is no tendency for the pulley to turn. What does he conclude about the motion of the elevator?

PROBLEMS

1(5). Two blocks, mass m_1 and m_2 are connected by a light spring on a horizontal frictionless table. Find the ratio of their accelerations a_1 and a_2 after they are pulled apart and then released. *Answer:* $a_1/a_2 = m_2/m_1$.

2(6). A book rests on a table that rests on the earth. Identify the reaction forces to (*a*) the force of gravity on the book, (*b*) the force of the table legs on the earth, (*c*) the force of gravity on the table, and (*d*) the force of the book on the tabletop.

3(6). Two blocks are in contact on a frictionless table. A horizontal force is applied to one block, as shown in Fig. 5–18. (*a*) If $m_1 = 2.0$ kg, $m_2 = 1.0$ kg, and $F = 3.0$ nt, find the force of contact between the two blocks. (*b*) Show that if the same force F is applied to m_2 rather than to m_1, the force of contact between the blocks in 2.0 nt, which is not the same as the value derived in (*a*). Explain. *Answer:* (*a*) 1.0 nt.

4(9). A certain particle has a weight of 20 nt at a point where the acceleration due to gravity is 9.8 meters/sec². (*a*) What are the weight and mass of the particle at a point where the acceleration due to gravity is 4.9 meters/sec²? (*b*) What are the weight and mass of the particle if it is moved to a point in space where the gravitational force is zero?

5(9). A space traveler whose mass is 75 kg leaves the earth. Compute his weight (*a*) on the earth, (*b*) 400 miles above the earth (where $g = 8.1$ meters/sec²), and (*c*) in interplanetary space. What is his mass at each of these locations?
Answer: (*a*) 740 nt. (*b*) 610 nt. (*c*) Zero.

6(9). What is the weight in newtons and the mass in kilograms of (*a*) a five-pound bag of sugar, (*b*) a 240-lb fullback, and (*c*) a 1.8-ton automobile?

7(11). A 400-lb motorcycle accelerates from rest to 50 miles/hr in 60 sec. (*a*) What average force acts on it? (*b*) What body provides the force?
Answer: (*a*) 15 lb. (*b*) The road.

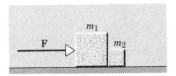

FIGURE 5–18   Problem 3(6).

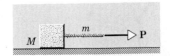

FIGURE 5–19   Problem 10(11).

8(11). A car moving initially at a speed of 50 miles/hr and weighing 3000 lb is brought to a stop in a distance of 200 ft. (*a*) Find the braking force and the time required to stop. (*b*) Assuming the same braking force, find the distance and time required to stop if the car was going 25 miles/hr initially.

9(11). An experimental rocket sled can be accelerated from rest to 1600 km/hr in 1.8 sec. What constant force would be required if the sled has a mass of 500 kg?
Answer: 1.2×10^5 nt.

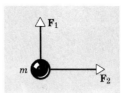

FIGURE 5-20 Problem 11(11).

10(11). A block of mass M is pulled along a horizontal frictionless surface by a rope of mass m, as in Fig. 5–19. A force **P** is applied to one end of the rope. (*a*) Find the acceleration of the block and the rope. (*b*) Find the force that the rope exerts on the block M.

11(11). A body of mass m is acted on by two forces **F**$_1$ and **F**$_2$, as shown in Fig. 5–20. If $m = 5.0$ kg, $F_1 = 3.0$ nt, and $F_2 = 4.0$ nt, find the vector acceleration of the body.
Answer: 1.0 meters/sec^2, 37° from **F**$_2$ toward **F**$_1$.

12(11). An electron is projected horizontally at a speed of 1.2×10^7 meters/sec into an electric field which exerts a constant vertical force of 4.5×10^{-16} nt on it. The mass of the electron is 9.1×10^{-31} kg. Determine the vertical distance the electron is deflected during the time it has moved forward 3.0 cm horizontally.

13(11). A body of mass 2.0 slugs is acted on by the downward force of gravity and a horizontal force of 130 lb. Find (*a*) its acceleration and (*b*) its velocity as a function of time, assuming it starts from rest.
Answer: (*a*) $a_{\mathrm{hor}} = 65$ ft/sec^2; $a_{\mathrm{vert}} = 32$ ft/sec^2. (*b*) $v_{\mathrm{hor}} = 65t$, ft/sec; $v_{\mathrm{vert}} = 32t$, ft/sec.

14(11). (*a*) Compute the initial upward acceleration of a rocket of mass 1.3×10^4 kg if the initial upward thrust of its engine is 2.6×10^5 nt. (*b*) Can you neglect the weight of the rocket (the downward pull of the earth on it)?

15(11). A charged sphere of mass 3.0×10^{-4} kg is suspended from a string. An electric force acts horizontally on the sphere so that the string makes an angle of 37° with the vertical when at rest. Find (*a*) the magnitude of the electric force and (*b*) the tension in the string.
Answer: (*a*) 2.2×10^{-3} nt. (*b*) 3.7×10^{-3} nt.

16(11). A 0.50-kg puck rests on a frictionless surface at the origin of a coordinate system. Two strings are attached. One is pulled with a force of 2.5 nt along the y-axis and the other with a force of 6.0 nt along the x-axis. (*a*) Where is the puck 10 sec later? (*b*) What is the puck's velocity at this time? The strings remain parallel to the axes.

17(11). An electron travels in a straight line from the cathode of a vacuum tube to its anode, which is exactly 1.0 cm away. It starts with zero speed and reaches the anode with a speed of 6.0×10^6 meters/sec. (*a*) Assume constant acceleration and compute the force on the electron. Take the electron's mass to be 9.1×10^{-31} kg. This force is electrical in origin. (*b*) Compare it with the gravitational force on the electron, which we neglected when we assumed straight-line motion.
Answer: (*a*) 1.6×10^{-15} nt. (*b*) 8.9×10^{-30} nt.

18(11). An 8.5-kg object passes through the origin with a velocity of 30 meters/sec parallel to the x-axis. It experiences a constant 17-nt force in the direction of the positive y-axis. (*a*) Describe the resulting motion. Calculate (*b*) the velocity and (*c*) the position of the particle after 15 sec have elapsed.

19(11). (*a*) Neglecting gravitational forces, what force would be required to accelerate a 1.0-kiloton spaceship from rest to $\frac{1}{10}$ the speed of light in 3 days? In two months? (*b*) Assuming the engines are shut down at the speed of $\frac{1}{10}c$, what would be the time required to complete a five light-month journey for each of these two cases?
Answer: (*a*) 1.0×10^8 nt; 5.2×10^6 nt. (*b*) 4.2 yr; 4.3 yr.

20(11). A man of mass 80 kg jumps down to a concrete patio from a window ledge only 0.50 meter above the ground. He neglects to bend his knees on landing, so that his motion is arrested in a distance of about 2.0 cm. (*a*) What is the average acceleration of the man from the time his feet first touch the patio to the time he is brought fully to rest? (*b*) With what average force does this jump jar his bone structure?

21(11). A 4.0-kg mass has a velocity, in meters/sec, of $\mathbf{v}_1 = 48\mathbf{i} + 36\mathbf{j}$ at one instant. After 12 sec the velocity has changed to $\mathbf{v}_2 = 18\mathbf{i} - 6.0\mathbf{j}$. If the force acting on the object is constant, what are (*a*) the *x*- and *y*-components of the force and (*b*) its magnitude?
Answer: (*a*) -10 nt; -14 nt. (*b*) 17 nt.

22(11). A neutron travels at a speed of about 1.4×10^7 meters/sec. Nuclear forces are of very short range, being essentially zero outside of a nucleus but very strong inside. If the neutron is captured by a nucleus whose diameter is 1.0×10^{-14} meters, what is the minimum force acting on this neutron?

23(11). Let the only forces acting on two bodies be their mutual interactions. If both bodies start from rest, show that the distances traveled by each are inversely proportional to the respective masses of the bodies.

24(11). Three blocks are connected, as shown in Fig. 5–21, on a horizontal frictionless table and pulled to the right with a force $T_3 = 60$ nt. If $m_1 = 10$ kg, $m_2 = 20$ kg, and $m_3 = 30$ kg, find the tensions T_1 and T_2. Draw an analogy to bodies being pulled in tandem, such as an engine pulling a train of coupled cars.

25(11). How could a 100-lb object be lowered from a roof using a cord with a breaking strength of 87 lb without breaking the cord?
Answer: Lower it with an acceleration of 4.2 ft/sec² or greater.

26(11). A 5.0-kg lead ball is dropped into a swimming pool from a diving board that is 3.0 meters above the water. It reaches the bottom of the pool, 4.0 meters below the surface, 0.75 sec after reaching the water surface. With what constant force did the water act on the ball?

27(11). An elevator and its load have a combined mass of 1600 kg. Find the tension in the supporting cable when the elevator, originally moving downward at 20 meters/sec, is brought to rest with constant acceleration in a distance of 50 meters.
Answer: 2.2×10^4 nt.

28(11). Refer to Fig. 5–5. Let the mass of the block be 2.0 slugs and the angle θ equal 30°. (*a*) Find the tension in the string and (*b*) the normal force acting on the block. (*c*) If the string is cut, find the acceleration of the block.

29(11). A block of mass $m_1 = 3.0$ slugs on a smooth inclined plane of angle 30° is connected by a cord over a small frictionless pulley to a second block of mass $m_2 = 2.0$ slugs hanging vertically (Fig. 5–22). (*a*) What is the acceleration of each body? (*b*) What is the tension in the cord?
Answer: (*a*) 3.2 ft/sec². (*b*) 58 lb.

30(11). An elevator weighing 6000 lb is pulled upward by a cable with an acceleration of 4.0 ft/sec². (*a*) What is the tension in the cable? (*b*) What is the tension when the elevator is accelerating downward at 4.0 ft/sec²?

31(11). A fireman weighing 160 lb slides down a vertical pole with an average acceleration of 10 ft/sec². What is the average vertical force he exerts on the pole?
Answer: 110 lb.

32(11). Refer to Fig. 5–6. Let $m_1 = 4.0$ slugs and $m_2 = 2.0$ slugs. Find (*a*) the tension in the string and (*b*) the acceleration of the two blocks.

33(11). Two blocks slide together down an inclined plane without friction (Fig. 5–23). The mass of each block is 30 kg. What is the magnitude of the normal component of the force between the two blocks?
Answer: 190 nt.

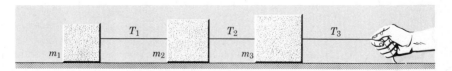

FIGURE 5-21 Problem 24(11).

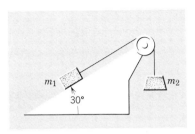

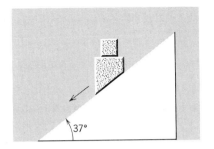

FIGURE 5-22 Problem 29(11).

FIGURE 5-23 Problem 33(11).

34(11). An 80-kg man is parachuting and experiencing a downward acceleration of 2.5 meters/sec². The mass of the parachute is 5.0 kg. (*a*) What is the value of the upward force exerted on the parachute by the air? (*b*) What is the value of the downward force exerted by the man on the parachute?

35(11). A block is projected up a frictionless inclined plane with a speed v_0. The angle of incline is θ. (*a*) How far up the plane does it go? (*b*) How long does it take to get there? (*c*) What is its speed when it gets back to the bottom? Find numerical answers for $\theta = 30°$ and $v_0 = 8.0$ ft/sec.
Answer: (*a*) $v_0^2/2g \sin \theta$; 2.0 ft. (*b*) $v_0/g \sin \theta$; 0.50 sec. (*c*) v_0; 8.0 ft/sec.

36(11). An elevator consists of the elevator cage (*A*), the counterweight (*B*), the driving mechanism (*C*), and the cable and pulleys as shown in Fig. 5–24. The mass of the cage is 1100 kg and the mass of the counterweight is 1000 kg. Neglect friction and the mass of the cable and pulleys. The elevator accelerates upward at 2.0 meters/sec² and the counterweight accelerates downward at the same rate. (*a*) What are the values of the tension in the two parts of the cable? (*b*) What force is exerted on the cable by the driving mechanism?

37(11). A lamp hangs vertically from a cord in a descending elevator. The elevator has a deceleration of 8.0 ft/sec² before coming to a stop. (*a*) If the tension in the cord is 20 lbs, what is the mass of the lamp? (*b*) What is the tension in the cord when the elevator ascends with an acceleration of 8.0 ft/sec²?
Answer: (*a*) 0.50 slugs. (*b*) 20 lbs.

38(11). An object is hung from a spring balance attached to the ceiling of an elevator. The balance reads 60 lbs when the elevator is standing still. (*a*) What is the reading when the elevator is moving upward with a constant speed of 24 ft/sec? (*b*) What is the reading of the balance when the elevator is accelerating downward at 24 ft/sec²?

39(11). A 100-kg man lowers himself to the ground from a height of 10 meters by means of a rope passed over a frictionless pulley and attached to a 70-kg sandbag. (*a*) What is the tension in the rope? (*b*) With what speed does the man hit the ground? (*c*) Is there anything he can do to reduce the speed with which he hits the ground?
Answer: (*a*) 810 nt. (*b*) 5.9 meters/sec. (*c*) Yes; he can climb the rope while falling.

40(11). A 10-kg monkey is climbing a massless rope attached to a 15-kg mass over a (fric-

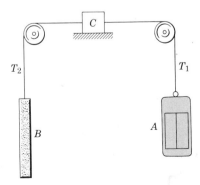

FIGURE 5-24 Problem 36(11).

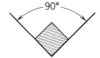

FIGURE 5-25 Problem 43(11).

tionless!) tree limb. (*a*) Explain quantitatively how the monkey can climb up the rope so that he can raise the 15-kg mass off the ground. (*b*) If, after the mass has been raised off the ground, the monkey stops climbing and holds on to the rope, what will now be (*b*) his acceleration and (*c*) the tension in the rope?

41(11). A plumb bob hanging from the ceiling of a railroad car acts as an accelerometer. (*a*) Derive the general expression relating the horizontal acceleration *a* of the car to the angle θ made by the bob with the vertical. (*b*) Find *a* when $\theta = 20°$. Find θ when $a = 5.0$ ft/sec². *Answer:* (*a*) $a = g \tan \theta$. (*b*) 12 ft/sec²; 8.9°.

42(11). A uniform flexible chain of length *l*, with weight per unit length λ, passes over a small, frictionless, massless pulley. It is released from a rest position with a length of chain *x* hanging from one side and a length $l - x$ from the other side. Find the acceleration *a* as a function of *x*.

43(11). A block of mass *m* slides in an inclined right-angled trough as in Fig. 5-25. If the coefficient of kinetic friction between the block and the material composing the trough is μ_k, find the acceleration of the block. *Answer:* $g(\sin \theta - \sqrt{2} \ \mu_k \cos \theta)$.

44(12). A wire will break under tensions exceeding 200 lb. (*a*) If the wire, not necessarily horizontal, is used to drag a box across the floor, what is the greatest weight that can be moved if the coefficient of static friction is 0.35? (*b*) If the wire is used to lift a box, what is the greatest weight that can be lifted with an upward acceleration of 3.0 ft/sec²?

45(12). A hockey puck weighing 0.25 lb slides on the ice for 50 ft before it stops. (*a*) If its initial speed was 20 ft/sec, what is the force of friction between puck and ice? (*b*) What is the coefficient of kinetic friction? *Answer:* (*a*) 0.031 lb. (*b*) 0.13.

46(12). The coefficient of kinetic friction in Fig. 5-26 is 0.20. What is the acceleration of the block?

47(12). Two blocks are connected over a massless pulley as shown in Fig. 5-27. The mass of block *A* is 10 kg and the coefficient of kinetic friction is 0.20. Block *A* slides down the incline at constant speed. What is the mass of block *B*? *Answer:* 3.3 kg.

48(12). A 5.0-kg block on an inclined plane is acted upon by a horizontal force of 50 nt (Fig. 5-28). The coefficient of friction between block and plane is 0.30. (*a*) What is the acceleration of the block if it is moving up the plane? (*b*) How far up the plane will the block go if it has an initial upward speed of 4.0 meters/sec? (*c*) What happens to the block after it reaches the highest point?

49(12). Frictional heat generated by the moving ski is the chief factor promoting sliding in skiing. The ski sticks at the start, but once in motion will melt the snow beneath it. Waxing the ski makes

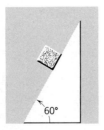

FIGURE 5-26 Problem 46(12).

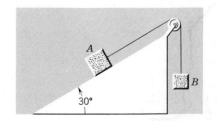

FIGURE 5-27 Problem 47(12).

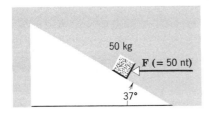

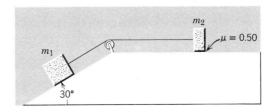

FIGURE 5–28 Problem 48(12).   **FIGURE 5–29** Problem 55(12).

it water repellent and reduces friction with the film of water. A magazine reports that a new type of plastic ski is even more water repellent and that on a gentle 700-ft slope in the Alps, a skier reduced his time from 61 to 42 sec with new skis. (*a*) Determine the average accelerations for each pair of skis. (*b*) Assuming a 3°-slope compute the coefficient of kinetic friction for each case. *Answer:* (*a*) 0.38 ft/sec²; 0.79 ft/sec². (*b*) 0.041; 0.028.

50(12). A 150-lb crate is dragged across a floor by pulling on a rope that inclined 15° above the horizontal. (*a*) If the coefficient of static friction is 0.50, what tension in the rope is required to start the crate moving? (*b*) If $\mu_k = 0.35$, what is the initial acceleration of the crate?

51(12). The handle of a floor mop of mass m makes an angle θ with the vertical direction. Let μ_k be the coefficient of kinetic friction between mop and floor, and μ_s be the coefficient of static friction between mop and floor. Neglect the mass of the handle. (*a*) Find the magnitude of the force F directed along the handle required to slide the mop with uniform velocity across the floor. (*b*) Show that if θ is smaller than a certain angle θ_0, the mop cannot be made to slide across the floor no matter how great a force is directed along the handle. (*c*) What is the angle θ_0? *Answer:* (*a*) $\mu_k mg/(\sin \theta - \mu_k \cos \theta)$. (*c*) $\theta_0 = \tan^{-1} \mu_s$.

52(12). A piece of ice slides down a 45°-incline in twice the time it takes to slide down a frictionless 45°-incline. What is the coefficient of kinetic friction between the ice and the incline?

53(12). A block slides down an inclined plane of slope angle φ with constant velocity. It is then projected up the same plane with an initial speed v_0. (*a*) How far up the incline will it move before coming to rest? (*b*) Will it slide down again? *Answer:* (*a*) $v_0^2/4g \sin \varphi$. (*b*) No.

54(12). A student wants to determine the coefficients of static friction and kinetic friction between a box and a plank. He places the box on the plank and gradually raises the plank. When the angle of inclination with the horizontal reaches 30°, the box starts to slip and slides 4.0 meters down the plank in 4.0 sec. What are the coefficients of friction?

55(12). Block m_1 in Fig. 5–29 has a mass of 4.0 kg and m_2 has a mass of 2.0 kg. The coefficient of friction between m_2 and the horizontal plane is 0.50. The inclined plane is smooth. Find (*a*) the tension in the string and (*b*) the acceleration of the blocks. *Answer:* (*a*) 13 nt. (*b*) 1.6 meters/sec².

56(12). A horizontal force F of 12 lb pushes a block weighing 5.0 lb against a vertical wall (Fig. 5–30). The coefficient of static friction between the wall and the block is 0.60 and the coefficient of kinetic friction is 0.40. Assume the block is not moving initially. (*a*) Will the block start moving? (*b*) What is the force exerted on the block by the wall?

57(12). A 10-lb block of steel is at rest on a horizontal table. The coefficient of static friction between block and table is 0.50. (*a*) What is the magnitude of the horizontal force that will just start the block moving? (*b*) What is the magnitude of a force acting upward 60° from the horizontal that will just start the block moving? (*c*) If the force acts down at 60° from the horizontal, how large can it be without causing the block to move? *Answer:* (*a*) 5.0 lb. (*b*) 5.4 lb. (*c*) 75 lb.

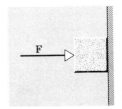

FIGURE 5–30 Problem 56(12).

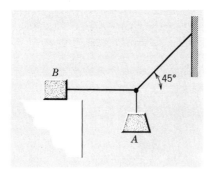

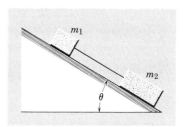

FIGURE 5-31   Problem 58(12).   **FIGURE 5-32**   Problem 59(12).

58(12). Block B in Fig. 5–31 weighs 160 lb. The coefficient of static friction between block and table is 0.25. Find the maximum weight of block A for which the system will be in equilibrium.

59(12). Two masses, $m_1 = 1.65$ kg and $m_2 = 3.30$ kg, attached by a massless rod parallel to the incline on which both slide, as shown in Fig. 5–32, travel down along the plane with m_1 trailing m_2. The angle of incline is $\theta = 30°$. The coefficient of kinetic friction between m_1 and the incline is $\mu_1 = 0.226$; between m_2 and the incline the corresponding coefficient is $\mu_2 = 0.113$. Compute (*a*) the tension in the rod linking m_1 and m_2 and (*b*) the common acceleration of the two masses. (*c*) Would the answers to (*a*) and (*b*) be changed if m_2 trails m_1?
Answer: (*a*) 1.06 nt, in tension. (*b*) 3.62 meters/sec². (*c*) 1.06 nt, in compression; 3.62 meters/sec².

60(12). A 4.0-kg block is put on top of a 5.0-kg block. In order to cause the top block to slip on the bottom one, a horizontal force of 12 nt must be applied to the top block (Fig. 5–33). Assume a frictionless table and find (*a*) the maximum horizontal force F which can be applied to the lower block so that the blocks will move together and (*b*) the resulting acceleration of the blocks.

61(12). Body B weighs 100 lb and body A weighs 32 lb (Fig. 5–34). Given $\mu_s = 0.56$ and $\mu_k = 0.25$, (*a*) find the acceleration of the system if B is initially at rest, and (*b*) find the acceleration if B is initially moving up the plane.
Answer: (*a*) Zero. (*b*) 14 ft/sec² down the plane.

62(12). A railroad flatcar is loaded with crates having a coefficient of static friction 0.25 with the floor. If the train is moving at 30 miles/hr, in how short a distance can the train be stopped without letting the crates slide?

63(12). A 40-kg slab rests on a frictionless floor. A 10-kg block rests on top of the slab (Fig. 5–35). The static coefficient of friction between the block and the slab is 0.60 while the kinetic coefficient is 0.40. The 10-kg block is acted upon by a horizontal force of 100 nt. What are the resulting accelerations of (*a*) the block, and (*b*) the slab?
Answer: (*a*) 6.1 meters/sec². (*b*) 0.98 meters/sec².

64(12). In Fig. 5–36, A is a 10-lb block and B is a 5.0-lb block. (*a*) Determine the minimum

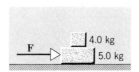

FIGURE 5-33   Problem 60(12).

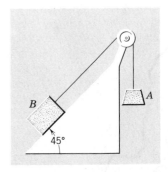

FIGURE 5-34   Problem 61(12).

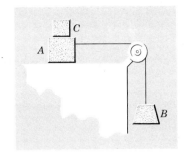

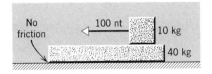

FIGURE 5-35 Problem 63(12).

FIGURE 5-36 Problem 64(12).

weight (block C) which must be placed on A to keep it from sliding, if μ_s between A and the table is 0.20. (b) The block C is suddenly lifted off A. What is the acceleration of block A, if μ_k between A and the table is 0.20?

65(12). An 8.0-lb block and a 16-lb block connected together by a string slide down a 30° inclined plane. The coefficient of kinetic friction between the 8.0-lb block and the plane is 0.10; between the 16-lb block and the plane it is 0.20. Find (a) the acceleration of the blocks and (b) the tension in the string, assuming that the 8.0-lb block leads. (c) Describe the motion if the blocks are reversed.
Answer: (a) 11 ft/sec². (b) 0.46 lb. (c) Same acceleration but blocks in contact.

66(13). What is the radial acceleration of (a) the tip of a 6.0-in. second hand on a clock; (b) of an 1800-miles/hr airplane making a 5.0-mile radius turn?

67(13). A small coin is placed on a flat, horizontal turntable. The turntable is observed to make three revolutions in 3.14 sec. (a) What is the speed of the coin when it rides without slipping at a distance 5.0 cm from the center of the turntable? (b) What is the acceleration (magnitude and direction) of the coin in part (a)? (c) What is the frictional force acting on the coin in part (a) if the coin has a mass m? (d) What is the coefficient of static friction between the coin and the turntable if the coin is observed to slide off the turntable when it is more than 10 cm from the center of the turntable?
Answer: (a) 30 cm/sec. (b) 180 cm/sec², radially inward. (c) 1.8m in newtons, for m in kilograms. (d) 0.37.

68(13). A small object is placed 10 cm from the center of a phonograph turntable. It is observed to remain on the table when it rotates at 45 rev/min but slides off if the table is caused to turn any faster than this. What is the coefficient of friction between the object and the surface of the turntable?

69(13). A mass m on a frictionless table is attached to a hanging mass M by a cord through a hole in the table (Fig. 5-37). Find the condition (v and r) with which m must spin for M to stay at rest.
Answer: $v^2/r = Mg/m$.

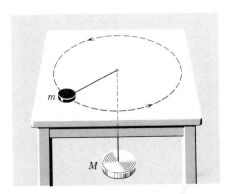

FIGURE 5-37 Problem 69(13).

70(13). A block at the end of a string is whirled around in a vertical circle with a radius of 1.5 meters. Find the critical speed v below which the string would become slack at the highest point.

71(13). If a spaceship can withstand the stresses of a 20-g acceleration, (*a* what is the minimum turning radius of such a craft moving at a speed of $\frac{1}{10}$ the speed of light? (*b*) How long would it take to complete a 90° turn at this speed?
Answer: (*a*) 4.6×10^{12} meters. (*b*) 2.8 days.

72(13). In the Bohr model of the hydrogen atom, the electron revolves in a circular orbit around the nucleus. If the radius is 5.3×10^{-11} meters and the electron makes 6.6×10^{15} rev/sec, find (*a*) the acceleration (magnitude and direction) of the electron and (*b*) the centripetal force acting on the electron. (This force is due to the attraction between the positively charged nucleus and the negatively charged electron.) The mass of the electron is 9.1×10^{-31} kg.

73(13). (*a*) What is the smallest radius of a circle at which a bicyclist can travel if his speed is 18 miles/hr and the coefficient of static friction between the tires and the road is 0.32? (*b*) Under these conditions what is the largest angle of inclination to the vertical at which the bicyclist can ride without falling?
Answer: (*a*) 68 ft. (*b*) 18°.

74(13). A block of mass m at the end of a string is whirled around in a vertical circle of radius R. Find the critical speed below which the string would become slack at the highest point.

75(13). Assume that the standard kilogram would weigh exactly 9.80 nt at sea level on the earth's equator if the earth did not rotate about its axis. Then take into account the fact that the earth does rotate so that this mass moves in a circle of radius 6.40×10^6 meters (earth's radius) at a constant speed of 465 meters/sec. (*a*) Determine the centripetal force needed to keep the standard moving in its circular path. (*b*) Determine the force exerted by the standard kilogram on a spring balance from which it is suspended at the equator (its weight).
Answer: (*a*) 0.0338 nt. (*b*) 9.77 nt.

76(13). Because of the rotation of the earth, a plumb bob may not hang exactly along the direction of the earth's gravitational pull (its weight) but deviates slightly from that direction. Calculate the deviation (*a*) at 40° latitude, (*b*) at the poles, and (*c*) at the equator.

77(13). A circular curve of highway is designed for traffic moving at 40 miles/hr. (*a*) If the radius of the curve is 400 ft, what is the correct angle of banking of the road? (*b*) If the curve is not banked, what is the minimum coefficient of friction between tires and road that would keep traffic from skidding at this speed?
Answer: (*a*) 14°. (*b*) 0.27.

78(13). A particle moves in a clockwise circular path of radius 0.60 meter centered at the origin of the *x-y* plane with a period of revolution of 2.0 sec. (*a*) What are the *x*- and *y*-components of the velocity when $x = 0.48$ meter in the first quadrant? (*b*) What will be the *x*- and *y*-components of the radial acceleration at this same instant?

79(13). An old streetcar rounds a corner on unbanked tracks. If the radius of the tracks is 30 ft and the car's speed is 10 miles/hr, what angle with the vertical will be made by the loosely hanging hand straps?
Answer: 13°.

80(13). A 150-lb student on a steadily rotating Ferris wheel has an apparent weight of 125 lb at his highest point. (*a*) What is his apparent weight at the lowest point? (*b*) What would be his apparent weight at the highest point if the speed of the Ferris wheel were doubled?

81(13). A banked circular highway curve is designed for traffic moving at 60 km/hr. The radius of the curve is 200 meters. Traffic is moving along the highway at 40 km/hr on a stormy day. What is the minimum coefficient of friction between tires and road that will allow cars to negotiate the turn without sliding off the road?
Answer: 0.080.

82(13). A man is driving a car at a speed of 60 miles/hr when he notices a barrier across the road exactly 200 feet ahead. (*a*) What is the minimum static coefficient of friction between his tires and the road that will allow him to stop before striking the barrier? (*b*) Suppose he is driving on a large empty parking lot. What is the minimum coefficient of static friction which would allow him to turn the car in a 200-ft radius circle and, in this way, avoid collision with the barrier?

83(13). A driver's manual states that a driver traveling at 30 miles/hr and desiring to stop as quickly as possible travels 33 ft before his foot reaches the brake. He travels an additional 68 ft before coming to rest. (*a*) What coefficient of friction is assumed in these calculations? (*b*) What is the minimum radius for turning a corner at 30 miles/hr without skidding?

Answer: (*a*) 0.45. (*b*) 140 ft.

84(13). A 5000-lb airplane loops at a speed of 200 miles/hr. Find (*a*) the radius of the largest circular loop possible and (*b*) the force on the plane at the bottom of this loop.

CHAPTER 6

Work and Energy

6–1 Introduction

A fundamental problem of particle dynamics is to find how a particle will move when we know the forces that act on it. By "how a particle will move" we mean how its position varies with time. If the motion is one-dimensional, the problem is to find x as a function of time, $x(t)$. In the previous two chapters we solved this problem for the special case of a constant force. The method used is this. We find the resultant force $\mathbf{F}$ acting on the particle from the appropriate force law. We then substitute $\mathbf{F}$ and the particle mass m into Newton's second law of motion. This gives us the acceleration $\mathbf{a}$ of the particle; or

$$\mathbf{a} = \mathbf{F}/m.$$

If the force $\mathbf{F}$ and the mass m are constant, the acceleration $\mathbf{a}$ must be constant. Let us choose the x-axis to be along the direction of this constant acceleration. We can then find the speed of the particle from Eq. 3–12,

$$v = v_0 + at,$$

and the position of the particle from Eq. 3–15 (with $x_0 = 0$), or

$$x = v_0 t + \tfrac{1}{2}at^2;$$

note that, for simplicity and convenience, we have dropped the subscript x in these equations. The last equation gives us directly what we usually want to know, namely $x(t)$, the position of the particle as a function of time.

The problem is more difficult, however, when the force acting on a particle is not constant. In such a case we still obtain the acceleration of the particle, as before, from

Newton's second law of motion. However, in order to get the speed or position of the particle, we can no longer use the formulas previously developed for constant acceleration because the acceleration now is not constant. To solve such problems, we use the mathematical process of integration, which we consider in this chapter.

We confine our attention to forces that vary with the position of the particle in its environment. This type of force is common in physics. Some examples are the gravitational forces between bodies, such as the sun and earth or earth and moon, and the force exerted by a stretched spring on a body to which it is attached. The procedure used to determine the motion of a particle subject to such a force leads us to the concepts of work and kinetic energy and to the development of the work-energy theorem, which is the central feature of this chapter. In Chapter 7 we consider a broader view of energy, embodied in the law of conservation of energy, a concept which has played a major role in the development of physics.

6–2 Work Done by a Constant Force

Consider a particle acted on by a force. In the simplest case the force **F** is the only force acting, is constant, and the motion takes place in a straight line in the direction of the force. We define the work done by this force on the particle as the product of the magnitude F and the distance d through which the particle moves. We write this as

$$W = Fd.$$

However, a constant force acting on a particle may not act in the direction in which the particle moves. In this case we define the work done by the force on the particle as the product of the component of the force along the line of motion by the distance d the body moves along that line. In Fig. 6–1 a constant force **F** makes an angle ϕ with the x-axis and acts on a particle whose displacement along the x-axis is **d.** If W represents the work done by **F** during this displacement, then according to our definition

$$W = (F \cos \phi)d. \tag{6–1}$$

Of course, other forces must act on a particle that moves in this way (its weight and the frictional force exerted by the plane, to name two). A particle acted on by only a single force may have a displacement in a direction other than that of this single force, as in projectile motion. But it cannot move in a straight line unless the line has the same direction as that of the single force applied to it. Equation 6–1 refers only to the work done on the particle by the particular force **F.** The work done on the particle by the other forces must be calculate separately. The total work done on the particle is the sum of the works done by the separate forces.

When ϕ is zero, the work done by **F** is simply Fd, in agreement with our previous equation. When ϕ is 90°, the force has no component in the direction of motion. That force then does no work on the body. For instance, the vertical force holding a body

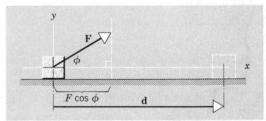

FIGURE 6–1 A force **F** makes the block undergo a displacement **d.** The component of **F** that does the work has magnitude $F \cos \phi$; the work done is $Fd \cos \phi$ ($= \mathbf{F} \cdot \mathbf{d}$).

a fixed distance off the ground does no work on the body, even if the body is moved horizontally over the ground. Also, the centripetal force acting on a body in motion does no work on that body because the force is always at right angles to the direction in which the body is moving. Of course, a force does no work on a body that does not move, for its displacement is then zero.

Notice that we can write Eq. 6–1 either as $(F \cos \phi)d$ or $F(d \cos \phi)$. This suggests that the work can be calculated in two different ways: Either we multiply the magnitude of the displacement by the component of the force in the direction of the displacement or we multiply the magnitude of the force by the component of the displacement in the direction of the force. These two methods always give the same result.

Work is a scalar, although the two quantities involved in its definition, force and displacement, are vectors. In Section 2–4 we defined the scalar product of two vectors as the scalar quantity that we find when we multiply the magnitude of one vector by the component of a second vector along the direction of the first. We promised in that section that we would soon run across physical quantities that behave like scalar products. Equation 6–1 shows that work is such a quantity. In the terminolgy of vector algebra we can write this equation as

$$W = \mathbf{F} \cdot \mathbf{d}, \tag{6–2}$$

where the dot indicates a scalar (or dot) product. Equation 6–2 for $\mathbf{F}$ and $\mathbf{d}$ corresponds to Eq. 2–11 for $\mathbf{a}$ and $\mathbf{b}$.

Work can be either positive or negative. If the particle on which a force acts has a component of motion opposite to the direction of the force, the work done by that force is negative. For example, when a person lowers an object to the floor, the work done on the object by the upward force of his hand holding the object is negative. In this case ϕ is 180°, for $\mathbf{F}$ points up and $\mathbf{d}$ points down.

Work as we have defined it (Eq. 6–2) proves to be a very useful concept in physics. Our special definition of the word "work" does not correspond to the colloquial usage of the term. This may be confusing. A person holding a heavy weight at rest in the air may say that he is doing hard work—and he may work hard in the physiological sense—but from the point of view of physics we say that he is not doing any work. We say this because the applied force causes no displacement. The word "work" is used only in the strict sense of Eq. 6–2. In many scientific fields words are borrowed from our everyday language and are used to name a very specific concept.

The unit of work is the work done by a unit force in moving a body a unit distance in the direction of the force. In the mks system the unit of work is 1 *newton-meter,* called 1 *joule.* In the British engineering system the unit of work is the *foot-pound.* In cgs systems the unit of work is 1 *dyne-centimeter,* called 1 *erg.* Using the relations between the newton, the dyne and the pound, and the meter, the centimeter, and foot, we obtain 1 joule = 10^7 ergs = 0.7376 ft-lb. A unit of work which is often convenient in atomic and nuclear physics is the *electron volt* (eV), which is equal to 1.602×10^{-19} joule.

Example 1. A block of mass 10.0 kg is to be raised from the bottom to the top of an incline 5.00 meters long and 3.00 meters off the ground at the top. Assuming frictionless surfaces, how much work must be done by a force parallel to the incline pushing the block up at constant speed at a place where $g = 9.80$ meters/sec^2.

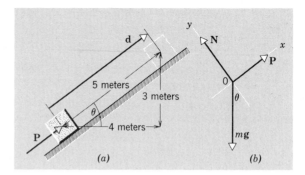

FIGURE 6-2   Example 1. (*a*) A force **P** displaces a block a distance **d** up an inclined plane which makes an angle θ with the horizontal. (*b*) A free-body force diagram for the block.

The situation is shown in Fig. 6–2*a*. The forces acting on the block are shown in Fig. 6–2*b*. We must first find *P*, the magnitude of the force pushing the block up the incline. Because the motion is not accelerated, the resultant force parallel to the plane must be zero. Thus

$$P - mg \sin \theta = 0,$$

or

$$P = mg \sin \theta = (10.0 \text{ kg})(9.80 \text{ meters/sec}^2)(\tfrac{3}{5}) = 58.8 \text{ nt}.$$

Then the work done by **P,** from Eq. 6–1 with $\phi = 0°$, is

$$W = \mathbf{P} \cdot \mathbf{d} = Pd \cos 0° = Pd = (58.8 \text{ nt})(5.00 \text{ meters}) = 294 \text{ joules}.$$

If a man were to raise the block vertically without using the incline, the work he would do would be the vertical force *mg* times the vertical distance or

$$(98.0 \text{ nt})(3.00 \text{ meters}) = 294 \text{ joules},$$

the same as before. The only difference is that with the incline he could apply a smaller force ($P = 58.8$ nt) to raise the block than is required without the incline ($mg = 98.0$ nt); on the other hand, he had to push the block a greater distance (5.00 meters) up the incline than he had to raise the block directly (3.00 meters).

Example 2. A boy pulls a 10-lb sled 30 ft along a horizontal surface at a constant speed. What work does he do on the sled if the coefficient of kinetic friction is 0.20 and his pull makes an angle of 45° with the horizontal?

The situation is shown in Fig. 6–3*a* and the forces acting on the sled are shown in Fig. 6–3*b*. **P** is the boy's pull, **w** the sled's weight, **f** the frictional force, and **N** the normal force exerted by the surface on the sled. The work done by the boy on the sled is

$$W = \mathbf{P} \cdot \mathbf{d} = Pd \cos \phi.$$

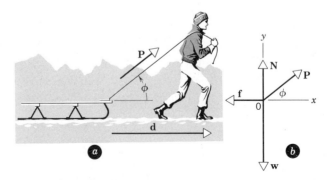

FIGURE 6-3   Example 2. (*a*) A boy displaces a sled an amount **d** by pulling with a force **P** on a rope that makes an angle ϕ with the horizontal. (*b*) A free-body force diagram for the sled.

To evaluate this we first must determine P, whose value has not been given. To obtain P we refer to the force diagram.

The sled is unaccelerated, so that from the second law of motion we obtain

$$P \cos \phi - f = 0,$$

and

$$P \sin \phi + N - w = 0.$$

We know also that f and N are related by

$$f = \mu_k N.$$

These three equations contain three unknown quantities, P, f, and N. To find P we eliminate f and N from these equations and solve the remaining equation for P. You should verify that

$$P = \mu_k w / (\cos \phi + \mu_k \sin \phi).$$

With $\mu_k = 0.20$, $w = 10$ lb, and $\phi = 45°$ we obtain

$$P = (0.20)(10 \text{ lb})/(0.707 + 0.141) = 2.4 \text{ lb.}$$

Then with $d = 30$ ft, the work done by the boy on the sled is

$$W = Pd \cos \phi = (2.4 \text{ lb})(30 \text{ ft})(0.707) = 51 \text{ ft-lb.}$$

The vertical component of the boy's pull **P** does no work on the sled. Notice, however, that it reduces the normal force between the sled and the surface ($N = w - P \sin \phi$) and thereby reduces the magnitude of the force of friction ($f = \mu_k N$).

Would the boy do more work, less work, or the same amount of work on the sled if he pulled horizontally instead of a 45° from the horizontal? Do any of the other forces acting on the sled do work on it?

6–3   Work Done by a Variable Force—One-Dimensional Case

Let us now consider the work done by a force that is not constant. We consider first a force that varies in magnitude only. Let the force be given as a function of position $F(x)$ and assume that the force acts in the x-direction. Suppose a body is moved along the x-direction by this force. What is the work done by this variable force in moving the body from x_1 to x_2?

In Fig. 6–4 we plot F versus x. Let us divide the total displacement into a large number of small equal intervals Δx (Fig. 6–4a). Consider the small displacement Δx from x_1 to $x_1 + \Delta x$. During this small displacement the force F has a nearly constant value and the work it does, ΔW, is approximately

$$\Delta W = F \Delta x, \tag{6–3}$$

where F is the value of the force at x_1. Likewise, during the small displacement from $x_1 + \Delta x$ to $x_1 + 2\Delta x$, the force F has a nearly constant value and the work it does is approximately $\Delta W = F \Delta x$, where F is the value of the force at $x_1 + \Delta x$. The total work done by F in displacing the body from x_1 to x_2, W_{12}, is approximately the sum of a large number of terms like that of Eq. 6–3, in which F has a different value for each term. Hence

$$W_{12} = \sum_{x_1}^{x_2} F \Delta x, \tag{6–4}$$

where the Greek letter sigma (Σ) stands for sum over all intervals from x_1 to x_2.

To make a better approximation we can divide the total displacement from x_1 to x_2

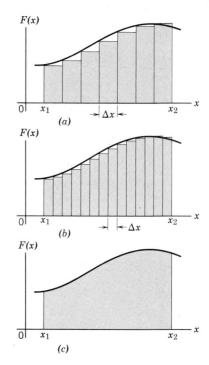

FIGURE 6–4 Computing $\int_{x_1}^{x_2} F(x)\,dx$ amounts to finding the area under the curve $F(x)$ between the limits x_1 and x_2 as discussed in the text.

into a larger number of equal intervals, as in Fig. 6–4b, so that Δx is smaller and the value of F at the beginning of each interval is more typical of its values within the interval. Can you see that better and better approximations can be obtained by taking Δx smaller and smaller so as to have a larger and larger number of intervals? We can obtain an exact result for the work done by F if we let Δx go to zero and the number of intervals go to infinity. Hence the exact result is

$$W_{12} = \lim_{\Delta x \to 0} \sum_{x_1}^{x_2} F\,\Delta x. \tag{6–5}$$

The relation

$$\lim_{\Delta x \to 0} \sum_{x_1}^{x_2} F\,\Delta x = \int_{x_1}^{x_2} F\,dx,$$

as you may have learned in your calculus course, defines the integral of F with respect to x from x_1 to x_2. Numerically, this quantity is equal to the area between the force curve and the x-axis between the limits x_1 and x_2 (Fig. 6–4c). Hence, graphically an integral can be interpreted as an area. The symbol $\int$ is a distorted S (for *sum*) and symbolizes the integration process. We can write the total work done by F in displacing a body from x_1 to x_2 as

$$W_{12} = \int_{x_1}^{x_2} F(x)\,dx. \tag{6–6}$$

As an example, consider a spring attached to a wall. Let the (horizontal) axis of the spring be chosen as an x-axis, and let the origin, $x = 0$, coincide with the endpoint of the spring in its normal, unstretched state. We assume that the positive x-direction points away from the wall. In what follows we imagine that we stretch the spring so slowly that it is essentially in equilibrium at all times ($\mathbf{a} = 0$).

If we stretch the spring so that its endpoint moves to a position x, the spring will exert a force on the agent doing the stretching

$$F = -kx, \tag{6-7}$$

where k is a constant called the *force constant* of the spring. Equation 6-7 is the *force law* for springs. The direction of the force is always opposite to the displacement of the endpoint from the origin. When the spring is stretched, $x > 0$ and F is negative; when the spring is compressed, $x < 0$ and F is positive. The force exerted by the spring is a *restoring force* in that it always points toward the origin. Real springs will obey Eq. 6-7, known as *Hooke's law*, if we do not stretch them beyond a limited range. We can think of k as the magnitude of the force per unit elongation. Thus very stiff springs have large values of k.

To stretch a spring we must exert a force F' on it equal but opposite to the force F exerted by the spring on us. The applied force is therefore $F' = kx$ and the work done by the applied force in stretching the spring so that its endpoint moves from x_1 to x_2 is

$$W_{12} = \int_{x_1}^{x_2} F'(x)\, dx = \int_{x_1}^{x_2} (kx)\, dx = \tfrac{1}{2}kx_2^2 - \tfrac{1}{2}kx_1^2.$$

If we let $x_1 = 0$ and $x_2 = x$, we obtain

$$W = \int_0^x (kx)\, dx = \tfrac{1}{2}kx^2. \tag{6-8}$$

This is the work done in stretching a spring so that its endpoint moves from its unstretched position to x. Note that the work to compress a spring by x is the same as that to stretch it by x because the displacement x is squared in Eq. 6-8; either sign for x gives a positive value for W.

We can also evaluate this integral by computing the area under the force-displacement curve and the x-axis from $x = 0$ to $x = x$. This appears as the white area in Fig. 6-5. The area is a triangle of base x and altitude kx. The white area is therefore

$$\tfrac{1}{2}(x)(kx) = \tfrac{1}{2}kx^2.$$

in agreement with Eq. 6-8.

6-4 Work Done by a Variable Force—Two-Dimensional Case

The force **F** acting on a particle may vary in direction as well as in magnitude, and the particle may move along a curved path. To compute the work in this general case we divide the path up into a large number of small displacements $\Delta\mathbf{r}$, each pointing along the path in the direction of motion. Figure 6-6 shows two selected displacements for a particular situation; it also shows the value of **F** and the angle ϕ between

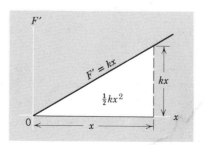

FIGURE 6-5 The force exerted in stretching a spring is $F' = kx$. The area under the force curve is the work done in stretching the spring a distance x and can be found by integrating or by using the formula for the area of a triangle.

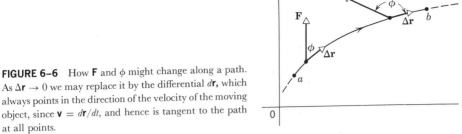

FIGURE 6–6   How **F** and ϕ might change along a path. As $\Delta\mathbf{r} \to 0$ we may replace it by the differential $d\mathbf{r}$, which always points in the direction of the velocity of the moving object, since $\mathbf{v} = d\mathbf{r}/dt$, and hence is tangent to the path at all points.

F and $\Delta\mathbf{r}$ at each location. We can find the amount of work done on the particle during a displacement $\Delta\mathbf{r}$ from

$$dW = \mathbf{F} \cdot \Delta\mathbf{r} = F \cos \phi \, \Delta r \qquad (6\text{–}9)$$

The work done by the variable force **F** on the particle as the particle moves, say, from a to b in Fig. 6–6 is found very closely by adding up (summing) the elements of work done over each of the line segments that make it up. As the line segments $\Delta\mathbf{r}$ become smaller they may be replaced by differentials $d\mathbf{r}$ and the sum over the line segments may be replaced by an integral, as in Eq. 6–6. The work is then found from

$$W_{ab} = \int_a^b \mathbf{F} \cdot d\mathbf{r} = \int_a^b F \cos \phi \, dr. \qquad (6\text{–}10)$$

We cannot evaluate this integral until we are able to say how F and ϕ in Eq. 6–10 vary from point to point along the path; both are functions of the x- and y-coordinates of the particle in Fig. 6–6.

6–5   Kinetic Energy and the Work-Energy Theorem

In our previous examples of work done by forces, we dealt with unaccelerated objects. In such cases the resultant force acting on the object is zero. Let us suppose now that the resultant force acting on an object is not zero, so that the object is accelerated. The conditions are the same in all respects to those that exist when a single unbalanced force acts on the object.

The simplest situation to consider is that of a constant resultant force **F**. Such a force, acting on a particle of mass m, will produce a constant acceleration **a**. Let us choose the x-axis to be in the common direction of **F** and **a**. What is the work done by this force on the particle in causing a displacement x? We have (for constant acceleration) the relations

$$a = \frac{v - v_0}{t}$$

and

$$x = \frac{v + v_0}{2} \cdot t,$$

which are Eqs. 3–12 and 3–14 respectively (in which we have dropped the subscript x, for convenience, and chosen $x_0 = 0$ in the last equation). Here v_0 is the particle's speed at $t = 0$ and v its speed at the time t. Then the work done is

$$W = Fx = max$$

$$= m\left(\frac{v - v_0}{t}\right)\left(\frac{v + v_0}{2}\right)t = \tfrac{1}{2}mv^2 - \tfrac{1}{2}mv_0^2. \tag{6–11}$$

We call one-half the product of the mass of a body and the square of its speed the *kinetic energy* of the body. If we represent kinetic energy by the symbol K, then

$$K = \tfrac{1}{2}mv^2. \tag{6–12}$$

We may then state Eq. 6–11 in this way: The work done by the resultant force acting on a particle is equal to the change in the kinetic energy of the particle.

Although we have proved this result for a constant force only, it holds whether the resultant force is constant or variable. Let the resultant force vary in magnitude (but not in direction), for example. Take the displacement to be in the direction of the force. Let this direction be the x-axis. The work done by the resultant force in displacing the particle from x_0 to x is

$$W = \int \mathbf{F} \cdot d\mathbf{r} = \int_{x_0}^{x} F \, dx.$$

But from Newton's second law we have $F = ma$, and the acceleration a can be written as

$$a = \frac{dv}{dt} = \frac{dv}{dx} \cdot \frac{dx}{dt} = \frac{dv}{dx} v = v \frac{dv}{dx}.$$

Hence

$$W = \int_{x_0}^{x} F \, dx = \int_{x_0}^{x} mv \frac{dv}{dx} \, dx = \int_{v_0}^{v} mv \, dv = \tfrac{1}{2}mv^2 - \tfrac{1}{2}mv_0^2. \tag{6–13}$$

A more general case is that in which the force varies both in direction and magnitude and the motion is along a curved path, as in Fig. 6–6. (See Problem 11.) Once again we find that the work done on a particle by the resultant force is equal to the change in the kinetic energy of the particle.

The work done on a particle by the resultant force is always equal to the change in the kinetic energy of the particle:

$$W \text{ (of the resultant force)} = K - K_0 = \Delta K. \tag{6–14}$$

Equation 6–14 is known as the _work-energy theorem_ for a particle.

Notice that when the speed of the particle is constant, there is no change in kinetic energy and the work done by the resultant force is zero. With uniform circular motion, for example, the speed of the particle is constant and the centripetal force does no work on the particle. A force at right angles to the direction of motion merely changes the direction of the velocity and not its magnitude.

If the kinetic energy of a particle decreases, the work done on it by the resultant force is negative. The displacement and the component of the resultant force along the line of motion are oppositely directed. The work done *on* the particle by the force is the negative of the work *by* the particle on whatever produced the force. This is a consequence of Newton's third law of motion. Hence Eq. 6–14 can be interpreted to say that the kinetic energy of a particle decreases by an amount just equal to the amount of work which the particle does. A body is said to have energy stored in it because of its motion; as it does work it slows down and loses some of this energy. Therefore, the kinetic energy of a body in motion is equal to the work it can do in being brought to rest. This result holds whether the applied forces are constant or variable.

The units of kinetic energy and of work are the same. Kinetic energy, like work, is a scalar quantity. The kinetic energy of a group of particles is simply the (scalar) sum of the kinetic energies of the individual particles in the group.

Example 3. Assume the force of gravity to be constant for small distances above the surface of the earth. A body is dropped from rest at a height h above the earth's surface. What will its kinetic energy be just before it strikes the ground?

The gain in kinetic energy is equal to the work done by the resultant force, which here is the force of gravity. This force is constant and directed along the line of motion, so that the work done by gravity is

$$W = \mathbf{F} \cdot \mathbf{d} = mgh.$$

Initially the body has a speed $v_0 = 0$ and finally a speed v. The gain in kinetic energy of the body is

$$\tfrac{1}{2}mv^2 - \tfrac{1}{2}mv_0^2 = \tfrac{1}{2}mv^2 - 0.$$

Equating these two equivalent terms we obtain

$$K = \tfrac{1}{2}mv^2 = mgh$$

as the kinetic energy of the body just before it strikes the ground.

The speed of the body is then

$$v = \sqrt{2gh}.$$

You should show that in falling from a height h_1 to a height h_2 a body will increase its kinetic energy from $\tfrac{1}{2}mv_1^2$ to $\tfrac{1}{2}mv_2^2$, where

$$\tfrac{1}{2}mv_2^2 - \tfrac{1}{2}mv_1^2 = mg(h_1 - h_2).$$

In this example we are dealing with a constant force and a constant acceleration. The methods developed in previous chapters should be useful here too. Can you show how the results obtained by energy considerations could be obtained directly from the laws of motion for uniformly accelerated bodies?

Example 4. A block weighing 8.0 lb slides on a horizontal frictionless table with a speed of 4.0 ft/sec. It is brought to rest in compressing a spring in its path. By how much is the spring compressed if its force constant is 0.25 lb/ft?

The kinetic energy of the block is

$$K = \tfrac{1}{2}mv^2 = \tfrac{1}{2}(w/g)v^2.$$

This kinetic energy is equal to the work W that the block can do before it is brought to rest. The work done in compressing the spring a distance x beyond its unstretched length is

$$W = \tfrac{1}{2}kx^2,$$

so that

$$\tfrac{1}{2}kx^2 = \tfrac{1}{2}(w/g)v^2$$

or

$$x = \sqrt{\frac{w}{gk}}\, v = \sqrt{\frac{8.0}{(32)(0.25)}}\, 4.0 \text{ ft} = 4.0 \text{ ft}.$$

6–6 Significance of the Work-Energy Theorem

The work-energy theorem does not represent a new, independent law of classical mechanics. We have simply defined work and kinetic energy and derived the relation between them directly from Newton's second law. The work-energy theorem is useful, however, for solving problems in which the work done by the resultant force is easily computed and in which we are interested in finding the particle's speed at certain positions. Of greater significance, perhaps, is the fact that the work-energy theorem is the starting point for a sweeping generalization in physics. It has been emphasized that the work-energy theorem is valid when W is interpreted as the work done by the *resultant* force acting on the particle. However, it is helpful in many problems to compute separately the work done by certain types of force and give special names to the work done by each type. This leads to the concepts of different types of energy and the principle of the conservation of energy, which is the subject of the next chapter.

6–7 Power

Let us now consider the time involved in doing work. The same amount of work is done in raising a given body through a given height whether it takes one second or one year to do so. However, the rate at which work is done is often more interesting to us than the total work performed.

We define *power* as the time rate at which work is done. The average power delivered by an agent is the total work done by the agent divided by the total time interval, or

$$\bar{P} = W/t.$$

The instantaneous power delivered by an agent is

$$P = dW/dt. \tag{6–15}$$

If the power is constant in time, then $P = \bar{P}$ and

$$W = Pt.$$

In the mks system the unit of power is 1 joule/sec, which is called 1 *watt*. This unit of power is named in honor of James Watt whose steam engine is the predecessor of today's more powerful engines. In the British engineering system, the unit of power is 1 ft-lb/sec. Because this unit is quite small for practical purposes, a larger unit, called the *horsepower*, has been adopted. Actually Watt himself suggested as a unit of power the power delivered by a horse as an engine. One horsepower was chosen to equal 550 ft-lb/sec. One horspower is equal to about 746 watts or about three-fourths of a kilowatt. A horse would not last very long at that rate.

Work can also be expressed in units of power × time. This is the origin of the term *kilowatt-hour*, for example. One kilowatt-hour is the work done in 1 hr by an agent working at a constant rate of 1 kw.

▮ **Example 5.** An automobile uses 100 hp and moves at a uniform speed of 60 miles/hr (= 88 ft/sec). What is the forward thrust exerted by the engine?

$$P = \frac{W}{t} = \frac{\mathbf{F} \cdot \mathbf{d}}{t} = \mathbf{F} \cdot \mathbf{v}.$$

The forward thrust **F** is in the direction of motion given by **v,** so that

$$P = Fv,$$

and
$$F = \frac{P}{v} = \left(\frac{100 \text{ hp}}{88 \text{ ft/sec}}\right)\left(\frac{550 \text{ ft-lb/sec}}{1 \text{ hp}}\right) = 630 \text{ lb.}$$

Why doesn't the car accelerate?

QUESTIONS

1. Can you think of other words like "work" whose colloquial meanings are often different from their scientific meanings?

2. In a tug of war one team is slowly giving way to the other. What work is being done and by whom?

3. The inclined plane (see Example 1) is a simple machine which enables us to do work with the application of a smaller force than is otherwise necessary. The same statement applies to a wedge, a lever, a screw, a gear wheel, and a pulley. Do such machines save us work?

4. Springs A and B are identical except that A is stiffer than B, that is, $k_A > k_B$. On which spring is more work expended if (a) they are stretched by the same distance, (b) they are stretched by the same force?

5. A man rowing a boat upstream is at rest with respect to the shore. (a) Is he doing any work? (b) If he stops rowing and moves down with the stream, is any work being done on him?

6. The work done by the resultant force is always equal to the change in kinetic energy. Can it happen that the work done by one of the component forces alone will be greater than the change in kinetic energy? If so, give examples.

7. When two children play catch on a train, does the kinetic energy of the ball depend on the speed of the train? Does the reference frame chosen affect your answer? If so, would you call kinetic energy a scalar quantity?

8. Does the work done in raising a box onto a platform depend on how fast it is raised?

9. A mass m is whirled in a horizontal circle of radius r at a uniform speed v. How much work is done on the particle during each revolution, assuming that there is no friction?

PROBLEMS

1(2). A man pushes a 60-lb block 30 ft along a level floor at constant speed with a force directed 45° below the horizontal. If the coefficient of kinetic friction is 0.20, how much work does the man do on the block?
Answer: 450 ft-lb.

2(2). A block of mass $m = 3.57$ kg is drawn at constant speed a distance $d = 4.06$ meters along a horizontal floor by a rope exerting a constant force of magnitude $F = 7.68$ nt making an angle $\theta = 15.0°$ with the horizontal. Compute (a) the total work done on the block, (b) the work done by the rope on the block, (c) the work done by friction on the block, and (d) the coefficient of kinetic friction between block and floor.

3(2). A 50-kg trunk is pushed 6.0 meters at constant speed up a 30° incline by a constant horizontal force. The coefficient of sliding friction between the trunk and the incline is 0.20. Calculate the work done by (a) the applied force, (b) the frictional force, and (c) the gravitational force.
Answer: (a) 2200 joules. (b) −770 joules. (c) −1500 joules.

4(2). A 100-lb block of ice slides down an incline 5.0 ft long and 3.0 ft high. A man pushes up on

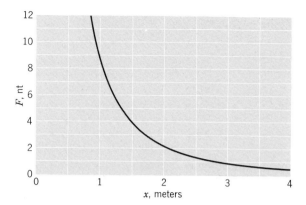

FIGURE 6-7 Problem 7(3).

the ice parallel to the incline so that it slides down at constant speed. The coefficient of friction between the ice and the incline is 0.10. Find (a) the force exerted by the man, (b) the work done by the man on the block, (c) the work done by gravity on the block, (d) the work done by the surface of the incline on the block, (e) the work done by the resultant force on the block, and (f) the change in kinetic energy of the block.

5(2). A crate weighing 500 lb is suspended from the end of a rope 40 ft long. The crate is then pushed aside 4.0 ft from the vertical and held there. (a) What is the force needed to keep the crate in this position? (b) Is work being done in holding it there? (c) Was work done in moving it aside? If so, how much? (d) Does the tension in the rope perform any work on the crate?
Answer: (a) 50 lb. (b) No. (c) Yes; 100 ft-lb. (d) No.

6(2). A cord is used to lower vertically a block of mass M a distance d at a constant downward acceleration of $g/4$. Find the work done by the cord on the block.

7(3). (a) Estimate the work done by the force shown on the graph (Fig. 6–7) in displacing a particle from $x = 1$ to $x = 3$ meters. Refine your method to see how close you can come to the exact answer of 6 joules. (b) The curve is given analytically by $F = a/x^2$ where $a = 9$ nt-meters^2. Show how to get the work done by the rules of integration.

8(3). A 5.0-kg block moves in a straight line on a horizontal frictionless surface under the influence of a force that varies with position as shown in Fig. 6–8. How much work is done by the force as the block moves from the origin to $x = 8.0$ meters?

9(3). A single force acts on a body in rectilinear motion. A plot of velocity versus time for the body is shown in Fig. 6–9. Find the sign (positive or negative) of the work done *by* the force *on* the body in each of the intervals AB, BC, CD, and DE.
Answer: AB, BC, CD, DE.
  +   0   −   +

10(3). The force exerted on an object is $F = F_0(x/x_0 - 1)$. Find the work done in moving the object from $x = 0$ to $x = 2x_0$ evaluating the integral both analytically and also by plotting $F(x)$ and finding the area under the curve.

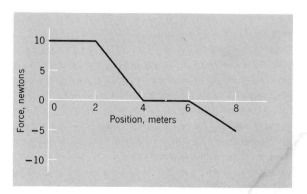

FIGURE 6-8 Problem 8(3).

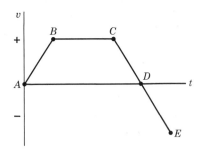

FIGURE 6–9 Problem 9(3).

11(4). When the force **F** varies both in direction and magnitude and the motion is along a curved path, the work done by **F** is obtained from $dW = \mathbf{F} \cdot d\mathbf{r}$, the subsequent integration being taken along the curved path. Notice that both F and ϕ, the angle between **F** and $d\mathbf{r}$, may vary from point to point (see Fig. 6–6). Show that for two-dimensional motion

$$W = \tfrac{1}{2}mv^2 - \tfrac{1}{2}mv_0^2,$$

where v is the final speed and v_0 the initial speed.

12(5). From what height would an automobile have to fall to gain the kinetic energy equivalent to what it would have when going 60 miles/hr?

13(5). A running man has half the kinetic energy that a boy of half his mass has. The man speeds up by 1.0 meter/sec and then has the same kinetic energy as the boy. What were the original speeds of man and boy?
Answer: Man, 2.4 meters/sec. Boy, 4.8 meters/sec.

14(5). A 30-gm bullet initially traveling 500 meters/sec penetrates 12 cm into a wooden block. What average force does it exert?

15(5). If a 2.9×10^5-kg Saturn V rocket with an Apollo spacecraft attached must achieve an escape velocity of 11.2 km/sec (25,000 miles/hr) near the surface of the earth, how much energy must the fuel contain? Would the system actually need as much or would it need more? Why?
Answer: 1.8×10^{13} joules.

16(5). A proton starting from rest is accelerated in a cyclotron to a final speed of 3.0×10^7 meters/sec (about one-tenth the speed of light). How much work, in electron volts, is done on the proton by the electrical force of the cyclotron that accelerates it? One eV $= 1.6 \times 10^{-19}$ joule.

17(5). A 1000-kg car is traveling at 60 km/hr on a level road. The brakes are applied long enough to do 7.0×10^4 joules of work. (a) What is the final speed of the car? (b) How much more work must be done by the brakes to stop the car?
Answer: (a) 42 km/hr. (b) 7.0×10^4 joules.

18(5). An outfielder throws a baseball with an initial speed of 60 ft/sec. An infielder at the same level catches the ball when its speed is reduced to 40 ft/sec. What work was done in overcoming the resistance of the air? The weight of a baseball is 9.0 oz.

19(5). A force acts on a 3.0-kg particle in such a way that the position of the particle as a function of time is given by $x = 3t - 4t^2 + t^3$, where x is in meters and t is in seconds. Find the work done by the force during the first 4.0 sec.
Answer: 530 joules.

20(5). A proton (nucleus of the hydrogen atom) is being accelerated in a linear accelerator. In each stage of such an accelerator the proton is accelerated along a straight line at 3.6×10^{15} meters/sec². If a proton enters such a stage moving initially with a speed of 2.4×10^7 meters/sec and the stage is 3.5 cm long, compute (a) its speed at the end of the stage and (b) the gain in kinetic energy resulting from the acceleration. Take the mass of the proton to be 1.67×10^{-27} kg and express the energy in electron volts. One eV $= 1.6 \times 10^{-19}$ joule.

21(5). A 2.0-kg block is forced against a horizontal spring of negligible mass, compressing the spring by 15 cm. When the block is released, it moves 60 cm across a horizontal tabletop before coming to rest. The force constant of the spring is 200 nt/meter. What is the coefficient of sliding friction between the block and the table?

Answer: 0.19.

22(5). A helicopter is used to lift a 160-lb astronaut 50 ft vertically from the ocean by means of a cable. The acceleration of the astronaut is $g/10$. (*a*) How much work is done by the helicopter on the astronaut? (*b*) How much work is done by the gravitational force on the astronaut? (*c*) What is the kinetic energy of the astronaut just before he reaches the helicopter?

23(5). If the velocity components of a 0.50-kg object moving in the *x-y* plane change from $v_x = 3.0$ meters/sec, $v_y = 5.0$ meters/sec to $v_x = 0.0$ meters/sec, $v_y = 7.0$ meters/sec in 3.0 sec, what is the work done on the body?

Answer: 3.8 joules.

24(5). Show from considerations of work and kinetic energy that the minimum stopping distance for a car of mass m moving with speed v along a level road is $v^2/2\mu_s g$, where μ_s is the coefficient of static friction between tires and road. (See Example 8, Chapter 5.)

25(7). (*a*) Show that the fundamental units of the derived quantity work are ML^2T^{-2}. (*b*) What are the fundamental units of kinetic energy? (*c*) of power?

Answer: (*b*) ML^2T^{-2}. (*c*) ML^2T^{-3}.

26(7). What power is developed by a grinding machine whose wheel has a radius of 8.0 in. and runs at 2.5 rev/sec when the tool to be sharpened is held against the wheel with a force of 40 lb? The coefficient of friction between the tool and the wheel is 0.32.

27(7). A horse pulls a cart with a force of 40 lb at an angle of 30° with the horizontal and moves along at a speed of 6.0 miles/hr. (*a*) How much work does the horse do in 10 min? (*b*) What is the power output of the horse?

Answer: (*a*) 1.8×10^5 ft-lb. (*b*) 0.55 hp.

28(7). A 2.0-kg object accelerates uniformly from rest to a speed of 10 meters/sec in 3.0 sec. (*a*) How much work is done on the object? (*b*) What is the instantaneous power delivered to the body 1.5 sec after it starts?

29(7). A net force of 5.0 nt acts on a 15-kg body initially at rest. Compute (*a*) the work done by the force in the first, second, and third second and (*b*) the instantaneous power exerted by the force at the end of the third second.

Answer: (*a*) 0.83, 2.5, and 4.2 joules. (*b*) 5.0 watts.

30(7). A boy whose mass is 51.0 kg climbs, with constant speed, a vertical rope 6.00 meters long in 10.0 sec. (*a*) How much work does the boy perform? (*b*) What is the boy's power output during the climb?

31(7). A 3200-lb automobile starts from rest on a horizontal road and gains a speed of 45 miles/hr in 30 sec. (*a*) What is the kinetic energy of the auto at the end of the 30 sec? (*b*) What is the average net power delivered to the car during the 30-sec interval? (*c*) What is the instantaneous power at the end of the 30-sec interval?

Answer: (*a*) 2.2×10^5 ft-lb. (*b*) 13 hp. (*c*) 26 hp.

32(7). A 1.5-ton block of granite is pulled up a plane with constant speed (3.0 ft/sec) by a steam winch (Fig. 6–10). The coefficient of sliding friction between the block and plane is 0.40. (*a*) How

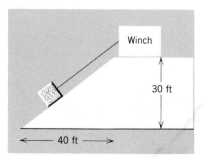

FIGURE 6–10   Problem 32(7).

much work is done by each of the forces which act on the block as it moves 30 ft up the incline? (*b*) How many horsepower must be supplied by the winch?

33(7). The force required to tow a boat at constant velocity is proportional to the speed. If it takes 10 hp to tow a certain boat at a speed of 2.5 miles/hr, how much horsepower does it take to tow it at a speed of 7.5 miles/hr?

Answer: 90 hp.

34(7). A truck can move up a road having a grade of 1.0-ft rise every 50 ft with a speed of 15 miles/hr. The resisting force is equal to one-twenty-fifth the weight of the truck. How fast will the same truck move down the hill with the same horsepower? Assume that the resisting force is proportional to the speed.

35(7). A satellite rocket weighing 100,000 lb acquires a speed of 4000 miles/hr in 1.0 min after launching. (*a*) What is its kinetic energy at the end of the first minute? (*b*) What is the average power expended during this time, neglecting frictional and gravitational forces?

Answer: (*a*) 5.4×10^{10} ft-lb. (*b*) 1.6×10^6 hp.

36(7). A body of mass m accelerates uniformly from rest to a speed v_f in time t_f. (*a*) Show that the work done on the body as a function of time t, in terms of v_f and t_f, is

$$\frac{1}{2} m \frac{v_f^2}{t_f^2} t^2.$$

(*b*) As a function of time t, what is the instantaneous power delivered to the body? (*c*) What is the instantaneous power at the end of 10 sec delivered to a 3200-lb body which accelerates to 60 miles/hr in 10 sec?

CHAPTER 7

The Conservation of Energy

7–1 Introduction

In Chapter 6 we derived the work-energy theorem from Newton's second law of motion. This theorem says that the work W done by the resultant force $\mathbf{F}$ acting on a particle as it moves from one point to another is equal to the change ΔK in the kinetic energy of the particle, or

$$W = \Delta K. \qquad (7\text{–}1)$$

Often several forces act on a particle, the resultant force $\mathbf{F}$ being their vector sum, that is, $\mathbf{F} = \mathbf{F}_1 + \mathbf{F}_2 + \cdots \mathbf{F}_n$, in which we assume that n forces act. The work done by the resultant force $\mathbf{F}$ is the algebraic sum of the work done by these individual forces, or $W = W_1 + W_2 + \cdots W_n$. Thus we can write the work-energy theorem (Eq. 7–1) as

$$W_1 + W_2 + \cdots + W_n = \Delta K. \qquad (7\text{–}2)$$

In this chapter we shall consider systems in which a single particle is acted upon by various kinds of forces and we shall compute W_1, W_2, etc., for these forces; this will lead us to define different kinds of energy such as potential energy and internal energy. The process culminates in the formulation of one of the great principles of science, the *conservation of energy principle*.

7–2 Conservative Forces

Let us first distinguish between two types of forces, *conservative* and *nonconservative*. We shall consider an example of each type and then discuss each example from several different, but related, points of view.

Imagine a spring fastened at one end to a rigid wall as in Fig. 7–1. Let us slide a

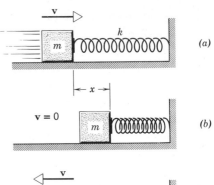

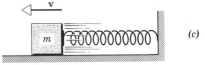

FIGURE 7–1   (*a*) A block of mass *m* is projected with speed *v* against a spring. (*b*) The block is brought to rest by the action of the spring force. (*c*) The block has regained its initial speed *v* as it returns to its starting point.

block of mass *m* with velocity **v** directly toward the spring; we assume that the horizontal plane is frictionless and that the spring is ideal, that is, that it obeys Hooke's law (Eq. 6–7)

$$F = -kx, \tag{7–3}$$

where F is the force exerted by the spring when its free end is displaced through a distance x; we assume further that the mass of the spring is so small compared to that of the block that we can neglect the kinetic energy of the spring. Thus, in the system (mass + spring), all the kinetic energy is concentrated in the mass.

After the block touches the spring, the speed and hence the kinetic energy of the block decrease until finally the block is brought to rest by the action of the spring force, as in Fig. 7–1*b*. The block now reverses its motion as the compressed spring expands. It gains speed and kinetic energy and, when it comes once again to its position of initial contact with the spring, we find that it has the same speed and kinetic energy as it had originally; only the direction of motion has changed. The block loses kinetic energy during one part of its motion but gains it all back during the other part of its motion as it returns to its starting point (Fig. 7–1*c*).

We have interpreted the kinetic energy of a body as its ability to do work by virtue of its motion. It is clear that at the completion of a round trip the ability of the block in Fig. 7–1 to do work remains the same; it has been *conserved*. The elastic force exerted by an ideal spring, and other forces that act in this same way, are called *conservative*. The force of gravity is also conservative; if we throw a ball vertically upward, it will (if we assume air resistance to be negligible) return to our hand with the same kinetic energy that it had when it left our hand.

If, however, a particle on which one or more forces act returns to its initial position with either more or less kinetic energy than it had initially, then in a round trip its ability to do work has been changed. In this case the ability to do work has not been conserved and at least one of the forces acting is labeled *nonconservative*.

To illustrate a nonconservative force let us assume that the surfaces of the block and the plane in Fig. 7–1 are not frictionless but rather that a force of friction **f** is exerted by the plane on the block. The frictional force opposes the motion of the block no matter which way the block is moving and we find that the block returns to its starting point with less kinetic energy than it had initially. Since we showed in our first experiment that the spring force was conservative, we must attribute this new result to the action of the friction force. We say that this force, and other forces that act in this same way, are nonconservative. The induction force in a betatron (Section 31–6) is also a nonconservative force. Instead of dissipating kinetic energy, however, it generates it, so that an electron moving in the circular betatron orbit will return to its initial position with more kinetic energy than it had there originally. In a round trip the electron gains kinetic energy, as it must do if the betatron is to be effective.

We can define conservative force from another point of view, that of the work done by the force on the particle. In our first example above, the work done by the elastic spring force on the block while the spring was being compressed was negative, because the force exerted on the block by the spring (to the left in Fig. 7–1a) was directed opposite to the displacement of the block (to the right in Fig. 7–1a). While the spring was being extended the work that the spring force did on the block was positive (force and displacement in the same direction). In our first example the net work done on the block by the spring force during a complete cycle or round trip, is zero.

In our second example we considered the effect of the frictional force. The work done on the block by this force was negative for each portion of the cycle because the frictional force always opposed the motion. Hence the work done by friction in a round trip cannot be zero. In general, then: *A force is conservative if the work done by the force on a particle that moves through any round trip is zero. A force is nonconservative if the work done by the force on a particle that moves through any round trip is not zero.* *(p. 587)*

The work-energy theorem shows that this second way of defining conservative and nonconservative forces is fully equivalent to our first definition. If there is no change in the kinetic energy of a particle moving through any round trip then $\Delta K = 0$ and, from Eq. 7–1, $W = 0$ and the resultant force acting must be conservative. Similarly, if $\Delta K \neq 0$ then, from Eq. 7–1, $W \neq 0$ and at least one of the forces acting must be nonconservative. *NEXT PAGE*

We can consider the difference between conservative and nonconservative forces in still a third way. Suppose a particle goes from a to b along path 1 and back from b to a along path 2 as in Fig. 7–2a. Several forces may act on the particle during this round trip; we consider each force separately. If the force being considered is conservative, the work done on the particle by that particular force for the round trip is zero, or

$$W_{ab,1} + W_{ba,2} = 0,$$

which we can write as

$$W_{ab,1} = -W_{ba,2}.$$

That is, the work in going from a to b along path 1 is the negative of the work in going from b to a along path 2. However, if we cause the particle to go from a to b along path 2, as shown in Fig. 7–2b, we merely reverse the direction of the previous motion along 2, so that

$$W_{ab,2} = -W_{ba,2}.$$

FIGURE 7–2   (a)   (b)

Hence

$$W_{ab,1} = W_{ab,2},$$

which tells us that the work done on the particle by a conservative force in going from *a* to *b* is the same for either path.

Paths 1 and 2 can be any paths at all as long as they go from *a* to *b*; and *a* and *b* can be chosen to be any two points at all. We always find the same result if the force is conservative. Hence, we have another equivalent definition of conservative and nonconservative forces: *A force is conservative if the work done by it on a particle that moves between two points depends only on these points and not on the path followed. A force is nonconservative if the work done by that force on a particle that moves between two points depends on the path taken between those points.*

To illustrate this third (equivalent) definition of conservative forces, let us consider a second kind of conservative force, that due to gravity. Suppose that we take a stone of mass *m* in our hand and raise it to a height *h* above the ground, going from *a* to *b* by several different paths as in Fig. 7–3. We already know that in a round trip the total work done by a conservative force is zero and that the gravitational force is conservative. The work done on the stone by gravity along the return path *bca* is simply *mgh*. Hence, because gravity is a conservative force, the work done by gravity on the stone along any of the paths from *a* to *b* must be $-mgh$, for only if this is true can the total work done by gravity in a round trip be zero. This means that gravity does negative work on the stone as it moves from *a* to *b*, or, to put it another way, work must be done against gravity along any of the paths *ab*. You can compute directly the result that the work done by gravity along any path *ab* equals $-mgh$. For any of these paths can be decomposed into infinitesimal displacements which are alternately horizontal and vertical; no work is done by gravity in horizontal displacements, and the net vertical displacement is the same in all cases. Hence the work done by gravity on the stone moving from *a* to *b* depends only on the positions of *a* and *b* and not at all on the path taken.

For a nonconservative force, such as friction, the work done is not independent of the path taken between two fixed points. We need only point out that as we push a block over a (rough) table between any two points *a* and *b* by various paths, the distance traversed varies and so does the work done by the frictional force. It depends on the path.

The definitions of conservative force which we have given are equivalent to one another. Which one we use depends only on convenience. The round-trip approach

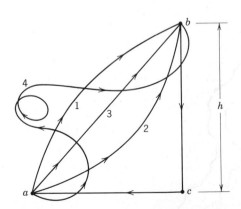

FIGURE 7–3   A stone is raised from *a* to *b* via various paths 1, 2, 3, and 4.

shows clearly that kinetic energy is conserved when conservative forces act. To develop the idea of potential energy, however, the path independence statement is preferable.

7–3 Potential Energy

In this section we shall focus attention not on the moving block of Fig. 7–1 but rather on the (isolated) system (block + spring). Instead of saying that the block is moving we prefer, from this point of view, to say that the configuration of the system is changing. We measure both the position of the block and the configuration of the system at any instant by the same parameter x, namely, the displacement of the free end of the spring from its normal position, corresponding to an unstretched spring. The kinetic energy of the system is the same as that of the block because we have assumed the spring to be massless.

We have seen that the kinetic energy of the system of Fig. 7–1 decreases during the first half of the motion, becomes zero, and then increases during the second half of the motion. If there is no friction, the kinetic energy of the system when it has regained its initial configuration returns to its initial value.

Under these circumstances (conservative forces acting) it makes sense to introduce the concept of *energy of configuration,* or *potential energy U,* and to say that if K for the system changes by ΔK as the configuration of the system changes (that is, as the block moves in the system of Fig. 7–1), then U for the system must change by an equal but opposite amount so that the sum of the two changes is zero, or

$$\Delta K + \Delta U = 0. \tag{7–4a}$$

Alternatively, we can say that any change in kinetic energy K of the system is compensated for by an equal but opposite change in the potential energy U of the system so that their sum remains constant throughout the motion, or

$$K + U = \text{a constant}. \tag{7–4b}$$

The potential energy of a system represents a form of stored energy which can be fully recovered and converted into kinetic energy. We cannot associate a potential energy with a nonconservative force such as the force of friction because the kinetic energy of a system in which such forces act does not return to its initial value when the system returns to its initial configuration.

Equations 7–4 apply to a closed system of interacting objects, such as the (mass + spring) system of Fig. 7–1. In this example, because we have taken the spring to be effectively massless, the kinetic energy may be associated with the moving mass alone. The block slows down (or speeds up) because a force is exerted on it by the spring; it is appropriate, then, to associate the potential energy of the system with this force, that is to say, with the spring. Thus in this simple case we say that kinetic energy, localized in the mass, decreases during the first part of the motion while potential energy, localized in the spring, increases during this same time.

Equations 7–4 are essentially bookkeeping statements about energy. However, they, and the concept of potential energy, have no real value until we have shown how to calculate U as a function of the configuration of the system within which the conservative forces act; in the example of Fig. 7–1 this means that we must be able to calculate $U(x)$, where x is the displacement of the end of the spring.

To refine our concept of potential energy U let us consider the work-energy theorem, $W = \Delta K$, in which W is the work done by the resultant force on a particle as it moves from a to b. For simplicity let us assume that only a single force $\mathbf{F}$ is acting on the particle; this is effectively true in the system of Fig. 7–1. If $\mathbf{F}$ is conservative we can combine the work-energy theorem (Eq. 7–1) with Eq. 7–4a, obtaining

$$W = \Delta K = -\Delta U. \qquad (7\text{–}5a)$$

The work W done by a conservative force depends only on the starting and the end points of the motion and not on the path followed between them. Such a force can depend only on the position of a particle; it does not depend on the velocity of the particle or on the time, for example.

For motion in one dimension, Eq. 7–5a becomes

$$\Delta U = -W = -\int_{x_0}^{x} F(x)\, dx, \qquad (7\text{–}5b)$$

the particle moving from x_0 to x. Equation 7–5b shows how to calculate the change in potential energy ΔU when a particle, acted on by a conservative force $F(x)$, moves from point a, described by x_0, to point b, described by x. The equation shows that we can only calculate ΔU if the force $\mathbf{F}$ depends only on the position of the particle (that is, on the system configuration), which is equivalent to saying that potential energy has meaning only for conservative forces.

Now that we know that the potential energy U depends on the position of the particle only, we can write Eq. 7–4b as

$$\tfrac{1}{2}mv^2 + U(x) = E \qquad \text{(one-dimension)} \qquad (7\text{–}6a)$$

in which E, which remains constant as the particle is moving, is called the *mechanical energy*. Suppose that the particle moves from point a (where its position is x_0 and its speed is v_0) to point b (where its position is x and its speed is v); the total mechanical energy E must be the same for each system configuration when the force is conservative, or, from Eq. 7–6a,

$$\tfrac{1}{2}mv^2 + U(x) = \tfrac{1}{2}mv_0^2 + U(x_0). \qquad (7\text{–}6b)$$

The quantity on the right depends only on the initial position x_0 and the initial speed v_0, which have definite values; it is, therefore, constant during the motion. This is the constant total mechanical energy E. Notice that force and acceleration do not appear in this equation, only position and speed. Equations 7–6 are often called the *law of conservation of mechanical energy* for conservative forces.

In many problems we find that although some of the individual forces are not conservative, they are so small that we can neglect them. In such cases we can use Eqs. 7–6 as a good approximation. For example, air resistance may be present but may have so little effect on the motion that we can ignore it.

Notice that, instead of starting with Newton's laws, we can simplify problem solving when conservative forces alone are involved by starting with Eqs. 7–6. This relation is derived from Newton's laws, of course, but it is one step closer to the solution (the so-called first integral of the motion). We often solve problems without analyzing the forces or writing down Newton's laws by looking instead for something in the motion that is constant; here the mechanical energy is constant and we can write down Eqs. 7–6 as the first step.

For one-dimensional motion we can also write the relation between force and potential energy (Eq. 7–5b) as

$$F(x) = -\frac{dU(x)}{dx}. \qquad (7-7)$$

To show this, substitute this expression for $F(x)$ into Eq. 7–5b and observe that you get an identity. Equation 7–7 gives us another way of looking at potential energy. The potential energy is a function of position whose negative derivative gives the force.

You may have noticed that we wrote down the quantity $U(x)$ in Eqs. 7–6 although we are only able to calculate changes in U (from Eq. 7–5b) and not U itself. Let us imagine that a particle moves from a to b along the x-axis and that a single conservative force $F(x)$ acts on it. To assign a value to U_b, the potential energy at point b, let us write

$$\Delta U = U_b - U_a,$$

or (see Eq. 7–5b),

$$U_b = \Delta U + U_a = -\int_{x_a}^{x_b} F(x)\, dx + U_a. \qquad (7-8)$$

We cannot assign a value to U_b until we have assigned one to U_a. If point b is any arbitrary position x, so that $U_b = U(x)$, we give meaning to $U(x)$ by choosing point a to be some convenient reference position, described by $x_a = x_0$, and by arbitrarily assigning a value to the potential energy $U_a = U(x_0)$ when the body is at that point. Thus Eq. 7–8 becomes

$$U(x) = -\int_{x_0}^{x} F(x)\, dx + U(x_0). \qquad (7-9)$$

The potential energy when the body is at the reference position, that is, $U(x_0)$, is usually given the arbitrary value zero.

It is often convenient to choose the reference position x_0 to be that at which the force acting on the particle is zero. Thus the force exerted by a spring is zero when the spring has its normal unstretched length; we usually say that the potential energy is also zero for this condition. Also, the attraction of the earth on a body decreases as the body moves away from the earth, becoming zero at an infinite distance. We usually take infinity as our reference position and assign the value zero to the potential energy associated with the gravitational force at that position (see Chapter 14). So far, however, we have been more concerned with the gravitational pull on bodies such as baseballs, etc., which, in comparison to the earth's radius, never move very far from the earth's surface. Here the gravitational force ($= m\mathbf{g}$) is essentially constant and we find it convenient to take the zero of potential energy to be, not at infinity, but at the surface of the earth.

The effect of changing the coordinate of the standard reference position x_0, or of the arbitrary value assigned to $U(x_0)$, is simply to change the value of $U(x)$ by an added constant. The presence of an arbitrary added constant in the potential energy expression (Eq. 7–9) makes no difference to the equations that we have written so far. This simply adds the same constant term to each side of Eq. 7–6b, for example, leaving that equation unchanged. Furthermore, changing $U(x)$ by an added constant does not change the force calculated from Eq. 7–7 because the derivative of a constant is zero. All this simply means that the choice of a reference point for potential energy is immaterial because we are always concerned with differences in potential energy, rather than with any absolute value of potential energy at a given point.

There is a certain arbitrariness in specifying kinetic energy also. In order to determine speed, and hence kinetic energy, we must specify a reference frame. The speed of a man sitting on a train is zero if we take the train as a reference frame, but it is not zero to an observer on the ground who sees the man move by with uniform velocity. The value of the kinetic energy depends on the reference frame used by the observer. Hence the important thing about mechanical energy E, which is the sum of the kinetic and the potential energies, is not its actual value during a given motion (this depends on the observer) but the fact that this value does not change during the motion for any particular observer when the forces are conservative.

7–4 One-Dimensional Conservative Systems

Let us now calculate the potential energy in one-dimensional motion for two examples of conservative forces, the force of gravity for motions near the earth's surface and the elastic restoring force of an (ideal) stretched spring.

For the force of gravity we take the one-dimensional motion to be vertical, along the y-axis. We take the positive direction of the y-axis to be upward; the force of gravity is then in the negative y-direction, or downward. We have $F(y) = -mg$, a constant. The potential energy at position y is found from Eq. 7–9, or

$$U(y) = -\int_0^y F(y)\, dy + U(0) = -\int_0^y (-mg)\, dy + U(0) = mgy + U(0).$$

The potential energy can be taken as zero where $y = 0$, so that $U(0) = 0$, and

$$U(y) = mgy. \tag{7–10}$$

The gravitational potential energy is then mgy. The relation $F(y) = -dU/dy$ (Eq. 7–7) is satisfied, for $-d(mgy)/dy = -mg$. We choose $y = 0$ to be at the surface of the earth for convenience, so that the gravitational potential energy is zero at the earth's surface and increases linearly with altitude y.

If we compare points y and $y = 0$, the conservation of kinetic plus potential energy, Eq. 7–6b, gives us the relation

$$\tfrac{1}{2}mv^2 + mgy = \tfrac{1}{2}mv_0{}^2.$$

This is equivalent mathematically to the well-known result (see Eq. 3–17),

$$v^2 = v_0{}^2 - 2gy.$$

If our particle moves from a height h_1 to a height h_2, we can use Eq. 7–6b to obtain

$$\tfrac{1}{2}mv_1{}^2 + mgh_1 = \tfrac{1}{2}mv_2{}^2 + mgh_2.$$

This result is equivalent to the result of Example 3, Chapter 6. The mechanical energy E is constant and is conserved during the motion, even though the kinetic energy and the potential energy vary as the configuration of the system (particle + earth) changes.

A second example of a conservative force is that exerted by an elastic spring on a body of mass m attached to it moving on a horizontal frictionless surface. If we take $x_0 = 0$ as the position of the end of the spring when unextended, the force exerted on the mass when the spring is stretched a distance x from its unextended length is $F = -kx$. The potential energy is obtained from Eq. 7–9,

$$U(x) = -\int_0^x F(x)\,dx + U(0) = -\int_0^x (-kx)\,dx + U(0).$$

If we choose $U(0) = 0$, the potential energy, as well as the force, is zero when the spring is unextended, and

$$U(x) = -\int_0^x (-kx)\,dx = \tfrac{1}{2}kx^2.$$

The result is the same whether we stretch or compress the spring, that is, whether x is plus or minus.

The relation $F(x) = -dU/dx$ (Eq. 7–7) is satisfied, for $-d(\tfrac{1}{2}kx^2)/dx = -kx$. The elastic potential energy of the spring is then

$$U(x) = \tfrac{1}{2}kx^2. \tag{7–11}$$

The body of mass m will undergo a motion in which mechanical energy E is conserved (Fig. 7–4). From Eq. 7–6b we have

$$\tfrac{1}{2}mv^2 + \tfrac{1}{2}kx^2 = \tfrac{1}{2}mv_0{}^2.$$

Here v_0 is the speed of the particle for $x = 0$. Physically we achieve such a result by stretching the spring with an applied force to some position, x_m, and then releasing the spring. Notice that at $x = 0$ the energy of the system (particle + spring) is all kinetic. At $x = x_m$ (the maximum value of x), v must be zero, so that here the system energy is all potential. At $x = x_m$, we have

$$\tfrac{1}{2}kx_m{}^2 = \tfrac{1}{2}mv_0{}^2$$

or

$$x_m = \sqrt{m/k}\,v_0.$$

For positions x_1 and x_2 between 0 and x_m, Eq. 7–6b gives

$$\tfrac{1}{2}kx_1{}^2 + \tfrac{1}{2}mv_1{}^2 = \tfrac{1}{2}kx_2{}^2 + \tfrac{1}{2}mv_2{}^2.$$

We have seen that the kinetic energy of a body is the work that a body can do by virtue of its motion. We express the kinetic energy by the formula $K = \tfrac{1}{2}mv^2$. We cannot give a similar universal formula by which potential energy can be expressed. The potential energy of a system of bodies is the work that the system of bodies can do by virtue of the relative position of its parts, that is, by virtue of its configuration. In each case we must determine how much work the system can do in passing from one configuration to another and then take this as the difference in potential energy of the system between these two configurations.

The potential energy of the spring depends on the relative position of the parts of the spring. Work can be obtained by allowing the spring to return from its extended to its unextended length, during which time it exerts a force through a distance. If a mass is attached to the spring, as in our example, the mass will be accelerated by this force and the potential energy will be converted to kinetic energy. In the gravitational case an object occupies a position with respect to the earth. The potential energy is a property of the object and the earth, considered as a system of bodies. It is the relative position of the parts of this system that determines its potential energy. The

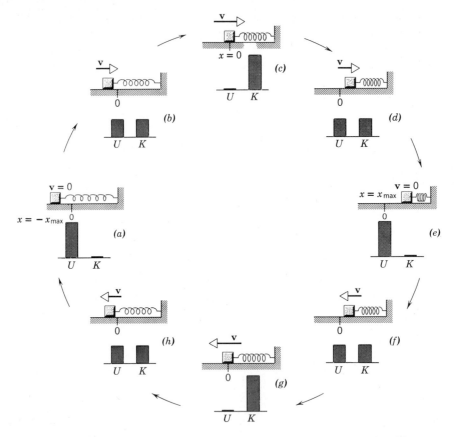

FIGURE 7-4   A mass attached to a spring slides back and forth on a frictionless surface. The system is called a harmonic oscillator. The motion of the mass through one cycle is illustrated. Starting at the left (9 o'clock) the mass is in its extreme left position and momentarily at rest: $K = 0$. The spring is extended to its maximum length: $U = U_{max}$. (K and U are illustrated in the bar graphs below each sketch.) An eighth-cycle later (next drawing) the mass has gained kinetic energy, but the spring is no longer so elongated; K and U have here the same value, $K = U = U_{max}/2$. At the top the spring is neither elongated nor compressed and the speed is a maximum: $U = 0$, $K = K_{max} = U_{max}$. The cycle continues, with mechanical energy $E = K + U$ always the same: $E = K_{max} = U_{max}$. The harmonic oscillator will be analyzed more closely in Chapter 13.

potential energy is greater when the parts are far apart than when they are close together. The loss of potential energy is equal to the work done in this process. This work is converted into kinetic energy of the bodies. In our example we ignored the kinetic energy acquired by the earth itself as an object fell toward it. In principle, this object exerts a force on the earth and causes it to acquire an acceleration, relative to some inertial frame. The resulting change in speed, however, is extremely small, and in spite of the enormous mass of the earth, its additional kinetic energy is negligible compared to that acquired by the falling object. This will be proved in a later chapter. In other cases, such as in planetary motion where the masses of the objects in our system may be comparable, we cannot ignore any part of the system. In general, potential energy is not assigned to either body separately but is considered a joint property of the system.

Example 1. What is the change in gravitational potential energy when a 1600-lb elevator moves from street level to the top of the Empire State Building, 1250 ft above street level?

The gravitational potential energy of the system (elevator + earth) is $U = mgy$. Then

$$\Delta U = U_2 - U_1 = mg(y_2 - y_1).$$

But $\qquad\qquad mg = W = 1600 \text{ lb} \qquad \text{and} \qquad y_2 - y_1 = 1250 \text{ ft},$

so that $\qquad\qquad \Delta U = 1600 \times 1250 \text{ ft-lb} = 2.00 \times 10^6 \text{ ft-lb}.$

Example 2. As an example of the simplicity and usefulness of the energy method of solving dynamical problems, consider the problem illustrated in Fig. 7–5. A block of mass m slides down a curved frictionless surface. The force exerted by the surface on the block is always normal to the surface and to the direction of the motion of the block, so that this force does no work. Only the gravitational force does work on the block and that force is conservative. The mechanical energy E is, therefore, conserved and we write at once

$$\tfrac{1}{2}mv_1{}^2 + mgy_1 = \tfrac{1}{2}mv_2{}^2 + mgy_2.$$

This gives

$$v_2{}^2 = v_1{}^2 + 2g(y_1 - y_2).$$

The speed at the bottom of the curved surface depends only on the initial speed and the change in vertical height but does not depend at all on the shape of the surface. In fact, if the block is initially at rest at $y_1 = h$, and if we set $y_2 = 0$, we obtain

$$v_2 = \sqrt{2gh}.$$

At this point you should recall the independence of path feature of work done by conservative forces and be able to justify applying the ideas developed for one-dimensional motion to this two-dimensional example.

In this problem the value of the force depends on the slope of the surface at each point. Hence, the acceleration is not constant but is a function of position. To obtain the speed by starting with Newton's laws we would first have to determine the acceleration at each point and then integrate the acceleration over the path. We avoid all this labor by starting at once from the fact that the mechanical energy is constant throughout the motion.

Example 3. The spring in a spring gun has a force constant of 4.0 lb/in. It is compressed 2.0 in. from its natural length, and a ball weighing 0.030 lb is put into the barrel against it. Assuming no friction and a horizontal gun barrel, with what speed will the ball leave the gun when released?

The force is conservative so that mechanical energy is conserved in the process. The initial mechanical energy is the elastic potential energy of the spring, $\tfrac{1}{2}kx^2$, and the final mechanical energy is the kinetic energy of the ball, $\tfrac{1}{2}mv^2$. Hence,

$$\tfrac{1}{2}kx^2 = \tfrac{1}{2}mv^2$$

or

$$v = \sqrt{\frac{k}{m}}\, x = \sqrt{\frac{48 \text{ lb/ft}}{(0.030 \text{ lb})/(32 \text{ ft/sec}^2)}}\,(\tfrac{1}{6} \text{ ft}) = 38 \text{ ft/sec.}$$

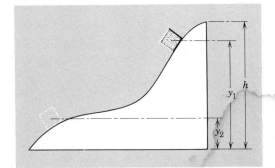

FIGURE 7–5   A block sliding down a frictionless curved surface.

7–5 Mechanical Energy and the Potential Energy Curve

A large amount of information can be found simply from a qualitative consideration
of a system by the use of the potential energy curve and the conservation of mechanical energy. Generally, one would like to solve the equations of motion which follow
from the definition of mechanical energy

$$\tfrac{1}{2}mv^2 + U(x) = E. \qquad (7\text{–}6a)$$

A complete solution would give the position of the particle as a function of time, $x(t)$
in the one-dimensional case.

For most systems, however, this solution is impossible to find without using difficult
computer techniques. This is because either the mechanical energy, E, does not have
a constant value with time (conservation of mechanical energy) or the potential function, $U(x)$ is not sufficiently simple.

However, even when $x(t)$ cannot be found explicitly, if E is constant a knowledge
of $U(x)$ and the value of E can yield a great amount of useful information through the
consideration of Eq. 7–6a.

For example, for a given mechanical energy E, Eq. 7–6a tells us that the particle is
restricted to those regions on the x-axis where $E > U(x)$. Physically we cannot have
an imaginary speed or a negative kinetic energy, so that $E - U(x)$ must be zero or
greater. Furthermore, we can obtain a good qualitative description of the types of
motion possible by plotting $U(x)$ versus x. This description depends on the fact that
the speed is proportional to the square root of the difference between E and U.

To take a specific example, consider the potential energy function shown in Fig.
7–6. This could be thought of as an actual profile of a frictionless roller coaster, but
in general it can represent the potential energy of a nongravitational system. Since
we must have $E \geqq U(x)$ for real motion, the lowest value of E possible is E_0. At this
value of mechanical energy, $E_0 = U$ and the kinetic energy must be zero. The particle
must be at rest at the point x_0. At a slightly higher energy E_1 the particle can move
between x_1 and x_2 only. As it moves from x_0 its speed decreases on approaching either
x_1 or x_2. At x_1 or x_2 the particle stops and reverses its direction. These points x_1 and x_2
are, therefore, called *turning points* of the motion. At an energy like E_2 there are four
turning points, and the particle can oscillate in either one of the two potential valleys.
At an energy like E_3 there is only one turning point of the motion, at x_3. If the par-

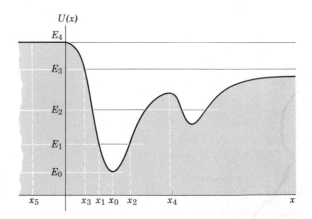

FIGURE 7–6   A potential energy
curve.

ticle is initially moving in the negative x-direction, it will stop at x_3 and then move in the positive x-direction. It will speed up as U decreases and slow down as U increases. At energies above E_4 there are no turning points, and the particle will not reverse direction. Its speed will change according to the value of the potential at each point.

At a point where $U(x)$ has a minimum value, such as at $x = x_0$, the slope of the curve is zero so that the force is zero, that is, $F(x_0) = -(dU/dx)_{x=x_0} = 0$. A particle at rest at this point will remain at rest. Furthermore, if the particle is displaced slightly in either direction, the force, $F(x) = -dU/dx$, will tend to return it, and it will oscillate about the equilibrium point. This equilibrium point is, therefore, called a point of *stable-equilibrium*.

At a point where $U(x)$ has a maximum value, such as at $x = x_4$, the slope of the curve is zero so that the force is again zero, that is, $F(x_4) = -(dU/dx)_{x=x_4} = 0$. A particle at rest at this point will remain at rest. However, if the particle is displaced even the slightest distance from this point, the force, $F(x) = -dU/dx$, will tend to push it farther away from the equilibrium position. Such an equilibrium point is, therefore, called a point of *unstable equilibrium*.

In an interval in which $U(x)$ is constant, such as near $x = x_5$, the slope of the curve is zero so that the force is zero, that is, $F(x_5) = -(dU/dx)_{x=x_5} = 0$. Such an interval is called one of *neutral equilibrium*, since a particle can be displaced slightly without experiencing either a repelling or a restoring force.

From this it is clear that if we know the potential energy function for the region of x in which the body moves, we know a great deal about the motion of the body.

7–6 Two- and Three-Dimensional Conservative Systems

So far we have discussed potential energy and energy conservation for one-dimensional systems in which the force was directed along the line of motion. We can easily generalize the discussion to three-dimensional motion.

If the work done by the force **F** depends only on the end points of the motion and is independent of the path taken between these points, the force is conservative. We define the potential energy U by analogy with the one-dimensional system and find that it is a function of three space coordinates, that is, $U = U(x,y,z)$. Again we obtain an expression for the conservation of mechanical energy.

The generalization of Eq. 7–5b to motion in three dimensions is

$$\Delta U = - \int_{\mathbf{r}_0}^{\mathbf{r}} \mathbf{F}(\mathbf{r}) \cdot d\mathbf{r} \qquad (7\text{–}5c)$$

in which ΔU is the change in potential energy for the system as the particle moves from the point (x_0,y_0,z_0), described by the position vector $\mathbf{r}_0$, to the point (x,y,z), described by the position vector $\mathbf{r}$. F_x, F_y, and F_z are the components of the conservative force $\mathbf{F}(\mathbf{r}) = \mathbf{F}(x,y,z)$.

The generalization of Eq. 7–6b to three-dimensional motion is

$$\tfrac{1}{2}mv^2 + U(x,y,z) = \tfrac{1}{2}mv_0^2 + U(x_0,y_0,z_0) \qquad (7\text{–}6c)$$

which can be written in vector notation as

$$\tfrac{1}{2}m\mathbf{v} \cdot \mathbf{v} + U(\mathbf{r}) = \tfrac{1}{2}m\mathbf{v}_0 \cdot \mathbf{v}_0 + U(\mathbf{r}_0) \qquad (7\text{–}6d)$$

in which $\mathbf{v} \cdot \mathbf{v} = v_x^2 + v_y^2 + v_z^2 = v^2$ and $\mathbf{v}_0 \cdot \mathbf{v}_0 = v_{0x}^2 + v_{0y}^2 + v_{0z}^2 = v_0^2$. Likewise Eq. 7–6$a$ becomes

$$\tfrac{1}{2}mv^2 + U(x,y,z) = E$$

in three dimensions, E being the constant mechanical energy.

Finally, the generalization of Eq. 7–7 to three dimensions is

$$\mathbf{F(r)} = -\mathbf{i}\,\frac{\partial U}{\partial x} - \mathbf{j}\,\frac{\partial U}{\partial y} - \mathbf{k}\,\frac{\partial U}{\partial z} \; = \; -\text{grad } U = (-\nabla U)$$

If we substitute this expression for **F** into Eq. 7–5c we again obtain an identity. In vector language the conservative force **F** is said to be the negative of the *gradient* of the potential energy $U(x,y,z)$.

Show that all these expressions reduce to the correct one-dimensional equation for motion along the x-axis.

Example 4. An important two-dimensional conservative system that will be encountered several times in this text is the simple pendulum as shown in Fig. 7–7. The forces on the bob are the gravitational attraction of the earth and the tension in the supporting cord. In the absence of friction there are no dissipative forces, so we can define potential energy. The mechanical energy will be constant.

If the cord does not stretch, the motion of the bob will always be at right angles to the direction of the tension so that this force does no work. After the bob is released only the gravitational force exerted on the bob by the earth does work on it. Since the gravitational force is conservative, we can use the equation of energy conservation in two dimensions,

$$\tfrac{1}{2}mv^2 + U(x,y) = E.$$

We previously showed that the gravitational potential energy in one dimension is $U(y) = mgy$ (see Eq. 7–10). Since the force of gravity is entirely in the y-direction, it will do no work by virtue of motion in the x-direction. Therefore, when the pendulum moves the gravitation potential will be dependent only upon the y coordinate; specifically $U(x,y)$ equals mgy, where y is taken as zero at the lowest point of the arc ($\phi = 0°$). Then,

$$\tfrac{1}{2}mv^2 + mgy = E.$$

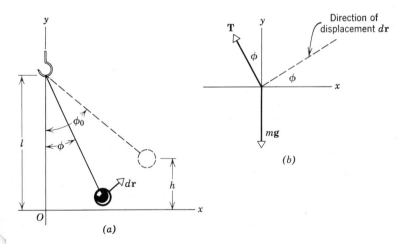

FIGURE 7–7   (*a*) A simple pendulum. The bob, a mass point m, is suspended on a string of length l. Its maximum displacement is ϕ_0. (*b*) A free-body force diagram for the mass subjected to an applied horizontal force.

The particle is pulled through an angle ϕ_0 before being released. The potential energy there is *mgh*, corresponding to a height $y = h$ above the reference point. At the release point ($\phi = \phi_0$) the speed and the kinetic energy are zero so that the potential energy equals the total mechanical energy at that point.

Hence,

$$E = mgh$$

and

$$\tfrac{1}{2}mv^2 + mgy = mgh,$$

or

$$\tfrac{1}{2}mv^2 = mg(h - y).$$

The maximum speed occurs at $y = 0$, where $v = \sqrt{2gh}$.

The minimum speed occurs at $y = h$, where $v = 0$.

At $y = 0$ the energy is entirely kinetic, the potential energy being zero.

At $y = h$ the energy is entirely potential, the kinetic energy being zero.

At intermediate positions the energy is partly kinetic and partly potential.

Notice that $U \leqq E$ at all points of the motion; the pendulum cannot rise higher than $y = h$, its initial release point.

7–7 Nonconservative Forces

So far we have considered only the action of a single conservative force on a particle. Starting from the work-energy theorem, or

$$W_1 + W_2 + \cdots + W_n = \Delta K \qquad (7\text{–}2)$$

we saw that, if only one force, say $\mathbf{F}_1$, was acting and if it was conservative, then we could represent the work W_1 that it did on the particle as a decrease in potential energy ΔU_1 of the system (see Eq. 7–5a), or

$$W_1 = -\Delta U_1.$$

Combining this with Eq. 7–2 yielded

$$\Delta K + \Delta U_1 = 0.$$

If several conservative forces such as gravity, an elastic spring force, an electrostatic force, etc., are acting, we can easily extend these two equations to

$$\Sigma W_c = -\Sigma \Delta U \qquad (7\text{–}12a)$$

and

$$\Delta K + \Sigma \Delta U = 0 \qquad (7\text{–}12b)$$

in which ΣW_c is the sum of the work done by the various (conservative) forces and the ΔU's are the changes in the potential energy of the system associated with these forces. The quantity on the left of Eq. 7–12b is simply ΔE, the change in the total mechanical energy, for the case in which several conservative forces are acting on a particle. We can write this equation then as

$$\Delta E = 0 \qquad \text{(conservative forces)}, \qquad (7\text{–}13)$$

which tells us that, as the system configuration changes the total mechanical energy E for the system remains constant.

Let us now suppose that, in addition to the several conservative forces, a single nonconservative force due to friction acts on the particle. We can then write Eq. 7–2 as

$$W_f + \Sigma\, W_c = \Delta K,$$

where $\Sigma\, W_c$ is again the sum of the work done by the conservative forces and W_f is the work done by friction. We can recast this (see Eq. 7–12b) as

$$\Delta K + \Sigma\, \Delta U = W_f. \tag{7–14}$$

Equation 7–14 shows that, if a frictional force acts, the mechanical energy is not constant, but changes by the work done by the frictional force. We can write Eq. 7–14 as

$$\Delta E = E - E_0 = W_f. \tag{7–15}$$

Since W_f, the work done by friction on the particle, is always negative, it follows from Eq. 7–15 that the final mechanical energy $E\,(= K + \Sigma\, U)$ is less than the initial mechanical energy $E_0\,(= K_0 + \Sigma\, U_0)$.

Friction is an example of a dissipative force, one which does negative work on a body and tends to diminish the mechanical energy of the system. If we had used another nonconservative force, then W_f in Eqs. 7–14 and 7–15 would be replaced by a term W_{nc}, showing again that the mechanical energy E of the system is not constant, but decreases by the amount of work done by the nonconservative force. Hence, only when there are no nonconservative forces, or when we can neglect the work they do, can we assume conservation of mechanical energy.

What happened to the "lost" mechanical energy in the case of friction? It is transformed into a form of energy called *internal energy* (U_{int}) which we may regard as stored within the system. If, for example, we slide a block over a rough horizontal surface, its mechanical energy—in this case entirely kinetic—decreases. There is a corresponding increase in the internal energy of the block, which manifests itself by a rise in temperature. We will discuss internal energy further in Chapter 19.

Just as the work done by a conservative force *on* an object is the negative of the potential energy gain, so the work done by a frictional force *on* an object is the negative of the internal energy gain. In other words, the internal energy produced is equal to the work done *by* the object. We can replace W_f in Eq. 7–15 by $-\Delta U_{int}$, or

$$\Delta E + \Delta U_{int} = 0. \tag{7–16}$$

This asserts that there is no change in the sum of the mechanical and internal energy of the system when only conservative and frictional forces act on the system. Writing this equation as $\Delta U_{int} = -\Delta E$ we see that the loss of mechanical energy equals the gain in internal energy.

Example 5. A 10-lb block is thrust up a 30° inclined plane with an initial speed of 16 ft/sec. It is found to travel 5.0 ft along the plane, stop, and slide back to the bottom. Compute the force of friction f (assumed to have a constant magnitude) acting on the block and find the speed v of the block when it returns to the bottom of the inclined plane.

Consider first the upward motion. At the top, where this motion ends,

$$E = K + U = 0 + (10\ \text{lb})(5.0\ \text{ft})(\sin 30°) = 25\ \text{ft-lb}.$$

At the bottom, where this motion begins,

$$E_0 = K_0 + U_0 = \frac{1}{2}\left(\frac{10\ \text{lb}}{32\ \text{ft/sec}^2}\right)(16\ \text{ft/sec})^2 + 0 = 40\ \text{ft-lb},$$

but

$$W_f = -fs = -f(5.0\ \text{ft}).$$

and

$$E - E_0 = W_f,$$

so that

$$25 \text{ ft-lb} - 40 \text{ ft-lb} = -f(5.0 \text{ ft})$$

and

$$f = 3.0 \text{ lb.}$$

Now consider the downward motion. The block returns to the bottom of the inclined plane with a speed v. Then, at the bottom, where this motion ends,

$$E = K + U = \frac{1}{2} \left(\frac{10 \text{ lb}}{32 \text{ ft/sec}^2} \right) v^2 + 0 = (\tfrac{5}{32} \text{ lb sec}^2/\text{ft}) v^2.$$

At the top, where this motion begins,

$$E_0 = K_0 + U_0 = 0 + (10 \text{ lb})(5.0 \text{ ft})(\sin 30°) = 25 \text{ ft-lb,}$$

but

$$W_f = -(3.0 \text{ lb})(5.0 \text{ ft}) = -15 \text{ ft-lb}$$

and

$$E - E_0 = W_f,$$

so that

$$(\tfrac{5}{32} \text{ lb sec}^2/\text{ft}) v^2 - 25 \text{ ft-lb} = -15 \text{ ft-lb,}$$

and

$$v = 8.0 \text{ ft/sec.}$$

7–8   The Conservation of Energy

We can extend the discussion of the previous section by considering not only conservative forces and the force of friction but also other, nonfrictional, nonconservative forces. Let us regroup the work-energy theorem (Eq. 7–2)

$$W_1 + W_2 + \cdots + W_n = \Delta K$$

as

$$\Sigma W_c + W_f + \Sigma W_{nc} = \Delta K \tag{7–17}$$

in which ΣW_c is the total work done on the particle by conservative forces, W_f is the work done by friction, and ΣW_{nc} is the total work done by nonconservative forces other than friction. We have seen that each conservative force can be associated with a potential energy U and that friction is associated with internal energy U_{int}, or

$$\Sigma W_c = -\Sigma \Delta U$$

and

$$W_f = -\Delta U_{\text{int}},$$

so that Eq. 7–17 becomes

$$\Sigma W_{nc} = \Delta K + \Sigma \Delta U + \Delta U_{\text{int}}.$$

Now whatever the W_{nc} are, it has always been possible to find new forms of energy which correspond to this work. We can then represent ΣW_{nc} by another change of energy term on the right-hand side of the equation, with the result that we can always write the work-energy theorem as

$$0 = \Delta K + \Sigma \Delta U + \Delta U_{\text{int}} + (\text{change in other forms of energy}).$$

In other words, the total energy—kinetic plus potential plus internal plus other forms —does not change. *Energy may be transformed from one kind to another, but it cannot be created or destroyed; the total energy is constant.*

This statement is a generalization from our experience, so far not contradicted by observation of nature. It is called the *principle of the conservation of energy.* Often in the history of physics this principle seemed to fail. But its apparent failure stimulated the search for the reasons. Experimentalists searched for physical phenomena besides motion that accompany the forces that act between bodies. Such phenomena have always been found. With work done against friction, internal energy is produced; in other interactions energy in the form of sound, light, electricity, etc., may be produced. Hence the concept of energy was generalized to include forms other than kinetic and potential energy of directly observable bodies. This procedure, which relates the mechanics of bodies observed to be in motion to phenomena which are not mechanical or in which motion is not directly detected, has linked mechanics to all other areas of physics. The energy concept now permeates all of physical science and has become one of the unifying ideas of physics.*

In subsequent chapters we shall study various transformations of energy—from mechanical to internal, mechanical to electrical, nuclear to internal, etc. It is during such transformations that we measure the energy changes in terms of work, for it is during these transformations that forces arise and do work.

Although the principle of the conservation of kinetic plus potential energy is often useful, we see that it is a restricted case of the more general principle of the conservation of energy. Kinetic and potential energy is conserved only when conservative forces act. Total energy is always conserved.

7–9 Mass and Energy

Historically, one of the great conservation laws of science has been the law of conservation of matter. Enunciated at least as early as the time of the Roman poet Lucretius (born 94 BC) who stated the principle in his writings, the concept was not established as a firm scientific principle for a long time. When it finally was, it proved extremely fruitful in chemistry and physics.

Serious doubts as to the general validity of this principle, however, were raised by Albert Einstein in his papers introducing the theory of relativity. Subsequent experiments on fast-moving electrons and on nuclear matter confirmed his conclusions.

Einstein's finding suggested that, if certain physical laws were to be retained, the mass of a particle had to be redefined as

$$m = \frac{m_0}{\sqrt{1 - v^2/c^2}} \tag{7–18}$$

Here m_0 is the mass of the particle when at rest with respect to the observer, called the *rest mass; m* is the mass of the particle measured as it moves at a speed v relative to the observer; and c is the speed of light, having a constant value of approximately 3×10^8 meters/sec. Experimental checks of this equation can be made, for example, by deflecting high-speed electrons in magnetic fields and measuring the radii of curvature of their paths. The paths are circular and the magnetic force a centripetal

* See for example, "Concept of Energy in Mechanics" by R. B. Lindsay, in *The Scientific Monthly,* October 1957.

one ($F = mv^2/r$, F and v being known). At ordinary speeds the difference between m and m_0 is too small to be detectable. Electrons, however, can be emitted from radio-active nuclei with speeds greater than nine-tenths that of light. In such cases the results (Fig. 7–8) confirm Eq. 7–18. Thus, in this formulation, mass is not a conserved quantity, contrary to the earlier belief.

In Einstein's formulation the concepts of kinetic energy and total energy also have to be redefined. Since mass is now a function of the square of the velocity, increasing as velocity increases, it appears that mass and kinetic energy are related. In relativity, kinetic energy, K, is defined as

$$K = (m - m_0)c^2 \qquad (7\text{--}19)$$

where m is given by Eq. 7–18. When the velocity is small this approaches the classical expression, $\frac{1}{2}m_0v^2$, and when $v = 0$, $m = m_0$, and $K = 0$, as expected.

The basic idea that energy is equivalent to mass can be extended to include energies other than kinetic. For example, when we compress a spring and give it elastic potential energy, U, its mass increases from m_0 to $m_0 + U/c^2$. When we add internal energy in amount U_{int} to an object, its mass increases by an amount Δm, where Δm is U_{int}/c^2. Equation 7–19 can be modified to include these types of energy,

$$K + U + U_{\text{int}} + \cdots = (m - m_0)c^2, \qquad (7\text{--}20)$$

where m now is increased over that in Eq. 7–18 by the amount $(U + U_{\text{int}} + \cdots)/c^2$.

One might think that the total energy could now be defined as equal to $(m - m_0)c^2$. Classically, in a closed system (one acted on by no external forces and having no loss or gain of mass or energy through the sides) the sum of the energies on the left of Eq. 7–20 would indeed be a constant, the total energy. However, by introducing expressions containing mass where the classical total energy appeared, Einstein raised the possibility of converting rest-mass energy itself into other forms, just as kinetic and potential energies may be converted from one to the other. If this is the case, then the conserved quantity could be mc^2 which would then be quite properly called the total energy. It would certainly reduce to that in the classical limit. We would then have

$$K + U + U_{\text{int}} \cdots + m_0c^2 = mc^2 = E_{\text{tot}}. \qquad (7\text{--}21)$$

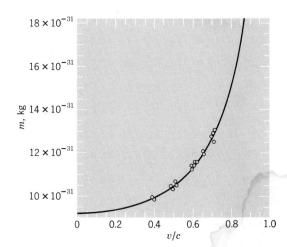

FIGURE 7–8 The way an electron's mass increases as its speed relative to the observer increases. The solid line is a plot of $m = m_0(1 - v^2/c^2)^{-1/2}$, and the circles are adapted from experimental values obtained by Bucherer and Neumann in 1914. The curve tends toward infinity as $v \to c$.

These predictions of the theory were dramatically verified when nuclear fission was discovered in 1939 by Otto Hahn and Lise Meitner, and this was followed by the construction of the atomic bomb in 1945. In the fission bomb, uranium atoms break into several fragments. The sum of the rest masses of the fragments is less than that of the parent atom. The mass difference, Δm, appears as energy, ε, of various forms. The total energy remains constant so, from Eq. 7–21

$$\varepsilon + \Delta m_0 c^2 = 0$$

and

$$\varepsilon = -\Delta m_0 c^2. \tag{7-22}$$

The minus sign shows that, for a gain of energy in the system the rest mass must decrease.

Example 6. Consider a quantitative example. On the atomic mass scale the unit of mass is 1.66×10^{-27} kg approximately. On this scale the mass of the proton (the nucleus of a hydrogen atom) is 1.00728 and the mass of the neutron (a neutral particle, one of the constituents of all nuclei except hydrogen) is 1.00867. A deuteron (the nucleus of heavy hydrogen) is known to consist of a neutron and a proton; the mass of the deuteron is found to be 2.01360. The mass of the deuteron is *less than* the combined masses of the neutron and proton by 0.00239 atomic mass units. The discrepancy is equivalent in energy to

$$\varepsilon = -\Delta m c^2 = (0.00239 \times 1.66 \times 10^{-27} \text{ kg})(3.00 \times 10^8 \text{ meters/sec})^2$$

$$= 3.58 \times 10^{-13} \text{ joules} = 2.23 \times 10^6 \text{ eV}.$$

When a neutron and a proton combine to form a deuteron, this exact amount of energy is given off in the form of γ-radiation. Similarly, it is found that the same amount of energy must be added to the deuteron to break it up into a proton and a neutron. This energy is therefore called the *binding energy* of the deuteron.

QUESTIONS

1. Mountain roads rarely go straight up the slope but wind up gradually. Explain why.

2. Is any work being done on a car moving with constant speed along a straight level road?

3. What happens to the potential energy an elevator loses in coming down from the top of a building to a stop at the ground floor?

4. In Example 2 (see Fig. 7–5) we asserted that the speed at the bottom does not depend at all on the shape of the surface. Would this still be true if friction were present?

5. Give physical examples of unstable equilibrium. Of neutral equilibrium. Of stable equilibrium.

6. Explain, using work and energy ideas, how a child pumps a swing up to large amplitudes from a rest position.

7. Two disks are connected by a stiff spring. Can one press the upper disk down enough so that when it is released it will spring back and raise the lower disk off the table (see Fig. 7–9)? Can mechanical energy be conserved in such a case?

8. In the case of work done against friction, the amount of internal energy generated is independent of the velocity (or inertial reference frame) of the observer. That is, different observers would assign the same quantity of mechanical energy transformed into internal energy due to friction. How can this be explained, considering that such observers measure different quantities of total work done and different changes in kinetic energy in general?

9. An object is dropped and observed to bounce to one and one-half time its original height. What conclusion can you draw from this observation?

10. The driver of an automobile traveling at speed v suddenly sees a brick wall at a distance d directly in front of him. To avoid crashing, is it better for him to slam on the brakes or to turn the car sharply away from the wall? (Hint: consider the force required in each case.)

FIGURE 7–9

11. A spring is kept compressed by tying its ends together tightly. It is then placed in acid and dissolved. What happened to its stored potential energy?

PROBLEMS

1(3). A chain is held on a frictionless table with one-fifth of its length hanging over the edge. If the chain has a length l and a mass m, how much work is required to pull the hanging part back on the table?
Answer: mgl/50.

2(3). A body moving along the x-axis is subject to a force repelling it from the origin, given by $F = kx$. (*a*) Find the potential energy function $U(x)$ for the motion and write down the conservation of energy condition. (*b*) Describe the motion of the system and show that this is the kind of motion we would expect near a point of unstable equilibrium.

3(3). If the magnitude of the force of attraction between a particle of mass m_1 and one of mass m_2 is given by

$$F = k \frac{m_1 m_2}{x^2}$$

where k is a constant and x is the distance between the particles, find (*a*) the potential energy function and (*b*) the work required to increase the separation of the masses from $x = x_1$ to $x = x_1 + d$.
Answer: (*a*) $U(x) = -km_1m_2/x$, if $U(\infty) \to 0$. (*b*). $\dfrac{km_1m_2d}{x_1(d + x_1)}$.

4(4). A 5.0-gram bullet is fired vertically upward by a spring gun. It is found that the spring must be compressed at least 10 cm if the bullet is to reach a 20-meter high target. What is the force constant of the spring?

5(4). A 10-kg projectile is fired straight up with an initial velocity of 500 meters/sec. (*a*) What is the projectile's potential energy at the top of its trajectory? (*b*) What would be the maximum potential energy if the projectile were fired at a 45° angle rather than straight up?
Answer: (*a*) 1.3×10^6 joules. (*b*) 6.3×10^5 joules.

6(4). A 2.0-gram penny is pushed down on a vertical spring, compressing the spring by 1.0 cm. The force constant of the spring is 40 nt/meter. How far above this original position will the penny fly if it is released?

7(4). A 200-lb man jumps out a window into a fire net 30 ft below. The net stretches 6.0 ft before bringing him to rest and tossing him back into the air. What is the potential energy of the stretched net if no energy is dissipated by nonconservative forces?
Answer: 7200 ft-lb.

8(4). Approximately 1 million kg of water drops 50 meters over Niagara Falls every second. (*a*) How much potential energy is lost every second by the falling water? (*b*) What would be the power output of an electric generating plant if it could convert all of the water's potential energy to electrical energy? (*c*) How many kilowatt-hours of electrical energy could be produced each year?

9(4). A 2.0-kg block is dropped from a height of 0.40 meter onto a spring of force constant $k = 1960$ nt/meter. Find the maximum distance the spring will be compressed (neglect friction).
Answer: 10 cm.

10(4). Show that for the same initial speed v_0, the speed v of a projectile will be the same at all points at the same elevation, regardless of the angle of projection.

11(4). A certain peculiar spring is found *not* to conform to Hooke's law. The force (in newtons) it exerts when stretched a distance x (in meters) is found to have magnitude $52.8x + 38.4x^2$ in the direction opposing the stretch. (*a*) Compute the total work required to stretch the spring from $x = 0.500$ to $x = 1.00$ meter. (*b*) With one end of the spring fixed, a particle of mass 2.17 kg is attached to the other end of the spring when it is extended by an amount $x = 1.00$ meter. If the particle is then released from rest, compute its speed at the instant the spring has returned to the configuration in which the extension is $x = 0.500$ meter. (*c*) Is the force exerted by spring conservative or nonconservative? Explain.
Answer: (*a*) 31.0 joules. (*b*) 5.33 meters/sec. (*c*) Conservative.

12(4). A 1.0-kg object is acted upon by a conservative force given by $F = -3.0x - 5.0x^2$ where F is in newtons and x in meters. (*a*) What is the potential energy of the object at $x = 2.0$ meters? (*b*) If the object has a speed of 4.0 meters/sec in the negative x-direction when it is at $x = 5.0$ meters, describe the subsequent motion.

13(4). It is claimed that large trees can evaporate as much as one ton of water per day. (*a*) Assuming the average rise of water to be 30 ft from the ground, how much energy (in kw-hr) must be supplied to do this? (*b*) What is the average power if the evaporation is assumed to occur during 12 hours of the day?
Answer: (*a*) 2.2×10^{-2} kw-hr. (*b*) 1.9 watts.

14(4). An object is attached to a vertical spring and slowly lowered to its equilibrium position. This stretches the spring by an amount d. If the same object is attached to the same vertical spring but permitted to fall instead, through what maximum distance does it stretch the spring?

15(5). A particle moves along a line in a region in which its potential energy varies as in Fig. 7–10. (*a*) Sketch, with the same scale of the abscissa, the force $F(x)$ acting on the particle. Indicate on the graph the approximate numerical scale for $F(x)$. (*b*) If the particle has a constant mechanical energy of 4.0 joules, sketch the graph of its kinetic energy. Indicate the numerical scale on the $K(x)$ axis.

16(6). The string in Fig. 7–11 has a length $l = 4.0$ ft. When the ball is released, it will swing down the dotted arc. How fast will it be going when it reaches the lowest point in its swing?

17(6). A 50-gm ball is thrown from a window with an initial velocity of 8.0 meters/sec at an angle of 30° above the horizontal. Using energy methods determine (*a*) the kinetic energy of the ball at the top of its flight and (*b*) its speed when it is 3.0 meters below the window.
Answer: (*a*) 1.2 joule. (*b*) 11 meters/sec.

18(6). A 2.0-kg block is placed against a compressed spring on a frictionless incline (Fig. 7–12). The spring, whose force constant is 1960 nt/meter, is compressed 20 cm, after which the block is released. How far up the incline will it go before coming to rest? Assume that the spring has negligible mass and that the spring and block are not attached to each other. Measure the final position of the block with respect to its position just before being released.

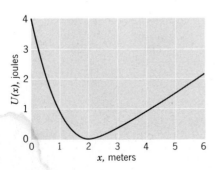

FIGURE 7–10   Problem 15(5).

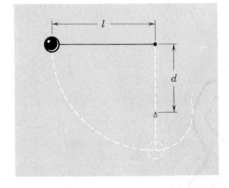

FIGURE 7–11   Problems 16(6) and 19(6).

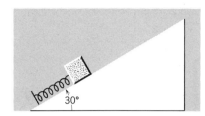

FIGURE 7–12   Problem 18(6).

19(6). The nail in Fig. 7–11 is located a distance d below the point of suspension. Show that d must be at least $0.6l$ if the ball is to swing completely around in a circle centered on the nail.

20(6). A frictionless roller coaster of mass m starts at point A with speed v_0, as in Fig. 7–13. Assume that the roller coaster can be considered as a particle and that it always remains on the track. (*a*) What will be the speed of the roller coaster at points B and C? (*b*) What constant deceleration is required to stop it at point E if the brakes are applied at point D?

21(6). Two children are playing a game in which they attempt to hit a small box on the floor using a spring-loaded marble gun placed horizontally on a frictionless table (Fig. 7–14). The first child compresses the spring 1.0 cm and the marble falls 20 cm short of the target, which is 2.0 meters horizontally from the edge of the table. How far should the second child compress the spring so that the same marble falls into the box?
Answer: 1.1 cm.

22(6). What force corresponds to a potential energy $U = ax^2 + bxy + z$?

23(6). A 2.0-lb block is given an initial speed of 4.0 ft/sec up an inclined plane starting from a point 2.0 ft from the bottom as measured along the plane. If the plane makes an angle of 30° with the horizontal and the coefficient of friction is 0.20, (*a*) how far up the plane will the block travel? (*b*) With what speed will it reach the bottom of the plane?
Answer: (*a*) 4.5 in. (*b*) 7.1 ft/sec.

24(6). An ideal massless spring S can be compressed 1.0 meter by a force of 1000 nt. This same spring is placed at the bottom of a frictionless inclined plane which makes an angle of $\theta = 30°$ with the horizontal (see Fig. 7–15). A 10-kg mass M is released from rest at the top of the incline and is

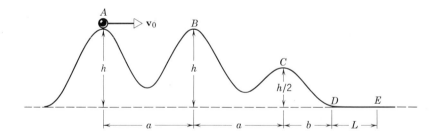

FIGURE 7–13   Problem 20(6).

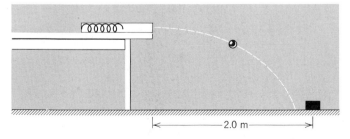

FIGURE 7–14   Problem 21(6).

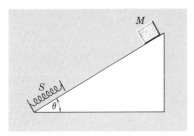

FIGURE 7-15   Problem 24(6).

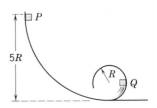

FIGURE 7-16   Problem 28(6).

brought to rest momentarily after compressing the spring 2.0 meters. (*a*) Through what distance does the mass slide before coming to rest? (*b*) What is the speed of the mass just before it reaches the spring?

25(6). The spring of a spring gun has a force constant of 4.0 lb/in. When the gun is inclined at an angle of 30°, a 2.0-oz ball is projected to a height of 6.0 ft. By how much must the spring have been compressed initially?

Answer: 15 in.

26(6). (*a*) A light rigid rod of length *l* has a mass *m* attached to its end, forming a simple pendulum. It is inverted and then released. What is its speed *v* at the lowest point and what is the tension *T* in the suspension at that instant? (*b*) The same pendulum is next put in a horizontal position and released from rest. At what angle from the vertical will the tension in the suspension equal the weight in magnitude?

27(6). An escalator joins one floor with another one 25 ft above. The escalator is 40 ft long and moves along its length at 2.0 ft/sec. (*a*) What power must its motor deliver if it is required to carry a maximum of 100 persons per minute, of average mass 5.0 slugs? (*b*) A 160-lb man walks up the escalator in 10 sec. How much work does the motor do on him? (*c*) If this man turned around at the middle and walked down the escalator so as to stay at the same level in space, would the motor do work on him? If so, what power does it deliver for this purpose? (*d*) Is there any (other?) way the man could walk along the escalator without consuming power from the motor?

Answer: (*a*) 6700 ft-lb/sec. (*b*) 2000 ft-lb. (*c*) No.

28(6). A small block of mass *m* slides along the frictionless loop-the-loop track shown in Fig. 7–16. (*a*) If it starts from rest at *P*, what is the resultant force acting on it at *Q*? (*b*) At what height above the bottom of the loop should the block be released so that the force exerted on it by the track at the top of the loop is equal to its weight?

29(6). The potential energy corresponding to a certain two-dimensional force field is given by $U(x,y) = \frac{1}{2}k(x^2 + y^2)$. (*a*) Derive F_x and F_y and describe the vector force at each point in terms of its coordinates *x* and *y*. (*b*) Derive F_r and F_θ and describe the vector force at each point in terms of the polar coordinates *r* and *θ* of the point. (*c*) Can you think of a physical model of such a force?

Answer: (*a*) $F_x = -kx$; $F_y = -ky$; **F** points toward the origin. (*b*) $F_r = -kr$; $F_\theta = 0$.

30(6). A boy is seated on the top of a hemispherical mound of ice (Fig. 7–17). He is given a very small push and starts sliding down the ice. Show that he leaves the ice at a point whose height is 2*R*/3 if the ice is frictionless.

31(6). The magnitude of the force of attraction between the positively charged nucleus and the

FIGURE 7-17   Problem 30(6).

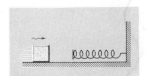

FIGURE 7-18   Problem 32(7).

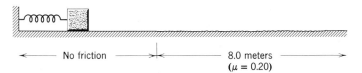

FIGURE 7-19   Problem 33(7).

negatively charged electron in the hydrogen atom is given by

$$F = k\frac{e^2}{r^2}$$

where e is the charge of the electron, k is a constant, and r is the separation between electron and nucleus. Assume that the nucleus is fixed. The electron, initially moving in a circle of radius R_1 about the nucleus, jumps suddenly into a circular orbit of smaller radius R_2. (a) Calculate the change in kinetic energy of the electron, using Newton's second law. (b) Using the relation between force and potential energy, calculate the change in potential energy of the atom. (c) Show by how much the mechanical energy of the atom has changed in this process. (This energy is given off in the form of radiation.)

Answer: (a) $+\frac{1}{2}ke^2\left(\frac{1}{R_2}-\frac{1}{R_1}\right)$. (b) $-ke^2\left(\frac{1}{R_2}-\frac{1}{R_1}\right)$. (c) $-\frac{1}{2}ke^2\left(\frac{1}{R_2}-\frac{1}{R_1}\right)$.

32(7). A 1.0-kg block collides with a horizontal weightless spring of force constant 2.0 nt/meter (Fig. 7-18). The block compresses the spring 4.0 meters from the rest position. Assuming that the coefficient of kinetic friction between block and horizontal surface is 0.25, what was the speed of the block at the instant of collision?

33(7). A 3.0-kg object is released from a compressed spring whose force constant is 120 nt/meter. After leaving the spring it travels over a horizontal surface, with coefficient of friction 0.20, for a distance of 8.0 meters before coming to rest (Fig. 7-19). (a) What was its maximum kinetic energy? (b) How far was the spring compressed before being released?
Answer: (a) 47 joules. (b) 89 cm.

34(7). A body of mass m starts from rest down a plane of length l inclined at an angle θ with the horizontal. (a) Take the coefficient of friction to be μ and find the body's speed at the bottom. (b) How far will it slide horizontally on a similar surface after reaching the bottom of the incline. Solve by using energy methods and solve again using Newton's laws directly.

35(7). A 4.0-kg block starts up a 30° incline with 128 joules of kinetic energy. How far will it slide up the plane if the coefficient of friction is 0.30?
Answer: 4.3 meters.

36(7). A 40-lb body is pushed up a frictionless 30° inclined plane 10 ft long by a horizontal force **F**. (a) If the speed at the bottom is 2.0 ft/sec and at the top is 10 ft/sec, how much work is done by **F**? (b) Suppose the plane is not frictionless and that $\mu_k = 0.15$. What work will this same force do? (c) How far up the plane does the body go?

37(7). A particle slides along a track with elevated ends and a flat central part, as shown in Fig. 7-20. The flat part has a length $l = 2.0$ meters. The curved portions of the track are frictionless. For the flat part the coefficient of kinetic friction is $\mu_k = 0.20$. The particle is released at point A which is a height $h = 1.0$ meter above the flat part of the track. Where does the particle finally come to rest?
Answer: In the center of the flat part.

FIGURE 7-20   Problem 37(7).

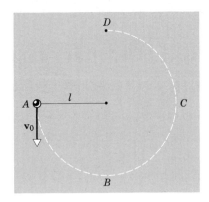

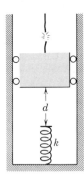

FIGURE 7-21 Problem 38(7).

FIGURE 7-22 Problem 39(7).

38(7). A very light rigid rod whose length is *l* has a ball of mass *m* attached to one end (Fig. 7–21). The other end is pivoted frictionlessly in such a way that the ball moves in a vertical circle. The system is launched from the horizontal position *A* with downward initial velocity **v₀**. The ball just reaches point *D* and then stops. (*a*) Derive an expression for v_0 in terms of *l*, *m*, and *g*. (*b*) What is the tension in the rod when the ball is at *B*? (*c*) A little sand is placed on the pivot, after which the ball just reaches *C* when launched from *A* with the same speed as before. How much work is done by friction during this motion? (*d*) How much total work is done by friction before the ball finally comes to rest at *B* after oscillating back and forth several times?

39(7). The cable of a 4000-lb elevator in Fig. 7–22 snaps when the elevator is at rest at the first floor so that the bottom is a distance *d* = 12 ft above a cushioning spring whose spring constant is *k* = 10,000 lb/ft. A safety device clamps the guide rails so that a constant friction force of 1000 lb opposes the motion of the elevator. (*a*) Find the speed of the elevator just before it hits the spring. (*b*) Find the distance *s* that the spring is compressed. (*c*) Find the distance that the elevator will "bounce" back up the shaft. (*d*) Using the conservation of energy principle, find the total distance that the elevator will move before coming to rest.
Answer: (*a*) 24 ft/sec. (*b*) 3.0 ft. (*c*) 9.0 ft. (*d*) 49 ft.

40(7). Show that when friction is present in an otherwise conservative mechanical system, the rate at which mechanical energy is dissipated equals the frictional force times the speed at that instant, or

$$\frac{d}{dt}(K+U) = -fv$$

41(9). The United States consumed about 1.6×10^{12} kw-hr of electrical energy in 1970. How many kilograms of matter would have to be completely destroyed to yield this energy?
Answer: 65 kg.

42(9). An electron (rest mass 9.1×10^{-31} kg) is moving with a speed 0.99*c*. (*a*) What is its total energy? (*b*) Find the ratio of the Newtonian kinetic energy to the relativistic kinetic energy in this case.

43(9). How much matter would have to be converted into energy in order to accelerate a 1.0-kiloton spaceship from rest to a speed of $\frac{1}{16}c$?
Answer: 4.5×10^3 kg (nonrelativistic expression for kinetic energy).

44(9). (*a*) The rest mass of a body is 10 gm. What is its mass when it moves at a speed of $3.0 = 10^7$ meters/sec relative to the observer? At 2.7×10^8 meters/sec? (*b*) Compare the classical and relativistic kinetic energies for these cases. (*c*) What if the observer, or measuring apparatus, is riding on the body?

45(9). It is believed that the sun obtains its energy by a fusion process in which four hydrogen nuclei are transformed, after a series of steps, into a helium nucleus and two electrons, accompanied by the release of energy. Calculate the energy released in each fusion event. The rest masses of the hydrogen atom (not nucleus), the helium atom, and the electron are 1.00783 amu,

4.00260 amu, and 0.000545 amu, respectively. See Example 6, in which, however, the masses are nuclear rather than atomic masses.

Answer: 27 MeV.

46(9). A vacuum diode consists of a cylindrical anode enclosing a cylindrical cathode. An electron with a potential energy relative to the anode of 4.8×10^{-16} joule leaves the surface of the cathode with zero initial speed. Assume that the electron does not collide with any air molecules and that the gravitational force is negligible. (*a*) What kinetic energy would the electron have when it strikes the anode? (*b*) Take 9.1×10^{-31} kg as the mass of the electron and find its final speed. (*c*) Were you justified in using classical relations for kinetic energy and mass rather than the relativistic ones?

CHAPTER 8

Conservation of Linear Momentum

8–1 Center of Mass

So far we have treated objects as though they were particles, having mass but no size. In translational motion each point on a body experiences the same displacement as any other point as time goes on, so that the motion of one particle represents the motion of the whole body. But even when a body rotates or vibrates as it moves, there is one point on the body, called the *center of mass*, that moves in the same way that a single particle subject to the same external forces would move. Figure 8–1 shows the simple parabolic motion of the center of mass of an Indian club thrown from one performer to another; no other point in the club moves in such a simple way. Note that, if the club were moving in pure translation, then every point in it would experience the same displacements as does the center of mass in Fig. 8–1. For this reason the motion of the center of mass of a body is called the translational motion of the body.

When the system with which we deal is not a rigid body like the Indian club, a center of mass, whose motion can also be described in a relatively simple way, can be assigned, even though the particles that make up the system may be changing their positions with respect to each other in a relatively complicated way as the motion proceeds. In this section we define the center of mass and show how to calculate its position. In the next section we discuss the properties that make it useful to describe the motion of extended objects or systems of particles.

Consider first the simple case of a system of two particles m_1 and m_2 at distances x_1 and x_2 respectively, from some origin O as in Fig. 8–2. We define a point C, the center of mass of the system, at a distance x_{cm} from the origin O, where x_{cm} is defined by

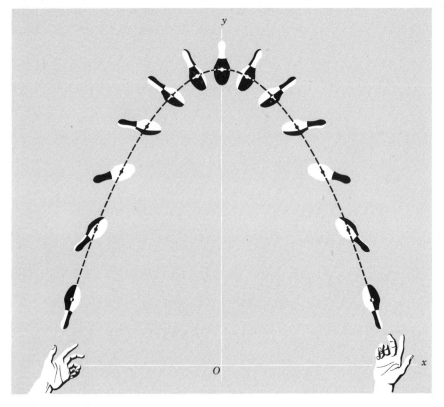

FIGURE 8–1 An Indian club is thrown from one performer to another. Even though it rotates and spins around its axis, as shown, there is one point on its axis, the center of mass, that follows a simple parabolic path.

$$x_{\text{cm}} = \frac{m_1 x_1 + m_2 x_2}{m_1 + m_2}. \tag{8–1}$$

The center of mass lies between the two masses and on the line joining them. Note that if all of the mass is concentrated at x_1 ($m_2 = 0$), then $x_{\text{cm}} = x_1$. If the masses are equal then the center of mass is equidistant from the two masses.

Although the usefulness of this definition will not become clear until later we can get some feeling for the concept at this point. Figure 8–2 shows two masses and the center of mass at point C. Notice from Eq. 8–1 that the product of the total mass M ($= m_1 + m_2$) and the center of mass distance is the sum of similar products for each mass element of the system; that is

$$M x_{\text{cm}} = m_1 x_1 + m_2 x_2.$$

FIGURE 8–2 The center of mass of the two masses m_1 and m_2 lies on the line joining m_1 and m_2 at C, a distance x_{cm} from the origin.

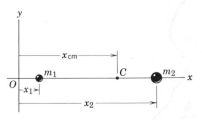

When we extend the definition to systems having more particles we will simply add the products for the additional particles. Thus, knowing the center of mass will be useful where the mechanical equations involve adding for each mass element in a system the mass times the distance from some point (such as from an axis of rotation). The operation is then simplified greatly. Why such products arise will be discussed later.

If we have n particles, $m_1, m_2, \ldots, m_n$, *along a straight line*, by definition the center of mass of these particles relative to some origin is located at

$$x_{cm} = \frac{m_1 x_1 + m_2 x_2 + \cdots + m_n x_n}{m_1 + m_2 + \cdots + m_n} = \frac{\Sigma\, m_i x_i}{\Sigma\, m_i}, \tag{8-2a}$$

where $x_1, x_2, \ldots, x_n$ are the distances of the masses from the origin from which x_{cm} is measured. The symbol Σ represents a summation operation, in this case over all n particles. The sum

$$\Sigma\, m_i = M$$

is the total mass of the system. We can then rewrite Eq. 8-2a in the form

$$M x_{cm} = \Sigma\, m_i x_i. \tag{8-2b}$$

For a large number of particles distributed in space, the center of mass is at x_{cm}, y_{cm}, z_{cm}, where

$$x_{cm} = \frac{1}{M} \Sigma\, m_i x_i, \quad y_{cm} = \frac{1}{M} \Sigma\, m_i y_i, \quad z_{cm} = \frac{1}{M} \Sigma\, m_i z_i. \tag{8-3a}$$

In vector notation each particle in the system can be described by a position vector $\mathbf{r}_i$ in a particular reference frame and the center of mass can be located by a position vector $\mathbf{r}_{cm}$. These vectors are related to x_i, y_i, z_i and x_{cm}, y_{cm}, z_{cm} in Eq. 8-3a by

$$\mathbf{r}_i = \mathbf{i} x_i + \mathbf{j} y_i + \mathbf{k} z_i$$

and

$$\mathbf{r}_{cm} = \mathbf{i} x_{cm} + \mathbf{j} y_{cm} + \mathbf{k} z_{cm}.$$

Thus the three scalar equations of Eq. 8-3a can be replaced by a single vector equation

$$\mathbf{r}_{cm} = \frac{1}{M} \Sigma\, m_i \mathbf{r}_i \tag{8-3b}$$

in which the sum is a vector sum. You can prove that Eq. 8-3b is true by substituting the expressions given for $\mathbf{r}_i$ and $\mathbf{r}_{cm}$ just above into Eq. 8-3b.

Equations 8-3 treat the most general case for a collection of particles. Equations 8-1 and 8-2 are special instances of this one. The location of the center of mass is independent of the reference frame used to locate it. The center of mass of a system of particles depends only on the masses of the particles and the positions of the particles relative to one another.

A rigid body, such as the Indian club, can be thought of as a system of closely packed particles; this would allow us to find a center of mass for it from the particle point of view (Eqs. 8-3). The number of particles (atoms) in the body is so large and their spacing so small, however, that we can treat the body as though it had a continuous distribution of mass. To obtain the expression for the center of mass of a continuous body, let us begin by subdividing the body into n small elements of mass Δm_i located approximately at the points x_i, y_i, z_i. The coordinates of the center of mass are then given approximately, by

$$x_{cm} \simeq \frac{\Sigma \, \Delta m_i x_i}{\Sigma \, \Delta m_i} \qquad y_{cm} \simeq \frac{\Sigma \, \Delta m_i y_i}{\Sigma \, \Delta m_i} \qquad z_{cm} \simeq \frac{\Sigma \, \Delta m_i z_i}{\Sigma \, \Delta m_i}.$$

Now let the elements of mass be further subdivided so that the number of elements n tends to infinity. The points x_i, y_i, z_i will locate the mass elements more precisely as n is increased. The coordinates of the center of mass can now be given precisely as

$$x_{cm} = \lim_{\Delta m_i \to 0} \frac{\Sigma \, \Delta m_i x_i}{\Sigma \, \Delta m_i} = \frac{\int x \, dm}{\int dm} = \frac{1}{M} \int x \, dm,$$

$$y_{cm} = \lim_{\Delta m_i \to 0} \frac{\Sigma \, \Delta m_i y_i}{\Sigma \, \Delta m_i} = \frac{\int y \, dm}{\int dm} = \frac{1}{M} \int y \, dm, \qquad (8\text{–}4a)$$

$$z_{cm} = \lim_{\Delta m_i \to 0} \frac{\Sigma \, \Delta m_i z_i}{\Sigma \, \Delta m_i} = \frac{\int z \, dm}{\int dm} = \frac{1}{M} \int z \, dm.$$

In these expressions dm is the differential element of mass at the point x, y, z, and $\int dm$ equals M, where M is the total mass of the object. For a continuous body the summation of Eq. 8–3a is replaced by the integration of Eq. 8–4a.

The vector expression that is equivalent to the three scalar expressions of Eq. 8–4a is

$$\mathbf{r}_{cm} = \frac{1}{M} \int \mathbf{r} \, dm \qquad (8\text{–}4b)$$

As before, the summation of Eq. 8–3b has been replaced by an integration.

Often we deal with homogeneous objects having a point, a line, or a plane of symmetry. Then the center of mass will lie at the point, on the line, or in the plane of symmetry. For example, the center of mass of a homogeneous sphere (which has a point of symmetry) will be at the center of the sphere, the center of mass of a cone (which has a line of symmetry) will be on the axis of the cone, etc. We can understand that this is so because, from symmetry, an origin of coordinates can be found such that $\int \mathbf{r} \, dm$ in Eq. 8–4b is zero (at the center of the sphere, somewhere along the axis of the cone, etc.). From this it follows that $\mathbf{r}_{cm} = 0$ for this point, which identifies it as the center of mass.

Example 1. Locate the center of mass of three particles of mass $m_1 = 1.0$ kg, $m_2 = 2.0$ kg, and $m_3 = 3.0$ kg at the corners of an equilateral triangle 1.0 meter on a side.

Choose the x-axis along one side of the triangle as shown in Fig. 8–3. Then,

$$x_{cm} = \frac{\Sigma \, m_i x_i}{\Sigma \, m_i} = \frac{(1.0 \text{ kg})(0) + (2.0 \text{ kg})(1.0 \text{ meter}) + (3.0 \text{ kg})(\frac{1}{2} \text{ meter})}{(1.0 + 2.0 + 3.0) \text{ kg}} = \frac{7}{12} \text{ meter,}$$

$$y_{cm} = \frac{\Sigma \, m_i y_i}{\Sigma \, m_i} = \frac{(1.0 \text{ kg})(0) + (2.0 \text{ kg})(0) + (3.0 \text{ kg})(\sqrt{3}/2 \text{ meter})}{(1.0 + 2.0 + 3.0) \text{ kg}} = \frac{\sqrt{3}}{4} \text{ meter.}$$

The center of mass C is shown in the figure. Why is it not at the geometric center of the triangle?

Example 2. Find the center of mass of the triangular plate of Fig. 8–4.

If a body can be divided into parts such that the center of mass of each part is known, the

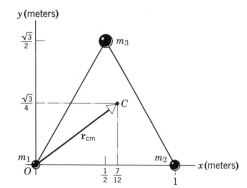

FIGURE 8-3 Example 1. Finding the center of mass C of three unequal masses forming an equilateral triangle.

center of mass of the body can sometimes be found simply. The triangular plate may be divided into narrow strips parallel to one side. The center of mass of each strip lies on the line which joins the middle of that side to the opposite vertex. But we can divide up the triangle in three different ways, using this process for each of three sides. Hence the center of mass lies at the intersection of the three lines which join the middle of each side with the opposite vertices. This is the only point that is common to the three lines.

8-2   Motion of the Center of Mass

Now we can discuss the physical importance of the center-of-mass concept. Consider the motion of a group of particles whose masses are $m_1, m_2, \ldots m_n$ and whose total mass is M. For the time being we will assume that mass neither enters nor leaves the system so that the total mass M of the system remains constant with time. In Section 8-6 we shall consider systems in which M is not constant; a familiar example is a rocket, which expels hot gases as its fuel burns, thus reducing its mass.

From Eq. 8-3b we have, for our fixed system of particles,

$$M\mathbf{r}_{cm} = m_1\mathbf{r}_1 + m_2\mathbf{r}_2 + \cdots + m_n\mathbf{r}_n,$$

where $\mathbf{r}_{cm}$ is the position vector identifying the center of mass in a particular reference frame. Differentiating this equation with respect to time, we obtain

$$M\mathbf{v}_{cm} = m_1\mathbf{v}_1 + m_2\mathbf{v}_2 + \cdots + m_n\mathbf{v}_n, \qquad (8-5)$$

where $\mathbf{v}_1$ is the velocity of the first particle, etc., and $d\mathbf{r}_{cm}/dt \, (= \mathbf{v}_{cm})$ is the velocity of the center of mass.

Differentiating Eq. 8-5 with respect to time, we obtain

$$M\frac{d\mathbf{v}_{cm}}{dt} = m_1\mathbf{a}_1 + m_2\mathbf{a}_2 + \cdots + m_n\mathbf{a}_n, \qquad (8-6)$$

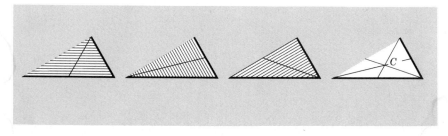

FIGURE 8-4   Example 2. Finding the center of mass C of a triangular plate.

where $\mathbf{a}_1$ is the acceleration of the first particle, etc., and $d\mathbf{v}_{cm}/dt\ (=\mathbf{a}_{cm})$ is the acceleration of the center of mass of the system. Now, from Newton's second law, the force $\mathbf{F}_1$ acting on the first particle is given by $\mathbf{F}_1 = m_1\mathbf{a}_1$ Likewise, $\mathbf{F}_2 = m_2\mathbf{a}_2$, etc. We can then write Eq. 8–6 as

$$M\mathbf{a}_{cm} = \mathbf{F}_1 + \mathbf{F}_2 + \cdots + \mathbf{F}_n. \qquad (8\text{–}7)$$

Hence *the total mass of the group of particles times the acceleration of its center of mass is equal to the vector sum of all the forces acting on the group of particles.*

Among all these forces will be internal forces exerted by the particles on each other. However, from Newton's third law, these internal forces will occur in equal and opposite pairs, so that they contribute nothing to the sum. Hence the internal forces can be removed from the problem. The right-hand side of Eq. 8–7 represents the sum of only the external forces acting on all the particles. We can then rewrite Eq. 8–7 as simply

$$M\mathbf{a}_{cm} = \mathbf{F}_{ext}. \qquad (8\text{–}8)$$

This states that *the center of mass of a system of particles moves as though all the mass of the system were concentrated at the center of mass and all the external forces were applied at that point.*

Notice that we obtain this simple result without specifying the nature of the system of particles. The system can be a rigid body in which the particles are in fixed positions with respect to one another, or it can be a collection of particles in which there may be all kinds of internal motion.

We have now found how to describe the translational motion of a system of particles and how to describe the translational motion of a body which may be rotating as well. In this chapter and the next we apply this result to the linear motion of a system of particles. In later chapters we shall see how it simplifies the analysis of rotational motion.

Example 3. Consider three particles of different masses acted on by external forces, as shown in Fig. 8–5. Find the acceleration of the center of mass of the system.

First we find the coordinates of the center of mass. From Eq. 8–3a,

$$x_{cm} = \frac{(8.0 \times 4) + (4.0 \times -2) + (4.0 \times 1)}{16} \text{ meters} = 1.8 \text{ meters},$$

$$y_{cm} = \frac{(8.0 \times 1) + (4.0 \times 2) + (4.0 \times -3)}{16} \text{ meters} = 0.25 \text{ meters}.$$

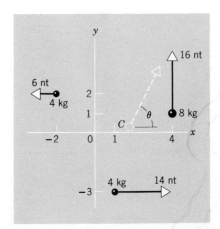

FIGURE 8–5 Example 3. Finding the motion of the center of mass of three masses, each subjected to a different (external) force. The forces lie in the plane defined by the particles. The distances along the axes are in meters.

These are shown as C in Fig. 8–5.

To obtain the acceleration of the center of mass, we first determine the resultant external force acting on the system consisting of the three particles. The x-component of this force is

$$F_x = 14 \text{ nt} - 6.0 \text{ nt} = 8.0 \text{ nt},$$

and the y-component is

$$F_y = 16 \text{ nt}.$$

Hence the resultant external force has a magnitude

$$F = \sqrt{(8.0)^2 + (16)^2} \text{ nt} = 18 \text{ nt},$$

and makes an angle θ with the x-axis given by

$$\tan \theta = \frac{16 \text{ nt}}{8.0 \text{ nt}} = 2.0 \quad \text{or} \quad \theta = 63°.$$

Then, from Eq. 8–8, the magnitude of the acceleration of the center of mass is

$$a_{cm} = \frac{F}{M} = \frac{18 \text{ nt}}{16 \text{ kg}} = 1.1 \text{ meters/sec}^2,$$

making an angle of 63° with the x-axis.

Although the three particles will change their relative positions as time goes on, the center of mass will move, as shown, with this constant acceleration.

8–3 Linear Momentum of a Particle

The *linear momentum* of a particle is a vector **p** defined as the product of its mass m and its velocity **v**. That is,

$$\mathbf{p} = m\mathbf{v}. \tag{8–9}$$

Momentum, being the product of a scalar by a vector, is itself a vector. Because it is proportional to **v**, the momentum **p** of a particular particle depends on the reference frame of the observer; we must always specify this frame.

Newton, in his famous *Principia*, expressed the second law of motion in terms of momentum (which he called "quantity of motion"). Expressed in modern terminology Newton's second law reads: *The rate of change of momentum of a body is proportional to the resultant force acting on the body and is in the direction of that force.* In symbolic form this becomes

$$\mathbf{F} = \frac{d\mathbf{p}}{dt}. \tag{8–10}$$

If our system is a single particle of (constant) mass m, this formulation of the second law is equivalent to the form $\mathbf{F} = m\mathbf{a}$, which we have used up to now. That is, if m is a constant, then

$$\mathbf{F} = \frac{d\mathbf{p}}{dt} = \frac{d}{dt}(m\mathbf{v}) = m\frac{d\mathbf{v}}{dt} = m\mathbf{a}.$$

The relations $\mathbf{F} = m\mathbf{a}$ and $\mathbf{F} = d\mathbf{p}/dt$ for single particles are completely equivalent in classical mechanics.

In relativity theory the second law for a single particle in the form $\mathbf{F} = m\mathbf{a}$ is not valid. However, it turns out that Newton's second law in the form $\mathbf{F} = d\mathbf{p}/dt$ is still a valid law if the momentum **p** for a single particle is defined not as $m_0\mathbf{v}$ but as

$$\mathbf{p} = \frac{m_0\mathbf{v}}{\sqrt{1 - v^2/c^2}}. \qquad (8\text{–}11)$$

This result suggested a new definition of mass (compare Eqs. 8–9 and 8–11)

$$m = \frac{m_0}{\sqrt{1 - v^2/c^2}},$$

so that the momentum could still be written as $\mathbf{p} = m\mathbf{v}$; see Section 7–9. In this equation v is the speed of the particle, c is the speed of light, and m_0 is the "rest mass" of the body (its mass when $v = 0$). From this definition we must expect the mass of a particle to increase with its speed. In atomic and nuclear systems particles may acquire enormous speeds, comparable to the speed of light. This concept can be put to a direct test in such systems because the increase in mass over the rest mass for such particles is large enough to measure accurately. Results of all such experiments indicate that this effect is real and given exactly by the equation above. (See for example, Fig. 7–8.)

8–4   Linear Momentum of a System of Particles

Suppose that instead of a single particle we have a system of n particles, with masses m_1, m_2, etc. We shall continue to assume, as we did in Section 8–2, that no mass enters or leaves the system, so that the mass $M\,(= \Sigma\, m_i)$ of the system remains constant with time. The particles may interact with each other and external forces may act on them as well. Each particle will have a velocity and a momentum. Particle 1 of mass m_1 and velocity $\mathbf{v}_1$ will have a momentum $\mathbf{p}_1 = m_1\mathbf{v}_1$, for example. The system as a whole will have a total momentum **P** in a particular reference frame, which is defined to be simply the vector sum of the momenta of the individual particles in that same frame, or

$$\mathbf{P} = \mathbf{p}_1 + \mathbf{p}_2 + \cdots + \mathbf{p}_n \qquad (8\text{–}12)$$

$$= m_1\mathbf{v}_1 + m_2\mathbf{v}_2 + \cdots + m_n\mathbf{v}_n.$$

If we compare this relation with Eq. 8–5 we see at once that

$$\mathbf{P} = M\mathbf{v}_{\text{cm}}, \qquad (8\text{–}13)$$

which is an equivalent definition for the momentum of a system of particles. In words, Eq. 8–13 states: *The total momentum of a system of particles is equal to the product of the total mass of the system and the velocity of its center of mass.*

We have seen (Eq. 8–8) that Newton's second law for a system of particles can be written as

$$\mathbf{F}_{\text{ext}} = M\mathbf{a}_{\text{cm}} \qquad (8\text{–}8)$$

in which $\mathbf{F}_{\text{ext}}$ is the vector sum of all the external forces acting on the system; we recall that the internal forces acting between particles cancel in pairs because of Newton's third law. If we differentiate Eq. 8–13 with respect to time we obtain, for an assumed constant mass M,

$$\frac{d\mathbf{P}}{dt} = M\frac{d\mathbf{v}_{\text{cm}}}{dt} = M\mathbf{a}_{\text{cm}}. \qquad (8\text{–}14)$$

Comparison of Eqs. 8–8 and 8–14 allows us to write Newton's second law for a system of particles in the form

$$\mathbf{F}_{ext} = \frac{d\mathbf{P}}{dt}. \tag{8–15}$$

This equation is the generalization of the single-particle equation $\mathbf{F} = d\mathbf{p}/dt$ (Eq. 8–10) to a system of many particles, no mass entering or leaving the system. Equation 8–15 reduces to Eq. 8–10 for the special case of a single particle, there being only external forces on a one-particle system.

8–5 Conservation of Linear Momentum

Suppose that the sum of the external forces acting on a system is zero. Then, from Eq. 8–15,

$$\frac{d\mathbf{P}}{dt} = 0 \quad \text{or} \quad \mathbf{P} = \text{constant.}$$

When the resultant external force acting on a system is zero, the total linear momentum of the system remains constant. This simple but quite general result is called *the principle of the conservation of linear momentum.* We shall see that it is applicable to many important situations.

The conservation of linear momentum principle is the second of the great conservation principles that we have met so far, the first being the conservation of energy principle. Later we shall meet several others, among them the conservation of electric charge and of angular momentum. Conservation principles are of theoretical and practical importance in physics because they are simple and universal. They are all cast in the form: While the system is changing there is one aspect of the system that remains unchanged. Different observers, each in his own reference frame, would all agree, if they watched the same changing system, that the conservation laws applied to the system. For the conservation of linear momentum, for example, observers in different reference frames would assign different values of $\mathbf{P}$ to the linear momentum of the system, but each would agree (assuming $\mathbf{F}_{ext} = 0$) that his own value of $\mathbf{P}$ remained unchanged as the particles that make up the system move about.

The total momentum of a system can only be changed by external forces acting on the system. The internal forces, being equal and opposite, produce equal and opposite changes in momentum which annul one another. For a system of particles

$$\mathbf{p}_1 + \mathbf{p}_2 + \cdots + \mathbf{p}_n = \mathbf{P},$$

so that when the total momentum $\mathbf{P}$ is constant we have

$$\mathbf{p}_1 + \mathbf{p}_2 + \cdots + \mathbf{p}_n = \text{constant} = \mathbf{P}_0. \tag{8–16}$$

The momenta of the individual particles may change, but their sum remains constant if there is no net external force.

Momentum is a vector quantity. Equation 8–16 is therefore equivalent to three scalar equations, one for each coordinate direction. Hence the conservation of linear momentum supplies us with three conditions on the motion of a system to which it applies. The conservation of energy on the other hand supplies us with only one condition on the motion of a system to which it applies, because energy is a scalar.

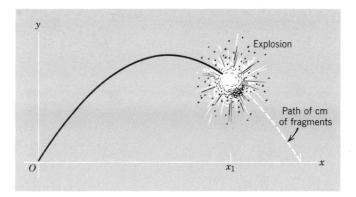

FIGURE 8–6   Example 4. A projectile, following the usual parabolic trajectory, bursts at x_1. The center of mass of the fragments continues along the same parabolic path.

The law of the conservation of linear momentum holds true even in atomic and nuclear physics, although Newtonian mechanics does not. Hence this conservation law must be more fundamental than the Newtonian principles.

8–6   Some Applications of the Momentum Principle

Example 4. Consider first a problem in which an external force acts on a system of particles. Recall our previous discussion of projectile motion (Chapter 4). Now let us imagine that our projectile is a shell that explodes while in flight. The path of the shell is shown in Fig. 8–6. We assume that the air resistance is negligible. The system is the shell, the earth is our reference frame, and the external force is that of gravity. At the point x_1 the shell explodes and shell fragments are blown in all directions. What can we say about the motion of this system thereafter?

The forces of the explosion are all internal forces; they are forces exerted by part of the sytsem on other parts of the system. These forces may change the momenta of all the individual fragments from the values they had when they made up the shell, but they cannot change the total vector momentum of the system. Only an external force can change the total momentum. The external force, however, is simply that due to gravity. Since a system of particles as a whole moves as though all its mass were concentrated at the center of mass with the external force applied there, the center of mass of the fragments will continue to move in the parabolic trajectory that the unexploded shell would have followed. The change in the total momentum of the system attributable to gravity is the same whether the shell explodes or not.

What can you say about the mechanical energy of the system before and after the explosion?

Example 5. Consider now two blocks A and B, of masses m_A and m_B, coupled by a spring and resting on a horizontal frictionless table. Let us pull the blocks apart and stretch the spring, as in Fig. 8–7, and then release the blocks. Describe the subsequent motion.

If the system consists of the two blocks and spring, then after we have released the blocks there is no net external force acting on the system. We can therefore apply the conservation of linear momentum to the motion. The momentum of the system before the blocks were released was zero

FIGURE 8–7   Example 5. Two blocks A and B, resting on a frictionless surface, are connected by a spring. If they are held apart and then released, the sum of their momenta remains zero.

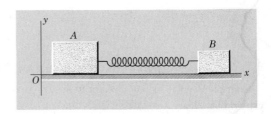

in the reference frame shown attached to the table, so the momentum must remain zero thereafter. The total momentum can be zero even though the blocks move because momentum is a vector quantity. One block will have positive momentum (A moves in the $+x$-direction) and the other block will have negative momentum (B moves in the $-x$-direction). From the conservation of momentum we have

$$\text{initial momentum} = \text{final momentum.}$$

$$0 = m_B\mathbf{v}_B + m_A\mathbf{v}_A.$$

Therefore

$$m_B\mathbf{v}_B = -m_A\mathbf{v}_A$$

or

$$\mathbf{v}_A = -\frac{m_B}{m_A}\mathbf{v}_B.$$

For example, if m_A is 2.0 slugs and m_B is 1.0 slug, then $\mathbf{v}_A$ will always be one-half $\mathbf{v}_B$ in magnitude and oppositely directed as the blocks move.

The kinetic energy of block A is $\frac{1}{2}m_Av_A^2$ and can be written as $(m_Av_A)^2/2m_A$ and that of block B is $\frac{1}{2}m_Bv_B^2$ and can be written as $(m_Bv_B)^2/2m_B$. But

$$\frac{K_A}{K_B} = \frac{2m_B(m_Av_A)^2}{2m_A(m_Bv_B)^2} = \frac{m_B}{m_A},$$

in which m_Av_A equals m_Bv_B because of momentum conservation. The kinetic energies of the blocks at any instant are inversely proportional to their respective masses. Because mechanical energy is conserved also, the blocks will continue to oscillate back and forth, the energy being partly kinetic and partly potential. What is the motion of the center of mass of this system?

If mechanical energy is not conserved, as would be true if friction were present, the motion will die out as the energy is dissipated. Can we apply the conservation of linear momentum in this case? Explain.

Example 6. As an example of recoil, consider radioactive decay. An α-particle (the nucleus of a helium atom) is emitted from a uranium-238 nucleus, originally at rest, with a speed of 1.4×10^7 meters/sec and a kinetic energy of 4.1 MeV (million electron volts). Find the recoil speed of the residual nucleus (thorium-234).

We think of the system (thorium + α-particle) as initially bound and forming the uranium nucleus. The system then fragments into two separate parts. The momentum of the system before fragmentation is zero. In the absence of external forces, the momentum after fragmentation is also zero. Hence,

$$\text{initial momentum} = \text{final momentum,}$$

$$0 = M_\alpha\mathbf{v}_\alpha + M_{\text{Th}}\mathbf{v}_{\text{Th}},$$

$$\mathbf{v}_{\text{Th}} = -\frac{M_\alpha}{M_{\text{Th}}}\mathbf{v}_\alpha.$$

The ratio of the α-particle mass to the thorium nucleus mass, M_α/M_{Th}, is 4/234 and $v_\alpha = 1.4 \times 10^7$ meters/sec. Hence,

$$v_{\text{Th}} = -(4/234)(1.4 \times 10^7 \text{ meters/sec}) = -2.4 \times 10^5 \text{ meters/sec.}$$

The minus sign indicates that the residual thorium nucleus recoils in a direction exactly opposite to the motion of the α-particle, so as to give a resultant vector momentum of zero.

How can we compute the kinetic energy of the recoiling nucleus (see previous example)? Where does the energy of the fragments come from?

Example 7. With manned flight to the moon now an accomplished fact and interplanetary exploration contemplated, the application of the conservation of momentum to rocket problems becomes timely and important. Rocket dynamics is a complicated subject, but we can handle a rather simple example.

Consider a rocket weighing 30,000 lb and drifting in gravity-free space along a straight line which we may identify as the *x*-axis of an inertial reference frame. The rocket is fired for 30 sec and, during this period, ejects gases at the rate of 10 slugs/sec with a speed of 5000 ft/sec relative to the rocket (exhaust velocity), both quantities being assumed to be constant during the "burn." Find the acceleration as a function of time during the burn. Note that there are no external forces, such as gravity or air resistance, in this problem.

This is an example of the more general problem of finding the acceleration of a body where the acceleration occurs because mass is either being ejected from the body or added to it, with a relative velocity $\mathbf{v}_{rel}$. We can solve such problems by including in the system both the body whose acceleration is to be calculated and the mass Δm ejected or added during a small time interval Δt.

Figure 8–8*a* shows the situation at time *t*. The rocket and fuel have total mass M and the combination is moving with velocity $\mathbf{v}$ as seen from a particular reference frame. At a time Δt later the configuration has changed to that shown in Fig. 8–8*b*. A mass ΔM has been ejected from the rocket, its center of mass moving with velocity $\mathbf{u}$ as seen by our observer. The rocket mass is reduced to $M - \Delta M$ and the velocity $\mathbf{v}$ of the center of mass of the rocket is changed to $\mathbf{v} + \Delta\mathbf{v}$. Because there are no external forces, $\mathbf{F}_{ext} = 0$ in Eq. 8–15, or,

$$\frac{d\mathbf{P}}{dt} = 0.$$

We can write, for the time interval Δt

$$0 = \frac{\Delta\mathbf{P}}{\Delta t} = \frac{\mathbf{P}_f - \mathbf{P}_i}{\Delta t}$$

in which $\mathbf{P}_f$ is the (final) system momentum in Fig. 8–8*b* and $\mathbf{P}_i$ is the (initial) system momentum for Fig 8–8*a*. But $\mathbf{P}_f = (M - \Delta M)(\mathbf{v} + \Delta\mathbf{v}) + \Delta M\mathbf{u}$ and $\mathbf{P}_i = M\mathbf{v}$. This leads to

$$0 = \frac{[(M - \Delta M)(\mathbf{v} + \Delta\mathbf{v}) + \Delta M\mathbf{u}] - [M\mathbf{v}]}{\Delta t} \tag{8–17}$$

Now, if we let Δt approach zero, $\Delta\mathbf{v}/\Delta t$ approaches $d\mathbf{v}/dt$, the acceleration of the body. The quantity ΔM is the mass ejected in Δt; this leads to a decrease in the mass M of the original body. Since dM/dt, the change in mass of the body with time, is intrinsically negative in this case, in the limit the positive quantity $\Delta M/\Delta t$ is replaced by $-dM/dt$.

The quantity $\mathbf{u} - (\mathbf{v} + \Delta\mathbf{v})$ in Eq. 8–17 is just $\mathbf{v}_{rel}$, the relative velocity of the ejected mass with respect to the rocket. With these changes, Eq. 8–17 may be written as

$$M\frac{d\mathbf{v}}{dt} = (\mathbf{u} - \mathbf{v})\frac{dM}{dt} \tag{8–18a}$$

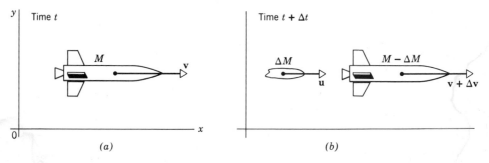

(a)   *(b)*

FIGURE 8–8

or

$$M \frac{d\mathbf{v}}{dt} = \mathbf{v}_{\text{rel}} \frac{dM}{dt}. \tag{8–18b}$$

The right-hand term in Eq. 8–18b depends on the characteristics of the rocket and—like the left-hand term—has the dimensions of a force. We call it the *thrust,* and we interpret it as the reaction force exerted on the rocket by the mass that leaves it. The rocket designer wants to make the thrust as large as possible. He can do this by designing the rocket to eject mass as rapidly as possible (dM/dt large) and with the highest possible relative speed ($\mathbf{v}_{\text{rel}}$ large).

Equation 8–18b is general and also holds for bodies gaining mass. In our present example, however,

$$\frac{dM}{dt} = -10 \text{ slugs/sec,}$$

$$M = \left(\frac{30{,}000}{32} - 10t \right) \text{slugs,}$$

and

$$v_{\text{rel}} = -5000 \text{ ft/sec,}$$

where t is the time interval, in seconds, between ignition and the time at which the acceleration is to be calculated. Equation 8–18b yields for the acceleration

$$a = \frac{dv}{dt} = \frac{v_{\text{rel}}(dM/dt)}{M} = \frac{(-5000 \text{ ft/sec})(-10 \text{ slugs/sec})}{(940 - 10t) \text{ slugs}}$$

$$= \frac{+5000}{(94 - t)} \text{ ft/sec}^2.$$

This expression holds during the burn only, yielding $a = 53$ ft/sec^2 at ignition ($t = 0$) and $a = 78$ ft/sec^2 at burnout ($t = 30$ sec). Both before and after the burn we have $\mathbf{a} = 0$. We can calculate the thrust from

$$\text{Thrust} = v_{\text{rel}}(dM/dt)$$
$$= (-5000 \text{ ft/sec})(-10 \text{ slugs/sec})$$
$$= +50{,}000 \text{ lb}$$

The thrust is constant during the burn in this example.

QUESTIONS

1. Must there necessarily be any mass at the center of mass of a system?

2. Does the center of mass of a solid necessarily lie within the body? If not, give examples.

3. How is the center of mass concept related to the concept of geographic center of the country? To the population center of the country? What can you conclude from the fact that the geographic center differs from the population center?

4. A sculptor decides to portray a bird (Fig. 8–9). Luckily the final model is actually able to stand upright. The model is formed of a single sheet of metal of uniform thickness. Of the points shown, which is most likely to be the center of mass?

5. If only an external force can change the state of motion of the center of mass of a body, how does it happen that the internal force of the brakes can bring a car to rest?

6. Can a body have energy without having momentum? Explain. Can a body have momentum without having energy? Explain.

7. A light and a heavy body have equal kinetic energies of translation. Which one has the larger momentum?

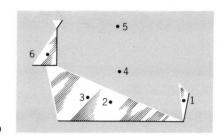

FIGURE 8-9

8. A bird is in a wire cage hanging from a spring balance. Is the reading of the balance when the bird is flying about greater than, less than, or the same as that when the bird sits in the cage?

9. Can a sailboat be propelled by air blown at the sails from a fan attached to the boat?

10. A man stands still on a large sheet of slick ice; in his hand he holds a lighted firecracker. He throws the firecracker into the air. Describe briefly, but as exactly as you can, the motion of the center of mass of the firecracker and the motion of the center of mass of the system consisting of man and firecracker. It will be most convenient to describe each motion during each of the following periods: (*a*) after he throws the firecracker, but before it explodes; (*b*) between the explosion and the first piece of firecracker hitting the ice; (*c*) between the first fragment hitting the ice and the last fragment landing; (*d*) during the time when all fragments have landed but none has reached the edge of the ice.

11. The final velocity of the final stage of a multistage rocket is much greater than the final velocity of a single-stage rocket of the same total weight and fuel supply. Explain this fact.

12. Two bodies, *A* and *B*, have the same translational kinetic energy. The mass of *A* is four times that of *B*. How do the magnitudes of their linear momenta compare?

PROBLEMS

1(1). The mass of the moon is about 0.013 times the mass of the earth, and the distance from the center of the moon to the center of the earth is about 60 times the radius of the earth. How far is the center of mass of the earth-moon system from the center of the earth? Take the earth's radius to be 4000 miles.

Answer: 3100 miles (within the earth).

2(1). Experiments using the diffraction of electrons show that the distance between the centers of the carbon (C) and oxygen (O) atoms in the carbon monoxide gas molecule is 1.130×10^{-10} meter. Locate the center of mass of a CO molecule relative to the carbon atom.

3(1). Where is the center of mass of the three particles shown in Figure 8–10?

Answer: $1.1\mathbf{i} + 1.3\mathbf{j}$, meters.

4(1). In the ammonia (NH_3) molecule, the three hydrogen (H) atoms form an equilateral trian-

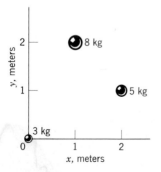

FIGURE 8-10   Problem 3(1).

FIGURE 8-11   Problem 4(1).

gle, the distance between centers of the atoms being 1.628×10^{-10} meter, so that the center of the triangle is 9.39×10^{-11} meter from each hydrogen atom. The nitrogen (N) atom is at the apex of a pyramid, the three hydrogens constituting the base (see Fig. 8–11). The hydrogen-nitrogen distance is 1.014×10^{-10} meter. Locate the center of mass relative to the nitrogen atom.

5(1). Show that the center of mass of two particles is on the line joining them at a point whose distance from each particle is inversely proportional to the mass of that particle.

6(1). Find the center of mass of a homogeneous semicircular plate. Let a be the radius of the circle.

7(2). The masses and coordinates of four particles are as follows: 5.0 kg, $x = y = 0.0$ cm; 3.0 kg, $x = y = 8.0$ cm; 2.0 kg, $x = 3.0$ cm, $y = 0.0$ cm; 6.0 kg, $x = -2.0$ cm, $y = -6.0$ cm. Find the coordinates of the center of mass of this collection of particles.
Answer: $x_{cm} = +1.1$ cm; $y_{cm} = -0.75$ cm.

8(2). (*a*) For the collection of masses in problem 7(2), if a 2.0-nt force acts on the 5.0-kg mass in the negative x-direction where will this object be at the end of 2.0 sec? (*b*) Where will the center of mass now be located? (*c*) If this same force acted on a single mass equal to the total mass of the collection and initially located at the point calculated in the preceding problem, where would this mass be at the end of 2.0 sec?

9(2). Two blocks of masses 1.0 kg and 3.0 kg connected by a spring rest on a frictionless surface. If the two are given velocities such that the first travels at 1.7 meters/sec toward the center of mass, which remains at rest, what is the velocity of the second?
Answer: 0.57 meter/sec, toward the center of mass.

10(2). Two particles P and Q are initially at rest 1.0 meter apart. P has a mass of 0.10 kg and Q a mass of 0.30 kg. P and Q attract each other with a constant force of 1.0×10^{-2} nt. No external forces act on the system. Describe the motion of the center of mass. At what distance from P's original position do the particles collide?

11(2). A cannon and a supply of cannon balls are inside a sealed railroad car as in Fig. 8–12. The cannon fires to the right, the car recoiling to the left. The cannon balls remain in the car after hitting the far wall. Show that no matter how the cannon balls are fired the railroad car cannot travel more than its length L, assuming it starts from rest.

12(2). A dog, weighing 10 lb is standing on a flatboat so that he is 20 ft from the shore. He walks 8.0 ft on the boat toward shore and then halts. The boat weighs 40 lb, and one can assume there is no friction between it and the water. How far is he from the shore at the end of this time? (Hint: The center of mass of boat + dog does not move. Why?) The shoreline is to the left in Fig. 8–13.

13(2). A ball of mass m and radius R is placed inside a larger hollow sphere with the same mass and inside radius $2R$. The combination is at rest on a frictionless surface in the position shown in Fig. 8–14. The smaller ball is released, rolls around the inside of the hollow sphere, and finally comes to rest at the bottom. How far will the larger sphere have moved during this process?
Answer: $\frac{1}{2}R$ to left.

14(3). What would be the speed of a 75-gm bullet if it had the same momentum as a 1000-kg car with a speed of 50 km/hr?

15(3). Find (*a*) the momentum and (*b*) the kinetic energy of a 10-gm bullet with a speed of 760 meters/sec. (*c*) How fast must a 450-kg deer move to have the same momentum?
Answer: (*a*) 7.6 kg meters/sec. (*b*) 2900 joules. (*c*) 1.7 cm/sec.

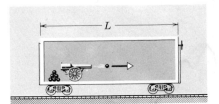

FIGURE 8–12   Problem 11(2).

FIGURE 8–13   Problem 12(2).

16(3). A 2000-kg truck traveling north at 40 km/hr turns east and accelerates to 50 km/hr. (*a*) What is the change in kinetic energy of the truck? (*b*) What is the magnitude and direction of the change in the truck's momentum?

17(3). How fast must a 1800-lb Volkswagen travel (*a*) to have the same momentum as a 5850-lb Cadillac going 10 miles/hr? (*b*) To have the same kinetic energy? (*c*) Make the same calculations using a 10-ton truck instead of a Cadillac.
Answer: (*a*) 33 miles/hr. (*b*) 18 miles/hr. (*c*) 110 miles/hr; 33 miles/hr.

18(3). A 50-gm ball is thrown into the air with an initial speed of 15 meters/sec at an angle of 45°. (*a*) What are the values of the kinetic energy of the ball initially and just before it hits the ground? (*b*) Find the corresponding values of the momentum (magnitude and direction). (*c*) Show that the change in momentum is just equal to the weight of the ball multiplied by the time of flight.

19(3). A 5.0-kg object with a speed of 30 meters/sec strikes a steel plate at an angle of 45° and rebounds with the same speed and angle (Fig. 8–15). What is the change (magnitude and direction) of the linear momentum of the object?
Answer: 210 kg meters/sec, perpendicular to the plate.

20(3). Two bodies, each made up of weights from a set, are connected by a light cord which passes over a light, frictionless pulley with a diameter of 5.0 cm. The two bodies are at the same level. Each originally has a mass of 500 gm. (*a*) Locate their center of mass. (*b*) Twenty grams are transferred from one body to the other, but the bodies are prevented from moving. Locate the center of mass. (*c*) The two bodies are now released. Describe the motion of the center of mass and determine its acceleration.

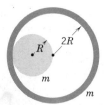

FIGURE 8–14   Problem 13(2).

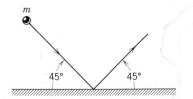

FIGURE 8–15   Problem 19(3).

21(4). A pellet gun fires 10 2.0-gm pellets per second with a speed of 500 meters/sec. The pellets are stopped by a rigid wall. (*a*) What is the momentum of each pellet? (*b*) What is the kinetic energy of each pellet? (*c*) What is the average force exerted by the pellets on the wall? *Answer:* (*a*) 1.0 kg meter/sec. (*b*) 250 joules. (*c*) 10 nt.

22(4). An 80-kg man is standing at the rear of a 400-kg iceboat that is moving at 4.0 meters/sec across ice that may be considered to be frictionless. He decides to walk to the front of the 18 meters-long boat and does so at a speed of 2.0 meters/sec with respect to the boat. How far did the boat move across the ice while he was walking?

23(5). A body of mass 8.0 kg is traveling at 2.0 meters/sec under the influence of no external force. At a certain instant an internal explosion occurs, splitting the body into two chunks of 4.0 kg mass each; 16 joules of translational kinetic energy are imparted to the two-chunk system by the explosion. Neither chunk leaves the line of the original motion. Determine the speed and direction of motion of each of the chunks after the explosion. *Answer:* One chunk comes to rest. The other moves ahead with a speed of 4.0 meters/sec.

24(5). A space vehicle is traveling at 4,000 km/hr with respect to the earth when the exhausted rocket motor is disengaged and sent backwards with a speed of 80 km/hr with respect to the command module. The mass of the motor is four times the mass of the module. What is the speed of the command module after the separation?

25(5). The last stage of a rocket is traveling at a speed of 25,000 ft/sec. This last stage is made up of two parts which are clamped together, namely, a rocket case with a mass of 20 slugs and a payload capsule with a mass of 10 slugs. When the clamp is released, a compressed spring causes the two parts to separate with a relative speed of 3000 ft/sec. (*a*) What are the speeds of the two parts after they have separated? Assume that all velocities are along the same line. (*b*) Find the total kinetic energy of the two parts before and after they separate and account for the difference, if any. *Answer:* (*a*) Case: 24,000 ft/sec; capsule: 27,000 ft/sec. (*b*) Before: 9.38×10^9 ft-lb; after: 9.40×10^9 ft-lb.

26(5). A stream of 40-gm bullets, fired horizontally with a speed of 1000 meters/sec, strikes a 10-kg wooden block that is free to move on a horizontal frictionless tabletop. What is the speed of the block after it has absorbed 15 bullets?

27(5). A hunter has a rifle that can fire 60-gm bullets with a muzzle velocity of 900 meters/sec. A 40-kg leopard springs at him with a speed of 10 meters/sec. How many bullets must the hunter fire into the leopard in order to stop him in his tracks? *Answer:* 8.

28(5). A machine gun fires 50-gm bullets at a speed of 1000 meters/sec. The gunner, holding the machine gun in his hands, can exert an average force of 180 nt against the gun. Determine the maximum number of bullets he can fire per minute.

29(5). A radioactive nucleus, initially at rest, decays by emitting an electron and a neutrino at right angles to one another. The momentum of the electron is 1.2×10^{-22} kg meter/sec and that of the neutrino is 6.4×10^{-23} kg meter/sec. (*a*) Find the direction and magnitude of the momentum of the recoiling nucleus. (*b*) The mass of the residual nucleus is 5.8×10^{-26} kg. What is its kinetic energy of recoil? *Answer:* (*a*) 1.4×10^{-22} kg meter/sec, 150° from the electron track and 120° from the neutrino track. (*b*) 1.0 eV.

30(5). A vessel at rest explodes, breaking into three pieces. Two pieces, having equal mass, fly off perpendicular to one another with the same speed of 30 meters/sec. The third piece has three times the mass of each other piece. What is the direction and magnitude of its velocity immediately after the explosion?

31(5). A projectile is fired from a gun at an angle of 45° with the horizontal and with a muzzle speed of 1500 ft/sec. At the highest point in its flight the projectile explodes into two fragments of equal mass. One fragment, whose speed immediately after the explosion is zero, falls vertically. How far from the gun does the other fragment land, assuming a level terrain? *Answer:* 1.1×10^5 ft.

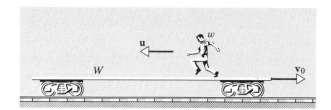

FIGURE 8-16   Problem 34(5).

32(5). Two 2.0-kg masses, *A* and *B*, collide. The velocities before the collision are $\mathbf{v}_A = 15\mathbf{i} + 30\mathbf{j}$, meters/sec and $\mathbf{v}_B = -10\mathbf{i} + 5.0\mathbf{j}$, meters/sec. After the collision $\mathbf{v}'_A = -5.0\mathbf{i} + 20\mathbf{j}$, meters/sec. (*a*) What is the final velocity of *B*? (*b*) How much energy was gained or lost in the collision?

33(5). A 6000-kg rocket is set for a vertical firing. If the exhaust speed is 1000 meters/sec, how much gas must be ejected per second to supply the thrust needed (*a*) to overcome the weight of the rocket, (*b*) to give the rocket an initial upward acceleration of 20 meters/sec²?
Answer: (*a*) 59 kg/sec. (*b*) 180 kg/sec.

34(5). A railroad flat car of weight *W* can roll without friction along a straight horizontal track as shown. Initially a man of weight *w* is standing on the car which is moving to the right with speed v_0. What is the change in velocity of the car if the man runs to the left (Fig. 8–16) so that his speed relative to the car is v_{rel} just before he jumps off at the left end?

35(5). Assume that the car in Problem 34(5) is initially at rest. It holds *n* men each of weight *w*. If each man in succession runs with a relative velocity v_{rel} and jumps off the end, do they impart to the car a greater velocity than if they all run and jump at the same time?
Answer: Yes.

36(5). A fireman plays a horizontal stream of water on the flat vertical back of a 2000-kg truck. The water hitting the truck splashes out along the back of the truck without rebounding and then runs down to the ground. The hose delivers 600 kg of water every 10 sec in a horizontal stream moving with a speed of 30 meters/sec. (*a*) What will be the initial acceleration of the truck if it is free to move and frictional forces are negligible? (*b*) How will the acceleration change as the truck begins to move? (*c*) How would your answers to (*a*) and (*b*) change if the water went through an open door on the back and remained inside the truck?

37(5). Two long barges are floating in the same direction in still water, one with a speed of 10 km/hr and the other with a speed of 20 km/hr. While they are passing each other, coal is shoveled from the slower to the faster one at a rate of 1000 kg/min. How much additional force must be provided by the driving engines of each barge if neither is to change speed? Assume that the shoveling is always perfectly sideways and that the frictional forces between the barges and the water do not depend on the weight of the barges.
Answer: Fast barge: 46 nt more; slow barge: no change.

38(5). A jet airplane is traveling 600 ft/sec. The engine takes in 2400 ft³ of air having a mass of 4.8 slugs each second. The air is used to burn 0.20 slug of fuel each second. The energy is used to compress the products of combustion and to eject them at the rear of the plane at 1600 ft/sec relative to the plane. Find (*a*) the thrust of the jet engine and (*b*) the delivered horsepower.

39(5). A block of mass *m* rests on a wedge of mass *M* which, in turn, rests on a horizontal table, as shown in Fig. 8–17. All surfaces are frictionless. If the system starts at rest with point *P* of the block a distance *h* above the table, find the velocity of the wedge the instant point *P* touches the table.

Answer: $\sqrt{\dfrac{2m^2gh\,\cos^2\alpha}{(M + m)^2 - m(M + m)\cos^2\alpha}}$.

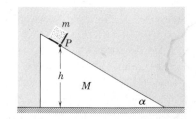

FIGURE 8-17   Problem 39(5).

CHAPTER 9

Collisions

9–1 What is a Collision?

We learn much about atomic, nuclear, and elementary particles experimentally by observing collisions between them. On a larger scale we can better interpret such things as the properties of gases in terms of particle collisions. In this chapter we apply the principles of conservation of energy and conservation of momentum to the collisions of particles.

In a *collision* a relatively large force acts on each colliding particle for a relatively short time. The basic idea of a collision is that the motion of the colliding particles (or of at least one of them) changes rather abruptly and that we can make a relatively clean separation of times that are "before the collision" and those that are "after the collision."

When a bat strikes a baseball for example, the beginning and the end of the collision can be determined fairly precisely. The bat is in contact with the ball for an interval that is quite short in comparison to the time during which we are watching the ball. During the collision the bat exerts a large force on the ball (Fig. 9–1). This force varies with time in a complex way that we can measure only with difficulty. Both the ball and the bat are deformed during the collision. Forces that act for a time that is short compared to the time of observation of the system are called *impulsive* forces.

When an alpha particle (He^4) "collides" with a nucleus of gold (Au^{197}), the force acting between them may be the well-known repulsive electrostatic force associated with the charges on the particles. The particles may not "touch," but we still may speak of a "collision" because a relatively strong force, acting for a time that is short in comparison to the time that the alpha particle is under observation, has a marked effect on the motion of the alpha particle.

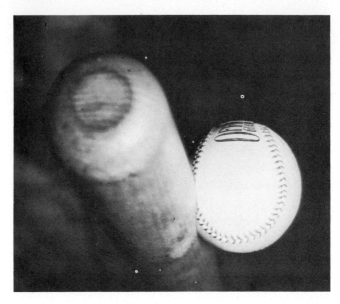

FIGURE 9–1 A high-speed flash photograph of a bat striking a baseball. Notice the deformation of the ball, indicating the enormous magnitude of the impulsive force at this instant. (Courtesy Harold E. Edgerton, Massachusetts Institute of Technology, Cambridge, Mass.)

The concept of collision may be broadened to include events in which the identities of the interacting particles change during the event. For instance, when a proton (H^1 or p) with energy of 25 MeV "collides" with a nucleus of a silver isotope (perhaps Ag^{107}), the particles may in a sense actually "touch," the predominant force then acting between them being, not the electrostatic repulsive force, but the strong, short-range, attractive nuclear force. The proton may enter the silver nucleus, forming a compound structure. A short time later—the "collision interval" may be 10^{-18} sec—this compound structure may break up into two different particles, according to a scheme such as

$$p + Ag^{107} \rightarrow \alpha + Pd^{104},$$

in which α ($=He^4$) is an alpha particle. This type of collision is called a *reaction*. The conservation principles are applicable to all these examples.

We may, if we wish, broaden our definition of collision even further to include the spontaneous decay of a single particle into two or more other particles. An example is the decay of the elementary particle called the *sigma particle* into two other particles, the *pion* and the *neutron* or

$$\Sigma^- \rightarrow \pi^- + n.$$

Although two bodies do not come in contact in this process (unless we consider it in reverse), it has many features in common with collisions: (1) there is a clean distinction between "before the event" and "after the event," and (2) the laws of conservation of momentum and energy permit us to learn much about such processes by studying the before and after situations, even though we may know little about the force laws that operate during the event itself.

In studying collisions in this chapter our aim will be this: given the initial motions of the colliding particles, what can we learn about their final motions from the principles of conservation of momentum and energy, assuming that we know nothing about the forces acting during the collision?

9–2 Impulse and Momentum

Let us assume that Fig. 9–2 shows the magnitude of the force exerted on a body during a collision. We assume that the force has a constant direction. The collision begins at time t_i and ends at time t_f, the force being zero before and after collision. From Eq. 8–10 we can write the change in momentum $d\mathbf{p}$ of a body in a time dt during which a force $\mathbf{F}$ acts on it as

$$d\mathbf{p} = \mathbf{F}\,dt \tag{9–1}$$

We can find the change in momentum of the body during a collision by integrating over the time of collision, that is,

$$\mathbf{p}_f - \mathbf{p}_i = \int_{\mathbf{p}_i}^{\mathbf{p}_f} d\mathbf{p} = \int_{t_i}^{t_f} \mathbf{F}\,dt \tag{9–2}$$

in which the subscripts i ($=$ *initial*) and f ($=$ *final*) refer to the times before and after the collision, respectively. The integral of a force over the time interval during which the force acts is called the *impulse* $\mathbf{J}$ of the force. Hence the change in momentum of a body acted on by an impulsive force is equal to the impulse. Both impulse and momentum are vectors and both have the same units and dimensions.

The impulsive force represented in Fig. 9–2 is assumed to have a constant direction. The impulse of this force, $\int_{t_i}^{t_f} \mathbf{F}\,dt$, is represented in magnitude by the area under the force-time curve.

9–3 Conservation of Momentum during Collisions

Consider now a collision between two particles, such as those of masses m_1 and m_2, shown in Fig. 9–3. During the brief collision these particles exert large forces on one another. At any instant $\mathbf{F}_1$ is the force exerted on particle 1 by particle 2 and $\mathbf{F}_2$ is the force exerted on particle 2 by particle 1. By Newton's third law these forces at any instant are equal in magnitude but opposite in direction.

The change in momentum of particle 1 resulting from the collision is

$$\Delta\mathbf{p}_1 = \int_{t_i}^{t_f} \mathbf{F}_1\,dt = \bar{\mathbf{F}}_1\,\Delta t$$

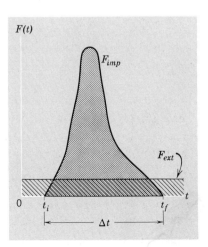

FIGURE 9–2   How an impulsive force $F(t)$ might vary with time during a collision starting at time t_i and ending at t_f. During a collision, the impulsive force F_{imp} is much greater (by definition of the term collision) than any external forces, F_{ext}, which may act on the system.

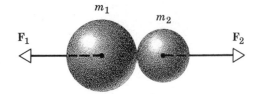

FIGURE 9-3 Two "particles" m_1 and m_2, in collision, experience equal and opposite forces along their line of centers, according to Newton's third law; $\mathbf{F}_2(t) \equiv -\mathbf{F}_1(t)$.

in which $\overline{\mathbf{F}_1}$ is the average value of the force $\mathbf{F}_1$ during the time interval of the collision $\Delta t = t_f - t_i$.

The change in momentum of particle 2 resulting from the collision is

$$\Delta\mathbf{p}_2 = \int_{t_i}^{t_f} \mathbf{F}_2 \, dt = \overline{\mathbf{F}_2} \, \Delta t$$

in which $\overline{\mathbf{F}_2}$ is the average value of the force $\mathbf{F}_2$ during the time interval of the collision $\Delta t = t_f - t_i$.

If no other forces act on the particles, then $\Delta\mathbf{p}_1$ and $\Delta\mathbf{p}_2$ give the total change in momentum for each particle. But we have seen that at each instant $\mathbf{F}_1 = -\mathbf{F}_2$, so that $\overline{\mathbf{F}_1} = -\overline{\mathbf{F}_2}$, and therefore

$$\Delta\mathbf{p}_1 = -\Delta\mathbf{p}_2.$$

If we consider the two particles as an isolated system, the total momentum of the system is

$$\mathbf{P} = \mathbf{p}_1 + \mathbf{p}_2,$$

and the total change in momentum of the system as a result of the collision is zero, that is,

$$\Delta\mathbf{P} = \Delta\mathbf{p}_1 + \Delta\mathbf{p}_2 = 0.$$

Hence, if there are no external forces the total momentum of the system is not changed by the collision. The impulsive forces acting during the collision are internal forces which have no effect on the total momentum of the system.

We have defined a collision as an interaction which occurs in a time Δt that is negligible compared to the time during which we are observing the system. We can also characterize a collision as an event in which the external forces that may act on the system are negligible compared to the impulsive collision forces. When a bat strikes a baseball, a golf club strikes a golf ball, or one billiard ball strikes another, external forces act on the system. Gravity or friction exerts forces on these bodies, for example; these external forces may not be the same on each colliding body nor are they necessarily cancelled by other external forces. Even so it is quite safe to neglect these external forces during the collision and to assume momentum conservation provided, as is almost always true, that the external forces are negligible compared to the impulsive forces of collision. As a result the change in momentum of a particle during a collision arising from an external force is negligible compared to the change in momentum of that particle arising from the impulsive collisional force (Fig. 9–2).

For example, when a bat strikes a baseball, the collision lasts only a small fraction of a second. Since the change in momentum is large and the time of collision is small, it follows from

$$\Delta \mathbf{p} = \overline{\mathbf{F}} \, \Delta t$$

that the average impulsive force $\overline{\mathbf{F}}$ is relatively large. Compared to this force, the external force of gravity is negligible. During the collision we can safely ignore this external force in determining the change in motion of the ball; the shorter the duration of the collision the more likely this is to be true.

In practice, therefore, we can apply the principle of momentum conservation during collisions if the time of collision is small enough. We can then say that the momentum of a system of particles just before the particles collide is equal to the momentum of the system just after the particles collide.

9–4 Collisions in One Dimension

We can always calculate the motions of bodies after collision from their motions before collision if we know the forces that act during the collision, and if we can solve the equations of motion. Often we do not know these forces. Nevertheless, we may be able to predict the results of the collision by using the principle of conservation of momentum and the principle of conservation of total energy, both of which still hold.

Collisions are usually classified according to whether or not kinetic energy is conserved in the collision. When kinetic energy is conserved, the collision is said to be *elastic*. Otherwise, the collision is said to be *inelastic*. Collisions between atomic, nuclear, and fundamental particles are sometimes elastic. These are, in fact, the only truly elastic collisions known. Collisions between gross bodies are always inelastic to some extent. We can often treat such collisions as approximately elastic, however, as, for example, collisions between ivory or glass balls. When two bodies stick together after collisions, the collision is said to be *completely inelastic*. For example, the collision between a bullet and its target is completely inelastic when the bullet remains embedded in the target. The term completely inelastic does not mean that all the initial kinetic energy is lost; as we shall see, it means rather that the loss is as great as is consistent with momentum conservation.

Even if the forces of collision are not known, we can find the motions of the particles after collision from the motions before collision, provided the collision is completely inelastic, or, if the collision is elastic, provided the collision is a one-dimensional one. For a one-dimensional collision the relative motion after collision is along the same line as the relative motion before collision. We restrict ourselves to one-dimensional motion for the present.

Consider first an *elastic* one-dimensional collision. We can imagine two smooth non-rotating spheres moving initially along the line joining their centers, then colliding head-on and moving along the same straight line without rotation after collision (see Fig. 9–4). These bodies exert forces on each other during the collision that are along the initial line of motion, so that the final motion is also along this same line.

The masses of the spheres are m_1 and m_2, the (scalar) velocity components being v_{1i} and v_{2i} before collision and v_{1f} and v_{2f} after collision. We take the positive direction of the momentum and velocity to be to the right. We assume, unless we specify otherwise, that the speeds of the colliding particles are not so high as to require the use of the relativistic expressions for momentum and kinetic energy. Then from conservation of momentum we obtain

$$m_1 v_{1i} + m_2 v_{2i} = m_1 v_{1f} + m_2 v_{2f}.$$

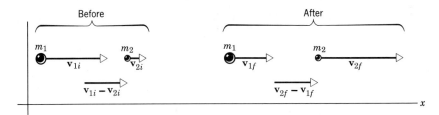

FIGURE 9–4 Two spheres before and after an elastic collision. The velocity, $\mathbf{v}_{1i} - \mathbf{v}_{2i}$, of m_1 relative to m_2 before collision is equal to the velocity, $\mathbf{v}_{2f} - \mathbf{v}_{1f}$, of m_2 relative to m_1 after collision.

Because the collision is elastic, kinetic energy is conserved and we obtain

$$\tfrac{1}{2}m_1v_{1i}^2 + \tfrac{1}{2}m_2v_{2i}^2 = \tfrac{1}{2}m_1v_{1f}^2 + \tfrac{1}{2}m_2v_{2f}^2.$$

It is clear at once that, if we know the masses and the initial velocities, we can calculate the two final velocities v_{1f} and v_{2f} from these two equations.

The momentum equation can be written as

$$m_1(v_{1i} - v_{1f}) = m_2(v_{2f} - v_{2i}), \tag{9–3}$$

and the energy equation can be written as

$$m_1(v_{1i}^2 - v_{1f}^2) = m_2(v_{2f}^2 - v_{2i}^2). \tag{9–4}$$

Dividing Eq. 9–4 by Eq. 9–3, and assuming $v_{2f} \neq v_{2i}$ and $v_{1f} \neq v_{1i}$ (see Question 5), we obtain

$$v_{1i} + v_{1f} = v_{2f} + v_{2i}$$

and, after rearrangement,

$$v_{1i} - v_{2i} = v_{2f} - v_{1f} \tag{9–5}$$

This tells us that in an elastic one-dimensional collision, the relative velocity of approach before collision is equal to the relative velocity of separation after collision.

To find the velocity components v_{1f} and v_{2f} after collision from the velocity components v_{1i} and v_{2i} before collision, we can use any two of the three previous numbered equations. Thus from Eq. 9–5

$$v_{2f} = v_{1i} + v_{1f} - v_{2i}.$$

Inserting this into Eq. 9–3 and solving for v_{1f}, we find that

$$v_{1f} = \left(\frac{m_1 - m_2}{m_1 + m_2}\right)v_{1i} + \left(\frac{2m_2}{m_1 + m_2}\right)v_{2i}.$$

Likewise, inserting $v_{1f} = v_{2f} + v_{2i} - v_{1i}$ (from Eq. 9–5) into Eq. 9–3 and solving for v_{2f}, we obtain

$$v_{2f} = \left(\frac{2m_1}{m_1 + m_2}\right)v_{1i} + \left(\frac{m_2 - m_1}{m_1 + m_2}\right)v_{2i}.$$

There are several cases of special interest. For example, when the colliding particles have the same mass, m_1 equals m_2 so that the two previous equations become simply

$$v_{1f} = v_{2i} \qquad \text{and} \qquad v_{2f} = v_{1i}.$$

That is, in a one-dimensional elastic collisions of two particles of equal mass, the particles simply exchange velocities during collision.

Another case of interest is that in which one particle m_2 is initially at rest. Then v_{2i} equals zero and

$$v_{1f} = \left(\frac{m_1 - m_2}{m_1 + m_2}\right) v_{1i}, \qquad v_{2f} = \left(\frac{2m_1}{m_1 + m_2}\right) v_{1i}.$$

Of course, if $m_1 = m_2$ also, then $v_{1f} = 0$ and $v_{2f} = v_{1i}$ as we expect. The first particle is "stopped cold" and the second one "takes off" with the velocity the first one originally had. If, however, m_2 is very much greater than m_1, we obtain

$$v_{1f} \cong -v_{1i} \qquad \text{and} \qquad v_{2f} \cong 0.$$

That is, when a light particle collides with a very much more massive particle at rest, the velocity of the light particle is approximately reversed and the massive particle remains approximately at rest. For example, suppose that we drop a ball vertically onto a horizontal surface attached to the earth. This is in effect a collision between the ball and the earth. If the collision is elastic, the ball will rebound with a reversed velocity and will reach the same height from which it fell.

If, finally, m_2 is very much smaller than m_1, we obtain

$$v_{1f} \cong v_{1i} \qquad v_{2f} \cong 2v_{1i}.$$

This means that the velocity of the massive incident particle is virtually unchanged by the collision with the light stationary particle, but that the light particle rebounds with approximately twice the velocity of the incident particle. The motion of a bowling ball is hardly affected by collision with an inflated beach ball of the same size, but the beach ball bounces away quickly.

If a collision is *inelastic* then, by definition, the kinetic energy is not conserved. The final kinetic energy may be less than the initial value, the difference being ultimately converted to internal or to potential energy of deformation in the collision, for example; or the final kinetic energy may exceed the initial value, as when potential energy is released in the collision. In any case, the conservation of momentum still holds, as does the conservation of *total* energy.

Let us consider finally a *completely inelastic* collision. The two particles stick together after collision, so that there will be a final common velocity $\mathbf{v}_f$. It is not necessary to restrict the discussion to one-dimensional motion. Using only the conservation of momentum principle, we find

$$m_1\mathbf{v}_{1i} + m_2\mathbf{v}_{2i} = (m_1 + m_2)\mathbf{v}_f. \tag{9–6}$$

This determines $\mathbf{v}_f$ when $\mathbf{v}_{1i}$ and $\mathbf{v}_{2i}$ are known.

Example 1. A baseball weighing 0.35 lb is struck by a bat while it is in horizontal flight with a speed of 90 ft/sec. After leaving the bat the ball travels with a speed of 110 ft/sec in a direction opposite to its original motion. Determine the impulse of the collision.

We cannot calculate the impulse from the definition $\mathbf{J} = \int \mathbf{F}\,dt$ because we do not know the force exerted on the ball as a function of time. However, we have seen (Eq. 9–2) that the change in momentum of a particle acted on by an impulsive force is equal to the impulse. Hence

$$\mathbf{J} = \text{change in momentum} = \mathbf{p}_f - \mathbf{p}_i = m\mathbf{v}_f - m\mathbf{v}_i = \left(\frac{W}{g}\right)(\mathbf{v}_f - \mathbf{v}_i).$$

Assuming arbitrarily that the direction of $\mathbf{v}_i$ is positive, the impulse is then

$$J = \left(\frac{0.35 \text{ lb}}{32 \text{ ft/sec}^2}\right)(-110 \text{ ft/sec} - 90 \text{ ft/sec}) = -2.2 \text{ lb-sec.}$$

The minus sign shows that the direction of the impulse acting on the ball is opposite that of the original velocity of the ball.

We cannot determine the force of the collision from the data we are given. Actually, any force whose impulse is -2.2 lb-sec will produce the same change in momentum. For example, if the bat and ball were in contact for 0.0010 sec, the average force during this time would be

$$\overline{F} = \frac{\Delta p}{\Delta t} = \frac{-2.2 \text{ lb-sec}}{0.0010 \text{ sec}} = -2200 \text{ lb.}$$

For shorter contact time the average force would be greater. The actual force would have a maximum value greater than this average value.

How far would gravity cause the baseball to fall during its collision time?

Example 2. (*a*) By what fraction is the kinetic energy of a neutron (mass m_1) decreased in a head-on elastic collision with an atomic nucleus (mass m_2) initially at rest?

The initial kinetic energy of the neutron K_i is $\frac{1}{2}m_1v_{1i}^2$. Its final kinetic energy K_f is $\frac{1}{2}m_1v_{1f}^2$. The fractional decrease in kinetic energy is

$$\frac{K_i - K_f}{K_i} = \frac{v_{1i}^2 - v_{1f}^2}{v_{1i}^2} = 1 - \frac{v_{1f}^2}{v_{1i}^2}.$$

But, for such a collision,

$$v_{1f} = \left(\frac{m_1 - m_2}{m_1 + m_2}\right)v_{1i},$$

so that

$$\frac{K_i - K_f}{K_i} = 1 - \left(\frac{m_1 - m_2}{m_1 + m_2}\right)^2 = \frac{4m_1m_2}{(m_1 + m_2)^2}.$$

(*b*) Find the fractional decrease in the kinetic energy of a neutron when it collides in this way with a lead nucleus, a carbon nucleus, and a hydrogen nucleus. The ratio of nuclear mass to neutron mass ($= m_2/m_1$) is 206 for lead, 12 for carbon, and 1 for hydrogen.

For lead, $m_2 = 206m_1$,

$$\frac{K_i - K_f}{K_i} = \frac{4 \times 206}{(207)^2} = 0.02 \qquad \text{or} \quad 2\%.$$

For carbon, $m_2 = 12m_1$,

$$\frac{K_i - K_f}{K_i} = \frac{4 \times 12}{(13)^2} = 0.28 \qquad \text{or} \quad 28\%.$$

For hydrogen, $m_2 = m_1$,

$$\frac{K_i - K_f}{K_i} = \frac{4 \times 1}{(2)^2} = 1 \qquad \text{or} \quad 100\%.$$

These results explain why paraffin, which is rich in hydrogen, is far more effective in slowing down neutrons than is lead.

Example 3. *The ballistic pendulum.* The ballistic pendulum is used to measure bullet speeds. The pendulum is a large wooden block of mass M hanging vertically by two cords. A bullet of mass m, traveling with a horizontal speed v_i, strikes the pendulum and remains embedded in it (Fig. 9–5). If the collision time (the time required for the bullet to come to rest with respect to the block) is very small compared to the time of swing of the pendulum, the supporting cords remain approxi-

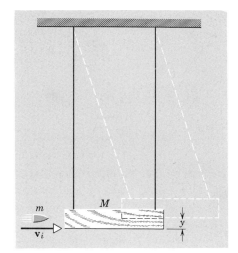

FIGURE 9–5 Example 3. A ballistic pendulum consisting of a large wooden block of mass M suspended by cords. When a bullet of mass m and velocity $\mathbf{v}_i$ is fired into it, the block swings, rising a maximum distance y.

mately vertical during the collision. Therefore, no external horizontal force acts on the system (bullet + pendulum) during collision, and the horizontal component of momentum is conserved. The speed of the system after collision v_f is much less than that of the bullet before collision. This final speed can be easily determined, so that the original speed of the bullet can be calculated from momentum conservation.

The initial momentum of the system is that of the bullet mv_i, and the momentum of the system just after collision is $(m + M)v_f$, so that

$$mv_i = (m + M)v_f.$$

After the collision is over, the pendulum and bullet swing up to a maximum height y, where the kinetic energy left after impact is converted into gravitational potential energy, Then, using the conservation of mechanical energy for this part of the motion, we obtain

$$\tfrac{1}{2}(m + M)v_f^2 = (m + M)gy.$$

Solving these two equations for v_i, we obtain

$$v_i = \frac{m + M}{m}\sqrt{2gy}.$$

Hence, we can find the initial speed of the bullet by measuring m, M, and y.

The kinetic energy of the bullet initially is $\tfrac{1}{2}mv_i^2$ and the kinetic energy of the system (bullet + pendulum) just after collision is $\tfrac{1}{2}(m + M)v_f^2$. The ratio is

$$\frac{\tfrac{1}{2}(m + M)v_f^2}{\tfrac{1}{2}mv_i^2} = \frac{m}{m + M}.$$

For example, if the bullet has a mass $m = 5$ gm and the block has a mass $M = 2000$ gm, only about one-fourth of 1% of the original kinetic energy remains; over 99% is converted into internal energy, resulting in a local temperature rise in the block.

9–5 Collisions in Two and Three Dimensions

In two or three dimensions (except for a completely inelastic collision) the conservation laws alone cannot tell us the motion of particles after a collision if we know the motion before the collision. For example, for a two-dimensional elastic collision, which is the simplest case, we have four unknowns, namely the two components of velocity for each of two particles after collision; but we have only three known relations between them, one for the conservation of kinetic energy and a conservation of momentum relation for each of the two dimensions. Hence we need more information than just the initial conditions. When we do not know the actual forces of interaction, as is often the case, the additional information must be obtained from experiment. It is simplest to specify the angle of recoil of one of the colliding particles.

Let us consider what happens when one particle is projected at a target particle which is at rest. This case is not as restrictive as it may seem, for we can always pick our reference frame to be one in which the target particle is at rest before the collision. Much experimental work in nuclear physics involves projecting nuclear particles at a target which is stationary in the laboratory reference frame. In such collisions, because of momentum conservation, the motion is in a plane determined by the lines of recoil of the colliding particles. The initial motion need not be along the line joining the centers of the two particles. The force of interaction may be electromagnetic, gravitational, or nuclear. The particles need not "touch"; strong forces, which act at relatively close distances of approach and for a time short compared to the observation time, deflect the particles from their initial courses.

A typical situation is shown in Fig. 9-6. The distance b between the initial line of motion and a line parallel to it through the center of the target particle is called the *impact parameter*. This is a measure of the directness of the collision, $b = 0$, corresponding to a head-on collision. The direction of motion of the incident particle m_1 after collision makes an angle θ_1 with the initial direction, and the target projectile m_2, initially at rest, moves in a direction after collision making an angle θ_2 with the initial direction of the incident projectile. Applying the conservation of momentum, which is a vector relation, we obtain two scalar equations; for the x-component of motion we have

$$m_1 v_{1i} = m_1 v_{1f} \cos \theta_1 + m_2 v_{2f} \cos \theta_2,$$

and for the y-component

$$0 = m_2 v_{2f} \sin \theta_2 - m_1 v_{1f} \sin \theta_1.$$

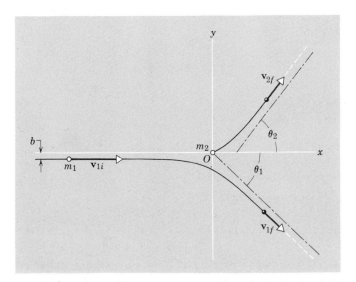

FIGURE 9–6 Two particles, m_1 and m_2, undergoing a collision. The open circles indicate their positions before collision, the shaded ones after collision. Initially m_2 is at rest. The impact parameter b is the distance by which the collision misses being head-on.

Let us now assume that the collision is *elastic*. Here the conservation of kinetic energy applies and we obtain a third equation,

$$\tfrac{1}{2}m_1 v_{1i}^2 = \tfrac{1}{2}m_1 v_{1f}^2 + \tfrac{1}{2}m_2 v_{2f}^2.$$

If we know the initial conditions (m_1, m_2, and v_{1i}), we are left with four unknowns (v_{1f}, v_{2f}, θ_1, and θ_2) but only three equations relating them. We can determine the motion after collision only if we specify a value for one of these quantities, such as θ_1.

▨ **Example 4.** A gas molecule having a speed of 300 meters/sec collides elastically with another molecule of the same mass which is initially at rest. After the collision the first molecule moves at an angle of 30° to its initial direction. Find the speed of each molecule after collision and the angle made with the incident direction by the recoiling target molecule.

   This example corresponds exactly to the situation just discussed, with $m_1 = m_2$, $v_{1i} = 300$ meters/sec, and $\theta_1 = 30°$. Setting m_1 equal to m_2, we have the relations

$$v_{1i} = v_{1f} \cos \theta_1 + v_{2f} \cos \theta_2,$$

$$v_{1f} \sin \theta_1 = v_{2f} \sin \theta_2,$$

and

$$v_{1i}^2 = v_{1f}^2 + v_{2f}^2.$$

We must solve for v_{1f}, v_{2f}, and θ_2. To do this we square the first equation (rewriting it as $v_{1i} - v_{1f} \cos \theta_1 = v_{2f} \cos \theta_2$), and add this to the square of the second equation (noting that $\sin^2 \theta + \cos^2 \theta = 1$); we obtain

$$v_{1i}^2 + v_{1f}^2 - 2v_{1i}v_{1f} \cos \theta_1 = v_{2f}^2.$$

Combining this with the third equation, we obtain

$$2v_{1f}^2 = 2v_{1i}v_{1f} \cos \theta_1$$

or (since $v_{1f} \neq 0$)

$$v_{1f} = v_{1i} \cos \theta_1 = (300 \text{ meters/sec})(\cos 30°)$$

or

$$v_{1f} = 260 \text{ meters/sec}.$$

From the third equation

$$v_{2f}^2 = v_{1i}^2 - v_{1f}^2 = (300 \text{ meters/sec})^2 - (260 \text{ meters/sec})^2,$$

or

$$v_{2f} = 150 \text{ meters/sec}.$$

Finally, from the second equation

$$\sin \theta_2 = (v_{1f}/v_{2f}) \sin \theta_1$$

$$= (260/150)(\sin 30°) = 0.866$$

or

$$\theta_2 = 60°.$$

The two molecules move apart at right angles ($\theta_1 + \theta_2 = 90°$ in Fig. 9–6).
   Can you show that in an elastic collision between particles of equal mass, one of which is initially at rest, the recoiling particles always move off at right angles to one another? ▨

   In Figure 9–7 a photograph of nuclear particle tracks in a Wilson cloud chamber is shown. The tracks are actually trails made up of small droplets of liquid that collect on charged particles (ions) left in the wake of the moving particles.
   A nuclear collision is evident and in this case the incident particle is an α-particle

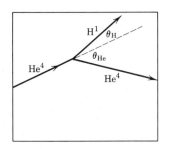

FIGURE 9–7   Photographs of trajectories of particles undergoing collisions in a cloud chamber, a device that makes these paths visible. The chamber contains saturated water vapor. If the vapor is slightly compressed and then allowed to expand quickly, the water vapor will condense in droplets along the trajectory. The incident particle is a helium nucleus (He^4, or α). The target is a hydrogen nucleus (H^1, or p).

(He^4). The target nucleus, a proton (H^1), is essentially at rest before collision. The tracks after the collision of both the incident and recoiling target particles are evident.

9–6   Cross Section

When we deal with particles of atomic or subatomic dimensions, we can seldom define the track of the incident particle or the location of the target particle precisely enough to apply to each collision the powerful method using conservation of linear momentum. In practice, as when we bombard a thin target foil with a beam of deuterons from a cyclotron, we must deal in a statistical way with a large number of collisions between deuterons and the nuclei in the target; the impact parameters for individual collisions cannot be determined.

The situation is much the same as if we were firing a machine gun at random (in the dark, say) at the side of a distant barn of area A on which someone had hung a number of small dinner plates, each of area σ, in random (but not overlapping) positions. If the number of plates is q and if the rate at which bullets strike the barn is R_0, what is the rate R at which plates are broken? It is, on the basis of the random character of the events,

$$R = R_0(\sigma q/A), \qquad (9–7a)$$

where σq is the total area of all the plates. We could, in fact, use this relation to measure σ, the geometrical area of a single plate. Solving for σ yields

$$\sigma = RA/R_0 q \qquad (9–7b)$$

which permits us to find σ from measured values of R, A, R_0, and q. We may call σ the *cross section* for the event consisting of the impact of a bullet on a plate.

Similarly, in nuclear physics we often bombard targets with nuclear projectiles, measure the rate at which events of a selected type occur, and assign a cross section to those events. While the details of individual collisions remain unknown, much information is obtained which is independent of the number or density of target atoms or the flux of incident particles. For example, if we bombarded a thin gold foil (Au^{197}) with deuterons (H^2, or d) whose energy is 30 MeV, many events would occur including scattering of the deuteron beam, the nuclear reaction $d + Au^{197} \rightarrow p + Au^{198}$, and the reaction $d + Au^{197} \to n + Hg^{198}$, where d represents a deuteron, n a neutron, and p a proton. Each of these events (and many others) has its own cross section σ_x.

Let the area of the foil exposed to the beam be A and the thickness of the foil be x. If there are n target particles per unit volume in the foil, the total number of available target particles is nAx. If the *effective area* (that is, the cross section) for the event we are concerned with is σ_x, the *total effective area* of all the nuclei is $nAx\sigma_x$. If R_0 is the rate at which projectiles strike the target and R_x is the rate at which the events in which we are interested occur, we have, because of the random nature of the events (see Eq. 9–7a),

$$\frac{R_x}{R_0} = \frac{(nAx\sigma_x)}{A}$$

or

$$R_x = R_0 nx\sigma_x. \tag{9-8}$$

Thus we can measure σ_x for the event by measuring R_x, R_0, n, and x and substituting into Eq. 9–8. Cross sections are commonly expressed in *barns* or submultiples thereof; one barn $= 10^{-28}$ meter^2.

In atomic and nuclear physics the cross section only rarely has anything to do with the geometrical area of the target. It is an *effective* target area, measuring the probability that a given event will occur. Cross sections for particular events generally vary with the energy of the incident particle, often reaching very high values at sharply defined energies.

9–7 Reactions and Decay Processes

We stated in Section 9–1 that reactions and radioactive decay processes, for atoms, nuclei, and elementary particles, can be treated by the same methods used in collision studies, namely: We can apply the principles of conservation of linear momentum and energy to the (well-defined) periods "before the event" and "after the event." For these processes we must use the conservation of total energy because kinetic energy is not conserved. In this section we only consider examples in which the speeds of the particles are negligible with respect to the speed of light. This means that we may use the classical expressions for momentum and energy and need not use the relativistic expressions.

Example 5. *Nuclear reactions.* A thin film containing a fluorine (F^{19}) compound is bombarded by a beam of protons (p) which have been accelerated to an energy of 1.85 MeV (million electron volts; 1 MeV $= 1.60 \times 10^{-13}$ joule) in a Van de Graaff accelerator. Some of the protons interact with the fluorine nuclei to produce the following nuclear reaction:

$$F^{19} + p \to O^{16} + \alpha.$$

It is observed that the α-particles (which are helium nuclei) that emerge at right angles to the incident proton beam (see Fig. 9–8) have speeds of 1.95×10^7 meters/sec. What can you learn about the reaction by applying the laws of conservation of linear momentum and of total energy? The masses involved are, to a precision good enough for our purposes,

$$m_p = 1.01 \text{ amu} \qquad m_O = 16.0 \text{ amu}$$

$$m_F = 19.0 \text{ amu} \qquad m_\alpha = 4.00 \text{ amu},$$

in which 1 amu (*atomic mass unit*) $= 1.66 \times 10^{-27}$ kg.

The x- and y-components of linear momentum are conserved, which means that they have the

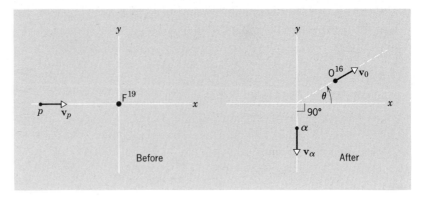

FIGURE 9–8   The nuclear reaction $p + F^{19} \rightarrow \alpha + O^{16}$, showing the situation before and after the event in the laboratory reference frame.

same values before and after the reaction. In the laboratory reference frame of Fig. 9–8, then

$$m_p v_p = m_O v_O \cos \theta \qquad \text{(x-component)} \qquad (9\text{–}9)$$

and

$$0 = m_O v_O \sin \theta - m_\alpha v_\alpha \qquad \text{(y-component)} \qquad (9\text{–}10)$$

For the conservation of total energy we write

$$Q + \tfrac{1}{2} m_p v_p^2 = \tfrac{1}{2} m_O v_O^2 + \tfrac{1}{2} m_\alpha v_\alpha^2 \qquad (9\text{–}11)$$

in which it is clear that Q is the amount by which the kinetic energy of the system after the reaction exceeds the kinetic energy of the system before the reaction. Note that we have assumed that the particles are moving slowly enough so that we may use the classical expression for kinetic energy ($\tfrac{1}{2}mv^2$) rather than the relativistic one for total energy,

$$\frac{m_0 c^2}{\sqrt{1 - v^2/c^2}} .$$

If Q is positive, kinetic energy must be generated by the reaction.

The energy represented by Q can only come from differences in the rest energies of the particles before or after the reaction, according to Einstein's well-known relation $E = \Delta m c^2$ (see Section 7–9) Thus (if Q is positive) we expect that the rest mass of the system after the reaction would be slightly less than its rest mass before the reaction and that Q would indeed be given by the Einstein relation

$$Q = \Delta m c^2$$
$$= [(m_p + m_F) - (m_\alpha + m_O)]c^2. \qquad (9\text{–}12)$$

Note that Eqs. 9–11 and 9–12 are independent relations for Q, being connected through Einstein's mass-energy relation (see Problem 37).

The three conservation equations contain just three unknowns, v_O, θ, and Q. To find Q from them let us first eliminate θ between the first two equations by squaring and adding (recalling that $\cos^2 \theta + \sin^2 \theta = 1$). We obtain

$$m_p^2 v_p^2 + m_\alpha^2 v_\alpha^2 = m_O^2 v_O^2.$$

We can now eliminate v_O between this relation and Eq. 9–11. After a little rearrangement, as you can easily show, we obtain

$$Q = K_\alpha(1 + m_\alpha/m_O) - K_p(1 - m_p/m_O). \qquad (9\text{–}13)$$

From the data given we know that $K_p \; (= \tfrac{1}{2} m_p v_p^2) = 1.85$ MeV and

$$K_\alpha = \tfrac{1}{2}m_\alpha v_\alpha{}^2$$

$$= \tfrac{1}{2}(4.00 \text{ amu} \times 1.66 \times 10^{-27} \text{ kg/amu})(1.95 \times 10^7 \text{ meters/sec})^2$$

$$= (1.26 \times 10^{-12} \text{ joule})(1 \text{ MeV}/1.60 \times 10^{-13} \text{ joule})$$

$$= 7.88 \text{ MeV}.$$

We may now calculate Q from Eq. 9–13 as

$$Q = (7.88 \text{ MeV})(1 + 4.00/16.0) - (1.85 \text{ MeV})(1 - 1.01/16.0) = 8.13 \text{ MeV}.$$

Thus, by using the principles of conservation of linear momentum and total energy, we can calculate Q for the reaction without making any observations on the recoiling O^{16} nucleus. If we want to know v_0 and θ for this nucleus we can easily calculate them from Eqs. 9–9 and 9–10.

The result $Q = 8.13 \text{ MeV}$ is an important bit of information about the reaction. From Eq. 9–12, which is a relation for Q independent of Eq. 9–13, we can now calculate that the decrease in rest mass during the reaction is given by

$$\Delta m = Q/c^2$$

$$= (8.13 \text{ MeV} \times 1.60 \times 10^{-13} \text{ joule/MeV})/(3.00 \times 10^8 \text{ meters/sec})^2$$

$$= (1.44 \times 10^{-29} \text{ kg})(1 \text{ amu}/1.66 \times 10^{-27} \text{ kg})$$

$$= 0.00873 \text{ amu}.$$

We can verify this result by calculating $\Delta m \; [= (m_p + m_F) - (m_\alpha + m_O)]$ from very precise measurements of the four separate masses made in a mass spectrometer. The excellent agreement that we get shows once again the essential validity of Einstein's mass-energy relationship.

QUESTIONS

1. Explain how conservation of momentum applies to a handball bouncing off a wall.

2. How can you reconcile the sailing of a sailboat into the wind with the conservation of momentum principle?

3. An hour glass is being weighed on a sensitive balance, first when sand is dropping in a steady stream from the upper to the lower part and then again after the upper part is empty. Are the two weights the same or not? Explain your answer.

4. The blades of a turbine are usually curved rather than flat in shape so that the fluid striking them follows a path resembling a u-turn. Explain the advantage of the curved shape over the flat one.

5. It is obvious from inspection of Eqs. 9–3 and 9–4 that a valid solution to the problem of finding the final velocities of two particles in a one-dimensional elastic collision is $v_{1f} = v_{1i}$ and $v_{2f} = v_{2i}$. What does this mean physically? Explain.

6. Consider a one-dimensional elastic collision between a given incoming body A and a body B initially at rest. How would you choose the mass of B, in comparison to the mass of A, in order that B should recoil with (a) the greatest speed, (b) the greatest momentum, and (c) the greatest kinetic energy?

7. Fast neutrons produced in a reactor are slowed down by allowing them to collide with particles in a *moderator*. Which particles would moderate the neutrons most effectively, lead nuclei, hydrogen nuclei or electrons? What material (i.e. oil, carbon tetrachloride, or such) would you suggest using?

8. When dealing with atoms, nuclei, or elementary particles, what does it mean to say that such bodies "touch" during a collision?

9. When the forces of interaction between two particles have an infinite range, such as the mutual gravitational attraction between two bodies, can the cross section for "collision" be finite? Is it at all useful to regard this interaction as a collision?

10. Could we determine in principle the cross section for a collision by using only one bombarding particle and one target particle? In practice?

PROBLEMS

1(2). A 150-gram baseball is moving at a speed of 40 meters/sec when it is struck by a bat that reverses its direction and gives it a speed of 60 meters/sec. What average force was exerted by the bat if it was in contact with the ball for 5.0 msec?
Answer: 3000 nt.

2(2). A ball of mass m and speed v strikes a wall perpendicularly and rebounds with undiminished speed. If the time of collision is t, what is the average force exerted by the ball on the wall?

3(2). A 1.0-kg ball drops vertically onto the floor with a speed of 25 meters/sec. It rebounds with an initial speed of 10 meters/sec. (*a*) What impulse acts on the ball during contact? (*b*) If the ball is in contact for 0.020 sec, what is the average force exerted on the floor?
Answer: (*a*) 35 nt-sec. (*b*) 1800 nt.

4(2). A cue strikes a billiard ball, exerting an average force of 50 nt over a time of 10 milliseconds. If the ball has mass 0.20 kg, what speed does it have after impact?

5(2). A croquet ball (mass 0.50 kg) is struck by a mallet, receiving the impulse shown in the graph (Fig. 9–9). What is the ball's velocity just after the force has become zero?
Answer: 8.8 meters/sec.

6(2). The force on a 10-kg object increases uniformly from zero to 50 nt in 4.0 sec. What is the object's final speed if it started at rest?

7(2). A 300-gm ball with a speed of 6.0 meters/sec strikes a wall at an angle of 30° and then rebounds with the same speed and angle (Fig. 9–10). It is in contact with the wall for 10 msec. (*a*) What impulse was experienced by the ball? (*b*) What was the average force exerted by the ball on the wall?
Answer: (*a*) 1.8 kg meters/sec. (*b*) 180 nt.

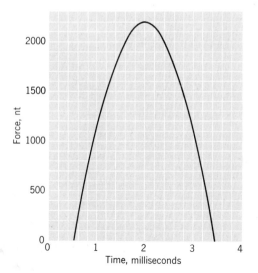

FIGURE 9–9 Problem 5(2).

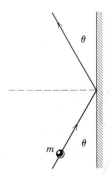

FIGURE 9–10 Problem 7(2).

8(2). A ball having a mass of 50 gm strikes a wall with a velocity of 5.0 meters/sec and rebounds with only 50% of its initial kinetic energy. (*a*) What is the final velocity of the ball after rebounding? (*b*) What was the impulse delivered by the ball to the wall? (*c*) If the ball was in contact with the wall for 0.035 sec, what was the average force exerted by the wall on the ball during this time interval?

9(2). A golfer hits a golf ball imparting to it an initial velocity of magnitude 5.0×10^3 cm/sec directed 30° above the horizontal. Assuming that the mass of the ball is 25 gm and the club and ball are in contact for 0.010 sec find (*a*) the impulse imparted to the ball; (*b*) the impulse imparted to the club; (*c*) the average force exerted on the ball by the club; (*d*) the work done on the ball. *Answer:* (*a*) 1.3 kg meters/sec, along the direction of the initial velocity. (*b*) 1.3 kg meters/sec, in the opposite direction. (*c*) 130 nt. (*d*) 31 joules.

10(2). If the velocity components of a 2.0-kg body change from $v_x = 3.0$ meters/sec, $v_y = 5.0$ meters/sec to $v_x = 0.0$ meters/sec, $v_y = 7.0$ meters/sec in a time interval of 3.0 sec, (*a*) what is the average force that acts on the body (magnitude and direction)? (*b*) What is the impulse delivered to the body?

11(2). A 50-gm particle having a velocity $\mathbf{v} = 7.0\mathbf{i} - 3.0\mathbf{j}$ meters/sec receives an impulse $\mathbf{J} = 1.5\mathbf{i} + 2.5\mathbf{j}$ nt-sec. (*a*) What is the final velocity? (*b*) How much work is done on the particle? *Answer:* (*a*) $37\mathbf{i} + 47\mathbf{j}$, meters/sec. (*b*) 88 joules.

12(2). A stream of water impinges on a stationary "dished" turbine blade, as shown in Fig. 9–11. The speed of the water is v, both before and after it strikes the curved surface of the blade, and the mass of water striking the blade per unit time is constant at the value μ. Find the force exerted by the water on the blade.

13(2). A stream of water from a hose is sprayed on a wall. If the velocity of the water is 5.0 meters/sec and the hose sprays 300 cm³/sec, what is the average force exerted on the wall by the stream of water? Assume that the water does not spatter back appreciably. *Answer:* 1.5 nt.

14(4). A body of 2.0 kg mass makes an elastic collision with another body at rest and afterwards continues to move in the original direction but with one-fourth of its original speed. What is the mass of the struck body?

15(4). A 5.0-kg particle with a velocity of 3.0 meters/sec collides with a 10-kg particle that has a velocity of 2.0 meters/sec in the same direction. After the collision, the 10-kg particle is observed to be traveling the original direction with a speed of 4.0 meters/sec. (*a*) What is the velocity of the 5.0-kg particle immediately after the collision? (*b*) By how much does the total kinetic energy of the system of two particles change because of the collision? *Answer:* (*a*) −1.0 meter/sec. (*b*) +40 joules.

16(4). A steel ball weighing 1.0 lb is fastened to a cord 27 in. long and is released when the cord is horizontal. At the bottom of its path the ball strikes a 5.0-lb steel block initially at rest on a frictionless surface (Fig. 9–12). The collision is elastic. Find (*a*) the speed of the ball and (*b*) the speed of the block, just after the collision.

17(4). A bullet weighing 1.0×10^{-2} lb is fired horizontally into a 4.0-lb wooden block at rest on a

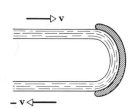

FIGURE 9–11   Problem 12(2).

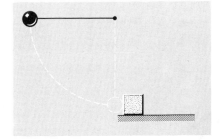

FIGURE 9–12   Problem 16(4).

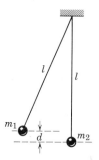

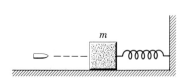

FIGURE 9–13   Problem 19(4).   **FIGURE 9–14**   Problem 20(4).

horizontal surface. The coefficient of kinetic friction between block and surface is 0.20. The bullet comes to rest in the block which moves 6.0 ft. Find the speed of the bullet.
Answer: 3500 ft/sec.

18(4). A bullet of mass 10 gm strikes a ballistic pendulum of mass 2.0 kg. The center of mass of the pendulum rises a vertical distance of 12 cm. Assuming the bullet remains embedded in the pendulum, calculate its initial speed.

19(4). Two pendulums each of length l are initially situated as in Fig. 9–13. The first pendulum is released and strikes the second. Assume that the collision is completely inelastic and neglect the mass of the strings and any frictional effects. How high does the center of mass rise after the collision?

Answer: $d \left(\dfrac{m_1}{m_1 + m_2} \right)^2$.

20(4). A 50-gram bullet embeds itself in a 500-gram wooden block, causing the block to compress a spring by 50 cm. (Fig. 9–14). The block moves on a horizontal, frictionless surface, and the force constant of the spring is 1000 nt/meter. What was the initial speed of the bullet?

21(4). Two particles, one having twice the mass of the other, are held together with a compressed spring between them. The energy stored in the spring is 60 joules. How much kinetic energy does each particle have after they are released?
Answer: 20 joules for the heavy particle; 40 joules for the light particle.

22(4). A railroad freight car weighing 32 tons and traveling 5.0 ft/sec overtakes one weighing 24 tons traveling 3.0 ft/sec in the same direction. (*a*) Find the speed of the cars after collision and the loss of kinetic energy during collision if the cars couple together. (*b*) If the collision is elastic, the freight cars do not couple but separate after collision. What are their speeds?

23(4). An alpha particle (atomic mass 4.0 units) experiences an elastic head-on collision with a gold nucleus (atomic mass 197 units) that is originally at rest. What is the fractional loss of kinetic energy for the alpha particle?
Answer: 7.8%.

24(4). An electron collides elastically with a hydrogen atom initially at rest. The initial and final motions are along the same straight line. What fraction of the electron's initial kinetic energy is transferred to the hydrogen atom? The mass of the hydrogen atom is 1840 times the mass of the electron.

25(4). An object with mass m and speed v explodes into two equal pieces in a gravity-free space. One piece comes to rest. How much kinetic energy was added to the system?
Answer: $\frac{1}{2}mv^2$.

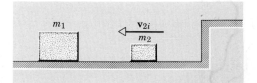

FIGURE 9–15   Problem 26(4).

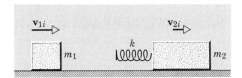

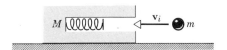

FIGURE 9-16   Problem 28(4).   **FIGURE 9-17**   Problem 29(4).

26(4). A block of mass $m_1 = 100$ kg is at rest on a long frictionless table, one end of which is terminated in a wall. Another block of mass m_2 is placed between the first block and the wall and set in motion to the left with constant speed v_{2i}, as in Figure 9–15. Assuming that all collisions are completely elastic, find the value of m_2 for which both blocks move with the same velocity after m_2 has collided once with m_1 and once with the wall. The wall has infinite mass effectively.

27(4). An electron, mass m, collides head-on with an atom, mass M, initially at rest. As a result of the collision a characteristic amount of energy E is stored internally in the atom. What is the minimum initial speed v_0 that the electron must have? (Hint: Conservation principles lead to a quadratic equation for the final electron velocity v and a quadratic equation for the final atom velocity V. The minimum value v_0 follows from the requirement that the radical in the solutions for v and V be real.)

Answer: $v_0 = \left(2E\,\dfrac{M+m}{Mm}\right)^{1/2}$.

28(4). A block of mass $m_1 = 2.0$ kg slides along a frictionless table with a speed of 10 meters/sec. Directly in front of it, and moving in the same direction, is a block of mass $m_2 = 5.0$ kg moving at 3.0 meters/sec. A massless spring with a spring constant of $k = 1120$ nt/meter is attached to the backside of m_2 as shown in Fig. 9–16. When the blocks collide, what is the maximum compression of the spring? Assume that the spring does not bend and always obeys Hooke's law.

29(4). A ball of mass m is projected with speed v_i into the barrel of a spring-gun of mass M initially at rest of a frictionless surface; see Fig. 9–17. The mass m sticks in the barrel at the point of maximum compression of the spring. No energy is lost in friction. What fraction of the initial kinetic energy of the ball is stored in the spring?

Answer: $M/(M + m)$.

30(4). The two masses on the right of Fig. 9–18 are slightly separated and initially at rest; the left mass is incident with speed v_0. Assuming head-on elastic collisions, (a) if $M \leq m$, show that there are two collisions and find all final velocities; (b) if $M > m$, show that there are three collisions and find all final velocities.

31(4). An elevator is moving up at 6.0 ft/sec in a shaft. At the instant the elevator is 60 ft from the top, a ball is dropped from the top of the shaft. The ball rebounds elastically from the elevator roof. (a) To what height can it rise relative to the top of the shaft? (b) Do the same problem assuming the elevator is moving down at 6.0 ft/sec.

Answer: (a) 23 ft above top of shaft. (b) 23 ft below top of shaft.

32(4). A scale is adjusted to read zero. Particles fall from a height of 9.0 ft colliding with the balance pan on the scale; the collisions are elastic, that is, the particles rebound upward with the same speed. If each particle has a mass of $\frac{1}{128}$ slug and collisions occur at the rate of 32 particles/sec, what is the scale reading in pounds?

33(4). A box is put on a scale that is adjusted to read zero when the box is empty. A stream of

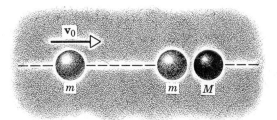

FIGURE 9-18   Problem 30(4).

pebbles is then poured into the box from a height h above its bottom at a rate of μ (pebbles per second). Each pebble has a mass m. If the collisions between the pebbles and box are completely inelastic, find the scale reading at time t after the pebbles begin to fill the box. Determine a numerical answer when $\mu = 100$ sec^{-1}, $h = 25$ ft, $mg = 0.010$ lb, and $t = 10$ sec.
Answer: 11 lb.

34(4). A 6.0-kg box sled is traveling across the ice at a speed of 9.0 meters/sec when a 12-kg package is dropped into it vertically. Describe the subsequent motion of the sled.

35(5). Two vehicles A and B are traveling west and south, respectively, toward the same intersection where they collide and lock together. Before the collision A (total weight, 900 lb) is moving with a speed of 40 miles/hr, and B (total weight, 1200 lb) has a speed of 60 miles/hr. Find the magnitude and direction of the velocity of the (interlocked) vehicles immediately after collision.
Answer: 38 miles/hr; 63° south of west.

36(5). Two skaters collide and embrace. One weighs 160 lb and is initially moving east at 4.0 miles/hr. The other has an initial velocity of 5.0 miles/hr north and weighs 96 lb. (*a*) What is the final velocity of the couple? (*b*) What fraction of the initial kinetic energy was lost in the collision?

37(5). A billiard ball moving at a speed of 2.2 meters/sec strikes an identical stationary ball a glancing blow. After the collision one ball is found to be moving at a speed of 1.1 meters/sec in a direction making a 60° angle with the original line of motion. (*a*) Find the velocity of the other ball. (*b*) Can the collision be inelastic, given these data?
Answer: (*a*) 1.9 meters/sec, 30° to initial direction. (*b*) No.

38(5). A proton (atomic mass 1.0 unit) with a speed of 500 meters/sec collides elastically with another proton at rest. The original proton is scattered 60° from its initial direction. (*a*) What are the speeds of the two protons after the collision? (*b*) What is the direction of the velocity of the target proton after the collision?

39(5). An α-particle collides with an oxygen nucleus, initially at rest. The α-particle is scattered at an angle of 64° from its initial direction of motion and the oxygen nucleus recoils at an angle of 51° on the other side of this initial direction. What is the ratio of the speeds of these particles? The mass of the oxygen nucleus is four times that of the α-particle.
Answer: 2.8.

40(5). A deuteron is a nuclear particle made up of one proton and one neutron. Its mass is about 4×10^{-24} gm. A deuteron, accelerated by a cyclotron to a speed of 10^9 cm/sec, collides with another deuteron at rest. (*a*) If the two particles stick together head on to form a helium nucleus, find the speed of the resultant nucleus. (*b*) The helium nucleus then breaks up into a neutron with a mass of about 2×10^{-24} gm and a helium isotope of mass 6×10^{-24} gm. If the neutron is given off at right angles to the direction of the original velocity with a speed of 5×10^8 cm/sec, find the magnitude and direction of the velocity of the helium isotope.

41(5). A 20-kg particle is moving in the direction of the positive x-axis with a speed of 200 meters/sec when, due to an internal explosion, it breaks into three parts. One part, whose mass is 10 kg, moves away from the point of explosion with a speed of 100 meters/sec along the positive y-axis. A second fragment, with a mass of 4.0 kg, moves along the negative x-axis with a speed of 500 meters/sec. (*a*) What is the velocity of the third (6.0-kg) particle? (*b*) How much energy was released in the explosion? Ignore effects due to gravity.
Answer: (*a*) $v_x = 1000$ meters/sec; $v_y = 170$ meters/sec. (*b*) 3.6×10^6 joules.

42(5). A certain nucleus, at rest, disintegrates into three particles. Two of them are detected, with masses and velocities as shown in Fig. 9–19. (*a*) What is the momentum of the third particle, which is known to have a mass of 12×10^{-27} kg? (*b*) How much energy was involved in the disintegration process?

43(5). A ball with an initial speed of 10 meters/sec collides elastically with two identical balls whose centers are on a line perpendicular to the initial velocity and which are originally in contact with each other (Fig. 9–20). The first ball is aimed directly at the contact point and all the balls are frictionless. Find the velocities of all three balls after the collision. (Hint: The directions of the two originally stationary balls can be found by considering the direction of the impulse they receive during the collision.)
Answer: $\mathbf{v}_2$ and $\mathbf{v}_3$ will be at 30° to $\mathbf{v}_0$ and will have a magnitude of 6.9 meters/sec. $\mathbf{v}_1$ will be in the opposite direction to $\mathbf{v}_0$ and will have magnitude 2.0 meters/sec.

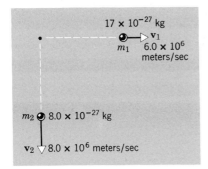

FIGURE 9-19 Problem 42(5).

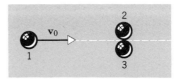

FIGURE 9-20 Problem 43(5).

44(5). Show that, in the case of an elastic collision between a particle of mass m_1 with a particle of mass m_2 initially at rest, (a) the maximum angle θ_m through which m_1 can be deflected by the collision is given by $\cos^2 \theta_m = 1 - m_2^2/m_1^2$, so that $0 \leqq \theta_m \leqq \pi/2$, when $m_1 > m_2$; (b) $\theta_1 + \theta_2 = \pi/2$, when $m_1 = m_2$; (c) θ_1 can take on all values between 0 and π, when $m_1 < m_2$.

45(6). A beam of slow neutrons strikes an aluminum foil 1.0×10^{-5} meters thick. Some neutrons are captured by the aluminum that becomes radioactive and decays by emitting an electron (β^-) forming silicon:

$$n + Al^{27} \rightarrow Al^{28} \rightarrow Si^{28} + \beta^-.$$

Suppose that the neutron flux is 3.0×10^{16}/meters^2 sec and the neutron capture cross section is 0.23 barn. How many transmutations per unit area will occur each second?

Answer: 4.2×10^{11} transmutations/meter^2 sec.

46(6). A beam of fast neutrons impinges on a 5.0-mg sample of Cu^{65}, a stable isotope of copper. A possibility exists that the copper nucleus may capture a neutron to form Cu^{66}, which is radioactive and decays to Zn^{66}, which is again stable. If a study of the electron emission of the copper sample implies that 4.6×10^{11} neutron captures occur each second, what is the neutron capture cross section in barns for this process? The intensity of the neutron beam is 1.1×10^{18} neutrons/meter^2 sec.

47(7). The precise masses in the reaction

$$p + F^{19} \rightarrow \alpha + O^{16}$$

have been determined by mass spectrometer measurements and are

$$m_p = 1.00783 \text{ amu} \qquad m_\alpha = 4.00260 \text{ amu}$$
$$m_F = 18.99840 \text{ amu} \qquad m_O = 15.99491 \text{ amu}$$

Calculate the Q of the reaction from these data and compare with the Q calculated in Example 5 from reaction studies.

Answer: 8.12 MeV.

48(7). An elementary particle called Σ^-, at rest in a certain reference frame, decays spontaneously into two other particles according to

$$\Sigma^- \rightarrow \pi^- + n$$

The masses are

$$m_{\Sigma^-} = 2340.5 m_e$$
$$m_{\pi^-} = 273.2 m_e$$
$$m_n = 1838.65 m$$

where m_e is the electron mass. (a) How much kinetic energy is generated in this process? (b) Which of the decay products (π^- and n) gets the larger share of this kinetic energy? Of the momentum?

CHAPTER 10

Rotational Kinematics

10–1 Rotational Motion

So far we have dealt mostly with the translational motion of single particles or of rigid bodies; that is, of bodies whose parts all have a fixed relationship to each other. No real body is truly rigid, but many bodies, such as molecules, steel beams, and planets, are rigid enough so that, in many problems, we can ignore the fact that they warp, bend or vibrate. *A rigid body moves in pure translation if each particle of the body undergoes the same displacement as every other particle in any given time interval.*

In this chapter we are interested in the rotation of rigid bodies. We shall not consider such rotational motions as those of the solar system or of water in a spinning beaker. We shall also deal only with rotation about axes that remain fixed in the reference frame in which we observe the rotation.

Figure 10–1 shows the rotational motion of a rigid body about a fixed axis, in this case the z-axis of our reference frame. Let P represent a particle in the rigid body, arbitrarily selected and described by the position vertor **r**. We then say that: *A rigid*

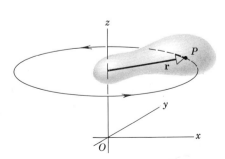

FIGURE 10–1 A rigid body rotating about the z-axis. Each point in the body, such as P, describes a circle about this axis.

body moves in pure rotation if every particle of the body (such as *P* in Fig. 10–1) *moves in a circle, the center of which lies on a straight line called the axis of rotation* (the *z*-axis in Fig. 10–1). If we draw a perpendicular from any point in the body to the axis, each such line will sweep through the same angle in any given time interval as another such line. Thus we can describe the pure rotation of a rigid body by considering the motion of any one of the particles (such as *P*) that make it up. (We must rule out, however, particles that are on the axis of rotation. Why?)

The general motion of a rigid body is a combination of translation and rotation, however, rather than one of pure rotation. As we saw in Chapter 8 we can describe the translational motion of any system of particles—whether rigid or not—whether rotating or not—by imagining that all of the mass *M* of the body is concentrated at the center of mass and that $\mathbf{F}_{ext}$, the resultant of the external forces acting on the body, acts at this point. The acceleration of the center of mass is then given by Eq. 8–8 or $\mathbf{F}_{ext} = M\mathbf{a}_{cm}$. It is very helpful to be able to represent the translational motion of a rigid body by the motion of a single point—its center of mass; all that is left is to determine its rotational motion. To do this, we must first describe the rotational motion. We call this description rotational kinematics; we must define the variables of angular motion and relate them to each other, just as in particle kinematics we defined the variables of translational motion and related them to each other. The next part of our program is to relate the rotational motion of a body to the properties of the body and of its environment. This is rotational dynamics. In this chapter we study the kinematics of rotation. We develop the dynamics of rotation in the next chapter.

10–2   Rotational Kinematics—the Variables

In Fig. 10–1 let us pass a plane through *P* at right angles to the axis of rotation. This plane, which cuts through the rotating body, contains the circle in which particle *P* moves. Figure 10–2 shows this plane, as we look downward on it from above, along the *z*-axis in Fig. 10–1.

We can tell exactly where the entire rotating body is in our reference frame if we know the location of any single particle (*P*) of the body in this frame. Thus, for the kinematics of this problem, we need only consider the (two-dimensional) motion of a particle in a circle.

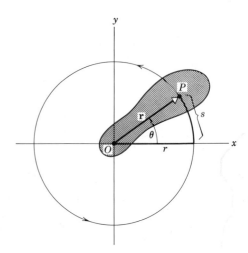

FIGURE 10–2   A cross sectional view of the rigid body of Fig. 10–1, showing point *P* and vector **r** of that figure. Point *P*, which is fixed in the rotating body, rotates counterclockwise about the origin in a circle of radius *r*.

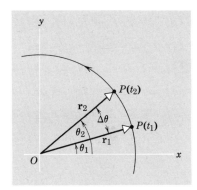

FIGURE 10–3 The reference line r ($= OP$), fixed in the body of Figs. 10–1 and 10–2, is displaced through angle $\Delta\theta (=\theta_2 - \theta_1)$ in time Δt ($= t_2 - t_1$).

The angle θ in Fig. 10–2 is the angular position of particle P with respect to the reference position. We arbitrarily choose the positive sense of rotation in Fig. 10–2 to be counterclockwise, so that θ increases for counterclockwise rotation and decreases for clockwise rotation.

It is convenient to measure θ in radians* rather than in degrees. By definition θ is given in radians by the relation

$$\theta = s/r,$$

in which s is the arc length shown in Fig. 10–2.

Let the body of Fig. 10–2 be rotating counterclockwise. At time t_1 the angular position of P is θ_1 and at a later time t_2 its angular position is θ_2. This is shown in Fig. 10–3, which gives the positions of P and of the position vector $\mathbf{r}$ at these times; the outline of the body itself has been omitted in that figure for simplicity. The *angular displacement* of P will be $\theta_2 - \theta_1 = \Delta\theta$ during the time interval $t_2 - t_1 = \Delta t$. The *average angular speed* $\bar{\omega}$ of particle P in this time interval is defined as

$$\bar{\omega} = \frac{\theta_2 - \theta_1}{t_2 - t_1} = \frac{\Delta\theta}{\Delta t}.$$

The *instantaneous angular speed* ω is defined as the limit approached by this ratio as Δt approaches zero:

$$\omega = \lim_{\Delta t \to 0} \frac{\Delta\theta}{\Delta t} = \frac{d\theta}{dt}. \tag{10–1}$$

For a rigid body all radial lines fixed in it perpendicular to the axis of rotation rotate through the same angle in the same time, so that the angular speed ω about this axis is the same for each particle in the body. Thus ω is characteristic of the body as a whole. Angular speed has the dimensions of an inverse time (T^{-1}); its units are commonly taken to be radians/sec or rev/sec.

If the angular speed of P is not constant, then the particle has an angular acceleration. Let ω_1 and ω_2 be the instantaneous angular speeds at the times t_1 and t_2 respectively; then the *average angular acceleration* $\bar{\alpha}$ of the particle P is defined as

$$\bar{\alpha} = \frac{\omega_2 - \omega_1}{t_2 - t_1} = \frac{\Delta\omega}{\Delta t}.$$

---

* The radian is a pure number, having no physical dimension since it is the ratio of two lengths. Since the circumference of a circle of radius r is $2\pi r$, there are 2π radians in a complete circle, that is, $\theta = 2\pi r/r = 2\pi$. Therefore 2π radians $= 360°$, π radians $= 180°$, and 1 radian $\cong 57.3°$.

The *instantaneous angular acceleration* is the limit of this ratio at Δt approaches zero, or

$$\alpha = \lim_{\Delta t \to 0} \frac{\Delta \omega}{\Delta t} = \frac{d\omega}{dt}. \tag{10-2}$$

Because ω is the same for all particles in the rigid body, it follows from Eq. 10–2 that α must be the same for each particle and thus α, like ω, is a characteristic of the body as a whole. Angular acceleration has the dimensions of an inverse time squared (T^{-2}); its units are commonly taken to be radians/sec^2 or rev/sec^2.

The rotation of a particle (or a rigid body) about a fixed axis has a formal correspondence to the translational motion of a particle (or a rigid body) along a fixed direction. The kinematical variables are θ, ω, and α in the first case and x, v, and a in the second. These quantities correspond in pairs: θ to x, ω to v, and α to a. Note that the angular quantities differ dimensionally from the corresponding linear quantities by a length factor. Note, too, that all six quantities may be treated as scalars in this special case. For example, a particle at any instant can be moving in one direction or the other along its straight-line motion, corresponding to a positive or a negative value for v; similarly a particle at any instant can be rotating in one direction or another about its fixed axis, corresponding to a positive or a negative value for ω.

10–3 Rotation with Constant Angular Acceleration

For translational motion of a particle or a rigid body along a fixed direction, such as the x-axis, we have seen (in Chapter 3) that the simplest type of motion is that in which the acceleration a is zero. The next simplest type corresponds to $a = a$ constant (other than zero); for this motion we derived the equations of Table 3–1, which connect the kinematic variables x, v, a, and t in all possible combinations.

For the rotational motion of a particle or a rigid body around a fixed axis the simplest type of motion is that in which the angular acceleration α is zero (such as uniform circular motion). The next simplest type of motion, in which $\alpha = a$ constant (other than zero), corresponds exactly to linear motion with $a = a$ constant (other than zero). As before, we can derive four equations linking the four kinematic variables θ, ω, α, and t in all possible combinations. You can either derive these angular equations by the methods used to derive the linear equations or you may write them down at once by substituting corresponding angular quantities for the linear quantities in the linear equations.

We list both sets of equations in Table 10–1, having chosen $x_0 = 0$ and $\theta_0 = 0$ in these relations for simplicity. Here ω_0 is the angular speed at the time $t = 0$. You might check these equations dimensionally before verifying them. Both sets of equations hold not only for particles but also for rigid bodies.

Table 10–1   Motion with Constant Linear or Angular Acceleration

	Translational Motion (Fixed Direction)	Rotational Motion (Fixed Axis)	
(3–12)	$v = v_0 + at$	$\omega = \omega_0 + \alpha t$	(10–3)
(3–14)	$x = \dfrac{v_0 + v}{2} t$	$\theta = \dfrac{\omega_0 + \omega}{2} t$	(10–4)
(3–15)	$x = v_0 t + \tfrac{1}{2} a t^2$	$\theta = \omega_0 t + \tfrac{1}{2} \alpha t^2$	(10–5)
(3–16)	$v^2 = v_0^2 + 2ax$	$\omega^2 = \omega_0^2 + 2\alpha\theta$	(10–6)

For the angular quantities, we arbitrarily select one of the two possible directions of rotation about the fixed axis as the direction in which θ is increasing. From Eq. 10–1 ($\omega = d\theta/dt$) we see that if θ is increasing with time ω is positive. Similarly, from Eq. 10–2 ($\alpha = d\omega/dt$), we see that if ω is increasing with time α is positive. There are corresponding sign conventions for the linear quantities.

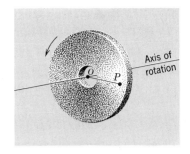

FIGURE 10–4   Example 1. The line *OP* is attached to a grindstone rotating as shown about an axis through O that is fixed in the reference frame of the observer.

Example 1. A grindstone has a constant angular acceleration α of 3.0 radians/sec². Starting from rest a line, such as *OP* in Fig. 10–4, is horizontal. Find (*a*) the angular displacement of the line *OP* (and hence of the grindstone) and (*b*) the angular speed of the grindstone 2.0 sec later.

(*a*) α and t are given; we wish to find θ. Hence, we use Eq. 10–5,

$$\theta = \omega_0 t + \frac{1}{2}\alpha t^2.$$

At $t = 0$, we have $\omega = \omega_0 = 0$ and $\alpha = 3.0$ radians/sec². Therefore, after 2.0 sec,

$$\theta = (0)(2.0 \text{ sec}) + \tfrac{1}{2}(3.0 \text{ radians/sec}^2)(2.0 \text{ sec})^2 = 6.0 \text{ radians} = 0.96 \text{ rev}.$$

(*b*) α and t are given; we wish to find ω. Hence we use Eq. 10–3

$$\omega = \omega_0 + \alpha t,$$

and

$$\omega = 0 + (3.0 \text{ radians/sec}^2)(2.0 \text{ sec}) = 6.0 \text{ radians/sec}.$$

Using Eq. 10–6 as a check, we have

$$\omega^2 = \omega_0{}^2 + 2\alpha\theta,$$

$$\omega^2 = 0 + (2)(3.0 \text{ radians/sec}^2)(6.0 \text{ radians}) = 36 \text{ radians}^2/\text{sec}^2,$$

$$\omega = 6.0 \text{ radians/sec}.$$

10–4   Relation between Linear and Angular Kinematics for a Particle in Circular Motion

In Section 4–4 we discussed the linear velocity and acceleration of a particle moving in a circle. When a rigid body rotates about a fixed axis, every particle in the body moves in a circle. Hence we can describe the motion of such a particle either in linear variables or in angular variables. The relation between the linear and angular variables enables us to pass back and forth from one description to another and is very useful.

Consider a particle at P in the rigid body, a distance r from the axis through O. This particle moves in a circle of radius r as the body rotates, as in Fig. 10–5a. The reference position is Ox. The particle moves through a distance s along the arc when the body rotates through an angle θ, such that

$$s = \theta r \tag{10–7}$$

where θ is in radians.

Differentiating both sides of this equation with respect to the time, and noting that r is constant, we obtain

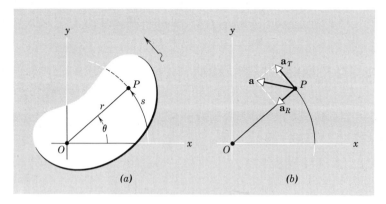

FIGURE 10–5 (*a*) A rigid body rotates about a fixed axis through *O* perpendicular to the page. The point *P* sweeps out an arc *s* which subtends an angle θ. (*b*) The acceleration **a** of point *P* has components $\mathbf{a}_T$ (tangential) where $a_T = \alpha r$ and $\mathbf{a}_R$ (radial) where $a_R = v^2/r = \omega^2 r$ (ω = angular speed).

$$\frac{ds}{dt} = \frac{d\theta}{dt} r.$$

But ds/dt is the linear speed of the particle at *P* and $d\theta/dt$ is the angular speed ω of the rotating body so that

$$v = \omega r. \tag{10–8}$$

This is a relation between the magnitudes of the linear velocity and the angular velocity; the linear speed of a particle in circular motion is the product of the angular speed and the distance *r* of the particle from the axis of rotation.

Differentiating Eq. 10–8 with respect to the time, we have

$$\frac{dv}{dt} = \frac{d\omega}{dt} r.$$

dv/dt is the magnitude of the *tangential* component of acceleration of the particle and $d\omega/dt$ is the magnitude of the angular acceleration of the rotating body, so that

$$a_T = \alpha r. \tag{10–9}$$

Hence the magnitude of the tangential component of the linear acceleration of a particle in circular motion is the product of the magnitude of the angular acceleration and the distance *r* of the particle from the axis of rotation.

We have seen that the radial component of acceleration is v^2/r for a particle moving in a circle. This can be expressed in terms of angular speed by use of Eq. 10–8. We have

$$a_R = \frac{v^2}{r} = \omega^2 r. \tag{10–10}$$

The resultant acceleration of point *P* is shown in Fig. 10–5*b*.

Equations 10–7 through 10–10 enable us to describe the motion of one point on a rigid body rotating about a fixed axis either in angular variables or in linear variables. We might ask why we need the angular variables when we are already familiar with the equivalent linear variables. The answer is that the angular description offers a distinct advantage over the linear description when various points on the same rotating body must be considered. Different points on the same rotating body do not have the same linear displacement, speed, or acceleration, but all points on a rigid body rotating about a fixed axis do have the same angular displacement, speed, or acceleration at any instant. By the use of angular variables we can describe the motion of the body as a whole in a simple way.

Example 2. If the radius of the grindstone of Example 1 is 0.50 meter, calculate (*a*) the linear or tangential speed of a particle on the rim, (*b*) the tangential acceleration of a particle on the rim, and (*c*) the centripetal acceleration of a particle on the rim, at the end of 2.0 sec.

We have $\alpha = 3.0$ radians/sec^2, $\omega = 6.0$ radians/sec after 2.0 sec, and $r = 0.50$ meter. Then,

(*a*)
$$v = \omega r$$
$$= (6.0 \text{ radians/sec})(0.50 \text{ meter})$$
$$= 3.0 \text{ meter/sec} \quad \text{(linear speed)};$$

(*b*)
$$a_T = \alpha r$$
$$= (3.0 \text{ radians/sec}^2)(0.50 \text{ meter})$$
$$= 1.5 \text{ meters/sec}^2 \quad \text{(tangential acceleration)};$$

(*c*)
$$a_R = v^2/r = \omega^2 r$$
$$= (6.0 \text{ radians/sec})^2(0.50 \text{ meter})$$
$$= 18 \text{ meters/sec}^2 \quad \text{(centripetal acceleration)}.$$

(*d*) Are the results the same for a particle halfway in from the rim, that is, at $r = 0.25$ meter? The angular variables are the same for this point as for a point on the rim. That is, once again

$$\alpha = 3.0 \text{ radians/sec}^2, \qquad \omega = 6.0 \text{ radians/sec}.$$

But now $r = 0.25$ meter, so that for this particle

$$v = 1.5 \text{ meters/sec}, \qquad a_T = 0.75 \text{ meter/sec}^2, \qquad a_R = 9.0 \text{ meters/sec}^2.$$

QUESTIONS

1. Could the angular quantities θ, ω, and α be expressed in terms of degrees instead of radians in the kinematical equations?

2. Explain why the radian measure of angle is equally satisfactory for all systems of units. Is the same true for degrees?

3. What is the angular speed of the second hand of a watch? Of the minute hand? Could you conveniently express the motion of these hands in terms of linear variables?

4. How could you express simply the relationship between the angular velocities of a pair of gears which are coupled?

5. A wheel is rotating about an axis through its center perpendicular to the plane of the wheel. Consider a point on the rim. When the wheel rotates with constant angular velocity, does the point have a radial acceleration? A tangential acceleration? When the wheel rotates with constant angular acceleration, does the point have a radial acceleration? A tangential acceleration? Do the magnitudes of these accelerations change?

6. Suppose you were asked to determine the equivalent distance traveled by a phonograph needle in playing, say, a 12-in., $33\frac{1}{3}$ rmp record. What information do you need? Discuss from the points of view of reference frames (*a*) fixed in the room, and (*b*) fixed on the rotating record.

7. In a centrifuge particles will tend to separate from the fluid in which they are suspended if their density (mass/volume) differs from that of the fluid. Discuss the dynamical principles upon which the operation of a centrifuge depends.

PROBLEMS

1(2). A phonograph record on a turntable rotates at 33 rev/min. What is the linear speed of a point on the record at the needle at (*a*) the beginning and (*b*) the end of the recording? The distances of the needle from the turntable axis are 5.9 and 2.9 in., respectively, at these two positions. *Answer:* (*a*) 1200 in./min. (*b*) 600 in./min.

2(2). (*a*) What is the angular speed about the polar axis of a point on the earth's surface at a latitude of 45°N? (*b*) What is the linear speed? (*c*) How do these compare with the similar values for a point at the equator?

3(2). One method of measuring the speed of light makes use of a rotating toothed wheel. A beam of light passes through a slot at the outside edge of the wheel, as in Fig. 10–6, travels to a distant mirror, and returns to the wheel just in time to pass through the next slot in the wheel. One such toothed wheel has a radius of 5.0 cm and 500 teeth at its edge. Measurements taken when the mirror was 500 meters from the wheel indicated a speed of light of 3.0×10^5 km/sec. (*a*) What was the (constant) angular speed of the wheel? (*b*) What was the linear speed of a point on its edge? *Answer:* (*a*) 3.8×10^3 radians/sec. (*b*) 190 meters/sec.

4(2). The earth's orbit about the sun is almost a circle. (*a*) What is the angular velocity of the earth (regarded as a particle) about the sun and (*b*) its average linear speed in its orbit? (*c*) What is the centripetal acceleration of the earth with respect to the sun?

5(2). If an airplane propeller of radius 5.0 ft rotates at 2000 rev/min and the airplane is propelled at a ground speed of 300 miles/hr, what is the speed of a point on the tip of the propeller, as seen by (*a*) the pilot and (*b*) an observer on the ground? *Answer:* (*a*) 6.3×10^4 ft/min. (*b*) 6.9×10^4 ft/min.

6(2). The angular position of a point on the rim of a rotating wheel is described by $\theta = 4.0t - 3.0t^2 + t^3$, where θ is in radians and t in seconds. (*a*) What is the total angular displacement during the time intervals which begin at $t = 0$ and end at $t = 1.0$, 2.0 and 3.0 sec? (*b*) What is the average angular velocity for the time interval which begins at $t = 2.0$ and ends at $t = 4.0$ sec?

7(2). The angle turned through by the flywheel of a generator during a time interval t is given by

$$\theta = at + bt^3 - ct^4,$$

where a, b, and c are constants. What is the expression for its angular acceleration? *Answer:* $6bt - 12ct^2$.

8(2). A wheel rotates with an angular acceleration α given by

$$\alpha = 4at^3 - 3bt^2,$$

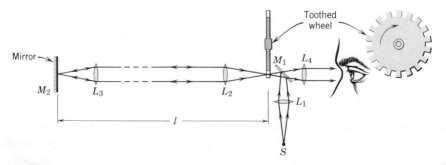

FIGURE 10–6 Problem 3(2).

where t is the time and a and b are constants. If the wheel has an initial angular speed ω_0, write the equations for (a) the angular speed and (b) the angle turned through, as functions of time.

9(3). The angular speed of an automobile engine is increased from 1200 rev/min to 3000 rev/min in 12 sec. (a) What is its angular acceleration, assuming it to be uniform? (b) How many revolutions does the engine make during this time?
Answer: (a) 16 radians/sec². (b) 420.

10(3). A phonograph turntable rotating at 78 rev/min slows down and stops in 30 sec after the motor is turned off. (a) Find its (uniform) angular acceleration. (b) How many revolutions did it make in this time?

11(3). A heavy flywheel rotating on its axis is slowing down because of friction in its bearings. At the end of the first minute its angular velocity is 0.90 of its angular velocity ω_0 at the start. Assuming constant frictional forces, find its angular velocity at the end of the second minute.
Answer: 0.80 ω_0.

12(3). A wheel has a constant angular acceleration of 3.0 radians/sec². In a 4.0-sec interval, it turns through an angle of 120 radians. Assuming the wheel started from rest, how long had it been in motion at the start of this 4.0-sec interval?

13(3). A certain wheel turns through 90 revolutions in 15 sec, its angular speed at the end of the period being 10 rev/sec. (a) What was the angular speed of the wheel at the beginning of the 15-sec interval, assuming constant angular acceleration? (b) How much time had elapsed between the time the wheel was at rest and the beginning of the 15-sec interval?
Answer: (a) 2.0 rev/sec. (b) 3.8 sec.

14(3). A uniform disk rotates about a fixed axis starting from rest and accelerating with constant angular acceleration. At one time it is rotating at 10 rev/sec. After completing 60 more complete revolutions its angular speed is 15 rev/sec. Calculate (a) the angular acceleration, (b) the time required to complete the 60 revolutions mentioned, (c) the time required to attain the 10 rev/sec angular speed, and (d) the number of revolutions from rest until the time the disk attained the 10 rev/sec angular speed.

15(3). A flywheel completes 40 revolutions as it slows from an angular speed of 1.5 radians/sec to a complete stop. Assuming uniform acceleration, (a) what is the time required for it to come to rest? (b) What is the angular acceleration? (c) How much time is required for it to complete one-half of the 40 revolutions?
Answer: (a) 330 sec. (b) -4.5×10^{-3} radians/sec². (c) 98 sec.

16(3). An automobile traveling 60 miles/hr has wheels of 30-in. diameter. (a) What is the angular speed of the wheels about the axle? (b) If the wheels are brought to a stop uniformly in 30 turns, what is the angular acceleration? (c) How far does the car advance during this braking period?

17(3). Derive the equation $\omega = \omega_0 + \alpha t$ for constant angular acceleration.

18(4). What is the centripetal acceleration of a point on the rim of a 12-in. diameter record rotating at 33.3 rev/min?

19(4). What is the angular speed of a car rounding a circular turn of radius 360 ft at 30 miles/hr?
Answer: 0.12 radians/sec.

20(4). What are (a) the angular speed, (b) the radial acceleration, and (c) the tangential acceleration of a spaceship negotiating a circular turn of radius 2000 miles at a constant speed of 18,000 miles/hr?

21(4). What is the ratio of the centripetal acceleration, associated with the earth's rotation, of a point on the equator, to the centripetal acceleration of the earth itself, associated with its motion around the sun? Assume a circular orbit.
Answer: 5.7.

22(4). The flywheel of a steam engine runs with a constant angular speed of 150 rev/min. When steam is shut off, the friction of the bearings and of the air brings the wheel to rest in 2.2 hr. (a) What is the average angular acceleration of the wheel? (b) How many rotations will the wheel make before coming to rest? (c) What is the tangential linear acceleration of a particle distant 50 cm from the axis of rotation when the flywheel is turning at 75 rev/min? (d) What is the magnitude of the total linear acceleration of the particle in part (c)?

23(4). A coin of mass M is placed a distance R from the center of a phonograph turntable. The coefficient of static friction is μ_s. The rotational speed ω of the turntable is increased to a value ω_0 at which time the coin just flies off. (*a*) Find ω_0 in terms of the quantities M, R, g and μ_s. (*b*) Make a sketch showing the path of the coin as it flies off the turntable.

Answer: (*a*) $\omega_0 = \sqrt{\mu_s g / R}$.

24(4). (*a*) What are the acceleration and velocity of a point on the top of a 26-in. diameter automobile tire if the automobile is traveling at 60 miles/hr on a level road? (*b*) What are the acceleration and velocity of a point on the bottom of the tire?

CHAPTER 11

Rotational Dynamics and the Conservation of Angular Momentum

11–1 Introduction

In Chapter 10 we considered the kinematics of rotation. In this chapter, following the pattern of our study of translational motion, we study the causes of rotation, a subject called *rotational dynamics*. Rotating systems are made up of particles and we have already learned how to apply the laws of classical mechanics to the motion of particles. For this reason rotational dynamics should contain no features that are fundamentally new. In the same way rotational kinematics contained no basic new features, the rotational parameters θ, ω, and α being related to corresponding translational parameters x, v, and a for the particles that make up the rotating system. As in Chapter 10, however, it is very useful to recast the concepts of translational motion into a new form, especially chosen for its convenience in describing rotating systems.

We restricted our kinematical studies in Chapter 10 to a single but important special case, the rotation of a rigid body about an axis that is fixed in the reference frame in which we make our measurements. In studying rotational dynamics we start from a more fundamental point of view, that of a single particle viewed from an inertial reference frame. Later we shall generalize to systems of many particles, including the special case of a rigid body rotating about a fixed axis.

Finally we shall consider systems on which no external torques act and shall introduce the important principle of *conservation of angular momentum*.

11–2 Torque Acting on a Particle

In translational motion we associate a force with the linear acceleration of a body. In rotational motion, what quantity shall we associate with the angular accelera-

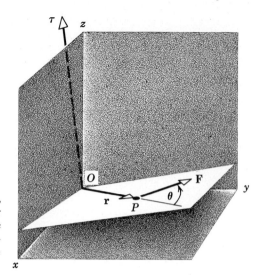

FIGURE 11-1   A force **F** is applied to a particle P, displaced **r** relative to the origin. The force vector makes an angle θ with the radius vector **r**. The torque τ about O is shown. Its direction is perpendicular to the plane formed by **r** and **F** with the sense given by the right-hand rule.

tion of a body? It cannot be simply force because, as experiment with a heavy revolving door teaches us, a given force (vector) can produce various angular accelerations of the door depending on where the force is applied and how it is directed; a force applied to the hinge line cannot produce any angular acceleration, whereas a force of given magnitude applied at right angles to the door at its outer edge produces a maximum acceleration.

We shall call the rotational analogue of force *torque* and shall now define it for the special case of a single particle observed from an inertial reference frame. Later we shall extend the torque concept to systems of particles (including rigid bodies) and shall show that torque is intimately associated with angular acceleration.

If a force **F** acts on a single particle at a point P whose position with respect to the origin O of the inertial reference frame is given by the position vector **r** (Fig. 11–1), the *torque* τ acting on the particle with respect to the origin $\dot{O}$ is defined as

$$\tau = r \times F. \tag{11–1}$$

Torque is a vector quantity. Its magnitude is given by

$$\tau = rF \sin \theta, \tag{11–2a}$$

where θ is the angle between **r** and **F**; its direction is normal to the plane formed by **r** and **F**. The sense is given by the right-hand rule for the vector product of two vectors, namely, one swings **r** into **F** through the smaller angle between them with the curled fingers of the right hand; the direction of the extended thumb then gives the direction of τ.

Torque has the same dimensions as force times distance, or in terms of our assumed fundamental dimensions, M, L, T, it has the dimensions ML^2T^{-2}. These are the same as the dimensions of work. However, torque and work are very different physical quantities. Torque is a vector and work is a scalar, for example. The unit of torque may be the nt-meter or lb-ft, among other possibilities.

Notice (Eq. 11–1) that the torque produced by a force depends not only on the magnitude and on the direction of the force but also on the point of application of the force relative to the origin, that is, on the vector **r**. In particular, when particle

P is at the origin, so that the line of action of **F** passes through the origin, **r** is zero and the torque τ about the origin is zero.

We can also write the magnitude of τ (Eq. 11–2a) either as

$$\tau = (r \sin \theta)F = Fr_\perp, \tag{11–2b}$$

or as

$$\tau = r(F \sin \theta) = rF_\perp, \tag{11–2c}$$

in which, as Fig. 11–2a shows, $r_\perp$ ($= r \sin \theta$) is the component of **r** at right angles to the line of action of **F**, and $F_\perp$ ($= F \sin \theta$) is the component of **F** at right angles to **r**. Torque is often called the *moment of force* and $r_\perp$ in Eq. 11–2b is called the *moment arm*. Equation 11–2c shows that only the component of **F** perpendicular to **r** contributes to the torque. In particular, when θ equals 0 or 180°, there is no perpendicular component ($F_\perp = F \sin \theta = 0$); then the line of action of the force passes through the origin and the moment arm $r_\perp$ about the origin is also zero. In this case both Eq. 11–2b and Eq. 11–2c show that the torque τ is zero.

If, as in Fig. 11–2b, we reverse the direction of **F**, the magnitude of τ remains unchanged but the direction of τ is reversed. Similarly, if, as in Fig. 11–2c, we reverse **r**, thereby changing the point of application of **F**, the magnitude of τ remains unchanged but the direction of τ is again reversed.

If, as in Fig. 11–2d, we reverse both **r** and **F**, then both the magnitude and the direction of τ remain unchanged. These results follow formally from the facts that: (1) $\sin \theta = \sin (180° - \theta)$, so that Eq. 11–2a for the magnitude of τ is unaffected;

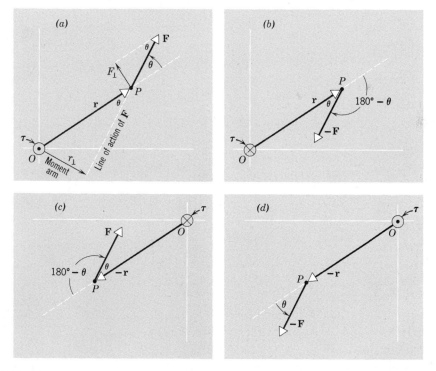

FIGURE 11–2 The plane shown is that defined by **r** and **F** in Fig. 11–1. (a) The magnitude of τ is given by $Fr_\perp$ (Eq. 11–2b) or by $rF_\perp$ (Eq. 11–2c). (b) Reversing **F** reverses the direction of τ. (c) Reversing **r** reverses the direction of τ. (d) Reversing **F** and **r** leaves the direction of τ unchanged. The directions of τ are represented by $\odot$ (perpendicularly out of the figure, the symbol representing the tip of an arrow) and by $\otimes$ (perpendicularly into the figure, the symbol representing the tail of an arrow).

(2) reversing the direction of one vector in a vector product (either **r** *or* **F**) reverses the direction of the product; and (3) reversing the directions of both vectors in a vector product (both **r** and **F**) leaves the direction of the product unchanged. Look at the directions of τ in Fig. 11–2 and verify them using the right-hand rule.

11–3 Angular Momentum of a Particle

We have found linear momentum to be useful in dealing with the translational motion of single particles or of systems of particles (including rigid bodies). For example, linear momentum is conserved in collisions. For a single particle the linear momentum is $\mathbf{p} = m\mathbf{v}$ (Eq. 8–9); for a system of particles it is $\mathbf{P} = M\mathbf{v}_{cm}$ (Eq. 8–13) in which M is the total system mass and $\mathbf{v}_{cm}$ is the velocity of the center of mass. In rotational motion, what is the analog of linear momentum? We call it *angular momentum* and we define it below for the special case of a single particle. Later we shall broaden the definition to include systems of particles and shall show that angular momentum, as we define it, is as useful a concept in rotational motion as linear momentum is in translational motion.

Consider a particle of mass m and linear momentum **p** at a position **r** relative to the origin O of an inertial reference frame (Fig. 11–3). We define the *angular momentum* **l** of the particle with respect to the origin O to be

$$\mathbf{l} = \mathbf{r} \times \mathbf{p}. \tag{11–3}$$

Angular momentum is a vector. Its magnitude is given by

$$l = rp \sin \theta, \tag{11–4a}$$

where θ is the angle between **r** and **p;** its direction is normal to the plane formed by **r** and **p**. The sense is given by the right-hand rule, namely, one swings **r** into **p,** through the smaller angle between them, with the curled fingers of the right hand; the extended right thumb then points in the direction of **l.**

We can also write the magnitude of **l** either as

$$l = (r \sin \theta)p = pr_{\perp}, \tag{11–4b}$$

or as

$$l = r(p \sin \theta) = rp_{\perp}, \tag{11–4c}$$

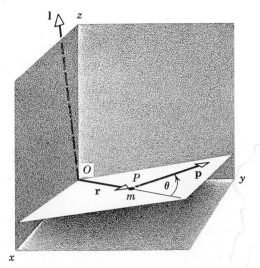

FIGURE 11–3 A particle of mass m is at point P displaced **r** relative to the origin, and has linear momentum **p**. The vector **p** makes an angle θ with the radius vector **r**. The angular momentum **l** of the particle with respect to origin O is shown. Its direction is perpendicular to the plane formed by **r** and **p** with the sense given by the right-hand rule.

in which $r_\perp$ ($= r \sin \theta$) is the component of **r** at right angles to the line of action of **p** and $p_\perp$ ($= p \sin \theta$) is the component of **p** at right angles to **r**. Angular momentum is often called *moment of* (linear) *momentum* and $r_\perp$ in Eq. 11–4b is often called the *moment arm*. Equation 11–4c shows that only the component of **p** perpendicular to **r** contributes to the angular momentum. When the angle θ between **r** and **p** is 0 or 180°, there is no perpendicular component ($p_\perp = p \sin \theta = 0$); then the line of action of **p** passes through the origin and $r_\perp$ is also zero. In this case both Eqs. 11–4b and 11–4c show that the angular momentum **l** is zero.

We now derive an important relation between torque and angular momentum. We have seen that $\mathbf{F} = d(m\mathbf{v})/dt = d\mathbf{p}/dt$ for a particle. Let us take the vector product of **r** with both sides of this equation, obtaining

$$\mathbf{r} \times \mathbf{F} = \mathbf{r} \times \frac{d\mathbf{p}}{dt}.$$

But $\mathbf{r} \times \mathbf{F}$ is the torque, or moment of a force, about O. We can then write

$$\boldsymbol{\tau} = \mathbf{r} \times \frac{d\mathbf{p}}{dt}. \tag{11–5}$$

Next we differentiate Eq. 11–3 and obtain

$$\frac{d\mathbf{l}}{dt} = \frac{d}{dt}(\mathbf{r} \times \mathbf{p}).$$

Now the derivative of a vector product is taken in the same way as the derivative of an ordinary product, except that we must not change the order of the terms. We have

$$\frac{d\mathbf{l}}{dt} = \frac{d\mathbf{r}}{dt} \times \mathbf{p} + \mathbf{r} \times \frac{d\mathbf{p}}{dt}.$$

But $d\mathbf{r}$ is the vector displacement of the particle in the time dt so that $d\mathbf{r}/dt$ is the instantaneous velocity **v** of the particle. Also, **p** equals $m\mathbf{v}$, so that the equation can be rewritten as

$$\frac{d\mathbf{l}}{dt} = (\mathbf{v} \times m\mathbf{v}) + \mathbf{r} \times \frac{d\mathbf{p}}{dt}.$$

Now $\mathbf{v} \times m\mathbf{v} = 0$, because the vector product of two parallel vectors is zero. Therefore,

$$\frac{d\mathbf{l}}{dt} = \mathbf{r} \times \frac{d\mathbf{p}}{dt}. \tag{11–6}$$

Inspection of Eqs. 11–5 and 11–6 shows that

$$\boldsymbol{\tau} = d\mathbf{l}/dt, \tag{11–7}$$

which states that the time rate of change of the angular momentum of a particle is equal to the torque acting on it. This result is the rotational analog of Eq. 8–10, which stated that the time rate of change of the linear momentum of a particle is equal to the force acting on it, that is, that $\mathbf{F} = d\mathbf{p}/dt$.

Equation 11–7, like all vector equations, is equivalent to three scalar equations, namely

$$\tau_x = dl_x/dt, \qquad \tau_y = dl_y/dt, \qquad \tau_z = dl_z/dt. \tag{11–8}$$

Hence, the x-component of the applied torque is given by the time derivative of the x-component of the angular momentum. Similar results hold for the y- and z-directions.

■ **Example 1.** A particle of mass m is released from rest at point a in Fig. 11–4, falling parallel to the (vertical) y-axis. (*a*) Find the torque acting on m at any time t, with respect to origin O. (*b*) Find the angular momentum of m at any time t, with respect to this same origin. (*c*) Show that the relation $\boldsymbol{\tau} = d\mathbf{l}/dt$ (Eq. 11–7) yields a correct result when applied to this familiar problem.

(*a*) The torque is given by Eq. 11–1 or $\boldsymbol{\tau} = \mathbf{r} \times \mathbf{F}$, its magnitude being given by

$$\tau = rF \sin \theta.$$

In this example $r \sin \theta = b$ and $F = mg$ so that

$$\tau = mgb = \text{a constant}.$$

Note that the torque is simply the product of the force (mg) times the moment arm (b). The right-hand rule shows that $\boldsymbol{\tau}$ is directed perpendicularly into the figure.

(*b*) The angular momentum is given by Eq. 11–3 or $\mathbf{l} = \mathbf{r} \times \mathbf{p}$, its magnitude being given by

$$l = rp \sin \theta.$$

In this example $r \sin \theta = b$ and $p = mv = m(gt)$ so that

$$l = mgbt.$$

The right-hand rule shows that $\mathbf{l}$ is directed perpendicularly into the figure, which means that $\mathbf{l}$ and $\boldsymbol{\tau}$ are parallel vectors. The vector $\mathbf{l}$ changes with time in magnitude only, its direction always remaining the same in this case.

(*c*) Since $d\mathbf{l}$, the change in $\mathbf{l}$, and $\boldsymbol{\tau}$ are parallel we can replace the vector relation $\boldsymbol{\tau} = d\mathbf{l}/dt$ by the scalar relation

$$\tau = dl/dt.$$

Using the expressions for τ and l from (*a*) and (*b*) above we have

$$mgb = \frac{d}{dt}(mgbt) = mgb,$$

which is an identity. Thus the relation $\boldsymbol{\tau} = d\mathbf{l}/dt$ yields correct results in this simple case. Indeed, if we cancel the constant b out of the first two terms above and if we substitute for gt the equivalent quantity v, we have

$$mg = \frac{d}{dt}(mv).$$

Since $mg = F$ and $mv = p$, this is the familiar result $F = dp/dt$. Thus, as we indicated earlier, relations such as $\boldsymbol{\tau} = d\mathbf{l}/dt$, though often vastly useful, are not new basic postulates of classical mechanics but are rather the reformulation of the Newtonian laws for rotational motion.

Note that the values of τ and l depends on our choice of origin, that is, on b. In particular, if $b = 0$, then $\tau = 0$ and $l = 0$.   ■

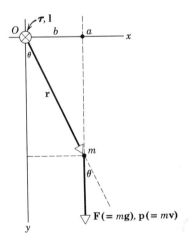

FIGURE 11–4 A particle of mass m drops vertically from point a. The torque and the angular momentum about O are directed perpendicularly into the figure, as shown by the symbol $\otimes$ at O.

11–4 Systems of Particles

So far we have talked only about single particles. Let us now consider a system of many particles. To calculate the total angular momentum **L** of a system of particles about a given point, we must add vectorially the angular momenta of all the individual particles of the system about this same point. For a system containing n particles we have, then,

$$\mathbf{L} = \mathbf{l}_1 + \mathbf{l}_2 + \cdots + \mathbf{l}_n = \sum_{i=1}^{i=n} \mathbf{l}_i,$$

in which the (vector) sum is taken over all particles in the system.

As time goes on, the total angular momentum **L** of the system about a fixed reference point (which we choose, as in our basic definition of **l** in Eq. 11–3, to be the origin of an inertial reference frame) may change. This change, $d\mathbf{L}/dt$, can arise from two sources: (1) torques exerted on the particles of the system by internal forces between the particles and (2) torques exerted on the particles of the system by external forces.

If Newton's third law holds in its so-called strong form, that is, if the forces between any two particles not only are equal and opposite but are also directed along the line joining the two particles, then the total internal torque is zero because the torque resulting from each internal action-reaction force pair is zero.

Hence the first source contributes nothing. For our reference point, therefore, only the second source remains, and we can write

$$\boldsymbol{\tau}_{\text{ext}} = d\mathbf{L}/dt, \tag{11–9}$$

where $\boldsymbol{\tau}_{\text{ext}}$ stands for the sum of all the external torques acting on the system. In words, the time rate of change of the total angular momentum of a system of particles about the origin of an inertial reference frame is equal to the sum of the external torques acting on the system. Later, for convenience, in situations in which no confusion is likely to arise, we shall drop the subscript on $\boldsymbol{\tau}_{\text{ext}}$.

Equation 11–9 is the generalization of Eq. 11–7 to many particles. When we have only one particle, there are no internal forces or torques. This relation (Eq. 11–9) holds whether the particles that make up the system are in motion relative to each other or whether they have fixed spatial relationships, as in a rigid body.

Equation 11–9 is the rotational analog of Eq. 8–15

$$\mathbf{F}_{\text{ext}} = d\mathbf{P}/dt \tag{8–15}$$

which tells us that for a system of particles (rigid body or not) the resultant external force acting on the system equals the time rate of change of the linear momentum of the system.

As we have derived it, Eq. 11–9 holds when $\boldsymbol{\tau}$ and **L** are measured with respect to the origin of an inertial reference frame. We may well ask whether it still holds if we measure these two vectors with respect to an arbitrary point (a particular particle, say) in the moving system. In general, such a point would move in a complicated way as the body or system of particles translated, tumbled, and changed its configuration and Eq. 11–9 would not apply to such a reference point. However, if the reference point is chosen to be the center of mass of the system, even though this point is not fixed in our reference frame, then Eq. 11–9 does hold. This is another remarkable property of the center of mass. Thus we can separate the general motion of a system of particles into the translational motion of its center of mass (Eq. 8–15) and rotational motion about its center of mass (Eq. 11–9).

11–5 Kinetic Energy of Rotation and Rotational Inertia

We shall now confine our attention to an important special case of a system of particles, a rigid body. In a rigid body the particles in the system always maintain the same positions with respect to one another. In studying the rotation of a rigid body we shall restrict our attention to the special case, often encountered, in which the axis of rotation is fixed in an inertial reference frame.

Let us now imagine a rigid body rotating with angular speed ω about an axis that is fixed in a particular inertial frame, as in Fig. 10–1. Each particle in such a rotating body has a certain amount of kinetic energy. A particle of mass m at a distance r from the axis of rotation moves in a circle of radius r with an angular speed ω about this axis and has a linear speed $v = \omega r$. Its kinetic energy therefore is $\frac{1}{2}mv^2 = \frac{1}{2}mr^2\omega^2$. The total kinetic energy of the body is the sum of the kinetic energies of its particles.

If the body is rigid, as we assume in this section, ω is the same for all particles. The radius r may be different for different particles. Hence the total kinetic energy K of the rotating body can be written as

$$K = \tfrac{1}{2}(m_1r_1{}^2 + m_2r_2{}^2 + \cdots)\omega^2 = \tfrac{1}{2}(\Sigma\, m_ir_i{}^2)\omega^2.$$

The term $\Sigma\, m_ir_i{}^2$ is the sum of the products of the masses of the particles by the squares of their respective distances from the axis of rotation. If we denote this quantity by I, then

$$I = \Sigma\, m_ir_i{}^2 \tag{11–10}$$

is called the *rotational inertia,* or moment of inertia, of the body with respect to the particular axis of rotation. Note that the rotational inertia of a body depends on the particular axis about which it is rotating as well as on the shape of the body and the manner in which its mass is distributed. Rotational inertia has the dimensions ML^2 and is usually expressed in kg-m^2 or slug-ft^2.

In terms of rotational inertia we can now write the kinetic energy of the rotating rigid body as

$$K = \tfrac{1}{2}I\omega^2. \tag{11–11}$$

This is analogous to the expression for the kinetic energy of translation of a body, $K = \frac{1}{2}Mv^2$. We have already seen that the angular speed ω is analogous to the linear speed v. Now we see that the rotational inertia I is analogous to the mass, or the translational inertia M. Although the mass of a body does not depend on its location, the rotational inertia of a body does depend on the axis about which it is rotating.

We should understand that the rotational kinetic energy given by Eq. 11–11 is simply the sum of the ordinary translational kinetic energy of all the parts of the body and not a new kind of energy. Rotational kinetic energy is simply a convenient way of expressing the kinetic energy for a rotating rigid body.

Example 2. Consider a body consisting of two spherical masses of 5.0 kg each connected by a light rigid rod 1.0 meter long (Fig. 11–5). Treat the spheres as point particles and neglect

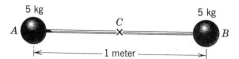

FIGURE 11–5 Example 2. Calculating the rotational inertia of a dumbell.

the mass of the rod. Determine the rotational inertia (or moment of inertia) of the body (*a*) about an axis normal to it through its center at C, and (*b*) about an axis normal to it through one sphere.

   (*a*) If the axis is normal to the page through C, we have

$$I_C = \Sigma m_i r_i^2 = m_a r_a^2 + m_b r_b^2$$

$$= (5.0 \text{ kg})(0.50 \text{ meter})^2 + (5.0 \text{ kg})(0.50 \text{ meter})^2 = 2.5 \text{ kg-m}^2.$$

   (*b*) If the axis is normal to the page through A or B, we have

$$I_A = m_a r_a^2 + m_b r_b^2 = (5.0 \text{ kg})(0 \text{ meter})^2 + (5.0 \text{ kg})(1.0 \text{ meter})^2 = 5.0 \text{ kg-m}^2,$$

$$I_B = m_a r_a^2 + m_b r_b^2 = (5.0 \text{ kg})(1.0 \text{ meter})^2 + (5.0 \text{ kg})(0 \text{ meter})^2 = 5.0 \text{ kg-m}^2.$$

Hence the rotational inertia of this rigid dumbbell model is twice as great about an axis through an end as it is about an axis through the center.

For a body that is not composed of discrete point masses but is instead a continuous distribution of matter, the summation in $I = \Sigma m_i r_i^2$ becomes an integration. We imagine the body to be subdivided into infinitesimal elements, each of mass dm. We let r be the distance from such an element to the axis of rotation. Then the rotational inertia is obtained from

$$I = \int r^2 \, dm, \tag{11–12}$$

where the integral is taken over the whole body. The procedure by which the summation Σ of a discrete distribution is replaced by the integral $\int$ for a continuous distribution is the same as that discussed for the center of mass in Section 8–1.

   For bodies of irregular shape the integrals may be hard to evaluate. For bodies of simple geometrical shape the integrals are relatively easy when an axis of symmetry is chosen as the axis of rotation.

   The rotational inertias about certain axes of some common solids (of uniform density) are listed in Table 11–1. Each of these results can be derived by integration. The total mass of the body is denoted by M in each equation.

   There is a simple and very useful relation between the rotational inertia I of a body about any axis and its rotational inertia I_{cm} with respect to a parallel axis through the center of mass. If M is the total mass of the body and h the distance between the two axes, the relation is

$$I = I_{cm} + Mh^2. \tag{11–13}$$

   The proof of this relation, which is called the *parallel axis theorem*, follows. Let C be the center of mass of the arbitrarily shaped body whose cross section is shown in Fig. 11–6. The center of mass has coordinates x_{cm} and y_{cm}. We choose the x-y plane to include C, so that z_{cm} equals zero. Consider an axis through C at right angles to the plane of the paper and another axis parallel to it through P at $(x_{cm} + a)$ and $(y_{cm} + b)$. The distance between the axes is $h = \sqrt{a^2 + b^2}$. Then the square of the distance of a particle from the axis through C is $x_i^2 + y_i^2$, where x_i and y_i measure the coordinates of a mass element m_i relative to the axis through C. The square of its distance from an axis through P is $(x_i - a)^2 + (y_i - b)^2$. Hence the rotational

Table 11-1

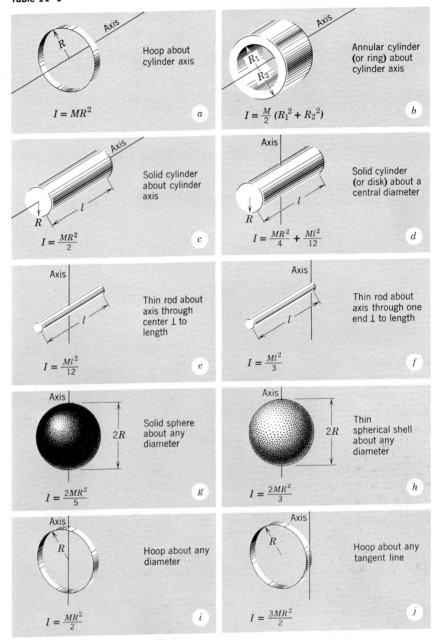

inertia about an axis through P is

$$I = \Sigma \, m_i[(x_i - a)^2 + (y_i - b)^2]$$
$$= \Sigma \, m_i(x_i{}^2 + y_i{}^2) - 2a \, \Sigma \, m_ix_i - 2b \, \Sigma \, m_iy_i + (a^2 + b^2) \, \Sigma \, m_i.$$

From the definition of center of mass,

$$\Sigma \, m_ix_i = \Sigma \, m_iy_i = 0,$$

so that the two middle terms are zero. The first term is simply the rotational inertia about an axis through the center of mass I_{cm} and the last term is Mh^2. Hence it follows that $I = I_{\text{cm}} + Mh^2$.

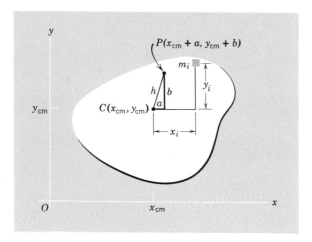

FIGURE 11–6 Derivation of the parallel-axis theorem. Knowing the rotational inertia about an axis through C, we can find its value about a parallel axis through P.

With the aid of this formula several of the results of Table 11–1 can be deduced from previous results. For example, (f) follows from (e), and (j) follows from (i) with the aid of Eq. 11–13.

11–6 Rotational Dynamics of a Rigid Body

In this section we continue to study the special case of a rigid body confined to rotate about an axis that is fixed in an inertial reference frame. First we shall review the concept of torque as applied to such a rigid body; then we shall show how the torque is related to the angular acceleration of the body about this axis.

Suppose that we apply a torque τ to one of the particles in a rigid body. Since all the particles of a truly rigid body maintain a fixed spatial relationship to all the other particles that make up the body, the torque may be said to act on the rigid body as a whole. In general, the vector τ will not lie along the axis around which the body is free to rotate. We are not concerned in this section with the actual torques applied to the body but only with the components of these torques that lie along (or as we sometimes say, *around*) the axis. Only these components can cause the body to rotate about this axis. Torque components perpendicular to the axis. tend to turn the axis from its fixed position. We have specifically assumed, however, that the axis maintains a fixed direction. The body may, for example, be attached to a shaft that is held in a fixed position by bearings at each end; if an applied torque has a component at right angles to the shaft, tending to turn it, the bearings will automatically apply an equal and opposite counter-torque to the shaft, canceling the effect of this component.

In Fig. 11–7 (compare Fig. 10–2) we show a section through a rigid body that is free to rotate about the z-axis of an inertial reference frame. A force **F**, taken for convenience to be in (or parallel to) the x-y plane of the section, acts on a particle at point P in the body, the position of P with respect to the rotational axis (the z-axis) being defined by the vector **r**. The torque acting on the particle at P may be said to act on the rigid body as a whole and is given by Eq. 11–1, or

$$\tau = \mathbf{r} \times \mathbf{F}.$$

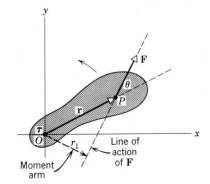

FIGURE 11–7 A force **F** acts on the particle P in a rigid body, exerting a torque $\tau = \mathbf{r} \times \mathbf{F}$ on the body, with respect to an axis through O at right angles to the plane of the figure. The moment arm $r_\perp$ is also shown, as is the torque $\boldsymbol{\tau}$, which is a vector emerging perpendicularly from the page.

Because we have chosen **r** and **F** to lie in a plane parallel to the x-y plane, the torque $\boldsymbol{\tau}$ will point along the z axis. The right-hand rule shows that it points perpendicularly out of the plane of Fig. 11–7. If **r** and **F** did not lie in the plane of the figure, $\boldsymbol{\tau}$ would not be parallel to the z-axis and we would concern ourselves here only with the component of $\boldsymbol{\tau}$ along this axis. The magnitude of $\boldsymbol{\tau}$ is given by Eq. 11–2 or

$$\tau = rF \sin \theta$$

which, as we have seen, can also be written as $\tau = rF_\perp$ or $\tau = Fr_\perp$.

■   **Example 3.** A wagon wheel is free to rotate about a horizontal axis through O. A force of 10 lb is applied to a spoke at the point P, 1.0 ft from the center. OP makes an angle of 30° with the horizontal (x-axis) and the force is in the plane of the wheel making an angle of 45° with the horizontal (x-axis). What is the torque on the wheel?

The angle between the displacement vector **r** from O to P and the applied force **F** (Fig. 11–8) is θ, where

$$\theta = 45° - 30° = 15°.$$

Then the magnitude of the torque is

$$\tau = rF \sin \theta$$

$$= (1.0 \text{ ft})(10 \text{ lb})(\sin 15°) = 2.6 \text{ lb-ft}.$$

It is clear that we can obtain this same result from $\tau = rF_\perp$ or $\tau = Fr_\perp$ as well (see Eqs. 11–2). The torque ($\boldsymbol{\tau} = \mathbf{r} \times \mathbf{F}$) is a vector pointing out $\odot$ along the axis through O having a magnitude 2.6 lb-ft.   ■

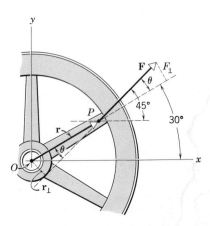

FIGURE 11–8   Example 3.

We now investigate the relationship between the torque applied to the rigid body of Fig. 11–7 and the angular acceleration of this body. Let us observe the rigid body for an infinitesimal time dt, during which it will rotate through an infinitesimal angle $d\theta$. We have seen earlier that we can describe the rotation of a rigid body about a fixed axis by examining the motion of any single point fixed in the body, such as P in Fig. 11–7. For convenience, then, we ignore the body itself in Fig. 11–9 and focus our attention on the representative point P and on the vector **r** which locates point P with respect to the axis of rotation.

During the time dt, the point P will move an infinitesimal distance ds along a circular path of radius r as the body rotates through an infinitesimal angle $d\theta$, where

$$ds = r\,d\theta.$$

The work dW done by this force during this small rotation is

$$dW = \mathbf{F}\cdot d\mathbf{s} = F\cos\phi\,ds = (F\cos\phi)(r\,d\theta),$$

where $F\cos\phi$ is the component of **F** in the direction of $d\mathbf{s}$.

The term $(F\cos\phi)r$, however, is the magnitude of the instantaneous torque exerted by **F** on the rigid body about the axis perpendicular to the page through O, so that

$$dW = \tau\,d\theta. \tag{11–14}$$

This differential expression for the work done in rotation (about a fixed axis) is equivalent to the expression $dW = F\,dx$ for the work done in translation (along a straight line).

To obtain the rate at which work is done in rotational motion (about a fixed axis), we divide both sides of Eq. 11–14 by the infinitesimal time interval dt during which the body is displaced through $d\theta$, obtaining

$$\frac{dW}{dt} = \tau\frac{d\theta}{dt}$$

or

$$P = \tau\omega,$$

giving the instantaneous power P. This last expression is the rotational analog of $P = Fv$ for translational motion (along a straight line).

If now a number of forces $\mathbf{F}_1$, $\mathbf{F}_2$, etc., are applied to the body in the plane normal to its axis of rotation, the work done by these forces on the body in a small rotation $d\theta$ will be

$$dW = F_1\cos\phi_1 r_1\,d\theta + F_2\cos\phi_2 r_2\,d\theta + \cdots,$$

$$= (\tau_1 + \tau_2 + \cdots)\,d\theta = \tau\,d\theta,$$

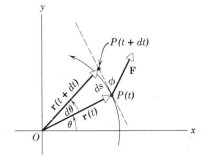

FIGURE 11–9   In time dt point P in the rigid body of Fig. 11–7 moves a distance ds along the arc of a circle of radius r. The rigid body (not shown) and the vector **r** that locates point P in it each rotate through an angle $d\theta$ during this interval.

where $r_1\,d\theta$ equals ds_1, the displacement of the point at which $\mathbf{F}_1$ is applied, and ϕ_1 is the angle between $\mathbf{F}_1$ and $d\mathbf{s}_1$, etc., and where τ is now the magnitude of the component of the resultant torque along the axis through O. In computing this sum each torque is considered positive or negative according to the sense in which it alone would tend to rotate the body about its axis. We can arbitrarily call the torque associated with a force positive if the effect of the force, acting alone, is to produce a counterclockwise rotation; then the torque is negative if the effect is to produce a clockwise rotation.

There is no internal motion of particles within a truly rigid body. The particles always maintain a fixed position relative to one another and move only with the body as a whole. Hence there can be no dissipation of energy within a truly rigid body. We can therefore equate the rate at which work is being done on the body to the rate at which its kinetic energy is increasing. The rate at which work is being done on the rigid body is

$$\frac{dW}{dt} = \tau\frac{d\theta}{dt} = \tau\omega. \tag{11-15}$$

The rate at which the kinetic energy of the rigid body is increasing is

$$\frac{d}{dt}\left(\tfrac{1}{2}I\omega^2\right).$$

But I is constant because the body is rigid and the axis is fixed. Hence

$$\frac{d}{dt}\left(\tfrac{1}{2}I\omega^2\right) = \tfrac{1}{2}I\frac{d}{dt}\left(\omega^2\right) = I\omega\frac{d\omega}{dt} = I\omega\alpha. \tag{11-16}$$

Equating the right-hand members of Eqs. 11–15 and 11–16, we obtain

$$\tau\omega = I\alpha\omega,$$

or

$$\tau = I\alpha. \tag{11-17}$$

In deriving Eq. 11–17 ($\tau = I\alpha$), we have simply transformed Newton's second law ($F = Ma$), written in scalar form suitable to describe rectilinear motion, into rotational terms. This suggests that just as we associate a force with the linear acceleration of a body, so we may associate a torque with the angular acceleration of a body about a given axis. The rotational inertia I is a measure of the resistance a body offers to having its rotational motion changed by a given torque just as the translational inertia, or mass, M is a measure of the resistance a body offers to having its translational motion changed by a given force.

In Table 11–2 we compare the translational motion of a rigid body along a straight line with the rotational motion of a rigid body about a fixed axis.

The rotation of a rigid body about a fixed axis (to which $\tau = I\alpha$ applies) is not the most general kind of rotary motion in that the body may not be rigid and the axis may not be fixed in an inertial reference frame. In this general case Eq. 11–9, or $\tau_{\text{ext}} = d\mathbf{L}/dt$, applies. As we have already pointed out, this is equivalent to Newton's second law for the general translational motion of a system of particles, namely, Eq. 8–15, or $\mathbf{F}_{\text{ext}} = d\mathbf{P}/dt$.

In the rest of this chapter we confine ourselves to the rotations of rigid bodies about fixed axes.

Table 11–2

Rectilinear Motion		Rotation about a Fixed Axis	
Displacement	x	Angular displacement	θ
Velocity	$v = \dfrac{dx}{dt}$	Angular velocity	$\omega = \dfrac{d\theta}{dt}$
Acceleration	$a = \dfrac{dv}{dt}$	Angular acceleration	$\alpha = \dfrac{d\omega}{dt}$
Mass	M	Rotational inertia	I
Force	$F = Ma$	Torque	$\tau = I\alpha$
Work	$W = \int F\, dx$	Work	$W = \int \tau\, d\theta$
Kinetic energy	$\frac{1}{2}Mv^2$	Kinetic energy	$\frac{1}{2}I\omega^2$
Power	$P = Fv$	Power	$P = \tau\omega$
Linear momentum	$P = Mv$	Angular momentum	$L = I\omega$

Example 4. A uniform disk of radius R and mass M is mounted on an axle supported in fixed frictionless bearings, as in Fig. 11–10. A light cord is wrapped around the rim of the wheel and a steady downward pull **T** is exerted on the cord. Find the angular acceleration of the wheel and the tangential acceleration of a point on the rim.

The torque about the central axis is $\tau = TR$, and the rotational inertia of the disk about its central axis is $I = \frac{1}{2}MR^2$. From

$$\tau = I\alpha,$$

we have

$$TR = (\tfrac{1}{2}MR^2)\alpha,$$

or

$$\alpha = \frac{2T}{MR}.$$

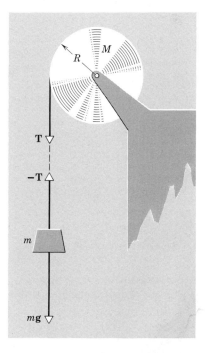

FIGURE 11–10 Example 4. A steady downward force **T** produces rotation of the disk. Example 5. Here **T** is supplied by the falling mass m.

If the mass of the disk is taken to be $M = 0.20$ slug, its radius $R = 0.50$ ft, and the force $T = 1.0$ lb, then

$$\alpha = \frac{(2)(1.0 \text{ lb})}{(0.20 \text{ slug})(0.50 \text{ ft})} = 20 \text{ radians/sec}^2.$$

The tangential acceleration of a point on the rim is given by

$$a = R\alpha = (20 \text{ radians/sec}^2)(0.50 \text{ ft}) = 10 \text{ ft/sec}^2.$$

Example 5. Suppose that we hang a body of mass m from the cord in the previous problem. Find the angular acceleration of the disk and the tangential acceleration of a point on the rim in this case.

Now, let T be the tension in the cord. Since the suspended body will accelerate downward, the magnitude of the downward pull of gravity on it, mg, must exceed the magnitude of the upward pull of the cord on it, T. The acceleration a of the suspended body is the same as the tangential acceleration of a point on the rim of the disk. From Newton's second law

$$mg - T = ma.$$

The resultant torque on the disk is TR and its rotational inertia is $\frac{1}{2}MR^2$, so that from

$$\tau = I\alpha$$

we obtain

$$TR = \frac{1}{2}MR^2\alpha.$$

Using the relation $a = R\alpha$, we can write this last equation as

$$2T = Ma.$$

Solving the first and last equations simultaneously leads to

$$a = \left(\frac{2m}{M + 2m}\right)g,$$

and

$$T = \left(\frac{Mm}{M + 2m}\right)g.$$

If now we let the disk have a mass $M = 0.20$ slug and a radius $R = 0.50$ ft as before, and we let the suspended body weigh 1.0 lb, we obtain

$$a = \frac{2mg}{M + 2m} = \frac{(2)(1.0 \text{ lb})}{(0.20 \text{ slug}) + (2)(\frac{1}{32} \text{ slug})} = 7.6 \text{ ft/sec}^2,$$

$$\alpha = \frac{a}{R} = \frac{(7.6 \text{ ft/sec}^2)}{0.50 \text{ ft}} = 15 \text{ radians/sec}^2.$$

Notice that the accelerations are less for a suspended 1.0-lb body than they were for a steady 1.0-lb pull on the string (Example 4). This corresponds to the fact that the tension in the string supplying the torque is now less than 1.0 lb, namely

$$T = \frac{Mmg}{M + 2m} = \frac{(0.20 \text{ slug})(1.0 \text{ lb})}{(0.20 + 2.0/32) \text{ slug}} = 0.76 \text{ lb}.$$

The tension in the string must be less than the weight of the suspended body if the body is to accelerate downward.

Example 6. Derive the relation $L = I\omega$, shown in Table 11–2, for the angular momentum of a rigid body confined to rotate about a fixed axis.

Starting from the scalar relation $\tau = I\alpha$ and the definition of α $(= d\omega/dt)$, we may write

$$\tau = I\alpha = I(d\omega/dt) = d(I\omega)/dt,$$

in which the last step is justified by the fact that I is a constant for a given rigid body and a specified (fixed) axis of rotation.

Next we use the vector relation $\tau_{\text{ext}} = d\mathbf{L}/dt$ (Eq. 11–9) and write the corresponding relation for the scalar components, τ and dL, of τ_{ext} and $d\mathbf{L}$ along the fixed axis of rotation, obtaining

$$\tau = dL/dt.$$

Simply by comparing the two equations above we obtain the relation sought, namely

$$L = I\omega. \tag{11–18}$$

Like Eq. 11–17 ($\tau = I\alpha$), this is a scalar relation holding for the rotation of a rigid body about a fixed axis. L is the component along that axis of the vector angular momentum $\mathbf{L}$ of the rigid body and I must refer to that same axis.

Equation 11–18 is the rotational analog of the expression $P = Mv$ for the linear momentum of a rigid body of mass M in pure translational motion with linear speed v. It gives the angular momentum about a fixed axis of a rigid body having rotational inertia I and angular speed ω about that same axis.

11–7   Conservation of Angular Momentum

In Section 11–4, we found that the time rate of change of the total angular momentum of a system of particles about a point fixed in an inertial reference frame (or about the center of mass) is equal to the sum of the external torques acting on the system, that is,

$$\tau_{\text{ext}} = d\mathbf{L}/dt. \tag{11–9}$$

Suppose now that $\tau_{\text{ext}} = 0$; then $d\mathbf{L}/dt = 0$ so that $\mathbf{L} = $ a constant.

When the resultant external torque acting on a system is zero, the total vector angular momentum of the system remains constant. This is the *principle of the conservation of angular momentum.*

For a system of n particles, the total angular momentum $\mathbf{L}$ about some point is

$$\mathbf{L} = \mathbf{l}_1 + \mathbf{l}_2 + \cdots + \mathbf{l}_n.$$

When the resultant external torque on the system is zero, we have

$$\mathbf{L} = \text{a constant} = \mathbf{L}_0, \tag{11–19}$$

where $\mathbf{L}_0$ is the constant total angular momentum vector. The angular momenta of the individual particles may change, but their vector sum $\mathbf{L}_0$ remains constant in the absence of a net external torque.

Angular momentum is a vector quantity so that Eq. 11–19 is equivalent to three scalar equations, one for each coordinate direction through the reference point. The conservation of angular momentum therefore supplies us with three conditions on the motion of a system to which it applies.

For a system consisting of a rigid body rotating about an axis (the z-axis, say) that is fixed in an inertial reference frame, we have

$$L_z = I\omega, \tag{11–20}$$

where L_z is the component of the angular momentum along the rotation axis and I is the rotational inertia for this same axis. It is possible for the rotational inertia I of a rotating body to change by rearrangement of its parts. If no net external torque acts, then L_z must remain constant and, if I does change, there must be a

FIGURE 11-11   A diver leaves the diving board with arms and legs outstretched and with some initial angular velocity. Since no torques are exerted on him about his center of mass, $L (= I\omega)$ is constant while he is in the air. When he pulls his arms and legs in, since I decreases, ω increases. When he again extends his limbs, his angular velocity drops back to its initial value. Notice the parabolic motion of his center of mass, common to all two-dimensional motion under the influence of gravity.

compensating change in ω. The principle of conservation of angular momentum in this case is expressed as

$$I\omega = I_0\omega_0 = \text{a constant.} \tag{11-21}$$

Equation 11–21 holds not only for rotation about a fixed axis but also about an axis through the center of mass of the system that moves so that it always remains parallel to itself.

Acrobats, divers, ballet dancers, ice skaters, and others often use this principle. Because I depends on the square of the distance of the parts of the body from the axis of rotation, a large variation is possible by extending or pulling in the limbs. Consider the diver in Fig. 11–11. Let us assume that as he leaves the diving board he has a certain angular speed ω_0 about a horizontal axis through the center of mass, such that he would rotate through half a turn before he strikes water. If he wishes to make a one and one-half turn somersault instead, in the same time, he must triple his angular speed. Now there are no external forces acting on him except gravity, and gravity exerts no torque about his center of mass. His angular momentum therefore remains constant, and $I_0\omega_0 = I\omega$. Since $\omega = 3\omega_0$, the diver must change his rotational inertia about the horizontal axis through the center of mass from the initial value I_0 to a value I, such that I equals $\frac{1}{3}I_0$. This he does by pulling in his arms and legs toward the center of his body. The greater his initial angular speed and the more he can reduce his rotational inertia, the greater the number of revolutions he can make in a given time.

We should notice that the rotational kinetic energy of the diver is not constant. In fact, in our example, since

$$I\omega = I_0\omega_0$$

and

$$I < I_0,$$

it follows that

$$\frac{1}{2}\frac{(I\omega)^2}{I} = \tfrac{1}{2}I\omega^2 > \tfrac{1}{2}I_0\omega_0^2,$$

and the diver's rotational kinetic energy increases. This increase in energy is supplied by the diver, who does work when he pulls the parts of his body together.

In a similar way the ice skater and ballet dancer can increase or decrease the angular speed of a spin about a vertical axis.

Example 7. A small object of mass m is attached to a light string which passes through a hollow tube. The tube is held by one hand and the string by the other. The object is set into rotation in a circle of radius r_1 with a speed v_1. The string is then pulled down, shortening the radius of the path to r_2 (Fig. 11–12). Find the new linear speed v_2 and the new angular speed ω_2 of the object in terms of the initial values v_1 and ω_1 and the two radii.

The downward pull on the string is transmitted as a radial force on the object. Such a force exerts a zero torque on the object about the center of rotation. Since no torque acts on the object about its axis of rotation, its angular momentum in that direction is constant. Hence

initial angular momentum = final angular momentum,

$$mv_1r_1 = mv_2r_2,$$

and

$$v_2 = v_1\left(\frac{r_1}{r_2}\right).$$

Since $r_1 > r_2$, the object speeds up when it is pulled in.

In terms of angular speed, since v_1 equals $\omega_1 r_1$ and v_2 equals $\omega_2 r_2$,

$$mr_1{}^2\omega_1 = mr_2{}^2\omega_2$$

and

$$\omega_2 = \left(\frac{r_1}{r_2}\right)^2\omega_1,$$

so that there is an even greater increase in angular speed over the initial value (see Problem 31). What effect does the force of gravity (the object's weight) have on this analysis?

Example 8. A student sits on a stool that is free to rotate about a vertical axis. He holds his arms extended horizontally with an 8.0-lb weight in each hand. The instructor sets him rotating with an angular speed of 0.50 rev/sec. Assume that friction is negligible and exerts no torque about the vertical axis of rotation. Assume that the rotational inertia of the student remains

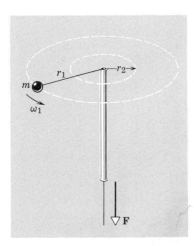

FIGURE 11–12 Example 7. A mass at the end of a cord moves in a circle of radius r_1 with angular speed ω_1. The cord passes down through a tube. **F** supplies the centripetal force.

constant at 4.0 slug-ft^2 as he pulls his hands to his sides and that the change in rotational inertia is due only to pulling the weights in. Take the original distance of the weights from the axis of rotation to be 3.0 ft and their final distance 0.50 ft. Find the final angular speed of the student.

The only external force is gravity acting through the center of mass, and that exerts no torque about the axis of rotation. Hence the angular momentum is conserved about this axis and

initial angular momentum = final angular momentum,

$$I_0\omega_0 = I\omega.$$

We have

$$I = I_{\text{student}} + I_{\text{weights}},$$

$$I_0 = 4.0 + 2\left(\frac{8.0}{32}\right)(3.0)^2 = 8.5 \text{ slug-ft}^2,$$

$$I = 4.0 + 2\left(\frac{8.0}{32}\right)\left(\frac{1}{2}\right)^2 = 4.1 \text{ slug-ft}^2,$$

$$\omega_0 = 0.50 \text{ rev/sec} = \pi \text{ radians/sec}.$$

Therefore

$$\omega = \frac{I_0}{I}\omega_0 = \frac{8.5}{4.1}\pi \text{ radians/sec} = 2.1\pi \text{ radians/sec} \cong 1.0 \text{ rev/sec}.$$

The final angular speed is approximately double.

If we had allowed for the decrease in I caused by the arms being pulled in, the final angular speed would have been much greater.

What change would friction make? Is kinetic energy conserved as the student pulls in his arms and then puts them out again, assuming there is no friction? Explain.

Example 9. A classsroom demonstration that illustrates the vector nature of the law of conservation of angular momentum is worth considering.

A student stands in a platform that can rotate only about a vertical axis. In his hand he holds the axle of a rim-loaded bicycle wheel with its axis vertical. The wheel is spinning about this vertical axis with an angular speed ω_0, but the student and platform are at rest. The student tries to change the direction of rotation of the wheel. What happens?

Let us choose as the system the student plus platform plus wheel. The initial total angular momentum of this system is $I_0\omega_0$, arising from the spinning wheel, I_0 being the rotational inertia of the wheel about its axis and ω_0 pointing vertically upward. Figure 11–13a shows the initial condition.

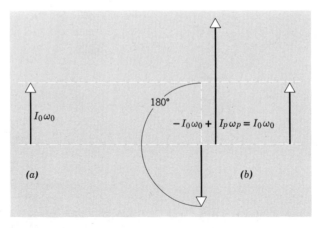

(a)

180°

$-I_0\omega_0 + |I_p\omega_p = I_0\omega_0$

(b)

FIGURE 11–13   Example 9. (a) The initial angular momentum of the system is shown. In (b) the wheel has been tilted 180°. Its contribution to the vertical component of angular momentum is reduced by $2I_0\omega_0$. The deficit is made up by the student and platform.

The student next turns the axis of the wheel through an angle of 180° from the vertical (to do this he must supply a torque. This torque, however, is internal to the system as we have defined it). Since there is no external component of torque on the system about the vertical axis, the vertical component of angular momentum of the system must be conserved. The wheel, however, now contributes a vertical component of angular momentum of $-I_0\omega_0$ to the system. Hence the student and platform must supply the additional angular momentum about the vertical axis, and they begin to rotate. This extra vertical angular momentum $I_p\omega_p$ when added to $-I_0\omega_0$ must equal the initial vertical angular momentum of the system $I_0\omega_0$. That is,

$$-I_0\omega_0 + I_p\omega_p = I_0\omega_0$$

This is shown in Fig. 11–13*b*. I_p is the rotational inertia of student and platform with respect to the vertical axis, and ω_p is their angular speed about this axis.

Thus, when the student turns the wheel through an angle $\theta = 180°$, the student and platform acquire a vertical angular momentum of $2I_0\omega_0$. The total vertical angular momentum of the system is still being conserved at the initial value $I_0\omega_0$.

Consider now the angular momentum of the wheel alone. As the student turns the axis of the wheel he exerts a torque on it which lasts for the time Δt that it takes to reorient the shaft. The vertical component of the reaction to this "torque-impulse" acts on the student and accounts for the vertical angular momentum acquired by him and the platform.

11–8 Rotational Dynamics—A Review

The subject of the rotary motions of particles and rigid bodies is rather complicated, so much so in fact that a completely general treatment is beyond our scope here. It seems advisable to collect in one place all equations dealing with rotational dynamics and to comment on the conditions under which they can be used. We do this in Table 11–3.

Table 11–3 SUMMARY OF EQUATIONS FOR ROTARY MOTION

Eq. No.	Equation	Remarks
		I. Defining Equations
11–1	$\boldsymbol{\tau} = \mathbf{r} \times \mathbf{F}$	Torque on a particle about a point O, due to a resultant force $\mathbf{F}$ applied at $\mathbf{r}$
	$\boldsymbol{\tau}_{\text{ext}} = \Sigma\boldsymbol{\tau}_i = \Sigma(\mathbf{r}_i \times \mathbf{F}_i)$	Resultant external torque on a system of particles about a point O
11–3	$\mathbf{l} = \mathbf{r} \times \mathbf{p}$	Angular momentum of a particle about a point O
	$\mathbf{L} = \Sigma\mathbf{l}_i = \Sigma(\mathbf{r}_i \times \mathbf{p}_i)$	Resultant angular momentum of a system of particles about a point O
		II. General Relations
11–7	$\boldsymbol{\tau} = d\mathbf{l}/dt$	The law of motion for a single particle acted on by a torque. It is the rotational analog of $\mathbf{F} = d\mathbf{p}/dt$ (Eq. 8–10). Equation 11–7 holds only if $\boldsymbol{\tau}$ and $\mathbf{l}$ are measured with respect to any point O fixed in an inertial reference frame
11–9	$\boldsymbol{\tau}_{\text{ext}} = d\mathbf{L}/dt$	The law of motion for a system of particles acted on by a resultant external torque $\boldsymbol{\tau}_{\text{ext}}$. It is the rotational analog of $\mathbf{F} = d\mathbf{P}/dt$ (Eq. 8–15). Equation 11–9 holds only if $\boldsymbol{\tau}_{\text{ext}}$ and $\mathbf{L}$ are measured with respect to (*a*) any point O fixed in an inertial reference frame or (*b*) the center of mass of the system

Table 11-3   SUMMARY OF EQUATIONS FOR ROTARY MOTION (*Continued*)

Eq. No.	Equation	Remarks
	III. Special Case of a Rigid Body Rotating about an Axis Fixed in an Inertial Reference Frame	
11–17	$\tau = I\alpha$	*I* must refer to the fixed axis and τ must be the scalar component of τ_{ext} directed along this same axis. It is the rotational analog of $F = Ma$ for rectilinear motion
11–18	$L = I\omega$	*I* must refer to the fixed axis and *L* must be the scalar component of the total angular momentum along this axis. It is the rotational analog of $P = Mv$ for rectilinear motion

QUESTIONS

1. What are the dimensions of angular momentum? Can you find any significance in the fact that they are the same as those of energy multiplied by time?

2. Can the mass of a body be considered as concentrated at its center of mass for purposes of computing its rotational inertia?

3. About what axis would a uniform cube have its minimum rotational inertia?

4. If two circular disks of the same weight and thickness are made from metals having different densities, which disk, if either, will have the larger rotational inertia about its central axis?

5. The rotational inertia of a body of rather complicated shape is to be determined. The shape makes a mathematical calculation from $\int r^2 \, dm$ exceedingly difficult. Suggest ways in which the rotational inertia could be measured experimentally.

6. Five solids are shown in cross section (Fig. 11–14). The cross sections have equal heights and maximum widths. The axes of rotation are perpendicular to the sections through the points shown. The solids have equal masses. Which one has the largest rotational inertia about a perpendicular axis through the center of mass? Which the smallest?

7. In Fig. 11–15a a meter stick, half of which is wood—the other half steel—is pivoted at the

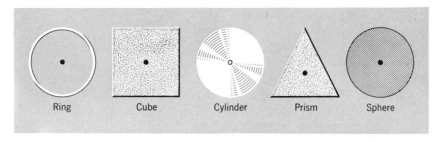

Ring   Cube   Cylinder   Prism   Sphere

FIGURE 11–14

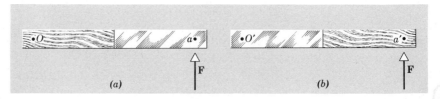

(a)   (b)

FIGURE 11–15

wooden end at O and a force is applied to the steel end at a. In Fig. 11–15b the stick is pivoted at the steel end at O' and the same force is applied at the wooden end at a'. Does one get the same angular acceleration in each case? Explain.

8. State Newton's three laws of motion in words suitable for rotating bodies.

9. Explain, in terms of angular momentum and rotational inertia, exactly how one "pumps up" a swing.

10. In Chapter 1 the melting of the polar icecaps was cited as a possible cause of the variation in the earth's time of rotation. Explain.

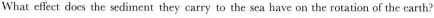

FIGURE 11–16

11. Many great rivers flow toward the equator. What effect does the sediment they carry to the sea have on the rotation of the earth?

12. A man turns on a rotating table with an angular speed ω. He is holding two equal masses at arm's length. Without moving his arms, he drops the two masses. What change, if any, is there in his angular speed? Is the angular momentum conserved? Explain.

13. In Example 7, if the string is released suddenly back to where the object can travel in a circle of radius r_1, will the object return to its original speed? What happens if one repeatedly pulls down on and suddenly releases the string? Explain the behavior in terms of work-energy and torque-angular momentum considerations.

14. A circular turntable rotates at constant angular velocity about a vertical axis. There is no friction and no driving torque. A circular pan rests on the turntable and rotates with it: see Fig. 11–16. The bottom of the pan is covered with a layer of ice of uniform thickness, which is, of course, also rotating with the pan. The ice melts but none of the water escapes from the pan. Is the angular velocity now greater than, the same as, or less than the original velocity? Give reasons for your answer.

PROBLEMS

1(2). Given a particle at $\mathbf{r} = \mathbf{i}x + \mathbf{j}y + \mathbf{k}z$ and given that a force acts on it given by $\mathbf{F} = \mathbf{i}F_x + \mathbf{j}F_y + \mathbf{k}F_z$, find the torque $\boldsymbol{\tau}(= \mathbf{r} \times \mathbf{F}.)$
Answer: $\mathbf{i}(yF_z - zF_y) + \mathbf{j}(zF_x - xF_z) + \mathbf{k}(xF_y - yF_x)$.

2(3). A 2.0-kg object moves in a plane with velocity components $v_x = 30$ meters/sec and $v_y = 60$ meters/sec as it passes through the point $(x,y) = (3.0,-4.0)$ meters. (a) What is its angular momentum relative to the origin at this moment? (b) What is its angular momentum relative to the point $(-2.0,-2.0)$ meters at this same moment?

3(3). A 3.0-kg particle is at $x = 3.0$ meters, $y = 8.0$ meters with a velocity of $\mathbf{v} = 5\mathbf{i} - 6\mathbf{j}$, meters/sec. It is acted on by a 7.0-newton force in the negative x-direction. (a) What is the angular momentum of the particle? (b) What torque acts on the particle? (c) At what rate is the angular momentum of the particle changing with time?
Answer: (a) $-170\mathbf{k}$, kg meter^2/sec. (b) $+56\mathbf{k}$, nt-meter. (c) $+56\mathbf{k}$ kg meter^2/sec^2.

4(3). A particle P with mass 2.0 kg has position $\mathbf{r}$ and velocity $\mathbf{v}$ as shown in Fig. 11–17. It is acted on by the force $\mathbf{F}$. All three vectors lie in a common plane. Presume that $r = 3.0$ meters, $v = 4.0$ meters/sec and $F = 2.0$ nt. Compute (a) the angular momentum of the particle and (b) the torque acting on the particle. What are the directions of these two vectors?

5(3). If we are given r, p, and θ, we can calculate the angular momentum of a particle from Eq. 11–4a. Sometimes, however, we are given the components (x,y,z) of $\mathbf{r}$ and (p_x,p_y,p_z) of $\mathbf{p}$ instead. (a) Show that the components of $\mathbf{l}$ along the x-, y-, and z-axes are then given by

$$l_x = yp_z - zp_y,$$
$$l_y = zp_x - xp_z,$$
$$l_z = xp_y - yp_x.$$

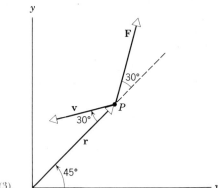

FIGURE 11-17   Problem 4(3).

(*b*) Show that if the particle moves only in the *x–y* plane the resultant angular momentum vector has only a *z*-component.

6(3). Show that the angular momentum about any point of a single particle moving with constant velocity remains constant throughout the motion.

7(3). (*a*) In Example 1, express **F** and **r** in terms of unit vectors and compute **τ**. Do the same in Example 3. (*b*) In Example 1, express **p** and **r** in terms of unit vectors and compute **l**.
Answer: (*a*) $\boldsymbol{\tau} = -\mathbf{k}mgb$; 2.6**k**, lb-ft. (*b*) $\mathbf{l} = -\mathbf{k}mgbt$.

8(4). Two particles, each of mass *m* and speed *v*, travel in opposite directions along parallel lines separated by a distance *d*. Show that the vector angular momentum of this system of particles is the same about any point taken as origin.

9(4). Three particles, each of mass *m*, are fastened to each other and to a rotational axis by three lengths of string each with length *l* as shown in Fig. 11–18. The combination rotates around the rotational axis with angular velocity ω in such a way that the particles remain in a straight line. (*a*) What is the angular momentum of the middle particle? (*b*) What is the total angular momentum of the three particles? Express your answers in terms of *m*, *l*, and ω.
Answer: (*a*) $4ml^2\omega$. (*b*) $14ml^2\omega$.

10(4). Starting from Newton's third law, prove that the resultant internal torque on a system of particles is zero.

11(5). Assume the earth to be a sphere of uniform density. (*a*) What is its rotational kinetic energy? Take the radius of the earth to be 6.4×10^3 km and the mass of the earth to be 6.0×10^{24} kg. (*b*) Suppose this energy could be harnessed for our use. For how long could the earth supply 1 kw of power to each of the 3.5×10^9 persons on earth?
Answer: (*a*) 2.6×10^{29} joules. (*b*) 2.4×10^9 yr.

12(5). The masses and coordinates of four particles are as follows: 50 gm, *x* = 2.0 cm, *y* = 2.0 cm; 25 gm, *x* = 0, *y* = 4.0 cm; 25 gm, *x* = −3.0 cm, *y* = −3.0 cm; 30 gm, *x* = −2.0 cm, *y* = 4.0 cm. What is the rotational inertia of this collection with respect to the *x*, *y*, and *z* axes?

13(5). An oxygen molecule has a total mass of 5.30×10^{-26} kg and a rotational inertia of 1.94×10^{-46} kg meter² about an axis through the center perpendicular to the line joining the two

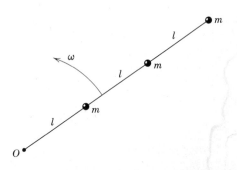

FIGURE 11-18   Problem 9(4).   O

atoms. Suppose that such a molecule in a gas has a mean speed of 500 meters/sec and that its rotational kinetic energy is two-thirds of its translational kinetic energy. Find its average angular velocity.

Answer: 6.75×10^{12} radians/sec.

14(5). What is the rotational kinetic energy of the system of particles described in Problem 9(4)?

15(5). Presume that the particles of Problem 9(4) are rigidly attached to each other by thin rigid rods rather than strings. Calculate the rotational inertia of the combination about O assuming that the mass of the rods is so small in comparison to the mass of the particles that their contribution to the rotational inertia may be neglected.

Answer: $14ml^2$.

16(5). Presume that the rods of Problem 15(5) each have mass M. What is the total rotational inertia of the system about O?

17(5). A thin rod of length l and mass m is suspended freely at its end. It is pulled aside and swung about a horizontal axis, passing through its lowest position with an angular speed ω. How high does its center of mass rise above its lowest position? Neglect friction and air resistance.

Answer: $l^2\omega^2/6g$.

18(5). A uniform spherical shell rotates about a vertical axis on frictionless bearings (Fig. 11-19). A light cord passes around the equator of the shell, over a pulley, and is attached to a small object that is otherwise free to fall under the influence of gravity. What is the speed of the object after it has fallen a distance h from rest?

19(5). (*a*) Prove that the rotational inertia of a thin rod of length l about an axis through its center perpendicular to its length is $I = \frac{1}{12}Ml^2$. (See Table 11-1.) (*b*) Use the parallel-axis theorem to show that $I = \frac{1}{3}Ml^2$ when the axis of rotation is through one end perpendicular to the length of the rod.

20(5). (*a*) Show that a solid cylinder of mass M and radius R is equivalent to a thin hoop of mass M and radius $R/\sqrt{2}$, for rotation about a central axis. (*b*) The radial distance from a given axis at which the mass of a body could be concentrated without altering the rotational inertia of the body about that axis is called the radius of gyration. Let k represent radius of gyration and show that

$$k = \sqrt{I/M}\,.$$

This gives the radius of the "equivalent hoop" in the general case.

21(5). A uniform disk of radius r rolls down a loop-the-loop (Fig. 11-20). What minimum height h will keep the disk on the track at the top of the loop? Assume that r is small compared to h and R.

Answer: $2.75R$.

22(5). A rigid body is made of three identical thin rods fastened together in the form of a letter H (Fig. 11-21). The body is free to rotate about a horizontal axis that passes through one of the legs of the H. The body is allowed to fall from rest from a position in which the plane of the H is horizontal. What is the angular speed of the body when the plane of the H is vertical?

23(6). If $R = 12$ cm, $M = 400$ gm and $m = 50$ gm in Fig. 11-10, find the speed of m after it has descended 50 cm starting from rest. Solve the problem using the results of Example 5 and by application of energy conservation principles.

Answer: 140 cm/sec.

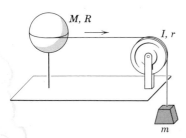

FIGURE 11-19   Problem 18(5).

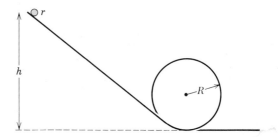

FIGURE 11-20   Problem 21(5).

FIGURE 11-21   Problem 22(5).

24(6). Assuming that the disk of Example 5 starts from rest, compute the work done by the applied torque on the disk in 2.1 sec. Also compute the increase in rotational kinetic energy of the disk.

25(6). The angular momentum of a flywheel having a rotational inertia of 0.125 kg meter² decreases from 3.0 to 2.0 kg meter²/sec in a period of 1.5 sec. (*a*) What is the average torque acting on the flywheel during this period? (*b*) Assuming a uniform angular acceleration, through how many revolutions will the flywheel have turned? (*c*) How much work was done? (*d*) What was the average power supplied by the flywheel?
Answer: (*a*) −0.67 nt-meter. (*b*) 4.8. (*c*) 20 joules. (*d*) 13 watts.

26(6). In an Atwood's machine (Fig. 5–7) one block has a mass of 500 gm and the other a mass of 460 gm. The pulley, which is mounted in horizontal frictionless bearings, has a radius of 5.0 cm. When released from rest the heavier block is observed to fall 75 cm in 5.0 sec. What is the rotational inertia of the pulley?

27(6). A uniform steel rod of length 1.20 meters and mass 6.40 kg has attached to each end a small ball of mass 1.06 kg. The rod is constrained to rotate in a horizontal plane about a vertical axis through its midpoint. At a certain instant it is observed to be making 39.0 rev/sec. Because of axle friction it comes to rest 32.0 sec later. Compute, assuming a constant frictional torque, (*a*) the angular acceleration, (*b*) the retarding torque exerted by axle friction, (*c*) the total work done by the axle friction, and (*d*) the number of revolutions executed during the 32.0 sec. (*e*) Suppose, however, that the frictional torque is known not to be constant. Which, if any, of the quantities (*a*), (*b*), (*c*), or (*d*) can still be computed without requiring any additional information? If such exists, give its value.
Answer: (*a*) −7.66 radians/sec². (*b*) −11.7 nt-meter. (*c*) 4.58 × 10⁴ joules. (*d*) 624. (*e*) "*c*" can be computed and has the same value.

28(6). A cylinder having a mass of 2.0 kg rotates about an axis through its center (see Fig. 11–22). If $F_1 = 6.0$ nt, $F_2 = 4.0$ nt and $F_3 = 2.0$ nt, and if $R_1 = 5.0$ cm and $R_2 = 12$ cm find the magnitude and direction of the angular acceleration of the cylinder.

29(6). A 6.0-lb block is put on a plane inclined 30° to the horizontal and is attached by a cord parallel to the plane over a pulley at the top to a hanging block weighing 18 lb. The pulley weighs 2.0 lb and has a radius of 0.33 ft. The coefficient of kinetic friction between block and plane is 0.10. Find the acceleration of the hanging block and the tension in the cord on each side of the pulley. Assume the pulley to be a uniform disk.
Answer: $a = 19$ ft/sec²; $T_{18} = 7.6$ lb; $T_6 = 7.0$ lb.

30(6). An automobile engine develops 100 hp when rotating at a speed of 1800 rev/min. What torque does it deliver?

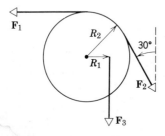

FIGURE 11-22   Problem 28(6).

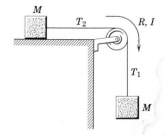

FIGURE 11-23   Problem 33(6).

31(6). Calculate (a) the torque, (b) the energy, and (c) the average power required to accelerate the earth from rest to its present angular speed about its axis in one day.

Answer: (a) 8.2×10^{28} nt-meter. (b) 2.6×10^{29} joules. (c) 3.0×10^{21} kw.

32(6). A 1000-kg car has four 10-kg wheels. What fraction of the total kinetic energy of the car is due to rotation of the wheels? Assume that the wheels have the same rotational inertia as disks of the same mass and size.

33(6). Two identical blocks, each of mass M, are connected by a light string over a frictionless pulley of radius R and rotational inertia I (Fig. 11–23). The string does not slip on the pulley, and it is not known whether or not there is friction between the plane and the sliding block. When this system is released, it is found that the pulley turns through an angle θ in time t and the acceleration of the blocks is constant. (a) What is the angular acceleration of the pulley? (b) What is the acceleration of the two blocks? (c) What is the tension in the upper and lower sections of the string. All answers are to be expressed in terms of M, I, R, θ, g and t.

Answer: (a) $2\theta/t^2$. (b) $2R\theta/t^2$. (c) $T_1 = M\,(g - 2R\theta/t^2)$; $T_2 = Mg - (2\theta/t^2)(MR - I/R)$.

34(6). A pulley having a rotational inertia of 1.0×10^4 gm-cm² and a radius of 10 cm is acted upon by a force, applied at its rim, that varies in time as

$$F = 0.50t + 0.30t^2$$

where F is in newtons and t in seconds. (a) What is the torque acting on the pulley at $t = 1.0$ sec and at 3.0 sec? (b) What is the magnitude of the linear acceleration of a point on the rim at $t = 2.0$ sec?

35(7). A wheel is rotating with an angular speed of 800 rev/min on a shaft whose rotational inertia is negligible. A second wheel, initially at rest and with twice the rotational inertia of the first, is suddenly coupled to the same shaft. (a) What is the angular speed of the resultant combination of the shaft and two wheels? (b) Account for any changes in rotational kinetic energy experienced by this system.

Answer: (a) 270 rev/min. (b) The system loses $\frac{2}{3}$ of its kinetic energy.

36(7). In Example 7 compare the kinetic energies of the object in two different orbits. Use the work-energy theorem to explain the difference quantitatively.

37(7). A man stands on a frictionless rotating platform that is rotating with a speed of 1.0 rev/sec; his arms are outstretched and he holds a weight in each hand. With his hands in this position the rotational inertia of the man, the weights and the platform is 6.0 kg meters². If by drawing in the weights the man decreases the rotational inertia to 2.0 kg meters², (a) what is the resulting angular speed of the platform? (b) By how much is the kinetic energy increased?

Answer: (a) 3.0 rev/sec. (b) By a factor of 3, to 240 joules.

38(7). With center and spokes of negligible mass, a certain bicycle wheel has a thin rim of radius 1.14 ft and weight 8.36 lb; it can turn on its axle with negligible friction. A man holds the wheel above his head with the axis vertical while he stands on a turntable free to rotate without friction; the wheel rotates clockwise, as seen from above, with an angular speed 57.7 radians/sec, and the turntable is initially at rest. The rotational inertia of wheel-plus-man-plus-turntable about the common axis of rotation is 1.54 slug ft². (a) The man's hand suddenly stops the rotation of the wheel (relative to the turntable). Determine the resulting angular velocity (magnitude and direction) of the system. (b) The experiment is repeated with noticeable friction introduced into the axle of the wheel, which, starting from the same initial angular speed (57.7 radians/sec) gradually comes to rest (relative to the turntable) while the man holds the wheel as described above. (The turntable is still free to rotate without friction.) Describe what happens to the system, giving as much quantitative information as the data permit.

39(7). A girl (mass M) stands on the edge of a frictionless merry-go-round (mass $10\,M$, radius R, rotational inertia I) that is not moving. She throws a rock (mass m) in a horizontal direction that is tangent to the outer edge of the merry-go-round. The speed of the rock, relative to the ground, is v. What are (a) the angular speed of the merry-go-round and (b) the linear speed of the girl after the rock is thrown?

Answer: (a) $mvR/(I + MR^2)$. (b) $vmR^2/(I + mR^2)$.

40(7). Two wheels, A and B, are arranged by a belt system as in Fig. 11–24. The radius of B is

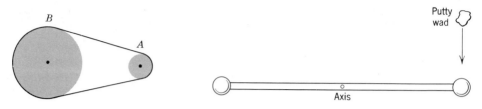

FIGURE 11-24   Problem 40(7).   **FIGURE 11-25**   Problem 43(7).

three times the radius of *A*. What would be the ratio of the rotational inertias I_A/I_B if (*a*) both wheels have the same angular momenta? (*b*) both wheels have the same rotational kinetic energy?

41(7). Two girls, each of mass *M*, sit on opposite ends of a thin rod with length *L* and mass *M* (the same as the mass of each girl). The rod is pivoted at its center and is free to rotate in a horizontal circle without friction. (*a*) What is the rotational inertia of the rod plus the girls about a vertical axis through the center of the rod? (*b*) What is the angular momentum of the system if it is rotating with an angular speed ω_0 in a clockwise direction as seen from above? What is the direction of the angular momentum? (*c*) While the system is rotating, the girls pull themselves toward the center of the rod until they are half as far from the center as before. What is the resulting angular speed in terms of ω_0? (*d*) What is the change in kinetic energy of the system due to the girls changing their positions? How is energy conserved during this motion?
Answer: (*a*) $(7/12)ML^2$. (*b*) $(7/12)ML^2\omega_0$; down. (*c*) $(14/5)\omega_0$. (*d*) $K_{\text{final}} = 2.8\,K_{\text{initial}}$.

42(7). In a playground there is a small merry-go-round of radius 4.0 ft and mass 12 slugs. The radius of gyration (see Problem 20(5)) is 3.0 ft. A child of mass 3.0 slugs runs at a speed of 10 ft/sec tangent to the rim of the merry-go-round when it is at rest and then jumps on. Neglect friction and find the angular velocity of the merry-go-round and child.

43(7). Two 2.0-kg masses are attached to the end of a thin rod of negligible mass 5.0 cm long. The rod is free to rotate without friction about a horizontal axis through its center. A 50-gm putty wad drops onto one of the masses with a speed of 3.0 meters/sec. and sticks to it. See Fig. 11-25. (*a*) What is the angular speed of the system just after the putty wad hits? (*b*) What is the ratio of the kinetic energy of the entire system after the collision to that of the putty wad just before? (*c*) Will the system subsequently rotate through a complete revolution? If not how far will it rotate?
Answer: (*a*) 1.5 radians/sec. (*b*) 0.012. (*c*) It will rotate through 190°.

44(7). Two skaters, each of mass 50 kg, approach each other along parallel paths separated by 3.0 meters. They have equal and opposite velocities of 10 meters/sec. The first skater carries a long light pole, 3.0 meters long, and the second skater grabs the end of it as he passes. (Assume frictionless ice.) (*a*) Describe quantitatively the motion of the skaters after they are connected by the pole. (*b*) By pulling on the pole, the skaters reduce their distance apart to 1.0 meter. What is their motion then? (*c*) Compare the kinetic energy of the system in parts (*a*) and (*b*). Where does the change come from?

45(7). A cockroach, mass *m*, runs counterclockwise around the rim of a lazy Susan (a circular dish mounted on a vertical axle) of radius *R* and rotational inertia *I* with frictionless bearings. The

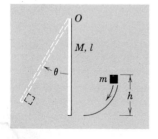

FIGURE 11-26   Problem 46(7).

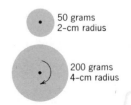

FIGURE 11-27   Problem 47(7).

cockroach's speed (relative to the earth) is v, whereas the lazy Susan turns clockwise with angular speed ω_0. The cockroach finds a bread crumb on the rim and of course, stops. (*a*) What is the angular speed of the lazy Susan after the cockroach stops? (*b*) Is energy conserved?

Answer: (*a*) $(I\omega_0 - mRv)/(I + mR^2)$ (*b*) No.

46(7). Particle m in Fig. 11–26 slides down the frictionless surface and collides with the uniform vertical rod, sticking to it. The rod pivots about O and rotates through the angle θ before coming to rest. Find θ in terms of the other parameters given in the figure.

47(7). Two uniform disks are mounted on two parallel, frictionless axes (Fig. 11–27). The larger disk is initially rotating about its axis with an angular speed ω_0 of 2000 rev/min, and the smaller disk is initially at rest. The axis of the smaller disk is then moved down so that the two disks touch each other. (*a*) What is the angular speed of the two disks when there is no further sliding at their point of contact? (*b*) What fraction of the initial energy is lost in this "collision." (Warning: Angular momentum is not conserved in this interaction. Can you see why?)

Answer: (*a*) 3200 rev/min. (*b*) 28%.

CHAPTER 12

Equilibrium of Rigid Bodies

12–1 The Equilibrium of a Rigid Body

The towers supporting a suspension bridge must be strong enough so that they do not collapse under the weight of the bridge and its traffic load; the landing gear of an aircraft must not collapse if the pilot makes a poor landing; the tines of a fork must not bend when we cut a tough steak. In all such problems the engineer is concerned that these presumed rigid structures do indeed remain rigid under the forces, and the associated torques, that act on them.

In such problems the engineer must ask two questions. (1) What forces and torques act on the presumed rigid body? (2) Considering its design and the materials used, will the body remain rigid under the action of these forces and torques? In this chapter we are concerned only with the first of these questions; the engineering student will deal at length with the second question in later courses.

We note that the presumed rigid bodies are in *mechanical equilibrium*. A rigid body is in mechanical equilibrium if, as viewed from an inertial reference frame, (1) the linear acceleration a_{cm} of its center of mass is zero and (2) its angular acceleration α about any axis fixed in this reference frame is zero.

This definition does not require the body to be at rest with respect to the observer but only to be unaccelerated. Its center of mass, for example, may be moving with constant velocity v_{cm} and the body may be rotating about a fixed axis with constant angular velocity ω. If the body is actually at rest (so that $v_{cm} = 0$ and $\omega = 0$), we often speak of *static equilibrium*. However, as we shall see, the restrictions imposed on the forces and torques are the same whether or not the equilibrium is static. Furthermore, we can transform any case of (nonstatic) equilibrium to one of static equilibrium by choosing an appropriate new reference frame.

The translational motion of a rigid body of mass M is governed by Eq. 8–8, or

$$\mathbf{F}_{\text{ext}} = M\mathbf{a}_{\text{cm}},$$

in which $\mathbf{F}_{\text{ext}}$ is the vector sum of all the external forces acting on the body. Because $\mathbf{a}_{\text{cm}}$ must be zero for equilibrium, the first condition of equilibrium (static or otherwise) is: *The vector sum of all the external forces acting on a body in equilibrium must be zero.*

We can write condition (1) as

$$\mathbf{F} = \mathbf{F}_1 + \mathbf{F}_2 + \cdots = 0, \tag{12–1}$$

in which we have dropped the subscript on $\mathbf{F}_{\text{ext}}$ for convenience. This vector equation leads to three scalar equations.

$$F_x = F_{1x} + F_{2x} + \cdots = 0,$$
$$F_y = F_{1y} + F_{2y} + \cdots = 0, \tag{12–2}$$
$$F_z = F_{1z} + F_{2z} + \cdots = 0,$$

which state that the sum of the components of the forces along each of any three mutually perpendicular directions is zero.

The second requirement for equilibrium is that $\alpha = 0$ for any axis. Since the angular acceleration of a rigid body is associated with torque—recall that $\tau = I\alpha$ for a fixed axis—we can state this second condition of equilibrium (static or otherwise) as: *The vector sum of all the external torques acting on a body in equilibrium must be zero.*

We can write condition (2) as

$$\boldsymbol{\tau} = \boldsymbol{\tau}_1 + \boldsymbol{\tau}_2 + \cdots = 0. \tag{12–3}$$

This vector equation leads to three scalar equations

$$\tau_x = \tau_{1x} + \tau_{2x} + \cdots = 0,$$
$$\tau_y = \tau_{1y} + \tau_{2y} + \cdots = 0, \tag{12–4}$$
$$\tau_z = \tau_{1z} + \tau_{2z} + \cdots = 0,$$

which state that, at equilibrium, the sum of the components of the torques acting on a body, along each of any three mutually perpendicular directions, is zero.

The resultant torque $\boldsymbol{\tau}$ in Eq. 12–3, which must be zero for mechanical equilibrium, is defined with respect to a particular origin O. The quantities τ_x, τ_y, and τ_z in Eq. 12–4 are the scalar components of $\boldsymbol{\tau}$ and refer to any set of three mutually perpendicular axes whose origin is at O, no matter how these axes are oriented in space. This follows because, if a vector is zero, its scalar components must be zero no matter how we orient the axes of the reference frame. You may wonder whether the choice of a specific origin is essential. The answer—as we shall show below— is that it is not, because (for a body in translational equilibrium), if $\tau = 0$ for any single origin O it is also zero for any other origin in the reference frame. The substance of this paragraph then is that condition 2 is satisfied for a body in translational equilibrium if we can show either that $(a)\,\tau = 0$ with respect to any one point (Eq. 12–3) or that (b) the torque components along any three mutually perpendicular axes are zero (Eq. 12–4).

Hence we have six independent conditions on our forces for a body to be in equilibrium. These conditions are the six algebraic relations of Eqs. 12–2 and 12–4. These six conditions are a condition on each of the six degrees of freedom of a rigid body, three translational and three rotational.

Often we deal with problems in which all the forces lie in a plane. Then we have only three conditions on our forces: The sum of their components must be zero for each of any two perpendicular directions in the plane, and the sum of their torques about any one axis perpendicular to the plane must be zero. These conditions correspond to the three degrees of freedom for motion in a plane, two of translation and one of rotation.

We shall limit ourselves henceforth mostly to planar problems in order to simplify the calculations. Also, as a matter of convenience, we shall consider only the case of static equilibrium, in which bodies are actually at rest.

12–2 Center of Gravity

One of the forces encountered in rigid-body motions is the force of gravity. Actually, for an extended body, this is not just one force but the resultant of a great many forces. Each particle in the body is acted on by a gravitational force. If the body of mass M is imagined to be divided into a large number of particles, say n, the gravitational force exerted by the earth on the ith particle of mass m_i is $m_i\mathbf{g}$. This force is directed down toward the earth. If the acceleration due to gravity $\mathbf{g}$ is the same everywhere in a region, we say that a uniform gravitational field exists there; that is, $\mathbf{g}$ has the same magnitude and direction everywhere in that region. For a rigid body in a uniform gravitational field, $\mathbf{g}$ must be the same for each particle in the body and the weight forces on the particles must be parallel to one another. If we assume that the earth's gravitational field is uniform, we can show that all the individual weight forces acting on a body can be replaced by a single force $M\mathbf{g}$ acting down at the center of mass of the body. This is equivalent to showing that the individual weight forces, acting downward, can be counteracted in their accelerating effects by a single force $\mathbf{F}$ $(= -M\mathbf{g})$ acting upward, provided this force $\mathbf{F}$ is applied at the center of mass of the body.

Figure 12–1 shows two typical particles or mass elements m_1 and m_2, selected from the n such elements into which the rigid body has been divided. An upward force $\mathbf{F}$ $(= -M\mathbf{g})$ is applied at a certain point O. It remains to show that the body is in

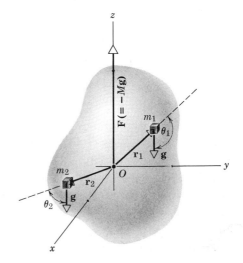

FIGURE 12–1 An irregular body is divided into n mass elements of which two typical elements m_1 and m_2 are shown. In the text we prove that the body can be held in translational and rotational equilibrium by a single force $\mathbf{F}$ $(= -M\mathbf{g})$ directed upward and applied at the center of mass of the body.

mechanical equilibrium if (and only if) point O is the center of mass. Condition 1 for equilibrium (Eq. 12–1) has already been satisfied by our choice of the magnitude and direction of **F**. That is,

$$\mathbf{F} + m_1\mathbf{g} + m_2\mathbf{g} + \cdots + m_n\mathbf{g} = 0,$$

or

$$\mathbf{F} = -(m_1 + m_2 + \cdots + m_n)\mathbf{g} = -M\mathbf{g},$$

which corresponds to our assumption.

It remains to prove that $\tau = 0$ for any single point in the body, such as O. This is the second condition for equilibrium. By choosing O as our origin we insure that the torque of **F** about this point is zero, because the moment arm of **F** is zero for this point. The torque about O due to the gravitational pull on the mass elements is

$$\tau = \mathbf{r}_1 \times m_1\mathbf{g} + \mathbf{r}_2 \times m_2\mathbf{g} + \cdots + \mathbf{r}_n \times m_n\mathbf{g}$$

which (because m_1, m_2, etc., are scalars) we can write as

$$\tau = m_1\mathbf{r}_1 \times \mathbf{g} + m_2\mathbf{r}_2 \times \mathbf{g} + \cdots + m_n\mathbf{r}_n \times \mathbf{g}.$$

Since **g** is the same in each term, we can factor it out, obtaining

$$\tau = (m_1\mathbf{r}_1 + m_2\mathbf{r}_2 + \cdots + m_n\mathbf{r}_n) \times \mathbf{g}$$

$$= \left(\sum_1^n m_i\mathbf{r}_i\right) \times \mathbf{g},$$

in which the sum is taken over all the mass elements that make up the body.

Now if point O is the center of mass of the body, the sum above is zero. This follows from the definition of the center of mass (see Eq. 8–3 and the discussion following it). We conclude then that if (and only if) point O is the center of mass, then $\tau = 0$ and the second condition for mechanical equilibrium is satisfied.

Thus the gravitational forces acting on the individual mass elements that make up a rigid body are equivalent in their translational and rotational effects to a single force equal to $M\mathbf{g}$, the total weight of the body, acting at the center of mass. We can obtain the same result if the body is continuous and divided into an infinite number of particles. You should be able to do this by the methods of integral calculus (see Section 8–1). The point of application of the equivalent resultant gravitational force is often called the *center of gravity*.

Because almost all problems in mechanics involve objects having dimensions small compared to the distances over which **g** changes appreciably, we can usually assume that **g** is uniform over the body. The center of mass and the center of gravity can then be taken as the same point. In fact, we can use this coincidence to determine experimentally the center of mass in irregularly shaped objects. For example, let us locate the center of mass of a thin plate of irregular shape, as in Fig. 12–2. We suspend the body by a cord from some point A on its edge. When the body is at rest, the center of gravity must lie directly under the point of support somewhere on the line Aa, for only then can the torque resulting from the cord and the weight add to zero. We next suspend the body from another point B on its edge. Again, the center of gravity must lie somewhere on Bb. The only point common to the lines Aa and Bb is O, the point of intersection, so that this point must be the

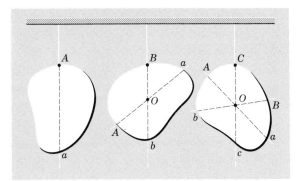

FIGURE 12-2 Since the center of mass O always hangs directly below the point of suspension, hanging a plate from two different points determines O.

center of gravity. If now we suspend the body from any other point on its edge, as C, the vertical line Cc will pass through O. Since we have assumed a uniform field, the center of gravity coincides with the center of mass, which is therefore located at O.

12–3   Examples of Equilibrium

In applying the conditions for equilibrium (zero resultant force and zero resultant torque about any axis), we can clarify and simplify the procedure in many ways.

First, we draw an imaginary boundary around the system under consideration. This assures that we see clearly just what body or system of bodies it is to which we are applying the laws of equilibrium. This process is called isolating the system.

Second, we draw vectors representing the magnitude, direction, and point of application of all external forces. An external force is one that acts from outside the boundary which was drawn earlier. Examples of external forces often encountered are gravitational forces and forces transmitted by strings, wires, rods, and beams which cross the boundary. A question sometimes arises about the direction of a force. In this case make an imaginary cut through the member transmitting the force at the point where it crosses the boundary. If the ends of this cut tend to pull apart, the force acts outward. If you are in doubt, choose the direction arbitrarily. A negative value for a force in the solution means that the force acts in the direction opposite to that assumed. Note that only external forces acting on the system need be considered; all internal forces cancel one another in pairs.

Third, we choose a convenient reference frame along whose axes we resolve the external forces before applying the first condition of equilibrium (Eq. 12–2). The object here is to simplify the calculations. The preferable reference frame is usually obvious.

Fourth, we choose a convenient reference frame along whose axes we resolve the external torques before applying the second condition of equilibrium (Eq. 12–4). The object again is to simplify caculations and we may use different reference frames in applying the two conditions for static equilibrium if this proves to be convenient. Suppose that an axis passes through the point at which two forces concur and is at right angles to the plane formed by these forces; these forces will automatically have no torque component along (or about) this axis. The torque components resulting from all external forces must be zero about any axis for equilibrium. Internal torques will cancel in pairs and need not be considered.

Example 1. A uniform steel meter bar, weighing 4.0 lb, rests on two scales at its ends. A 6.0-lb block is placed at the 25-cm mark. Find the reading on the scales.

Our system is the bar and block. The forces acting on the bar are **W** and **w**, the gravitational forces acting down at the centers of gravity of the bar and block, and **F₁** and **F₂**, the forces exerted upward on the bar at its ends by the scales. These are shown in Fig. 12–3. By Newton's third law, the force exerted by a scale on the bar is equal and opposite to that exerted by the bar on the scale. Therefore, to obtain the readings on the scales, we must determine the magnitudes of **F₁** and **F₂**.

For translational equilibrium (Eq. 12–1) the condition is

$$\mathbf{F}_1 + \mathbf{F}_2 + \mathbf{W} + \mathbf{w} = 0$$

All forces act vertically, so that if we choose the y-axis to be vertical, no other axes need be considered. Then we get the scalar equation

$$F_1 + F_2 - W - w = 0$$

For rotational equilibrium, the component of the resultant torque on the bar must be zero about any axis. We have seen that it is enough to show that the torque components are zero for any set of three mutually perpendicular axes. These components are certainly zero for any two perpendicular axes that lie in the plane of Fig. 12–3 (Why?). It remains to require that the resultant torque is zero about any one axis at right angles to the plane of the figure. Let us choose an axis through the center of the bar. Then, taking clockwise rotation as positive and counterclockwise rotation as negative, the condition for rotational equilibrium (Eq. 12–4) is

$$F_1\left(\frac{l}{2}\right) - F_2\left(\frac{l}{2}\right) + W(0) - w\left(\frac{l}{4}\right) = 0$$

or

$$F_1 - F_2 - w/2 = 0$$

Adding this equation to the one obtained for translational equilibrium and solving for F_1

$$F_1 = \frac{W}{2} + \frac{3w}{4} = 6.5 \text{ lb.}$$

Similarly,

$$F_2 = F_1 - w/2 = 3.5 \text{ lb.}$$

Notice, in the equation from which F_1 is obtained, if $w = 0$, $F_1 = W/2 = 2.0$ lb and consequently $F_2 = 2.0$ lb also as we would have expected. If we had chosen an axis through one end of the bar, we would have obtained the same result.

Example 2. (*a*) A 60-ft ladder weighing 100 lb rests against a wall at a point 48 ft above the ground. The center of gravity of the ladder is one-third the way up the ladder. A 160-lb man climbs halfway up the ladder. Assuming that the wall is frictionless, find the forces exerted by the system on the ground and the wall.

The forces acting on the ladder are shown in Fig. 12–4. **W** is the weight of the man standing

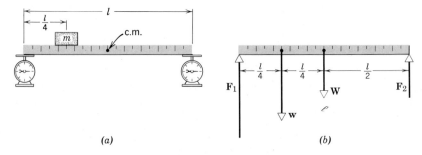

(*a*)   (*b*)

FIGURE 12–3   Example 1. (a) A weight is placed a quarter of the way from one end of a uniform steel bar resting on two spring scales. (b) The force diagram.

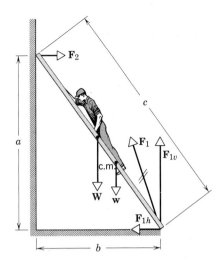

FIGURE 12–4 Example 2.

on the ladder, and **w** is the weight of the ladder itself. A force **F**$_1$ is exerted by the ground on the ladder. **F**$_{1v}$ is the vertical component and **F**$_{1h}$ is the horizontal component of this force (due to friction). This wall, being frictionless, can exert only a force normal to its surface, called **F**$_2$. We are given the following data:

$$W = 160 \text{ lb}, \qquad w = 100 \text{ lb},$$
$$a = 48 \text{ ft}, \qquad c = 60 \text{ ft}.$$

From the geometry we conclude that $b = 36$ ft. The line of action of **W** intersects the ground at a distance $b/2$ from the wall and the line of action of **w** intersects the ground at a distance $2b/3$ from the wall.

We choose the x-axis to be along the ground and the y-axis along the wall. Then, the conditions on the forces for translational equilibrium (Eq. 12–2) are

$$F_2 - F_{1h} = 0,$$
$$F_{1v} - W - w = 0.$$

For rotational equilibrium (Eq. 12–4) choose an axis through the point of contact with the ground and obtain

$$F_2(a) - W\left(\frac{b}{2}\right) - w\left(\frac{b}{3}\right) = 0.$$

Using the data given, we obtain

$$F_2(48 \text{ ft}) - (160 \text{ lb})(18 \text{ ft}) - (100 \text{ lb})(12 \text{ ft}) = 0,$$
$$F_2 = 85 \text{ lb},$$
$$F_{1h} = F_2 = 85 \text{ lb},$$
$$F_{1v} = 160 \text{ lb} + 100 \text{ lb} = 260 \text{ lb}.$$

By Newton's third law the forces exerted by the ground and the wall on the ladder are equal but opposite to the forces exerted by the ladder on the ground and the wall respectively. Therefore, the normal force on the wall is 85 lb, and the force on the ground has components of 260 lb down and 85 lb to the right.

(*b*) If the coefficient of static friction between the ground and the ladder is $\mu_s = 0.40$, how high up the ladder can the man go before it starts to slip?

Let x be the fraction of the total length of the ladder the man can climb before slipping begins. Then our equilibrium conditions are

$$F_2 - F_{1h} = 0,$$

$$F_{1v} - W - w = 0,$$

and

$$F_2 a - Wbx - w\left(\frac{b}{3}\right) = 0.$$

Now we obtain

$$F_2(48 \text{ ft}) = (160 \text{ lb})(36 \text{ ft})x + (100 \text{ lb})(12 \text{ ft}),$$

$$F_2 = (120x + 25) \text{ lb.}$$

Hence

$$F_{1h} = (120x + 25) \text{ lb,}$$

and, as before,

$$F_{1v} = 260 \text{ lb.}$$

The maximum force of static friction is given by

$$F_{1h} = \mu_s F_{1v} = (0.40)(260 \text{ lb}) = 104 \text{ lb.}$$

Therefore,

$$F_{1h} = (120x + 25) \text{ lb} = 104 \text{ lb}$$

and

$$x = \tfrac{79}{120},$$

so that the man can climb up the ladder

$$60x \text{ ft} = 39.5 \text{ ft}$$

before slipping begins.

In this example the ladder is treated as a one-dimensional object, with only one point of contact at the wall and ground. You might reflect on how this limits consideration of the less artificial case of two contact points at each end.

In the preceding examples we have been careful to limit the number of unknown forces to the number of independent equations relating the forces. When all the forces act in a plane, we can have only three independent equations of equilibrium, one for rotational equilibrium about any axis normal to the plane and two others for translational equilibrium in the plane. However, we often have more than three unknown forces. For example, in the ladder problem of Example 2a, if we drop the artificial assumption of a frictionless wall, we have four unknown scalar quantities, namely, the horizontal and vertical components of the force acting on the ladder at the wall and the horizontal and vertical components of the force acting on the ladder at the ground. Since we have only three scalar equations, these forces cannot be determined. For any value assigned to one unknown force, the other three forces can be determined. But if we have no basis for assigning any particular value to an unknown force, there are an infinite number of solutions possible mathematically. We must therefore find another independent relation between the unknown forces if we hope to solve the problem uniquely.

Another simple example of such underdetermined structures is the automobile. In this case we wish to determine the forces exerted by the ground on each of the four tires when the car is at rest on a horizontal surface. If we assume that these forces are normal to the ground, we have four unknown scalar quantities. All other forces, such as the weight of the car and passengers, act normal to the ground. Therefore we have only three independent equations giving the equilibrium conditions, one for translational equilibrium in the single direction of all the forces and two for rotational equilibrium about the two axes perpendicular to each other in a horizontal plane. Again the solution of the problem is indeterminate, mathematically. A four-legged table with all its legs in contact with the floor is a similar example.

Of course, since there is actually a unique solution to any real physical problem, we must find a physical basis for the additional independent relation between the forces that enable us to solve the problem. The difficulty is removed when we realize that structures are never perfectly rigid, as we have tacitly assumed throughout. Actually our structures will be somewhat deformed. For example, the automobile tires and the ground will be deformed, as will the ladder and wall. The laws of elasticity and the elastic properties of the structure determine the nature of the deformation and will provide the necessary additional relation between the four forces. A complete analysis therefore requires not only the laws of rigid body mechanics but also the laws of elasticity. In courses in civil and mechanical engineering, many such problems are encountered and analyzed in this way. We shall not consider the matter further here.

QUESTIONS

1. Are Eqs. 12–1 and 12–3 both necessary and sufficient conditions for mechanical equilibrium? For static equilibrium?

2. A wheel rotating at constant angular velocity ω about a fixed axis is in mechanical equilibrium because no net external force or torque acts on it. However, the particles that make up the wheel undergo a centripetal acceleration $\mathbf{a}$ directed toward the axis. Since $\mathbf{a} \neq 0$ how can the wheel be said to be in equilibrium?

3. Give several examples of a body which is not in equilibrium, even though the resultant of all the forces acting on it is zero.

4. If a body is not in translational equilibrium, will the torque about any point be zero if the torque about some particular point is zero?

5. Which is more likely to break in use, a hammock stretched tightly between two trees or one that sags quite a bit? Prove your answer.

6. A ladder is at rest with its upper end against a wall and the lower end on the ground. Is it more likely to slip when a man stands on it at the bottom or at the top? Explain.

7. In Example 2, if the wall were rough, would the empirical laws of friction supply us with the extra condition needed to determine the extra (vertical) force exerted by the wall on the ladder?

8. A picture hangs from a wall by two wires. What orientation should the wires have to be under minimum tension? Explain how equilibrium is possible with any number of orientations and tensions, even though the picture has a definite mass.

9. Show how to use a spring balance to weigh objects well beyond the maximum reading of the balance.

10. Do a center of mass and the center of gravity coincide for a building? For a lake? Under what conditions does the difference between the center of mass and the center of gravity of a body become significant?

11. If a rigid body is thrown into the air without spinning, it does not spin during its flight, provided air resistance can be neglected. What does this simple result imply about the location of the center of gravity?

12. A uniform block, in the shape of a rectangular parallelepiped of sides in the ratio $1:2:3$, lies on a horizontal surface. In which position, that is, on which of its three different faces, can it be said to be most stable, if any?

13. A virus particle in a rotating liquid-filled centrifuge tube is in uniform circular motion (that is, in accelerated motion) as viewed by an observer in the laboratory. An observer rotating with the centrifuge, however, would declare the particle to be unaccelerated. Explain how the virus particle can be in equilibrium for this second observer but not for the first.

14. In Chapter 5 we defined force in terms of acceleration from the relation **F** = m**a**. For a body in equilibrium, however, there are no accelerations. How, then, can we give meaning to the forces acting on such a body?

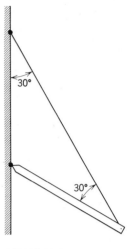

FIGURE 12–5

PROBLEMS

1(1). A 10-lb book remains at rest on a table. Identify the forces on it.

2(1). A 10-gm cork floats in a stream flowing west with constant velocity. Identify the forces on the cork and show that they add up to zero.

3(1). Prove that when only three forces act on a body in equilibrium, they must be coplanar and their lines of action must meet at a point or at infinity.

4(2). A nonuniform bar of weight W is suspended at rest in a horizontal position by two light cords as shown in Fig. 12–5; the angle one cord makes with the vertical is $\theta = 36.9°$; the other makes the angle $\phi = 53.1°$ with the vertical. If the length l of the bar is 20.0 ft, compute the distance x from the left-hand end to the center of gravity.

5(2). A circular section of radius r is cut out of a uniform disk of radius R, the center of the hole being $R/2$ from the center of the original disk. Locate the center of gravity of the resulting body. *Answer:* Along a line from the center of the hole through the center of the disk, beyond the latter point by a distance $Rr^2/2(R^2 - r^2)$.

6(3). A beam is carried by three men, one man at one end and the other two supporting the beam between them on a crosspiece so placed that the load is equally divided among the three men. Find where the crosspiece is placed. Neglect the mass of the crosspiece.

7(3). In Fig. 12–6 a man is trying to get his car out of the mud on the shoulder of a road. He ties one end of a rope tightly around the front bumper and the other end tightly around a telephone pole 60 ft away. He then pushes sideways on the rope at its midpoint with a force of 125 lb, displacing the center of the rope 1.0 ft from its previous position and the car almost moves. What force does the rope exert on the car? (The rope stretches somewhat under the tension.) *Answer:* 1900 lb.

8(3). Forces **F₁**, **F₂**, and **F₃** act on the structure of Fig. 12–7 as shown. We wish to put the structure in equilibrium by applying a force, at a point such as P, whose vector components are **F**$_h$ and **F**$_v$. We are given that $a = 2.0$ ft, $b = 3.0$ ft, $c = 1.0$ ft, $F_1 = 20$ lb, $F_2 = 10$ lb, and $F_3 = 5.0$ lb. Find (a) F_h, (b) F_v, and (c) d.

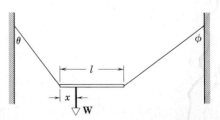

FIGURE 12–5   Problem 4(2).

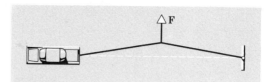

FIGURE 12–6   Problem 7(3).

9(3). What force F applied horizontally at the axle of the wheel is necessary to raise the wheel over an obstacle of height h? Take r as the radius of the wheel and W as its weight (Fig. 12–8).
Answer: $\dfrac{W\sqrt{h(2r-h)}}{r-h}$.

10(3). One end of a uniform beam weighing 50 lb and 3.0-ft long is attached to a wall with a hinge. The other end is supported by a wire (see Fig. 12–9). (*a*) Find the tension in the wire. (*b*) What are the horizontal and vertical components of the force at the hinge?

11(3). A trap door in a ceiling is 3.0 ft square, weighs 25 lb, and is hinged along one side with a catch at the opposite side. If the center of gravity of the door is 4.0 in. from the door's center and closer to the hinged end, what forces must (*a*) the catch and (*b*) the hinge sustain?
Answer: (*a*) 9.7 lb; up. (*b*) 15 lb; up.

12(3). A meter stick balances on a knife edge at the 50.0-cm mark. When two nickels are stacked over the 12.0-cm mark, the loaded stick is found to balance at the 45.5-cm mark. A nickel has a mass of 5.0 gm. What is the mass of the meter stick? Try this technique and check your answer experimentally.

13(3). A 160-lb man is walking across a level bridge and stops $\frac{3}{4}$ of the way from one end. The bridge is uniform and weighs 600 lb. What are the values of the vertical forces exerted on each end of the bridge by its supports?
Answer: 340 lb and 420 lb.

14(3). The system shown in Fig. 12–10 is in equilibrium. Find (*a*) the tension T in the cable and (*b*) the force exerted on the strut S by the pivot P. The strut weighs 100 lb.

15(3). The system in Fig. 12–11 is in equilibrium with the string in the center exactly horizontal. Find (*a*) the angle θ and (*b*) the tension in each string.
Answer: (*a*) 29°. (*b*) $T_1 = 49$ lb; $T_2 = 28$ lb; $T_3 = 57$ lb.

16(3). A door 7.0 ft high and 3.0 ft wide weighs 60 lb. A hinge 1.0 ft from the top and another 1.0 ft from the bottom each support half the door's weight. Assume that the center of gravity is at the geometrical center of the door and determine the horizontal and vertical force components exerted by each hinge on the door.

17(3). Two uniform beams are attached to a wall with hinges and then loosely bolted together as in Fig. 12–12. Find (*a*) the forces on each hinge and (*b*) the force exerted by the bolt on each beam.
Answer: (*a*) For the lower hinge, $F_{vert} = 210$ lb and $F_{hor} = 180$ lb; for the upper hinge, $F_{vert} = 60$ lb and $F_{hor} = 180$ lb. (*b*) $F_{vert} = 60$ lb; $F_{hor} = 180$ lb.

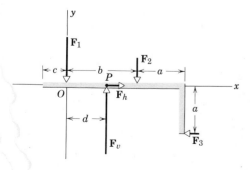

FIGURE 12–7   Problem 8(3).

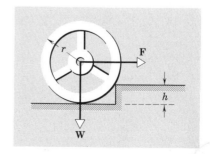

FIGURE 12–8   Problem 9(3).

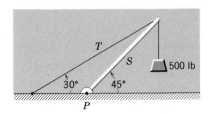

FIGURE 12-10   Problem 14(3).

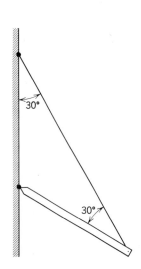

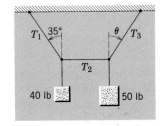

FIGURE 12-9   Problem 10(3).

FIGURE 12-11   Problem 15(3).

18(3). An automobile weighing 3000 lb has a wheel base of 120 in. Its center of gravity is located 70 in. behind the front axle. Determine (*a*) the force exerted on each of the front wheels (assumed the same), and (*b*) the force exerted on each of the back wheels (assumed the same) by the level ground.

19(3). A 100-lb plank, of length $l = 20$ ft, rests on the ground and on a frictionless roller (not shown) at the top of a wall of height $h = 10$ ft (see Fig. 12–13). The center of gravity of the plank is at its center. The plank remains in equilibrium for any value of $\theta \geqslant 70°$, but slips if $\theta < 70°$. (*a*) Draw a diagram showing all forces acting on the plank. (*b*) Find the coefficient of friction between the plank and the ground.
Answer: (*b*) 0.34.

20(3). A thin horizontal bar *AB* of negligible weight and length *l* is pinned to a vertical wall at *A* and supported at *B* by a thin wire *BC* that makes an angle θ with the horizontal. A weight *W* can be moved anywhere along the bar as defined by the distance *x* from the wall (Fig. 12–14). (*a*) Find the tensile force *T* in the thin wire as a function of *x*. (*b*) Find the horizontal and vertical components of the force exerted on the bar by the pin at *A*.

21(3). In the stepladder shown in Fig. 12–15, *AC* and *CE* are 8.0 ft long and hinged at *C*. *BD* is a tie rod 2.5 ft long, halfway up. A man weighing 192 lb climbs 6.0 ft along the ladder. Assuming that the floor is frictionless and neglecting the weight of the ladder, find (*a*) the tension in the tie

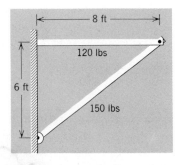

FIGURE 12-12   Problem 17(3).

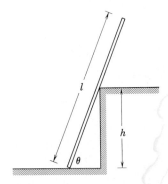

FIGURE 12-13   Problem 19(3).

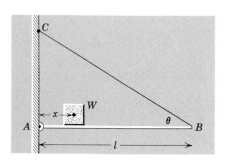

FIGURE 12-14   Problem 20(3).

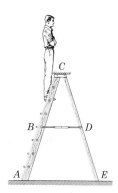

FIGURE 12-15   Problem 21(3).

rod and (*b*) the forces exerted on the ladder by the floor. (Hint: it will help to isolate parts of the ladder in applying the equilibrium conditions.)

Answer: (*a*) 47 lb. (*b*) $F_A = 120$ lb; $F_E = 72$ lb.

22(3). The system in Fig. 12–16 is in equilibrium, but it begins to slip if any additional mass is added to the 5.0-kg object. What is the coefficient of static friction between the 10-kg block and the plane on which it rests?

23(3). A certain skydiver free falls until she reaches a speed of 70 ft/sec. She then opens her parachute and experiences an initial upwards acceleration of 30 ft/sec². What will be her terminal speed (the speed with which she will fall with no acceleration) if the retarding force of her parachute is proportional to her speed of descent?

Answer: 36 ft/sec.

24(3). A balance is made up of a rigid rod free to rotate about a point not at the center of the rod. It is balanced by unequal weights placed in the pans at each end of the rod. When an unknown mass *m* is placed in the left-hand pan, it is balanced by a mass m_1 placed in the right-hand pan, and similarly when the mass *m* is placed in the right-hand pan, it is balanced by a mass m_2 in the left-hand pan. Show that

$$m = \sqrt{m_1 m_2}.$$

25(3). By means of a turnbuckle *G*, a tension force **T** is produced in bar *AB* of the square frame *ABCD* in Fig. 12–17. Determine the forces produced in the other bars. The diagonals *AC* and *BD* pass each other freely at *E*. Symmetry considerations can lead to considerable simplification in this and similar problems.

Answer: Bars *AD*, *BC*, and *DC* are in tension (force *T*); diagonals *AC* and *BD* are in compression (force $\sqrt{2}\ T$).

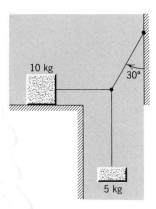

FIGURE 12-16   Problem 22(3).

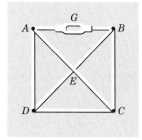

FIGURE 12-17   Problem 25(3).

26(3). A particle is acted upon by forces in newtons, $F_1 = 10i - 4j$ and $F_2 = 17i + 2j$. (*a*) Find a force F_3 that would keep it in equilibrium. (*b*) What direction does F_3 have relative to the *x*-axis?

27(3). A crate in the form of a 4.0-ft cube contains a piece of machinery whose design is such that the center of gravity of the crate and its contents is located one foot above its geometrical center. (*a*) If the crate is to be slid down a ramp without tipping over, what is the maximum angle which the ramp may make with the horizontal? (*b*) What is the maximum value for the coefficient of static friction between the crate and the ramp in this case such that the crate will just begin to slide?

Answer: (*a*) 34°. (*b*) 0.67.

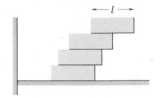

FIGURE 12–18   Problem 28(3).

28(3). Four bricks, each of length *l*, are put on top of one another (see Fig. 12–18) in such a way that part of each extends beyond the one beneath. Show that the largest equilibrium extensions are (*a*) top brick overhanging the one below by *l*/2, (*b*) second brick from top overhanging the one below by *l*/4, and (*c*) third brick from top overhanging the bottom one by *l*/6.

29(3). A cubical box is filled with sand and weighs 200 lb. It is desired to "roll" the box by pushing horizontally on one of the upper edges. (*a*) What minimum force is required? (*b*) What minimum coefficient of friction is required? (*c*) Is there a more efficient way to roll the box? If so, find the lowest possible force that would be required to be applied directly to the box.

Answer: (*a*) 100 lb. (*b*) 0.50. (*c*) Yes. Push 45° up. $F = 71$ lb.

CHAPTER 13

Oscillations

13–1 Oscillations

Any motion that repeats itself in equal intervals of time is called *periodic motion*. As we shall see, the displacement of a particle in periodic motion can always be expressed in terms of sines and cosines. Because the term harmonic is applied to expressions containing these functions, periodic motion is often called *harmonic motion*.

If a particle in periodic motion moves back and forth over the same path, we call the motion *oscillatory* or *vibratory*. The world is full of oscillatory motions. Some examples are the oscillations of the balance wheel of a watch, a violin string, a mass attached to a spring, atoms in molecules or in a solid lattice, and air molecules as a sound wave passes by.

Many oscillating bodies do not move back and forth between precisely fixed limits because frictional forces dissipate the energy of motion. Thus a violin string soon stops vibrating and a pendulum stops swinging. We call such motions *damped* harmonic motions. Although we cannot eliminate friction from the periodic motions of gross objects, we can often cancel out its damping effect by feeding energy into the oscillating system so as to compensate for the energy dissipated by friction. The main spring of a watch and the falling weight in a pendulum clock supply external energy in this way, so that the oscillating system, that is, the balance wheel or the pendulum, moves as if it were undamped.

Not only mechanical systems can oscillate. Radio waves, microwaves, and visible light are oscillating magnetic and electric field vectors. Thus a tuned circuit in a radio and a closed metal cavity in which microwave energy is introduced can oscillate electromagnetically. The analogy is close, being based on the fact that mechanical and electromagnetic oscillations are described by the same basic mathematical equations. We will make the most of this analogy in later chapters.

The *period* T of a harmonic motion is the time required to complete one round trip of the motion, that is, one complete oscillation or *cycle*. The *frequency* of the motion ν is the number of oscillations (or cycles) per unit of time. The frequency is therefore the reciprocal of the period, or

$$\nu = 1/T. \qquad (13\text{--}1)$$

The mks unit of frequency is the cycle per second, $(\sec^{-1})$. The position at which no net force acts on the oscillating particle is called its *equilibrium* position. The *displacement* (linear or angular) is the distance (linear or angular) of the oscillating particle from its equilibrium position at any instant.

Let us focus attention on a particle oscillating back and forth along a straight line between fixed limits. Its displacement **x** changes periodically in both magnitude and direction. Its velocity **v** and acceleration **a** also vary periodically in magnitude and direction and, in view of the relation **F** = m**a,** so does the force **F** acting on the particle.

In terms of energy, we can say that a particle undergoing harmonic motion passes back and forth through a point (its equilibrium position) at which its potential energy is a minimum. A swinging pendulum is a good example, its potential energy being a minimum at the bottom of the swing, that is, at the equilibrium position. Figure 13–1a shows a particle oscillating between the limits x_1 and x_2, O being the equilibrium position. Figure 13–1b shows the corresponding potential

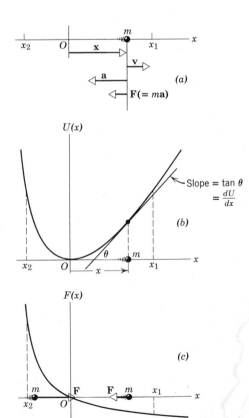

FIGURE 13–1 (a) A particle of mass m oscillates between points x_1 and x_2, O being the equilibrium position. (b) The potential energy of the particle as a function of position. The force acting on the particle at position x is given by $F = -dU/dx$. (c) The force acting on the particle as a function of position x; note that the force is directed toward the equilibrium position.

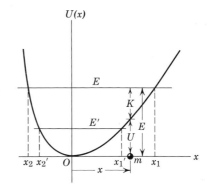

FIGURE 13–2 The mechanical energy E for the motion of Fig. 13–1 is shown. If the mechanical energy of the oscillating particle is reduced to E', the limits of oscillation are reduced to x_1' and x_2' respectively.

energy curve, which has a minimum value at that position. The force acting on the particle at any position is derivable from the potential energy function; it is given by Eq. 7–7,

$$F = -dU/dx, \qquad (7\text{–}7)$$

and is displayed in Fig. 13–1c. The force is zero at the equilibrium position O, points to the right (that is, has a positive value) when the particle is to the left of O, and points to the left (that is, has a negative value) when the particle is to the right of O. The force is a *restoring force* because it always acts to accelerate the particle in the direction of its equilibrium position. Hence in harmonic motion the position of equilibrium is always one of stable equilibrium.

The mechanical energy E for an oscillating particle is the sum of its kinetic energy and its potential energy, or

$$E = K + U \qquad (13\text{–}2)$$

in which E remains constant if no nonconservative forces, such as the force of friction, are acting. Figure 13–2 shows E for the motion of Fig. 13–1. Note how Eq. 13–2 is satisfied for the particle in the typical position shown. The particle cannot move outside the limits x_1 and x_2 because, in these regions, U exceeds E. This, as Eq. 13–2 shows, would require the kinetic energy to be negative, an impossibility.

For a given environment, that is, for a given function $U(x)$, an oscillating particle can have various total energies, depending on how it is set into motion initially. Thus the total energy may be E', rather than E, in which case the limits of oscillation would be x_1' and x_2', as Fig. 13–2 shows, rather than x_1 and x_2.

13–2 The Simple Harmonic Oscillator

Let us consider an oscillating particle (Fig. 13–3a) moving back and forth about an equilibrium position through a potential that varies as

$$U(x) = \tfrac{1}{2}kx^2 \qquad (13\text{–}3)$$

in which k is a constant; see Fig. 13–3b. The force acting on the particle is given by Eq. 7–7, or

$$F(x) = -dU/dx = -d(\tfrac{1}{2}kx^2)/dx = -kx; \qquad (13\text{–}4)$$

see Fig. 13–3c. Such an oscillating particle is called a *simple harmonic oscillator* and its motion is called *simple harmonic motion*. In such motion, as Eq. 13–3 shows, the potential energy curve varies as the square of the displacement, and, as Eq. 13–4

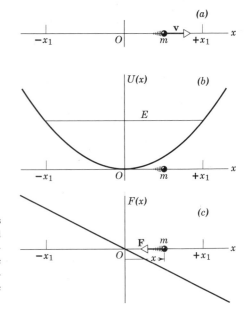

FIGURE 13-3   (*a*) A particle of mass *m* oscillates with simple harmonic motion between points $+x_1$ and $-x_1$, *O* being the equilibrium position. (*b*) The potential energy and the mechanical energy *E*. (*c*) The force acting on the particle. Compare this figure carefully with Fig. 13–1, which illustrates the general case of harmonic motion.

shows, the force acting on the particle is proportional to the displacement but is opposite to it in direction. In simple harmonic motion the limits of oscillation are equally spaced about the equilibrium position. This is not true for the more general motion of Fig. 13–1 which, although harmonic, is not simple harmonic. The magnitude of the maximum displacement, that is, the quantity x_1 in Fig. 13–3, always taken as positive, is called the *amplitude* of the simple harmonic motion.

You might have recognized Eq. 13–3 $[U(x) = \frac{1}{2}kx^2]$ as the expression for the potential energy of an "ideal" spring, compressed or extended by a distance *x*; see Section 7–4. In this same section an ideal spring was defined as one in which the force exerted by the stretched or compressed spring is given by $F(x) = -kx$ (see Eq. 13–4), *k* being called the *force constant*.

Hence, a body of mass *m* attached to an ideal spring of force constant *k* and free to move over a frictionless horizontal surface is an example of a simple harmonic oscillator (see Fig. 13–4). Note that there is a position (the equilibrium position;

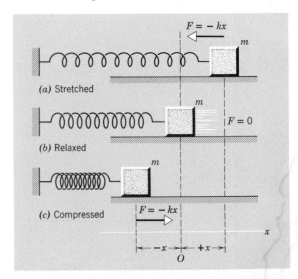

FIGURE 13-4   A simple harmonic oscillator. The force exerted by the spring is shown in each case. The block slides on a frictionless horizontal table.

see Fig. 13–4*b*) in which the spring exerts no force on the body. If the body is displaced to the right (as in Fig. 13–4*a*), the force exerted by the spring on the body points to the left and is given by $F = -kx$. If the body is displaced to the left (as in Fig. 13–4*c*), the force points to the right and is also given by $F = -kx$. In each case the force is a restoring force. The motion of the oscillating mass is simple harmonic motion.

Let us apply Newton's second law, $F = ma$, to the motion of Fig. 13–4. For F we substitute $-kx$ (from Eq. 13–4) and for the acceleration a we put in d^2x/dt^2 ($= dv/dt$). This gives us

$$-kx = m\frac{d^2x}{dt^2}$$

or

$$\frac{d^2x}{dt^2} + \frac{k}{m}x = 0. \tag{13–5}$$

This equation involves derivatives and is, therefore, called a *differential equation.* To solve this equation means to determine how the displacement x of the particle must depend on the time t in order that the equation be satisfied. When we know how x depends on time, we know the motion of the particle; thus, Eq. 13–5 is called the *equation of motion* of a simple harmonic oscillator. We shall solve this equation and describe the motion in detail in the next section.

The simple harmonic oscillator problem is important for two reasons. First, most problems involving mechanical vibrations reduce to that of the simple harmonic oscillator at small amplitudes of vibration, or to a combination of such vibrations. This is equivalent to saying that if we consider a small enough portion of the restoring force curve of Fig. 13–1*c* (around the origin), it becomes arbitrarily close to a straight line which, as Fig. 13–3*c* shows, is characteristic of simple harmonic motion. Or, in other words, the potential energy curve of Fig. 13–1*b* for general oscillatory motion reduces to that of Fig. 13–3*b* for simple harmonic oscillation when the amplitude of vibration is made sufficiently small about the equilibrium position *O*.

Second, as we have indicated, equations like Eq. 13–5 turn up in many physical problems in acoustics, in optics, in mechanics, in electrical circuits, and even in atomic physics. The simple harmonic oscillator exhibits features common to many physical systems.

Equation 13–4 ($F = -kx$) is an empirical relation known as *Hooke's law*. It is a special case of a more general relation, dealing with the deformation of elastic bodies, discovered by Robert Hooke (1635–1703). It is obeyed by springs and other elastic bodies provided the deformation is not too great. If the solid is deformed beyond a certain point, called its *elastic limit,* it will not return to its original shape when the applied force is removed (Fig. 13–5). It turns out that Hooke's law holds almost up to the elastic limit for many common materials. The range of applied forces over which Hooke's law is valid is called the "proportional region." Beyond the elastic limit, the force can no longer be specified by a potential energy function, because the force then depends on many factors including the speed of deformation and the previous history of the solid.

Notice that the restoring force and potential energy function of the simple harmonic oscillator are the same as that of a solid deformed in one dimension in the proportional region. If the deformed solid is released, it will vibrate, just as the simple harmonic oscillator does. Therefore, as long as the amplitude of the vibra-

FIGURE 13-5 Typical graph of applied force F versus resulting elongation of an aluminum bar under tension. The sample was a foot long and a square inch in cross section. Notice that we may write $F = kx$ only for the portion Oa, since beyond this point the slope is no longer constant but varies in a complicated way with x. At some point such as b (the *elastic limit*) the sample does not return to its original length when the applied force is removed. Between b and b' the elongation increases, even though the force is held constant; the material flows like a viscous fluid. At c, the sample can be stretched no farther; any increase in elongation results in the sample's breaking in two. The applied force is equal in magnitude to the restoring force so that no minus sign appears in the relation $F = kx$.

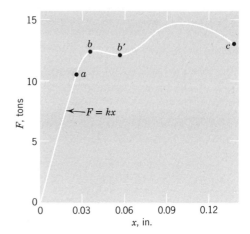

tion is small enough, that is, as long as the deformation remains in the proportional region, mechanical vibrations behave exactly like simple harmonic oscillators. It is easy to generalize this discussion to show that any problem involving mechanical vibrations of small amplitude in three dimensions reduces to a combination of simple harmonic oscillators.

The vibrating string or membrane, sound vibrations, the vibrations of atoms in solids, and electrical or acoustical oscillations in a cavity can all be described in a form which is identical mathematically to a system of harmonic oscillators. The analogy enables us to solve problems in one area by using the techniques developed in other areas.

13-3   Simple Harmonic Motion

Let us now solve the equation of motion of the simple harmonic oscillator,

$$\frac{d^2x}{dt^2} + \frac{k}{m}x = 0. \tag{13-6}$$

Recall that any system of mass m upon which a force $F = -kx$ acts will be governed by this equation. In the case of a spring, the proportionality constant k is the force constant of the spring determined by the stiffness of the spring. In other oscillating systems the proportionality constant k may be related to other physical features of the system, as we shall see later. We can use the oscillating spring as our prototype.

Equation 13-6 is a differential equation. It gives a relation between a function of the time $x(t)$ and its second time derivative d^2x/dt^2. To find the position of the particle as a function of the time, we must find a function $x(t)$ which satisfies this relation.

We can rewrite Eq. 13-6 as

$$\frac{d^2x}{dt^2} = -\frac{k}{m}x. \tag{13-7}$$

Equation 13-7 then requires that $x(t)$ be some function whose second derivative is the negative of the function itself, except for a constant factor k/m. We know from the calculus, however, that the sine function or the cosine function has this property. For example,

$$\frac{d}{dt}\cos t = -\sin t \quad \text{and} \quad \frac{d^2}{dt^2}\cos t = -\frac{d}{dt}\sin t = -\cos t.$$

This property is not affected if we multiply the cosine function by a constant A. We can allow for the fact that the sine function will do as well, and for the fact that Eq. 13–7 contains a constant factor, by writing as a tentative solution of Eq. 13–7,

$$x = A\cos(\omega t + \delta). \tag{13–8}$$

Here since

$$\cos(\omega t + \delta) = \cos\delta\cos\omega t - \sin\delta\sin\omega t = a\cos\omega t + b\sin\omega t,$$

the constant δ allows for any combination of sine and cosine solutions. Hence, with the (as yet) unknown constants A, ω, and δ, we have written as general a solution to Eq. 13–7 as we can. In order to determine these constants such that Eq. 13–8 is actually the solution of Eq. 13–7, we differentiate Eq. 13–8 twice with respect to the time. We have

$$\frac{dx}{dt} = -\omega A\sin(\omega t + \delta)$$

and

$$\frac{d^2x}{dt^2} = -\omega^2 A\cos(\omega t + \delta).$$

Putting this into Eq. 13–7, we obtain

$$-\omega^2 A\cos(\omega t + \delta) = -\frac{k}{m}A\cos(\omega t + \delta).$$

Therefore, if we choose the constant ω such that

$$\omega^2 = \frac{k}{m}, \tag{13–9}$$

then

$$x = A\cos(\omega t + \delta)$$

is in fact a solution of the equation of motion of a simple harmonic oscillator.

The constants A and δ are still undetermined and, therefore, still completely arbitrary. This means that any choice of A and δ whatsoever will satisfy Eq. 13–7, so that a large variety of motions is possible for the oscillator. Actually, this is characteristic of a differential equation of motion, for such an equation does not describe just one single motion but a group or family of possible motions which have some features in common but differ in other ways. In this case ω is common to all the allowed motions, but A and δ may differ among them. We shall see later that A and δ are determined for a particular harmonic motion by how the motion starts.

Let us find the physical significance of the constant ω. If the time t in Eq. 13–8 is increased by $2\pi/\omega$, the function becomes

$$x = A\cos[\omega(t + 2\pi/\omega) + \delta],$$
$$= A\cos(\omega t + 2\pi + \delta),$$
$$= A\cos(\omega t + \delta).$$

That is, the function merely repeats itself after a time $2\pi/\omega$. Therefore, $2\pi/\omega$ is the period of the motion T. Since $\omega^2 = k/m$, we have

$$T = \frac{2\pi}{\omega} = 2\pi \sqrt{\frac{m}{k}}. \tag{13–10}$$

Hence, all motions given by Eq. 13–7 have the same period of oscillation, and this is determined only by the mass m of the vibrating particle and the force constant k. The *frequency* ν of the oscillator is the number of complete vibrations per unit time and is given by

$$\nu = \frac{1}{T} = \frac{\omega}{2\pi} = \frac{1}{2\pi} \sqrt{\frac{k}{m}}. \tag{13–11}$$

Hence,

$$\omega = 2\pi\nu = \frac{2\pi}{T}. \tag{13–12}$$

The quantity ω is called the *angular frequency;* it differs from the frequency ν by a factor 2π. It has the dimension of reciprocal time (the same as angular speed) and its unit is the radian/sec. In Section 13–6 we shall give a geometric meaning to this angular frequency.

The constant A has a simple physical meaning. The cosine function takes on values from -1 to 1. The displacement x from the central equilibrium position $x = 0$, therefore, has a maximum value of A. Hence, A ($= x_{max}$) is the amplitude of the motion. Since A is not fixed by our differential equation, motions of various amplitudes are possible, but all have the same frequency and period. *The frequency of a simple harmonic motion is independent of the amplitude of the motion.*

The quantity $(\omega t + \delta)$ is called the *phase* of the motion. The constant δ is called the *phase constant.* Two motions may have the same amplitude and frequency but differ in phase. If $\delta = -\pi/2$, for example,

$$x = A \cos (\omega t + \delta) = A \cos (\omega t - 90°)$$
$$= A \sin \omega t$$

so that the displacement is zero at the time $t = 0$. When $\delta = 0$, the displacement $x = A \cos \omega t$ is a maximum at the time $t = 0$. Other initial conditions correspond to other phase constants.

The amplitude A and the phase constant δ of the oscillation are determined by the initial position and speed of the particle. These two initial conditions will specify A and δ. Of course, δ may be increased by any integral multiple of 2π (or $360°$) without changing x. Once the motion has started the particle will continue to oscillate with a constant amplitude and phase constant at a fixed frequency, unless other forces disturb the system.

In Fig. 13–6, we plot the displacement x versus the time t for several simple harmonic motions. Three comparisons are made. In Fig. 13–6a, I and II have the same amplitude and frequency but differ in phase by $\delta = \pi/4$ or $45°$. In Fig. 13–6b, I and III have the same frequency and phase constant but differ in amplitude by a factor of 2. In Fig. 13–6c, I and IV have the same amplitude and phase constant but differ in frequency by a factor $\frac{1}{2}$ or in period by a factor 2. You should study these curves carefully to become familiar with the terminology used in simple harmonic motion.

Another distinctive feature of simple harmonic motion is the relation between the displacement, the velocity, and the acceleration of the oscillating particle. Let us compare these quantities for curve I of Fig. 13–6, which is typical. In Fig. 13–7

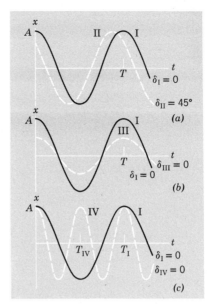

FIGURE 13–6 Several solutions of the simple harmonic oscillator equation. (*a*) Both solutions have the same amplitude and period but differ in phase by 45°. (*b*) Both have the same period and phase constant but differ in amplitude by a factor of 2. (*c*) Both have the same phase constant and amplitude but differ in frequency by a factor of 2.

we plot separately the displacement x versus the time t, the velocity $v = dx/dt$ versus the time t, and the acceleration $a = d^2x/dt^2$ versus the time t. The equations of these curves are

$$x = A \cos (\omega t + \delta),$$

$$v = \frac{dx}{dt} = -\omega A \sin (\omega t + \delta), \qquad (13\text{–}13)$$

$$a = \frac{dv}{dt} = -\omega^2 A \cos (\omega t + \delta).$$

For the case plotted we have taken $\delta = 0$. The units and scale of displacement, velocity, and acceleration are omitted for simplicity of comparison. Notice that the

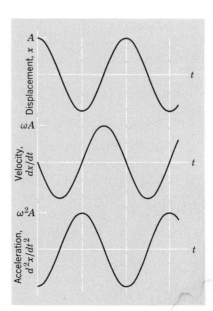

FIGURE 13–7 The relations between displacement, velocity, and acceleration in simple harmonic motion. The phase constant δ is zero in this particular case since the displacement is maximum at $t = 0$; see Eq. 13–8.

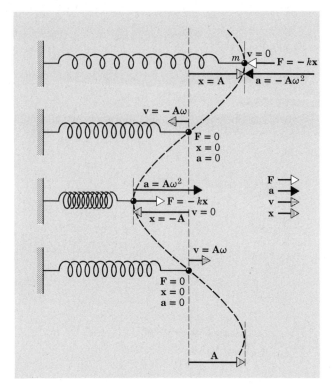

FIGURE 13-8   The force acting on, and the acceleration, velocity and displacement of a mass m undergoing simple harmonic motion. Compare carefully with Fig. 13-7.

maximum displacement is A, the maximum speed is ωA, and the maximum acceleration is $\omega^2 A$.

When the displacement is a maximum in either direction, the speed is zero because the velocity must now change its direction. The acceleration at this instant, like the restoring force, has a maximum value but is directed opposite to the displacement. When the displacement is zero, the speed of the particle is a maximum and the acceleration is zero, corresponding to a zero restoring force. The speed increases as the particle moves toward the equilibrium position and then decreases as it moves out to the maximum displacement, just as for a pendulum bob.

In Fig. 13-8 we show the instantaneous values of **x, v,** and **a** at four instants in the motion of a particle oscillating at the end of a spring.

13-4   Energy Considerations in Simple Harmonic Motion

Equation 13-2 tells us that for harmonic motion, including simple harmonic motion, in which no dissipative forces are present the mechanical energy $E\ (= K + U)$ is conserved. We can now study this in more detail for the special case of simple harmonic motion, for which the displacement is given by

$$x = A \cos (\omega t + \delta). \qquad (13\text{-}8)$$

The potential energy U at any instant is given by

$$U = \tfrac{1}{2}kx^2$$
$$= \tfrac{1}{2}kA^2 \cos^2 (\omega t + \delta) \qquad (13\text{-}14)$$

The potential energy has a maximum value of $\frac{1}{2}kA^2$. During the motion the potential energy varies between zero and this maximum value, as the curves in Fig. 13–9a and 13–9b show.

The kinetic energy K at any instant is $\frac{1}{2}mv^2$. Using the relations

$$v = dx/dt = -\omega A \sin(\omega t + \delta)$$

and

$$\omega^2 = k/m,$$

we obtain

$$K = \tfrac{1}{2}mv^2,$$

$$= \tfrac{1}{2}m\omega^2 A^2 \sin^2(\omega t + \delta),$$

$$= \tfrac{1}{2}kA^2 \sin^2(\omega t + \delta). \tag{13–15}$$

The kinetic energy, therefore, has a maximum value of $\frac{1}{2}kA^2$ or $\frac{1}{2}m(\omega A)^2$, in agreement with the maximum speed ωA noted earlier. During the motion the kinetic energy varies between zero and this maximum value, as shown by the curves in Fig. 13–9a and 13–9b.

The mechanical energy is just the sum of the kinetic and the potential energy. Using Eqs. 13–14 and 13–15 we obtain

$$E = K + U = \tfrac{1}{2}kA^2 \sin^2(\omega t + \delta) + \tfrac{1}{2}kA^2 \cos^2(\omega t + \delta) = \tfrac{1}{2}kA^2. \tag{13–16}$$

We see that the mechanical energy is constant, as we expect, and that it has the value $\frac{1}{2}kA^2$. At the maximum displacement the kinetic energy is zero, but the potential energy has the value $\frac{1}{2}kA^2$. At the equilibrium position the potential energy is zero, but the kinetic energy has the value $\frac{1}{2}kA^2$. At other positions the kinetic and potential energies each contribute energy whose sum is always $\frac{1}{2}kA^2$. This constant total

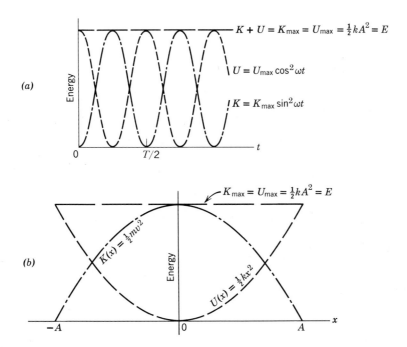

FIGURE 13–9 Energies of a simple harmonic oscillator. (*a*) Potential energy (– – – – –), kinetic energy (—– ·– —–), and mechanical energy (—— ——) plotted as a function of time. (*b*) Potential, kinetic, and mechanical energy plotted as a function of displacement from the equilibrium position. Compare with Fig. 7–4.

energy is shown in Fig. 13–9a and 13–9b. The total energy of a particle executing simple harmonic motion is proportional to the square of the amplitude of the motion. It is clear from Fig. 13–9a that the average kinetic energy for the motion during one period is exactly equal to the average potential energy and that each of these average quantities is $\frac{1}{4}kA^2$.

Equation 13–16 can be written quite generally as

$$K + U = \tfrac{1}{2}mv^2 + \tfrac{1}{2}kx^2 = \tfrac{1}{2}kA^2. \tag{13–17}$$

From this relation we obtain $v^2 = (k/m)(A^2 - x^2)$ or

$$v = \frac{dx}{dt} = \pm\sqrt{\frac{k}{m}(A^2 - x^2)}. \tag{13–18}$$

This relation shows clearly that the speed is a maximum at the equilibrium position $x = 0$ and zero at the maximum displacement $x = A$. In fact, we can start from the conservation of energy principle, Eq. 13–17 (in which $\frac{1}{2}kA^2 = E$), and by integration of Eq. 13–18 obtain the displacement as a function of time. The result is identical with Eq. 13–8 which we deduced from the differential equation of the motion, Eq. 13–6. (See Problem 22.)

Example 1. The horizontal spring of Fig. 13–4 is found to be stretched 3.0 in. from its equilibrium position when a force of 0.75 lb acts on it. Then a 1.5-lb body is attached to the end of the spring and is pulled 4.0 in. along a horizontal frictionless table from the equilibrium position. The body is then released and executes simple harmonic motion.

(a) What is the force constant of the spring?

A force of 0.75 lb on the spring produces a displacement of 0.25 ft. Hence,

$$k = F/x = 0.75 \text{ lb}/0.25 \text{ ft} = 3.0 \text{ lb/ft}.$$

Why didn't we use $k = -F/x$ here?

(b) What is the force exerted by the spring on the 1.5-lb body just before it is released?

The spring is stretched 4.0 in. or $\frac{1}{3}$ ft. Hence, the force exerted by the spring is

$$F = -kx = -(3.0 \text{ lb/ft})(\tfrac{1}{3} \text{ ft}) = -1.0 \text{ lb}.$$

The minus sign indicates that the force is directed opposite to the displacement.

(c) What is the period of oscillation after release?

$$T = 2\pi\sqrt{\frac{m}{k}} = 2\pi\sqrt{\frac{1.5/32}{3.0}} \text{ sec} = 0.79 \text{ sec}.$$

This corresponds to a frequency $\nu(= 1/T)$ of 1.3 cycles/sec and to an angular frequency $\omega(= 2\pi\nu)$ of 8.0 rad/sec.

(d) What is the amplitude of the motion?

The maximum displacement corresponds to zero kinetic energy and a maximum potential energy. This is the initial condition before release, so that the amplitude is the initial displacement of 4.0 in. Hence, $A = \frac{1}{3}$ ft.

(e) What is the maximum speed of the vibrating body?

From Eq. 13–13, $v_{max} = \omega A = (2\pi/T)A$,

$$v_{max} = \left(\frac{2\pi}{0.79 \text{ sec}}\right)\left(\frac{1}{3} \text{ ft}\right) = 2.7 \text{ ft/sec}.$$

The maximum speed occurs at the equilibrium position, where $x = 0$. This value is achieved twice in each period, the velocity being -2.7 ft/sec when the body passes through $x = 0$ first after release and $+2.7$ ft/sec when the body passes through $x = 0$ on the return trip.

(f) What is the maximum acceleration of the body?

From Eq. 13–13, $a_{max} = \omega^2 A = (k/m)A$,

$$a_{max} = \left(\frac{3.0}{1.5/32}\right)\left(\frac{1}{3}\right) \text{ ft/sec}^2 = 21 \text{ ft/sec}^2.$$

The maximum acceleration occurs at the ends of the path where $x = \pm A$ and $v = 0$. Hence, $a = -21$ ft/sec^2 at $x = +A$ and $a = +21$ ft/sec^2 at $x = -A$, the acceleration and displacement being oppositely directed.

(g) Compute the velocity, the acceleration, and the kinetic and potential energies of the body when it has moved in halfway from its initial position toward the center of motion.

At this point, $$x = \frac{A}{2} = \tfrac{1}{6} \text{ ft},$$

so that from Eq. 13–18,

$$v = -\frac{2\pi}{T}\sqrt{A^2 - x^2}$$

$$= -\frac{2\pi}{\pi/4}\sqrt{\left(\tfrac{1}{3}\right)^2 - \left(\tfrac{1}{6}\right)^2} \text{ ft/sec} = -2.3 \text{ ft/sec},$$

$$a = -\frac{k}{m}x = \frac{-3.0}{1.5/32}\left(\tfrac{1}{6}\right) \text{ ft/sec}^2 = -11 \text{ ft/sec}^2,$$

$$K = \tfrac{1}{2}mv^2 = \left(\tfrac{1}{2}\right)\left(\frac{1.5}{32}\right)\left(\frac{4}{\sqrt{3}}\right)^2 \text{ ft-lb} = \tfrac{1}{8} \text{ ft-lb}$$

$$U = \tfrac{1}{2}kx^2 = \left(\tfrac{1}{2}\right)(3)\left(\tfrac{1}{6}\right)^2 \text{ ft-lb} = \tfrac{1}{24} \text{ ft-lb}.$$

(h) Compute the mechanical energy of the oscillating system.

Since the mechanical energy is conserved, we can compute it at any stage of the motion. Using previous results, we obtain

$$E = K + U = \tfrac{1}{8} + \tfrac{1}{24} = \tfrac{1}{6} \text{ ft-lb}, \quad \text{(particle at } x = A/2\text{)}$$

$$E = U_{max} = \tfrac{1}{2}kx_{max}^2 = \left(\tfrac{1}{2}\right)(3)\left(\tfrac{1}{3}\right)^2 \text{ ft-lb} = \tfrac{1}{6} \text{ ft-lb}, \quad \text{(particle at } x = A\text{)}$$

$$E = K_{max} = \tfrac{1}{2}mv_{max}^2 = \left(\tfrac{1}{2}\right)\left(\frac{1.5}{32}\right)\left(\tfrac{8}{3}\right)^2 \text{ ft-lb} = \tfrac{1}{6} \text{ ft-lb}, \quad \text{(particle at } x = 0\text{)}$$

(i) What is the displacement of the body as a function of time?

In general, we have

$$x = A\cos(\omega t + \delta).$$

We have already found that $A = \tfrac{1}{3}$ ft. We must now determine ω and δ. We obtain

$$\omega = \frac{2\pi}{T} = \frac{2\pi}{0.79 \text{ sec}} = 8.0 \text{ radians/sec},$$

so that, with our particular units,

$$x = \tfrac{1}{3}\cos(8t + \delta).$$

At the time $t = 0$, $x = \tfrac{1}{3}$ ft, so that at that instant

$$x = \tfrac{1}{3}\cos\delta = \tfrac{1}{3}$$

or $$\delta = 0 \text{ radian}.$$

Therefore, with $A = \tfrac{1}{3}$ ft, $\omega = 8.0$ radians/sec, and $\delta = 0$ radian, we obtain

$$x = \tfrac{1}{3}\cos 8.0t.$$

This describes the motion of the body, where x is in feet, t is in seconds, and the angle $8.0t$ is in radians.

13–5   Applications of Simple Harmonic Motion

A few physical systems that move with simple harmonic motion are considered here. We will discuss others from time to time throughout the text.

The Simple Pendulum. A simple pendulum is an idealized body consisting of a point mass, suspended by a light inextensible cord. When pulled to one side of its equilibrium position and released, the pendulum swings in a vertical plane under the influence of gravity. The motion is periodic and oscillatory. We wish to determine the period of the motion.

Figure 13–10 shows a pendulum of length l, particle mass m, making an angle θ with the vertical. The forces acting on m are $m\mathbf{g}$, the gravitational force, and $\mathbf{T}$, the tension in the cord. Choose axes tangent to the circle of motion and along the radius. Resolve $m\mathbf{g}$ into a radial component of magnitude $mg \cos \theta$, and a tangential component of magnitude $mg \sin \theta$. The radial components of the forces supply the necessary centripetal acceleration to keep the particle moving on a circular arc. The tangential component is the restoring force acting on m tending to return it to the equilibrium position. Hence, the restoring force is

$$F = -mg \sin \theta.$$

Notice that the restoring force is not proportional to the angular displacement θ but to $\sin \theta$ instead. The resulting motion is, therefore, not simple harmonic. However, if the angle θ is small, $\sin \theta$ is very nearly equal to θ in radians. For example, if $\theta = 5° = 0.0873$ radian, $\sin \theta = 0.0872$; even at $\theta = 15°$ the two figures differ only by 1.1%. The displacement along the arc is $x = l\theta$, and for small angles this is nearly straight-line motion. Hence, assuming

$$\sin \theta \cong \theta,$$

we obtain
$$F = -mg\theta = -mg \frac{x}{l} = -\frac{mg}{l} x.$$

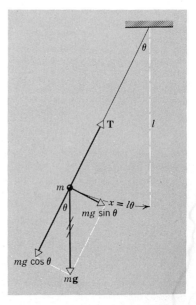

FIGURE 13–10   The forces acting on a simple pendulum are the tension **T** in the string and the weight $m\mathbf{g}$ of mass m. The magnitudes of the radial and tangential components of $m\mathbf{g}$ are labeled.

For small displacements, therefore, the restoring force is proportional to the displacement and is oppositely directed. This is exactly the criterion for simple harmonic motion. The constant mg/l represents the constant k in $F = -kx$. Check the dimensions of k and mg/l. The period of a simple pendulum when its amplitude is small is

$$T = 2\pi \sqrt{\frac{m}{k}} = 2\pi \sqrt{\frac{m}{mg/l}} \quad \text{or} \quad T = 2\pi \sqrt{\frac{l}{g}}. \quad (13\text{–}19)$$

Notice that the period is independent of the mass of the suspended particle.

The Torsional Pendulum. In Fig. 13–11 we show a disk suspended by a wire attached to the center of mass of the disk. The wire is securely fixed to a solid support and to the disk. At the equilibrium position of the disk a radial line is drawn from its center to P, as shown. If the disk is rotated in a horizontal plane to the radial position Q, the wire will be twisted. The twisted wire will exert a torque on the disk tending to return it to the position P. This is a restoring torque. For small twists the restoring torque is found to be proportional to the amount of twist, or the angular displacement (Hooke's law), so that

$$\tau = -\kappa\theta. \quad (13\text{–}20)$$

Here κ is a constant that depends on the properties of the wire and is called the *torsional* constant. The minus sign shows that the torque is directed opposite to the angular displacement θ. Equation 13–20 is the condition for *angular simple harmonic motion*.

The equation of motion for such a system is

$$\tau = I\alpha = I\frac{d^2\theta}{dt^2},$$

so that, on using Eq. 13–20, we obtain

$$-\kappa\theta = I\frac{d^2\theta}{dt^2}$$

or

$$\frac{d^2\theta}{dt^2} = -\frac{\kappa}{I}\theta. \quad (13\text{–}21)$$

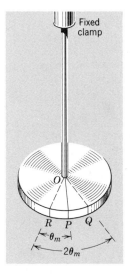

Notice the similarity between Eq. 13–21 for simple angular harmonic motion and Eq. 13–7 for simple linear harmonic motion. In fact, the equations are identical mathematically. We have simply substituted angular displacement θ for linear displacement x, rotational inertia I for mass m, and torsional constant κ for force constant k. The solution of Eq. 13–21 is, therefore, a simple harmonic oscillation in the angle coordinate θ, namely

$$\theta = \theta_m \cos(\omega t + \delta). \quad (13\text{–}22)$$

Here, θ_m is the maximum angular displacement, that is, the amplitude of the angular oscillation. In Fig. 13–11 the disk oscillates

FIGURE 13–11 The torsional pendulum. The line drawn from the center to P oscillates between Q and R, sweeping out an angle $2\theta_m$ where θ_m is the (angular) amplitude of the motion.

about the equilibrium position $\theta = 0$ (line OP), the total angular range being $2\theta_m$ (from OQ to OR).

The period of the oscillation by analogy with Eq. 13–10 is

$$T = 2\pi \sqrt{\frac{I}{\kappa}}. \tag{13–23}$$

If κ is known and T is measured, the rotational inertia I about the axis of rotation of any oscillating rigid body can be determined. If I is known and T is measured, the torsional constant κ of any sample of wire can be determined.

Many laboratory instruments involve torsional oscillations, notably the galvanometer. The Cavendish balance is a torsional pendulum (Chapter 14). The balance wheel of a watch is another example of angular harmonic motion, the restoring torque here being supplied by a spiral hairspring.

Example 2. A thin rod of mass 0.10 kg and length 0.10 meter is suspended by a wire which passes through its center and is perpendicular to its length. The wire is twisted and the rod set oscillating. The period is found to be 2.0 sec. When a flat body in the shape of an equilateral triangle is suspended similarly through its center of mass, the period is found to be 6.0 sec. Find the rotational inertia of the triangle about this axis.

The rotational inertia of the rod is $Ml^2/12$ (see Table 11–1). Hence

$$I_{\text{rod}} = \frac{(0.10 \text{ kg})(0.10 \text{ meter})^2}{12} = 8.3 \times 10^{-5} \text{ kg-m}^2.$$

From Eq. 13–23,

$$\frac{T_{\text{rod}}}{T_{\text{triangle}}} = \left(\frac{I_{\text{rod}}}{I_{\text{triangle}}}\right)^{1/2} \quad \text{or} \quad I_{\text{triangle}} = I_{\text{rod}}\left(\frac{T_t}{T_r}\right)^2,$$

so that

$$I_{\text{triangle}} = (8.3 \times 10^{-5} \text{ kg-m}^2)\left(\frac{6.0 \text{ sec}}{2.0 \text{ sec}}\right)^2 = 7.5 \times 10^{-4} \text{ kg-m}^2.$$

Does the amplitude of the oscillation affect the period in these cases?

13–6   Relation between Simple Harmonic Motion and Uniform Circular Motion

Let us consider the relation between simple harmonic motion along a straight line and uniform circular motion. This relation is useful in describing many features of simple harmonic motion. It also gives a simple geometric meaning to the angular frequency ω and the phase constant δ. Uniform circular motion is also an example of a combination of simple harmonic motions, a phenomenon we deal with rather often in wave motion.

In Fig. 13–12 Q is the point moving around a circle of radius A with a constant angular speed of ω, expressed, say, in radians/sec. P is the perpendicular projection of Q on the horizontal diameter, along the x-axis. Let us call Q the *reference point* and the circle on which it moves the *reference circle*. As the reference point revolves, the projected point P moves back and forth along the horizontal diameter. The x-component of Q's displacement is always the same as the displacement of P; the x-component of the velocity of Q is always the same as the velocity of P; and the x-component of the acceleration of Q is always the same as the acceleration of P.

Let the angle between the radius OQ and the x-axis at the time $t = 0$ be called δ. At any later time t, the angle between OQ and the x-axis is $(\omega t + \delta)$, the point Q moving with constant angular speed. The x-coordinate of Q at any time is, therefore,

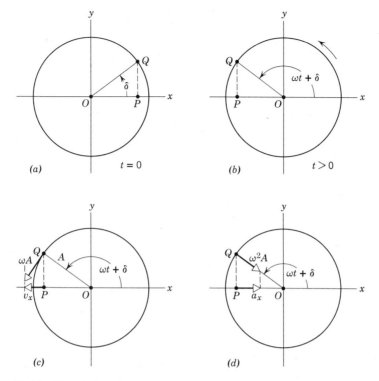

FIGURE 13–12 The relation of simple harmonic motion to uniform circular motion. Q moves in uniform circular motion and P in simple harmonic motion. Q has angular speed ω, P angular frequency ω. (a, b) The x-component of Q's displacement is always equal to P's displacement. (c) The x-component of Q's velocity is always equal to P's velocity. (d) The x-component of Q's acceleration is always equal to P's acceleration.

$$x = A \cos (\omega t + \delta). \tag{13–24}$$

Hence, the projected point P moves with simple harmonic motion along the x-axis. Therefore, simple harmonic motion can be described as the projection along a diameter of uniform circular motion.

The angular frequency ω of simple harmonic motion is the same as the angular speed of the reference point. The frequency of the simple harmonic motion is the same as the number of revolutions per unit time of the reference point. Hence, $\nu = \omega/2\pi$ or $\omega = 2\pi\nu$. The time for a complete revolution of the reference point is the same as the period T of the simple harmonic motion. Hence, $T = 2\pi/\omega$ or $\omega = 2\pi/T$. The phase of the simple harmonic motion, $\omega t + \delta$, is the angle that OQ makes with the x-axis at any time t (Fig. 13–12b). The angle that OQ makes with the x-axis at the time $t = 0$ (Fig. 13–12a) is δ, the phase constant or initial phase of the motion. The amplitude of the simple hormonic motion is the same as the radius of the reference circle.

The tangential velocity of the reference point Q has a magnitude of ωA. Hence, the x-component of this velocity (Fig. 13–12c) is

$$v_x = -\omega A \sin (\omega t + \delta).$$

This relation gives a negative v_x when Q and P are moving to the left and a positive v_x when Q and P are moving to the right. Notice that v_x is zero at the end points of the simple harmonic motion, where $\omega t + \delta$ is zero and π, as required.

The acceleration of point Q in uniform circular motion is directed radially inward and has a magnitude of $\omega^2 A$. The acceleration of the projected point P is the x-component of the acceleration of the reference point Q (Fig. 13–12d). Hence,

$$a_x = -\omega^2 A \cos(\omega t + \delta)$$

gives the acceleration of the point executing simple harmonic motion. Notice that a_x is zero at the midpoints of the simple harmonic motion, where $\omega t + \delta = \pi/2$ or $3\pi/2$, as required.

These results are all identical with those of simple harmonic motion along the x-axis; see Eqs. 13–13.

If we had taken the perpendicular projection of the reference point onto the y-axis, instead, we would have obtained for the motion of the y-projected point

$$y = A \sin(\omega t + \delta). \tag{13–25}$$

This is again a simple harmonic motion. It differs only in phase from Eq. 13–24, for if we replace δ by $\delta - \pi/2$, then $\cos(\omega t + \delta)$ becomes $\sin(\omega t + \delta)$. It is clear that the projection of uniform circular motion along any diameter gives a simple harmonic motion.

Conversely, uniform circular motion can be described as a combination of two simple harmonic motions. It is that combination of two simple harmonic motions, occurring along perpendicular lines, which have the same amplitude and frequency but differ in phase by 90°. When one component is at the maximum displacement, the other component is at the equilibrium point. If we combine these components (Eqs. 13–24 and 13–25), we obtain at once the relation

$$r = \sqrt{x^2 + y^2} = A.$$

By writing the relations for v_y and a_y (you should do this) and combining corresponding quantities, we obtain also the relations

$$v = \sqrt{v_x^2 + v_y^2} = \omega A,$$
$$a = \sqrt{a_x^2 + a_y^2} = \omega^2 A.$$

These relations correspond respectively to the magnitudes of the displacement, the velocity, and the acceleration in uniform circular motion.

It will be possible for us to analyze many complicated motions as combinations of individual simple harmonic motions. Circular motion is a particularly simple combination. In the next section we shall consider other combinations of simple harmonic motions.

Example 3. In Example 1 we considered a body executing a horizontal simple harmonic motion. The equation of that motion (units?) was

$$x = \tfrac{1}{3} \cos 8.0t.$$

This motion can also be represented as the projection of uniform circular motion along a horizontal diameter.

(*a*) Give the properties of the corresponding uniform circular motion.

The x-component of the circular motion is given by

$$x = A \cos(\omega t + \delta).$$

Therefore, the reference circle must have a radius $A = \tfrac{1}{3}$ ft, the initial phase or phase constant must be $\delta = 0$, and the angular velocity must be $\omega = 8.0$ radians/sec, in order to obtain the equation $x = \frac{1}{3}\cos 8.0t$ for the horizontal projection.

   (*b*) From the motion of the reference point determine the time required for the body to come halfway in toward the center of motion from its initial position.

   As the body moves halfway in, the reference point moves through an angle of $\omega t = 60°$ (Fig. 13–13). The angular velocity is constant at 8.0 radians/sec so that the time required to move through $60°$ is

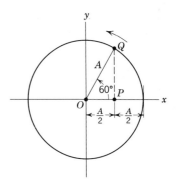

$$t = \frac{60°}{\omega} = \frac{\pi/3 \text{ radians}}{8.0 \text{ radians/sec}} = \frac{\pi}{24}\text{ sec} = 0.13 \text{ sec}.$$

FIGURE 13–13   Example 3. The particles Q and P of Fig. 13–12 are shown for $\omega t = 60°$. Since ω is known, t may be found.

The time may also be computed directly from the equation of motion. Thus,

$$x = \tfrac{1}{3}\cos 8.0t \qquad \text{and} \qquad x = \frac{A}{2} = \tfrac{1}{6},$$

Hence

$$\tfrac{1}{6} = \tfrac{1}{3}\cos 8.0t \qquad \text{or} \qquad 8.0t = \cos^{-1}\left(\tfrac{1}{2}\right) = \frac{\pi}{3}.$$

Therefore,

$$t = \frac{\pi}{24}\text{ sec} = 0.13 \text{ sec}.$$

13–7   Combinations of Harmonic Motions

Often two linear simple harmonic motions act on a particle at right angles. The resulting motion is the sum of two independent oscillations. Consider first the case in which the frequencies of the vibrations are the same, such as

$$x = A_x \cos(\omega t + \delta),$$
$$y = A_y \cos(\omega t + \alpha). \tag{13–26}$$

The *x*- and *y*-motions have different amplitudes and different phase constants, however.

   If the phase constants are the same so that $\delta = \alpha$, the resulting motion is a straight line. This can be shown analytically, for when we eliminate t from the equations

$$x = A_x \cos(\omega t + \delta) \qquad y = A_y \cos(\omega t + \delta),$$

we obtain $\qquad\qquad\qquad y = (A_y/A_x)x.$

This is the equation of a straight line, whose slope is A_y/A_x. In Fig. 13–14a and *b* we show the resultant motion for two cases, $A_y/A_x = 1$ and $A_y/A_x = 2$. In these cases both the *x*- and *y*-displacements reach a maximum at the same time and reach a minimum at the same time. They are in phase.

   If the phase constants are different, the resulting motion will not be a straight line. For example, if the phase constants differ by $\pi/2$, the maximum *x*-displacement occurs when the *y*-displacement is zero and vice versa. When the amplitudes are equal, the resulting motion is circular; when the amplitudes are unequal, the resulting motion is elliptical. Two cases, $A_y/A_x = 1$ and $A_y/A_x = 2$, are shown in Fig. 13–14c and *d*, for $\delta = \alpha + \pi/2$. The cases $A_y/A_x = 1$ and $A_y/A_x = 2$, for $\delta = \alpha - \pi/4$, are shown in Fig. 13–14e and *f*.

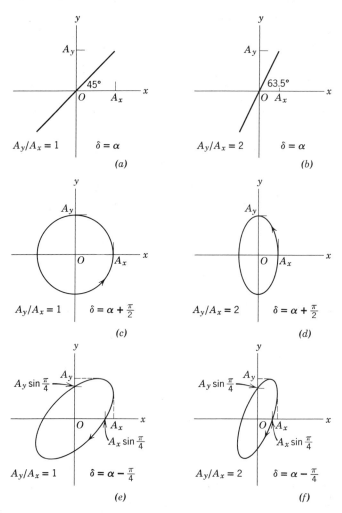

FIGURE 13-14 Simple harmonic motions in two dimensions. (*a*) The amplitudes of x and y (namely A_x and A_y) are the same, as are their phase constants. (*b*) y's amplitude is twice x's, but their phase constants are the same. (*c*) Their amplitudes are equal, but x leads y in phase by 90°. (*d*) Same as (*c*) except that y's amplitude is twice x's. (*e*) Equal amplitudes, but x lags y in phase by 45°. (*f*) Same as (*e*) except that y's amplitude is twice x's.

All possible combinations of two simple harmonic motions at right angles having the same frequency correspond to elliptical paths, the circle and straight line being special cases of an ellipse. This can be shown analytically by combining Eqs. 13–26 and eliminating the time; you should show that the resulting equation is that of an ellipse. The shape of the ellipse depends only on the ratio of the amplitudes, A_y/A_x, and the difference in phase between the two oscillations, $\delta - \alpha$. The actual motion can be either clockwise or counterclockwise, depending on which component leads in phase.

A simple way to produce such patterns is by means of an oscilloscope. In this, electrons are deflected by each of two electric fields at right angles to one another. The field strengths alternate sinusoidally with the same frequency, but their phases and amplitudes can be varied. In this way the electrons can be made to trace out the various patterns discussed above on a fluorescent screen. We can also produce these patterns mechanically by means of a pendulum swinging with small amplitude but not confined to one vertical plane. Such combinations of two simple harmonic motions at right angles having the same frequency are particularly important in the study of polarized light and alternating current circuits.

Combinations of simple harmonic motions of the same frequency in the same direction, but with different amplitudes and phases, are of special interest in the study of diffraction and interference of light, sound, and electromagnetic radiation. We will discuss this later in the text.

If two oscillations of different frequencies are combined at right angles, the resulting motion is more complicated. The motion is not even periodic unless the two component frequencies ω_1 and ω_2 are the ratio of two integers (see Problem 31). Oscillations of different frequencies in the same direction may also be combined. The treatment of this motion is particularly important in the case of sound vibrations and will be discussed in Chapter 17.

QUESTIONS

1. Give some examples of motions that are approximately simple harmonic. Why are motions that are exactly simple harmonic rare?

2. A spring has a force constant k, and a mass m is suspended from it. The spring is cut in half and the same mass is suspended from one of the halves. Is the frequency of vibration the same before and after the spring is cut? How are the frequencies related?

3. An unstressed spring has a force constant k. It is stretched, by a weight hung from it, to an equilibrium length well within the elastic limit. Does the spring have the same force constant k for displacements from this new equilibrium position?

4. Any real spring has mass. If this mass is taken into account, explain qualitatively how this will change our expressions for the period of oscillation of a mass-spring system.

5. Suppose we have a block of unknown mass and a spring of unknown force constant. Show how we can predict the period of oscillation of this block-spring system simply by measuring the extension of the spring produced by attaching the block to it.

6. Can one have an oscillator which even for small amplitudes is not simple harmonic? That is, can one have a nonlinear restoring force in an oscillator even at arbitrarily small amplitudes?

7. Could we ever construct a true simple pendulum?

8. Could standards of mass, length, and time be based on properties of a pendulum? Explain.

9. What would happen to the motion of an oscillating system if the sign of the force term, $-kx$ in Eq. 13–4, were changed?

10. Predict by qualitative arguments whether a pendulum oscillating with large amplitude will have a period longer or shorter than the period for oscillations with small amplitude. (Consider extreme cases.)

11. How is the period of a pendulum affected when its point of suspension is (a) moved horizontally in the plane of the oscillation with acceleration a; (b) moved vertically upward with acceleration a; (c) moved vertically downward with acceleration $a < g$. Which case, if any, applies to a pendulum mounted on a cart rolling down an inclined plane?

12. A hollow sphere is filled with water through a small hole in it. It is hung by a long thread and, as the water slowly flows out of the hole at the bottom, one finds that the period of oscillation first increases and then decreases. Explain.

13. Two pendula, each consisting of a disk attached to a light bar, are identical except for the coupling between disk and bar. In one the bar is rigidly mounted to the disk; in the other ball-bearings are used so that the disk would be free to spin about the end of the bar, for example. Both pendula are hung, pulled aside to the same height, and released. Which has the greater period? Explain.

14. How can a pendulum be used so as to trace out a sinusoidal curve?

15. What combination of simple harmonic motions would give a figure 8 as the resultant motion?

16. Is there a connection between the F vs. x relation at the molecular level and the macroscopic relation between F and x in a spring?

PROBLEMS

1(1). An oscillating mass-spring system takes 0.75 sec to begin repeating its motion. Find (*a*) the period, (*b*) the frequency in cycles/sec, and (*c*) the frequency in radians/sec.
Answer: (*a*) 0.75 sec. (*b*) 1.3 cycles/sec. (*c*) 8.4 radians/sec.

2(2). A uniform spring whose unstressed length is l has a force constant k. The spring is cut into two pieces of unstressed lengths l_1 and l_2, where $l_1 = nl_2$ and n is an integer. What are the corresponding force constants k_1 and k_2 in terms of n and k? Does your result seem reasonable for $n = 1$?

3(3). A 4.0-kg block extends a spring 16 cm from its unstretched position. The block is removed and a 0.50-kg body is hung from the same spring. If the spring is then stretched and released, what is its period of oscillation?
Answer: 0.28 sec.

4(3). A 2.0-kg mass hangs from a spring. A 300-gm body hung below the mass stretches the spring 2.0 cm farther. If the 300-gm body is removed and the mass is set into oscillation, find the period of motion.

5(3). Find the maximum displacement of a 1.0×10^{-20}-kg particle vibrating with simple harmonic motion with a period of 1.0×10^{-5} sec and a maximum speed of 1.0×10^3 meters/sec.
Answer: 1.6 mm.

6(3). The scale of a spring balance reading from 0 to 32 lb is 4.0 in. long. A package suspended from the balance is found to oscillate vertically with a frequency of 2.0 oscillations per second. How much does the package weigh?

7(3). The piston in the cylinder head of a locomotive has a stroke of 2.5 ft. What is the maximum speed of the piston if the drive wheels make 180 rev/min and the piston moves with simple harmonic motion?
Answer: 24 ft/sec.

8(3). An automobile can be considered to be mounted on a spring as far as vertical oscillations are concerned. The springs of a certain car are adjusted so that the vibrations have a frequency of 3.0 per second. (*a*) What is the spring's force constant if the car weighs 3200 lb? (*b*) What will the vibration frequency be if five passengers, averaging 160 lb each, ride in the car?

9(3). A particle executes linear harmonic motion about the point $x = 0$. At $t = 0$ it has displacement $x = 0.37$ cm and zero velocity. The frequency of the motion is 0.25 vib/sec. Determine (*a*) the period, (*b*) the angular frequency, (*c*) the amplitude, (*d*) the displacement at time t, (*e*) the velocity at time t, (*f*) the maximum speed, (*g*) the maximum acceleration, (*h*) the displacement at $t = 3.0$ sec, and (*i*) the speed at $t = 3.0$ sec.
Answer: (*a*) 4.0 sec. (*b*) $\pi/2$ radians/sec. (*c*) 0.37 cm. (*d*) $0.37 \cos{(\pi t/2)}$, in centimeters. (*e*) $-0.58 \sin{(\pi t/2)}$, in centimeters per second. (*f*) 0.58 cm/sec. (*g*) 0.91 cm/sec². (*h*) Zero. (*i*) 0.58 cm/sec.

10(3). A 50-gm mass is attached to the bottom of a vertical spring and set vibrating. If the maximum speed of the mass is 15 cm/sec and the period is 0.50 sec find (*a*) the spring constant k of the spring, (*b*) the amplitude of the motion, and (*c*) the frequency of oscillation.

11(3). A small body of mass 0.10 kg is undergoing simple harmonic motion of amplitude 1.0

meter and period 0.20 sec. (*a*) What is the maximum value of the force acting on it? (*b*) If the oscillations are produced by a spring, what is the force constant of the spring?

Answer: (*a*) 99 nt. (*b*) 99 nt/meter.

12(3). A block is on a piston which is moving vertically with a simple harmonic motion of period 1.0 sec. (*a*) At what amplitude of motion will the block and piston separate? (*b*) If the piston has an amplitude of 5.0 cm, what is the maximum frequency for which the block and piston will be in contact continuously?

13(3). A body is vibrating with simple harmonic motion of amplitude 15 cm and frequency 4.0 cycles/sec. Calculate (*a*) the maximum values of its velocity and acceleration, (*b*) the acceleration and speed when the body is 9.0 cm from the equilibrium position, and (*c*) the time required for the body to move from the equilibrium position to a point 12 cm away.

Answer: (*a*) 3.8 meter/sec; 94 meter/sec^2. (*b*) 57 meters/sec^2; 3.0 meters/sec. (*c*) 3.7×10^{-2} sec.

14(3). A massless spring hangs from the ceiling with a small object attached to its lower end. The object is initially held at rest in such a position that the spring is not stretched. The object is then released from this position and oscillates up and down with its lowest position being 10 cm below the initial position. (*a*) What is the frequency of the oscillation? (*b*) What is the speed of the object when it is 8.0 cm below the initial position? (*c*) A 300-gram weight is attached to the first object, after which the system oscillates with half the original frequency. What is the mass of the first object? (*d*) Where is the new equilibrium position with both objects attached to the spring?

15(3). The end of one of the prongs of a tuning fork that executes simple harmonic motion of frequency 1000 per second has an amplitude of 0.40 mm. Find (*a*) the maximum acceleration and maximum speed of the end of the prong and (*b*) the acceleration and the speed of the end of the prong when it has a displacement 0.20 mm.

Answer: (*a*) 1.6×10^4 meters/sec^2; 2.5 meters/sec. (*b*) 7.9×10^3 meters/sec^2; 2.2 meters/sec.

16(3). A body oscillates with simple harmonic motion according to the equation

$$x = 6.0 \cos (3\pi t + \pi/3)$$

where *x* is in meters, *t* is in seconds, and the numbers in the parentheses are in radians. What is (*a*) the displacement, (*b*) the velocity, (*c*) the acceleration, and (*d*) the phase at the time $t = 2.0$ sec. Find also (*e*) the frequency ν, and (*f*) the period of the motion.

17(3). A body oscillates with simple harmonic motion according to the equation

$$x = 5.0 \cos \left(3t + \frac{\pi}{2}\right)$$

where *x* is in meters and *t* is in seconds. Calculate (*a*) the displacement from equilibrium when $t = \pi$ sec, (*b*) the speed of the body when $t = \pi/2$ sec and (*c*) the phase of the motion when $t = \pi/6$ sec.

Answer: (*a*) Zero. (*b*) Zero. (*c*) π.

18(3). A 0.10-kg block slides back and forth along a straight line on a smooth horizontal surface. Its displacement from the origin is given by

$$x = 10 \cos \left(10t + \frac{\pi}{2}\right)$$

with *x* is in cm and *t* in seconds. (*a*) What is the oscillation frequency? (*b*) What is the maximum speed acquired by the block? At what value of *x* does this occur? (*c*) What is the maximum acceleration of the block? At what value of *x* does this occur? (*d*) What force must be applied to the block to give it this motion?

19(3). A loudspeaker produces a musical sound by the oscillation of a diaphram. If the amplitude of oscillation is limited to 1.0×10^{-3} mm, what frequencies will result in the acceleration of the diaphram exceeding *g*?

Answer: $\nu > 500$ cycles/sec.

20(3). The vibration frequencies of atoms in solids at normal temperatures are of the order 10^{13}/sec. Imagine the atoms to be connected to one another by "springs." Suppose that a single sil-

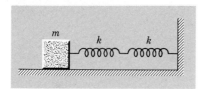

FIGURE 13-15   Problem 21(3).

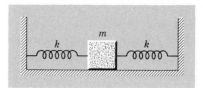

FIGURE 13-16   Problem 22(3).

ver atom vibrates with this frequency and that all the other atoms are at rest. Compute the force constant of a single spring. One mole of silver has a mass of 108 gm and contains 6.02×10^{23} atoms. Assume that the atoms are in a simple cubic lattice and that the only significant interaction of each atom is with its nearest neighbors.

21(3). Two springs are joined and connected to a mass m as shown in Fig. 13–15. The surfaces are frictionless. If the springs each have force constant k, show that the frequency of oscillation of m is

$$\nu = \frac{1}{2\pi} \sqrt{\frac{k}{2m}}.$$

22(3). The springs are now attached to m and to fixed supports as shown in Fig. 13–16. Show that the frequency of oscillation in this case is

$$\nu = \frac{1}{2\pi} \sqrt{\frac{2k}{m}}.$$

23(3). Two particles execute simple harmonic motion of the same amplitude and frequency along the same straight line. They pass one another when going in opposite directions each time their displacement is half their amplitude. What is the phase difference between them?
Answer: 120°.

24(3). Two particles oscillate in simple harmonic motion along a common straight line segment of length A. Each particle has a period of 1.5 seconds but they differ in phase by 30°. (*a*) How far apart are they (in terms of A) 0.50 sec after the lagging particle leaves one end of the path? (*b*) Are they moving in the same direction, towards each other, or away from each other at this time?

25(3). A block is on a horizontal surface which is moving horizontally with a simple harmonic motion of frequency two oscillations per second. The coefficient of static friction between block and plane is 0.50. How great can the amplitude be if the block does not slip along the surface?
Answer: 3.1 cm.

26(4). A massless spring of force constant 19 nt/meter hangs vertically. A body of mass 0.20 kg is attached to its free end and then released. Assume that the spring was unstretched before the body was released. Find (*a*) how far below the initial position the body descends, (*b*) the frequency, and (*c*) the amplitude of the resulting motion, assumed to be simple harmonic.

27(4). An oscillating mass-spring system has a mechanical energy of 1.0 joule, amplitude of 0.10 meter and maximum speed of 1.0 meter/sec. Find (*a*) the force constant of the spring. (*b*) the mass, and (*c*) the frequency of oscillation.
Answer: (*a*) 200 nt/meter. (*b*) 2.0 kg. (*c*) 1.6 cycles/sec.

28(4). Find the mechanical energy of a mass-spring system having a force constant of 1.3 nt/cm, and an amplitude of 2.4 cm.

29(4). (*a*) When the displacement is one-half the amplitude A, what fraction of the total energy is kinetic and what fraction is potential in simple harmonic motion? (*b*) At what displacement is the energy half kinetic and half potential?
Answer: (*a*) $\frac{3}{4}$; $\frac{1}{4}$. (*b*) $A/\sqrt{2}$.

30(4). A 1.0×10^{-2}-kg particle is undergoing simple harmonic motion with an amplitude of 2.0×10^{-3} meters. The maximum acceleration experienced by the particle is 8.0×10^{3} meters/sec². (*a*) Write an equation for the force on the particle as a function of time. (*b*) What is the period of the motion? (*c*) What is the maximum speed of the particle? (*d*) What is the total mechanical energy of this simple harmonic oscillator?

31(4). A 3.0-kg particle is in simple harmonic motion in one dimension and moves according to the equation

$$x = 5.0 \cos \left(\frac{\pi}{3} t - \frac{\pi}{4} \right),$$

in which *x* is in meters and *t* is in seconds. (*a*) At what value of *x* is the potential energy equal to half the total energy? (*b*) How long does it take the particle to move to this position from the equilibrium position?

Answer: (*a*) 3.5 meters. (*b*) 0.75 sec.

32(4). A 5.0-kg object moves on a horizontal frictionless surface under the influence of a spring with force constant 1.0×10^3 nt/meter. The object is displaced 50 cm and given an initial velocity of 10 meters/sec back towards the equilibrium position. (*a*) What is the frequency of the motion? (*b*) How much energy is associated with the motion? (*c*) What is the amplitude of the oscillation?

33(4). Start from Eq. 13–17 for the conservation of energy (with $\frac{1}{2}kA^2 = E$) and obtain the displacement as a function of the time by integration of Eq. 13–18. Compare with Eq. 13–8.

34(4). An 8.0-lb block is suspended from a spring with a force constant of 3.0 lb/in. A bullet weighing 0.10 lb is fired into the block from below with a speed of 500 ft/sec and comes to rest in the block. (*a*) Find the amplitude of the resulting simple harmonic motion. (*b*) What fraction of the original kinetic energy of the bullet is stored in the harmonic oscillator?

35(5). What is the length of a simple pendulum whose period is 1.00 sec at a point where $g = 32.2$ ft/sec^2?

Answer: 9.80 in.

36(5). A simple pendulum of length 1.00 meter makes 100 complete oscillations in 204 sec at a certain location. What is the acceleration of gravity at this point?

37(5). A long uniform rod of length *l* and mass *m* is free to rotate in a horizontal plane about a vertical axis through its center. A spring with force constant *k* is connected horizontally between the end of the rod and a fixed wall as shown in Fig. 13–17. What is the period of the small oscillations that result when the rod is pushed slightly to one side and released?

Answer: $2\pi \sqrt{m/3k}$.

38(5). A pendulum is formed by pivoting a long thin rod of length *l* and mass *m* about a point on the rod which is a distance *d* above the center of the rod. Find the small amplitude period of this pendulum in terms of *d*, *l*, *m*, and *g*.

39(5). An 8.0-kg solid sphere with a 15-cm radius is suspended by a vertical wire attached to the ceiling of a room. A torque of 0.20 nt-meter is required to twist the sphere through an angle of one radian. What is the period of oscillation when the sphere is released from this position?

Answer: 3.8 sec.

40(5). The balance wheel of a watch vibrates with an angular amplitude of π radians and a period of 0.50 sec. Find (*a*) the maximum angular speed of the wheel, (*b*) the angular speed of the wheel when its displacement is $\pi/2$ radians, and (*c*) the angular acceleration of the wheel when its displacement is $\pi/4$ radians.

41(5). (*a*) What is the frequency of a simple pendulum 2.0 meters long? (*b*) Assuming small amplitudes, what would its frequency be in an elevator accelerating upward at a rate of 2.0 meters/sec^2? (*c*) What would its frequency be in free fall?

Answer: (*a*) 0.35 cycles/sec. (*b*) 0.39 cycles/sec. (*c*) Zero.

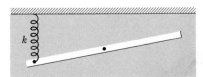

FIGURE 13–17 Problem 37(5).

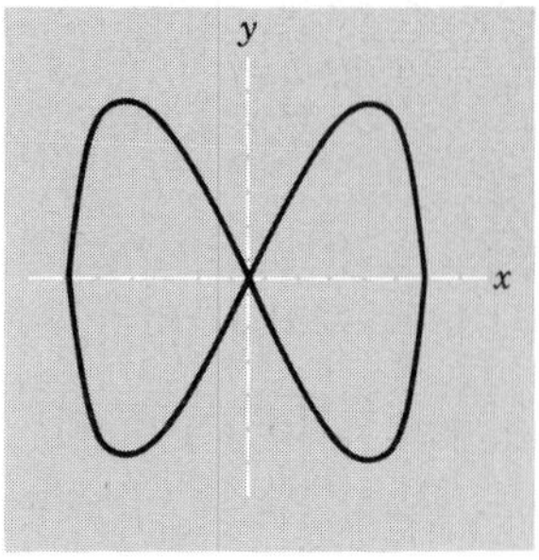

FIGURE 13-18   Problem 47(7).

42(5). A simple pendulum of length l and mass m is suspended in a car that is traveling with a constant speed v around a circle of radius R. If the pendulum undergoes small oscillations in a radial direction about its equilibrium position, what will its frequency of oscillation be?

43(5). (*a*) Show that the general relations for the period and frequency of any simple harmonic motion are

$$T = 2\pi \sqrt{-\frac{x}{a}} \quad \text{and} \quad \nu = \frac{1}{2\pi} \sqrt{-\frac{a}{x}}.$$

(*b*) Show that the general relations for the period and frequency of any simple angular harmonic motion are

$$T = 2\pi \sqrt{-\frac{\theta}{\alpha}} \quad \text{and} \quad \nu = \frac{1}{2\pi} \sqrt{-\frac{\alpha}{\theta}}.$$

44(7). *Lissajous Figures.* When oscillations at right angles are combined, the frequencies for the motion of the particle in the x- and y-directions need not be equal, so that in the general case Eqs. 13–26 become

$$x = A_x \cos(\omega_x t + \delta) \quad \text{and} \quad y = A_y \cos(\omega_y t + \alpha).$$

The path of the particle is no longer an ellipse but is called a *Lissajous curve*, after Jules Antoine Lissajous who first demonstrated such curves in 1857. (*a*) If ω_x/ω_y is a rational number, so that the angular frequencies ω_x and ω_y are "commensurable," then the curve is closed and the motion repeats itself at regular intervals of time. Assume $A_x = A_y$ and $\delta = \alpha$ and draw the Lissajous curve for $\omega_x/\omega_y = \frac{1}{2}, \frac{1}{3}$, and $\frac{2}{3}$. (*b*) Let ω_x/ω_y be a rational number, either $\frac{1}{2}, \frac{1}{3}$, or $\frac{2}{3}$ say, and show that the shape of the Lissajous curve depends upon the phase difference $\alpha - \delta$. Draw curves for $\alpha - \delta = 0$, $\pi/4$, and $\pi/2$ radians. (*c*) If ω_x/ω_y is not a rational number, then the curve is "open." Convince yourself that after a long time the curve will have passed through every point lying in the rectangle bounded by $x = \pm A_x$ and $y = \pm A_y$, the particle never passing twice through a given point with the same velocity.

45(7). Sketch the path of a particle which moves in the x-y plane according to the equations $x = A \cos(\omega t - \pi/2)$, $y = 2A \cos(\omega t)$, in which x and y are in meters and t is in seconds.

46(7). Electrons in an oscilloscope are deflected by two mutually perpendicular electric fields in such a way that at any time t the displacement is given by

$$x = A \cos \omega t, \qquad y = A \cos(\omega t + \alpha).$$

(*a*) Describe the path of the electrons and determine its equation when $\alpha = 0°$. (*b*) When $\alpha = 30°$. (*c*) When $\alpha = 90°$.

47(7). The Lissajous figure shown in Fig. 13–18 is the result of combining the two simple harmonic motions $x = A_x \cos \omega_x t$ and $y = A_y \cos(\omega_y t + \delta)$. (*a*) What is the value of A_x/A_y? (*b*) What is the value of ω_x/ω_y? (*c*) What is the value of δ?
Answer: (*a*) 1.0. (*b*) 0.50. (*c*) $\pm\frac{1}{2}\pi$.

48(7). Sketch the motion of a mass attached to two identical springs which are aligned along the x- and y-axes when at rest if the amplitude of the x-motion is twice that of the y and the mass reaches maximum x-position and minimum y-position simultaneously. Find the phase difference $(\delta - \alpha)$ in Eq. 13–26. Why should you assume small vibrations?

CHAPTER 14

Gravitation

14–1 The Law of Universal Gravitation

Until the seventeenth century the tendency of a body to fall toward the earth, that is, its weight, was regarded as an inherent property of all bodies needing no further explanation. That the weight of a body should be regarded as a force of attraction between the earth and that body was an idea that occurred to Newton and some of his contemporaries, notably Robert Hooke.

The laws governing celestial motions were regarded as quite different from those governing the motions of bodies on the earth. The motion of the heavenly bodies, particularly the planets and the sun, was a subject of much active interest at this time. This subject was discussed by students of natural philosophy at Cambridge in 1664. In 1665, the plague broke out. School was suspended and the students were sent home. One of them was Isaac Newton, then a 23-year-old "scholar of the college."

At home in Woolsthorpe, Newton continued to think about these questions. According to his friend and biographer Stukeley, he was inspired as he saw an apple fall to the earth from a tree. It occurred to him that the same force of gravitation which attracts the apple to the earth might also attract the moon to the earth. Newton thought that the centripetal acceleration of the moon in its orbit and the downward acceleration of a body on the earth might have the same origin (Fig. 14–1). The very idea that celestial motions and terrestrial motions were subject to similar laws was a break with tradition.

The acceleration of the moon toward the earth can be computed from its period of revolution and the radius of its orbit. We obtain 0.0089 ft/sec² (see Example 3, Chapter 4). This value is about 3600 times smaller than g, the acceleration due to gravity on the surface of the earth. Newton sought to account for this difference by

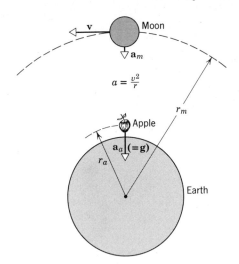

FIGURE 14–1   Both the moon and the apple are accelerated toward the center of the earth. The difference in their motions arises because the moon has a tangential velocity **v** whereas the apple does not.

assuming that the acceleration of a falling body is inversely proportional to the square of its distance from the earth.

The question of what we mean by "distance from the earth" arises immediately. Newton regarded every particle of the earth as contributing to the gravitational attraction it had on other bodies. The distance and direction of particles of the earth from some object are different for each particle. Newton made the daring assumption that the earth could be treated for such purposes as if all the mass were concentrated at its center.

We could treat the earth as a particle with respect to the sun, for example. It is not obvious at all, however, that we could safely treat the earth as a particle with respect to an apple located only a few feet above its surface. On this assumption, however, a falling body near the earth's surface is a distance of one earth radius from the effective center of attraction of the earth, or 4000 miles. The moon is about 240,000 miles away. The inverse square of the ratio of these distances is $(4000/240,000)^2 = 1/3600$, in agreement with the ratio of the accelerations of the moon and apple.

It is believed that Newton made these calculations in 1666. To quote him,

> And in the same year (1666) I began to think of gravity extending to the orb of the moon. . . . I deduced that the forces which keep the planets in their Orbs must [be] reciprocally as the squares of their distances from the centers about which they revolve: and thereby compared the force requisite to keep the Moon in her Orb with the force of gravity at the surface of the earth and found them answer pretty nearly.

His results were not published until 1687, however, when his *Principia Mathematica* appeared. The reason is thought to be that he was not satisfied at first that he could prove his basic assumption about the earth's acting as a mass particle for objects outside it.* Before he could solve this problem exactly, Newton had to invent the calculus. His proof will be given in Section 14–4.

The force on the moon and on the apple depends on the mass of the moon and the

---

* Some students of history believe that he merely wished to avoid controversy. The publication of his *Theory of Light and Colors* had involved him in bitter arguments. Newton was a shy and introspective man. Bertrand Russell writes of Newton, "If he had encountered the sort of opposition with which Galileo had to contend, it is probable that he would never have published a line." It was Halley, a devoted friend with a great interest in celestial mechanics, who persuaded Newton to publish the *Principia.*

mass of the apple, respectively, as well as on that of the earth. Hence, Newton assumed that the gravitational force depended on the masses of the attracting bodies as well as inversely on the square of their distance of separation. He then generalized his concept of gravitational attraction into a law of universal gravitation. He thought that all bodies, no matter where they were located, exerted forces of gravitational attraction upon one another. In order to discover the exact nature of this attractive force, he had to consider bodies of various different masses at significantly different distances from one another. He could not change the distance between the center of the earth and a body on the earth very appreciably, however. It was for this reason that he first compared the motion of the moon and a body on earth. The force between different macroscopic bodies on the earth was so small that it was not detected in Newton's time. Newton apparently realized that this force was small and easily masked by frictional or other forces. Hence, he focused his attention on the motion of the planets in an attempt to confirm his ideas.

Newton's law of universal gravitation, which he developed from these early studies may be expressed as follows:

The force between any two particles having masses m_1 and m_2 separated by a distance r is an attraction acting along the line joining the particles and has the magnitude

$$F = G \frac{m_1 m_2}{r^2}, \tag{14–1}$$

where G is a universal constant having the same value for all pairs of particles. It is important to stress at once many features of this law in order that we understand it clearly.

First, the gravitational forces between two particles are an action-reaction pair. The first particle exerts a force on the second particle that is directed toward the first particle along the line joining the two. Likewise, the second particle exerts a force on the first particle that is directed toward the second particle along the line joining the two. These forces are equal in magnitude but oppositely directed.

The universal constant G must not be confused with the **g** which is the acceleration of a body arising from the earth's gravitational pull on it. The constant G has the dimensions L^3/MT^2 and is a scalar; **g** has the dimensions L/T^2, is a vector, and is neither universal nor constant.

Notice that Newton's law of universal gravitation is not a defining equation for any of the physical quantities (force, mass, or length) contained in it. According to our program for classical mechanics in Chapter 5, force is defined from Newton's second law, $\mathbf{F} = m\mathbf{a}$. The essence of this law, however, is the assumption that the force on a particle, so defined, can be related in a simple way to measurable properties of the particle and of its environment, that is, the existence of simple force laws is assumed. The law of universal gravitation is such a simple law. The constant G must be found from experiment. Once G is determined for a given pair of bodies, we can use that value in the law of gravitation to determine the gravitational forces between any other pairs of bodies.

Notice also that Eq. 14–1 expresses the force between mass particles. If we want to determine the force between extended bodies, as for example the earth and the moon, we must regard each body as decomposed into particles. Then the interaction between all particles must be computed. Integral calculus makes such a calculation possible. Newton's motive in developing the calculus arose in part from a desire to solve such problems. In general, it is incorrect to assume that all the mass of a body can be concentrated as its center of mass for gravitational purposes. This assumption is correct for uniform spheres, however, a result that we shall use often and shall prove in Section 14–4.

Implicit in the law of universal gravitation is the idea that the gravitational force between two particles is independent of the presence of other bodies or the properties of the intervening space. This is consistent with all experimental evidence and has been used by some to rule out the possible existence of "gravity screens."

We can express the law of universal gravitation in vector form. Let the displacement vector $\mathbf{r}_{12}$ point from the particle of mass m_1 to the particle of mass m_2, as Fig. 14–2a shows. The gravitational force $\mathbf{F}_{21}$, exerted on m_2 by m_1, is given in direction and magnitude by the vector relation

$$\mathbf{F}_{21} = -G\frac{m_1 m_2}{r_{12}{}^3}\,\mathbf{r}_{12} \qquad (14\text{--}2a)$$

in which r_{12} is the magnitude of $\mathbf{r}_{12}$. The minus sign in Eq. 14–2a shows that $\mathbf{F}_{21}$ points in a direction opposite to $\mathbf{r}_{12}$; that is, the gravitational force is attractive, m_2 feeling a force directed toward m_1 (see Fig. 14–2). That Eq. 14–2a is indeed an inverse square law can be seen by writing it as $\mathbf{F}_{21} = -(Gm_1 m_2/r_{12}{}^2)(\mathbf{r}_{12}/r_{12})$; here the displacement vector divided by its own magnitude, $\mathbf{r}_{12}/r_{12}$, is simply a unit vector $\mathbf{u}_r$ in the direction of the displacement. If we express the relation in scalar form by equating the magnitudes of each side, a factor r_{12} in the numerator cancels one of the factors of $r_{12}{}^3$ in the denominator and the inverse square relation of Eq. 14–1 results.

The force exerted on m_1 by m_2 is clearly

$$\mathbf{F}_{12} = -G\frac{m_2 m_1}{r_{21}{}^3}\,\mathbf{r}_{21}. \qquad (14\text{--}2b)$$

Note, in Eqs. 14–2, that $\mathbf{r}_{21} = -\mathbf{r}_{12}$ (see Fig. 14–2a,b) so that, as we expect, $\mathbf{F}_{12} = -\mathbf{F}_{21}$ (see Fig. 14–2c); that is, the gravitational forces acting on the two bodies form an action-reaction pair.

14–2 The Constant of Universal Gravitation, G

To determine the value of G it is necessary to measure the force of attraction between two known masses. The first accurate measurement was made by Lord Cavendish in 1798. Significant improvements were made by Poynting and Boys in the nineteenth century. The present accepted value of G was obtained by P. R. Heyl and P. Chizanowski at the U. S. National Bureau of Standards in 1942. This value is

$$G = 6.6732 \times 10^{-11}\ \text{nt-m}^2/\text{kg}^2,$$

accurate to within 0.0031×10^{-11} nt-m^2/kg^2. In the British engineering system this value is 3.436×10^{-8} lb-ft^2/slug^2.

The constant G can be determined by the maximum deflection method illustrated

FIGURE 14–2   The force exerted on m_2 (by m_1), $\mathbf{F}_{21}$, is directed opposite to the displacement, $\mathbf{r}_{12}$, of m_2 from m_1. The force exerted on m_1 (by m_2), $\mathbf{F}_{12}$, is directed opposite to the displacement, $\mathbf{r}_{21}$, of m_1 from m_2. $\mathbf{F}_{21} = -\mathbf{F}_{12}$, the forces being an action-reaction pair.

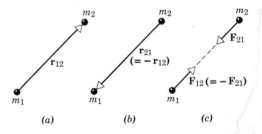

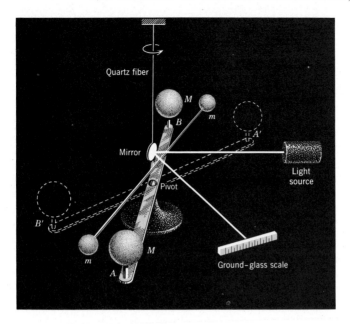

FIGURE 14–3 The Cavendish balance, used for experimental verification of Newton's law of universal gravitation. Masses m,m are suspended from a fiber. Masses M,M can rotate on a stationary support. An image of the lamp filament is reflected by the mirror attached to m,m onto the scale so that any rotation of m,m can be measured.

in Fig. 14–3. Two small balls, each of mass m, are attached to the ends of a light rod. This rigid "dumbbell" is suspended, with its axis horizontal, by a fine vertical fiber. Two large balls each of mass M are placed near the ends of the dumbbell on opposite sides. When the large masses are in the positions A and B, the small masses are attracted, by the law of gravitation, and a torque is exerted on the dumbbell rotating it counterclockwise, as viewed from above. When the large masses are in the positions A' and B', the dumbbell rotates clockwise. The fiber opposes these torques as it is twisted. The angle θ through which the fiber is twisted when the balls are moved from one position to the other is measured by observing the deflection of a beam of light reflected from the small mirror attached to it. If the masses and their distances of separation and the torsional constant of the fiber are known, we can calculate G from the measured angle of twist. The force of attraction is very small so that the fiber must have an extremely small torsion constant if we are to obtain a detectable twist. In Example 1 at the end of this section some data are given from which G can be calculated.

The masses in the Cavendish balance of Fig. 14–3 are, of course, not particles but extended objects. Since they are uniform spheres, however, they act gravitationally as though all their mass were concentrated at their centers (Section 14–4).

Because G is so small, the gravitational forces between bodies on the earth's surface are extremely small and can be neglected for ordinary purposes. For example, two spherical objects each having a mass of 100 kg (about 220-lb weight) and separated by 1.0 meter at their centers attract each other with a force

$$F = \frac{Gm_1 m_2}{r^2} = \frac{(6.67 \times 10^{-11}) \times (100) \times (100)}{(1.0)^2} \text{ nt}$$

$$= 6.7 \times 10^{-7} \text{ nt}$$

or about 1.5×10^{-7} lb! The Cavendish experiment must be a very delicate one indeed.

The large gravitational force which the earth exerts on all bodies near its surface is due to the extremely large mass of the earth. In fact, we can determine the mass of the earth from the law of universal gravitation and the value of G calculated from the Cavendish experiment. For this reason Cavendish is said to have been the first person to "weigh" the earth. Conside the earth, mass M_e, and an object on its surface of mass m. The force of attraction is given both by

$$F = mg$$

and

$$F = \frac{GmM_e}{R_e{}^2}.$$

Here R_e is the radius of the earth, which is the separation of the two bodies, and g is the acceleration due to gravity at the earth's surface. Combining these equations we obtain

$$M_e = \frac{g\,R_e^2}{G} = \frac{(9.80 \text{ meters/sec}^2)(6.37 \times 10^6 \text{ meters})^2}{6.67 \times 10^{-11} \text{ nt-m}^2/\text{kg}^2} = 5.97 \times 10^{24} \text{ kg}$$

or 6.6×10^{21} tons   "weight."

If we were to divide the total mass of the earth by its total volume, we would obtain the average density of the earth. This turns out to be 5.5 gm/cm^3, or about 5.5 times the density of water. The average density of the rock on the earth's surface is much less than this value. We conclude that the interior of the earth contains material of density greater than 5.5 gm/cm^3. From the Cavendish experiment we have obtained information about the nature of the earth's core. (Scc Question 4.)

▪ **Example 1.** Let the small spheres of Fig. 14–3 each have a mass of 10.0 gm and let the light rod be 50.0 cm long. The period of torsional oscillation of this system is found to be 769 sec. Then two large fixed spheres each of mass 10.0 kg are placed near each suspended sphere so as to produce the maximum torsion. The angular deflection of the suspended rod is then 3.96×10^{-3} radian and the distance between centers of the large and small spheres is 10.0 cm. Calculate the universal constant of gravitation G from these data.

The period of torsional oscillation is given by Eq. 13–23,

$$T = 2\pi \sqrt{\frac{I}{\kappa}}.$$

For the rigid dumbbell, if we neglect the contribution of the light rod,

$$I = \Sigma mr^2 = (10.0 \text{ gm})(25.0 \text{ cm})^2 + (10.0 \text{ gm})(25.0 \text{ cm})^2$$

or $I = 1.25 \times 10^{-3}$ kg-m^2.

With $T = 769$ sec, we can obtain the torsional constant κ as

$$\kappa = \frac{4\pi^2 I}{T^2} = \frac{(4\pi^2)(1.25 \times 10^{-3} \text{ kg-m}^2)}{(769 \text{ sec})^2} = 8.34 \times 10^{-8} \text{ kg-m}^2/\text{sec}^2.$$

The relation between the applied torque and the angle of twist is $\tau = \kappa\theta$. We now know κ and the value of θ at maximum deflection.

The torque arises from the gravitational forces exerted by the large spheres on the small ones. This torque will be a maximum for a given separation when the line joining the centers of these spheres is at right angles to the rod. The force on each small sphere is

$$F = \frac{GMm}{r^2},$$

and the moment arm for each force is half the length of the rod ($l/2$). Then,

$$\text{torque} = \text{force} \times \text{moment arm}$$

or

$$\tau = 2\frac{GMm}{r^2}\frac{l}{2}.$$

Combining this with

$$\tau = \kappa\theta,$$

we obtain

$$G = \frac{\kappa\theta r^2}{Mml} = \frac{(8.34 \times 10^{-8} \text{ kg-m}^2/\text{sec}^2)(3.96 \times 10^{-3} \text{ radian})(0.100 \text{ meter})^2}{(10.0 \text{ kg})(0.0100 \text{ kg})(0.500 \text{ meter})}$$

$$= 6.63 \times 10^{-11} \text{ nt-m}^2/\text{kg}^2.$$

Notice that this result is about 1% lower than the accepted value. What have we neglected in this calculation that might account for this difference?

14–3 Inertial and Gravitational Mass and the Principle of Equivalence

The gravitational force on a body is proportional to its mass, as Eq. 14–1 shows. This proportionality between gravitational force and mass is the reason we ordinarily consider the theory of gravitation to be a branch of mechanics, whereas theories of other kinds of force (electromagnetic, nuclear, etc.) may not be.

An important consequence of this proportionality is that we can measure a mass by measuring the gravitational force on it. This can be done by using a spring balance or by comparing the gravitational force on one mass with that on a standard mass, as in a balance; in other words, we can determine the mass of a body by weighing it. This gives us a more practical and more convenient method of measuring mass than is given by our original definition of mass (Section 5–5).

The question arises whether these two methods really measure the same property. The word mass has been used in two quite different experimental situations. For example, if we try to push a block that is at rest on a horizontal frictionless surface, we notice that it requires some effort to move it. The block seems to be inert and tends to stay at rest, or if it is moving it tends to keep moving. Gravity does not enter here at all. It would take the same effort to accelerate the block in gravity-free space. It is the mass of the block which makes it necessary to exert a force to change its motion. This is the mass occurring in $\mathbf{F} = m\mathbf{a}$ in our original experiments in dynamics. We call this mass m the *inertial mass*. Now there is a different situation which involves the mass of the block. For example, it requires effort just to hold the block up in the air at rest above the earth. If we do not support it, the block will fall to the earth with accelerated motion. The force required to hold up the block is equal in magnitude to the force of gravitational attraction between it and the earth. Here inertia plays no role whatever; the property of material bodies, that they are attracted to other objects such as the earth, does play a role. The force is given by

$$F = G\frac{m'M_e}{R_e^2},$$

where m' is the gravitational mass of the block. Are the gravitational mass m' and the inertial mass m of the block really the same?

Newton devised an experiment to test directly the apparent equivalence of inertial and gravitational mass. If we go back (Section 13–5) and look up the derivation of the period of a simple pendulum, we find that the period (for small angles) was given by

$$T = 2\pi \sqrt{\frac{ml}{m'g}},$$

where m is the numerator refers to the inertial mass of the pendulum bob and m' in the denominator is the gravitational mass of the pendulum bob, such that $m'g$ gives the gravitational pull on the bob. Only if we assume that m equals m', as we did there implicitly, do we obtain the expression

$$T = 2\pi \sqrt{\frac{l}{g}}$$

for the period. Newton made a pendulum bob in the form of a tin shell. Into this hollow bob he put different substances, being careful always to have the same weight of substance as determined by a balance. Hence, in all cases the force on the pendulum was the same at the same angle. Because the external shape of the bob was always the same, even the air resistance on the moving pendulum was the same. As one substance replaced another inside the bob, any difference in acceleration could only be due to a difference in the inertial mass. Such a difference would show up by a change in the period of the pendulum. But in all cases Newton found the period of the pendulum to be the same, always given by $T = 2\pi\sqrt{l/g}$. Hence, he concluded that $m = m'$ and that inertial and gravitational masses are equivalent.

In 1909, Eötvös devised an apparatus which could detect a difference of 5 parts in 10^9 in gravitational force. He found that equal inertial masses always experienced equal gravitational forces within the accuracy of his apparatus. A refined version of the Eötvös experiment was reported in 1964 by R. H. Dicke and his collaborators, who improved the accuracy of the original experiment by a factor of several hundred.

In classical physics the equivalence of gravitational and inertial mass was looked upon as a remarkable accident having no deep significance. But in modern physics this equivalence is regarded as a clue leading to a deeper understanding of gravitation. This was, in fact, an important clue leading to the development of the general theory of relativity.

Consider two reference frames: (1) a nonaccelerating (inertial) reference frame in which there is a uniform gravitational field and (2) a reference frame which is accelerating uniformly with respect to an inertial frame but in which there is no gravitational field. In his general theory of relativity, Albert Einstein showed that two such frames are exactly equivalent physically. That is, experiments carried out under the same conditions in these two frames should give the same results. This is the *principle of equivalence.*

Suppose you were standing in an elevator at rest and you suddenly felt the sensation of being pulled toward the floor. You would probably conclude that the elevator was accelerating upward. However, could you not as well conclude that the gravitational force acting on you had increased? Two such reference frames would be equivalent. If the cable broke, the free-fall acceleration could be interpreted as a reduction of the gravitational field to zero.

The principle of equivalence says that no physical measurement can detect a difference between a situation where an observer is accelerating relative to an inertial frame in a region having no gravitational field and a situation in which one is not accelerating in an inertial frame where a uniform gravitational field exists. Of course, if it were found that gravitational mass and inertial mass were measurably different, then a way would indeed exist for telling the difference between the two frames and the principle of equivalence would be invalid.

14–4 Gravitational Effect of a Spherical Distribution of Mass

We have already used the fact that a large sphere attracts particles outside it just as though the mass of the sphere were concentrated at its center. Let us now prove this result.

Consider a uniformly dense shell whose thickness t is small compared to its radius r (Fig. 14–4). We seek the gravitational force it exerts on an external particle P of mass m.

We assume that each particle of the shell exerts on P a force which is proportional to the mass of the small part, inversely proportional to the square of the distance between that part of the shell and P, and directed along the line joining them. We must then obtain the resultant force on P attributable to all parts of the spherical shell.

The small part of the shell at A attracts m with a force $\mathbf{F}_1$. A small part of equal mass at B, equally far from m but diametrically opposite A, attracts m with a force $\mathbf{F}_2$. The resultant of these two forces on m is $\mathbf{F}_1 + \mathbf{F}_2$. Notice, however, that the vertical components of these two forces cancel one another and that the horizontal components, $F_1 \cos \alpha$ and $F_2 \cos \alpha$, are equal. By dividing the spherical shell into pairs of particles like these, we can see at once that all transverse forces on m cancel in pairs. A small mass in the upper hemisphere exerts a force having an upward component on m that will annul the downward component of force exerted on m by an equal symmetrically located mass in the lower hemisphere of the shell. To find the resultant force on m arising from the shell, we need consider only horizontal components.

Let us take as our element of mass of the shell a circular strip labeled dS in the figure. Its length is $2\pi(r \sin \theta)$, its width is $r\, d\theta$, and its thickness is t. Hence, it has a volume

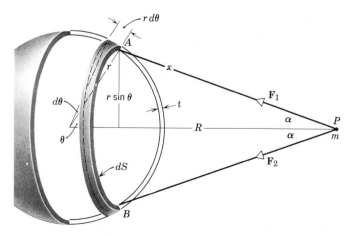

FIGURE 14–4   Gravitational attraction of a section dS of a spherical shell of matter on a particle of mass m.

$$dV = 2\pi t r^2 \sin \theta \, d\theta.$$

Let us call the density ρ, so that the mass within the strip is

$$dM = \rho \, dV = 2\pi t \rho r^2 \sin \theta \, d\theta.$$

The force exerted by dM on the particle of mass m at P is horizontal and has the value

$$dF = G \frac{m \, dM}{x^2} \cos \alpha$$

$$= 2\pi G t \rho m r^2 \frac{\sin \theta \, d\theta}{x^2} \cos \alpha.$$

(14–3)

The variables x, α, and θ are related. From the figure we see that

$$\cos \alpha = \frac{R - r \cos \theta}{x}.$$

(14–4)

Since, by the law of cosines,

$$x^2 = R^2 + r^2 - 2Rr \cos \theta,$$

(14–5)

we have

$$r \cos \theta = \frac{R^2 + r^2 - x^2}{2R}.$$

(14–6)

On differentiating Eq. 14–5, we obtain

$$2x \, dx = 2Rr \sin \theta \, d\theta$$

or

$$\sin \theta \, d\theta = \frac{x}{Rr} \, dx.$$

(14–7)

We now put Eq. 14–6 into Eq. 14–4 and then put Eqs. 14–4 and 14–7 into Eq. 14–3. As a result we eliminate θ and α and obtain

$$dF = \frac{\pi G t \rho m r}{R^2} \left(\frac{R^2 - r^2}{x^2} + 1 \right) dx.$$

This is the force exerted by the circular strip dS on the particle m.

We must now consider every element of mass in the shell and sum up over all the circular strips in the entire shell. This operation is an integration over the shell with respect to the variable x. But x ranges from a minimum value of $R - r$ to a maximum value $R + r$.

Since

$$\int_{R-r}^{R+r} \left(\frac{R^2 - r^2}{x^2} + 1 \right) dx = 4r,$$

we obtain the resultant force

$$F = \int_{R-r}^{R+r} dF = G \frac{(4\pi r^2 \rho t)m}{R^2} = G \frac{Mm}{R^2}$$

(14–8)

where

$$M = (4\pi r^2 t \rho)$$

is the total mass of the shell. This is exactly the same result we would obtain for the force between particles of mass M and m separated a distance R. We have proved, therefore, that *a uniformly dense spherical shell attracts an external mass point as if all its mass were concentrated at its center.*

A solid sphere can be regarded as composed of a large number of concentric shells. If each spherical shell has a uniform density, even though different shells may have different densities, the same result applies to the solid sphere. Hence, a body like the earth, the moon, or the sun, to the extent that they are such spheres, may be regarded gravitationally as point particles to bodies outside them.

Notice that our proof applies only to spheres and then only when the density is constant over the sphere or a function of radius alone.

An interesting result of some significance is the force exerted by a spherical shell on a particle inside it. This force is zero. In this case R is now smaller than r. The limits of our integration over x are now $r - R$ to $R + r$. But

$$\int_{r-R}^{R+r} \left(\frac{R^2 - r^2}{x^2} + 1 \right) dx = 0,$$

so that $F = 0$.

This last result, although not obvious, is plausible because the mass elements of the shell on opposite sides of m now exert forces of opposite directions on m inside. The total annulment depends on the fact that the force varies precisely as an inverse square of the separation distance of two particles. (See Problem 12.) Important consequences of this result will be discussed in the chapters on electricity. There we shall see that the electrical force between charged particles also depends inversely on the square of the distance between them. A consequence of interest in gravitation is that the gravitational force exerted by the earth on a particle decreases as the particle goes deeper into the earth, assuming a constant density for the earth, for the portions of matter in shells external to the position of the particle exert no force on it, the force becoming zero at the center of the earth. Hence, g would be a maximum at the earth's surface and decrease both outward and inward from that point if the earth had constant density. Can you imagine a spherically symmetric distribution of the earth's mass which would not give this result?

Example 2. Suppose a tunnel could be dug through the earth from one side to the other along a diameter, as shown in Fig. 14–5.

(*a*) Show that the motion of a particle dropped into the tunnel is simple harmonic motion. Neglect all frictional forces and assume that the earth has a uniform density.

The gravitational attraction of the earth for the particle at a distance r from the center of the earth arises entirely from that portion of matter of the earth in shells internal to the position of the particle. The external shells exert no force on the particle. Let us assume that the earth's density is uniform at the value ρ. Then the mass inside a sphere of radius r is

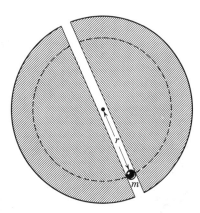

FIGURE 14–5 Example 2. Particle moving in a tunnel through the earth.

$$M' = \rho V' = \rho \frac{4\pi r^3}{3}.$$

This mass can be treated as though it were concentrated at the center of the earth for gravitational purposes. Hence, the force on the particle of mass m is

$$F = \frac{-GM'm}{r^2}.$$

The minus sign is used to indicate that the force is attractive and directed toward the center of the earth.

Substituting for M', we obtain

$$F = -G \frac{(\rho 4\pi r^3)m}{3r^2} = -\left(G\rho \frac{4\pi m}{3}\right) r = -kr.$$

Here $G\rho 4\pi m/3$ is a constant, which we have called k. The force is, therefore, proportional to the displacement r but oppositely directed. This is exactly the criterion for simple harmonic motion.

(*b*) If mail were delivered through this chute, how much time would elapse between deposit at one end and delivery at the other end?

The period of this simple harmonic motion is

$$T = 2\pi \sqrt{\frac{m}{k}} = 2\pi \sqrt{\frac{3m}{G\rho 4\pi m}} = \sqrt{\frac{3\pi}{G\rho}}.$$

Let us take $\rho = 5.51 \times 10^3$ kg/meter^3 and $G = 6.67 \times 10^{-11}$ nt-m^2/kg^2. This gives

$$T = \sqrt{\frac{3\pi}{G\rho}} = \sqrt{\frac{3\pi}{(6.67 \times 10^{-11})(5.51 \times 10^3)}} \text{ sec} = 5050 \text{ sec} = 84.2 \text{ min}.$$

The time for delivery is one-half period, or about 42 min. Notice that this time is independent of the mass of the mail.

The earth does not really have a uniform density. Suppose ρ were some function of r, rather than a constant. What effect would this have on our problem?

14–5 Gravitational Acceleration, g

In previous chapters we have taken the acceleration due to gravity **g** to be a constant. It is well known, however, that even on the surface of the earth it varies measurably depending upon one's position. We can see why by explicitly calculating g from the law of gravitation. Consider, as we did in Section 14–2, the force exerted by the earth (mass M_e) on an object of mass m as defined by Newton's second law of motion and also that given by the law of gravitation,

$$F = ma$$

$$= \frac{GmM_e}{r^2}.$$

We are treating both the objects as spheres. We are also assuming that the density of the earth varies only with respect to the distance from the center and that the size of the object is small compared with the radius of the earth. The measure of the separation of centers is r. When the acceleration of the object is that caused by the gravitational attraction of the earth alone, we define $a = g$. Equating the forces and solving for g we obtain

$$g = \frac{GM_e}{r^2}. \tag{14–9}$$

It should now be apparent that *g* is not a constant but depends upon several factors. First, it depends upon the distance from the center of the earth. Next, *g* varies because the earth deviates slightly from a sphere both in gross shape and in the roughness of the surface. Finally, the density criterion stated above does not hold exactly. In actual measurements yet another effect must be taken into account, the rotation of the earth. Let us discuss a few of these effects briefly.

Table 14–1 shows the variation of *g* with altitude. Even at altitudes of low artificial satellites (100 miles or 161,000 meters) *g* differs from the surface value by only about 2.5%.

Calculations made near the surface are facilitated by relating the fractional change in *g*, *dg/g*, with the fractional change in *r*, *dr/r*. From Eq. 14–9 we easily find

$$\frac{dg}{g} = -2\frac{dr}{r}. \tag{14–10}$$

The gross nonspherical shape of the earth causes significant changes in *g* measured as a function of latitude. To define the problem more closely we usually consider not the earth itself but an imaginary closed surface called the *geoid*. Over the oceans the geoid is defined to coincide with mean sea level, whereas over the continents it is defined as a continuation of this level; in principle the position of the geoid can be found by digging small sea-level canals across the continents and noting the mean water level.

The ancient Greeks believed the earth to be round and one of them, Eratosthenes (*c* 276–194 BC), measured the radius of the earth on the assumption that it is a sphere. He obtained a value of 7400 km, which is to be compared with the modern value of 6371 km. This basic information about the shape of the earth was gradually forgotten and was not rediscovered until the great voyages of exploration of the fifteenth century.

Later it was learned by measurement that, to a good second approximation, the geoid is not a sphere but is an ellipsoid of revolution, flattened along the earth's rotational axis and bulging at the equator. The equatorial radius, in fact, exceeds the polar radius by 21 km. This flattening is caused by centrifugal effects in the rotating, plastic earth. The geoidic surface is not exactly ellipsoidal, lying outside the ellipsoid of closest fit under mountain masses and inside it over the oceans.

The fact that the equator is farther from the center of the earth than are the poles means that there should be a steady increase in the measured value of *g* as one goes from the equator (latitude 0°) to either pole (latitude 90°).

In 1959, it was observed that the orbit of the Vanguard artificial earth satellite,

Table 14–1 VARIATION OF *g* WITH ALTITUDE AT 45° LATITUDE

Altitude (meters)	*g* (meters/sec²)	Altitude (meters)	*g* (meters/sec²)
0	9.806	32,000	9.71
1,000	9.803	100,000	9.60
4,000	9.794	500,000	8.53
8,000	9.782	1,000,000*	7.41
16,000	9.757	380,000,000†	0.00271

* Typical satellite orbit altitude (= 620 miles).
† Radius of moon's orbit (= 240,000 miles).

calculated using values of **g** based on an ellipsoidal geoid, did not agree exactly with the observed orbit. It was concluded that the geoid is best approximated not by an ellipsoid of revolution but by a slightly pear-shaped figure, the small end of the "pear" being in the northern hemisphere and extending about 15 meters above the reference ellipsoid. The motion of a satellite is governed at all times by the value of **g** at its position. Thus an artificial earth satellite forms a useful probe to explore the values of **g** near the surface of the earth and from this to deduce information about the shape of the geoid. These studies* are continuing as of this date.

The final effect we shall consider, that of the earth's rotation, is also dependent upon latitude. This will be introduced as an example.

Example 3. *Effect on g of the rotation of the earth.* Figure 14–6 is a schematic view of the earth looking down on the North Pole. In it we show an enlarged view of a body of mass m hanging from a spring balance at the equator. The forces on this body are the upward pull of the spring balance **w,** which is the apparent weight of the body, and the downward pull of the earth's gravitational attraction $F = GmM_e/R_e^2$. This body is not in equilibrium because it experiences a centripetal acceleration $\mathbf{a}_R$ as it rotates with the earth. There must, therefore, be a net force acting on the body toward the center of the earth. Consequently, the force **F** of gravitational attraction (the true weight of the body) must exceed the upward pull of the balance **w** (the apparent weight of the body).

From the second law of motion we obtain

$$F - w = ma_R,$$

$$\frac{GM_e m}{R_e^2} - mg = ma_R,$$

$$g = \frac{GM_e}{R_e^2} - a_R \qquad \text{at the equator.}$$

At the poles $a_R = 0$ so that

$$g = \frac{GM_e}{R_e^2} \qquad \text{at the poles.}$$

This is the value of g we would obtain anywhere (assuming a spherical earth) were the rotation of the earth to be neglected.

* See, for example, "Satellite Orbits and Their Geophysical Implications" by D. G. King-Hele, in *Contemporary Physics,* April 1961.

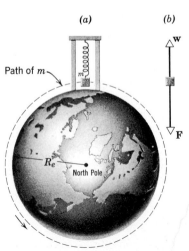

FIGURE 14–6   Example 3. Effect of the earth's rotation on the weight of a body as measured by a spring balance.

Actually the centripetal acceleration is not directed in toward the center of the earth other than at the equator. It is directed perpendicularly in toward the earth's axis of rotation at any given latitude. The detailed analysis is, therefore, really a two-dimensional one. However, the extreme case is at the equator. There

$$a_R = \omega^2 R_e = \left(\frac{2\pi}{T}\right)^2 R_e = \frac{4\pi^2 R_e}{T^2},$$

in which ω is the angular speed of the earth's rotation, T is the period, and R_e is the radius of the earth. Using the values

$$R_e = 6.37 \times 10^6 \text{ meters,}$$

$$T = 8.64 \times 10^4 \text{ sec,}$$

we obtain
$$a_R = 0.0336 \text{ meter/sec}^2.$$

Any measurement of g as a function of latitude will reveal the combined effects of the earth's asphericity and its rotation. Table 14–2 shows the total variation. Comparing the data in this table with the results of Example 3 reveals that more than half the difference between the observed values of g at low and high latitudes is due to the rotation effect.

The consideration of **g** as a function of space leads to the interesting concept of this quantity as a vector field.

14–6   The Gravitational Field

A basic fact of gravitation is that two masses exert forces on one another. We can think of this as a direct interaction between the two mass particles, if we wish. This point of view is called *action-at-a-distance,* the particles interacting even though they are not in contact. Another point of view is the field concept which regards a mass particle as modifying the space around it in some way and setting up a gravitational field. This field then acts on any other mass particle in it, exerting the force of gravitational attraction on it. The field, therefore, plays an intermediate role in our thinking about the forces between mass particles. According to this view we have two separate parts to our problem. First, we must determine the field established by a given distribution of mass particles; and secondly, we must calculate the force that this field exerts on another mass particle placed in it.

For example, consider the earth as an isolated mass. If a body is now brought in the vicinity of the earth, a force is exerted on it. This force has a definite direction and magnitude at each point in space. The direction is radially in toward the center of the earth and the magnitude is *mg*. We can, therefore, associate with each point near the earth a vector **g** which is the acceleration that a body would experience if it were released at this point. We call **g** the *gravitational field* at the point in ques-

Table 14–2   Variation of g with Latitude at Sea Level

Latitude	g (meters/sec^2)	Latitude	g (meters/sec^2)
0°	9.78039	50°	9.81071
10°	9.78195	60°	9.81918
20°	9.78641	70°	9.82608
30°	9.79329	80°	9.83059
40°	9.80171	90°	9.83217

tion. Since

$$\mathbf{g} = \frac{\mathbf{F}}{m},\qquad\qquad (14\text{–}11)$$

we may define the gravitational field at any point as the gravitational force per unit mass at that point. We calculate the force from the field simply by multiplying **g** by the mass m of the particle placed at any point.

The gravitational field is an example of a vector field, each point in this field having a vector associated with it. There are also scalar fields, such as the temperature field in a heat-conducting solid. The gravitational field arising from a fixed distribution of matter is also an example of a stationary field, because the value of the field at a given point does not change with time.

The field concept is particularly useful for understanding electromagnetic forces between moving electric charges. It has distinct advantages, both conceptually and in practice, over the action-at-a-distance concept. The field concept was not used in Newton's day. It was developed much later by Faraday for electromagnetism and only then applied to gravitation. Subsequently, this point of view was adopted for gravitation in the general theory of relativity. The chief purpose of introducing it here is to give the student an early familiarity with a concept that proves to be important in the development of physical theory.

14–7 The Motions of Planets and Satellites

Through a lifelong study of the motions of bodies in the solar system Johannes Kepler (1571–1630) was able to derive empirically the three basic laws of planetary motion bearing his name. Tycho Brahe (1546–1601), who was the last and greatest astronomer to make observations without the use of a telescope, compiled the data. Kepler, who had been Brahe's assistant, after years of laborious calculations found these regularities, now known as Kepler's three laws of planetary motion.

1. All planets move in elliptical orbits having the sun as one focus (the law of orbits).
2. A line joining any planet to the sun sweeps out equal areas in equal times (the law of areas).
3. The square of the period of any planet about the sun is proportional to the cube of the planet's mean distance from the sun (the law of periods).

These laws can be deduced from the laws of motion and the law of universal gravitation. Indeed, Newton used these as basic information in the formulation of his gravitational theory.

Though, as Kepler pointed out, all planets move in elliptical orbits, the sun being at one focus, we can learn a lot about planetary motion by considering the special case of circular orbits.

We shall neglect the forces between planets, considering only the interaction between the sun and a given planet. These considerations apply equally well to the motion of a satellite (natural or artificial) about a planet.

Consider two spherical bodies of masses M and m moving in circular orbits under the influence of each other's gravitational attraction. The center of mass of this system of two bodies lies along the line joining them at a point C such that $mr = MR$ (Fig. 14–7). If there are no external forces acting on this system, the center of mass

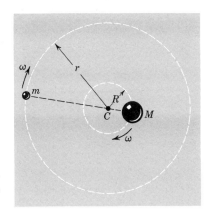

FIGURE 14–7 Two bodies moving in circular orbits under the influence of each other's gravitational attraction. They both have the same angular velocity ω.

has no acceleration. In this case we choose C to be the origin of our reference frame. The large body of mass M moves in an orbit of constant radius R and the small body of mass m in an orbit of constant radius r, both having the same angular velocity ω. In order for this to happen, the gravitational force acting on each body must provide the necessary centripetal acceleration. Since these gravitational forces are simply an action-reaction pair, the centripetal forces must be equal but oppositely directed. That is, $m\omega^2 r$ (the magnitude of the centripetal force exerted by M on m) must equal $M\omega^2 R$ (the magnitude of the centripetal force exerted by m on M). That this is so follows at once, for $mr = MR$ so that $m\omega^2 r = M\omega^2 R$. The specific requirement, then, is that the gravitational force on either body must equal the centripetal force needed to keep it moving in its circular orbit, that is,

$$\frac{GMm}{(R + r)^2} = m\omega^2 r. \qquad (14\text{–}12)$$

If one body has a much greater mass than the other, as in the case of the sun and a planet, its distance from the center of mass is much smaller than that of the other body. Let us therefore assume that R is negligible compared to r. Equation 14–12 then becomes

$$GM_s = \omega^2 r^3,$$

where M_s is the mass of the sun. If we express the angular velocity in terms of the period of the revolution, $\omega = 2\pi/T$, we obtain

$$GM_s = \frac{4\pi^2 r^3}{T^2}. \qquad (14\text{–}13)$$

This is a basic equation of planetary motion; it holds also for elliptical orbits if we define r to be the semi-major axis of the ellipse. Let us consider some of its consequences.

One immediate consequence of Eq. 14–13 is that it predicts Kepler's third law of planetary motion in the special case of circular orbits, for we can express Eq. 14–13 as

$$T^2 = \frac{4\pi^2}{GM_s} r^3.$$

Notice that the mass of the planet is not involved in this expression. Here, $4\pi^2/GM_s$ is a constant, the same for all planets.

When the period T and radius r of revolution are known for any planet, Eq. 14–13 can be used to determine the mass of the sun. For example, the earth's period is

$$T = 365 \text{ days} = 3.15 \times 10^7 \text{ sec,}$$

and its orbital radius is

$$r = 93 \text{ million miles} = 1.5 \times 10^{11} \text{ meters.}$$

Hence,

$$M_s = \frac{4\pi^2 r^3}{GT^2} = \frac{(4\pi^2)(1.5 \times 10^{11} \text{ meters})^3}{(6.67 \times 10^{-11} \text{ nt-m}^2/\text{kg}^2)(3.15 \times 10^7 \text{ sec})^2} \cong 2.0 \times 10^{30} \text{ kg.}$$

The mass of the sun is thus 300,000 times the mass of the earth. The error made in neglecting R compared to r is seen to be trivial, for

$$R = \frac{m}{M}r = \frac{1}{300,000}r \cong 300 \text{ miles;} \qquad \frac{R}{r} 100\% \cong \frac{1}{3000} \text{ of } 1\%.$$

In a similar manner we can determine the mass of the earth from the period and radius of the moon's orbit about the earth. (See Problem 24.)

If we know the mass of the sun M_s and the period of revolution T of any planet about it, we can determine the radius of the planet's orbit r from Eq. 14–13. Since the period is easily obtained from astronomical observations, this method of determining a planet's distance from the sun is fairly reliable.

Equation 14–13 holds also for the motion of artificial satellites about the earth; we need only substitute the mass of the earth M_e for M_s in that equation.

Kepler's second law of planetary motion (see page 262) must, of course, hold for circular orbits. In such orbits both ω and r are constant so that equal areas are swept out in equal times by the line joining a planet and the sun. For the exact elliptical orbits, however, or for any orbit in general, both r and ω will vary. Let us consider this case.

Figure 14–8 shows a particle revolving about C along some arbitrary path. The area swept out by the radius vector in a very short time interval Δt is shown shaded in the figure. This area, neglecting the small triangular region at the end, is one-half the base times the altitude or approximately $\frac{1}{2}(r\omega \, \Delta t) \cdot r$. This expression becomes more exact in the limit as $\Delta t \to 0$, the small triangle going to zero more rapidly than the large one. The rate at which area is being swept out instantaneously is therefore

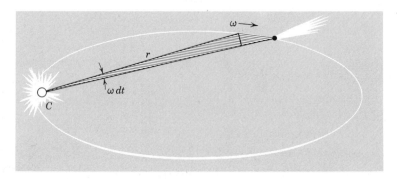

FIGURE 14–8   A comet moving along an elliptical path with the sun C at the focus of the ellipse. In time dt the comet sweeps out an angle $d\theta = \omega \, dt$.

$$\lim_{\Delta t \to 0} \frac{\frac{1}{2}(r\omega\,\Delta t)(r)}{\Delta t} = \frac{1}{2}\omega r^2.$$

But $m\omega r^2$ is simply the angular momentum of the particle about C. Hence, Kepler's second law, which requires that the rate of sweeping out of area $\frac{1}{2}\omega r^2$ be constant, is entirely equivalent to the statement that the angular momentum of any planet about the sun remains constant. The angular momentum of the particle about C cannot be changed by a force on it directed toward C. Kepler's second law would, therefore, be valid for any *central force,* that is, any force directed toward the sun. The exact nature of this force—how it depends on distance of separation or other properties of the bodies—is not revealed by this law.

It is Kepler's first law that requires the gravitational force to depend exactly on the inverse square of the distance between two bodies, that is, on $1/r^2$. Only such a force, it turns out, can yield planetary orbits which are elliptical with the sun at one focus.

14–8 Gravitational Potential Energy

In Chapter 7 we discussed the gravitational potential energy of a particle (mass m) and the earth (mass M). We considered only the special case in which the particle remains close to the earth so that we could assume the gravitational force acting on the particle to be constant for all positions of the particle. In this section we remove that restriction and consider particle-earth separations that may be appreciably greater than the earth's radius.

Equation 7–5b, which we may write as,

$$\Delta U = U_b - U_a = -W_{ab}, \tag{7–5b}$$

defines the change ΔU in the potential energy of any system, in which a conservative force (gravity, say) acts, as the system changes from configuration a to configuration b. W_{ab} is the work done by that conservative force as the system changes.

The potential energy of the system in any arbitrary configuration b is (see Eq. 7–5b)

$$U_b = -W_{ab} + U_a. \tag{14–14}$$

To give a value to U_b we must (arbitrarily) choose configuration a to be some agreed-upon reference configuration and we must assign to U_a some (arbitrarily) agreed-upon value, usually zero.

In Chapter 7 we chose as a reference configuration for the earth-particle system that in which the particle is resting on the surface of the earth and we assigned to his configuration the potential energy $U_a = 0$. When the particle is at a height y above the surface of the earth, the potential energy $U\,(= U_b)$ is given from Eq. 14–14 as

$$U = -W_{ab} + 0 = -(-mg)(y) = mgy.$$

The conservative force in question, gravity, points down and has the value $(-mg)$; the displacement of the particle $(+y)$ points up from the reference level; hence the difference in sign for these quantities.

For the more general case, in which the restriction $y \ll R$ (in which R is the radius of the earth) is not imposed, we find it convenient to select a different reference configuration, namely that in which the particle and the earth are infinitely far apart. We assign the value zero to the potential energy of the system in this configuration.

Thus the zero-potential-energy configuration is also the zero-force configuration. We made a similar choice when we defined the zero-energy configuration of a spring to be its normal unstressed state, for which the restoring force is zero.

When the particle of mass m is a distance r from the center of the earth, the system potential energy is given by Eq. 14–14 as

$$U(r) = -W_{\infty r} + 0 \tag{14–15}$$

in which $W_{\infty r}$ is the work done by the conservative force (gravity) on the particle as the particle moves in from infinity to a distance r from the center of the earth. For simplicity we assume for the present that the particle moves toward the earth along a radial line. The gravitational force $F(r)$ acting on the particle (assuming $r \gg R$) will then be $-GMm/r^2$, the minus sign indicating an attractive force, that is, a force that pulls the particle toward the earth. We may then find $U(r)$ from Eq. 14–15 as

$$
\begin{aligned}
U(r) &= -W_{\infty r} \\
&= -\int_{\infty}^{r} F(r')\, dr' \\
&= -\int_{\infty}^{r} \left(-\frac{GMm}{r'^2}\right) dr' = -\left. \frac{GMm}{r'} \right|_{\infty}^{r} \\
&= -\frac{GMm}{r}.
\end{aligned}
\tag{14–16}
$$

The minus sign indicates that the potential energy is negative at any finite distance; that is, the potential energy is zero at infinity and decreases as the separation distance decreases. This corresponds to the fact that the gravitational force exerted on the particle by the earth is attractive. As the particle moves in from infinity, the work $W_{\infty r}$ done by this force on the particle is positive, which means, from Eq. 14–15, that $U(r)$ is negative.

Equation 14–16 holds no matter what path is followed by the particle in moving in from infinity to radius r. We can show this by breaking up any arbitrary path into infinitesimal steplike portions, which are drawn alternately along the radius and perpendicular to it (Fig. 14–9). No work is done along perpendicular segments, like AB, because along them the force is perpendicular to the displacement. But the work done along the radial parts of the path, such as BC, adds up to the work done in going directly along a radial path, such as AE. The work done in moving the particle be-

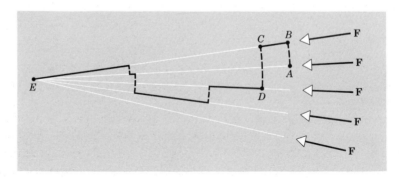

FIGURE 14–9   Work done in taking a mass from A to E is independent of the path.

tween two points in a gravitational field is, therefore, independent of the actual path connecting these points. Hence, the gravitational force is a conservative force.

Equation 14–16 shows that the potential energy of the particles M and m is a characteristic of the system $M + m$. The potential energy is a property of the system of bodies, rather than of either body alone. The potential energy changes whether M or m is displaced; each is in the gravitational field of the other. Nor does it make any sense to assign part of the potential energy to M and part of it to m. Often, however, we do speak of the potential energy of a body m (planet or stone, say) in the gravitational field of a much more massive body M (sun or earth, respectively). The justification for speaking as though the potential energy belongs to the planet or to the stone alone is this: When the potential energy of a system of two bodies changes into kinetic energy, the lighter body gets most of the kinetic energy. The sun is so much more massive than a planet that the sun receives hardly any of the kinetic energy; and the same is true for the earth in the earth-stone system.

We can derive the gravitational force from the potential energy. The relation for spherically symmetric potential energy functions is $F = -dU/dr$; (see Eq. 7–7). This relation is the converse of Eq. 14–16. From it we obtain

$$F = -\frac{dU}{dr} = -\frac{d}{dr}\left(-\frac{GMm}{r}\right) = -\frac{GMm}{r^2}. \tag{14–17}$$

The minus sign here shows that the force is an attractive one, directed inward along a radius opposite to the radial displacement vector.

▨ **Example 4.** *Escape velocity.* We can readily find the gravitational potential energy of a particle of mass m at the surface of the earth as (Eq. 14–16) $U(R) = -GM_em/R_e$. The amount of work required to move a body from the surface of the earth to infinity is GM_em/R_e, or about 6.0×10^7 joules/kg. If we could give a projectile more than this energy at the surface of the earth, then, neglecting the resistance of the earth's atmosphere, it would escape from the earth never to return. As it proceeds outward its kinetic energy decreases and its potential energy increases, but its speed is never reduced to zero. The critical initial speed, called the escape speed v_0, such that the projectile does not return, is given by

$$\tfrac{1}{2}mv_0^2 = \frac{GM_em}{R_e}$$

or

$$v_0 = \sqrt{2\frac{GM_e}{R_e}} = 7.0 \text{ miles/sec} = 25{,}000 \text{ miles/hr} = 11.2 \text{ km/sec}.$$

Should a projectile be given this intial speed, it would escape from the earth. For initial speeds less than this the projectile will return. Its kinetic energy becomes zero at some finite distance from the earth and the projectile falls back to earth. ▨

The lighter molecules in the earth's upper atmosphere can attain high enough speeds by thermal agitation to escape into outer space. Hydrogen gas, which must have been present in the earth's atmosphere a long time ago, has now disappeared from it. Helium gas escapes at a steady rate, much of it resupplied by radioactive decay from the earth's crust. The escape speed for the sun is much too great to allow hydrogen to escape from its atmosphere. On the other hand, for most gases the speed of escape on the moon is so small that it can hardly keep any atmosphere at all. (See Problem 25.)

14–9   Potential Energy for Many-Particle Systems

If two particles are separated by a distance r, their potential energy is given from Eq. 14–15 as

$$U(r) = -W_{\infty r} \qquad (14\text{–}15)$$

in which $W_{\infty r}$ is the work done by the gravitational force as the particles move from an infinite separation to separation r. We now give another interpretation to $U(r)$.

Let us balance out the gravitational force by an external force applied by some external agent and let us arrange it so that, at all times, this external force is equal and opposite to the gravitational force for each particle. The work done by the external force as the particles move from an infinite separation to separation r is not $W_{\infty r}$ but $-W_{\infty r}$; this follows because the displacements are the same but the forces are equal and opposite. Thus we may interpret Eq. 14–15 as follows: *The potential energy of a system of particles is equal to the work that must be done by an external agent to assemble the system, starting from the standard reference configuration.*

Thus, if you lift a stone of mass m a distance of y above the earth's surface, you are the external agent (separating earth and stone) and the work you do in "assembling the system" is $+mgy$, which is also the potential energy. Similarly, the work done by an external agent as a body of mass m moves in from infinity to a distance r from the earth is negative because the agent must exert a restraining force on the body; this is in agreement with Eq. 14–15.

These considerations hold for systems that contain more than two particles. Consider three bodies of masses m_1, m_2, and m_3. Let them initially be infinitely far from one another. The problem is to compute the work done by an external agent to bring them into the positions shown in Fig. 14–10. Let us first bring m_2 in toward m_1 from an infinite separation to the separation $\mathbf{r}_{12}$. The work done against the gravitational force exerted by m_1 on m_2 is $-Gm_1m_2/r_{12}$. Now let us bring m_3 in from infinity to the separation $\mathbf{r}_{13}$ from m and $\mathbf{r}_{23}$ from m_2. The work done against the gravitational force exerted by m_1 on m_3 is $-Gm_1m_3/r_{13}$ and that against the gravitational force exerted by m_2 on m_3 is $-Gm_2m_3/r_{23}$. The total work done in assembling this system is the total potential energy of the system

$$-\left(\frac{Gm_1m_2}{r_{12}} + \frac{Gm_1m_3}{r_{13}} + \frac{Gm_2m_3}{r_{23}} \right)$$

Notice that no vector operations are needed in this procedure.

No matter how we assemble the system, that is, regardless of which bodies are moved or which paths are taken, we always find this same amount of work required to bring the bodies into the configuration of Fig. 14–10 from an initial infinite separa-

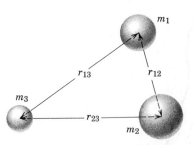

FIGURE 14–10   Three masses m_1, m_2, and m_3 brought together from infinity.

tion. The potential energy must, therefore, be associated with the system rather than with any one or two bodies. If we wanted to separate the system into three isolated masses once again, we would have to supply an amount of energy

$$+ \left(\frac{Gm_1m_2}{r_{12}} + \frac{Gm_1m_3}{r_{13}} + \frac{Gm_2m_3}{r_{23}} \right)$$

This energy may be regarded as a sort of *binding energy* holding the mass particles together in the configuration shown.

Just as we can associate elastic potential energy with the compressed or stretched configuration of a spring holding a mass particle, so we can associate gravitational potential energy with the configuration of a system of mass particles held together by gravitational forces. Similarly, if we want to think of the elastic potential energy of a particle as being stored in the spring, so we can think of the gravitational potential energy as being stored in the gravitational field of the system of particles. A change in either configuration results in a change of potential energy.

These concepts occur again when we meet forces of electric or magnetic origin, or, in fact, of nuclear origin. Their application is rather broad in physics. The advantage of the energy method over the dynamical method is derived from the fact that the energy method uses scalar quantities and scalar operations rather than vector quantities and vector operations. When the actual forces are not known, as is often the case in nuclear physics, the energy method is essential.

14–10 Energy Considerations in the Motions of Planets and Satellites

Consider again the motion of a body of mass m (planet or satellite, say) about a massive body of mass M (sun or earth, say). We shall consider M to be at rest in an inertial reference frame with the body m moving about it in a circular orbit. The potential energy of the system is

$$U(r) = -\frac{GMm}{r},$$

where r is the radius of the circular orbit. The kinetic energy of the system is

$$K = \tfrac{1}{2}m\omega^2r^2$$

the sun being at rest. From the equation preceding Eq. 14–13 we obtain

$$\omega^2 r^2 = \frac{GM}{r},$$

so that $$K = \frac{1}{2}\frac{GMm}{r}.$$

The mechanical energy is

$$E = K + U = \frac{1}{2}\frac{GMm}{r} - \frac{GMm}{r} = -\frac{GMm}{2r}. \tag{14–18}$$

This energy is constant and is negative. Now the kinetic energy can never be negative, but from Eq. 14–18 we see that it must go to zero as the separation goes to infinity. The potential energy is always negative, except for its zero value at infinite separa-

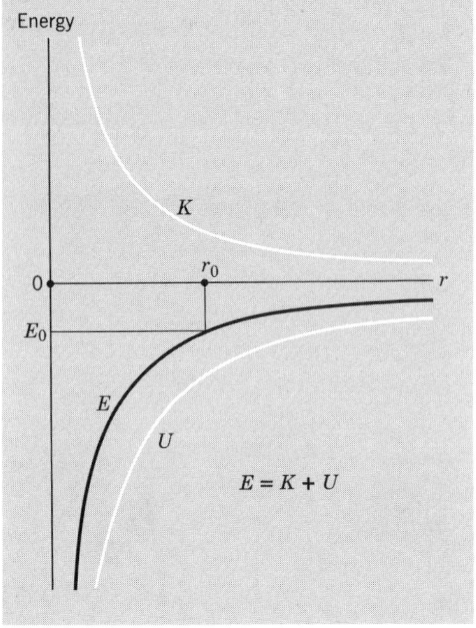

FIGURE 14–11 Kinetic energy K, potential energy U, and mechanical energy $E = U + K$ of a body in circular planetary motion. A planet with total energy $E_0 < 0$ will remain in an orbit of radius r_0. The farther the planet is from the sun, the greater (that is, less negative) its (constant) total energy E. To escape from the center of force and still have kinetic energy at infinity, it would need positive total energy.

tion. The meaning of the negative mechanical energy, then, is that the system is a closed one, the planet m always being bound to the attracting solar center M and never escaping from it (Fig. 14–11).

Even when we consider elliptical orbits, in which r and ω vary, the mechanical energy is negative. It is also constant, corresponding to the fact that gravitational forces are conservative. Hence, both the mechanical energy and the total angular momentum are constant in planetary motion. These quantities are often called *constants of the motion.* We obtain the actual orbit of a planet with respect to the sun by starting with these conservation relations and eliminating the time variable by use of the laws of dynamics and gravitation. The result is that planetary orbits are elliptical.

In the earlier theories of the atom, as in the Bohr theory of the hydrogen atom, these identical mechanical relations are used in describing the motion of an electron about an attracting nuclear center. These same relations are used for open orbits (total energy positive) as in the experiments of Rutherford on the scattering of charged nuclear particles. Central forces, and particularly inverse square forces, are encountered often in physical systems.

Example 5. What is the binding energy of the earth-sun system? Neglect the presence of other planets or satellites.

For simplicity assume that the earth's orbit about the sun is circular at a radius r_{es}. The work done against the gravitational force to bring the earth and sun from an infinite separation to a separation r_{es} is

$$-G\frac{M_s M_e}{r_{es}} \cong -5.0 \times 10^{33} \text{ joules,}$$

where we take $M_s \cong 330{,}000 M_e$, $M_e = 6.0 \times 10^{24}$ kg, $r_{es} = 150 \times 10^9$ meters. The minus sign indicates that the force is attractive, so that work is done by the gravitational force. It would take an equivalent amount of work by an outside agent to separate these bodies completely from rest. Because the kinetic energy of the earth in its orbit is half the magnitude of the potential energy of the earth-sun system, only half of this work is needed to break up the system, so that the effective binding energy, assuming that the earth-sun-system is at rest after breakup, is about 2.5×10^{33} joules.

What effect does the presence of the moon and other planets have on the energy binding the earth to the solar system?

QUESTIONS

1. If the force of gravity acts on all bodies in proportion to their masses, why doesn't a heavy body fall faster than a light body?

2. How does the weight of a body vary en route from the earth to the moon? Would its mass change?

3. Would we have more sugar to the pound at the pole or the equator?

4. Does the concentration of the earth's mass near its center change the variation of g with height compared with a homogeneous sphere? How?

5. Because the earth bulges near the equator, the source of the Mississippi River, although high above sea level, is nearer to the center of the earth than is its mouth. How can the river flow "uphill"?

6. The earth is an oblate spheroid because of the "flattening" effect of the earth's rotation. Is a degree of latitude larger or smaller near either pole than near the equator? Why?

7. Why can we learn more about the shape of the earth by studying the motion of an artificial satellite than by studying the motion of the moon?

8. How can one determine the mass of the moon?

9. One clock is based on an oscillating spring, the other on a pendulum. Both are taken to Mars. Will they keep the same time there that they kept on Earth? Will they agree with each other? Explain. Mars has a mass 0.1 that of Earth and a radius half as great.

10. From Kepler's second law and observations of the sun's motion as seen from the earth, we can conclude that the earth is closer to the sun during winter in the Northern hemisphere than during summer. Explain.

11. Does the law of universal gravitation require the planets of our solar system to have the actual orbits observed? Would planets of another star, similar to our Sun, have the same orbits? Suggest factors that might have determined the special orbits observed.

12. The gravitational force exerted by the sun on the Moon is greater than (about twice as great as) the gravitational force exerted by the earth on the moon. Why then doesn't the moon escape from the earth (during a solar eclipse, for example)?

13. Explain why the following reasoning is wrong. "The sun attracts all bodies on the earth. At midnight, when the sun is directly below, it pulls on an object in the same direction as the pull on the earth on that object; at noon, when the sun is directly above, it pulls on an object in a direction opposite to the pull of the earth. Hence, all objects should be heavier at midnight (or night) than they are at noon (or day)."

14. The gravitational attraction of the sun and the moon on the earth produces tides. The sun's tidal effect is about half as great as that of the moon's. The direct pull of the sun on the earth, however, is about 175 times that of the moon. Why is it then that the moon causes the larger tides?

15. If lunar tides slow down the rotation of the earth (owing to friction), the angular momentum of the earth decreases. What happens to the motion of the moon as a consequence of the conservation of angular momentum? Does the sun (and solar tides) play a role here?

16. Would you expect the total energy of the solar system to be constant? The total angular momentum? Explain your answers.

17. Does a rocket really need the escape speed of 25,000 miles/hr initially to escape from the earth?

18. Objects at rest on the earth's surface move in circular paths with a period of 24 hr. Are they "in orbit" in the sense that an earth satellite is in orbit? Why not? What would the length of the "day" have to be to put such objects in true orbit?

19. Neglecting air friction and technical difficulties, can a satellite be put into an orbit by being fired from a huge cannon at the earth's surface? Explain.

20. Can a satellite move in a stable orbit in a plane not passing through the earth's center? Explain.

21. As measured by an observer on earth would there be any difference in the periods of two satellites, each in a circular orbit near the earth in an equatorial plane, but one moving eastward and the other westward?

22. After Sputnik I was put into orbit we were told that it would not return to earth but would burn up in its descent. Considering the fact that it did not burn up in its ascent, how is this possible?

23. Show that a satellite may speed down: that is, show that if frictional forces cause a satellite to lose total energy, it will move into an orbit closer to the earth and may have increased kinetic energy.

24. An artificial satellite is in a circular orbit about the earth. How will its orbit change if one of its rockets is momentarily fired (*a*) toward the earth, (*b*) away from the earth, (*c*) in a forward direction, (*d*) in a backward direction, (*e*) at right angles to the plane of the orbit?

25. Inside a space ship what difficulties would you encounter in walking? In jumping? In drinking?

26. If a planet of given density were made larger, its force of attraction for an object on its surface would increase because of the planet's greater mass but would decrease because of the greater distance from the object to the center of the planet. Which effect predominates?

27. Consider a hollow spherical shell. How does the gravitational potential inside compare with that on the surface? What is the gravitational field strength inside?

28. A stone is dropped along the center of a deep vertical mine shaft. Assume no air resistance but consider the earth's rotation. Will the stone continue along the center of the shaft? If not, describe its motion.

PROBLEMS

1(1). How far from the earth must a body be along a line toward the sun so that the sun's gravitational pull balances the earth's? The sun is 9.3×10^7 miles away and its mass is $3.24 \times 10^5 \, M_e$. *Answer:* 1.6×10^5 miles.

2(1). A spaceship is going from the earth to the moon in a trajectory along the line joining the centers of the two bodies. At what distance from the earth will the net gravitational force on the spaceship be zero? See Appendix *C*.

3(1). An 800-kg mass and a 600-kg mass are separated by 0.25 meter. What is the net gravitational force due to these masses acting on a 1.0-kg mass at a point 0.20 meter from the 800-kg mass and 0.15 meter from the 600-kg mass?
Answer: 2.2×10^{-6} nt, at right angles toward a line joining the centers.

4(4). Two concentric shells of uniform density of mass M_1 and M_2 are situated as shown in Fig. 14–12. Find the force on a particle of mass m when the particle is located at (*a*) $r = a$, (*b*) $r = b$, and (*c*) $r = c$. The distance r is measured from the center of the shells.

5(4). With what speed would mail pass through the center of the earth if it were delivered by the chute of Example 2?
Answer: 1.1×10^4 meters/sec.

6(4). (*a*) Show that in a chute through the earth along a chord line, rather than along a diameter, the motion of an object will be simple harmonic; assume a uniform earth density. (*b*) Find the period. (*c*) Will the object attain the same maximum speed along a chord as it does along a diameter?

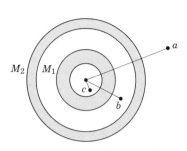

FIGURE 14–12   Problem 4(4).

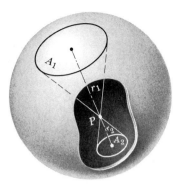

FIGURE 14–13   Problem 7(4).

7(4). Consider a mass particle at a point P anywhere inside a spherical shell of matter. Assume the shell is of uniform thickness and density. Construct a narrow double cone with apex at P intercepting areas A_1 and A_2 on the shell (Fig. 14–13). (*a*) Show that the resultant gravitational force exerted on the particle at P by the intercepted mass elements is zero. (*b*) Show then that the resultant gravitational force of the entire shell on an internal particle is zero everywhere.

8(4). The following problem is from the 1946 "Olympic" examination of Moscow State University (see Fig. 14–14): A spherical hollow is made in a lead sphere of radius R, such that its surface touches the outside surface of the lead sphere and passes through its center. The mass of the sphere before hollowing was M. With what force, according to the law of universal gravitation, will the lead sphere attract a small sphere of mass m, which lies at a distance d from the center of the lead sphere on the straight line connecting the centers of the spheres and of the hollow?

9(5). At what altitude above the earth's surface would the acceleration of gravity be 4.9 meters/sec². The mass of the earth is 6.0×10^{24} kg and its mean radius is 6.4×10^6 meters. *Answer:* 2.7×10^6 meters.

10(5). (*a*) What is the period of a "seconds pendulum" (period $= 2$ sec on earth) on the surface of the moon? The moon's mass is 7.35×10^{22} kg and its radius is 1,720 km. (*b*) Why should a "seconds pendulum" have a period of two seconds rather than one second?

11(5). A certain hypothetical planet has a radius of 500 km and a surface gravity of 3.0 meters/sec². (*a*) What is the gravitational acceleration 100 km above the surface of the planet? (*b*) What is the mass of the planet? *Answer:* (*a*) 2.1 meters/sec². (*b*) 1.1×10^{22} kg.

12(5). (*a*) Calculate the acceleration due to gravity on the surface of the moon if g_{earth} is 9.8 meters/sec², the moon's radius is 27% of the earth's radius and the moon's mass is 1.2% that of the earth. (*b*) What will an object weigh on the moon's surface if it weighs 100 nt on the earth's surface? (*c*) How many earth radii must this same object be from the center of the earth if it is to weigh the same as it does on the moon? (*d*) Is the gravitational force of the earth on the moon greater than, less than, or the same as the gravitational force of the moon on the earth?

13(5). Certain neutron stars (extremely dense stars) are believed to be rotating at about one revolution per second. If such a star has a radius of 20 km, what must be its mass so that objects on its surface will be attracted to the star and not "thrown off" by the rapid rotation? *Answer:* 4.7×10^{24} kg.

14(5). The fact that g varies from place to place over the earth's surface drew attention when

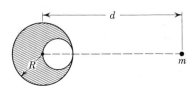

FIGURE 14–14   Problem 8(4).

Jean Richer in 1672 took a pendulum clock from Paris to Cayenne, French Guiana, and found that it lost 2.5 min/day. If $g = 9.81$ meters/sec² in Paris, what is g in Cayenne?

15(5). (*a*) If g is to be determined by dropping an object through a distance of (exactly) 10 meters, how accurately must the time be measured in order to obtain a result good to 0.1%? (*b*) Compare the accuracy of this method to that of timing a pendulum through a large number of swings.

Answer: (*a*) To 0.2%, or to 2.8×10^{-3} sec.

16(5). A scientist is making a precise measurement of g at a certain point in the Indian Ocean (on the equator) by timing the swings of a pendulum of accurately known construction. To provide a stable base the measurements are conducted in a submerged submarine. It is observed that a slightly different result for g is obtained when the submarine is moving eastward through the point than when it is moving westward, the speed in each case being 10 miles/hr. Account for this difference and calculate the fractional error $\Delta g/g$ in g.

17(5). A body is suspended on a spring balance in a ship sailing along the equator with a speed v. (*a*) Show that the scale reading will be very close to $W_0(1 \pm 2\omega v/g)$, where ω is the angular speed of the earth and W_0 is the scale reading when the ship is at rest. (*b*) Explain the plus or minus.

18(5). Masses m, assumed equal, hang from strings of different lengths on a balance at the surface of the earth, as shown in Fig. 14–15. If the strings have negligible mass and differ in length by h, (*a*) show that the error in weighing, associated with the fact that W' is closer to the earth than W, is $W' - W = 8\pi G\rho mh/3$ in which ρ is the mean density of the earth (5.5 gm/cm³). (*b*) Find the difference in length which will give an error of one part in a million.

19(5). An object at rest on the earth's equator is accelerated (*a*) toward the center of the earth because the earth rotates, (*b*) toward the sun because the earth revolves around the sun in an almost circular orbit, and (*c*) toward the center of our galaxy; take the period of the sun's rotation about the galactic center to be 2.5×10^8 yr and the sun's distance from this center to be 2.4×10^{20} meters. Compare these three accelerations.

Answer: (*a*) 3.4×10^{-2} meters/sec². (*b*) 5.9×10^{-3} meters/sec². (*c*) 1.5×10^{-10} meters/sec².

20(7). Calculate the shortest possible period for an earth satellite from the value of G and the values of the earth's mass (6.0×10^{24} kg) and radius (6.4×10^6 meters).

21(7). (*a*) Can a satellite be sent out to a distance where it will revolve about the earth with an angular velocity equal to that at which the earth rotates, so that it remains always above the same point on the earth? (*b*) What would be the radius of such an orbit?

Answer: (*a*) Yes. The plane of the orbit must be equatorial. (*b*) 4.2×10^4 km.

22(7). (*a*) With what horizontal speed must a satellite be projected at 100 miles above the surface of the earth so that it will have a circular orbit about the earth? Take the earth's radius as 4000 miles? (*b*) What will be the period of rotation?

23(7). The planet Mars has a satellite, Phobos, which travels in an orbit of radius 9.4×10^6 meters with a period of 7 hr 39 min. Calculate the mass of Mars from this information.

Answer: 6.5×10^{23} kg.

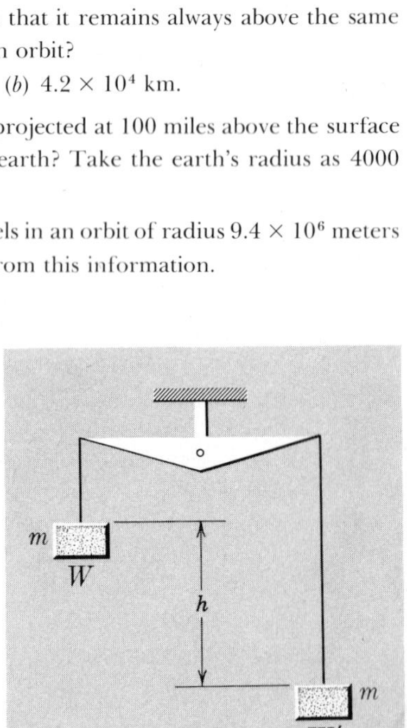

FIGURE 14–15 Problem 18(5).

24(7). What is the shortest period for any satellite revolving about a hypothetical planet whose radius is 500 km and whose surface gravity is 3.0 meters/sec².

25(7). The mean distance of Mars from the sun is 1.52 times that of Earth from the sun. Find the number of years required for Mars to make one revolution about the sun.
Answer: 1.88 yr.

26(7). Determine the mass of the earth from the period T and the radius r of the moon's orbit about the earth, $T = 27.3$ days and $r = 2.39 \times 10^5$ miles.

27(7). A hypothetical planet has the same mean density as the earth and a diameter twice as large. (*a*) What is the mass of a man on the planet if his mass on the earth is 100 kg? (*b*) What is the value of g at the surface of the planet? (*c*) What is the weight of a man on the planet if his mass on the earth is 100 kg? (*d*) What is the speed of a satellite in a circular orbit just above the surface of the planet if the corresponding speed for the earth is 8000 meters/sec?
Answer: (*a*) 100 kg. (*b*) 20 meters/sec². (*c*) 2.0×10^3 nt. (*d*) 1.6×10^4 meters/sec.

28(7). Three identical bodies of mass M are located at the vertices of an equilateral triangle with side L. At what speed must they move if they all revolve under the influence of one another's gravity in a circular orbit circumscribing the triangle while still preserving the equilateral triangle?

29(7). (*a*) Satellite A is in a circular earth orbit with radius R and satellite B is in a circular earth orbit with radius $4R$. Calculate the ratio of the periods of revolution, T_A/T_B. (*b*) A pendulum and a mass-spring system oscillate with approximately the same frequency on the earth's surface. How do their frequencies compare if they are mounted first in satellite A and then in satellite B?
Answer: (*a*) $T_A/T_B = \frac{1}{8}$. (*b*) The pendulum frequency is zero; the mass-spring frequency is unchanged.

30(7). A projectile is fired vertically from the earth's surface with an initial speed of 10 km/sec. Neglecting atmospheric friction, how far above the surface of the earth would it go? Take the earth's radius as 6400 km.

31(8). (*a*) What is the escape velocity on a hypothetical planet whose radius is 500 km and whose surface gravity is 3.0 meters/sec²? (*b*) How high will a particle rise if it leaves the surface of the planet with a vertical velocity of 1000 meters/sec? (*c*) With what speed will an object hit the planet if it is dropped from a height of 1000 km (1500 km from the center of the planet)?
Answer: (*a*) 1700 meters/sec. (*b*) 250 km. (*c*) 1400 meters/sec.

32(8). (*a*) What is the speed necessary for a 1.0-kiloton space ship at a distance from the sun equal to the radius of Jupiter's orbit (7.8×10^8 km) to escape from the sun's gravitational field? (*b*) What average power would the ship's engines need to supply in order to achieve this speed in 60 days? (*c*) How does this compare with the power outputs of terrestrial power stations?

33(8). Mars has a mean diameter of 6900 km, Earth one of 1.3×10^4 km. The mass of Mars is $0.11M_e$. (*a*) How does the mean density of Mars compare with that of Earth? (*b*) What is the value of g on Mars? (*c*) What is the escape velocity on Mars?
Answer: (*a*) $\rho_M = 0.73\, \rho_E$. (*b*) 3.8 meters/sec.² (*c*) 5.1 km/sec.

34(8). (*a*) Show that to escape from the atmosphere of a planet a necessary condition for a molecule is that it have a speed such that $v^2 > 2GM/r$, where M is the mass of the planet and r is the distance of the molecule from the center of the planet. (*b*) Determine the escape speed from the earth for an atmospheric particle 1000 km above the earth's surface. (*c*) Do the same for the moon and the sun.

35(8). Masses of 200 and 800 gm are 12 cm apart. (*a*) Find the gravitational force per unit mass on an object at a point on the line joining the masses 4.0 cm from the 200-gm mass. (*b*) Find the gravitational potential energy per unit mass at that point. (*c*) How much work per unit mass is needed to move this object to a point 4.0 cm from the 800-gm mass along the line of centers?
Answer: (*a*) Zero. (*b*) -1.0×10^{-9} joule/kg. (*c*) -5.0×10^{-10} joules/kg.

36(8). (*a*) Write an expression for the potential energy of a body of mass m in the gravitational field of the earth and moon. Let M_e be the earth's mass, M_m the moon's mass (where $M_e = 81M_m$), R the distance from the earth's center, and r the distance from the moon's center. The distance between earth and moon is about 240,000 miles. (*b*) At what point between the earth and the moon will the total gravitational field strength attributable to the earth and moon be zero? (*c*) What will the gravitational potential energy and the gravitational field strength be for m when it is on the earth's surface? (*d*) Answer for the moon's surface.

37(8). Two identical stars are separated by a distance of 10^{10} meters. They each have a mass of 10^{30} kg and a radius of 10^5 meters. They are initially at rest with respect to one another. (*a*) How fast are they moving when their separation has decreased to one-half its initial value? (*b*) How fast are they moving just before they collide?
Answer: (*a*) 82 km/sec. (*b*) 1.8×10^4 km/sec.

38(8). Two particles of mass m and M are initially at rest an infinite distance apart. Show that at any instant their relative velocity of approach attributable to gravitational attraction is $\sqrt{2G(M + m)/d}$, where d is their separation at that instant.

39(8). In a double star, two stars of mass 3×10^{30} kg each rotate about their common center of mass, 10^{11} meters away. (*a*) What is their common angular speed? (*b*) Suppose that a meteorite passes through this center of mass moving at right angles to the orbital plane of the stars. What must its speed be if it is to escape from the gravitational field of the double star?
Answer: (*a*) 2×10^{-7} radians/sec. (*b*) 9×10^4 meters/sec.

40(8). For interstellar travel, a space ship must overcome the sun's gravitational field as well as that of the earth. (*a*) What is the total amount of energy required for a 1.0-kiloton space ship to free itself from the combined earth-sun gravitational field starting from an orbit 300 mi above the earth's surface? Neglect all other bodies in the solar system. (*b*) What fraction of this energy is used to overcome the sun's field?

41(9). The masses and coordinates of three spheres are as follows: 20 kg, $x = 0.50$ m, $y = 1.0$ m; 40 kg, $x = -1.0$ m, $y = -1.0$ m; 60 kg, $x = 0.0$ m, $y = -0.50$ m. What is (*a*) the gravitational force and (*b*) the gravitational potential energy of a 20-kg sphere located at the origin?
Answer: (*a*) 3.2×10^{-7} nt, at $-92°$. (*b*) -2.2×10^{-7} joule.

42(10). An asteroid, whose mass is 2.0×10^{-4} times the mass of the earth, revolves in a circular orbit around the sun at a distance which is twice the earth's distance from the sun. (*a*) Calculate the period of revolution of the asteroid in earth years. (*b*) What is the ratio of the kinetic energy of the asteroid to that of the earth?

43(10). (*a*) Does it take more energy to get a satellite up to 1000 miles above the earth than to put it in orbit there? (*b*) What about 2000 miles? (*c*) What about 3000 miles? Take the earth's radius to be 4000 miles.
Answer: (*a*) No. (*b*) The same. (*c*) Yes.

44(10). Consider two satellites A and B of equal mass m, moving in the same circular orbit of radius r around the earth E but in opposite senses of rotation and therefore on a collision course (see Fig. 14–16). (*a*) In terms of G, M_e, m, and r, find the total mechanical energy $E_A + E_B$ of the two-satellite-plus-earth system before collision. (*b*) If the collision is completely inelastic so that wreckage remains as one piece of tangled material (mass $= 2m$), find the total mechanical energy immediately after collision. (*c*) Describe the subsequent motion of the wreckage.

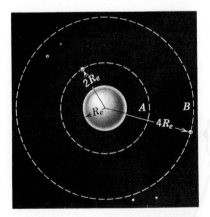

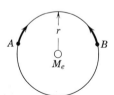

FIGURE 14–16   Problem 44(10).   **FIGURE 14–17**   Problem 45(10).

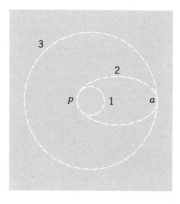

FIGURE 14–18 Problem 48(10).

45(10). Two earth satellites, A and B, each of mass m, are to be launched into (nearly) circular orbits about the earth's center. Satellite A is to orbit at an altitude of 4000 miles. Satellite B is to orbit at an altitude of 12,000 miles. The radius of the earth R_e is 4000 miles (Fig. 14–17). (a) What is the ratio of the potential energy of satellite B to that of satellite A, in orbit? (Explain the result in terms of the work required to get each satellite from its orbit to infinity.) (b) What is the ratio of the kinetic energy of satellite B to that of satellite A, in orbit? (c) Which satellite has the greater total energy if each has a mass of 1.0 slug? By how much?
Answer: (a) $\frac{1}{2}$. (b) $\frac{1}{2}$. (c) B, by 8.4×10^7 ft-lb.

46(10). A satellite is to be launched so that it will have an elliptical orbit about the sun. Its closest distance to the sun is to be equal to the radius of Venus's orbit and the largest distance from the sun is to be equal to the radius of the earth's orbit. (a) What is its speed, with respect to the sun, at its largest distance from the sun? (b) What is its speed, with respect to the earth, at the same distance? (c) How should such a satellite be launched from the earth? Hint: Use appropriate conservation laws.

47(10). A satellite travels initially in an approximately circular orbit 400 miles above the surface of the earth; its mass is 220 kg. (a) Determine its speed. (b) Determine its period. (c) For various reasons the satellite loses mechanical energy at the (average) rate of 1.4×10^5 joules per complete orbital revolution. Adopting the reasonable approximation that the trajectory is a "circle of slowly diminishing radius," determine the distance from the surface of the earth, the speed, and the period of the satellite at the end of its 1500th orbital revolution. (d) What is the magnitude of the average retarding force? (e) Is angular momentum conserved?
Answer: (a) 7.5×10^3 meters/sec. (b) 5.8×10^3 sec. (c) 420 km; 7.7×10^3 meters/sec; 5.6×10^3 sec. (d) 3.3×10^{-3} nt. (e) For the satellite: No. For the system Satellite and Earth: Yes.

48(10). Three identical artificial earth satellites have been launched into the three orbits shown in Fig. 14–18. The important properties of the three orbits can be ordered, using inequality and/or equality signs. For example, the three apogees (the furthest distance from the earth) would be ordered as $a_1 < a_2 = a_3$. (a) Order the three orbital angular momenta L. (b) Order the total energies E of the three satellites. (c) Order the four potential energies V_1, V_{2p}, V_{2a}, V_3 where p refers to perigee (the closest distance to the earth) and a refers to apogee. (d) Order the four kinetic energies K_1, K_{2p}, K_{2a}, K_3.

CHAPTER 15

Fluid Mechanics

15–1 Fluids

We usually classify matter, viewed macroscopically, into solids and fluids. A *fluid* is something that can flow. Hence the term fluid includes liquids and gases. Such classifications are not always clearcut. Some fluids, such as glass or pitch, flow so slowly that they behave like solids for the time intervals that we usually work with them. Even the distinction between a liquid and a gas is not clearcut because, by changing the pressure and temperature in the right way, it is possible to change a liquid (water, say) into a gas (steam, say) without the appearance of a meniscus and without boiling; the density and viscosity change in a continuous manner throughout the process.

In this chapter we will apply Newton's laws of motion, including the appropriate force laws, to fluids. We will find, just as we did for rigid bodies, that it will be useful to develop a special formulation of Newton's laws. However, we introduce no new principles.

15–2 Pressure and Density

There is a difference in the way a surface force acts on a fluid and on a solid. For a solid there are no restrictions on the direction of such a force, but for a fluid at rest the surface force must always be directed at right angles to the surface. A fluid at rest cannot sustain a tangential force; the fluid layers would simply slide over one another. Indeed, it is the inability of fluids to resist such tangential forces (or shearing stresses) that permits them to change their shape or to flow.

It is convenient, therefore, to describe the force acting on a fluid by specifying the *pressure p,* which is defined as the magnitude of the normal force per unit surface

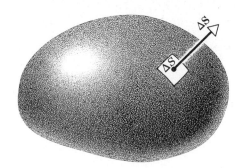

FIGURE 15–1 An element of surface ΔS can be represented by a vector $\Delta \mathbf{S}$, equal to its area in magnitude and outwardly normal to it in direction.

area. Pressure is transmitted to solid boundaries or across arbitrary sections of fluid at right angles to these boundaries or sections at every point. Pressure is a scalar quantity. Some common units are lb/in.^{2}, nt/meter^2, dynes/cm^2, bars (1 bar $= 10^6$ dynes/cm^2), atmospheres (1 atm $= 14.7$ lb/in.^{2}), and mm-Hg (760 mm-Hg $= 1$ atm).

A fluid under pressure exerts a force on any surface in contact with it. Consider a closed surface containing a fluid (Fig. 15–1). We can represent an element of the surface by a vector $\Delta \mathbf{S}$ whose magnitude gives the area of the element and whose direction is taken to be the outward normal to the surface of the element. Then the force $\Delta \mathbf{F}$ exerted by the fluid against this surface element is

$$\Delta \mathbf{F} = p \, \Delta \mathbf{S}.$$

Since $\Delta \mathbf{F}$ and $\Delta \mathbf{S}$ have the same direction, the pressure p is

$$p = \frac{\Delta F}{\Delta S}.$$

The pressure so defined may depend on the size of the element of area ΔS that we choose. To avoid this difficulty we refine our definition by taking a small element of surface containing a point in question and considering this quotient as the element shrinks to the point. Then the pressure at the point is

$$p = \lim_{\Delta S \to 0} \frac{\Delta F}{\Delta S}.$$

The pressure may vary from point to point on the surface.

The density ρ of a homogeneous fluid may depend on factors such as its temperature and the pressure to which it is subjected. For liquids the density varies very little over wide ranges in pressure and temperature, and we can safely treat it as a constant for our present purposes; the density of a gas, however, is very sensitive to changes in temperature and pressure; see entries under "Air" and "Water" in Table 15–1. This table shows the range of densities that occur in nature.

15–3   The Variation of Pressure in a Fluid at Rest

If a fluid is in equilibrium, every part of the fluid is in equilibrium. Let us consider a small volume element that is entirely submerged within the body of the fluid. Let this element be a thin disk and be a distance y above some reference level, as shown in Fig. 15–2a. The thickness of the disk is dy and each face has an area A. The mass of this element is $\rho A \, dy$ and its weight is $\rho g A \, dy$. The forces exerted on the element by the surrounding fluid are perpendicular to its surface at each point (Fig. 15–2b).

The fluid element is unaccelerated in the vertical direction, so that the resultant

Table 15–1 DENSITIES OF SOME MATERIALS AND OBJECTS IN KG/METER^3

Interstellar space	10^{-18} to 10^{-21}
Best laboratory vacuum	$\sim 10^{-16}$
Hydrogen: at 0° C and 1.0 atm	9.0×10^{-2}
Air: at 0° C and 1.0 atm	1.3
at 100° C and 1.0 atm	0.95
at 0° C and 50 atm	6.5
Styrofoam	$\sim 1 \times 10^2$
Ice	0.92×10^3
Water: at 0° C and 1.0 atm	1.000×10^3
at 100° C and 1.0 atm	0.958×10^3
at 0° C and 50 atm	1.002×10^3
Aluminum	2.7×10^3
Mercury	1.4×10^4
Platinum	2.1×10^4
The earth: average density	5.5×10^3
density of core	9.5×10^3
density of crust	2.8×10^3
The sun: average density	1.4×10^3
density at center	$\sim 1.6 \times 10^5$
White dwarf stars (central densities)	10^8 to 10^{15}
A uranium nucleus	$\sim 10^{17}$

vertical force on it must be zero. The vertical forces are due not only to the pressure of the fluid on its faces but also to the weight of the element. If we let p be the pressure on the lower face and $p + dp$ the pressure on its upper face, the upward force is pA (exerted on the lower face) and the downward force is $(p + dp)A$ plus the weight of the element $dw (= \rho g A\, dy)$. Hence, for vertical equilibrium

$$pA = (p + dp)A + \rho g A\, dy,$$

and

$$\frac{dp}{dy} = -\rho g. \qquad (15\text{--}1)$$

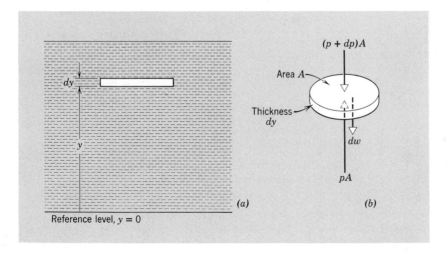

FIGURE 15–2 (*a*) A small volume element of fluid at rest. (*b*) The forces on the element.

This equation tells us how the pressure varies with elevation above some reference level in a fluid in static equilibrium. As the elevation increases (dy positive), the pressure decreases (dp negative). The cause of this pressure variation is the weight per unit cross-sectional area of the layers of fluid lying between the points whose pressure difference is being measured. The quantity ρg is often called the *weight density* of the fluid. For water, for example, the weight density is 62.4 lb/ft^3.

If p_1 is the pressure at elevation y_1 and p_2 the pressure at elevation y_2 above some reference level, integration of Eq. 15–1 gives

$$\int_{p_1}^{p_2} dp = -\int_{y_1}^{y_2} \rho g \, dy$$

or

$$p_2 - p_1 = -\int_{y_1}^{y_2} \rho g \, dy. \tag{15–2}$$

For liquids ρ is practically constant because liquids are nearly incompressible; and differences in level are rarely so great that any change in g need be considered. Hence, taking ρ and g as constants, we obtain

$$p_2 - p_1 = -\rho g(y_2 - y_1) \tag{15–3}$$

for a homogeneous liquid.

If a liquid has a free surface, we take y_2 to be the elevation of that surface, where the pressure is p_0, and we take the elevation y_1 to be at any level with pressure p, and $y_2 - y_1$ is the depth h (see Fig. 15–3). Equation 15–3 then becomes

$$p = p_0 + \rho g h. \tag{15–4}$$

This shows that the pressure is the same at all points at the same depth. Often p_0 is the atmospheric pressure.

For gases ρ is comparatively small and the difference in pressure between two points is usually negligible (see Eq. 15–3). In a vessel containing a gas we can therefore take the pressure as the same everywhere. However, this is not the case if $y_2 - y_1$ is very great. The pressure of the atmosphere varies greatly as we ascend to great heights. In such cases the density ρ varies with altitude and we must know ρ (and also g) as a function of y before we can integrate Eq. 15–2.

Equation 15–3 gives the relation between the pressures at any two points in a fluid, regardless of the shape of the containing vessel. For no matter what the shape of the containing vessel, two points in the fluid can be connected by a path made up of vertical and horizontal steps. For example, consider points A and B in the homogeneous liquid contained in the U-tube of Fig. 15–4a. Along the stepped path from A to B there is a difference in pressure $\rho g y'$ for each vertical segment of length y', whereas along each horizontal segment there is no change in pressure. Hence, the difference in pressure $p_B - p_A$ is ρg times the algebraic sum of the vertical segments from A to B, or $\rho g(y_2 - y_1)$.

If the U-tube contains two immiscible liquids,

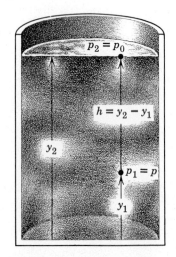

FIGURE 15–3   A liquid whose top surface is open to the atmosphere.

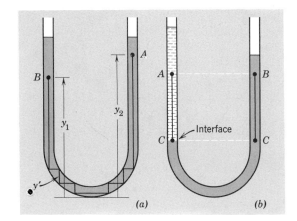

FIGURE 15–4 (*a*) The difference in pressure between two points *A* and *B* in a homogeneous liquid depends only on their difference in elevation $y_2 - y_1$. (*b*) Two points *A* and *B* at the same elevation can be at different pressures if the densities there differ.

say a dense liquid in the right tube and a less dense one in the left tube, as Fig. 15–4*b* shows, the pressure can be different at the same level on different sides. In the figure the liquid surface is higher in the left tube than in the right. The pressure at *A* will be greater than at *B*. The pressure at *C* is the same on both sides, but the pressure falls less from *C* to *A* than from *C* to *B*, for a column of liquid of unit cross-sectional area connecting *A* and *C* will weigh less than a corresponding column connecting *B* and *C*.

Example 1. A U-tube is partly filled with water. Another liquid, which does not mix with water, is poured into one side until it stands a distance *d* above the water level on the other side (Fig. 15–5). Find the density of the liquid relative to that of water.

In Fig. 15–5 points *C* are at the same pressure. Hence, the pressure drop from *C* to each surface is the same, for each surface is at atmospheric pressure.

The pressure drop on the water side is $\rho_w g l$; that on the other side is $\rho g(d + l)$, where ρ is the density of the unknown liquid. Hence,

$$\rho_w g l = \rho g(d + l)$$

and

$$\frac{\rho}{\rho_w} = \frac{l}{(l + d)}.$$

The ratio of the density of a substance to the density of water is called the *relative density* (or the *specific gravity*) of that substance.

15–4 Pascal's Principle and Archimedes' Principle

Figure 15–6 shows a liquid in a cylinder that is fitted with a piston to which we may apply an external pressure p_0. The pressure p at any arbitrary point *P* a distance *h* below the upper surface of the liquid is given by Eq. 15–4, or

$$p = p_0 + \rho g h.$$

Let us increase the external pressure by an arbitrary amount Δp_0 which need not be small compared to p_0. Since liquids are nearly incompressible, the density ρ remains essentially constant. The equation shows that, to this extent, the change in pressure Δp at the arbitrary point *P* is equal to Δp_0. This result was stated by Blaise

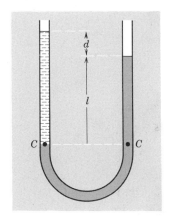

FIGURE 15–5 Example 1.

Pascal (1623–1662) and is called *Pascal's principle.* We usually say it this way: *Pressure applied to an enclosed fluid is transmitted undiminished to every portion of the fluid and to the walls of the containing vessel.* This is a necessary consequence of the laws of mechanics, and not an independent principle.

Actually, slight changes in density do occur and provide the means by which a wave propagates through a fluid (see Section 17–2). This means that a change of pressure applied to one portion of a liquid propagates through the liquid as a wave at the speed of sound in that liquid. Pascal's principle also holds for gases with slight complications of interpretation caused by the large volume changes that may occur when the pressure on a confined gas is changed.

Archimedes' principle is also a necessary consequence of the laws of fluid mechanics. When a body is wholly or partly immersed in a fluid (either a liquid or a gas) at rest, the fluid exerts pressure on every part of the body's surface in contact with the fluid. The pressure is greater on the parts immersed more deeply. The resultant of all the forces is an upward force called the buoyant force of the immersed body. We can find the magnitude and direction of this resultant force as follows.

The pressure on each part of the surface of the body certainly does not depend on the material the body is made of. Suppose, then, that the body, or as much of it as is immersed, is replaced by fluid like the surroundings. This fluid will experience the pressures that acted on the immersed body (Fig. 15–7) and will be at rest. Hence **R,** the resultant upward force on it, will equal its weight and will act vertically upward through its center of gravity. From this follows *Archimedes' Principle: A body wholly or partly immersed in a fluid is buoyed up with a force equal to the weight of the fluid displaced by the body.* We have seen that the force acts vertically up through the center of gravity of the fluid before its displacement. The corresponding point in the immersed body is called its *center of buoyancy.*

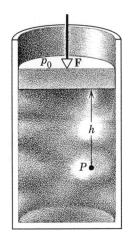

FIGURE 15–6 A fluid in a cylinder fitted with a movable piston. The pressure at any point P is due not only to the weight of the fluid above the level of P but also to the force exerted by the piston.

■ **Example 2.** What fraction of the volume of an iceberg is exposed? The density of ice is $\rho_i = 0.92$ gm/cm^3 and that of sea water is $\rho_w = 1.03$ gm/cm^3.

The weight of the iceberg is

$$W_i = \rho_i V_i g,$$

where V_i is the volume of the iceberg; the weight of the volume V_w of sea water displaced is the buoyant force

$$B = \rho_w V_w g.$$

But B equals W_i, for the iceberg is in equilibrium, so that

$$\rho_w V_w g = \rho_i V_i g,$$

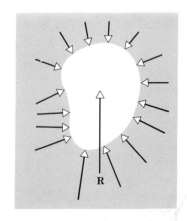

FIGURE 15–7 Archimedes' principle.

and
$$\frac{V_w}{V_i} = \frac{\rho_i}{\rho_w} = \frac{0.92}{1.03} = 89\%.$$

The volume of water displaced, V_w, is the volume of the submerged portion of the iceberg, so that 11% of the iceberg is exposed.

15–5　Measurement of Pressure

Evangelista Torricelli (1608–1647) invented the mercury barometer in 1643. This is a long glass tube that has been filled with mercury and then inverted in a dish of mercury, as in Fig. 15–8. The space above the mercury column contains only mercury vapor, whose pressure is so small at ordinary temperatures that it can be neglected. You can easily show (see Eq. 15–3) that the atmospheric pressure p_0 is

$$p_0 = \rho g h.$$

Most pressure gauges use atmospheric pressure as a reference level and measure the difference between the actual pressure and atmospheric pressure, called the *gauge pressure*. The actual pressure at a point in a fluid is called the *absolute pressure*. Gauge pressure is given either above or below atmospheric pressure.

The pressure of the atmosphere at any point is equal to the weight per unit area of a column of air extending from that point to the top of the atmosphere. The atmospheric pressure, therefore, decreases with altitude. There are variations in atmospheric pressure from day to day since the atmosphere is not static. The mercury column in the barometer will have a height of about 76 cm at sea level, varying with the atmospheric pressure. A pressure equivalent to that exerted by exactly 76 cm of mercury at exactly $0°$ C under standard gravity, $g = 32.174$ ft/sec^2 = 980.665 cm/sec^2, is defined to be *one atmosphere* (1 atm). The density of mercury at this temperature is 13.5950 gm/cm^3. Hence, one atmosphere is equal to

1 atm = (13.5950 gm/cm^3)(980.665 cm/sec^2)

$$\times \text{ (76.00 cm)}$$

$$= 1.013 \times 10^5 \text{ nt/meter}^2$$

$$= 2116 \text{ lb/ft}^2$$

$$= 14.70 \text{ lb/in.}^2$$

Often pressures are specified by giving the height of a mercury column, at $0°$ C under standard gravity, which exerts the same pressure. This is the origin of the expression "centimeters of mercury (cm-Hg)" or "inches of mercury (in-Hg)" pressure. Pressure is the ratio of force to area, however, and not a length.

The open-tube manometer (Fig. 15–9) measures gauge pressure. It consists of a U-shaped tube containing a liquid, one end of the tube being open to the atmosphere and the other end being connected to the system (tank) whose pressure p we want to measure. From Eq. 15–4 we obtain

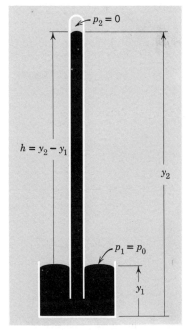

FIGURE 15–8　The Torricelli barometer.

$$p - p_0 = \rho g h.$$

Thus the gauge pressure, $p - p_0$, is proportional to the difference in height of the liquid columns in the U-tube.

Example 3. An open-tube mercury manometer (Fig. 15–9) is connected to a gas tank. The mercury is 39.0 cm higher on the right side than on the left when a barometer nearby reads 75.0 cm-Hg. What is the absolute pressure of the gas? Express the answer in cm-Hg, atm, and lb/in².

The gas pressure is the pressure at the top of the left mercury column. This is the same as the pressure at the same horizontal level in the right column. The pressure at this level is the atmospheric pressure (75.0 cm-Hg) plus the pressure exerted by the extra 39.0-cm column of Hg, or (assuming standard values of mercury density and gravity) a total of 114 cm-Hg. Therefore, the absolute pressure of the gas is

$$114 \text{ cm-Hg} = \tfrac{114}{76} \text{ atm} = 1.50 \text{ atm} = (1.50)(14.7) \text{ lb/in.}^2 = 22.1 \text{ lb/in.}^2.$$

What is the gauge pressure of the gas?

15–6   Fluid Dynamics

Now we consider fluids that are flowing and on which definable forces act. One way of describing the motion of a fluid is to divide it into infinitesimal volume elements, which we may call fluid particles, and to follow the motion of each of these particles. This is a formidable task. We could give coordinates x, y, z to each such fluid particle and then specify these as functions of the time t and the initial position of the particle x_0, y_0, and z_0. This procedure is a direct generalization of the concepts of particle mechanics and was first developed by Joseph Louis Lagrange (1736–1813).

There is, however, a treatment, developed by Leonhard Euler (1707–1783), which is much more convenient. In it we give up the attempt to specify the history of each fluid particle and instead specify the density and the velocity of the fluid at each point in space at each instant of time. This is the method we shall follow here.

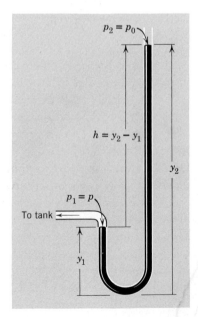

FIGURE 15–9   The open-tube manometer, as used to measure the gauge pressure in a tank.

We describe the motion of the fluid by specifying the density $\rho(x,y,z,t)$ and the velocity
$\mathbf{v}(x,y,z,t)$ at the point (x,y,z) at the time t. We
thus focus attention on what is happening at a
particular point in space at a particular time,
rather than on what is happening to a particular fluid particle. Any quantity used in describing the state of the fluid, for example the pressure p, will have a definite value at each point
in space and at each instant of time. Although
this description of fluid motion focuses attention on a point in space rather than on a fluid
particle, we cannot avoid following the fluid

FIGURE 15–10   A small paddle wheel
placed in a flowing liquid rotates in
rotational flow and does not rotate in
irrotational flow.

particles themselves, at least for short time intervals dt. For it is the particles, after
all, and not the space points, to which the laws of mechanics apply.

Let us first consider some general characteristics of fluid flow and specify a number of restrictions we shall make in order to simplify the treatment.

Fluid flow can be *steady* or *nonsteady*. When the fluid velocity $\mathbf{v}$ at every point
is constant in time, we say that the fluid motion is steady. That is, at any given
point in a steady flow the velocity of each passing fluid particle is always the same.
At some other point a particle may travel with a different velocity, but every
other particle which passes this second point behaves there just as this particle did
when it passed this point. These conditions can be achieved at low flow speeds; a
gently flowing stream is an example. In nonsteady flow, as in a tidal bore, the
velocities $\mathbf{v}$ *are* a function of the time at any given point.

Fluid flow can be *rotational* or *irrotational*. If the element of fluid at each point has
no net angular velocity about that point, the fluid flow is irrotational. We can
imagine a small paddle wheel immersed in the moving fluid (Fig. 15–10). If
the wheel does not rotate the motion is irrotational. Rotational flow includes
vortex motion, such as whirlpools or eddies, and motion in which the velocity
vector varies in the transverse direction.

Fluid flow can be *compressible* or *incompressible*. Liquids can usually be considered
as flowing incompressibly. But even a highly compressible gas may sometimes
undergo unimportant changes in density. Its flow is then practically incompressible.
In flight at speeds much lower than the speed of sound in air (subsonic aerodynamics), the motion of the air relative to the wings is one of nearly incompressible
flow. In such cases the density ρ is a constant, independent of x, y, z, and t, and the
mathematical treatment of fluid flow is thereby greatly simplified.

Finally, fluid flow can be *viscous* or *nonviscous*. Viscosity in fluid motion is the
analog of friction in the motion of solids. In many cases, such as in lubrication problems, it is extremely important. Sometimes, however, it is negligible. Viscosity
introduces tangential forces between layers of fluid in relative motion and results in
dissipation of mechanical energy.

We shall confine our discussion of fluid dynamics for the most part to steady,
irrotational, incompressible, nonviscous flow. The mathematical simplifications resulting should be obvious. We run the danger, however, of making so many
simplifying assumptions that we are no longer talking about a meaningfully real
fluid. Furthermore, we sometimes find it difficult to decide whether a given property
of a fluid—its viscosity, say—can be neglected in a particular situation. In spite of

all this the restricted analysis that we are going to give has wide application in practice.

15–7   Streamlines and the Equation of Continuity

In steady flow the velocity **v** at a given point is constant in time. Consider the point P (Fig. 15–11) within the fluid. Since **v** at P does not change in time, every particle arriving at P will pass on with the same speed in the same direction. The same is true about the points Q and R. Hence, if we trace out the path of the particle, as is done in the figure, that curve will be the path of every particle arriving at P. Such a curve is called a *streamline*. A streamline is tangent to the velocity of the fluid particles at every point. No two streamlines can cross, for if they did, an oncoming fluid particle could go either one way or the other, and the flow could not be steady. In steady flow the pattern of streamlines in a flow is stationary with time.

In principle we can draw a streamline through every point in the fluid. Let us assume steady flow and select a finite number of streamlines to form a bundle, like the streamline pattern of Fig. 15–12. This tubular region is called a *tube of flow*. Since the boundary of such a tube consists of streamlines no fluid can cross the boundary of a tube of flow and the tube behaves somewhat like a pipe of the same shape. The fluid that enters at one end must leave at the other.

In Fig. 15–12 we have drawn a thin tube of flow. The velocity of the fluid inside, although parallel to the tube at any point, may have different magnitudes at different points. Let the speed be v_1 for fluid particles at P and v_2 for fluid particles at Q. Let A_1 and A_2 be the cross-sectional areas of the tubes perpendicular to the streamlines at the points P and Q, respectively, and assume that the speed is essentially constant over each surface separately. In the time interval Δt a fluid element travels approximately the distance $v\,\Delta t$. Then the mass of fluid Δm_1 crossing A_1 in the time interval Δt is approximately

$$\Delta m_1 = \rho_1 A_1 v_1 \,\Delta t$$

or the *mass flux*, $\Delta m_1/\Delta t$, is approximately $\rho_1 A_1 v_1$. We must take Δt small enough so that in this time interval neither v nor A varies appreciably over the distance the fluid travels. In the limit as $\Delta t \to 0$, we obtain the precise definitions:

$$\text{mass flux at } P = \rho_1 A_1 v_1, \qquad \text{and}$$

$$\text{mass flux at } Q = \rho_2 A_2 v_2,$$

where ρ_1 and ρ_2 are the fluid densities at P and Q, respectively.

We will assume that there are no "sources" or "sinks" wherein fluid is created or destroyed in the tube. Since in steady flow no fluid can leave through the walls nor can fluid particles temporarily bunch up, the mass crossing each section of the tube per unit time must be the same. In particular, the mass flux at P must equal that at Q:

FIGURE 15–11   A particle passing through points P, Q, and R traces out a streamline, assuming steady flow. Any other particle passing through P must be traveling along the same streamline in steady flow.

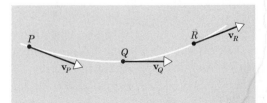

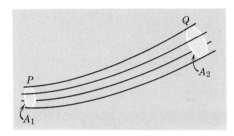

FIGURE 15–12   A tube of flow used in proving the equation of continuity.

$$\rho_1 A_1 v_1 = \rho_2 A_2 v_2,$$

or
$$\rho A v = \text{constant.} \tag{15–5}$$

This result (Eq. 15–5) expresses the law of conservation of mass in fluid dynamics.

Would you expect Eq. 15–5 to hold when the flow is viscous? When it is rotational?

If the fluid is incompressible, as we shall henceforth assume, $\rho_1 = \rho_2$ and Eq. 15–5 takes on the simpler form

$$A_1 v_1 = A_2 v_2,$$

or
$$A v = \text{constant.} \tag{15–6}$$

The product Av gives the *volume flux* or flow rate, as it is often called. Notice that it predicts that in steady incompressible flow the speed of flow varies inversely with the cross-sectional area, being larger in narrower parts of the tube. The fact that the product Av remains constant along a tube of flow allows us to interpret the streamline picture somewhat. In a narrow part of the tube the streamlines must crowd closer together than in a wide part. Hence, as the distance between streamlines decreases, the fluid speed must increase. Therefore, we conclude that widely spaced streamlines indicate regions of lower speed, and closely spaced streamlines indicate regions of higher speed.

We can obtain another interesting result by applying the second law of motion to the flow of fluid. In moving from P to Q (Fig. 15–12) the fluid is decelerated from v_1 to the lower speed v_2. The deceleration can come about from a difference in pressure acting on the fluid particle flowing from P to Q or from the action of gravity. In a horizontal tube of flow the gravitational force does not change. Hence we can conclude that in steady horizontal flow the pressure is greatest where the speed is least. These results will be presented quantitatively in the next section.

15–8 Bernoulli's Equation

Bernoulli's equation is a fundamental relation in fluid mechanics. Like all equations in fluid mechanics it is not a new principle but is derivable from Newtonian mechanics. We will derive it from the work-energy theorem (see Section 6–5), for it is essentially an application of this theorem to fluid flow.

Consider the nonviscous, steady, incompressible flow of a fluid through the pipeline or tube of flow in Fig. 15–13. The portion of pipe shown in the figure has a uniform cross section A_1 at the left. It is horizontal there at an elevation y_1 above some reference level. It gradually widens and rises until at the right it has a uniform cross section A_2. It is horizontal there also at an elevation y_2. Let us concentrate our attention on the shaded portion of fluid between A_1 and A_2 and call this fluid the "system." Consider then the motion of the system from the position shown in (*a*) to

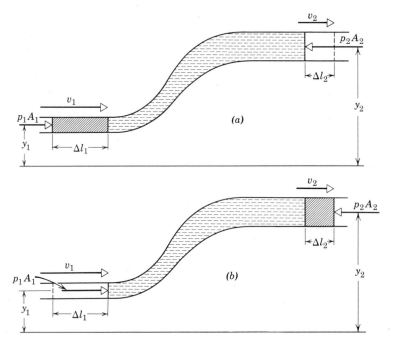

FIGURE 15–13   A portion of fluid (cross-shading *and* horizontal shading) moves through a section of pipeline from the position shown in (*a*) to that shown in (*b*).

that in (*b*). At all points in the narrow part of the pipe the pressure is p_1 and the speed v_1; at all points in the wide portion the pressure is p_2 and the speed v_2.

The work energy theorem (see Eq. 6–14) states: The work done by the resultant force acting on a system is equal to the change in kinetic energy of the system. In Fig. 15–13 the forces that do work on the system, assuming that we can neglect viscous forces, are the pressure forces p_1A_1 and p_2A_2 that act on the left- and right-hand ends of the system, respectively, and the force of gravity. As fluid flows through the pipe the net effect, as a comparison of Figs. 15–13*a* and *b* shows, is to raise an amount of fluid represented by the cross-shaded area in Fig. 15–13*a* to the position shown in Fig. 15–13*b*. The amount of fluid represented by the horizontal shading is unchanged by the flow.

We can find the work W done on the system by the resultant force as follows:

1. The work done on the system by the pressure force p_1A_1 is $p_1A_1 \, \Delta l_1$.
2. The work done on the system by the pressure force p_2A_2 is $-p_2A_2 \, \Delta l_2$.

Note that it is negative, which means that positive work is done by the system.

3. The work done on the system by gravity is associated with lifting the cross-shaded fluid from height y_1 to height y_2 and is $-mg(y_2 - y_1)$ in which m is the mass of fluid in either cross-shaded area. It too is negative because work is done by the system against the gravitational force.

$$W = p_1A_1 \, \Delta l_1 - p_2A_2 \, \Delta l_2 - mg(y_2 - y_1).$$

Now $A_1 \, \Delta l_1 \; (= A_2 \, \Delta l_2)$ is the volume of the cross-shaded fluid element, which we can write as m/ρ in which ρ is the (constant) fluid density. Recall that the two fluid elements have the same mass, so that in setting $A_1 \, \Delta l_1 = A_2 \, \Delta l_2$ we have assumed the fluid to be incompressible. With this assumption we have

$$W = (p_1 - p_2)(m/\rho) - mg(y_2 - y_1). \tag{15–7a}$$

The change in kinetic energy of the fluid element is

$$\Delta K = \tfrac{1}{2}mv_2^2 - \tfrac{1}{2}mv_1^2. \tag{15–7b}$$

From the work-energy theorem (Eq. 6–14) we then have

$$W = \Delta K$$

or

$$(p_1 - p_2)(m/\rho) - mg(y_2 - y_1) = \tfrac{1}{2}mv_2^2 - \tfrac{1}{2}mv_1^2, \tag{15–8a}$$

which can be rearranged to read

$$p_1 + \tfrac{1}{2}\rho v_1^2 + \rho g y_1 = p_2 + \tfrac{1}{2}\rho v_2^2 + \rho g y_2. \tag{15–8b}$$

Since the subscripts 1 and 2 refer to any two locations along the pipeline, we can drop the subscripts and write

$$p + \tfrac{1}{2}\rho v^2 + \rho g y = \text{constant.} \tag{15–9}$$

Equation 15–9 is called *Bernoulli's equation* for steady, nonviscous, incompressible flow. It was first presented by Daniel Bernoulli (1700–1782) in 1738.

Bernoulli's equation is strictly applicable only to steady flow, the quantities involved being evaluated along a streamline. In our figure the streamline used is along the axis of the pipeline. If the flow is irrotational, however, we can show that the constant in Eq. 15–9 is the same for all streamlines.

Just as the statics of a particle is a special case of particle dynamics, so fluid statics is a special case of fluid dynamics. It should come as no surprise, therefore, that the law of pressure change with height in a fluid at rest is included in Bernoulli's equation as a special case. For let the fluid be at rest; then $v_1 = v_2 = 0$ and Eq. 15–8b becomes

$$p_1 + \rho g y_1 = p_2 + \rho g y_2$$

or

$$p_2 - p_1 = -\rho g(y_2 - y_1),$$

which is the same as Eq. 15–3.

15–9 Applications of Bernoulli's Equation and the Equation of Continuity

Bernoulli's equation can be used to find fluid speeds by making pressure measurements. The principle generally used in such measuring devices is the following: The equation of continuity requires that the speed of the fluid at a constriction increase; Bernoulli's equation then shows that the pressure must fall there. That is, for a horizontal pipe $\tfrac{1}{2}\rho v^2 + p$ equals a constant; if v increases and the fluid is incompressible, p must decrease.

1. The Venturi Meter. This (Fig. 15–14) is a gauge put in a flow pipe to measure the flow speed of a liquid. A liquid of density ρ flows through a pipe of cross-sectional area A. At the throat the area is reduced to a and a manometer tube is attached, as shown. Let the manometer liquid, such as mercury, have a density ρ'. By applying Bernoulli's equation and the equation of continuity at points 1 and 2, you can show that the speed of flow at A (see Problem 23) is

$$v = a\sqrt{\frac{2(\rho' - \rho)gh}{\rho(A^2 - a^2)}}.$$

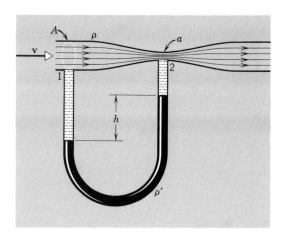

FIGURE 15–14   The Venturi meter, used to measure the speed of flow of a fluid.

If we want the volume flux or flow rate R, which is the volume of liquid transported past any point per second, we simply compute

$$R = vA.$$

2. The Pitot Tube. This device (Fig. 15–15) is used to measure the flow speed of a gas. Consider the gas, say air, flowing past the opening at a. These openings are parallel to the direction of flow and are set far enough back so that the velocity and pressure outside the openings have the free-stream values. The pressure in the left arm of the manometer, which is connected to these openings, is then the static pressure in the gas stream, p_a. The opening of the right arm of the manometer is at right angles to the stream. The velocity is zero at b and the pressure is p_b.

Applying Bernoulli's equation to points a and b, we obtain

$$p_a + \tfrac{1}{2}\rho v^2 = p_b,$$

where, as the figure shows, p_b is greater than p_a. If h is the difference in height of the liquid in the manometer arms and ρ' is the density of the manometer liquid, then

$$p_a + \rho'gh = p_b.$$

Comparing these two equations, we find

$$\tfrac{1}{2}\rho v^2 = \rho'gh$$

or

$$v = \sqrt{\frac{2gh\rho'}{\rho}},$$

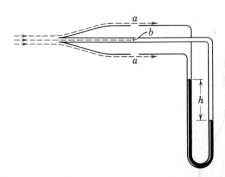

FIGURE 15–15   Cross-sectional diagram of a Pitot tube.

which gives the gas speed. This device can be calibrated to read v directly and is then known as an air-speed indicator. You will see them on airplanes.

3. Dynamic Lift. *Dynamic lift* is the force that acts on a body, such as an airplane wing, a hydrofoil, or a spinning baseball, because of its motion through a fluid. We must distinguish it from *static lift*, which is the buoyant force that acts on an object, such as a balloon, in accord with Archimedes' principle.

Figure 15–16*a* shows the streamlines around a (nonspinning) baseball as it moves through the air. For convenience, we use a reference frame in which the baseball is at rest and the air moves past it; this reference frame can be realized in practice by mounting a baseball in a wind tunnel. From the symmetry of the streamlines it is clear that the velocity of the air is the same at corresponding points above and below the ball, such as 1 and 2 in Fig. 15–16*a*. From Bernoulli's equation we then deduce that the pressures at such corresponding points are equal and that the air exerts no upward or downward force on the ball by virtue of its motion; the dynamic lift is zero.

In another experiment let us spin a resting baseball about an axis that is perpendicular to the plane of Fig. 15–16*b*. Since the ball is not perfectly smooth, it drags some air around with it as illustrated by the streamlines in the figure.

Finally, let us combine both motions by throwing a baseball and spinning it at the same time. Figure 15–16*c* shows the resulting streamlines from a reference frame in which the center of mass of the ball is at rest. These streamlines represent a distribution of velocities found by adding (vectorially) at every point the velocities in Figs. 15–16*a* and *b*. The velocities at point 1 add numerically while those at point 2 subtract. Thus the speed at point 1 in Fig. 15–16*c* is greater than the speed at point 2 as, indeed, the crowding of the streamlines suggests. Since from Bernoulli's principle the pressure at point 1 is less than the pressure at point 2, can you see that there is a net upward force (a dynamic lift) on the spinning baseball? Such forces on actual spinning baseballs and tennis balls are well known in practice.

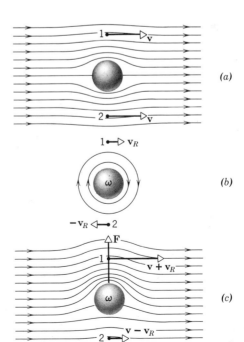

FIGURE 15–16 (*a*) Streamlines for air moving past a (nonrotating) baseball. The velocity **v** is shown for typical corresponding points 1 and 2. (*b*) Streamlines for air dragged around a rotating baseball, showing the velocities $\mathbf{v}_R$ and $-\mathbf{v}_R$ at points 1 and 2 respectively. (*c*) The super-position of (*a*) and (*b*). The velocities at points 1 and 2 are shown along with the dynamic lift **F**.

When dynamic lift on an object occurs it is always associated with an un-symmetrical set of streamlines relatively close together on one side and relatively far apart on the other, like those of Fig. 15–16*c*, that correspond, as Fig. 15–16*b* shows, to circulation of fluid around the object. For the spinning baseball the circulation is obtained by spinning the object; in other cases of dynamic lift, of which the airplane wing is a good example, streamline patterns that contain the necessary circulation are obtained by properly shaping the object and properly orienting it in the moving fluid. Figure 15–17 shows streamlines around an airplane wing. As required, they are closer together above the wing than they are below so that (compare Fig. 15–16*c*) Bernoulli's principle predicts the observed upward dynamic lift.

4. Thrust on a Rocket. As our final example let us compute the thrust on a rocket produced by the escape of its exhaust gases. Consider a chamber (Fig. 15–18) of cross-sectional area A filled with a gas of density ρ at a pressure p. Let there be a small orifice of cross-sectional area A_0 at the bottom of the chamber. We wish to find the speed v_0 with which the gas escapes through the orifice.

Let us write Bernoulli's equation (Eq. 15–8*b*) as

$$p_1 - p_2 = \rho g(y_2 - y_1) + \tfrac{1}{2}\rho(v_2{}^2 - v_1{}^2).$$

For a gas the density is so small that we can neglect the variation in pressure with height in a chamber (see Section 15–3). Hence, if p represents the pressure p_1 in the chamber and p_0 represents the atmospheric pressure p_2 just outside the orifice, we have

$$p - p_0 = \tfrac{1}{2}\rho(v_0{}^2 - v^2)$$

or
$$v_0{}^2 = \frac{2(p - p_0)}{\rho} + v^2, \tag{15-10}$$

where v is the speed of the flowing gas inside the chamber and v_0 is the *speed of efflux* of the gas through the orifice. Although a gas is compressible and the flow may become turbulent, we can treat the flow as steady and incompressible for pressure and efflux speeds that are not too high.

Now let us assume continuity of mass flow (in a rocket engine this is achieved when the mass of escaping gas equals the mass of gas created by burning the fuel), so that (for an assumed constant density)

$$Av = A_0 v_0.$$

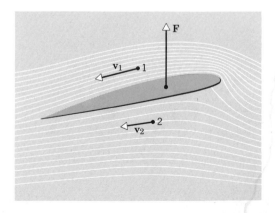

FIGURE 15–17   Streamlines about an airfoil.

If the orifice is very small so that $A_0 \ll A$, then $v_0 \gg v$, and we can neglect v^2 compared to v_0^2 in Eq. 15–10. Hence, the speed of efflux is

$$v_0 = \sqrt{\frac{2(p - p_0)}{\rho}}. \qquad (15\text{–}11)$$

If our chamber is the exhaust chamber of a rocket, the thrust on the rocket (Section 8–6) is $v_0 \, dM/dt$. But the mass of gas flowing out in time dt is $dM = \rho A_0 v_0 \, dt$, so that

$$v_0 \frac{dM}{dt} = v_0 \rho A_0 v_0 = \rho A_0 v_0^2,$$

and from Eq. 15–11 the thrust is

$$2A_0(p - p_0). \qquad (15\text{–}12)$$

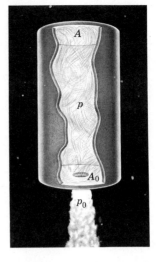

FIGURE 15–18 Fluid streaming out of a chamber.

QUESTIONS

1. (*a*) Two bodies (for example, balls) have the same shape and size but one is denser than the other. Assuming the air resistance to be the same on each, show that when they are released simultaneously from the same height the heavier body will get to the ground first. (*b*) Two bodies (for example, raindrops) have the same shape and density but one is larger than the other. Assuming the air resistance to be proportional to the body's speed through the air, which body will fall faster?

2. Water is poured to the same level in each of the vessels shown, all of the same base area (Fig. 15–19). If the pressure is the same at the bottom of each vessel, the force experienced by the base of each vessel is the same. Why then do the three vessels have different weights when put on a scale? This apparently contradictory result is commonly known as the "hydrostatic paradox."

3. (*a*) An ice cube is floating in a glass of water. When the ice melts, will the water level rise? Explain. (*b*) If the ice cube contains a piece of lead, the water level will fall when the ice melts. Explain.

4. Does Archimedes' law hold in a vessel in free fall? In a satellite moving in a circular orbit? Explain.

5. A spherical bob made of cork floats half submerged in a pot of tea at rest on the earth. Will the cork float or sink aboard a spaceship coasting in free space? On the surface of Jupiter?

6. Two bodies of equal weight and volume and having the same shape, except that one has an opening at the bottom and the other is sealed, are immersed to the same depth in water. Is less work required to immerse one than the other? If so, which one and why?

7. A ball floats on the surface of water in a container exposed to the atmosphere. Will the ball remain immersed at its former depth or will it sink or rise somewhat if (*a*) the container is covered and the air is removed, (*b*) the container is covered and the air is compressed?

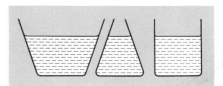

FIGURE 15–19

8. A soft plastic bag weighs the same when empty as when filled with air at atmospheric pressure. Why? Would the weights be the same if measured in a vacuum?

9. A leaky tramp steamer that is barely able to float in the North Sea steams up the Thames estuary toward the London docks. It sinks before it arrives. Why?

10. Is it true that a floating object will only be in stable equilibrium if its center of buoyancy lies above its center of gravity? Illustrate with examples.

11. Very often a sinking ship will turn over as it becomes immersed in water. Why?

12. A barge filled with scrap iron is in a canal lock. If the iron is thrown overboard, what happens to the water level in the lock?

13. A bucket of water is suspended from a spring balance. Does the balance reading change when a piece of iron suspended from a string is immersed in the water? When a piece of cork is put in the water?

14. Explain why a uniform wooden stick which will float horizontally if it is not loaded will float vertically if enough weight is added to one end. See Problem 8.

15. Explain how it can be that pressure is a scalar quantity when forces, which are vectors, can be produced by the action of pressures.

16. Can you assign a coefficient of static friction between two surfaces, one of which is a fluid surface?

17. Describe the forces acting on an element of fluid as it flows through a pipe of non-uniform cross section.

18. The height of the liquid in the standpipes indicates that the pressure drops along the channel, even though the channel has a uniform cross section and the flowing liquid is incompressible (Fig. 15–20). Explain.

19. In a lecture demonstration a ping-pong ball is kept in midair by a vertical jet of air. Is the equilibrium stable, unstable, or neutral? Explain.

20. (*a*) Explain how a pitcher can make a baseball curve to his right or left. Justify your answer by drawing a diagram of the streamlines and applying Bernoulli's equation. (*b*) Why is it easier to throw a curve with a tennis ball than with a baseball?

21. Two rowboats moving parallel to one another in the same direction are pulled toward one another. Two automobiles moving parallel are also pulled together. Explain such phenomena using Bernoulli's equation.

22. Can the action of a parachute in retarding free fall be explained by Bernoulli's equation?

23. Liquid is flowing inside a horizontal pipe which has a constriction along its length. Vertical tube manometers are attached at both the wide portion and the narrow portion of the pipe. If a stopcock at the exit end is closed, will the liquid in the manometer tubes rise or fall? Explain.

24. Why does water flow in a continuous stream down a vertical pipe, whereas it breaks into drops when falling freely?

25. Why does an object falling from a great height reach a steady terminal speed?

26. On take off would it be better for an airplane to move into the wind or with the wind? On landing . . . ?

27. Does the difference in pressure between the lower and upper surfaces of an airplane wing depend on the altitude of the moving plane? Explain.

28. The accumulation of ice on an airplane wing may change its shape in such a way that its lift is greatly reduced. Explain.

29. How is an airplane able to fly upside down?

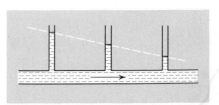

FIGURE 15–20

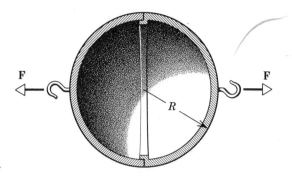

FIGURE 15–21

30. Why does the factor "2" appear in Eq. 15–12, rather than "1"? One might expect that the thrust would simply be the pressure difference times the area, that is, $A_0(p - p_0)$.

31. The destructive effect of a tornado (twister) is greater near the center of the disturbance than near the edge. Explain.

32. In steady flow the velocity vector **v** at any point is constant. Can there then be accelerated motion of the fluid particles? Discuss.

PROBLEMS

1(2). In 1654 Otto von Guericke, burgomeister of Magdeburg and inventor of the air pump, gave a demonstration before the Imperial Diet in which two teams of eight horses could not pull apart two evacuated brass hemispheres. (a) Show that the force F required to pull apart the hemispheres is $F = \pi R^2 P$ where R is the (outside) radius of the hemispheres and P is the difference in pressure outside and inside the sphere (Fig. 15–21). (b) Taking R equal to 1.0 ft and the inside pressure as 0.10 atm, what force would the team of horses have had to exert to pull apart the hemispheres? (c) Why were two teams of horses used? Would not one team prove the point just as well?
Answer: (b) 6000 lb.

2(3). What is the pressure at the bottom of a 10-ft deep swimming pool? Assume normal atmospheric pressure at the surface. Express your answer in lb/ft² and in nt/m².

3(3). Find the total pressure, in lb/in.², 500 ft below the surface of the ocean. The relative density of sea water is 1.03.
Answer: 240 lb/in.²

4(3). What would be the height of the atmosphere if the air density (a) were constant and (b) decreased linearly to zero with height? Assume a sea-level density of 1.3 kg/meter³.

5(3). What is the magnitude and direction of the force on a square millimeter area of the side of a container filled with water at a depth of 6.0 ft. Consider only the water and not the atmosphere.
Answer: 4.0×10^{-3} lb, normal to the surface.

6(3). A swimming pool has the dimensions 80 ft × 30 ft × 8.0 ft. (a) When it is filled with water, what is the force (due to the water alone) on the bottom? On the ends? On the sides? (b) If you are

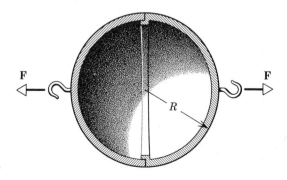

FIGURE 15–21   Problem 1(2).

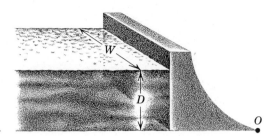

FIGURE 15-22   Problem 9(3).

concerned with whether or not the concrete walls and floor will collapse, is it appropriate to take the atmospheric pressure into account?

7(3). Three liquids that will not mix are poured into a cylindrical container 20 cm in diameter. The amounts and densities of the liquids are 0.50 liter, 2.6 gm/cm³; 0.25 liter, 1.0 gm/cm³; and 0.40 liter, 0.80 gm/cm³. What is the total fluid force acting on the bottom of the container? *Answer:* 18 nt.

8(3). A simple U-tube contains mercury. When 13.6 cm of water is poured into the right arm, how high does the mercury rise in the left arm from its initial level?

9(3). Water stands at a depth D behind the vertical upstream face of a dam, as shown in Fig. 15–22. Let W be the width of the dam. (a) Find the resultant horizontal force exerted on the dam by the water and (b) the net torque due to the water exerted about a line through O parallel to the width of the dam.
Answer: (a) $\frac{1}{2}\rho g D^2 W$, acting $D/3$ up from bottom. (b) $\frac{1}{6}\rho g D^3 W$.

10(3). A U-tube is filled with a single homogeneous liquid. The liquid is temporarily depressed in one side by a piston. The piston is removed and the level of the liquid in each side oscillates. Show that the period of oscillation is $\pi\sqrt{2L/g}$ where L is the total length of the liquid in the tube.

11(3). Two identical cylindrical vessels with their bases at the same level each contain a liquid of density ρ. The area of either base is A, but in one vessel the liquid height is h_1 and in the other h_2. Find the work done by gravity in equalizing the levels when the two vessels are connected.
Answer: $\frac{1}{4}\rho g A(h_2 - h_1)^2$.

12(4). A piston of small cross-sectional area a is used in the hydraulic press to exert a small force f on the enclosed liquid. A connecting pipe leads to a larger piston of cross-sectional area A (Fig. 15–23). (a) What force F will the larger piston sustain? (b) If the small piston has a diameter of 1.5 in. and the large piston one of 21 in., what weight on the small piston will support 2.0 tons on the large piston?

13(4). A cubical object of dimensions L (2.0 ft) on a side and weight W (1000 lb.) in a vacuum is suspended by a rope in an open tank of liquid of density ρ (2.0 slugs/ft³) as in Fig. 15-24. (a) Find the total downward force exerted by the liquid and the atmosphere on the top of the object of area A (4.0 ft²). (b) Find the total upward force on the bottom of the object. (c) Find the tension in the rope.
Answer: (a) 8700 lb. (b) 9300 lb. (c) 490 lb.

14(4). (a) What is the minimum area of a block of ice 1.0 ft thick floating on water that will hold up an automobile weighing 2500 lb? (b) Does it matter where the car is placed on the block of ice?

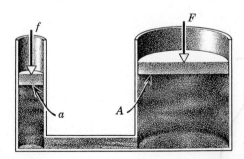

FIGURE 15-23   Problem 12(4).

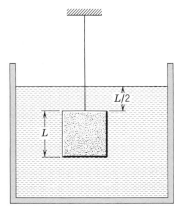

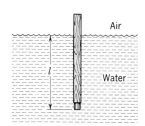

FIGURE 15–24 Problem 13(4).   **FIGURE 15–25** Problem 22(4).

15(4). Three boys each of weight W (80 lb) make a log raft by lashing together logs of diameter D (1.0 ft) and length L (6.0 ft). How many logs will be needed to keep them afloat? Take the specific gravity of wood to be 0.80.
Answer: Five.

16(4). A block of wood floats in water with two-thirds of its volume submerged. In oil it has 0.90 of its volume submerged. Find the density of (a) the wood and (b) the oil.

17(4). A block of wood weighs 8.0 lb and has a relative density of 0.60. It is to be loaded with lead so that it will float in water with 0.90 of its volume immersed. What weight of lead is needed (a) if the lead is on top of the wood? (b) if the lead is attached below the wood? The density of lead is 11.3 gm/cm³.
Answer: (a) 4.0 lb. (b) 4.4 lb.

18(4). Assume the density of brass weights to be 8.0 gm/cm³ and that of air to be 0.0012 gm/cm³. What percent error arises from neglecting the buoyancy of air in weighing an object of mass m and density ρ on a beam balance?

19(4). A hollow sphere of inner radius 8.0 cm and outer radius 9.0 cm floats half submerged in a liquid of specific gravity 0.80. Calculate the density of the material of which the sphere is made.
Answer: 1.3×10^3 kg/meter³.

20(4). A hollow spherical iron shell floats almost completely submerged in water. If the outer diameter is 2.00 ft and the relative density of iron is 7.80, find the inner diameter.

21(4). An iron casting weighs 60 lb in air and 40 lb in water. What is the volume of cavities in the casting? Assume the relative density of iron to be 7.8.
Answer: 0.20 ft³.

22(4). A cylindrical wooden rod is loaded with lead at one end so that it floats upright in water as in Fig. 15–25. The length of the submerged portion is $l = 8.0$ ft. The rod is set into vertical oscillation. (a) Show that the oscillation is simple harmonic. (b) Find the period of the oscillation. Neglect the fact that the water has a damping effect on the motion.

23(4). A long uniform wooden bar with square cross section floats on water. Show that, if the relative density of the wood is 0.50, the bar will float with all four faces making an angle of 45° with the water surface. (Hint: The bar floats in the equilibrium position for which the potential energy is a minimum.)

24(7). A ¾-in. I.D. water pipe is coupled to three ½-in. I.D. pipes. (a) If the flow rates in the three pipes are 7.0, 5.0, and 3.0 gal/min, what is the flow rate in the ¾-in. pipe? (b) How will the speed of the water in the ¾-in. pipe compare with that carrying 7.0 gal/min?

25(7). A garden hose having an internal diameter of 0.75 in. is connected to a lawn sprinkler that consists merely of an enclosure with 24 holes, each 0.050 in. in diameter. If the water in the hose has a speed of 3.0 ft/sec, at what speed does it leave the sprinkler holes?
Answer: 28 ft/sec.

26(8). How much work is done by pressure in forcing 50 ft³ of water through a 0.50-in. pipe if the difference in pressure at the two ends of the pipe is 15 lb/in.²?

27(8). In a horizontal oil pipeline of constant cross-sectional area the pressure decrease between two points 1000 ft apart is 5.0 lb/in.². What is the energy loss per cubic foot of oil per unit distance? *Answer:* 0.72 ft-lb/ft³, per foot.

28(8). Water falls from a height of 60 ft at the rate of 500 ft³/min and drives a water turbine. What is the maximum power that can be developed by this turbine?

29(8). Water is moving with a speed of 5.0 meters/sec through a pipe with a cross-sectional area of 4.0 cm². The water gradually descends 10 meters as the pipe increases in area to 8.0 cm². (a) What is the speed of flow at the lower level? (b) If the pressure at the upper level is 1.50×10^5 nt/meter², what is the pressure at the lower level? *Answer:* (a) 2.5 meter/sec. (b) 2.6×10^5 nt/meter².

30(8). A water pipe having a 1.0-in. inside diameter carries water into the basement of a house at a velocity of 3.0 ft/sec at a pressure of 25 lb/in². If the pipe tapers to $\frac{1}{2}$ in. and rises to the second floor 25 ft above the input point, what are (a) the velocity and (b) water pressure there?

31(8). If a person blows with a speed of 15 meters/sec across the top of one side of a U-tube containing water, what will be the difference between the water levels on the two sides? *Answer:* 1.4 cm.

32(8). Figure 15–26 shows liquid discharging from an orifice in a large tank at a distance h below the water level. (a) Apply Bernoulli's equation to a streamline connecting points 1, 2, and 3, and show that the speed of efflux is

$$v = \sqrt{2gh}.$$

This is known as Torricelli's law. (b) If the orifice were curved directly upward, how high would the liquid stream rise? (c) How would viscosity or turbulence affect the analysis?

33(8). Suppose that two tanks, each with a large opening at the top, contain different liquids. A small hole is made in the side of each tank at the same depth h below the liquid surface, but one hole has twice the cross-sectional area of the other. (a) What is the ratio of the densities of the fluids if it is observed that the mass flux is the same for each hole? (b) How do the flow rates (volume flux) from each hole compare? (c) Could the flow rates be made equal? How? *Answer:* Let the subscript 1 refer to the tank with the small hole; subscript 2 to the tank with the large hole. (a) $\rho_1 = 2\rho_2$. (b) $A_1 v_1 = \frac{1}{2}(A_2 v_2)$. (c) If $h_1 = 4h_2$ the flow rates would be the same.

34(8). A tank is filled with water to a height H. A hole is punched in one of the walls at a depth h below the water surface (Fig. 15–27). (a) Show that the distance x from the foot of the wall at which the stream strikes the floor is given by $x = 2\sqrt{h(H-h)}$. (b) Could a hole be punched at another depth so that this second stream would have the same range? If so, at what depth?

35(8). The upper surface of water in a standpipe is a height H above level ground. (a) At what

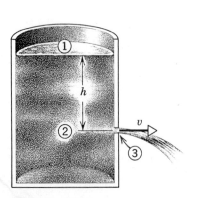

FIGURE 15–26 Problem 32(8).

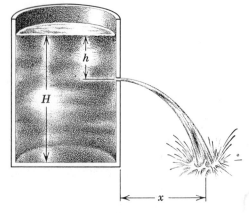

FIGURE 15–27 Problem 34(8).

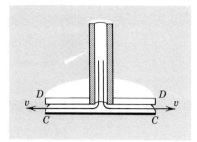

FIGURE 15-28 Problem 45(9).

depth h should a small hole be put to make the emerging horizontal water stream strike the ground at the maximum distance from the base of the standpipe? (*b*) What is this maximum distance? *Answer:* (*a*) $\frac{1}{2}H$. (*b*) H.

36(8). A tank of large area is filled with water to a depth D (1.0 ft). A hole of cross section A (1.0 in.²) in the bottom of the tank allows water to drain out. (*a*) What is the rate at which water flows out in ft³/sec? (*b*) At what distance below the bottom of the tank is the cross-sectional area of the stream equal to one-half the area of the hole?

37(9). By applying Bernoulli's equation and the equation of continuity to points 1 and 2 of Fig. 15–14, show that the speed of flow at the entrance is

$$v = a \sqrt{\frac{2(\rho' - \rho)gh}{\rho(A^2 - a^2)}}.$$

38(9). A Venturi meter has a pipe diameter of 10 in. and a throat diameter of 5.0 in. If the water pressure in the pipe is 8.0 lb/in.² and in the throat is 6.0 lb/in.², determine the rate of flow of water in ft³/sec (volume flux).

39(9). Consider the Venturi tube of Fig. 15–14 without the manometer. Let A equal $5a$. Suppose the pressure at A is 2.0 atm. (*a*) Compute the values of v at A and v' at a that would make the pressure p' at a equal to zero. (*b*) Compute the corresponding volume flow rate if the diameter at A is 5.0 cm. The phenomenon at a when p' falls to nearly zero is known as cavitation. The water vaporizes into small bubbles. *Answer:* (*a*) $v = 4.1$ meters/sec; $v' = 21$ meters/sec. (*b*) 8.0×10^{-3} meters³/sec.

40(9). A Pitot tube is mounted on an airplane wing to determine the speed of the plane relative to the air. The tube contains alcohol and indicates a level difference of 4.9 in. What is the plane's speed relative to the air? The density of alcohol is 0.81×10^3 kg/meter³.

41(9). In what direction will a baseball thrown north with the leading edge of the ball spinning east curve? Is it more likely that the pitcher is a right-hander or a left-hander? *Answer:* East; left-handed.

42(9). A 12-in.² plate weighing 1.0 lb is hinged along one side. If air is blown over the upper surface only, what speed must the air have to hold the plate horizontal?

43(9). An airplane has a wing area (each wing) of 100 ft². At a certain air speed, air flows over the upper wing surface at 160 ft/sec and over the lower wing surface at 130 ft/sec. What is the weight of the plane? Assume that the plane travels at constant velocity and that lift effects associated with the fuselage and tail assembly are small. Discuss the lift if the airplane, flying at the same air speed, is (*a*) in level flight, (*b*) climbing at 15°, and (*c*) descending at 15°. *Answer:* 2180 lb. Lift is the same in all three cases.

44(9). If the speed of flow past the lower surface of a wing is 350 ft/sec, what speed of flow over the upper surface will give a lift of 20 lb/ft²?

45(9). A hollow tube has a disk DD attached to its end. When air is blown through the tube, the disk attracts the card CC. Let the area of the card be A and let v be the average airspeed between CC and DD (Fig. 15–28); calculate the resultant upward force on CC. Neglect the card's weight. *Answer:* $\frac{1}{2}\rho v^2 A$.

CHAPTER 16

Waves in Elastic Media

16–1 Mechanical Waves

Wave motion appears in almost every branch of physics. We are all familiar with water waves. There are also sound waves, as well as light waves, radio waves, and other electromagnetic waves. One formulation of the mechanics of atoms and sub-atomic particles is called wave mechanics (matter waves).

In this chapter and the next we confine our attention to waves in deformable or elastic media. These waves, among which ordinary sound waves in air are an example, are called *mechanical waves*. They originate in the displacement of some portion of an elastic medium from its normal position, causing it to oscillate about an equilibrium position. Because of the elastic forces on adjacent layers, the disturbance is transmitted from one layer to the next through the medium. The medium itself does not move as a whole; rather, the various parts oscillate in limited paths. For example, with surface waves in water, small floating objects like corks show that the actual motion of the water molecules is elliptical, slightly up and down and back and forth. Yet the water waves move steadily along the water. As they reach floating objects, they set them in motion, thus transferring energy to them. The energy in the waves is in the form of both kinetic and potential energy and its transmission comes about by its being passed from one part of the matter to the next, not by any long-range motion of the matter itself. Thus, mechanical waves are characterized by the transport of energy through matter by the motion of a disturbance in that matter without any corresponding bulk motion of the matter itself.

A material medium is necessary for the transmission of mechanical waves. Such a medium, however, is not needed to transmit electromagnetic waves. For example, light from stars comes to us through the near vacuum of space. These waves will be discussed later (Chapter 35).

16–2   Types of Waves

In listing water waves, light waves, and sound waves as examples of wave motion, we are classifying waves according to their broad physical properties. Waves can be classified in other ways.

We can distinguish different kinds of waves by considering how the motions of the particles of matter are related to the direction of propagation of the waves themselves. If the motions of the matter particles conveying the wave are perpendicular to the direction of propagation of the wave itself, we then have a *transverse* wave. For example, when a vertical string under tension is set oscillating back and forth at one end, a transverse wave travels down the string; the disturbance moves along the string but the string particles vibrate at right angles to the direction of propagation of the disturbance (Fig. 16–1*a*).

While it is true that light waves are not mechanical, they are nevertheless transverse. Just as material particles move perpendicular to the direction of propagation in some mechanical waves, electric and magnetic fields are perpendicular to the direction of propagation of light waves.

If the motion of the particles conveying a mechanical wave is back and forth along the direction of propagation, we then have a *longitudinal* wave. For example, when a vertical spring under tension is set oscillating up and down at one end, a longitudinal wave travels along the spring; the coils vibrate back and forth in the direction in which the disturbance travels along the spring (Fig. 16–1*b*). Sound waves in a gas are longitudinal waves. We shall discuss them in greater detail in Chapter 17.

Some waves are neither purely longitudinal nor purely transverse. For example, in waves on the surface of water the particles of water move both up and down and back and forth, tracing out elliptical paths as the water waves move by.

Waves can also be classified as one-, two-, and three-dimensional waves, according to the number of dimensions in which they propagate energy. Waves moving along the string or the spring of Fig. 16–1 are one-dimensional. Surface waves or ripples on water, caused by dropping a pebble into a quiet pond, are two-dimensional. Sound waves and light waves which emanate radially from a small source are three-dimensional.

Waves may be classified further according to the behavior of a particle of the matter conveying the wave during the course of time the wave propagates. For example, we can produce a *pulse* or a *single wave* traveling down a taut rope by applying a single sidewise movement at its end. Each particle remains at rest until the pulse reaches it, then it moves during a short time, and then it again remains at rest. If we continue to move the end of the rope back and forth (Fig. 16–1*a*), we produce a *train of waves* traveling along the rope. If our motion is periodic, we produce a *periodic train of waves* in which each particle of the rope has a periodic motion. The simplest special case of a periodic wave is a *simple harmonic wave* which gives each particle a simple harmonic motion.

Consider a three-dimensional pulse. We can draw a surface through all points undergoing a similar disturbance at a given instant. As time goes on, this surface moves along showing how the pulse propagates. We can draw similar surfaces for subsequent pulses. For a periodic wave we can generalize the idea by drawing in surfaces, all of whose points are in the same phase of motion. These surfaces are called *wavefronts*. If the medium is homogeneous and isotropic, the direction of propagation

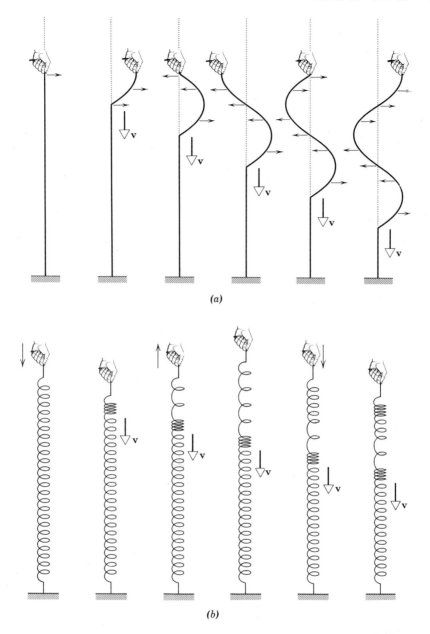

FIGURE 16–1 (*a*) In a transverse wave the particles of the medium (string) vibrate at right angles to the direction in which the wave itself is propagated. (*b*) In a longitudinal wave the particles of the medium (spring) vibrate in the same direction as that in which the wave itself is propagated. In both (*a*) and (*b*) we imagine that all the energy of the wave is absorbed by the device at the bottom. Thus we do not have to worry about waves reflected back up the string or spring.

is always at right angles to the wavefront. A line normal to the wavefronts, indicating the direction of motion of the waves, is called a *ray*.

Wavefronts can have many shapes. If the disturbances are propagated in a single direction, the waves are called *plane waves*. At a given instant conditions are the same everywhere on any plane perpendicular to the direction of propagation. The wavefronts are planes and the rays are parallel straight lines (Fig. 16–2*a*). Another simple case is that of *spherical waves*. Here the disturbance is propagated out in all directions from a point source of waves. The wavefronts are spheres and the rays are radial lines leaving the point source in all directions (Fig. 16–2b). Far from the source the spherical wavefronts have very small curvature, and over a limited region they can often be regarded as planes. Of course, there are many other possible shapes for wavefronts.

We shall refer to all these wave types as we progress through the wave phenomena of physics. In this chapter we often use the transverse wave in a string to illustrate the general properties of waves. In the next chapter we shall see the consequences of these properties for sound, a longitudinal mechanical wave. Later in the text we will discuss the properties of nonmechanical waves such as light and matter waves.

16–3   Traveling Waves

Let us consider a long string stretched in the x-direction along which a transverse wave is traveling. At some instant of time, say $t = 0$, the shape of the string can be represented by

$$y = f(x) \qquad t = 0, \qquad (16–1)$$

where y is the transverse displacement of the string at the position x. In Fig. 16–3a we show a possible waveform (a pulse) on the string at $t = 0$. Experiment shows that as time goes on such a wave travels along the string without changing its form, provided internal frictional losses are small enough. At some time t later the wave has traveled a distance vt to the right, where v is magnitude of the wave velocity, assumed constant. The equation of the curve at the time t is therefore

$$y = f(x - vt) \qquad \textit{for all times } t \qquad (16–2)$$

This gives us the same waveform about the point $x = vt$ at time t as we had about $x = 0$ at the time $t = 0$ (Fig. 16–3b). Equation 16–2 is the general equation representing a wave of any shape traveling to the right. To describe a particular shape we must specify exactly what the function f is. The variables x and t, of course, can only appear in the combination $x - vt$. For example, $\sin k(x - vt)$ and $(x - vt)^3$ are appropriate functions; $x^2 - v^2 t^2$ is not.

Let us look more carefully at this equation. If we wish to follow a particular part (or phase) of the wave as time goes on, then in the equation we look at a particular value of y (say, the top of the pulse just described). Mathematically this means we look at how x changes with t when $(x - vt)$ has some particular fixed value. We see at once that as t increases x must increase in order to keep $(x - vt)$ fixed. Hence, Eq.

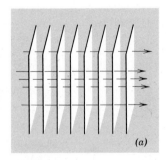
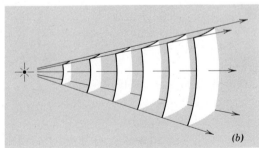

(a)   (b)

FIGURE 16–2   (a) A plane wave. The planes represent wavefronts spaced a wavelength apart, and the arrows represent rays. (b) A spherical wave. The rays are radial and the wavefronts, spaced a wavelength apart, form spherical shells. Far out from the source, however, small portions of the wavefronts become nearly plane.

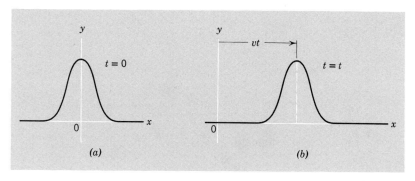

FIGURE 16–3 (*a*) The shape of a string (in this case a pulse) at $t = 0$. (*b*) At a later time t the pulse has traveled to the right a distance $x = vt$.

16–2 does in fact represent a wave traveling to the right (increasing x as time goes on). If we wished to represent a wave traveling to the left, we would write

$$y = f(x + vt), \qquad (16\text{–}3)$$

for here the position x of some fixed phase $(x + vt)$ of the wave decreases as time goes on. The velocity of a particular phase of the wave is easily obtained. For a particular phase of a wave traveling in the $+x$-direction we require that

$$x - vt = \text{constant}.$$

Then differentiation with respect to time gives

$$\frac{dx}{dt} - v = 0 \qquad \text{or} \qquad \frac{dx}{dt} = v, \qquad (16\text{–}4)$$

so that v is really the *phase velocity* of the wave. For a wave traveling in the $-x$-direction we obtain $-v$, in the same way, as its phase velocity.

The general equation of a wave can be interpreted further. Note that for any fixed value of the time t the equation gives y as a function of x. This defines a curve, and this curve represents the actual shape of the string at this chosen time. It gives us a snapshot of the wave at this time. Suppose, on the other hand, we wish to focus our attention on one point of the string, that is, a fixed value of x. Then the equation gives us y as a function of the time t. This describes how the transverse position of this point on the string changes with time.

The argument just presented holds for longitudinal waves as well as for transverse waves. The analogous longitudinal example is that of a long straight tube of gas whose axis is taken as the x-axis, and the wave or pulse is a pressure change traveling along the tube. Then the same reasoning leads us to an equation, having the form of Eqs. 16–2 and 16–3, which gives the pressure variations with time at all points of the tube. (See Section 17–3.)

Let us now consider a particular waveform, whose importance will soon become clear. Suppose that at the time $t = 0$ we have a wavetrain along the string given by

$$y = y_m \sin \frac{2\pi}{\lambda} x. \qquad (16\text{–}5)$$

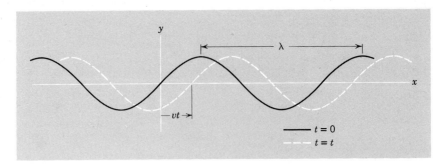

FIGURE 16–4 At $t = 0$, the string has a shape $y = y_m \sin 2\pi x / \lambda$ (solid line). At a later time t the sine wave has moved to the right a distance $x = vt$, and the string has a shape given by $y = y_m \sin 2\pi (x - vt) / \lambda$.

The wave shape is a sine curve (Fig. 16–4). The maximum displacement y_m is the *amplitude* of the sine curve. The value of the transverse displacement y is the same at x as it is at $x + \lambda$, $x + 2\lambda$, etc. The symbol λ is called the *wavelength* of the wavetrain and represents the distance between two adjacent points in the wave having the same phase. As time goes on let the wave travel to the right with a phase velocity v. Hence, the equation of the wave at the time t is

$$y = y_m \sin \frac{2\pi}{\lambda} (x - vt). \tag{16–6}$$

Notice that this has the form required for a traveling wave (Eq. 16–2).

The *period* T is the time required for the wave to travel a distance of one wavelength λ, so that

$$\lambda = vT. \tag{16–7}$$

Putting this relation into the equation of the wave, we obtain

$$y = y_m \sin 2\pi \left(\frac{x}{\lambda} - \frac{t}{T} \right). \tag{16–8}$$

From this form it is clear that y, at any given time, has the same value at $x + \lambda$, $x + 2\lambda$, etc., as it does at x, and that y, at any given position, has the same value at the time $t + T$, $t + 2T$, etc., as it does at the time t.

To reduce Eq. 16–8 to a more compact form, we define two quantities, the *wave number* k and the *angular frequency* ω (see Eq. 13–12). They are given by

$$k = \frac{2\pi}{\lambda} \quad \text{and} \quad \omega = \frac{2\pi}{T}. \tag{16–9}$$

In terms of these quantities, the equation of a sine wave traveling to the right is

$$y = y_m \sin (kx - \omega t). \tag{16–10a}$$

For a sine wave traveling to the left, we have

$$y = y_m \sin (kx + \omega t). \tag{16–10b}$$

Comparing Eqs. 16–7 and 16–9, we see that the phase velocity v of the wave is given by

$$v = \frac{\lambda}{T} = \frac{\omega}{k}. \qquad (16\text{–}11)$$

In the traveling waves of Eqs. 16–10a and 16–10b we have assumed that the displacement y is zero at the position $x = 0$ at the time $t = 0$. This, of course, need not be the case. The general expression for a sinusoidal wave traveling in the $+x$-direction is

$$y = y_m \sin (kx - \omega t - \phi),$$

where ϕ is called the phase constant. For example, if $\phi = -90°$, the displacement y at $x = 0$ and $t = 0$ is y_m. This particular example is

$$y = y_m \cos (kx - \omega t),$$

since the cosine function is displaced by 90° from the sine function.

If we fix our attention on a given point of the string, say $x = \pi/k$, the displacement y at that point can be written as

$$y = y_m \sin (\omega t + \phi).$$

This is similar to Eq. 13–25 for simple harmonic motion. Hence, any particular element of the string undergoes simple harmonic motion about its equilibrium position as this wavetrain travels along the string.

16–4 Wave Speed in a Stretched String

We will now continue with the example of waves in a stretched string and derive the wave velocity by a mechanical analysis.

In Fig. 16–5 we show a wave pulse proceeding from right to left in the string with a speed v. For convenience in the derivation imagine the entire string to be moving from left to right with this same speed so that the wave pulse remains fixed in space, whereas the particles composing the string successively pass through the pulse. Instead of taking our reference frame to be the walls between which the string is stretched, we are choosing a reference frame which is in uniform motion with respect to the walls and at rest with respect to the pulse. Because Newton's laws involve only accelerations, which are the same in both frames, we can use them in either frame, but this one is more convenient.

Consider a small section of the pulse of length Δl to form an arc of a circle of radius R, as shown in the diagram. If μ is the mass per unit length of the string, the so-called linear density, then $\mu \, \Delta l$ is the mass of this element. The tension F in the string is a tangential pull at each end of this small segment of the string. The horizontal components cancel and the vertical components are each equal to $F \sin \theta$. Hence, the total vertical force is $2F \sin \theta$. Since θ is small, we can take $\sin \theta \cong \theta$ and

$$2F \sin \theta \cong 2F\theta = 2F \frac{(\Delta l/2)}{R} = F \frac{\Delta l}{R}.$$

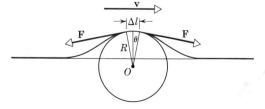

FIGURE 16–5   Derivation of wave speed by considering the forces on a section of string of length Δl.

This gives the force supplying the centripetal acceleration of the string particles directed toward O. Now the centripetal force acting on a mass $\mu \, \Delta l$ moving in a circle of radius R with speed v is $\mu \, \Delta l \, v^2/R$; see Section 5–13. Notice that the tangential velocity v of this mass element along the top of the arc is horizontal and is the same as the pulse phase velocity. Combining the equivalent expressions just given we obtain

$$F \frac{\Delta l}{R} = \frac{\mu \, \Delta l \, v^2}{R}$$

or

$$v = \sqrt{\frac{F}{\mu}}.$$

If the amplitude of the pulse were very large compared to the length of the string, we would not have been able to use the approximation $\sin \theta \cong \theta$. Furthermore, the tension F in the string would be changed by the presence of the pulse, whereas we assumed F to be unchanged from the original tension in the stretched string. Therefore, our result holds only for relatively small transverse displacements of the string— which case, however, is widely applicable in practice. Notice also that the wave speed is independent of the shape of the wave, for no particular assumption about the shape of the pulse was used in the proof.

The frequency of a wave is naturally determined by the frequency of the source. The speed with which the wave travels through a medium is determined by the properties of the medium, as illustrated before. Once the frequency ν and speed v of the wave are determined, the wavelength λ is fixed. In fact, from Eq. 16–7 and the relation, $\nu = 1/T$, we have

$$\lambda = \frac{v}{\nu}. \tag{16–12}$$

Example 1. A transverse sinusoidal wave is generated at one end of a long horizontal string by a bar which moves the end up and down through a distance of 0.20 in. The motion is continuous and is repeated regularly 120 times per second.

(*a*) If the string has a linear density of 0.0050 slug/ft and is kept under a tension of 20 lb, find the speed, amplitude, frequency, and wavelength of the wave motion.

The end moves 0.10 in. away from the equilibrium position, first above it, then below it; therefore, the amplitude y_m is 0.10 in.

The entire motion is repeated 120 times each second so that the frequency is 120 vibrations per second.

The wave speed is given by $v = \sqrt{F/\mu}$. But $F = 20$ lb and $\mu = 0.0050$ slug/ft, so that

$$v = \sqrt{\frac{20 \text{ lb}}{0.0050 \text{ slug/ft}}} = 63 \text{ ft/sec}$$

The wavelength is given by $\lambda = v/\nu$, so that

$$\lambda = \frac{63 \text{ ft/sec}}{120 \text{ vib/sec}} = 0.53 \text{ ft.}$$

(*b*) Assuming the wave moves in the $+x$-direction and that, at $t = 0$, the end of the string described by $x = 0$ is in its equilibrium position $y = 0$, write the equation of the wave.

The general expression for a transverse sinusoidal wave moving in the $+x$-direction is

$$y = y_m \sin (kx - \omega t - \phi).$$

Requiring that $y = 0$ for the conditions $x = 0$ and $t = 0$ yields

$$0 = y_m \sin(-\phi),$$

which means that the phase constant ϕ may be taken to be zero. You should show that integral multiples of π yield the same final results. Hence for this wave

$$y = y_m \sin(kx - \omega t),$$

and with the values just found,

$$y_m = 0.10 \text{ in.} = 8.3 \times 10^{-3} \text{ ft},$$

$$\lambda = 0.53 \text{ ft} \quad \text{or} \quad k = \frac{2\pi}{\lambda} = \frac{2\pi}{0.53 \text{ ft}} = 12 \text{ ft}^{-1}$$

$$v = 63 \text{ ft/sec} \quad \text{or} \quad \omega = vk = (63 \text{ ft/sec})(12 \text{ ft}^{-1}) = 740 \text{ sec}^{-1}$$

we obtain as the equation for the wave

$$y = (8.3 \times 10^{-3}) \sin(12x - 740t)$$

where x and y are in feet and t is in seconds.

Example 2. As this wave passes along the string, each particle of the string moves up and down at right angles to the direction of the wave motion. Find the velocity and acceleration of a particle 2.0 ft from the end.

The general form of this wave is

$$y = y_m \sin(kx - \omega t) = y_m \sin k(x - vt).$$

The v in this equation is the constant horizontal velocity of the wavetrain. What we are after now is the velocity of a particle in the string through which this wave moves; this particle velocity is neither horizontal nor constant. In fact, each particle moves vertically, that is, in the y-direction. In order to determine the particle velocity, which we shall designate by the symbol u, let us fix our attention on a particle at a particular position x—that is, x is now a constant in this equation— and ask how the particle displacement y changes with time. With x constant we obtain

$$u = \frac{\partial y}{\partial t} = -y_m \omega \cos(kx - \omega t),$$

in which the *partial derivative* $\partial y/\partial t$ reminds us that although in general y is a function of both x and t, we here assume that x remains constant so that t becomes the only variable. The acceleration a of the particle at this (constant) value of x is

$$a = \frac{\partial^2 y}{\partial t^2} = \frac{\partial u}{\partial t} = -y_m \omega^2 \sin(kx - \omega t) = -\omega^2 y.$$

This shows that for each particle through which this transverse sinusoidal wave passes we have precisely SHM (simple harmonic motion), for the acceleration a is proportional to the displacement y, but oppositely directed.

For a particle at $x = 2.0$ ft with the wave of Example 1, in which

$$y_m = 8.3 \times 10^{-2} \text{ ft}, \quad k = 12 \text{ ft}^{-1}, \quad \omega = 740 \text{ sec}^{-1},$$

we obtain

$$u = -y_m \omega \cos(kx - \omega t)$$

or

$$u = -(8.3 \times 10^{-3})(740) \cos[(12)(2.0) - (740)t] = -6.2 \cos(24 - 740\,t)$$

and

$$a = -\omega^2 y$$

or

$$a = -(740)^2(8.3 \times 10^{-3}) \sin[(12)(2.0) - (740)t] = -4.6 \times 10^3 \sin(24 - 740\,t)$$

where t is expressed in seconds u in ft/sec and a in ft/sec^2.

16–5 Power and Intensity in Wave Motion

An important quantity associated with one-dimensional waves is the amount of energy transferred past a given point per unit time. This is a measure of the power transfer. In the case of three-dimensional waves the corresponding quantity is the energy transferred across a given area per unit time or *intensity*.

   We will derive the power transfer in the case of a one-dimensional string. Consider one wavelength of the wave shown in Fig. 16–6. In one period, *T*, the one wavelength shown will move past the point at *D*. If we can find the energy contained in this one wavelength, the power can be obtained simply by dividing by the period *T*.

   At a given time each particle of equal mass on the string has the same energy. At point *A* it is all in potential energy, at *B* it is all in kinetic, and at *C* it is shared. We will find the kinetic energy at *B* for the mass element *dm*; then, inferring that all elements have the same total energy, we will sum over all elements. This sum is accomplished by replacing *dm* by the total mass contained in one wavelength which is $\mu\lambda$ where μ is the mass per unit length. The kinetic energy of the element is

$$dK = \tfrac{1}{2}dm \left(\frac{\partial y}{\partial t}\right)^2_{\text{max}}$$

The power follows easily by replacing *dm* by $\mu\lambda$ and dividing by *T*.

$$P = \tfrac{1}{2}\mu\lambda \left(\frac{\partial y}{\partial t}\right)^2_{\text{max}} \times \frac{1}{T} = \tfrac{1}{2}\mu v \left(\frac{\partial y}{\partial t}\right)^2_{\text{max}}$$

where we see that the velocity *y* at point *B* is the maximum value that can be obtained. From the equation for the displacement,

$$y = y_m \sin (kx - \omega t), \tag{16–10a}$$

we see that

$$\frac{\partial y}{\partial t} = -\omega y_m \cos (kx - \omega t)$$

and

$$\left(\frac{\partial y}{\partial t}\right)_{\text{max}} = \omega y_m.$$

Therefore, the power is

$$P = \tfrac{1}{2}\mu v \omega^2 y_m{}^2. \tag{16–13}$$

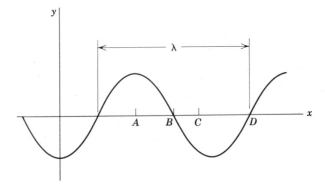

FIGURE 16–6

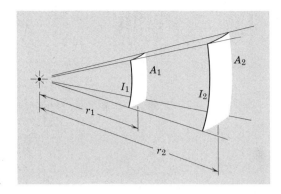

FIGURE 16–7   Example 3.

The fact that the rate of transfer of energy depends on the square of the wave amplitude and the square of the wave frequency is true in general, holding for all types of waves.

   Example 3. Spherical waves travel from a source of waves whose power output is P; see Fig. 16–7. Find how the wave intensity depends on the distance from the source. We assume that the medium is isotropic and that the source radiates uniformly in all directions, that is, that its emission is spherically symmetric.

   The intensity of a 3-dimensional wave is the power transmitted across a unit area normal to the direction of propagation. As the wave front expands from a distance r_1 from the source at the center to a distance r_2, its surface area increases from $4\pi r_1{}^2$ to $4\pi r_2{}^2$. If there is no absorption of energy, the total energy transported per second by the wave remains constant at the value P, so that

$$P = 4\pi r_1{}^2 I_1 = 4\pi r_2{}^2 I_2,$$

where I_1 and I_2 are the wave intensities at r_1 and r_2 respectively. Hence,

$$\frac{I_1}{I_2} = \frac{r_2{}^2}{r_1{}^2}$$

and the wave intensity varies inversely as the square of its distance from the source. Since the intensity is proportional to the square of the amplitude, the amplitude of the wave must vary inversely as the distance from the source.

16–6   The Superposition Principle

It is an experimental fact that, for many kinds of waves, two or more waves can traverse the same space independently of one another. The fact that waves act independently of one another means that the displacement of any particle at a given time is simply the sum of the displacements that the individual waves alone would give it. This process of vector addition of the displacements of a particle is called *superposition*. For example, radio waves of many frequencies pass through a radio antenna; the electric currents set up in the antenna by the superposed action of all these waves are very complex. Nevertheless, we can still tune to a particular station, the signal that we receive from it being in principle the same as that which we would receive if all other stations were to stop broadcasting. Likewise, in sound we can listen to notes played by individual instruments in an orchestra, even though the sound wave reaching our ears from the full orchestra is very complex.

   For waves in deformable media the superposition principle holds whenever the mathematical relation between the deformation and the restoring force is one of simple proportionality. Such a relation is expressed mathematically by a linear equation. For electromagnetic waves the superposition principle holds because the mathematical relations between the electric and magnetic fields are linear.

The importance of the superposition principle physically is that, where it holds, it makes it possible to analyze a complicated wave motion as a combination of simple waves. In fact, as was shown by the French mathematician J. Fourier (1786–1830), all that we need to build up the most general form of periodic wave are simple harmonic waves. Fourier showed that any periodic motion of a particle can be represented as a combination of simple harmonic motions. The general expression of such a combination is called a Fourier series. An example of the analysis of a sawtooth wave as a combination of simple harmonic waves is shown in Fig. 16–8. Six terms of the Fourier series representing the sawtooth wave give a rather good fit.

16–7 Interference of Waves

Interference refers to the physical effects of superimposing two or more wave trains. Let us consider two waves of equal frequency and amplitude traveling with the same speed in the same direction ($+x$) but with a phase difference ϕ between them. The

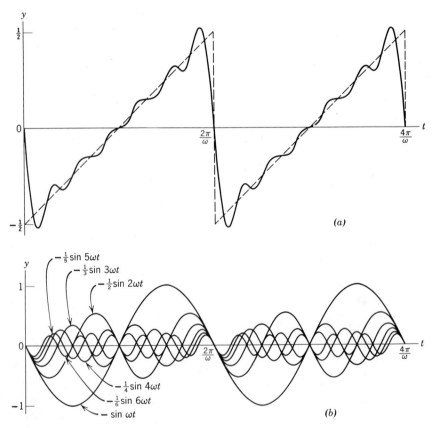

FIGURE 16–8 (*a*) The dashed line is a sawtooth "wave" commonly encountered in electronics. The Fourier series for this function is $y(t) = -\sin \omega t - \frac{1}{2}\sin 2\omega t - \frac{1}{3}\sin 3\omega t - \ldots$. The solid line is the sum of the first six terms of this series and can be seen to approximate the sawtooth quite closely. As more terms of the series are included, the approximation becomes better and better. (*b*) Here we show the first six terms of the Fourier series which, when added together, yield the solid curve in (*a*).

equations of the two waves will be

$$y_1 = y_m \sin (kx - \omega t - \phi) \tag{16–14}$$

and
$$y_2 = y_m \sin (kx - \omega t). \tag{16–15}$$

We can rewrite the first equation in two equivalent forms

$$y_1 = y_m \sin\left[k\left(x - \frac{\phi}{k}\right) - \omega t\right] \tag{16–14a}$$

or
$$y_1 = y_m \sin\left[kx - \omega\left(t + \frac{\phi}{\omega}\right)\right]. \tag{16–14b}$$

Equations 16–14a and 16–15 suggest that if we take a "snapshot" of the two waves at any time t, we will find them displaced from one another along the x-axis by the constant distance ϕ/k. Equations 16–14b and 16–15 suggest that if we station ourselves at any position x the two waves will give rise to two simple harmonic motions having a constant time difference ϕ/ω. This gives some insight into the meaning of the phase difference ϕ.

Now let us find the resultant wave, which, on the assumption that superposition occurs, is the sum of Eqs. 16–14 and 16–15 or

$$y = y_1 + y_2 = y_m[\sin (kx - \omega t - \phi) + \sin (kx - \omega t)].$$

From the trigonometric equation for the sum of the sines of two angles

$$\sin B + \sin C = 2 \sin \tfrac{1}{2}(B + C) \cos \tfrac{1}{2}(C - B), \tag{16–16}$$

we obtain
$$y = y_m \left[2 \sin\left(kx - \omega t - \frac{\phi}{2}\right) \cos \frac{\phi}{2}\right],$$

$$= \left(2y_m \cos \frac{\phi}{2}\right) \sin\left(kx - \omega t - \frac{\phi}{2}\right). \tag{16–17}$$

This resultant wave corresponds to a new wave having the same frequency but with an amplitude $2y_m \cos (\phi/2)$. If ϕ is very small compared to 180°, the resultant amplitude will be nearly $2y_m$. That is, when ϕ is very small, $\cos (\phi/2) \cong \cos 0° = 1$. When ϕ is zero, the two waves have the same phase everywhere. The crest of one corresponds to the crest of the other and likewise for the troughs. The waves are then said to interfere constructively. The resultant amplitude is just twice that of either wave alone. If ϕ is near 180°, on the other hand, the resultant amplitude will be nearly zero. That is, when $\phi \cong 180°$, $\cos (\phi/2) \cong \cos 90° = 0$. When ϕ is exactly 180°, the crest of one wave corresponds exactly to the trough of the other. The waves are then said to interfere destructively. The resultant amplitude is zero.

In Fig. 16–9a we show the superposition of two wavetrains almost in phase (ϕ small) and in Fig. 16–9b the superposition of two wavetrains almost 180° out of phase ($\phi \cong 180°$). Notice that in these figures the algebraic sum of the ordinates of the thin (component) curves at any value of x equals the ordinate of the thick (resultant) curve. The sum of two waves can, therefore, have different values, depending on their phase relations.

The resultant wave will be a sine wave, even when the amplitudes of the component sine waves are unequal. Figure 16–10, for example, illustrates the addition of two sine waves of the same frequency and velocity but different amplitudes. The

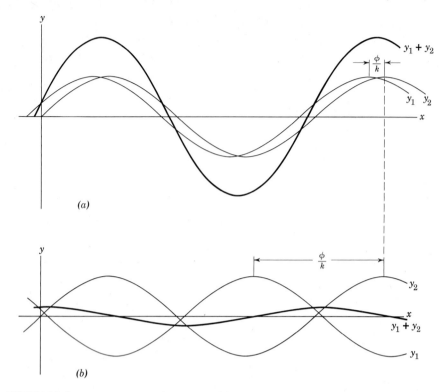

(a)

(b)

FIGURE 16–9 (*a*) The superposition of two waves of equal frequency and amplitude that are almost in phase results in a wave of almost twice the amplitude of either component. (*b*) The superposition of two waves of equal frequency and amplitude and almost 180° out of phase results in a wave whose amplitude is nearly zero. Note that in both the resultant frequency is unchanged. (The drawings correspond to the instant $t = 0$.)

resultant amplitude depends on the phase difference, which is taken as zero in this figure. The result for other phase differences could be obtained by shifting one of the component waves sideways with respect to the other and would give a smaller resultant amplitude. The smallest resultant amplitude would be the difference in the amplitudes of the components, obtained when the phases differ by 180°. However, the resultant is always a sine wave. The addition of any number of sine waves having the same frequency and velocity gives a similar result. The resultant waveform will always have a constant amplitude because the component waves (and their resultant) all move with the same velocity and maintain the same relative position. The actual state of affairs can be pictured by having all the waves in Figs. 16–9 and 16–10 move toward the right with the same speed.

In practice, interference effects are obtained from wavetrains which originate in the same source (or in sources having a fixed phase relationship to one another) but

FIGURE 16–10 The addition of two waves of same frequency and phase but differing amplitudes (light lines) yields a third wave of the same frequency and phase (heavy line).

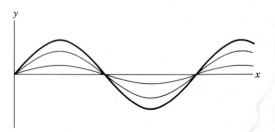

which follow different paths to the point of interference. The phase difference ϕ between the waves arriving at a point can be calculated by finding the difference between the paths traversed by them from the source to the point of interference. The phase difference is $k\,\Delta L$ or $2\pi\,\Delta L/\lambda$ if the path difference is ΔL. When the path difference is $0, \lambda, 2\lambda, 3\lambda$, etc., so that $\phi = 0, 2\pi, 4\pi$, etc., the two waves interfere constructively. For path differences of $\frac{1}{2}\lambda, \frac{3}{2}\lambda, \frac{5}{2}\lambda$, etc., ϕ is $\pi, 3\pi, 5\pi$, etc., and the waves interfere destructively. We shall return to these matters later in more detail.

16–8 Standing Waves

In a one-dimensional body of finite size, such as a taut string held by two clamps a distance l apart, traveling waves in the string are reflected from the boundaries of the body, that is, from the clamps. Each such reflection gives rise to a wave traveling in the string in the opposite direction. The reflected waves add to the incident waves according to the principle of superposition.

Consider two wavetrains of the same frequency, speed, and amplitude which are traveling in opposite directions along a string. Two such waves may be represented by the equations

$$y_1 = y_m \sin (kx - \omega t),$$

$$y_2 = y_m \sin (kx + \omega t).$$

We can write the resultant as

$$y = y_1 + y_2 = y_m \sin (kx - \omega t) + y_m \sin (kx + \omega t) \qquad (16\text{–}18a)$$

or, making use of the trigonometric relation of Eq. 16–16, as

$$y = 2y_m \sin kx \cos \omega t. \qquad (16\text{–}18b)$$

Equation 16–18b is the equation of a *standing* wave. Notice that a particle at any particular point x executes simple harmonic motion as time goes on, and that all particles vibrate with the same frequency. In a traveling wave each particle of the string vibrates with the same amplitude. Characteristic of a standing wave, however, is the fact that the amplitude is not the same for different particles but varies with the location x of the particle. In fact, the amplitude, $2y_m \sin kx$, has a maximum value of $2y_m$ at positions where

$$kx = \frac{\pi}{2}, \frac{3\pi}{2}, \frac{5\pi}{2}, \text{ etc.}$$

or

$$x = \frac{\lambda}{4}, \frac{3\lambda}{4}, \frac{5\lambda}{4}, \text{ etc.}$$

These points are called *antinodes* and are spaced one-half wavelength apart. The amplitude has a minimum value of zero at positions where

$$kx = \pi, 2\pi, 3\pi, \text{ etc.}$$

or

$$x = \frac{\lambda}{2}, \lambda, \frac{3\lambda}{2}, 2\lambda, \text{ etc.}$$

These points are called *nodes* and are spaced one-half wavelength apart. The separation between a node and an adjacent antinode is one-quarter wavelength.

It is clear that energy is not transported along the string to the right or to the left, for energy cannot flow past the nodal points in the string which are permanently at rest. Hence, the energy remains "standing" in the string, although it alternates between vibrational kinetic energy and elastic potential energy. We call the motion a wave motion because we can think of it as a superposition of waves traveling in opposite directions (Eq. 16–18*a*). We can equally well regard the motion as an oscillation of the string as a whole (Eq. 16–18*b*), each particle oscillating with SHM of angular frequency ω and with an amplitude that depends on its location. Each small part of the string has inertia and elasticity, and the string as a whole can be thought of as a collection of coupled oscillators. Hence, the vibrating string is the same in principle as a spring-mass system, except that a spring-mass system has only one natural frequency, and a vibrating string has a large number of natural frequencies (Section 16–9).

In Fig. 16–11, in (*a*), (*b*), (*c*), and (*d*), we show a standing wave pattern separately at intervals of one-quarter of a period in the lower figures, 3. The traveling waves, one moving in the positive *x*-direction and the other moving in the negative *x*-direction, whose superposition can be considered to give rise to the standing wave, are shown for the same quarter-period intervals in the upper figures 2 and 1. Standing waves can also be produced with electromagnetic waves and with sound waves.

In Fig. 16–12 we show how the energy associated with the oscillating string shifts back and forth between kinetic energy of motion K and potential energy of deformation U during one cycle. You should compare this with Fig. 7–4, which shows the same thing for a mass-spring oscillator. Oscillating strings often vibrate so rapidly that the eye perceives only a blur whose shape is that of the envelope of the motion; see Fig. 16–13.

The superposition of an incident wave and a reflected wave, being the sum of two waves traveling in opposite directions, will give rise to a standing wave. We shall now consider the process of reflection of a wave more closely. Suppose a pulse travels down a stretched string which is fixed at one end, as shown in Fig. 16–14*a*. When the pulse arrives at that end, it exerts an upward force on the support. The support is rigid, however, and does not move. By Newton's third law the support exerts an equal but oppositely directed force on the string. This reaction force generates a pulse at the support, which travels back along the string in a direction opposite to that of the incident pulse. We say that the incident pulse has been *reflected* at the fixed end

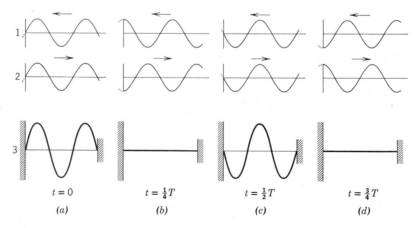

FIGURE 16–11   Standing waves as the superposition of left- and right-going waves; 1 and 2 are the components, 3 the resultant.

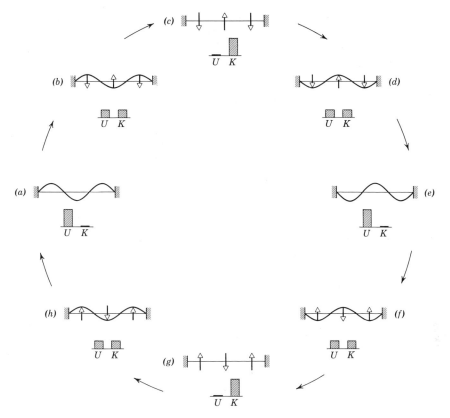

FIGURE 16–12   A standing wave in a stretched string, showing one cycle of oscillation. At (*a*) the string is momentarily at rest and the energy of the system is all potential energy of elastic deformation associated with the transverse displacement of the string. (*b*) An eighth-cycle later the displacement is reduced and the string is in motion. The three arrows show the velocities of the string particles at the positions shown. *K* and *U* have the same value. (*c*) The string is not displaced, but its particles have their maximum speeds; the energy is all kinetic. The motion continues until the initial condition (*a*) is reached after which the cycle continues to repeat itself.

point of the string. Notice that the reflected pulse returns with its transverse displacement reversed. If a wavetrain is incident on the fixed end point, a reflected wavetrain is generated at that point in the same way. The displacement of any point along the string is the sum of the displacements caused by the incident and reflected wave. Since the end point is fixed, these two waves must always interfere destructively at that point so as to give zero displacement there. Hence, the reflected wave is always 180° out of phase with the incident wave at a fixed boundary. We say that on reflection from a fixed end a wave undergoes a phase change of 180°.

Let us now consider the reflection of a pulse at a free end of a stretched string, that is, at an end that is free to move transversely. This can be achieved by attaching the

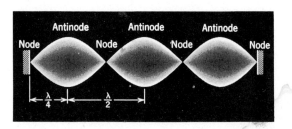

FIGURE 16–13   The envelope of a standing wave, corresponding to a time exposure of the motion, and showing the patterns of nodes and antinodes.

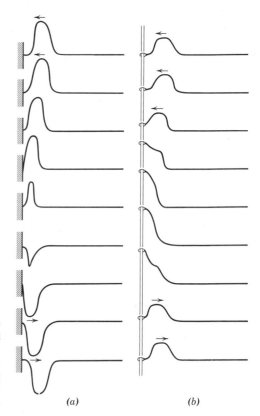

FIGURE 16–14 (*a*) Reflection of a pulse at the fixed end of a string. The drawings are spaced uniformly in time. The phase is changed by 180° on reflection. (*b*) Reflection of a pulse at an end free to move in a transverse direction. (The string is attached to a ring which slides vertically without friction.) The phase is unchanged on reflection.

(*a*)   (*b*)

end to a very light ring free to slide without friction along a transverse rod, or to a long and very much lighter string. When the pulse arrives at the free end, it exerts a force on the element of string there. This element is accelerated and its inertia carries it past the equilibrium point; it "overshoots" and exerts a reaction force on the string. This generates a pulse which travels back along the string in a direction opposite to that of the incident pulse. Once again we get reflection, but now at a free end. The free end will obviously suffer the maximum displacement of the particles on the string; an incident and a reflected wavetrain must interfere constructively at that point if we are to have a maximum there. Hence, the reflected wave is always in phase with the incident wave at that point (see Fig. 16–14*b*). We say that at a free end a wave is reflected without change of phase.

Hence, when we have a standing wave in a string, there will be a node at a fixed end and an antinode at a free end. Later, we will apply these ideas to sound waves and electromagnetic waves.

16–9 Resonance

In general, whenever a system capable of oscillating is acted on by a periodic series of impulses having a frequency equal or nearly equal to one of the natural frequencies of oscillation of the system, the system is set into oscillation with a relatively large amplitude. This phenomenon is called *resonance* and the system is said to resonate with the applied impulses.

Consider a string fixed at both ends. Oscillations or standing waves can be established in the string. The only requirement we have to satisfy is that the end points be nodes. There may be any number of nodes in between or none at all, so that the wavelength associated with the standing waves can take on many different values. The distance between adjacent nodes is $\lambda/2$, so that in a string of length l there must be exactly an integral number n of half wavelengths, $\lambda/2$. That is,

$$\frac{n\lambda}{2} = l$$

or
$$\lambda = \frac{2l}{n}, \qquad n = 1, 2, 3, \ldots.$$

But $\lambda = v/\nu$ and $v = \sqrt{F/\mu}$, so that the natural frequencies of oscillation of the system are

$$\nu = \frac{n}{2l}\sqrt{\frac{F}{\mu}}, \qquad n = 1, 2, 3, \ldots. \tag{16–19}$$

If the string is set vibrating and left to itself, the oscillations gradually die out. The motion is damped by dissipation of energy through the elastic supports at the ends and by the resistance of the air to the motion. We can pump energy into the system by applying a driving force. If the driving frequency is near that of any natural frequency, the string will vibrate at that frequency with a large amplitude. Because the string has a large number of natural frequencies, resonance can occur at many different frequencies. A mass-spring, by contrast, has only one resonant frequency.

Resonance in a string is often demonstrated by attaching a string to a fixed end, by means of a weight attached to it over a pulley, and connecting the other end to a vibrator, as shown in Fig. 16–15. The transverse oscillations of the vibrator set up a traveling wave in the string which is reflected back from the fixed end. The frequency of the waves is that of the vibrator, and the wavelength is determined by $\lambda = v/\nu$. The fixed end P is a node, but the end Q vibrates and is not. If we now vary the tension in the string by changing the hanging weight, for example, we can change the wavelength. Changing the tension changes the wave velocity, and the wavelength changes in proportion to the velocity, the frequency being constant. Whenever the wavelength becomes nearly equal to $2l/n$, where l is the length of the string, we obtain standing waves of great amplitude. The string now vibrates in one of its natural modes and resonates with the vibrator. The vibrator does work on the string to maintain these oscillations against the losses due to damping. The amplitude builds up only to the point at which the vibrator expends all its energy input against damping losses. The point Q is almost a node because the amplitude of the vibrator is small compared to that of the string.

If the frequency of the vibrator is much different from a natural frequency of the

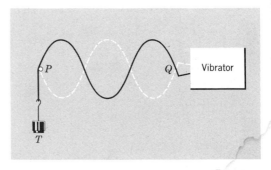

FIGURE 16–15 Standing waves in a driven string when the natural and driving frequencies are very nearly equal.

system, as given by Eq. 16–19, the wave reflected at P on returning to Q may be out of phase with the vibrator, and it can do work on the vibrator. That is, the string can give up some energy to the vibrator just as well as receive energy from it. The "standing" wave pattern is not fixed in form but wiggles about. On the average the amplitude is small and not much different from that of the vibrator. This situation is analogous to the erratic motion of a swing being pushed periodically with a frequency other than its natural one. The displacement of the swing is rather small.

Hence, the string absorbs peak energy from the vibrator at resonance. Tuning a radio is an analogous process. By tuning a dial the natural frequency of an alternating current in the receiving circuit is made equal to the frequency of the waves broadcast by the station desired. The circuit resonates with the transmitted signals and absorbs peak energy from the signal. We shall encounter resonance conditions again in sound, in electromagnetism, in optics, and in atomic and nuclear physics. In these areas, as in mechanics, the system will absorb peak energy from the source at resonance and relatively little energy off resonance.

Example 4. In a demonstration with the apparatus just described, the vibrator has a frequency $v = 20$ cycles/sec, and the string has a linear density $\mu = 1.56 \times 10^{-4}$ slug/ft and a length $l = 24$ ft. The tension F is varied by pulling down on the end of the string over the pulley. If the demonstrator wants to show resonance, starting with one loop and then with two, three, and four loops, what force must he exert on the string?

At resonance,

$$v = \frac{n}{2l} \sqrt{\frac{F}{\mu}}.$$

Hence, the tension F is given by

$$F = \frac{4l^2 v^2 \mu}{n^2}.$$

For one loop, $n = 1$, so that

$$F_1 = 4l^2 v^2 \mu = 4(24 \text{ ft})^2 (20 \text{ sec}^{-1})^2 (1.56 \times 10^{-4} \text{ slug/ft}) = 144 \text{ lb}.$$

For two loops, $n = 2$, and

$$F_2 = \frac{4l^2 v^2 \mu}{4} = \frac{F_1}{4} = 36 \text{ lb}.$$

Likewise, for three and four loops

$$F_3 = \frac{F_1}{(3)^2} = 16 \text{ lb},$$

$$F_4 = \frac{F_1}{(4)^2} = 9 \text{ lb}.$$

Hence, the demonstrator gradually relaxes the tension to obtain resonance with an increasing number of loops. Although the resonant frequency is always the same under these circumstances, the speed of propagation and the wavelength at resonance decrease proportionately.

QUESTIONS

1. How could you prove experimentally that energy is associated with a wave?

2. Energy can be transferred by particles as well as by waves. How can we distinguish experimentally between these methods of energy transfer?

3. Can a wave motion be generated in which the particles of the medium vibrate with angular simple harmonic motion? If so, explain how and describe the wave.

4. Are torsional waves transverse or longitudinal? Can they be considered as a superposition of two waves, which are either transverse or longitudinal?

5. How can one create plane waves? Spherical waves?

6. The following functions, in which A is a constant, are of the form $f(x \pm vt)$:

$$y = A(x - vt), \qquad y = A(x + vt)^2,$$
$$y = A\sqrt{x - vt}, \qquad y = A \ln (x + vt).$$

Explain why these functions are not useful in wave motion.

7. How do the amplitude and the intensity of surface water waves vary with the distance from the source?

8. The inverse square law does not apply exactly to the decrease in intensity of sounds with distance. Why not?

9. When two waves interfere, does one alter the progress of the other?

10. When waves interfere, is there a loss of energy? Explain your answer.

11. Why don't we observe interference effects between the light beams emitted from two flashlights or between the sound waves emitted by two violins.

12. If two waves differ only in amplitude and are propagated in opposite directions through a medium, will they produce standing waves? Is energy transported? Are there any nodes?

13. Is an oscillation a wave? Explain.

14. Consider the standing waves in a string to be a superposition of traveling waves and explain, using superposition ideas, why there are no true nodes in the resonating string of Fig. 16–15, even at the "fixed" end. (*Hint:* Consider damping effects.)

15. In the discussion of transverse waves in a string we have dealt only with displacements in a single plane, the *x-y* plane. If all displacements lie in one plane the wave is said to be *plane polarized*. Can there be displacements in a plane other than the single plane dealt with? If so, can two differently plane-polarized waves be combined? What appearance would such a combined wave have?

PROBLEMS

1(3). The speed of electromagnetic waves in vacuum is 3×10^8 meters/sec. (*a*) Wavelengths in the visible part of the spectrum (light) range from about 4×10^{-7} meter in the violet to about 7×10^{-7} meter in the red. What is the range of frequencies of light waves? (*b*) The range of frequencies for shortwave radio (for example, FM radio and VHF television) is 1.5 megacycles/sec to 300 megacycles/sec. What is the corresponding wavelength range? (*c*) X rays are also electromagnetic. Their wavelength range extends from about 5×10^{-9} meter to about 1.0×10^{-11} meter. What is the frequency range for x rays?

Answer: (*a*) 4×10^{14} to 8×10^{14} cycles/sec. (*b*) 1 to 200 meters. (*c*) 6×10^{16} to 3×10^{19} cycles/sec.

2(3). Show that $y = y_m \sin (kx - \omega t)$ may be written in the alternative forms

$$y = y_m \sin k(x - vt), \qquad y = y_m \sin 2\pi \left(\frac{x}{\lambda} - vt \right),$$

$$y = y_m \sin \omega \left(\frac{x}{v} - t \right), \qquad y = y_m \sin 2\pi \left(\frac{x}{\lambda} - \frac{t}{T} \right).$$

3(3). The equation of a transverse wave traveling along a very long string is given by $y = 6.0 \sin (0.020\pi x + 4.0\pi t)$, where x and y are expressed in cm and t in seconds. Calculate (*a*) the amplitude, (*b*) the wavelength, (*c*) the frequency, (*d*) the speed, (*e*) the direction of propagation of the wave, and (*f*) the maximum transverse speed of a particle in the string.

Answer: (*a*) 6.0 cm. (*b*) 100 cm. (*c*) 2.0 cycles/sec. (*d*) 200 cm/sec. (*e*) $-x$. (*f*) 75 cm/sec.

4(3). The equation of a transverse wave traveling in a string is given by

$$y = 0.20 \sin (2.0x - 600t),$$

where y and x are expressed in centimeters and t in seconds. (*a*) Find the amplitude, frequency, velocity, and wavelength of the wave. (*b*) Find the maximum transverse speed of a particle in the string.

5(3). A sinusoidal wave travels along a string. If the time for a particular point to move from maximum displacement to zero displacement is 0.17 sec, what are (*a*) the period, and (*b*) frequency? (*c*) If the wavelength is 1.4 meters what is the velocity?
Answer: (*a*) 0.68 sec. (*b*) 1.5 cycles/sec. (*c*) 2.1 meters/sec.

6(3). A wave of frequency 500 cycles/sec has a velocity of 350 meters/sec. (*a*) How far apart are two points 60° out of phase? (*b*) What is the phase difference between two displacements at a certain point at times 10^{-3} sec apart?

7(3). (*a*) Write an expression describing a transverse wave traveling on a cord in the $+y$-direction with a wave number of 60 cm^{-1}, a period of 0.20 sec and having an amplitude of 3.0 cm. Take the transverse direction to be the z-direction. (*b*) What is the maximum transverse velocity of a point on the cord?
Answer: (*a*) $z = 3.0 \sin (60y - 10\pi t)$, with y and z in centimeters and t in seconds. (*b*) 94 cm/sec.

8(3). Write the equation for a wave traveling in the negative direction along the x-axis and having an amplitude 0.010 meter, a frequency 550 vib/sec, and a speed 330 meters/sec.

9(3). (*a*) Write an expression describing a transverse wave traveling on a cord in the $+x$ direction with a wavelength of 10 cm, a frequency of 400 cycles/sec and an amplitude of 2.0 cm. (*b*) What is the maximum speed of a point on the cord? (*c*) What is the velocity of the wave?
Answer: (*a*) $y = 2.0 \sin 2\pi(0.10x - 400t)$, with x and y in centimeters and t in seconds. (*b*) 5000 cm/sec. (*c*) 4000 cm/sec.

10(3). (*a*) A continuous sinusoidal longitudinal wave is sent along a coiled spring from a vibrating source attached to it. The frequency of the source is 25 vib/sec, and the distance between successive rarefactions in the spring is 24 cm. Find the wave speed. (*b*) If the maximum longitudinal displacement of a particle in the spring is 0.30 cm and the wave moves in the $-x$-direction, write the equation for the wave. Let the source be at $x = 0$ and the displacement $x = 0$ and $t = 0$ be zero.

11(4). What is the speed of a transverse wave in a rope of length 2.0 meters and mass 0.060 kg under a tension of 500 nt?
Answer: 130 meters/sec.

12(4). If the string in Problem 4(3) is under a tension of 10.0 nt, what is the linear density of the string?

13(4). The linear density of a vibrating string is 1.3×10^{-4} kg/meter. A transverse wave is propagating on the string and is described by the equation $y = 0.021 \sin (x + 30t)$, where x and y are measured in meters and t in seconds. What is the tension in the string?
Answer: 0.12 nt.

14(4). A continuous sinusoidal wave is traveling on a string with velocity 80 cm/sec. The displacement of the particles of the string at $x = 10$ cm is found to vary with time according to the equation $y = 5.0 \sin (1.0 - 4.0t)$ in cm. The linear density of the string is 4.0 gm/cm. (*a*) What is the frequency of the wave? (*b*) What is the wavelength of the wave? (*c*) Write the general equation giving the transverse displacement of the particles of the string as a function of position and time. (*d*) Calculate the tension in the string.

15(4). A simple harmonic transverse wave is propagating along a string toward the left (or $-x$) direction. Fig. 16–16 shows a plot of the displacement as a function of position at time $t = 0$. The string tension is 3.6 nt and its linear density is 25 grams/meter. Calculate (*a*) the amplitude, (*b*) the period, (*c*) the wavelength, (*d*) the wave speed, and (*e*) the maximum speed of a particle in the string. (*f*) Write an equation describing the traveling wave.
Answer: (*a*) 5.0 cm. (*b*) 0.033 sec. (*c*) 40 cm. (*d*) 12 meters/sec. (*e*) 9.4 meters/sec. (*f*) $y = 5.0 \sin (0.16x + 190t + 0.79)$ in which x and y are in centimeters and t is in seconds.

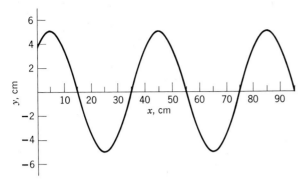

FIGURE 16–16 Problem 15(4).   **FIGURE 16–17** Problem 18(4).

16(4). A wire 10.0 meters long and having a mass of 100 gm is stretched under a tension of 250 nt. If two disturbances, separated in time by 0.030 sec, are generated one at each end of the wire where will the disturbances meet?

17(4). Prove that the slope of a string at any point x is numerically equal to the ratio of the particle speed to the wave speed at that point.

18(4). A uniform circular hoop of string is rotating clockwise in the absence of gravity (see Fig. 16–17). The tangential speed is v_0. Find the speed of waves traveling on this string. (Remark: The answer is independent of the radius of the circle and the mass per unit length of the string!)

19(5). Spherical waves are emitted from a 1.0-watt source in an isotropic nonabsorbing medium. What is the wave intensity 1.0 meter from the source?
Answer: 0.080 watts/meter².

20(5). (*a*) Show that the intensity I (the energy crossing unit area per unit time) is the product of the energy per unit volume u and the speed of propagation v of a wave disturbance. (*b*) Radio waves travel at a speed of 3.0×10^8 meters/sec. Find the energy density in a radio wave 300 miles from a 50,000-watt source, assuming the waves to be spherical and the propagation to be isotropic.

21(5). A line source emits a cylindrical expanding wave. Assuming the medium absorbs no energy, find how (*a*) the amplitude and (*b*) the intensity of the wave depend on the distance from the source.
Answer: (*a*) Proportional to $r^{-1/2}$. (*b*) Proportional to r^{-1}.

22(5). (*a*) From Example 2 show that the maximum speed of a particle in a string through which a sinusoidal wave is passing is $u = y_m\omega$. (*b*) In Example 2 we saw that the particles in the string oscillate with simple harmonic motion. The mechanical energy of each particle is the sum of its potential and kinetic energies and is always equal to the maximum value of its kinetic energy. Consider an element of string of mass $\mu\Delta x$ and show that the energy per unit length of the string is given by

$$E_l = 2\pi^2\mu\nu^2 y_m{}^2.$$

(*c*) Show finally that the average power or average rate of transfer of energy is the product of the energy per unit length and the wave speed. (*d*) Do these results hold only for a sinusoidal wave?

23(5). A wave travels out uniformly in all directions from a point source. (*a*) Justify the following expression for the displacement y of the medium at any distance r from the source:

$$y = \frac{Y}{r} \sin k(r - vt).$$

Consider the speed, direction of propagation, periodicity, and intensity of the wave. (*b*) What are the dimensions of the constant Y?
Answer: (*b*) Length².

24(7). Determine the amplitude of the resultant motion when two sinusoidal motions having the

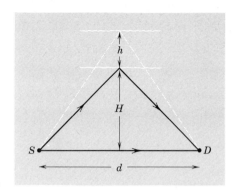

FIGURE 16-18 Problem 26(7).

same frequency and traveling in the same direction are combined, if their amplitudes are 3.0 cm and 4.0 cm and they differ in phase by $\pi/2$ radians.

25(7). Two waves are propagating on the same very long string. A generator at the left end of the string creates a wave given by $y = 6.0 \cos \dfrac{\pi}{2} (0.020x + 8.0t)$, and one at the right end of the string creates the wave $y = 6.0 \cos \dfrac{\pi}{2} (0.020x - 8.0t)$. In both cases, y and x are in centimeters and t is in seconds. (a) Calculate the frequency, wavelength, and speed of each wave. (b) Find the points at which there is no motion (the nodes). (c) At which points is the motion a maximum?
Answer: (a) 2.0 cycles/sec; 200 cm; 400 cm/sec. (b) $x = 50$ cm, 150 cm, 250 cm, etc. (c) $x = 0$, 100 cm, 200 cm, etc.

26(7). A source S and a detector D of high-frequency waves are a distance d apart on the ground. The direct wave from S is found to be in phase at D with the wave from S that is reflected from a horizontal layer at an altitude H (Fig. 16–18). The incident and reflected rays make the same angle with the reflecting layer. When the layer rises a distance h, no signal is detected at D. Neglect absorption in the atmosphere and find the relation between d, h, H, and the wavelength λ of the waves.

27(7). Four component sine waves have frequencies in the ratio 1, 2, 3, and 4 and amplitudes in the ratio 1, $\frac{1}{2}$, $\frac{1}{3}$, and $\frac{1}{4}$, respectively. The first and third components are 180° out of phase with the second and fourth components. Plot the resultant waveform and discuss its nature.

28(7). Two pulses are traveling along a string in opposite directions, as shown in Fig. 16–19. (a) If the wave velocity is 2.0 m/sec and the pulses are 6.0 cm apart, sketch the patterns after 0.5, 1.0, 1.5, 2.0, 2.5 msec, where 1 msec = 10^{-3} sec. (b) What has happened to the energy at $t = 1.5$ msec?

29(7). Three component sinusoidal waves have the same period, but their amplitudes are in the ratio 1, $\frac{1}{2}$, and $\frac{1}{3}$ and their phase angles are 0, $\pi/2$, and π respectively. Plot the resultant waveform and discuss its nature.

30(8). A string vibrates according to the equation

$$y = 2.0 \sin 0.16x \cos 750t$$

where x and y are in centimeters and t is in seconds. (a) What are the amplitude and velocity of the component waves whose superposition give rise to this vibration? (b) What is the distance between nodes? (c) What is the velocity of a particle of the string at the position $x = 5.0$ cm when $t = 2.0 \times 10^{-3}$ sec?

31(8). The equation of a transverse wave traveling in a string is given by

$$y = 10 \cos (0.0079x - 13t - 0.89),$$

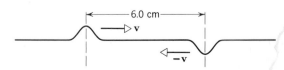

FIGURE 16-19 Problem 28(7).

in which x and y are expressed in centimeters and t in seconds. Write down the equation of a wave which, when added to the given one, would produce standing waves on the rope.

Answer: $y = 10 \cos (0.0079x + 13t - 0.89)$.

32(8). Two waves on a string are described by the equations

$$y_1 = 0.10 \sin 2\pi \left(\frac{x}{2} + 20t\right)$$

and

$$y_2 = 0.20 \sin \pi \left(\frac{x}{2} - 20t\right)$$

where x and y are in meters and t in seconds. Sketch the total response for the point on the string at $x = 3.0$ m; that is, plot y vs. t for that value of x.

33(8). Two transverse sinusoidal waves travel in opposite directions along a string. Each has an amplitude of 0.30 cm and a wavelength of 6.0 cm. The speed of a transverse wave in the string is 1.5 m/sec. Plot the shape of the string at each of the following times: $t = 0$ (arbitrary), $t = 1.0$, $t = 2.0$, $t = 3.0$, $t = 4.0$, and $t = 5.0$ msec, where 1 msec $= 10^{-3}$ sec.

34(8). What are the three lowest frequencies for standing waves on the wire of Problem 16(4)?

35(8). When played in a certain manner the lowest frequency of vibration of a certain violin string is concert A (440 cycles/sec). What two higher frequencies could also be found on that string if the length isn't changed?

Answer: 880 cycles/sec and 1320 cycles/sec.

36(9). In a laboratory experiment on standing waves a string 3.0 ft long is attached to the prong of an electrically driven tuning fork which vibrates perpendicular to the length of the string at a frequency of 60 vib/sec. The weight of the string is 0.096 lb. (*a*) What tension must the string be under (weights are attached to the other end) if it is to vibrate in four loops? (*b*) What would happen if the tuning fork is turned so as to vibrate parallel to the length of the string?

37(9). A 1.0-meter long wire has a mass of 10 grams and is held under a tension of 100 nt. The wire is rigidly held at both ends and set into vibration. Calculate (*a*) the velocity of waves on the wire, (*b*) the wavelengths of the waves that produce one- and two-loop standing waves on the string, and (*c*) the frequencies of the waves that produce one- and two-loop standing waves. *Answer:* (*a*) 100 meters/sec. (*b*) 2.0 and 1.0 meters. (*c*) 50 and 100 cycles/sec.

38(9). Vibrations from a 600-cycle/sec tuning fork set up standing waves in a string clamped at both ends. The wave speed for the string is 400 meters/sec. The standing wave has four loops and an amplitude of 2.0 mm. (*a*) What is the length of the string? (*b*) Write an equation for the displacement of the string as a function of position and time.

39(9). A 3.0-meter long string is vibrating as a 3-loop standing wave whose amplitude is 1.0 cm. The wave speed is 100 meters/sec. (*a*) What is the frequency? (*b*) Write equations for two waves that, when combined, will result in this standing wave.

Answer: (*a*) 50 cycles/sec. (*b*) $y = 0.50 \sin \pi \left(\dfrac{x}{100} \pm 100t\right)$, with x and y in centimeters and t in seconds.

CHAPTER 17

Sound Waves

17–1 Audible, Ultrasonic, and Infrasonic Waves

Sound waves are longitudinal mechanical waves. They can be propagated in solids, liquids, and gases. The material particles transmitting such a wave oscillate in the direction of propagation of the wave itself. There is a large range of frequencies within which longitudinal mechanical waves can be generated, sound waves being confined to the frequency range which can stimulate the human ear and brain to the sensation of hearing. This range is from about 20 cycles/sec to about 20,000 cycles/sec and is called the *audible* range. A longitudinal mechanical wave whose frequency is below the audible range is called an *infrasonic* wave, and one whose frequency is above the audible range is called an *ultrasonic* wave.

Infrasonic waves of interest are usually generated by large sources, earthquake waves being an example.* The high frequencies associated with ultrasonic waves may be produced by elastic vibrations of a quartz crystal induced by resonance with an applied alternating electric field (piezoelectric effect). It is possible to produce ultrasonic frequencies as high as 6×10^8 cycles/sec in this way; the corresponding wavelength in air is about 5×10^{-5} cm, the same as the length of visible light waves.

Audible waves originate in vibrating strings (violin, human vocal cords), vibrating air columns (organ, clarinet), and vibrating plates and membranes (xylophone, loudspeaker, drum). All of these vibrating elements alternately compress the surrounding air on a forward movement and rarefy it on a backward movement. The air transmits these disturbances outward from the source as a wave. Upon

---

* See "Long Earthquake Waves" by Jack Oliver, in *Scientific American*, March 1959.

entering the ear, these waves produce the sensation of sound. Waveforms which are approximately periodic or consist of a small number of approximately periodic components often create a pleasant sensation (if the intensity is not too high), as, for example, musical sounds. Sound whose waveform is nonperiodic is heard as noise. Noise can be represented as a superposition of periodic waves, but the number of components is very large.

In this chapter we deal with the properties of longitudinal mechanical waves, using sound waves as the prototype.

17–2 Propagation and Speed of Longitudinal Waves

Sound waves, if unimpeded, will spread out in all directions from a source. It is simpler to deal with one-dimensional propagation, however, than with three-dimensional propagation, so that we consider first the transmission of longitudinal waves in a tube.

Figure 17–1 shows a piston at one end of a long tube filled with a compressible medium. The vertical lines divide the compressional (fluid) medium into thin slices, each of which contains the same mass of fluid. Where the lines are relatively close together the fluid pressure and density are greater than they are in the normal undisturbed fluid, and conversely. We shall treat the fluid as a continuous medium and ignore for the time being the fact that it is made up of molecules that are in continual random motion.

If we push the piston of Fig. 17–1 forward, the fluid in front of it is compressed, the fluid pressure and density rising above their normal undisturbed values. The compressed fluid moves forward, compressing the fluid layers next to it, and a compressional pulse travels down the tube. If we then withdraw the piston, the fluid in front of it expands, its pressure and density falling below their normal undisturbed values; a pulse of rarefaction travels down the tube. These pulses are similar to transverse pulses traveling along a string, except that the oscillating fluid elements are displaced along the direction of propagation (longitudinal) instead of at right angles to this direction (transverse). If the piston oscillates back and forth, a continuous train of compressions and rarefactions will travel along the tube (Fig. 17–1). As for transverse waves in a string (see Section 16–4) we should be able, using Newton's laws of motion, to express the speed of propagation of this longitudinal wave in terms of an elastic and an inertial property of the medium. We now do so.

For the moment, let us assume that the tube is very long so that we can ignore

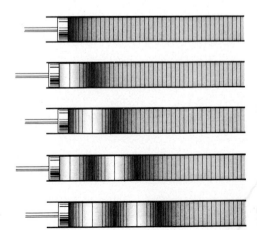

FIGURE 17–1 Sound waves generated in a tube by an oscillating piston. The vertical lines divide the compressible medium in the tube into layers of equal mass. We assume the tube to be infinitely long so that there is no confusion caused by reflection of the wave at the end of the tube.

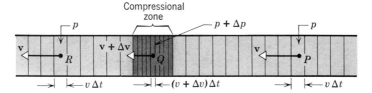

FIGURE 17–2 A compressional pulse travels along a gas-filled tube. In a reference frame in which the undisturbed gas is at rest the pulse moves from left to right with speed v. We view the pulse, however, from a reference frame in which the pulse is stationary; in such a frame the gas outside the pulse streams through the tube from right to left with speed v, as shown. Note that Δv is negative.

reflections from the other end. As for the string of Fig. 16–5, we will consider not an extended wave but a single (compressional) pulse that we might generate by giving the piston in Fig. 17–1 a short, rapid, inward stroke.

Figure 17–2 shows such a pulse (labeled "compressional zone") traveling at speed v along the tube from left to right. For simplicity we have assumed this pulse to have sharply defined leading and trailing edges and to have a uniform fluid pressure and density in its interior. When we analyzed the motion of a transverse pulse in a string, we found it convenient to choose a reference frame in which the pulse remained stationary; we will do this here also. In Fig. 17–2, then, the compressional zone remains stationary in our reference frame while the fluid moves through it from right to left with speed v, as shown.

Let us follow the motion of the element of fluid contained between the vertical lines at P in Fig. 17–2. This element moves forward at speed v until it strikes the compressional zone. While it is entering this zone it encounters a difference of pressure Δp between its leading and its trailing edges. The element is compressed and decelerated, moving with a lower speed $v + \Delta v$ within the zone, the quantity Δv being negative. The element eventually emerges from the left face of the zone where it expands to its original volume and the pressure differential Δp acts to accelerate it to its original speed v. The figure shows the element at point R, having passed through the compressional zone and moving again with speed v, as at P.

Let us apply Newton's laws to the fluid element while it is entering the compressional zone. The resultant force acting during entry points is to the right in Fig. 17–2 and has magnitude

$$F = (p + \Delta p)A - pA = \Delta p A$$

in which A is the cross-sectional area of the tube.

The length of the element outside the compressional zone (at P, say) is $v\Delta t$, where Δt is the time required for the element to move past any given point. The volume of the element is thus $vA\,\Delta t$ and its mass is $\rho_0 vA\,\Delta t$, where ρ_0 is the density of the fluid outside the compressional zone. The acceleration a experienced by the element as it enters the zone is $-\Delta v/\Delta t$; because Δv is inherently negative; a is positive, which means that, like the force $\Delta p A$ in Fig. 17–2, it points to the right. Thus Newton's second law

$$F = ma$$

yields

$$\Delta p A = (\rho_0 vA\,\Delta t)\frac{-\Delta v}{\Delta t},$$

which we may write as

$$\rho_0 v^2 = \frac{-\Delta p}{\Delta v/v}.$$

Now the fluid that would occupy a volume $V (= Av\,\Delta t)$ at P is compressed by an amount $A(\Delta v)\,\Delta t = \Delta V$ on entering the compressional zone. Hence,

$$\frac{\Delta V}{V} = \frac{A\,\Delta v\,\Delta t}{Av\,\Delta t} = \frac{\Delta v}{v}$$

and we obtain

$$\rho_0 v^2 = \frac{-\Delta p}{\Delta V/V}.$$

The ratio of the change in pressure on a body, Δp, to the resulting fractional change in volume, $-\Delta V/V$, is called the *bulk modulus of elasticity* B of the body. That is, $B = -V\,\Delta p/\Delta V$. B is positive because an increase in pressure causes a decrease in volume. In terms of B, the speed of the longitudinal pulse in the medium of Fig. 17–2 is

$$v = \sqrt{B/\rho_0}. \tag{17-1}$$

A more extended analysis than given above shows that Eq. 17–1 applies not only to rectangular pulses of the type displayed in Fig. 17–2 but also to pulses of any shape and to extended wave trains. Notice that the speed of the wave is determined by the properties of the medium through which it propagates, and that an elastic property B and an inertial property ρ_0 are involved. Table 17–1 gives the speed of longitudinal (sound) waves in various media.

If the medium is a gas, such as air, it is possible to express B in terms of the undisturbed gas pressure p_0. For a sound wave in a gas we obtain

$$v = \sqrt{\gamma p_0/\rho_0},$$

where γ is a constant called the ratio of specific heats for the gas (see Section 20–6).

If the medium is a solid, for a thin rod the bulk modulus is replaced by a stretch modulus (called Young's modulus). If the solid is extended, we must allow for the fact that, unlike a fluid, a solid offers elastic resistance to tangential or shearing forces and the speed of longitudinal waves will depend on the shear modulus as well as the bulk modulus.

Table 17–1   SPEED OF SOUND

Medium	Tempera-ture, °C	Speed meters/sec	ft/sec
Air	0	331.3	1,087
Hydrogen	0	1,286	4,220
Oxygen	0	317.2	1,041
Water	15	1,450	4,760
Lead	20	1,230	4,030
Aluminum	20	5,100	16,700
Copper	20	3,560	11,700
Iron	20	5,130	16,800
Granite	—	6,000	19,700
Vulcanized rubber	—	54	177

17–3 Traveling Longitudinal Waves

Consider again the continuous train of compressions and rarefactions traveling down the tube of Fig. 17–1. As the wave advances along the tube, each small volume element of fluid oscillates about its equilibrium position. The displacement is to the right or left along the x-direction of propagation of the wave. For convenience let us represent the displacement of any such volume element (or layer of elements that move in the same way) from its equilibrium position at x by the letter y. You should understand that the displacement y is along the direction of propagation for a longitudinal wave, whereas for a transverse wave the displacement y is at right angles to the direction of propagation. Then the equation of a longitudinal wave traveling to the right in Fig. 17–1 may be written as

$$y = f(x - vt).$$

For the particular case of a simple harmonic oscillation we may have

$$y = y_m \cos \frac{2\pi}{\lambda} (x - vt).$$

In this equation v is the speed of the longitudinal wave, y_m is its amplitude, and λ is its wavelength; y gives the displacement of a particle at time t from its equilibrium position at x. As before, we may write this more compactly as

$$y = y_m \cos (kx - \omega t). \tag{17–2}$$

It is usually more convenient to deal with pressure variations in a sound wave than with the actual displacements of the particles conveying the wave. Let us therefore write the equation of the wave in terms of the pressure variation rather than in terms of the displacement.

From the relation

$$B = -\frac{\Delta p}{\Delta V / V},$$

we have

$$\Delta p = -B \frac{\Delta V}{V}.$$

Just as we let y represent the displacement from the equilibrium position x, so we now let p represent the change from the undisturbed pressure p_0. Then p replaces Δp, and

$$p = -B \frac{\Delta V}{V}.$$

If a layer of fluid at pressure p_0 has a thickness Δx and cross-sectional area A, its volume is $V = A \, \Delta x$. When the pressure changes, its volume will change by $A \, \Delta y$, where Δy is the amount by which the thickness of the layer changes during compression or rarefaction. Hence,

$$p = -B \frac{\Delta V}{V} = -B \frac{A \, \Delta y}{A \, \Delta x}.$$

As we let $\Delta x \to 0$ so as to shrink the fluid layer to infinitesimal thickness, we obtain

$$p = -B \frac{\partial y}{\partial x}. \tag{17–3}$$

We have used partial derivative notation because (see Eq. 17–2) y is a function of both x and t and we take the latter quantity as constant in this discussion. If the particle displacement is simple harmonic, then, from Eq. 17–2, we obtain

$$\frac{\partial y}{\partial x} = -ky_m \sin (kx - \omega t),$$

and from Eq. 17–3 $$p = Bky_m \sin (kx - \omega t). \qquad (17\text{–}4)$$

Hence, the pressure variation at each position x is also simple harmonic.

Since $v = \sqrt{B/\rho_0}$, we can write Eq. 17–4 more conveniently as

$$p = [k\rho_0 v^2 y_m] \sin (kx - \omega t).$$

Recall that p represents the change from standard pressure p_0. The term in brackets represents the maximum change in pressure and is called the *pressure amplitude*. If we denote this by P, then

$$p = P \sin (kx - \omega t), \qquad (17\text{–}5)$$

where $$P = k\rho_0 v^2 y_m. \qquad (17\text{–}6)$$

Hence, a sound wave may be considered either as a displacement wave or as a pressure wave. If the former is written as a cosine function, the latter will be a sine function and vice versa. The displacement wave is thus 90° out of phase with the pressure wave. That is, when the displacement from equilibrium at a point is a maximum or a minimum, the excess pressure there is zero; when the displacement at a point is zero, the excess or deficiency of pressure there is a maximum. Equation 17–6 gives the relation between the pressure amplitude (maximum variation of pressure from equilibrium) and the displacement amplitude (maximum variation of position from equilibrium). Check the dimension of each side of Eq. 17–6 for consistency. What units may the pressure amplitude have?

Example 1. (*a*) The maximum pressure variation P that the ear can tolerate in loud sounds is about 28 nt/meter². Normal atmospheric pressure is about 100,000 nt/meter². Find the corresponding maximum displacement for a sound wave in air having a frequency of 1000 cycles/sec.

From Eq. 17–6 we have

$$y_m = \frac{P}{k\rho_0 v^2}.$$

From Table 17–1, $v = 331$ meters/sec so that

$$k = \frac{2\pi}{\lambda} = \frac{2\pi\nu}{v} = \frac{2\pi \times 10^3}{331}\ \text{meter}^{-1} = 19\ \text{meter}^{-1}.$$

The density of air ρ_0 is 1.22 kg/meter³. Hence, for $P = 28$ nt/meter² we obtain

$$y_m = \frac{28}{(19)(1.22)(331)^2}\ \text{meter} = 1.1 \times 10^{-5}\ \text{meter}.$$

The displacement amplitudes for the loudest sounds are about 10^{-5} meter, a very small value indeed.

(*b*) In the faintest sound that can be heard at 1000 cycles/sec the pressure amplitude is about 2.0×10^{-5} nt/meter². Find the corresponding displacement amplitude.

From $y_m = P/k\rho_0 v^2$, using these values for k, v, and ρ_0, we obtain, with $P = 2.0 \times 10^{-5}$ nt/meter²,

$$y_m \cong 8 \times 10^{-12} \text{ meter} \cong 10^{-11} \text{ meter}.$$

This is smaller than the radius of an atom, which is about 10^{-10} meter! How can it be that the ear responds to such a small displacement?

17–4 Vibrating Systems and Sources of Sound

If a string fixed at both ends is bowed, transverse vibrations travel along the string; these disturbances are reflected at the fixed ends, and a standing wave pattern is formed. The natural modes of vibration of the string are excited and these vibrations give rise to longitudinal waves in the surrounding air which transmits them to our ears as a musical sound.

We have seen (Section 16–9) that a string of length l, fixed at both ends, can resonate at frequencies given by

$$\nu_n = \frac{n}{2l} v = \frac{n}{2l} \sqrt{\frac{F}{\mu}}, \qquad n = 1, 2, 3, \dots . \tag{17-7}$$

Here v is the speed of the transverse waves in the string whose superposition can be thought of as giving rise to the vibrations; the speed $v \, (= \sqrt{F/\mu})$ is the same for all frequencies. At any one of these frequencies the string will contain a whole number n of loops between its ends, and the condtion that the ends be nodes is met (Fig. 17–3).

The lowest frequency, $\sqrt{F/\mu}/2l$, is called the *fundamental* frequency ν_1 and the others are called *overtones*. Overtones whose frequencies are integral multiples of the fundamental are said to form a harmonic series. The fundamental is the first harmonic. The frequency $2\nu_1$ is the first overtone or the second harmonic, the frequency $3\nu_1$ is the second overtone or the third harmonic, and so on.

If the string is initially distorted so that its shape is the same as any one of the possible harmonics, it will vibrate at the frequency of that particular harmonic, when released. The initial conditions usually arise from striking or bowing the string, however, and in such cases not only the fundamental but many of the overtones are present in the resulting vibration. We have a superposition of several natural modes of oscillation. The actual displacement is the sum of the several harmonics with various amplitudes. The impulses that are sent through the air to the ear and brain give rise to one net effect which is characteristic of the particular stringed instrument. The quality of the sound of a particular note (fundamental frequency) played by an instrument is determined by the number of overtones present and their respective intensities. Figure 17–4 shows the sound spectra and corresponding waveforms for the violin and piano.*

---

* See "The Physics of the Piano" by E. Donnell Blackham, in *Scientific American*, December 1965.

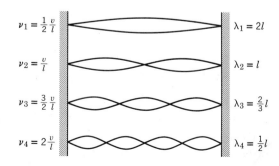

$$\nu_1 = \frac{1}{2} \frac{v}{l} \qquad\qquad \lambda_1 = 2l$$

$$\nu_2 = \frac{v}{l} \qquad\qquad \lambda_2 = l$$

$$\nu_3 = \frac{3}{2} \frac{v}{l} \qquad\qquad \lambda_3 = \frac{2}{3} l$$

FIGURE 17–3   The first four modes of vibration of a string fixed at both ends. Note that $\nu_n \lambda_n = v = \sqrt{F/\mu}.$

$$\nu_4 = 2 \frac{v}{l} \qquad\qquad \lambda_4 = \frac{1}{2} l$$

Longitudinal waves traveling along a tube are reflected at the ends of the tube, just as transverse waves in a string are reflected at its ends. Interference between the waves traveling in opposite directions gives rise to standing longitudinal waves.

If the end of the tube is closed, the reflected wave is 180° out of phase with the incident wave. This result is a necessary consequence of the fact that the displacement of the small volume elements at a closed end must always be zero. Hence, a closed end is a displacement *node*. If the end of the tube is open, the fluid elements there are free to move. However, the nature of the reflection there depends on whether the tube is wide or narrow compared to the wavelength. If the tube is narrow compared to the wavelength, as in most musical instruments, the reflected wave has nearly the same phase as the incident wave. Then the open end is almost a displacement *antinode*. The exact antinode is usually somewhere near the opening, but the effective length of the air columns of a wind instrument, for example, is not as definite as the length of a string fixed at both ends.

An organ pipe is a simple example of sound originating in a vibrating air column. If both ends of a pipe are open and a stream of air is directed against an edge, standing longitudinal waves can be set up in the tube. The air column will then resonate at its natural frequencies of vibration, given by

$$\nu_n = \frac{n}{2l}v, \qquad n = 1, 2, 3, \ldots.$$

Here v is the speed of the longitudinal waves in the column whose superposition can be thought of as giving rise to the vibrations, and n is the number of half wavelengths in the length l of the column. As with the bowed string, the fundamental and overtones are excited at the same time.

In an open pipe the fundamental frequency corresponds (approximately) to a displacement antinode at each end and a displacement node in the middle, as shown in Fig. 17–5*a*. The succeeding drawings of Fig. 17–5*a* show three of the overtones, the second, third, and fourth harmonics. Hence, in an open pipe the fundamental frequency is $v/2l$ and all harmonics are present.

In a closed pipe the closed end is a displacement node. Figure 17–5*b* shows the modes of vibration of a closed pipe. The fundamental frequency is $v/4l$ (approximately), which is one-half that of an open pipe of the same length. The only overtones present are those that give a displacement node at the closed end and an antinode (approximately) at the open end. Hence, as is shown in Fig. 17–5*b*, the

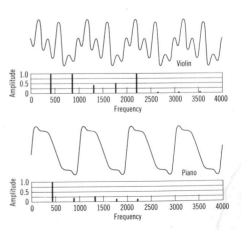

FIGURE 17–4 Waveform and sound spectrum for two stringed instruments, the violin and the piano. The fundamental frequency in both cases is 440 cycles/sec (concert A). In each diagram we show only four cycles of the wave. The sound spectrum shows the relative amplitude of the various harmonic components of the wave. Notice the presence of loud higher harmonics (especially the fifth) in the violin spectrum.

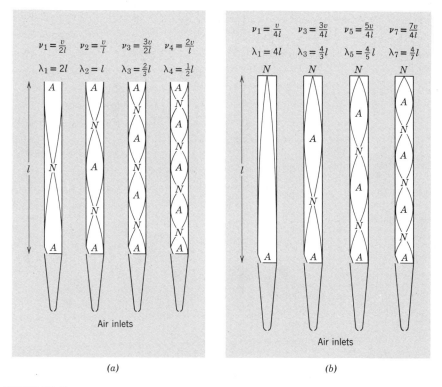

$$\nu_1 = \frac{v}{2l} \quad \nu_2 = \frac{v}{l} \quad \nu_3 = \frac{3v}{2l} \quad \nu_4 = \frac{2v}{l}$$

$$\lambda_1 = 2l \quad \lambda_2 = l \quad \lambda_3 = \tfrac{2}{3}l \quad \lambda_4 = \tfrac{1}{2}l$$

$$\nu_1 = \frac{v}{4l} \quad \nu_3 = \frac{3v}{4l} \quad \nu_5 = \frac{5v}{4l} \quad \nu_7 = \frac{7v}{4l}$$

$$\lambda_1 = 4l \quad \lambda_3 = \tfrac{4}{3}l \quad \lambda_5 = \tfrac{4}{5}l \quad \lambda_7 = \tfrac{4}{7}l$$

Air inlets

Air inlets

(a) *(b)*

FIGURE 17–5 (*a*) The first four modes of an open organ pipe. The distance from the center line of the pipe to the light lines drawn inside the pipe represents the displacement amplitude at each place. N and A mark the locations of the displacement nodes and antinodes. Note that both ends of the pipe are open. (*b*) The first four modes of vibration of a closed organ pipe. Notice that the even-numbered harmonics are absent and the upper end of the pipe is closed.

second, fourth, etc., harmonics are missing. In a closed pipe the fundamental frequency is $v/4l$, and only the odd harmonics are present. The quality of the sounds from an open pipe is therefore different from that from a closed pipe.

Vibrating rods, plates, and stretched membranes also give rise to sound waves. Consider a stretched flexible membrane, such as a drumhead. If it is struck a blow, a two-dimensional pulse travels outward from the struck point and is reflected again and again at the boundary of the membrane. If some point of the membrane is forced to vibrate periodically, continuous trains of waves travel out along the membrane. Just as in the one-dimensional case of the string, so here too standing waves can be set up in the two-dimensional membrane. Each of these standing waves has a certain frequency natural to (or characteristic of) the membrane. Again the lowest frequency is called the fundamental and the others are overtones. Generally, a number of overtones are present along with the fundamental when the membrane is vibrating. These vibrations may excite sound waves of the same frequency.

The nodes of a vibrating membrane are lines rather than points (as in a vibrating string) or planes (as in a pipe). Since the boundary of the membrane is fixed, it must be a nodal line. For a circular membrane fixed at its edge, possible modes of vibration along with their nodal lines are shown in Fig. 17–6. The natural frequency of each mode is given in terms of the fundamental ν_1. Notice that the frequencies of the overtones are not harmonics, that is, they are not integral

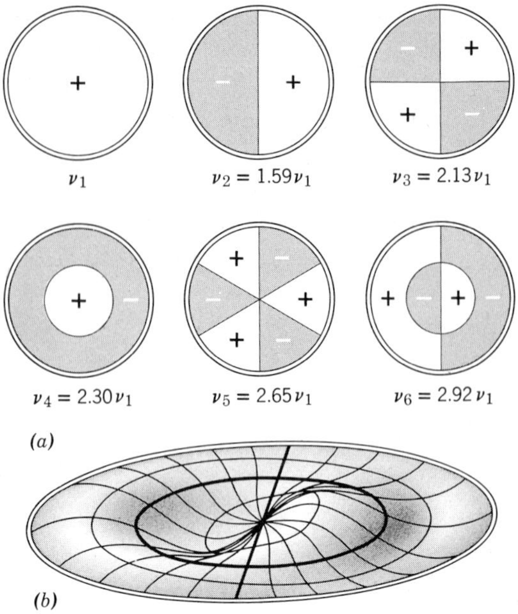

FIGURE 17-6 (*a*) The first six modes of vibration of a circular drumhead clamped around its periphery. The lines represent nodes, the circumference being a node in every case. The + and − signs represent opposite displacements; at an instant when the + areas are raised, the − areas will be depressed. Note that the frequency of each mode is not an integral multiple of the fundamental ν_1 as is the case for strings and tubes. (*b*) A sketch of a drum-head vibrating in mode ν_6. The displacement shown here is exaggerated for clarity.

multiplies of ν_1. Vibrating rods also have a nonharmonic set of natural frequencies. Rods and plates have limited use as musical instruments for this reason.

■ **Example 2.** A tuning fork with a frequency $\nu = 1080$ cycles/sec sets up a resonance in a tube open on one end, closed on the other, and having a variable length (such as by using a tube partially filled with water). The shortest effective length l for which resonance occurs is 7.65 cm. Find the speed of sound in air.

Referring to Fig. 17–5*b* the relation between the wavelength λ of the sound and the effective length of the tube is

$$l = \lambda/4.$$

Knowing the frequency and the wavelength above we obtain the velocity

$$v = \nu\lambda = (1080/\text{sec})(4)(7.65\ \text{cm}) = 330\ \text{meters/sec}.$$

What series of measurements would you make to eliminate the edge effect? Hint: Consider measuring the distance of the water level from the actual physical edge for several different resonances using the same frequency. Then see whether or not you can infer the wavelength without using the position of the physical edge. ■

17–5 Beats

When two wavetrains of the same frequency travel along the same line in opposite directions, standing waves are formed in accord with the principle of superposition. We may characterize these waves by drawing a plot of the amplitude of oscillation as a function of distance, as in Fig. 17–3. This illustrates a type of interference that we can call *interference in space*.

The same principle of superposition leads us to another type of interference, which we can call *interference in time*. It occurs when two wavetrains of slightly different frequency travel through the same region. With sound such a condition exists when, for example, two adjacent piano keys are struck simultaneously.

Consider some one point in space through which the waves are passing. In Fig.

17–7a we plot the displacements produced at such a point by the two waves separately as a function of time. For simplicity we have assumed that the two waves have equal amplitude, although this is not necessary. The resultant vibration at that point as a function of time is the sum of the individual vibrations and is plotted in Fig. 17–7b. We see that the amplitude of the resultant wave at the given point is not constant but varies in time. In the case of sound the varying amplitude gives rise to variations in loudness which are called *beats*. Two strings may be tuned to the same frequency by tightening one of them while sounding both until the beats disappear.

Let us represent the displacement at the point produced by one wave as

$$y_1 = y_m \cos 2\pi\nu_1 t,$$

and the displacement at the point produced by the other wave of equal amplitude as

$$y_2 = y_m \cos 2\pi\nu_2 t,$$

By the superposition principle, the resultant displacement is

$$y = y_1 + y_2 = y_m(\cos 2\pi\nu_1 t + \cos 2\pi\nu_2 t),$$

and since

$$\cos a + \cos b = 2 \cos \frac{a-b}{2} \cos \frac{a+b}{2},$$

this can be written as

$$y = \left[2y_m \cos 2\pi \left(\frac{\nu_1 - \nu_2}{2}\right) t\right] \cos 2\pi \left(\frac{\nu_1 + \nu_2}{2}\right) t. \tag{17–8}$$

The resulting vibration may then be considered to have a frequency

$$\bar{\nu} = \frac{\nu_1 + \nu_2}{2},$$

which is the average frequency of the two waves, and an amplitude given by the expression in brackets. Hence, the amplitude itself varies with time with a frequency

$$\nu_{\text{amp}} = \frac{\nu_1 - \nu_2}{2}.$$

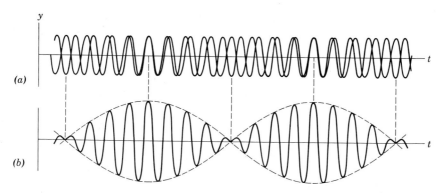

FIGURE 17–7 The beat phenomenon. Two waves of slightly different frequencies, shown in (*a*), combine in (*b*) to give a wave whose envelope (dashed line) varies periodically with time.

If ν_1 and ν_2 are nearly equal, this term is small and the amplitude fluctuates slowly. This phenomenon is a form of amplitude modulation which has a counterpart (side bands) in AM radio receivers.

A beat, that is, a maximum of amplitude, will occur whenever

$$\cos 2\pi \left(\frac{\nu_1 - \nu_2}{2}\right) t$$

equals 1 or -1. Since each of these values occurs once in each cycle, the number of beats per second is twice the frequency ν_{amp} or $\nu_1 - \nu_2$. Hence, the number of beats per second equals the difference of the frequencies of the component waves. Beats between two tones can be detected by the ear up to a frequency of about seven per second. At higher frequencies individual beats cannot be distinguished in the sound produced.

17–6 The Doppler Effect

When a listener is in motion toward a stationary source of sound, the pitch (frequency) of the sound heard is higher than when he is at rest. If the listener is in motion away from the stationary source, he hears a lower pitch than when he is at rest. We obtain similar results when the source is in motion toward or away from a stationary listener. The pitch of the whistle of the locomotive is higher when the source is approaching the hearer than when it has passed and is receding.

Christian Johann Doppler (1803–1853), an Austrian, in a paper of 1842, called attention to the fact that the color of a luminous body, just as the pitch of a sounding body, must be changed by relative motion of the body and the observer. This *Doppler effect,* as it is called, applies to waves in general. Let us apply it now to sound waves. We consider only the special case in which the source and observer move along the line joining them.

Let us consider a reference frame at rest in the medium through which the sound travels. Figure 17–8 shows a source of sound S at rest in this frame and an observer O (note the ear) moving toward the source at a speed v_0. The circles represent wavefronts, spaced one wavelength apart, traveling through the medium. If the observer were at rest in the medium he would receive vt/λ waves in time t, where v is the speed of sound in the medium and λ is the wavelength. Because of his motion toward the source, however, he receives $v_0 t/\lambda$ additional waves in this same time t. The frequency ν' that he hears is the number of waves received per unit time or

$$\nu' = \frac{vt/\lambda + v_0 t/\lambda}{t} = \frac{v + v_0}{\lambda} = \frac{v + v_0}{v/\nu}.$$

That is,

$$\nu' = \nu \frac{v + v_0}{v} = \nu \left(1 + \frac{v_0}{v}\right). \tag{17–9a}$$

The frequency ν' heard by the observer is the ordinary frequency ν heard at rest plus the increase $\nu(v_0/v)$ arising from the motion of the observer. When the observer is in motion away from the stationary source, there is a decrease in frequency $\nu(v_0/v)$ corresponding to the waves that do not reach the observer each unit of time because of his receding motion. Then

$$\nu' = \nu \left(\frac{v - v_0}{v}\right) = \nu \left(1 - \frac{v_0}{v}\right). \tag{17–9b}$$

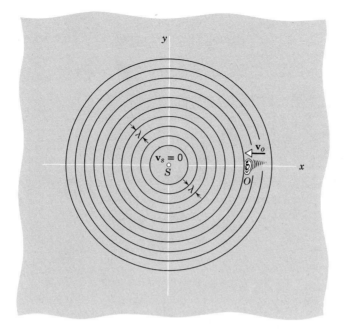

FIGURE 17–8   The Doppler effect due to motion of the observer (ear). The source is at rest.

Hence, the general relation holding when the source is at rest with respect to the medium but the observer is moving through it is

$$\nu' = \nu \left(\frac{v \pm v_0}{v} \right).$$ (17–9)

where the plus sign holds for motion toward the source and the minus sign holds for motion away from the source. Notice that the cause of the change here is the fact that the observer intercepts more or fewer waves each second because of his motion through the medium.

When the source is in motion toward a stationary observer, the effect is a shortening of the wavelength (see Fig. 17–9), for the source is following after the approaching waves and the crests therefore come closer together. If the frequency of the source is ν and its speed is v_s, then during each vibration it travels a distance v_s/ν and each wavelength is shortened by this amount. Hence, the wavelength of the sound arriving at the observer is not $\lambda = v/\nu$ but $\lambda' = v/\nu - v_s/\nu$. Therefore, the frequency of the sound heard by the observer is increased, being

$$\nu' = \frac{v}{\lambda'} = \frac{v}{(v - v_s)/\nu} = \nu \left(\frac{v}{v - v_s} \right).$$ (17–10a)

If the source moves away from the observer, the wavelength emitted is v_s/ν greater than λ, so that the observer hears a decreased frequency, namely

$$\nu' = \frac{v}{(v + v_s)/\nu} = \nu \left(\frac{v}{v + v_s} \right).$$ (17–10b)

Hence, the general relation holding when the observer is at rest with respect to the medium but the source is moving through it is

$$\nu' = \nu \left(\frac{v}{v \mp v_s} \right),$$ (17–10)

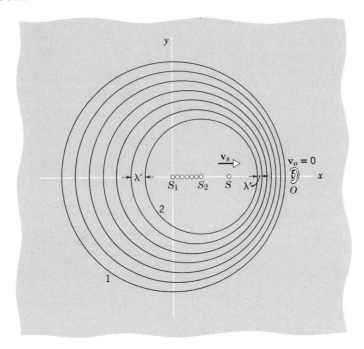

FIGURE 17–9 The Doppler effect due to motion of the source. The observer is at rest. Wavefront 1 was emitted by the source when it was at S_1, wavefront 2 was emitted when it was at S_2, etc. At the instant the "snapshot" was taken, the source was at S.

where the minus sign holds for motion toward the observer and the plus sign holds for motion away from the observer. Notice that the cause of the change here is the fact that the motion of the source through the medium shortens or increases the wavelength transmitted through the medium.

If both source and observer move through the transmitting medium, you should be able to show that the observer hears a frequency

$$\nu' = \nu \left(\frac{v \pm v_o}{v \mp v_s} \right). \tag{17–11}$$

where the upper signs (+ numerator, − denominator) correspond to the source and observer moving along the line joining the two in the direction toward the other, and the lower signs in the direction away from the other. Notice that Eq. 17–11 reduces to Eq. 17–9 when $v_s = 0$ and to Eq. 17–10 when $v_o = 0$, as it must.

If a vibrating tuning fork on its resonating box is moved rapidly toward a wall, the observer will hear two notes of different frequency. One is the note heard directly from the receding fork and is lowered in pitch by the motion. The other note is due to the waves reflected from the wall, and this is raised in pitch. The superposition of these two wave trains produces beats.

When v_o or v_s becomes comparable in magnitude to v, the formulas just given for the Doppler effect must be modified. The modification is required because the linear relation between restoring force and displacement assumed up until now no longer holds in the medium. The speed of wave propagation is no longer the normal phase velocity, and the wave shapes change in time. Components of the motion at right angles to the line joining source and observer also contribute to the Doppler effect at these high speeds. When v_o or v_s exceeds v, the Doppler formula clearly has no meaning.

There are many instances in which the source moves through a medium at a speed greater than the speed of the wave in that medium. In such cases the wavefront takes the shape of a cone with the moving body at its apex. Some examples are the bow wave from a speedboat on the water and the "shock wave" from an airplane or projectile moving through the air at a speed greater than the velocity of sound in that medium (supersonic speeds). Cerenkov radiation consists of light waves emitted by charged particles which move through a medium with a speed greater than the speed of light in that medium.

In Fig. 17–10 we show the present positions of the spherical waves which originated at various positions of the source during its motion. The radius of each sphere at this time is the product of the wave speed v and the time t which has elapsed since the source was at its center. The envelope of these waves is a cone whose surface makes an angle θ with the direction of motion of the source. From the figure we obtain the result

$$\sin \theta = \frac{v}{v_s}.$$

For water waves the cone reduces to a pair of intersecting lines. In aerodynamics the ratio v_s/v is called the Mach number.

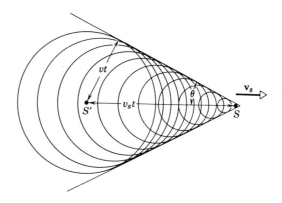

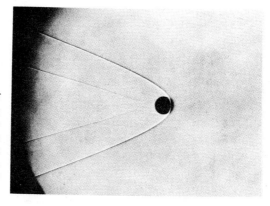

FIGURE 17–10 Above, right, a group of wavefronts associated with a projectile moving with supersonic speed. The wavefronts are spherical and their envelope is a cone. Notice the relation between this figure and the previous one. Below, right, a spark photograph of a projectile undergoing this motion. (U.S. Navy Photograph.)

QUESTIONS

1. List some sources of infrasonic waves. Of ultrasonic waves.

2. What experimental evidence is there for assuming that the speed of sound is the same for all wavelengths?

3. What quantity, if any, for transverse waves in a string corresponds to the pressure amplitude for longitudinal waves in a tube?

4. A bell is rung for a short time in a school. After a while its sound is inaudible. Trace the sound waves and the energy they transfer from the time of emission until they become inaudible.

5. How can we experimentally locate the positions of nodes and antinodes in a string? In an air column? On a vibrating surface?

6. Explain how a stringed instrument is "tuned."

7. Discuss the factors that determine the range of frequencies in your voice and the quality of your voice.

8. The bugle has no valves. How then can we sound different notes on it? To what notes is the bugler limited? Why?

9. Would a plucked violin oscillate for a longer or shorter time if the violin had no sounding board? Explain.

10. Two ships with steam whistles of the same pitch sound off in the harbor. Would you expect this to produce an interference pattern with regions of high and low intensity?

11. Suppose that, in the Doppler effect for sound, the source and receiver are at rest in some reference frame but the transmitting medium is moving with respect to this frame. Will there be a change in wavelength, or in frequency, received?

12. Is there a Doppler effect for sound when the observer or the source moves at right angles to the line joining them? How then can we determine the Doppler effect when the motion has a component at right angles to this line?

13. A satellite emits radio waves of constant frequency. These waves are picked up on the ground and made to beat against some standard frequency. The beat frequency is then sent through a loudspeaker and one "hears" the satellite signals. Describe how the sound changes as the satellite approaches, passes overhead, and recedes from the detector on the ground.

14. Two identical tuning forks emit notes of the same frequency. Explain how you might hear beats between them.

15. Transverse waves in a string can be polarized (see, for example, Question 15 of Chapter 16). Can sound waves be polarized?

PROBLEMS *

1(2). The lowest pitch detectable as sound by the average human ear is about 20 vib/sec and the highest is about 20,000 vib/sec. What is the wavelength of each in air?
Answer: 17 meters; 1.7 cm.

2(2). Bats emit ultrasonic waves. The shortest wavelength emitted in air by a bat is about 0.13 in. What is the highest frequency a bat can emit?

3(2). A sound wave has a frequency of 440 vib/sec. (*a*) What is the wavelength of this sound in air? (*b*) In water?
Answer: (*a*) 0.75 meter. (*b*) 3.3 meters.

4(2). What is the value for the bulk modulus of oxygen at standard temperature and pressure if one mole of oxygen occupies 22.4 liters under these conditions and the speed of sound in oxygen is 317 meters/sec?

5(2). (*a*) A conical loudspeaker has a diameter of 6.0 in. At what frequency will the wavelength of the sound it emits in air be equal to its diameter? Be ten times its diameter? Be one-tenth its

---

* Unless otherwise specified, take the speed of sound in air to be 331 meters/sec.

diameter? (b) Make the same calculations for a speaker of diameter 12 in. (Note: if the wavelength is large compared to the diameter of the speaker, the sound waves spread out almost uniformly in all directions from the speaker, but when the wavelength is small compared to the diameter of the speaker, the wave energy is propagated mostly in the forward direction.)

Answer: (a) 2.2 kc/sec; 0.22 kc/sec; 22 kc/sec. (b) 1.1 kc/sec; 0.11 kc/sec; 11 kc/sec.

6(2). An experimenter wishes to measure the speed of sound in an aluminum rod 10 cm long by measuring the time it takes for a sound pulse to travel the length of the rod. If results good to four significant figures are desired, how precisely must the length of the bar be known and how closely must he be able to resolve time intervals?

7(2). A rule for finding your distance from a lightning flash is to count seconds from the time you see the flash until you hear the thunder and then divide the count by five. The result is supposed to give the distance in miles. Explain this rule and determine the percent error in it at standard conditions.

Answer: About 3%.

8(2). A stone is dropped into a well. The sound of the splash is heard 3.0 sec later. What is the depth of the well?

9(2). The speed of sound in a certain metal is V. One end of a long pipe of that metal of length l is struck a hard blow. A listener at the other end hears two sounds, one from the wave that has traveled along the pipe and the other from the wave that has traveled through the air. (a) If v is the speed of sound in air, what time interval t elapses between the two sounds? (b) Suppose $t = 1.0$ sec and the metal is iron. Find the length l.

Answer: (a) $\dfrac{l(V-v)}{Vv}$. (b) 1200 ft.

10(2). Two spectators at a soccer game in a large stadium see, and a moment later hear, the ball being kicked on the playing field. If the time delay for one spectator is 0.90 sec and for the other 0.60 sec, and lines through each spectator and the player kicking the ball meet at an angle of 90°, (a) how far is each spectator from the player? (b) How far are the spectators from each other?

11(3). The pressure in a traveling sound wave is given by the equation

$$p = 1.5 \sin \pi(x - 330t).$$

where x is in meters, t in seconds, and p in nt/meter². Find (a) the pressure amplitude, (b) the frequency, (c) the wavelength, and (d) the speed of the wave.

Answer: (a) 1.5 nt/meter². (b) 170 cycles/sec. (c) 2.0 meters. (d) 330 meters/sec.

12(3). A hi-fi engineer has designed a speaker that is spherical in shape and emits sound isotropically (the same intensity in all directions). The speaker emits 10 watts of acoustic power into a room with completely absorbent walls, floor, and ceiling. (a) What is the intensity (watts/meters²) of the sound waves at 3.0 meters from the center of the source? (b) How would the amplitude of the waves at 4.0 meters compare with that 3.0 meters from the center of the source? (Hint: see Example 3, Chapter 16, and note that the intensity of a wave is proportional to the square of its amplitude.)

13(3). Two waves give rise to pressure variations at a certain point in space given by

$$p_1 = P \sin 2\pi \nu t,$$

$$p_2 = P \sin 2\pi(\nu t - \phi).$$

What is the pressure amplitude of the resultant wave at this point when $\phi = 0$, $\phi = \frac{1}{4}$, $\phi = \frac{1}{6}$, $\phi = \frac{1}{8}$? All ϕ's are measured in radians.

Answer: 2.00P; 1.41P; 1.73P; 1.85P.

14(3). Two sound waves with the same frequency, 540 vib/sec, travel at a speed of 330 meters/sec. What is the phase difference of the waves at a point that is 4.40 meters from one source and 4.00 meters from the other if the sources are in phase?

15(3). Two sources of sound are separated by a distance of 10 meters. They both emit sound at the same amplitude and frequency, 300 vib/sec, but they are 180° out of phase. At what points along the line between them will the sound intensity be the largest?

Answer: At ±0.28, 0.83, 1.38, 1.93, 2.48, 3.03, 3.58, 4.13, and 4.68 meters from the center.

16(3). Two loudspeakers are located 11 ft apart on the stage of an auditorium. A listener is seated 60 ft from one and 64 ft from the other. A signal generator drives the two speakers in phase with the same amplitude and frequency. The frequency is swept through the audio range. (*a*) What are the three lowest frequencies for which the listener will hear minimum intensity because of destructive interference? (*b*) What are the three lowest frequencies for which he will hear maximum intensity?

17(3). Two stereo loudspeakers are separated by a distance of 6.0 ft. Assume the amplitude of the sound from each speaker is approximately the same at the position of a listener, who is 10 ft directly in front of one of the speakers. (*a*) For what frequencies in the audible range (20 − 20,000 cycles/sec) will there be a minimum signal? (*b*) For what frequencies is the sound a maximum? *Answer:* (*a*) $330(1 + 2n)$ cycles/sec, with *n* being an integer from 0 to 29. (*b*) $660n$ cycles/sec, with *n* being an integer from 1 to 30.

18(3). In Fig. 17-11 we show an acoustic interferometer, used to demonstrate the interference of sound waves. *S* is a diaphragm that vibrates under the influence of an electromagnet. *D* is a sound detector, such as the ear or a microphone. Path *SBD* can be varied in length, but path *SAD* is fixed. The interferometer contains air, and it is found that the sound intensity has a minimum value of 100 units at one position of *B* and continuously climbs to a maximum value of 900 units at a second position 1.65 cm from the first. Find (*a*) the frequency of the sound emitted from the source, and (*b*) the relative amplitudes of the two waves arriving at the detector. This device is called "König's trombone." (Note: The intensity of a wave is proportional to the square of its amplitude.)

19(3). A spherical sound source is placed at P_1 near a reflecting wall *AB* and a microphone is located at point P_2, as shown in Fig. 17-12. The frequency of the sound source P_1 is variable. Find two different frequencies for which the sound intensity, as observed at P_2, will be a maximum. The speed of sound in air is 1100 ft/sec.
Answer: 32 and 95 cycles/sec.

20(3). Two loudspeakers, S_1 and S_2, each emit sound of frequency 200 vib/sec uniformly in all directions. S_1 has an acoustic output of 1.2×10^{-3} watt and S_2 one of 1.8×10^{-3} watt. S_1 and S_2 vibrate in phase. Consider a point *P* which is 4.0 meters from S_1 and 3.0 meters from S_2. (*a*) How are the phases of the two waves arriving at *P* related? (*b*) What is the intensity of sound at *P* with both S_1 and S_2 on? (*c*) What is the intensity of sound at *P* if S_1 is turned off (S_2 on)? (*d*) What is the intensity of sound at *P* if S_2 is turned off (S_1 on)? (Note: The intensity of a wave is proportional to the square of its amplitude.)

21(3). The intensity of a sound wave (watts/meter²), when expressed in terms of the pressure amplitude *P* is

$$I = P^2/2\rho_0 v.$$

When expressed in terms of the displacement amplitude y_m it is

$$I = 2\pi^2 \rho_0 v \nu^2 y_m{}^2.$$

(*a*) Can you prove these relations? (*b*) Two sound waves, one in water and one in air, are equal in intensity. What is the ratio of the pressure amplitude *P* of the wave in water to that of the wave in air? (*c*) If the pressure amplitudes are equal instead, what is the ratio of the intensities *I* of the waves?
Answer: (*b*) 58. (*c*) 2.9×10^{-4}.

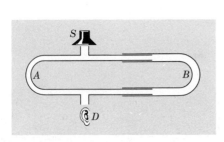

FIGURE 17–11   Problem 18(3).

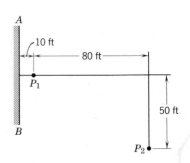

FIGURE 17–12   Problem 19(3).

22(3). A certain loudspeaker produces a sound with a frequency of 2000 cycles/sec and an intensity of 1.0×10^{-3} watts/meter² at a distance of 20 ft. Presume that there are no reflections and that the loudspeaker emits the same in all directions. (*a*) What would be the intensity at 100 ft? (*b*) What is the displacement amplitude at 20 ft? (*c*) What is the pressure amplitude at 20 ft? (Hint: see Example 3, Chapter 16 and the preceding problem.)

23(3). A note of frequency 300 vib/sec has an intensity of 1.0 microwatt/meter². What is the amplitude of the air vibrations caused by this sound? (Hint: See Problem 21–3.)
Answer: 3.6×10^{-8} meters.

24(4). If a violin string is tuned to a certain note, by how much must the tension in the string be increased if it is to emit a note of double the original frequency (that is, a note one octave higher in pitch).

25(4). An open organ pipe has a fundamental frequency of 300 vib/sec. The first overtone of a closed organ pipe has the same frequency as the first overtone of the open pipe. How long is each pipe?
Answer: 55 cm; 41 cm.

26(4). A 10-ft skip rope is used essentially in its fundamental mode of oscillation. If the rope has a mass of 1.0 kg and the children are pulling back with a force of 10 nt, what is the frequency of oscillation?

27(4). A certain violin string is 50 cm long between its fixed points and has a mass of 2.0 gm. The string sounds an *A* note (440 vib/sec) when played without fingering. Where must one put one's finger to play a *C* (528 vib/sec)?
Answer: 8.4 cm from the end.

28(4). The strings of a cello have a length *L*. (*a*) By what length *l* must they be shortened by fingering to change the pitch by a frequency ratio *r*? (*b*) Find *l*, if $L = 0.80$ meter and $r = 6/5$, $5/4$, $4/3$, and $3/2$.

29(4). A tuning fork placed over an open vertical tube partly filled with water causes strong resonances when the water surface is 8 cm and 28 cm from the top of the tube and for no other positions. The speed of sound in the air in the room is 330 meters/sec. What is the frequency of the tuning fork?
Answer: 830 cycles/sec.

30(4). The water level in a vertical glass tube 1.0 meter long can be adjusted to any position in the tube. A tuning fork vibrating at 660 vib/sec is held just over the open top end of the tube. At what positions of the water level will there be resonance?

31(4). A sound wave in a fluid medium is reflected at a barrier so that a standing wave is formed. The distance between nodes is 3.8 cm and the velocity of propagation is 1500 meters/sec. Find the frequency.
Answer: 2.0×10^4 cycles/sec.

32(4). In Fig. 17–13 a rod *R* is clamped at its center and a disk *D* at its end projects into a glass tube, which has cork filings spread over its interior. A plunger *P* is provided at the other end of the tube. The rod is set into longitudinal vibration and the plunger is moved until the filings form a pattern of nodes and antinodes (the filings form well-defined ridges at the pressure antinodes). If we know the frequency v of the longitudinal vibrations in the rod, a measurement of the average distance *d* between successive antinodes determines the speed of sound *v* in the gas in the tube. Show that

$$v = 2vd.$$

This is Kundt's method for determining the speed of sound in various gases.

33(4). *S* in Fig. 17–14 is a small loudspeaker driven by an audio oscillator and amplifier, adjust-

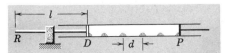

FIGURE 17–13   Problem 32(4).

able in frequency from 1000 to 2000 cycles/sec only. *D* is a piece of cylindrical sheetmetal pipe 18.0 in. long. (*a*) If the speed of sound in air is 1130 ft/sec at the existing temperature, at what frequencies will resonance occur when the frequency emitted by the speaker is varied from 1000 to 2000 cycles/sec? (*b*) Sketch the displacement nodes for each. Neglect end effects.
Answer: (*a*) 1130, 1500, and 1880 cycles/sec.

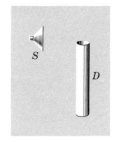

FIGURE 17–14   Problem 33(4).

34(4). A well with vertical sides and water at the bottom resonates at 7.0 vib/sec and at no lower frequency. The air in the well has a density of 1.1 kg/m³, a pressure of 9.5×10^4 nt/m², and a ratio of specific heats equal to 7/5. How deep is the well?

35(4). A tube 1.0 meter long is closed at one end. A stretched wire is placed near the open end. The wire is 0.30 meter long and has a mass of 0.010 kg. It is held fixed at both ends and vibrates in its fundamental mode. It sets the air column in the tube into vibration at its fundamental frequency by resonance. Find (*a*) the frequency of oscillation of the air column and (*b*) the tension in the wire.
Answer: (*a*) 83 cycles/sec. (*b*) 82 nt.

36(4). An aluminum wire of length $l_1 = 60.0$ cm and of cross-sectional area 1.00×10^{-2} cm² is connected to a steel wire of the same cross-sectional area. The compound wire, loaded with a block *m* of mass 10.0 kg, is arranged as shown in Fig. 17–15, so that the distance l_2 from the joint to the supporting pulley is 88.0 cm. Transverse waves are set up in the wire by using an external source of variable frequency. (*a*) Find the lowest frequency of excitation for which standing waves are observed such that the joint in the wire is a node. (*b*) What is the total number of nodes observed at this frequency, excluding the two at the ends of the wire? The density of aluminum is 2.70 gm/cm³, and that of steel is 7.80 gm/cm³.

37(4). A 31.6-cm violin string with linear density 0.65 gram/meter is placed near a loudspeaker that is fed by an audio-oscillator of variable frequency. It is found that the string is set in oscillation only at the frequencies 880 and 1320 cycles/sec as the frequency of the oscillator is varied continuously over the range 500 to 1500 cycles/sec. What is the tension in the string?
Answer: 50 nt.

38(5). Two identical piano wires have a fundamental frequency of 600 vib/sec when kept under the same tension. What fractional increase in the tension of one wire will lead to the occurrence of six beats per second when both wires vibrate simultaneously?

39(5). A tuning fork of unknown frequency makes three beats per second with a standard fork of frequency 384 vib/sec. The beat frequency decreases when a small piece of wax is put on a prong of the first fork. What is the frequency of this fork?
Answer: 387 vib/sec.

40(6). In 1845 Buys Ballot first tested the Doppler effect for sound. He put a trumpet player on a flatcar drawn by a locomotive and another player near the tracks. If each player blows a 440 cycles/sec note, and if there are 4.0 beats/sec as they approach each other, what is the speed of the flatcar?

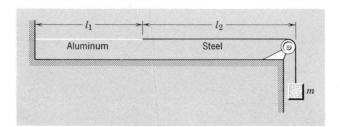

FIGURE 17–15   Problem 36(4).

41(6). A whistle of frequency 540 vib/sec rotates in a circle of radius 2.00 ft at an angular speed of 15.0 radians/sec. What is (*a*) the lowest and (*b*) the highest frequency heard by a listener a long distance away at rest with respect to the center of the circle?
Answer: (*a*) 525 vib/sec. (*b*) 555 vib/sec.

42(6). Two trains are traveling toward each other at 100 ft/sec relative to the ground. One train is blowing a whistle at 500 cycles/sec. (*a*) What frequency will be heard on the other train in still air? (*b*) What frequency will be heard on the other train if the wind is blowing at 100 ft/sec toward the whistle and away from the listener? (*c*) What frequency will be heard if the wind direction is reversed?

43(6). A bat is flittering about in a cave, navigating very effectively by the use of ultrasonic bleeps (short emissions lasting a millisecond or less and repeated several times a second). Assume that the sound emission frequency of the bat is 39,000 cycles/sec. During one fast swoop directly toward a flat wall surface, the bat is moving at 1/40 of the speed of sound in air. What frequency does he hear reflected off the wall?
Answer: 41,000 cycles/sec.

44(6). A source of sound waves of frequency 1080 vib/sec moves to the right with a speed of 108 ft/sec relative to the ground. To its right is a reflecting surface moving to the left with a speed of 216 ft/sec relative to the ground. Take the speed of sound in air to be 1080 ft/sec and find (*a*) the wavelength of the sound emitted in air by the source, (*b*) the number of waves per second arriving at the reflecting surface, (*c*) the speed of the reflected waves, (*d*) the wavelength of the reflected waves.

45(6). A siren emitting a sound of frequency 1000 vib/sec moves away from you toward a cliff at a speed of 10 meters/sec. (*a*) What is the frequency of the sound you hear coming directly from the siren? (*b*) What is the frequency of the sound you hear reflected off the cliff? (*c*) Could you hear the beat frequency? Take the speed of sound in air as 330 meters/sec.
Answer: (*a*) 970 vib/sec. (*b*) 1030 vib/sec. (*c*) No. It is too high.

46(6). A girl is leaning out of the window of a train that is moving at a velocity of 10.0 meters/sec to the east. The girl's uncle stands on the tracks and watches the train move away. The locomotive whistle vibrates at 500 cycles/sec. The air is still. (*a*) What frequency does the uncle hear? (*b*) What frequency does the girl hear?
  A wind begins to blow from the west at 10 meters/sec. (*c*) What frequency does the uncle now hear? (*d*) What frequency does the girl now hear?

47(6). Trooper *B* is chasing speeder *A* along a straight stretch of road. Both are moving at a top speed of about 100 miles/hr, which is about 150 ft/sec. Trooper *B*, failing to catch up, sounds his siren again. Take the speed of sound in air to be 1100 ft/sec and the frequency of the source to be 500 cycles/sec. What, if any, will be the Doppler shift in the frequency heard by speeder *A*?
Answer: There is no Doppler shift.

48(6). A bullet is fired with a speed of 2200 ft/sec. Find the angle made by the shock wave with the line of motion of the bullet.

49(6). Calculate the speed of the projectile illustrated in the photograph in Fig. 17–10. Assume the speed of sound in the medium through which the projectile is traveling to be 380 meters/sec.
Answer: 1100 meters/sec.

50(6). A plane flies with 5/4 the speed of sound. The sonic boom reaches a man on the ground exactly one minute after the plane passed directly overhead. What is the altitude of the plane? Assume the speed of sound to be 330 meters/sec.

51(6). A jet plane passes overhead at a height of 5000 meters and a speed of Mach 1.5 (that is, 1.5 times the speed of sound). (*a*) Find the angle made by the shock wave with the line of motion of the jet. (*b*) How long after the jet has passed directly overhead will the shock wave reach the ground?
Answer: (*a*) 42°. (*b*) 11 sec.

52(6). The speed of light in water is about three-fourths the speed of light in vacuum. A beam of high-speed electrons from a betatron emits Cerenkov radiation in water, the wavefront being a cone of angle 60°. Find the speed of the electrons in the water.

CHAPTER 18

Temperature

18–1 Macroscopic and Microscopic Descriptions

In analyzing physical situations we usually focus our attention on some portion of matter which we separate, in our minds, from the enviroment external to it. We call such a portion the *system*. Everything outside the system which has a direct bearing on its behavior we call the *environment*. We then seek to determine the behavior of the system by finding how it interacts with its environment. For example, a ball can be the system and the environment can be the air and the earth. In free fall we seek to find how the air and the earth affect the motion of the ball. Or the gas in a container can be the system, and a movable piston and a Bunsen burner can be the environment. We seek to find how the behavior of the gas is affected by the action of the piston and burner. In all such cases we must choose suitable observable quantities to describe the behavior of the system. We classify these quantities, which are gross properties of the system measured by laboratory operations, as *macroscopic*. For processes in which heat is involved the laws relating the appropriate macroscopic quantities (which include pressure, volume, temperature, internal energy, and entropy, among others) form the basis for the science of *thermodynamics*. Many of the macroscopic quantities (pressure, volume, and temperature, for example) are directly associated with our sense perceptions. We can also adopt a *microscopic* point of view. Here we consider quantities that describe the atoms and molecules that make up the system, their speeds, energies, masses, angular momenta, behavior during collisions, etc. These quantities, or mathematical formulations based on them, form the basis for the science of *statistical mechanics*. The microscopic properties are not directly associated with our sense perceptions.

For any system the macroscopic and the microscopic quantities must be related because they are simply different ways of describing the same situation. In particular, we should be able to express the former in terms of the latter. The pressure of a gas, viewed macroscopically, is measured operationally using a manometer (Fig. 15–9). Viewed microscopically it is related to the average rate per unit area at which the molecules of the gas deliver momentum to the manometer fluid as they strike its surface. In Section 20–4 we will make this microscopic definition of pressure quantitative. Similarly (see Section 20–5), the temperature of a gas may be related to the average kinetic energy of translation of the molecules.

If the macroscopic quantities can be expressed in terms of the microscopic quantities, we should be able to express the laws of thermodynamics quantitatively in the language of statistical mechanics. We can indeed do this. In the words of R. C. Tolman:

> The explanation of the complete science of thermodynamics in terms of the more abstract science of statistical mechanics is one of the greatest achievements of physics. In addition, the more fundamental character of statistical mechanical considerations makes it possible to supplement the ordinary principles of thermodynamics to an important extent.

We begin our examination of heat phenomena in this chapter with a study of temperature. As we progress we shall try to gain a deeper understanding of these phenomena by interweaving the microscopic and the macroscopic description—statistical mechanics and thermodynamics. The interweaving of the microscopic and the macroscopic points of view is characteristic of modern physics.

18–2   Thermal Equilibrium—The Zeroth Law of Thermodynamics

The sense of touch is the simplest way to distinguish hot bodies from cold bodies. By touch we can arrange bodies in the order of their hotness, deciding that A is hotter than B, B than C, etc. We speak of this as our *temperature* sense. This is a very subjective procedure for determining the temperature of a body and certainly not very useful for purposes of science. A simple experiment, suggested in 1690 by John Locke, shows the unreliability of this method. Let a person immerse his hands, one in hot water, the other in cold. Then let him put both hands in water of intermediate hotness. This will seem cooler to the first hand and warmer to the second hand. Our judgment of temperature can be rather misleading. Further, the range of our temperature sense is limited. What we need is an objective, numerical, measure of temperature.

To begin with, we should try to understand the meaning of temperature. Let an object A which feels cold to the hand and an identical object B which feels hot be placed in contact with each other. After a sufficient length of time, A and B give rise to the same temperature sensation. Then A and B are said to be in *thermal equilibrium* with each other. We can generalize the expression "two bodies are in thermal equilibrium" to mean that the two bodies are in states such that, if the two were connected, the combined systems would be in thermal equilibrium. The logical and operational test for thermal equilibrium is to use a third or test body, such as a thermometer. This is summarized in a postulate often called *the zeroth law of thermodynamics: If A and B are in thermal equilibrium with a third body C (the "thermometer"), then A and B are in thermal equilibrium with each other.*

This discussion expresses the idea that the temperature of a system is a property which eventually attains the same value as that of other systems when all these systems are put in contact. This concept agrees with the everyday idea of temperature as the measure of the hotness or coldness of a system, because as far as our temperature sense can be trusted, the hotness of all objects becomes the same after they have been in contact long enough. The idea contained in the zeroth law, although simple, is not obvious. For example, Jones and Smith each know Green, but they may or may not know each other. Two pieces of iron attract a magnet but they may or may not attract each other.

18–3 Measuring Temperature

There are many measurable physical properties that vary as our physiological perception of temperature varies. Among these are the volume of a liquid, the length of a rod, the electrical resistance of a wire, the pressure of a gas kept at constant volume, the volume of a gas kept at constant pressure, and the color of a lamp filament. Any of these properties can be used in the construction of a thermometer—that is, in the setting up of a particular "private" temperature scale. Such a temperature scale is established by choosing a particular thermometric substance and a particular thermometric property of this substance. We then define the temperature scale by an assumed continuous monotonic relation between the chosen thermometric property of our substance and the temperature as measured on our (private) scale. For example, the thermometric substance may be a liquid in a glass capillary tube and the thermometric property can be the length of the liquid column; or the thermometric substance may be a gas kept in a container at constant volume and the thermometric property can be the pressure of the gas; and so forth. We must realize that each choice of thermometric substance and property—along with the assumed relation between property and temperature—leads to an individual temperature scale whose measurements need not necessarily agree with measurements made on any other independently defined temperature scale.

This apparent chaos in the definition of temperature is removed by universal agreement, within the scientific community, on the use of a particular thermometric substance, a particular thermometric property, and a particular functional relation between measurements of that property and a universally accepted temperature scale. A private temperature scale defined in any other way can then always be calibrated against the universal scale. We describe such a universal scale in Section 18–4.

The *constant volume gas thermometer* illustrates the techniques by which an arbitrary temperature scale might be defined. If the volume of a gas is kept constant, its pressure depends on the temperature and increases steadily with rising temperature. The constant-volume gas thermometer uses the pressure at constant volume as the thermometric property.

The thermometer is shown diagrammatically in Fig. 18–1. It consists of a bulb of glass, porcelain, quartz, platinum, or platinum-iridium (depending on the temperature range over which it is to be used), connected by a capillary tube to a mercury manometer. The bulb containing some gas is put into the bath or environment whose temperature is to be measured; by raising or lowering the mercury reservoir the mercury in the left branch of the U-tube can be made to coincide with a fixed reference mark, thus keeping the confined gas at a constant volume. Then we read the height of the mercury in the right branch. The pressure of the confined gas is the difference of the heights of the mercury columns (times ρg) plus the atmospheric pressure, as indicated by the barometer. In practice the apparatus is very elaborate and we must

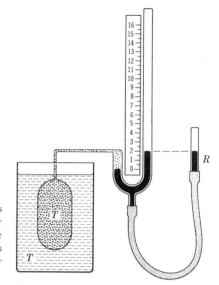

FIGURE 18–1   A representation of a constant-volume gas thermometer. As long as the mercury in the left manometer tube remains at the same position on the scale (zero) the volume of the confined gas will be constant. The meniscus can always be brought to the zero position by raising or lowering reservoir R.

make many corrections, for example, (1) to allow for the small volume change owing to slight contraction or expansion of the bulb and (2) to allow for the fact that not all the confined gas (such as that in the capillary) has been immersed in the bath. Assume that all corrections have been made, and let P be the corrected value of the pressure at the temperature of the bath.

The temperature may be defined most simply to be linearly related to the pressure. Algebraically we write

$$T = aP \tag{18–1}$$

where a is a constant which may be defined arbitrarily.

To determine the constant a, and hence to calibrate the thermometer, we specify a *standard fixed point* at which all thermometers must give the same reading for temperature T. This fixed point is chosen to be that at which ice, liquid water, and water vapor coexist in equilibrium and is called the *triple point of water*. This state can be achieved only at a definite pressure and is unique. The water vapor pressure at the triple point is 4.58 mm-Hg. The temperature at this standard fixed point is arbitrarily* set at 273.16 degrees Kelvin and is abbreviated as 273.16° K. The Kelvin degree is a unit temperature interval.

We indicate values at the triple point by the subscript *tr*. The thermometer may now be calibrated (that is, the value of a may be determined) by measuring the pressure P_{tr} at the triple point temperature $T_{tr} = 273.16°$ K. Inserting this value into Eq. 18–1 we obtain

$$T_{tr} = aP_{tr}$$
$$a = T_{tr}/P_{tr}$$
$$= 273.16° \text{ K}/P_{tr}.$$

The temperature at the measured pressure P is then provisionally (see below) defined to be

* Adopted in 1954 at the Tenth General Conference on Weights and Measures in Paris.

$$T(P) = 273.16° \text{ K } \frac{P}{P_{tr}} \qquad \text{(constant } V\text{).} \qquad (18\text{–}2)$$

The constant-volume thermometer, used as described below, is the thermometer which serves to establish the temperature scale used universally in scientific work today.

18–4 Ideal Gas Temperature Scale

Let a certain amount of gas be put into the bulb of a constant-volume gas thermometer so that the bulb is surrounded by water at the triple point. The pressure P_{tr} is equal to a definite value, say 80 cm-Hg. Now surround the bulb with steam condensing at 1-atm pressure and, with the volume kept constant at its previous value, measure the gas pressure P_s, the pressure at the steam point, in this case, P_{s80}. Then calculate the temperature provisionally from $T(P_{s80}) = 273.16°$ K $(P_{s80}/80 \text{ cm-Hg})$. Next remove some of the gas so that P_{tr} has a smaller value, say 40 cm-Hg. Then measure the new value of P_s and calculate another provisional temperature from $T(P_{s40}) = 273.16°$ K $(P_{s40}/40 \text{ cm-Hg})$. Continue this same procedure, reducing the amount of gas in the bulb again, and at this new lower value of P_{tr} calculating the temperature at the steam point $T(P_s)$. If we plot the values $T(P_s)$ against P_{tr} and have enough data, we can extrapolate the resulting curve to the intersection with the axis where $P_{tr} = 0$.

In Fig. 18–2, we plot curves obtained from such a procedure for constant-volume thermometers of some different gases. These curves show that the temperature read-

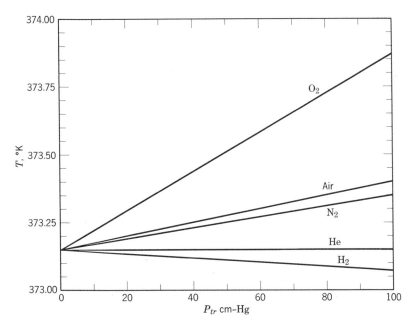

FIGURE 18–2   The readings of a constant-volume gas thermometer for the temperature T of condensing steam as a function of P_{tr}, when different gases are used. As the amount of gas in the thermometer is reduced its pressure P_{tr} at the triple point decreases. Note that at a particular P_{tr} the values of T given by different gas thermometers differ. The discrepancy is small but measurable, being about 0.2 per cent in the most extreme cases shown (O_2 and H_2 at 100 cm-Hg). Helium gives nearly the same T at all pressures (the curve is almost horizontal) so that its behaviour is the most similar to that of an ideal gas over the entire range shown.

ings of a constant-volume gas thermometer depend on the gas used at ordinary values of the reference pressure. However, as the reference pressure is decreased, the temperature readings of constant-volume gas thermometers using different gases approach the same value. Therefore, the extrapolated value of the temperature depends only on the general properties of gases and not on any particular gas. We therefore define an *ideal gas temperature scale* by the relation

$$T = 273.16° \text{ K} \lim_{P_{tr} \to 0} \left(\frac{P}{P_{tr}}\right) \quad \text{(constant } V). \qquad (18\text{--}3)$$

Our standard thermometer is therefore chosen to be a constant-volume gas thermometer using a temperature scale defined by Eq. 18–3.

Although our temperature scale is independent of the properties of any one particular gas, it does depend on the properties of gases in general (that is, on the properties of an ideal gas). Therefore, to measure a temperature, a gas must be used at that temperature. The lowest temperature that can be measured with any gas thermometer is about 1° K. To obtain this temperature we must use low-pressure helium, for helium becomes a liquid at a temperature lower than any other gas. Therefore we cannot give experimental meaning to temperatures below about 1° K, by means of a gas thermometer.

We would like to define a temperature scale in a way that is independent of the properties of any particular substance. An *absolute thermodynamic temperature scale,* called the Kelvin scale, is such a scale. It can be shown that the ideal gas scale and the Kelvin scale are identical in the range of temperatures in which a gas thermometer may be used. For this reason we can write °K after an ideal gas temperature, as we have already done.

It can also be shown that the Kelvin scale has an *absolute zero* of 0° K and that temperatures below this do not exist. The absolute zero of temperature has defied all attempts to reach it experimentally, although it is possible to come arbitrarily close. The existence of the absolute zero is inferred by extrapolation. You should not think of absolute zero as a state of zero energy and no motion. The conception that all molecular action would cease at absolute zero is incorrect. This notion assumes that the purely macroscopic concept of temperature is strictly connected to the microscopic concept of molecular motion. When we try to make such a connection, we find in fact that as we approach absolute zero the kinetic energy of the molecules approaches a finite value, the so-called zero-point energy. The molecular energy is a minimum, but not zero, at absolute zero.

In Table 18–1 we list the temperatures, on the Kelvin scale, of various bodies and processes.

18–5   The Celsius and Fahrenheit Scales

Two temperature scales in common use are the Celsius (formerly called "centigrade") and the Fahrenheit scales. These are defined in terms of the Kelvin scale, which is the fundamental temperature scale in science.

The Celsius temperature scale uses a degree (the unit of temperature) which has the same magnitude as the degree on the Kelvin scale. If we let T_C represent the Celsius temperature, then

$$T_C = T - 273.15° \qquad (18\text{--}4)$$

Table 18–1 SOME TEMPERATURES* (°K)

Carbon thermonuclear reaction	5×10^8
Helium thermonuclear reaction	10^8
Solar interior	10^7
Solar corona	10^6
Shock wave in air at Mach 20	2.5×10^4
Luminous nebulae	10^4
Solar surface	6×10^3
Tungsten melts	3.6×10^3
Lead melts	6.0×10^2
Water freezes	2.7×10^2
Oxygen boils (1 atm)	9.0×10^1
Hydrogen boils (1 atm)	2.0×10^1
Helium (He^4) boils at 1 atm	4.2
He^3 boils at attainable low pressure	3.0×10^{-1}
Adiabatic demagnetization of paramagnetic salts	10^{-3}
Adiabatic demagnetization of nuclei	10^{-6}

* See *Scientific American*, September 1954; special issue on heat.

relates the Celsius temperature t (°C) and the Kelvin temperature T (°K). We see that the triple point of water ($= 273.16°$ K by definition) corresponds to 0.01° C. By experiment the temperature at which ice and air-saturated water are in equilibrium at atmospheric pressure—the so-called ice point—proves to be 0.00° C and the temperature at which steam and liquid water are in equilibrium at 1-atm pressure—the so-called steam point—proves to be 100.00° C.

The Fahrenheit scale, in common use in English-speaking countries (except in England itself, which adopted the Celsius scale for commercial and civil use in 1964) is not used in scientific work. The relationship between the Fahrenheit and Celsius scales is defined to be

$$T_F = 32° \text{ F} + \tfrac{9}{5} T_C.$$

From this relation we can conclude that the ice point (0.00° C) equals 32.0° F, that the steam point (100.0° C) equals 212.0° F, and that one Fahrenheit degree is exactly $\tfrac{5}{9}$ as large as one Celsius degree. In Fig. 18–3 we compare the Kelvin, Celsius, and Fahrenheit scales.

18–6 The International Practical Temperature Scale

Let us now summarize the ideas of the last few sections. The standard fixed point in thermometry is the triple point of water which is arbitrarily assigned a value 273.16° K. The constant-volume gas thermometer is the standard thermometer. The extrapolated gas scale is used to define the ideal gas temperature from

$$T = 273.16° \text{ K} \lim_{P_{tr} \to 0} (P/P_{tr}).$$

This scale is identical with the (absolute thermodynamic) Kelvin scale in the range in which a gas thermometer can be used.

By using the standard thermometer in this way, we can experimentally determine other reference points for temperature measurements, called fixed points. We list the basic fixed points adopted for experimental reference in Table 18–2. The temperatures can be expressed on the Celsius scale, with the use of Eq. 18–4, once the Kelvin temperature is determined.

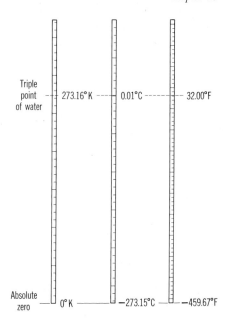

FIGURE 18-3 The Kelvin, Celsius, and Fahrenheit temperature scales.

Determining ideal gas temperatures is a painstaking job. It would not make sense to use this procedure to determine temperatures for all work. Hence, an International Practical Temperature Scale (IPTS) was adopted in 1927 (revised in 1948 and again in 1954 and 1960) to provide a scale that can be used easily for practical purposes, such as for calibration of industrial or scientific instruments. This scale consists of a set of recipes for providing in practice the best possible approximations to the Kelvin scale. A set of fixed points, the basic points in Table 18–2, is adopted, and a set of instruments is specified to be used in interpolating between these fixed points and in extrapolating beyond the highest fixed point. Formulas are specified for correcting the basic temperatures according to the barometer reading. The IPTS departs from the Kelvin scale at temperatures between the fixed points, but the difference is usually negligible. The IPTS has become the legal standard in nearly all countries.

18-7 Temperature Expansion

Common effects of temperature changes are changes in size and changes of state of materials. Let us consider changes in size which occur without changes of state. Consider a simple model of a crystalline solid. The atoms are held together in a regular

Table 18-2 FIXED POINTS ON THE INTERNATIONAL PRACTICAL TEMPERATURE SCALE (1960)*

| | | Temperature | |
Substance	Designation	°C	°K
Oxygen	Normal boiling point	− 182.97	90.18
Water	Triple point	0.01	273.16
Water	Normal boiling point	100.00	373.15
Sulfur†	Normal boiling point	444.60	717.75
Silver	Normal melting point	960.80	1233.95
Gold	Normal melting point	1063.00	1336.15

* All temperatures assumed exact for the purposes of establishing the scale.
† The normal melting point of zinc (419.505° C) may be substituted.

array by forces of electrical origin. The forces between atoms are like those that would be exerted by a set of springs connecting the atoms, so that we can visualize the solid body as a microscopic bedspring (Fig. 18–4). These "springs" are quite stiff (Problem 14, Chapter 13), and there are about 10^{22} of them per cubic centimeter. At any temperature the atoms of the solid are vibrating. The amplitude of vibration is about 10^{-9} cm and the frequency about 10^{13}/sec.

When the temperature is increased the average distance between atoms increases. This leads to an expansion of the whole solid body as the temperature is increased. The change in any linear dimension of the solid, such as its length, width, or thickness, is called a linear expansion. If the length of this linear dimension is l, the change in length, arising from a change in temperature ΔT, is Δl. We find from experiment that, if ΔT is small enough, this change in length Δl is proportional to the temperature change ΔT and to the original length l. Hence, we can write

$$\Delta l = \alpha l \, \Delta T, \tag{18–5}$$

where α, called the *coefficient of linear expansion,* has different values for different materials. Rewriting this formula we obtain

$$\alpha = \frac{1}{l} \frac{\Delta l}{\Delta T},$$

so that α has the meaning of a fractional change in length per degree temperature change.

Strictly speaking, the value of α depends on the actual temperature and the reference temperature chosen to determine l. However, its variation is usually negligible compared to the accuracy with which engineering measurements need to be made. We can safely take it as a constant for a given material, independent of the temperature. In Table 18–3 we list the experimental values for the average coefficient of linear expansion $\bar{\alpha}$ of several common solids. For all the substances listed, the change in size consists of an expansion as the temperature rises, for $\bar{\alpha}$ is positive. The order of magnitude of the expansion is about 1 millimeter per meter length per 100 Celsius degrees.*

*One Celsius degree (1 C°) means a temperature interval (ΔT) of one unit measured on a Celsius scale. One degree Celsius (1° C) means a specific temperature reading (T) on that scale.

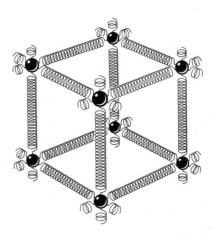

FIGURE 18–4   A solid behaves in many ways as if it is a microscopic "bed-spring" in which the molecules are held together by elastic forces.

Table 18–3   SOME VALUES* OF $\overline{\alpha}$

Substance	$\overline{\alpha}$ (per C°)	Substance	$\overline{\alpha}$ (per C°)
Aluminum	23×10^{-6}	Hard rubber	80×10^{-6}
Brass	19×10^{-6}	Ice	51×10^{-6}
Copper	17×10^{-6}	Invar	0.7×10^{-6}
Glass (ordinary)	9×10^{-6}	Lead	29×10^{-6}
Glass (Pyrex)	3.2×10^{-6}	Steel	11×10^{-6}

* For the range 0° C to 100° C; except −10° C to 0° C for ice.

Example 1. A steel metric scale is to be ruled so that the millimeter intervals are accurate to within about 5×10^{-5} mm at a certain temperature. What is the maximum temperature variation allowable during the ruling?

From Eq. 18–5,

$$\Delta l = \alpha l \, \Delta T,$$

we have

$$5 \times 10^{-5} \text{ mm} = (11 \times 10^{-6}/\text{C}°)(1.0 \text{ mm}) \, \Delta T$$

in which we have used $\overline{\alpha}$ for steel, taken from Table 18–3. This yields $\Delta T \cong 5$ C°. The temperature maintained during the ruling process must be maintained when the scale is being used and it must be held constant to within about 5 C°.

Note (see Table 18–3) that if the alloy invar is used instead of steel, then for the same required tolerance one can permit a temperature variation of about 75 C°; or for the same temperature variation ($\Delta T = 5$ C°) the tolerance achieved would be more than an order of magnitude better.

For many solids, called *isotropic,* the per cent change in length for a given temperature change is the same for all lines in the solid. The expansion is quite analogous to a photographic enlargement, except that a solid is three-dimensional. Thus, if you have a flat plate with a hole punched in it, $\Delta l/l \, (= \alpha \, \Delta T)$ for a given ΔT is the same for the length, the thickness, the face diagonal, the body diagonal, and the hole diameter. Every line, whether straight or curved, lengthens in the ratio α per degree temperature rise. If you scratch your name on the plate, the line representing your name has the same fractional change in length as any other line. The analogy to a photographic enlargement is shown in Fig. 18–5.

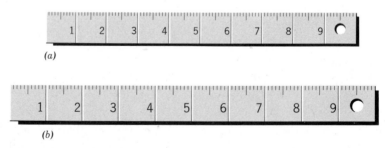

(a)

(b)

FIGURE 18–5   The same steel rule at two different temperatures. On expansion every dimension is increased by the same proportion: the scale, the numbers, the hole, and the thickness are all increased by the same factor. (The expansion shown, from (a) to (b), is obviously exaggerated, for it would correspond to an imaginary temperature rise of about 100,000 C°!)

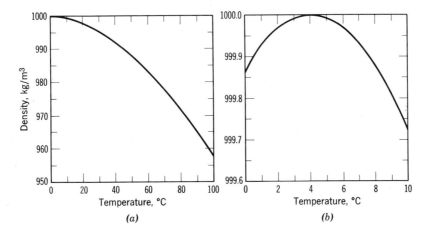

FIGURE 18–6 (*a*) The variation with temperature of density of water under atmospheric pressure. (*b*) The variation between 0 and 10° C in more detail.

With these ideas in mind, you should be able to show [(see Problems 15(7) and 17(7)] that to a high degree of accuracy the fractional change in area A per degree temperature change for an isotropic solid is 2α, that is,

$$\Delta A = 2\alpha A\,\Delta T,$$

and the fractional change in volume V per degree temperature change for an isotropic solid is 3α, that is,

$$\Delta V = 3\alpha V\,\Delta T.$$

Because the shape of a fluid is not definite, only the change in volume with temperature is significant. Gases respond strongly to temperature or pressure changes, whereas the change in volume of liquids with changes in temperature or pressure is very small. If we let β represent the coefficient of volume expansion for a liquid, that is,

$$\beta = \frac{1}{V}\frac{\Delta V}{\Delta T},$$

we find that β is relatively independent of the temperature. Liquids typically expand with increasing temperature, their volume expansion being generally about ten times greater than that of solids.

However, the most common liquid, water, does not behave like other liquids. In Fig. 18–6 we show the expansion curve for water. Notice that above 4° C water expands as the temperature rises, although not linearly. As the temperature is lowered from 4 to 0° C, however, water expands instead of contracting. Such an expansion with decreasing temperature is not observed in any other common liquid; it is observed in rubberlike substances and in certain crystalline solids over limited temperature intervals. The density of water is a maximum at 4° C, where its value is 1000 kg/meter³ or 1.000 gm/cm³. At all other temperatures its density is less. This behavior of water is the reason why lakes freeze first at their upper surface.

QUESTIONS

1. Is temperature a microscopic or macroscopic concept?

2. Does our "temperature sense" have a built-in sense of direction; that is, does hotter necessarily mean higher temperature, or is this just an arbitrary convention? Celsius, by the way, originally chose the steam point as $0°$ C and the ice point as $100°$ C.

3. How would you suggest measuring the temperature of (*a*) the sun, (*b*) the earth's upper atmosphere, (*c*) an insect, (*d*) the moon, (*e*) the ocean floor, and (*f*) liquid helium?

4. Is one gas any better than another for purposes of a standard constant-volume gas thermometer? What properties are desirable in a gas for such purposes?

5. State some objections to using water-in-glass as a thermometer. Is mercury-in-glass an improvement?

6. Can you explain why the column of mercury first descends and then rises when a mercury-in-glass thermometer is put in a flame?

7. What are the dimensions of α, the coefficient of linear expansion? Does the value of α depend on the unit of length used? When $F°$ are used instead of $C°$ as a unit of temperature change, does the numerical value of α change? If so, how?

8. A metal ball can pass through a metal ring. When the ball is heated, however, it gets stuck in the ring. What would happen if the ring, rather than the ball, were heated?

9. A bimetallic strip, consisting of two different metal strips riveted together, is used as a control element in the common thermostat. Explain how it works.

10. Explain how the period of a pendulum clock can be kept constant with temperature by attaching tubes of mercury to the bottom of the pendulum. (See Problem 17).

11. Explain why the apparent expansion of a liquid in a bulb does not give the true expansion of the liquid.

12. Does the change in volume of a body when its temperature is raised depend on whether the body has cavities inside, other things being equal? Consider a solid sphere and a hollow sphere, for example.

13. What difficulties would arise if you defined temperature in terms of the density of water?

14. Explain why lakes freeze first at the surface.

PROBLEMS

1(3). A *resistance thermometer* is a thermometer in which the thermometric property is electrical resistance. We are free to define temperatures measured by such a thermometer in $°K$ to be directly proportional to the resistance R, measured in ohms. A certain resistance thermometer is found to have a resistance R of 90.35 ohms when its bulb is placed in water at the triple point temperature ($273.16°$ K). What temperature is indicated by the thermometer if the bulb is placed in an environment such that its resistance is 96.28 ohms?
Answer: $291.1°$ K.

2(3). The amplification or "gain" of a transistor amplifier may depend upon the temperature. The gain for a certain amplifier at room temperature ($20°$ C) is 30.0 while at $55°$ C it is 35.2. What would the gain be at $30°$ C if the gain depends linearly upon temperature, over a limited range?

3(3). It is an everday observation that hot and cold objects cool down or warm up to the temperature of their surroundings. If the temperature difference ΔT between an object and its surroundings is not too great, the rate of cooling or warming is approximately proportional to the temperature difference between the object and its surroundings; that is,

$$\frac{d\,\Delta T}{dt} = -K\,\Delta T,$$

where K is a constant. The minus sign appears because ΔT decreases with time if ΔT is positive and vice versa. This is known as *Newton's law of cooling*. (a) On what factors does K depend? What are its dimensions? (b) If at some instant the temperature difference is ΔT_0, show that it is

$$\Delta T = \Delta T_0 e^{-Kt}$$

at a time t later.

Answer: (a) Materials, shape, absolute temperature, air currents, etc; dimensions are $\sec^{-1}$.

4(3). A mercury-in-glass thermometer is placed in boiling water for a few minutes and then removed. The temperature readings at various times after removal are as follows:

t, sec	T, °C	t, sec	T, °C	t, sec	T, °C	t, sec	T, °C
0	98.4	25	65.1	100	50.3	700	26.5
5	76.1	30	63.9	150	43.7	1000	26.1
10	71.1	40	61.6	200	38.8	1400	26.0
15	67.7	50	59.4	300	32.7	2000	26.0
20	66.4	70	55.4	500	27.8	3000	26.0

Plot K as a function of time, assuming Newton's law of cooling to apply [see Problem 3(3)]. To what extent are you justified in assuming that Newton's law of cooling applies here?

5(4). If the ideal gas temperature at the steam point is 373.15° K, what is the limiting value of the ratio of the pressures of a gas at the steam point and at the triple point of water when the gas is kept at constant volume?

Answer: 1.366.

6(4). Let p_{tr} be the pressure in the bulb of a constant-volume gas thermometer when the bulb is at the triple-point temperature of 273.16° K and p the pressure when the bulb is at room temperature. Given three constant-volume gas thermometers: For No. 1 the gas is oxygen and $p_{tr} = 20$ cm-Hg; for No. 2 for the gas is also oxygen but $p_{tr} = 40$ cm-Hg; for No. 3 the gas is hydrogen and $p_{tr} = 30$ cm-Hg. The measured values of p for the three thermometers are p_1, p_2, and p_3. (a) An approximate value of the room temperature T can be obtained with each of the thermometers using

$$T_1 = 273.16° \text{ K } \frac{p_1}{20 \text{ cm-Hg}}; \ T_2 = 273.16° \text{ K } \frac{p_2}{40 \text{ cm-Hg}}; \ T_3 = 273.16° \text{ K } \frac{p_3}{30 \text{ cm-Hg}}.$$

Mark each of the following statements "true" or "false": (1) With the method described, all three thermometers will give the same value of T. (2) The two oxygen thermometers will agree with each other but not with the hydrogen thermometer. (3) Each of the three will give a different value of T. (b) In the event that there is disagreement among the three thermometers, explain how you would change the method of using them to cause all three to give the same value of T.

7(4). (a) The temperature of the surface of the sun is about 6000° K. Express this on the Fahrenheit scale. (b) Express normal human body temperature, 98.6° F, on the Celsius scale. (c) In the continental United States, the highest recorded temperature is 134° F at Death Valley, California, and the lowest is −70° F at Rogers Pass, Montana. Express these extremes on the Celsius scale. (d) Express the normal boiling point of oxygen, −183° C, on the Fahrenheit scale. (e) At what Celsius temperature would you find a room to be uncomfortably warm?

Answer: (a) 10,000° F. (b) 37.0° C. (c) 56.7° C; −57° C. (d) −297° F. (e) 25° C = 77° F, for example.

8(5). (a) At what temperature do the Fahrenheit and Celsius scales give the same reading? (b) The Fahrenheit and the Kelvin scales?

9(6). In the interval between 0 and 600° C, a platinum resistance thermometer of definite specifications is used for interpolating temperatures on the International Practical Temperature Scale. The temperature T_C is given by a formula for the variation of resistance with temperature:

$$R = R_0(1 + AT_C + BT_C{}^2).$$

R_0, A, and B are constants determined by measurements at the ice point, the steam point, and the

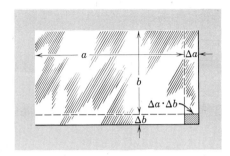

FIGURE 18–7 Problem 15(7).

sulphur point. (*a*) If *R* equals 10.000 ohms at the ice point, 13.946 ohms at the steam point, and 24.817 ohms at the sulphur point, find R_0, *A*, and *B*. (*b*) Plot *R* versus T_C in the temperature range from 0 to 600° C.

Answers: (*a*) 10.000 ohms; 4.124×10^{-3}/°C; -1.779×10^{-6}/°C².

10(7). A steel rod has a length of exactly 20 cm at 30° C. How much longer is it at 50° C?

11(7). An aluminum-alloy rod has a length of 10.000 cm at room temperature (20° C) and a length of 10.015 cm at the boiling point of water. (*a*) What is the length of the rod at the freezing point of water? (*b*) What is the temperature if the length of the rod is 10.009 cm?

Answer: (*a*) 9.996 cm. (*b*) 68° C.

12(7). The Pyrex glass mirror in the telescope at the Mount Palomar Observatory has a diameter of 200 in. The temperature ranges from −10 to 50° C on Mount Palomar. Determine the maximum change in the diameter of the mirror.

13(7). A circular hole in an aluminum plate is 1.000 in. in diameter at 0° C. What is its diameter when the temperature of the plate is raised to 100° C?

Answer: 1.002 in.

14(7). A steel rod is 3.000 cm in diameter at 25° C. A brass ring has an interior diameter of 2.992 cm at 25° C. At what common temperature will the ring just slide onto the rod?

15(7). The area *A* of a rectangular plate is *ab*. Its coefficient of linear expansion is α. After a temperature rise ΔT, side *a* is longer by Δa and side *b* is longer by Δb. Show that if we neglect the small area $\Delta a \cdot \Delta b$, shown cross-hatched and greatly exaggerated in size in Fig. 18–7, then $\Delta A = 2\alpha A \, \Delta T$.

16(7). A glass window is exactly 20 cm by 30 cm at 10° C. By how much has its area increased when its temperature is 40° C?

17(7). Prove that, if we neglect extremely small quantities, the change in volume of a solid on expansion through a temperature rise ΔT is given by $\Delta V = 3\alpha V \, \Delta T$ where α is the coefficient of linear expansion.

18(7). What is the volume of a lead ball at 30° C if its volume at 60° C is (exactly) 50 cm³?

19(7). Find the change in volume of an aluminum sphere of 10.0-cm radius when it is heated from 0 to 100° C.

Answer: 29 cm³.

20(7). When the temperature of a "copper" penny is raised by 100 C°, its diameter increases by 0.18%. To two significant figures give the per cent increase in the (*a*) area of a face, (*b*) thickness, (*c*) volume, and (*d*) mass of the penny. (*e*) What is the coefficient of linear expansion?

21(7). Density is mass per unit volume. If the volume *V* is temperature dependent, so is the density ρ. Show that the change in density $\Delta \rho$ with change in temperature ΔT is given by

$$\Delta \rho = -\beta \rho \, \Delta T$$

where β is the volume coefficient of expansion. Explain the minus sign.

22(7). When the temperature of a metal cylinder is raised from 0° C to 100° C, its length increases by 0.23%. (*a*) Find the percentage change in density. (*b*) What is the metal?

23(7). A pendulum clock with a pendulum made of brass is designed to keep accurate time at 20° C. What will be the error, in seconds per hour, if the clock operates at 0° C?

Answer: +0.68 sec/hr.

24(7). A clock pendulum made of Invar has a period of 0.500 sec at 20° C. If the clock is used in a climate where the temperature averages 30° C, what correction (approximately) is necessary at the end of 30 days to the time given by the clock?

25(7). The timing of a certain electric watch is governed by a small tuning fork. The frequency of the fork is inversely proportional to the square root of the length of the fork. What is the percentage gain or loss in time for a steel tuning fork 8.0 mm long at (a) −40° F and (b) at +120° F if it keeps perfect time at 25° C?

Answer: (a) +0.036%. (b) −0.013%.

26(7). A rod is measured to be exactly 20.000 cm long using a steel ruler at room temperature (20° C). Both the rod and the ruler are placed in an oven at 270° C, where the rod now measures 20.1 cm using the same ruler. What is the coefficient of thermal expansion for the material of which the rod is made?

27(7). A 1.0-meter long vertical glass tube is half-filled with a liquid at 20° C. How much will the liquid column change when the tube is heated to 30° C? Take $\bar{\alpha}_{glass} = 1.0 \times 10^{-5}/C°$ and $\bar{\beta}_{liquid} = 4 \times 10^{-5}/C°$.

Answer: Increases by 0.05 mm.

28(7). (a) Show that if the lengths of two rods of different solids are inversely proportional to their respective coefficients of linear expansion at some initial temperature, the difference in length between them will be constant at all temperatures. (b) What should be the lengths of a steel and a brass rod at 0° C so that at all temperatures their difference in length is 0.30 meter?

29(7). The distance between the towers of the main span of the Golden Gate Bridge at San Francisco is 4200 ft. The sag of the cable halfway between the towers at 50° F is 470 ft. Take $\alpha = 6.5 \times 10^{-6}/F°$ for the cable and compute (a) the change in length of the cable, and (b) the change in sag for a temperature change from −20 to 110° F. Assume no bending or separation of the towers and a parabolic shape for the cable.

Answer: (a) 3.7 ft. (b) 6.3 ft.

CHAPTER 19

Heat and the First Law of Thermodynamics

19–1 Heat, a Form of Energy

When two systems at different temperatures are placed together, the final temperature reached by both systems is somewhere between the two starting temperatures. This is a common observation. Man has long sought for a deeper understanding of such phenomena. Until the beginning of the nineteenth century, they were explained by postulating that a material substance, *caloric,* existed in every body. It was believed that a body at high temperature contained more caloric than one at a low temperature. When the two bodies were put together, the body rich in caloric lost some to the other until both bodies reached the same temperature. The caloric theory was able to describe many processes, such as heat conduction or the mixing of substances in a calorimeter, in a satisfactory way. However, the concept of heat as a substance, whose total amount remained constant, eventually could not stand the test of experiment. Nevertheless, we still describe many common temperature changes as the transfer of "something" from one body at a higher temperature to one at the lower, and this "something" we call heat. A useful but nonoperational definition, is: *Heat is that which is transferred between a system and its surroundings as a result of temperature differences only.*

Eventually it became generally understood that heat is a form of energy rather than a substance. The first conclusive evidence that heat could not be a substance was given by Benjamin Thompson (1753–1814), an American who later became Count Rumford of Bavaria.

Rumford made his discovery while supervising the boring of cannon for the Bavarian government. To prevent overheating, the bore of the cannon was kept full of water. The water was replenished as it boiled away during the boring process. It was accepted that caloric had to be supplied to water to boil it. The continuous production of caloric was explained by assuming that when a substance was more finely subdivided, as in boring, its capacity for retaining caloric became smaller, and that the caloric released in this way was what caused the water to boil. Rumford observed, however, that the water boiled away even when his boring tools became so dull that they were no longer cutting or subdividing matter. He was able to rule out, by experiment, all possible caloric interpretations and to conclude that the mechanical motion of the boring process produced the same effect on the water (a temperature rise and then boiling) as would be produced by putting the water near a flame.

Here we have the germ of the idea that doing mechanical work on a system, such as a can of water, and adding heat to it from an external source may be *equivalent* in their effects, both work and heat being forms of energy.

Although the concept of energy and its conservation seems self-evident today, it was a novel idea as late as the 1850's and had eluded such men as Galileo and Newton. Throughout the subsequent history of physics this conservation idea led men to new discoveries. Its early history was remarkable in many ways. Several thinkers arrived at this great concept at about the same time; at first, all of them either met with a cold reception or were ignored. The principle of the conservation of energy was established independently by Julius von Mayer (1814–1878) in Germany, James Joule (1818–1889) in England, Hermann von Helmholtz (1821–1894) in Germany, and L. A. Colding (1815–1888) in Denmark.

It was Joule who showed by experiment that a given amount of mechanical work done on a system was *quantitatively equivalent* in its effects to a specific quantity of heat added to the system from an external source at a higher temperature. Thus, the equivalence of heat and mechanical work as two forms of energy was definitely established.

Helmholtz first expressed clearly the idea that not only heat and mechanical energy but all forms of energy are equivalent, and that a given amount of one form cannot disappear without an equal amount appearing in some of the other forms.

19–2 Quantity of Heat and Specific Heat

The unit of heat Q is defined quantitatively in terms of a specified change produced in a body during a specified process. Thus, if the temperature of one kilogram of water is raised from 14.5 to 15.5° C by heating, we say that one *kilocalorie* (kcal) of heat has been added to the system. The *calorie* ($= 10^{-3}$ kcal) is also used as a heat unit. (Incidentally, the "calorie" used to measure the energy content of foods is actually a kilocalorie.) In the engineering system the unit of heat is the *British thermal unit* (Btu), which is defined as the heat necessary to raise the temperature of one pound of water from 63 to 64° F.

The reference temperatures are stated because, near room temperature, there is a slight variation in the heat needed for a one-degree temperature rise with the interval chosen. We will neglect this variation for most practical purposes. The heat units are related as follows:

$$1.000 \text{ kcal} = 1000 \text{ cal} = 3.968 \text{ Btu}$$

Substances differ from one another in the quantity of heat needed to produce a given rise of temperature in a given mass. The ratio of the heat ΔQ supplied to a body to its corresponding temperature rise ΔT is called the *heat capacity* C of the body; that is,

$$C = \text{heat capacity} = \frac{\Delta Q}{\Delta T}.$$

The word "capacity" may be misleading because it suggests the essentially meaningless statement "the amount of heat a body can hold," whereas what is meant is simply the heat added per unit temperature rise.

The heat capacity per unit mass of a body, called *specific heat,* is characteristic of the material of which the body is composed:

$$c = \frac{\text{heat capacity}}{\text{mass}} = \frac{\Delta Q}{m\,\Delta T}. \tag{19–1}$$

We properly speak, on the one hand, of the heat capacity of a penny but, on the other, of the specific heat of copper.

Neither the heat capacity of a body nor the specific heat of a material is constant but depends on the location of the temperature interval. At ordinary temperatures and over ordinary temperature intervals, however, specific heats can be considered to be constants. For example, the specific heat of water varies less than 1% from its value of 1.000 cal/gm C° over the temperature range from 0° C to 100° C.

Equation 19–1 does not define specific heat uniquely. We must also specify the conditions under which the heat ΔQ is added to the specimen. We have implied that the condition is that the specimen remain at normal (constant) atmospheric pressure while we add the heat. This is a common condition, but there are many other possibilities, each leading, in general, to a different value for c. To obtain a unique value for c we must specify the conditions, such as specific heat at constant pressure c_p, specific heat at constant volume c_v, etc.

Table 19–1 (second column) shows the specific heats at constant pressure of some solid elements; we will discuss the specific heats of gases later. You should realize from the way the calorie and the Btu are defined that 1 cal/gm C° = 1 kcal/kg C° = 1 Btu/lb F°, exactly. Note that the specific heat of water, equal to 1.00 cal/gm C°, is large compared to that of most substances.

Example 1. A 75-gm block of copper, taken from a furnace, is dropped into a 300-gm glass beaker containing 200 gm of water. The temperature of the water rises from 12 to 27° C. What was the temperature of the furnace?

Table 19–1   VALUES FOR c_p FOR SOME SOLIDS   (AT ROOM TEMPERATURE AND FOR $p = 1.0$ ATM)

Substance	Specific heat (cal/gm C°)	Molecular weight (gm/mole)	Molar heat capacity* (cal/mole C°)
Aluminum	0.215	27.0	5.82
Carbon	0.121	12.0	1.46
Copper	0.0923	63.5	5.85
Lead	0.0305	207	6.32
Silver	0.0564	108	6.09
Tungsten	0.0321	184	5.92

*The molecular weight and the mole are defined on page 376.

This is an example of two systems originally at different temperatures reaching thermal equilibrium after contact. No mechanical energy is involved, only heat exchange. Hence,

$$\text{heat lost by copper} = \text{heat gained by (beaker + water)},$$

$$m_C c_C (T_C - T_e) = (m_G c_G + m_W c_W)(T_e - T_W).$$

The subscript C stands for copper, G for glass, and W for water. The initial copper temperature is T_C, the initial beaker water temperature is T_W, and T_e is the final equilibrium temperature. Substituting the given values, with $c_C = 0.092$ cal/gm C°, $c_G = 0.12$ cal/gm C°, and $c_W = 1.0$ cal/gm C°, we obtain

$$(75 \text{ gm})(0.092 \text{ cal/gm C}°)(T_C - 27° \text{ C}) = [(300 \text{ gm})(0.12 \text{ cal/gm C}°)$$
$$+ (200 \text{ gm})(1.0 \text{ cal/gm C}°)](27° \text{ C} - 12° \text{ C})$$

or, solving for T_C,
$$T_C = 530° \text{ C}.$$

What approximations, both experimental and theoretical, were used implicitly to arrive at this answer?

19–3   Heat Conduction

The transfer of energy arising from the temperature difference between adjacent parts of a body is called *heat conduction*. Consider a slab of material of cross-sectional area A and thickness Δx, whose faces are kept at different temperatures. We measure the heat ΔQ that flows perpendicular to the faces for a time Δt. Experiment shows that ΔQ is proportional to Δt and to the cross-sectional area A for a given temperature difference ΔT, and that ΔQ is proportional to $\Delta T/\Delta x$ for a given Δt and A, providing both ΔT and Δx are small. That is,

$$\frac{\Delta Q}{\Delta t} \propto A \frac{\Delta T}{\Delta x} \qquad \text{approximately.}$$

In the limit of a slab of infinitesimal thickness dx, across which there is a temperature difference dT, we obtain the fundamental law of heat conduction

$$\frac{dQ}{dt} = -kA \frac{dT}{dx}. \tag{19–2}$$

Here dQ/dt is the time rate of heat transfer across the area A, dT/dx is called the *temperature gradient,* and k is a constant of proportionality called the *thermal conductivity.* We choose the direction of heat flow to be the direction in which x increases; since heat flows in the direction of decreasing T, we introduce a minus sign in Eq. 19–2 (that is, we wish dQ/dt to be positive when dT/dx is negative).

A substance with a large thermal conductivity k is a good heat conductor; one with a small thermal conductivity k is a poor heat conductor, or a good thermal insulator. The value of k depends on the temperature, increasing slightly with increasing temperature, but k can be taken to be practically constant throughout a substance if the temperature difference between its parts is not too great. In Table 19–2 we list values of k for various substances; we see that metals as a group are better heat conductors than nonmetals, and that gases are poor heat conductors.

Let us apply Eq. 19–2 to a rod of length L and constant cross sectional area A in which a steady state has been reached (Fig. 19–1). In a steady state the temperature at each point is constant in time. Hence, dQ/dt is the same at all cross-sections. (Why?) But $dQ/dt = -kA \ (dT/dx)$, so that, for a constant k and A, the temperature

Table 19–2 THERMAL CONDUCTIVITIES, KCAL/SEC METER C°
(GASES AT 0° C; OTHERS AT ABOUT ROOM TEMPERATURE)

Metals		Others	
Aluminum	4.9×10^{-2}	Asbestos	2×10^{-5}
Brass	2.6×10^{-2}	Concrete	2×10^{-4}
Copper	9.2×10^{-2}	Cork	4×10^{-5}
Lead	8.3×10^{-3}	Glass	2×10^{-4}
Silver	9.9×10^{-2}	Ice	4×10^{-4}
Steel	1.1×10^{-2}	Wood	2×10^{-5}
Gases			
Air	5.7×10^{-6}		
Hydrogen	3.3×10^{-5}		
Oxygen	5.6×10^{-6}		

gradient dT/dx is the same at all cross-sections. Hence, T decreases linearly along the rod so that $-dT/dx = (T_2 - T_1)/L$. Therefore, the heat ΔQ transferred in time Δt is

$$\frac{\Delta Q}{\Delta t} = kA \frac{T_2 - T_1}{L}. \tag{19–3}$$

The phenomenon of heat conduction also shows that the concepts of heat and temperature are distinctly different. Different rods, having the same temperature difference between their ends, may transfer entirely different quantities of heat in the same time.

Example 2. Consider a compound slab, consisting of two materials having different thicknesses, L_1 and L_2, and different thermal conductivities, k_1 and k_2. If the temperatures of the outer surfaces are T_2 and T_1, find the rate of heat transfer through the compound slab (Fig. 19–2) in a steady state.

Let T_x be the temperature at the interface between the two materials. Then

$$\frac{\Delta Q_2}{\Delta t} = \frac{k_2 A (T_2 - T_x)}{L_2}$$

and

$$\frac{\Delta Q_1}{\Delta t} = \frac{k_1 A (T_x - T_1)}{L_1}.$$

In a steady state $\Delta Q_1/\Delta t = \Delta Q_2/\Delta t$, so that

$$\frac{k_2 A (T_2 - T_x)}{L_2} = \frac{k_1 A (T_x - T_1)}{L_1}.$$

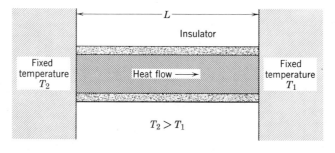

FIGURE 19–1   Conduction of heat through an insulated conducting bar.

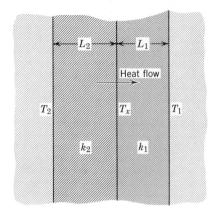

FIGURE 19–2   Example 2. Conduction of heat through two layers of matter with different thermal conductivities.

Let $\Delta Q/\Delta t$ be the rate of heat transfer (the same for all sections). Then, solving for T_x and substituting into either of these equations, we obtain

$$\frac{\Delta Q}{\Delta t} = \frac{A(T_2 - T_1)}{(L_1/k_1) + (L_2/k_2)}.$$

19–4   The Mechanical Equivalent of Heat

If heat is just another form of energy, any energy unit could be a heat unit. The calorie and Btu originated before it was generally accepted that heat is energy. It was Joule who first carefully measured the mechanical energy equivalent of heat energy, that is, the number of joules equivalent to 1 calorie, or the number of foot-pounds equivalent to 1 Btu.

The relative size of the "heat units" and the "mechanical units" can be found from experiments in which a measured quantity of mechanical energy—in joules, say—is added to a system such as a bucket of water; from the measured temperature rise we calculate, using the definition of Q in section 19–2, how much heat—in calories, say—we would have to add to the water sample to produce this same effect. Thus we can calculate the ratio of the joule to the calorie, the so-called *mechanical equivalent of heat*. Joule originally used an apparatus in which falling weights rotated a set of paddles in a water container (Fig. 19–3). The loss of mechanical energy was computed from a knowledge of the weights and the heights through which they fell and the equivalent heat energy by determining the mass of water and its rise in temperature. Joule wanted to show that the same amount of heat energy would be obtained from a given expenditure of work regardless of the method used to produce the work. He did so by stirring mercury, by rubbing together iron rings in a mercury bath, by adding electrical energy using a hot wire immersed in water, and in other ways. Always the constant of proportionality between the equivalent heat and work performed agreed within his experimental error of 5%. Joule did

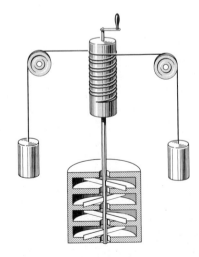

FIGURE 19–3   Joule's arrangement for measuring the mechanical equivalent of heat. The falling weights turn paddles which stir the water in the container, thus raising its temperature.

not have at his disposal the accurately standardized thermometers of today, nor could he make such reliable corrections for heat losses from the system as are possible now. His pioneer experiments are noteworthy not only for the skill and ingenuity he showed but also for the influence they had in convincing scientists everywhere of the correctness of the concept that heat is a form of energy.

The accepted results are

$$1 \text{ kcal} = 1000 \text{ cal} = 4186 \text{ joules}$$

$$1 \text{ Btu} = 252.0 \text{ cal} = 777.9 \text{ ft lb};$$

that is, 4186 joules of mechanical energy will raise the temperature of 1 kg of water from 14.5 to 15.5° C, just as will 1000 calories of heat.

19–5 Heat and Work

We have seen that heat is the energy that flows from one body to another because of a temperature difference between them. Heat is not a property of a body in the sense that we can assign a value to the amount of heat "contained" in the body.

Work is similar to heat in that it is a measure of energy being transmitted from one body to another. In fact, *work* may be defined as *energy that is transmitted from one system to another in such a way that a difference of temperature is not directly involved.* Work is not a specific property of a body in the sense that we can assign a value to the work "contained" within the body. In Joule's experiment we can imagine that a steady state has been achieved, in which we do work on the system at a constant rate and the temperature of the system rises to such a steady value that heat energy is lost to the (cooler) surroundings at this same constant rate. We speak of work and heat only as they enter or leave the system.

Work is a measure of energy transfer by mechanical means, such as by gravitational, electrical, or magnetic forces. Heat is a measure of energy transfer by means of temperature differences.

Thermodynamics is concerned with the energy transfers that occur when a system undergoes any thermodynamic process from one state of the system to another. It is also concerned with the relationship of these energy transfers to changes in the measurable properties of the system.

For example, suppose we wish to discuss a system composed of a certain number of gas molecules contained in a cylinder with a moveable piston, as in Fig. 19–4. The environment with which the system of molecules interacts would be the cylinder walls and the piston. The properties which describe the system would be the pressure, temperature, volume, and chemical state of the gas. A thermodynamic process in such a system might involve a process in which an amount of heat energy Q is transferred to the gas because of a difference in temperature between the system (the gas) and its environment (the cylinder walls). The gas might do work W on its environment (the piston) by causing the piston to move because of an imbalance between the forces on the two sides of the piston. During the process the gas might change from a configuration described by initial parameters p_i, V_i, and T_i to a final configuration with parameters p_f, V_f, and T_f. (Chemical reactions or changes of state might also be involved and these would also be described by appropriate system parameters.)

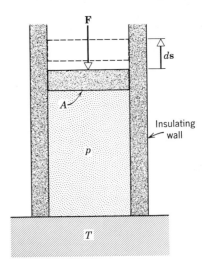

FIGURE 19–4 Work is done by the gas at pressure p as it expands against the piston. Heat may enter or leave the system from the heat reservoir on which the cylinder rests.

The laws of thermodynamics allow us to calculate certain relationships between the initial and final parameters of the system on the one hand and the energy transfers Q and W on the other.

Let us now compute W for a process of this type. In Fig. 19-4 we show the gas expanding against the piston. The work done by the gas in displacing the piston through an infinitesimal distance ds is

$$dW = \mathbf{F} \cdot d\mathbf{s} = pA \, ds = p \, dV$$

where dV is the differential change in the volume of the gas. In general, the pressure will not be constant during a displacement. To obtain the total work W done on the piston by the gas in a large displacement, we must know how p varies with the displacement. Then we compute the integral

$$W = \int dW = \int_{V_i}^{V_f} p \, dV.$$

This integral can be evaluated graphically as the area under the curve in a p–V diagram, as shown for a special case in Fig. 19–5.

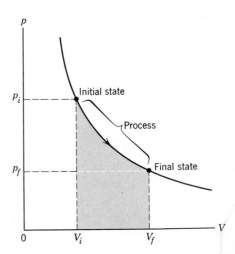

FIGURE 19–5 The work done by a gas is equal to the area under a p-V curve.

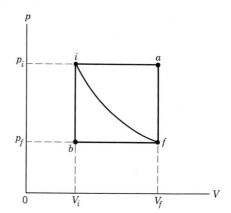

FIGURE 19–6 The work done by a system depends not only on the initial state (i) and the final state (f) but on the intermediate path as well.

There are many different ways in which the system can be taken from the initial state i to the final state f. For example (Fig. 19–6), the pressure may be kept constant from i to a and then the volume kept constant from a to f. Then the work done by the expanding gas is equal to the area under the line ia. Another possibility is the path ibf, in which case the work done by the gas is the area under the line bf. The continuous curve from i to f is another possible path in which the work done by the gas is still different from the previous two paths. We can see, therefore, that the work done by a system depends not only on the initial and final states but also on the intermediate states, that is, on the path of the process.

A similar result follows if we compute the flow of heat during the process. State i is characterized by a temperature T_i and state f by a temperature T_f. The heat flowing into the system, say, depends on how the system is heated. We can heat it at a constant pressure p_i, for example, until we reach the temperature T_f, and then change the pressure at constant temperature to the final value p_f. Or we can first lower the pressure to p_f and then heat it at that pressure to the final temperature T_f. Or we can follow many other paths. Each path gives a different result for the heat flowing into the system. Hence, the heat lost or gained by a system depends not only on the initial and final states but also on the intermediate states, that is, on the path of the process. This is an experimental fact.

Both heat and work "depend on the path" taken; neither one is independent of the path, and neither one can be conserved alone.

19–6   The First Law of Thermodynamics

We can now tie all these ideas together. Let a system change from an initial equilibrium state i to a final equilibrium state f in a definite way, the heat absorbed by the system being Q and the work done by the system being W. Then we compute the value of $Q - W$. Now we start over and change the system from the same state i to the same state f, but this time by a different path. We do this over and over again, using different paths each time. We find that in every case the quantity $Q - W$ is the same. That is, although Q and W separately depend on the path taken, $Q - W$ does not depend at all on how we took the system from state i to state f but only on the initial and final (equilibrium) states.

You will remember from mechanics that when an object is moved from an initial point i to a final point f in a gravitational field in the absence of friction, the work done depends only on the positions of the two points and not at all on the path through which the body is moved. From this we concluded that there is a function of the space coordinates of the body whose final value minus its initial value equals the work done in displacing the body. We called it the potential energy function. Now in thermodynamics we find from experiment that when a system has its state changed from state i to state f, the quantity $Q - W$ depends only on the initial and final coordinates and not at all on the path taken between these end points. We conclude that there is a function of the thermodynamic coordinates whose final value minus its initial value equals the change $Q - W$ in the process. We call this function the *internal energy function.*

Let us represent the internal energy function by the letter U. Then the internal energy of the system in state f, U_f, minus the internal energy of the system in state i, U_i, is simply the change in internal energy of the system, and this quantity has a definite value independent of how the system went from state i to state f. We have

$$U_f - U_i = \Delta U = Q - W. \tag{19-4}$$

Just as for potential energy, so for internal energy too it is the change that matters. If some arbitrary value is chosen for the internal energy in some standard reference state, its value in any other state can be given a definite value. Equation 19-4 is known as the *first law of thermodynamics.* In applying Eq. 19-4 we must remember that Q is considered positive when heat enters the system and W is positive when work is done by the system.

If our system undergoes only an infinitesimal change in state, only an infinitesimal amount of heat dQ is absorbed and only an infinitesimal amount of work dW is done, so that the internal energy change dU is also infinitesimal. Although dW and dQ are not true differentials*, we may write the first law in differential form as

$$dU = dQ - dW. \tag{19-5}$$

We may express the first law in words by saying: *Every thermodynamic system in an equilibrium state possesses a state variable called the internal energy U whose change dU in a differential process is given by Eq. 19-5.*

The first law of thermodynamics applies to every process in nature that starts in one equilibrium state and ends in another. We say that a system is in an equilibrium state when we can describe it by an appropriate set of constant system parameters such as pressure, volume, temperature, magnetic field, etc. The first law still holds if the states through which the system passes from its initial (equilibrium) state to its final (equilibrium) state are not themselves equilibrium states. For example, we can apply the first law of thermodynamics to the explosion of a firecracker in a closed steel drum.

There are some important questions that the first law cannot answer. For example, although it tells us that energy is conserved in every process, it does not tell us whether any particular process can actually occur. An entirely different generalization, called the second law of thermodynamics, gives us this information, and much of the subject matter of thermodynamics depends on this second law (Chapter 21).

19-7   Some Applications of the First Law of Thermodynamics

We have seen that when a gas expands the work it does on its environment is

$$W = \int p \, dV,$$

---

* See an advanced text in thermodynamics.

where p is the pressure exerted on or by the gas and dV is the differential change in volume of the gas. Consider a special case in which the pressure remains constant while the volume changes by a finite amount, say from V_i to V_f. Then

$$W = \int_{V_i}^{V_f} p \, dV = p \int_{V_i}^{V_f} dV = p(V_f - V_i) \qquad \text{(constant pressure).}$$

A process taking place at constant pressure is called an *isobaric* process. For example, water is heated in the boiler of a steam engine up to its boiling point and is vaporized to steam; then the steam is superheated, all processes proceeding at constant pressure.

In Fig. 19-7 we show an isobaric process. The system is H_2O in a cylindrical container. A frictionless airtight piston is loaded with sand to produce the desired pressure on the H_2O and to maintain it automatically. Heat can be transferred from the environment to the system by a Bunsen burner. If the process continues long enough, the water boils and some is converted to steam; we assume that this occurs. The system may expand, very slowly (quasi-statically) but the pressure it exerts on the piston is automatically always the same, for this pressure must be equal to the constant pressure which the piston exerts on the system. If we wedged the piston so that it could not move, or if we added or took away some sand during the heating process, the process would not be isobaric.

Let us consider the boiling process. We know that substances will change their phase from liquid to vapor at a definite combination of values of pressure and temperature. Water will vaporize at 100° C and atmospheric pressure, for example. For a system to undergo a change of phase heat must be added to it, or taken from it, quite apart from the heat necessary to bring its temperature to the required value. Consider the change of phase of a mass m of liquid to a vapor occurring at

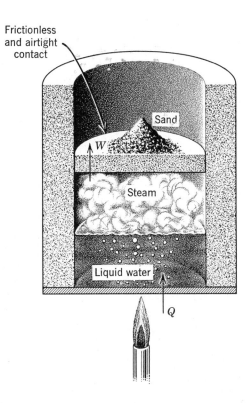

FIGURE 19–7 Water boiling at constant pressure (isobarically). The pressure is kept constant by the weight of the sand and the piston.

constant temperature and pressure. Let V_l be the volume of liquid and V_v the volume of vapor. The work done by this substance in expanding from V_l to V_v at constant pressure is

$$W = p(V_v - V_l).$$

Let L represent the heat of vaporization, that is, the heat needed per unit mass to change a substance from liquid to vapor at constant temperature and pressure. Then the heat absorbed by the mass m during the change of state is

$$Q = mL.$$

From the first law of thermodynamics, we have

$$\Delta U = Q - W$$

so that

$$\Delta U = mL - p(V_v - V_l)$$

for this process.

Example 3. At atmospheric pressure 1.00 gm of water, having a volume of 1.00 cm³, becomes 1671 cm³ of steam when boiled. The heat of vaporization of water is 539 cal/gm at 1 atm. Hence, if $m = 1.00$ gm,

$$Q = mL = 539 \text{ cal.}$$

This quantity, which represents heat added to the system from the environment, is positive.

$$W = p(V_v - V_l) = (1.013 \times 10^5 \text{ nt/meter}^2)[(1671 - 1) \times 10^{-6} \text{ meter}^3]$$

$$= 169.5 \text{ joules.}$$

This quantity, which represents work done by the system on the environment, is positive.

Since 1 cal equals 4.186 joules, $W = 41$ cal. Then,

$$\Delta U = U_v - U_l = mL - p(V_v - V_l) = (539 - 41) \text{ cal}$$

$$= 498 \text{ cal.}$$

This quantity is positive; the internal energy of the system increases during this process. Hence, of the 539 cal needed to boil 1 gm of water (at 100° C and 1 atm), 41 cal go into external work of expansion and 498 cal go into internal energy added to the system. This energy represents the internal work done in overcoming the strong attraction of H_2O molecules for one another in the liquid state.

How would you expect the 80 cal that are needed to melt 1 gm of ice to water (at 0° C and 1 atm) to be shared by the external work and the internal energy?

A process that takes place in such a way that no heat flows into or out of the system is called an *adiabatic process*. Experimentally such processes are achieved either by sealing the system off from its surroundings with heat insulating material or by performing the process quickly. Because the flow of heat is somewhat slow, any process can be made practically adiabatic if it is performed quickly enough.

For an adiabatic process Q equals zero, so that from the first law we obtain

$$\Delta U = U_f - U_i = -W.$$

Hence, the internal energy of a system increases exactly by the amount of work done on the system in an adiabatic process. If work is done by the system in an adiabatic process, the internal energy of the system decreases by exactly the amount of external work it performs. An increase of internal energy usually raises the system's temperature and conversely, a decrease of internal energy usually lowers the system's temperature. A gas that expands adiabatically does external work and its internal energy decreases; such a process is used to attain low temperatures. The increase of temperature during an adiabatic compression of air is well known from the heating of a bicycle pump.

In Fig. 19–8 we show a simple adiabatic process. The system is a gas inside a cylinder made of heat-insulating material. Heat cannot enter the system from its environment or leave the system to the environment. Again we have a pile of sand on a frictionless airtight piston. The only interaction permitted between system and environment is through the performance of work. Such a process can occur when sand is added or removed from the piston, so that the gas can be compressed or can expand against the piston.

Among the many engineering examples of adiabatic processes are the expansion of steam in the cylinder of a steam engine, the expansion of hot gases in an internal combustion engine, and the compression of air in a Diesel engine or in an air compressor. These processes all take place rapidly enough so that only a very small amount of heat can enter or leave the system through its walls during that short time. The compressions and rarefactions in a sound wave are so rapid that the behavior of the transmitting gas is adiabatic.

The most important reason for studying adiabatic processes however, is that ideal engines use processes that are exactly adiabatic. These ideal engines determine the theoretical limits to the operation and capabilities of real engines. We shall look further into this in Chapter 21.

A process of much theoretical interest is that of *free expansion*. This is an adiabatic process in which no work is performed on or by the system. Something like this can be achieved by connecting one vessel which contains a gas to another evacuated vessel with a stopcock connection, the whole system being enclosed with thermal insulation (Fig. 19–9). If the stopcock is suddenly opened, the gas rushes into the vacuum and expands freely. Because of the heat insulation this process is adiabatic,

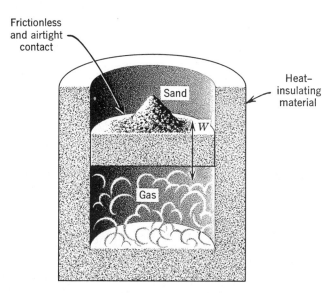

FIGURE 19-8   In an adiabatic process there is no flow of heat to or from the system. Here the walls are insulated and, as sand is removed or added, the volume of the gas changes adiabatically.

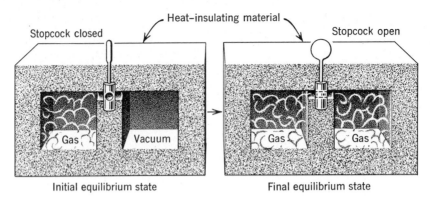

FIGURE 19–9   Free expansion. There is no change of internal energy U since there is no flow of heat Q and no external work W is done.

and because the walls of the vessels are rigid no external work is done on the system. Hence, in the first law we have $Q = 0$ and $W = 0$, so that $U_i = U_f$ for this process. The initial and final internal energies are equal in free expansion.

A free expansion differs from the other examples that we have given in that there is no way to carry it out very slowly (quasi-statically). After we open the stopcock we have no further control over the process. At intermediate states the pressure, volume and temperature do not have unique values characteristic of the system as a whole, that is, the system passes through non-equilibrium states so that we cannot plot the course of the process by a curve on a p-V diagram. We can plot the initial and final states as points on such plots because they are well-defined, equilibrium states. The free expansion is a good example of an irreversible process; see Section 21–2.

QUESTIONS

1. Give examples to distinguish clearly between temperature and heat.

2. (*a*) Show how heat conduction and calorimetry could be explained by the caloric theory. (*b*) List some heat phenomena that cannot be explained by the caloric theory.

3. Can heat be considered a form of stored (or potential) energy? Would such an interpretation contradict the concept of heat as energy in process of transfer because of a temperature difference?

4. Apply Eq. 19–1 to boiling water.

5. Can heat be added to a substance without causing the temperature of the substance to rise? If so, does this contradict the concept of heat as energy in process of transfer because of a temperature difference?

6. Explain the fact that the presence of a large body of water nearby, such as a sea or ocean, tends to moderate the temperature extremes of the climate on adjacent land.

7. Give an example of a process in which no heat is transferred to or from the system but the temperature of the system changes.

8. Both heat conduction and wave propagation involve the transfer of energy. Is there any difference in principle between these two phenomena?

9. When a hot body warms a cool one are their temperature changes equal in magnitude? Give examples. Can one then say that temperature passes from one to the other?

10. What connection is there between an object's feeling hot or cold and its heat capacity? Between this and its thermal conductivity?

11. A block of wood and a block of metal are at the same temperature. When the blocks feel cold the metal feels colder than the wood; when the blocks feel hot the metal feels hotter than the wood. Explain. At what temperature will the blocks feel equally cold or hot?

12. On a winter day the temperature of the inside surface of a wall is much lower than room temperature and that of the outside surface is much higher than the outdoor temperature. Explain.

13. What requirements for thermal conductivity, specific heat capacity, and coefficient of expansion would you want a material to be used as a cooking utensil to satisfy?

14. Is the mechanical equivalent of heat, J, a physical quantity or merely a conversion factor for converting energy from heat units to mechanical units and vice versa?

15. Is the temperature of an isolated system (no interaction with the environment) conserved?

16. Can one distinguish between whether the internal energy of a body was acquired by heat transfer or acquired by performance of work?

17. If the pressure and volume of a system are given is the temperature always uniquely determined?

18. Does a gas do any work when it expands adiabatically? If so, what is the source of the energy needed to do this work?

19. A quantity of gas occupies an initial volume V_0 at a pressure p_0 and a temperature T_0. It expands to a volume V (*a*) at constant temperature and (*b*) at constant pressure. In which case does the gas do more work?

20. Discuss the process of the freezing of water from the point of view of the first law of thermodynamics. Remember that ice occupies a greater volume than an equal mass of water.

21. A thermos bottle contains coffee. The thermos bottle is vigorously shaken. Consider the coffee as the system. (*a*) Does its temperature rise? (*b*) Has heat been added to it? (*c*) Has work been done on it? (*d*) Has its internal energy changed?

PROBLEMS

1(2). A certain substance has a molecular weight of 50 grams/mole. 75 calories of heat are added to a 30-gram sample of this material and its temperature rises from 25° C to 45° C. (*a*) What is the specific heat of this substance? (*b*) What is the molar heat capacity of the substance?
Answer: (*a*) 0.13 cal/gm C°. (*b*) 6.3 cal/mole C°.

2(2). Calculate the specific heat of a metal from the following data. A container made of the metal weighs 8.0 lb and contains 30 lb of water. A 4.0-lb piece of the metal initially at a temperature of 350° F is dropped into the water. The water and container initially have a temperature of 60° F and the final temperature of the entire system is 65° F.

3(2). Two 50-gm ice cubes are dropped into 200 gm of water in a glass. If the water was initially at a temperature of 25° C, and if the ice came directly from a freezer operating at a temperature of −15° C, what will be the final temperature of the drink? The specific heat of ice is approximately 0.50 cal/gm C° in this temperature range and the heat required to melt ice to water is approximately 80 cal/gm.
Answer: 0.0° C.

4(2). A 300-gram copper object is dropped into a 150-gram copper calorimeter containing 220 grams of water at 20° C. This causes the water to boil, 5.0 gm being converted to steam. The heat of vaporization of water is 540 cal/gram. What was the original temperature of the copper?

5(2). A thermometer of mass 0.0550 kg and of specific heat 0.200 cal/gm C° reads 15.0° C. It is then completely immersed in 0.300 kg of water and it comes to the same final temperature as the

water. If the thermometer reads 44.4° C and is accurate, what was the temperature of the water before insertion of the thermometer, neglecting other heat losses?
Answer: 45.5° C.

6(2). By means of a heating coil, energy is transferred at a constant rate to a substance in a thermally insulated container. The temperature of the substance is measured as a function of the time. Show how we can deduce the way in which the heat capacity of the body depends on the temperature from this information.

7(2). An aluminum ring has a diameter of exactly 1.00000 in. at its temperature of 0° C. A copper sphere has a diameter of exactly 1.00200 in. at its temperature of 100° C. The sphere is placed on top of the ring (Fig. 19–10), and the two are allowed to come to thermal equilibrium, no heat being lost to the surroundings. The sphere just passes through the ring at the equilibrium temperature. What is the ratio of the mass of the ring to the mass of the sphere? See Table 18–3.
Answer: 0.43.

8(3). Consider the rod shown in Fig. 19–1. Suppose $L = 25$ cm, $A = 1.0$ cm², and the material is copper. If $T_2 = 125°$ C, $T_1 = 0°$ C, and a steady state is reached, find (*a*) the temperature gradient, (*b*) the rate of heat transfer, and (*c*) the temperature at a point in the rod 10 cm from the high-temperature end.

9(3). Compare the heat flow through two storm doors 2.0 meters high and 0.75 meter wide. (*a*) One door is made with aluminum panels 1.5 mm thick and a 3.0-mm glass pane that covers 75% of its surface (the structural frame is considered to have a negligible area). (*b*) The second door is made entirely of wood averaging 2.5 cm in thickness. Take the temperature drop across the doors to be 60° F. See Table 19–2.
Answer: (*a*) 410 kcal/sec. (*b*) 0.040 kcal/sec.

10(3). (*a*) What is the heat loss in watts/ft² through a glass window 3.0 mm thick if the outside temperature is −20° F and the inside temperature is +72° F? (*b*) If a storm window is installed having the same thickness of glass but with an air gap of 7.5 cm between the two windows, what will be the corresponding heat loss?

11(3). Two identical square rods of metal are welded end-to-end as shown in Fig. 19–11*a*. Assume that 10 cal of heat flows through the rods in 2 min. How long would it take for 10 cal to flow through the rods if they are welded as shown in Fig. 19–11*b*?
Answer: 0.50 min.

12(3). Assume that the thermal conductivity of copper is twice that of aluminum and four times that of brass. Three metal rods, made of copper, aluminum, and brass, respectively, are each 6.0 in. long and 1.0 in. in diameter. These rods are placed end-to-end, with the aluminum between the other two. The free ends of the copper and brass rods are maintained at 100 and 0° C, respectively. Find the equilibrium temperatures of the copper-aluminum junction and the aluminum-brass junction.

13(3). A tank of water has been outdoors in cold weather until a 5.0-cm thick slab of ice has formed on its surface (Fig. 19–12). The air above the ice is at −10° C. Calculate the rate of formation of ice (in cm/hr) on the bottom surface of the ice slab. Take the thermal conductivity, density and heat of fusion of ice to be 0.0040 cal/sec cm C°, 0.92 gm/cm³ and 80 cal/gm respectively. Assume that no heat enters or leaves the water through the walls of the tank.
Answer: 0.39 cm/hr.

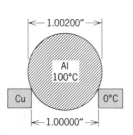

FIGURE 19–10 Problem 7(2).

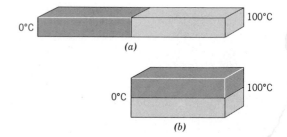

(a)

(b)

FIGURE 19–11 Problem 11(3).

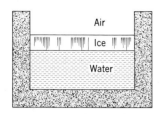

FIGURE 19-12 Problem 13(3).

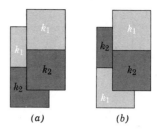

FIGURE 19-13 Problem 15(3).

14(3). Show that in a compound slab the temperature gradient in each portion is inversely proportional to the thermal conductivity.

15(3). Four square pieces of insulation of two different materials, all with the same thickness and area A, are available to cover an opening of area $2A$. This can be done in either of the two ways shown in Fig. 19–13. Which arrangement would give the lower heat flow if $k_2 \neq k_1$?
Answer: Arrangement (*b*).

16(4). In a Joule experiment, a mass of 6.00 kg falls through a height of 50.0 meters and rotates a paddle wheel that stirs 0.600 kg of water. The water is initially at 15.0° C. By how much does its temperature rise?

17(4). An energetic athlete dissipates all the energy in a diet of 4000 kcal per day. If he were to release this energy as heat at a steady rate, how would this heat output compare with the energy output of a 100-watt bulb? (Note: The calorie of nutrition is really a kilocalorie, as we have defined it.)
Answer: 190 watts.

18(4). What quantity of butter (6000 cal/gm) would supply the energy needed for a 160-lb man to ascend to the top of Mt. Everest, elevation 29,000 ft, from sea level?

19(4). Power is supplied at the rate of 0.40 hp for 2.0 min in drilling a hole in a 1.0-lb brass block. (*a*) How much heat is generated? (*b*) What is the rise in temperature of the brass if only 75% of the power warms the brass? (*c*) What happens to the other 25%?
Answer: (*a*) 34 Btu. (*b*) 280 F°.

20(4). Compute the possible increase in temperature for water going over Niagara Falls, 162 ft high. What factors would tend to prevent this possible rise?

21(4). A 2.0-gm lead bullet moving at a speed of 200 meters/sec becomes embedded in a 2.0-kg wooden block suspended as a pendulum bob (a ballistic pendulum). Calculate the rise in temperature of the bullet, assuming that all the absorbed energy raises the bullet's temperature.
Answer: 160 C°.

22(4). A block of ice at 0° C whose mass is initially 50.0 kg slides along a horizontal surface, starting at a speed of 5.38 meters/sec and finally coming to rest after traveling 28.3 meters. Compute the mass of ice melted as a result of the friction between the block and the surface.

23(4). A "flow calorimeter" is used to measure the specific heat of a liquid. Heat is added at a known rate to a stream of the liquid as it passes through the calorimeter at a known rate. Then a measurement of the resulting temperature difference between the inflow and the outflow points of the liquid stream enables us to compute the specific heat of the liquid.

A liquid of density 0.85 gm/cm³ flows through a calorimeter at the rate of 8.0 cm³/sec. Heat is added by means of a 250-watt electric heating coil, and a temperature difference of 15 C° is established in steady-state conditions between the inflow and outflow points. Find the specific heat of the liquid.
Answer: 0.59 cal/gm C°.

24(6). Determine the value of J, the mechanical equivalent of heat, from the following data: 2000 cal of heat are supplied to a system; the system does 3350 joules of external work during that time; the increase in internal energy during the process is 5030 joules.

25(7). Gas within a chamber passes through the cycle shown in Fig. 19–14. Determine the net

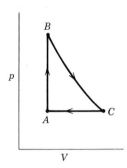

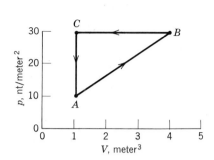

FIGURE 19-14   Problem 25(7).   **FIGURE 19-15**   Problem 26(7).

heat added to the system during process *CA* if $Q_{AB} = 4.77$ cal, $Q_{BC} = 0$ and $W_{BCA} = 15.0$ joules. *Answer:* 1.19 cal.

26(7). Gas within a chamber passes through the processes shown in the p-*V* diagram of Fig. 19-15. Calculate the net heat added to the system during one complete cycle.

27(7). A thermodynamic system is taken from an initial state *A* to another *B* and back again to *A*, via state *C*, as shown by the path *A-B-C-A* in the p-*V* diagram of Fig. 19-16*a*. (*a*) Complete the table in Fig. 19-16*b* by filling in appropriate + or − indications for the signs of the thermodynamic quantities associated with each process. (*b*) Calculate the numerical value of the work done by the system for the complete cycle *A-B-C-A*.

Answer: (*a*) $A \rightarrow B, + + +; B \rightarrow C, + 0 +; C \rightarrow A, − − −.$ (*b*) −20 joules.

28(7). Figure 19-17*a* shows a cylinder containing gas and closed by a movable piston. The cylin-

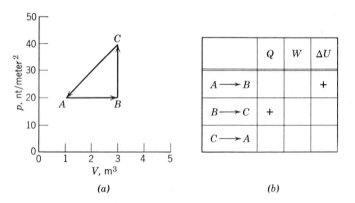

	Q	W	ΔU
$A \longrightarrow B$			+
$B \longrightarrow C$	+		
$C \longrightarrow A$			

(*a*)   (*b*)

FIGURE 19-16   Problem 27(7).

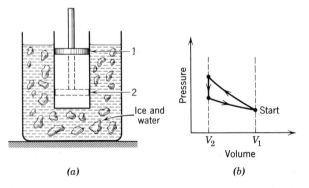

(*a*)   (*b*)

FIGURE 19-17   Problem 28(7).

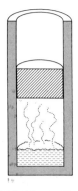

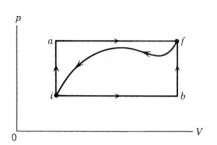

FIGURE 19-18 Problem 29(7).   **FIGURE 19-19** Problem 30(7).

der is submerged in an ice-water mixture. The piston is quickly pushed down from position (1) to position (2). The piston is held at position (2) until the gas is again at 0° C and then is slowly raised back to position (1). Figure 19–17b is a p-V diagram for the process. If 100 gm of ice are melted during the cycle, how much work has been done on the gas?

29(7). A cylinder has a well-fitted 2.0-kg metal piston whose cross-sectional area is 2.0 cm² (Fig. 19–18). The cylinder contains water and steam at 100° C. The piston is observed to fall slowly at a rate of 0.30 cm/sec because heat flows out of the cylinder through the cylinder walls. As this happens, some steam condenses in the chamber. The density of the steam inside the chamber is 6.0×10^{-4} gm/cm³. (a) Calculate the rate of condensation of steam. (b) What is the rate of change of internal energy of the steam and water inside the chamber? (c) At what rate is heat leaving the chamber?

Answer: (a) 3.6×10^{-4} gm/sec. (b) -0.18 cal/sec. (c) 0.19 cal/sec.

30(7). When a system is taken from state i to state f along the path iaf in Fig. 19–19 it is found that $Q = 50$ cal and $W = 20$ cal. Along the path ibf, $Q = 36$ cal. (a) What is W along the path ibf? (b) If $W = -13$ cal for the curved return path fi, what is Q for this path? (c) Take $U_i = 10$ cal. What is U_f? (d) If $U_b = 22$ cal, what is Q for the process ib? For the process bf?

31(7). An iron ball is dropped onto a concrete floor from a height of 10 meters. On the first rebound it rises to a height of 0.50 meter. Assume that all the macroscopic mechanical energy lost in the collision with the floor goes into the ball. The specific heat of iron is 0.12 cal/gm C°. During the collision (a) has heat been added to the ball? (b) Has work been done on it? (c) Has its internal energy changed? If so, by how much? (d) How much has the temperature of the ball risen after the first collision?

Answer: (a) No. (b) Yes. (c) Yes, by $+93$ joules/kg. (d) 0.20 C°.

CHAPTER 20

Kinetic Theory of Gases

20–1 Introduction

Thermodynamics deals only with macroscopic variables, such as pressure, temperature, and volume. Its basic laws, expressed in terms of such quantities, say nothing at all about the fact that matter is made up of atoms. *Statistical mechanics,* however, which deals with the same areas of science that thermodynamics does, presupposes the existence of atoms. Its basic laws are the laws of mechanics, which are applied to the atoms that make up the system.

No existing electronic computer could solve the problem of applying the laws of mechanics individually to every atom in a bottle of oxygen, for instance. Even if the problem could be solved, the results of such calculations would be too voluminous to be useful. Fortunately, the detailed life histories of individual atoms in a gas are not important if we want to calculate only the macroscopic behavior of the gas. We apply the laws of mechanics statistically, then, and we find that we are able to express all the thermodynamic variables as certain averages of atomic properties. For example, the pressure exerted by a gas on the wall of the containing vessel is the average rate per unit area at which the atoms of the gas transfer momentum to the wall as they collide with it. The number of atoms in a macroscopic system is usually so large that such averages are very sharply defined quantities indeed.

We can apply the laws of mechanics statistically to assemblies of atoms at two different levels. At the level called *kinetic theory* we proceed in a rather physical way, using relatively simple mathematical averaging techniques. In this chapter we will use these methods to enlarge our understanding of pressure, temperature, specific heat, and internal energy at the atomic level.[*] In this book we apply kinetic theory

[*] See "John James Waterston and the Kinetic Theory of Gases" by S. G. Brush, in *American Scientist,* June 1961, for an interesting aspect of the history of kinetic theory.

to gases only, because the interactions between atoms in gases are much weaker than in liquids and solids; this greatly simplifies the mathematical difficulties.

At another level, we can apply the laws of mechanics statistically, using techniques that are more formal and abstract than those of kinetic theory. This approach, developed by J. Willard Gibbs (1839–1903) and by Ludwig Boltzmann (1844–1906) among others, is called *statistical mechanics,* a term that includes kinetic theory as a sub-branch. Using these methods one can derive the laws of thermodynamics, thus establishing that science as a branch of mechanics. The fullest flowering of statistical mechanics (*quantum statistics*) involves the statistical application of the laws of quantum mechanics—rather than those of classical mechanics—to many-atom systems.*

20–2   Ideal Gas—A Macroscopic Description

Let a mass $\mathfrak{M}$ of a gas be confined in a container of volume V. The density ρ of the gas is $\mathfrak{M}/V$ and it is clear that we can reduce ρ either by removing some gas from the container (reducing $\mathfrak{M}$) or by putting the gas in a larger container (increasing V). We find from experiment that, at low enough densities, all gases tend to show a certain simple relationship among the thermodynamic variables p, V, and T. This suggests the concept of an *ideal gas,* one that would have the same simple behavior under all conditions of temperature and pressure. In this section we give a macroscopic or thermodynamic definition of an ideal gas. In Section 20–3 we will define an ideal gas microscopically, from the standpoint of kinetic theory, and we will see what we can learn by comparing these two approaches.

Given a mass $\mathfrak{M}$ of any gas in a state of thermal equilibrium we can measure its pressure p, its temperature T, and its volume V. For low enough values of the density experiment shows that (1) for a given mass of gas held at a constant temperature, the pressure is inversely proportional to the volume (Boyle's law), and (2) for a given mass of gas held at a constant pressure, the volume is directly proportional to the temperature (law of Charles and Gay-Lussac). We can summarize these two experimental results by the relation

$$\frac{pV}{T} = \text{a constant} \quad \text{(for a fixed mass of gas)}. \tag{20–1}$$

The volume occupied by a gas (real or ideal) at a given pressure and temperature is proportional to its mass. Thus the constant in Eq. 20–1 must also be proportional to the mass of the gas. We therefore write the constant in Eq. 20–1 as μR, where μ is the mass of the gas in moles† and R is a constant that must be determined for each gas by experiment. Experiment shows that, at low enough densities, R has the same value for all gases, namely

$$R = 8.314 \text{ joule/mole K}° = 1.986 \text{ cal/mole K}°.$$

---

* See *Thermal Physics* by Philip M. Morse, W. A. Benjamin, Inc., New York, 1962, for a fuller treatment of thermodynamics, kinetic theory, and (particularly) statistical mechanics proper, then we can give here.

† A *mole* of any substance is that mass of the substance that contains a specified number of molecules, namely, 6.02217×10^{23}, called Avogadro's number. This number is the result of the defining relation that one mole of carbon (actually, of the isotope C^{12}) shall have a mass of 12 gm, exactly. The *molecular weight M* of a substance is a dimensionless quantity expressing the number of grams per mole of that substance. Thus the molecular weight of oxygen is 32.0 gm/mole. Although the mole is a unit of mass, we cannot translate it into, say, grams, until we know the chemical composition of the substance; for this reason we find it convenient to use a special symbol (μ) for masses expressed in moles.

R is called the *universal gas constant.* We then write Eq. 20–1 as

$$pV = \mu RT \qquad (20\text{–}2)$$

and we define an ideal gas as one that obeys this relation under all conditions. There is no such thing as a truly ideal gas, but it remains a useful and simple concept connected with reality by the fact that all real gases approach the ideal gas abstraction in their behavior if the density is low enough. Equation 20–2 is called the *equation of state* of an ideal gas.

If we could fill the bulb of an (ideal) constant-volume gas thermometer with an ideal gas, we see from Eq. 20–2 that we could define temperature in terms of its pressure readings, that is,

$$T = 273.16°\ \text{K} \frac{p}{p_{tr}} \quad \text{(ideal gas)}.$$

Here p_{tr} is the gas pressure at the triple point of water, at which the temperature T_{tr} is 273.16° K by definition. In practice we must fill our thermometer with a real gas and measure the temperature by extrapolating to zero density using Eq. 18–3,

$$T = 273.16°\ \text{K} \lim_{p_{tr}\to 0} \frac{p}{p_{tr}} \quad \text{(real gas)}.$$

Example 1. A cylinder contains oxygen at a temperature of 20°C and a pressure of 15 atm in a volume of 100 liters. A piston is lowered into the cylinder, decreasing the volume occupied by the gas to 80 liters and raising the temperature to 25° C. Assuming oxygen to behave like an ideal gas under these conditions, what then is the gas pressure?

From Eq. 20–1, since the mass of gas remains unchanged, we may write

$$\frac{p_i V_i}{T_i} = \frac{p_f V_f}{T_f}.$$

Our initial conditions are

$$p_i = 15\ \text{atm}, \qquad T_i = 293°\ \text{K}, \qquad V_i = 100\ \text{liters}.$$

Our final conditions are

$$p_f = ?, \qquad T_f = 298°\ \text{K}, \qquad V_f = 80\ \text{liters}.$$

Hence,

$$p_f = \left(\frac{T_f}{V_f}\right)\left(\frac{p_i V_i}{T_i}\right) = \left(\frac{298°\ \text{K}}{80\ \text{liters}}\right)\left(\frac{15\ \text{atm} \times 100\ \text{liters}}{293°\ \text{K}}\right) = 19\ \text{atm}.$$

Example 2. Calculate the work per mole done by an ideal gas which expands isothermally, that is, at constant temperature, from an initial volume V_i to a final volume V_f.

The work done may be represented as

$$W = \int_{V_i}^{V_f} p\ dV.$$

From the ideal gas law we have

$$p = \frac{\mu RT}{V},$$

so that W/μ, the work per mole, is

$$\frac{W}{\mu} = \frac{RT}{V}\ dV.$$

The temperature is constant so that

$$\frac{W}{\mu} = RT \int_{V_i}^{V_f} \frac{dV}{V} = RT \ln \frac{V_f}{V_i}$$

is the work per mole done by an ideal gas in an isothermal expansion at temperature T from an initial volume V_i to a final volume V_f.

Notice that when the gas expands, so that $V_f > V_i$, the work done by the gas is positive; when the gas is compressed, so that $V_f < V_i$, the work done by the gas is negative. This is consistent with the sign convention adopted for W in the first law of thermodynamics. The work done is shown as the shaded area in Fig. 20–1. The curved line is an isotherm, that is, a curve giving the relation of p to V at a constant temperature.

In practice, how can we keep an expanding or contracting gas at constant temperature?

20–3 Ideal Gas—A Microscopic Description

From the microscopic point of view we define an ideal gas by making the following assumptions; it will then be our task to apply the laws of classical mechanics statistically to the gas atoms and to show that our microscopic definition is consistent with the macroscopic definition of the preceding section:

1. *A gas consists of particles called molecules.* Depending on the gas, each molecule will consist of one atom or a group of atoms. If the gas is an element or a compound and is in a stable state, we consider all its molecules to be identical.

2. *The molecules are in random motion and obey Newton's laws of motion.* The molecules move in all directions and with various speeds. In computing the properties of the motion, we assume that Newtonian mechanics works at the microscopic level. As for all our assumptions, this one will stand or fall depending on whether or not the experimental facts it predicts are correct.

3. *The total number of molecules is large.* The direction and speed of motion of any one molecule may change abruptly on collision with the wall or another molecule. Any particular molecule will follow a zigzag path because of these collisions. However, because there are so many molecules we assume that the resulting large number of collisions maintains the over-all distribution of molecular velocities and the randomness of the motion.

4. *The volume of the molecules is a negligibly small fraction of the volume occupied by the gas.* Even though there are many molecules, they are extremely small. We know that the volume occupied by a gas can be changed through a large range of values with little difficulty, and that when a gas condenses the volume occupied by the

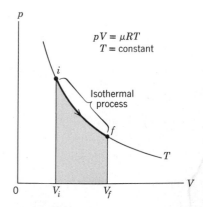

FIGURE 20–1   Example 2. The shaded area represents the work done by μ moles of gas in expanding from V_i to V_f with the temperature held constant.

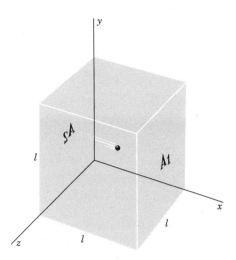

FIGURE 20-2   A cubical box of side l, containing an ideal gas. A molecule is shown moving toward A_1.

liquid may be thousands of times smaller than that of the gas. Hence, our assumption is plausible.

5. *No appreciable forces act on the molecules except during a collision.* To the extent that this is true a molecule moves with uniform velocity between collisions. Because we have assumed the molecules to be so small, the average distance between molecules is large compared to the size of a molecule. Hence, we assume that the range of molecular forces is comparable to the molecular size.

6. *Collisions are elastic and are of negligible duration.* Collisions between molecules and with the walls of the container conserve momentum and (we assume) kinetic energy. Because the collision time is negligible compared to the time spent by a molecule between collisions, the kinetic energy which is converted to potential energy during the collision is available again as kinetic energy after such a brief time that we can ignore this exchange entirely.

20–4   Kinetic Calculation of the Pressure

Let us now calculate the pressure of an ideal gas from kinetic theory. To simplify matters, we consider a gas in a cubical vessel whose walls are perfectly elastic. Let each edge be of length l. Call the faces normal to the x-axis (Fig. 20–2) A_1 and A_2, each of area l^2. Consider a molecule which has a velocity **v**. We can resolve **v** into components v_x, v_y, and v_z in the directions of the edges. If this particle collides with A_1, it will rebound with its x-component of velocity reversed. There will be no effect on v_y or v_z, so that the change Δp in the particle's momentum will be

$$\Delta p = p_f - p_i = (-mv_x) - (mv_x) = -2mv_x,$$

normal to A_1. Hence, the momentum imparted to A_1 will be $2mv_x$, since the total momentum is conserved.

Suppose that this particle reaches A_2 without striking any other particle on the way. The time required to cross the cube will be l/v_x. At A_2 it will again have its x-component of velocity reversed and will return to A_1. Assuming no collisions in between, the round trip will take a time $2l/v_x$. Hence, the number of collisions per unit time this particle makes with A_1 is $v_x/2l$, so that the rate at which it transfers momentum to A_1 is

$$2mv_x \frac{v_x}{2l} = \frac{mv_x^2}{l}.$$

To obtain the total force on A_1, that is, the rate at which momentum is imparted to A_1 by all the gas molecules, we must sum up mv_x^2/l for all the particles. Then, to find the pressure, we divide this force by the area of A_1, namely l^2.

If m is the mass of each molecule, we have

$$p = \frac{m}{l^3}(v_{x1}^2 + v_{x2}^2 + \cdots),$$

where v_{x1} is the x-component of the velocity of particle 1, v_{x2} is that of particle 2, etc. If N is the total number of particles in the container and n is the number per unit volume, then $N/l^3 = n$ or $l^3 = N/n$. Hence,

$$p = mn\left(\frac{v_{x1}^2 + v_{x2}^2 + \cdots}{N}\right).$$

But mn is simply the mass per unit volume, that is, the density ρ. The quantity $(v_{x1}^2 + v_{x2}^2 + \cdots)/N$ is the average value of v_x^2 for all the particles in the container. Let us call this $\overline{v_x^2}$. Then

$$p = \rho\overline{v_x^2}.$$

For any particle $v^2 = v_x^2 + v_y^2 + v_z^2$. Because we have many particles and because they are moving entirely at random, the average values of v_x^2, v_y^2, and v_z^2 are equal and the value of each is exactly one-third the average value of v^2. There is no preference among the molecules for motion along any one of the three axes. Hence, $\overline{v_x^2} = \tfrac{1}{3}\overline{v^2}$, so that

$$p = \rho\overline{v_x^2} = \tfrac{1}{3}\rho\overline{v^2}. \tag{20-3}$$

Although we derived this result by neglecting collisions between particles, the result is true even when we consider collisions. Because of the exchange of velocities in an elastic collision between identical particles, there will always be some one molecule that will collide with A_2 with momentum mv_x corresponding to the one that left A_1 with this same momentum. Also, the time spent during collisions is negligible compared to the time spent between collisions. Hence, our neglect of collisions is merely a convenient device for calculation. Likewise, we could have chosen a container of any shape—the cube merely simplifies the calculation. Although we have calculated the pressure exerted only on the side A_1, if we neglect the weight of the gas, it follows from Pascal's law that the pressure is the same on all sides and everywhere in the interior.

The square root of $\overline{v^2}$ is called the *root-mean-square* speed of the molecules and is a kind of average molecular speed (see Section 20–9). Using Eq. 20–3, we can calculate this root-mean-square speed from measured values of the pressure and density of the gas. Thus,

$$v_{\text{rms}} = \sqrt{\overline{v^2}} = \sqrt{\frac{3p}{\rho}}. \tag{20-4}$$

In Eq. 20–3 we relate a macroscopic quantity (the pressure p) to an average value of a microscopic quantity (that is, to v^2 or v_{rms}^2. However, averages can be taken over short times or over long times, over small regions of space or large regions of space. The average computed in a small region for a short time might depend on the time or region chosen, so that the values obtained in this way may fluctuate.

This could happen in a gas of very low density, for example. We can ignore fluctuations, however, when the number of particles in the system is large enough.

Example 3. Calculate the root-mean-square speed of hydrogen molecules at 0.00° C and 1.00-atm pressure, assuming hydrogen to be an ideal gas. Under these conditions hydrogen has a density ρ of 8.99×10^{-2} kg/meter^3. Then, since $p = 1.00$ atm $= 1.01 \times 10^5$ nt/meter^2,

$$v_{\text{rms}} = \sqrt{\frac{3p}{\rho}} = 1840 \text{ meters/sec.}$$

This is of the order of a mile per second, or 3600 miles/hr.

Table 20–1 gives the results of similar calculations for some gases at 0°C. These molecular speeds are of the same order as the speed of sound at the same temperature. For example, in air at 0° C, $v_{\text{rms}} = 485$ meters/sec and the speed of sound is 331 meters/sec; in hydrogen $v_{\text{rms}} = 1838$ meters/sec and sound travels at 1286 meters/sec; in oxygen $v_{\text{rms}} = 461$ meters/sec and sound travels at 317 meters/sec. These results are to be expected in terms of our model of a gas; see Problem 7. We must distinguish between the speeds of individual molecules, described by v_{rms}, and the very much lower speed at which one gas diffuses into another. Thus, if we open a bottle of ammonia in one corner of a room, we smell ammonia in the opposite corner only after an easily measurable time lag. The diffusion speed is slow because large numbers of collisions with air molecules greatly reduce the tendency of ammonia molecules to spread themselves uniformly throughout the room.

Example 4. Assume that the speed of sound in a gas is the same as the root-mean-square speed of the molecules (this is only roughly true; see above), and show how the speed of sound for an ideal gas depends on the temperature.

The density of a gas is

$$\rho = \frac{\mathfrak{M}}{V} = \frac{\mu M}{V}$$

Table 20–1

Gas	v_{rms} (at 0° C), (meters/sec)	Molecular weight (gm/mole)	Translational kinetic energy per mole (at 0° C),* $\frac{1}{2}Mv_{\text{rms}}^2$ (joules/mole)
O_2	461	32	3400
N_2	493	28	3390
Air	485	28.8	3280
CO	493	28	3390
H_2	1838	2.02	3370
He	1311	4.0	3430
CO_2	393	44	3400
H_2O	615	18	3400
Ne	584	20.1	3420

* The molecular weight and the mole are defined on page 376. We will discuss the last column in this table in the next section.

in which $\mathfrak{M}$ is the mass of the gas, M is the molecular weight (grams/mole), and μ is the mass in moles. Combining this with the ideal gas law

$$pV = \mu RT$$

yields

$$\frac{p}{\rho} = \frac{RT}{M}.$$

We obtain from Eq. 20–4

$$v_{\mathrm{rms}} = \sqrt{\frac{3p}{\rho}} = \sqrt{\frac{3RT}{M}},$$

so that the speed of sound v_1 at a temperature T_1 is related to the speed of sound v_2 in the same gas at a temperature T_2 by

$$\frac{v_1}{v_2} = \sqrt{\frac{T_1}{T_2}}.$$

For example, if the speed of sound at 273° K is 332 meters/sec in air, its speed in air at 300° K will be

$$\sqrt{\tfrac{300}{273}} \times 332 \text{ meters/sec} = 348 \text{ meters/sec}.$$

Would our result change if the speed of sound were proportional to, rather than equal to, the root-mean-square speed of the molecules of a gas? See Problem 7.

20–5   Kinetic Interpretation of Temperature

If we multiply each side of Eq. 20–3 by the volume V, we obtain

$$pV = \tfrac{1}{3}\rho V v_{\mathrm{rms}}^2,$$

where ρV is simply the total mass $\mathfrak{M}$ of gas, ρ being the density. We can also write the mass of gas as μM, in which μ is the mass in moles and M is the molecular weight. Making this substitution yields

$$pV = \tfrac{1}{3}\mu M v_{\mathrm{rms}}^2.$$

The quantity $\tfrac{1}{3}\mu M v_{\mathrm{rms}}^2$ is two-thirds the total kinetic energy of translation of the molecules, that is, $\tfrac{2}{3}(\tfrac{1}{2}\mu M v_{\mathrm{rms}}^2)$. We can then write

$$pV = \tfrac{2}{3}(\tfrac{1}{2}\mu M v_{\mathrm{rms}}^2).$$

The equation of state of an ideal gas is

$$pV = \mu RT.$$

Combining these two expressions, we obtain

$$\tfrac{1}{2}M v_{\mathrm{rms}}^2 = \tfrac{3}{2}RT. \qquad (20\text{–}5)$$

That is, *the total translational kinetic energy per mole of the molecules of an ideal gas is proportional to the temperature.* We may say that this result, Eq. 20–5, is necessary to fit the kinetic theory to the equation of state of an ideal gas, or we may consider Eq. 20–5 as a definition of gas temperature on a kinetic theory or microscopic basis. In either case, we gain some insight into the meaning of temperature for gases.

The temperature of a gas is related to the total translational kinetic energy measured with respect to the center of mass of the gas. The kinetic energy associated with the motion of the center of mass of the gas has no bearing on the gas temperature. The temperature of a gas in a container does not increase when we put the container on a moving train.

Let us now divide each side of Eq. 20–5 by Avogadro's number, N_0, which (see page 376, footnote) is the number of molecules per mole of a gas. Thus $M/N_0 \; (= m)$ is the mass of a single molecule and we have

$$\tfrac{1}{2}(M/N_0)v_{\mathrm{rms}}^2 = \tfrac{1}{2}mv_{\mathrm{rms}}^2 = \tfrac{3}{2}(R/N_0)T.$$

Now $\tfrac{1}{2}mv_{\mathrm{rms}}^2$ is the average translational kinetic energy per molecule. The ratio R/N_0—which we call k, the *Boltzmann constant*—plays the role of the gas constant per molecule. We have

$$\tfrac{1}{2}mv_{\mathrm{rms}}^2 = \tfrac{3}{2}kT \qquad\qquad (20\text{–}6)$$

in which

$$k = \frac{R}{N_0} = \frac{8.317 \text{ joules/mole K}^\circ}{6.022 \times 10^{23} \text{ molecules/mole}} = 1.381 \times 10^{-23} \text{ joule/molecule K}^\circ.$$

We shall return to Boltzmann's constant in Section 20–9.

In the last column of Table 20–1 we list calculated values of $\tfrac{1}{2}Mv_{\mathrm{rms}}^2$. As Eq. 20–5 predicts, this quantity (the translational kinetic energy per mole) has (closely) the same value for all gases at the same temperatures, 0° C in this case. From Eq. 20–6 we conclude that at the same temperature T the ratio of the root-mean-square speeds of molecules of two different gases is equal to the square root of the inverse ratio of their masses. That is, from

$$T = \frac{2}{3k}\frac{m_1 v_{1\mathrm{rms}}^2}{2} = \frac{2}{3k}\frac{m_2 v_{2\mathrm{rms}}^2}{2}$$

we obtain

$$\frac{v_{1\mathrm{rms}}}{v_{2\mathrm{rms}}} = \sqrt{\frac{m_2}{m_1}}. \qquad\qquad (20\text{–}7)$$

We can apply Eq. 20–7 to the diffusion of two different gases in a container with porous walls placed in an evacuated space. The lighter gas, whose molecules move more rapidly on the average, will escape faster than the heavier one. The ratio of the numbers of molecules of the two gases which find their way through the porous walls for a short time interval is equal to the square root of the inverse ratio of their masses, $\sqrt{m_2/m_1}$. This diffusion process is one method of separating (readily fissionable) U^{235} (0.7% abundance) from a normal sample of uranium containing mostly U^{238} (99.3% abundance).

20–6 Specific Heats of an Ideal Gas

We picture the molecules in an ideal gas as hard elastic spheres; that is, we assume that there are no forces between the molecules except during collisions and that the molecules are not deformed by collisions. If this is so there is no internal potential energy and the internal energy of an ideal gas is entirely kinetic. We have already found that the average translational kinetic energy per molecule is $\tfrac{3}{2}kT$, so that the internal energy U of an ideal monatomic gas containing N molecules is

$$U = \tfrac{3}{2}NkT = \tfrac{3}{2}\mu RT. \qquad\qquad (20\text{–}8)$$

This prediction of kinetic theory says that *the internal energy of an ideal gas is proportional to the Kelvin temperature and depends only on the temperature,* being independent of pressure and volume. With this result we can now obtain information about the specific heats of an ideal gas.

The specific heat of a substance is the heat required per unit mass per unit temperature change. A convenient unit of mass is the mole. The corresponding specific heat is called the *molar heat capacity* and is represented by C. Only two varieties of molar heat capacity are important for gases, namely, that at constant volume, C_v, and that at constant pressure, C_p.

Let us confine a certain number of moles of an ideal gas in a piston-cylinder arrangement as in Fig. 20–3a. The cylinder rests on a heat reservoir whose temperature can be raised or lowered at will, so that we may add heat to the system or remove it, as we wish. The gas has a pressure p such that its upward force on the (frictionless) piston just balances the weight of the piston and its sand load. The state of the system is represented by point a in the p-V diagram of Fig. 20–3d; this diagram shows two isothermal lines, all points on one corresponding to a temperature T and all points on the other to a (higher) temperature $T + \Delta T$.

Now let us raise the temperature of the system by ΔT, by slowly increasing the reservoir temperature. As we do this let us add sand to the piston so that the volume V does not change. This constant-volume process carries the system from the initial state of Fig. 20–3a to the final state of Fig. 20–3c. Equivalently, it goes from point a to point c in Fig. 20–3d. Let us apply the first law of thermodynamics

$$\Delta Q = \Delta U + \Delta W$$

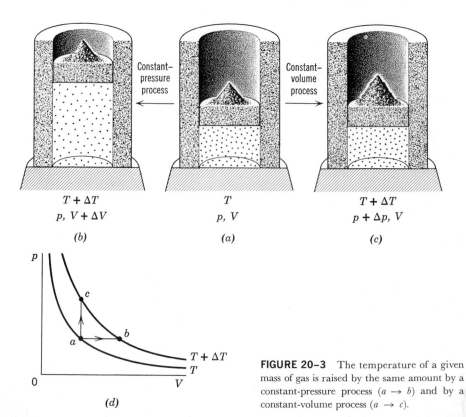

$T + \Delta T$
$p,\ V + \Delta V$

(b)

Constant–pressure process

T
$p,\ V$

(a)

Constant–volume process

$T + \Delta T$
$p + \Delta p,\ V$

(c)

(d)

FIGURE 20–3 The temperature of a given mass of gas is raised by the same amount by a constant-pressure process ($a \rightarrow b$) and by a constant-volume process ($a \rightarrow c$).

to this process. By definition of C_v we have $\Delta Q = \mu C_v \, \Delta T$. Also, $\Delta W (= p \, \Delta V) = 0$ because $\Delta V = 0$. Thus

$$\Delta U = \mu C_v \, \Delta T. \tag{20–9}$$

Let us restore the system to its original state and again raise its temperature by ΔT, this time leaving the sand load undisturbed so that the pressure p does not change. This constant-pressure process carries the system from the initial state of Fig. 20–3a to the final state of Fig. 20–3b. Equivalently, it goes from point a to point b in Fig. 20–3d. Let us apply the first law to this process. By definition of C_p we have $\Delta Q = \mu C_p \, \Delta T$. Also, $\Delta W = p \, \Delta V$. Now for an ideal gas, U depends only on the temperature. Since processes $a \rightarrow b$ and $a \rightarrow c$ in Fig. 20–3 involve the same change ΔT in temperature, they must also involve the same change ΔU in internal energy, namely, that given by Eq. 20–9. Thus for the constant-pressure process the first law yields

$$\mu C_p \, \Delta T = \mu C_v \, \Delta T + p \, \Delta V.$$

Let us apply the equation of state $pV = \mu RT$ to the constant-pressure process $a \rightarrow b$. For p constant we have, by taking differences,

$$p \, \Delta V = \mu R \, \Delta T.$$

Combining these equations yields

$$C_p - C_v = R. \tag{20–10}$$

This shows that the molar heat capacity of an ideal gas at constant pressure is always larger than that at constant volume by an amount equal to the universal gas constant $R(= 8.31$ joules/mole K° or 1.99 cal/mole K°). Although Eq. 20–10 is exact only for an ideal gas, it is nearly true for real gases at moderate pressure (see Table 20–2). Notice that in obtaining this result we did not use the specific relation $U = \frac{3}{2} \mu RT$, but only the fact that U depends on temperature alone.

If we can compute C_v, then Eq. 20–10 will give us C_p and vice versa. We can obtain C_v by combining Eq. 20–9 with the kinetic theory result for the internal energy of an ideal gas, $U = \frac{3}{2} \mu RT$ (Eq. 20–8). Thus, in the limit of differential changes,

$$C_v = \frac{dU}{\mu \, dT} = \frac{d}{\mu \, dT} [\tfrac{3}{2} \mu RT] = \tfrac{3}{2} R. \tag{20–11}$$

This result (about 3 cal/mole K°) turns out to be rather good for monatomic gases. However, it is in serious disagreement with values obtained for diatomic and polyatomic gases (see Table 20–2). This suggests that Eq. 20–8 is not generally correct. Since that relation followed directly from the kinetic theory model, we conclude that we must change the model if kinetic theory is to survive as a useful approximation to the behavior of real gases.

■ **Example 5.** Show that for an ideal gas undergoing an adiabatic process $pV^\gamma = $ a constant, where $\gamma = C_p/C_v$.

Let us apply the first law of thermodynamics

$$\Delta Q = \Delta U + \Delta W.$$

For an adiabatic process $\Delta Q = 0$. For ΔW we put $p\,\Delta V$. Since the gas is assumed to be ideal, U depends only on temperature and, from Eq. 20–9, $\Delta U = \mu C_v\,\Delta T$. With these substitutions we have

$$0 = \mu C_v\,\Delta T + p\,\Delta V$$

or

$$\Delta T = -\frac{p\,\Delta V}{\mu C_v}.$$

For an ideal gas $pV = \mu RT$, so that, if p, V, and T are allowed to take on small variations,

$$p\,\Delta V + V\,\Delta p = \mu R\,\Delta T$$

or

$$\Delta T = \frac{p\,\Delta V + V\,\Delta p}{\mu R}.$$

Equating these two expressions and using Eq. 20–10 ($C_p - C_v = R$), we obtain, after some re-arrangement,

$$p\,\Delta V C_p + V\,\Delta p C_v = 0$$

Dividing by pVC_v and recalling that, by definition, $C_p/C_v = \gamma$, we get

$$\frac{\Delta p}{p} + \gamma\frac{\Delta V}{V} = 0.$$

In the limiting case of differential changes this reduces to

$$\frac{dp}{p} + \gamma\frac{dV}{V} = 0,$$

which (assuming γ to be constant) we can integrate as

$$\ln p + \gamma \ln V = \text{a constant}$$

or

$$pV^\gamma = \text{a constant.} \tag{20–12}$$

The value of the constant is proportional to the quantity of gas. In Fig. 20–4 we compare the isothermal and adiabatic behaviors of an ideal gas.

20–7 Equipartition of Energy

A modification of the kinetic theory model designed to explain the specific heats of gases was first suggested by Clausius in 1857. Recall that in our model we assumed a molecule to behave like a hard elastic sphere and we treated its kinetic energy as purely translational. The specific heat prediction was satisfactory for monatomic molecules. Further, because of the success of this simple model in other respects in predicting the correct behavior of gases of all kinds over wide temperature ranges, we feel confident that it is the average kinetic energy of translation which determines what we measure as the temperature of a gas.

However, in the case of specific heats we are concerned with all possible ways of absorbing energy and we must ask whether or not a molecule can store energy internally, that is, in a form other than kinetic energy of translation. This would certainly be so if we pictured a molecule, not as a rigid particle, but as an object with internal structure. For then a molecule could rotate and vibrate as well as move with translational motion. In collisions, the rotational and vibrational modes of motion could be excited, and this would contribute to the internal energy of the gas. Here then is a model which enables us to modify the kinetic theory formula for the internal energy of a gas.

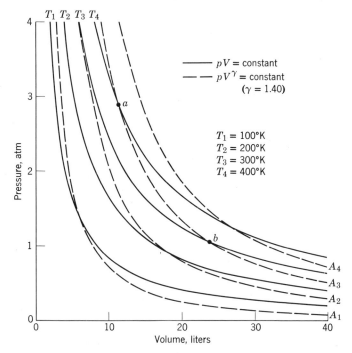

FIGURE 20–4 T_1, T_2, T_3, and T_4 show how the pressure of one mole of an ideal gas changes as its volume is changed, the temperature being held constant (isothermal process). A_1, A_2, A_3, and A_4 show how the pressure of an ideal gas changes as its volume is changed, no heat being allowed to flow to or from the gas (adiabatic process). An adiabatic increase in volume (for example going from a to b along A_3) is always accompanied by a decrease in temperature, since at a, $T = 400°$ K, whereas at b, $T = 300°$ K.

Let us now find the total energy of a system containing a large number of such molecules, where each molecule is thought of as an object having internal structure. The energy will consist of kinetic energy of translation, kinetic energy of vibration of the atoms in a molecule, and potential energy of vibration of the atoms in a molecule. Although other kinds of energy contributions exist, such as magnetic, for gases we can describe the total energy quite accurately by terms such as these. We can show from statistical mechanics that when the number of particles is large and Newtonian mechanics holds, all these terms have the same average value, and this average value depends only on the temperature. In other words, the available energy depends only on the temperature and distributes itself in equal shares to each of the independent ways in which the molecules can absorb energy. This theorem, stated here without proof, is called the *equipartition of energy* and was deduced by Clerk Maxwell. Each such independent mode of energy absorption is called a *degree of freedom.*

From Eq. 20–8 we know that the kinetic energy of translation per mole of gaseous molecules is $\frac{3}{2}RT$. The kinetic energy of translation per mole is the sum of three terms, however, namely $\frac{1}{2}M\overline{v_x^2}$, $\frac{1}{2}M\overline{v_y^2}$, and $\frac{1}{2}M\overline{v_z^2}$. The theorem of equipartition requires that each such term contribute the same amount to the total energy per mole, or $\frac{1}{2}RT$ per degree of freedom.

For monatomic gases the molecules have only translational motion (no internal structure in kinetic theory), so that $U = \frac{3}{2}\mu RT$. It follows from Eq. 20–11 that $C_v = \frac{3}{2}R \cong 3$ cal/mole K°. Then from Eq. 20–10, $C_p = \frac{5}{2}R$, and the ratio of specific heat is

$$\gamma = \frac{C_p}{C_v} = \frac{5}{3} = 1.67.$$

For a diatomic gas we can think of each molecule as having a dumbbell shape (two spheres joined by a rigid rod). Such a molecule can rotate about any one of three mutually perpendicular axes. However, the rotational inertia about an axis along the rigid rod should be negligible compared to that about axes perpendicular to the rod. This is consistent with our assumption that a monatomic molecule could not rotate. Implicitly, we have adopted a point mass model of the atom. The rotational energy, therefore, should consist of only two terms, such as $\frac{1}{2}I\omega_y{}^2$ and $\frac{1}{2}I\omega_z{}^2$. Each rotational degree of freedom is required by equipartition to contribute the same energy as each translational degree, so that for a diatomic gas having both rotational and translational motion,

$$U = 3\mu(\tfrac{1}{2}RT) + 2\mu(\tfrac{1}{2}RT) = \tfrac{5}{2}\mu RT,$$

or
$$C_v = \frac{dU}{\mu\,dT} = \tfrac{5}{2}R \cong 5 \text{ cal/mole K}°$$

and
$$C_p = C_v + R = \tfrac{7}{2}R,$$

or
$$\gamma = \frac{C_p}{C_v} = \frac{7}{5} = 1.40.$$

For polyatomic gases, each molecule contains three or more spheres (atoms) joined together by rods in our model, so that the molecule is capable of rotating energetically about each of three mutually perpendicular axes. Hence, for a polyatomic gas having both rotational and translational motion,

$$U = 3\mu(\tfrac{1}{2}RT) + 3\mu(\tfrac{1}{2}RT) = 3\mu RT,$$

or
$$C_v = \frac{dU}{\mu\,dT} = 3R = 6 \text{ cal/mole K}°,$$

and
$$C_p = 4R,$$

or
$$\gamma = \frac{C_p}{C_v} = 1.33.$$

Let us now turn to experiment to test these ideas. In Table 20–2 we list the experimentally determined molar heat capacities for common gases at 20° C and 1.0 atm. Notice that for monatomic and diatomic gases the values of C_v, C_p, and γ are close to the ideal gas predictions. In some diatomic gases, like chlorine, and in most polyatomic gases the specific heats are larger than the predicted values. Even γ shows no simple regularity for polyatomic gases. This suggests that our model is not yet close enough to reality.*

We have not yet considered energy contributions from the vibrations of the atoms in diatomic and polyatomic molecules. That is, we can modify the dumbbell model and join the spheres instead by springs. This new model will greatly improve our results in some cases. However, instead of having a theoretical model for all gases,

---

* The last column of Table 19–1 suggests that similar considerations would pertain to the specific heats of solids. This is indeed the case, but we will leave the details for more advanced texts (or your imagination).

Table 20–2

Type of Gas	Gas	C_p (cal/mole K°)	C_v (cal/mole K°)	$C_p - C_v$	$\gamma = C_p/C_v$
Monatomic	He	4.97	2.98	1.99	1.67
	A	4.97	2.98	1.99	1.67
Diatomic	H_2	6.87	4.88	1.99	1.41
	O_2	7.03	5.03	2.00	1.40
	N_2	6.95	4.96	1.99	1.40
	Cl_2	8.29	6.15	2.14	1.35
Polyatomic	CO_2	8.83	6.80	2.03	1.30
	SO_2	9.65	7.50	2.15	1.29
	C_2H_6	12.35	10.30	2.05	1.20
	NH_3	8.80	6.65	2.15	1.31

we now require an empirical model which differs from gas to gas. We can obtain a reasonably good picture of molecular behavior this way and the empirical model is therefore useful; however it ceases to be fundamental.

To see this more clearly, let us consider Fig. 20–5, which shows the variation of the molar heat capacity of hydrogen with temperature. The value of 5 cal/mole K°, which is predicted for diatomic molecules by our model, is characteristic of hydrogen only in the temperature range from about 250 to 750° K. Above 750° K, C_v increases steadily to 7 cal/mole K° and below 250° K, C_v decreases steadily to 3 cal/mole K°. Other gases show similar variations of molar heat with temperature.

Here is a possible explanation. At low temperatures apparently (see Example 6) the hydrogen molecule has translational energy only and, for some reason, cannot rotate. As the temperature rises rotation becomes possible so that at "ordinary" temperatures a hydrogen molecule acts like our dumbbell model. At high temperatures the collisions between molecules cause the atoms in the molecule to vibrate and the molecule ceases to behave as a rigid body. Different gases, because of their different molecular structure, may show these effects at different temperatures. Thus a chlorine molecule appears to vibrate at room temperature.

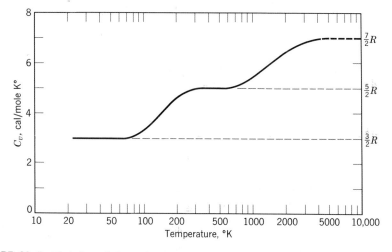

FIGURE 20–5 Variation of the molar heat C_v of hydrogen with temperature. Note that T is drawn on a logarithmic scale. Hydrogen dissociates below 3200° K and liquifies at 20° K. The dashed curve is for a diatomic molecule that does not dissociate below 10,000° K.

Although this description is essentially correct, and we have obtained much insight into the behavior of molecules, this behavior contradicts classical kinetic theory. For kinetic theory is based on Newtonian mechanics applied to a large collection of particles, and the equipartition of energy is a necessary consequence of this classical statistical mechanics. But if equipartition of energy holds, then, no matter what happens to the total internal energy as the temperature changes, each part of the energy—translational, rotational, and vibrational—must share equally in the change. There is no classical mechanism for changing one mode of mechanical energy at a time in such a system. Kinetic theory requires that the specific heats of gases be independent of the temperature.

Hence, we have come to the limit of validity of classical mechanics when we try to explain the structure of the atom (or molecule). Just as Newtonian principles break down at very high speeds (near the speed of light), so here in the region of very small dimensions they also break down. Relativity theory modifies Newtonian ideas to account for the behavior of physical systems in the region of high speeds. It is quantum physics that modifies Newtonian ideas to account for the behavior of physical systems in the region of small dimensions. Both relativity theory and quantum physics are generalizations of classical theory in the sense that they give the (correct) Newtonian results in the regions in which Newtonian physics has accurately described experimental observations. We shall confine our attention to the very fruitful application of thermodynamics and the kinetic theory to "classical" systems.

Example 6. According to quantum theory the internal energy of an atom (or molecule) is "quantized"; that is, the atom cannot have any of a continuous set of internal energies but only certain discrete ones. After being raised from its lowest energy state to some higher one the atom can give up this energy by emitting radiation whose energy equals the difference in energy between the upper and lower internal energy states of the atom.

When two atoms collide, some of their translational kinetic energy may be converted into internal energy of one or both of the atoms. In such a case the collision is inelastic, for translational kinetic energy is not conserved. In a gas, the average translational kinetic energy of an atom is $\frac{3}{2}kT$. When the temperature is raised to a value where $\frac{3}{2}kT$ is about equal to some allowed internal excitation energy of the atom, then an appreciable number of the atoms can absorb enough energy through inelastic collisions to be raised to this higher internal energy state. We can detect this because, after an interval, radiation corresponding to the absorbed energy will be emitted.

(*a*) Compute the average translational kinetic energy per molecule in a gas at room temperature.

We have, for $T = 300°$ K,

$$\frac{3}{2}kT = \frac{3}{2}(1.38 \times 10^{-23} \text{ joule/molecule K°})(300° \text{ K})$$

$$= 6.21 \times 10^{-21} \text{ joule/molecule}$$

$$= 3.88 \times 10^{-2} \text{ eV/molecule}$$

This is about $\frac{1}{25}$ eV per molecule. Some molecules will have larger energies and some smaller energies than this average value.

(*b*) The first allowed (internal) excited state of a hydrogen atom is 10.2 eV above its lowest (ground) state. What temperature is needed to excite a "large" number of hydrogen atoms to emit radiation of this energy?

We require

$$\frac{3}{2}kT = 10.2 \text{ eV}$$

and we have from above

$$\tfrac{3}{2}k(300° \text{ K}) = \tfrac{1}{25} \text{ eV}.$$

Hence

$$T = 300° \text{ K} \times 10.2/(\tfrac{1}{25}) \simeq 7.5 \times 10^{4°} \text{ K}.$$

Actually, because many molecules have energies much greater than the average value, appreciable excitation may occur at somewhat lower temperatures.

We can now appreciate why the kinetic theory assumption, that molecules can be regarded as having no internal structure and collide elastically with one another, holds true at ordinary temperatures. Only at temperatures high enough to give the molecules an average translational kinetic energy comparable to the energy difference between the lowest and first allowed excited state of the molecule will the internal structure of the molecule change and the collisions become inelastic. Indeed, in retrospect one may say that early evidence that the internal energy of an atom is quantized existed in experiments with gas collisions and that the seeds of quantum theory lay in the kinetic theory of gases.

20–8 Mean Free Path

Between successive collisions a molecule in a gas moves with constant speed along a straight line. The average distance between such successive collisions is called the *mean free path* (Fig. 20–6). If molecules were points, they would not collide at all and the mean free path would be infinite. Molecules, however, are not points and hence collisions occur. If they were so numerous that they completely filled the space available to them, leaving no room for translational motion, the mean free path would be zero. Thus the mean free path is related to the size of the molecules and to their number per unit volume.

Consider the molecules of a gas to be spheres of diameter d. The cross section for a collision is then πd^2. That is, a collision will take place when the centers of two molecules approach within a distance d of one another. An equivalent description of collisions made by any one molecule is to regard that molecule as having a diameter $2d$ and all other molecules as point particles (see Fig. 20–7).

Imagine a typical molecule of equivalent diameter $2d$ moving with speed v through a gas of equivalent point particles and let us assume, for the time being, that the molecule and the point particles exert no forces on each other. In time t our

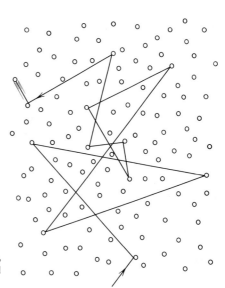

FIGURE 20–6   A molecule traveling through a gas, colliding with other molecules in its path. Of course, all the other molecules are moving in a similar fashion.

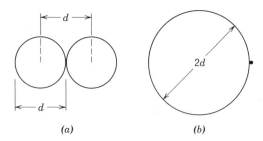

FIGURE 20-7 (*a*) If a collision occurs when two molecules come within a distance *d* of each other, the process can be treated equivalently (*b*) by thinking of one molecule as having an effective diameter 2*d* and the other as being a point mass.

molecule will sweep out a cylinder of cross-sectional area πd^2 and length vt. If n is the number of molecules per unit volume the cylinder will contain $(\pi d^2 vt)\,n$ particles (see Fig. 20–8). Since our molecule and the point particles do exert forces on each other, this will be the number of collisions experienced by the molecule in time t. The cylinder of Fig. 20–8 will, in fact, be a broken one, changing direction with every collision and v will not be a constant.

The mean free path l is the average distance between successive collisions. Hence, l is the total distance vt covered in time t divided by the number of collisions that take place in this time, or

$$\bar{l} = \frac{vt}{\pi d^2 nvt} = \frac{1}{\pi n d^2}.$$

This equation is based on the picture of a molecule hitting stationary targets. Actually the molecule hits moving targets. When the target molecules are moving, the two v's in the first equation above are not the same. The one in the numerator ($= \bar{v}$) is the mean molecular speed measured with respect to the container. The one in the denominator ($= \bar{v}_{\text{rel}}$) is the mean relative speed with respect to other molecules; it is this relative speed that determines the collision rate.

We can see qualitatively that $\bar{v}_{\text{rel}} > v$. Thus two molecules of speed v moving toward each other have a relative speed of $2v(> v)$; two molecules with speed v moving at right angles on a collision course have a relative speed of $\sqrt{2}v$ (also $> v$); two molecules moving with speed v in the same direction have a relative speed of zero ($< v$). Thus molecules arriving from all of the forward hemisphere and part of the backward hemisphere have $v_{\text{rel}} > v$. The molecules arriving from the rest of the backward hemisphere have $v_{\text{rel}} < v$ but, since their numbers are smaller, the average over both hemispheres yields $v_{\text{rel}} > v$. A quantitative calculation, taking into account the actual speed distribution of the molecules, gives $v_{\text{rel}} = \sqrt{2}v$. With this change, the mean free path is reduced to

$$l = \frac{1}{\pi \sqrt{2}\, n d^2}. \tag{20–13}$$

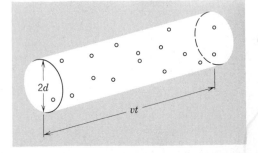

FIGURE 20-8 A molecule of equivalent diameter 2*d* traveling with speed *v* sweeps out a cylinder of base πd^2 and length *vt* in a time *t*. It suffers a collision with every other molecule whose center lies within this cylinder.

Example 7. Let us calculate the magnitude of the mean free path and the collision frequency for air molecules at 0° C and 1-atm pressure.

We take 2×10^{-8} cm as an effective molecular diameter d. For the conditions stated, the average speed of air molecules is about 1×10^5 cm/sec and there are about 3×10^{19} molecules/cm³. The mean free path is then

$$l = \frac{1}{\pi \sqrt{2} \, nd^2} = \frac{1}{\pi \sqrt{2}(3 \times 10^{19}/\text{cm}^3)(2 \times 10^{-8}\ \text{cm})^2}$$

$$= 2 \times 10^{-5}\ \text{cm}.$$

This is about a thousand molecular diameters.

The corresponding collision frequency is

$$\frac{v}{l} = (1 \times 10^5\ \text{cm/sec})/(2 \times 10^{-5}\ \text{cm})$$

$$= 5 \times 10^9/\text{sec}.$$

Thus, on the average, each molecule makes five billion collisions per second!

In the earth's atmosphere the mean free path of air molecules at sea level (760 mm-Hg) is, as we have seen, 2×10^{-5} cm. At 100 km above the earth (10^{-3} mm-Hg) the mean free path is 2 mm. At 300 km (10^{-6} mm-Hg) it is 15 om, and yet there are about 10^8 molecules/cm³ in this region. This emphasizes that molecules are indeed small. At great enough heights the mean free path concept fails because the upward-directed molecules follow ballistic paths and may escape from the atmosphere.

In the laboratory the mean free path concept is useful in situations such as that of Example 7. In even modest laboratory vacuums, however, it loses some of its meaning because nearly all the collisions are with the wall of the containing vessel rather than with other molecules. Consider a box 10 cm on edge containing air at 10^{-6} mm-Hg pressure. The mean free path (see above) is 15 om, so that collisions between molecules are rare indeed. And yet this box contains about 10^{12} molecules!

Even in a finite "box," however, there are some conditions in which particles can travel great distances without striking the walls. In a typical proton synchrotron, used to accelerate protons to the billion-electron-volt range of energies, the protons are constrained by a magnetic field to move in a circular path and may travel several hundred thousand miles during the acceleration process. Mean free path considerations are important if the accelerating protons are to have essentially no collisions with residual air molecules. In this case the effective cross section of the proton is so much smaller than that of the air molecules that if we have a vacuum of about 10^{-6} mm-Hg, there is essentially no beam loss by proton scattering from gas molecules inside the vacuum chamber.

20–9 Distribution of Molecular Speeds

In earlier sections we discussed the root-mean-square speed of the molecules of a gas. However, the speeds of individual molecules vary over a wide range of magnitude; there is a characteristic distribution of molecular speeds of a given gas which depends, as we will see below, on the temperature. If all the molecules of a gas had the same speed v, this situation would not persist for very long because the molecular speeds would be changed by collisions. However, we do not expect many molecules to have speeds $\ll v_{\text{rms}}$ (that is, near zero) or $\gg v_{\text{rms}}$ because such extreme speeds would require an unlikely sequence of preferential collisions.

Clerk Maxwell first solved the problem of the most probable distribution of speeds in a large number of molecules of a gas. His molecular speed distribution law, for a sample of gas containing N molecules, is

$$N(v) = 4\pi N(m/2\pi kT)^{3/2}v^2 e^{-mv^2/2kT}. \tag{20-14}$$

In this equation $N(v)\,dv$ is the number of molecules in the gas sample having speeds between v and $v + dv$. T is the absolute temperature, k is Boltzmann's constant, and m is the mass of a molecule. Note that for a given gas the speed distribution depends only on the temperature. We find N, the total number of molecules in the sample, by adding up (that is, by integrating) the number present in each differential speed interval from zero to infinity, or

$$N = \int_0^\infty N(v)\,dv. \tag{20-15}$$

The unit of $N(v)$ is, say, molecules/(cm/sec).

In Fig. 20–9 we plot the Maxwell distribution of speeds for molecules of oxygen at two different temperatures. The number of molecules having a speed between v_1 and v_2 equals the area under the curve between the vertical lines at v_1 and v_2. As Eq. 20–15 shows, the area under the speed distribution curve, which is the integral in that equation, is equal to the total number of molecules in the sample. At any temperature the number of molecules in a given speed interval Δv increases as the speed increases up to a maximum (the most probable speed v_p) and then decreases asymptotically toward zero. The distribution curve is not symmetrical about the most probable speed because the lowest speed must be zero, whereas there is no classical limit to the upper speed a molecule can attain. In this case the average speed $\bar{v}$ is somewhat larger than the most probable value. The root-mean-square value, v_{rms}, being the square root of the average of the sum of the squares of the speeds, is still larger.

As the temperature increases, the root-mean-square speed v_{rms} (and $\bar{v}$ and v_p as well) increases, in accord with our microscopic interpretation of temperature. The range of typical speeds is now greater, so that the distribution broadens. Since the

FIGURE 20–9   The Maxwellian distribution of speeds of 10^6 oxygen molecules at two different temperatures. The number of molecules within a certain range of speeds (say, 300 to 600 meters/sec) is the area under this section of the curve. The complete area under either curve is the total number of molecules (equals 10^6); this area must be the same for each temperature if, as in this case, the curves refer to a given number of molecules. The pressure is lower than atmospheric because oxygen is in liquid form at 1.0 atm and 73° K.

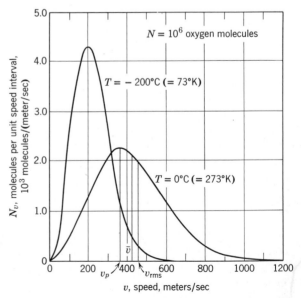

area under the distribution curve (which is the total number of molecules in the sample) remains the same, the distribution must also flatten as the temperature rises. Hence the number of molecules which have speeds greater than some given speed increases as the temperature increases (see Fig. 20–9). This explains many phenomena, such as the increase in the rates of chemical reactions with rising temperature.

The distribution of speeds of molecules in a liquid also resembles the curves of Fig. 20–9. This explains why some molecules in a liquid (the fast ones) can escape through the surface (evaporate) at temperatures well below the normal boiling point. Only these molecules can overcome the attraction of the molecules in the surface and escape by evaporation. The average kinetic energy of the remaining molecules drops correspondingly, leaving the liquid at a lower temperature. This explains why evaporation is a cooling process.

From Eq. 20–14 we see that the distribution of molecular speeds depends on the mass of the molecule as well as on the temperature. The smaller the mass, the larger the proportion of high-speed molecules at any given temperature. Hence hydrogen is more likely to escape from the atmosphere at high altitudes than oxygen or nitrogen. The moon may have a tenuous atmosphere. For the molecules in this atmosphere not to have a great probability of escaping from the weak gravitational pull of the moon, even at the low temperatures there, we would expect them to be molecules or atoms of the heavier elements. Evidence points to the heavy inert gases, such as krypton and xenon, which were produced largely by radioactive decay early in the moon's history. The atmospheric pressure on the moon cannot be larger than about 10^{-13} of the earth's atmospheric pressure.

Example 8. The speeds of ten particles in meters/sec are 0, 1.0, 2.0, 3.0, 3.0, 3.0, 4.0, 4.0, 5.0, and 6.0. Find (a) the average speed, (b) the root-mean-square speed, and (c) the most probable speed of these particles.

(a) The average speed is

$$v = \frac{0 + 1.0 + 2.0 + 3.0 + 3.0 + 3.0 + 4.0 + 4.0 + 5.0 + 6.0}{10} = 3.1 \text{ meters/sec.}$$

(b) The mean-square speed is

$$\overline{v^2} = \frac{0 + (1.0)^2 + (2.0)^2 + (3.0)^2 + (3.0)^2 + (3.0)^2 + (4.0)^2 + (4.0)^2 + (5.0)^2 + (6.0)^2}{10}$$

$$= 12.5 \text{ meters}^2/\text{sec}^2$$

and the root-mean-square speed is

$$v_{\text{rms}} = \sqrt{12.5 \text{ meters}^2/\text{sec}^2} = 3.5 \text{ meters/sec.}$$

(c) Of the ten particles three have speeds of 3.0 meters/sec, two have speeds of 4.0 meters/sec, and the other five each have a different speed. Hence, the most probable speed of a particle v_p is

$$v_p = 3.0 \text{ meters/sec.}$$

QUESTIONS

1. In discussing the fact that it is impossible to apply the laws of mechanics individually to atoms in a macroscopic system, Mayer and Mayer state: "The very complexity of the problem [that is, the fact that the number of atoms is large] is the secret of its solution." Discusss this sentence.

2. In kinetic theory we assume that there are a large number of molecules in a gas. Real gases behave like an ideal gas at low densities. Are these statements contradictory? If not, what conclusion can you draw from them?

3. We have assumed that the walls of the container are elastic for molecular collisions. Actually, the walls may be inelastic. In practice this makes no difference as long as the walls are at the same temperature as the gas. Explain.

4. In large-scale inelastic collisions mechanical energy is lost through internal friction resulting in a rise in temperature owing to increased internal molecular agitation. Is there a loss of mechanical energy to heat in an inelastic collision between molecules?

5. What justification is there in neglecting the change in gravitational potential energy of molecules in a gas?

6. We have assumed that the force exerted by molecules on the wall of a container is steady in time. How is this justified?

7. The average velocity of the molecules in a gas must be zero if the gas as a whole and the container are not in translational motion. Explain how it can be that the average speed is not zero.

8. By considering quantities which must be conserved in an elastic collision, show that in general molecules of a gas cannot have the same speeds after a collision as they had before. Is it possible, then, for a gas to consist of molecules which all have the same speed?

9. Justify the fact that the pressure of a gas depends on the square of the speed of its particles by explaining the dependence of pressure on the collision frequency and the momentum transfer of the particles.

10. The gas kinetic temperature in the upper atmosphere (see Eq. 20–5) is of the order of $1000°$ K. It is quite cold up there. Explain this paradox.

11. Explain how we might keep a gas at a constant temperature during a thermodynamic process.

12. Explain why the temperature of a gas drops in an adiabatic expansion.

13. If hot air rises, why is it cooler at the top of a mountain than near sea level?

14. A sealed rubber balloon contains a very light gas. The balloon is released and it rises high into the atmosphere. Describe and explain the thermal and mechanical behavior of the balloon.

15. Explain why the specific heat at constant pressure is greater than the specific heat at constant volume.

16. It is more common to excite radiation from gaseous atoms by use of electrical discharge than by thermal methods. Why?

17. Give a qualitative explanation of the connection between the mean free path of ammonia molecules in air and the time it takes to smell the ammonia when a bottle is opened across the room.

18. The two opposite walls of a container of gas are kept at different temperatures. Describe the mechanism of heat conduction through the gas.

19. A gas can transmit only those sound waves whose wavelength is long compared with the mean free path. Can you explain this? Where might this limitation arise?

20. If molecules are not spherical, what meaning can we give to d in Eq. 20–13 for the mean free path? In which gases would the molecules act the most nearly as rigid spheres?

21. Suppose we dispense with the hypothesis of elastic collisions in kinetic theory and consider the molecules as centers of force acting at a distance. Does the concept of mean free path have any meaning under these circumstances?

22. Justify qualitatively the statement that, in a mixture of molecules of different kinds in complete equilibrium, each kind of molecule has the same Maxwellian distribution in speed that it would have if the other kinds were not present.

23. The fraction of molecules within a given range Δv of the root-mean-square speed decreases as the temperature of a gas rises. Explain why.

24. (*a*) Do half the molecules in a gas in thermal equilibrium have speeds greater than v_p? Than $\bar{v}$? Than v_{rms}? (*b*) Which speed, v_p, $\bar{v}$, or v_{rms}, corresponds to a molecule having average kinetic energy?

25. Keeping in mind that internal energy of a body consists of kinetic energy and potential energy of its particles how would you distinguish between the internal energy of a body and its temperature?

PROBLEMS

1(2). An air bubble of 20 cm³ volume is at the bottom of a lake 40 meters deep where the temperature is 4° C. The bubble rises to the surface which is at a temperature of 20° C. Take the temperature of the bubble to be the same as that of the surrounding water and find its volume just before it reaches the surface.
Answer: 100 cm³.

2(2). Oxygen gas having a volume of 1.0 liter at 40° C and a pressure of 76 cm-Hg expands until its volume is 1.5 liters and its pressure is 80 cm-Hg. Find (*a*) the mass in moles of oxygen in the system and (*b*) its final temperature.

3(2). An automobile tire has a volume of 1000 in.³ and contains air at a gauge pressure of 24 lb/in.² when the temperature is 0° C. What is the gauge pressure of the air in the tires when its temperature rises to 27° C and its volume increases to 1020 in.³?
Answer: 27 lb/in².

4(2). Two containers, each with the same volume *V*, are connected by an open pipe and filled with gas to a gauge pressure of 30 lb/in² while one is held at 100° C and the other is held at −100° C (Fig. 20-10). (*a*) The high temperature container contains *x* times as many moles of gas as does the low temperature container. What is the value of *x*? (*b*) The pipe connecting the two containers is closed with a valve and both containers are brought to 20° C. What is the gauge pressure in each container at this temperature?

5(2). What is the coefficient of volume expansion for an ideal gas at constant pressure and at 100° C? (See Section 18–7).
Answer: 2.68×10^{-3}/°K.

6(2). Consider a given mass of an ideal gas. Compare curves representing constant-pressure, constant-volume, and isothermal processes on (*a*) a *p-V* diagram, (*b*) a *p-T* diagram and (*c*) a *V-T* diagram. (*d*) How do these curves depend on the mass of gas chosen?

7(2). Compute the number of molecules in a gas contained in a volume of 1.00 cm³ at a pressure of 1.00×10^{-3} atm and a temperature of 200° K.
Answer: 3.67×10^{16}.

8(2). An excellent laboratory vacuum is 10^{-10} mm-Hg. How many molecules/cm³ of gas remain at 20° C in this "vacuum"?

9(2). If the water molecules in 1.0 gm of water were distributed uniformly over the surface of the earth, how many such molecules would there be on 1.0 cm² of the earth's surface?
Answer: 6600.

10(2). Calculate the work done in compressing 1.00 mole of oxygen from a volume of 22.4 liters at 0° C and 1.00-atm pressure to 16.8 liters at the same temperature.

11(2). Air that occupies 5.0 ft³ at 15 lb/in² gauge pressure is expanded isothermally to atmospheric pressure and then cooled at constant pressure until it reaches its initial volume. Compute the work done by the gas.
Answer: 4200 ft-lb.

12(2). The equation of state for a certain material is given by

$$p = \frac{AT - BT^2}{V}.$$

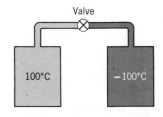

FIGURE 20–10 Problem 4(2).

Find an expression for the work done if the temperature changes from T_1 to T_2 while the pressure remains constant $p = p_0$.

13(4). The mass of a H_2 molecule is 3.32×10^{-24} gm. If 1.0×10^{23} hydrogen molecules per second strike 2.0 cm^2 of wall at an angle of 45° with the normal when moving with a speed of 10^5 cm/sec, what pressure do they exert on the wall?

Answer: 2400 nt/meter^2.

14(5). At 273° K. and 1.00×10^{-2} atm the density of a gas is 1.24×10^{-5} gm/cm^3. (*a*) Find v_{rms} for the gas molecules. (*b*) Find the molecular weight of the gas and identify it.

15(5). Calculate the root-mean-square speed of helium atoms at 1000° K.

Answer: 2500 meters/sec.

16(5). (*a*) Compute the root-mean-square speed of an argon atom at room temperature (20° C). (*b*) At what temperature will the root-mean-square speed be half that value? Twice that value?

17(5). What is the mean kinetic energy of individual nitrogen molecules at 1500° K in joules; in electron volts?

Answer: 3.1×10^{-20} joules; 0.19 eV.

18(5). (*a*) Determine the average value of the kinetic energy of the molecules of an ideal gas at 0.0° C and at 100° C. (*b*) What is the kinetic energy per mole of an ideal gas at these temperatures?

19(5). At what temperature is the average translational kinetic energy of a molecule equal to the kinetic energy of an electron accelerated from rest through a potential difference of one volt (that is, an energy of 1.0 eV)?

Answer: 7700° K.

20(5). (*a*) At what temperature does the root-mean-square speed of hydrogen approach $0.10c$, where c is the speed of light (3.0×10^8 meters/sec)? (*b*) What is the rms speed of interstellar hydrogen if its temperature is 3.2° K?

21(5). Oxygen gas at 273° K and 1.00-atm pressure is confined to a cubical container 10 cm on a side. Compare the change in gravitational potential energy of an oxygen molecule falling the height of the box with its mean translational kinetic energy.

Answer: Ratio of the mean translational kinetic energy to the change in gravitational potential energy is 1.1×10^5.

22(5). Find the root-mean-square speeds of helium and argon molecules at 40° C from that of oxygen molecules (460 meters/sec at 0.00° C). The molecular weight of oxygen is 32 gm/mole, of argon 40, of helium 4.0.

23(5). (*a*) Compute the temperature at which the root-mean-square speed is equal to the speed of escape from the surface of the earth for molecular hydrogen. For molecular oxygen. (*b*) Do the same for the moon, assuming gravity on its surface to be 0.16 *g*. (*c*) The temperature high in the earth's upper atmosphere is about 1000° K. Would you expect to find much hydrogen there? Much oxygen?

Answer: (*a*) H_2: 1.0×10^4 °K; O_2: 16×10^4 °K. (*b*) H_2: 450° K: O_2: 7200° K.

24(5). Water standing in the open at 27° C evaporates due to the escape of some of the surface molecules. The heat of vaporization (540 cal/gm) may be found approximately from ϵn, where ϵ is the average energy of the escaping molecules and n is the number of molecules per gram. (*a*) Find ϵ. (*b*) How many times greater is ϵ than the average kinetic energy of H_2O molecules, assuming that the kinetic energy is related to temperature in the same way as it is for gases?

25(6). One mole of an ideal gas undergoes an isothermal expansion. Find the heat flow into the gas in terms of the initial and final volumes and the temperature.

Answer: $RT \ln (V_f/V_i)$.

26(6). One mole of an ideal gas expands adiabatically from an initial temperature T_1 to a final temperature T_2. Prove that the work done by the gas is $C_v(T_1 - T_2)$, where C_v is the molar heat capacity.

27(6). Let 5.0 cal of heat be added to a particular substance. As a result, its volume changes from 50 to 100 cm^3 while the pressure remains constant at 1.0 atm. (*a*) By how much did the internal energy of the substance change? (*b*) If the mass of the substance is 2.0×10^{-3} moles, find the molar heat capacity at constant volume. (*c*) Find the molar heat capacity at constant pressure.

Answer: (*a*) 3.8 cal. (*b*) 6.2 cal/mole-K°. (*c*) 8.2 cal/mole-K°.

28(6). C_v for a certain ideal gas is 6.00 cal/mole-K°. The temperature of 3.0 moles of the gas is raised 50 K° by each of three different processes; at constant volume, at constant pressure, and by an adiabatic compression. Complete the following table, showing for each process the heat added (or subtracted), the work done on or by the gas, the change in internal energy of the gas, and the change in total translational kinetic energy of the gas.

Process	Heat Added	Work Done by Gas	Change in Internal Energy	Change in Kinetic Energy
Constant volume	_____	_____	_____	_____
Constant pressure	_____	_____	_____	_____
Adiabatic	_____	_____	_____	_____

29(6). A sample of ideal monatomic gas occupies a volume V_i at pressure p_i. The gas is caused to expand to a volume V_f in three different ways: isothermal, isobaric (p = constant), and adiabatic. (a) List, for each process, the algebraic signs (+, −, or 0) for ΔU, Δp, ΔT and Q. (b) Which process causes the most work to be done by the gas? Which process the least?

Answer: (a)

	ΔU	Δp	ΔT	Q
Isothermal	0	−	0	+
Isobaric	+	0	+	+
Adiabatic	−	−	−	0

(b) In descending order: isobaric, isothermal, adiabatic.

30(6). A reversible heat engine carries 1.00 mole of an ideal monatomic gas around the cycle shown in Fig. 20–11. Process 1–2 takes place at constant volume, process 2–3 is adiabatic, and process 3–1 takes place at a constant pressure. (a) Compute the heat Q, the change in internal energy ΔU, and the work done W, for each of the three processes and for the cycle as a whole. (b) If the initial pressure at point 1 is 1.00 atm, find the pressure and the volume at points 2 and 3.

31(6). The mass of a gas molecule can be computed from the specific heat at constant volume. Take $C_v = 0.075$ kcal/kg K° for argon and calculate (a) the mass of an argon atom and (b) the atomic weight of argon.

Answer: (a) 6.6×10^{-23} gm. (b) 40 gm/mole.

32(6). Take the mass of a helium atom to be 6.66×10^{-27} kg. Compute the specific heat at constant volume for helium gas.

33(6). Show that the speed of sound in a gas v_s and the molecular speed v_{rms} are related by

$$v_{\text{rms}}/v_s = \sqrt{3/\gamma}.$$

Test this prediction using the data given following Example 3; see also Section 17–2.

34(6). A mass of gas occupies a volume of 4.0 liters at a pressure of 1.0 atm and a temperature of 300° K. It is compressed adiabatically to a volume of 1.0 liter. Determine (a) the final pressure and (b) the final temperature, assuming it to be an ideal gas for which $\gamma = 1.5$.

35(6). (a) A liter of gas with $\gamma = 1.3$ is at 273° K and 1.0-atm pressure. It is suddenly compressed to half its original volume. Find its final pressure and temperature. (b) The gas is now cooled back to 0° C at constant pressure. What is its final volume?

Answer: (a) 2.5 atm; 340° K. (b) 0.40 liter.

36(7). An ideal diatomic gas (4.0 moles) at high temperature experiences a temperature increase of 60° K under constant pressure conditions. (a) How much heat was added to the gas? (b) By how much did the internal energy of the gas increase? (c) How much work was done by the gas? (d) By how much did the internal translational kinetic energy of the gas increase?

37(7). One mole of oxygen is heated at a constant pressure starting at 0.00° C. How much heat energy must be added to the gas to double its volume?

Answer: 1900 cal.

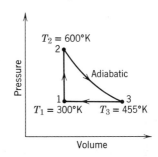

FIGURE 20–11 Problem 30(6).

38(7). Ten grams of oxygen are heated at constant atmospheric pressure from 27.0 to 127° C. (*a*) How much heat is transferred to the oxygen? (*b*) What fraction of the heat is used to raise the internal energy of the oxygen?

39(7). Calculate the mechanical equivalent of heat from the value of R and the values of C_v and γ for oxygen from Table 20–2.
Answer: 4.13 joules/cal.

40(7). An ideal gas experiences an adiabatic compression from $p = 1.0$ atm, $V = 1.0 \times 10^6$ liters, $T = 0°$ C to $p = 1.0 \times 10^5$ atm, $V = 1.0 \times 10^3$ liters. (*a*) Is this a monatomic, a diatomic or a polyatomic gas? (*b*) What is the final temperature? (*c*) How many moles of the gas are present? (*d*) What is the total translational kinetic energy per mole before and after the compression? (*e*) What is the ratio of the squares of the rms speeds before and after the compression?

41(7). (*a*) An ideal gas, initially at pressure p_0 undergoes a free expansion (adiabatic, no external work) until its final volume is 3.0 times its initial volume. What is the pressure of the gas after the free expansion? (*b*) The gas is then slowly and adiabatically compressed back to its original volume. The pressure after compression is $(3.0)^{1/3}p_0$. Determine whether the gas is monatomic, diatomic or polyatomic. (*c*) How does the average kinetic energy per molecule in this final state compare with that of the initial state?
Answer: (*a*) $p_0/3$. (*b*) Polyatomic (ideal). (*c*) $K_f/K_i = 1.4$.

42(7). (*a*) A monatomic ideal gas initially at 17° C is suddenly compressed to one-tenth its original volume. What is its temperature after compression? (*b*) Make the same calculation for a diatomic gas.

43(7). How would you explain the observed value of $C_v = 7.50$ cal/mole K° for gaseous SO_2 at 15.0° C and 1.00 atm?

44(7). *Avogadro's law* states that under the same condition of temperature and pressure equal volumes of gas contain equal numbers of molecules. Derive this law from kinetic theory using Eq. 20–3 and the equipartition of energy assumption.

45(7). A room of volume V is filled with a diatomic ideal gas (air) at temperature T_1 and pressure p_0. The air is heated to a higher temperature T_2, the pressure remaining constant at p_0 because the walls of the room are not air-tight. Show that the internal energy content of the air remaining in the room is the same at T_1 and T_2, and that the energy supplied by the furnace to heat the air has all gone to heat the air *outside* the room. If we add no energy to the air, why bother to light the furnace? (Ignore the furnace energy used to raise the temperature of the walls, and consider only the energy used to raise the air temperature.)

46(8). The mean free path of nitrogen molecules at 0° C and 1.0 atm is 0.80×10^{-5} cm. At this temperature and pressure there are 2.7×10^{19} molecules/cm³. What is the molecular diameter?

47(8). In a certain particle accelerator the protons travel around a circular path of diameter 75 ft in a chamber of 10^{-6} mm-Hg pressure. (*a*) Estimate the number of gas molecules per cubic centimeter at this pressure. (*b*) What is the mean free path of the gas molecules under these conditions if the molecular diameter is 2.0×10^{-8} cm?
Answer: (*a*) 3.5×10^{10} molecules/cm³. (*b*) 160 meters.

48(8). At what frequency would the wavelength of sound be of the order of the mean free path in oxygen at 1.0 atm pressure and 0.0° C? Take the diameter of the oxygen molecule to be 3.00×10^{-8} cm.

49(8). What is the mean free path for 15 spherical jelly beans in a bag that is vigorously shaken? Take the volume of the bag to be 1.0 liter and the diameter of a jelly bean to be 1.0 cm.
Answer: 15 cm.

50(8). (*a*) What is the molar volume (the volume per mole) of an ideal gas at standard conditions (0° C, 1.0 atm)? (*b*) Calculate the ratio of the root-mean-square speeds of helium and neon under these conditions. (*c*) What would be the mean-free-path of helium atoms under these conditions? Assume the molecular diameter d to be 1.0×10^{-8} cm. (*d*) What would be the mean-free-path of neon atoms under these conditions? (*e*) Comment on the results of parts (*c*) and (*d*) in view of the fact that the helium atoms are traveling faster than the neon atoms.

51(8). At 2500 km above the earth's surface the density is about one molecule/cm³. What mean free path is predicted by Eq. 20–13, and what is its significance under these conditions?

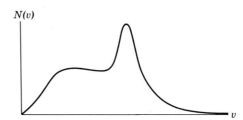

FIGURE 20–12   Problem 57(9).

Answer: Equation 20–13 yields $\bar{l} \approx 6 \times 10^9$ km., but this has little significance because, at this altitude, nearly all molecules would follow collisionless ballistic paths in the earth's gravitational field, and many would escape from the atmosphere.

52(8). The mean free path $\bar{l}$ of the molecules of a gas may be determined from measurement (for example, from measurement of the viscosity of the gas). At 20° C and 75 cm-Hg pressure such measurements yield values of $\bar{l}_A$ (argon) $= 9.9 \times 10^{-6}$ cm and $\bar{l}_{N_2}$ (nitrogen) $= 27.5 \times 10^{-6}$ cm. (a) Find the ratio of the effective cross-section diameters of argon and nitrogen. (b) What would be the value of the mean free path of argon at 20° C and 15 cm-Hg? (c) What would be the value of the mean free path of argon at −40° C and 75 cm-Hg?

53(8). A molecule of hydrogen (diameter 1.0×10^{-8} cm) escapes from a furnace ($T = 4000°$ K) with the root-mean-square speed into a chamber containing atoms of cold argon (diameter 3.0×10^{-8} cm) at a density of 4.0×10^{19} atoms/cm³. (a) What is the speed of the hydrogen molecule? (b) If the molecule and an argon atom collide, what is the closest distance between their centers, considering each as spherical? (c) What is the initial number of collisions per unit time experienced by the hydrogen molecule?
Answer: (a) 7000 meters/sec. (b) 2.0×10^{-8} cm. (c) 5.0×10^{10} collisions/sec.

54(8). For a gas in which all molecules travel with the same speed v, show that $\bar{v}_{rel} = \frac{4}{3}v$ rather than $\sqrt{2}\ \bar{v}$ (which is the result obtained when we consider the actual distribution of molecular speeds). See p. 392.

55(9). You are given the following group of particles (N_i represents the number of particles which have a speed v_i).

N_i	v_i(cm/sec)
2	1.00
4	2.00
6	3.00
8	4.00
2	5.00

(a) Compute the average speed $\bar{v}$. (b) Compute the root-mean-square speed v_{ms}. (c) Among the five speeds shown, which is the most probable speed v_p for the entire group?
Answer: (a) 3.2 cm/sec. (b) 3.4 cm/sec. (c) 4.0 cm/sec.

56(9). The speeds of a group of ten molecules are 2.0, 3.0, 4.0, . . . , 11.0 km/sec. (a) What is the average speed of the group? (b) What is the root-mean-square speed for the group?

57(9). Consider the distribution of speeds shown in Fig. 20–12. (a) List v_{rms}, v, and v_p in the order of increasing speed. (b) How does this compare with the Maxwellian distribution?
Answer: (a) $\bar{v}$, v_{rms}, v_p. (b) v_p, $\bar{v}$, v_{rms}.

CHAPTER 21

Entropy and the Second Law of Thermodynamics

21–1 Introduction

The first law of thermodynamics states that energy is conserved. However, we can think of many thermodynamic processes which conserve energy but which actually never occur. For example, when a hot body and a cold body are put into contact, it simply does not happen that the hot body gets hotter and the cold body colder. Or again, a pond does not suddenly freeze on a hot summer day by giving up heat to its environment. And yet neither of these processes violates the first law of thermodynamics. Similarly, the first law does not restrict our ability to convert work into heat or heat into work, except that energy must be conserved in the process. And yet in practice, although we can convert a given quantity of work completely into heat, we have never been able to find a scheme that converts a given amount of heat completely into work. The second law of thermodynamics deals with this question of whether processes, assumed to be consistent with the first law, do or do not occur in nature. Although the ideas contained in the second law may seem subtle or abstract, in application they prove to be extremely practical.

21–2 Reversible and Irreversible Processes

Consider a typical system in thermodynamic equilibrium, say a mass $\mathfrak{M}$ of a (real) gas confined in a cylinder-piston arrangement of volume V, the gas having a pressure p and a temperature T. In an equilibrium state these thermodynamic variables remain constant with time. Suppose that the cylinder, whose walls are an (ideal) heat insulator but whose base is an (ideal) heat conductor is placed on a large heat reservoir maintained at this same temperature T, as in Fig. 19–4. Now let us change the system to another equilibrium state in which the temperature T is the same but the volume V is reduced by one-half. Of the many ways in which we could do this we discuss two extreme cases.

I. We depress the piston very rapidly; we then wait for equilibrium with the reservoir to be re-established. During this process the gas is turbulent and its pressure and temperature are not well defined; we cannot plot the process as a continuous line on a p-V diagram because we would not know what value of pressure (or temperature) to associate with a given volume. The system passes from one equilibrium state i to another f through a series of nonequilibrium states (Fig. 21–1a).

II. We depress the piston (assumed to be frictionless) exceedingly slowly—perhaps by adding sand to the top of the piston—so that the pressure, volume, and temperature of the gas are, at all times, well-defined quantities. We first drop a few grains of sand on the piston. This will reduce the volume of the system a little and the temperature will tend to rise; the system will depart from equilibrium, but only slightly. A small amount of heat will be transferred to the reservoir and in a short time the system will reach a new equilibrium state, its temperature again being that of the reservoir. Then we drop a few more grains of sand on the piston, reducing the volume further. Again we wait for a new equilibrium state to be established, and so forth. By many repetitions of this procedure we finally reduce the volume by one-half. During this entire process the system is never in a state differing much from an equilibrium state. If we imagine carrying out this procedure with still smaller successive increases in pressure, the intermediate states will depart from equilibrium even less. By indefinitely increasing the number of changes and correspondingly decreasing the size of each change, we arrive at an ideal process in which the system passes through a continuous succession of equilibrium states, which we can plot as a continuous line on a p-V diagram (Fig. 21–1b). During this process a certain amount of heat Q is transferred from the system to the reservoir.

Processes of type I are called *irreversible* and those of type II are called *reversible. A reversible process is one that, by a differential change in the environment, can be made to retrace its path.* Thus as we cause the piston to move slowly downward, in II, the external pressure on the piston exceeds the pressure exerted on it by the gas by only a differential amount dp. If at any instant we reduce the external pressure ever so slightly (by removing a few sand grains), so that it is less than the internal gas pressure by dp, the gas will expand instead of contracting and the system will retrace the equilib-

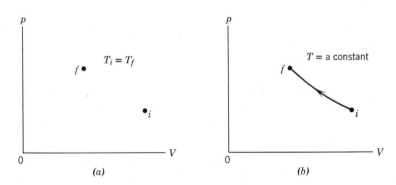

FIGURE 21–1   We cause a real gas to go from an initial equilibrium state i described by p_i, V_i, T_i to a final equilibrium state f described by p_f, $V_f (= \frac{1}{2}V_i)$, and $T_f (= T_i)$. We carry out the process (a) irreversibly, and (b) reversibly.

rium states through which it has just passed. In practice all processes are irreversible, but we can approach reversibility arbitrarily closely by making appropriate experimental refinements. The strictly reversible process is a simple and useful abstraction that bears a similar relation to real processes that the ideal gas abstraction does to real gases.

The process described in II is not only reversible but *isothermal,* because we have assumed that the temperature of the gas differs at all times by only a differential amount dT from the (constant) temperature of the reservoir on which the cylinder rests.

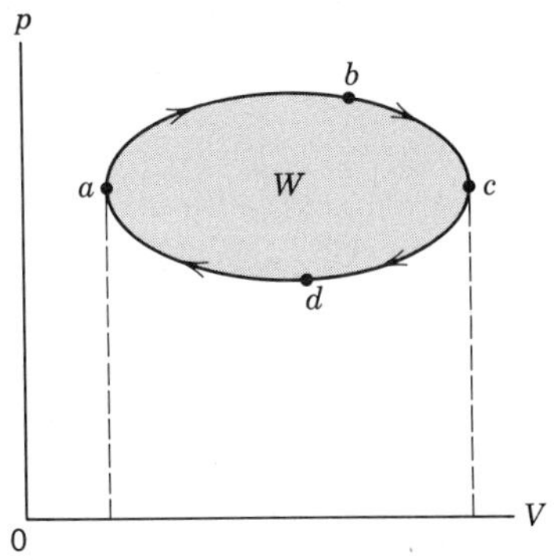

FIGURE 21–2 A p-V diagram of a gas undergoing a reversible cycle. The shaded area W represents the net work done by the gas in the cycle.

We could also reduce the volume *adiabatically* by removing the cylinder from the heat reservoir and putting it on a nonconducting stand. In an adiabatic process no heat is allowed to enter or to leave the system. An adiabatic process can be either reversible or irreversible—the definition does not exclude either. In a reversible adiabatic process we move the piston exceedingly slowly—perhaps using the sand-loading technique; in an irreversible adiabatic process we shove the piston down quickly.

The temperature of the gas will rise during an adiabatic compression because, from the first law, with $Q = 0$, the work W done in pushing down the piston must appear as an increase ΔU in the internal energy of the system. W will have different values for different rates of pushing down the piston, being given by $\int p\,dV$—that is, by the area under a curve on a p-V diagram—only for reversible processes (for which p has a well-defined value). Thus ΔU and the corresponding temperature change ΔT will not be the same for reversible and irreversible adiabatic processes.

21–3 The Carnot Cycle

Suppose that we have a system (a real gas, say) in an equilibrium state in a cylinder-piston arrangement. By using our ability to make changes in the environment of the system we can carry out, at our pleasure, a wide variety of processes. We can let the gas expand or we can compress it; we can add or subtract energy in the form of heat; we can do these things and others irreversibly or reversibly. We can also choose to carry out a sequence of processes such that the system returns to its original equilibrium state; we call this a *cycle.* If the processes involved are all reversible, we call it a *reversible cycle.*

Figure 21–2 shows a reversible cycle on a p-V diagram. Along the curve *abc* we allow the system to expand, and the area under this curve represents the work done by the system during the expansion. Along the curve *cda,* which returns the system to its original state, we compress the system, and the area under this curve represents the work we must do on the system during the compression. Hence, the net work done by the system is represented by the area enclosed by the curve and is positive. If we decided to traverse the cycle in the opposite sense, that is, expanding along *adc* and compressing along *cba,* the net work done by the system would be the negative of that of the previous case.

An important reversible cycle is the *Carnot cycle,* introduced by Sadi Carnot in 1824.

We shall see later that this cycle will determine the limit of our ability to convert heat into work. The system consists of a "working substance," such as a gas, and the cycle is made up of two isothermal and two adiabatic reversible processes. The working substance, which we can think of as an ideal gas for concreteness, is contained in a cylinder with a heat-conducting base and nonconducting walls and piston. We also provide, as part of the environment, a heat reservoir in the form of a body of large heat capacity at a temperature T_1, another reservoir of large heat capacity at a temperature T_2, and two nonconducting stands. We carry out the Carnot cycle in four steps, as shown in Fig. 21–3. The cycle is shown on the p-V diagram of Fig. 21–4.

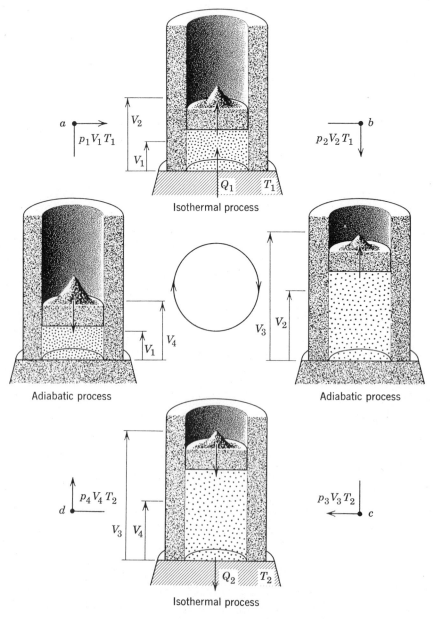

FIGURE 21–3 A Carnot cycle. The points a, b, c and d correspond to the points so labelled in Fig. 21–4. The cylinder-piston arrangements show intermediate steps in the processes that connect adjacent points. The arrows on the pistons suggest expansions (caused by removing sand) and compressions (caused by adding sand).

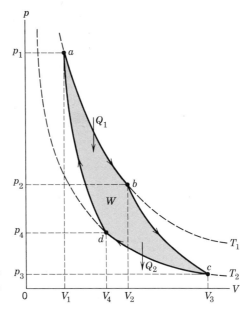

FIGURE 21–4 The Carnot cycle illustrated in the previous figure, plotted on a *p-V* diagram for an ideal gas.

Step 1. The gas is in an initial equilibrium state represented by p_1, V_1, T_1 (*a*, Fig. 21–4). We put the cylinder on the heat reservoir at temperature T_1, and allow the gas to expand slowly to p_2, V_2, T_1 (*b*, Fig. 21–4). During the process heat energy Q_1 is absorbed by the gas by conduction through the base. The expansion is isothermal at T_1 and the gas does work in raising the piston and its load.

Step 2. We put the cylinder on the nonconducting stand and allow the gas to expand slowly to p_3, V_3, T_2 (*c*, Fig. 21–4). The expansion is adiabatic because no heat can enter or leave the system. The gas does work in raising the piston and its temperature falls to T_2.

Step 3. We put the cylinder on the (colder) heat reservoir T_2 and compress the gas slowly to p_4, V_4, T_2 (*d*, Fig. 21–4). During the process heat energy Q_2 is transferred from the gas to the reservoir by conduction through the base. The compression is isothermal at T_2 and work is done on the gas by the piston and its load.

Step 4. We put the cylinder on a nonconducting stand and compress the gas slowly to the initial condition p_1, V_1, T_1. The compression is adiabatic because no heat can enter or leave the system. Work is done on the gas and its temperature rises to T_1.

The net work W done by the system during the cycle is represented by the area enclosed by path *abcd* of Fig. 21–4. The net amount of heat energy received by the system in the cycle is $Q_1 - Q_2$, where Q_1 is the heat absorbed in Step 1 and Q_2 is that given up in Step 3. The initial and final states are the same so that there is no net change in the internal energy U of the system. Hence, from the first law of thermodynamics,

$$W = Q_1 - Q_2 \qquad (21\text{–}1)$$

for the cycle, in which Q_1 and Q_2 are taken as positive quantities. The result of the cycle is that heat has been converted into work by the system. Any required amount of work can be obtained by simply repeating the cycle. Hence, the system acts like a *heat engine*.

We have used an ideal gas as an example of a working substance. The working substance can be anything at all, although the *p-V* diagrams for other substances would be different. Common heat engines use steam or a mixture of fuel and air, or fuel and oxygen as their working substance. Heat may be obtained from the combustion of a fuel such as gasoline or coal, or from the annihilation of mass in nuclear fission processes in nuclear reactors. Heat may be discharged at the exhaust or to a condenser. Although real heat engines do not operate on a reversible cycle, the Carnot cycle, which is reversible, gives useful information about the behavior of any heat engine.

The efficiency e of a heat engine is the ratio of the net work done by the engine during one cycle to the heat taken in from the high temperature source in one cycle. Hence,

$$e = \frac{W}{Q_1} = \frac{Q_1 - Q_2}{Q_1} = 1 - \frac{Q_2}{Q_1}. \tag{21-2}$$

Equation 21–2 shows that the efficiency of a heat engine is less than one (100%) so long as the heat Q_2 delivered to the exhaust is not zero. Experience shows that every heat engine rejects some heat during the exhaust stroke. This represents the heat absorbed by the engine that is not converted to work in the process.

We may choose to carry out the Carnot cycle by starting at any point, such as a in Fig. 21–4, and traversing each process in a direction opposite to that of the arrowheads in that figure. Then an amount of heat Q_2 is removed from the lower temperature reservoir at T_2, and an amount of heat Q_1 is delivered to the higher temperature reservoir at T_1; work must be done on the system by an outside agency. In this reversed cycle work must be done on the system which extracts heat from the lower temperature reservoir. Any amount of heat can be removed from this reservoir by simply repeating the reverse cycle. Hence, the system acts like a *refrigerator*, transferring heat from a body at a lower temperature (the freezing compartment) to one at a higher temperature (the room) by means of work supplied to it (the electric power input).

Example 1. Show that the efficiency of a Carnot engine using an ideal gas as the working substance is $e = (T_1 - T_2)/T_1$.

Along the isothermal path ab, the temperature, and hence the internal energy of an ideal gas, remains constant. From the first law, the heat Q_1 absorbed by the gas in its expansion must be equal to the work W_1 done in this expansion. From Example 2, Chapter 20, we have

$$Q_1 = W_1 = \mu R T_1 \ln (V_2/V_1).$$

Likewise, in the isothermal compression along the path cd, we have

$$Q_2 = W_2 = \mu R T_2 \ln (V_3/V_4).$$

On dividing the first equation by the second, we obtain

$$\frac{Q_1}{Q_2} = \frac{T_1 \ln (V_2/V_1)}{T_2 \ln (V_3/V_4)}.$$

From the equation describing an isothermal process for an ideal gas we obtain for the paths ab and cd

$$p_1 V_1 = p_2 V_2,$$
$$p_3 V_3 = p_4 V_4.$$

From the equation describing an adiabatic process for an ideal gas we have for paths bc and da

$$p_2 V_2{}^\gamma = p_3 V_3{}^\gamma,$$

$$p_4 V_4{}^\gamma = p_1 V_1{}^\gamma.$$

Multiplying these four equations together and canceling the factor $p_1 p_2 p_3 p_4$ appearing on both sides, we obtain

$$V_1 V_2{}^\gamma V_3 V_4{}^\gamma = V_2 V_3{}^\gamma V_4 V_1{}^\gamma,$$

from which

$$(V_2 V_4)^{\gamma-1} = (V_3 V_1)^{\gamma-1}$$

and

$$V_2/V_1 = V_3/V_4.$$

Using this result in our expression for Q_1/Q_2, we see that

$$Q_1/Q_2 = T_1/T_2, \tag{21–3}$$

so that

$$e = 1 - Q_2/Q_1 = 1 - T_2/T_1$$

or

$$e = \frac{Q_1 - Q_2}{Q_1} = \frac{T_1 - T_2}{T_1}.$$

The temperatures T_1 and T_2 are those measured on the ideal gas scale described in Chapter 18.

21–4 The Second Law of Thermodynamics

The first heat engines constructed were very inefficient devices. Only a small fraction of the heat absorbed at the high-temperature source could be converted to useful work. Even as engineering design improved, a sizable fraction of the absorbed heat was still discharged at the lower-temperature exhaust of the engine, remaining un-converted to mechanical energy. It remained a hope to devise an engine that could take heat from an abundant reservoir, like the ocean, and convert it completely into useful work. Then it would not be necessary to provide a source of heat at a higher temperature than the outside environment by burning fuels (Fig. 21–5). Likewise, we might hope to be able to devise a refrigerator that simply transfers heat from a cold body to a hot body, without requiring the expense of outside work (Fig. 21–6). Neither of these hopeful ambitions violates the first law of thermodynamics. The heat engine would simply convert heat energy completely into mechanical energy, the

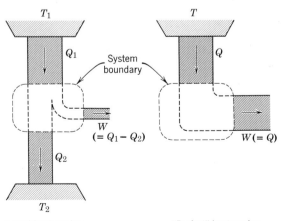

FIGURE 21–5 In an actual heat engine, some of the heat Q_1 taken in by the engine is converted into work W, but the rest is rejected as heat Q_2. In a "perfect" heat engine all the heat input would be converted into work output.

Actual heat engine   "Perfect" heat engine

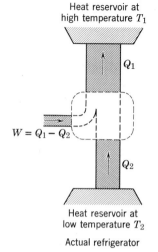
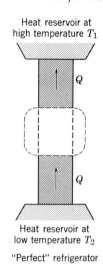

Actual refrigerator   "Perfect" refrigerator

FIGURE 21–6 In an actual refrigerator, work W is needed to transfer heat from a low-temperature to a high-temperature reservoir. In a "perfect" refrigerator, heat would flow from the low-temperature to the high-temperature reservoir without any work being done on the engine.

total energy being conserved in the process. In the refrigerator, the heat energy would simply be transferred from cold body to hot body without any loss of energy in the process. Nevertheless neither of these ambitions has ever been achieved, and there is reason to believe they never will be.

The *second law of thermodynamics,* which is a generalization of experience, is an assertion that such devices do not exist. There have been many statements of the second law, each emphasizing another facet of the law, but all can be shown to be equivalent to one another. Clausius stated it as follows: *It is impossible for any cyclical machine to produce no other effect than to convey heat continuously from one body to another at a higher temperature.* This statement rules out our ambitious refrigerator, for it implies that to convey heat continuously from a cold to a hot object it is necessary to supply work by an outside agent. We know from experience that when two bodies are in contact, heat energy flows from the hot body to the cold body. The second law rules out the possibility of heat energy flowing from cold to hot body in such a case and so determines the direction of transfer of heat. The direction can be reversed only by an expenditure of work.

Kelvin (with Planck) stated the second law in words equivalent to these: *A transformation whose only final result is to transform into work heat extracted from a source which is at the same temperature throughout is impossible.* This statement rules out our ambitious heat engine, for it implies that we cannot produce mechanical work by extracting heat from a single reservoir without returning any heat to a reservoir at a lower temperature.

To show that the two statements are equivalent we need to show that, if either statement is false, the other statement must be false also. Suppose Clausius' statement were false so that we could have a refrigerator operating without needing a work input. We could use an ordinary engine to remove heat from a hot body, to do work and to return part of the heat to a cold body. But by connecting our "perfect" refrigerator into the system, this heat would be returned to the hot body without expenditure of work and would become available again for use by the heat engine. Hence, the combination of an ordinary engine and the "perfect" refrigerator would constitute a heat engine which violates the Kelvin-Planck statement. Or we can reverse the argument. If the Kelvin-Planck statement were incorrect, we could have a heat engine which simply takes heat from a source and converts it completely into work. By connecting this "perfect" heat engine to an ordinary refrigerator, we could extract heat from the hot body, convert it completely to work, use this work to run the ordinary refrigerator, extract heat from the cold body, and deliver it plus the work converted to heat by the refrigerator to the hot body. The net result is a transfer of heat from cold to hot body without expenditure of work and this violates Clausius' statement.

The second law tells us that many processes are irreversible. For example, Clausius' statement specifically rules out a simple reversal of the process of heat transfer from hot body to cold body. Not only will some processes not run backward by themselves, but no combination of processes can undo the effect of an irreversible process without causing another corresponding change elsewhere. In later sections we shall develop these ideas more fully and formulate the second law quantitatively.

21–5   The Efficiency of Engines

Carnot first wrote scientifically on the theory of heat engines. In 1824 he published *Reflections on the Motive Power of Heat.* By then the steam engine was commonly used in industry. Carnot wrote:

> In spite of labor of all sorts expended on the steam engine, and in spite of the perfection to which it has been brought, its theory is very little advanced. . . .
>
> The production of motion in the steam engine is always accompanied by a circumstance which we should particularly notice. This circumstance is the passage of caloric from one body where the temperature is more or less elevated to another where it is lower. . . .
>
> The motive power of heat is independent of the agents employed to develop it; its quantity is determined solely by the temperature of the bodies between which, in the final result, the transfer of the caloric occurs.

Hence, Carnot directed attention to the facts that the difference in temperature was the real source of "motive power," that the transfer of heat played a significant role, and that the choice of working substance was of no theoretical importance.

Carnot's achievement was remarkable when we recall that the mechanical equivalence of heat and the conservation of energy principle were not known in 1824. In his later papers, published posthumously in 1872, it became clear that Carnot had foreseen the principle of the conservation of energy and had made an accurate determination of the mechanical equivalent of heat. He had planned a program of research which included all the important developments in the field made by other investigators during the following several decades. However, he died during a cholera epidemic in 1832 at the age of 36, leaving it to others to extend his work. It was William Thomson (later Lord Kelvin) who modified Carnot's reasoning to bring it into accord with the mechanical theory of heat, and who, together with Clausius, successfully developed the science of thermodynamics.

Carnot developed the concept of a reversible engine and the reversible cycle named after him. He stated a theorem of great practical importance: *The efficiency of all reversible engines operating between the same two temperatures is the same, and no irreversible engine working between the same two temperatures can have a greater efficiency than this.* Clausius and Kelvin showed that this theorem was a necessary consequence of the second law of thermodynamics. Notice that nothing is said about the working substance, so that the efficiency of a reversible engine is independent of the working substance and depends only on the temperatures. Furthermore, a reversible engine operates at the maximum efficiency possible for any engine working between the same two temperature limits. The proof of this theorem follows.

Let us call the two reversible engines H and H'. They operate between the temperatures T_1 and T_2 where $T_1 > T_2$. They may differ, say, in their working substance or in their initial pressures and lengths of stroke. We choose H to run forward and H' to run backward (as a refrigerator). The forward-running engine H takes in heat energy Q_1 at T_1 and gives out heat energy Q_2 at T_2. The backward-running engine (refrigerator) H' takes in heat Q_2' at T_2 and gives out heat Q_1' at T_1. We now connect the engines mechanically and adjust the stroke lengths so that the work done per cycle by H is just sufficient to operate H' (Fig. 21–7). Suppose the efficiency e of H were greater than the efficiency e' of H'. Then

$$e > e', \qquad \text{(assumption)}$$

or

$$\frac{Q_1 - Q_2}{Q_1} > \frac{Q_1' - Q_2'}{Q_1'}$$

Since the work per cycle done by one engine equals the work per cycle done on the other engine,

$$W = W',$$

or

$$Q_1 - Q_2 = Q_1' - Q_2'.$$

Comparing these relations, we see that (since $Q_1 - Q_2 > 0$)

$$\frac{1}{Q_1} > \frac{1}{Q_1'}$$

or

$$Q_1 < Q_1'.$$

Hence (from the work equality),

$$Q_2 < Q_2'.$$

Thus, the hot source gains heat $Q_1' - Q_1$ (positive) and the cool source loses heat $Q_2' - Q_2$ (positive). But no work is done in the process by the combined system

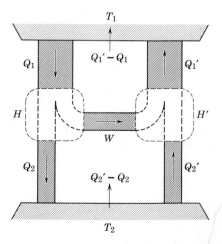

FIGURE 21–7   Proof of Carnot's theorem.

$H + H'$ so that we have transferred heat from a body at one temperature to a body at a higher temperature without performing work—in direct contradiction to Clausius' statement of the second law. Hence, we conclude that e cannot be greater than e'. Likewise, by reversing the engines we can use the same reasoning to prove that e' cannot be greater than e. Hence,

$$e = e',$$

proving the first part of Carnot's theorem.

Now suppose that H is an irreversible engine. Then by the exact same procedure we can prove that e_{ir} cannot be greater than e'. But H cannot be reversed, so we cannot prove that e' cannot be greater than e_{ir}. Therefore, e_{ir} is either equal to or less than e'. Since $e' = e = e_{reversible}$, we have

$$e_{irreversible} \leqslant e_{reversible},$$

thus proving the second part of Carnot's theorem.

Example 2. A steam engine takes steam from the boiler at 200° C (225 lb/in² pressure) and exhausts directly into the air (14 lb/in² pressure) at 100° C. What is its maximum possible efficiency?

Using the result of Example 1 (which applies to this case by virtue of Carnot's theorem, which we have just proved) we have

$$e_{max} = e_{reversible} = \frac{T_1 - T_2}{T_1} = \frac{473° \text{ K} - 373° \text{ K}}{473° \text{ K}} \times 100\% = 21.1\%.$$

Actual efficiencies of about 15% are usually realized. Energy is lost by friction, turbulence, and heat conduction. Lower exhaust temperatures on more complicated steam engines may raise the maximum possible efficiency to 35% and the actual efficiency to 20%. The efficiency of an ordinary automobile engine is about 22% and that of a large diesel oil engine about 40%.

21–6 Entropy—Reversible Processes

The zeroth law of thermodynamics is related to the concept of temperature T and the first law is related to the concept of internal energy U. In this and the following sections we show that the second law of thermodynamics is related to a thermodynamic variable called *entropy*, S, and that we can express the second law quantitatively in terms of this variable. We start by considering a Carnot cycle. For such a cycle we have seen (Eq. 21–3) that

$$\frac{Q_1}{T_1} = \frac{Q_2}{T_2},$$

in which the Q's were taken as positive quantities, that is, we dealt only with the magnitudes or absolute values of the Q's. If we now interpret them again as algebraic quantities, Q being positive when heat enters the system and negative when heat leaves the system, we can write this relation as

$$\frac{Q_1}{T_1} + \frac{Q_2}{T_2} = 0.$$

This equation states that the sum of the algebraic quantities Q/T is zero for a Carnot cycle.

As a next step, we assert that any reversible cycle is equivalent, to as close an approximation as we wish, to an assembly of Carnot cycles. Figure 21–8a shows an arbitrary reversible cycle superimposed on a family of isotherms. We can approximate the actual cycle by connecting the isotherms by suitably chosen adiabatic lines (Fig. 21–8b), thus forming an assembly of Carnot cycles. You should convince yourself that traversing the individual Carnot cycles in Fig. 21–8b is exactly equivalent, in terms of heat transferred and work done, to traversing the jagged sequence of isotherms and adiabatic lines that approximates the actual cycle. This is so because adjacent Carnot cycles have a common isotherm and the two traversals, in opposite directions, cancel each other in the region of overlap as far as heat transfer and work done are concerned. By making the temperature interval between the isotherms in Fig. 21–8b small enough we can approximate the actual cycle as closely as we wish by an alternating sequence of isotherms and adiabatic lines.

We can write, then, for the isothermal-adiabatic sequence of lines in Fig. 21–8b,

$$\sum \frac{Q}{T} = 0,$$

or, in the limit of infinitesimal temperature differences between the isotherms of Fig. 21–8b,

$$\oint \frac{dQ}{T} = 0, \tag{21–4}$$

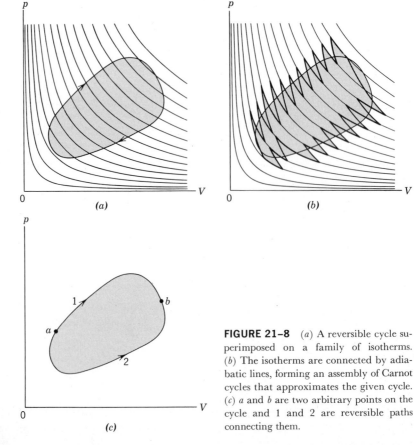

FIGURE 21–8   (a) A reversible cycle superimposed on a family of isotherms. (b) The isotherms are connected by adiabatic lines, forming an assembly of Carnot cycles that approximates the given cycle. (c) a and b are two arbitrary points on the cycle and 1 and 2 are reversible paths connecting them.

in which $\oint$ indicates that the integral is evaluated for a complete traversal of the cycle, starting (and ending) at any arbitrary point of the cycle.

If the integral of a quantity around any closed path is zero, that quantity is called a state variable, that is, it has a value that is characteristic only of the state of the system, regardless of how that state was arrived at. We call the variable in this case the *entropy* S and we have, from Eq. 21–4,

$$dS = \frac{dQ}{T} \quad \text{and} \quad \oint dS = 0. \tag{21-5}$$

Common units for entropy are joules/K° or cal/K°.

Gravitational potential energy U_g, internal energy U, pressure p, and temperature T are other state variables and equations of the form $\oint dX = 0$ hold for each of them, where for X we substitute the appropriate symbol. Heat Q and work W are not state variables and we know that, in general $\oint dQ \neq 0$ and $\oint dW \neq 0$, as you can easily show for the special case of a Carnot cycle.

The property of a state variable expressed by $\oint dX = 0$ can also be expressed by saying that $\int dX$ between any two equilibrium states has the same value for all (reversible) paths connecting those states. Let us prove this for the state called entropy. We can write Eq. 21–5 (see Fig. 21–8c) as

$$_1\int_a^b dS + {}_2\int_b^a dS = 0 \tag{21-6}$$

where a and b are arbitrary points and 1 and 2 describe the paths connecting these points. Since the cycle is reversible, we can write Eq. 21–6 as

$$_1\int_a^b dS - {}_2\int_a^b dS = 0$$

or

$$_1\int_a^b dS = {}_2\int_a^b dS \tag{21-7}$$

In Eq. 21–7 we have simply decided to traverse path 2 in the opposite direction, that is, from a to b rather than from b to a. We do this by changing the order of the limits in the second integral of Eq. 21–6, which requires that we also change the sign of the integral, thus yielding Eq. 21–7. This latter equation tells us that the quantity $\int_a^b dS$ between any two equilibrium states of the system, such as a and b, is independent of the path connecting those states, for 1 and 2 are quite arbitrary paths. Recall our almost identical discussion in Section 7–2, where we introduced the concept of a conservative force.

The change in entropy between a and b in Fig. 21–8c is, then

$$S_b - S_a = \int_a^b dS = \int_a^b \frac{dQ}{T} \quad \text{(reversible process)}, \tag{21-8}$$

where the integral is evaluated over any reversible path connecting these two states.

21–7   Entropy—Irreversible Processes

In Section 21–6 we spoke only of reversible processes. However, entropy, like all state variables, depends only on the state of the system and we must be able to calculate the change in entropy for irreversible processes, provided only that they begin and end in equilibrium states. Let us consider two examples.

1. Free Expansion. As in Section 19–7 (see Fig. 19–9) let a gas double its volume by expanding into an evacuated enclosure. Since no work is done against the vacuum, $W = 0$ and, since the gas is enclosed by nonconducting walls, $Q = 0$. From the first law, then $\Delta U = 0$ or

$$U_i = U_f \tag{21–9}$$

where i and f refer to the initial and final (equilibrium) states. If the gas is an ideal gas, then U depends on temperature alone and not on the pressure or the volume so that Eq. 21–9 implies $T_i = T_f$.

The free expansion is certainly irreversible because we lose control of the environment once we turn the stopcock in Fig. 19–9. There is, however, an entropy difference $S_f - S_i$ between the initial and final equilibrium states, but we cannot calculate it from Eq. 21–8 because that relation applies only to reversible paths; if we tried to use that equation we would have the immediate difficulty that $Q = 0$ for the free expansion and—further—we would not know how to assign meaningful values of T to the intermediate, nonequilibrium states.

How, then, do we calculate $S_f - S_i$ for a free expansion? We do so by finding a reversible path (any reversible path) that connects the states i and f and we calculate the entropy change for that path. In the free expansion a convenient reversible path (assuming an ideal gas) is an isothermal expansion from V_i to V_f ($= 2V_i$). This corresponds to the isothermal expansion carried out between the points a and b of the Carnot cycle of Fig. 21–4. It represents quite a different set of operations from the free expansion and has in common with it only the fact that it connects the same set of equilibrium states, i and f. From Eq. 21–8 and Example 1 we have

$$S_f - S_i = \int_i^f \frac{dQ}{T} = \mu R \ln \left(V_f / V_i \right)$$

$$= \mu R \ln 2.$$

This is positive so that the entropy of the system increases in this irreversible, adiabatic process.

2. Heat Conduction. For another example consider two bodies that are similar in every respect except that one is at a temperature T_1 and the other at temperature T_2, where $T_1 > T_2$. If we put both objects in contact inside a box with nonconducting walls, they will eventually reach a common temperature T_m, approximately half-way between T_1 and T_2. Like the free expansion, the process is irreversible because we lose control of the environment once we put the two bodies in the box. Like the free expansion this process is also (irreversibly) adiabatic because no heat enters or leaves the system during the process.

To calculate the entropy change for the system during this process we must again find a reversible process connecting the same initial and final states and calculate the system entropy change by applying Eq. 21–8 to that process. We can do so if we imagine that we have at our disposal a heat reservoir of large heat capacity whose temperature T is at our control, by turning a knob, say. We first adjust the reservoir temperature to T_1 and put the first (hotter) object in contact with the reservoir. We then slowly (reversibly) lower the reservoir temperature from T_1 to T_m, extracting heat from the hot body as we do so. The hot body *loses* entropy in this process, the amount being approximately

$$\Delta S_1 \cong - \frac{Q}{T_{1,m}}$$

where $T_{1,m}$ is the average of T_1 and T_m and Q is the heat extracted.

We then adjust our reservoir temperature to T_2 and place it in contact with the second (cooler) object. We then slowly (reversibly) raise the reservoir temperature from T_2 to T_m, adding heat to the cool body as we do so. The cool body gains entropy in this process, the amount being approximately

$$\Delta S_2 \cong + \frac{Q}{T_{2,m}},$$

where $T_{2,m}$ is the average of T_2 and T_m and Q is the heat added.

The two bodies are now at the same temperature T_m and the system, which consists of these two bodies, is now in its final equilibrium state. The change in entropy for the complete system is

$$S_f - S_i = \Delta S_1 + \Delta S_2$$

$$= - \frac{Q}{T_{1,m}} + \frac{Q}{T_{2,m}}.$$

Since $T_{1,m} > T_{2,m}$ we have $S_f > S_i$. Again, as for the free expansion, the entropy of the system has increased in this irreversible, adiabatic process.

In each of these examples we must distinguish carefully between the actual (irreversible) process (free expansion or heat conduction) and the reversible process that we introduce just so that we can calculate the entropy change in the actual process. We can choose any reversible process, as long as it connects the same initial and final states as the actual process; all such reversible processes will yield the same entropy change because this depends only on the initial and final states and not on the process connecting them—be it reversible or irreversible.

21–8 Entropy and the Second Law

We are now ready to formulate the second law of thermodynamics in terms of entropy. Since this law is a generalization from experience we cannot *prove* it but can only write it down and show that our statement is in agreement with experiment and is equivalent to other formulations of the second law that we have given earlier. In this spirit we assert that the second law is: *A natural process that starts in one equilibrium state and ends in another will go in the direction that causes the entropy of the system plus environment to increase.*

The two experiments of Section 21–7 (free expansion and heat conduction) are consistent with the second law. The entropy of the system increased in each of these irreversible processes. Note that the entropy of the environment in these two cases remains unchanged because, both being carried out in adiabatic enclosures, there was no interchange of heat with the environment. Thus, as required by our statement of the second law, the entropy of the system plus environment increased for each of these (natural) processes.

In the form that we have written it the second law applies only to irreversible processes because only such processes have a "natural direction." Indeed (see Section 21–1) the understanding of the natural direction of such processes is the main concern of the second law. Reversible processes can go equally well in either direction, however, and *for reversible processes the entropy of the system plus environment remains unchanged.* This is so because if heat dQ is transferred from the environment to the system the entropy of the environment *decreases* by dQ/T while that of the system *increases* by dQ/T, the net change for the system plus environment being zero. The fact that the process is reversible means that the environment and the system can differ in temperature by only a differential amount dT when the heat transfer takes place; this is in sharp contrast to our (irreversible) heat conduction problem of the previous section, in which the temperature difference of the two bodies placed in contact was large.

Another class of processes of particular interest are adiabatic processes (reversible or irreversible); they involve no transfer of heat with the environment so that the only entropy change possible is that of the system. From our statement of the second law and from our remarks about reversible processes in the paragraph above, we conclude that

$$S_f = S_i \qquad \text{(reversible adiabatic process)}$$

and

$$S_f > S_i \qquad \text{(irreversible adiabatic process),}$$

where S_f and S_i are the final and intial entropies of the system.

Our statement of the second law is consistent with the Clausius statement (page 408) which declares that there is no such thing as a "perfect" refrigerator (see Fig. 21–6). If there were, the entropy of the lower temperature reservoir would decrease by Q/T_2; that of the upper temperature reservoir would increase by Q/T_1; that of the system would remain unchanged because the system traverses a cycle, returning to its starting point. Thus the net change in the entropy of the system plus environment is a decrease, because $T_2 < T_1$. This violates the statement of the second law that we have just given and, if we wish to retain the statement, we must conclude (with Clausius) that there is no such thing as a "perfect" refrigerator.

Our statement of the second law is also consistent with the Kelvin-Planck statement (page 408) which declares that there is no such thing as a "perfect" heat engine (see Fig. 21–5). If there were, the entropy of the reservoir at temperature T would decrease by Q/T; that of the system would remain unchanged because the system traverses a cycle, returning to its starting point. Thus the net change of entropy of the system plus environment is a decrease. This violates the statement of the second law that we have just given and, if we wish to retain the statement, we must conclude (with Kelvin) that there is no such thing as a "perfect" heat engine.

Example 3. Compute the entropy change of a system consisting of 1.00 kg of ice at 0° C which melts (reversibly) to water at that same temperature. The latent heat of melting is 79.6 cal/gm.

The requirement that we melt the ice reversibly means that we must put it in contact with a heat reservoir whose temperature exceeds 0° C by only a differential amount. (If we lower the reservoir temperature until it is a differential amount below 0° C, the melted ice will begin to freeze.) Since the process is reversible, we can use Eq. 21–8 to compute the entropy change of the system. The temperature remains constant at 273° K. Therefore,

$$S_{\text{water}} - S_{\text{ice}} = \int_0^Q \frac{dQ}{T} = \frac{1}{T} \int_0^Q dQ = \frac{Q}{T}.$$

But

$$Q = 1.00 \times 10^3 \text{ gm} \times 79.6 \text{ cal/gm} = 7.96 \times 10^4 \text{ cal}$$

or

$$S_{\text{water}} - S_{\text{ice}} = \frac{7.96 \times 10^4}{273} \text{ cal/K}° = 292 \text{ cal/K}°$$

$$= 1220 \text{ joules/K}°.$$

In this example of reversible melting the entropy change of the system plus environment is zero, as it must be for all reversible processes. The entropy change calculated above is the increase in entropy of the system; there is an exactly equal decrease in entropy of the environment (-1220 joules/K°) associated with the heat that leaves the reservoir (environment), at 273° K, to melt the ice.

In practice, melting is likely to be irreversible, as when we put an ice cube in a glass of water at room temperature. This process has only one natural direction—the ice will melt. The entropy of the system plus environment will increase in this process as required by the second law. The (irreversible) heat conduction example of the previous section should make this understandable.

Example 4. Calculate the entropy change that an ideal gas undergoes in a reversible isothermal expansion from a volume V_i to a volume V_f.

From the first law

$$dU = dQ - p \, dV.$$

But $dU = 0$, since U depends only on temperature for an ideal gas and the temperature is constant. Hence,

$$dQ = p \, dV$$

and

$$dS = \frac{dQ}{T} = \frac{p \, dV}{T}.$$

But

$$pV = \mu RT,$$

so that

$$dS = \mu R \frac{dV}{V}$$

and

$$S_f - S_i = \int_{V_i}^{V_f} \mu R \frac{dV}{V} = \mu R \ln \frac{V_f}{V_i}. \tag{21–10}$$

Since $V_f > V_i$, $S_f > S_i$ and the entropy of the gas increases.

In order to carry out this process we must have a reservoir at temperature T which is in contact with the system and supplies the heat to the gas. Hence, the entropy of the reservoir decreases by $\int dQ/T \ [= \mu R \ln (V_f/V_i)]$, so that in this process the entropy of system plus environment does not change. As in the previous example, this is characteristic of a reversible process.

QUESTIONS

1. What requirements should a system meet in order to be in thermodynamic equilibrium?

2. In the irreversible process of Fig. 21–1a can we calculate the work done in terms of an area on a p-V diagram? Is any work done?

3. Can a given amount of mechanical energy be converted completely into heat energy? If so, give an example.

4. Can you suggest a reversible process whereby heat can be added to a system? Would adding heat by means of a Bunsen burner be a reversible process?

5. Give some examples of irreversible processes in nature.

6. Give a qualitative explanation of how frictional forces between moving surfaces produce heat energy. Does the reverse process (heat energy producing relative motion of these surfaces) occur? Can you give a plausible explanation?

7. A block returns to its initial position after dissipating mechanical energy to heat through friction. Is this process reversible thermodynamically?

8. To carry out a Carnot cycle need we start at point a in Fig. 21–4? May we equally well start at points b, c, or d, or indeed at any intermediate point?

9. If a Carnot engine is independent of the working substance, then perhaps real engines should be similarly independent, to a certain extent. Why then, for real engines, are we so concerned to find suitable fuels such as coal, gasoline, or fissionable material? Why not use stones as a fuel?

10. Couldn't we just as well define the efficiency of an engine as $e = W/Q_2$ rather than as $e = W/Q_1$? Why don't we?

11. What factors reduce the efficiency of a heat engine from its ideal value?

12. In order to increase the efficiency of a Carnot engine most effectively, would you increase T_1, keeping T_2 constant, or would you decrease T_2, keeping T_1 constant?

13. Can a kitchen be cooled by leaving the door of an electric refrigerator open? Explain.

14. Is there a change in entropy in purely mechanical motions?

15. Two samples of a gas initially at the same temperature and pressure are compressed from a volume V to a volume $(V/2)$, one isothermally, the other adiabatically. In which sample is the final pressure greater? Does the entropy of the gas change in either process?

16. Suppose we had chosen to represent the state of a system by its entropy and its absolute temperature rather than by its pressure and volume. What would a Carnot cycle look like on a *T-S* diagram?

17. Show that the total entropy increases when work is converted into heat by friction between sliding surfaces.

18. Discuss the following comment of Panofsky and Phillips: "From the standpoint of formal physics there is only one concept which is asymmetric in the time, namely entropy. But this makes it reasonable to assume that the second law of thermodynamics can be used to ascertain the sense of time independently in any frame of reference; that is, we shall take the positive direction of time to be that of increasing . . . entropy. . . ."

PROBLEMS

1(3). An ideal gas heat engine operates in a Carnot cycle between 227 and 127° C. It absorbs 6.0×10^4 cal at the higher temperature. How much work per cycle is this engine capable of performing?

Answer: 5.0×10^4 joules.

2(3). In a Carnot cycle, the isothermal expansion of an ideal gas takes place at 400° K and the isothermal compression at 300° K. During the expansion 500 cal of heat energy are transferred to the gas. Determine (*a*) the work performed by the gas during the isothermal expansion, (*b*) the heat rejected from the gas during the isothermal compression, (*c*) the work done on the gas during the isothermal compression.

3(3). If the Carnot cycle is run backward, we have an ideal refrigerator. A quantity of heat Q_2 is taken in at the lower temperature T_2 and a quantity of heat Q_1 is given out at the higher temperature T_1. The difference is the work W that must be supplied to run the refrigerator. (*a*) Show that

$$W = Q_2 \frac{T_1 - T_2}{T_2} .$$

(b) The *coefficient of performance* K of a refrigerator is defined as the ratio of the heat extracted from the cold source to the work needed to run the cycle. Show that ideally

$$K = \frac{T_2}{T_1 - T_2} .$$

In actual refrigerators K has a value of 5 or 6.

4(3). How is the efficiency of a reversible heat engine related to the coefficient of performance of the reversible refrigerator obtained by running the engine backward?

5(3). (a) A Carnot engine operates between a hot reservoir at 320° K and a cold reservoir at 260° K. If it absorbs 500 joules of heat at the hot reservoir, how much work does it deliver? (b) If the same engine, working in reverse, functions as a refrigerator between the same two reservoirs, how much work must be supplied to remove 1000 joules of heat from the cold reservoir?
Answer: (a) 94 joules. (b) 230 joules.

6(3). How much work must be done to transfer 1.0 joule of heat from a reservoir (a) at 7° C to one at 27° C by means of a refrigerator using a Carnot cycle? (b) From one at −73° C to one at 27° C? (c) From one at −173° C to one at 27° C? (d) From one at −223° C to one at 27° C?

7(3). (a) Plot accurately a Carnot cycle on a p-V diagram for 1.0 mole of an ideal gas. Let point a correspond to $p = 1.0$ atm, $T = 300°$ K, and let b correspond to $p = 0.50$ atm, $T = 300°$ K; take the low temperature reservoir to be at 100° K. Let $\gamma = 1.5$. (b) Compute graphically the work done in this cycle.
Answer: (b) About 1100 joules.

8(3). A Carnot engine works between temperatures T_1 and T_2. It drives a Carnot refrigerator that works between two different temperatures T_3 and T_4 (see Fig. 21–9). Find the ratio Q_3/Q_1 in terms of the four temperatures.

9(3). In a two-stage heat engine a quantity of heat Q_1 is absorbed at a temperature T_1, work W_1 is done, and a quantity of heat Q_2 is expelled at a lower temperature T_2 by the first stage. The second stage absorbs the heat expelled by the first, does work W_2, and expels a quantity of heat Q_3 at a lower temperature T_3. Prove that the efficiency of the combination engine is $(T_1 - T_3)/T_1$.

10(5). A combination mercury-steam turbine takes saturated mercury vapor from a boiler at 876° F and exhausts it to heat a steam boiler at 460° F. The steam turbine receives steam at this temperature and exhausts it to a condenser at 100° F. What is the maximum efficiency of the combination?

11(5). Suppose a deep shaft were drilled in the earth's crust near one of the poles where the surface temperature is −40° C to a depth where the temperature is 800° C. (a) What is the theoretical limit to the efficiency of an engine operating between these temperatures? (b) If all of the heat released into the low temperature reservoir were used to melt ice that was initially at −40° C, at what rate could water at 0° C be produced by a power plant having an output of 100 megawatts? The specific heat of ice is 0.50 cal/gm C°; its heat of fusion is 80 cal/gm.
Answes: (a) 78%. (b) 18 gal/sec.

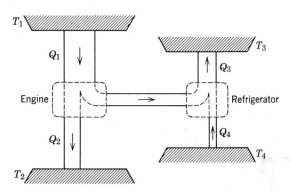

FIGURE 21–9   Problem 8(3).

12(5). In a mechanical refrigerator the low-temperature coils are at a temperature of $-13°$ C, and the compressed gas in the condenser has a temperature of $27°$ C. What is the theoretical maximum coefficient of performance? [See Problem 3(3).]

13(5). The motor in a refrigerator has a power output of 200 watts. If the freezing compartment is at $270°$ K and the outside air is at $300°$ K, assuming ideal efficiency, what is the maximum amount of heat that can be extracted from the freezing compartment in 10 min?
Answer: 1.1×10^6 joules.

14(5). In a heat pump, heat Q_2 is extracted from the outside atmosphere at T_2 and a larger quantity of heat Q_1 is delivered to the inside of the house at T_1, requiring the performance of work W. (*a*) Draw a schematic diagram of a heat pump. (*b*) How does it differ in principle from a refrigerator? (*c*) How are Q_1, Q_2 and W related to one another? (*d*) Can a heat pump be reversed for use in summer? Explain. (*e*) What advantages does such a pump have over other heating devices?

15(5). In a heat pump, heat from the outdoors at $-5°$ C is transferred to a room at $17°$ C, energy being supplied by an electric motor. How many joules of heat will be delivered to the room for each joule of electric energy consumed, ideally?
Answer: 13 joules.

16(5). Compute the efficiency of the cycle shown in Figure 20–11. See Problem 30(6) of Chapter 20.

17(5). One mole of a monatomic ideal gas initially at a volume of 10 liters and a temperature of $300°$ K is heated at constant volume to a temperature of $600°$ K, allowed to expand isothermally to its initial pressure and finally compressed isobarically to its original volume, pressure and temperature. (*a*) Compute the heat input to the system during one cycle. (*b*) What is the net work done by the gas during one cycle? (*c*) What is the efficiency of this cycle?
Answer: (*a*) 7200 joules. (*b*) 960 joules. (*c*) 13%.

18(5). One mole of a monatomic ideal gas is caused to go through the cycle shown in Fig. 21–10. Process *bc* is a reversible adiabatic expansion. $p_b = 10$ atm, $V_b = 1$ m³ and $V_c = 8$ m³. Calculate (*a*) the heat added to the gas, (*b*) the heat leaving the gas, (*c*) the net work done by the gas and (*d*) the efficiency of the cycle.

19(5). A gasoline internal combustion engine can be approximated by the cycle shown in Figure 21–11. Assume an ideal gas and use a compression ratio of 4:1 ($V_4 = 4V_1$). Assume $p_2 = 3p_1$. (*a*) Determine the pressure and temperature of each of the vertex points of the *p-V* diagram in terms of p_1, T_1 and the ratio of specific heats of the gas. (*b*) What is the efficiency of this cycle?
Answer: (*a*) $T_2 = 3T_1$
$$T_3 = 3(4)^{1-\gamma}T_1$$
$$T_4 = (4)^{1-\gamma}T_1$$
$$p_2 = 3p_1$$
$$p_3 = (3)(4)^{-\gamma}p_1$$
$$p_4 = (4)^{-\gamma}p_1$$
(*b*) $1 - (4)^{1-\gamma}$

20(6). One mole of an ideal monatomic gas is caused to go through the cycle shown in Fig.

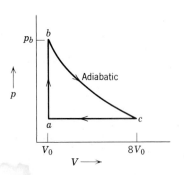

FIGURE 21–10   Problem 18(5).

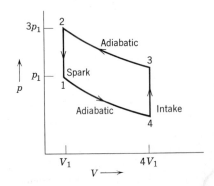

FIGURE 21–11   Problem 19(5).

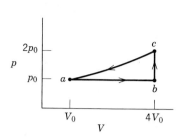

FIGURE 21-12   Problem 20(6).

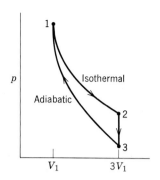

FIGURE 21-13   Problem 22(6).

21–12. (a) How much work is done in expanding the gas from a to c along path abc? (b) What is the change in internal energy and entropy in going from b to c? (c) What is the change in internal energy and entropy in going through one complete cycle? Express all answers in terms of the pressure p_0, volume V_0 and temperature T_0 at point a in the diagram.

21(6). A mole of a monatomic ideal gas is taken from an initial state of pressure p and volume V to a final state of pressure $2p$ and volume $2V$ by two different processes. (I) It expands isothermally until its volume is doubled, and then its pressure is increased at constant volume to the final state. (II) It is compressed isothermally until its pressure is doubled, and then its volume is increased at constant pressure to the final state.

Show the path of each process on a p-V diagram. For each process calculate in terms of p and V (a) the heat absorbed by the gas in each part of the process; (b) the work done by the gas in each part of the process; (c) the change in internal energy of the gas $U_f - U_i$; (d) the change in entropy of the gas $S_f - S_i$.

Answer: (a) Path I: $Q_T = pV \ln (2)$ Path II: $Q_T = -pV \ln (2)$

$\qquad\qquad Q_V = \frac{9}{2}pV$ $\qquad\qquad Q_p = \frac{15}{2}pV.$

$\qquad$(b) Path I: $W_T = pV \ln (2)$ Path II: $W_T = -pV \ln (2)$

$\qquad\qquad W_V = 0$ $\qquad\qquad W_V = 3pV.$

$\qquad$(c) $\frac{9}{2}pV$ for either case.

$\qquad$(d) $4R \ln (2)$ for either case.

22(6). An ideal diatomic gas is caused to pass through the cycle shown on the pV diagram in Figure 21–13. $V_2 = 3V_1$. Determine, in terms of p_1, V_1, T_1, and R, (a) p_2, p_3, and T_3; (b) W, Q, ΔU, and ΔS per mole for all three processes.

23(6). Find (a) the heat absorbed and (b) the change in entropy of a 1.0-kg block of copper whose temperature is increased reversibly from 25° C to 100° C. The specific heat for copper is 9.2×10^{-2} cal/gmC°.

Answer: 6900 cal. (b) 21 cal/°K.

24(7). Heat can be removed from water at 0° C and atmospheric pressure without causing the water to freeze, if done with little disturbance of the water. Suppose the water is cooled to −5.0° C before ice begins to form. What is the change to entropy per unit mass occurring during the sudden freezing that then takes place?

25(7). In a specific heat experiment 100 gm of lead ($c_p = 0.0345$ cal/gmC°) at 100° C is mixed with 200 gm of water at 20° C. Find the difference in entropy of the system at the end from its value before mixing.

Answer: +0.11 cal/°K.

26(7). A 50-gram block of copper having a temperature of 400° K is placed in an insulating box with a 100-gram block of lead having a temperature of 200° K. (a) What is the equilibrium temperature of this two-block system? (b) What is the change in the internal energy of the two-block system as it changes from the initial condition to the equilibrium condition? (c) What is the change in the entropy of the two-block system? (See Table 19–1, page 359.)

27(7). An 8.00-gm ice cube at −10.0° is dropped into a thermos flask containing 100 cm³ of water at 20.0° C. What is the change in entropy of the system when a final equilibrium state is reached? The specific heat of ice is 0.52 cal/gm C°.

Answer: +0.18 cal/°K.

28(7). A 10-gram ice cube at −10° C is placed in a lake whose temperature is +15° C. Calculate the change in entropy of the system as the ice cube comes to thermal equilibrium with the lake.

29(8). (*a*) Show that when a substance of mass m having a constant specific heat c is heated from T_1 to T_2 the entropy change is

$$S_2 - S_1 = mc \ln \frac{T_2}{T_1}.$$

(*b*) Does the entropy of the substance decrease on cooling? (*c*) If so, does the total entropy of the universe decrease in such a process?

Answer: (*b*) Yes. (*c*) No.

30(8). Four moles of an ideal gas are caused to expand from a volume V_1 to a volume V_2 ($=2V_1$). (*a*) If the expansion is isothermal at the temperature $T = 400°$ K, find the work done by the expanding gas. (*b*) Find the change in entropy, if any. (*c*) If the expansion were reversibly adiabatic instead of isothermal, would the change in entropy be positive, negative, or zero?

31(8). A brass rod is in contact thermally with a heat reservoir at 127° C at one end and a heat reservoir at 27° C at the other end. (*a*) Compute the total change in the entropy arising from the process of conduction of 1200 cal of heat through the rod. (*b*) Does the entropy of the rod change in the process?

Answer: (*a*) 1.0 cal/°K. (*b*) No.

32(8). A mixture of 1773 grams of water and 227 grams of ice at 0° C is, in a reversible process, brought to a final equilibrium state where the water-ice ratio is 1:1 at 0° C. (*a*) Calculate the entropy change of the system during this process. (The heat of fusion for water is approximately 80 cal/gm.) (*b*) The system is then returned to the first equilibrium state, but in an irreversible way (by using a bunsen burner, for instance). Calculate the entropy change of the system during this process. (*c*) Is your answer consistent with the second law of thermodynamics?

33(8). Two pieces of molding clay of equal mass are moving in opposite directions with equal speed. They strike and stick together. Treat the two pieces as a single system and state whether each of the following quantities is positive, negative, or zero for this process: ΔU, W, Q, and ΔS. Justify your answers.

Answer: $\Delta U = 0$. $W = 0$. $Q = 0$. $\Delta S > 0$.

CHAPTER 22

Charge and Matter

22–1 Electromagnetism

The science of electricity has its roots in the observation, known in 600 B.C., that a rubbed piece of amber will attract bits of straw. The study of magnetism goes back to the observation that naturally occurring "stones" (that is, magnetite) will attract iron. These two sciences developed quite separately until 1820, when Hans Christian Oersted observed a connection between them, namely, that an electric current in a wire can affect a magnetic compass needle (Section 29–1).

The new science of electromagnetism was developed further by many workers, of whom one of the most important was Michael Faraday (1791–1867). James Clerk Maxwell (1831–1879) put the laws of electromagnetism in the form in which we know them today. These laws, often called *Maxwell's equations,* are displayed in Table 34–2. These laws play the same role in electromagnetism that Newton's laws of motion and of gravitation do in mechanics.

Although Maxwell's synthesis of electromagnetism rests heavily on the work of his predecessors, his own contribution was vital. Maxwell deduced that light is electromagnetic in nature and that its speed can be found by making purely electric and magnetic measurements. Thus the science of optics was connected with those of electricity and of magnetism. The scope of Maxwell's equations is remarkable, including as it does the principles of all large-scale electromagnetic and optical devices such as motors, cyclotrons, electronic computers, radio, television, radar, microscopes, and telescopes.

The English physicist Oliver Heaviside (1850–1925) and especially the Dutch physicist H. A. Lorentz (1853–1928) also contributed substantially to the clarification

of electromagnetic theory. Heinrich Hertz (1857–1894)* took a great step forward
when, more than twenty years after Maxwell set up his theory, he produced in the
laboratory electromagnetic "Maxwellian waves" of a kind that we now call short
radio waves. It remained for Marconi and others to exploit this application of the
electromagnetic waves of Maxwell and Hertz.

At the level of engineering applications Maxwell's equations are used constantly in
the solution of practical problems.

The science of electromagnetism, based on Maxwell's equations, is called *classical
electromagnetism*. It gives correct answers for all problems involving electric and magnetic fields down to dimensions of about 10^{-10} cm. This is much smaller than an atom
and is well into the region in which Newtonian mechanics must be replaced by quantum mechanics. A blending of classical electromagnetism, quantum mechanics, and
relativity theory, called *quantum electrodynamics,* gives correct answers down to very
much smaller dimensions. F. J. Dyson has said of this theory: "It is the only field in
which we can choose a hypothetical experiment and predict the results to five places
of decimals, confident that the theory takes into account all the factors that are
involved." This book deals only with classical electromagnetism.

22–2 Electric Charge

The rest of this chapter deals with electric charge and its relationship to matter.
There are two kinds of charge. Let us rub a glass rod with silk and hang it from a long
thread as in Fig. 22–1. If a second glass rod is rubbed with silk and held near the
rubbed end of the first rod, the rods will repel each other. On the other hand, a Lucite

* "Heinrich Hertz" by P. and E. Morrison, *Scientific American,* December 1957.

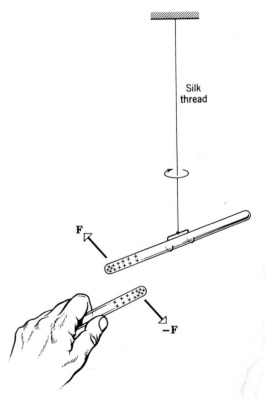

FIGURE 22–1 Two positively charged glass
rods repel each other.

rod rubbed with fur will attract the glass rod. Two Lucite rods rubbed with fur will repel each other. We explain these facts by saying that rubbing a rod gives it an electric charge and that the charges on the two rods exert forces on each other. Clearly the charges on the glass and on the Lucite must be different.

Benjamin Franklin (1706–1790), who, among his other achievements, was the first American physicist, named the kind of electricity that appears on the glass *positive* and the kind that appears on the Lucite (hard rubber in his day) *negative;* these names have remained to this day. We can sum up these experiments by saying that *like charges repel and unlike charges attract.*

Electric effects are not limited to glass rubbed with silk or to Lucite rubbed with fur. Any substance rubbed with any other under suitable conditions will become charged to some extent; by comparing the unknown charge with a charged glass rod or a charged Lucite rod, it can be labeled as either positive or negative.

The modern view of bulk matter is that, in its normal or neutral state, it contains equal amounts of positive and negative electricity. If two bodies like glass and silk are rubbed together, a small amount of charge is transferred from one to the other, upsetting the electric neutrality of each. In this case the glass would become positive, the silk negative.

22–3 Conductors and Insulators

A metal rod held in the hand and rubbed with fur will not seem to develop a charge. It is possible to charge such a rod, however, if it is furnished with a glass handle and if the metal is not touched with the hands while rubbing it. The explanation is that metals, the human body, and the earth are *conductors* of electricity and that glass, plastics, etc., are *insulators* (also called *dielectrics*).

In conductors electric charges are free to move through the material, whereas in insulators they are not. Although there are no perfect insulators, the insulating ability of fused quartz is about 10^{25} times as great as that of copper, so that for many practical purposes some materials behave as if they were perfect insulators.

In metals a fairly subtle experiment called the Hall effect (see Section 29–5) shows that only negative charge is free to move. Positive charge is as immobile as it is in glass or in any other dielectric. The actual charge carriers in metals are the free electrons. When isolated atoms are combined to form a metal, the outer electrons of the atom do not remain attached to individual atoms but become free to move throughout the volume of the solid. For some conductors, such as electrolytes, both positive and negative charges can move.

P. 546

Semiconductors are intermediate between conductors and insulators in their ability to conduct electricity. Among the elements, silicon and germanium are well-known examples. In semiconductors the electrical conductivity can often be greatly increased by adding small amounts of other elements; traces of arsenic or boron are often added to silicon for this purpose. Semiconductors have many practical applications, among which is their use in the construction of transistors. The mode of action of semiconductors cannot be described adequately without some understanding of quantum physics.

22–4 Coulomb's Law

Charles Augustin de Coulomb in 1785 first measured electrical attractions and repulsions quantitatively and deduced the law that governs them. His apparatus, shown in

Fig. 22–2, resembles the hanging rod of Fig. 22–1, except that the charges in Fig. 22–2 are confined to small spheres *a* and *b*.

If *a* and *b* are charged, the electric force on *a* will tend to twist the suspension fiber. Coulomb canceled out this twisting effect by turning the suspension head through the angle θ needed to keep the two charges at the particular distance apart in which he was interested. The angle θ is then a relative measure of the electric force acting on charge *a*. The device of Fig. 22–2 is called a torsion balance; a similar arrangement was used later by Cavendish to measure gravitational attractions (Section 14–2).

Coulomb's first experimental results can be represented by

$$F \propto \frac{1}{r^2} \,.$$

F is the magnitude of the force that acts on each of the two charges *a* and *b*; *r* is their distance apart. These forces, as Newton's third law requires, act along the line joining the charges but point in opposite directions. Note that the magnitude of the force on each charge is the same, even though the charges may be different.

Coulomb also studied how the electrical force varied with the relative size of the charges on the spheres of his torsion balance. For example, if we touch a charged conducting sphere to an identical but uncharged conducting sphere, the original charge must divide equally between the spheres. By such techniques Coulomb extended the inverse square relationship to

$$\vec{F} \propto \frac{q_1 q_2}{r^2} \hat{r} \tag{22-1}$$

where q_1 and q_2 are relative measures of the charges on spheres *a* and *b*. Equation 22–1, which is called *Coulomb's law*, holds only for charged objects whose sizes are much smaller than the distance between them. We often say that it holds only for point charges.

Coulomb's law resembles the inverse square law of gravitation which was already more than 100 years old at the time of Coulomb's experiments; *q* plays the role of *m* in that law. In gravity, however, the forces are always attractive, this corresponds to the fact that there are two kinds of electricity but (apparently) only one kind of mass.

Our belief in Coulomb's law does not rest quantitatively on Coulomb's experiments. Torsion balance measurements are difficult to make to an accuracy of better than a few per cent. Such measurements could not, for example, convince us that the exponent in Eq. 22–1 is exactly 2.00 and not, say, 2.01. In Section 24–5 we show that Coulomb's law can also be deduced from an indirect experiment which shows that the exponent in Eq. 22–1 lies between the limits of approximately 2.000000002

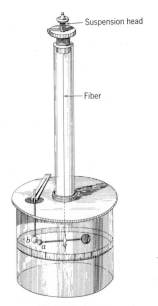

FIGURE 22–2   Coulomb's torsion balance from his 1785 memoir to the French Academy of Sciences.

and 1.999999998. Small wonder that we usually assume the exponent to be exactly 2.

Although we have established the physical concept of electric charge, we have not yet defined a unit in which it may be measured. It is possible to do so by putting equal charges q on the spheres of a torsion balance and by measuring the magnitude F of the force that acts on each when the charges are a measured distance r apart. We could then define q to have a unit value if a unit force acts on each charge when the charges are separated by a unit distance and we can give a name to the unit of charge so defined. This scheme is the basis for the definition of the unit of charge called the *statcoulomb*. However, in this book we do not use this unit or the systems of units of which it is a part.

For practical reasons having to do with the accuracy of measurements, the unit of charge in the mks system is defined in terms of the unit of electric current. If the ends of a long wire are connected to the terminals of a battery, it is common knowledge that an electric current i is set up in the wire. We visualize this current as a flow of charge. The mks unit of current is the *ampere* (abbr. *amp*). In Section 30–4 we describe the operational procedures in terms of which the ampere is defined.

The mks unit of charge is the *coulomb* (abbr. *coul*). *A coulomb is defined as the amount of charge that flows through a given cross section of a wire in 1 second if there is a steady current of 1 ampere in the wire.* In symbols

$$q = it, \tag{22–2}$$

where q is in coulombs if i is in amperes and t is in seconds. Thus, if a wire is connected to an insulated metal sphere, a charge of 10^{-6} coul can be put on the sphere if a current of 1.0 amp exists in the wire for 10^{-6} sec.

Example 1. A copper penny has a mass of 3.1 gm. Being electrically neutral, it contains equal amounts of positive and negative electricity. What is the magnitude q of these charges? A copper atom has a positive nuclear charge of 4.6×10^{-18} coul and a negative electronic charge of equal magnitude.

The number N of copper atoms in a penny is found from the ratio

$$\frac{N}{N_0} = \frac{m}{M},$$

where N_0 is Avogadro's number, m the mass of the coin, and M the atomic weight of copper. This yields, solving for N,

$$N = \frac{(6.0 \times 10^{23} \text{ atoms/mole})(3.1 \text{ gm})}{64 \text{ gm/mole}} = 2.9 \times 10^{22} \text{ atoms}.$$

The charge q is

$$q = (4.6 \times 10^{-18} \text{ coul/atom})(2.9 \times 10^{22} \text{ atoms}) = 1.3 \times 10^5 \text{ coul}.$$

In a 100-watt, 110-volt light bulb the current is 0.91 amp. Verify that it would take 40 hr for a charge of this amount to pass through this bulb.

Equation 22–1 can be written as an equality by inserting a constant of proportionality. Instead of writing this simply as, say, k, it is usually written in a more complex way as $1/4\pi\epsilon_0$ or

$$\vec{F} = \frac{1}{4\pi\epsilon_0} \frac{q_1 q_2}{r^2} \hat{r} \tag{22–3}$$

Certain equations that are derived from Eq. 22–3, but are used more often than it is, will be simpler in form if we do this.

In the mks system we can measure q_1, q_2, r, and F in Eq. 22–3 in ways that do not depend on Coulomb's law. Numbers with units can be assigned to them. There is no choice about the so-called *permittivity constant* ϵ_0; it must have that value which makes the right-hand side of Eq. 22–3 equal to the left-hand side. This (measured) value turns out to be

$$\epsilon_0 = 8.85418 \times 10^{-12} \text{ coul}^2/\text{nt-m}^2.$$

In this book the value 8.9×10^{-12} coul²/nt-m² will be accurate enough for all problems. For direct application of Coulomb's law or in any problem in which the quantity $1/4\pi\epsilon_0$ occurs we may use, with sufficient accuracy for this book,

$$1/4\pi\epsilon_0 = 9.0 \times 10^9 \text{ nt-m}^2/\text{coul}^2.$$

For practical reasons this value is not actually measured by direct application of Eq. 22–3 but in an equivalent although more circuitous way that is described in Section 26–2.

If more than two charges are present, Eq. 22–3 holds for every pair of charges. Let the charges be q_1, q_2, and q_3, etc.; we calculate the force exerted on any one (say q_1) by all the others from the vector equation

$$\mathbf{F}_1 = \mathbf{F}_{12} + \mathbf{F}_{13} + \mathbf{F}_{14} + \cdots, \tag{22–4}$$

SEE

p. 309

SUPERPOSITION
PRINCIPLE where $\mathbf{F}_{12}$, for example, is the force exerted on q_1 by q_2.

Example 2. Figure 22–3 shows three charges q_1, q_2, and q_3. What force acts on q_1? Assume that $q_1 = -1.0 \times 10^{-6}$ coul, $q_2 = +3.0 \times 10^{-6}$ coul, $q_3 = -2.0 \times 10^{-6}$ coul, $r_{12} = 15$ cm, $r_{13} = 10$ cm, and $\theta = 30°$.

From Eq. 22–3, ignoring the signs of the charges, since we are interested only in the magnitudes of the forces,

$$F_{12} = \frac{1}{4\pi\epsilon_0}\frac{q_1 q_2}{r_{12}{}^2}$$

$$= \frac{(9.0 \times 10^9 \text{ nt-m}^2/\text{coul}^2)(1.0 \times 10^{-6} \text{ coul})(3.0 \times 10^{-6} \text{ coul})}{(1.5 \times 10^{-1} \text{ meter})^2}$$

$$= 1.2 \text{ nt}$$

and
$$F_{13} = \frac{(9.0 \times 10^9 \text{ nt-m}^2/\text{coul}^2)(1.0 \times 10^{-6} \text{ coul})(2.0 \times 10^{-6} \text{ coul})}{(1.0 \times 10^{-1} \text{ meter})^2}$$

$$= 1.8 \text{ nt.}$$

The directions of $\mathbf{F}_{12}$ and $\mathbf{F}_{13}$ are shown in the figure. The components of the resultant force $\mathbf{F}_1$ acting on q_1 (see Eq. 22–4) are

$$F_{1x} = F_{12x} + F_{13x} = F_{12} + F_{13}\sin\theta$$

$$= 1.2 \text{ nt} + (1.8 \text{ nt})(\sin 30°) = 2.1 \text{ nt}$$

and
$$F_{1y} = F_{12y} + F_{13y} = 0 - F_{13}\cos\theta$$

$$= -(1.8 \text{ nt})(\cos 30°) = -1.6 \text{ nt.}$$

Find the magnitude of $\mathbf{F}_1$ and the angle it makes with the x-axis.

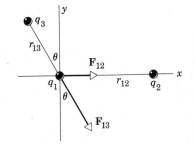

FIGURE 22–3   Example 3. Showing the forces exerted on q_1 by q_2 and q_3.

22–5   Charge Is Quantized

In Franklin's day electric charge was thought of as a continuous fluid, an idea that was useful for many purposes. The atomic theory of matter, however, has shown that fluids themselves, such as water and air, are not continuous but are made up of atoms. Experiment shows that the "electric fluid" is not continuous either but that it is made up of integral multiples of a certain minimum electric charge. This <u>fundamental charge</u>, to which we give the symbol e, has the magnitude $1.6021917 \times 10^{-19}$ coul. Any existing charge q, no matter what its origin, can be written as ne where n is a positive or a negative integer.

When a physical property such as charge exists in discrete "packets" rather than in continuous amounts, the property is said to be *quantized*. The existence of atoms and of particles like the electron and the proton indicates that mass is quantized also. Later we will see that several other properties prove to be quantized when suitably examined on the atomic scale; among them are energy and angular momentum.

The quantum of charge e is so small that the "graininess" of electricity does not show up in large-scale experiments, just as we do not realize that the air we breathe is made up of atoms. In an ordinary 110-volt, 100-watt light bulb, for example, 6×10^{18} elementary charges enter and leave the filament every second.

There exists today no theory that predicts the quantization of charge (or the quantization of mass, that is, the existence of fundamental particles such as protons, electrons, pions, etc.).

22–6   Charge and Matter

Matter as we ordinarily experience it can be regarded as composed of three kinds of elementary particles, the proton, the neutron, and the electron. Table 22–1 shows their masses and charges. Note that the masses of the neutron and the proton are approximately equal but that the electron is lighter by a factor of about 1836.

Atoms are made up of a dense, positively charged nucleus, surrounded by a cloud of electrons; see Fig. 22–4. The nucleus varies in radius from about 1×10^{-15} meter for hydrogen to about 7×10^{-15} meter for the heaviest atoms. The outer diameter of the electron cloud, that is, the diameter of the atom, lies in the range 1–3×10^{-10} meter, about 10^5 times larger than the nuclear diameter.

Table 22–1   PROPERTIES OF THE PROTON, THE NEUTRON, AND THE ELECTRON

Particle	Symbol	Charge	Mass
Proton	p	$+e$	1.67261×10^{-27} kg
Neutron	n	0	1.67492×10^{-27} kg
Electron	e^-	$-e$	9.10956×10^{-31} kg

Example 3. The distance r between the electron and the proton in the hydrogen atom is about 5.3×10^{-11} meter. What are the magnitudes of (a) the electrical force and (b) the gravitational force between these two particles?

From Coulomb's law,

$$F_e = \frac{1}{4\pi\epsilon_0} \frac{q_1 q_2}{r^2}$$

$$= \frac{(9.0 \times 10^9 \text{ nt-m}^2/\text{coul}^2)(1.6 \times 10^{-19} \text{ coul})^2}{(5.3 \times 10^{-11} \text{ meter})^2}$$

$$= 8.1 \times 10^{-8} \text{ nt.}$$

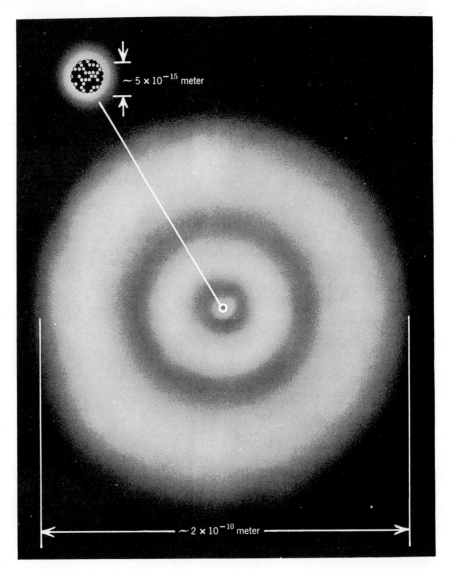

FIGURE 22-4   An atom, suggesting the electron cloud and, above, an enlarged view of the nucleus.

The gravitational force is

$$F_g = G \frac{m_1 m_2}{r^2}$$

$$= \frac{(6.7 \times 10^{-11} \text{ nt-m}^2/\text{kg}^2)(9.1 \times 10^{-31} \text{ kg})(1.7 \times 10^{-27} \text{ kg})}{(5.3 \times 10^{-11} \text{ meter})^2}$$

$$= 3.7 \times 10^{-47} \text{ nt.}$$

Thus the electrical force is about 10^{39} times stronger than the gravitational force.

The significance of Coulomb's law goes far beyond the description of the forces acting between charged balls or rods. This law, when incorporated into the structure of quantum physics, correctly describes (*a*) the electric forces that bind the electrons of an atom to its nucleus, (*b*) the forces that bind atoms together to form molecules, and (*c*) the forces that bind atoms or molecules together to form solids or liquids. Most of the forces of our daily experience that are not gravitational in nature are electrical. A force transmitted by a steel cable is an electrical force because, if we pass an imaginary plane through the cable at right angles to it, it is only the attractive electrical interatomic forces acting between atoms on opposite sides of the plane that keep the cable from parting. In fact, each of us is an assembly of nuclei and electrons bound together in a stable configuration by Coulomb forces.

In the atomic nucleus we encounter a new force which is neither gravitational nor electrical. This strong attractive force, which binds together the protons and neutrons that make up the nucleus, is called simply *the nuclear force*. If this force were not present, the nucleus would fly apart at once because of the strong Coulomb repulsion force that acts between its protons.

Example 4. What repulsive Coulomb force exists between two protons in a nucleus of iron? Assume a separation of 4.0×10^{-15} meter.

From Coulomb's law,

$$F = \frac{1}{4\pi\epsilon_0} \frac{q_1 q_2}{r^2}$$

$$= \frac{(9.0 \times 10^9 \text{ nt-m}^2/\text{coul}^2)(1.6 \times 10^{-19} \text{ coul})^2}{(4.0 \times 10^{-15} \text{ meter})^2}$$

$$= 14 \text{ nt.}$$

This enormous repulsive force must be more than compensated for by the strong attractive nuclear forces. This example, combined with Example 3, shows that nuclear binding forces are much stronger than electrical forces. Electric binding forces are, in turn, much stronger than gravitational forces for the same particles separated by the same distance.

22–7 Charge Is Conserved

When a glass rod is rubbed with silk, a positive charge appears on the rod. Measurement shows that a negative charge of equal magnitude appears on the silk. This suggests that rubbing does not create charge but merely transfers it from one object to another, disturbing slightly the electrical neutrality of each. This hypothesis of the conservation of charge has stood up under close experimental scrutiny both for large-scale events and at the atomic and nuclear level; no exceptions have been found.

An interesting example of charge conservation comes about when an electron (charge $= -e$) and a positron (charge $= +e$) are brought close to each other. The two particles simply disappear, converting all their rest mass into energy according to the $E = mc^2$ relationship. The energy appears in the form of two oppositely directed *gamma rays*, which are similar to x-rays. The net charge is zero both before and after the event so that charge is conserved. Rest mass is *not* conserved, being turned completely into energy.

Another example of charge conservation is found in radioactive decay, of which the following process is typical:

$$U^{238} \rightarrow Th^{234} + He^4. \tag{22–5}$$

The radioactive parent nucleus, U^{238}, contains 92 protons (that is, its atomic number Z is 92). It disintegrates spontaneously by emitting an α-particle (He^4; $Z = 2$) transmuting itself into the nucleus Th^{234}, with $Z = 90$. Thus the charge present before disintegration ($+92e$) is the same as that present after the disintegration.

Still another example of charge conservation is found in nuclear reactions, of which the bombardment of Ca^{44} with cyclotron-accelerated protons is typical. In a particular collision the proton may be absorbed and a neutron may emerge, leaving Sc^{44} as a residual nucleus:

$$Ca^{44} + p \rightarrow Sc^{44} + n.$$

The sum of the atomic numbers before the reaction $(20 + 1)$ is exactly equal to the sum of the atomic numbers after the reaction $(21 + 0)$. Again charge is conserved.

QUESTIONS

1. You are given two metal spheres mounted on portable insulating supports. Find a way to give them equal and opposite charges. You may use a glass rod rubbed with silk but may not touch it to the spheres. Do the spheres have to be of equal size for your method to work?

2. A charged rod attracts bits of dry cork dust which, after touching the rod, often jump violently away from it. Explain.

3. If a charged glass rod is held near one end of an insulated uncharged metal rod as in Fig. 22–5, electrons are drawn to one end, as shown. Why does the flow of electrons cease? There is an almost inexhaustible supply of them in the metal rod.

4. In Fig. 22–5 does any net electrical force act on the metal rod? Explain.

5. An insulated rod is said to carry an electric charge. How could you verify this and determine the sign of the charge?

6. Why do electrostatic experiments not work well on humid days?

7. A person standing on an insulated stool touches a charged, insulated conductor. Is the conductor discharged completely?

8. (*a*) A positively charged glass rod attracts a suspended object. Can we conclude that the object is negatively charged? (*b*) A positively charged glass rod repels a suspended object. Can we conclude that the object is positively charged?

9. Is the Coulomb force that one charge exerts on another changed if other charges are brought nearby?

10. The quantum of charge is 1.60×10^{-19} coul. Is there a corresponding single quantum of mass?

11. What does it mean to say that a physical quantity is (*a*) quantized or (*b*) conserved? Give some examples.

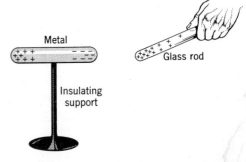

Metal

Glass rod

Insulating
support

FIGURE 22–5

PROBLEMS

1(4). Protons in the cosmic rays strike the earth's upper atmosphere at a rate, averaged over the earth's surface, of 1.0 proton/cm²-sec. What total current does the earth receive from beyond its atmosphere in the form of incident cosmic ray protons? The earth's radius is 6.4×10^6 meters. *Answer:* 0.82 amp.

2(4). Estimate roughly the number of coulombs of positive charge in a glass of water. Assume the volume to be 500 cm³.

3(4). (*a*) How many electrons would have to be removed from a penny to leave it with a charge of $+10^{-7}$ coul? (*b*) To what fraction of the electrons in the penny does this correspond? (*c*) What would be the force between two such pennies placed 1.0 meter apart? *Answer:* (*a*) 6.3×10^{11}. (*b*) 7.3×10^{-13}. (*c*) 9.0×10^{-5} nt.

4(4). What would be the force of attraction between two 1.0-coul charges separated by a distance of 1.0 meter? By 1.0 mile?

5(4). A point charge of $+3.0 \times 10^{-6}$ coul is 12 cm distant from a second point charge of -1.5×10^{-6} coul. Calculate the magnitude and direction of the force on each charge. *Answer:* 2.8 nt; attractive.

6(4). What is the force of attraction between a single sodium ion and an adjacent chlorine ion in a salt crystal if their separation is 2.82×10^{-10} meter?

7(4). The electrostatic force between two like ions that are separated by a distance of 5.0×10^{-10} m is 3.7×10^{-9} nt. (*a*) What is the charge on each ion? (*b*) How many electrons are missing from each ion? *Answer:* (*a*) 3.2×10^{-9} coul. (*b*) Two.

8(4). What equal positive charges would have to be placed on the earth and on the moon to neutralize their gravitational attraction?

9(4). How far apart must two protons be if the electrical repulsive force acting on either one is equal to its weight on the earth's surface? The mass of a proton is 1.7×10^{-27} kg. *Answer:* 12 cm.

10(4). Each of two small spheres is charged positively, the combined charge being 5.0×10^{-5} coul. If each sphere is repelled from the other by a force of 1.0 nt when the spheres are 2.0 meters apart, how is the total charge distributed between the spheres?

11(4). A certain charge Q is to be divided into two parts, q and $Q - q$. What is the relationship of Q to q if the two parts, placed a given distance apart, are to have a maximum Coulomb repulsion? *Answer:* $q = \frac{1}{2}Q$.

12(4). In the radioactive decay of U^{238} (see Eq. 22–5) the center of the emerging α-particle is, at a certain instant, 9.0×10^{-15} meter from the center of the residual nucleus Th^{234}. At this instant (*a*) what is the force on the α-particle and (*b*) what is its acceleration?

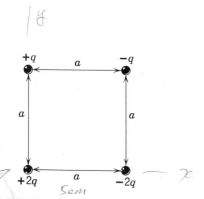

FIGURE 22–6 Problem 13(4).

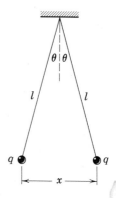

FIGURE 22–7 Problem 14(4).

13(4). In Fig. 22–6 what is the force on the charge in the lower left corner of the square? Assume that $q = 1.0 \times 10^{-7}$ coul and $a = 5.0$ cm.

Answer: $F_x = +0.17$ nt; $F_y = -0.046$ nt.

14(4). Two similar balls of mass m are hung from silk threads of length l and carry similar charges q as in Fig. 22–7. Assume that θ is so small that $\tan \theta$ can be replaced by $\sin \theta$. (a) To this approximation show that

$$x = \left(\frac{q^2 l}{2\pi\epsilon_0 mg} \right)^{1/3}$$

where x is the separation between the balls. (b) If $l = 120$ cm, $m = 10$ gm, and $x = 5.0$ cm, what is q?

15(4). The balls in Problem 14(4) carry positive charges q and q'. (a) Without assuming that x is small, find an expression from which θ can be found. (b) Are the two angles marked θ in Fig. 22–7 equal if $q \neq q'$? (c) Are they equal if the masses are unequal?

Answer: (a) $\sin^3\theta / \cos \theta = qq' / (16\pi\epsilon_0 l^2 mg)$. (b) Yes. (c) No.

16(4). The charges and coordinates of two charged particles located in the x-y plane are: $q_1 = +3.0 \times 10^{-6}$ coul; $x = 3.5$ cm, $y = 0.50$ cm, and $q_2 = -4.0 \times 10^{-6}$ coul; $x = -2.0$ cm, $y = 1.5$ cm. (a) Find the magnitude and direction of the force on q_2. (b) Where would you locate a third charge $q_3 = +4.0 \times 10^{-6}$ coul such that the total force on q_2 is zero?

17(4). A charge Q is placed at each of two opposite corners of a square. A charge q is placed at each of the other two corners. (a) If the resultant electrical force on Q is zero, how are Q and q related? (b) Could q be chosen to make the resultant force on every charge zero?

Answer: (a) $Q = -2\sqrt{2}\, q$. (b) No.

18(4). Two equally charged particles, 3.2×10^{-3} meters apart, are released from rest. The acceleration of the first particle is observed to be 7.0 meters/sec² and that of the second to be 9.0 meters/sec². If the mass of the first particle is 6.3×10^{-7} kg, what are (a) the mass, and (b) the charge of the second particle?

19(4). Two free point charges $+q$ and $+4q$ are a distance l apart. A third charge is so placed that the entire system is in equilibrium. Find the location, magnitude, and sign of the third charge.

Answer: $-4q/9$, located $l/3$ from $+q$.

20(4). Two equal positive charges, q, are a distance $2a$ apart. The force on a small positive test charge midway between them is zero. If the test charge is displaced a short distance either (a) toward one of the charges or (b) at right angles to the line joining the charges, find the direction of the force on it. Is the equilibrium stable or unstable in each case?

21(4). Three charged particles lie on a straight line and are separated by a distance d as shown in Fig. 22–8. Charges q_1 and q_2 are held fixed. If q_3 is free to move but in fact remains stationary, then how are q_1 and q_2 related?

Answer: $q_1 = -4q_2$.

22(4). Two identical conducting spheres, having charges of opposite sign, attract each other with a force of 0.108 nt when separated by 0.500 meter. The spheres are connected by a conducting wire, which is then removed, and thereafter repel each other with a force of 0.0360 nt. What were the initial charges on the spheres?

23(4). Find the frequency of oscillation if the test charge Q in Problem 20(4)a has mass m. Assume that the displacement is small compared to a and set the origin of coordinates at the midpoint.

Answer: $v = \dfrac{1}{2\pi} \sqrt{\dfrac{qQ}{\pi\epsilon_0 a^3 m}}$.

FIGURE 22–8   Problem 20(4).

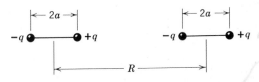

FIGURE 22–9   Problem 23(4).

24(4). If the balls of Fig. 22–7 are conducting, what happens to them after one is discharged? Find the new equilibrium separation.

25(4). A "dipole" is formed from a rod of length $2a$ and two charges, $+q$ and $-q$. Two such dipoles are oriented as shown in Fig. 22–9, their centers being separated by the distance R. (a) Calculate the force exerted on the left dipole. (b) For $R \gg a$, show that the magnitude of the force exerted on the left dipole is approximately given by $F = 3p^2/2\pi\epsilon_0 R^4$, where $p = 2qa$ is the "dipole moment."

Answer: (a) $\dfrac{q^2}{2\pi\epsilon_0}\left[\dfrac{R^2 + 4a^2}{(R^2 - 4a^2)^2} - \dfrac{1}{R^2}\right]$.

26(4). An electron is projected with an initial speed of 3.24×10^5 meter/sec directly toward a proton that is essentially at rest. If the electron is initially a great distance from the proton, at what distance from the proton is its speed instantaneously equal to twice its initial value? (Hint: Use the work-energy theorem.)

CHAPTER 23

The Electric Field

23–1 The Electric Field

With every point in space near the earth we can associate a gravitational field **g** (see Eq. 14–11). This is the gravitational acceleration that a test body, placed at that point and released, would experience. If m is the mass of the body and **F** the gravitational force acting on it, **g** is given by

$$\mathbf{g} = \mathbf{F}/m. \qquad (23\text{–}1)$$

This is an example of a vector field. For points near the surface of the earth the field is often taken as uniform; that is, **g** is the same for all points.

The flow of water in a river provides another example of a vector field, called a flow field. Every point in the water has associated with it a vector, the velocity **v** with which the water flows past the point. If **g** and **v** do not change with time, the corresponding fields are described as stationary. In the case of the river note that even though the water is moving the vector **v** at any point does not change with time for what are described as steady-flow conditions.

The space surrounding a charged rod is affected by the presence of the rod, and we speak of an *electric field* in this space. In the same way we speak of a *magnetic field* in the space around a bar magnet. In the classical theory of electromagnetism the electric and magnetic fields are central concepts.

Before Faraday's time, the force acting between charged particles was thought of as a direct and instantaneous interaction between the two particles. This action-at-a-distance view was also held for magnetic and for gravitational forces. Today we prefer to think in terms of electric fields as follows:

1. Charge q_1 in Fig. 23–1 sets up an electric field in the space around itself. This field is suggested by the shading in the figure.

2. The field acts on charge q_2; this shows up in the force **F** that q_2 experiences.

The field plays an intermediary role in the forces between charges. There are two separate problems: (*a*) calculating the fields that are set up by given distributions of charge and (*b*) calculating the forces that given fields will exert on charges placed in them. We think in terms of

$$\text{charge} \rightleftharpoons \text{field} \rightleftharpoons \text{charge}$$

and not, as in the action-at-a-distance point of view, in terms of

$$\text{charge} \rightleftharpoons \text{charge}.$$

In Fig. 23–1 we can also imagine that q_2 sets up a field and that this field acts on q_1, producing a force $-\mathbf{F}$ on it. The situation is completely symmetrical, each charge being immersed in a field associated with the other charge.

If the only problem in electromagnetism was that of the forces between stationary charges, the field and the action-at-a-distance points of view would be equivalent. Suppose, however, that q_1 in Fig. 23–1 suddenly moves to the right. How quickly does the charge q_2 learn that q_1 has moved and that the force which it (q_2) experiences must increase? Electromagnetic theory predicts that q_2 learns about q_1's motion by a field disturbance that emanates from q_1, traveling with the speed of light. The action-at-a-distance point of view requires that information about q_1's acceleration be communciated instantaneously to q_2; this is not in accord with experiment. Accelerating electrons in the antenna of a radio transmitter influence electrons in a distant receiving antenna only after a time l/c where l is the separation of the antennas and c is the speed of light.

23–2   The Electric Field E

To define the electric field operationally, we place a small test body carrying a test charge q_0 (assumed positive for convenience) at the point in space that is to be examined, and we measure the electric force **F** (if any) that acts on this body. The *electric field* **E** at the point is defined as

$$\mathbf{E} = \mathbf{F}/q_0. \tag{23–2}$$

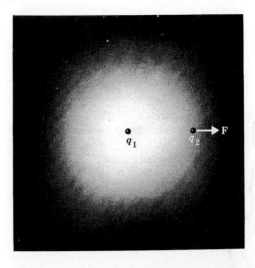

FIGURE 23–1   Charge q_1 sets up a field that exerts a force **F** on charge q_2.

Here **E** is a vector because **F** is one, q_0 being a scalar. The direction of **E** is the direction of **F**, that is, it is the direction in which a resting positive charge placed at the point would tend to move.

The definition of gravitational field **g** is much like that of electric field, except that the mass of the test body rather than its charge is the property of interest. Although the units of **g** are usually written as meters/sec², they could also be written as nt/kg (Eq. 23–1); those for **E** are nt/coul (Eq. 23–2). Thus both **g** and **E** are expressed as a force divided by a property (mass or charge) of the test body.

Example 1. What is the magnitude of the electric field **E** such that an electron, placed in the field, would experience an electrical force equal to its weight near the surface of the earth?

From Eq. 23–2, replacing q_0 by e and F by mg, where m is the electron mass, we have

$$E = \frac{F}{q_0} = \frac{mg}{e}$$

$$= \frac{(9.1 \times 10^{-31} \text{ kg})(9.8 \text{ meters/sec}^2)}{1.6 \times 10^{-19} \text{ coul}}$$

$$= 5.6 \times 10^{-11} \text{ nt/coul.}$$

This is a very weak electric field. Which way will **E** have to point if the electric force is to cancel the gravitational force?

In applying Eq. 23–2 we must use a test charge as small as possible. A large test charge might disturb the primary charges that are responsible for the field, thus changing the very quantity that we are trying to measure. Equation 23–2 should, strictly, be replaced by

$$\mathbf{E} = \lim_{q_0 \to 0} \frac{\mathbf{F}}{q_0}. \tag{23–3}$$

This equation instructs us to use a smaller and smaller test charge q_0, evaluating the ratio $\mathbf{F}/q_0$ at every step. The electric field **E** is then the limit of this ratio as the size of the test charge approaches zero.

23–3 Lines of Force

The concept of the electric field as a vector was not appreciated by Michael Faraday, who always thought in terms of *lines of force*. The lines of force still form a convenient way of visualizing electric-field patterns. We shall use them for this purpose but we shall not employ them quantitatively.

The relationship between the (imaginary) lines of force and the electric field is this:

1. The tangent to a line of force at any point gives the direction of **E** at that point.

2. The lines of force are drawn so that the number of lines per unit cross-sectional area is proportional to the magnitude of **E**. Where the lines are close together E is large and where they are far apart E is small.

It is not obvious that it is possible to draw a continuous set of lines to meet these requirements. Indeed, it turns out that if Coulomb's law were not true it would not be possible to do so; see Problem 4.

Figure 23–2 shows the lines of force for a uniform sheet of positive charge. We

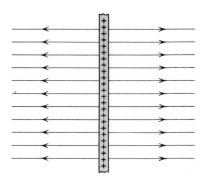

FIGURE 23–2   Lines of force for a section of an infinitely large sheet of positive charge.

assume that the sheet is infinitely large, which, for a sheet of finite dimensions, is equivalent to considering only those points whose distance from the sheet is small compared to the distance to the nearest edge of the sheet. A positive test charge, released in front of such a sheet, would move away from the sheet along a perpendicular line. Thus the electric field at any point near the sheet must be at right angles to the sheet. The lines of force are uniformly spaced, which means that **E** has the same magnitude for all points near the sheet.

Figure 23–3 shows the lines of force for a negatively charged sphere. From symmetry, the lines must lie along radii. They point inward because a free positive charge would be accelerated in this direction. The electric field E is not constant but decreases with increasing distance from the charge. This is evident in the lines of force, which are farther apart at greater distances. From symmetry, E is the same for all points that lie a given distance from the center of the charge.

▧ **Example 2.** In Fig. 23–3 how does E vary with the distance r from the center of the uniformly charged sphere?

Suppose that N lines originate on the sphere. Draw an imaginary concentric sphere of radius r; the number of lines per unit cross-sectional area at every point on the sphere is $N/4\pi r^2$. Since E is proportional to this, we can write that

$$E \propto 1/r^2.$$

We derive an exact relationship in Section 23–4. How should E vary with distance from an infinitely long uniform cylinder of charge? ▧

Figures 23–4 and 23–5 show the lines of force for two equal like charges and for two equal unlike charges, respectively. Michael Faraday, as we have said, used lines

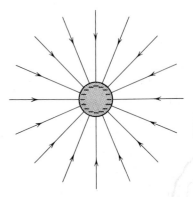

FIGURE 23–3   Lines of force for a negatively charged sphere.

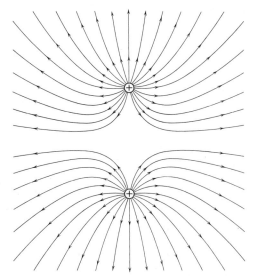

FIGURE 23-4　Lines of force for two equal positive charges.

of force a great deal in his thinking. They were more real for him than they are for most scientists and engineers today. It is possible to sympathize with Faraday's point of view. Can't you almost "see" the charges being pushed apart in Fig. 23–4 and pulled together in Fig. 23–5 by the lines of force?

23–4　Calculation of E

Let a test charge q_0 be placed a distance r from a point charge q. The magnitude of the force acting on q_0 is given by Coulomb's law, or

$$F = \frac{1}{4\pi\epsilon_0} \frac{q q_0}{r^2}.$$

The magnitude of the electric field at the site of the test charge is given by Eq. 23–2, or

$$E = \frac{F}{q_0} = \frac{1}{4\pi\epsilon_0} \frac{q}{r^2} \left(\frac{q_0}{q_0} \right) \; nt/coul \tag{23-4}$$

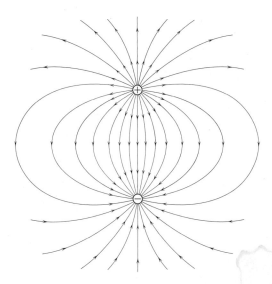

FIGURE 23-5　Lines of force for equal but opposite charges.

The direction of **E** is on a radial line from q, pointing outward if q is positive and inward if q is negative.

To find **E** for a group of point charges: (*a*) Calculate $\mathbf{E}_n$ due to each charge at the given point as if it were the only charge present. (*b*) Add these separately calculated fields vectorially to find the resultant field **E** at the point. In equation form,

$$\mathbf{E} = \mathbf{E}_1 + \mathbf{E}_2 + \mathbf{E}_3 + \cdots = \Sigma\,\mathbf{E}_n \qquad n = 1, 2, 3, \ldots . \qquad (23\text{--}5)$$

p. 309

The sum is a vector sum, taken over all the charges.

If the charge distribution is a continuous one, the field it sets up at any point P can be computed by dividing the charge into infinitesimal elements dq. We then calculate the field $d\mathbf{E}$ due to each element at the point in question, treating the elements as point charges. The magnitude of $d\mathbf{E}$ (see Eq. 23–4) is given by

$$dE = \frac{1}{4\pi\epsilon_0}\frac{dq}{r^2}, \qquad (23\text{--}6)$$

where r is the distance from the charge element dq to the point P. We find the resultant field at P by adding (that is, integrating) the field contributions due to all the charge elements, or, in the language of the calculus,

$$\mathbf{E} = \int d\mathbf{E}. \qquad (23\text{--}7)$$

The integration, like the sum in Eq. 23–5, is a vector operation; in Example 5 we will see how such an integral is handled in a simple case.

Example 3. *An electric dipole.* Figure 23–6 shows a positive and a negative charge of equal magnitude q placed a distance $2a$ apart, a configuration called an electric dipole. The pattern of lines of force is that of Fig. 23–5, which also shows an electric dipole. What is the field **E** due to these charges at point P, a distance r along the perpendicular bisector of the line joining the charges? Assume $r \gg a$.

Equation 23–5 gives the vector equation

$$\mathbf{E} = \mathbf{E}_1 + \mathbf{E}_2,$$

where, from Eq. 23–4,*

$$E_1 = E_2 = \frac{1}{4\pi\epsilon_0}\frac{q}{a^2 + r^2}.$$

The vector sum of $\mathbf{E}_1$ and $\mathbf{E}_2$ points vertically downward and has the magnitude

$$E = 2E_1 \cos\theta.$$

From the figure we see that

$$\cos\theta = \frac{a}{\sqrt{a^2 + r^2}}.$$

Substituting the expressions for E_1 and for $\cos\theta$ into that for E yields

$$E = \frac{2}{4\pi\epsilon_0}\frac{q}{(a^2 + r^2)}\frac{a}{\sqrt{a^2 + r^2}} = \frac{1}{4\pi\epsilon_0}\frac{2aq}{(a^2 + r^2)^{3/2}}.$$

If $r \gg a$, we can neglect a in the denominator; this equation then reduces to

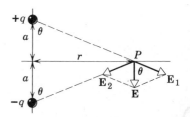

FIGURE 23–6   Example 3.

---

* Note that the r's in Eq. 23–4 and in this equation have different meanings.

$$E \cong \frac{1}{4\pi\epsilon_0} \frac{(2a)(q)}{r^3}. \qquad (23\text{–}8a)$$

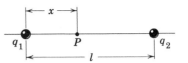

FIGURE 23–7   Example 4.

The essential properties of the charge distribution in Fig. 23–6, the magnitude of the charge q and the separation $2a$ between the charges, enter Eq. 23–8a only as product. This means that, if we measure **E** at various distances from the electric dipole (assuming $r \gg a$), we can never deduce q and $2a$ separately but only the product $2aq$; if q were doubled and a simultaneously cut in half, the electric field at large distances from the dipole would not change.

The product $2aq$ is called the *electric dipole moment* p. Thus we can rewrite this equation for E, for distant points along the perpendicular bisector, as

$$E = \frac{1}{4\pi\epsilon_0} \frac{p}{r^3}. \qquad (23\text{–}8b)$$

The result for distant points along the dipole axis (see Problem 10) and the general result for any distant point (see Problem 17) also contain the quantities $2a$ and q only as the product $2aq\,(= p)$. The variation of E with r in the general result for distant points is also as $1/r^3$, as in Eq. 23–8b.

The dipole of Fig. 23–6 is two equal and opposite charges placed close to each other so that their separate fields at distant points almost, but not quite, cancel. On this point of view it is easy to understand that $E(r)$ for a dipole varies as $1/r^3$ (Eq. 23–8b), whereas for a point charge $E(r)$ drops off more slowly, namely as $1/r^2$ (Eq. 23–4).

Example 4. Figure 23–7 shows a charge $q_1\,(= +1.0 \times 10^{-6}$ coul) 10 cm from a charge $q_2\,(= +2.0 \times 10^{-6}$ coul). At what point on the line joining the two charges is the electric field zero?

The point must lie between the charges because only here do the forces exerted by q_1 and q_2 on a test charge oppose each other. If $\mathbf{E}_1$ is the electric field due to q_1 and $\mathbf{E}_2$ that due to q_2, we must have

$$E_1 = E_2$$

or (see Eq. 23–4)

$$\frac{1}{4\pi\epsilon_0}\frac{q_1}{x^2} = \frac{1}{4\pi\epsilon_0}\frac{q_2}{(l-x)^2}$$

where x is the distance from q_1 and l equals 10 cm. Solving for x yields

$$x = \frac{l}{1 \pm \sqrt{q_2/q_1}} = \frac{10 \text{ cm}}{1 + \sqrt{2}} = 4.1 \text{ cm}.$$

On what basis was the second root of the resulting quadratic equation discarded?

Example 5. Figure 23–8 shows a ring of charge q and of radius a. Calculate **E** for points on the axis of the ring a distance x from its center.

Consider a differential element of the ring of length ds, located at the top of the ring in Fig. 23–8. It contains an element of charge dq which sets up a differential electric field $d\mathbf{E}$ at point P.

The resultant field **E** at P is found by adding, vectorially, the effects of all the elements that make up the ring. This looks like a hard job because the vectors do not even lie in the same plane. However, note that the charge has cylindrical symmetry about the axis of the ring. This must mean that the resultant electric field must lie along this axis.

Thus, all the components of $d\mathbf{E}$ that are perpendicular to the axis may be neglected and the resultant field may be obtained by adding only those components of $d\mathbf{E}$ that lie along the axis. These are all colinear, so the vector addition becomes a scalar addition of parallel axial components.

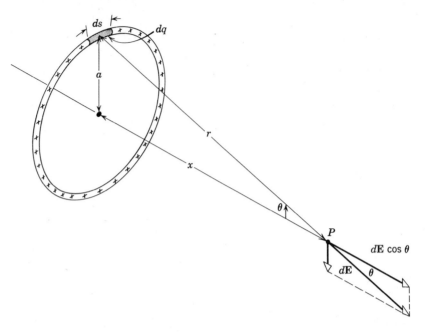

FIGURE 23–8   Example 5.

The quantity dE follows from Eq. 23–6, or

$$dE = \frac{1}{4\pi\epsilon_0}\frac{dq}{r^2} = \frac{1}{4\pi\epsilon_0}\frac{dq}{(a^2 + x^2)}.$$

From Fig. 23–8 we have

$$\cos\theta = \frac{x}{\sqrt{a^2 + x^2}}$$

The axial component of $d\mathbf{E}$ is thus

$$dE\cos\theta = \frac{dq}{4\pi\epsilon_0}\left[\frac{x}{(a^2 + x^2)^{3/2}}\right].$$

To add the various contributions, we need only add the dq's, since all the other factors in $dE\cos\theta$ are the same for all dq's. Thus, the addition yields

$$E = \frac{1}{4\pi\epsilon_0}\left[\frac{qx}{(a^2 + x^2)^{3/2}}\right].$$

Verify that this expression for E reduces to an expected result for $x = 0$.

For $x \gg a$ we can neglect a in the denominator of this equation, yielding

$$E \cong \frac{1}{4\pi\epsilon_0}\frac{q}{x^2}.$$

This is an expected result (compare Eq. 23–4) because at great enough distances the ring behaves like a point charge q.

23–5   A Point Charge in an Electric Field

An electric field will exert a force on a charged particle given by (see Eq. 23–2)

$$\mathbf{F} = \mathbf{E}q.$$

This force will produce an acceleration

$$\mathbf{a} = \mathbf{F}/m,$$

where m is the mass of the particle. We will consider two examples of the acceleration of a charged particle in a uniform electric field. Such a field can be produced by connecting the terminals of a battery to two parallel metal plates which are otherwise insulated from each other. If the spacing between the plates is small compared with the dimensions of the plates, the field between them will be fairly uniform except near the edges. Note that in calculating the motion of a particle in a field set up by external charges the field due to the particle itself (that is, its self-field) is ignored. Similarly, the earth's gravitational field can have no effect on the earth itself but only on a second object, say a stone, placed in that field.

Example 6. A particle of mass m and charge q is placed at rest in a uniform electric field (Fig. 23–9) and released. Describe its motion.

The motion resembles that of a body falling in the earth's gravitational field. The (constant) acceleration is given by

$$a = \frac{F}{m} = \frac{qE}{m}.$$

The equations for uniformly accelerated motion (Table 3–1) then apply. With $v_0 = 0$, they are

$$v = at = \frac{qEt}{m},$$

$$y = \tfrac{1}{2}at^2 = \frac{qEt^2}{2m},$$

and

$$v^2 = 2ay = \frac{2qEy}{m}.$$

The kinetic energy attained after moving a distance y is found from

$$K = \tfrac{1}{2}mv^2 = \tfrac{1}{2}m\left(\frac{2qEy}{m}\right) = qEy.$$

This result also follows directly from the work-energy theorem because a constant force qE acts over a distance y.

Example 7. *Deflecting an electron beam.* Figure 23–10 shows an electron of mass m and charge e projected with speed v_0 at right angles to a uniform field **E**. Describe its motion.

The motion is like that of a projectile fired horizontally in the earth's gravitational field. The considerations of Section 4–3 apply, the horizontal (x) and vertical (y) motions being

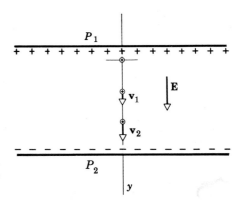

FIGURE 23–9 A charge is released from rest in a uniform electric field set up between two oppositely charged metal plates P_1 and P_2.

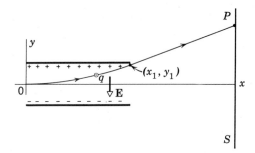

FIGURE 23–10   Example 7. An electron is projected into a uniform electric field.

given by

$$x = v_0 t$$

and

$$y = \tfrac{1}{2}at^2 = \frac{eE}{2m} t^2.$$

Eliminating t yields

$$y = \frac{eE}{2mv_0{}^2} x^2 \tag{23–9}$$

for the equation of the trajectory (a parabola).

When the electron emerges from the plates in Fig. 23–10, it travels (neglecting gravity) in a straight line tangent to the parabola at the exit point. We can let it fall on a fluorescent screen S placed some distance beyond the plates. Together with other electrons following the same path, it will then make itself visible as a small luminous spot; this is the principle of the electrostatic cathode-ray oscilloscope.

Example 8. The electric field between the plates of a cathode-ray oscilloscope is 1.2×10^4 nt/coul. What deflection will an electron experience if it enters at right angles to the field with a kinetic energy of 2000 eV ($= 3.2 \times 10^{-16}$ joule), a typical value? The deflecting assembly is 1.5 cm long.

Recalling that $K_0 = \tfrac{1}{2}mv_0{}^2$, we can rewrite Eq. 23–9 as

$$y = \frac{eEx^2}{4K_0}.$$

If x_1 is the horizontal position of the far edge of the plate, y_1 will be the corresponding deflection (see Fig. 23–10), or

$$y_1 = \frac{eEx_1{}^2}{4K_0}$$

$$= \frac{(1.6 \times 10^{-19} \text{ coul})(1.2 \times 10^4 \text{ nt/coul})(1.5 \times 10^{-2} \text{ meter})^2}{(4)(3.2 \times 10^{-16} \text{ joule})}$$

$$= 3.4 \times 10^{-4} \text{ meter} = 0.34 \text{ mm}.$$

The deflection measured at the fluorescent screen is much larger.

23–6   A Dipole in an Electric Field

An electric dipole moment can be regarded as a vector **p** whose magnitude p, for a dipole like that described in Example 3, is the product $2aq$ of the magnitude of either charge q and the distance $2a$ between the charges. The direction of **p** for such a dipole is from the negative to the positive charge. The vector nature of the electric dipole moment permits us to cast many expressions involving electric dipoles into concise form, as we shall see.

Figure 23–11a shows an electric dipole formed by placing two charges $+q$ and $-q$ a fixed distance $2a$ apart. The arrangement is placed in a uniform external electric field **E,** its dipole moment **p** making an angle θ with this field. Two equal and opposite forces **F** and $-$**F** act as shown, where

$$F = q\mathbf{E}.$$

The net force is clearly zero, but there is a net torque about an axis through O (see Eq. 11–2) given by

$$\tau = 2F(a \sin \theta) = 2aF \sin \theta.$$

Combining these two equations and recalling that $p = (2a)(q)$, we obtain

$$\tau = 2aqE \sin \theta = pE \sin \theta. \tag{23–10}$$

Thus an electric dipole placed in an external electric field **E** experiences a torque tending to align it with the field. Equation 23–10 can be written in vector form as

$$\tau = \mathbf{p} \times \mathbf{E}, \tag{23–11}$$

the appropriate vectors being shown in Fig. 23–11b.

Work (positive or negative) must be done by an external agent to change the orientation of an electric dipole in an external field. This work is stored as potential energy U in the system consisting of the dipole and the arrangement used to set up the external field. If θ in Fig. 23–11a has the initial value θ_0, the work required to turn the dipole axis to an angle θ is given (see Table 11–2) from

$$W = \int dW = \int_{\theta_0}^{\theta} \tau \, d\theta = U,$$

where τ is the torque exerted by the agent that does the work. Combining this equation with Eq. 23–10 yields

$$U = \int_{\theta_0}^{\theta} pE \sin \theta \, d\theta = pE \int_{\theta_0}^{\theta} \sin \theta \, d\theta$$

$$= pE \left| -\cos \theta \right|_{\theta_0}^{\theta}.$$

NOTE: WE DEFINE $U = 0$
FOR $\theta_0 = \frac{\pi}{2}$

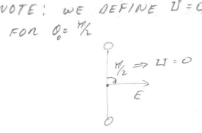

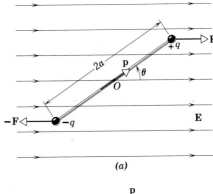

(a)

(b)

FIGURE 23–11 (a) An electric dipole in a uniform external field. (b) An oblique view, illustrating $\tau = $ **p** $\times$ **E.**

Since we are interested only in changes in potential energy, we may choose the reference orientation θ_0 to have any convenient value, in this case $90°$. This gives

$$U = -pE \cos \theta \qquad (23\text{--}12)$$

or, in vector form,

$$U = -\mathbf{p} \cdot \mathbf{E}. \qquad (23\text{--}13)$$

QUESTIONS

1. Name as many scalar fields and vector fields as you can.

2. In the gravitational attraction between the earth and a stone, can we say that the earth lies in the gravitational field of the stone? How is the gravitational field due to the stone related to that due to the earth?

3. A positively charged ball hangs from a long thread. We wish to measure $\mathbf{E}$ at a point in the same horizontal plane as that of the hanging charge. To do so, we put a positive test charge q_0 at the point and measure $\mathbf{F}/q_0$. Will F/q_0 be less than, equal to, or greater than E at the point in question?

4. Taking into account the quantization of electric charge (the single electron providing the basic charge unit), how can we justify the procedure suggested by Eq. 23–3?

5. In Fig. 23–5 the force on the lower charge points up and is finite. The crowding of the lines of force, however, suggests that E is infinitely great at the site of this (point) charge. A charge immersed in an infinitely great field should have an infinitely great force acting on it. What is the solution to this dilemma?

6. Electric lines of force never cross. Why?

7. In Fig. 23–4 why do the lines of force around the edge of the figure appear, when extended backwards, to radiate uniformly from the center of the figure?

8. Figure 23–2 shows that $\mathbf{E}$ has the same value for all points in front of an infinite uniformly charged sheet. Is this reasonable? One might think that the field should be stronger near the sheet because the charges are so much closer.

9. Two point charges of unknown magnitude and sign are a distance d apart. The electric field is zero at one point between them, on the line joining them. What can you conclude about the charges?

10. Compare the way E varies with r for (*a*) a point charge (Eq. 23–4) and (*b*) a dipole (Eq. 23–8*a*).

11. If a point charge q of mass m is released from rest in a nonuniform field, will it follow a line of force?

12. An electric dipole is placed in a *nonuniform* electric field. Is there a net force on it?

13. An electric dipole is placed at rest in a uniform external electric field, as in Fig. 23–11*a*, and released. Discuss its motion.

PROBLEMS

1(2) What are the magnitude and direction of an electric field that will balance the weight of (*a*) an electron and (*b*) an alpha particle?
Answer: (*a*) 5.6×10^{-11} nt/coul, down. (*b*) 2.0×10^{-7} nt/coul, up.

2(2). A particle having a charge of -2.0×10^{-9} coul is acted on by a downward electric force of 3.0×10^{-6} nt in a uniform electric field. (*a*) What is the magnitude of the electric field? (*b*) What are the magnitude and direction of the electric force exerted on a proton placed in this field? (*c*) What is the gravitational force on the proton? (*d*) What is the ratio of the electric to the gravitational forces in this case?

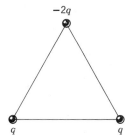

FIGURE 23-12   Problem 5(3).

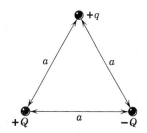

FIGURE 23-13   Problem 6(3).

3(2). An electric field **E** with an average magnitude of about 150 nt/coul points downward in the earth's atmosphere. We wish to "float" a sulfur sphere weighing 1.0 lb in this field by charging it. (*a*) What charge (sign and magnitude) must be used? (*b*) Why is the experiment not practical? Give a qualitative reason supported by a very rough numerical calculation to prove your point. *Answer:* (*a*) −0.030 coul. (*b*) Sphere would blow up because of mutual Coulomb repulsion.

4(3). Sketch qualitatively the lines of force associated with two separated point charges q and $-2q$.

5(3). Sketch qualitatively the lines of force associated with three point charges arranged in an equilateral triangle as shown in Fig. 23–12. The charges in this figure are spread uniformly over long parallel rods that are perpendicular to the plane of the figure.

6(3). Three charges are arranged in an equilateral triangle as in Fig. 23–13. What is the direction of the electric field acting on the charge $+q$?

7(3). Sketch qualitatively the lines of force associated with a thin, circular, uniformly charged disk of radius R. (Hint: Consider as limiting cases points very close to the surface and points far from it.) Show the lines only in a plane containing the axis of the disk.

8(3). (*a*) Sketch qualitatively the lines of force associated with three long parallel lines of positive charge, in a perpendicular plane. Assume that the intersections of the lines of charge with such a plane form an equilateral triangle and that each line of charge has the same linear charge density (coul/meter). (*b*) Discuss the nature of the equilibrium of a positive test charge placed on the central axis of the charge assembly.

9(3). Sketch qualitatively the lines of force between two concentric conducting spherical shells, charge $+q_1$ being placed on the inner sphere and $-q_2$ on the outer. Consider the cases $q_1 > q_2$, $q_1 = q_2$, $q_1 < q_2$.

10(3). (*a*) Sketch qualitatively the lines of force associated with a positively charged circular ring of radius R. Show the lines in a plane containing the axis of the ring. (*b*) Describe the motion of an electron moving toward the ring along its axis.

11(3). Assume that the exponent in Coulomb's law is not "two" but n. Show that for $n \neq 2$ it is impossible to construct lines that will have the properties listed for lines of force in Section 23–3. For simplicity, treat an isolated point charge.

12(4). What is the magnitude of a point charge chosen so that the electric field 52 cm away has the magnitude 2.2 nt/coul?

13(4). Two equal and opposite charges of magnitude 2.0×10^{-7} coul are 15 cm apart. (*a*) What are the magnitude and direction of **E** at a point midway between the charges? (*b*) What force (magnitude and direction) would act on an electron placed there? *Answer:* (*a*) 6.4×10^5 nt/coul, toward the negative charge. (*b*) 1.0×10^{-13} nt, toward the positive charge.

14(4). Two point charges of magnitude $+2.0 \times 10^{-7}$ coul and $+8.5 \times 10^{-8}$ coul are 12 cm apart. (*a*) What electric field does each produce at the site of the other? (*b*) What force acts on each?

15(4). An electron is placed at each corner of an equilateral triangle having sides of length a (20 cm). (*a*) What is the electric field at the midpoint of one of the sides? (*b*) What force would another electron placed there experience?

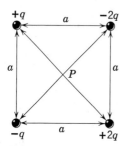

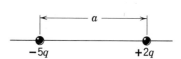

FIGURE 23-14 Problem 16(4).

FIGURE 23-15 Problem 18(4).

Answer: (a) 4.8×10^{-8} nt/coul, at right angles to the side, pointing in. (b) 7.7×10^{-27} nt, at right angles to the side, pointing out.

16(4) What is **E** in magnitude and direction at the center of the square of Fig. 23–14? Assume that $q = 1.0 \times 10^{-8}$ coul and $a = 5.0$ cm.

17(4). Two charges q_1 (2.1×10^{-8} coul), and q_2 ($= -4q_1$) are placed at a distance a (50 cm) apart. Find the point along the straight line joining the two charges at which the electric field is zero. *Answer:* 50 cm from q_1 and 100 cm from q_2.

18(4). (a) In Fig. 23–15 locate the point (or points) at which the electric field is zero. (b) Sketch qualitatively the lines of force. Take $a = 50$ cm.

19(4). Two point charges are a distance d apart (Fig. 23–16). Plot $E(x)$, assuming $x = 0$ at the left-hand charge. Consider both positive and negative values of x. Plot E as positive if **E** points to the right and negative if **E** points to the left. Assume $q_1 = +1.0 \times 10^{-6}$ coul, $q_2 = +3.0 \times 10^{-6}$ coul, and $d = 10$ cm.

20(4). Two point charges of unknown magnitude and sign are placed a distance d apart. (a) If it is possible to have **E** $= 0$ at any point not between the charges but on the line joining them, what are the necessary conditions and where is the point located? (b) Is it possible, for any arrangement of two point charges, to find two points (neither at infinity) at which **E** $= 0$; if so, under what conditions?

21(4). Make a quantitative plot of the electric field on the axis of a charged ring having a diameter of 6.0 cm if the ring carries a uniform charge of 1.0×10^{-6} coul. (See Example 5.)

22(4). An electron is constrained to move along the axis of the ring of charge in Example 5. Show that the electron can perform oscillations whose frequency is given by

$$\omega = \sqrt{\frac{eq}{4\pi\epsilon_0 ma^3}}.$$

This formula holds only for small oscillations, that is, for $x \ll a$ in Fig. 23–8. (Hint: Show that the motion is simple harmonic and use Eq. 13–11.)

23(4). *Axial field due to an electric dipole.* In Fig. 23–6, consider a point a distance r from the center of the dipole along its axis. (a) Show that, at large values of r, the electric field is

$$E = \frac{1}{2\pi\epsilon_0} \frac{p}{r^3},$$

which is twice the value given for the conditions of Example 3. (b) What is the direction of **E**? *Answer:* (b) Parallel to **p**.

24(4). In Fig. 23–6 assume that both charges are positive. (a) Show that E at point P in that figure, assuming $r \gg a$, is given by

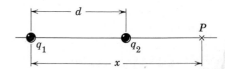

FIGURE 23-16 Problem 19(4).

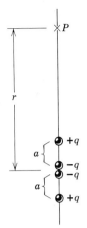

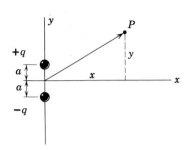

FIGURE 23-17   Problem 25(4).

FIGURE 23-18   Problem 26(4).

$$E = \frac{1}{4\pi\epsilon_0} \frac{2q}{r^2} .$$

(*b*) What is the direction of **E**? (*c*) Is it reasonable that E should vary as r^{-2} here and as r^{-3} for the dipole of Fig. 23-6?

25(4). Fig. 23-17 shows an electric quadrupole consisting of two electric dipoles. Show that if $r \gg a$ the value of E at point P is $E = 3Q/4\pi\epsilon_0 r^4$ where $Q = 2qa^2$. (Q is called the quadrupole moment of the charge distribution.)

26(4). *Field due to an electric dipole.* (*a*) Show that the components of **E** due to a dipole are given at distant points by

$$E_x = \frac{1}{4\pi\epsilon_0} \frac{3pxy}{(x^2 + y^2)^{5/2}}$$

$$E_y = \frac{1}{4\pi\epsilon_0} \frac{p(2y^2 - x^2)}{(x^2 + y^2)^{5/2}},$$

where x and y are coordinates of a point in Fig. 23-18. (*b*) Show that this general result includes the special results of Eq. 23-8*b* and of Problem 23(4).

27(4). A thin glass rod is bent into a semicircle of radius R. A charge $+Q$ is uniformly distributed along the upper half and a charge $-Q$ is uniformly distributed along the lower half, as shown in Fig. 23-19. Find the electric field **E** at P, the center of the semicircle.
Answer: $E = Q/(\pi^2\epsilon_0 R^2)$, vertically down.

28(4). Find an expression for the magnitude of the electric field at a distance x along the perpendicular bisector of a uniformly charged rod of length a and total charge q; see Fig. 23-20.

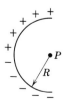

FIGURE 23-19   Problem 27(4).

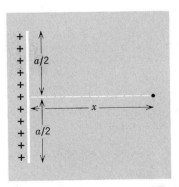

FIGURE 23-20   Problem 28(4).

FIGURE 23-21 Problem 29(4).

29(4). A "semi-infinite" insulating rod (Fig. 23–21) carries a constant charge per unit length of λ. Show that (*a*) the electric field at the point P makes an angle of 45° with the rod and (*b*) that this result is independent of the distance R.

30(5). A uniform electric field exists in a region between two oppositely charged plates. An electron is released from rest at the surface of the negatively charged plate and strikes the surface of the opposite plate, 2.0 cm away in a time 1.5×10^{-8} sec. (*a*) What is the speed of the electron as it strikes the second plate? (*b*) What is the magnitude of the electric field **E**?

31(5). *Oil drop experiment.* R. A. Millikan set up an apparatus (Fig. 23–22) in which a tiny, charged oil drop, placed in an electric field **E**, could be "balanced" by adjusting E until the electric force on the drop was equal and opposite to its weight. If the radius of the drop is 1.64×10^{-4} cm and E at balance is 1.92×10^5 nt/coul, (*a*) what charge is on the drop? (*b*) Why did Millikan not try to balance electrons in his apparatus instead of oil drops? The density of the oil is 0.851 gm/cm³. (Millikan first measured the electronic charge in this way. He measured the drop radius by observing the limiting speed that the drops attained when they fell in air with the electric field turned off. He charged the oil drops by irradiating them with x rays.)
Answer: (*a*) $5.0e$. (*b*) Cannot see electrons; also **E** at balance would be inconveniently small.

32(5). In an early run (1911), Millikan observed that the following measured charges, among others, appeared at different times on a single drop:

6.563×10^{-19} coul	13.13×10^{-19} coul	19.71×10^{-19} coul
8.204×10^{-19} coul	16.48×10^{-19} coul	22.89×10^{-19} coul
$11.50 \ \times 10^{-19}$ coul	18.08×10^{-19} coul	26.13×10^{-19} coul

What value for the elementary charge e can be deduced from these data?

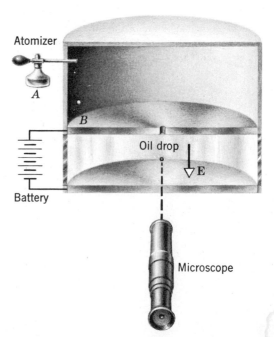

FIGURE 23-22 Problem 31(5) Millikan's oil drop apparatus. Charged oil drops from atomizer A fall through the hole in plate B.

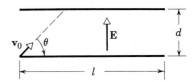

FIGURE 23–23   Problem 37(5).

33(5). (*a*) What is the acceleration of an electron in a uniform electric field of 1.0×10^6 nt/coul? (*b*) How long would it take for the electron, starting from rest, to attain one-tenth the speed of light? (*c*) What considerations limit the applicability of Newtonian mechanics to such problems? *Answer:* (*a*) 1.8×10^{17} meters/sec². (*b*) 1.7×10^{-10} sec.

34(5). An electron moving with a speed of 5.0×10^8 cm/sec is shot parallel to an electric field of 1.0×10^3 nt/coul arranged so as to retard its motion. (*a*) How far will the electron travel in the field before coming (momentarily) to rest, and (*b*) how much time will elapse? (*c*) If the electric field ends abruptly after 0.80 cm, what fraction of its initial kinetic energy will the electron lose in traversing it?

35(5). An object having a mass of 10 gm and a charge of $+8.0 \times 10^{-5}$ coul is placed in an electric field given, in nt/coul, by $\mathbf{E} = \mathbf{i}3.0 \times 10^3 - \mathbf{j}6.0 \times 10^2$. (*a*) What are the magnitude and direction of the force on the object? (*b*) If the object starts from rest at the origin, what will be its coordinates after 3.0 sec? *Answer:* (*a*) 0.24 nt, at $-11°$ from the *x*-axis. (*b*) $x = 110$ meters; $y = -22$ meters.

36(5). At some instant the velocity components of an electron moving between two charged parallel plates are $v_x = 1.5 \times 10^5$ m/sec and $v_y = 0.30 \times 10^4$ m/sec. If the electric field between the plates is given by $\mathbf{E} = \mathbf{j}1.2 \times 10^4$ nt/coul, (*a*) what is the acceleration of the electron? (*b*) When the *x*-coordinate of the electron has changed by 2.0 cm what will be the velocity of the electron?

37(5). An electron is projected as in Fig. 23–23 at a speed of 6.0×10^6 meters/sec and at an angle θ of 45°; $E = 2.0 \times 10^3$ nt/coul (directed upward), $d = 2.0$ cm, and $l =$ cm. (*a*) Will the electron strike either of the plates? (*b*) If it strikes a plate, where does it do so? *Answer:* (*a*) Yes. (*b*) It strikes the upper plate 2.7 cm from its left edge.

38(5). A uniform vertical field $\mathbf{E}$ is established in the space between two large parallel plates. In this field one suspends a small conducting sphere of mass m from a string of length l. Find the period of this pendulum when the sphere is given a charge $+q$ if (*a*) the lower plate is charged positively and (*b*) if it is charged negatively.

39(6). A charge $q = 3.0 \times 10^{-6}$ coul is 30 cm from a small dipole along its perpendicular bisector. The magnitude of the force on the charge is 5.0×10^{-6} nt. Show on a diagram (*a*) the direction of the force on the charge, (*b*) the direction of the force on the dipole, and (*c*) find the magnitude of the force on the dipole. *Answer:* (*a*) Antiparallel to $\mathbf{p}$ of dipole. (*b*) Parallel to $\mathbf{p}$ of dipole. (*c*) 5.0×10^{-6} nt.

40(6). An electric dipole consists of two opposite charges of magnitude $q = 1.0 \times 10^{-6}$ coul separated by $d = 2.0$ cm. The dipole is placed in an external field of 1.0×10^5 nt/coul. (*a*) What maximum torque does the field exert on the dipole? (*b*) How much work must an external agent do to turn the dipole end for end, starting from a position of alignment ($\theta = 0$)?

CHAPTER 24

Gauss's Law

24–1 Flux of the Electric Field

Phi

Flux (symbol Φ) is a property of any vector field; it refers to a hypothetical surface in the field, which may be closed or open. For a flow field the flux (Φ_v) is measured by the number of streamlines that cut through such a surface. For an electric field the flux (Φ_E) is measured by the number of lines of force that cut through such a surface.

SEE THOMAS p. 597 –

For closed surfaces we shall see that Φ_E is positive if the lines of force point outward everywhere and negative if they point inward. Figure 24–1 shows two equal and opposite charges and their lines of force. Curves S_1, S_2, S_3, and S_4 are the intersections with the plane of the figure of four hypothetical closed surfaces. From the statement just given, Φ_E is positive for surface S_1 and negative for S_2. The flux of the electric field is important because Gauss's law, one of the four basic equations of electromagnetism (see Table 34–2), is expressed in terms of it. Although the concept of flux may seem a little abstract at first, you will soon see its value in solving problems.

To define Φ_E precisely, consider Fig. 24–2, which shows an arbitrary closed surface immersed in an electric field. Let the surface be divided into elementary squares ΔS, each of which is small enough so that it may be considered to be plane. Such an element of area can be represented as a vector $\Delta \mathbf{S}$, whose magnitude is the area ΔS; the direction of $\Delta \mathbf{S}$ is taken as the outward-drawn normal to the surface.

At every square in Fig. 24–2 we can also construct an electric field vector $\mathbf{E}$. Since the squares have been taken to be arbitrarily small, $\mathbf{E}$ may be taken as constant for all points in a given square.

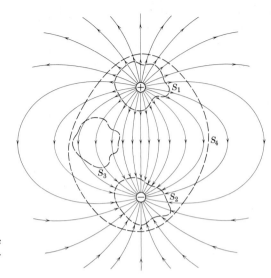

FIGURE 24–1 Two equal and opposite charges. The dashed lines represent hypothetical closed surfaces.

The vectors **E** and **ΔS** that characterize each square make an angle θ with each other. Figure 24–2*b* shows an enlarged view of the three squares on the surface of Fig. 24–2*a* marked *x*, *y*, and *z*. Note that at *x*, $\theta > 90°$; at *y*, $\theta = 90°$; and at *z*, $\theta < 90°$.

A semiquantitative definition of flux is

$$\Phi_E \cong \Sigma \, \mathbf{E} \cdot \mathbf{\Delta S}, \tag{24–1}$$

which instructs us to add up the scalar quantity $\mathbf{E} \cdot \mathbf{\Delta S}$ for all elements of area into which the surface has been divided. For points such as *x* in Fig. 24–2 the contribution to the flux is negative; at *y* it is zero and at *z* it is positive. Thus if **E** is every-

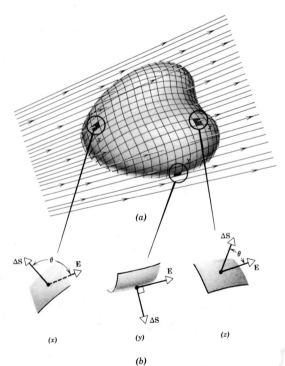

FIGURE 24–2 (*a*) A hypothetical surface immersed in an electric field. (*b*) Three elements of area on this surface, shown enlarged.

where outward, $\theta < 90°$, $\mathbf{E} \cdot \Delta \mathbf{S}$ will be positive, and Φ_E for the entire surface will be positive; see Fig. 24–1, surface S_1. If $\mathbf{E}$ is everywhere inward, $\theta > 90°$, $\mathbf{E} \cdot \Delta \mathbf{S}$ will be negative, and Φ_E for the surface will be negative; see Fig. 24–1, surface S_2. From Eq. 24–1 we see that the appropriate mks unit for Φ_E is the newton-meter²/coul.

The exact definition of electric flux is found in the differential limit of Eq. 24–1. Replacing the sum over the surface by an integral over the surface yields

$$\Phi_E = \int \mathbf{E} \cdot d\mathbf{S}. \qquad (24\text{--}2)$$

SEE p. 538

This *surface integral* indicates that the surface in question is to be divided into infinitesimal elements of area $d\mathbf{S}$ and that the scalar quantity $\mathbf{E} \cdot d\mathbf{S}$ is to be evaluated for each element and the sum taken for the entire surface.

Flux may be evaluated in this way for any surface in the field whether it is open or closed. When the surface is closed, this fact is sometimes emphasized by placing a circle on the integral sign. From Eq. 24–2 we see that the mks unit for Φ_E is the nt m²/coul.

Example 1. Figure 24–3 shows a hypothetical cylinder of radius R immersed in a uniform electric field $\mathbf{E}$, the cylinder axis being parallel to the field. What is Φ_E for this closed surface?

The flux Φ_E can be written as the sum of three terms, an integral over (*a*) the left cylinder cap, (*b*) the cylindrical surface, and (*c*) the right cap. Thus

$$\Phi_E = \oint \mathbf{E} \cdot d\mathbf{S}$$

$$= \int_{(a)} \mathbf{E} \cdot d\mathbf{S} + \int_{(b)} \mathbf{E} \cdot d\mathbf{S} + \int_{(c)} \mathbf{E} \cdot d\mathbf{S}.$$

For the left cap, the angle θ for all points is $180°$, $\mathbf{E}$ has a constant value, and the vectors $d\mathbf{S}$ are all parallel. Thus

$$\mathbf{E} \cdot d\mathbf{S} = E \cos 180 \, dS$$

$$= -E \, dS$$

for all elements dS of surface a. Thus

$$\int_{(a)} \mathbf{E} \cdot d\mathbf{S} = -ES$$

where $S \, (= \pi R^2)$ is the cap area. Similarly, for the right cap,

$$\int_{(c)} \mathbf{E} \cdot d\mathbf{S} = +ES,$$

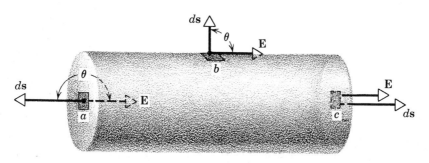

FIGURE 24–3   Example 1. A cylindrical surface immersed in a uniform field $\mathbf{E}$ parallel to its axis.

the angle θ for all points being zero here. Finally, for the cylinder wall, $\mathbf{E} \cdot d\mathbf{S} = 0$ for all points on the cylindrical surface because $\theta = 90°$. Thus

$$\int_{(b)} \mathbf{E} \cdot d\mathbf{S} = 0,$$

and

$$\Phi_E = -ES + 0 + ES = 0.$$

24–2 Gauss's Law

Gauss's law, which applies to any closed hypothetical surface (called a *Gaussian surface*), gives a connection between Φ_E for the surface and the net charge q enclosed by the surface. It is

$$\epsilon_0 \Phi_E = q \tag{24–3}$$

or, using Eq. 24–2,

$$\epsilon_0 \oint \mathbf{E} \cdot d\mathbf{S} = q. = \int_V \rho \, dv \tag{24–4}$$

The fact that Φ_E proves to be zero in Example 1 is predicted by Gauss's law because no charge is enclosed by the Gaussian surface in Fig. 24–3 ($q = 0$).

Note that q in Eq. 24–3 (or in Eq. 24–4) is the net charge, taking its algebraic sign into account. If a surface encloses equal and opposite charges, the flux Φ_E is zero. Charge outside the surface makes no contribution to the value of q, nor does the exact location of the inside charges affect this value.

Gauss's law can be used to evaluate $\mathbf{E}$ if the charge distribution is so symmetric that by proper choice of a Gaussian surface we can easily evaluate the integral in Eq. 24–4. Conversely, if $\mathbf{E}$ is known for all points on a given closed surface, Gauss's law can be used to compute the charge inside. If $\mathbf{E}$ has an outward component for every point on a closed surface, Φ_E, as Eq. 24–2 shows, will be positive and, from Eq. 24–4, there must be a net positive charge within the surface (see Fig. 24–1, surface S_1). If $\mathbf{E}$ has an inward component for every point on a closed surface, there must be a net negative charge within the surface (see Fig. 24–1, surface S_2). Surface S_3 in Fig. 24–1 encloses no charge, so that Gauss's law predicts that $\Phi_E = 0$. This is consistent with the fact that lines of $\mathbf{E}$ pass directly through surface S_3, the contribution to the integral on one side canceling that on the other. What would be the value of Φ_E for surface S_4 in Fig. 24–1, which encloses both charges?

24–3 Gauss's Law and Coulomb's Law

Coulomb's law can be deduced from Gauss's law and symmetry considerations. To do so, let us apply Gauss's law to an isolated point charge q as in Fig. 24–4. Although Gauss's law holds for any surface whatever, information can most readily be extracted for a spherical surface of radius r centered on the charge. The advantage of this surface is that, from symmetry, $\mathbf{E}$ must be normal to it and must have the same (as yet unknown) magnitude for all points on the surface.

In Fig. 24–4 both $\mathbf{E}$ and $d\mathbf{S}$ at any point on the Gaussian surface are directed radially outward. The angle between them is zero and the quantity $\mathbf{E} \cdot d\mathbf{S}$ becomes simply $E \, dS$. Gauss's law (Eq. 24–4) thus reduces to   SINCE cos θ = 1

$$\epsilon_0 \oint \mathbf{E} \cdot d\mathbf{S} = \epsilon_0 \oint E \, dS = q.$$

Because E is constant for all points on the sphere, it can be factored from inside the integral sign, leaving

$$\epsilon_0 E \oint dS = q,$$

where the integral is simply the area of the sphere. This equation gives

$$\epsilon_0 E (4\pi r^2) = q$$

or
$$E = \frac{1}{4\pi\epsilon_0} \frac{q}{r^2}. \qquad (24\text{–}5)$$

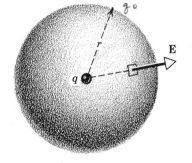

FIGURE 24–4 A spherical Gaussian surface of radius r surrounding a point charge q.

Equation 24–5 gives the magnitude of the electric field **E** at any point a distance r from an isolated point charge q. The direction of **E** is already known from symmetry.

Let us put a second point charge q_0 at the point at which **E** is calculated. The magnitude of the force that acts on it (see Eq. 23–2) is

$$F = Eq_0.$$

Combining with Eq. 24–5 gives

$$F = \frac{1}{4\pi\epsilon_0} \frac{qq_0}{r^2},$$

which is precisely Coulomb's law. Thus we have deduced Coulomb's law from Gauss's law. Note how vital symmetry arguments are in this derivation.

Gauss's law is one of the fundamental equations of electromagnetic theory and is displayed in Table 34–2 as one of Maxwell's equations. Coulomb's law is not listed in that table because, as we have just proved, it can be deduced from Gauss's law and from simple assumptions about the symmetry of **E** due to a point charge. The usefulness of Gauss's law depends on our ability to find a surface over which, from symmetry, both E and θ (see Fig. 24–2) have constant values. Then $E \cos \theta$ can be factored out of the integral and E can be found simply, as in this example.

It is interesting to note that writing the proportionality constant in Coulomb's law as $1/4\pi\epsilon_0$ (see Eq. 22–3) permits a particularly simple form for Gauss's law (Eq. 24–3). If we had written the Coulomb law constant simply as k, Gauss's law would have to be written as $(1/4\pi k)\Phi_E = q$. We prefer to leave the factor 4π in Coulomb's law so that it will not appear in Gauss's law or in other much used relations that will be derived later.

24–4 An Insulated Conductor

Gauss's law can be used to make an important prediction, namely: *An excess charge, placed on an insulated conductor, resides entirely on its outer surface.* This hypothesis was shown to be true by experiment (see Section 24–5) before either Gauss's law or Coulomb's law were advanced. Indeed, the experimental proof of the hypothesis is the foundation upon which both laws rest: We have already pointed out that Coulomb's torsion balance experiments, although direct and convincing, are not capable of great accuracy. In showing that the italicized hypothesis is predicted by Gauss's law, we are simply reversing the historical situation.

Figure 24–5 is a cross section of an insulated conductor of arbitrary shape carrying an excess charge q. The dashed lines show a Gaussian surface that lies a small distance below the actual surface of the conductor. Although the Gaussian surface can be as close to the actual surface as we wish, it is important to keep in mind that the Gaussian surface is inside the conductor.

An initially uncharged insulated conductor must have $\mathbf{E} = 0$ for all internal points. If this were not so, electrical forces would act on the charge carriers, producing internal currents. There are, in fact, no such currents in, say, a

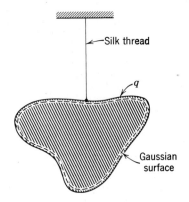

FIGURE 24–5 An insulated conductor.

coin resting on a table top. Electrostatic conditions prevail. We must also have $\mathbf{E} = 0$ for all internal points even if we put a charge on the conductor. It is true that random electric fields will appear inside a conductor at the instant that a charge is placed on it. However, these fields will not last long. Internal currents will act to redistribute the added charge until the electric fields vanish and we again have electrostatic conditions. What can we say about the distribution of the excess charge when such electrostatic conditions have been achieved?

If, at electrostatic equilibrium, $\mathbf{E}$ is zero everywhere inside the conductor, it must be zero for every point on the Gaussian surface. This means that the flux Φ_E for this surface must be zero. Gauss's law then predicts (see Eq. 24–3) that there must be no net charge inside the Gaussian surface. If the excess charge q is not inside this surface, it can only be outside it, that is, it must be on the actual surface of the conductor.

24–5 Experimental Proof of Gauss's and Coulomb's Laws

Let us turn to the experiments that prove that the hypothesis of Section 24–4 is true. For a simple test, charge a metal ball and lower it with a silk thread deep into a metal can as in Fig. 24–6. Touch the ball to the inside of the can; when the ball is removed from the can, all its charge will have vanished. When the metal ball touches the can, the ball and can together form an "insulated conductor" to which the hypothesis of Section 24–4 applies. That the charge moves entirely to the out-

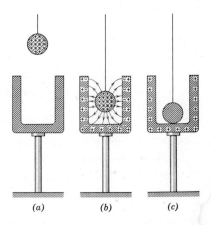

FIGURE 24–6 The *entire* charge on the ball is transferred to the *outside* of the can.

(a)   (b)   (c)

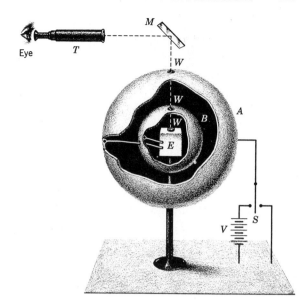

FIGURE 24–7 The apparatus of
Plimpton and Lawton.

side surface of the can can be shown by touching a small insulated metal object to
the can; only on the outside of the can will it be possible to pick up a charge.

Henry Cavendish (1731–1810) carried out an improved version of the experiment
of Fig. 24–6. With the instruments available to him, Cavendish proved experimentally that the exponent in the force law lay, with high probability, between
2.02 and 1.98. Cavendish, however, did not publish his results so that almost no-
body knew about them at the time. Maxwell repeated Cavendish's experiment
with more accuracy and set these limits as 2.00005 and 1.99995. In 1936 Plimpton
and Lawton repeated the experiment again; they set the probability limits as
2.000000002 and 1.999999998. In 1971, Williams, Faller, and Hill, set the limits
as 2.0000000000000003 and 1.9999999999999997.

Figure 24–7 is an idealized sketch of the apparatus of Plimpton and Lawton. It
consists in principle of two concentric metal shells, A and B, the former being
5 ft in diameter. The inner shell contains a sensitive electrometer E connected so
that it will indicate whether any charge moves between shells A and B.

By throwing switch S to the left, a substantial charge can be placed on the
sphere assembly. If any of this charge moves to shell B, it will have to pass through
the electrometer and will cause a deflection, which can be observed optically using
telescope T, mirror M, and windows W.

However, when the switch S is thrown alternately from left to right, thus connecting the shell assembly alternately to the battery or to the ground, no effect is
observed on the galvanometer. This is the strongest experimental evidence to date
that the hypothesis of Section 24–4 is correct. Knowing the sensitivity of their electrometer, Plimpton and Lawton calculated that the exponent in Coulomb's law
lies, with high probability, between the limits already stated.

24–6 Gauss's Law—Some Applications

Gauss's law can be used to calculate **E** if the symmetry of the charge distribution is
high. One example of this, the calculation of **E** for a point charge, has already been
discussed (Eq. 24–5). Here are some other examples.

Example 2. *Spherically symmetric charge distribution*. Figure 24–8 shows a spherical distribution of charge of radius R. The charge density ρ (that is, the charge per unit volume) at any point depends only on the distance of the point from the center and not on the direction, a condition called spherical symmetry. Find an expression for E for points (a) outside and (b) inside the charge distribution. Note that the object in Fig. 24–8 cannot be a conductor or, as we have seen, the excess charge would reside on its surface.

Applying Gauss's law to a spherical Gaussian surface of radius r in Fig. 24–8a (see Section 24–3) leads exactly to Eq. 24–5, or

$$E = \frac{1}{4\pi\epsilon_0} \frac{q}{r^2}, \quad = \frac{k\,\theta}{r^2} \tag{24–5}$$

where q is the total charge. Thus for points outside a spherically symmetric distribution of charge, the electric field has the value that it would have if the charge were concentrated at its center. This reminds us that a sphere of mass m behaves gravitationally, for outside points, as if the mass were concentrated at its center. At the root of this similarity lies the fact that both Coulomb's law and the law of gravitation are inverse square laws. The gravitational case was proved in detail in Section 14–5; the proof using Gauss's law in the electrostatic case is certainly much simpler.

Figure 24–8b shows a spherical Gaussian surface of radius r drawn *inside* the charge distribution. Gauss's law (Eq. 24–4) gives

$$\epsilon_0 \oint \mathbf{E} \cdot d\mathbf{S} = \epsilon_0 E(4\pi r^2) = q'$$

or

$$E = \frac{1}{4\pi\epsilon_0} \frac{q'}{r^2},$$

in which q' is that part of q contained within the sphere of radius r. The part of q that lies outside this sphere makes no contribution to $\mathbf{E}$ at radius r. This corresponds, in the gravitational case (Section 14–4), to the fact that a spherical shell of matter exerts no gravitational force on a body inside it. Actually, if the charged assembly of Fig. 24–8b is made up of atoms, polarization effects (see Chapter 26) will reduce the value E inside the distribution by a factor κ, the dielectric constant. Our derivation is correct, however, for charge assemblies made up of charges of only one sign, as in the atomic nucleus.

 An interesting special case of a spherically symmetric charge distribution is a uniform sphere of charge. For such a sphere, the charge density ρ would have a constant value for all points within a sphere of radius R and would be zero for all points outside this sphere. For points inside

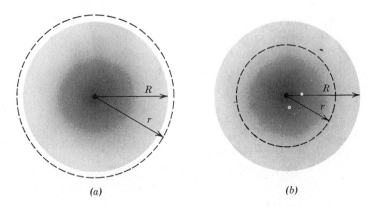

(a)   (b)

FIGURE 24–8   Example 2. A spherically symmetric charge distribution, showing two Gaussian surfaces. The density of charge, as the shading suggests, varies with distance from the center but not with direction.

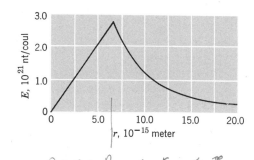

FIGURE 24-9 Example 2. The electric field due to the positive charge in a gold nucleus.

such a uniform sphere of charge we can put

$$0 \leqslant r \leqslant R \qquad \leqslant r \qquad \leqslant \infty$$

$$q' = q \frac{\frac{4}{3}\pi r^3}{\frac{4}{3}\pi R^3}$$

or

$$q' = q \left(\frac{r}{R}\right)^3$$

where $\frac{4}{3}\pi R^3$ is the volume of the spherical charge distribution. The expression for E then becomes

$$E = \frac{1}{4\pi\epsilon_0} \frac{qr}{R^3}. \tag{24-6}$$

This equation becomes zero, as it should, for $r = 0$. Note that Eqs. 24–5 and 24–6 give the same result, as they must, for points on the surface of the charge distribution (that is, if $r = R$). Note that Eq. 24–6 does not apply to the charge distribution of Fig. 24–8b because the charge density, suggested by the shading, is not constant in that case.

Figure 24–9 shows $E(r)$ for a uniform sphere of charge, using Eq. 24–5 for points outside the sphere and Eq. 24–6 for internal points. The "sphere" actually represents the nucleus of a gold atom, with $R = 6.6 \times 10^{-15}$ m and $q = Ze = (79)(1.6 \times 10^{-19} \text{ coul}) = 1.3 \times 10^{-17}$ coul.

Example 3. *Line of charge.* Figure 24–10 shows a section of an infinite rod of charge, the linear charge density λ (that is, the charge per unit length) being constant for all points on the line. Find an expression for E at a distance r from the line.

From symmetry, **E** due to a uniform linear charge can only be radially directed. As a Gaussian surface we choose a circular cylinder of radius r and length h, closed at each end by plane caps normal to the axis. E is constant over the cylindrical surface and the flux of **E** through this surface is $E(2\pi rh)$ where $2\pi rh$ is the area of the surface. There is no flux through the circular caps because **E** here lies in the surface at every point.

The charge enclosed by the Gaussian surface of Fig. 24–10 is λh. Gauss's law (Eq. 24–4),

$$\epsilon_0 \oint \mathbf{E} \cdot d\mathbf{S} = q,$$

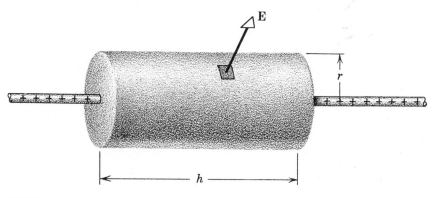

FIGURE 24-10 Example 3. An infinite rod of charge, showing a cylindrical Gaussian surface.

then becomes $\qquad\qquad\qquad\qquad \epsilon_0 E(2\pi r h) = \lambda h,$

whence $\qquad\qquad\qquad\qquad\qquad E = \dfrac{\lambda}{2\pi\epsilon_0 r}. \qquad\qquad\qquad\qquad\qquad (24\text{--}7)$

The direction of **E** is radially outward for a line of positive charge.

Note that the solution using Gauss's law is possible only if we choose our Gaussian surface to take full advantage of the radial symmetry of the electric field set up by a long line of charge. We are free to choose any surface, such as a cube or a sphere, for a Gaussian surface. Even though Gauss's law holds for all such surfaces, they are not useful for the problem at hand; only the cylindrical surface of Fig. 24–10 is appropriate in this case.

Gauss's law has the property that it provides a useful technique for calculation only in problems that have a certain degree of symmetry, but in these problems the solutions are strikingly simple.

You may wonder about the usefulness of solving a problem involving an infinite rod of charge when any actual rod must have a finite length. However, for points close enough to finite rods and not near their ends, the equation that we have just derived yields results that are so close to the correct values that the difference can be ignored in many practical situations. It is usually unnecessary to solve exactly every geometry encountered in practical problems. Indeed, if idealizations or approximations are not made, the vast majority of significant problems of all kinds in physics and engineering cannot be solved at all.

Example 4. *A sheet of charge.* Figure 24–11 shows a portion of a thin, nonconducting, infinite sheet of charge, the charge density σ (that is, the charge per unit area) being constant. What is **E** at a distance r in front of the plane?

A convenient Gaussian surface is a closed circular cylinder of cross-sectional area A and height $2r$, arranged to pierce the plane as shown. From symmetry, **E** points at right angles to the end caps and away from the plane. Since **E** does not pierce the cylindrical surface, there is no contribution to the flux from this source. Thus Gauss's law,

$$\epsilon_0 \oint \mathbf{E} \cdot d\mathbf{S} = q$$

becomes $\qquad\qquad\qquad\qquad \epsilon_0(EA + EA) = \sigma A$

where σA is the enclosed charge. This gives

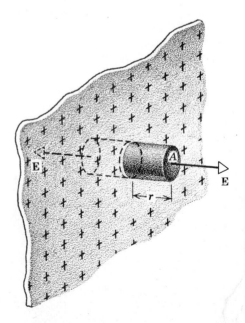

FIGURE 24–11   Example 4. An infinite sheet of charge pierced by a cylindrical Gaussian surface. The cross section of the cylinder need not be circular, as shown, but can have an arbitrary shape.

$$E = \frac{\sigma}{2\epsilon_0}. \qquad (24\text{--}8)$$

Note that E is the same for all points on each side of the plane; compare Fig. 23–2. Although an infinite sheet of charge cannot exist, this derivation is still useful in that Eq. 24–8 yields substantially correct results for real charge sheets if we consider only points not near the edges whose distance from the sheet is small compared to the dimensions of the sheet.

Example 5. *A charged conductor*. Figure 24–12 shows a conductor carrying on its surface a charge whose surface charge density at any point is σ; in general σ will vary from point to point. What is **E** for points a short distance above the surface?

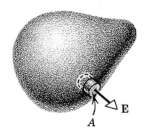

FIGURE 24–12   Example 5. A charged insulated conductor, showing a Gaussian surface. The cross section of the surface need not be circular, as shown, but can have an arbitrary shape.

·' *S U R F A C E   I S   C U R V E D*

The direction of **E** for points close to the surface is at right angles to the surface, pointing away from the surface if the charge is positive. If **E** were not normal to the surface, it would have a component lying in the surface. Such a component would act on the charge carriers in the conductor and set up surface currents. Since there are no such currents under the assumed electrostatic conditions, **E** must be normal to the surface.

The magnitude of **E** can be found from Gauss's law using a small cylinder of cross section A as a Gaussian surface. Since **E** equals zero everywhere inside the conductor (see Section 24–4), the only contribution to Φ_E is through the plane cap of area A that lies outside the conductor. Gauss's law

$$\epsilon_0 \oint \mathbf{E} \cdot d\mathbf{S} = q$$

becomes

$$\epsilon_0(EA) = \sigma A$$

where σA is the net charge within the Gaussian surface. This yields

$$E = \frac{\sigma}{\epsilon_0}. \qquad (24\text{--}9)$$

Comparison with Eq. 24–8 shows that the electric field is twice as great near a conductor carrying a charge whose surface charge density is σ as that near a nonconducting sheet with the same surface charge density. Compare the Gaussian surfaces in Figs. 24–11 and 24–12 carefully. In Fig. 24–11 lines of force leave the surface through each end cap, an electric field existing on both sides of the sheet. In Fig. 24–12 the lines of force leave only through the outside end cap, the inner end cap being inside the conductor where no electric field exists. If we assume the same surface charge density and cross-sectional area A for the two Gaussian surfaces, the enclosed charge ($= \sigma A$) will be the same. Since, from Gauss's law, the flux Φ_E must then be the same in each case, it follows that E ($= \Phi_E/A$) must be twice as large in Fig. 24–12 as in Fig. 24–11. It is helpful to note that in Fig. 24–11 half the flux emerges from one side of the surface and half from the other, whereas in Fig. 24–12 all the flux emerges from the outside surface.

QUESTIONS

1. A point charge is placed at the center of a spherical Gaussian surface. Is Φ_E changed (*a*) if the surface is replaced by a cube of the same volume, (*b*) if the sphere is replaced by a cube of one-tenth the volume, (*c*) if the charge is moved off-center in the original sphere, still remaining inside, (*d*) if the charge is moved just outside the original sphere, (*e*) if a second charge is placed near, and outside, the original sphere, and (f) if a second charge is placed inside the Gaussian surface?

2. By analogy with Φ_E, how would you define the flux Φ_g of a gravitational field? What is the flux of the earth's gravitational field through the boundaries of a room, assumed to contain no matter?

3. In Gauss's law,

$$\epsilon_0 \oint \mathbf{E} \cdot d\mathbf{S} = q,$$

is $\mathbf{E}$ the electric field attributable to the charge q?

4. A surface encloses an electric dipole. What can you say about Φ_E for this surface?

5. Suppose that a Gaussian surface encloses no net charge. Does Gauss's law require that $\mathbf{E}$ equal zero for all points on the surface? Is the converse of this statement true, that is, if $\mathbf{E}$ equals zero everywhere on the surface, does Gauss's law require that there be no net charge inside?

6. Would Gauss's law hold if the exponent in Coulomb's law were not exactly two?

7. Does Gauss's law, as applied in Section 24–4, require that all the conduction electrons in an insulated conductor reside on the surface?

8. In Section 24–4 we assumed that $\mathbf{E}$ equals zero everywhere inside a conductor. However, there are certainly very large electric fields inside the conductor, at points close to the electrons or to the nuclei. Does this invalidate the proof of Section 24–4?

9. Is Gauss's law useful in calculating the field due to three equal charges located at the corners of an equilateral triangle? Explain.

10. The use of line, surface, and volume densities of charge to calculate the charge contained in an element of a charged object implies a continuous distribution of charge, whereas, in fact, charge on the microscopic scale is discontinuous. How, then, is this procedure justified?

11. Is $\mathbf{E}$ necessarily zero inside a charged rubber balloon if the balloon is (a) spherical, or (b) sausage-shaped? For each shape assume the charge to be distributed uniformly over the surface.

12. A spherical rubber balloon carries a charge that is uniformly distributed over its surface. How does E vary for points (a) inside the balloon, (b) at the surface of the balloon, and (c) outside the balloon, as the balloon is blown up?

13. As you penetrate a uniform sphere of charge, E should decrease because less charge lies inside a sphere drawn through the observation point. On the other hand, E should increase because you are closer to the center of this charge. Which effect predominates and why?

14. Given a spherically symmetric charge distribution (not of uniform density of charge), is E necessarily a maximum at the surface? Comment on various possibilities.

PROBLEMS

1(1). In Example 1 compute Φ_E for the cylinder if it is turned so that its axis is perpendicular to the electric field.
Answer: Zero.

2(1). A plane surface of area A is inclined so that its normal makes an angle θ with a uniform field of $\mathbf{E}$. Calculate Φ_E for this surface.

3(1). Calculate Φ_E through a hemisphere of radius R. The field of $\mathbf{E}$ is uniform and is parallel to the axis of the hemisphere.
Answer: $\pi R^2 E$.

4(1). A butterfly net is in a uniform electric field as shown in Fig. 24–13. The rim, a circle of radius a, is aligned perpendicular to the field. Find the electric flux through the netting.

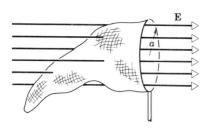

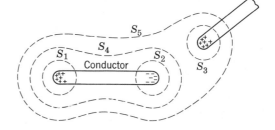

FIGURE 24-13 Problem 4(1).

FIGURE 24-14 Problem 5(2).

5(2). Charge on an originally uncharged insulated conductor is separated by holding a positively charged rod nearby, as in Fig. 24–14. What can you learn from Gauss's law about the flux for the five Gaussian surfaces shown?

Answer: For S_1, $\phi_1 > 0$; for S_2, $\phi_2 = -\phi_1$; for S_3, $\phi_3 = \phi_1$; for S_4, $\phi_4 = 0$; for S_5, $\phi_5 = \phi_1$.

6(2). A uniformly charged conducting sphere of 1.0 meter diameter has a surface charge density of 8 coul/m². What is the total electric flux leaving the surface of the sphere?

7(2). A point charge of 1.0×10^{-6} coul is at the center of a cubical Gaussian surface 0.50 meter on edge. What is Φ_E for the surface?

Answer: 1.1×10^5 nt meters/coul.

8(2). Suppose that an electric field in some region is found to have a constant direction but to be decreasing in strength in that direction. What do you conclude about the charge in the region?

9(2). It is found experimentally that the electric field in a large region of the earth's atmosphere is directed vertically down. At an altitude of 300 meters the field is 60 nt/coul and at an altitude of 200 meters it is 100 nt/coul. Find the net amount of charge contained in a cube 100 meters on edge located between 200 and 300 meters altitude. Neglect the curvature of the earth.

Answer: $+3.5 \times 10^{-6}$ coul.

10(3). "Gauss's law for gravitation" is

$$\frac{1}{4\pi G}\, \Phi_g = \frac{1}{4\pi G} \oint \mathbf{g} \cdot d\mathbf{S} = -m.$$

where m is the enclosed mass and G is the universal gravitation constant (Section 14–2). Derive Newton's law of gravitation from this.

11(6). Fig. 24–15 shows a point charge of 1.0×10^{-7} coul at the center of a spherical cavity of radius 3.0 cm in a piece of metal. (*a*) Use Gauss's law to find the electric field at point *a*, halfway from the center to the surface. (*b*) What is the electric field at *b*?

Answer: (*a*) 4.0×10^6 nt/coul. (*b*) Zero.

12(6). An uncharged spherical thin metallic shell has a point charge q at its center. Give expres-

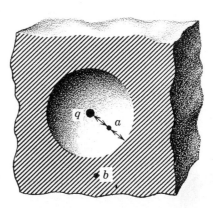

FIGURE 24-15 Problem 11(6).

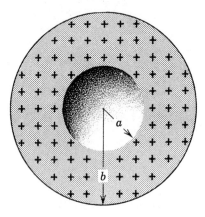

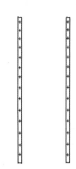

FIGURE 24-16   Problem 18(6).   **FIGURE 24-17**   Problem 20(6).

sions for the electric field (a) inside the shell, and (b) outside the shell, using Gauss's law. (c) Has the shell any effect on the field due to q? (d) Has the presence of q any effect on the shell? (e) If a second point charge is held outside the shell, does this outside charge experience a force? (f) Does the inside charge experience a force? (g) Is there a contradiction with Newton's third law here?

13(6). A thin-walled metal sphere has a radius of 25 cm and carries a charge of 2.0×10^{-7} coul. Find E for a point (a) inside the sphere, (b) just outside the sphere, and (c) 3.0 meters from the center of the sphere.
Answer: (a) Zero. (b) 2.9×10^4 nt/coul. (c) 200 nt/coul.

14(6). A proton orbits with a speed v (3.0×10^5 meters/sec) just outside a charged sphere of radius r (10^{-6} meter). What is the charge on the sphere?

15(6). A thin metallic spherical shell of radius a carries a charge q_a. Concentric with it is another thin metallic spherical shell of radius b ($b > a$) carrying a charge q_b. Use Gauss's law to find the electric field at radial points r where (a) $r < a$; (b) $a < r < b$; (c) $r > b$. (d) How is the charge on each shell distributed between the inner and outer surfaces of that shell?
Answer: (a) Zero. (b) $q_a/(4\pi\epsilon_0 r^2)$. (c) $(q_a + q_b)/(4\pi\epsilon_0 r^2)$. (d) Inner shell: zero and q_a; outer shell: $-q_a$, $q_a + q_b$.

16(6). Two charged concentric spheres have radii of 10 cm and 15 cm. The charge on the inner sphere is 4.0×10^{-8} coul and that on the outer sphere 2.0×10^{-8} coul. Find the electric field (a) at $r = 12$ cm, and (b) at $r = 20$ cm.

17(6). Two concentric conducting spherical shells have radii $R_1 = 0.145$ meter and $R_2 = 0.207$ meter. The inner sphere bears a charge -6.00×10^{-2} coul. An electron escapes from the inner sphere with negligible speed. Assuming that the region between the spheres is a vacuum, compute the speed with which the electron strikes the outer sphere.
Answer: 1.99×10^7 meters/sec.

18(6). Figure 24–16 shows a spherical nonconducting shell of charge of uniform charge density ρ. Plot E for distances r from the center of the shell ranging from zero to 30 cm. Assume that $\rho = 1.0 \times 10^{-6}$ coul/meter^3, $a = 10$ cm, and $b = 20$ cm.

19(6). A metal plate 8.0 cm on a side carries a total charge of 6.0×10^{-6} coul. (a) Estimate the electric field 0.50 cm above the surface of the plate near the plate's center. (b) Estimate the field at a distance of 3.0 meters.
Answer: (a) 5.3×10^7 nt/coul. (b) 6.0×10^3 nt/coul.

20(6). Two large nonconducting sheets of positive charge face each other as in Fig. 24-17. What is **E** at points (a) to the left of the sheets, (b) between them, and (c) to the right of the sheets? Assume the same surface charge density σ for each sheet. The separation of the sheets is much less than their dimensions. Do not consider points near the edges of the sheets. (Hint: **E** at any point is the vector sum of the separate electric fields set up by each sheet.)

21(6). Two large metal plates face each other as in Fig. 24–18 and carry charges with surface charge density $+\sigma$ and $-\sigma$, respectively, on their inner surfaces. What is **E** at points (a) to the left

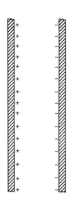

FIGURE 24-18 Problem 21(6).

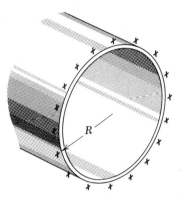

FIGURE 24-19 Problem 26(6).

of the sheets, (b) between them, and (c) to the right of the sheets. The separation of the sheets is much less than their dimensions. Do not consider points near the edges of the sheets.

Answer: (a) Zero. (b) $E = \sigma/\epsilon_0$, to the right. (c) Zero.

22(6). Two large metal plates of area 1.0 meter^2 face each other. They are 5.0 cm apart and carry equal and opposite charges on their inner surfaces. If E between the plates is 55 nt/coul, what is the charge on the plates? Neglect edge effects. See Problem 21(6).

23(6). A 100-eV electron is fired directly toward a large metal plate that has a surface charge density of -2.0×10^{-6} coul/meter^2. From what distance must the electron be fired if it is to just fail to strike the plate? One eV $= 1.60 \times 10^{-19}$ joule.

Answer: 0.45 mm.

24(6). A nonconducting plane slab of thickness d has a uniform volume charge density ρ. Find the magnitude of the electric field at all points in space both (a) inside and (b) outside the slab.

25(6). An infinite line of charge produces a field of 4.5×10^4 nt/coul at a distance of 2.0 meters. Calculate the linear charge density.

Answer: 5.0×10^{-6} coul/meter.

26(6). Fig. 24–19 shows a section through a long, thin-walled metal tube of radius R, carrying a charge per unit length λ on its surface. Derive expressions for E for various distances r from the tube axis, considering both (a) $r > R$ and (b) $r < R$. (c) Plot your results for the range $r = 0$ to $r = 5.0$ cm, assuming that $\lambda = 2.0 \times 10^{-8}$ coul/meter and $R = 3.0$ cm.

27(6). Fig. 24–20 shows a section through two long concentric cylinders of radii a and b. The cylinders carry equal and opposite charges per unit length λ. Using Gauss's law, prove (a) that $E = 0$ for $r > b$ and for $r < a$ and (b) that between the cylinders E is given by

$$E = \frac{1}{2\pi\epsilon_0} \frac{\lambda}{r}.$$

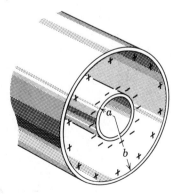

FIGURE 24-20 Problem 27(6).

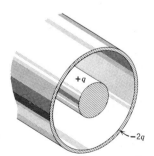

FIGURE 24-21 Problem 29(6).

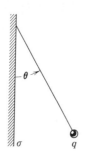

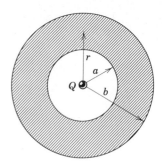

FIGURE 24-22   Problem 33(6).

FIGURE 24-23   Problem 34(6).

28(6). In Problem 27(6) a positive electron revolves in a circular path of radius r, between and concentric with the cylinders. What must be its kinetic energy K? Assume $a = 2.0$ cm, $b = 3.0$ cm, and $\lambda = 3.0 \times 10^{-8}$ coul/meter.

29(6). A long conducting cylinder of length l, carrying a total charge $+q$, is surrounded by a conducting cylindrical shell of total charge $-2q$, as shown in cross section in Fig. 24–21. Use Gauss's law to find (a) the electric field at points outside the conducting shell, (b) the distribution of the charge on the conducting shell, and (c) the electric field in the region between the cylinders. *Answer:* (a) $E = q/(2\pi\epsilon_0 rl)$, radially in. (b) $-q$ on both inner and outer surfaces. (c) $E = q/(2\pi\epsilon_0 rl)$, radially out.

30(6). Two charged concentric cylinders have radii of 3.0 cm and 6.0 cm. The charge per unit length on the inner cylinder is 5.0×10^{-6} coul/m and that on the outer cylinder is -7.0×10^{-6} coul/m. Find the electric field at (a) $r = 4.0$ cm and (b) at $r = 8.0$ cm.

31(6). Charge is distributed uniformly throughout an infinitely long cylinder of radius R. (a) Show that E at a distance r from the cylinder axis ($r < R$) is given by

$$E = \frac{\rho r}{2\epsilon_0}$$

where ρ is the density of charge. (b) What result do you expect for $r > R$?
Answer: (b) $E = \rho R^2/2\epsilon_0 r$.

32(6). Equation 24–9 ($E = \sigma/\epsilon_0$) gives the electric field at points near a charged conducting surface. Show that this equation leads to a familiar result when applied to a conducting sphere of radius r, carrying a charge q.

33(6). A small charged sphere of mass m and charge q hangs from a silk thread that makes an angle θ with a large, flat charged conducting surface, as in Fig. 24–22. Find the surface charge density σ.
Answer: $\sigma = (\epsilon_0 mg \tan \theta)/q$.

34(6). The spherical region $a < r < b$ carries a charge per unit volume of $\rho = A/r$, where A is constant (Fig. 24–23). At the center ($r = 0$) of the enclosed cavity is a point charge Q. What should the value of A be so that the electric field in the region $a < r < b$ has constant magnitude?

35(6). A solid insulating sphere carries a uniform charge per unit volume of ρ. Let r be the vector from the center of the sphere to a general point P within the sphere. (a) Show that the electric field at P is given by $\mathbf{E} = \rho\mathbf{r}/3\epsilon_0$. (b) A spherical "cavity" is removed from the above sphere, as shown in Fig. 24–24. Using superposition concepts show that the electric field at all points within the cavity is $\mathbf{E} = \rho\mathbf{a}/3\epsilon_0$ (uniform field), where $\mathbf{a}$ is the vector connecting the center of the sphere with the center of the cavity. Note that both these results are independent of the radii of the sphere and the cavity.

FIGURE 24-24   Problem 35(6).

CHAPTER 25

Electric Potential

25–1 Electric Potential

The electric field around a charged rod can be described not only by a (vector) electric field **E** but also by a scalar quantity, the *electric potential V*. These quantities are intimately related, and often it is only a matter of convenience which is used in a given problem.

To find the electric potential difference between two points A and B in an electric field, we move a test charge q_0 from A to B, always keeping it in equilibrium, and we measure the work W_{AB} that must be done by the agent moving the charge. The electric potential difference* is defined from

$$V_B - V_A = \frac{W_{AB}}{q_0}.\qquad(25\text{--}1)$$

The work W_{AB} may be positive, negative, or zero. In these cases the electric potential at B will be higher, lower, or the same as the electric potential at A, respectively.

The mks unit of potential difference that follows from Eq. 25–1 is the joule/coul. This combination occurs so often that a special unit, the *volt*, is used to represent it; that is,

$$1 \text{ volt} = 1 \text{ joule/coul.}$$

Usually point A is chosen to be at a large distance from all charges, and the electric potential V_A at this infinite distance is arbitrarily taken as zero. This allows us to

---

* This definition of potential difference, though suitable for our present purpose, is rarely carried out in practice because of technical difficulties. Equivalent and more technically feasible methods are usually adopted.

define the electric potential at a point. Putting $V_A = 0$ in Eq. 25–1 and dropping the subscripts leads to

$$V = \frac{W}{q_0},\tag{25–2}$$

where W is the work that an external agent must do to move the test charge q_0 from infinity to the point in question. Keep in mind that *potential differences* are of fundamental concern and that Eq. 25–2 depends on the arbitrary assignment of the value zero to the potential V_A at the reference position (infinity); this reference potential could equally well have been chosen as any other value, say -137 volts. Similarly any other agreed-upon point could be chosen as a reference position. In many circuit problems the earth is taken as a reference of potential and assigned the value zero.

Bearing in mind the assumptions made about the reference position, we see from Eq. 25–2 that V near an isolated positive charge is positive because positive work must be done by an outside agent to push a (positive) test charge in from infinity. Similarly, the potential near an isolated negative charge is negative because an outside agent must exert a restraining force on (that is, must do negative work on) a (positive) test charge as it comes in from infinity. Electric potential as defined in Eq. 25–2 is a scalar because W and q_0 in that equation are scalars.

Both W_{AB} and $V_B - V_A$ in Eq. 25–1 are independent of the path followed in moving the test charge from point A to point B. If this were not so, point B would not have a unique electric potential (with respect to point A as a defined reference position) and the concept of potential would have limited usefulness.

We can easily prove that potential differences are path-independent for the special case shown in Fig. 25–1. This figure illustrates the case in which the two points A and B are in a field set up by a spherical charge q; the two points are further chosen, for simplicity, to lie along a radial line. Although our path-independence proof applies only to this special case, it illustrates the general principles involved.

Point A in Fig. 25–1 may be taken as a defined reference point, and we imagine a positive test charge q_0 moved by an external agent from A to B. We consider two paths, path I being a radial line between A and B and path II being a completely arbitrary path between these two points. The open arrows on path II show the electric force per unit charge that would act at various points on a test charge q_0.

Path II may be approximated by a broken path made up of alternating elements of arc and of radius. Since these elements can be arbitrarily small, the broken path can be made arbitrarily close to the actual path. On path II the external agent does work only along the radial segments because along the arcs the force **F** and the dis-

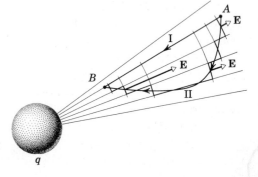

FIGURE 25–1   A test charge q_0 is moved from A to B in the field of charge q along either of two paths. The open arrows show **E** at three points on path II. For simplicity, we have chosen path II to lie in the plane of the figure; our analysis still holds if it does not.

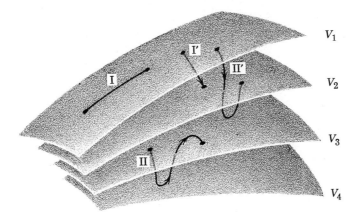

FIGURE 25-2 Portions of four equipotential surfaces. The heavy lines show four paths along which a test charge is moved.

placement $d\mathbf{l}$ are at right angles, $\mathbf{F} \cdot d\mathbf{l}$ being zero in such cases. The sum of the work done on the radial segments that make up path II is the same as the work done on path I because each path has the same array of radial segments. Since path II is arbitrary, we have proved that the work done is the same for all paths connecting A and B. Although this proof holds only for the special case of Fig. 25–1, the potential difference is path-independent for any two points in any electrostatic field. We discussed path independence in Section 7–2 for the general class of conservative forces; electrostatic forces, like gravitational forces, are conservative.

The locus of points, all of which have the same electric potential, is called an *equipotential surface*. A family of equipotential surfaces, each surface corresponding to a different value of the potential, can be used to give a general description of the electric field in a certain region of space. We have seen earlier (Section 23–3) that electric lines of force can also be used for this purpose; in later sections (see, for example, Fig. 25–15) we explore the intimate connection between these two ways of describing the electric field.

No work is required to move a test charge between any two points on an equipotential surface. This follows from Eq. 25–1,

$$V_B - V_A = \frac{W_{AB}}{q_0},$$

because W_{AB} must be zero if $V_A = V_B$. This is true, because of the path independence of potential difference, even if the path connecting A and B does not lie entirely on the equipotential surface.

Figure 25–2 shows an arbitrary family of equipotential surfaces. The work to move a charge along paths I and II is zero because all these paths begin and end on the same equipotential surface. The work to move a charge along paths I' and II' is not zero but is the same for each path because the initial and the final potentials are identical; paths I' and II' connect the same pair of equipotential surfaces.

From symmetry, the equipotential surfaces for a spherical charge are a family of concentric spheres. For a uniform field they are a family of planes at right angles to the field. In all cases (including these two examples) the equipotential surfaces are at right angles to the lines of force and thus to $\mathbf{E}$ (see Fig. 25–15). If $\mathbf{E}$ was not at right angles to the equipotential surface, it would have a component lying in that surface.

Then work would have to be done in moving a test charge about on the surface. Work cannot be done if the surface is an equipotential, so **E** must be at right angles to the surface.

25–2   Potential and the Electric Field

Let A and B in Fig. 25–3 be two points in a uniform electric field **E,** set up by an arrangement of charges not shown, and let A be a distance d in the field direction from B. Assume that a positive test charge q_0 is moved, by an external agent and without acceleration, from A to B along the straight line connecting them.

The electric force on the charge is q_0**E** and points down. To move the charge in the way we have described we must counteract this force by applying an external force **F** of the same magnitude but directed upward. The work W done by the agent that supplies this force is

$$W_{AB} = Fd = q_0Ed. \tag{25–3}$$

Substituting this into Eq. 25–1 yields

$$V_B - V_A = \frac{W_{AB}}{q_0} = Ed. \tag{25–4}$$

This equation shows the connection between potential difference and electric field for a simple special case. Note from this equation that another mks unit for **E** is the volt/meter. You should prove that a volt/meter is identical with a nt/coul; this latter unit is the one first presented for **E** in Section 23–2.

In Fig. 25–3 B has a higher potential than A. This is reasonable because an external agent would have to do positive work to push a positive test charge from A to B. Figure 25–3 could be used as it stands to illustrate the act of lifting a stone from A to B in the uniform gravitational field near the earth's surface.

What is the connection between V and **E** in the more general case in which the field is not uniform and in which the test body is moved along a path that is not straight, as in Fig. 25–4? The electric field exerts a force q_0**E** on the test charge, as

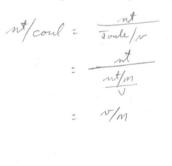

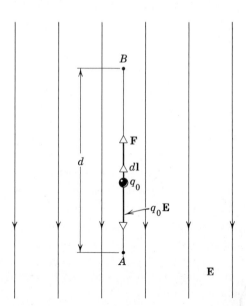

FIGURE 25–3   A test charge q_0 is moved from A to B in a uniform electric field **E** by an external agent that exerts a force **F** on it.

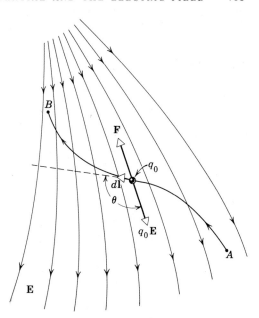

FIGURE 25–4 A test charge q_0 is moved from A to B in a nonuniform electric field by an external agent that exerts a force **F** on it. Again, as in Fig. 25–1, the path connecting A and B need not lie in a single plane.

shown. To keep the test charge from accelerating, an external agent must apply a force **F** chosen to be exactly equal to $-q_0\mathbf{E}$ for all positions of the test body.

If the external agent causes the test body to move through a displacement $d\mathbf{l}$ along the path from A to B, the element of work done by the external agent is $\mathbf{F} \cdot d\mathbf{l}$. To find the total work W_{AB} done by the external agent in moving the test charge from A to B, we add up (that is, integrate) the work contributions for all the infinitesimal segments into which the path is divided. This leads to

$$W_{AB} = \int_A^B \mathbf{F} \cdot d\mathbf{l} = -q_0 \int_A^B \mathbf{E} \cdot d\mathbf{l}.$$

Such an integral is called a *line integral*. Note that we have substituted $-q_0\mathbf{E}$ for its equal, **F**.

Substituting this expression for W_{AB} into Eq. 25–1 leads to

$$V_B - V_A = \frac{W_{AB}}{q_0} = -\int_A^B \mathbf{E} \cdot d\mathbf{l}. \tag{25–5}$$

If point A is taken to be infinitely distant and the potential V_A at infinity is taken to be zero, this equation gives the potential V at point B, or, dropping the subscript B,

$$V = -\int_\infty^B \mathbf{E} \cdot d\mathbf{l}. \tag{25–6}$$

These two equations allow us to calculate the potential difference between any two points, or the potential at any point, if **E** is known at various points in the field.

Example 1. In Fig. 25–3 calculate $V_B - V_A$ using Eq. 25–5. Compare the result with that obtained by direct analysis of this special case (Eq. 25–4).

In moving the test charge, the element of path $d\mathbf{l}$ always points in the direction of motion; this is upward in Fig. 25–3. The electric field **E** in this figure points down so that the angle θ between **E** and $d\mathbf{l}$ is $180°$.

Equation 25–5 then becomes

$$V_B - V_A = -\int_A^B \mathbf{E} \cdot d\mathbf{l} = -\int_A^B E \cos 180° \, dl = \int_A^B E \, dl.$$

E is constant for all parts of the path in this problem and can thus be taken outside the integral sign, giving

$$V_B - V_A = E \int_A^B dl = Ed,$$

which agrees with Eq. 25–4, as it must.

Example 2. In Fig. 25–5 let a test charge q_0 be moved without acceleration from A to B over the path shown. Compute the potential difference between A and B.

For the path AC we have $\theta = 135°$ and, from Eq. 25–5,

$$V_C - V_A = -\int_A^C \mathbf{E} \cdot d\mathbf{l} = -\int_A^C E \cos 135° \, dl = \frac{E}{\sqrt{2}} \int_A^C dl.$$

The integral is the length of the line AC which is $\sqrt{2}d$. Thus

$$V_C - V_A = \frac{E}{\sqrt{2}} (\sqrt{2}d) = Ed.$$

Points B and C have the same potential because no work is done in moving a charge between them, $\mathbf{E}$ and $d\mathbf{l}$ being at right angles for all points on the line CB. In other words, B and C lie on the same equipotential surface at right angles to the lines of force. Thus

$$V_B - V_A = V_C - V_A = Ed.$$

This is the same value derived for a direct path connecting A and B, a result to be expected because the potential difference between two points is path independent.

25–3 Potential Due to a Point Charge

Figure 25–6 shows two points A and B near an isolated point charge q. For simplicity we assume that A, B, and q lie on a straight line. Let us compute the potential differ-

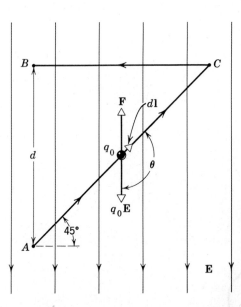

FIGURE 25–5 Example 2. A test charge q_0 is moved along path ACB in a uniform electric field by an external agent.

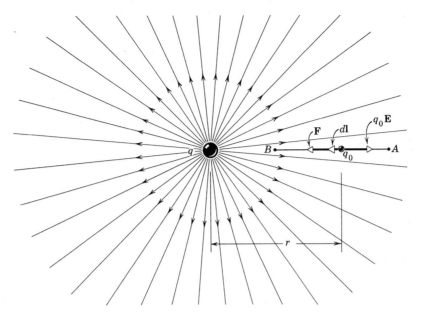

FIGURE 25–6 A test charge q_0 is moved by an external agent from A to B in the field set up by a point charge q. Points A and B need not lie on a radial line and the path connecting them can be quite arbitrary. We have assumed the conditions of the figure for simplicity.

ence between points A and B, assuming that a test charge q_0 is moved without acceleration along a radial line from A to B.

Along this path **E** points to the right and $d\mathbf{l}$, which is always in the direction of motion, points to the left. Therefore, in Eq. 25–5,

$$\mathbf{E} \cdot d\mathbf{l} = E \cos 180° \, dl = -E \, dl.$$

However, as we move a distance dl to the left, we are moving in the direction of decreasing r because r is measured from q as an origin. Thus

$$dl = -dr.$$

Combining yields 	$$\mathbf{E} \cdot d\mathbf{l} = E \, dr.$$

Substituting this into *Eq.* 25–5 gives

$$V_B - V_A = -\int_A^B \mathbf{E} \cdot d\mathbf{l} = -\int_{r_A}^{r_B} E \, dr.$$

Combining with Eq. 23–4,

$$E = \frac{1}{4\pi\epsilon_0} \frac{q}{r^2}$$

leads to 	$$V_B - V_A = -\frac{q}{4\pi\epsilon_0} \int_{r_A}^{r_B} \frac{dr}{r^2} = \frac{q}{4\pi\epsilon_0}\left(\frac{1}{r_B} - \frac{1}{r_A}\right). \tag{25–7}$$

Choosing reference position A to be at infinity (that is, letting $r_A \to \infty$), choosing $V_A = 0$ at this position, and dropping the subscript B leads to

$$V = \frac{1}{4\pi\epsilon_0} \frac{q}{r}. \tag{25–8}$$

This shows that equipotential surfaces for an isolated point charge are spheres concentric with the point charge (see Fig. 25–15a). A study of the derivation will show that this relation also holds for points external to spherically symmetric charge distributions.

Example 3. What must the magnitude of an isolated positive point charge be for the electric potential at 10 cm from the charge to be $+100$ volts?

Solving Eq. 25–8 for q yields

$$q = V4\pi\epsilon_0 r = (100 \text{ volts})(4\pi)(8.9 \times 10^{-12} \text{ coul}^2/\text{nt-m}^2)(0.10 \text{ meter})$$

$$= 1.1 \times 10^{-9} \text{ coul}.$$

This charge is comparable to charges that can be produced by friction.

Example 4. What is the electric potential at the surface of a gold nucleus? The radius is 6.6×10^{-15} meter and the atomic number $Z = 79$.

The nucleus, assumed spherically symmetrical, behaves electrically for external points as if it were a point charge, equal to that of 79 protons. Thus we can use Eq. 25–8, or, recalling that the proton charge is 1.6×10^{-19} coul,

$$V = \frac{1}{4\pi\epsilon_0} \frac{q}{r} = \frac{(9.0 \times 10^9 \text{ nt-m}^2/\text{coul}^2)(79)(1.6 \times 10^{-19} \text{ coul})}{6.6 \times 10^{-15} \text{ meter}}$$

$$= 1.7 \times 10^7 \text{ volts}.$$

25–4 A Group of Point Charges

The potential at any point due to a group of point charges is found by calculating the potential V_n due to each charge, as if the other charges were not present, and adding the quantities so obtained, or (see Eq. 25–8)

SUPERPOSITION
PRINCIPLE

$$V = \sum_n V_n = \frac{1}{4\pi\epsilon_0} \sum_n \frac{q_n}{r_n}, \qquad (25-9)$$

where q_n is the value of the nth charge and r_n is the distance of this charge from the point in question. The sum used to calculate V is an algebraic sum and not a vector sum like the one used to calculate $\mathbf{E}$ for a group of point charges (see Eq. 23–5). Herein lies an important computational advantage of potential over electric field.

If the charge distribution is a continuous one, rather than a collection of points, the sum in Eq. 25–9 must be replaced by an integral, or

$$V = \int dV = \frac{1}{4\pi\epsilon_0} \int \frac{dq}{r}, \qquad (25-10)$$

where dq is a differential element of the charge distribution, r is its distance from the point at which V is to be calculated, and dV is the potential it establishes at that point.

Example 5. What is the potential at the center of the square of Fig. 25–7? Assume that $q_1 = +1.0 \times 10^{-8}$ coul, $q_2 = -2.0 \times 10^{-8}$ coul, $q_3 = +3.0 \times 10^{-8}$ coul, $q_4 = +2.0 \times 10^{-8}$ coul, and $a = 1.0$ meter.

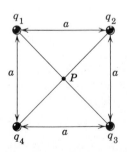

FIGURE 25–7   Example 5.

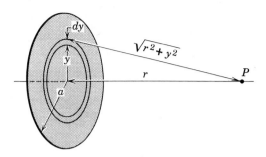

FIGURE 25–8 Example 6. A point P on the axis of a uniformly charged circular disk of radius a.

The distance r of each charge from P is $a/\sqrt{2}$ or 0.71 meter. From Eq. 25–9

$$V = \sum_n V_n = \frac{1}{4\pi\epsilon_0} \frac{q_1 + q_2 + q_3 + q_4}{r}$$

$$= \frac{(9.0 \times 10^9 \text{ nt-m}^2/\text{coul}^2)(1.0 - 2.0 + 3.0 + 2.0) \times 10^{-8} \text{ coul}}{0.71 \text{ meter}} \quad ; \quad nt\text{-}m/coul = V$$

$$= 500 \text{ volts.}$$

Is the potential constant within the square? Does any point inside have a negative potential? Sketch roughly the intersection of the plane of Fig. 25–7 with the equipotential surface corresponding to zero volts.

Example 6. *A charged disk.* Find the electric potential for points on the axis of a uniformly charged disk whose surface charge density is σ (see Fig. 25–8). $\quad \sigma = \dfrac{dq}{dA} = \dfrac{dq}{2\pi y\, dy}$

Consider a charge element dq consisting of a flat circular strip of radius y and width dy. We have

$$dq = \sigma(2\pi y)(dy),$$

where $(2\pi y)(dy)$ is the area of the strip. All parts of this charge element are the same distance $r' (= \sqrt{y^2 + r^2})$ from axial point P so that their contribution dV to the electric potential at P is given by Eq. 25–8, or

$$dV = \frac{1}{4\pi\epsilon_0} \frac{dq}{r'} = \frac{1}{4\pi\epsilon_0} \frac{\sigma 2\pi y\, dy}{\sqrt{y^2 + r^2}}.$$

The potential V is found by integrating over all the strips into which the disk can be divided (Eq. 25–10) or

$$V = \int dV = \frac{\sigma}{2\epsilon_0} \int_0^a (y^2 + r^2)^{-1/2} y\, dy$$

$$= \frac{\sigma}{2\epsilon_0}(\sqrt{a^2 + r^2} - r).$$

This general result is valid for all values of r. In the special case of $r \gg a$ the quantity $\sqrt{a^2 + r^2}$ can be approximated as

$$\sqrt{a^2 + r^2} = r\left(1 + \frac{a^2}{r^2}\right)^{1/2} = r\left(1 + \frac{1}{2}\frac{a^2}{r^2} + \cdots\right) \cong r + \frac{a^2}{2r},$$

in which the quantity in parentheses in the second member of this equation has been expanded by the binomial theorem (see Appendix G). This equation means that V becomes

$$V \cong \frac{\sigma}{2\epsilon_0}\left(r + \frac{a^2}{2r} - r\right) = \frac{\sigma\pi a^2}{4\pi\epsilon_0 r} = \frac{1}{4\pi\epsilon_0}\frac{q}{r},$$

where $q (= \sigma\pi a^2)$ is the total charge on the disk. This limiting result is expected because the disk behaves like a point charge for $r \gg a$.

25–5 Potential Due to a Dipole

Two equal charges, q, of opposite sign, separated by a distance $2a$, constitute an electric dipole; see Example 3, Chapter 23. The electric dipole moment **p** has the magnitude $2aq$ and points from the negative charge to the positive charge. Here we derive an expression for the electric potential V at any point of space due to a dipole, provided only that the point is not too close to the dipole.

A point P is specified by giving the quantities r and θ in Fig. 25–9. From symmetry, it is clear that the potential will not change as point P rotates about the z-axis, r and θ being fixed. Thus we need only find $V(r,\theta)$ for any plane containing this axis; the plane of Fig. 25–9 is such a plane. Applying Eq. 25–9 gives

$$V = \sum_n V_n = V_1 + V_2 = \frac{1}{4\pi\epsilon_0}\left(\frac{q}{r_1} - \frac{q}{r_2}\right) = \frac{q}{4\pi\epsilon_0}\frac{r_2 - r_1}{r_1 r_2},$$

which is an exact relationship.

We now limit consideration to points such that $r \gg 2a$. These approximate relations then follow from Fig. 25–9:

$$r_2 - r_1 \cong 2a\cos\theta \qquad \text{and} \qquad r_1 r_2 \cong r^2,$$

and the potential reduces to

$$V = \frac{q}{4\pi\epsilon_0}\frac{2a\cos\theta}{r^2} = \frac{1}{4\pi\epsilon_0}\frac{p\cos\theta}{r^2}, \tag{25–11}$$

in which $p\,(= 2aq)$ is the dipole moment. Note that V vanishes everywhere in the equatorial plane ($\theta = 90°$). This reflects the fact that it takes no work to bring a test charge in from infinity along the perpendicular bisector of the dipole. For a given radius, V has its greatest positive value for $\theta = 0$ and its greatest negative value for $\theta = 180°$. Note that the potential does not depend separately on q and $2a$ but only on their product p.

It is convenient to call any assembly of charges, for which V at distant points is given by Eq. 25–11, an *electric dipole*. Two point charges separated by a small distance behave this way, as we have just proved. However, other charge configurations also obey Eq. 25–11. Suppose that by measurement at points outside an imaginary box (Fig. 25–10) we find a pattern of lines of force that can be described quantitatively by Eq. 25–11. We then declare that the object inside the box is an electric dipole, that its axis is the line zz', and that its dipole moment **p** points vertically upward.

Many molecules have electric dipole moments. That for H_2O in its vapor state is 6.2×10^{-30} coul-m. Figure 25–11 is a representation of this molecule, showing the three nuclei and the surrounding electron cloud. The dipole moment **p** is represented by the arrow on the axis of

SINCE
$\theta = 90° \Rightarrow$ $q_{net} = 0$
FOR $r \gg 2a$

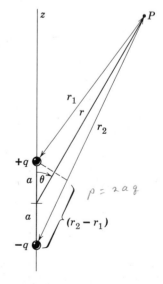

FIGURE 25–9 A point P in the field of an electric dipole.

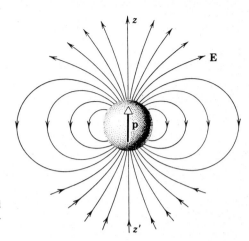

FIGURE 25-10 If an object inside the spherical box sets up the electric field shown (described quantitatively by Eq. 25-11), it is an electric dipole.

symmetry of the molecule. In this molecule the effective center of positive charge does not coincide with the effective center of negative charge. It is precisely because of this separation that the dipole moment exists.

Atoms, and many molecules, do not have permanent dipole moments. However, dipole moments may be induced by placing any atom or molecule in an external electric field. The action of the field (Fig. 25-12) is to separate the centers of positive and of negative charge. We say that the atom becomes *polarized* and acquires an *induced electric dipole moment*. Induced dipole moments disappear when the electric field is removed, except under very special circumstances.

Electric dipoles are important in situations other than atomic and molecular ones. Radio and radar antennas are often in the form of a metal wire or rod in which electrons surge back and forth periodically. At a certain time one end of the wire or rod will be negative and the other end positive. Half a cycle later the polarity of the ends is exactly reversed. This is an *oscillating* electric dipole. It is so named because its dipole moment changes in a periodic way with time.

25-6 Electric Potential Energy

If we raise a stone from the earth's surface, the work that we do against the earth's gravitational attraction is stored as potential energy in the system earth + stone.

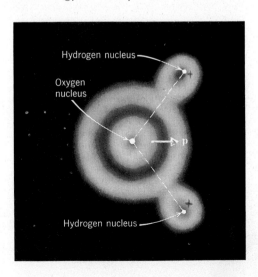

FIGURE 25-11 A schematic representation of a water molecule, showing the three nuclei, the electron cloud, and the orientation of the dipole moment.

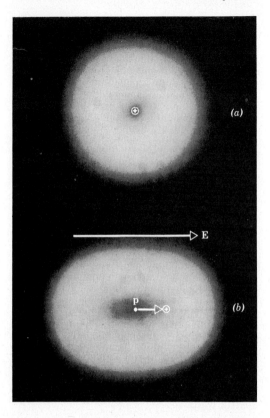

FIGURE 25-12   (*a*) An atom, showing the nucleus and the electron cloud. The center of negative charge coincides with the center of positive charge, that is, with the nucleus. (*b*) If an external field **E** is applied, the electron cloud is distorted so that the center of negative charge, marked by the dot, and the center of positive charge no longer coincide. An electric dipole appears.

If we release the stone, the stored potential energy changes steadily into kinetic energy as the stone drops. After the stone comes to rest on the earth, this kinetic energy, equal in magnitude just before the time of contact to the originally stored potential energy, is transformed into thermal energy in the system earth + stone.

A similar situation exists in electrostatics. Consider two charges q_1 and q_2 a distance r apart, as in Fig. 25–13. If we increase the separation between them, an external agent must do work that will be positive if the charges are opposite in sign and negative otherwise. The energy represented by this work can be thought of as stored in the system $q_1 + q_2$ as *electric potential energy*. This energy, like all varieties of potential energy, can be transformed into other forms. If q_1 and q_2, for example, are charged masses of opposite sign and we release them, they will accelerate toward each other, transforming the stored potential energy into kinetic energy of the accelerating masses. The analogy to the earth + stone system is exact, save for the fact that electric forces may be either attractive or repulsive whereas gravitational forces are always attractive.

We define the electric potential energy of a system of point charges as the work required to assemble this system of charges by bringing them in from an infinite distance. We assume that the charges are all at rest when they are infinitely separated, that is, they have no initial kinetic energy.

In Fig. 25–13 let us imagine q_2 removed to infinity and at rest. The electric potential* at the original site of q_2, caused by q_1, is given by Eq. 25–8, or

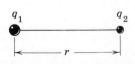

FIGURE 25–13

---

* *Potential* and *potential energy* are very different quantities, measured in different units. Don't confuse them.

$$V = \frac{1}{4\pi\epsilon_0} \frac{q_1}{r}.$$

If q_2 is moved in from infinity to the original distance r, the work required is, from the definition of electric potential (Eq. 25–2),

$$W = Vq_2. \qquad (25\text{–}12)$$

Combining these two equations and recalling that this work W is precisely the electric potential energy U of the system $q_1 + q_2$ yields

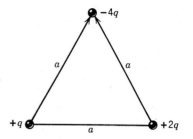

FIGURE 25–14   Example 8. Three charges are fixed rigidly, as shown, by external forces.

$$U (= W) = \frac{1}{4\pi\epsilon_0} \frac{q_1 q_2}{r_{12}}. \qquad (25\text{–}13)$$

$$U_B - U_A = W_{AB} = -\int_A^B \vec{F} \cdot d\vec{l}$$

The subscript of r emphasizes that the distance involved is that between the point charges q_1 and q_2.

For systems containing more than two charges the procedure is to compute the potential energy for every pair of charges separately and to add the results algebraically. This procedure rests on a physical picture in which (*a*) charge q_1 is brought into position, (*b*) q_2 is brought from infinity to its position near q_1, (*c*) q_3 is brought from infinity to its position near q_1 and q_2, etc.

The potential energy of continuous charge distributions (an ellipsoid of charge, for example) can be found by dividing the distribution into infinitesimal elements dq, treating each such element as a point charge, and using the procedures of the preceding paragraph, with the summation process replaced by an integration.

Example 7. Two protons in a nucleus of U^{238} are 6.0×10^{-15} meter apart. What is their mutual electric potential energy?

From Eq. 25–13

$$U = \frac{1}{4\pi\epsilon_0} \frac{q_1 q_2}{r} = \frac{(9.0 \times 10^9 \text{ nt-m}^2/\text{coul}^2)(1.6 \times 10^{-19} \text{ coul})^2}{6.0 \times 10^{-15} \text{ meter}}$$

$$= 3.8 \times 10^{-14} \text{ joule} = 2.4 \times 10^5 \text{ eV}.$$

Example 8. Three charges are arranged as in Fig. 25–14. What is their mutual potential energy? Assume that $q = 1.0 \times 10^{-7}$ coul and $a = 10$ cm.

The total energy of the configuration is the sum of the energies of each pair of particles. From Eq. 25–13,

$$U = U_{12} + U_{13} + U_{23}$$

$$= \frac{1}{4\pi\epsilon_0} \left[\frac{(+q)(-4q)}{a} + \frac{(+q)(+2q)}{a} + \frac{(-4q)(+2q)}{a} \right]$$

$$= -\frac{10}{4\pi\epsilon_0} \frac{q^2}{a}$$

$$= -\frac{(9.0 \times 10^9 \text{ nt-m}^2/\text{coul}^2)(10)(1.0 \times 10^{-7} \text{ coul})^2}{0.10 \text{ meter}} = -9.0 \times 10^{-3} \text{ joule}.$$

The fact that the total energy is negative means that negative work would have to be done to assemble this structure, starting with the three charges separated and at rest at infinity. Expressed otherwise, 9.0×10^{-3} joule of work must be done to dismantle this structure, removing the charges to an infinite separation from one another.

25–7 Calculation of E from V

We have stated that V and **E** are equivalent descriptions and have determined (Eq. 25–6) how to calculate V from **E.** Let us now consider how to calculate **E** if V is known throughout a certain region.

This problem has already been solved graphically. If **E** is known at every point in space, the lines of force can be drawn; then a family of equipotentials can be sketched in by drawing surfaces at right angles. These equipotentials describe the behavior of V. Conversely, if V is given as a function of position, a set of equipotential surfaces can be drawn. The lines of force can then be found by drawing lines at right angles, thus describing the behavior of **E.** It is the mathematical equivalent of this second graphical process that we seek here. Figure 25–15 shows some examples of lines of force and of the corresponding equipotential surfaces.

Figure 25–16 shows the intersection with the plane of the figure of a family of equipotential surfaces. The figure shows that **E** at any point P is at right angles to the equipotential surface through P, as it must be.

Let us move a test charge q_0 from P along the path marked $\Delta \mathbf{l}$ to the equipotential surface marked $V + \Delta V$. The work that must be done by the agent exerting the force **F** (see Eq. 25–1) is $q_0 \, \Delta V$.

FOR A SPACIAL CURVE dl

$$\vec{E} = - \operatorname{Grad} V$$

$$= - \nabla V$$

$$\left(\text{SEE THOMAS p. 510} \right)$$

$$\vec{E} = - \frac{\partial V}{\partial x} \hat{i} - \frac{\partial V}{\partial y} \hat{j} - \frac{\partial V}{\partial z} \hat{k}$$

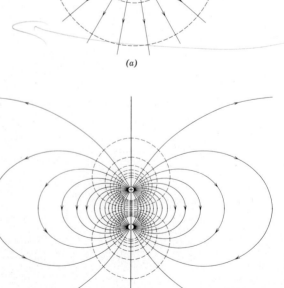

(a)

FIGURE 25–15 Equipotential surfaces (dashed lines) and lines of force (solid lines) for (a) a point charge, (b) an electric dipole. In both figures there is a constant difference of potential ΔV between adjacent equipotential surfaces. Thus from Eq. 25–14, written for the case of $\theta = 180°$ as $\Delta l = -\Delta V/E$, the surfaces will be relatively close together where E is relatively large and relatively far apart where E is small. Similarly (see Section 23–3) the lines of force are relatively close together where E is large and far apart where E is small.

(b)

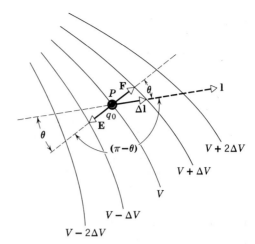

FIGURE 25–16 A test charge q_0 is moved from one equipotential surface to another along an arbitrarily selected direction marked **l**.

From another point of view we can calculate the work from*

$$\Delta W = \mathbf{F} \cdot \Delta \mathbf{l},$$

where **F** is the force that must be exerted on the charge to overcome exactly the electrical force $q_0\mathbf{E}$. Since **F** and $q_0\mathbf{E}$ have opposite signs and are equal in magnitude,

$$\Delta W = -q_0\mathbf{E} \cdot \Delta \mathbf{l} = -q_0 E \cos(\pi - \theta)\,\Delta l = q_0 E \cos\theta\,\Delta l.$$

These two expressions for the work must be equal, which gives

$$q_0\,\Delta V = q_0 E \cos\theta\,\Delta l$$

or
$$E \cos\theta = \frac{\Delta V}{\Delta l}. \qquad (25\text{–}14)$$

Now $E \cos\theta$ is the component of **E** in the direction $-\mathbf{l}$ in Fig. 25–16; the quantity $-E \cos\theta$, which we call E_l, would then be the component of **E** in the $+\mathbf{l}$ direction. In the differential limit Eq. 25–14 can then be written as

$$E_l = -\frac{dV}{dl}. \qquad (25\text{–}15)$$

In words, this equation says: If we travel through an electric field along a straight line and measure V as we go, the rate of change of V with distance that we observe, when changed in sign, is the component of **E** in that direction. The minus sign implies that **E** points in the direction of decreasing V, as in Fig. 25–16. It is particularly clear from Eq. 25–15 that appropriate units for **E** are the volts/meter.

Example 9. Calculate $E(r)$ for a point charge q, using Eq. 25–15 and assuming that $V(r)$ is given by Eq. 25–8

$$V = \frac{1}{4\pi\epsilon_0}\frac{q}{r}.$$

From symmetry, **E** must be directed radially outward for a (positive) point charge. Consider a point P in the field a distance r from the charge. Thus, from Eq. 25–15,

---

* We assume that the equipotentials are so close together that **F** is constant for all parts of the path $\Delta \mathbf{l}$. In the limit of a differential path ($d\mathbf{l}$) there will be no difficulty.

$$E = -\frac{dV}{dr} = -\frac{d}{dr}\left(\frac{1}{4\pi\epsilon_0}\frac{q}{r}\right)$$

$$= -\frac{q}{4\pi\epsilon_0}\frac{d}{dr}\left(\frac{1}{r}\right) = \frac{1}{4\pi\epsilon_0}\frac{q}{r^2}.$$

This result agrees exactly with Eq. 23–4, as it must.

25–8   An Insulated Conductor

We proved in Section 24–4, using Gauss's law, that after a steady state is reached an excess charge q placed on an insulated conductor will move to its outer surface. We now present a completely equivalent statement, namely, that this charge q will distribute itself on this surface so that all points of the conductor, including those on the surface and those inside, have the same potential.

Consider any two points A and B in or on the conductor. If they were not at the same potential, the charge carriers in the conductor near the point of lower potential would tend to move toward the point of higher potential. We have assumed, however, that a steady-state situation, in which such currents do not exist, has been reached; thus all points, both on the surface and inside it, must have the same potential. Since the surface of the conductor is an equipotential surface, **E** for points on the surface must be at right angles to the surface.

Figure 25–17a is a plot of potential against radial distance for an isolated spherical conducting shell of 1.0-meter radius carrying a positive charge of 1.0×10^{-6} coul. For points outside the shell $V(r)$ can be calculated from Eq. 25–8 because the charge q behaves, for such points, as if it were concentrated at the center of the sphere. Equation 25–8 is correct right up to the surface of the shell. Now let us push the test charge through the surface, assuming that there is a small hole, and into the interior. No extra work is needed because no electrical forces act on the test charge once it is inside the shell. Thus the potential everywhere inside is the same as that on the surface, as Fig. 25–17a shows.

Figure 25–17b shows the electric field for this same sphere. Note that E equals zero inside. The lower of these curves can be derived from the upper by differentiation with respect to r, using Eq. 25–15; the upper can be derived from the lower by integration with respect to r, using Eq. 25–6. Figure 25–17 holds without change if the conductor is a solid sphere rather than a spherical shell as we have assumed.

Finally we remark that, for isolated conductors, the charge density is high where the radius of curvature is small (points) and low where the radius of curvature is large (flat spots). The electric field E at points immediately above a charged surface is proportional to the charge density σ so that E may also reach very high values near sharp points. Glow discharges from sharp points during thunderstorms are a familiar example. The lightning rod acts in this way to neutralize charged clouds and thus prevent lightning strokes.

The fact that the surface charge density σ, and thus E, can become very large near sharp points is important in the design of high-voltage equipment. *Corona discharge* can result from such points if the conducting object is raised to high potential and surrounded by air. Normally air is thought of as a nonconductor. However, it contains a small number of ions produced, for example, by the cosmic rays. A positively charged conductor will attract negative ions from the surrounding air and thus will slowly neutralize itself.

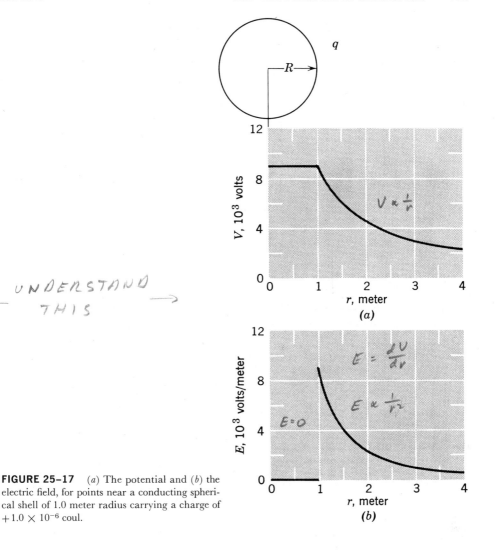

UNDERSTAND
THIS

FIGURE 25–17   (*a*) The potential and (*b*) the electric field, for points near a conducting spherical shell of 1.0 meter radius carrying a charge of $+1.0 \times 10^{-6}$ coul.

If the charged conductor has sharp points, E in the air near the points can be very high. If E is high enough, the ions, as they are drawn toward the conductor, will receive such large accelerations that, by collision with air molecules, they will produce many more ions. The air is thus made much more conducting, and the discharge of the conductor by this corona discharge may be very rapid. The air surrounding sharp conducting points may even glow visibly because of light emitted from the air molecules during these collisions.

25–9   The Electrostatic Generator

The electrostatic generator was conceived by Lord Kelvin in 1890 and put into useful practice by R. J. Van de Graaff in 1931. It is a device for producing electric potential differences of the order of several millions of volts. Its chief application in physics is the use of this potential difference to accelerate charged particles to high energies. Beams of energetic particles made in this way can be used in many different "atom-smashing" experiments. The technique is to let a charged particle "fall" through a potential difference V, gaining kinetic energy as it does so.

Let a particle of (positive) charge q move in a vacuum under the influence of an electric field from one position A to another position B whose electric potential is lower by V. The electric potential energy of the system is reduced by qV because this is the work that an external agent would have to do to restore the system to its original condition. This decrease in potential energy appears as kinetic energy of the particle, or

$$K = qV. \tag{25-16}$$

K is in joules if q is in coulombs and V in volts. If the particle is an electron or a proton, q will be the quantum of charge e.

If we adopt the quantum of charge e as a unit in place of the coulomb, we arrive at another unit for energy, the *electron volt*, which is used extensively in atomic and nuclear physics. By substituting into Eq. 25–16,

$$1 \text{ electron volt} = (1 \text{ quantum of charge})(1 \text{ volt})$$
$$= (1.60 \times 10^{-19} \text{ coul})(1.00 \text{ volt})$$
$$= 1.60 \times 10^{-19} \text{ joule}.$$

The electron volt can be used interchangeably with any other energy unit. Thus a 10-gm object moving at 1000 cm/sec can be said to have a kinetic energy of 3.1×10^{18} eV. Most physicists would prefer to express this result as 0.50 joule, the electron volt being inconveniently small. In atomic and nuclear problems, however, the electron volt (eV) and its multiples the KeV ($= 10^3$ eV), the MeV ($= 10^6$ eV), and the GeV ($= 10^9$ eV) are the usual units of choice.

▨   **Example 10.** *The electrostatic generator.* Figure 25–18, which illustrates the basic operating principle of the electrostatic generator, shows a small sphere of radius r placed inside a large spherical shell of radius R. The two spheres carry charges q and Q, respectively. Calculate their potential difference.

The potential of the large sphere is caused in part by its own charge and in part because it lies in the field set up by the charge q on the small sphere. From Eq. 25–8,

$$V_R = \frac{1}{4\pi\epsilon_0}\left(\frac{Q}{R} + \frac{q}{R}\right).$$

The potential of the small sphere is caused in part by its own charge and in part because it is inside the large sphere; see Fig. 25–17a. From Eq. 25–8,

$$V_r = \frac{1}{4\pi\epsilon_0}\left(\frac{q}{r} + \frac{Q}{R}\right).$$

The potential difference is

$$V_r - V_R = \frac{q}{4\pi\epsilon_0}\left(\frac{1}{r} - \frac{1}{R}\right).$$

Thus, assuming q is positive, the inner sphere will always be higher in potential than the outer sphere. If the spheres are connected by a fine wire, the charge q will flow *entirely* to the outer sphere, regardless of the charge Q that may already be present. From another point of view, we note that since the spheres when electrically connected form a single conductor at electrostatic equilibrium there can be only a single potential. This means that $V_r - V_R = 0$, which can occur only if $q = 0$.

In actual electrostatic generators charge is carried into the shell on rapidly moving belts made of insulating material. Charge is "sprayed" onto the belts outside the shell by corona discharge

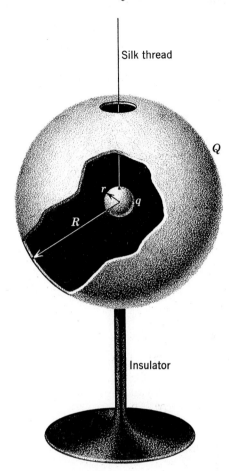

Silk thread

Q

r

q

R

Insulator

FIGURE 25–18   Example 10. A small charged sphere of radius r is suspended inside a charged spherical shell whose outer surface has radius R.

from a series of sharp metallic points connected to a source of moderately high potential difference. Charge is removed from the belts inside the shell by a similar series of points connected to the shell. Electrostatic generators are built commercially to accelerate protons to energies up to at least 10 MeV, using a single acceleration.

QUESTIONS

1. Are we free to call the potential of the earth $+100$ volts instead of zero? What effect would such an assumption have on measured values of (*a*) potentials and (*b*) potential differences?

2. What would happen to a person on an insulated stand if his potential was increased by 10,000 volts?

3. Do electrons tend to go to regions of high potential or of low potential?

4. Suppose that the earth has a net charge that is not zero. Is it still possible to adopt the earth as a standard reference point of potential and to assign the potential $V = 0$ to it?

5. Does the potential of a positively charged insulated conductor have to be positive? Give an example to prove your point.

6. Can two different equipotential surfaces intersect?

7. If **E** equals zero at a given point, must V equal zero for that point?

8. If you know **E** at a given point, can you calculate V at that point?

9. If V equals a constant throughout a given region of space, what can you say about **E** in that region?

10. In Section 14–4 we saw that the gravitational field is zero inside a spherical shell of matter. The electrical field is zero not only inside an isolated charged spherical conductor but inside an isolated conductor of any shape. Is the gravitational field inside, say, a cubical shell of matter zero? If not, in what respect is the analogy not complete?

11. How can you insure that the electric potential in a given region of space will have a constant value?

12. An isolated conducting spherical shell carries a negative charge. What will happen if a positively charged metal object is placed in contact with the shell interior? Assume that the positive charge is (*a*) less than, (*b*) equal to, and (*c*) greater than the negative charge in magnitude.

13. A charge is placed on an insulated conductor in the form of a perfect cube. What will be the relative charge density at various points on the cube (surfaces, edges, corners); what will happen to the charge if the cube is in air?

14. Is the uniformly charged (non-conducting) disk of Example 6 a surface of constant potential.

PROBLEMS

1(2). An infinite charged sheet has a surface charge density σ of 1.0×10^{-7} coul/meter^2. How far apart are the equipotential surfaces whose potentials differ by 5.0 volts?
Answer: 0.89 mm.

2(2). Two identical metal plates a distance d apart carry equal and opposite charges on their inner surfaces. Plot (*a*) the magnitude of the electric field **E** and (*b*) the potential V as a function of the distance x ($0 \leqslant x \leqslant d$) from the positive plate. Neglect edge effects.

3(2). In the Millikan oil drop experiment (see Fig. 23–22) an electric field of 1.92×10^5 nt/coul is maintained at balance across two plates separated by 1.50 cm. Find the potential difference between the plates.
Answer: 2900 volts.

4(2). An oil drop of mass m (3.27×10^{-10} gm) remains stationary between two horizontal parallel plates of separation d (2.00 cm) and potential V (2.00×10^4 volts), the electric force balancing the gravitational force. What is the charge on the oil drop?

5(2). As shown in Fig. 25-19 two large, flat, horizontal metal plates are a distance a (2.0 cm) apart and are maintained at a potential difference V (500 volts), the lower plate being positive. A beam of electrons is introduced midway between the plates moving parallel to them at speed v (2.0×10^7 meters/sec). At what horizontal distance d will the beam hit the positive plate?
Answer: 4.3 cm.

6(2). Two line charges are parallel to the z-axis. One, of charge per unit length $+\lambda$, is a distance a to the right of this axis. The other, of charge per unit length $-\lambda$, is a distance a to the left of this axis (the lines and the z-axis being in the same plane). Sketch some of the equipotential surfaces.

7(2). A Geiger counter has a metal cylinder 2.0 cm in diameter along whose axis is stretched a wire 1.3×10^{-2} cm in diameter. If 850 volts are applied between them, what is the electric field at the surface of (*a*) the wire, and (*b*) the cylinder?
Answer: (*a*) 2.7×10^6 volts/meter. (*b*) 1.7×10^4 volts/meter.

8(3). A thin conducting spherical shell of outer radius R carries a positive charge $+Q$. Sketch

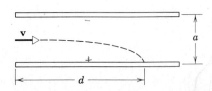

FIGURE 25–19   Problem 5(2).

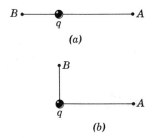

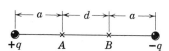

FIGURE 25-20   Problem 13(3).   **FIGURE 25-21**   Problem 17(4).

(a) the magnitude of the electric field **E** and (b) the potential V, versus the distance r from the center of the shell.

9(3). A charge of 10^{-8} coul can be produced by simple rubbing. To what potential would such a charge raise an insulated conducting sphere of 10-cm radius?
Answer: 900 volts.

10(3). A charged sphere of radius 1.5 meters contains a total charge of 3.0×10^{-6} coul. (a) What is the electric field at the sphere's surface? (b) What is the electric potential at the sphere's surface? (c) At what distance from the sphere's surface has the electric potential decreased by 5000 volts?

11(3). Consider a point charge with $q = 1.5 \times 10^{-8}$ coul. (a) What is the radius of an equipotential surface having a potential of 30 volts? (b) Are surfaces whose potentials differ by a constant amount (say 1.0 volt) evenly spaced in radius?
Answer: (a) 4.5 meters. (b) No.

12(3). What is the charge density on the surface of a conducting sphere of radius 0.15 meter whose potential is 200 volts?

13(3). A point charge has $q = +1.0 \times 10^{-6}$ coul. Consider point A, which is 2.0 meters distant, and point B, which is 1.0 meter distant, in a direction diametrically opposite, as in Fig. 25-20a. (a) What is the potential difference $V_A - V_B$? (b) Repeat if points A and B are located as in Fig. 25-20b.
Answer: (a) -4500 volts. (b) -4500 volts.

14(3). A spherical drop of water carrying a charge of 3×10^{-11} coul has a potential of 500 volts at its surface. (a) What is the radius of the drop? (b) If two such drops, of the same charge and radius, combine to form a single spherical drop, what is the potential at the surface of the new drop so formed?

15(3). A positive charge q is distributed uniformly throughout a nonconducting spherical volume of radius R. Show that the potential a distance a from the center, where $a < R$, is given by

$$V = \frac{q(3R^2 - a^2)}{8\pi\epsilon_0 R^3}.$$

16(3). The space between two concentric spheres of radii r_1 and r_2 is filled with a nonconducting material having a uniform charge density ρ. Find the electric potential V as a function of the distance r from the center of the spheres, considering the regions (a) $r > r_2$, (b) $r_2 > r > r_1$, and (c) $r < r_1$. (d) Do these solutions agree at $r = r_2$ and at $r = r_1$?

17(4). (a) In Fig. 25-21 derive an expression for $V_A - V_B$. (b) Does your result reduce to the expected answer when $d = 0$? When $q = 0$?
Answer: (a) $V_A - V_B = qd/2\pi\epsilon_0 a(a + d)$. (b) Yes; $V_A - V_B = 0$ in each case.

18(4). Two metal spheres are 3.0 cm in radius and carry charges of $+1.0 \times 10^{-8}$ coul and -3.0×10^{-8} coul, respectively, assumed to be uniformly distributed. If their centers are 2.0 meters apart, calculate (a) the potential of the point halfway between their centers and (b) the potential of each sphere.

19(4). Two identical conducting spheres of radius r (0.15 meter) are separated by a distance a

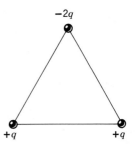

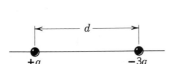

FIGURE 25-22   Problem 20(4).   **FIGURE 25-23**   Problem 22(4).

(10 meters). What is the charge on each sphere if the potential of one is $+1500$ volts and the potential of the other is -1500 volts?

Answer: $+2.5 \times 10^{-8}$ coul and -2.5×10^{-8} coul.

20(4). In Fig. 25–22, locate the points (*a*) where $V = 0$ and (*b*) where $\mathbf{E} = 0$. Consider only points on the axis and choose $d = 1.0$ meter.

21(4). In Fig. 25–22 [see Problem 20(4)] sketch qualitatively (*a*) the lines of force and (*b*) the intersections of the equipotential surfaces with the plane of the figure. (Hint: Consider the behavior close to each point charge and at considerable distances from the pair of charges.)

22(4). In Fig. 25–23 draw the lines of force and the intersections of the constant potential surfaces with the plane of the figure. The charges in the figure are spread uniformly over long parallel rods that are perpendicular to the plane of the figure.

23(5). Calculate the dipole moment of a water vapor molecule under the assumption that all 10 electrons in the molecule circulate symmetrically about the oxygen atom, that the OH distance is 0.96×10^{-8} cm, and that the angle between the two OH bonds is $104°$. Compare with the value quoted on p. 474: see Fig. 25–11.

Answer: 1.9×10^{-29} coul-meter. The text value (0.61×10^{-29} coul-meter) is correct because the assumptions made in this problem are oversimplified.

24(6). In the rectangle shown in Fig. 25–24 the sides have lengths 5.0 cm and 15 cm, $q_1 = -5.0 \times 10^{-6}$ coul and $q_2 = +2.0 \times 10^{-6}$ coul. (*a*) What is the electric potential at B? At A? (*b*) How much work is involved in moving a third charge $q_3 = +3.0 \times 10^{-6}$ coul from B to A along a diagonal of the rectangle? (*c*) In this process, is external work converted into electrostatic potential energy or vice versa? Explain.

25(6). The charges and coordinates of two charges located on the *x-y* plane are: $q_1 = +3.0 \times 10^{-6}$ coul; $x = +3.5$ cm, $y = +0.50$ cm, and $q_2 = -4.0 \times 10^{-6}$ coul; $x = -2.0$ cm, $y = +1.5$ cm. (*a*) Find the electric potential at the origin. (*b*) How much work must be done to locate these charges at their given positions, starting from infinity?

Answer: (*a*) -6.8×10^5 volts. (*b*) -1.9 joules.

26(6). Three charges of $+0.10$ coul are placed on the corner of an equilateral triangle, 1.0 meter on a side. If energy is supplied at the rate of 1.0 kw, how many days would be required to move one of the charges onto the midpoint of the line joining the other two?

27(6). What is the energy required to completely separate a sodium ion from its six nearest chlorine ion neighbors in a salt crystal? The chlorine ions may be considered to lie on the faces of a cube with the sodium ion at the center of the cube. The sodium-chlorine distance is 2.82×10^{-10} meter. In a real crystal, how would the presence of other ions affect your result?

Answer: 4.9×10^{-18} joules.

28(6). In a linear accelerator using a constant electric field, a proton acquires a kinetic energy K (10 MeV) in traveling a distance d(5.0 meters). What is the magnitude of the field?

29(6). An alpha particle is accelerated through a potential difference of 1.0×10^6 volts in an electrostatic generator. (*a*) What kinetic energy does it acquire? (*b*) What kinetic energy would a proton acquire under these same circumstances? (*c*) Which particle would acquire the greater speed, starting from rest?

Answer: (*a*) 3.2×10^{-13} joules. (*b*) 1.6×10^{-13} joules. (*c*) The proton.

30(6). Between two parallel, flat, conducting surfaces of spacing d (1.0 cm) and potential dif-

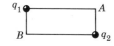

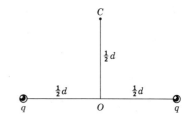

FIGURE 25-24   Problem 24(6).   **FIGURE 25-25**   Problem 39(6).

ference V (1.0×10^4 volts), an electron is projected from one plate directly toward the second. What is the initial velocity of the electron if it comes to rest just at the surface of the second plate?

31(6). A particle of charge Q is kept in a fixed position at a point P and a second particle of mass m, having the same charge Q, is initially held at rest a distance r_1 from P. The second particle is then released and is repelled from the first one. Find its speed at the instant it is a distance r_2 from P. Let $Q = 3.1 \times 10^{-6}$ coul, $m = 2.0 \times 10^{-5}$ kg, $r_1 = 9.0 \times 10^{-4}$ meter, and $r_2 = 25 \times 10^{-4}$ meter. *Answer:* 2.5×10^3 meters/sec.

32(6). A gold nucleus contains a positive charge equal to that of 79 protons. An α-particle ($Z = 2$) has a kinetic energy K at points far from this charge and is traveling directly toward the charge. The particle just touches the surface of the charge (assumed spherical) and is reversed in direction. Calculate K, assuming a nuclear radius of 5.0×10^{-15} meter.

33(6). A metal sphere having a radius r (2.0 cm) carries a charge q (1.0×10^{-8} coul). A point P is at a distance d (3.0 cm) from the center of the sphere. (*a*) Relative to infinity, what is the potential of the sphere and the potential at the point P? (*b*) Now suppose that an electron is released from rest at P. Neglecting relativistic effects what is its speed as it strikes the sphere? *Answer:* (*a*) $V_{sphere} = +4500$ volts; $V_P = +3000$ volts. (*b*) 2.3×10^7 meters/sec.

34(6). Two electrons are 2.0 meters apart. Another electron is shot from infinity and comes to rest midway between the two. What must its initial velocity be?

35(6). Two small metal spheres of mass m_1 (5.0 gm) and mass m_2 (10 gm) carry equal positive charges q (5.0×10^{-6} coul). The spheres are connected by a massless string of length d (1.0 meter), which is much greater then the sphere radius. (*a*) What is the electrostatic potential energy of the system? (*b*) You cut the string. At that instant what is the acceleration of each of the spheres? (*c*) A long time after you cut the string, what is the velocity of each sphere? Ignore gravitational effects. *Answer:* (*a*) 0.23 joule. (*b*) $a_1 = 45$ meters/sec²; $a_2 = 22.5$ meters/sec². (*c*) $v_1 = 7.7$ meters/sec; $v_2 = 3.9$ meters/sec.

36(6). Calculate (*a*) the electric potential established by the nucleus of a hydrogen atom at the mean distance of the circulating electron ($r = 5.3 \times 10^{-11}$ meter), (*b*) the electric potential energy of the atom when the electron is at this radius, and (*c*) the kinetic energy of the electron, assuming it to be moving in a circular orbit of this radius centered on the nucleus. (*d*) How much energy is required to ionize the hydrogen atom? Express all energies in electron volts.

37(6). A particle of mass m, positive charge q, and initial kinetic energy K is projected from infinity directly toward a heavy nucleus of charge Q. How close to the center of the nucleus is the particle when it comes instantaneously to rest? *Answer:* $qQ/4\pi\epsilon_0 K$.

38(6). A particle of (positive) charge Q is assumed to have a fixed position at P. A second particle of mass m and (negative) charge $-q$ moves at constant speed in a circle of radius r_1, centered at P. Derive an expression for the work W that must be done by an external agent on the second particle in order to increase the radius of the circle of motion, centered at P, to r_2. Express W in terms of quantities chosen from among m, r_1, r_2, q, Q, and ϵ_0 only.

39(6). Two charges q ($= +2.0 \times 10^{-6}$ coul) are fixed in space a distance d (2.0 cm) apart, as shown in Fig. 25-25. (*a*) What is the electric potential at point C? (*b*) You bring a third charge q ($+2.0 \times 10^{-6}$ coul) very slowly from infinity to C. How much work must you do? (*c*) What is the potential energy U of the configuration when the third charge is in place? *Answer:* (*a*) 2.5×10^6 volts. (*b*) 5.1 joules. (*c*) 6.9 joules.

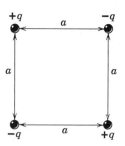

FIGURE 25-26   Problem 40(6).   **FIGURE 25-27**   Problem 42(6).

40(6). Figure 25–26 shows an idealized representation of a U^{238} nucleus $(Z = 92)$ on the verge of fission. Calculate (*a*) the repulsive force acting on each fragment and (*b*) the mutual electric potential energy of the two fragments. Assume that the fragments are equal in size and charge, spherical, and just touching. The radius of the initially spherical U^{238} nucleus is 8.0×10^{-15} meters. Assume that the material out of which nuclei are made has a constant density, both of mass and charge.

41(6). What is the electric potential energy of the charge configuration of Fig. 25–7? Use the numerical values of Example 5.
Answer: -6.4×10^{-7} joules.

42(6). Derive an expression for the work required to put the four charges together as indicated in Fig. 25–27.

43(6). Devise an arrangement of three point charges, separated by finite distances, that has zero electric potential energy.

44(6). If the earth had a net charge equivalent to 1.0 electron/meter² of surface area, (*a*) what would the earth's potential be? (*b*) What would the electric field due to the earth be just outside its surface?

45(7). The electric potential varies along the *x*-axis as shown in the graph of Fig. 25–28. For each of the intervals shown (ignore the behavior at the end points of the intervals), find the *x*-component of the electric field and plot E_x versus *x*.
Answer: (*a–b*) -6.0 volts/meter; (*b–c*) 0.0 volts/meter; (*c–e*) $+3.0$ volts/meter; (*e–f*) $+15.0$ volts/meter; (*f–g*) 0.0 volts/meter; (*g–h*) -3.0 volts/meter.

46(7). (*a*) Starting from Eq. 25–11, find the magnitude E_r of the radial component of the electric field due to a dipole. (*b*) For what values of θ is E_r zero?

47(7). In Example 6 the potential at an axial point for a charged disk was shown to be

$$V = \frac{\sigma}{2\epsilon_0} \, (\sqrt{a^2 + r^2} - r).$$

From this result show that E for axial points is given by

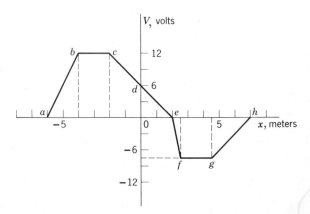

FIGURE 25-28   Problem 45(7).

$$E = \frac{\sigma}{2\epsilon_0} \left(1 - \frac{r}{\sqrt{a^2 + r^2}} \right).$$

Does this expression for E reduce to an expected result for (a) $r \gg a$ and (b) for $r \ll a$?
Answer: (a) Yes; the disk appears as a point charge. (b) Yes; the disk appears as an infinite sheet of charge; see Eq. 24–8.

48(7). (a) Show that the electric potential at a point on the axis of a ring of charge of radius a, computed directly from Eq. 25–10, is given by

$$V = \frac{1}{4\pi\epsilon_0} \frac{q}{\sqrt{x^2 + a^2}}.$$

(b) From this result derive an expression for E at axial points; compare with the direct calculation of E in Example 5, Chapter 23.

49(8). Can a conducting sphere 10 cm in radius hold a charge of 4×10^{-6} coul in air without breakdown? The dielectric strength (minimum field required to produce breakdown) of air at 1 atm is 3×10^6 volts/meter.
Answer: No.

50(8). Two thin insulated concentric conducting shells of radii R_1 and R_2 carry charges q_1 and q_2, respectively. Find $V(r)$ and $E(r)$, where r is the distance from the center of the shells, for (a) $r < R_1$, (b) $R_1 < r < R_2$, and (c) $r > R_2$. (d) Do these equations agree for $r = R_1$ and $r = R_2$?

51(8). Two conducting spheres, one of radius 6.0 cm and the other of radius 12 cm, each have a charge of 3.0×10^{-8} coul and are very far apart. If the spheres are connected by a conducting wire, find (a) the direction of motion and the magnitude of the charge transferred, and (b) the final charge on and potential of each sphere.
Answer: (a) 1.0×10^{-8} coul, from the smaller to the larger sphere. (b) Smaller sphere: 2.0×10^{-8} coul and 3000 volts; larger sphere: 4.0×10^{-8} coul and 3000 volts.

52(9). Let the potential difference between the shell of an electrostatic generator and the point at which charges are sprayed onto the moving belt be 3.0×10^6 volts. If the belt transfers charge to the shell at the rate of 3.0×10^{-3} coul/sec, what power must be provided to drive the belt, considering only electrical forces?

53(9). (a) How much charge is required to raise an isolated metallic sphere of 1.0-meter radius to a potential of 1.0×10^6 volts? Repeat for a sphere of 1.0-cm radius. (b) Why use a large sphere in an electrostatic generator since the same potential can be achieved for a smaller charge with a small sphere?
Answer: (a) 1.1×10^{-4} coul; 1.1×10^{-6} coul. (b) Due to larger E at the surface of the smaller sphere charge will leak off more rapidly.

54(9). The high-voltage electrode of an electrostatic generator is a charged spherical metal shell having a potential V ($+9.0 \times 10^6$ volts). (a) Electrical breakdown occurs in the gas in this machine at a field E (1.0×10^8 volt/meter). In order to prevent such breakdown, what restriction must be made on the radius r of the shell? (b) A long moving rubber belt transfers charge to the shell at 3.0×10^{-4} coul/sec, the potential of the shell remaining constant because of leakage. What minimum power is required to transfer the charge? (c) The belt is of width w (0.50 meters) and travels at speed v (30 meters/sec). What is the surface charge density on the belt?

CHAPTER 26

Capacitors and Dielectrics

26–1 Capacitance

In Section 25–3 we showed that the potential of a charged conducting sphere, assumed to be completely isolated with no other bodies (conducting or nonconducting) nearby, is given by

$$V_+' = \frac{1}{4\pi\epsilon_0}\frac{q}{R},$$ (26–1)

in which q is the charge on the sphere and R is the sphere radius. The subscript on V indicates that we assume the charge to be positive. We represent this potential in Fig. 26–1 by the line marked V_+'. The line marked V_∞ in that figure represents the potential of an infinitely distant reference position; it has been assigned the value zero, following the usual convention.

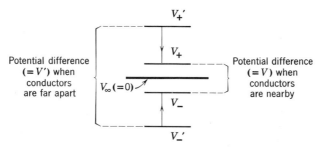

FIGURE 26–1 The potential difference between two conductors that carry constant, equal, and opposite charges is reduced as the conductors are brought closer together.

Let us now imagine a second sphere of radius R, carrying a negative charge $-q$ and located a large distance ($\gg R$) from the first sphere so that each may still be considered to be electrically isolated. The potential of the second sphere is given by

$$V_-' = -\frac{1}{4\pi\epsilon_0}\frac{q}{R},\tag{26–2}$$

and this quantity also is represented in Fig. 26–1.

The potential difference V' between the two spheres is

$$V' = V_+' - V_-' = \frac{1}{4\pi\epsilon_0}\frac{2q}{R}.$$

This shows that V', the potential difference, and q, the magnitude of the charge on either sphere, are proportional to each other. We may rewrite the equation as

$$q = (2\pi\epsilon_0 R)V' = C'V',\tag{26–3}$$

in which the proportionality constant in parentheses is called the *capacitance* of the two spheres and assigned the symbol C'.

Let us move the two spheres close together. The presence of each will now spoil the spherical symmetry of the lines of force emanating from the other. Lines from a given sphere, which, for large sphere separations, radiated uniformly in all directions to infinity, now terminate, in part, on the other sphere. Under these conditions Eqs. 26–1 and 26–2 no longer apply, since they were derived (see Section 25–3) on the assumption that spherical symmetry existed, permitting the useful application of Gauss's law.

A positive charge brought near an isolated object serves to raise the potential of that object and a negative charge serves to lower it, as you can see by considering the work required to move a positive test charge from infinity to points near such charges. Thus the potential of the positively charged sphere will be lowered by the presence nearby of the negatively charged sphere, from V_+' to some lower value V_+. Similarly the potential of the negative sphere will be raised from V_-' to a higher value V_-. These new potentials are shown in Fig. 26–1, the potential changes for each sphere being suggested by the vertical arrows.

From Fig. 26–1, although the charges on the spheres have not changed, it is clear that the potential difference between the spheres has been considerably reduced. Put another way, the capacitance of the system of two spheres (see Eq. 26–3), defined from

$$q = CV \qquad C = \frac{q}{V}\tag{26–4}$$

has been made considerably larger than its initial value C' by bringing the spheres closer together.

It is also possible to use Eq. 26–4 to define the capacitance of a single isolated conductor such as a sphere. In such cases we may imagine that the second "object" carrying an equal and opposite charge, is a conducting sphere of very large—essentially infinite—radius centered about the conductor. The potential of this infinitely distant sphere, according to the usual convention for potential measurements, is zero. The capacitance of an isolated sphere of radius R is given from Eqs. 26–4 and 26–1 as

$$C = \frac{q}{V} = 4\pi\epsilon_0 R.$$

The mks unit of capacitance that follows from Eq. 26–4 is the coul/volt. A special unit, the *farad,* is used to represent it. It is named in honor of Michael Faraday who, among other contributions, developed the concept of capacitance. Thus

$$1 \text{ farad} = 1 \text{ coul/volt.}$$

The submultiples of the farad, the *microfarad* (1 μf $= 10^{-6}$ farad) and the *picofarad* (1 pf $= 10^{-12}$ farad), are more convenient units in practice.

Figure 26–2 shows a more general case of two nearby conductors, which are now permitted to be of any shape, carrying equal and opposite charges. Such an arrangement is called a *capacitor,* the conductors being called *plates.* The equal and opposite charges might be established by connecting the plates momentarily to opposite poles of a battery. The capacitance C of any capacitor is defined from Eq. 26–4 in which V is the potential difference between the plates and q is the magnitude of the charge on either plate; q must not be taken as the net charge of the capacitor, which is zero. The capacitance of a capacitor depends on the geometry of each plate, their spatial relationship to each other, and the medium in which the plates are immersed. For the present, we take this medium to be a vacuum.

Capacitors are very useful devices, of great interest to engineers and physicists. For example:

1. In this book we stress the importance of fields to the understanding of natural phenomena. A capacitor can be used to establish desired electric field configurations for various purposes. In Section 23–5 we described the deflection of an electron beam in a uniform field set up by a capacitor, although we did not use this term in that section. In later sections we discuss the behavior of dielectric materials when placed in an electric field (provided conveniently by a capacitor) and we shall see how the laws of electromagnetism can be generalized to take the presence of dielectric bodies more readily into account.

2. A second important concept stressed in this book is energy. By analyzing a charged capacitor we show that electric energy may be considered to be stored in the electric field between the plates and indeed in any electric field, however generated. Because capacitors can confine strong electric fields to small volumes, they can serve as useful devices for storing energy. Many experiments and devices in plasma physics make use of bursts of energy stored in this way.

3. The electronic age could not exist without capacitors. They are used, in

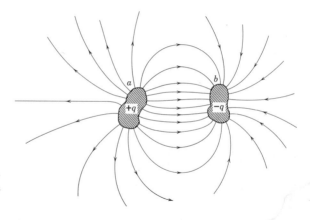

FIGURE 26–2 Two insulated conductors carrying equal and opposite charges form a capacitor.

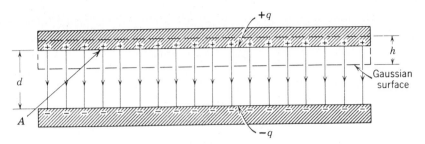

FIGURE 26–3   A parallel-plate capacitor with plates of area A. The dashed line represents a Gaussian surface whose height is h and whose top and bottom caps are the same shape and size as the capacitor plates.

conjunction with other devices, to reduce voltage fluctuations in electronic power supplies, to transmit pulsed signals, to generate or detect electromagnetic oscillations at radio frequencies, and to provide time delays. In most of these applications the potential difference between the plates will not be constant, as we assume in this chapter, but will vary with time, often in a sinusoidal or a pulsed fashion. In later chapters we consider some aspects of the capacitor used as a circuit element.

26–2   Calculating Capacitance

Figure 26–3 shows a *parallel-plate capacitor* formed of two parallel conducting plates of area A separated by a distance d. If we connect each plate to the terminal of a battery, a charge $+q$ will appear on one plate and a charge $-q$ on the other. If d is small compared with the plate dimensions, the electric field **E** between the plates will be uniform, which means that the lines of force will be parallel and evenly spaced. The laws of electromagnetism (see Problem 22, Chapter 31) require that there be some "fringing" of the lines at the edges of the plates, but for small enough d it can be neglected for our present purpose.

We can calculate the capacitance of this device using Gauss's law. Figure 26–3 shows (dashed lines) a Gaussian surface of height h closed by plane caps of area A that are the shape and size of the capacitor plates. The flux of **E** is zero for the part of the Gaussian surface that lies inside the top capacitor plate because the electric field inside a conductor carrying a static charge is zero. The flux of **E** through the wall of the Gaussian surface is zero because, to the extent that the fringing of the lines of force can be neglected, **E** lies in the wall.

This leaves only the face of the Gaussian surface that lies between the plates. Here **E** is constant and the flux Φ_E is simply EA. Gauss's law gives

$$\epsilon_0 \Phi_E = \epsilon_0 E A = q. \qquad (26\text{–}5)$$

The work required to carry a test charge q_0 from one plate to the other can be expressed either as $q_0 V$ (see Eq. 25–2) or as the product of a force $q_0 E$ times a distance d or $q_0 Ed$. These expressions must be equal, or

$$V = Ed. \qquad (26\text{–}6)$$

More formally, Eq. 26–6 is a special case of the general relation (Eq. 25–5; see also Example 1, Chapter 25)

$$V = -\int \mathbf{E} \cdot d\mathbf{l},$$

p. 468

where V is the difference in potential between the plates. The integral may be taken over any path that starts on one plate and ends on the other because each plate is an equipotential surface and the electrostatic potential difference is path independent. Although the simplest path between the plates is a perpendicular straight line, Eq. 26–6 follows no matter what path of integration we choose.

If we substitute Eqs. 26–5 and 26–6 into the relation $C = q/V$, we obtain

$$C = \frac{q}{V} = \frac{\epsilon_0 EA}{Ed} = \frac{\epsilon_0 A}{d}. \qquad \frac{k\,\epsilon_0\,A}{d} = \frac{\epsilon\,A}{d} \qquad (26\text{–}7)$$

$$k\,\epsilon_0 = \epsilon$$

Equation 26–7 holds only for capacitors of the parallel-plate type; different formulas hold for capacitors of different geometry.

In Section 22–4 we stated that ϵ_0, which we first met in connection with Coulomb's law, was not measured in terms of that law because of experimental difficulties. Equation 26–7 suggests that ϵ_0 might be measured by building a capacitor of accurately known plate area and plate spacing and determining its capacitance experimentally by measuring q and V in the relation $C = q/V$. Thus Eq. 26–7 can be solved for ϵ_0 and a numerical value found in terms of the measured quantities A, d, and C; ϵ_0 has been measured accurately in this way.

Example 1. The parallel plates of an air-filled capacitor are everywhere 1.0 mm apart. What must the plate area be if the capacitance is to be 1.0 farad? $k = 1.00054 \doteq vac,$
From Eq. 26–7

$$A = \frac{dC}{\epsilon_0} = \frac{(1.0 \times 10^{-3}\ \text{meter})(1.0\ \text{farad})}{8.9 \times 10^{-12}\ \text{coul}^2/\text{nt-m}^2} = 1.1 \times 10^8\ \text{meter}^2.$$

This is the area of a square sheet more than 6 miles on edge; the farad is indeed a large unit.

Example 2. *A cylindrical capacitor.* A cylindrical capacitor consists of two coaxial cylinders (Fig. 26–4) of radii a and b and length l. What is the capacitance of this device? Assume that the capacitor is very long (that is, that $l \gg b$) so that fringing of the lines of force at the ends can be ignored for the purpose of calculating the capacitance.

As a Gaussian surface construct a coaxial cylinder of radius r and length l, closed by plane caps. Gauss's law

$$\epsilon_0 \oint \mathbf{E} \cdot d\mathbf{S} = q$$

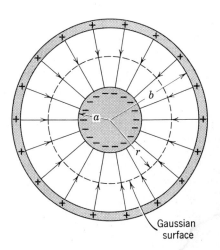

FIGURE 26–4 Example 2. A cross section of a cylindrical capacitor. The dashed circle is a cross section of a cylindrical Gaussian surface of radius r and length l.

Gaussian surface

gives
$$\epsilon_0 E(2\pi r)(l) = q,$$

the flux being entirely through the cylindrical surface and not through the end caps. Solving for E yields

$$E = \frac{q}{2\pi\epsilon_0 r l}.$$

The potential difference between the plates is given by Eq. 25–5 [note that $\mathbf{E}$ and $d\mathbf{l}\,(= d\mathbf{r})$ point in opposite directions] or

$$V = -\int_a^b \mathbf{E} \cdot d\mathbf{l} = \int_a^b E\,dr = \int_a^b \frac{q}{2\pi\epsilon_0 l}\frac{dr}{r} = \frac{q}{2\pi\epsilon_0 l}\ln\frac{b}{a}.$$

Finally, the capacitance is given by

$$\left(\ln b - \ln a = \ln \frac{b}{a} \right)$$

$$C = \frac{q}{V} = \frac{2\pi\epsilon_0 l}{\ln (b/a)}.$$

Like the relation for the parallel-plate capacitor (Eq. 26–7), this relation also depends only on geometrical factors.

Example 3. *Capacitors in series.* Figure 26–5 shows three capacitors connected in series. What single capacitance C is "equivalent" to this combination?

For capacitors connected as shown, the magnitude q of the charge on each plate must be the same. This is true because the net charge on the part of the circuit enclosed by the dashed line in Fig. 26–5 must be zero; that is, the charge present on these plates initially is zero and connecting a battery between a and b will only produce a charge separation, the net charge on these plates still being zero. Assuming that neither C_1 nor C_2 "sparks over," there is no way for charge to enter or leave the region enclosed by the dashed line.

Applying the relation $q = CV$ to each capacitor yields

$$V_1 = q/C_1; \qquad V_2 = q/C_2; \qquad \text{and} \qquad V_3 = q/C_3.$$

The potential difference for the series combination is

$$V = V_1 + V_2 + V_3$$

$$= q\left(\frac{1}{C_1} + \frac{1}{C_2} + \frac{1}{C_3}\right).$$

$$SERIES:$$
$$\frac{1}{C} = \sum_{i=1}^{n}\frac{1}{C_i}$$

The equivalent capacitance is

$$C = \frac{q}{V} = \frac{1}{\dfrac{1}{C_1} + \dfrac{1}{C_2} + \dfrac{1}{C_3}},$$

$$PARALLEL:$$
$$C = \sum_{i=1}^{n} C_i$$
$$(NOTE\ C < any\ C_i)$$

or

$$\frac{1}{C} = \frac{1}{C_1} + \frac{1}{C_2} + \frac{1}{C_3}.$$

The equivalent series capacitance is always less than the smallest capacitance in the chain.

26–3   Parallel-Plate Capacitor with Dielectric

Equation 26–7 holds only for a parallel-plate capacitor with its plates in a vacuum. Michael Faraday, in 1837, first investigated the effect of filling the space between the plates with a dielectric, say mica or oil.

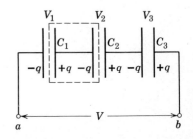

FIGURE 26–5   Example 3. Three capacitors in series.

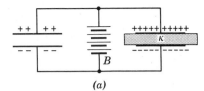

(a)

FIGURE 26–6   (a) Battery B supplies the same potential difference to each capacitor; the one on the right has the higher charge. (b) Both capacitors carry the same charge; the one on the right has the lower potential difference, as indicated by the meter readings.

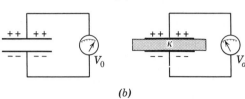

(b)

He found that q is larger, for the same V, if a dielectric is present; see Fig. 26–6a. It follows from the relation $C = q/V$ that *the capacitance of a capacitor increases if a dielectric is placed between the plates*. The ratio of the capacitance with the dielectric (assumed to fill completely the space between the plates) to that without is called the *dielectric constant* κ of the material; see Table 26–1.

Instead of maintaining the two capacitors at the same potential difference, we can place the same charge on them, as in Fig. 26–6b. Experiment then shows that the potential difference V_d between the plates of the right-hand capacitor is smaller than that for the left-hand capacitor by the factor $1/\kappa$, or

$$V_d = V_0/\kappa. \qquad k = \frac{V_o}{V_d} = \frac{C_d}{C_o} = \frac{\epsilon}{\epsilon_o} = \frac{E_o}{E}$$

We are led once again to conclude, from the relation $C = q/V$, that the effect of the dielectric is to increase the capacitance by a factor κ.

For a parallel-plate capacitor we can write, as an experimental result,

$$C = \frac{\kappa\epsilon_0 A}{d}. \qquad (26\text{–}8)$$

Table 26–1   PROPERTIES OF SOME DIELECTRICS

Material	Dielectric Constant	Dielectric Strength* (kV/mm)
Vacuum	1.00000	∞
Air	1.00054	0.8
Water	78	—
Paper	3.5	14
Ruby mica	5.4	160
Amber	2.7	90
Porcelain	6.5	4
Fused quartz	3.8	8
Pyrex glass	4.5	13
Polyethylene	2.3	50
Polystyrene	2.6	25
Teflon	2.1	60
Neoprene	6.9	12
Pyranol oil	4.5	12
Titanium dioxide	100	6

* This is the maximum potential gradient that may exist in the dielectric without the occurrence of electrical breakdown.

Equation 26–7 is a special case of this relation found by putting $\kappa = 1$, corresponding to a vacuum between the plates. Experiment shows that the capacitance of all types of capacitor is increased by the factor κ if the space between the plates is filled with a dielectric. Thus the capacitance of any capacitor can be written as

$$C = \kappa \epsilon_0 L,$$

where L depends on the geometry and has the dimensions of a length. For a parallel-plate capacitor (see Eq. 26–7) L is A/d; for a cylindrical capacitor (see Example 2) it is $2\pi l / \ln (b/a)$.

26–4   Dielectrics—An Atomic View

We now seek to understand, in atomic terms, what happens when a dielectric is placed in an electric field. There are two possibilities. The molecules of some dielectrics, like water, have permanent electric dipole moments. In such materials (called *polar*) the electric dipole moments **p** tend to align themselves with an external electric field, as in Fig. 26–7; see also Section 23–6. Because the molecules are in constant thermal agitation, the degree of alignment will not be complete but will increase as the applied electric field is increased or as the temperature is decreased.

Whether or not the molecules have permanent electric dipole moments, they acquire them by induction when placed in an electric field. In Section 25–5 we saw that the external electric field tends to separate the negative and the positive charge in the atom or molecule. This induced electric dipole moment is present only when the electric field is present, except under unusual circumstances. It is proportional to the electric field (for normal fields) and is created already lined up with the electric field as Fig. 25–12 suggests.

Let us use a parallel-plate capacitor, carrying a fixed charge q and not connected to a battery (see Fig. 26–6b), to provide a uniform external electric field $\mathbf{E}_0$ into which we place a dielectric slab. The over-all effect of alignment and induction is to separate the center of positive charge of the entire slab slightly from the center of negative charge. The slab, as a whole, although remaining electrically neutral, becomes *polarized*, as Fig. 26–8b suggests. The net effect is a pile-up of positive charge on the right face of the slab and of negative charge on the left face; within the slab no excess charge appears in any given volume element. Since the slab as a whole remains neutral, the positive induced surface charge must be equal in magnitude to the negative induced surface charge. Note that in this process electrons in the dielectric are displaced from their equilibrium positions by distances that are

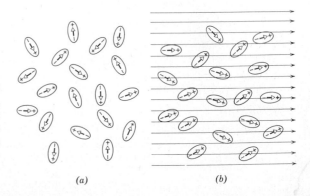

FIGURE 26–7 (*a*) Molecules with a permanent electric dipole moment, showing their random orientation in the absence of an external electric field. (*b*) An electric field is applied, producing partial alignment of the dipoles. Thermal agitation prevents complete alignment.

(*a*)   (*b*)

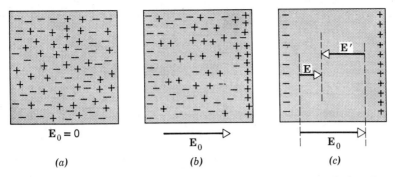

$$\mathbf{E}_0 = 0$$

$$\mathbf{E}_0$$

$$\mathbf{E}_0$$

(a)　　　　　　　　　　　　(b)　　　　　　　　　　　　(c)

FIGURE 26–8　(a) A dielectric slab, showing the random distribution of plus and minus charges. (b) An external field $\mathbf{E}_0$, established by putting the slab between the plates of a parallel-plate capacitor (not shown), separates the center of plus charge in the slab slightly from the center of minus charge, resulting in the appearance of surface charges. No net charge exists in any volume element located in the interior of the slab. (c) The surface charges set up a field $\mathbf{E}'$ which opposes the external field $\mathbf{E}_0$ associated with the charges on the capacitor plates. The resultant field $\mathbf{E}$ ($= \mathbf{E}_0 + \mathbf{E}'$) in the dielectric is thus less than $\mathbf{E}_0$.

considerably less than an atomic diameter. There is no transfer of charge over macroscopic distances such as occurs when a current is set up in a conductor.

Figure 26–8c shows that the induced surface charges will always appear in such a way that the electric field set up by them ($\mathbf{E}'$) opposes the external electric field $\mathbf{E}_0$. The resultant field in the dielectric $\mathbf{E}$ is the vector sum of $\mathbf{E}_0$ and $\mathbf{E}'$. It points in the same direction as $\mathbf{E}_0$ but is smaller. *If a dielectric is placed in an electric field, induced surface charges appear which tend to weaken the original field within the dielectric.*

This weakening of the electric field reveals itself in Fig. 26–6b as a reduction in potential difference between the plates of a charged isolated capacitor when a dielectric is introduced between the plates. The relation $V = Ed$ for a parallel-plate capacitor (see Eq. 26–6) holds whether or not dielectric is present and shows that the reduction in V described in Fig. 26–6b is directly connected to the reduction in E described in Fig. 26–8. More specifically, if a dielectric slab is introduced into a charged parallel-plate capacitor, then

$$\frac{E_0}{E} = \frac{V_0}{V_d} = \kappa \; = \; \frac{C_d}{C_v} \tag{26–9}$$

where the symbols on the left refer to Fig. 26–8 and the symbols V_0 and V_d refer to Fig. 26–6b.

Equation 26–9 does not hold if the battery remains connected while the dielectric slab is introduced. In this case V (hence E) could not change. Instead, the charge q on the capacitor plates would increase by a factor κ, as Fig. 26–6a suggests.

Induced surface charge is the explanation of the most elementary fact of static electricity, namely, that a charged rod will attract uncharged bits of paper, etc. Figure 26–9 shows a bit of paper in the field of a charged rod. Surface charges appear on the paper as shown. The negatively charged end of the paper will be pulled toward the rod and the positively charged end will be repelled. These two forces do not have the same magnitude because the negative end, being closer to the rod, is in a stronger field and experiences a stronger force. The net effect is an attraction. A dielectric body in a uniform electric field will not experience a net force.

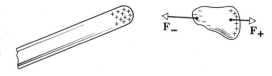

FIGURE 26-9   A charged rod attracts an uncharged piece of paper because unbalanced forces act on the induced surface charges.

26-5   Dielectrics and Gauss's Law

So far our use of Gauss's law has been confined to situations in which no dielectric was present. Now let us apply this law to a parallel-plate capacitor filled with a dielectric of dielectric constant κ.

Figure 26-10 shows the capacitor both with and without the dielectric. It is assumed that the charge q on the plates is the same in each case. Gaussian surfaces have been drawn after the fashion of Fig. 26-3.

If no dielectric is present (Fig. 26-10a), Gauss's law (see Eq. 26-5) gives

$$\epsilon_0 \oint \mathbf{E} \cdot d\mathbf{S} = \epsilon_0 E_0 A = q$$

or
$$E_0 = \frac{q}{\epsilon_0 A}. \tag{26-10}$$

If the dielectric is present (Fig. 26-10b), Gauss's law gives

$$\epsilon_0 \oint \mathbf{E} \cdot d\mathbf{S} = \epsilon_0 E A = q - q'$$

or
$$E = \frac{q}{\epsilon_0 A} - \frac{q'}{\epsilon_0 A}, \tag{26-11}$$

in which $-q'$, the induced surface charge, must be distinguished from q, the so-called free charge on the plates. These two charges, both of which lie within the Gaussian surface, are opposite in sign; $q - q'$ is the net charge within the Gaussian surface.

Equation 26-9 shows that in Fig. 26-10b

$$E = \frac{E_0}{\kappa}.$$

Combining this with Eq. 26-10, we have

$$E = \frac{E_0}{\kappa} = \frac{q}{\kappa \epsilon_0 A}. \tag{26-12}$$

Inserting this in Eq. 26-11 yields

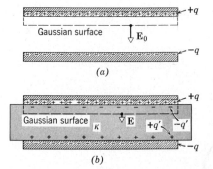

FIGURE 26-10   A parallel-plate capacitor (a) without and (b) with a dielectric. The charge q on the plates is assumed to be the same in each case.

$$\frac{q}{\kappa\epsilon_0 A} = \frac{q}{\epsilon_0 A} - \frac{q'}{\epsilon_0 A} \tag{26–13a}$$

or

$$q' = q\left(1 - \frac{1}{\kappa}\right). \tag{26–13b}$$

This shows correctly that the induced surface charge q' is always less in magnitude than the free charge q and is equal to zero if no dielectric is present, that is, if $\kappa = 1$.

Now we write Gauss's law for the case of Fig. 26–10b in the form

$$\epsilon_0 \oint \mathbf{E} \cdot d\mathbf{S} = q - q', \tag{26–14}$$

$q - q'$ again being the net charge within the Gaussian surface. Substituting from Eq. 26–13b for q' leads, after some rearrangement, to

$$\epsilon_0 \oint \kappa \mathbf{E} \cdot d\mathbf{S} = q. \tag{26–15}$$

This important relation, although derived for a parallel-plate capacitor, is true generally and is the form in which Gauss's law is usually written when dielectrics are present. Note the following:

1. The flux integral now contains a factor κ.

2. The charge q contained within the Gaussian surface is taken to be the free charge only. Induced surface charge is deliberately ignored on the right side of this equation, having been taken into account by the introduction of κ on the left side. Equations 26–14 and 26–15 are completely equivalent formulations.

26–6 Energy Storage in an Electric Field

In Section 25–6 we saw that all charge configurations have a certain electric potential energy U, equal to the work W (which may be positive or negative) that must be done to assemble them from their individual components, originally assumed to be infinitely far apart and at rest. This potential energy reminds us of the potential energy stored in a compressed spring or the gravitational potential energy stored in, say, the earth-moon system.

For a simple example, work must be done to separate two equal and opposite charges. This energy is stored in the system and can be recovered if the charges are allowed to come together again. Similarly, a charged capacitor has stored in it an electrical potential energy U equal to the work W required to charge it. This energy can be recovered if the capacitor is allowed to discharge. We can visualize the work of charging by imagining that an external agent pulls electrons from the positive plate and pushes them onto the negative plate, thus bringing about the charge separation; normally the work of charging is done by a battery, at the expense of its store of chemical energy.

Suppose that at a time t a charge $q'(t)$ has been transferred from one plate to the other. The potential difference $V(t)$ between the plates at that moment will be $q'(t)/C$. If an extra increment of charge dq' is transferred, the small amount of additional work needed will be

$$dW = V\,dq' = \left(\frac{q'}{C}\right)dq'.$$

If this process is continued until a total charge q has been transferred, the total work will be found from

$$W = \int dW = \int_0^q \frac{q'}{C} \, dq' = \frac{1}{2} \frac{q^2}{C}. \qquad (26\text{–}16)$$

From the relation $q = CV$ we can also write this as

$$W (= U) = \tfrac{1}{2} CV^2. \quad \text{JOULES} \qquad (26\text{–}17)$$

It is reasonable to suppose that the energy stored in a capacitor resides in the electric field. As q or V in Eqs. 26–16 and 26–17 increase, for example, so does the electric field E; when q and V are zero, so is E.

In a parallel-plate capacitor, neglecting fringing, the electric field has the same value for all points between the plates. Thus the *energy density u,* which is the stored energy per unit volume, should also be uniform; u (see Eq. 26–17) is given by

$$u = \frac{U}{Ad} = \frac{\tfrac{1}{2} CV^2}{Ad}, \quad \text{JOULES} \frac{}{m^3}$$

where Ad is the volume between the plates. Substituting the relation $C = \kappa\epsilon_0 A/d$ (Eq. 26–8) leads to

$$u = \frac{\kappa\epsilon_0}{2} \left(\frac{V}{d}\right)^2.$$

However, V/d is the electric field E, so that

$$u = \tfrac{1}{2} \kappa\epsilon_0 E^2. \qquad (26\text{–}18)$$

Although this equation was derived for the special case of a parallel-plate capacitor, it is true in general. If an electric field $\mathbf{E}$ exists at any point in space, we can think of that point as the site of stored energy in amount, per unit volume, of $\tfrac{1}{2}\kappa\epsilon_0 E^2$.

Example 4. A capacitor C_1 is charged to a potential difference V_0. This charging battery is then removed and the capacitor is connected as in Fig. 26–11 to an uncharged capacitor C_2.

(*a*) What is the final potential difference V across the combination?

The original charge q_0 is now shared by the two capacitors. Thus

$$q_0 = q_1 + q_2.$$

Applying the relation $q = CV$ to each of these terms yields

$$C_1 V_0 = C_1 V + C_2 V$$

or

$$V = V_0 \frac{C_1}{C_1 + C_2}.$$

This suggests a way to measure an unknown capacitance (C_2, say) in terms of a known one.

(*b*) What is the stored energy before and after the switch in Fig. 26–11 is thrown?

The initial stored energy is

$$U_0 = \tfrac{1}{2} C_1 V_0^2.$$

The final stored energy is

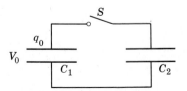

FIGURE 26–11 Example 4. C_1 is charged and then connected to C_2 by closing switch S.

$$U = \tfrac{1}{2}C_1V^2 + \tfrac{1}{2}C_2V^2 = \tfrac{1}{2}(C_1 + C_2)\left(\frac{V_0C_1}{C_1 + C_2}\right)^2 = \left(\frac{C_1}{C_1 + C_2}\right)U_0.$$

Thus U is less than U_0! The "missing" energy appears as thermal energy in the connecting wires as the charges move through them.

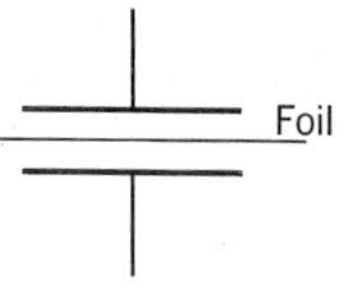

Example 5. A parallel-plate capacitor has plates with area A and separation d. A battery charges the plates to a potential difference V_0. The battery is then disconnected, and a dielectric slab of thickness d is introduced. Calculate the stored energy both before and after the slab is introduced and account for any difference.

The energy U_0 before introducing the slab is

$$U_0 = \tfrac{1}{2}C_0V_0{}^2.$$

After the slab is in place, we have

$$C = \kappa C_0 \quad \text{and} \quad V = V_0/\kappa$$

and thus

$$U = \tfrac{1}{2}CV^2 = \tfrac{1}{2}\kappa C_0 \left(\frac{V_0}{\kappa}\right)^2 = \frac{1}{\kappa}U_0.$$

The energy after the slab is introduced is less by a factor $1/\kappa$. The "missing" energy would be apparent to the person who inserted the slab. He would feel a "tug" on the slab and would have to restrain it if he wished to insert the slab without acceleration. This means that he would have to do negative work on it, or, alternatively, that the capacitor + slab system would do positive work on him. This positive work is

$$W = U_0 - U = \tfrac{1}{2}C_0V_0{}^2\left(1 - \frac{1}{\kappa}\right).$$

FIGURE 26–12

QUESTIONS

1. A capacitor is connected across a battery. (*a*) Why does each plate receive a charge of exactly the same magnitude? (*b*) Is this true if the plates are of different sizes?

2. Can there be a potential difference between two adjacent conductors that carry the same positive charge?

3. A sheet of aluminum foil of negligible thickness is placed between the plates of a capacitor as in Fig. 26–12. What effect has it on the capacitance if (*a*) the foil is electrically insulated and (*b*) the foil is connected to the upper plate.

4. Discuss what happens when (*a*) a dielectric slab and (*b*) a conducting slab are inserted between the plates of a parallel-plate capacitor. Assume the slab thicknesses to be one-half the plate separation.

5. An oil-filled parallel-plate capacitor has been designed to have a capacitance C and to operate safely at or below a certain maximum potential difference V_m without arcing over. However, the designer did not do a good job and the capacitor occasionally arcs over. What can be done to redesign the capacitor, keeping C and V_m unchanged and using the same dielectric?

6. Would you expect the dielectric constant, for substances containing permanent molecular electric dipoles, to vary with temperature?

7. For a given potential difference does a capacitor store more or less charge with a dielectric than it does without a dielectric (vacuum)? Explain in terms of the microscopic picture of the situation.

8. A dielectric slab is inserted in one end of a charged parallel-plate capacitor (the plates

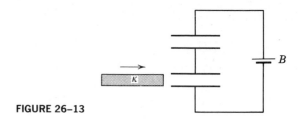

FIGURE 26-13

being horizontal and the charging battery having been disconnected) and then released. Describe what happens. Neglect friction.

9. A capacitor is charged by using a battery, which is then disconnected. A dielectric slab is then slipped between the plates. Describe qualitatively what happens to the charge, the capacitance, the potential difference, the electric field, and the stored energy.

10. While a capacitor remains connected to a battery, a dielectric slab is slipped between the plates. Describe qualitatively what happens to the charge, the capacitance, the potential difference, the electric-field, and the stored energy. Is work required to insert the slab?

11. Two identical capacitors are connected as shown in Fig. 26–13. A dielectric slab is slipped between the plates of one capacitor, the battery remaining connected. Describe qualitatively what happens to the charge, the capacitance, the potential difference, the electric field, and the stored energy for each capacitor.

PROBLEMS

1(1). Calculate the capacitance of the earth, viewed as a spherical conductor of radius 6400 km. *Answer:* 710 μf.

2(1). If we solve the equation on the bottom of page 490 for ϵ_0, we see that its mks units are farads/meter. Show that these units are equivalent to those obtained earlier for ϵ_0, namely coul^2/nt-m^2.

3(1). A spherical liquid drop of radius R has a capacitance given by $C = 4\pi\epsilon_0 R$ (see page 490). If two such drops combine to form a single larger drop, what is its capacitance? *Answer:* $C = 5.05\pi\epsilon_0 R$.

4(1). Two metallic spheres, radii a and b, are connected by a thin wire. Their separation is large compared with their dimensions. A charge Q is put onto this system. (*a*) How much charge resides on each sphere? (*b*) Show that the capacitance of this system is $C = 4\pi\epsilon_0(a + b)$.

5(2). A parallel-plate capacitor has circular plates of 8.0-cm radius and 1.0-mm separation. What charge will appear on the plates if a potential difference of 100 volts is applied? *Answer:* 1.8×10^{-8} coul.

6(2). *Capacitors in parallel.* Fig. 26–14 shows three capacitors connected in parallel. Show that the single capacitance C which is equivalent to this combination is $C = C_1 + C_2 + C_3$.

7(2). How many 1.0-μf capacitors would need to be connected in parallel in order to store a charge of 1.0 coul with a potential of 300 volts across the capacitors? *Answer:* 3300.

8(2). A 6.0-μf capacitor is connected in series with a 4.0-μf capacitor and a potential difference of 200 volts is applied across the pair. (*a*) What is the charge on each capacitor? (*b*) What is the potential difference across each capacitor?

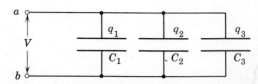

FIGURE 26–14 Problem 6(2).

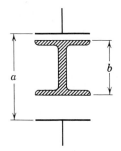

FIGURE 26-15 Problem 11(2).

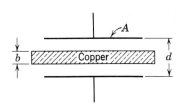

FIGURE 26-16 Problems 12(2) and 46(6).

9(2). Repeat the previous problem for the same two capacitors connected in parallel.
Answer: (*a*) $q_4 = 8.0 \times 10^{-4}$ coul; $q_6 = 1.2 \times 10^{-3}$ coul. (*b*) 200 volts.

10(2). You have material available to construct two parallel plate capacitors having a combined plate area A. How would you distribute this area between the two capacitors to obtain maximum total capacitance if you intend to connect them (*a*) in parallel and (*b*) in series?

11(2). Figure 26–15 shows two capacitors in series, the rigid center section of length b being movable vertically. Show that the equivalent capacitance of the series combination is independent of the position of the center section and is given by

$$C = \frac{\epsilon_0 A}{a - b}.$$

12(2). A slab of copper of thickness b is thrust into a parallel-plate capacitor as shown in Fig. 26–16. It is exactly halfway between the plates. What is the capacitance (*a*) before and (*b*) after the slab is introduced?

13(2). In Fig. 26–17 a variable air capacitor of the type sometimes used in tuning radios is shown. Alternate plates are connected together, one group being fixed in position, the other group being capable of rotation. Consider a pile of n plates of alternate polarity, each having an area A and separated from adjacent plates by a distance d. Show that this capacitor has a maximum capacitance of

$$C = \frac{(n - 1)\epsilon_0 A}{d}.$$

14(2). A potential difference of 300 volts is applied to a 2.0-μf capacitor and an 8.0-μf capacitor connected in series. (*a*) What are the charge and the potential difference for each capacitor? (*b*) The charged capacitors are reconnected with their positive plates together and their negative plates together, no external voltage being applied. What are the charge and the potential difference for each? (*c*) The charged capacitors in (*a*) are reconnected with plates of opposite sign together. What are the charge and the potential difference for each?

15(2). A 100-pf capacitor is charged to a potential difference of 50 volts, the charging battery then being disconnected. The capacitor is then connected, as in Fig. 26–11, to a second capacitor. If the measured potential difference drops to 35 volts, what is the capacitance of this second capacitor?
Answer: 43 pf.

FIGURE 26-17 Problem 13(2).

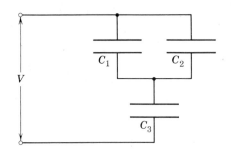

FIGURE 26-18   Problems 16(2), 17(2), and 44(6).   **FIGURE 26-19**   Problems 18(2) and 45(6).

16(2). In Fig. 26-18 find the equivalent capacitance of the combination. Assume that $C_1 = 10\ \mu f$, $C_2 = 5.0\ \mu f$, $C_3 = 4.0\ \mu f$, and $V = 100$ volts.

17(2). In Fig. 26-18 suppose that capacitor C_3 breaks down electrically, becoming equivalent to a conducting path. What *changes* in (*a*) the charge and (*b*) the potential difference occur for capacitor C_1? Use the numerical values of Problem 16(2).

Answer: (*a*) $+7.9 \times 10^{-4}$ coul. (*b*) $+79$ volts.

18(2). In Fig. 26-19 find the equivalent capacitance of the combination. Assume that $C_1 = 10\ \mu f$, $C_2 = 5.0\ \mu f$, $C_3 = 4.0\ \mu f$, and $V = 100$ volts.

19(2). Capacitors C_1 (1.0 μf) and C_2 (3.0 μf) are each charged to a potential V (100 volts) but with opposite polarity, so that points a and c are on the side of the respective positive plates of C_1 and C_2, and points b and d are on the side of the respective negative plates (see Fig. 26-20). Switches S_1 and S_2 are now closed. (*a*) What is the potential difference between points e and f? (*b*) What is the charge on C_1? (*c*) What is the charge on C_2?

Answer: (*a*) 50 volts. (*b*) 0.50×10^{-4} coul. (*c*) 1.5×10^{-4} coul.

20(2). A spherical capacitor consists of two concentric spherical shells of radii a and b, with $b > a$. Show that its capacitance is

$$C = 4\pi\epsilon_0 \frac{ab}{b-a}.$$

21(2). If the radii of the spherical capacitor in Problem 20(2) differ by only a small amount d where $d \ll a$, show that the expression for the capacitance reduces to that for a parallel plate capacitor having the same surface area.

22(2). The plate and cathode of a vacuum tube diode are in the form of two concentric cylinders with the cathode as the central cylinder. If the cathode diameter is 1.5 mm and the plate diameter 18 mm with both elements having a length of 2.3 cm, what is the capacitance of the diode?

23(2). If you have available a supply of 2.0-μf capacitors, each capable of withstanding 200 volts without breakdown, how would you assemble a combination having an equivalent capacitance of (*a*) 0.40 μf, or of (*b*) 1.2 μf, each capable of withstanding 1000 volts.

Answer: (*a*) Five in series. (*b*) Three arrays as in part (*a*), in parallel; there are other solutions.

24(2). In Fig. 26-21 the battery B supplies 12 volts. (*a*) Find the charge on each capacitor when

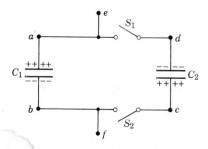

FIGURE 26-20   Problem 19(2).

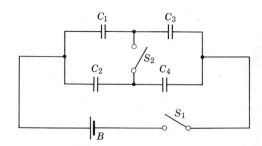

FIGURE 26-21   Problem 24(2).

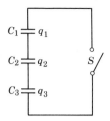

FIGURE 26-22   Problem 25(2).

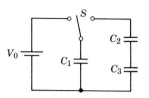

FIGURE 26-23   Problem 26(2).

switch S_1 is closed and (b) when switch S_2 is also closed. Take $C_1 = 1.0\ \mu f$, $C_2 = 2.0\ \mu f$, $C_3 = 3.0\ \mu f$, and $C_4 = 4.0\ \mu f$.

25(2). Charges q_1, q_2, and q_3 are placed on capacitors of capacitance C_1, C_2, C_3 respectively, arranged in series as shown in Fig. 26–22. Switch S is then closed. What are the final charges (a) q_1', (b) q_2', and (c) q_3', on the capacitors?

Answer: (a) $q_1' = \dfrac{(C_1C_2 + C_1C_3)q_1 - C_1C_3q_2 - C_1C_2q_3}{C_1C_2 + C_1C_3 + C_2C_3}$.

(b) $q_2' = \dfrac{(C_1C_2 + C_2C_3)q_2 - C_1C_2q_3 - C_2C_3q_1}{C_1C_2 + C_1C_3 + C_2C_3}$.

(c) $q_3' = \dfrac{(C_1C_3 + C_2C_3)q_3 - C_1C_3q_2 - C_2C_3q_1}{C_1C_2 + C_1C_3 + C_2C_3}$.

26(2). When switch S is thrown on the left in Fig. 26–23, the plates of the capacitor of capacitance C_1 acquire a potential difference V_0. C_2 and C_3 are initially uncharged. The switch is now thrown to the right. What are the final charges (a) q_1, (b) q_2, and (c) q_3, on the corresponding capacitors?

27(3). Two parallel plates of area 100 cm² are each given equal but opposite charges of 8.9×10^{-7} coul. The electric field within the dielectric material filling the space between the plates is 1.4×10^6 volts/meter. (a) Find the dielectric constant of the material. (b) Determine the magnitude on the charge induced on each dielectric surface.

Answer: (a) 7.1. (b) 7.7×10^{-7} coul.

28(3). A certain material has a dielectric constant of 2.8 and a dielectric strength of 18×10^6 volts/meter. If it is used as the dielectric material in a parallel-plate capacitor, what minimum area may the plates of the capacitor have in order that the capacitance be $7.0 \times 10^{-2}\ \mu f$ and that the capacitor be able to withstand a potential difference of 4000 volts?

29(3). A parallel-plate capacitor is filled with two dielectrics as in Fig. 26–24. Show that the capacitance is given by

$$C = \frac{\epsilon_0 A}{d}\left(\frac{\kappa_1 + \kappa_2}{2}\right).$$

Check this formula for all the limiting cases that you can think of. [Hint: Can you justify regarding this arrangement as two capacitors in parallel; see Problem 6(2)?]

30(3). A parallel-plate capacitor is filled with two dielectrics of equal thicknesses as in Fig. 26–25. Show that the capacitance is given by

FIGURE 26-24   Problem 29(3).

FIGURE 26-25   Problem 30(3).

$$C = \frac{2\epsilon_0 A}{d} \left(\frac{\kappa_1 \kappa_2}{\kappa_1 + \kappa_2} \right).$$

Check this formula for all the limiting cases that you can think of. (Hint: Can you justify regarding this arrangement as two capacitors in series?)

31(3). What is capacitance of the capacitor in Fig. 26–26? The plate area is A.

Answer: $C = \dfrac{\epsilon_0 A}{2d} \left[\dfrac{\kappa_1}{2} + \dfrac{\kappa_2 \kappa_3}{\kappa_2 + \kappa_3} \right].$

32(3). For making a capacitor you have available two plates of copper, a sheet of mica (thickness $= 0.10$ mm, $\kappa = 6.0$), a sheet of glass (thickness $= 2.0$ mm, $\kappa = 7.0$), and a slab of paraffin (thickness $= 1.0$ cm, $\kappa = 2.0$). To obtain the largest capacitance, which sheet (or sheets) should you place between the copper plates?

33(5). A Geiger counter is made of two long concentric metal cylinders with a gas of dielectric constant κ between them. Neglecting end effects, use Gauss's law to calculate the capacitance of this configuration. The center rod has a radius a, the surrounding tube a radius b, and the length $l \gg b$.

Answer: $2\pi\epsilon_0 \kappa l / \ln(b/a)$.

34(6). A parallel-plate air capacitor having area A (40 cm²) and spacing d (1.0 mm) is charged to a potential V (600 volts). Find (*a*) the capacitance, (*b*) the magnitude of the charge on each plate, (*c*) the stored energy, (*d*) the electric field between the plates, and (*e*) the energy density between the plates.

35(6). An isolated metal sphere whose diameter is 10 cm has a potential of 8000 volts. What is the energy density at the surface of the sphere?

Answer: 0.11 joule/meter³.

36(6). Two capacitors (2.0 μf and 4.0 μf) are connected in parallel across a 300-volt potential difference. Calculate the total energy stored in the system.

37(6). (*a*) If the potential difference across a cylindrical capacitor is doubled the energy stored in the capacitor is changed by what factor? (*b*) If the radii of the inner and outer cylinders are each doubled, keeping the stored charge constant, how does the stored energy change?

Answer: (*a*) Factor of four. (*b*) No change.

38(6). A parallel-plate capacitor has plates of area A and separation d and is charged to a potential difference V. The charging battery is then disconnected and the plates are pulled apart until their separation is $2d$. Derive expressions in terms of A, d, and V for (*a*) the final potential difference, (*b*) the initial and the final stored energy, and (*c*) the work required to separate the plates.

39(6). What would be the capacitance required to store energy U (10 kw-hr) at voltage V (1000 volts)?

Answer: 72f.

40(6). A parallel-connected bank of 2000 5.0-μf capacitors is used to store electric energy. What does it cost to charge this bank to 50,000 volts, assuming a rate of 2c/kw-hr?

41(6). A parallel-plate air capacitor has a capacitance of 100 pf. (*a*) What is the stored energy if the applied potential difference is 50 volts? (*b*) Is it possible to calculate the energy density for points between the plates?

Answer: (*a*) 1.3×10^{-7} joules. (*b*) No.

[handwritten in left margin:]
$a \quad 35.4\,pf$
$b. \quad 2.12 \times 10^{-8}\,Coul$
$c. \quad 6.4 \times 10^{-6}\,Joules$
$d. \quad 6 \times 10^5\,v/m$
$e. \quad 1.59 \; J/m^3$

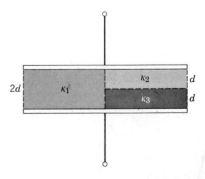

FIGURE 26–26 Problem 31(3).

$$U_c = \frac{Q^2}{2C}, \quad Q_1 = \frac{Q}{2}$$

42(6). One capacitor is charged until its stored energy is U (4.0 joules). A second uncharged $U_f = \frac{Q^2}{4C}$ capacitor is then connected to it in parallel. (a) If the charge distributes equally, what is now the total energy stored in the electric fields? (b) Where did the excess energy go?

$$\left(ex\ 4 \right) = \frac{U_i}{2}$$

43(6). For the capacitors of Problem 14(2) compute the energy stored for the three different connections of parts (a), (b), and (c). Compare your answers and explain any differences. *Answer:* (a) $U_2 = 5.8 \times 10^{-2}$ joules; $U_8 = 1.4 \times 10^{-2}$ joule. (b) $U_2 = 9.2 \times 10^{-3}$ joules; $U_8 = 37 \times 10^{-3}$ joule. (c) $U_2 = U_8 =$ zero.

44(6). In Fig. 26–18 find (a) the charge, (b) the potential difference, and (c) the stored energy for each capacitor. Assume the numerical values of Problem 16(2).

45(6). In Fig. 26–19 find (a) the charge, (b) the potential difference, and (c) the stored energy for each capacitor. Assume the numerical values of Problem 18(2). *Answer:* (a) $q_1 = q_2 = 3.3 \times 10^{-4}$ coul; $q_3 = 4.0 \times 10^{-4}$ coul. (b) $V_1 = 33$ volts; $V_2 = 67$ volts; $V_3 = 100$ volts. (c) $U_1 = 5.4 \times 10^{-3}$ joules; $U_2 = 1.1 \times 10^{-2}$ joules; $U_3 = 2.0 \times 10^{-2}$ joules.

46(6). In terms of the original capacitance C, find the work done in inserting a copper slab of thickness $d/2$ in Problem 12(2) if (a) the voltage is held constant, and (b) if the charge is held constant.

47(6). Show that the plates of a parallel-plate capacitor attract each other with a force given by

$$F = \frac{q^2}{2\epsilon_0 A} \quad .$$

Prove this by calculating the work necessary to increase the plate separation from x to $x + dx$ with the battery disconnected.

48(6). A parallel plate capacitor has plates of area 0.12 meter^2 and a separation of 1.2 cm. A battery charges the plates to a potential difference of 120 volts and is then disconnected. A dielectric slab of thickness 0.40 cm and dielectric constant 4.8 is then placed symmetrically between the plates. (a) Find the capacitance before the slab is inserted. (b) What is the capacitance with the slab in place? (c) What is the free charge q before and after the slab is inserted? (d) Find the electric field in the space between the plates and dielectric. (e) What is the electric field in the dielectric? (f) With the slab in place what is the potential difference across the plates? (g) How much external work is involved in the process of inserting the slab?

CHAPTER 27

Current and Resistance

27–1 Current and Current Density

The free electrons in an isolated metallic conductor, such as a length of copper wire, are in random motion like the molecules of a gas confined to a container. They have no net directed motion along the wire. If a hypothetical plane is passed through the wire, the rate at which electrons pass through it from right to left is the same as the rate at which they pass through from left to right; the net rate is zero.

If the ends of the wire are connected to a battery, an electric field will be set up at every point within the wire. If the potential difference maintained by the battery is 10 volts and if the wire (assumed uniform) is 5 meters long, this field will have the value 2 volts/meter at every point. This field **E** will act on the electrons and will give them a resultant motion in the direction of −**E**. We say that an *electric current i* is established; if a net charge q passes through any cross section of the conductor in time t the current, assumed constant, is

$$i = q/t. \qquad amp = \frac{coul}{sec} \qquad (27\text{--}1)$$

def of amp
p. 564

The appropriate mks units are amperes for i, coulombs for q, and seconds for t. Recall (Section 22–4) that Eq. 27–1 is the defining equation for the coulomb and that we have not yet given an operational definition of the ampere; we do so in Section 30–4.

If the rate of flow of charge with time is not constant, the current varies with time and is given by the differential limit of Eq. 27–1, or

$$i = dq/dt. \qquad (27\text{--}2)$$

def

507

In the rest of this chapter we consider only constant currents.

The current i is the same for all cross sections of a conductor, even though the cross-sectional area may be different at different points. In the same way the rate at which water (assumed incompressible) flows past any cross section of a pipe is the same even if the cross section varies. The water flows faster where the pipe is smaller and slower where it is larger, so that the volume rate, measured perhaps in gal/min, remains unchanged. This constancy of the electric current follows because charge must be conserved; it does not pile up steadily or drain away steadily from any point in the conductor under the assumed steady-state conditions.

The existence of an electric field inside a conductor does not contradict Section 24–4, in which we asserted that $E = 0$ inside a conductor. In that section, which dealt with a state in which all net motion of charge had stopped (electrostatics), we assumed that the conductor was insulated and that no potential difference was deliberately maintained between any two points on it, as by a battery. In this chapter, which deals with charges in motion, we relax this restriction.

The electric field that acts on the electrons in a conductor does not produce a net acceleration because the electrons keep colliding with the atoms (strictly, ions) that make up the conductor. This array of ions, coupled together by strong spring-like forces of electric origin, is called the *lattice* (see Fig. 18–4). The over-all effect of these collisions is to transfer kinetic energy from the accelerating electrons into vibrational energy of the lattice. The electrons acquire a constant average *drift speed* v_d in the direction $-\mathbf{E}$. The analogy is to a marble rolling down a long flight of stairs and not to a marble falling freely from the same height. In the first case the acceleration caused by the (gravitational) field is effectively canceled by the decelerating effects of collisions with the stair treads so that, under proper conditions, the marble rolls down the stairs with zero average acceleration, that is, at constant average speed.

Although in metals the charge carriers are electrons, in electrolytes or in gaseous conductors they may also be positive or negative ions or both. A convention for labeling the directions of currents is needed because charges of opposite sign move in opposite directions in a given field. A positive charge moving in one direction is equivalent in nearly all external effects to a negative charge moving in the opposite direction. Hence, for simplicity we assume that all charge carriers are positive and we draw the current arrows in the direction that such charges would move. If the charge carriers are negative, they simply move opposite to the direction of the current arrow (see Fig. 27–1). When we encounter a case (as in the Hall effect; see Section 29–5) in which the sign of the charge carriers makes a difference in the external effects, we will disregard the convention and take the actual situation into account.

Current i is a characteristic of a particular conductor. It is a macroscopic quantity, like the mass of an object, the volume of an object, or the length of a rod. A related microscopic quantity is the current density **j**. It is a vector and is charac-

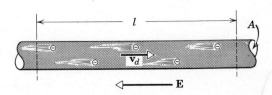

FIGURE 27–1   Electrons drift in a direction opposite to the electric field in a conductor.

teristic of a point inside a conductor rather than of the conductor as a whole. If the current is distributed uniformly across a conductor of cross-sectional area A, the magnitude of the <u>current density</u> for all points on that cross section is

$i = \int \vec{j} \cdot d\vec{s}$

$$j = i/A. \qquad \text{amp}/m^2 \tag{27-3}$$

$\mathcal{DEF}$

The vector **j** at any point is oriented in the direction that a positive charge carrier would move at that point. An electron at that point would move in the direction $-\mathbf{j}$.

The arrow often associated with the current in a wire does not indicate that current is a vector but merely shows the sense of charge flow. Positive charge carriers either move in a certain direction along the wire or in the opposite direction, these two possibilities being represented by $+$ or $-$ in algebraic equations. Note that (*a*) the current in a wire remains unchanged if the wire is bent, tied into a knot, or otherwise distorted, and (*b*) the arrows representing the sense of currents do not in any way obey the laws of vector addition.

The drift speed v_d of charge carriers in a conductor can be computed from the current density j. Figure 27–1 shows the conduction electrons in a wire moving to the right at an assumed constant drift speed v_d. The number of conduction electrons in the wire is nAl where n is the number of conduction electrons per unit volume and Al is the volume of the wire. A charge of magnitude

$$q = (nAl)e$$

passes out of the wire, through its right end, in a time t given by

$$t = \frac{l}{v_d}.$$

The current i is given by

$$i = \frac{q}{t} = \frac{nAle}{l/v_d} = nAev_d.$$

Solving for v_d and recalling that $j = i/A$ (Eq. 27–3) yields

$$v_d = \frac{i}{nAe} = \frac{j}{ne}. \qquad \vec{j} = ne\,\vec{v_d} \tag{27-4}$$

Example 1. An aluminum wire whose diameter is 0.10 in. is welded end to end to a copper wire with a diameter of 0.064 in. The composite wire carries a steady current of 10 amp. What is the current density in each wire?

The current is distributed uniformly over the cross section of each conductor, except near the junction, which means that the current density may be taken as constant for all points within each wire. The cross-sectional area of the aluminum wire is 0.0079 in.² Thus, from Eq. 27–3,

$$j_{Al} = \frac{i}{A} = \frac{10 \text{ amp}}{0.0079 \text{ in.}^2} = 1300 \text{ amp/in.}^2$$

The cross-sectional area of the copper wire is 0.0032 in.² Thus

$$j_{Cu} = \frac{i}{A} = \frac{10 \text{ amp}}{0.0032 \text{ in.}^2} = 3100 \text{ amp/in.}^2$$

The fact that the wires are of different materials does not enter into consideration here.

Example 2. What is v_d for the copper wire in Example 1?

We can write the current density for the copper wire as 480 amp/cm². To compute n we start from the assumption that there is one free electron per atom in copper. The number of atoms per unit volume is $N_0 d/M$ where d is the density, N_0 is Avogadro's number, and M is the atomic weight. The number of free electrons per unit volume is then

$$n = \frac{N_0 d}{M} = \frac{(6.0 \times 10^{23} \text{ atoms/mole})(9.0 \text{ gm/cm}^3)(1 \text{ electron/atom})}{64 \text{ gm/mole}}$$

$$= 8.4 \times 10^{22} \text{ electrons/cm}^3.$$

Finally, v_d is, from Eq. 27–4

$$v_d = \frac{j}{ne} = \frac{480 \text{ amp/cm}^2}{(8.4 \times 10^{22} \text{ electrons/cm}^3)(1.6 \times 10^{-19} \text{ coul/electron})}$$

$$= 3.6 \times 10^{-2} \text{ cm/sec.}$$

It takes 28 sec for the electrons in this wire to drift 1.0 cm. Would you have guessed that v_d was so low? The drift speed of electrons must not be confused with the speed at which changes in the electric field configuration travel along wires, a speed which approaches that of light. When pressure is applied to one end of a long water-filled tube, a pressure wave travels rapidly along the tube. The speed at which water moves through the tube is much lower, however. ▪

27–2 Resistance, Resistivity, and Conductivity

If the same potential difference is applied between the ends of a rod of copper and of a rod of wood, very different currents result. The characteristic of the conductor that enters here is its *resistance*. We define the resistance of a conductor (often called a resistor; symbol ‑ⱲⱲⱲⱲ‑) between two points by applying a potential difference V between those points, measuring the current i, and dividing:

$$R = V/i. \qquad (27\text{–}5)$$

If V is in volts and i in amperes, the resistance R will be in *ohms*.

The flow of charge through a conductor is often compared with the flow of water through a pipe, which occurs because there is a difference in pressure between the ends of the pipe, established perhaps by a pump. This pressure difference can be compared with the potential difference established between the ends of a resistor by a battery. The flow of water (ft³/sec, say) is compared with the current (coul/sec or amp). The rate of flow of water for a given pressure difference is determined by the nature of the pipe. Is it long or short? Is it narrow or wide? Is it empty or filled, perhaps with gravel? These characteristics of the pipe are analogous to the resistance of a conductor.

Related to resistance is the *resistivity ρ*, which is a characteristic of a material rather than of a particular specimen of a material; it is defined, for isotropic materials (that is, materials whose properties—electrical in this case—do not vary with direction), from

$$\rho = \frac{E}{j}. \qquad (27\text{–}6)$$

The resistivity of copper is 1.7×10^{-8} ohm-m; that of fused quartz is about 10^{16} ohm-m. Few physical properties are measurable over such a range of values; Table 27–1 lists some values for common metals.

The macroscopic quantities V, i, and R are related to the microscopic quantities

Table 27-1 PROPERTIES OF METALS AS CONDUCTORS

	RESISTIVITY			
Aluminum	2.8×10^{-8}	3.9×10^{-3}	2.7	659
Copper	1.7×10^{-8}	3.9×10^{-3}	8.9	1080
Carbon (amorphous)	3.5×10^{-5}	-5×10^{-4}	1.9	3500
Iron	1.0×10^{-7}	5.0×10^{-3}	7.8	1530
Manganin	4.4×10^{-7}	1×10^{-5}	8.4	910
Nickel	6.8×10^{-8}	6×10^{-3}	8.9	1450
Silver	1.6×10^{-8}	3.8×10^{-3}	10.5	960
Steel	1.8×10^{-7}	3×10^{-3}	7.7	1510
Wolfram (tungsten)	5.6×10^{-8}	4.5×10^{-3}	19	3400

(annotations: "RESISTIVITY", "Ω/m")

* This quantity is the fractional change in resistivity $(d\rho/\rho)$ per unit change in temperature (dT). The resistivity ρ at any temperature T is given by

$$\rho = \rho_0[1 + \alpha(T - T_0)]$$

where, in this case, ρ_0 is the resistivity at T_0 (= 20° C). The coefficient α varies slightly with temperature; the value given here is for 20° C. *(annotation: COEF OF RESISTIVITY)*

E, j, and ρ. To see this in a particular simple case, consider a cylindrical conductor, of cross-sectional area A and length l, carrying a steady current i. Let us apply a potential difference V between its ends. If the cross sections of the cylinder at each end are equipotential surfaces, the electric field and the current density will be constant for all points in the cylinder and will have the values

$$E = \frac{V}{l} \quad \text{and} \quad j = \frac{i}{A}.$$

The resistivity ρ may then be written as

$$\rho = \frac{E}{j} = \frac{V/l}{i/A}.$$

But V/i is the resistance R which leads to

$$R = \rho \frac{l}{A}. \qquad \Omega \cdot \frac{m \cdot m}{m^2} = \Omega \qquad (27\text{–}8)$$

Figure 27–2 shows how the resistivity of copper varies with temperature. It does not go to zero at the absolute zero of temperature, even though it appears to do so, the residual resistivity at this temperature being 0.02×10^{-8} ohm-m. For many substances the resistance does become zero at some low temperature. Figure 27–3 shows the resistance of a specimen of mercury for temperatures below 6° K. In the space of about 0.05 K° the resistance drops abruptly to an immeasurably low value. This phenomenon, called *superconductivity*,† was discovered by Kamerlingh Onnes in the Netherlands in 1911. The resistance of materials in the superconducting state seems to be truly zero; currents, once established in closed superconducting circuits,

† See "Superconductivity" by B. T. Matthias in *Scientific American*, p. 92, November 1957.

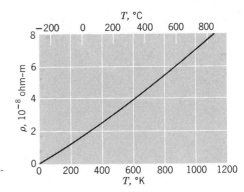

FIGURE 27-2 The resistivity of copper as a function of temperature.

persist for weeks without diminution, even though there is no battery in the circuit. If the temperature is raised slightly above the superconducting point, such currents drop rapidly to zero.

27-3 Ohm's Law

Let us apply a variable potential difference V between the ends of a 100-foot coil of #18 copper wire. For each applied potential difference, let us measure the current i and plot it against V as in Fig. 27–4. The straight line that results means that the resistance of this conductor is the same no matter what applied voltage is used to measure it. This important result, which holds for metallic conductors, is known as *Ohm's law*. We assume that the temperature of the conductor is essentially constant throughout the measurements.

Many conductors do not obey Ohm's law. Figure 27–5, for example, shows a V–i plot for a type 2A3 vacuum tube. The plot is not straight and the resistance depends on the voltage used to measure it. Also, the current for this device is almost vanishingly small if the polarity of the applied potential difference is reversed. For metallic conductors the current reverses direction when the potential difference is reversed, but its magnitude does not change.

Figure 27–6 shows a typical V–i plot for another nonohmic device, a *thermistor*. This is a semiconductor with a large and negative temperature coefficient of resistivity α (see Table 27–1) that varies greatly with temperature. We note that two different currents through the thermistor can correspond to the same potential difference between its ends. Thermistors are often used to measure the rate of energy flow in microwave beams by allowing the microwave beam to fall on the thermistor and heat it. The relatively small temperature rise so produced results in a relatively large change in resistance, which serves as a measure of the microwave power.

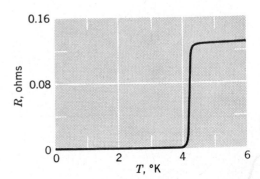

FIGURE 27-3 The resistance of mercury disappears below about 4° K.

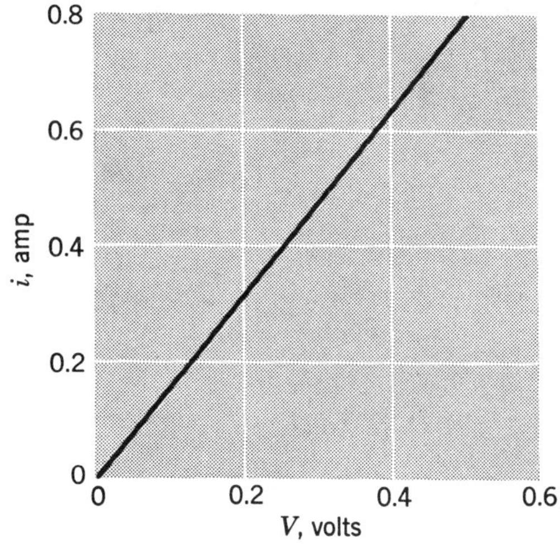

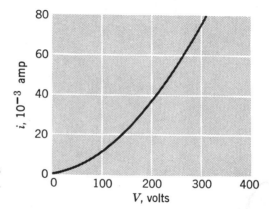

FIGURE 27-4 The current in a particular copper conductor as a function of potential difference. This conductor obeys Ohm's law.

Modern electronics, and therefore much of the character of our present technological civilization, depends in a fundamental way on the fact that many conductors, such as vacuum tubes, crystal rectifiers, thermistors, and transistors, do *not* obey Ohm's law.

We stress that the relationship $V = iR$ is not a statement of Ohm's law. A conductor obeys Ohm's law only if its V–i curve is linear, that is, if R is independent of V and i. The relationship $R = V/i$ remains as the definition of the resistance of a conductor whether or not the conductor obeys Ohm's law.

The microscopic equivalent of the relationship $V = iR$ is Eq. 27-6, or $E = j\rho$. A conducting material is said to obey Ohm's law if a plot of E versus j is linear, that is, if the resistivity ρ is independent of E and j. Ohm's law is a specific property of certain materials and is not a general law of electromagnetism, for example, like Gauss's law.

A close analogy exists between the flow of charge because of a potential difference and the flow of heat because of a temperature difference. Consider a thin electrically conducting slab of thickness Δx and area A. Let a potential difference ΔV be maintained between opposing faces. The current i is given by Eqs. 27-5 ($i = V/R$) and 27-8 ($R = \rho l/A$), or

$$i = \frac{\Delta V}{R} = \frac{A\Delta V}{\rho \Delta x}.$$

FIGURE 27-5 The current in a type 2A3 vacuum tube as a function of potential difference. This conductor does not obey Ohm's law.

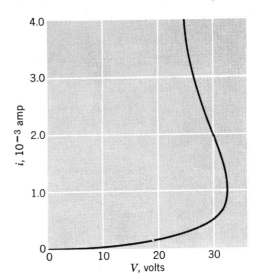

FIGURE 27-6 A plot of current as a function of potential difference in a Western Electric 1–B thermistor. The curve shows how the voltage across the thermistor varies as the current through it is increased. The shape of the curve can be accounted for in terms of the large negative temperature coefficient of resistivity of the material of which the device is made.

In the limiting case of a slab of thickness dx this becomes

$$i = \frac{1}{\rho} A \frac{dV}{dx}$$

or

$$i = \frac{dq}{dt} = -\sigma A \frac{dV}{dx}, \tag{27–9}$$

where $\sigma \ (= 1/\rho)$ is the *conductivity* of the material. Since positive charge flows in the direction of decreasing V, we introduce a minus sign into Eq. 27–9, that is, dq/dt is positive when dV/dx is negative.

The analogous heat flow equation (see Section 19–3) is

$$\frac{dQ}{dt} = -kA \frac{dT}{dx}, \tag{27–10}$$

which shows that k, the thermal conductivity, corresponds to σ and dT/dx, the temperature gradient, corresponds to dV/dx, the potential gradient. There is more than a formal mathematical analogy between Eqs. 27–9 and 27–10. Both heat energy and charge are carried by the free electrons in a metal; empirically, a good electrical conductor (silver, say) is also a good heat conductor and conversely.

27–4 Resistivity—An Atomic View

We can understand why metals obey Ohm's law on the basis of simple classical ideas. If these ideas are modified when necessary by the requirements of quantum physics, it is possible to go further and to calculate theoretical values of the resistivity ρ for various metals. These calculations are not simple, but when they have been carried out the agreement with the experimental value of ρ has usually been good.

In a metal the valence electrons are not attached to individual atoms but are free to move about within the lattice and are called *conduction electrons*. In copper there is one such electron per atom, the other 28 remaining bound to the copper nuclei to form ionic cores.

The speed distribution of conduction electrons can be described correctly only in terms of quantum physics. For our purposes, however, it suffices to consider only a suitably defined average speed $\bar{v}$; for copper $\bar{v} = 1.6 \times 10^8$ cm/sec. In the absence of an electric field, the directions in which the electrons move are completely random, like those of the molecules of a gas confined to a container.

In this classical picture the electrons collide constantly with the ionic cores of the conductor, that is, they interact with the lattice, often suffering sudden changes in speed and direction. Collisions between electrons occur only rarely and have little effect on the resistivity. As in the case of molecular collisions, we can describe electron-lattice collisions by a *mean free path* $\bar{l}$, where $\bar{l}$ is the average distance that an electron travels between collisions. In copper the mean free path is about 4×10^{-6} cm (about 200 atomic diameters) and the mean time between collisions ($= \bar{l}/\bar{v}$) is about 3×10^{-14} sec.

When an electric field is applied to a metal, the electrons modify their random motion in such a way that they drift slowly, in the opposite direction to that of the field, with an average drift speed v_d. This drift speed is much less than the effective average speed $\bar{v}$ mentioned above (see Example 2). Figure 27–7 suggests the relationship between these two speeds. The solid lines suggest a possible random path followed by an electron in the absence of an applied field; the electron proceeds from x to y, making six collisions on the way. The dashed curves show how this same event might have occurred if an electric field **E** had been applied. Note that the electron drifts steadily to the right, ending at y' rather than at y. In preparing Fig. 27–7, it has been assumed that the drift speed v_d is $0.02\bar{v}$; actually, it is more like $10^{-10}\bar{v}$, so that the "drift" exhibited in the figure is greatly exaggerated.

The drift speed v_d can be calculated in terms of the applied electric field E and of $\bar{v}$ and $\bar{l}$. When a field is applied to an electron in the metal it will experience a force eE which will impart to it an acceleration a given by Newton's second law,

$$a = \frac{eE}{m}.$$

Consider an electron that has just collided with an ion core. The collision, in general, will momentarily destroy the tendency to drift and the electron will have a random direction after the collision. At its next collision the electron's velocity component in the $-\mathbf{E}$ direction will have changed, on the average, by $a(\bar{l}/\bar{v})$ where $\bar{l}/\bar{v}$ is the mean time between collisions. We call this the drift speed v_d, or

$$v_d = a \left(\frac{\bar{l}}{\bar{v}} \right) = \frac{eE\bar{l}}{m\bar{v}}. \tag{27–11}$$

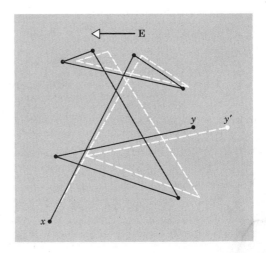

FIGURE 27–7 The solid lines show an electron moving from x to y, making six collisions. The dashed curves show what the electron path *might* have been in the presence of an electric field **E**. Note the steady drift in the direction of $-\mathbf{E}$.

We may express v_d in terms of the current density (Eq. 27–4) and combine with Eq. 27–11 to obtain

$$v_d = \frac{j}{ne} = \frac{eE\bar{l}}{m\bar{v}}.$$

Combining this with Eq. 27–6 ($\rho = E/j$) leads finally to

$$\rho = \frac{m\bar{v}}{ne^2l}. \qquad (27\text{–}12)$$

Equation 27–12 can be taken as a statement that metals obey Ohm's law if we can show that $\bar{v}$ and $\bar{l}$ do not depend on the applied electric field E. In this case ρ will not depend on E, which (see Section 27–3) is the criterion that a material obey Ohm's law. The quantities $\bar{v}$ and $\bar{l}$ depend on the speed distribution of the conduction electrons. We have seen that this distribution is affected only slightly by the application of even a relatively large electric field, since $\bar{v}$ is of the order of 10^8 cm/sec and v_d (see Example 2) only of the order of 10^{-2} cm/sec, a ratio of 10^{10}. We may be sure that whatever the values of $\bar{v}$ and $\bar{l}$ are (for copper at 20°C, say) in the absence of a field they remain essentially unchanged when the field is applied. Thus the right side of Eq. 27–12 is independent of E and the material obeys Ohm's law.

27–5   Energy Transfers in an Electric Circuit

Figure 27–8 shows a circuit consisting of a battery B connected to a "black box." A steady current i exists in the connecting wires and a steady potential difference V_{ab} exists between the terminals a and b. The box might contain a resistor, a motor, or a storage battery, among other things.

Terminal a, connected to the positive battery terminal, is at a higher potential than terminal b. If a charge dq moves through the box from a to b, this charge will decrease its electric potential energy by $dq\, V_{ab}$ (see Section 25–6). The conservation-of-energy principle tells us that this energy is transferred in the box from electric potential energy to some other form. What that other form will be depends on what is in the box. In a time dt the energy dU transferred inside the box is then

$$dU = dq\, V_{ab} = i\, dt\, V_{ab}.$$

We find the rate of energy transfer, or the power P, by dividing by the time, or

$$P = \frac{dU}{dt} = iV_{ab}. \qquad (27\text{–}13)$$

If the device in the box is a motor, the energy appears largely as mechanical work done by the motor; if the device is a storage battery that is being charged, the energy appears largely as stored chemical energy in this second battery.

If the device is a resistor, we assert that the energy appears as internal energy in the resistor, resulting in an increase of temperature (just as if heat had been added from an external source

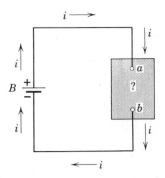

FIGURE 27–8   A battery B sets up a current in a circuit containing a "black box."

at a higher temperature). To see this, consider a stone of mass m that falls through a height h. It decreases its gravitational potential energy by mgh. If the stone falls in a vacuum or—for practical purposes—in air, this energy is transformed into kinetic energy of the stone. If the stone falls in water, however, its speed eventually becomes constant, which means that the kinetic energy no longer increases. The potential energy that is then steadily being made available as the stone falls at constant speed appears as internal energy in the stone and the surrounding water, causing a temperature rise.

The course of the electrons through the resistor is much like that of the stone through water. The electrons travel with a constant drift speed v_d and thus do not gain kinetic energy. The electric potential energy that they lose is transferred to the resistor as internal energy. On a microscopic scale this can be understood in that collisions between the electrons and the lattice increase the amplitude of the thermal vibrations of the lattice; on a macroscopic scale this corresponds to a temperature increase. This effect, which is thermodynamically irreversible, is called the *Joule effect.*

For a resistor we can combine Eqs. 27–13 and 27–5 ($R = V/i$) and obtain either

$$P = i^2R \qquad \qquad (27\text{–}14)$$

$$P = iV$$

or

$$P = \frac{V^2}{R}. \qquad \qquad (27\text{–}15)$$

Note that Eq. 27–13 applies to electrical energy transfer of all kinds; Eqs. 27–14 and 27–15 apply only to the transfer of electrical energy to internal energy in a resistor. Equations 27–14 and 27–15 are known as *Joule's law.* This law is a particular way of writing the conservation-of-energy principle for the special case in which electrical energy is transferred into internal energy.

The unit of power that follows from Eq. 27–13 is the volt-amp. It can be written as

$$= \frac{nt \cdot m}{coul} \cdot \frac{coul}{sec} = \frac{nt \cdot m}{sec} = Joule/sec$$

$$1 \text{ volt-amp} = 1 \text{ volt-amp} \left(\frac{1 \text{ joule}}{1 \text{ volt} \times 1 \text{ coul}}\right)\left(\frac{1 \text{ coul}}{1 \text{ amp} \times 1 \text{ sec}}\right)$$

$$= 1 \text{ joule/sec.} = 1 \text{ WATT}$$

The first conversion factor in parenthesis comes from the definition of the volt (Eq. 25–1); the second comes from the definition of the coulomb. The joule/sec is such a common unit that it is given a name of its own, the *watt* (see Section 6–7). Power is not an exclusively electrical concept, of course, and we can express in watts the power ($= \mathbf{F} \cdot \mathbf{v}$) expended by an agent that exerts a force $\mathbf{F}$ while it moves with a velocity $\mathbf{v}$.

Example 3. You are given a 20-ft length of "heating wire" made of the special alloy Nichrome; it has a resistance of 24 ohms. Can you obtain more power output by winding one coil or by cutting the wire in two and winding two separate coils? In each case the coils are to be connected individually across a 110-volt line.

The power P for the single coil is given by Eq. 27–15:

$$P = \frac{V^2}{R} = \frac{(110 \text{ volts})^2}{24 \text{ ohms}} = 500 \text{ watts.}$$

The power for a coil of half the length is given by

$$P' = \frac{(110 \text{ volts})^2}{12 \text{ ohms}} = 1000 \text{ watts.}$$

There are two "half-coils," so that the total power obtained by cutting the wire in half is 2000 watts, or four times that for the single coil. This would seem to suggest that we could buy a 500-watt heating coil, cut it in half, and rewind it to obtain 2000 watts. Why is this not a practical idea?

QUESTIONS

1. Name other physical quantities that, like current, are scalars having a sense represented by an arrow in a diagram.

2. A potential difference V is applied to a circular cylinder of carbon by clamping it between circular copper electrodes, as in Fig. 27–9. Discuss the difficulty of calculating the resistance of the carbon cylinder, using the relation $R = \rho l/A$.

3. How would you measure the resistance of a pretzel-shaped conductor? Give specific details to clarify the concept.

4. Discuss the difficulties of testing whether the filament of a light bulb obeys Ohm's law.

5. Does the relation $V = iR$ apply to nonohmic resistors?

6. A potential difference V is applied to a copper wire of diameter d and length l. What is the effect on the electron drift speed of (a) doubling V, (b) doubling l, and (c) doubling d?

7. If the drift speeds of the electrons in a conductor under ordinary circumstances are so slow (see Example 2), why do the lights in a room turn on so quickly after the switch is closed?

8. Can you think of a way to measure the drift speed for electrons by timing their travel along a conductor?

9. Why are the dashed white lines in Fig. 27–7 curved slightly?

10. A current i enters the top of a copper sphere of radius R and leaves at a diametrically opposite point. Are all parts of the sphere equally effective from the point of view of the Joule effect?

11. What special characteristics must (a) heating wire and (b) fuse wire have?

12. Equation 27–14 ($P = i^2R$) seems to suggest that the Joule effect in a resistor is reduced if the resistance is made less; Eq. 27–15 ($P = V^2/R$) seems to suggest just the opposite. How do you reconcile this apparent paradox?

13. Is the filament resistance lower or higher in a 500-watt light bulb than in a 100-watt bulb? Both bulbs are designed to operate on 110 volts.

14. Five wires of the same length and diameter are connected in turn between two points maintained at constant potential difference. Will the Joule effect be largest in the wire of the smallest or the largest resistance?

PROBLEMS

1(1). A current of 5.0 amp exists in a 10-ohm resistor for 4.0 min. (a) How many coulombs and (b) how many electrons pass through any cross section of the resistor in this time?
Answer: (a) 1.2×10^3 coul. (b) 7.5×10^{21} electrons.

2(1). A current is established in a gas discharge tube when a sufficiently high potential difference is applied across the two electrodes in the tube. The gas ionizes; electrons move toward the positive terminal and positive ions toward the negative terminal. What are the magnitude and sense of the current in a hydrogen discharge tube in which 3.1×10^{18} electrons and 1.1×10^{18} protons move past a cross-sectional area of the tube each second?

Copper

Carbon

Copper

FIGURE 27–9

3(1). The belt of an electrostatic generator is 50 cm wide and travels at 30 meters/sec. The belt carries charge into the sphere at a rate corresponding to 10^{-4} amp. What is the surface charge density on the belt?

Answer: 6.7×10^{-6} coul/meter^2.

4(1). A steady beam of alpha particles ($q = 2e$) traveling with constant kinetic energy 20 MeV carries a current 0.25×10^{-6} ampere. (*a*) If the beam is directed perpendicular to a plane surface, how many alpha particles strike the surface in 3.0 sec? (*b*) At any instant, how many alpha particles are there in a given 20-cm length of the beam? (*c*) Through what potential difference was it necessary to accelerate each alpha particle from rest to bring it to an energy of 20 MeV?

5(1). A small but measurable current of 1.0×10^{-10} amp exists in a copper wire whose diameter is 0.10 in. Calculate the electron drift speed.

Answer: 1.5×10^{-15} meters/sec.

6(1). The National Electric Code, which sets maximum safe currents for rubber insulated copper wires of various diameters, is given below. (*a*) Plot safe current density and drift speed as a function of diameter. (*b*) What wire gauge would you choose to obtain the maximum drift speed? (*c*) What is the ratio of your maximum drift speed to the speed of light (3.00×10^8 meters/sec)?

Gauge[1]	4	6	8	10	12	14	16	18
Diameter (mils)[2]	204	162	129	102	81	64	51	40
Safe current (amp)	70	50	35	25	20	15	6	3

[1] Just a way of identifying the wire.
[2] 1 mil = 10^{-3} in.

7(2). A rectangular carbon block has dimensions 1.0 cm $\times$ 1.0 cm $\times$ 50 cm. (*a*) What is the resistance measured between the two square ends? (*b*) Between two opposing rectangular faces? The resistivity of carbon at 20° C is 3.5×10^{-5} ohm-meter.

Answer: (*a*) 0.18 ohms. (*b*) 7.0×10^{-5} ohms.

8(2). Steel trolley-car rail has a cross-sectional area of 7.1 in.^{2} What is the resistance of 10 miles of single track? The resistivity of the steel is 6.0×10^{-7} ohm-meter.

9(2). A rod of a certain metal is 1.00 meter long and 0.550 cm in diameter. The resistance between its ends (at 20° C) is 2.87×10^{-3} ohm. (*a*) What is the metal? (*b*) A round disk is formed of this same material, 2.00 cm in diameter and 1.00 mm thick. What is the resistance between the opposing round faces?

Answer: (*a*) Nickel; see Table 27–1. (*b*) 2.2×10^{-7} ohms.

10(2). A coil is made by winding 15,000 turns of gauge 28 copper wire (diameter = 0.013 in.) at an average radius of 1.6 in. What is the resistance of the coil?

11(2). A square aluminum rod is 1.0 meter long and 5.0 mm on edge. (*a*) What is the resistance between its ends? (*b*) What must be the diameter of a circular 1.0-meter copper wire if its resistance is to be the same?

Answer: (*a*) 1.1×10^{-3} ohms. (*b*) 4.4 mm. 2,2 ?.

12(2). A wire with a resistance of 6.0 ohms is drawn out so that its new length is three times its original length. Find the resistance of the longer wire, assuming that the resistivity and density of the material are not changed during the drawing process.

13(2). A wire of length L (4.0 meters) and diameter D (6.0 mm) has resistance R (20 ohms). If a potential difference V (30 volts) is applied between its ends, (*a*) What is the current in the wire? (*b*) What is the current density? (*c*) What is the resistivity of the wire?

Answer: (*a*) 1.5 amp. (*b*) 5.3×10^4 amp/meter^2. (*c*) 1.4×10^{-4} ohm-meter.

14(2). A 6.0-volt battery is connected across the length of a square iron rod, 4.0 mm on a side and 0.50 meter long. Find (*a*) the resistance, (*b*) the current, (*c*) the current density, and (*d*) the electric field.

15(2). Wires A and B, each 40 meters long and 0.10 meter^2 in cross-sectional area, are connected in series. A potential of 60 volts is applied between the ends of the connected wires. The resistances of the wires are 40 ohms and 20 ohms, respectively. Find (*a*) the resistivities of the two wires, (*b*) the potential difference across each wire, (*c*) the magnitude of the electric field in each wire, and (*d*) the current density in each wire.

Answer: (*a*) $\rho_A = 0.10$ ohm-meter; $\rho_B = 0.050$ ohm-meter. (*b*) $V_A = 40$ volts; $V_B = 20$ volts. (*c*) $E_A = 1.0$ volt/meter; $E_B = 0.50$ volt/meter; (*d*) $j_A = j_B = 10$ amp/meter^2.

16(2). A copper wire and an iron wire of equal length *l* (10 meters) and diameter *d* (2.0 mm) are joined and a potential difference V (100 volts) is applied between the ends of the composite wire. Find (*a*) the potential difference across each wire, (*b*) the magnitude of the electric field in each wire, and (*c*) the current density in each wire.

17(2). A copper wire and an iron wire of the same length have the same potential difference applied to them. (*a*) What must be the ratio of their radii if the current is to be the same? (*b*) Can the current density be made the same by suitable choices of the radii?

Answer: (*a*) 2.4, iron being larger. (*b*) No.

18(2). Will a copper wire or an aluminum wire give the larger current to weight ratio for high voltage transmission lines if both wires are to have the same resistance between any two points?

19(2). (*a*) At what temperature would the resistance of a copper conductor be double its resistance at 0° C? (*b*) Does this same temperature hold for all copper conductors, regardless of shape or size? See Table 27-1.

Answer: (*a*) 260° C. (*b*) Yes.

20(2). It is desired to make a long cylindrical conductor whose temperature coefficient of resistance at 20° C will be close to zero. If such a conductor is made by assembling alternate disks of iron and carbon, what is the ratio of the thickness of a carbon disk to that of an iron disk? Assume that the temperature remains essentially the same in each disk. See Table 27-1.

21(2). A resistor is in the shape of a truncated right circular cone (Fig. 27-10). The end radii are *a* and *b*, the altitude is *l*. If the taper is small, we may assume that the current density is uniform across any cross section. (*a*) Calculate the resistance of this object. (*b*) Show that your answer reduces to $\rho(l/A)$ for the special case of zero taper (*a* = *b*). (Hint: Construct an equation that expresses the radius of a cross section at any altitude as a function of the altitude. Then integrate over the altitude.)

Answer: (*a*) $\rho l/\pi ab$.

22(3). (*a*) Using data from Fig. 27-5, plot the resistance of the vacuum tube as a function of applied potential difference. (*b*) Repeat for the thermistor of Fig. 27-6.

23(5). Heat is developed in a resistor at a rate of 100 watts when the current is 3.0 amp. What is the resistance?

Answer: 11 ohms.

24(5). If the resistance of a battery is 25 ohms, what is the current when it delivers a power of 75 watts?

25(5). A 1250-watt radiant heater is constructed to operate at 115 volts. (*a*) What is the current in the heater? (*b*) What is the resistance of the heating coil? (*c*) How many calories are radiated in one hour by the heater?

Answer: (*a*) 10.9 amp. (*b*) 10.6 ohms. (*c*) 1.08×10^6.

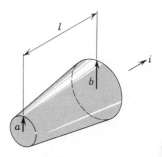

FIGURE 27-10 Problem 21(2).

26(5). A potential difference of 1.0 volt is applied to a 100-ft length of gauge 18 copper wire [see Problem 6(1)]. Calculate (a) the current, (b) the current density, (c) the electric field, and (d) the rate at which internal energy appears by the Joule effect.

27(5). For gauge 10 rubber-coated copper wire [see Problem 6(1)] the maximum safe current is 25 amp. At this current, find (a) the current density, (b) the electric field, (c) the potential difference for 1000 ft of wire, and (d) the power associated with the Joule effect for 1000 ft of wire. *Answer:* (a) 4.9×10^6 amp/meter². (b) 8.3×10^{-2} volt/meter. (c) 25 volts. (d) 630 watts.

28(5). A wire carries a current i (4.0 amps) when a potential V (60 volts) is applied to it. (a) What is the resistance of the wire? (b) How much electrical energy is converted into heat when operating for time t (3.0 sec)? (c) How many electrons pass through a cross section of the wire in that time?

29(5). (a) Derive the formulas $P = j^2\rho$ and $P = E^2/\rho$ where $P =$ power per unit volume in a resistor. (b) A cylindrical resistor of radius 0.50 cm and length 2.0 cm has a resistivity of 3.5×10^{-5} ohm-meter. What are the current density and potential difference when the power dissipation is 1.0 watt?
Answer: (b) $j = 1.3 \times 10^5$ amp/meter²; $V = 0.094$ volts.

30(5). A beam of 16-MeV deuterons (charge $= +1.60 \times 10^{-19}$ coul) from a cyclotron falls on a copper block. The beam is equivalent to a current of 1.5×10^{-5} amp. (a) At what rate do deuterons strike the block? (b) At what rate is internal energy produced in the block by the Joule effect?

31(5). A "500-watt" heating unit is designed to operate from a 115-volt line. (a) By what percentage will its heat output drop if the line voltage drops to 110 volts? Assume no change in resistance. (b) Taking the variation of resistance with temperature into account, would the actual heat output drop be larger or smaller than that calculated in (a)?
Answer: (a) 8.6%. (b) Smaller.

32(5). A 500-watt immersion heater is placed in a pot containing 2.0 liters of water at 20° C. (a) How long will it take to bring the water to boiling temperature, assuming that 80% of the available energy is absorbed by the water? (b) How much longer will it take to boil half the water away?

33(5). An electron linear accelerator produces a pulsed beam of electrons. The pulse current is 0.50 amp and the pulse duration 0.10 μsec. (a) How many electrons are accelerated per pulse? (b) What is the average current for a machine operating at 500 pulses/sec? (c) If the electrons are accelerated to an energy of 50 MeV, what are the average and peak power outputs of the accelerator?
Answer: (a) 3.1×10^{11}. (b) 25 μamp. (c) 1200 watts (average); 2.5×10^7 watts (peak).

CHAPTER 28

Electromotive Force and Circuits

28–1 Electromotive Force

There exist in nature certain devices such as batteries and electric generators which are able to maintain a potential difference between two points to which they are attached. Such devices are called seats of *electromotive force* (abbr. emf). In this chapter we do not discuss their construction or detailed mode of action but confine ourselves to describing their gross electrical characteristics and to exploring their usefulness in electric circuits.

Figure 28–1a shows a seat of emf B, represented by a battery, connected to a resistor R. The seat of emf maintains its upper terminal positive and its lower terminal negative. In the circuit external to B positive charge carriers would be driven in the direction shown by the arrows marked i. In other words, a clockwise current would be set up.

An emf is represented by an arrow which is placed next to the seat and points in the direction in which the seat, acting alone, would cause a positive charge carrier to move in the external circuit. A small circle is drawn on the tail of an emf arrow so that it will not be confused with a current arrow.

A seat of emf must be able to do work on charge carriers that enter it. In the circuit of Fig. 28–1a, for example, the seat acts to move positive charges from a point of low potential (the negative terminal) through the seat to a point of high potential (the positive terminal). This reminds us of a pump, which can cause water to move from a place of low gravitational potential to a place of high potential.

In Fig. 28–1a a charge dq passes through any cross section of the circuit in time dt. In particular, this charge enters the seat of emf $\mathcal{E}$ at its low-potential end and leaves

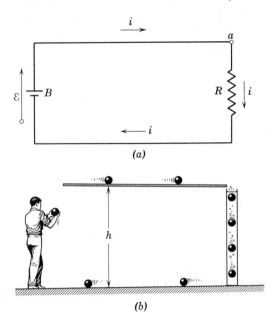

FIGURE 28-1   (*a*) A simple electric circuit and (*b*) its gravitational analog.

at its high-potential end. The seat must do an amount of work dW on the (positive) charge carriers to force them to go to the point of higher potential. The emf ε of the seat is defined from

$$\varepsilon = dW/dq. \qquad \text{Joule/coul} = \text{volt} \qquad (28\text{-}1)$$

The unit of emf is the joule/coul (see Eq. 25-1) which is the *volt*. We might be inclined to say that a battery has an emf of 1 volt if it maintains a difference of potential of 1 volt between its terminals. This is true only under certain conditions, which we describe in Section 28-4.

If a seat of emf does work on a charge carrier, energy must be transferred within the seat. In a battery, for example, chemical energy is transferred into electrical energy. Thus we can describe a seat of emf as a device in which chemical, mechanical, or some other form of energy is changed (reversibly) into electrical energy. The chemical energy provided by the battery in Fig. 28-1*a* is stored in the electric and the magnetic fields that surround the circuit. This stored energy does not increase because it is being drained away, by transfer to internal energy in the resistor, at the same rate at which it is supplied. The electric and magnetic fields play an intermediary role in the energy transfer process, acting as a storage reservoir.

Figure 28-1*b* shows a gravitational analog of Fig. 28-1*a*. In the top figure the seat of emf B does work on the charge carriers. This energy, stored temporarily as electromagnetic field energy, appears eventually as Joule energy in resistor R. In the lower figure the man, in lifting the bowling balls from the floor to the shelf, does work on them. This energy is stored temporarily as gravitational field energy. The balls roll slowly and uniformly along the shelf, dropping from the right end into a cylinder full of viscous oil. They sink to the bottom at constant speed, are removed by a trapdoor mechanism not shown, and roll back along the floor to the left. The energy put into the system by the man appears eventually as internal energy in the viscous fluid, causing a temperature rise. The energy supplied by the man comes from his own internal (chemical) energy. The circulation of charges in Fig. 28-1*a* will stop eventually if battery B is not charged; the circulation of bowling balls in Fig. 28–1b will stop eventually if the man does not replenish his store of internal energy by eating.

Figure 28–2 shows a circuit containing two (ideal) batteries, A and B, a resistor R, and an (ideal) electric motor employed in lifting a weight. The batteries are connected so that they tend to send charges around the circuit in opposite directions; the actual direction of the current is determined by B, which supplies the larger potential difference. The chemical energy in B is being steadily depleted, the energy appearing as Joule energy in the resistor, work done by the motor, and as chemical energy stored in A. Battery A is being "charged" while battery B is being discharged. Again, the electric and magnetic fields that surround the circuit act as an intermediary.

28–2 Calculating the Current

In a time dt an amount of energy given by $i^2R\,dt$ will appear in the resistor of Fig. 28–1a as Joule energy. During this same time a charge $dq\,(= i\,dt)$ will have moved through the seat of emf, and the seat will have done work on this charge (see Eq. 28–1) given by

$$dW = \mathcal{E}dq = \mathcal{E}i\,dt.$$

From the conservation of energy principle, the work done by the seat must equal the Joule energy, or

$$\mathcal{E}i\,dt = i^2R\,dt.$$

Solving for i, we obtain

$$i = \mathcal{E}/R. \qquad \text{or} \quad \mathcal{E} = iR \qquad\qquad (28\text{–}2)$$

We can also derive Eq. 28–2 by considering that if electric potential is to have any meaning a given point can have only one value of potential at any given time. If we start at any point in the circuit of Fig. 28–1a and, in imagination, go around the circuit in either direction, adding up algebraically the changes in potential that we encounter, we must arrive at the same potential when we return to our starting point. In other words, the *algebraic sum of the changes in potential encountered in a complete traversal of the circuit must be zero.*   LOOP THM

In Fig. 28–1a let us start at point a, whose potential is V_a, and traverse the circuit clockwise. Because we are interested only in changes in potential, the actual value of V_a is arbitrary. In going through the resistor, there is a change in potential of $-iR$. The minus sign shows that the top of the resistor is higher in potential than the bottom, which must be true, because positive charge carriers move of their own accord from high to low potential. As we traverse the battery from bottom to top,

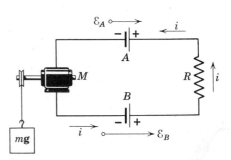

FIGURE 28–2 Two batteries, a resistor, and a motor, connected in a single-loop circuit. It is given that $\mathcal{E}_B > \mathcal{E}_A$.

there is an increase of potential $+\mathcal{E}$ because the battery does (positive) work on the charge carriers, that is, it moves them from a point of low potential to one of high potential. Adding the algebraic sum of the changes in potential to the initial potential V_a must yield the identical value V_a, or

$$V_a - iR + \mathcal{E} = V_a.$$

We write this as

$$-iR + \mathcal{E} = 0,$$

which is independent of the value of V_a and which asserts explicitly that the algebraic sum of the potential changes for a complete circuit traversal is zero. This relation leads directly to Eq. 28–2.

These two ways to find the current in single-loop circuits, based on the conservation of energy and on the concept of potential, are completely equivalent because potential differences are defined in terms of work and energy (Section 25–1). The statement that the sum of the changes in potential encountered in making a complete loop is zero is called the *loop theorem*. This theorem is simply a particular way of stating the law of conservation of energy for electric circuits.

To prepare for the study of more complex circuits, let us examine the rules for finding potential differences; these rules follow from the previous discussion. They are not meant to be memorized but rather to be so thoroughly understood that it becomes trivial to re-derive them on each application.

1. If a resistor is traversed in the direction of the current, the change in potential is $-iR$; in the opposite direction it is $+iR$.

2. If a seat of emf is traversed in the direction of the emf, the change in potential is $+\mathcal{E}$; in the opposite direction it is $-\mathcal{E}$.

28–3   Other Single-Loop Circuits

Figure 28–3a shows a circuit which emphasizes that all seats of emf have an intrinsic internal resistance r. This resistance cannot be removed—although we would usually like to do so—because it is an inherent part of the device. The figure shows the internal resistance r and the emf separately, although, actually, they occupy the same region of space.

If we apply the loop theorem, starting at b and going around clockwise, we obtain

$$V_b + \mathcal{E} - ir - iR = V_b$$

or

$$+\mathcal{E} - ir - iR = 0.$$

Compare this equation with Fig. 28–3b, which shows the changes in potential graphically. In writing this equation, note that we traversed r and R in the direction of the current and $\mathcal{E}$ in the direction of the emf. The same equation follows if we start at any other point in the circuit or if we traverse the circuit in a counterclockwise direction. Solving for i gives

$$i = \frac{\mathcal{E}}{R + r}. \tag{28–3}$$

■ **Example 1.** *Resistors in series.* Resistors in series are connected so that there is only one conducting path through them, as in Fig. 28–4. What is the equivalent resistance R of this series com-

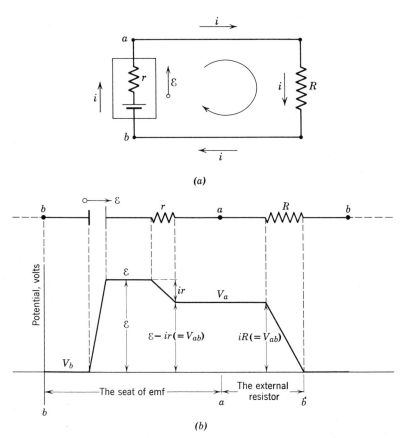

FIGURE 28–3   A single-loop circuit. The rectangular block is a seat of emf with internal resistance *r*. (*b*) The same circuit is drawn for convenience as a straight line. Directly below are shown the changes in potential that one encounters in traversing the circuit clockwise from point *b*.

bination? The equivalent resistance is the single resistance R which, substituted for the series combination between the terminals ab, will leave the current i unchanged.

Applying the loop theorem (going clockwise from a) yields

$$-iR_1 - iR_2 - iR_3 + \mathcal{E} = 0$$

or

$$i = \frac{\mathcal{E}}{R_1 + R_2 + R_3}.$$

For the equivalent resistance R

$$i = \frac{\mathcal{E}}{R}$$

or

$$R = R_1 + R_2 + R_3. \qquad (28\text{–}4)$$

The extension to more than three resistors is clear. ▤

28–4   Potential Differences

We often want to compute the potential difference between two points in a circuit. In Fig. 28–3*a* for example, what is the relationship between the potential difference $V_{ab}\,(= V_a - V_b)$ between points b and a and the fixed circuit

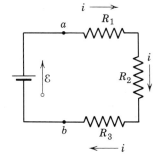

FIGURE 28–4   Example 1. Three resistors are connected in series between terminals *a* and *b*.

parameters $\mathcal{E}, r$, and R? To find this relationship, let us start from point b and traverse the circuit to point a, passing through resistor R against the current. If V_b and V_a are the potentials at b and a, respectively, we have

$$V_b + iR = V_a$$

because we experience an increase in potential in traversing a resistor against the current arrow. We rewrite this relation as

$$V_{ab} = V_a - V_b = +iR,$$

which tells us that V_{ab}, the potential difference between points a and b, has the magnitude iR and that point a is at a higher potential than point b. Combining this last equation with Eq. 28–3 yields

$$V_{ab} = \mathcal{E}\,\frac{R}{R+r}. \qquad\qquad 28\text{-}3 \qquad\qquad (28\text{–}5)$$
$$\text{TIMES } R$$

To sum up: To find the potential difference between any two points in a circuit start at one point and traverse the circuit to the other, following any path, and add up algebraically the potential changes encountered. This algebraic sum will be the potential difference. This procedure is similar to that for finding the current in a closed loop, except that here the potential differences are added up over part of a loop and not over the whole loop.

The potential difference between any two points can have only one value and we must obtain the same answer for all paths that connect these points. If we consider two points on the side of a hill, the measured difference in gravitational potential (that is, in altitude) between them is the same no matter what path is followed in going from one to the other. In Fig. 28–3a let us calculate V_{ab}, using a path passing through the seat of emf. We have

$$V_b + \mathcal{E} - ir = V_a$$

or (see also Fig. 28–3b)

$$V_{ab} = V_a - V_b = +\mathcal{E} - ir.$$

Again, combining with Eq. 28–3 leads to Eq. 28–5.

The terminal potential difference of the battery V_{ab}, as Eq. 28–5 shows, is less than $\mathcal{E}$ unless the battery has no internal resistance ($r = 0$) or if it is on open circuit ($R = \infty$); then V_{ab} is equal to $\mathcal{E}$. Thus the emf of a device is equal to its terminal potential difference when on open circuit.

Example 2. In Fig. 28–5a let $\mathcal{E}_1$ and $\mathcal{E}_2$ be 2.0 volts and 4.0 volts, respectively; let the resistances r_1, r_2, and R be 1.0 ohm, 2.0 ohms, and 5.0 ohms, respectively. What is the current?

Emfs $\mathcal{E}_1$ and $\mathcal{E}_2$ oppose each other, but because $\mathcal{E}_2$ is larger it controls the direction of the current. Thus i will be counterclockwise. The loop theorem, going clockwise from a, yields

$$-\mathcal{E}_2 + ir_2 + iR + ir_1 + \mathcal{E}_1 = 0.$$

Check that the same result is obtained by going around counterclockwise and compare this equation carefully with Fig. 28–5b, which shows the potential changes graphically.

Solving for i yields

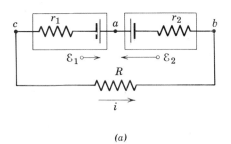

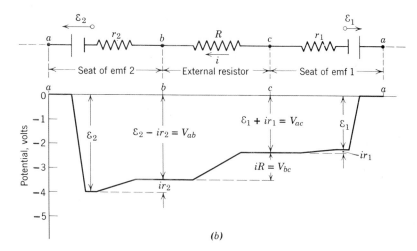

(a)

(b)

FIGURE 28–5   (*a*) A single-loop circuit. (*b*) The same circuit is shown schematically as a straight line, the potential differences encountered in traversing the circuit clockwise from point *a* being displayed directly below. In the lower figure the potential of point *a* was assumed to be zero for convenience.

$$i = \frac{\mathcal{E}_2 - \mathcal{E}_1}{R + r_1 + r_2} = \frac{4.0 \text{ volts} - 2.0 \text{ volts}}{5.0 \text{ ohms} + 1.0 \text{ ohm} + 2.0 \text{ ohms}}$$

$$= 0.25 \text{ amp.}$$

It is not necessary to know in advance what the actual direction of the current is. To show this, let us assume that the current in Fig. 28–5*a* is clockwise, an assumption that we know is incorrect. The loop theorem then yields (going clockwise from *a*)

$$-\mathcal{E}_2 - ir_2 - iR - ir_1 + \mathcal{E}_1 = 0$$

or

$$i = \frac{\mathcal{E}_1 - \mathcal{E}_2}{R + r_1 + r_2}.$$

Substituting numerical values (see above) yields -0.25 amp for the current. The minus sign tells us that the current is in the opposite direction from the one we have assumed.

In more complex circuit problems involving many loops and branches it is often impossible to know in advance the correct directions for the currents in all parts of the circuit. We can assume directions for the currents at random. Those currents for which positive numerical values are obtained will have the correct directions; those for which negative values are obtained will be exactly opposite to the assumed directions. In all cases the numerical values will be correct.

Example 3. What is the potential difference (*a*) between points *b* and *a* in Fig. 28–5*a*? (*b*) Between points *c* and *a*?

(*a*) For points *a* and *b* we start at *b* and traverse the circuit to *a*, obtaining

$$V_{ab}\,(= V_a - V_b) = -ir_2 + \mathcal{E}_2 = -(0.25 \text{ amp})(2.0 \text{ ohms}) + 4.0 \text{ volts}$$
$$= +3.5 \text{ volts.}$$

Thus *a* is more positive than *b* and the potential difference (3.5 volts) is less than the emf (4.0 volts); see Fig. 28–5*b*.

(*b*) For points *c* and *a*, we start at *c* and traverse the circuit to *a*, obtaining

$$V_{ac}\,(= V_a - V_c) = +\mathcal{E}_1 + ir_1 = +2.0 \text{ volts} + (0.25 \text{ amp})(1.0 \text{ ohm})$$
$$= +2.25 \text{ volts.}$$

This tells us that *a* is at a higher potential than *c*. The terminal potential difference of $\mathcal{E}_1$ (2.25 volts) is larger than the emf (2.0 volts); see Fig. 28–5*b*. Charge is being forced through $\mathcal{E}_1$ in a direction opposite to the one in which it would send charge if it were acting by itself; if $\mathcal{E}_1$ is a storage battery, it is being charged at the expense of $\mathcal{E}_2$.

Let us test the first result by proceeding from *b* to *a* along a different path, namely, through *R*, r_1, and $\mathcal{E}_1$. We have

$$V_{ab} = iR + ir_1 + \mathcal{E}_1 = (0.25 \text{ amp})(5.0 \text{ ohms})$$
$$+ (0.25 \text{ amp})(1.0 \text{ ohm}) + 2.0 \text{ volts} = +3.5 \text{ volts,}$$

which is the same as the earlier result.

28–5 Multiloop Circuits

Figure 28–6 shows a circuit containing two loops. For simplicity, we have neglected the internal resistances of the batteries. There are two *junctions, b* and *d*, and three *branches* connecting these junctions. The branches are the left branch *bad*, the right branch *bcd*, and the central branch *bd*. If the emfs and the resistances are given, what are the currents in the various branches?

We label the currents in the branches as i_1, i_2, and i_3, as shown. Current i_1 has the same value for any cross section of the left branch from *b* to *d*. Similarly, i_2 has the same value everywhere in the right branch and i_3 in the central branch. The directions of the currents have been chosen arbitrarily. Perhaps you have noted that i_3 must point in a direction opposite to the one we have shown. We have deliberately drawn it in wrong to show how the formal mathematical procedures will always indicate this to us.

The three currents i_1, i_2, and i_3 carry charge either toward junction *d* or away from it. Charge does not accumulate at junction *d*, nor does it drain away from this junction because the circuit is in a steady-state condition. Thus charge must be removed from the junction by the currents at the same rate that it is brought into it. If we arbitrarily call a current approaching the junction positive and the one leaving the junction negative, then

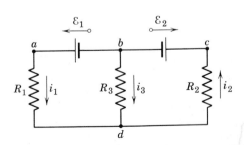

FIGURE 28–6   A multiloop circuit.

$$i_1 + i_3 - i_2 = 0.$$

This equation suggests a general principle, called the *junction theorem*, for the solution of multiloop circuits: *At any junction the algebraic sum of the currents must be zero.* This theorem is simply a statement of the conservation of charge. Thus our basic tools for solving circuits are (*a*) the conservation of energy (see Section 28–2) and (*b*) the conservation of charge.

For the circuit of Fig. 28–6, the junction theorem yields only one relationship among the three unknowns. Applying the theorem at junction *b* leads to exactly the same equation. To solve for the three unknowns, we need two more independent equations; they can be found from the loop theorem.

In single-loop circuits there is only one conducting loop around which to apply the loop theorem, and the current is the same in all parts of this loop. In multiloop circuits there is more than one loop, and the current in general will not be the same in all parts of any given loop.

If we traverse the left loop of Fig. 28–6 in a counterclockwise direction, the loop theorem gives

$$\mathcal{E}_1 - i_1 R_1 + i_3 R_3 = 0. \tag{28–6}$$

The right loop gives

$$-i_3 R_3 - i_2 R_2 - \mathcal{E}_2 = 0. \tag{28–7}$$

These two equations, together with the relation derived earlier with the junction theorem, are the three simultaneous equations needed to solve for the unknowns i_1, i_2, i_3. Doing so yields

$$i_1 = \frac{\mathcal{E}_1(R_2 + R_3) - \mathcal{E}_2 R_3}{R_1 R_2 + R_2 R_3 + R_1 R_3}, \tag{28–8a}$$

$$i_2 = \frac{\mathcal{E}_1 R_3 - \mathcal{E}_2(R_1 + R_3)}{R_1 R_2 + R_2 R_3 + R_1 R_3}, \tag{28–8b}$$

and

$$i_3 = \frac{-\mathcal{E}_1 R_2 - \mathcal{E}_2 R_1}{R_1 R_2 + R_2 R_3 + R_1 R_3}. \tag{28–8c}$$

Equation 28–8*c* shows that no matter what numerical values are given to the emfs and to the resistances the current i_3 will always have a negative value. This means that it will always point up in Fig. 28–6 rather than down, as we assumed. The currents i_1 and i_2 may be in either direction, depending on the particular numerical values given.

Verify that Eqs. 28–8 reduce to sensible conclusions in special cases. For $R_3 = \infty$, for example, we find

$$i_1 = i_2 = \frac{\mathcal{E}_1 - \mathcal{E}_2}{R_1 + R_2} \quad \text{and} \quad i_3 = 0.$$

What do these equations reduce to for $R_2 = \infty$?

The loop theorem can be applied to a large loop consisting of the entire circuit *abcda* of Fig. 28–6. This fact might suggest that there are more equations than we need, for there are only three unknowns and we already have three equations written in terms of them. However, the loop theorem yields for this loop

$$-i_1 R_1 - i_2 R_2 - \mathcal{E}_2 + \mathcal{E}_1 = 0,$$

which is nothing more than the sum of Eqs. 28–6 and 28–7. Thus this large loop does not yield another independent equation. It will never be found in solving multiloop circuits that there are more independent equations than variables.

Example 4. *Resistors in parallel.* Figure 28–7 shows three resistors connected across the same seat of emf. Resistances across which the identical potential difference is applied are said to be in parallel. What is the equivalent resistance R of this parallel combination? The equivalent resistance is that single resistance which, substituted for the parallel combination between terminals ab, would leave the current i unchanged.

The currents in the three branches are

$$i_1 = \frac{V}{R_1}, \qquad i_2 = \frac{V}{R_2}, \qquad \text{and} \qquad i_3 = \frac{V}{R_3},$$

where V is the potential difference that appears between points a and b. The total current i is found by applying the junction theorem to an extended junction consisting of the uppermost horizontal wire in Fig. 28–7, or

$$i = i_1 + i_2 + i_3 = V\left(\frac{1}{R_1} + \frac{1}{R_2} + \frac{1}{R_3}\right).$$

If the equivalent resistance is used instead of the parallel combination, we have

$$i = \frac{V}{R}.$$

Combining these two equations gives

$$\frac{1}{R} = \frac{1}{R_1} + \frac{1}{R_2} + \frac{1}{R_3}. \tag{28–9}$$

This formula can easily be extended to more than three resistances. Note that the equivalent resistance of a parallel combination is less than any of the resistances that make it up.

28–6   RC Circuits

The preceding sections dealt with circuits in which the circuit elements were resistors and in which the currents did not vary with time. Here we introduce the capacitor as a circuit element, which will lead us to the concept of time-varying currents. In Fig. 28–8 let switch S be thrown to position a. What current is set up in the single-loop circuit so formed? Let us apply conservation of energy principles.

In time dt a charge $dq \, (= i \, dt)$ moves through any cross section of the circuit. The work done by the seat of emf $(= \mathcal{E} \, dq$; see Eq. 28–1) must equal the energy that appears from the Joule effect in the resistor during time $dt \, (= i^2 R \, dt)$ plus the increase in the amount of energy U that is stored in the capacitor $[= dU = d(q^2/2C)$; see Eq. 26–16]. In equation form

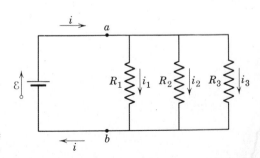

FIGURE 28–7   Example 4. Three resistors are connected in parallel between terminals a and b.

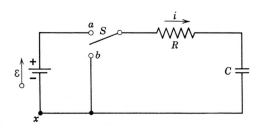

FIGURE 28–8 An *RC* circuit.

$$\mathcal{E} \, dq = i^2R \, dt + d\!\left(\frac{q^2}{2C}\right)$$

or

$$\mathcal{E} \, dq = i^2R \, dt + \frac{q}{C} \, dq.$$

Dividing by *dt* yields

$$\mathcal{E} \, \frac{dq}{dt} = i^2R + \frac{q}{C} \frac{dq}{dt}.$$

But *dq/dt* is simply *i*, so that this equation becomes

$$\mathcal{E} = iR + \frac{q}{C}. \tag{28–10}$$

This equation also follows from the loop theorem, as it must, since the loop theorem was derived from the conservation of energy principle. Starting from point *x* and traversing the circuit clockwise, we experience an increase in potential in going through the seat of emf and decreases in potential in traversing the resistor and the capacitor, or

BY DEF $C = \frac{q}{v}$

$$\mathcal{E} - iR - \frac{q}{C} = 0,$$

which is identical with Eq. 28–10.

We cannot immediately solve Eq. 28–10 because it contains two variables, *q* and *i*, which, however, are related by

$$i = \frac{dq}{dt}. \tag{28–11}$$

Substituting for *i* into Eq. 28–10 gives

$$\mathcal{E} = R \frac{dq}{dt} + \frac{q}{C}. \tag{28–12}$$

Our task now is to find the function *q(t)* that satisfies this differential equation. Although this particular equation is not difficult to solve, we choose to avoid mathematical complexity by simply presenting the solution, which is

$$q = C\mathcal{E}(1 - e^{-t/RC}). \tag{28–13}$$

We can easily test whether this function *q(t)* is really a solution of Eq. 28–12 by substituting it into that equation and seeing whether an identity results. Differentiating Eq. 28–13 with respect to time yields

$$\frac{dq}{dt} (= i) = \frac{\mathcal{E}}{R} e^{-t/RC}. \tag{28–14}$$

Substituting q (Eq. 28–13) and dq/dt (Eq. 28–14) into Eq. 28–12 yields an identity, as you should verify. Thus Eq. 28–13 is a solution of Eq. 28–12.

Figure 28–9 shows some plots of Eqs. 28–13 and 28–14 for a particular case. Study of these plots and of the corresponding equations shows that (*a*) at $t = 0$, $q = 0$ and $i = \mathcal{E}/R$, and (*b*) as $t \to \infty$, $q \to C\mathcal{E}$ and $i \to 0$: that is, the current is initially $\mathcal{E}/R$ and finally zero; the charge on the capacitor plates is initially zero and finally $C\mathcal{E}$.

The quantity RC in Eqs. 28–13 and 28–14 has the dimensions of time (since the exponent must be dimensionless) and is called the *capacitative time constant* of the circuit. It is the time at which the charge on the capacitor has increased to within a factor of $(1 - e^{-1})$ $(= 63\%)$ of its equilibrium value. To show this, we put $t = RC$ in Eq. 28–13 to obtain

$$q = C\mathcal{E}(1 - e^{-1}) = 0.63C\mathcal{E}.$$

Since $C\mathcal{E}$ is the equilibrium charge on the capacitor, corresponding to $t \to \infty$, the foregoing statement follows.

Example 5. After how many time constants will the energy stored in the capacitor in Fig. 28–8 reach one-half its equilibrium value?

The energy is given by Eq. 26–16, or

$$U = \frac{1}{2C}q^2,$$

the equilibrium energy U_∞ being $(1/2C)(C\mathcal{E})^2$. From Eq. 28–13, we can write the energy as

$$U = \frac{1}{2C}(C\mathcal{E})^2(1 - e^{-t/RC})^2$$

or
$$U = U_\infty(1 - e^{-t/RC})^2.$$

Putting $U = \frac{1}{2}U_\infty$ yields

$$\tfrac{1}{2} = (1 - e^{-t/RC})^2$$

and solving this relation for t yields finally

$$t = 1.22RC = 1.22 \text{ time constants.}$$

Figure 28–9 shows that if a resistance is included in the circuit the rate of increase of the charge of a capacitor toward its final equilibrium value is delayed in a way measured by the time constant RC. With no resistor present ($RC = 0$), the charge

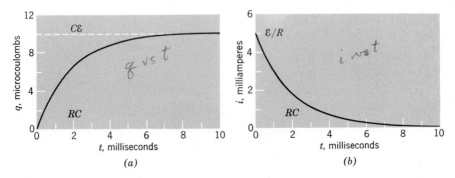

FIGURE 28–9   If, in Fig. 28–8, we assume that $R = 2000$ ohms, $C = 1.0$ μf, and $\mathcal{E} = 10$ volts, then (*a*) shows the variation of q with t during the charging process and (*b*) the variation of i with t. The time constant is $RC = 2.0 \times 10^{-3}$ sec.

would rise immediately to its equilibrium value. Although we have shown that this time delay follows from an application of the loop theorem to RC circuits, it is important to understand the causes of the delay.

When switch S in Fig. 28–8 is closed on a, the resistor experiences instantaneously an applied potential difference of $\mathcal{E}$, and an initial current of $\mathcal{E}/R$ is set up. Initially, the capacitor experiences no potential difference because its initial charge is zero, the potential difference always being given by q/C. The flow of charge through the resistor starts to charge the capacitor, which has several effects. First, the existence of a capacitor charge means that there must be a potential difference ($= q/C$) across the capacitor; this, in turn, means that the potential difference across the resistor must decrease by this amount, since the sum of the two potential differences must always equal $\mathcal{E}$. This decrease in the potential difference across R means that the charging current is reduced. Thus the charge of the capacitor builds up and the charging current decreases until the capacitor is fully charged. At this point the full emf $\mathcal{E}$ is applied to the capacitor, there being no potential drop ($i = 0$) across the resistor. This is precisely the reverse of the initial situation.

Assume now that the switch S in Fig. 28–8 has been in position a for a time t such that $t \gg RC$. The capacitor is then fully charged for all practical purposes. The switch S is then thrown to position b. How do the charge of the capacitor and the current vary with time?

With the switch S closed on b, there is no emf in the circuit and Eq. 28–10 for the circuit, with $\mathcal{E} = 0$, becomes simply

$$iR + \frac{q}{C} = 0. \tag{28–15}$$

Putting $i = dq/dt$ allows us to write, as the differential equation of the circuit (compare Eq. 28–12),

$$R\frac{dq}{dt} + \frac{q}{C} = 0. \tag{28–16a}$$

The solution, as you can readily verify by substitution, is

$$q = q_0 e^{-t/RC}, \tag{28–16b}$$

q_0 being the initial charge on the capacitor. The capacitative time constant RC appears in this expression for capacitor discharge as well as in that for the charging process (Eq. 28–13). We see that at a time such that $t = RC$ the capacitor charge is reduced to $q_0 e^{-1}$, which is 37% of the initial charge q_0.

The current during discharge follows from differentiating Eq. 28–16b, or

$$i = \frac{dq}{dt} = -\frac{q_0}{RC} e^{-t/RC}. \tag{28–17}$$

The negative sign shows that the current is in the direction opposite to that shown in Fig. 28–8. This is as it should be, since the capacitor is discharging rather than charging. Since $q_0 = C\mathcal{E}$, we can write Eq. 28–17 as

$$i = -\frac{\mathcal{E}}{R} e^{-t/RC},$$

in which $\mathcal{E}/R$ appears as the initial current, corresponding to $t = 0$. This is reasonable because the initial potential difference for the fully charged capacitor is $\mathcal{E}$.

QUESTIONS

1. Does the direction of the emf provided by a battery depend on the direction of current flow through the battery?

2. Discuss in detail the statement that the energy method and the loop theorem method for solving circuits are perfectly equivalent.

3. It is possible to generate a 10,000-volt potential difference by rubbing a pocket comb with wool. Why is this large voltage not dangerous when the much lower voltage provided by an ordinary electric outlet is very dangerous?

4. Devise a method for measuring the emf and the internal resistance of a battery.

5. A 25-watt, 110-volt bulb glows at normal brightness when connected across a bank of batteries. A 500-watt, 110-volt bulb glows only dimly when connected across the same bank. Explain.

6. Under what circumstances can the terminal potential difference of a battery exceed its emf?

7. What is the difference between an emf and a potential difference?

8. Compare and contrast the formulas for the equivalent values of (*a*) capacitors and (*b*) resistors, in series and in parallel.

9. Does the time required for the charge on a capacitor in an *RC* circuit to build up to a given fraction of its equilibrium value depend on the value of the applied emf?

10. Devise a method whereby an *RC* circuit can be used to measure very high resistances.

PROBLEMS

1(2). A 5.0-amp current is set up in an external circuit by a 6.0-volt storage battery for 6.0 min. By how much is the chemical energy of the battery reduced?
Answer: 1.1×10^4 joules.

2(2). A certain car battery (12 volt) carries an initial charge Q (120 amp-hr). Assuming that the potential across the terminals stays constant until the battery is completely discharged, for how many hours can it deliver power P (100 watts)?

3(2). The current in a simple series circuit is 5.0 amp. When an additional resistance of 2.0 ohms is inserted, the current drops to 4.0 amp. What was the resistance of the original circuit?
Answer: 8.0 ohms.

4(2). Two batteries having the same emf ε but different internal resistances r_1 and r_2 are connected in series to an external resistance R. Find the value of R that makes the potential difference zero between the terminals of the first battery.

5(3). In Fig. 28–3*a* put $\varepsilon = 2.0$ volts and $r = 100$ ohms. Plot (*a*) the current, and (*b*) the potential difference across R, as functions of R over the range 0 to 500 ohms. Make both plots on the same graph. (*c*) Make a third plot by multiplying together, for each value of R, the two curves plotted. What is the physical significance of this plot?

6(3). (*a*) In the circuit of Fig. 28–3*a* show that the power delivered to R by the Joule effect is a maximum when R is equal to the internal resistance r of the battery. (*b*) Show that this maximum power is $P = \varepsilon^2/4r$.

7(3). (*a*) In Fig. 28–10 what value must R have if the current in the circuit is to be 0.0010 amp? Take $\varepsilon_1 = 2.0$ volts, $\varepsilon_2 = 3.0$ volts, and $r_1 = r_2 = 3.0$ ohms. (*b*) What is the rate of generation of Joule energy in R?
Answer: (*a*) 990 ohms. (*b*) 9.4×10^{-4} watts.

8(3). A wire of resistance 5.0 ohms is connected to a battery whose emf ε is 2.0 volts and whose internal resistance is 1.0 ohm. In 2.0 min (*a*) how much energy is transferred from chemical to electric form? (*b*) How much energy appears in the wire from the Joule effect? (*c*) Account for the difference between (*a*) and (*b*).

9(3). Heat is to be generated in a 0.10-ohm resistor at the rate of 10 watts by connecting it to a

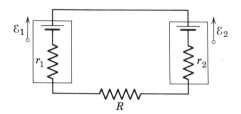

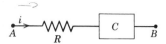

FIGURE 28-10 Problem 7(3).

FIGURE 28-11 Problem 11(4).

battery whose emf is 1.5 volts. (*a*) What is the internal resistance of the battery? (*b*) What potential difference exists across the resistor?
Answer: (*a*) 0.050 ohm. (*b*) 1.0 volt.

10(3). A light bulb designed to consume power P (100 watts) when connected to a source V (100 volts) is actually connected to a source V' (50 volts). What power does it consume?

11(4). A section of a circuit AB (see Fig. 28-11) absorbs power P (50 watts) and a current i (1.0 amp) passes through it in the direction shown. The resistance R is 2.0 ohms. (*a*) What is the potential difference between A and B? (*b*) What is the emf of element C, assuming that it has no internal resistance? (*c*) What is the polarity of C?
Answer: (*a*) 50 volts. (*b*) 48 volts. (*c*) B is the negative terminal.

12(4). In Example 2 an ammeter whose resistance is 0.050 ohm is inserted in the circuit. What percent change in the current results because of the presence of the meter?

13(4). In Fig. 28-5 calculate the potential difference between a and c by considering a path that contains R and $\mathcal{E}_2$. Compare your answer with that obtained in Example 3.
Answer: 2.25 volts.

14(4). A battery of emf $\mathcal{E}$ (2.0 volts) and internal resistance r (1.0 ohm) is driving a motor. The motor is lifting a weight W (2.0 nt) at constant speed v (0.50 meter/sec). Assuming no power losses, find (*a*) the current i in the circuit and (*b*) the voltage drop V across the terminals of the motor.

15(5). By using only two resistance coils—singly, in series, or in parallel—you are able to obtain resistances of 3.0, 4.0, 12, and 16 ohms. What are the separate resistances of the coils?
Answer: 4.0 ohms and 12 ohms.

16(5). Two resistors, R_1 and R_2, may be connected either in series or parallel across a (resistanceless) battery with emf $\mathcal{E}$. We desire the Joule energy for the parallel combination to be five times that for the series combination. If R_1 equals 100 ohms, what is R_2?

17(5). Two light bulbs, one of resistance R and the other of resistance r ($<R$), are connected (*a*) in parallel and (*b*) in series. Which bulb is brighter?
Answer: (*a*) r. (*b*) R.

18(5). In Fig. 28-12 find the current in each resistor and the potential difference between a and b. Put $\mathcal{E}_1 = 6.0$ volts, $_2 = 5.0$ volts, $\mathcal{E}_3 = 4.0$ volts, $R_1 = 100$ ohms, and $R_2 = 50$ ohms.

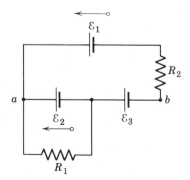

FIGURE 28-12 Problem 18(5).

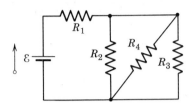

FIGURE 28-13   Problem 20(5).

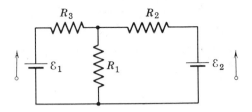

FIGURE 28-14   Problem 21(5).

19(5). In Fig. 28–6 calculate the potential difference between points c and d by as many paths as possible. Assume that $\mathcal{E}_1 = 4.0$ volts, $\mathcal{E}_2 = 1.0$ volt, $R_1 = R_2 = 10$ ohms, and $R_3 = 5.0$ ohms.
Answer: $V_d - V_c = +0.25$ volts.

20(5). (*a*) In Fig. 28–13 what is the equivalent resistance of the network shown? (*b*) What are the currents in each resistor? Put $R_1 = 100$ ohms, $R_2 = R_3 = 50$ ohms, $R_4 = 75$ ohms, and $\mathcal{E} = 6.0$ volts.

21(5). (*a*) In Fig. 28–14 what power appears as Joule energy in R_1? In R_2? In R_3? (*b*) What power is supplied by $\mathcal{E}_1$? By $\mathcal{E}_2$? (*c*) Discuss the energy balance in this circuit. Assume that $\mathcal{E}_1 = 3.0$ volts, $\mathcal{E}_2 = 1.0$ volt, $R_1 = 5.0$ ohms, $R_2 = 2.0$ ohms, and $R_3 = 4.0$ ohms.
Answer: (*a*) 0.35 watt (R_1); 0.049 watt (R_2); 0.71 watt (R_3). (*b*) 1.3 watts ($\mathcal{E}_1$); −0.16 watt ($\mathcal{E}_2$).

22(5). (*a*) Find the three currents in Fig. 28–15. (*b*) Find V_{ab}. Assume that $R_1 = 1.0$ ohm, $R_2 = 2.0$ ohms, $\mathcal{E}_1 = 2.0$ volts, and $\mathcal{E}_2 = \mathcal{E}_3 = 4.0$ volts.

23(5). You are given two batteries of emf $\mathcal{E}$ and internal resistance r. They may be connected either in series or in parallel and are used to establish a current in a resistor R, as in Fig. 28–16. (*a*) Derive expressions for the current in R for both methods of connection. (*b*) Which connection yields the larger current if $R > r$ and if $R < r$?
Answer: (*a*) Series, $i = 2\mathcal{E}/(r + 2R)$; Parallel, $i = 2\mathcal{E}/(2r + R)$. (*b*) Series, $R > r$; Parallel, $R < r$.

24(5). What current, in terms of $\mathcal{E}$ and R, does the meter M in Fig. 28–17 read?

25(5). Four 100-watt heating coils are to be connected in all possible series-parallel combinations of four coils and plugged into a 100-volt line. What different rates of heat dissipation are possible?
Answer: 400, 250, 133, 100, 75, 60, 40, and 25 watts.

26(5). What is the equivalent resistance between the terminal points x and y of the circuits shown in (*a*) Fig. 28–18*a*, (*b*) Fig. 28–18*b*, and (*c*) Fig. 28–18*c*? Assume that the resistance of each resistor is 10 ohms.

27(5). A voltmeter and an ammeter are used to measure two unknown resistances, R_1 by the method shown in Fig. 28–19*a*, and R_2 by that shown in Fig. 28–19*b*. The voltmeter resistance is 310 ohms and the ammeter resistance is 3.6 ohms. In method (*a*) the ammeter reads 0.32 amp and the voltmeter reads 28 volts, whereas in method (*b*) the ammeter reads 0.36 amp and the voltmeter reads 24 volts. What are R_1 and R_2?
Answer: $R_1 = 84$ ohms; $R_2 = 85$ ohms.

28(5). N identical batteries of emf $\mathcal{E}$ and internal resistance r may be connected all in series or all in parallel. Show that each arrangement will give the same current in an external resistor R if $R = r$.

29(5). A copper wire of radius a (0.25mm) has an aluminum jacket of outside radius b (0.38

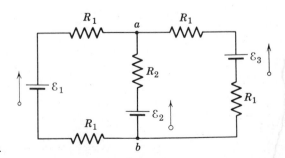

FIGURE 28-15   Problem 22(5).

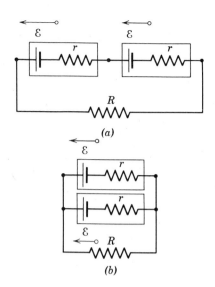

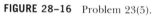

(a)

(b)

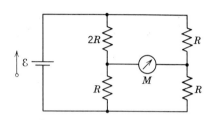

FIGURE 28–16 Problem 23(5).

FIGURE 28–17 Problem 24(5).

mm). (*a*) If there is a current *i* (2.0 amps) in the wire, find the current in each material. (*b*) What is the wire length if voltage *V* (12 volts) maintains the current?

Answer: (*a*) $i_{Cu} = 1.1$ amp; $i_{Al} = 0.86$ amp. (*b*) 130 meters.

30(5). A resistor is formed by filling the space between two thin spherical conducting shells of radii *a* and *b* with a material of resistivity ρ. Show that the resistance between the inner and outer

shells is $R = \dfrac{\rho}{4\pi}\left(\dfrac{1}{a} - \dfrac{1}{b}\right)$.

31(6). A capacitor *C* (1.0 μf) with initial stored energy *U* (0.50 joule) is discharged through a resistance *R* (1.0×10^6 ohms). (*a*) What was the initial charge on the capacitor? (*b*) What was the initial current through the resistor? (*c*) Find the potential difference V_C across the capacitor as a func-

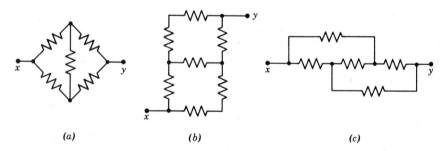

(a)

(b)

(c)

FIGURE 28–18 Problem 26(5).

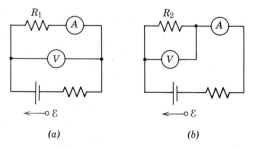

(a)

(b)

FIGURE 28–19 Problem 27(5).

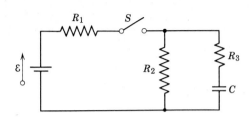

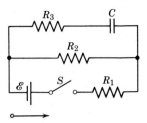

FIGURE 28-20   Problem 39(6).

FIGURE 28-21   Problem 41(6).

tion of time. (*d*) Do the same for V_R, the potential difference across the resistor (*e*) Express the rate of Joule heating in the resistor as a function of time.
Answer: (*a*) 1.0×10^{-3} coul. (*b*) 1.0×10^{-3} amp. (*c*) $10^3 e^{-t}$ volts, with *t* in seconds. (*d*) Same as (*c*). (*e*) e^{-2t} watts, with *t* in seconds.

32(6). After fully charging the capacitor in the circuit shown in Fig. 28–8 the switch is thrown to *b*. (*a*) Find the current in the resistor immediately after throwing the switch. (*b*) If the charge on the capacitor were to reduce to zero at this constant rate, how long would it take?

Answer: (*a*) $i = \dfrac{\mathcal{E}}{R}$; (*b*) $t = RC$.

33(6). Show that the units of *RC* are indeed time units, that is, that 1 ohm $\times$ 1 farad = 1 sec.

34(6). How many time constants must elapse before a capacitor in an *RC* circuit is charged to within 1.0 per cent of its equilibrium charge?

35(6). An *RC* circuit is discharged by closing a switch at time $t = 0$. The initial potential difference across the capacitor is 100 volts. If the potential difference has decreased to 1.0 volt after 10 sec, (*a*) what will the potential difference be 20 sec after $t = 0$? (*b*) What is the time constant of the circuit?
Answer: (*a*) 0.010 volts. (*b*) 2.2 sec.

36(6). A 10,000-ohm resistor and a capacitor are connected in series and a 10-volt potential is suddenly applied. If the potential across the capacitor rises to 5.0 volts in 1.0 μsec, what is the capacitance of the capacitor?

37(6). The potential difference between the plates of a leaky capacitance *C* (2.0 μf) drops from V_0 to V ($\frac{1}{4}V_0$) in time *t* (2.0 sec). What is the equivalent resistance between the capacitor plates?
Answer: 7.2×10^5 ohms.

38(6). A 3.0×10^6-ohm resistor and a 1.0-μf capacitor are connected in a single-loop circuit with a seat of emf with $\mathcal{E} = 4.0$ volts. At 1.0 sec after the connection is made, what are the rates at which (*a*) the charge of the capacitor is increasing, (*b*) energy is being stored in the capacitor, (*c*) Joule energy is appearing in the resistor, and (*d*) energy is being delivered by the seat of emf?

39(6). In the circuit of Fig. 28–20 let i_1, i_2, and i_3 be the currents through resistors R_1, R_2, and R_3, respectively, and let V_1, V_2, V_3, and V_C be the corresponding potential differences across the resistors and across the capacitor *C*. (*a*) Plot qualitatively as a function of time after switch *S* is closed the currents and voltages listed above. (*b*) After being closed for a large number of time constants, the switch *S* is now opened. Plot qualitatively as a function of time after the switch is opened the currents and voltages listed above.

40(6). In the circuit shown in Fig. 28–8 every *T* second the switch position is changed. As functions of time plot qualitatively the potential difference across the resistor and the capacitor. Sketch the plots for the cases $T \gg RC$ and $T \cong RC$.

41(6). In the circuit of Fig. 28–21, $\mathcal{E} = 1200$ volts, $C = 6.5$ μf, $R_1 = R_2 = R_3 = 7.3 \times 10^5$ ohms. With *C* uncharged, the switch *S* is suddenly closed ($t = 0$). (*a*) Find, for $t = 0$ and $t \to \infty$, the currents through each resistor. (*b*) Draw qualitatively a graph of the potential drop V_2 across R_2 from $t = 0$ to $t \to \infty$. (*c*) What are the numerical values of V_2 at $t = 0$ and $t \to \infty$?
Answer: (*a*) $t = 0$: $i_1 = 11 \times 10^{-4}$ amp, $i_2 = i_3 = 5.5 \times 10^{-4}$ amp; $t \to \infty$: $i_1 = i_2 = 8.2 \times 10^{-4}$ amp, $i_3 = 0$.
(*c*) $t = 0$: 400 volts; $t \to \infty$, 600 volts.

CHAPTER 29

The Magnetic Field

29–1 The Magnetic Field

The science of magnetism grew from the observation that certain "stones" (magnetite) would attract bits of iron. The word *magnetism* comes from the district of Magnesia in Asia Minor, which is one of the places where these stones were found. Another "natural magnet" is the earth itself, whose orienting action on a magnetic compass needle has been known since ancient times.

In 1820 Oersted first discovered that a current in a wire can also produce magnetic effects, namely, that it can change the orientation of a compass needle. We pointed out in Section 22–1 how this important discovery linked the then separate sciences of magnetism and electricity. The magnetic effect of a current in a wire can be intensified by forming the wire into a coil of many turns and by providing an iron core.

We define the space around a magnet or a current-carrying conductor as the site of a *magnetic field*, just as we defined the space near a charged rod as the site of an electric field. In the following section we will define the basic magnetic field vector **B**. It can be represented by lines of **B**, just as the electric field was represented by lines of **E**. As for the electric field (see Section 23–3), the magnetic field vector is related to its lines in this way:

1. The tangent to a line of **B** at any point gives the direction of **B** at that point.
2. The lines of **B** are drawn so that the number of lines per unit cross-sectional area is proportional to the magnitude of the magnetic field vector **B**. Where the lines are close together B is large and where they are far apart B is small.

As for the electric field, the field vector **B** is of fundamental importance, the lines of

B simply giving a graphic representation of the way **B** varies throughout a certain region of space.

The flux Φ_B for a magnetic field can be defined in exact analogy with the flux Φ_E for the electric field, namely

SEE
THOMAS
p. 597

$$WEBER \qquad \Phi_B = \int \mathbf{B} \cdot d\mathbf{S}, \qquad (29\text{–}1)$$

SEE p. 451   24-2

in which the integral is taken over the surface (closed or open) for which Φ_B is defined.

29–2 The Definition of B

This chapter is not concerned with the causes of the magnetic field; we seek to determine (*a*) whether a magnetic field exists at a given point P and (*b*) the action of this field on charges moving through it. As for the electric field, a particle of charge q_0 serves as a test body. We assume that there is no electric field present, which means that no electric force will act on the test body.

Let us fire a positive test charge with arbitrary velocity **v** through the point P. *If a sideways deflecting force* **F** *acts on it, we assert that a magnetic field is present at P* and we define **B** in terms of **F**, **v**, and q_0.

If we vary the direction of **v** through point P, keeping the magnitude of **v** unchanged, we find that although **F** will always remain at right angles to **v** its magnitude F will change. For a particular orientation of **v** (and also for the opposite orientation $-\mathbf{v}$) the force **F** becomes zero. We define this direction as the direction of **B,** the specification of the sense of **B** (that is, the way it points along this line) being left to the more complete definition of **B** that we give below.

Having found the direction of **B,** we are now able to orient **v** so that the test charge moves at right angles to **B.** We will find that the force **F** is now a maximum, and we define the magnitude of **B** from the measured magnitude of this maximum force $F_\perp$, or

$$B = \frac{F_\perp}{q_0 v}. \qquad (29\text{–}2)$$

Let us regard this definition of **B** (in which we have specified its magnitude and direction, but not its sense) as preliminary to the complete vector definition that we now give: *If a positive test charge q_0 is moving with velocity* **v** *through a point P and if a (sideways) force* **F** *acts on the moving charge, a magnetic field* **B** *is present at point P, where* **B** *is the vector that satisfies the relation*

$$\mathbf{F} = q_0 \mathbf{v} \times \mathbf{B}, \qquad (29\text{–}3a)$$

v, q_0, and **F** being measured quantities. The magnitude of the magnetic deflecting force **F,** according to the rules for vector products, is given by

$$F = q_0 v B \sin \theta, \qquad (29\text{–}3b)$$

in which θ is the angle between **v** and **B** (see Section 2–4).

Figure 29–1 shows the relations among the vectors. We see that **F,** being at right angles to the plane formed by **v** and **B,** will always be at right angles to **v** (and also to **B**) and thus will always be a sideways force. Equation 29–3 is consistent with the observed facts that (*a*) the magnetic force vanishes as $v \to 0$, (*b*) the magnetic force vanishes if **v** is either parallel or antiparallel to the direction of **B** (in these cases $\theta = 0$

$\theta = 90° \Rightarrow F_m = q_0 v B$

$\theta = 0° \Rightarrow F_m = 0$

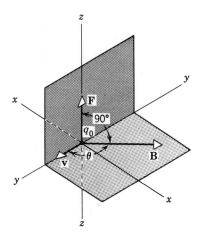

FIGURE 29–1   Illustrating $\mathbf{F} = q_0\mathbf{v} \times \mathbf{B}$ (Eq. 29–3a).

or 180° and $\mathbf{v} \times \mathbf{B} = 0$), and (c) if $\mathbf{v}$ is at right angles to $\mathbf{B}$ ($\theta = 90°$), the deflecting force has its maximum value, given by Eq. 29–2, that is, $F_\perp = q_0vB$.

This definition of $\mathbf{B}$ is similar in spirit, although more complex, than the definition of the electric field $\mathbf{E}$, which we can cast into this form: *If a positive test charge q_0 is placed at point P and if an (electric) force $\mathbf{F}$ acts on the stationary charge, an electric field $\mathbf{E}$ is present at P, where $\mathbf{E}$ is the vector satisfying the relation*

$$\mathbf{F} = q_0\mathbf{E}, \qquad \text{SEE } p. 434$$

q_0 and $\mathbf{F}$ being measured quantities. In defining $\mathbf{E}$, the only characteristic direction to appear is that of the electric force $\mathbf{F}$ which acts on the positive test body; the direction of $\mathbf{E}$ is taken to be that of $\mathbf{F}$. In defining $\mathbf{B}$, two characteristc directions appear, those of $\mathbf{v}$ and of the magnetic force $\mathbf{F}_B$.

In Fig. 29–2 a positive and a negative electron are created at point P in a bubble chamber. A magnetic field is perpendicular to the chamber, pointing out of the plane of the figure.* The relation $\mathbf{F} = q_0\mathbf{v} \times \mathbf{B}$ (Eq. 29–3a) shows that the deflecting forces acting on the two particles are as indicated in the figure. These deflecting forces would make the tracks deflect as shown.

The unit of $\mathbf{B}$ that follows from Eq. 29–3 is the (nt/coul)/(meter/sec). This is given the special name weber/meter², or tesla. Recalling that a coul/sec is an ampere,

$$1 \text{ tesla} = 1 \text{ weber/meter}^2 = \frac{1 \text{ nt}}{\text{coul (meter/sec)}} = \frac{1 \text{ nt}}{\text{amp-m}} \qquad w = \frac{Joule}{amp}$$

An earlier unit for $\mathbf{B}$, still in common use, is the *gauss;* the relationship is

$$1 \text{ tesla} = 1 \text{ weber/meter}^2 = 10^4 \text{ gauss.}$$

The *weber* is used to measure Φ_B, the flux of $\mathbf{B}$; see Eq. 29–1.

The fact that the magnetic force is always at right angles to the direction of motion means that (for steady magnetic fields) the work done by this force on the particle is zero. For an element of the path of the particle of length $d\mathbf{l}$, this work dW is $\mathbf{F}_B \cdot d\mathbf{l}$; dW is zero because $\mathbf{F}_B$ and $d\mathbf{l}$ are always at right angles. Thus a static magnetic field cannot change the kinetic energy of a moving charge; it can only deflect it sideways.

If a charged particle moves through a region in which both an electric field and a

---

* The symbol $\otimes$ indicates a vector into the page, the $\times$ being thought of as the tail of an arrow; the symbol $\odot$ indicates a vector out of the page, the dot being thought of as the tip of an arrow.

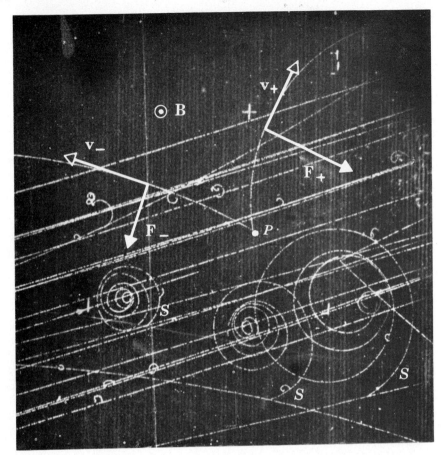

FIGURE 29–2   A *bubble chamber* is a device for rendering visible, by means of small bubbles, the tracks of charged particles that pass through the chamber. The figure is a photograph taken with such a chamber immersed in a magnetic field **B** and exposed to radiations from a large cyclotron-like accelerator. The curved *V* at point *P* is formed by a positive and a negative electron, which deflect in opposite directions in the magnetic field. The spirals *S* are the tracks of three low-energy electrons. (Courtesy E. O. Lawrence Radiation Laboratory, University of California.)

magnetic field are present, the resultant force is found by combining Eqs. 23–2 and 29–3a, or

$$\mathbf{F} = q_0\mathbf{E} + q_0\mathbf{v} \times \mathbf{B}. \; = q_0\left(\vec{E} + \vec{v} \times \vec{B}\right) \tag{29–4}$$

This is sometimes called the *Lorentz relation* in tribute to H. A. Lorentz who did so much to develop and clarify the concepts of the electric and magnetic fields.

■ **Example 1.** A uniform magnetic field **B** points horizontally from south to north; its magnitude is 1.5 webers/meter². If a 5.0-MeV proton moves vertically downward through this field, what force will act on it?

The kinetic energy of the proton is

$$K = (5.0 \times 10^6 \text{ eV})(1.6 \times 10^{-19} \text{ joule/eV}) = 8.0 \times 10^{-13} \text{ joule}.$$

Its speed can be found from the relation $K = \frac{1}{2}mv^2$, or

$$v = \sqrt{\frac{2K}{m}} = \sqrt{\frac{(2)(8.0 \times 10^{-13} \text{ joule})}{1.7 \times 10^{-27} \text{ kg}}} = 3.1 \times 10^7 \text{ meters/sec.}$$

Equation 29–3b gives

$$F = qvB \sin \theta = (1.6 \times 10^{-19}\ \text{coul})(3.1 \times 10^{7}\ \text{meters/sec})(1.5\ \text{webers/meter}^2)(\sin 90°)$$

$$= 7.4 \times 10^{-12}\ \text{nt.}$$

Show that this force is about 4×10^{14} times greater than the weight of the proton.

The relation $\mathbf{F} = q\mathbf{v} \times \mathbf{B}$ shows that the direction of the deflecting force is to the east. If the particle had been negatively charged, the deflection would have been to the west. This is predicted automatically by Eq. 29–3a if we substitute $-e$ for q_0.

29–3   Magnetic Force on a Current

A current is an assembly of moving charges. Because a magnetic field exerts a sideways force on a moving charge, we expect that it will also exert a sideways force on a wire carrying a current. Figure 29–3 shows a length l of wire carrying a current i and placed in a magnetic field $\mathbf{B}$. For simplicity we have oriented the wire so that the current density vector $\mathbf{j}$ is at right angles to $\mathbf{B}$.

The current i in a metal wire is carried by the free (or conduction) electrons, n being the number of such electrons per unit volume of the wire. The magnitude of the average force on one such electron is given by Eq. 29–3b, or, since $\theta = 90°$,

$$F' = q_0 vB \sin \theta = ev_d B$$

where v_d is the drift speed. From the relation $v_d = j/ne$ (Eq. 27–4), $j = nev_d$

$$F' = e\left(\frac{j}{ne}\right)B = \frac{jB}{n}.$$

The length l of the wire contains nAl free electrons, Al being the volume of the section of wire of cross section A that we are considering. The total force on the free electrons in the wire, and thus on the wire itself, is

$$F = (nAl)F' = nAl\frac{jB}{n}.$$ $j = \dfrac{i}{A}$

Since jA is the current i in the wire, we have

$$F = ilB. \tag{29–5}$$

The negative charges, which move to the right in the wire of Fig. 29–3, are equivalent to positive charges moving to the left, that is, in the direction of the current arrow. For such a positive charge the velocity $\mathbf{v}$ would point to the left and the force on the wire, given by Eq. 29–3a ($\mathbf{F} = q_0\mathbf{v} \times \mathbf{B}$) points out of the page. This same conclusion follows if we consider the actual negative charge carriers for which $\mathbf{v}$ points to the right as in Fig. 29–3 but q_0 has a negative sign. Thus by measuring the side-

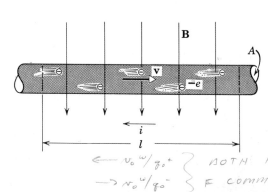

FIGURE 29–3   A wire carrying a current i is placed at right angles to $\mathbf{B}$.

$\leftarrow v_0\ ^w/q_0^+$ $\Big\}$ BOTH HAVE

$\rightarrow v_0\ ^w/q_0^-$ $\Big\}$ $\text{F COMMING OUT OF PAGE}$

ways magnetic force on a wire carrying a current and placed in a magnetic field we cannot tell whether the current carriers are negative charges moving in a given direction or positive charges moving in the opposite direction.

Equation 29–5 holds only if the wire is at right angles to **B**. We can express the more general situation in vector form as

θ BTWN l + B

$$\mathbf{F} = i\mathbf{l} \times \mathbf{B}, \tag{29–6}$$

where **l** is a (displacement) vector that points along the (straight) wire in the direction of the current. Equation 29–6 is equivalent to the relation $\mathbf{F} = q_0\mathbf{v} \times \mathbf{B}$ (Eq. 29–3a); either can be taken as a defining equation for **B**. Note that the vector **l** in Fig. 29–3 points to the left and that the magnetic force $\mathbf{F}$ $(= i\mathbf{l} \times \mathbf{B})$ points up, out of the page. This agrees with the conclusion obtained by analyzing the forces that act on the individual charge carriers.

SEE p. 184

29–4 Torque on a Current Loop

Figure 29–4 shows a rectangular loop of wire of length a and width b placed in a uniform magnetic field **B**, with sides 1 and 3 always normal to the field direction. The normal nn' to the plane of the loop makes an angle θ with the direction of **B**.

Assume the current to be as shown in the figure. Wires must be provided to lead the current into the loop and out of it. If these wires are twisted tightly together, there will be no net magnetic force on the twisted pair because the currents in the two wires are in opposite directions. Thus the lead wires may be ignored. Also, some way of supporting the loop must be provided. Let us imagine it to be suspended from a long string attached to the loop at its center of mass. In this way the loop will be free to turn, through a small angle at least, about any axis through the center of mass.

The net force on the loop is the resultant of the forces on the four sides of the loop. On side 2 the vector **l** points in the direction of the current and has the magnitude b. The angle between **l** and **B** for side 2 (see Fig. 29–4b) is $90° - \theta$. Thus the magnitude of the force on this side is

$$F_2 = ibB \sin (90° - \theta) = ibB \cos \theta.$$

From the relation $\mathbf{F} = i\mathbf{l} \times \mathbf{B}$ (Eq. 29–6), we find the direction of $\mathbf{F}_2$ to be out of the plane of Fig. 29–4b. Show that the force $\mathbf{F}_4$ on side 4 has the same magnitude as

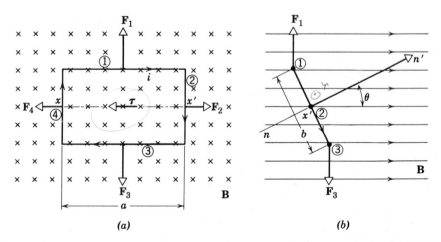

(a)   (b)

FIGURE 29–4   A rectangular coil carrying a current i is placed in a uniform external magnetic field.

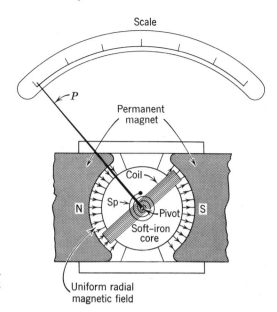

FIGURE 29-5 Example 2. The elements of a galvanometer, showing the coil, the helical spring *Sp*, and pointer *P*.

$\mathbf{F}_2$ but points in the opposite direction. Thus $\mathbf{F}_2$ and $\mathbf{F}_4$, taken together, have no effect on the motion of the loop. The net force they provide is zero, and, since they have the same line of action, the net torque due to these forces is also zero.

The common magnitude of $\mathbf{F}_1$ and $\mathbf{F}_3$ is iaB. These forces, too, are oppositely directed so that they do not tend to move the coil bodily. As Fig. 29–4*b* shows, however, they do not have the same line of action if the coil is in the position shown; there is a net torque, which tends to rotate the coil clockwise about the line *xx'*. The coil can be supported on a rigid axis that lies along *xx'*, with no loss of its freedom of motion. This torque can be represented in Fig. 29–4*b* by a vector pointing into the figure at point *x'* or in Fig. 29–4*a* by a vector pointing along the *xx'*-axis from right to left.

The magnitude of the torque $\boldsymbol{\tau}'$ is found by calculating the torque caused by $\mathbf{F}_1$ about axis *xx'* and doubling it, for $\mathbf{F}_3$ exerts the same torque about this axis that $\mathbf{F}_1$ does. Thus

$$\tau' = 2(iaB)\left(\frac{b}{2}\right)(\sin\theta) = iabB\sin\theta. \quad = iAB\sin\theta$$

This torque acts on every turn of the coil. If there are *N* turns, the torque on the entire coil is

$$\tau = N\tau' = NiabB\sin\theta = NiAB\sin\theta, \qquad (29\text{--}7)$$

in which *A*, the area of the coil, is substituted for *ab*.

This equation can be shown to hold for all plane loops of area *A*, whether they are rectangular or not. A torque on a current loop is the basic operating principle of the electric motor and of most electric meters used for measuring current or potential difference.

Example 2. *A galvanometer.* Figure 29–5 shows the rudiments of a galvanometer, which is a device used to measure currents. The coil is 2.0 cm high and 1.0 cm wide; it has 250 turns and is mounted so that it can rotate about a vertical axis in a uniform radial magnetic field with $B =$

2000 gauss. A spring Sp provides a countertorque that cancels out the magnetic torque, resulting in a steady angular deflection ϕ corresponding to a given current i in the coil. If a current of 1.0×10^{-4} amp produces an angular deflection of $30°$, what is the torsional constant κ of the spring (see Eq. 13–20)?

Equating the magnetic torque to the torque caused by the spring (see Eq. 29–7) yields

$$\tau = NiAB \sin \theta = \kappa\phi$$

or

$$\kappa = \frac{NiAB \sin \theta}{\phi}$$

$$= \frac{(250)(1.0 \times 10^{-4} \text{ amp})(2.0 \times 10^{-4} \text{ meter}^2)(0.20 \text{ weber/meter}^2)(\sin 90°)}{30°}$$

$$= 3.3 \times 10^{-8} \text{ nt-m/deg.}$$

Note that the normal to the plane of the coil (that is, the pointer P) is always at right angles to the (radial) magnetic field so that $\theta = 90°$.

A current loop orienting itself in an external magnetic field reminds us of the action of a compass needle in such a field. One face of the loop behaves like the north pole of the needle (that is, like the end of the needle that points toward the geographic north). The other face behaves like the south pole. Compass needles, bar magnets, and current loops can all be regarded as *magnetic dipoles*. We show this here for the current loop, reasoning entirely by analogy with electric dipoles.

A structure is called an electric dipole if (*a*) when placed in an external electric field it experiences a torque given by Eq. 23–11,

$$\tau = \mathbf{p} \times \mathbf{E,} \qquad \textit{SEE} \quad \textit{p.443} \qquad (29\text{–}8)$$

where $\mathbf{p}$ is the electric dipole moment, and (*b*) it sets up a field of its own at distant points, described qualitatively by the lines of force of Fig. 25–10 and quantitatively by Eq. 25–11. These two requirements are not independent; if one is fulfilled, the other follows automatically.

The magnitude of the torque described by Eq. 29–8 is

$$\tau = pE \sin \theta, \qquad (29\text{–}9)$$

where θ is the angle between $\mathbf{p}$ and $\mathbf{E}$. Let us compare this with Eq. 29–7, the expression for the torque on a current loop:

$$\tau = (NiA)B \sin \theta. \qquad (29\text{–}7)$$

In each case the appropriate field (E or B) appears, as does a term $\sin \theta$. Comparison suggests that NiA in Eq. 29–7 can be taken as the *magnetic dipole moment μ,* corresponding to p in Eq. 29–9, or

$$\mu = NiA. \qquad (29\text{–}10)$$

Equation 29–7 suggests that we write the torque on a current loop in vector form, in analogy with Eq. 29–8, or

$$\tau = \mu \times \mathbf{B.} \quad \textit{nt-m} \qquad (29\text{–}11)$$

The magnetic dipole moment of the loop μ must be taken to lie along the axis of the loop; its direction is given by the following rule: Let the fingers of the right hand curl around the loop in the direction of the current; the extended right thumb will then point in the direction of $\boldsymbol{\mu}$. Check carefully that, if $\boldsymbol{\mu}$ is defined by this rule and Eq. 29–10, Eq. 29–11 correctly describes in every detail the torque acting on a current loop in an external field (see Fig. 29–4).

Since a torque acts on a current loop, or other magnetic dipole, when it is placed in an external magnetic field, it follows that work (positive or negative) must be done by an external agent to change the orientation of such a dipole. Thus a magnetic dipole has potential energy associated with its orientation in an external magnetic field. This energy may be taken to be zero for any arbitrary position of the dipole. By analogy with the assumption made for electric dipoles in Section 23–6, we assume that the magnetic energy U is zero when $\boldsymbol{\mu}$ and **B** are at right angles, that is, when $\theta = 90°$. This choice of a zero-energy configuration for U is arbitrary because we are interested only in the changes in energy that occur when the dipole is rotated.

see p. 443

The magnetic potential energy in any position θ is defined as the work that an external agent must do to turn the dipole from its zero-energy position ($\theta = 90°$) to the given position θ. Thus

$$U = \int_{90°}^{\theta} \tau \, d\theta = \int_{90°}^{\theta} NiAB \sin\theta \, d\theta = \mu B \int_{90°}^{\theta} \sin\theta \, d\theta = -\mu B \cos\theta,$$

in which Eq. 29–7 is used to substitute for τ. In vector symbolism this relation can be written as

$$U = -\boldsymbol{\mu} \cdot \mathbf{B}, \tag{29–12}$$

which is in perfect correspondence with Eq. 23–13, the expression for the energy of an electric dipole in an external electric field,

$$U = -\mathbf{p} \cdot \mathbf{E}.$$

Example 3. A circular coil of N turns has an effective radius a and carries a current i. How much work is required to turn it in an external magnetic field **B** from a position in which θ equals zero to one in which θ equals $180°$? Assume that $N = 100$, $a = 5.0$ cm, $i = 0.10$ amp, and $B = 1.5$ webers/meter².

The work required is the difference in energy between the two positions, or, from Eq. 29–12,

$$W = U_{\theta=180°} - U_{\theta=0} = (-\mu B \cos 180°) - (-\mu B \cos 0) = 2\mu B.$$

But $\mu = NiA$, so that

$$W = 2NiAB = 2Ni(\pi a^2)B$$

$$= (2)(100)(0.10 \text{ amp})(\pi)(5.0 \times 10^{-2} \text{ meter})^2(1.5 \text{ webers/meter}^2) = 0.24 \text{ joule.}$$

29–5 The Hall Effect

In 1879 E. H. Hall devised an experiment that gives the sign of the charge carriers in a conductor. Figure 29–6 shows a flat strip of copper, carrying a current i in the direction shown. As usual, the direction of the current arrow, labeled i, is the direction in which the charge carriers would move if they were positive. The current arrow can represent either positive charges moving down (as in Fig. 29–6a) or negative charges moving up (as in Fig. 29–6b). The Hall effect can be used to decide between these two possibilities.

A magnetic field **B** is set up at right angles to the strip by placing the strip between the polefaces of an electromagnet. This field exerts a deflecting force **F** on the strip

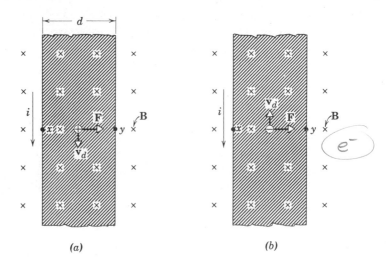

FIGURE 29–6   A current i is set up in a copper strip placed in a magnetic field **B,** assuming (*a*) positive carriers and (*b*) negative carriers.

(given by $i\mathbf{l} \times \mathbf{B}$), which points to the right in the figure. Since the sideways force on the strip is due to the sideways forces on the charge carriers (given by $q\mathbf{v} \times \mathbf{B}$), it follows that these carriers, whether they are positive or negative, will tend to drift toward the right in Fig. 29–6 as they drift along the strip, producing a *transverse Hall potential difference* V_{xy} between opposite points such as x and y. The sign of the charge carriers is determined by the sign of this Hall potential difference. If the carriers are positive, y will be at a higher potential than x; if they are negative, y will be at a lower potential than x. Experiment shows that in metals the charge carriers are negative.

$$V_H = \frac{i}{nqt} B \qquad t = THICKNESS$$

29–6   Circulating Charges

$SEE\ p.\ 230$

Figure 29–7 shows a negatively charged particle introduced with velocity **v** into a uniform magnetic field **B.** We assume that **v** is at right angles to **B** and thus lies entirely in the plane of the figure. The relation $\mathbf{F} = q\mathbf{v} \times \mathbf{B}$ (Eq. 29–3*a*) shows that the particle will experience a sideways deflecting force of magnitude qvB. This force will lie in the plane of the figure, which means that the particle cannot leave this plane.

This reminds us of a stone held by a rope and whirled in a horizontal circle on a

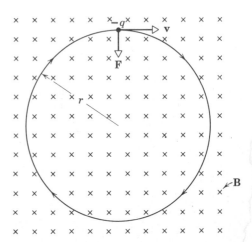

FIGURE 29–7   A charge $-q$ circulates at right angles to a uniform magnetic field.

smooth surface. Here, too, a force of constant magnitude, the tension in the rope, acts in a plane and at right angles to the velocity. The charged particle, like the stone, also moves with constant speed in a circular path. From Newton's second law we have

$$F = qvB = \frac{mv^2}{r} \quad \text{or} \quad r = \frac{mv}{qB}, \tag{29–13}$$

p. 82

which gives the radius of the path. The three spirals in Fig. 29–2 show relatively low-energy electrons in a bubble chamber. The paths are not circles because the electrons lose energy by collisions in the chamber as they move. Also, when viewed stereoptically, one sees that the spirals do not lie in a single plane but have a helical motion.

The angular velocity ω is given by v/r or, from Eq. 29–13,

$$\omega = \frac{v}{r} = \frac{qB}{m}.$$

The frequency ν measured, say, in rev/sec, is given by

$$\nu = \frac{\omega}{2\pi} = \frac{qB}{2\pi m}. \tag{29–14}$$

Note that ν does not depend on the speed of the particle. Fast particles move in large circles (Eq. 29–13) and slow ones in small circles, but all require the same time T to complete one revolution in the field.

The frequency ν is a characteristic frequency for the charged particle in the field and may be compared to the characteristic frequency of a swinging pendulum in the earth's gravitational field or to the characteristic frequency of an oscillating mass-spring system. It is sometimes called the *cyclotron frequency* of the particle in the field because particles circulate at this frequency in the cyclotron.

Example 4. A 10-eV electron is circulating in a plane at right angles to a uniform magnetic field of 1.0×10^{-4} weber/meter² ($= 1.0$ gauss).

(*a*) What is its orbit radius?

The velocity of an electron whose kinetic energy is K can be found from

$$v = \sqrt{\frac{2K}{m}}.$$

Show that this yields 1.9×10^6 meters/sec for v. Then, from Eq. 29–13,

$$r = \frac{mv}{qB} = \frac{(9.1 \times 10^{-31} \text{ kg})(1.9 \times 10^6 \text{ meters/sec})}{(1.6 \times 10^{-19} \text{ coul})(1.0 \times 10^{-4} \text{ weber/meter}^2)} = 0.11 \text{ meter} = 11 \text{ cm}.$$

(*b*) What is the cyclotron frequency?

From Eq. 29–14,

$$\nu = \frac{qB}{2\pi m} = \frac{(1.6 \times 10^{-19} \text{ coul})(1.0 \times 10^{-4} \text{ weber/meter}^2)}{(2\pi)(9.1 \times 10^{-31} \text{ kg})} = 2.8 \times 10^6 \text{ rev/sec}.$$

(*c*) What is the period of revolution T?

$$T = \frac{1}{\nu} = \frac{1}{2.8 \times 10^6 \text{ rev/sec}} = 3.6 \times 10^{-7} \text{ sec}.$$

Thus an electron requires 0.36 μsec to make 1.0 revolution in a 1.0-gauss field.

(*d*) What is the direction of circulation as viewed by an observer sighting along the field?

In Fig. 29–7 the magnetic force $q\mathbf{v} \times \mathbf{B}$ must point radially inward, since it provides the centripetal force. Since **B** points into the plane of the paper, **v** would have to point to the left at the position shown in the figure if the charge q were positive. However, the charge is an electron, with $q = -e$, which means that **v** must point to the right. Thus the charge circulates clockwise as viewed by an observer sighting in the direction of **B.**

29–7 The Cyclotron

The cyclotron, first put into operation by Ernest Lawrence (1902–1958) in 1932, accelerates charged particles, such as protons or deuterons, to high energies so that they can be used in atom-smashing experiments. Figure 29–8 shows a cyclotron formerly used at the University of Pittsburgh. Although devices like this are no longer used, they are the starting point for the high energy cyclotron-like accelerators that have replaced them.

In an ion source at the center of the cyclotron molecules of deuterium are bombarded with electrons whose energy is high enough (say 100 eV) so that many positive ions are formed during the collisions. Many of these ions are free deuterons, which enter the cyclotron proper through a small hole in the wall of the ion source and are available to be accelerated.

The cyclotron uses a modest potential difference for accelerating (say 10^5 volts), but it requires the ion to pass through this potential difference a number of times. To reach 10 MeV with 10^5 volts accelerating potential requires 100 passages. A magnetic field is used to bend the ions around so that they may pass again and again through the same accelerating potential.

Figure 29–9 is a top view of the part of the cyclotron that is inside the vacuum tank marked V in Fig. 29–8. The two D-shaped objects, called *dees*, are made of copper sheet and form part of an electric oscillator which establishes an accelerating poten-

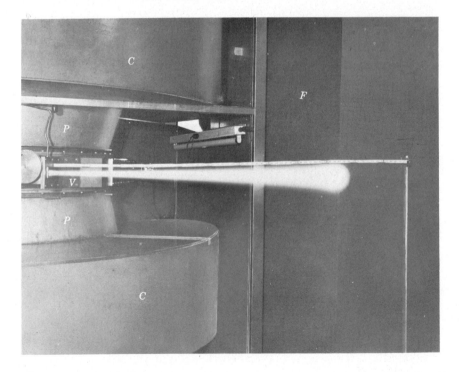

FIGURE 29–8 A cyclotron. Note vacuum chamber V, magnet frame F, magnetic pole faces P, magnet coils C, and the deuteron beam emerging into the air of the laboratory. The rule is 6 ft long.

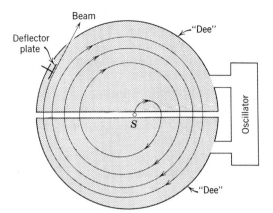

FIGURE 29–9 The elements of a cyclotron showing the ion source S and the dees. The deflector plate, held at a suitable negative potential, deflects the particles out of the dee system.

tial difference across the gap between the dees. The direction of this potential difference is made to change sign some millions of times per second.

The dees are immersed in a magnetic field ($B \cong 1.6$ webers/meters^2) whose direction is out of the plane of Fig. 29–9. The field is set up by a large electromagnet, marked F in Fig. 29–8. Finally, the space in which the ions move is evacuated to a pressure of about 10^{-6} mm-Hg. If this were not done, the ions would continually collide with air molecules.

Suppose that a deuteron, emerging from the ion source, finds the dee that it is facing to be negative; it will accelerate toward this dee and wil! enter it. Once inside, it is screened from electrical forces by the metal walls of the dee. The magnetic field is not screened by the dees so that the ion bends in a circular path whose radius, which depends on the speed, is given by Eq. 29–13, or

$$r = \frac{mv}{qB}.$$

After a time t_0 the ion emerges from the dee on the other side of the ion source. Let us assume that the accelerating potential has now changed sign. Thus the ion again faces a negative dee, is further accelerated, and again describes a semicircle, of somewhat larger radius (see Eq. 29–13), in the dee. The time of passage through this dee, however, is still t_0. This follows because the period of revolution T of an ion circulating in a magnetic field does not depend on the speed of the ion; see Eq. 29–14. This process goes on until the ion reaches the outer edge of one dee where it is pulled out of the system by a negatively charged deflector plate.

The key to the operation of the cyclotron is that the characteristic frequency ν at which the ion circulates in the field must be equal to the fixed frequency ν_0 of the electric oscillator, or

$$\nu = \nu_0.$$

This resonance condition says that if the energy of the circulating ion is to increase energy must be fed to it at a frequency ν_0 that is equal to the natural frequency ν at which the ion circulates in the field. In the same way we feed energy to a swing by pushing it at a frequency equal to the natural frequency of oscillation of the swing.

From Eq. 29–14 ($\nu = qB/2\pi m$), we can rewrite the resonance equation as

$$\frac{qB}{2\pi m} = \nu_0. \tag{29–15}$$

Once we have selected an ion to be accelerated, q/m is fixed; usually the oscillator is designed to work at a single frequency ν_0. We then "tune" the cyclotron by varying B until Eq. 29–15 is satisfied and an accelerated beam appears.

The energy of the particles produced in the cyclotron depends on the radius R of the dees. From Eq. 29–13 ($r = mv/qB$), the speed of a particle circulating at this radius is given by

$$v = \frac{qBR}{m}.$$

The kinetic energy is then

$$K = \tfrac{1}{2}mv^2 = \frac{q^2B^2R^2}{2m}. \tag{29–16}$$

Example 5. The cyclotron of Fig. 29–8 has an oscillator frequency of 12×10^6 cycles/sec and a dee radius of 21 in. (*a*) What magnetic field B is needed to accelerate deuterons?

From Eq. 29–15, $\nu_0 = qB/2\pi m$, so that

$$B = \frac{2\pi\nu_0 m}{q} = \frac{(2\pi)(12 \times 10^6/\text{sec})(3.3 \times 10^{-27}\ \text{kg})}{1.6 \times 10^{-19}\ \text{coul}} = 1.6\ \text{webers/meter}^2.$$

Note that the deuteron has the same charge as the proton but (very closely) twice the mass.

(*b*) What deuteron energy results?

From Eq. 29–16,

$$K = \frac{q^2B^2R^2}{2m} = \frac{(1.6 \times 10^{-19}\ \text{coul})^2(1.6\ \text{webers/meter}^2)^2(21 \times 0.0254\ \text{meters})^2}{(2)(3.3 \times 10^{-27}\ \text{kg})}$$

$$= (2.8 \times 10^{-12}\ \text{joule})\left(\frac{1\ \text{eV}}{1.6 \times 10^{-19}\ \text{joule}}\right) = 17\ \text{MeV}$$

The cyclotron fails to operate at high energies because one of its assumptions, that the frequency of rotation of an ion circulating in a magnetic field is independent of its speed, is true only for speeds much less than that of light. As the particle speed increases, we must use the relativistic mass m in Eq. 29–14. The relativistic mass increases with velocity (Eq. 7–18) so that at high enough speeds v decreases with velocity. Thus the ions get out of step with the electric oscillator, and eventually the energy of the circulating ion stops increasing.

Another difficulty associated with the acceleration of particles to high energies is that the size of the magnet that would be required to guide such particles in a circular path is very large. For a 30-GeV proton, for example, in a field of 15,000 gauss the radius of curvature is 65 meters. Incidentally, a 30-GeV proton has a speed equal to 0.99998 that of light.

These difficulties have been overcome and high energy cyclotron-like devices in the 50-GeV range are now in operation.

29–8 Thomson's Experiment

In 1897 J. J. Thomson, working at the Cavendish Laboratory in Cambridge, measured the ratio of the charge e of the electron to its mass m by observing its deflection in both electric and magnetic fields. The discovery of the electron is usually said to date from this historic experiment, although H. A. Lorentz and P. Zeeman, during the previous year, had measured this same quantity for electrons bound in atoms, using a method entirely different from Thomson's.

Figure 29–10 shows a modern version of Thomson's apparatus. Electrons are emitted from hot filament F and accelerated by an applied potential difference V. They then enter a region in which an electric field **E** and a magnetic field **B** are present; **E** and **B** are at right angles to each other. The beam is made visible as a spot of light when it strikes fluorescent screen S. The entire region in which the electrons move is highly evacuated so that collisions with air molecules will not occur.

The resultant force on a charged particle moving through an electric and a magnetic field is given by Eq. 29–4, or

$$\mathbf{F} = q_0\mathbf{E} + q_0\mathbf{v} \times \mathbf{B}.$$

Study of Fig. 29–10 shows that the electric field deflects the particle upward and the magnetic field deflects it downward. If these deflecting forces are to cancel (that is, if **F** = 0), this equation, for this problem, reduces to

$$eE = evB$$

or $$E = vB. \tag{29–17}$$

Thus for a given electron speed v the condition for zero deflection can be satisfied by adjusting E or B.

When Thomson's experiment is repeated today, usually as a student experiment, the procedure* is: (*a*) note the position of the undeflected beam spot, with **E** and **B** both equal to zero; (*b*) apply a fixed electric field **E,** measuring on the fluorescent screen the deflection so caused; and (*c*) apply a magnetic field and adjust its value until the beam deflection is restored to zero.

In Section 23–5 we saw that the deflection y of an electron in a purely electric field (step *b*), measured at the far edge of the deflecting plates, is given by Eq. 23–9, or, with small changes in notation,

$$y = \frac{eEl^2}{2mv^2},$$

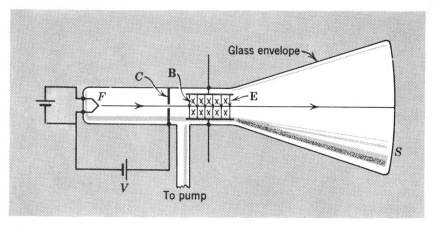

FIGURE 29–10   Electrons from the heated filament F are accelerated by a potential difference V and pass through a hole in the screen C. After passing through a region in which perpendicular electric and magnetic fields are present, they strike the fluorescent screen S.

---

*Thomson, no doubt for practical reasons, did not use this null method. He applied the fields **E** and **B** separately and adjusted them to obtain the same beam deflection. Equation 29–18 still holds. See his original paper: *Philosophical Magazine,* **44,** 293 (1897).

where v is the electron speed and l is the length of the deflecting plates; y is not measurable directly, but it may be calculated from the measured displacement of the spot on the screen if the geometry of the apparatus is known. Thus y, E, and l are known; the ratio e/m and the velocity v are unknown. We cannot calculate e/m until we have found the speed, which is the purpose of step c above.

If (step c) the electric force is set equal and opposite to the magnetic force, the net force is zero and we can write (Eq. 29–17)

$$v = \frac{E}{B}$$

Substituting this equation into the equation for y and solving for the ratio e/m leads to

$$\frac{e}{m} = \frac{2yE}{B^2 l^2}. \tag{29–18}$$

in which all the quantities on the right can be measured. Thomson's value for e/m was 1.7×10^{11} coul/kg, in as close agreement as could have been expected with the modern value of 1.7588028×10^{11} coul/kg.

QUESTIONS

1. Of the three vectors in the equation $\mathbf{F} = q\mathbf{v} \times \mathbf{B}$, which pairs are always at right angles? Which may have any angle between them?

2. Why do we not simply define the magnetic field $\mathbf{B}$ to point in the direction of the magnetic force that acts on the moving charge?

3. Imagine that you are sitting in a room with your back to one wall and that an electron beam, traveling horizontally from that toward the opposite wall, is deflected to your right. What is the direction of the magnetic field that exists in the room? Assume that no electric field is present (and that the room is evacuated!)

4. If an electron is not deflected in passing through a certain region of space, can we be sure that there is no magnetic field in that region?

5. If a moving electron is deflected sideways in passing through a certain region of space, can we be sure that a magnetic field exists in that region?

6. A beam of protons is deflected sideways. Could this deflection be caused by an electric field? By a magnetic field? If either could be responsible, how would you be able to tell which was present?

7. A conductor, even though it is carrying a current, has zero net charge. Why, then, does a magnetic field exert a force on it?

8. Equation 29–11 ($\tau = \boldsymbol{\mu} \times \mathbf{B}$) shows that there is no torque on a current loop in an external magnetic field if the angle between the axis of the loop and the field is (a) 0° or (b) 180°. Discuss the nature of the equilibrium (that is, is it stable, neutral, or unstable?) for these two positions.

9. In Example 3 we showed that the work required to turn a current loop end-for-end in an external magnetic field is $2\mu B$. Does this hold no matter what the original orientation of the loop was?

10. Imagine that the room in which you are seated is filled with a uniform magnetic field with $\mathbf{B}$ pointing vertically upward. A circular loop of wire has its plane horizontal and is mounted so that it is free to rotate about a horizontal axis. For what direction of current in the loop, as viewed from above, will the loop be in stable equilibrium with respect to forces and torques of magnetic origin?

11. A rectangular current loop is in an arbitrary orientation in an external uniform magnetic field. Is any work required to rotate the loop about an axis perpendicular to its plane?

12. In measuring Hall potential differences, why must we be careful that points x and y in Fig. 29–6 are exactly opposite to each other? If one of the contacts is movable, what procedure might we follow in adjusting it to make sure that the two points are properly located?

13. A uniform magnetic field fills a certain cubical region of space. Can an electron be fired into this cube from the outside in such a way that it will travel in a closed circular path inside the cube?

14. Imagine the room in which you are seated to be filled with a uniform magnetic field with **B** pointing vertically downward. At the center of the room two electrons are suddenly projected horizontally with the same speed but in opposite directions. Discuss their motions. Discuss their motions if one particle is an electron and one a positron. Assume that you are wearing a space suit and that the room is evacuated.

15. What are the primary functions of (*a*) the electric field and (*b*) the magnetic field in the cyclotron?

16. For Thomson's e/m experiment to work properly (Section 29–8), is it essential that all the electrons have the same speed?

PROBLEMS

1(2). Particles 1, 2, and 3 follow the paths shown in Fig. 29–11 as they pass through the magnetic field there. What can one conclude about each particle?
Answer: 1 is positively charged, 2 is neutral, and 3 is negatively charged.

2(2). What direction is a magnetic field if an electron moving south in this field experiences a force directed (*a*) down and (*b*) east.

3(2). Express magnetic field B and magnetic flux Φ in terms of the fundamental dimensions M, L, T, and Q (mass, length, time, and charge).
Answer: ML^2/QT.

4(2). An electron has a velocity given in meters/sec by $\mathbf{v} = 2.0 \times 10^6\mathbf{i} + 3.0 \times 10^6\mathbf{j}$. It enters a magnetic field given in webers/meters² by $\mathbf{B} = 0.03\mathbf{i} - 0.15\mathbf{j}$. (*a*) Find the magnitude and direction of the force on the electron. (*b*) Repeat your calculation for a deuteron having the same velocity.

5(2). The electrons in the beam of a television tube have an energy of 12 KeV. The tube is oriented so that the electrons move horizontally from south to north. The vertical component of the earth's magnetic field points down and has $B = 5.5 \times 10^{-5}$ weber/meter². (*a*) In what direction will the beam deflect? (*b*) What is the acceleration of a given electron? (*c*) How far will the beam deflect in moving 20 cm through the television tube?
Answer: (*a*) East. (*b*) 6.3×10^{14} meters/sec.² (*c*) 3.0 mm.

6(2). An electron in a uniform magnetic field **B** has a velocity $\mathbf{v} = 4.0 \times 10^5\mathbf{i} + 7.1 \times 10^5\mathbf{j}$, in meters/sec. It experiences a force $\mathbf{F} = -2.7 \times 10^{-13}\mathbf{i} + 1.5 \times 10^{-13}\mathbf{j}$, in nt. If $B_x = 0$, find the magnetic field.

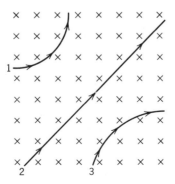

FIGURE 29-11 Problem 1(2).

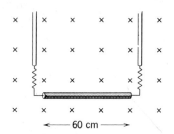

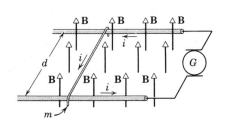

FIGURE 29–12 Problem 9(3).   **FIGURE 29–13** Problem 11(3).

7(3). A wire 1.0 meter long carries a current of 10 amp and makes an angle of 30° with a uniform magnetic field with $B = 1.5$ webers/meter². Calculate the magnitude and direction of the force on the wire.

Answer: 7.5 nt, perpendicular both to the wire and to **B**.

8(3). A wire 50 cm long lying along the *x*-axis carries a current of 0.50 amp in the positive *x*-direction. A magnetic field is present that is given in webers/meters² by **B** = 0.0030**j** + 0.010**k**. Find the components of the force on the wire.

9(3). A wire of 60 cm length and mass 10 gm is suspended as shown in Fig. 29–12 by a pair of flexible leads in a magnetic field of 0.40 weber/meter². What are the magnitude and direction of the current required to remove the tension in the supporting leads?

Answer: 0.41 amp from left to right.

10(3). A 1.0-kg copper rod rests on two horizontal rails 1.0 meter apart and carries a current of 50 amp from one rail to the other. The coefficient of static friction is 0.60. What is the smallest magnetic field (not necessarily vertical) that would cause the bar to slide?

11(3). A metal wire of mass *m* slides without friction on two rails spaced a distance *d* apart, as in Fig. 29–13. The track lies in a vertical uniform magnetic field **B**. A constant current *i* flows from generator *G* along one rail, across the wire, and back down the other rail. Find the velocity (speed and direction) of the wire as a function of time, assuming it to be at rest at $t = 0$.

Answer: $v = Bidt/m$, away from generator.

12(3). Consider the possibility of a new design for an electric train. The engine is driven by the force due to the vertical component of the earth's magnetic field on a conducting axle. Current is passed down one rail, into a conducting wheel, through the axle, through another conducting wheel and then back to the source via the other rail. (*a*) What current is needed to provide a modest 10,000 nt force? Take the vertical component of **B** to be 10^{-5} weber/meter² and the length of the axle to be 3.0 meters. (*b*) How much power would be lost for each ohm of resistance in the rails? (*c*) Is such a train totally unrealistic or marginally unfeasible?

13(4). Figure 29–14 shows a rectangular 20-turn loop of wire, 10 cm by 5.0 cm. It carries a current of 0.10 amp and is hinged at one side. What torque (direction and magnitude) acts on the loop if it is mounted with its plane at an angle of 30° to the direction of a uniform magnetic field $B = 0.50$ weber/meter²?

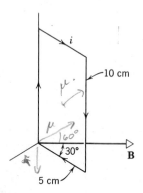

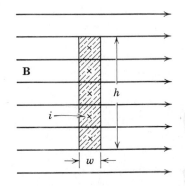

FIGURE 29–14 Problem 13(4).   **FIGURE 29–15** Problem 18(5).

Answer: 4.3×10^{-3} nt-meter. The torque vector is parallel to the long side of the coil and points down.

14(4). A circular loop of wire having a radius of 8.0 cm carries a current of 0.20 amp. A unit vector parallel to the dipole moment $\boldsymbol{\mu}$ of the loop is given by $0.60\mathbf{i} - 0.80\mathbf{j}$. If the loop is located in a magnetic field given in webers/meters^2 by $\mathbf{B} = 0.25\mathbf{i} + 0.30\mathbf{k}$, find (*a*) the magnitude and direction of the torque on the loop, and (*b*) the magnetic potential energy of the loop. Assume the same zero-energy configuration that we assumed in Section 29–4.

15(4). A wire loop of radius r (15 cm) carrying a current i (2.0 amp) is placed so that its plane makes an angle θ (30°) with a magnetic field. If the magnetic flux through the loop is ϕ (0.70 weber) what is the torque on the loop?
Answer: 0.81 nt-meter.

16(4). An N-turn circular coil of radius R is suspended in a uniform magnetic field $\mathbf{B}$ that points vertically downward. The coil can rotate about a horizontal axis through its center. A mass m hangs by a string from the bottom of the coil. When a current i is put through the coil it eventually assumes an equilibrium position in which the perpendicular to the plane of the coil makes an angle ϕ with the direction of $\mathbf{B}$. Find ϕ and draw a sketch of this equilibrium position. Take $B = 0.50$ tesla. $R = 10$ cm, $N = 10$, $m = 500$ gm, and $i = 1.0$ amp.

17(4). Prove that the relation $\tau = NAiB \sin \theta$ holds for closed loops of arbitrary shape. (Hint: Replace the loop of arbitrary shape by an assembly of adjacent long, thin—approximately rectangular—loops which are equivalent to it as far as the distribution of current is concerned.)

18(5). A current i, indicated by the crosses in Fig. 29–15, is established in a strip of copper of height h and width w. A uniform magnetic field $\mathbf{B}$ is applied at right angles to the strip. (*a*) Calculate the drift speed v_d for the electrons. (*b*) What are the magnitude and direction of the magnetic force $\mathbf{F}$ acting on each electron? (*c*) What would the magnitude and direction of a homogeneous electric field $\mathbf{E}$ have to be in order to counterbalance the effect of the magnetic field? (*d*) What is the voltage V necessary between two sides of the conductor in order to create this field $\mathbf{E}$? Between which sides of the conductor would this voltage have to be applied? (*e*) If no electric field is applied from the outside, the electrons will be pushed somewhat to one side and therefore will give rise to a uniform electric field $\mathbf{E}_H$ ("H" for "Hall effect") across the conductor until the forces of this electrostatic field $\mathbf{E}_H$ balance the magnetic forces encountered in part (*b*). What will be the magnitude and direction of the field $\mathbf{E}_H$? Assume that n, the number of conduction electrons per unit volume, is 1.1×10^{29}/meter^3 and that $h = 0.020$ meter, $w = 0.10$ cm, $i = 50$ amp, and $B = 2.0$ webers/meter^2.

19(5). In a Hall effect experiment a current of 3.0 amp lengthwise in a conductor 1.0 cm wide, 4.0 cm long, and 1.0×10^{-3} cm thick produced a transverse Hall voltage (across the width) of 1.0×10^{-5} volt when a magnetic field of 1.5 webers/meter^2 passed perpendicularly through the thin conductor. From these data, find (*a*) the drift velocity of the charge carriers and (*b*) the number of carriers per cubic centimeter. (*c*) Show on a diagram the polarity of the Hall voltage with a given current and magnetic field direction, assuming the charge carriers are (negative) electrons.
Answer: (*a*) 0.067 cm/sec. (*b*) 2.8×10^{23} electrons/cm^3.

20(5). Show that the ratio of the Hall electric field E_H to the electric field E responsible for the current is $E_H/E = B/ne\rho$, where ρ is the resistivity of the material.

21(6). An electron is accelerated through 15,000 volts and is then allowed to circulate at right angles to a uniform magnetic field with $B = 250$ gauss. What is its path radius?
Answer: 1.7 cm.

22(6). (*a*) In a magnetic field with $B = 0.50$ weber/meter^2, for what path radius will an electron circulate at 0.10 the speed of light? (*b*) What will its kinetic energy be?

23(6). An α-particle travels in a circular path of radius 0.45 meter in a magnetic field with $B = 1.2$ webers/meter^2. Calculate (*a*) its speed, (*b*) its period of revolution, (*c*) its kinetic energy, and (*d*) the potential difference through which it would have to be accelerated to achieve this energy.
Answer: (*a*) 2.6×10^7 meters/sec. (*b*) 1.1×10^{-7} sec. (*c*) 14 MeV. (*d*) 7.0×10^6 volts.

24(6). A 10-KeV electron is circulating in a plane at right angles to a uniform magnetic field. The orbit radius is 25 cm. Find (*a*) the magnetic field, (*b*) the cyclotron frequency, and (*c*) the period of the motion.

25(6). (*a*) What is the cyclotron frequency of an electron with an energy of 100 eV in the earth's magnetic field of 1.0×10^{-4} tesla? (*b*) What is the radius of curvature of the path of this electron if its velocity is perpendicular to the magnetic field?
Answer: (*a*) 2.8×10^6 rev/sec. (*b*) 0.34 meters.

26(6). What is the smallest magnetic field that can be set up at the equator to permit a proton of speed 1.0×10^7 meters/sec to circulate around the earth? What is its direction?

27(6). In a nuclear experiment a 1.0-MeV proton moves in a uniform magnetic field in a circular path. What energy must (*a*) an alpha particle, and (*b*) a deuteron have if they are to circulate in the same orbit?
Answer: (*a*) 1.0 MeV. (*b*) 0.50 MeV.

28(6). A proton, a deuteron, and an α-particle, accelerated through the same potential difference, enter a region of uniform magnetic field, moving at right angles to **B**. (*a*) Compare their kinetic energies. If the radius of the proton's circular path is 10 cm, what are the radii of (*b*) the deuteron and (*c*) the α-particle paths?

29(6). Show that the radius of curvature of a charged particle moving at right angles to a magnetic field is proportional to its momentum.

30(6). A deuteron in a large cyclotron is moving in a magnetic field with $B = 1.5$ webers/meter² and an orbit radius of 2.0 meters. Because of a grazing collision with a target, the deuteron breaks up, with a negligible loss of kinetic energy, into a proton and a neutron. Discuss the subsequent motions of each. Assume that the deuteron energy is shared equally by the proton and neutron at breakup.

31(6). A neutral particle, viewed in a reference frame in which it is at rest, lies in a homogeneous magnetic field of magnitude B. At time $t = 0$ it decays into two charged particles each of mass m. (*a*) If the charge of one of the particles is $+q$, what is the charge of the other? (*b*) The two particles move off in separate paths both of which lie in the plane perpendicular to **B**. What is the direction of the path of particle 1 relative to particle 2 immediately after decay? (*c*) Describe the subsequent motion of the particles. (*d*) At a later time the particles collide. Express the time from decay until collision, t, in terms of m, B, and q.
Answer: (*a*) $-q$. (*b*) $t = \pi m/qB$.

32(6). *Time-of-flight spectrometer.* S. A. Goudsmit has devised a method for measuring accurately the masses of heavy ions by timing their period of circulation in a known magnetic field. A singly charged ion of iodine makes 7.00 rev in a field of 4.50×10^{-2} weber/meter² in about 1.29×10^{-3} sec. What (approximately) is its mass in kilograms? Actually, the mass measurements are carried out to much greater accuracy than these approximate data suggest.

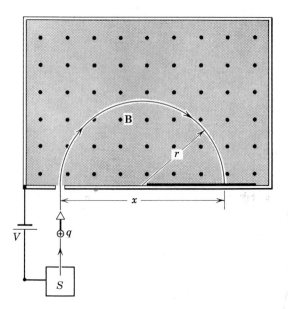

FIGURE 29-16 Problem 33(6).

4, 7, 9, 21, 26

33(6). *Mass spectrometer.* Figure 29–16 shows an arrangement used by Dempster to measure the masses of ions. An ion of mass M and charge $+q$ is produced essentially at rest in source S, a chamber in which a gas discharge is taking place. The ion is accelerated by potential difference V and allowed to enter a magnetic field $\mathbf{B}$. In the field it moves in a semicircle, striking a photographic plate at distance x from the entry slit and being recorded. Show that the mass M is given by

$$M = \frac{B^2 q}{8V} x^2.$$

34(6). Two types of singly ionized atoms having the same charge q, and mass differing by a small amount ΔM are introduced into the mass spectrometer described in Problem 33(6). (*a*) Calculate the difference in mass in terms of V, q, M (of either), B, and the distance Δx between the spots on the photographic plate. (*b*) Calculate Δx for a beam of singly ionized chlorine atoms of masses 35 and 37 amu if $V = 7.3 \times 10^3$ volts and $B = 0.50$ tesla. (1.00 amu $= 1.67 \times 10^{-27}$ kg).

35(6). Singly ionized chlorine atoms of 35 amu and 37 amu, traveling with speed 2.0×10^5 meters/sec, enter perpendicularly a uniform magnetic field of 0.50 tesla. (*a*) After bending through $180°$ the atoms strike a photographic film. What is the separation distance between the two spots on the film? (1.00 amu $= 1.67 \times 10^{-27}$ kg.) (*b*) What is the difference between this setup and that of Problem 34(6)?
Answer: (*a*) 1.7 cm.

36(6). A 2.0-KeV positron is projected into a uniform magnetic field $\mathbf{B}$ of 0.10 weber/meter² with its velocity vector making an angle of $89°$ with $\mathbf{B}$. (*a*) Convince yourself that the path will be a helix, its axis being the direction of $\mathbf{B}$. Find (*b*) the period, (*c*) the pitch p, and (*c*) the radius r of the helix; see Fig. 29–17.

37(7). In a certain cyclotron a proton moves in a circle of radius $r = 0.50$ meter. The magnitude of the $\mathbf{B}$ field is 1.2 webers/meter? (*a*) What is the cyclotron frequency? (*b*) What is the kinetic energy of the proton?
Answer: (*a*) 1.8×10^7 cycles/sec. (*b*) 17 MeV.

38(7). The cyclotron of Example 5 was normally adjusted to accelerate deuterons. (*a*) What energy of protons could it produce, using the same oscillator frequency as that used for deuterons? (*b*) What magnetic field would be required? (*c*) What energy of protons could be produced if the magnetic field was left at the value used for deuterons? (*d*) What oscillator frequency would then be required? (*e*) Answer the same questions for α-particles.

39(7). Estimate the total path length traversed by a deuteron in the cyclotron of Example 5 during the acceleration process. Assume an accelerating potential between the dees of 80,000 volts.
Answer: Approximately 240 meters.

40(8). A 10-KeV electron moving horizontally enters a region of space in which there is a

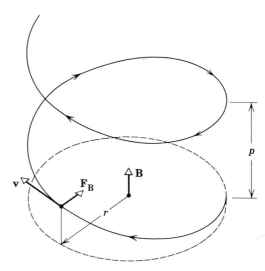

FIGURE 29–17 Problem 36(6).

downward-directed electric field of magnitude 100 volts/cm. What are the magnitude and direction of the (smallest) magnetic field that will allow the electron to continue to move horizontally? Ignore gravitational forces, which are rather small.

41(8). An electron is accelerated through a potential difference of 1000 volts and directed into a region between two parallel plates separated by 0.020 meter with a potential difference of 100 volts between them. If the electron enters moving perpendicular to the electric field between the plates, what magnetic field is necessary perpendicular to both the electron path and the electric field so that the electron travels in a straight line?

Answer: 2.7×10^{-4} webers/meter^2.

42(8). An electric field of 1500 volts/meter and a magnetic field of 0.40 weber/meter^2 act on a moving electron to produce no force. (*a*) Calculate the minimum electron speed v. (*b*) Draw the vectors **E**, **B**, and **v**.

43(8). A positive point charge Q travels in a straight line with constant speed through an evacuated region in which there is a uniform electric field **E** and a uniform magnetic field **B**. (*a*) If **E** is directed vertically up and the charge travels horizontally from north to south with speed v, determine the least value of the magnitude of **B** and the corresponding direction of **B**. (*b*) Explain why **B** is not uniquely determined when **E** and **v** alone are given. (*c*) Suppose the charge is a proton that enters the region after having been accelerated through a potential difference of 3.1×10^5 volts. If $E = 1.9 \times 10^5$ volt/meter, compute the value of B in part (*a*). (*d*) If in part (*c*) the electric field **E** is turned off, determine the radius r of the circle in which the proton now moves.

Answer: (*a*) $B = e/v$. (*b*) The angle between **B** and **V** or **B** and **E** is not uniquely determined by the force equation. (*c*) 2.5×10^{-2} weber/meter^2. (*d*) 3.2 meters.

CHAPTER 30

Ampère's Law

30–1 Ampère's Law

One class of problems involving magnetic fields, dealt with in Chapter 29, concerns the forces exerted by a magnetic field on a moving charge or on a current-carrying conductor and the torque exerted on a magnetic dipole. A second class concerns the production of a magnetic field by a current-carrying conductor or by moving charges. This chapter deals with problems of this second class.

The discovery that currents produce magnetic effects was made by Oersted in 1820. Figure 30–1, which shows a wire surrounded by a number of small magnets, illustrates a modification of his experiment. If there is no current in the wire, all the magnets are aligned with the earth's magnetic field. When a strong current is present, the magnets point so as to suggest that the lines of **B** form closed circles around the wire. This view is strengthened by the experiment of Fig. 30–2, which shows iron filings on a horizontal glass plate, through the center of which a current-carrying conductor passes.

Today we write the quantitative relationship between current i and the magnetic field **B** as

SEE p.633
34.13

$$\oint \mathbf{B} \cdot d\mathbf{l} = \mu_0 i, \qquad (30\text{–}1)$$

which is known as _Ampère's law_. Ampère, being an advocate of the action-at-a-distance point of view, did not formulate his results in terms of fields; this was first done by Maxwell. Ampère's law, including an important extension of it made later by Maxwell, is one of the basic equations of electromagnetism (see Table 34–2).

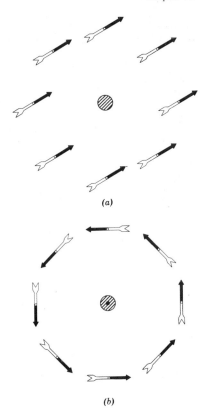

(a)

(b)

FIGURE 30-1   An array of compass needles near a wire. In (*a*) there is no current in the wire and the magnets line up with the earth's magnetic field. In (*b*) there is a strong current in the wire. The black ends of the compass needles are their north poles. The dot shows the current emerging from the page. As usual, the direction of a current is taken as the direction of flow of positive charge.

We can gain an appreciation of the way Ampère's law developed historically by considering a hypothetical experiment which has, in fact, much in common with experiments that were actually carried out. The experiment consists of measuring **B** at various distances r from a long straight wire of circular cross section and carrying a current i. This can be done by making quantitative the qualitative observation of Fig. 30–1*b*.

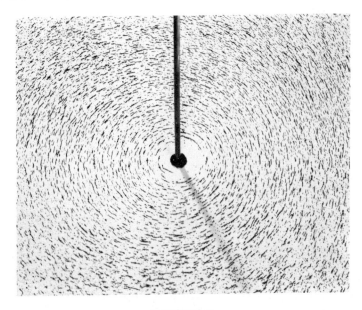

FIGURE 30-2

Let us put a small compass needle a distance r from the wire. Such a needle, a small magnetic dipole, tends to line up with an external magnetic field, with its north pole pointing in the direction of **B.** Figure 30–1b makes it clear that **B** at the site of the dipole is tangent to a circle of radius r centered on the wire.

If the current in the wire of Fig. 30–1b is reversed in direction, all the compass needles would reverse end-for-end. This experimental result leads to the "right-hand rule" for finding the direction of **B** near a wire carrying a current i: *Grasp the wire with the right hand, the thumb pointing in the direction of the current. The fingers will curl around the wire in the direction of* **B.**

Let us now turn the dipole through an angle θ from its equilibrium position. To do this, we must exert an external torque just large enough to overcome the restoring torque τ that will act on the dipole. τ, θ, and **B** are related by Eq. 29–11 ($\boldsymbol{\tau} = \boldsymbol{\mu} \times \mathbf{B}$), which can be written in terms of magnitudes as

$$\tau = \mu B \sin \theta \tag{30–2}$$

and in which μ is the magnitude of the magnetic moment of the dipole, θ being the angle between the vectors $\boldsymbol{\mu}$ and **B.** Even though we may not know the value of μ for the compass needle, we may take it to be a constant, independent of the position or orientation of the needle. Thus by measuring τ and θ in Eq. 30–2 we can obtain a relative measure of B for various distances r and for various currents i in the wire. The experimental results can be described by the proportionality

$$B \propto \frac{i}{r}. \tag{30–3}$$

We can convert this proportionality into an equality by inserting a proportionality constant. As for Coulomb's law, and for similar reasons (see Section 22–4), we do not write this constant simply as k but in a more complex form, namely $\mu_0/2\pi$, in which μ_0 (which has no connection with the dipole moment μ of Eq. 30–2) is called the *permeability constant.* Equation 30–3 then becomes

$$B = \frac{\mu_0 i}{2\pi r}, \tag{30–4}$$

which we choose to write in the form

$$(B)(2\pi r) = \mu_0 i. \tag{30–5}$$

The left side of Eq. 30–5 can easily be shown to be $\oint \mathbf{B} \cdot d\mathbf{l}$ for a path consisitng of a circle of radius r centered on the wire. For all points on this circle **B** has the same (constant) magnitude B and $d\mathbf{l}$, which is always tangent to the path of integration, points in the same direction as **B,** as Fig. 30–3 shows. Thus

$$\oint \mathbf{B} \cdot d\mathbf{l} = \oint B \, dl = B \oint dl = (B)(2\pi r),$$

$\oint dl$ being simply the circumference of the circle. In this special case, therefore, we can write the experimentally observed connection between the field and the current as

$$\oint \mathbf{B} \cdot d\mathbf{l} = \mu_0 i, \; = \mu_0 \int \vec{j} \cdot d\vec{s} \tag{30–1}$$

$$(i \; enclosed)$$

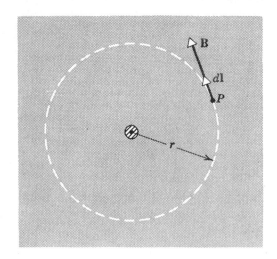

FIGURE 30-3   A circular path of integration surrounding a long straight wire. The central dot suggests a current i in the wire emerging from the page. Note that the angle between **B** and $d\mathbf{l}$ is zero so that $\mathbf{B} \cdot d\mathbf{l} = B\,dl$.

which is Ampère's law. A host of other experiments suggests that Eq. 30–1 is true in general (as long as time-varying electric fields are absent; see Section 34–4) for any magnetic field configuration, for any assembly of currents, and for any path of integration.

In applying Ampère's law in the general case, we construct a closed linear path in the magnetic field as shown in Fig. 30–4. This path is divided into elements of length $d\mathbf{l}$, and for each element the quantity $\mathbf{B} \cdot d\mathbf{l}$ is evaluated. Recall that $\mathbf{B} \cdot d\mathbf{l}$ has the magnitude $B\,dl\cos\theta$ and can be interpreted as the product of dl and the component of $\mathbf{B}$ ($= B\cos\theta$) parallel to $d\mathbf{l}$. The integral is the sum of the quantities $\mathbf{B} \cdot d\mathbf{l}$ for all path elements in the complete loop. The term i on the right of Eq. 30–1 is the net current (strictly, $i = \int \mathbf{j} \cdot d\mathbf{S}$) that passes through the area bounded by the closed path.

The permeability constant in Ampère's law has an assigned value of

$$\mu_0 = 4\pi \times 10^{-7} \text{ weber/amp-meter}$$

→ $p.564$

Both this and the permittivity constant (ϵ_0) occur in electromagnetic formulas when the mks system of units is used.

You may wonder why ϵ_0 in Coulomb's law is a measured quantity, whereas μ_0 in Ampère's law is an assigned quantity. The answer is that the ampere, which is the mks unit for the current i in Ampère's law, is defined by a laboratory technique (the

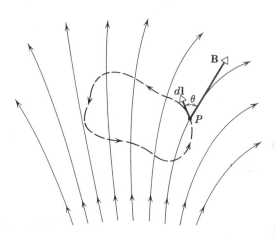

FIGURE 30-4   A path of integration in a magnetic field.

current balance) that involves forces exerted by magnetic fields and in which this same constant μ_0 appears. In effect, as we show in detail in Section 30–4, the size of the current that we agree to define as one ampere is adjusted so that μ_0 will have exactly the value assigned to it above. In Coulomb's law, on the other hand, the quantities **F,** q, and r are measured in ways in which the constant ϵ_0 plays no role. This constant must then take on the particular value that makes the left side of Coulomb's law equal to the right side; no arbitrary assignment is possible.

30–2 B Near a Long Wire

We have seen that the lines of **B** for a long straight wire carrying a current i are concentric circles centered on the wire and that B at a distance r from the wire is given by Eq. 30–4:

$$B = \frac{\mu_0 i}{2\pi r}. \tag{30–4}$$

We may regard this as an experimental result consistent with, and readily derivable from, Ampère's law.

It is interesting to compare Eq. 30–4 with the expression for the electric field near a long line of charge, or

$$E = \frac{1}{2\pi\epsilon_0} \frac{\lambda}{r}. \tag{24–7}$$

In each case there are multiplying constants, namely $\mu_0/2\pi$ and $1/2\pi\epsilon_0$, and factors describing the device responsible for the field, namely i and λ. Finally, each field varies as $1/r$.

Equation 24–7 may be derived from Gauss's law by relating the electric field at a Gaussian surface to the net charge within this surface. The (surface) integral in Gauss's law is evaluated for a closed cylindrical surface to which the lines of **E** are everywhere perpendicular.

Equation 30–4 may be derived from Ampère's law by relating the magnetic field at a path of integration to the net current that pierces this path. The (line) integral in Ampère's law is evaluated for a closed circular path to which the lines of **B** are everywhere tangent.

Example 1. Derive an expression for B at a distance r from the center of a long cylindrical wire of radius R, where $r < R$. The wire carries a current i_0, distributed uniformly over the cross section of the wire. Assume that the wire itself has no intrinsic magnetic effects.

Figure 30–5 shows a circular path of integration inside the wire. Symmetry suggests that **B** is tangent to the path as shown. Ampère's law,

$$\oint \mathbf{B} \cdot d\mathbf{l} = \mu_0 i,$$

gives
$$(B)(2\pi r) = \mu_0 i_0 \frac{\pi r^2}{\pi R^2},$$

since only the fraction of the current that passes through the path of integration is included in the factor i on the right. Solving for B and dropping the subscript on the current yields

$$B = \frac{\mu_0 i r}{2\pi R^2}.$$

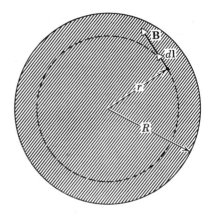

FIGURE 30–5 Example 1. A circular path of integration inside a wire. A current i_0, distributed uniformly over the cross section of the wire, emerges from the page.

At the surface of the wire ($r = R$) this equation reduces to the same expression as that found by putting $r = R$ in Eq. 30–4 ($B = \mu_0 i/2\pi R$).

30–3 Lines of B

Figure 30–6 shows the lines representing the field of **B** near a long straight wire. Note the increase in the spacing of the lines with increasing distance from the wire. This represents the $1/r$ decrease in B predicted by Eq. 30–4.

Figure 30–7 shows the resultant lines of **B** associated with a current in a wire that is oriented at right angles to a uniform external field $\mathbf{B}_e$. At any point the resultant **B** will be the vector sum of $\mathbf{B}_e$ and $\mathbf{B}_i$, where $\mathbf{B}_i$ is the magnetic field set up by the current in the wire. The fields $\mathbf{B}_e$ and $\mathbf{B}_i$ tend to cancel above the wire and to re-enforce each other below the wire. At point P in Fig. 30–7 $\mathbf{B}_e$ and $\mathbf{B}_i$ cancel exactly. Very near the wire the field is represented by circular lines and is essentially $\mathbf{B}_i$.

Michael Faraday, who originated the concept of lines of **B,** endowed them with more reality than they are currently given. He imagined that, like stretched rubber bands, they represent the site of mechanical forces. With this picture can we not visualize that the wire in Fig. 30–7 will be pushed up? Today we use lines of **B** largely for purposes of visualization. For quantitative calculations we use the field vectors, describing the force on the wire in Fig. 30–7, for example, from the relation $\mathbf{F} = i\mathbf{l} \times \mathbf{B}$.

In applying this relation to Fig. 30–7, we recall that **B** is always the external field in which the wire is immersed; that is, it is $\mathbf{B}_e$ and thus points to the right. Since **l** points out of the page, the magnetic force on the wire ($= i\mathbf{l} \times \mathbf{B}_e$) does indeed point up. It is necessary to use only the external field in such calculations because the field set up

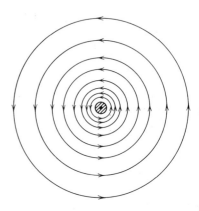

FIGURE 30–6 Lines of **B** near a long cylindrical wire. A current i, suggested by the central dot, emerges from the page.

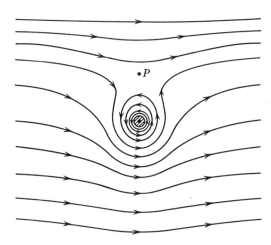

FIGURE 30–7 Lines of **B** near a long current-carrying wire immersed in a uniform external field $\mathbf{B}_e$ that points to the right. The current i is emerging from the page.

$\vec{B} = \vec{B}_e + \vec{B}_i$, NOTE $B_p = 0$

by the current in the wire cannot exert a force on the wire itself, just as the gravitational field of the earth cannot exert a force on the earth itself but only on another body. In Fig. 30–6, for example, there is no magnetic force on the wire because no external magnetic field is present.

30–4 Two Parallel Conductors

Figure 30–8 shows two long parallel wires separated by a distance d and carrying currents i_a and i_b. It is an experimental fact, noted by Ampère only one week after word of Oersted's experiments reached Paris, that two such conductors attract each other.

Wire a in Fig. 30–8 will produce a magnetic field $\mathbf{B}_a$ at all nearby points. The magnitude of $\mathbf{B}_a$, due to the current i_a, at the site of the second wire is, from Eq. 30–4,

$$B_a = \frac{\mu_0 i_a}{2\pi d}.$$

The right-hand rule shows that the direction of $\mathbf{B}_a$ at wire b is down, as shown in the figure.

Wire b, which carries a current i_b, finds itself immersed in an external magnetic field $\mathbf{B}_a$. A length l of this wire will experience a sideways magnetic force ($= i\mathbf{l} \times \mathbf{B}$)

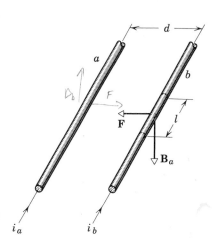

FIGURE 30–8 Two parallel wires that carry parallel currents attract each other.

and the magnitude of the force per unit length is

$$(\mathbf{F}_b/l) = i_b B_a = \frac{\mu_0 i_b i_a}{2\pi d}. \tag{30-6}$$

The vector rule of signs tells us that $\mathbf{F}_b/l$ lies in the plane of the wires and points to the left in Fig. 30–8.

We could have started with wire b, computed the magnetic field which it produces at the site of wire a, and then computed the force per unit length on wire a. $\mathbf{F}_a/l$ would, for parallel currents, point to the right. The forces that the two wires exert on each other are equal and opposite, as they must be according to Newton's law of action and reaction. For antiparallel currents the two wires repel each other.

This discussion reminds us of our discussion of the electric field between two point charges in Section 23–1. There we saw that the charges act on each other through the intermediary of the electric field. The conductors in Fig. 30–8 act on each other through the intermediary of the magnetic field. We think in terms of

$$\text{current} \rightleftharpoons \text{field} \rightleftharpoons \text{current}$$

and not, as in the action-at-a-distance point of view, in terms of

$$\text{current} \rightleftharpoons \text{current}.$$

The attraction between long parallel wires is used to define the ampere. Suppose that the wires are 1 meter apart ($d = 1.0$ meter) and that the two currents are equal ($i_a = i_b = i$). If this common current is adjusted until, by measurement, the force of attraction per unit length between the wires is 2×10^{-7} nt/meter, the current is defined to be 1 ampere. From Eq. 30–6,

$$\frac{F}{l} = \frac{\mu_0 i^2}{2\pi d} = \frac{(4\pi \times 10^{-7} \text{ weber/amp-m})(1 \text{ amp})^2}{(2\pi)(1 \text{ meter})}$$

$$= 2 \times 10^{-7} \text{ nt/meter}$$

as expected.

Note that μ_0 appears in this relation used to define the ampere. As we said in Section 30–1, μ_0 is assigned the (arbitrary) value of $4\pi \times 10^{-7}$ weber/amp-m, and the size of the current that we define as 1 ampere is adjusted to give the required force of attraction per unit length.

At the National Bureau of Standards primary measurements of current are made with a current balance. This consists of a carefully wound coil placed between two other coils, as in Fig. 30–9. The outer pair of coils is fastened to the table, and the inner one is hung from the arm of a balance. The coils are so connected that the current to be measured exists, as a common current, in all three of them.

The coils exert forces on one another—just as the parallel wires of Fig. 30–8 do—which can be measured by loading weights on the balance pan. The current is defined in terms of this measured force and the dimensions of the coils. The current balance is perfectly equivalent to the long parallel wires of Fig. 30–8 but is a much more practical arrangement. Current balance measurements are used primarily to standardize other, more convenient, secondary methods of measuring currents.

Example 2. Two parallel wires a distance d apart carry equal currents i in opposite directions. Find the magnetic field for points between the wires and at a distance x from one wire.

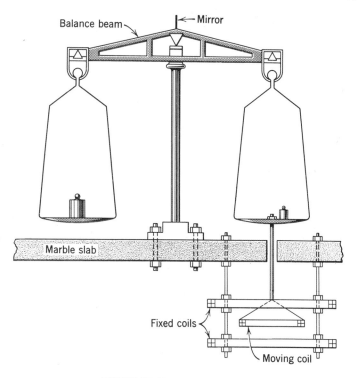

FIGURE 30–9   A current balance.

Study of Fig. 30–10 shows that $\mathbf{B}_a$ due to the current i_a and $\mathbf{B}_b$ due to the current i_b point in the same direction at P. Each is given by Eq. 30–4 ($B = \mu_0 i / 2\pi r$) so that

$$B = B_a + B_b = \frac{\mu_0 i}{2\pi} \left(\frac{1}{x} + \frac{1}{d-x} \right).$$

This relationship does not hold for points inside the wires because Eq. 30–4 is not valid there.

30–5 B for a Solenoid

A *solenoid* is a long wire wound in a close-packed helix and carrying a current i. We assume that the helix is very long compared with its diameter. What is the nature of the field of **B** that is set up?

For points close to a single turn of the solenoid, the observer is not aware that the wire is bent in an arc. The wire behaves magnetically almost like a long straight wire, and the lines of **B** due to this single turn are almost concentric circles.

The solenoid field is the vector sum of the fields set up by all the turns that make up the solenoid. Figure 30–11, which shows a "solenoid" with widely spaced turns, suggests that the fields tend to cancel between the wires. It suggests also that, at points inside the solenoid and reasonably far from the wires, **B** is parallel to the solenoid axis.

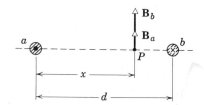

FIGURE 30–10   Example 2.

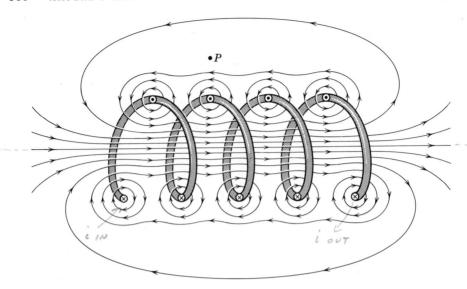

FIGURE 30–11 A loosely wound solenoid.

In the limiting case of adjacent square tightly packed wires, the solenoid becomes essentially a cylindrical current sheet and the requirements of symmetry then make the statement just given necessarily true. We assume that it is true in what follows.

For points such as P in Fig. 30–11 the field set up by the upper part of the solenoid turns (marked $\odot$) points to the left and tends to cancel the field set up by the lower part of the solenoid turns (marked $\otimes$), which points to the right. As the solenoid becomes more and more ideal, that is, as it approaches the configuration of an infinitely long cylindrical current sheet, the magnetic field at outside points approaches zero. Taking the external field to be zero is not a bad assumption for a practical solenoid if its length is much greater than its diameter and if we consider only external points near the central region of the solenoid, that is, away from the ends. Figure 30–12 shows the lines of **B** for a real solenoid, which is far from ideal in that the length is not much greater than the diameter. Even here the spacing of the lines of **B** in the central plane shows that the external field is much weaker than the internal field.

Let us apply Ampère's law,

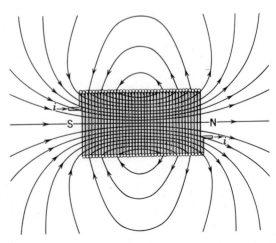

FIGURE 30–12 A solenoid of finite length. The right end, from which lines of **B** emerge, behaves like the north pole of a compass needle. The left end behaves like the south pole.

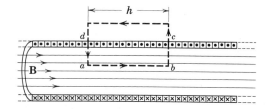

FIGURE 30–13 A section of an ideal solenoid, made of adjacent square turns, equivalent to an infinitely long cylindrical current sheet.

$$\oint \mathbf{B} \cdot d\mathbf{l} = \mu_0 i,$$

to the rectangular path $abcd$ in the ideal solenoid of Fig. 30–13. We write the integral $\oint \mathbf{B} \cdot d\mathbf{l}$ as the sum of four integrals, one for each path segment:

$$\oint \mathbf{B} \cdot d\mathbf{l} = \int_a^b \mathbf{B} \cdot d\mathbf{l} + \int_b^c \mathbf{B} \cdot d\mathbf{l} + \int_c^d \mathbf{B} \cdot d\mathbf{l} + \int_d^a \mathbf{B} \cdot d\mathbf{l}.$$

The first integral on the right is Bh, where h is the arbitrary length of the path from a to b. Note that path ab, though parallel to the solenoid axis, need not coincide with it.

The second and fourth integrals are zero because for every element of these paths $\mathbf{B}$ is at right angles to the path. This makes $\mathbf{B} \cdot d\mathbf{l}$ zero and thus the integrals are zero. The third integral, which includes the part of the rectangle that lies outside the solenoid, is zero because we have taken $\mathbf{B}$ as zero for all external points for an ideal solenoid.

Thus $\oint \mathbf{B} \cdot d\mathbf{l}$ for the entire rectangular path has the value Bh. The net current i that passes through the area bounded by the path of integration is not the same as the current i_0 in the solenoid because the path of integration encloses more than one turn. Let n be the number of turns per unit length; then

$$i = i_0(nh).$$

Ampère's law then becomes

$$Bh = \mu_0 i_0 nh$$

or

$$B = \mu_0 i_0 n. \tag{30–7}$$

Although Eq. 30–7 was derived for an infinitely long ideal solenoid, it holds quite well for actual solenoids for internal points near the center of the solenoid. It shows that B does not depend on the diameter or the length of the solenoid and that B is constant over the solenoid cross section. A solenoid is a practical way to set up a known uniform magnetic field for experimentation, just as a parallel-plate capacitor is a practical way to set up a known uniform electric field.

Example 3. A solenoid is 1.0 meter long and 3.0 cm in mean diameter. It has five layers of windings of 850 turns each and carries a current of 5.0 amp.

(a) What is B at its center? From Eq. 30–7,

$$B = \mu_0 i_0 n = (4\pi \times 10^{-7} \text{ weber/amp-m})(5.0 \text{ amp})(5 \times 850 \text{ turns/meter})$$

$$= 2.7 \times 10^{-2} \text{ weber/meter}^2.$$

We can use Eq. 30–7 even if the solenoid has more than one layer of windings because the diameter of the windings does not enter.

(b) What is the magnetic flux Φ_B for a cross section of the solenoid at its center? To the extent that $\mathbf{B}$ is constant, we can calculate the flux from

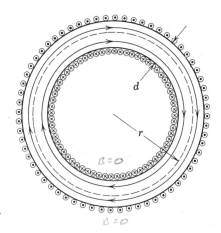

FIGURE 30–14 Example 4. A toroid.

$$\Phi_B = \int \mathbf{B} \cdot d\mathbf{S} = BA,$$

where A is the effective cross-sectional area. Let us assume that A represents the area of a circular disk whose diameter is the mean diameter of the windings (3.0 cm). The effective area can then be shown to be 7.1×10^{-4} meter², and

$$\Phi_B = BA = (2.7 \times 10^{-2} \text{ weber/meter}^2)(7.1 \times 10^{-4} \text{ meter}^2)$$

$$= 1.9 \times 10^{-5} \text{ weber.}$$

Example 4. *A toroid.* Figure 30–14 shows a toroid, which may be described as a solenoid of finite length bent into the shape of a doughnut. Calculate **B** at interior points.

From symmetry the lines of **B** form concentric circles inside the toroid, as shown in the figure. Let us apply Ampère's law to the circular path of integration of radius r:

$$\oint \mathbf{B} \cdot d\mathbf{l} = \mu_0 i$$

or

$$(B)(2\pi r) = \mu_0 i_0 N,$$

where i_0 is the current in the toroid windings and N is the total number of turns. This gives

$$B = \frac{\mu_0}{2\pi} \frac{i_0 N}{r}.$$

In contrast to the solenoid, B is not constant over the cross section of a toroid. Show from Ampère's law that B equals zero for points outside an ideal toroid.

30–6 The Biot-Savart Law

Ampère's law can be used to calculate magnetic fields only if the symmetry of the current distribution is high enough to permit the easy evaluation of the line integral $\oint \mathbf{B} \cdot d\mathbf{l}$. This requirement limits the usefulness of the law in practical problems. The law does not fail; it simply becomes difficult to apply in a useful way.

Similarly, in electrostatics, Gauss's law can be used to calculate electric fields only if the symmetry of the charge distribution is high enough to permit the easy evaluation of the surface integral $\oint \mathbf{E} \cdot d\mathbf{S}$. We can, for example, use Gauss's law to find the electric field due to a long uniformly charged rod but we cannot apply it usefully to an electric dipole, for the symmetry is not high enough in this case.

To compute **E** at a given point for an arbitrary charge distribution, we divided the distribution into charge elements dq and (see Section 23–4) we used Coulomb's law to calculate the field contribution $d\mathbf{E}$ due to each element at the point in question. We found the field $\mathbf{E}$ at that point by adding, that is, by integrating, the field contributions $d\mathbf{E}$ for the entire distribution.

We now describe a similar procedure for computing $\mathbf{B}$ at any point due to an arbitrary current distribution. We divide the current distribution into current elements and, using the law of Biot and Savart (which we describe below), we calculate the field contribution $d\mathbf{B}$ due to each current element at the point in question. We find the field $\mathbf{B}$ at that point by integrating the field contributions for the entire distribution.

Figure 30–15 shows an arbitrary current distribution consisting of a current i in a curved wire. The figure also shows a typical current element; it is a length $d\mathbf{l}$ of the conductor carrying a current i. Its direction is that of the tangent to the conductor (dashed line). A current element cannot exist as an isolated entity because a way must be provided to lead the current into the element at one end and out of it at the other. Nevertheless, we can think of an actual circuit as made up of a large number of current elements placed end to end.

Let P be the point at which we want to know the magnetic field $d\mathbf{B}$ associated with the current element. According to the Biot-Savart law, $d\mathbf{B}$ is given in magnitude by

$$dB = \frac{\mu_0 i}{4\pi} \frac{dl \sin \theta}{r^2},\qquad(30\text{--}8)$$

where $\mathbf{r}$ is a displacement vector from the element to P and θ is the angle between this vector and $d\mathbf{l}$. The direction of $d\mathbf{B}$ is that of the vector $d\mathbf{l} \times \mathbf{r}$. In Fig. 30–15, for example, $d\mathbf{B}$ at point P for the current element shown is directed into the page at right angles to the plane of the figure. Note that Eq. 30–8, being an inverse square law that describes the magnetic field due to a current element, may be viewed as the magnetic equivalent of Coulomb's law, which is an inverse square law that describes the electric field due to a charge element.

The law of Biot and Savart may be written in vector form as

$$d\mathbf{B} = \frac{\mu_0 i}{4\pi} \frac{d\mathbf{l} \times \mathbf{r}}{r^3}.\qquad(30\text{--}9)$$

This formulation reduces to that of Eq. 30–8 when expressed in terms of magnitudes; it also gives complete information about the direction of $d\mathbf{B}$, namely that it is the same as the direction of the vector $d\mathbf{l} \times \mathbf{r}$.

The resultant field at P is found by integrating Eq. 30–9, or

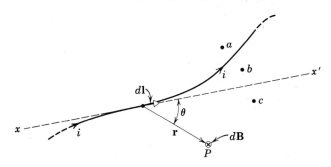

FIGURE 30–15   The current element $d\mathbf{l}$ establishes a magnetic field contribution $d\mathbf{B}$ at point P. The wire need not lie entirely in the plane of the page, although we have assumed that the selected tangent vector $d\mathbf{l}$ does.

$$\mathbf{B} = \int d\mathbf{B}, \tag{30-10}$$

where the integral is a vector integral.

Example 5. *A circular current loop.* Figure 30–16 shows a circular loop of radius R carrying a current i. Calculate $\mathbf{B}$ for points on the axis.

The vector $d\mathbf{l}$ for a current element at the top of the loop points perpendicularly out of the page. The angle θ between $d\mathbf{l}$ and $\mathbf{r}$ is 90°, and the plane formed by $d\mathbf{l}$ and $\mathbf{r}$ is normal to the page. The vector $d\mathbf{B}$ for this element is at right angles to this plane and thus lies in the plane of the figure and at right angles to $\mathbf{r}$, as the figure shows.

Let us resolve $d\mathbf{B}$ into two components, one, $d\mathbf{B}_{\parallel}$, along the axis of the loop and another, $d\mathbf{B}_{\perp}$, at right angles to the axis. Only $d\mathbf{B}_{\parallel}$ contributes to the total field $\mathbf{B}$ at point P. This follows because the components $d\mathbf{B}_{\parallel}$ for all current elements lie on the axis and add directly; however, the components $d\mathbf{B}_{\perp}$ point in different directions perpendicular to the axis, and their resultant for the complete loop is zero, from symmetry. Thus

$$\mathbf{B} = \int d\mathbf{B}_{\parallel},$$

where the integral is a simple scalar integration over the current elements.

For the current element shown in Fig. 30–16 we have, from the Biot-Savart law (Eq. 30–8),

$$dB = \frac{\mu_0 i}{4\pi} \frac{dl \sin 90°}{r^2}.$$

We also have

$$dB_{\parallel} = dB \cos \alpha.$$

Combining gives

$$dB_{\parallel} = \frac{\mu_0 i \cos \alpha \, dl}{4\pi r^2}.$$

Figure 30–16 shows that r and α are not independent of each other. Let us express each in terms of a new variable x, the distance from the center of the loop to the point P. The relationships are

$$r = \sqrt{R^2 + x^2}$$

and

$$\cos \alpha = \frac{R}{r} = \frac{R}{\sqrt{R^2 + x^2}}.$$

Substituting these values into the expression for $dB_{\parallel}$ gives

$$dB_{\parallel} = \frac{\mu_0 i R}{4\pi (R^2 + x^2)^{3/2}} \, dl.$$

Note that i, R, and x have the same values for all current elements. Integrating this equation, noting that $\int dl$ is simply the circumference of the loop ($= 2\pi R$), yields

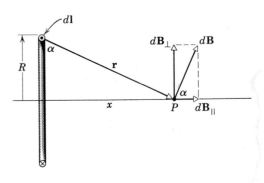

FIGURE 30–16   Example 5. A ring of radius R carrying a current i.

Table 30–1 SOME DIPOLE EQUATIONS

Property	Dipole Type	Equation
Torque in an external field	electric	$\tau = \mathbf{p} \times \mathbf{E}$
	magnetic	$\tau = \mu \times \mathbf{B}$
Energy in an external field	electric	$U = -\mathbf{p} \cdot \mathbf{E}$
	magnetic	$U = -\mu \cdot \mathbf{B}$
Field at distant points along axis	electric	$E = \dfrac{1}{2\pi\epsilon_0} \dfrac{p}{x^3}$
	magnetic	$B = \dfrac{\mu_0}{2\pi} \dfrac{\mu}{x^3}$
Field at distant points along perpendicular bisector	electric	$E = \dfrac{1}{4\pi\epsilon_0} \dfrac{p}{x^3}$
	magnetic	$B = \dfrac{\mu_0}{4\pi} \dfrac{\mu}{x^3}$

Handwritten margin notes:

$LINE$

$B = \dfrac{\mu_0 \, i}{2\pi r}$

$E = \dfrac{1}{2\pi\epsilon_0} \dfrac{\lambda}{r}$

$SOLINOID$

$B = \mu_0 i_0 n$

$TOROID$

$B = \dfrac{\mu_0}{2\pi} \dfrac{i_0 N}{r}$

$$B = \int dB_\parallel = \frac{\mu_0 i R}{4\pi (R^2 + x^2)^{3/2}} \int dl$$

$$= \frac{\mu_0 i R^2}{2(R^2 + x^2)^{3/2}}. \tag{30–11}$$

If we put $x \gg R$ in Example 5 so that points close to the loop are not considered, Eq. 30–11 reduces to

$$B = \frac{\mu_0 i R^2}{2x^3}.$$

Recalling that πR^2 is the area A of the loop and considering loops with N turns, we can write this equation as

$$B = \frac{\mu_0}{2\pi} \frac{(NiA)}{x^3} = \frac{\mu_0}{2\pi} \frac{\mu}{x^3},$$

where μ is the magnetic dipole moment of the current loop. This reminds us of the result derived in Problem 10, Chapter 23 [$E = (1/2\pi\epsilon_0)(p/x^3)$], which is the formula for the electric field on the axis of an electric dipole.

Thus we have shown in two ways that a current loop can be regarded as a magnetic dipole: It experiences a torque given by $\tau = \mu \times \mathbf{B}$ when placed in an external magnetic field (Eq. 29–11); it generates its own magnetic field given, for points on the axis, by the equation just developed.

Take 30–1 is a summary of the properties of electric and magnetic dipoles.

QUESTIONS

1. Can the path of integration around which we apply Ampère's law pass through a conductor?

2. Suppose we set up a path of integration around a cable that contains twelve wires with different currents (some in opposite directions) in each wire. How do we calculate i in Ampère's law in such a case?

3. Is **B** constant in magnitude for points that lie on a given line of **B**?

4. Discuss and compare Gauss's law and Ampère's law.

5. A current is set up in a long copper pipe. Is there a magnetic field (*a*) inside and (*b*) outside the pipe?

6. Equation 30–4 ($B = \mu_0 i / 2\pi r$) suggests that a strong magnetic field is set up at points near a long wire carrying a current. Since there is a current *i* and a magnetic field **B,** why is there not a force on the wire in accord with the equation $\mathbf{F} = i\mathbf{l} \times \mathbf{B}$?

7. In electronics, wires that carry equal but opposite currents are often twisted together to reduce their magnetic effect at distant points. Why is this effective?

8. A beam of 20-MeV protons emerges from a cyclotron. Is a magnetic field associated with these particles?

9. Comment on this statement: "The magnetic field outside a long solenoid cannot be zero, if only for the reason that the helical nature of the windings produces a field for external points like that of a straight wire along the solenoid axis."

10. A current is sent through a vertical spring from whose lower end a weight is hanging; what will happen?

11. Does Eq. 30–7 ($B = \mu_0 i_0 n$) hold for a solenoid of square cross section?

12. What is the direction of the magnetic fields at points *a*, *b*, and *c* in Fig. 30–15 set up by the particular current element shown?

13. In a circular loop of wire carrying a current *i*, is **B** uniform for all points within the loop?

14. Discuss analogies and differences between Coulomb's law and the Biot-Savart law.

15. Equation 30–9 gives the law of Biot and Savart in vector form. Write its electrostatic equivalent [that is, Eq. 23–6, or $dE = dq/(4\pi\epsilon_0 r^2)$] in vector form.

16. How might you measure the magnetic dipole moment of a compass needle?

17. What is the basis for saying that a current loop is a magnetic dipole?

PROBLEMS

1(1). Evaluate $\oint \mathbf{B} \cdot d\mathbf{l}$ for the three paths shown in Fig. 30–17. The current is 5.0 amperes. *Answer:* (*a*) -6.3×10^{-6} webers/meter. (*b*) Zero. (*c*) $+6.3 \times 10^{-6}$ webers/meter.

2(1). Eight wires cut the page perpendicularly at the points shown in Fig. 30–18. A wire labeled with the integer *k* ($k = 1, 2, \ldots$ 8) bears the current ki_0. For those with odd *k*, the current flows up out of the page; for those with even *k* it flows down into the page. Evaluate $\oint \mathbf{B} \cdot d\mathbf{l}$ along the closed path shown in the direction indicated by the single arrowhead.

3(1). Show that it is impossible for a uniform magnetic field **B** to drop abruptly to zero as one moves at right angles to it, as suggested by the horizontal arrow in Fig. 30–19 (see point *a*). In actual magnets fringing of the lines of force always occurs, which means that **B** approaches zero in a continuous and gradual way. (Hint: Apply Ampère's law to the rectangular path shown by the dashed lines.)

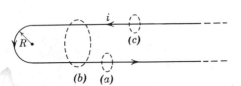

FIGURE 30–17 Problem 1(1).

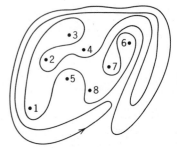

FIGURE 30–18 Problem 2(1).

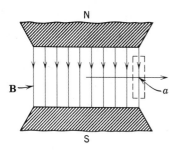

FIGURE 30-19 Problem 3(1).

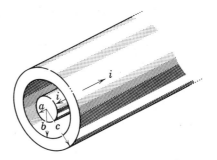

FIGURE 30-20 Problem 10(2).

4(2). A #10 gauge bare copper wire (0.10 in. in diameter) can carry a current of 50 amp without overheating. For this current, what is B at the surface of the wire?

5(2). A surveyor is using a compass 20 ft below a power line in which there is a steady current of 100 amp. Will this interfere seriously with the compass reading? The horizontal component of the earth's magnetic field at the site is 0.20 gauss.
Answer: Yes.

6(2). A long straight wire carries a current i (63 amps). A proton is projected parallel to the wire and in the direction of the current at a distance d (10 cm) and speed v (10^6 meters/sec). What force acts on the proton?

7(2). If a point charge of magnitude $+q$ and speed v is located a distance d from the axis of a long straight wire carrying a current i and is traveling perpendicular to the axis of the wire, what is the direction and magnitude of the force acting on it if the charge is moving (*a*) toward, or (*b*) away from the wire?

Answer: (*a*) $\dfrac{\mu_0}{2\pi}\dfrac{qvi}{r}$, antiparallel to i. (*b*) Same magnitude, parallel to i.

8(2). A long straight wire carries a current of 50 amp. An electron, traveling at 1.0×10^7 meters/sec, is 5.0 cm from the wire. What force acts on the electron if the electron velocity is directed (*a*) toward the wire, (*b*) parallel to the wire, and (*c*) at right angles to the directions defined by (*a*) and (*b*)?

9(2). A long solid cylindrical copper wire of radius R carries a current i distributed uniformly over the cross section of the wire. Sketch the magnitude of the magnetic field B as a function of the distance r from the axis of the wire for (*a*) $r < R$, and (*b*) $r > R$.

10(2). A long coaxial cable consists of two concentric conductors with the dimensions shown in Fig. 30–20. There are equal and opposite currents i in the conductors. (*a*) Find the magnetic field B at r within the inner conductor ($r < a$). (*b*) Find B between the two conductors ($a < r < b$). (*c*) Find B within the outer conductor ($b < r < c$). (*d*) Find B outside the cable ($r > c$).

11(2). Figure 30–21 shows a hollow cylindrical conductor of radii a and b that carries a current i uniformly spread over its cross section. (*a*) Show that the magnetic field B for points inside the body of the conductor (that is, $a < r < b$) is given by

$$B = \frac{\mu_0 i}{2\pi\,(b^2 - a^2)}\ \frac{r^2 - a^2}{r}.$$

Check this formula for the limiting case of $a = 0$. (*b*) Make a rough plot of $B(r)$ from $r = 0$ to $r \to \infty$.

12(2). A long copper wire carries a current

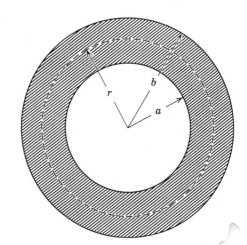

FIGURE 30-21 Problem 11(2).

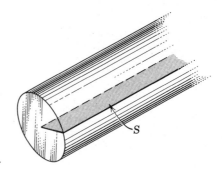

FIGURE 30-22 Problem 12(2).

of 10 amp. Calculate the magnetic flux per meter of wire for a plane surface S inside the wire, as in Fig. 30–22.

13(2). A conductor consists of an infinite number of adjacent wires, each infinitely long and carrying a current i. Show that the lines of **B** will be as represented in Fig. 30–23 and that B for all points in front of the infinite current sheet will be given by

$$B = \tfrac{1}{2}\mu_0 n i,$$

where n is the number of conductors per unit length.

14(2). A long straight conductor has a circular cross section of radius R and carries a current i. Inside the conductor there is a cylindrical hole of radius a whose axis is parallel to the conductor axis at a distance b from it. Use superposition ideas and find an expression for the magnetic field B inside the hole. (Hint: It is helpful to work with current densities.)

15(3). A long wire carrying a current of 100 amp is placed in a uniform external magnetic field of 50 gauss. The wire is at right angles to this external field. Locate the points at which the resultant magnetic field is zero.
Answer: **B** = 0 along a line parallel to the wire and 4.0 mm from it. If the current is horizontal and points toward the observer, and the external field points horizontally from left to right, the line is directly above the wire.

16(4). Two long straight wires pass near one another at right angles. If the wires are free to move, describe what happens when currents are sent through them.

17(4). A messy loop of limp wire is placed on a table and anchored at points a and b as shown in Fig. 30–24. If a current i is now passed through the wire will it try to form a circular loop or will it try to bunch up further?

18(4). Figure 30–25 shows a long wire carrying a current of 30 amp. The rectangular loop carries a current of 20 amp. Calculate the resultant force acting on the loop. Assume that $a = 1.0$ cm, $b = 8.0$ cm, and $l = 30$ cm.

19(4). Two long parallel wires of negligible radius are a distance d apart. Let there be a current i in each wire (a) in the same direction and (b) in opposite directions. Letting r be the perpendicular distance from the center of one wire, find the magnetic field in the region between the wires at points in the plane of the two wires.

Answer: (a) $\dfrac{\mu_0 i}{2\pi}\left[\dfrac{d-2r}{r(d-r)}\right]$. (b) $\dfrac{\mu_0 i}{2\pi}\left[\dfrac{d}{r(d-r)}\right]$.

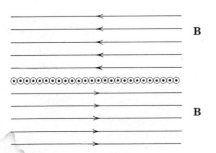

FIGURE 30-23 Problem 13(2).

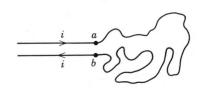

FIGURE 30-24 Problem 17(4).

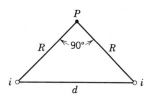

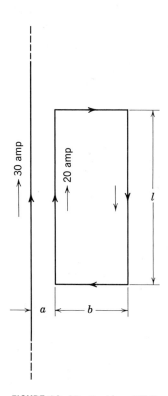

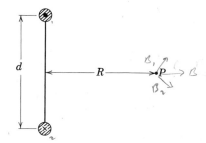

FIGURE 30-26   Problem 21(4).

FIGURE 30-25   Problem 18(4).

FIGURE 30-27   Problem 22(4).

20(4). Two long, parallel #10 gauge copper wires (diameter = 0.10 in.) carry currents of 10 amp in opposite directions. If their centers are 2.0 cm apart (a) calculate the flux per meter that exists in the space between the axes of the wires. (b) What fraction of the flux lies inside the wires?

21(4). Two long straight wires a distance d (10 cm) apart each carry a current i (100 amps). Figure 30–26 shows a cross section, with the wires running perpendicular to the page and point P lying on the perpendicular bisector of d. Find the magnitude and direction of the magnetic field at P when the current in the left-hand wire is out of the page and the current in the right-hand wire is (a) in the same direction and (b) in the opposite direction.
Answer: (a) 4.0×10^{-4} webers/meter², to the left. (b) 4.0×10^{-4} webers/meter², up.

22(4). Two long wires a distance d apart carry equal antiparallel currents i, as in Fig. 30–27. Show that B at point P, which is equidistant from the wires, is given by

$$B = \frac{2\mu_0 id}{\pi (4R^2 + d^2)}.$$

23(4). Four long #10 gauge copper wires are parallel to each other, their perpendicular cross section forming a square 20 cm on edge. A 20-amp current is set up in each wire in the direction shown in Fig. 30–28. What are (a) the magnitude and (b) the direction of **B** at the center of the square?
Answer: (a) 8.0×10^{-5} webers/meter². (b) Up.

24(4). In Problem 23(4) what are the components of the force per unit length (nt/meter) acting on the lower left wire?

25(4). Suppose, in Fig. 30–28 that the currents are all in the same direction. What is the force per unit length (magnitude and direction) on any one wire? In the analogous case of parallel motion of charged particles in a plasma this is known as the pinch effect.
Answer: $\dfrac{3\sqrt{2}\mu_0 i}{4\pi a}$, toward the center of the square.

26(5). A 200-turn solenoid having a length of 25 cm and a diameter of 10 cm carries a current

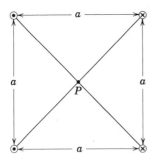

FIGURE 30-28 Problems 23(4), 24(4), 25(4).

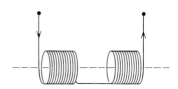

FIGURE 30-29 Problem 28(5).

of 0.30 amp. (*a*) What is the magnitude of the magnetic field **B** near the center of the solenoid? (*b*) What is the magnetic flux through an annular ring having an inside diameter of 2.0 cm and an outside diameter of 8.0 cm if the plane of the ring is perpendicular to the axis of the solenoid?

27(5). A toroid having a 5.0 cm $\times$ 5.0 cm cross section and an inside radius of 15 cm has 500 turns of wire and carries a current of 0.80 amp. (*a*) What is the magnitude of the magnetic field **B** in the center of the toroid (i.e. at a radius of 17.5 cm)? (*b*) What is the magnetic flux through the cross section?

Answer: (*a*) 4.6×10^{-4} webers/meter². (*b*) 1.2×10^{-6} webers.

28(5). Two identical short solenoids are located close together with their axes coinciding as shown in Fig. 30-29. They are connected in series so that the current in each goes around in the same sense. (*a*) Do the solenoids attract or repel each other? (*b*) How does the magnitude of the force depend upon the current?

29(6). A cylindrical bar of copper 50 cm long having a diameter of 2.0 cm is bent into a circular ring. An electric potential of 1.0 volt is applied to the ends of the bar. (*a*) What is the current in the bar? (*b*) What is the magnitude of the magnetic field at the center of the ring? (*c*) What is the power dissipated in the ring to establish this field?

Answer: (*a*) 3.7×10^4 amp. (*b*) 0.29 webers/meter². (*c*) 3.7×10^4 watts.

30(6). A circular copper loop of radius 10 cm carries a current of 15 amp. At its center is placed a second loop of radius 1.0 cm, having 50 turns and a current of 1.0 amp. (*a*) What magnetic field **B** does the large loop set up at its center? (*b*) What torque acts on the small loop? Assume that the planes of the two loops are at right angles and that the magnetic field **B** provided by the large loop is essentially uniform throughout the volume occupied by the small loop.

31(6). The wire shown in Fig. 30-30 carries a current *i*. What is the magnetic field **B** at the center *C* of the semicircle arising from (*a*) each straight segment of length *l*, (*b*) the semicircular segment of radius *R*, and (*c*) the entire wire?

Answer: (*a*) Zero. (*b*) $\dfrac{\mu_0 i}{4R}$, into the page. (*c*) $\dfrac{\mu_0 i}{4R}$, into the page.

32(6). A straight conductor is split into identical semicircular turns as shown in Fig. 30-31. What is the magnetic field at the center *C* of the circular loop so formed?

33(6). A long wire is bent into the shape shown in Fig. 30-32, without cross-contact at *P*. Determine the magnitude and direction of **B** at the center *C* at the circular portion when the current *i* flows as indicated.

Answer: $\dfrac{\mu_0 i}{2R} \left(1 + \dfrac{1}{\pi} \right)$, out of the page.

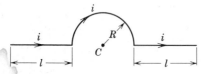

FIGURE 30-30 Problem 31(6).

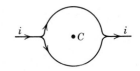

FIGURE 30-31 Problem 32(6).

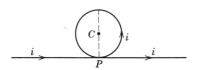

FIGURE 30-32   Problem 33(6).

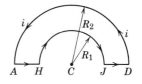

FIGURE 30-33   Problem 34(6).

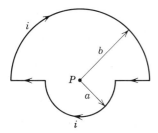

FIGURE 30-34   Problem 35(6).

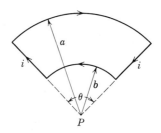

FIGURE 30-35   Problem 36(6).

34(6). Use the Biot-Savart law to calculate the magnetic field **B** at C, the common center of the semicircular arcs AD and HJ, of radii R_2 and R_1 respectively, forming part of the circuit $AHJDA$ and carrying current i, as shown in Fig. 30–33.

35(6). You are given a closed circuit with radii a and b, as shown in Fig. 30–34, carrying a current i. (a) What is the magnitude and direction of **B** at point P? (b) Find the dipole moment of the circuit.

Answer: (a) $\dfrac{\mu_0 i}{4}\left(\dfrac{1}{a}+\dfrac{1}{b}\right)$, into the page. (b) $\dfrac{i\pi}{2}\,(a^2+b^2)$, into the page.

[handwritten: HOW CAN $\mu \times B \parallel B$?]

36(6). Consider the circuit of Fig. 30–35. The curved segments are part of circles of radii a and b. The straight segments are along the radii. Find the magnetic field **B** at P, assuming a current i in the circuit.

37(6). A circular loop of radius R (see Fig. 30–36) is constructed from wire of resistivity ρ and cross section A. Two wires connect a battery of emf $\mathcal{E}$, located at the center, to the noop. Find **B** at the center of the loop. (Hint: Break up the circuit into two loops, find **B** for each loop, and then use the loop theorem to relate the two expressions.)

Answer: Zero.

38(6). A straight wire segment of length l carries a current i. (a) Show that the field **B** to be associated with this segment, at a distance R from the segment along a perpendicular bisector (see Fig. 30–37), is given in magnitude by

$$B = \frac{\mu_0 i}{2\pi R}\,\frac{l}{(l^2+4R^2)^{1/2}}.$$

(b) Does this expression reduce to an expected result as $l \to \infty$?

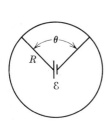

FIGURE 30-36   Problem 37(6).

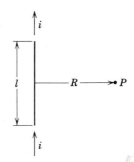

FIGURE 30-37   Problem 38(6).

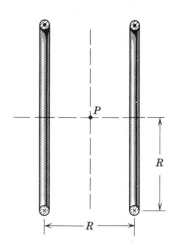

FIGURE 30-38   Problem 42(6).

39(6). A one-turn square loop of wire of edge *a* carries a current *i*. Show that *B* at the center is given by

$$B = \frac{2\sqrt{2}\,\mu_0 i}{\pi a}.$$

Use the result of Problem 38(6).

40(6). You are given a length *l* of wire in which a current *i* may be established. The wire may be formed into a circle or a square. Which yields the larger value for *B* at the central point? See Problem 39(6).

41(6). A one-turn square loop of wire of edge *a* carries a current *i*. (*a*) Show that *B* for a point on the axis of the loop and a distance *x* from its center is given by

$$B = \frac{4\mu_0 i a^2}{\pi (4x^2 + a^2)(4x^2 + 2a^2)^{1/2}}.$$

(*b*) Does this reduce to the result of Problem 39(6) for *x* = 0? (*c*) Does the square loop behave like a dipole for points such that *x* ≫ *a*? If so, what is its dipole moment?

Answer: (*b*) *ia²*.

42(6). *Hemholtz coils.* Two 300-turn coils are placed a distance apart equal to their radius, as in Fig. 30–38. For *R* = 5.0 cm and *i* = 50 amp, plot *B* as a function of distance *x* along the common axis over the range *x* = −5.0 cm to *x* = +5.0 cm, taking *x* = 0 at point *P*. (Such coils provide an especially uniform field of **B** near point *P*.)

43(6). A plastic disk of radius R has a charge *q* uniformly distributed over its surface. If the disk is rotated at an angular frequency *ω* about its axis, show that (*a*) the magnitude of the magnetic field at the center of the disk is

$$B = \frac{\mu_0 \omega q}{2\pi R}$$

and (*b*) the magnetic dipole moment of the disk is

$$\mu = \frac{\omega q R^2}{4}.$$

(Hint: The rotating disk is equivalent to an array of current loops; see Example 5.)

CHAPTER 31

Faraday's Law

31–1 Faraday's Experiments

Faraday's law of induction, which is one of the basic equations of electromagnetism (see Table 34–2) was deduced to explain a number of simple experiments carried out by Michael Faraday in England in 1831 and by Joseph Henry in the United States at about the same time.

Figure 31–1 shows the terminals of a coil connected to a galvanometer. Normally we would not expect this instrument to deflect because there seems to be no electromotive force in this circuit; but if we push a bar magnet toward the coil a remarkable thing happens. While the magnet is moving, the galvanometer deflects, showing that a current has been set up in the coil. If the magnet is held stationary with respect to the coil, the galvanometer does not deflect. If the magnet is moved away from the coil, the galvanometer again deflects, but in the opposite direction, which means

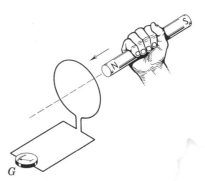

FIGURE 31–1 Galvanometer *G* deflects while the magnet is moving with respect to the coil.

that the current in the coil is in the opposite direction. If we turn the magnet end for end the experiment works as described but the deflections are reversed. Further experimentation shows that what matters is the relative motion of the magnet and the coil. It makes no difference whether the magnet is moved toward the coil or the coil toward the magnet.

The current that appears in this experiment is called an *induced current* and is said to be set up by an *induced electromotive force*. Faraday was able to deduce from experiments like this the law that gives their magnitude and direction. Such emfs are important in practice. It is almost certain that the lights in the room in which you are reading this book are operated from an induced emf produced in a commercial electric generator.

In another experiment the apparatus of Fig. 31–2 is used. The coils are placed close together but at rest with respect to each other. When the switch S is closed, thus setting up a steady current in the right-hand coil, the galvanometer deflects momentarily; when the switch is opened, thus interrupting this current, the galvanometer again deflects momentarily, but in the opposite direction.

Experiment shows that there will be an induced emf in the left coil of Fig. 31–2 whenever the current in the right coil is changing. It is the rate at which the current is changing and not the size of the current that counts.

31–2   Faraday's Law of Induction

Faraday had the insight to perceive that the change in the flux Φ_B of magnetic field for the left coil in the preceding experiments is the important common factor. This flux may be set up by a bar magnet or a current loop. Faraday's law says that the induced emf $\mathcal{E}$ in a circuit is equal to the negative rate at which the flux through the circuit is changing. If the rate of change of flux is in webers/sec, the emf $\mathcal{E}$ will be in volts. In equation form

p. 585   *volts* $$\mathcal{E} = -\frac{d\Phi_B}{dt}.$$ *wb/sec*   (31–1)

This equation is called <u>Faraday's law of induction.</u> The minus sign is an indication of the direction of the induced emf, a matter we discuss further in Section 31–3.

If Eq. 31–1 is applied to a coil of N turns, an emf appears in every turn and these emfs are to be added. If the coil is so tightly wound that each turn can be said to occupy the same region of space, the flux through each turn will then be the same. The flux through each turn is also the same for (ideal) toroids and solenoids (see Section 30–5). The induced emf in all such devices is given by

$$\mathcal{E} = -N\frac{d\Phi_B}{dt} = -\frac{d(N\Phi_B)}{dt},   \quad (31\text{–}2)$$

where $N\Phi_B$ measures the so-called *flux linkages* in the device.

Figures 31–1 and 31–2 suggest that there are at least two ways in which we can make the flux through a circuit change and thus induce an emf in that circuit. The coil that is connected to the galvanometer cannot tell in which of these ex-

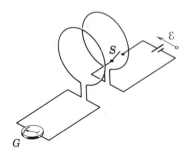

FIGURE 31–2 Galvanometer G deflects momentarily when switch S is closed or opened.

periments it is participating; it is aware only that the flux passing through its cross-sectional area is changing. The flux through a circuit can also be changed by changing its shape, that is, by squeezing or stretching it.

Example 1. A long solenoid has 200 turns/cm and carries a current of 1.5 amp; its diameter is 3.0 cm. At its center we place a 100-turn, close-packed coil of diameter 2.0 cm. This coil is arranged so that **B** at the center of the solenoid is parallel to its axis. The current in the solenoid is reduced to zero and then raised to 1.5 amp in the other direction at a steady rate over a period of 0.050 sec. What induced emf appears in the coil while the current is being changed?

The magnetic field B at the center of the solenoid is given by Eq. 30–7, or

$$B = \mu_0 n i_0 = (4\pi \times 10^{-7} \text{ weber/amp-m})(200 \times 10^2 \text{ turns/meter})(1.5 \text{ amp})$$

$$= 3.8 \times 10^{-2} \text{ weber/meter}^2.$$

The area of the coil (not of the solenoid) is 3.1×10^{-4} meter^2. The initial flux Φ_B through each turn of the coil is given by

$$\Phi_B = BA = (3.8 \times 10^{-2} \text{ weber/meter}^2)(3.1 \times 10^{-4} \text{ meter}^2) = 1.2 \times 10^{-5} \text{ weber}.$$

The flux goes from an initial value of 1.2×10^{-5} weber to a final value of -1.2×10^{-5} weber. The change in flux $\Delta\Phi_B$ for each turn of the coil during the 0.050-sec period is thus twice the initial value. The induced emf is given by

$$\mathcal{E} = -\frac{N\Delta\Phi_B}{\Delta t} = -\frac{(100)(2 \times 1.2 \times 10^{-5} \text{ weber})}{0.050 \text{ sec}} = -4.8 \times 10^{-2} \text{ volt} = -48 \text{ mV}.$$

The minus sign deals with the direction of the emf, as we explain below.

31–3 Lenz's Law

So far we have not specified the directions of the induced emfs. Although these directions can be found from a formal analysis of Faraday's law, we prefer to find them from the conservation-of-energy principle which, in this context, takes the form of Lenz's law, deduced by Heinrich Friedrich Lenz (1804–1865): *The induced current will appear in such a direction that it opposes the change that produced it.* The minus sign in Faraday's law suggests this opposition. In mechanics the energy principle often allows us to draw conclusions about mechanical systems without analyzing them in detail. We use the same approach here.

Lenz's law refers to induced currents, which means that it applies only to closed circuits. If the circuit is open, we can usually think in terms of what would happen if it were closed and in this way find the direction of the induced emf.

Consider the first of Faraday's experiments described in Section 31–1. Figure 31–3 shows the north pole of a magnet and a cross section of a nearby conducting loop. As we push the magnet toward the loop, an induced current is set up in the loop. What is its direction?

A current loop sets up a magnetic field at distant points like that of a magnetic dipole, one face of the loop being a north pole, the opposite face being a south pole. The north pole, as for bar magnets, is that face from which the lines of **B** emerge. If, as Lenz's law predicts, the loop in Fig. 31–3 is to oppose the motion of the magnet toward it, the face of the loop toward the magnet must become a north pole. The two north poles—one of the current loop and one of the magnet—will repel each other. The right-hand rule shows that for the magnetic field set up by the loop to emerge from the right face of the loop the induced current must be as shown. The current will be counterclockwise as we sight along the magnet toward the loop.

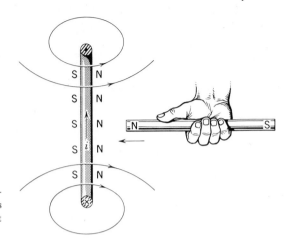

FIGURE 31-3 If the magnet is moved toward the loop, the induced current points as shown, setting up a magnetic field that opposes the motion of the magnet.

When we push the magnet toward the loop (or the loop toward the magnet), an induced current appears. In terms of Lenz's law this pushing is the "change" that produces the induced current, and, according to this law, the induced current will oppose the "push." If we pull the magnet away from the coil, the induced current will oppose the "pull" by creating a south pole on the right-hand face of the loop of Fig. 31-3. To make the right-hand face a south pole, the current must be opposite to that shown in Fig. 31-3. Whether we pull or push the magnet, its motion will always be automatically opposed.

The agent that causes the magnet to move, either toward the coil or away from it, will always experience a resisting force and will thus be required to do work. From the conservation-of-energy principle this work done on the system must be exactly equal to the Joule energy produced in the coil, since these are the only two energy transfers that occur in the system. If we move the magnet more rapidly, we will have to do work at a faster rate and the rate of Joule energy production will increase correspondingly. If we cut the loop and then perform the experiment, there will be no induced current, no Joule energy, no force on the magnet, and no work required to move it. There will still be an emf in the loop, but, like a battery connected to an open circuit, it will not set up a current.

If the current in Fig. 31-3 were in the opposite direction to that shown, the face of the loop toward the magnet would be a south pole, which would pull the bar magnet toward the loop. We would only need to push the magnet slightly to start the process and then the action would be self-perpetuating. The magnet would accelerate toward the loop, increasing its kinetic energy all the time. At the same time Joule energy would appear in the loop at a rate that would increase with time. This would indeed be a something-for-nothing situation! Needless to say, it does not occur.

Let us apply Lenz's law to Fig. 31-3 in a different way. Figure 31-4 shows the lines of **B** for the bar magnet.* From this point of view the "change" is the increase in Φ_B through the loop caused by bringing the magnet nearer. The induced current opposes this change by setting up a field that tends to oppose the increase in flux caused by the moving magnet. Thus the field due to the induced current must point from left to right through the plane of the coil, in agreement with our earlier conclusion.

* There are two magnetic fields in this problem—one connected with the current loop and one with the bar magnet. Be sure that you know which one is meant.

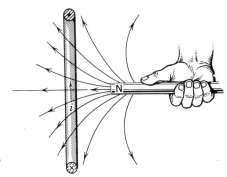

FIGURE 31–4   In moving the magnet toward the loop, we increase Φ_B through the loop.

It is not significant here that the induced field opposes the magnet field but rather that it opposes the change, which in this case is the increase in Φ_B through the loop. If we withdraw the magnet, we reduce Φ_B through the loop. The induced field will now oppose this decrease in Φ_B (that is, the change) by re-enforcing the magnet field. In each case the induced field opposes the change that gives rise to it.

31–4 Induction—A Quantitative Study

The example of Fig. 31–4, although easy to understand qualitatively, does not lend itself to quantitative calculations. Consider then Fig. 31–5, which shows a rectangular loop of wire of width l, one end of which is in a uniform field **B** pointing at right angles to the plane of the loop. The dashed lines show the assumed limits of the magnetic field. The experiment consists in pulling the loop to the right at a constant speed v.

Note that the situation described by Fig. 31–5 does not differ in any essential particular from that of Fig. 31–4. In each case a conducting loop and a magnet are in relative motion; in each case the flux of the field of the magnet through the loop is being caused to change with time.

The flux Φ_B enclosed by the loop in Fig. 31–5 is

$$\Phi_B = Blx,$$

where lx is the area of that part of the loop in which B is not zero. The emf $\mathcal{E}$ is found from Faraday's law, or

$$\mathcal{E} = -\frac{d\Phi_B}{dt} = -\frac{d}{dt}(Blx) = -Bl\frac{dx}{dt} = Blv, \qquad (31\text{–}3)$$

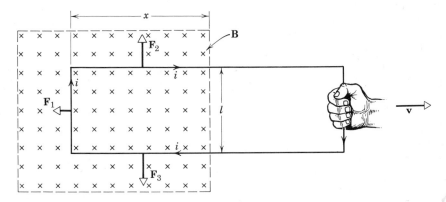

FIGURE 31–5   A rectangular loop is pulled out of a magnetic field with velocity **v**.

where we have set $-dx/dt$ equal to the speed v at which the loop is pulled out of the magnetic field. Note that the only dimension of the loop that enters into Eq. 31-3 is the length l of the left end conductor. As we shall see later, the induced emf in Fig. 31–5 may be regarded as localized here.

The emf Blv sets up a current in the loop given by

$$i = \frac{\mathcal{E}}{R} = \frac{Blv}{R},$$ (31–4)

where R is the loop resistance. From Lenz's law, this current (and thus $\mathcal{E}$) must be clockwise in Fig. 31–5; it opposes the "change" (the decrease in Φ_B) by setting up a field that is parallel to the external field within the loop.

The current in the loop will cause forces $\mathbf{F}_1$, $\mathbf{F}_2$, and $\mathbf{F}_3$ to act on the three conductors, in accord with Eq. 29–6, or

$$\mathbf{F} = i\mathbf{l} \times \mathbf{B}.$$ (31–5)

Because $\mathbf{F}_2$ and $\mathbf{F}_3$ are equal and opposite, they cancel each other; $\mathbf{F}_1$, which is the force that opposes our effort to move the loop, is given in magnitude from Eqs. 31–5 and 31–4 as

$$F_1 = ilB \sin 90° = \frac{B^2 l^2 v}{R}.$$

The agent that pulls the loop must do work at the steady rate of

$$\frac{dw}{dt} \quad P = F_1 v = \frac{B^2 l^2 v^2}{R}.$$ (31–6)

From the principle of the conservation of energy, Joule energy must appear in the resistor at this same rate. We introduced the conservation-of-energy principle into our derivation when we wrote down the expression for the current (Eq. 31–4); recall that the relation $i = \mathcal{E}/R$ for single-loop circuits is a direct consequence of this principle. Thus we should be able to write down the expression for Joule energy production in the loop with the expectation that we will obtain a result identical with Eq. 31–6. Recalling Eq. 31–4, we put

$$P_J = i^2 R = \left(\frac{Blv}{R}\right)^2 R = \frac{B^2 l^2 v^2}{R},$$

which is indeed the expected result. This example provides a quantitative illustration of the conversion of mechanical energy (the work done by an external agent) into electrical energy (the induced emf) and finally into internal energy (the Joule effect, which causes the temperature of the conductor to rise).

Figure 31–6 shows a side view of the coil in the field. In Fig. 31–6*a* the coil is stationary; in Fig. 31–6*b* we are moving it to the right; in Fig. 31–6*c* we are moving it to the left. The lines in these figures represent the resultant field produced by the vector addition of the field $\mathbf{B}_0$ due to the magnet and the field $\mathbf{B}_i$ due to the induced current, if any, in the coil. These lines suggest convincingly that the agent moving the coil always experiences an opposing force.

Example 2. Figure 31–7 shows a rectangular loop of resistance R, width l, and length a being pulled at constant speed v through a region of thickness d in which a uniform field $\mathbf{B}$ is set up by a magnet.

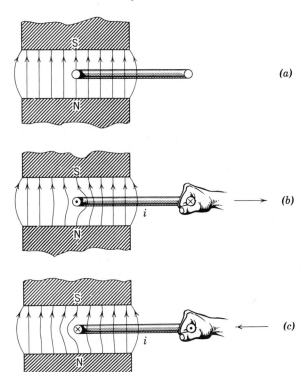

(a)

(b)

(c)

FIGURE 31–6 Side view of a rectangular loop in a magnetic field showing the loop (*a*) at rest, (*b*) being pulled out, and (*c*) being pushed in.

(*a*) Plot the flux Φ_B through the loop as a function of the coil position x. Assume that $l = 4$ cm, $a = 10$ cm, $d = 15$ cm, $R = 16$ ohms, $B = 2.0$ webers/meter^2, and $v = 1.0$ meter/sec.

The flux Φ_B is zero when the loop is not in the field; it is Bla when the loop is entirely in the field; it is Blx when the loop is entering the field and $Bl[a - (x - d)]$ when the loop is leaving the field. These conclusions, which you should verify, are shown graphically in Fig. 31–8*a*.

(*b*) Plot the induced emf $\mathcal{E}$.

The induced emf $\mathcal{E}$ is given by $\mathcal{E} = -d\Phi_B/dt$, which can be written as

$$\mathcal{E} = -\frac{d\Phi_B}{dt} = -\frac{d\Phi_B}{dx}\frac{dx}{dt} = -\frac{d\Phi_B}{dx}v,$$

where $d\Phi_B/dx$ is the slope of the curve of Fig. 31–8*a*. $\mathcal{E}(x)$ is plotted in Fig. 31–8*b*. Lenz's law, from the same type of reasoning as that used for Fig. 31–5, shows that when the coil is entering the field the emf $\mathcal{E}$ acts counterclockwise as seen from above. Note that there is no emf when the coil is entirely in the magnetic field because the flux Φ_B through the coil is not changing with time, as Fig. 31–8*a* shows.

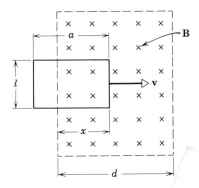

FIGURE 31–7 Example 2. A rectangular loop is caused to move with a velocity **v** through a magnetic field. The position of the loop is measured by x, the distance between the effective left edge of the field **B** and the right end of the loop.

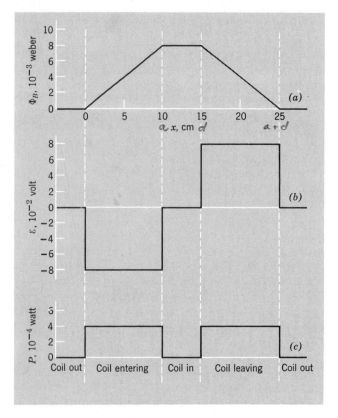

FIGURE 31–8 Example 2.

(c) Plot the rate P of production of Joule energy in the loop.

This is given by $P = \mathcal{E}^2/R$. It may be calculated by squaring the ordinate of the curve of Fig. 31–8b and dividing by R. The result is plotted in Fig. 31–8c.

If the fringing of the magnetic field, which cannot be avoided in practice (see Problem 30–28), is taken into account, the sharp bends and corners in Fig. 31–8 will be replaced by smooth curves. What changes would occur in the curves of Fig. 31–8 if the coil were open circuited?

31–5 Time-Varying Magnetic Fields

So far we have considered emfs induced by the relative motion of magnets and coils. In this section we assume that there is no physical motion of gross objects but that the magnetic field may vary with time. If a conducting loop is placed in such a time-varying field, the flux through the loop will change and an induced emf will appear in the loop. This emf will set the charge carriers in motion, that is, it will induce a current.

From a microscopic point of view we can say, equally well, that the changing flux of **B** sets up an induced electric field **E** at various points around the loop. These induced electric fields are just as real as electric fields set up by static charges and will exert a force **F** on a test charge q_0 given by $\mathbf{F} = q_0\mathbf{E}$. Thus we can restate Faraday's law of induction in a loose but informative way as: *A changing magnetic field produces an electric field.*

To fix these ideas, consider Fig. 31–9, which shows a uniform magnetic field **B** at right angles to the plane of the page. We assume that **B** is increasing in magnitude at the same constant rate dB/dt at every point. This could be done by causing the current in the windings of the electromagnet that establishes the field to increase with time in the proper way.

The circle of arbitrary radius r shown in Fig. 31–9 encloses, at any instant, a flux Φ_B. Because this flux is changing with time, an induced emf given by $\mathcal{E} = -d\Phi_B/dt$ will appear around the loop. The electric fields **E** induced at various points of the loop must, from symmetry and from experiment, be tangent to the loop. Thus the lines of **E** set up by the changing magnetic field are in this case concentric circles.

If we consider a test charge q_0 moving around the circle of Fig. 31–9, the work W $\mathcal{E} = \dfrac{dW}{dq}$ done on it per revolution is, in terms of the definition of an emf, simply $\mathcal{E}q_0$. From another point of view, it is $(q_0 E)(2\pi r)$, where $q_0 E$ is the force that acts on the charge and $2\pi r$ is the distance over which the force acts. Setting the two expressions for W equal and canceling q_0 yields

$$\mathcal{E} = E2\pi r. \tag{31–7}$$

In a more general case than that of Fig. 31–9 we must write

$$\mathcal{E} = \oint \mathbf{E} \cdot d\mathbf{l}. \qquad form\ 25,6 \tag{31–8}$$

If this integral is evaluated for the conditions of Fig. 31–9, we obtain Eq. 31–7 at once. If Eq. 31–8 is combined with Eq. 31–1 ($\mathcal{E} = -d\Phi_B/dt$), <u>Faraday's law of induction</u> can be written as

$$\oint \mathbf{E} \cdot d\mathbf{l} = -\frac{d\Phi_B}{dt}, \tag{31–9}$$

which is the form in which this law is expressed in Table 34–2.

Example 3. Let B in Fig. 31–9 be increasing at the rate dB/dt. Let R be the radius of the cylindrical region in which the magnetic field is assumed to exist. What is the magnitude of the electric field **E** at any radius r? Assume that $dB/dt = 0.10$ weber/m²-sec and $R = 10$ cm.

(a) For $r < R$, the flux Φ_B through the loop is

$$\Phi_B = B(\pi r^2). \qquad form\ 29,1\ p.\ 538$$

Substituting into Faraday's law (Eq. 31–9),

$$\oint \mathbf{E} \cdot d\mathbf{l} = -\frac{d\Phi_B}{dt}$$

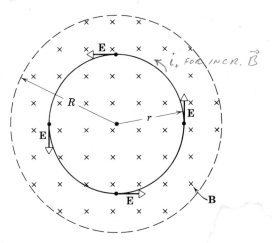

FIGURE 31–9 The induced electric fields at four points produced by an increasing magnetic field.

yields
$$(E)(2\pi r) = -\frac{d\Phi_B}{dt} = -(\pi r^2)\frac{dB}{dt}.$$

Solving for E yields
$$E = -\frac{1}{2}r\frac{dB}{dt}.$$

The minus sign is retained to suggest that the induced electric field **E** acts to oppose the change of the magnetic field. Note that $E(r)$ depends on dB/dt and not on B. Substituting numerical values, assuming $r = 5.0$ cm, yields, for the magnitude of E,

$$E = \frac{1}{2}r\frac{dB}{dt} = (\tfrac{1}{2})(5.0 \times 10^{-2} \text{ meter})\left(\frac{0.10 \text{ weber}}{\text{m}^2\text{-sec}}\right) = 2.5 \times 10^{-3} \text{ volt/meter.}$$

(*b*) For $r > R$ the flux through the loop is

$$\Phi_B = \int \mathbf{B} \cdot d\mathbf{S} = B(\pi R^2).$$

This equation is true because $\mathbf{B} \cdot d\mathbf{S}$ is zero for those points of the loop that lie outside the effective boundary of the magnetic field.

From Faraday's law (Eq. 31–9),

$$(E)(2\pi r) = -\frac{d\Phi_B}{dt} = -(\pi R^2)\frac{dB}{dt}.$$

Solving for E yields
$$E = -\frac{1}{2}\frac{R^2}{r}\frac{dB}{dt}.$$

These two expressions for $E(r)$ yield the same result, as they must, for $r = R$. Figure 31–10 is a plot of $E(r)$ for the numerical values given.

In applying Lenz's law to Fig. 31–9, imagine that a circular conducting loop is placed concentrically in the field. Since Φ_B through this loop is increasing, the induced current in the loop will tend to oppose this "change" by setting up a magnetic field of its own that points up within the loop. Thus the induced current i must be counterclockwise, which means that the lines of the induced electric field **E**, which is responsible for the current, must also be counterclockwise. If the magnetic field in Fig. 31–9 were decreasing with time, the induced current and the lines of force of the induced electric field **E** would be clockwise, again opposing the change in Φ_B.

(POS. CURRENT)

Figure 31–11 shows four of many possible loops to which Faraday's law may be applied. For loops 1 and 2, the induced emf $\mathcal{E}$ is the same because these loops lie entirely within the changing magnetic field and thus have the same value of $d\Phi_B/dt$. Note that even though the emf $\mathcal{E}$ ($= \oint \mathbf{E} \cdot d\mathbf{l}$) is the same for these two loops, the distribution of electric fields **E** around the perimeter of each loop, as indicated by the

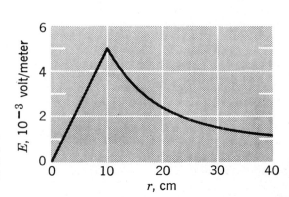

FIGURE 31–10   Example 3. If the necessary fringing of the field in Fig. 31–9 were to be taken into account, the result would be a rounding of the sharp cusp at $r = R$ ($= 10$ cm).

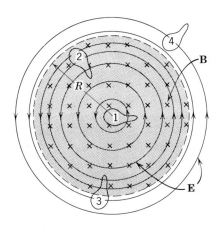

FIGURE 31–11 Showing the circular lines of **E** from an increasing magnetic field. The four loops are imaginary paths around which an emf can be calculated.

electric lines of force, is different. For loop 3 the emf is less because Φ_B and $d\Phi_B/dt$ for this loop are less, and for loop 4 the induced emf is zero.

The induced electric fields that are set up by the induction process are not associated with charges but with a changing magnetic flux. Although both kinds of electric fields exert forces on charges, there is a difference between them. The simplest manifestation of this difference is that lines of **E** associated with a changing magnetic flux can form closed loops (see Fig. 31–11); lines of **E** associated with charges cannot but can always be drawn to start on a positive charge and end on a negative charge.

Equation 25–5, which defined the potential difference between two points a and b, is

$$V_b - V_a = \frac{W_{ab}}{q_0} = -\int_a^b \mathbf{E} \cdot d\mathbf{l}.$$

We have insisted that if potential is to have any useful meaning this integral (and W_{ab}) must have the same value for every path connecting a with b. If a and b are the same point, the path connecting them is a closed loop and this equation reduces to

$$\oint \mathbf{E} \cdot d\mathbf{l} = 0.$$

SEE p. 111

However, when changing magnetic flux is present, $\oint \mathbf{E} \cdot d\mathbf{l}$ is precisely not zero but is, according to Faraday's law (see Eq. 31–9), $-d\Phi_B/dt$. Electric fields associated with stationary charges are conservative, but those associated with changing magnetic fields are nonconservative; see Section 7–2. Electric potential, which can be defined only for a conservative force, has no meaning for electric fields produced by induction.

31–6 The Betatron

The betatron is a device used to accelerate electrons to high speeds by allowing them to be acted upon by induced electric fields that are set up by a changing magnetic flux. It provides an excellent illustration of the "reality" of such induced fields. The energetic electrons can be used for fundamental research in physics or to produce penetrating x rays which are useful in cancer therapy and in industry.

Figure 31–12 shows a vertical cross section through the central part of a betatron. The magnetic field in the betatron has several functions: (*a*) it guides the electrons in a circular path; (*b*) it accelerates the electrons in this path; (*c*) it keeps the radius of the orbit in which the electrons are moving a constant; (*d*) it introduces the electrons

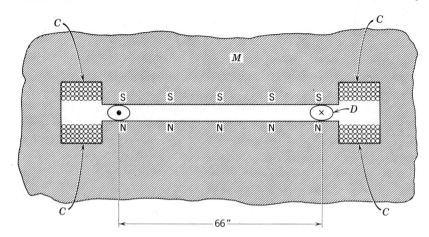

FIGURE 31–12   Cross section of a betatron, showing magnet *M*, coils *C*, and "doughnut" *D*. Electrons emerge from the page at the left and enter it at the right.

into the orbit initially and removes them from the orbit after they have reached full energy; and finally (*e*) it provides a restoring force that resists any tendency for the electrons to leave their orbit, either vertically or radially. It is remarkable that it is possible to do all these things by proper shaping and control of the magnetic field.

The object marked *D* in Fig. 31–12 is an evacuated glass "doughnut" inside which the electrons travel. Their orbit is a circle at right angles to the plane of the figure. The electrons emerge from the plane at the left (·) and enter it at the right (×).

The current in coils *C* is made to alter periodically, 60 times/sec, to produce a changing flux through the orbit, shown in Fig. 31–13. Here Φ_B is taken as positive when **B** is pointing up, as in Fig. 31–12. Show that if the electrons are to circulate in the direction shown, they must do so during the positive half-cycle, marked *ac* in Fig. 31–13 (see Section 29–6). The electrons are accelerated by electric fields set up by the changing flux. The direction of these induced fields depends on the sign of $d\Phi_B/dt$ and must be chosen to accelerate, and not to decelerate, the electrons. Thus only half the positive half-cycle in Fig. 31–13 can be used for acceleration; it will prove to be *ab*.

The average value of $d\Phi_B/dt$ during the quarter-cycle *ab* is the slope of the dashed line, or

$$\overline{\frac{d\Phi_B}{dt}} = \frac{1.8 \text{ weber}}{4.2 \times 10^{-3} \text{ sec}} = 430 \text{ volts}.$$

FIGURE 31–13   The flux through the orbit of a betatron, during one cycle. Rotation of the electrons in the desired direction (counterclockwise as viewed from above in Fig. 31–12) is possible only during half-cycle *ac*.

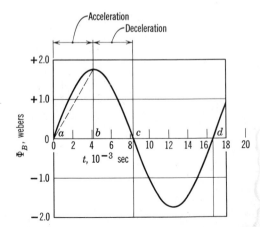

From Faraday's law (Eq. 31–1), this is also the emf in volts. The electron will thus increase its energy by 430 eV every time it makes a trip around the orbit in the changing flux. If the electron gains only 430 eV of energy per revolution, it must make about 230,000 rev to gain its full 100 MeV. For an orbit radius of 33 in., this corresponds to a length of path of some 750 miles.

FIGURE 31–14

The betatron provides a good example of the fact that electric potential has no meaning for electric fields produced by induction. If a potential exists, we must have $\oint \mathbf{E} \cdot d\mathbf{l} = 0$ for any closed path. In the betatron, however, this integral, evaluated around the orbit, is precisely not zero but is, in our example, 430 volts. This gain in kinetic energy of the circulating electron (430 eV/rev) must be supplied by an identifiable energy source, in this case the generator that energizes the magnet coils, thus providing the changing magnetic field.

QUESTIONS

1. The north pole of a magnet is moved away from a metallic ring, as in Fig. 31–14. In the part of the ring farthest from you, which way does the current point?

2. *Eddy currents.* A sheet of copper is placed in a magnetic field as shown in Fig. 31–15. If we attempt to pull it out of the field or push it further in, an automatic resisting force appears. Explain its origin. (Hint: Currents, called eddy currents, are induced in the sheet in such a way as to oppose the motion.)

3. *Electromagnetic shielding.* Consider a conducting sheet lying in a plane perpendicular to a magnetic field **B**, as shown in Fig. 31–16. If **B** changes periodically at high enough frequency and the conductor is made of a material of low enough resistivity, the region near P is almost completely shielded from the *changes* in flux. Explain. Is such a conductor useful as a shield from *static* magnetic fields?

4. *Magnetic damping.* A strip of copper is mounted as a pendulum about O in Fig. 31–17. It is free to swing through a magnetic field normal to the page. If the strip has slots cut in it as shown, it can swing freely through the field. If a strip without slots is substituted, the vibratory motion

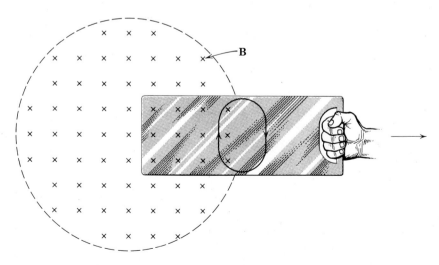

FIGURE 31–15

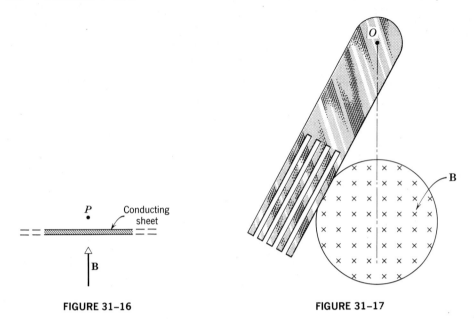

FIGURE 31–16   **FIGURE 31–17**

is strongly damped. Explain. (Hint: Use Lenz's law; consider the paths that the charge carriers in the strip must follow if they are to oppose the motion.)

5. Two conducting loops face each other a distance d apart (Fig. 31–18). An observer sights along their common axis in the direction shown. If a clockwise current i is suddenly established in the larger loop, what is the direction of the induced current in the smaller loop? What is the direction of the force (if any) that acts on the smaller loop?

6. What is the direction, if any, of the current through resistor R in Fig. 31–19 (*a*) immediately after switch S is closed, (*b*) some time after switch S was closed, and (*c*) immediately after switch S is opened? When switch S is held closed, which end of the coil acts as a north pole?

7. A current-carrying solenoid is moved toward a conducting loop as in Fig. 31–20. What is the direction of circulation of current in the loop as we sight toward it as shown?

8. If the resistance R in the left-hand circuit of Fig. 31–21 is increased, what is the direction of the induced current in the right-hand circuit?

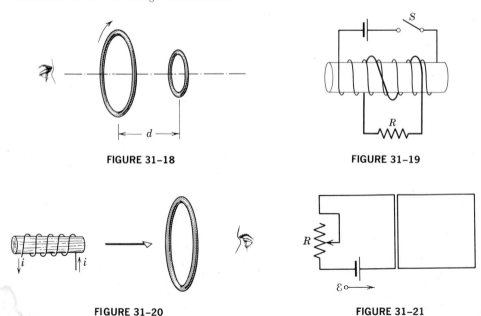

FIGURE 31–18   **FIGURE 31–19**

FIGURE 31–20   **FIGURE 31–21**

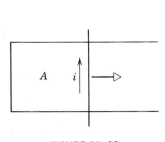

FIGURE 31–22

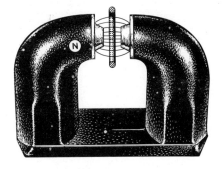

FIGURE 31–23

9. In Figure 31–22 the movable wire is moved to the right, causing an induced current as shown. What is the direction of **B** in region A?

10. A loop, shown in Fig. 31–23, is removed from the magnet by pulling it vertically upward. What is the direction of the induced current? Is a force required to remove the loop? Does the total amount of Joule energy produced in removing the loop depend on the time taken to remove it?

11. A bar magnet is dropped down a long vertical copper tube. Show that, even neglecting air resistance, the magnet will reach a constant terminal velocity.

12. A magnet is dropped from the ceiling along the axis of a copper loop lying flat on the floor. If the falling magnet is photographed with a time sequence camera, what differences, if any, will be noted if (a) the loop is at room temperature and (b) the loop is packed in dry ice?

13. A copper ring and a wooden ring of the same dimensions are placed so that there is the same changing magnetic flux through each. How do the induced electric fields in each ring compare?

14. In Fig. 31–11 how can the induced emfs around paths 1 and 2 be identical? The induced electric fields are much weaker near path 1 than near path 2, as the spacing of the lines of **E** shows. See also Fig. 31–10.

15. In a certain betatron the electrons rotate counterclockwise as seen from above. In what direction must the magnetic field point and how must it change with time while the electron is being accelerated?

16. Why can a betatron be used for acceleration only during one-quarter of a cycle?

17. To make the electrons in a betatron orbit spiral outward, would it be necessary to increase or to decrease the central flux? Assume that **B** at the orbit remains essentially unchanged.

PROBLEMS

1(2). The current in the solenoid of Example 1 changes according to the equation $i = 3.0t + 1.0t^2$, where i is in amperes and t in seconds. (a) Plot quantitatively the induced emf in the coil from $t = 0$ to $t = 4.0$ sec. (b) The resistance of the coil is 0.15 ohm. What is the instantaneous current in the coil at $t = 2.0$ sec?
Answer: (b) 3.6×10^{-2} amp.

2(2). Show that emf has the same dimensions as the time rate of change of magnetic flux, as Eq. 31–1 requires.

3(2). A small loop of area A is inside of, and has its axis in the same direction as, a long solenoid of n turns per unit length and current i. If $i = i_0 \sin \omega t$ find the emf $\mathcal{E}$ in the loop.
Answer: $-\mu_0 n A i_0 \omega \cos \omega t$.

4(2). A circular coil of radius r (10 cm) is made of heater wire of resistance R (10 ohms). Perpendicular to the plane of the coil is a uniform magnetic field **B**. (a) At what constant rate must B

increase so that there is a steady current i (0.010 amp) in the circuit? (*b*) What power is dissipated in the resistor?

5(2). A uniform magnetic field **B** is normal to the plane of a circular ring 10 cm in diameter made of #10 gauge copper wire (diameter = 0.10 in.). At what rate must B change with time if an induced current of 10 amp is to appear in the ring?

Answer: 1.3 webers/meter^2 sec.

6(2). You are given 50 cm of #18 gauge copper wire (diameter = 0.040 in.). It is formed into a circular loop and placed at right angles to a uniform magnetic field that is increasing with time at the constant rate of 100 gauss/sec. At what rate is Joule energy generated in the loop?

7(2). You are given a mass m of copper of resistivity ρ and density δ which is to be drawn into a wire of radius a and formed into a circular loop of radius b. Perpendicular to the plane of this loop a uniform magnetic field is changing in magnitude at a constant rate dB/dt. Find the induced current in the loop.

Answer: $\dfrac{m}{4\pi\rho\delta}\dfrac{dB}{dt}$.

8(2). (*a*) For the arrangement of Fig. 30–25, what would be the current induced around the rectangular loop if the current in the wire decreased uniformly from 30 amp to zero in 1.0 sec? Assume no initial current in the loop and a resistance for the loop of 0.020 ohm. Take $a = 1.0$ cm, $b = 8.0$ cm, and $l = 30$ cm. (*b*) How much energy would be transferred to the loop in the 1.0-sec interval?

9(3). In Fig. 31–24 the magnetic flux through the loop perpendicular to the plane of the coil and directed into the paper is varying according to the relation

$$\Phi_B = 6t^2 + 7t + 1,$$

where Φ_B is in milliwebers (1 milliweber = 10^{-3} weber) and t is in seconds. (*a*) What is the magnitude of the emf induced in the loop when $t = 2.0$ sec? (*b*) What is the direction of the current through R?

Answer: (*a*) 3.1×10^{-2} volts. (*b*) Left to right.

10(3). (*a*) Prove that if the flux of **B** through the coil of N turns of Fig. 31–25 changes in any way from Φ_1 to Φ_2, then the charge q that flows through the circuit of total resistance R is given by

$$q = \frac{N(\Phi_2 - \Phi_1)}{R}.$$

(*b*) Suppose the change in flux, $\Phi_2 - \Phi_1$, is zero. Does it then follow that no joule-heating occurred during this time interval?

11(3). A hundred turns of insulated copper wire are wrapped around an iron cylinder of cross-sectional area 0.0010 meter^2 and are connected to a resistor. The total resistance in the circuit is 10 ohms. If the longitudinal magnetic field **B** in the iron changes from 1.0 weber/meter^2 in one direction to 1.0 weber/meter^2 in the opposite direction, how much charge flows through the circuit?

Answer: 2.0×10^{-2} coul.

12(3). A wire is bent into three circular segments of radius r (10 cm) as shown in Fig. 31–26. Each segment is a quadrant of a circle, *ab* lying in the x-y plane, *bc* lying in the y-z plane, and *ca* lying

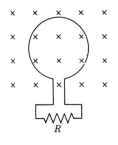

FIGURE 31-24   Problem 9(3).

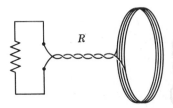

FIGURE 31-25   Problem 10(3).

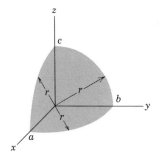

FIGURE 31-26 Problem 12(3).

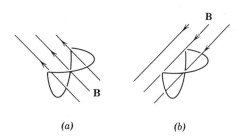

FIGURE 31-27 Problem 13(3).

in the z-x plane. (a) If a spatially uniform magnetic field **B** points in the x-direction, what is the magnitude of the emf ε developed in the wire when **B** increases at the rate of 3.0×10^{-3} webers/meter² sec? (b) What is the direction of the current in the segment bc?

13(3). A closed loop of wire consists of a pair of equal semicircles of radius r (3.7 cm) lying in mutually perpendicular planes. The loop was formed by folding a circular loop along a diameter until the two halves became perpendicular. A uniform magnetic field **B** of magnitude 760 gauss is directed perpendicular to the fold-diameter and makes equal angles (45°) with the planes of the semicircles as shown in Fig. 31–27a. (a) The magnetic field is reduced at a uniform rate to zero during a time interval 4.5×10^{-3} sec. Determine the induced emf ε and the sense of the induced current in the loop during this interval. (b) How would the answers change if **B** is directed as shown in Fig 31–27b perpendicular to the direction first given for it but still perpendicular to the fold-diameter?

Answer: (a) 5.1×10^{-2} volts, i clockwise, as viewed along the direction of **B**. (b) ε is zero.

14(4). A circular loop of wire 10 cm in diameter is placed with its perpendicular axis making an angle of 30° with the direction of a uniform 5000-gauss magnetic field. The loop is "wobbled" so that its axis rotates about the field direction at the constant rate of 100 rev/min; the angle between the axis and the field direction (30°) remains unchanged during this process. What emf ε appears in the loop?

15(4). A small bar magnet is pulled rapidly through a conducting loop, along its axis. Sketch qualitatively (a) the induced current and (b) the rate of Joule energy production as a function of the position of the center of the magnet. Assume that the north pole of the magnet enters the loop first and that the magnet moves at constant speed. Plot the induced current as positive if it is clockwise as viewed along the path of the magnet.

16(4). The perpendicular field B through a one-turn circular loop of wire of negligible resistance changes with time as shown in Fig. 31–28. The loop is of radius r (10 cm) and is connected to a resistor R (10 ohms). (a) Plot the emf appearing across the resistor. (b) Plot the current i through the resistor R. (c) Plot the rate of the joule-heating in the resistor.

17(4). *Alternating current generator.* A rectangular loop of N turns and of length a and width b is rotated at a frequency ν in a uniform magnetic field **B**, as in Fig. 31–29. (a) Show that an induced emf given by

$$\varepsilon = 2\pi\nu NbaB \sin 2\pi\nu t = \varepsilon_0 \sin 2\pi\nu t$$

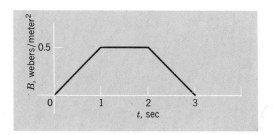

FIGURE 31-28 Problem 16(4).

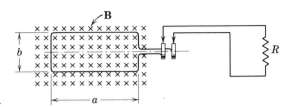

FIGURE 31-29   Problem 17(4).

appears in the loop. This is the principle of the commercial alternating-current generator. (*b*) Design a loop that will produce an emf with $\mathcal{E}_0 = 150$ volts when rotated at 60 rev/sec in a field of 5000 gauss.

Answer: (*b*) $NA = \dfrac{5}{2\pi}$ turn-meter².

18(4). A wire bent into a semicircle of radius R is rotated with a frequency ν in a uniform magnetic field **B**, as shown in Fig. 31-30. What is the amplitude of the induced emf $\mathcal{E}$?

19(4). In Fig. 31-31 a conducting rod AB makes contact with the metal rails AD and BC that are 50 cm apart in a uniform magnetic field of 1.0 weber/meter² perpendicular to the plane of the paper. The total resistance of the circuit $ABCD$ is 0.40 ohm (assumed constant). (*a*) What is the magnitude and sense of the emf induced in the rod when it is moved to the left with a speed of 8.0 meters/sec? (*b*) What force is required to keep the rod in motion? (*c*) Compare the rate at which mechanical work is done by the force **F** with the rate of increase of internal energy from the Joule effect.

Answer: (*a*) 4.0 volts (counterclockwise). (*b*) 5.0 nt. (*c*) Mechanical power = electrical power = 40 watts.

20(4). Figure 31-32 shows a copper rod moving on conducting rails with velocity **v** parallel to a long straight wire carrying a current i. Calculate the induced emf $\mathcal{E}$ in the rod, assuming that $v = 5.0$ meters/sec, $i = 100$ amp, $a = 1.0$ cm, and $b = 20$ cm.

21(4). A metal wire of mass m and resistance R slides without friction on two rails spaced a distance d apart, as in Fig. 31-33. The track lies in a vertical uniform field **B**. (*a*) A constant current i flows from generator G along one rail, across the wire, and back down the other rail. Find the velocity (speed and direction) of the wire as a function of time, assuming it to be at rest at $t = 0$. (*b*) The generator is replaced by a battery with constant emf $\mathcal{E}$. The velocity of the wire now approaches a constant final value. What is this terminal speed? (*c*) What is the current in part (*b*) when the terminal speed has been reached? Neglect the resistance of the rails.

Answer: (*a*) $Bidt/m$, away from G. (*b*) $\mathcal{E}/Bd$. (*c*) Zero.

22(4). A square wire of length l, mass m, and resistance R slides without friction down parallel conducting rails of negligible resistance, as in Fig. 31-34. The rails are connected to each other at the bottom by a resistanceless rail parallel to the wire, so that the wire and rails form a closed rec-

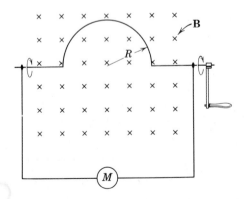

FIGURE 31-30   Problem 18(4).

FIGURE 31-31   Problem 19(4).

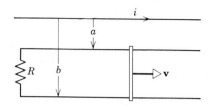

FIGURE 31-32   Problem 20(4).

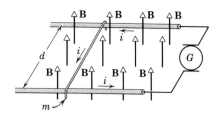

FIGURE 31-33   Problem 21(4).

tangular conducting loop. The plane of the rails makes an angle θ with the horizontal, and a uniform vertical magnetic field **B** exists throughout the region. (*a*) Show that the wire acquires a steady-state speed of magnitude

$$v = \frac{mgR \sin \theta}{B^2 l^2 \cos^2 \theta}.$$

(*b*) Prove that this result is consistent with the conservation of energy principle. (*c*) What change, if any, would there be if **B** were directed down instead of up?

23(4). Figure 31–35 shows two single turn loops of wire having the same axis. The smaller loop is above the larger one, a distance x that is large compared to the radius R of the larger loop. Hence, with current i as indicated in the larger loop, the consequent magnetic field is nearly constant throughout the plane area πr^2 bounded by the smaller loop. Suppose now that x is not constant but is changing at the constant rate $dx/dt = v$. (*a*) Find the magnetic flux Φ across the area bounded by the smaller loop as a function of x. (*b*) Compute the emf $\mathcal{E}$ generated in the smaller loop at the instant when $x = R$. (*c*) Find the direction of the induced current in the smaller loop if $v > 0$.

Answer: (*a*) $\dfrac{\mu_0 i \pi r^2 R^2}{2x^3}$. (*b*) $\dfrac{3\mu_0 \pi i r^2 v}{2R^2}$. (*c*) In the same sense as the current in the large loop.

24(4). A circular loop of radius r (10 cm) is placed in a uniform magnetic field **B** (0.80 webers/meter^2) normal to the plane of the loop. The radius of the loop begins shrinking at a constant rate dr/dt (80 cm/sec). (*a*) What is the emf $\mathcal{E}$ induced in the loop? (*b*) At what constant rate would the area have to shrink to induce this same emf?

25(4). A circular copper disk 10 cm in diameter rotates at 1800 rev/min about an axis through its center and at right angles to the disk. A uniform magnetic field **B** of 10,000 gauss is perpendicular to the disk. What potential difference develops between the axis of the disk and its rim?
Answer: 0.24 volts.

26(5). A long solenoid of radius r (2.5 cm) and turns per unit length n (100/cm) carries an initial current of i_0 (1.0 amp). A single loop of wire of diameter D (10 cm) is placed around the solenoid,

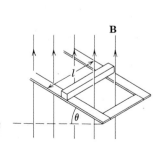

FIGURE 31-34   Problem 22(4).

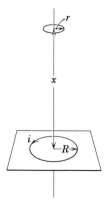

FIGURE 31-35   Problem 23(4).

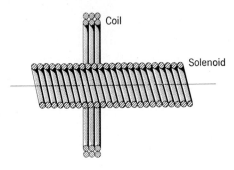

FIGURE 31-36   Problem 27(5).

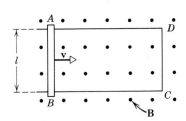

FIGURE 31-37   Problem 28(5).

the axes coinciding. The current in the solenoid is reduced uniformly to i (0.50 amp) over a period T (0.010 sec). While the current is changing what is the induced emf $\mathcal{E}$ in the surrounding loop?

27(5). In Fig. 31–36 a closed copper coil with 100 turns and a total resistance of 5.0 ohms is placed outside a solenoid like that of Example 1. If the current in the solenoid is changed as in that example, what current appears in the coil?

Answer: 2.1×10^{-2} amp.

28(5). In Fig. 31–37, $l = 2.0$ meters and $v = 50$ cm/sec. **B** is the earth's magnetic field, directed perpendicularly out of the page and having a magnitude 6.0×10^{-5} webers/m² at that place. The resistance R of the circuit $ADCB$, contained only in segment CD, (explain how this may be achieved approximately), is 1.2×10^{-5} ohm. (*a*) What is the emf $\mathcal{E}$ induced in the circuit? (*b*) What is the electric field **E** in the wire AB? (*c*) What force **F** does each electron in the wire experience due to the motion of the wire in the magnetic field? (*d*) What is the magnitude of the current i in the wire? (*e*) What force **F** must an external agency exert in order to keep the wire moving with this constant velocity? (*f*) Compute the rate at which the external agency is doing work. (*g*) Compute the rate at which electrical energy is being converted into Joule energy.

29(5). Figure 31–38 shows a uniform magnetic field **B** confined to a cylindrical volume of radius R. B is decreasing at a constant rate of 100 gauss/sec. What is the instantaneous acceleration (direction and magnitude) experienced by an electron placed at a, at b, and at c? Assume $r = 5.0$ cm.

Answer: (*a*) 4.4×10^7 meters/sec², to the right. (*b*) Zero. (*c*) 4.4×10^7 meters/sec², to the left.

30(5). For the arrangement of Example 3, let $dB/dt = 5.0$ weber/meters²-sec and $r = 25$ cm. (*a*) What is the electric field for $r = 20$ cm? (*b*) For $r = 40$ cm?

31(5). Prove that the electric field **E** in a charged parallel-plate capacitor cannot drop abruptly to zero as one moves at right angles to it, as suggested by the arrow in Fig. 31–39 (see point *a*). In actual capacitors fringing of the lines of force always occurs, which means that **E** approaches zero in a continuous and gradual way. See Problem 3(1) in Chapter 30. (Hint: Apply Faraday's law to the rectangular path shown by the dashed lines.)

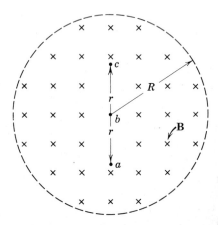

FIGURE 31-38   Problem 29(5).

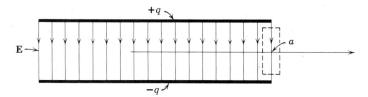

FIGURE 31–39   Problem 31(5).

32(5). An electromagnetic "eddy current" brake consists of a disk of conductivity σ and thickness t rotating about an axis through its center with a magnetic field **B** applied perpendicular to the plane of the disk over a small area a^2 (see Fig. 31–40). If the area a^2 is at a distance r from the axis, show that an approximate expression for the torque tending to slow down the disk at the instant its angular velocity equals ω is $r = B^2a^2r^2\omega\sigma t$.

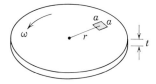

FIGURE 31–40   Problem 32(5).

CHAPTER 32

Inductance

32–1 Inductance

If two coils are near each other, a current i in one coil will set up a flux Φ_B through the second coil. If this flux is changed by changing the current, an induced emf will appear in the second coil according to Faraday's law. However, two coils are not needed to show an inductive effect. An induced emf appears in a coil if the current in that same coil is changed. This is called *self-induction* and the electromotive force produced is called a *self-induced emf*. It obeys Faraday's law of induction just as other induced emfs do.

Consider first a "close-packed" coil, a toroid, or the central section of a long solenoid. In all three cases the flux Φ_B set up in each turn by a current i is essentially the same for every turn. Faraday's law for such coils (Eq. 31–2)

$$\mathcal{E} = -\frac{d(N\Phi_B)}{dt} \tag{32–1}$$

shows that the number of flux linkages $N\Phi_B$ (N being the number of turns) is the important characteristic quantity for induction. For a given coil, provided no magnetic materials such as iron are nearby, this quantity is proportional to the current i, or

$$N\Phi_B = Li, \tag{32–2}$$

in which L, the proportionality constant, is called the *inductance* of the device.

From Faraday's law (see Eq. 32–1) the induced emf can be written as

$$\mathcal{E} = -\frac{d(N\Phi_B)}{dt} = -L\frac{di}{dt}. \tag{32–3a}$$

GEOMETRY

Written in the form   *henry* $L = -\dfrac{\mathcal{E}}{di/dt},$ $\dfrac{volt \cdot sec}{amp}$ (32–3b)

this relation may be taken as the defining equation for inductance for coils of all shapes and sizes, whether or not they are close-packed and whether or not iron or other magnetic material is nearby. It is analogous to the defining relation for capacitance, namely

$$C = \frac{q}{V}.$$

If no iron or similar materials are nearby, L depends only on the geometry of the device. In an *inductor* (symbol ⌇⌇⌇⌇) the presence of a magnetic field is the significant feature, corresponding to the presence of an electric field in a capacitor.

The unit of inductance, from Eq. 32–3b, is the volt-sec/amp. A special name, the *henry,* has been given to this combination of units, or

$$1 \text{ henry} = 1 \text{ volt-sec/amp.}$$

The unit of inductance is named after Joseph Henry (1797–1878), an American physicist and a contemporary of Faraday. Henry independently discovered the law of induction at about the same time Faraday did.

The direction of a self-induced emf can be found from Lenz's law. Suppose that a steady current i, produced by a battery, exists in a coil. Let us suddenly reduce the (battery) emf in the circuit to zero. The current i will start to decrease at once; this decrease in current, in the language of Lenz's law, is the "change" which the self-induction must oppose. To oppose the falling current, the induced emf must point in the same direction as the current, as in Fig. 32–1a. When the current in a coil is increased, Lenz's law shows that the self-induced emf points in the opposite direction to that of the current, as in Fig. 32–1b. In each case the self-induced emf acts to oppose the change in the current. The minus sign in Eq. 32–3 shows that $\mathcal{E}$ and di/dt are opposite in sign, since L is always a positive quantity.

32–2 Calculation of Inductance

It has proved possible to make a direct calculation of capacitance in terms of geometrical factors for a few special cases, such as the parallel-plate capacitor. In the same way, it is possible to calculate the self-inductance L, simply, for a few special cases.

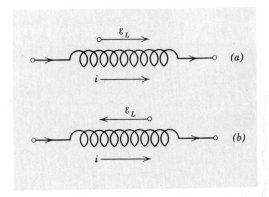

FIGURE 32–1

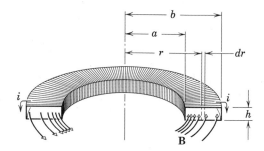

FIGURE 32–2   Example 1. A cross section of a toroid, showing the current in the windings and the lines of **B**.

For a close-packed coil with no iron nearby, we have, from Eq. 32–2,

$$L = \frac{N\Phi_B}{i}. \tag{32–4}$$

Let us apply this equation to calculate L for a section of length l near the center of a long solenoid. The number of flux linkages in the length l of the solenoid is

$$N\Phi_B = (nl)(BA),$$

where n is the number of turns per unit length, B is the magnetic field inside the solenoid, and A is the cross-sectional area. From Eq. 30–7, B is given by

$$B = \mu_0 ni.$$

Combining these equations gives

$$N\Phi_B = \mu_0 n^2 liA.$$

Finally, the inductance, from Eq. 32–4, is

$$L = \frac{N\Phi_B}{i} = \mu_0 n^2 lA. \tag{32–5}$$

The inductance of a length l of a solenoid is proportional to its volume (lA) and to the square of the number of turns per unit length. Note that it depends on geometrical factors only. The proportionality to n^2 is expected. If the number of turns per unit length is doubled, not only is the total number of turns N doubled but also the flux through each turn Φ_B is also doubled, an over-all factor of four for the flux linkages $N\Phi_B$, hence also a factor of four for the inductance (Eq. 32–4).

Example 1. Derive an expression for the inductance of a toroid of rectangular cross section as shown in Fig. 32–2. Evaluate for $N = 10^3$, $a = 5.0$ cm, $b = 10$ cm, and $h = 1.0$ cm.

The lines of **B** for the toroid are concentric circles. Applying Ampère's law,

$$\oint \mathbf{B} \cdot d\mathbf{l} = \mu_0 i,$$

to a circular path of radius r yields

$$(B)(2\pi r) = \mu_0 i_0 N,$$

where N is the number of turns and i_0 is the current in the toroid windings; recall that i in Ampère's law is the total current that passes through the path of integration. Solving for B yields

$$B = \frac{\mu_0 i_0 N}{2\pi r}.$$

The flux Φ_B for the cross section of the toroid is

$$\Phi_B = \int \mathbf{B} \cdot d\mathbf{S} = \int_a^b (B)(h\,dr) = \int_a^b \frac{\mu_0 i_0 N}{2\pi r} h\,dr$$

$$= \frac{\mu_0 i_0 Nh}{2\pi} \int_a^b \frac{dr}{r} = \frac{\mu_0 i_0 Nh}{2\pi} \ln \frac{b}{a},$$

where $h\,dr$ is the area of the elementary strip shown in the figure.

The inductance follows from Eq. 32–4, or

$$L = \frac{N\Phi_B}{i_0} = \frac{\mu_0 N^2 h}{2\pi} \ln \frac{b}{a}.$$

Substituting numerical values yields

$$L = \frac{(4\pi \times 10^{-7} \text{ weber/amp-m})(10^3)^2(1.0 \times 10^{-2} \text{ meter})}{2\pi} \ln \frac{10 \times 10^{-2} \text{ meter}}{5.0 \times 10^{-2} \text{ meter}}$$

$$= 1.4 \times 10^{-3} \text{ weber/amp} = 1.4 \text{ mh.}$$

32–3 An *LR* Circuit

In Section 28–6 we saw that if an emf $\mathcal{E}$ is suddenly introduced, perhaps by using a battery, into a single loop circuit containing a resistor R and a capacitor C the charge does not build up immediately to its final equilibrium value ($= C\mathcal{E}$) but approaches it in an exponential fashion described by Eq. 28–13, which we may write as

$$q = C\mathcal{E}\,(1 - e^{-t/\tau_C}). \qquad p.531 \qquad (32\text{--}6)$$

The delay in the rise of the charge is described by the capacitative time constant τ_C, defined from

$$\tau_C = RC. \qquad p.532 \qquad (32\text{--}7)$$

If in this same circuit the battery emf $\mathcal{E}$ is suddenly removed, the charge does not immediately fall to zero but approaches zero in an exponential fashion, described by Eq. 28–16b, or

$$q = C\mathcal{E}\,e^{-t/\tau_C}. \qquad p.533 \qquad (32\text{--}8)$$

The same time constant τ_c describes the fall of the charge as well as its rise.

An analogous delay in the rise or fall of the current occurs if an emf $\mathcal{E}$ is suddenly introduced into, or removed from, a single loop circuit containing a resistor R and an inductor L. When the switch S in Fig. 32–3 is closed on a, for example, the current in the resistor starts to rise. If the inductor were not present, the current would rise rapidly to a steady value $\mathcal{E}/R$. Because of the inductor, however, a self-induced emf $\mathcal{E}_L$ appears in the circuit and, from Lenz's law, this emf opposes the rise of current, which means that it opposes the battery emf $\mathcal{E}$ in polarity. Thus the resistor responds to the difference between two emfs, a constant one $\mathcal{E}$ due to the battery and a variable

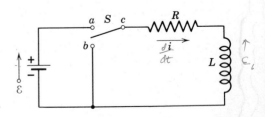

FIGURE 32–3 An *LR* circuit.

one $\mathcal{E}_L$ $(= -L\,di/dt)$ due to self-induction. As long as this second emf is present, the current in the resistor will be less than $\mathcal{E}/R$.

As time goes on, the rate at which the current increases becomes less rapid and the self-induced emf $\mathcal{E}_L$, which is proportional to di/dt, becomes smaller. Thus a time delay is introduced, and the current in the circuit approaches the value $\mathcal{E}/R$ asymptotically.

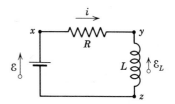

FIGURE 32–4 The circuit of Fig. 32–3 just after switch S is closed on *a*.

When the switch S in Fig. 32–3 is thrown to *a*, the circuit reduces to that of Fig. 32–4. Let us apply the loop theorem, starting at *x* in this figure and going clockwise around the loop. For the direction of current shown, *x* will be higher in potential than *y*, which means that we encounter a drop in potential of $-iR$ as we traverse the resistor. Point *y* is higher in potential than point *z* because, for an increasing current, the induced emf will oppose the rise of the current by pointing as shown. Thus as we traverse the inductor from *y* to *z* we encounter a drop in potential of $-L(di/dt)$. We encounter a rise in potential of $+\mathcal{E}$ in traversing the battery from *z* to *x*. The loop theorem thus gives

$$-iR - L\frac{di}{dt} + \mathcal{E} = 0$$

or
$$L\frac{di}{dt} + iR = \mathcal{E}. \tag{32–9}$$

Equation 32–9 is a differential equation involving the variable *i* and its first derivative di/dt. We seek the function $i(t)$ such that when it and its first derivative are substituted in Eq. 32–9 the equation is satisfied.

Although there are formal rules for solving various classes of differential equations (and Eq. 32–9 can, in fact, be easily solved by direct integration, after rearrangement) we often find it simpler to guess at the solution, guided by physical reasoning and by previous experience. Any proposed solution can be tested by substituting it in the differential equation and seeing whether this equation reduces to an identity.

The solution to Eq. 32–9 is, we assert,

$$i = \frac{\mathcal{E}}{R}(1 - e^{-Rt/L}). \tag{32–10}$$

To test this solution by substitution, we find the derivative di/dt, which is

$$\frac{di}{dt} = \frac{\mathcal{E}}{L}e^{-Rt/L}. \tag{32–11}$$

Substituting *i* and di/dt into Eq. 32–9 leads to an identity, as you can easily verify. Thus Eq. 32–10 is a solution of Eq. 32–9. Figure 32–5 shows how the potential difference V_R across the resistor $(= iR;$ see Eq. 32–10) and V_L across the inductor $(= L\,di/dt;$ see Eq. 32–11) vary with time for particular values of $\mathcal{E}$, *L*, and *R*. Compare this figure carefully with the corresponding figure for an *RC* circuit (Fig. 28–9).

We can rewrite Eq. 32–10 as

$$i = \frac{\mathcal{E}}{R}(1 - e^{-t/\tau_L}), \tag{32–12}$$

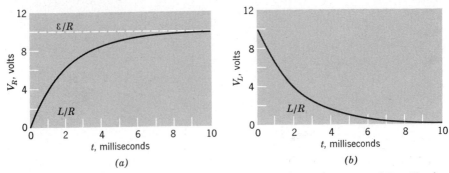

FIGURE 32–5 If in Fig. 32–3 we assume that $R = 2000$ ohms, $L = 4$ henrys and $\mathcal{E} = 10$ volts, then (*a*) shows the variation of i with t during the current buildup after switch S is closed on a, and (*b*) the variations of V_L with t. The time constant is $L/R = 2.0 \times 10^{-3}$ sec.

in which τ_L, the *inductive time constant,* is given by

$$\tau_L = L/R. \tag{32–13}$$

Note the correspondence between Eqs. 32–12 and 32–6.

To show that the quantity $\tau_L \,(= L/R)$ has the dimensions of time, we put

$$\frac{1 \text{ henry}}{\text{ohm}} = \frac{1 \text{ henry}}{\text{ohm}} \left(\frac{1 \text{ volt-sec}}{1 \text{ henry-amp}} \right)\left(\frac{1 \text{ ohm-amp}}{1 \text{ volt}} \right) = 1 \text{ sec.}$$

The first quantity in parenthesis is a conversion factor based on the defining equation for inductance $[L = -\mathcal{E}/(di/dt);$ Eq. 32–3*b*]. The second conversion factor is based on the relation $V = iR$.

The physical significance of the time constant follows from Eq. 32–12. If we put $t = \tau_L = L/R$ in this equation, it reduces to

$$i = \frac{\mathcal{E}}{R} (1 - e^{-1}) = (1 - 0.37)\frac{\mathcal{E}}{R} = 0.63 \frac{\mathcal{E}}{R}.$$

Thus the time constant τ_L is that time at which the current in the circuit will reach a value within $1/e$ (about 37%) of its final equilibrium value (see Fig. 32–5).

If the switch S in Fig. 32–3, having been left in position a long enough for the equilibrium current $\mathcal{E}/R$ to be established, is thrown to b, the effect is to remove the battery from the circuit. The differential equation that governs the subsequent decay of the current in the circuit can be found by putting $\mathcal{E} = 0$ in Eq. 32–9, or

$$L \frac{di}{dt} + iR = 0. \tag{32–14}$$

Show by the test of substitution that the solution of this differential equation is

$$i = \frac{\mathcal{E}}{R} e^{-t/\tau_L}. \tag{32–15}$$

▓ **Example 2.** A solenoid has an inductance of 50 henrys and a resistance of 30 ohms. If it is connected to a 100-volt battery, how long will it take for the current to reach one-half its final equilibrium value?

The equilibrium value of the current is reached as $t \to \infty$; from Eq. 32–12 it is $\mathcal{E}/R$. If the current has half this value at a particular time t_0, this equation becomes

$$\frac{1}{2}\frac{\mathcal{E}}{R} = \frac{\mathcal{E}}{R} (1 - e^{-t_0/\tau_L}).$$

Solving for t_0 yields

$$t_0 = \tau_L \ln 2 = 0.69 \frac{L}{R}.$$

Putting $\tau_L = L/R$ and using the values given, this reduces to

$$t_0 = 0.69\tau_L = 0.69 \left(\frac{50 \text{ henrys}}{30 \text{ ohms}} \right) = 1.2 \text{ sec.}$$

32–4 Energy and the Magnetic Field

In Section 26–6 we saw that the electric field could be viewed as the site of stored energy, the energy per unit volume being given, in a vacuum, by

$$u_E = \tfrac{1}{2}\epsilon_0 E^2,$$

p. 500

where E is the electric field at the point in question. Although this formula was derived for a parallel-plate capacitor, it holds for all kinds of electric field configurations.

Energy can also be stored in a magnetic field. For example, two parallel wires carrying currents in the same direction attract each other, and to pull them further apart work must be done. It is useful to think that this expended energy is stored in the magnetic field between and around the wires. The energy can be recovered from the field if the wires are allowed to move back to their original separation. In the electrostatic case the same argument was applied to pulling apart two unlike charges and in the gravitational case to pulling apart two masses.

To derive a quantitative expression for the storage of energy in the magnetic field, consider Fig. 32–4, which shows a source of emf $\mathcal{E}$ connected to a resistor R and an inductor L.

$$\mathcal{E} = iR + L\frac{di}{dt}, \tag{32–9}$$

is the differential equation that describes the growth of current in this circuit. We stress that this equation follows from the loop theorem and that the loop theorem in turn is an expression of the principle of conservation of energy for single-loop circuits. If we multiply each side of Eq. 32–9 by i, we obtain

$$\mathcal{E}i = i^2 R + Li\frac{di}{dt}. \tag{32–16}$$

which has the following physical interpretation in terms of work and energy:

1. If a charge dq passes through the seat of emf $\mathcal{E}$ in Fig. 32–4 in time dt, the seat does work on it in amount $\mathcal{E}\,dq$. The rate of doing work is $(\mathcal{E}\,dq)/dt$, or $\mathcal{E}i$. Thus the left term in Eq. 32–16 is the rate at which the seat of emf delivers energy to the circuit.

2. The second term in Eq. 32–16 is the rate at which energy appears as Joule energy in the resistor.

3. Energy that does not appear as Joule energy must, by our hypothesis, be stored in the magnetic field. Since Eq. 32–16 represents a statement of the conservation of energy for LR circuits, the last term must represent the rate dU_B/dt at which energy is stored in the magnetic field, or

$$\frac{dU_B}{dt} = Li\frac{di}{dt}. \tag{32–17}$$

We can write this as
$$dU_B = Li\,di.$$

Integrating yields

p. 500

$$JOULES \qquad U_B = \int_0^{U_B} dU_B = \int_0^i Li \, di = \tfrac{1}{2}Li^2, \qquad (32\text{–}18)$$

which represents the total stored magnetic energy in an inductance L carrying a current i.

This relation can be compared with the expression for the energy associated with a capacitor C carrying a charge q, namely

$$U_E = \frac{1}{2}\frac{q^2}{C} \cdot = \tfrac{1}{2}cv^2 =$$

Here the energy is stored in an electric field. In each case the expression for the stored energy was derived by setting it equal to the work that must be done to set up the field.

Example 3. A coil has an inductance of 5.0 henrys and a resistance of 20 ohms. If a 100-volt emf is applied, what energy is stored in the magnetic field after the current has built up to its maximum value $\mathcal{E}/R$?

The maximum current is given by

$$i = \frac{\mathcal{E}}{R} = \frac{100 \text{ volts}}{20 \text{ ohms}} = 5.0 \text{ amp.}$$

The stored energy is given by Eq. 32–18:

$$U_B = \tfrac{1}{2}Li^2 = \tfrac{1}{2}(5.0 \text{ henrys})(5.0 \text{ amp})^2 = 63 \text{ joules.}$$

Note that the time constant for this coil ($= L/R$) is 0.25 sec. After how many time constants will half of this equilibrium energy be stored in the field?

Example 4. A 3.0-henry inductor is placed in series with a 10-ohm resistor, an emf of 3.0 volts being suddenly applied to the combination. At 0.30 sec (which is one inductive time constant) after the contact is made, (*a*) what is the rate at which energy is being delivered by the battery?

The current is given by Eq. 32–12, or

$$i = \frac{\mathcal{E}}{R}(1 - e^{-t/\tau_L}),$$

which at $t = 0.30$ sec $(= \tau_L)$ has the value

$$i = \left(\frac{3.0 \text{ volts}}{10 \text{ ohms}}\right)(1 - e^{-1}) = 0.19 \text{ amp.}$$

The rate $P_\mathcal{E}$ at which energy is delivered by the battery is

$$P_\mathcal{E} = \mathcal{E}i$$
$$= (3.0 \text{ volts})(0.19 \text{ amp})$$
$$= 0.57 \text{ watt.}$$

(*b*) At what rate does energy appear as Joule energy in the resistor? This is given by

$$P_J = i^2R$$
$$= (0.19 \text{ amp})^2(10 \text{ ohms})$$
$$= 0.36 \text{ watt.}$$

(*c*) At what rate P_B is energy being stored in the magnetic field? This is given by the last term in Eq. 32–16, which requires that we know di/dt. Differentiating Eq. 32–12 yields

$$\frac{di}{dt} = \left(\frac{\mathcal{E}}{R}\right)\left(\frac{R}{L}\right)e^{-t/\tau_L}$$

$$= \frac{\mathcal{E}}{L}\,e^{-t/\tau_L}.$$

At $t = \tau_L$ we have

$$\frac{di}{dt} = \left(\frac{3.0 \text{ volts}}{3.0 \text{ henrys}}\right)e^{-1} = 0.37 \text{ amp/sec}.$$

From Eq. 32–17, the desired rate is

$$P_B = \frac{dU_B}{dt} = Li\frac{di}{dt}$$

$$= (3.0 \text{ henrys})(0.19 \text{ amp})(0.37 \text{ amp/sec})$$

$$= 0.21 \text{ watt}.$$

Note that as required by the principle of conservation of energy (see Eq. 32–16)

$$P_{\mathcal{E}} = P_J + P_B,$$

or

$$0.57 \text{ watt} = 0.36 \text{ watt} + 0.21 \text{ watt}$$

$$= 0.57 \text{ watt}.$$

32–5   Energy Density and the Magnetic Field

We now derive an expression for the density of energy u in a magnetic field. Consider a length l near the center of a long solenoid; Al is the volume associated with this length. The stored energy must lie entirely within this volume because the magnetic field outside such a solenoid is essentially zero. Moreover, the stored energy must be uniformly distributed throughout the volume of the solenoid because the magnetic field is uniform everywhere inside. Thus we can write

$$u_B = \frac{U_B}{Al}$$

or, since

$$U_B = \tfrac{1}{2}Li^2,$$

$$u_B = \frac{\tfrac{1}{2}Li^2}{Al}.$$

To express this in terms of the magnetic field, we can substitute for L in this equation, using the relation $L = \mu_0 n^2 lA$ (Eq. 32–5). Also we can solve Eq. 30–7 ($B = \mu_0 in$) for i and substitute in this equation. Doing so yields

JOULES/m^3

$$u_B = \frac{1}{2}\frac{B^2}{\mu_0}. \tag{32–19}$$

This equation gives the energy density stored at any point (in a vacuum or in a non-magnetic substance) where the magnetic field is **B**. The equation is true for all magnetic field configurations, even though it was derived by considering a special case, the solenoid. Equation 32–19 is to be compared with Eq. 26–18,

p. 500

$$u_E = \tfrac{1}{2}\epsilon_0 E^2, \qquad (\kappa = 1) \tag{32–20}$$

which gives the energy density (in a vacuum) at any point in an electric field. Note that both u_B and u_E are proportional to the square of the appropriate field quantity, B or E.

The solenoid plays a role with relationship to magnetic fields similar to the role the parallel-plate capacitor plays with respect to electric fields. In each case we have a simple device that can be used for setting up a uniform field throughout a well-defined region of space and for deducing, in a simple way, some properties of these fields.

Example 5. A long coaxial cable (Fig. 32–6) consists of two concentric cylinders with radii a and b. Its central conductor carries a steady current i, the outer conductor providing the return path. (a) Calculate the energy stored in the magnetic field for a length l of such a cable.

In the space between the two conductors Ampère's law

$$\oint \mathbf{B} \cdot d\mathbf{l} = \mu_0 i$$

leads to

$$(B)(2\pi r) = \mu_0 i$$

or

$$B = \frac{\mu_0 i}{2\pi r}.$$

Ampère's law shows further that the magnetic field is zero for points outside the outer conductor (why?). Magnetic fields exist inside each of the conductors; we choose to ignore them, on the assumption that the cable dimensions are chosen so that most of the stored magnetic energy is in the space between the conductors see, however, Problem 23.

The energy density for points between the conductors, from Eq. 32–19, is

$$u = \frac{1}{2\mu_0} B^2 = \frac{1}{2\mu_0} \left(\frac{\mu_0 i}{2\pi r} \right)^2 = \frac{\mu_0 i^2}{8\pi^2 r^2}.$$

Consider a volume element dV consisting of a cylindrical shell whose radii are r and $r + dr$ and whose length is l. The energy dU contained in it is

$$dU = u \, dV = \frac{\mu_0 i^2}{8\pi^2 r^2} (2\pi r l)(dr) = \frac{\mu_0 i^2 l}{4\pi} \frac{dr}{r}.$$

The total stored magnetic energy is found by integration, or

$$U = \int dU = \frac{\mu_0 i^2 l}{4\pi} \int_a^b \frac{dr}{r} = \frac{\mu_0 i^2 l}{4\pi} \ln \frac{b}{a},$$

which is the desired expression.

(b) What is the inductance of a length l of coaxial cable?

The inductance L can be found from Eq. 32–18 ($U = \frac{1}{2} L i^2$), which leads to

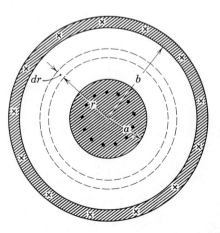

FIGURE 32–6   Example 5. Cross section of a coaxial cable, showing steady currents in the central and outer conductors.

$$L = \frac{2U}{i^2} = \frac{\mu_0 l}{2\pi} \ln \frac{b}{a}.$$

You can also derive this expression directly from the definition of inductance, using the procedures of Example 1.

Example 6. Compare the energy required to set up, in a cube 10 cm on edge, (a) a uniform electric field of 1.0×10^5 volts/meter and (b) a uniform magnetic field of 1.0 weber/meter^2 ($= 10^4$ gauss). Both these fields would be judged reasonably large but they are readily available in the laboratory.

(a) In the electric case we have, where V_0 is the volume of the cube,

$$U_E = u_E V_0 = \tfrac{1}{2}\epsilon_0 E^2 V_0$$

$$= (0.5)(8.9 \times 10^{-12} \text{ coul}^2/\text{nt-m}^2)(1.0 \times 10^5 \text{ volts/meter})^2(0.10 \text{ meter})^3$$

$$= 4.5 \times 10^{-5} \text{ joule.}$$

(b) In the magnetic case, from Eq. 32–19, we have

$$U_B = u_B V_0 = \frac{B^2}{2\mu_0} V_0 = \frac{(1.0 \text{ weber/meter}^2)^2(0.10 \text{ meter})^3}{(2)(4\pi \times 10^{-7} \text{ weber/amp-m})}$$

$$= 400 \text{ joules.}$$

In terms of fields normally available in the laboratory, much larger amounts of energy can be stored in a magnetic field than in an electric one, the ratio being about 10^7 in this example. Conversely, much more energy is required to set up a magnetic field of reasonable laboratory magnitude than is required to set up an electric field of similarly reasonable magnitude.

QUESTIONS

1. Two coils are connected in series. Does their equivalent inductance depend on their geometrical relationship to each other?

2. Is the inductance per unit length for a solenoid near its center the same as, less than, or greater than the inductance per unit length near its ends?

3. Two solenoids, A and B, have the same diameter and length and contain only one layer of windings, with adjacent turns touching, insulation thickness being negligible. Solenoid A contains many turns of fine wire and solenoid B contains fewer turns of heavier wire. Which solenoid has the larger inductance? Which solenoid has the larger inductive time constant?

4. If the flux passing through each turn of a coil is the same, the inductance of the coil may be computed from $L = N\Phi_B/i$ (Eq. 32–4). How might one compute L for a coil for which this assumption is not valid?

5. If a current in a source of emf is in the direction of the emf, the energy of the source decreases; if a current is in a direction opposite to the emf (as in charging a battery), the energy of the source increases. Do these statements apply to the inductor in Fig. 32–1a and 32–1b?

6. Show that the dimensions of the two expressions for L, $N\Phi_B/i$ (Eq. 32–4) and $\mathcal{E}/(di/dt)$ (Eq. 32–3b), are the same.

7. Does the time required for the current in a particular LR circuit to build up to any given fraction of its equilibrium value depend on the value of the applied emf?

8. A steady current is set up in a coil with a large inductive time constant. When the current is interrrupted with a switch, a heavy arc tends to appear at the switch blades. Explain.

9. In an LR circuit like that of Fig. 32–4 can the self-induced emf ever be larger than the battery emf?

10. In an *LR* circuit like that of Fig. 32–4 is the current in the resistance always the same as the current in the inductance?

11. In the circuit of Fig. 32–3 the self-induced emf is a maximum at the instant the switch is closed on *a*. How can this be since there is no current in the inductance at this instant?

12. Give some arguments to show that energy can be stored in a magnetic field.

13. The switch in Fig. 32–3 is thrown from *a* to *b*. What happens to the energy stored in the inductor?

14. In a toroid is the energy density larger near the inner radius or near the outer radius?

15. Each item (*a*) coulomb-ohm-meter/weber, (*b*) volt-second, (*c*) coulomb-ampere/farad, (*d*) kilogram-volt-meter^2/(henry-ampere)^2, (*e*) (henry/farad)$^{1/2}$ is equal to one of the items in the following list: meter, second, kilogram, dimensionless number, newton, joule, volt, ohm, watt, coulomb, ampere, weber, henry, farad. Give the equalities.

PROBLEMS

1(1). The inductance of a close-packed coil of 400 turns is 8.0 mh. What is the magnetic flux through the coil when the current is 5.0×10^{-3} amp?
Answer: 1.0×10^{-7} weber.

2(1). A 10-henry inductor carries a steady current of 2.0 amp. How can a 100-volt self-induced emf be made to appear in the inductor?

3(1). Each item (*a*) kilogram-volt-meter^2/(henry-ampere)^2, (*b*) (henry/farad)$^{1/2}$, (*c*) coulomb-ohm-meter/weber, (*d*) volt-second is equal to one of the items in the following list: meter, second, kilogram, dimensionless number, newton, joule, volt, ohm, watt, coulomb, ampere, weber, henry, farad. Give the equalities.
Answer: (*a*) Coulomb. (*b*) Ohm. (*c*) meter. (*d*) Weber.

4(2). What is the inductance of the toroid of Problem 27(5), Chapter 30?

5(2). A solenoid is wound with a single layer of #10 gauge copper wire (diameter, 0.10 in.). It is 4.0 cm in diameter and 2.0 meters long. What is the inductance per unit length for the solenoid near its center? Assume that adjacent wires touch and that insulation thickness is negligible.
Answer: 2.5×10^{-4} henry/meter.

6(2). Two parallel wires whose centers are a distance *d* apart carry equal currents in opposite directions. Show that, neglecting the flux within the wires themselves, the inductance of a length *l* of such a pair of wires is given by

$$L = \frac{\mu_0 l}{\pi} \ln \frac{d-a}{a},$$

where *a* is the wire radius. See Example 2, Chapter 30.

7(2). Calculate the self-inductance of two concentric hollow cylinders of radii *a* and *b*, and of length $l \gg a,b$. At one end the cylinders are connected by a flat conducting plate so that the current travels down in the inner cylinder and back in the outer.
Answer: $\frac{\mu_0 l}{2\pi} \ln \left(\frac{b}{a}\right)$.

8(2). A very wide copper strip of width *W* is bent into a piece of slender tubing of radius *R* with two plane extensions as shown in Fig. 32–7. A current *i* flows through the strip distributed uniformly over the width. In this way a "one-turn solenoid" has been formed. (*a*) Find the magnitude of the magnetic field **B** in the tubular part (far away from the edges). (Hint: Assume that the field outside this one-turn solenoid is negligibly small.) (*b*) Find the inductance of this one turn solenoid, neglecting the two plane extensions.

9(2). A long thin solenoid can be bent into a ring to form a toroid. Show that if the solenoid is long and thin enough the equation for the inductance of a toroid (see Example 1) reduces to that for a solenoid (Eq. 32–5).

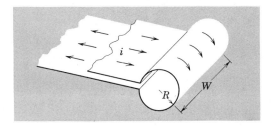

FIGURE 32-7 Problem 8(2).

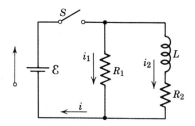

FIGURE 32-8 Problem 13(3).

10(3). A wooden toroidal core with a square cross section has an inner radius of 10 cm and an outer radius of 12 cm. It is wound with one layer of #18 gauge wire (diameter, 0.040 in.; resistance 0.0064 ohm/ft). What are (*a*) the inductance and (*b*) the inductive time constant? Ignore the thickness of the insulation.

11(3). (*a*) Two inductances L_1 and L_2 are connected in series and are separated by a large distance. Show that the equivalent inductance L is $L_1 + L_2$. (*b*) If two solenoids of inductance L are "close wound," show that the equivalent inductance is either zero or $4L$ depending on the direction of the windings. (Hint: They form a single solenoid.)

12(3). Show that if two inductors with equal inductance L are connected in parallel the equivalent inductance of the combination is $\frac{1}{2}L$. The inductors are separated by a large distance.

13(3). In the circuit shown in Fig. 32–8, $\mathcal{E} = 10$ volts, $R_1 = 5.0$ ohms, $R_2 = 10$ ohms and $L = 5.0$ henrys. For the two separate conditions (*I*) switch S just closed and (*II*) switch S closed for a very long time, calculate (*a*) the current i_1 through R_1, (*b*) the current i_2 through R_2, (*c*) the current i through the switch, (*d*) the potential difference across R_2, (*e*) the potential difference across L, and (*f*) di_2/dt.

Answer: $t \approx 0$; (*a*) 2.0 amp. (*b*) Zero. (*c*) 2.0 amp. (*d*) Zero. (*e*) 10 volts. (*f*) 2.0 amp/sec. $t \to \infty$; (*a*) 2.0 amp. (*b*) 1.0 amp. (*c*) 3.0 amp. (*d*) 10 volts. (*e*) Zero. (*f*) Zero.

14(3). An emf $\mathcal{E}$ is connected to a switch, two resistors and an inductor as shown in Fig. 32–9. After being closed for a long time at time $t = 0$ the switch is opened. (*a*) Find the current through R_2 as a function of the time t. (*b*) Find the potential difference across R_2 as a function of time.

15(3). In Figure 32–10, $\mathcal{E} = 100$ volts, $R_1 = 10$ ohms, $R_2 = 20$ ohms, $R_3 = 30$ ohms, and $L = 2$ henry. Find the value of i_1 and i_2 (*a*) immediately after S is closed, (*b*) a long time later, (*c*) immediately after S is opened again, and (*d*) a long time later.

Answer: (*a*) $i_1 = i_2 = 3.3$ amp. (*b*) $i_1 = 4.5$ amp; $i_2 = 2.8$ amp. (*c*) $i_1 = 0$; $i_2 = 1.8$ amp. (*d*) $i_1 = i_2 = 0$.

16(3). The current in an *LR* circuit drops from 1.0 amp at $t = 0$ to 0.010 amp one second later. If L is 10 henrys, find the resistance R in the circuit.

17(3). A solenoid having an inductance of 6.0×10^{-6} henrys is connected in series to a 1.0×10^3-ohm resistor. (*a*) If a 10-volt battery is switched across the pair, how long will it take for the current through the resistor to reach 80% of its final value? (*b*) What is the current through the resistor after one time constant?

Answer: (*a*) 9.7×10^{-9} sec. (*b*) 6.3×10^{-3} amp.

18(3). The current in an *LR* circuit builds up to one-third of its steady-state value in 5.0 sec. What is the inductive time constant?

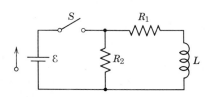

FIGURE 32-9 Problem 14(3).

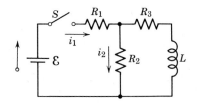

FIGURE 32-10 Problem 15(3).

19(3). How long would it take for the voltage across the resistor in an *LR* circuit ($L = 1.0$ henry, $R = 1.0$ ohm) to drop to 10% of its initial value?

Answer: 2.3 sec.

20(3). How many "time constants" must we wait for the current in an *LR* circuit to build up to within 0.10 per cent of its equilibrium value?

21(3). A 50-volt potential difference is suddenly applied to a coil with $L = 50$ mh and $R = 180$ ohms. At what rate is the current increasing after 0.0010 sec?

Answer: 27 amps/sec.

22(3). A uniform magnetic field **B** is normal to the plane of a one-turn loop of area A made of superconducting wire (the resistance is zero at a sufficiently low temperature). If the external magnetic field is turned off (*a*) what is the final magnetic flux through the loop? (*b*) What can be said about the final magnetic field? (*c*) How do these quantities depend on the rate the external field is decreased?

23(4). A coil with an inductance of 2.0 henrys and a resistance of 10 ohms is suddenly connected to a resistanceless battery with $\varepsilon = 100$ volts. (*a*) What is the equilibrium current? (*b*) How much energy is stored in the magnetic field when this current exists in the coil?

Answer: (*a*) 10 amps. (*b*) 100 joules.

24(4). A given coil is connected in series with a 10,000-ohm resistor. When a 50-volt battery is applied to the two, the current reaches a value of 2.0 milliamp after 5.0 milliseconds. (*a*) Find the value of the inductance of the coil. (*b*) What is the energy stored in the inductance at this same moment?

25(4). A coil with an inductance of 2.0 henrys and a resistance of 10 ohms is suddenly connected to a resistanceless battery with $\varepsilon = 100$ volts. At 0.10 sec after the connection is made, what are the rates at which (*a*) energy is being stored in the magnetic field, (*b*) Joule energy is appearing, and (*c*) energy is being delivered by the battery?

Answer: (*a*) 240 joules/sec. (*b*) 150 joules/sec, (*c*) 390 joules/sec.

26(5). What must be the magnitude of a uniform electric field if it is to have the same energy density as that possessed by a 5000-gauss magnetic field?

27(5). A length of #10 gauge copper wire carries a current of 10 amp. Calculate (*a*) the magnetic energy density and (*b*) the electric energy density at the surface of the wire. The wire diameter is 0.10 in. and its resistance per unit length is 1.0 ohm/1000 ft.

Answer: (*a*) 0.99 joules/meter³. (*b*) 4.8×10^{-15} joules/meter³.

28(5). What is the energy density in the magnetic field near the center of the solenoid of Problem 26(5), Chapter 30?

29(5). A circular loop of wire 5.0 cm in radius carries a current of 100 amp. What is the energy density at the center of the loop?

Answer: 0.63 joules/meter³.

30(5). (*a*) What is the magnetic energy density of the earth's magnetic field of 0.50 gauss? (*b*) Assuming this to be relatively constant over distances small compared with the earth's radius and neglecting the variations near the magnetic poles, how much energy would be stored in a shell between the earth's surface and 10 miles above the surface?

31(5). (*a*) Find an expression for the energy density as a function of the radius for the toroid of Problem 27(5), Chapter 30. (*b*) Integrating the energy density over the volume of the toroid, find the total energy stored in the field of the toroid. (*c*) Using the result of Problem 4(2) of this chapter evaluate the energy stored in the toroid directly from the inductance and compare with (*b*).

Answer: (*a*) $\dfrac{\mu_0 i^2 N^2}{8\pi^2 r^2}$. (*b*) 2.3×10^{-4} joule. (*c*) 2.3×10^{-4} joule.

32(5). A long wire carries a current of uniform density. Let i be the total current carried by the wire and show that the magnetic energy per unit length stored within the wire equals $\mu_0 i^2/16\pi$. Note that it does not depend on the wire diameter.

33(5). Show that the self-inductance for a length l of a long wire associated with the flux inside the wire is $\mu_0 l/8\pi$, independent of the diameter. Assume a uniform current density; see Problem 32(5).

34(5). The coaxial cable of Example 5 has $a = 1.0$ mm, $b = 4.0$ mm, and $c = 5.0$ mm (c is the radius of the outer surface of the outer conductor). It carries a current of 10 amp in the inner conductor and an equal but oppositely directed return current in the outer conductor. Calculate and compare the stored magnetic energy per meter of cable length (a) within the central conductor, (b) in the space between the conductors, and (c) within the outer conductor.

CHAPTER 33

Magnetic Properties of Matter

33–1 Poles and Dipoles

In electricity the isolated charge q is the simplest structure that can exist. If two such charges of opposite sign are placed near each other, they form an electric dipole, characterized by an electric dipole moment **p**. In magnetism isolated magnetic "poles," which would correspond to isolated electric charges, apparently do not exist. The simplest magnetic structure is the magnetic dipole, characterized by a magnetic dipole moment **μ**. Table 30–1 summarizes some characteristics of electric and magnetic dipoles.

A current loop, a bar magnet, and a solenoid of finite length are examples of magnetic dipoles. Their magnetic dipole moments can be measured by placing them in an external magnetic field **B**, measuring the torque **τ** that acts on them, and computing **μ** from Eq. 29–11, or

$$\tau = \mu \times \mathbf{B}. \tag{33–1}$$

Alternatively, we can measure B due to the dipole at a point along its axis a (large) distance r from its center and compute μ from the expression in Table 30–1, or

$$B = \frac{\mu_0}{2\pi} \frac{\mu}{r^3}. \tag{33–2}$$

Figure 33–1, which shows iron filings sprinkled on a sheet of paper under which there is a bar magnet, suggests that this dipole might be viewed as two "poles" separated by a distance d. However, all attempts to isolate these poles fail. If we break the magnet, as in Fig. 33–2, the fragments prove to be dipoles and not isolated poles. If we break up a magnet into the electrons and nuclei that make up its atoms, we will find

FIGURE 33–1 A bar magnet is a magnetic dipole. The iron filings suggest the lines of **B** in Fig. 33–4a. (Courtesy Physical Science Study Committee.)

that even these elementary particles are magnetic dipoles. Figure 33–3 contrasts the electric and the magnetic characteristics of the free electron.

All electrons have a characteristic *"spin"* angular momentum about a certain axis, which has the value of

$$L_s = 0.527596 \times 10^{-34} \text{ joule-sec.}$$

This is suggested by the vector $\mathbf{L}_s$ in Fig. 33–3b. We can view such a spinning charge classically as being made up of infinitesimal current loops. Each such loop is a tiny magnetic dipole, its moment being given by (Eq. 29–10)

$$\mu = NiA, \tag{33–3}$$

where i is the equivalent current in each infinitesimal loop and A is the loop area. The number of turns, N, is unity. The magnetic dipole moment of the spinning charge can be found by integrating over the moments of the infinitesimal current loops that make it up; see Problem 2.

Although this model of the spinning electron is too mechanistic and is not in accord

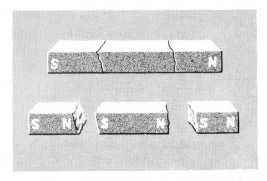

FIGURE 33–2 If a bar magnet is broken, each fragment becomes a dipole.

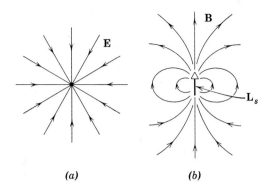

FIGURE 33–3 (*a*) The lines of **E** and (*b*) the lines of **B** for an electron. The magnetic dipole moment of the electron, μ_l, is directed opposite to the spin angular momentum vector, $\mathbf{L}_s$.

(a) (b)

with modern quantum physics, it remains true that the magnetic dipole moments of elementary particles are closely connected with their intrinsic angular momenta. Those particles and nuclei whose angular momentum is zero (the α-particle, the pion, the O^{16} nucleus, etc.) have no magnetic dipole moment. The "intrinsic" or "spin" magnetic moment of the electron must be distinguished from any additional magnetic moment it may have because of its orbital motion in an atom.

Example 1. Find a method for measuring μ for a bar magnet.

(*a*) Place the magnet in a uniform external magnetic field **B**, with μ making an angle θ with **B**. The magnitude of the torque acting on the magnet (see Eq. 33–1) is given by

$$\tau = \mu B \sin \theta.$$

Clearly μ can be learned if τ, B, and θ are measured.

(*b*) A second technique is to suspend the magnet from its center of mass and to allow it to oscillate about its stable equilibrium position in the external field **B**. For small oscillations, $\sin \theta$ can be replaced by θ and the equation just given becomes

$$\tau = -(\mu B)\theta = -\kappa\theta,$$

where κ is a constant. The minus sign has been inserted to show that τ is a restoring torque. Since τ is proportional to θ, the condition for simple angular harmonic motion is met. The frequency ν is given by the reciprocal of Eq. 13–23, or

$$\nu = \frac{1}{2\pi} \sqrt{\frac{\kappa}{I}} = \frac{1}{2\pi} \sqrt{\frac{\mu B}{I}}.$$

With this equation μ can be found from the measured quantities ν, B, and I.

Example 2. An electron in an atom circulating in an assumed circular orbit of radius r behaves like a current loop and has an orbital magnetic dipole moment* usually represented by μ_l. Derive a connection between μ_l and the orbital angular momentum L_l.

Newton's second law ($F = ma$) yields, if we substitute Coulomb's law for F,

$$\frac{1}{4\pi\epsilon_0} \frac{e^2}{r^2} = ma = \frac{mv^2}{r}$$

or

$$v = \sqrt{\frac{e^2}{4\pi\epsilon_0 mr}}. \tag{33–4}$$

The angular speed ω is given by

$$\omega = \frac{v}{r} = \sqrt{\frac{e^2}{4\pi\epsilon_0 mr^3}}.$$

* This must not be confused with the magnetic dipole moment μ_s of the electron spin, which is also present.

The current for the orbit is the rate at which charge passes any given point, or

$$i = ev = e\left(\frac{\omega}{2\pi}\right) = \sqrt{\frac{e^4}{16\pi^3\epsilon_0 mr^3}}.$$

The orbital dipole moment μ_l is given from Eq. 33–3 if we put $N = 1$ and $A = \pi r^2$, or

$$\mu_l = NiA = (1)\sqrt{\frac{e^4}{16\pi^3\epsilon_0 mr^3}}(\pi r^2) = \frac{e^2}{4}\sqrt{\frac{r}{\pi\epsilon_0 m}}. \tag{33–5}$$

The orbital angular momentum L_l is

$$L_l = (mv)r.$$

Combining with Eq. 33–4 leads to

$$L_l = \sqrt{\frac{e^2 mr}{4\pi\epsilon_0}}.$$

Finally, eliminating r between this equation and Eq. 33–5 yields

$$\mu_l = L_l\left(\frac{e}{2m}\right),$$

which shows that the orbital magnetic moment of an electron is proportional to its orbital angular momentum.

For $r = 5.1 \times 10^{-11}$ meter, which corresponds to hydrogen in its normal state, we have from Eq. 33–5

$$\mu_l = \frac{e^2}{4}\sqrt{\frac{r}{\pi\epsilon_0 m}}$$

$$= \frac{(1.6 \times 10^{-19}\ \text{coul})^2}{4}\sqrt{\frac{5.1 \times 10^{-11}\ \text{meter}}{(\pi)(8.9 \times 10^{-12}\ \text{coul}^2/\text{nt-m}^2)(9.1 \times 10^{-31}\ \text{kg})}}$$

$$= 9.1 \times 10^{-24}\ \text{amp-m}^2.$$

33–2   Gauss's Law for Magnetism

Gauss's law for magnetism, which is one of the basic equations of electromagnetism (see Table 34–2), is a formal way of stating a conclusion that seems to be forced on us by the facts of magnetism, namely, that *isolated magnetic poles do not exist*. This law asserts that the flux Φ_B through any closed Gaussian surface must be zero, or

$$\Phi_B = \oint \mathbf{B} \cdot d\mathbf{S} = 0, \tag{33–6}$$

where the integral is to be taken over the entire closed surface. We contrast this with Gauss's law for electricity, which is

$$\epsilon_0 \oint \mathbf{E} \cdot d\mathbf{S} = q. \tag{33–7}$$

The fact that a zero appears at the right of Eq. 33–6, but not at the right of Eq. 33–7, means that in magnetism there is no counterpart to the free charge q in electricity.

Figure 33–4*a* shows a Gaussian surface that encloses one end of a bar magnet. Note that the lines of **B** enter the surface for the most part inside the magnet and leave it for the most part outside the magnet. There is thus an inward (or negative) flux inside the magnet and an outward (or positive) flux outside it. The total flux for the whole surface is zero.

Figure 33–4*b* shows a similar surface for a solenoid of finite length which, like a bar

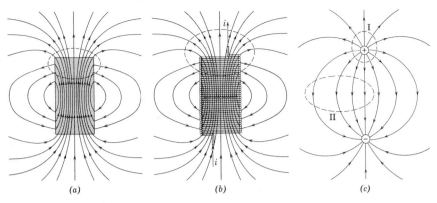

FIGURE 33–4   Lines of **B** (*a*) for a bar magnet and (*b*) for a short solenoid. (*c*) Lines of **E** for an electric dipole. At large enough distances all three fields vary like those for a dipole. The four dashed curves are intersections with the plane of the figure of closed Gaussian surfaces. Note that Φ_B equals zero for (*a*) or (*b*). Φ_E equals zero for surfaces like II in (*c*), which do not contain any charge, but Φ_E is not zero for surfaces like I.

magnet, is also a magnetic dipole. Here, too, Φ_B equals zero. Figures 33–4*a* and *b* show clearly that there are no "sources" of **B**; that is, there are no points from which lines of **B** emanate. Also, there are no "sinks" of **B**; that is, there are no points toward which **B** converges. In other words, there are no free magnetic poles.

Figure 33–4*c* shows a Gaussian surface (I) surrounding the positive end of an electric dipole. Here there *is* a net flux of the lines of **E**. There is a "source" of **E**; it is the charge q. If q is negative, we have a "sink" of **E** because the lines of **E** end on negative charges. For surfaces like surface II in Fig. 33–4*c* for which the charge inside is zero, the flux of **E** over the surface is also zero.

33–3 Paramagnetism

Magnetism as we know it in our daily experience is an important but special branch of the subject called *ferromagnetism;* we discuss this in Section 33–5. Here we discuss a weaker form of magnetism called *paramagnetism.*

For most atoms and ions, the magnetic effects of the electrons, including both their spins and orbital motions, exactly cancel so that the atom or ion is not magnetic. This is true for the rare gases such as neon and for ions such as Cu^+, which makes up ordinary copper. These materials do not exhibit paramagnetism. For other atoms or ions the magnetic effects of the electrons do not cancel, so that the atom as a whole has a magnetic dipole moment $\boldsymbol{\mu}$. Examples are found among the so-called transition elements, such as Mn^{++}, the rare earths, such as Gd^{+++}, and the actinide elements, such as U^{++++}.

If a sample of N atoms, each of which has a magnetic dipole moment $\boldsymbol{\mu}$, is placed in a magnetic field, the elementary atomic dipoles tend to line up with the field. For perfect alignment, the sample as a whole would have a magnetic dipole moment of $N\mu$. However, the aligning process is seriously interfered with by the collisions that take place between the atoms if the sample is a gas and by temperature vibrations if the sample is a solid. The importance of this thermal agitation effect may be measured by comparing two energies: one $(= \frac{3}{2}kT)$ is the mean kinetic energy of translation of a gas atom at temperature T; the other $(= 2\mu B)$ is the difference in energy between an atom lined up with the magnetic field and one pointing in the opposite direction. As Example 3 shows, the effect of the collisions at ordinary temperatures and fields is large. The sample acquires a magnetic moment when placed in an external magnetic field, but this moment is usually much smaller than the maximum possible moment $N\mu$.

Example 3. A paramagnetic gas, whose atoms (see Example 2) have a magnetic dipole moment of about 10^{-23} amp-m^2, is placed in an external magnetic field of magnitude 1.0 weber/meter^2. At room temperature ($T = 300°$ K) calculate and compare U_T, the mean kinetic energy of translation ($= \frac{3}{2}kT$), and U_B, the magnetic energy ($= 2\mu B$):

$$U_T = \tfrac{3}{2}kT = (\tfrac{3}{2})(1.38 \times 10^{-23} \text{ joule}/° \text{ K})(300° \text{ K}) = 6.0 \times 10^{-21} \text{ joule},$$

$$U_B = 2\mu B = (2)(10^{-23} \text{ amp-m}^2)(1.0 \text{ weber/meter}^2) = 2.0 \times 10^{-23} \text{ joule}.$$

Because U_T equals 300 U_B, we see that energy exchanges in collisions can interfere seriously with the alignment of the dipoles with the external field.

For any substance the *magnetization* **M** is defined as the magnetic moment per unit volume, or

$$\mathbf{M} = \frac{\mu}{V},$$

where V is the volume of the specimen. It is a vector because $\boldsymbol{\mu}$, the dipole moment of the specimen, is a vector.

In 1895 Pierre Curie (1859–1906) discovered experimentally that the magnetization M of a paramagnetic specimen is directly proportional to B, the effective magnetic field in which the specimen is placed, and inversely proportional to the temperature, or

$$M = C\frac{B}{T}, \tag{33–8}$$

in which C is a constant. This equation is known as *Curie's law*. The law is physically reasonable in that increasing B tends to align the elementary dipoles in the specimen, that is, to increase M, whereas increasing T tends to interfere with this alignment, that is, to decrease M. Curie's law is well verified experimentally, provided that the ratio B/T does not become too large.

M cannot increase without limit, as Curie's law implies, but must approach a value M_{max} ($= \mu N/V$) corresponding to the complete alignment of the N dipoles contained in the volume V of the specimen. Figure 33–5 shows this saturation effect for a sample of $CrK(SO_4)_2 \cdot 12H_2O$. The chromium ions are responsible for all the paramagnetism of this salt, all the other elements being paramagnetically inert. To achieve 99.5% saturation, it is necessary to use applied magnetic fields as high as 50,000 gauss and temperatures as low as 1.3° K. Note that for more readily achievable conditions, such as $B = 10,000$ gauss and $T = 10°$ K, the abscissa in Fig. 33–5 is only 1.0 so that Curie's law would appear to be well obeyed for this and for all lower values of B/T. The curve that passes through the experimental points in this figure is calculated from a theory based on modern quantum physics; it is in excellent agreement with experiment.

The magnetic dipoles of a paramagnetic substance tend to align themselves parallel to the external magnetic field. This is easily seen if we consider the atomic dipoles to be tiny bar magnets between the poles of a larger permanent magnet. Under such

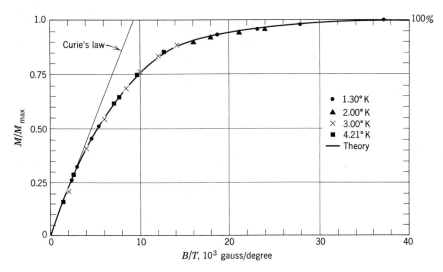

FIGURE 33–5   The ratio M/M_{max} for a paramagnetic salt (chromium potassium alum) in various magnetic fields and at various temperatures. The curve through the experimental points is a theoretical curve calculated from modern quantum physics. (From W. E. Henry.)

conditions, the north pole of the dipole will be attracted to the south pole of the external field as shown in Fig. 33–6. The dipole moment μ, which always points from the south pole to the north pole of the dipole, is thus parallel to the external field **B,** as shown in the figure.

Paramagnetism is the magnetic equivalent of the electric polarization of polar molecules in an external electric field, which we discussed in Section 26–4.

33–4 Diamagnetism

The magnetic equivalent of induced electric dipole moments (see Section 25–5 and Fig. 25–12) results in *diamagnetism*. It is present in all substances, but it is such a feeble effect that its presence is masked in substance made of atoms that have a permanent fixed magnetic dipole moment, that is, in paramagnetic or ferromagnetic substances.

Diamagnetic materials also differ from paramagnetic and ferromagnetic materials (and also dielectric materials) in that the atomic dipoles are aligned in a direction opposite to that of the external field. Such materials are thus repelled from the pole of a strong magnet, whereas paramagnetic and ferromagnetic materials would be attracted under similar circumstances. To understand this result, we must analyze the mechanism by which the magnetic dipoles are induced in diamagnetic materials.

Figures 33–7*a* and *b* each show an electron circulating in a diamagnetic atom at angular frequency ω_0 in an assumed circular orbit of radius *r*. Each electron is moving under the action of a centripetal force $\mathbf{F}_E$ of electrostatic origin where from Newton's second law,

$$F_E = ma = m\omega_0^2 r. \qquad (33\text{–}9)$$

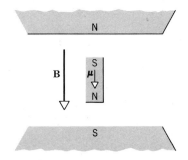

FIGURE 33–6   An atomic dipole, represented by a tiny bar magnet, in an external magnetic field set up by a large permanent magnet whose poles are shown. Note that μ, the atomic dipole moment, is parallel to **B.**

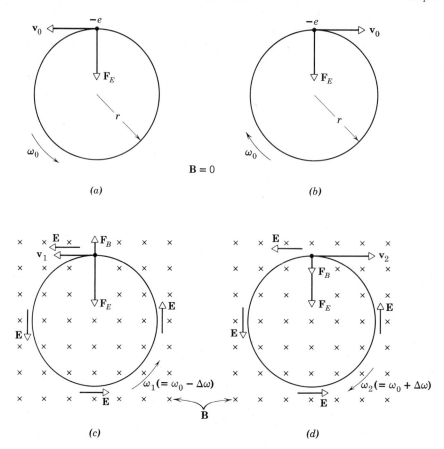

FIGURE 33–7   (*a*) An electron circulating in an atom. (*b*) An electron circulating in the opposite direction. (*c*) A magnetic field is introduced, *decreasing* the linear speed of the electron in (*a*), that is, $v_1 < v_0$. (*d*) The magnetic field *increases* the linear speed of the electron in (*b*), that is, $v_2 > v_0$.

Each rotating electron has an orbital magnetic moment, but for the atom as a whole the orbits are randomly oriented so that there is no *net* magnetic effect. In Fig. 33–7*a*, for example, the magnetic dipole moment μ_l points into the page; in Fig. 33–7*b* it points out and the net effect for the two orbits shown is cancellation. This cancellation is also shown at the left in Fig. 33–8.

If an external field **B** is applied as in Figs. 33–7*c* and *d*, these circulating electrons experience an additional acceleration, parallel to their motion, during the time that the field is increasing. This is due to the betatron effect (see Section 31–5, especially Fig. 31–9). The induced electric fields are shown in Figs. 33–7*c* and *d*. Since the electrons are negatively charged, they are accelerated in a direction opposite to that of **E.** Thus, in Fig. 33–7*c* the electron is slowed down whereas the electron in Fig. 33–7*d* is speeded up.

We would usually expect the changing electron speed to result in a new orbit. However, an additional radial force, given by $-e(\mathbf{v} \times \mathbf{B})$, also acts on the electron. This is shown as $\mathbf{F}_B$ in Fig. 33–7*c* where it subtracts from the original $\mathbf{F}_E$, and in Fig. 33–7*d* where it adds to $\mathbf{F}_E$. It can be shown that this additional magnetic force is exactly right to keep the electrons in their original circular orbits even though their speeds are changed.

Since the electron of Fig. 33–7*c* is slower than that of Fig. 33–7*a* and since it has the same orbital radius, its magnetic moment is reduced below the zero-field value. Similarly, the moment of the electron in Fig. 33–7*d* is greater than that in Fig. 33–7*b*. These results are shown in Fig. 33–8*b*. Note that the resultant magnetic moment is no longer zero, but that there is a net moment which is opposite to the direction of the external field.

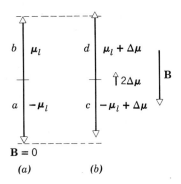

B = 0

(*a*)   (*b*)

33–5 Ferromagnetism

For five elements (Fe, Co, Ni, Gd, and Dy) and for many alloys of these and other elements a special effect occurs which permits a specimen to achieve a high degree of magnetic alignment in spite of the randomizing tendency of the thermal motions of the atoms. In such materials, described as *ferromagnetic,* a special form of interaction called *exchange coupling* occurs between adjacent atoms, coupling their magnetic

FIGURE 33–8 The magnetic moments of the two oppositely circulating electrons in an atom cancel when there is no external magnetic field, as in (*a*), but do not cancel when a field is applied, as in (*b*). Note that the resultant moment in (*b*) points in the opposite direction to **B**. Compare carefully with Fig. 37–7.

moments together in rigid parallelism. This is a purely quantum effect and cannot be explained in terms of classical physics. Modern quantum physics successfully predicts that this will occur only for the five elements listed. If the temperature is raised above a certain critical value, called the *Curie temperature,* the exchange coupling suddenly disappears and the materials become simply paramagnetic. For iron the Curie temperature is 1043° K. Ferromagnetism is evidently a property not only of the individual atom or ion but also of the interaction of each atom or ion with its neighbors in the crystal lattice of the solid.

Figure 33–9 shows a magnetization curve for a specimen of iron. To obtain such a curve, we form the specimen, assumed initially unmagnetized, into a ring and wind a toroidal coil around it as in Fig. 33–10, to form a so-called *Rowland ring*. When a current i is set up in the coil, if the iron core is not present, a magnetic field is set up within the toroid given by (see Eq. 30–7)

$$B_0 = \mu_0 ni, \tag{33–10}$$

where n is the number of turns per unit length for the toroid. Although this formula was derived for a long solenoid, it can be applied to a toroid if $d \ll r$ in Fig. 33–10. Because of the iron core, the actual value of **B** in the toroidal space will exceed **B₀**, by

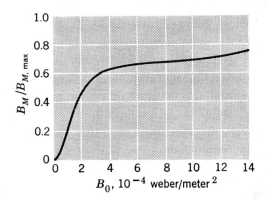

FIGURE 33–9   A magnetization curve for iron.

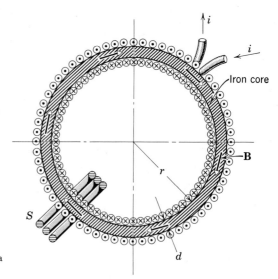

FIGURE 33-10   A Rowland ring, showing a secondary coil S.

a large factor in many cases, since the elementary atomic dipoles in the core line up with the applied field $\mathbf{B}_0$, thereby setting up their own magnetic field. Thus we can write

$$B = B_0 + B_M \tag{33-11}$$

where B_M is the magnetic field due to the specimen; it is proportional to the magnetization M of the specimen. Often $B_M \gg B_0$.

The field B_0 is proportional to the current in the toroid and can be calculated readily, using Eq. 33-10; B can be measured in a way that is described below. An experimental value for B_M can then be derived from Eq. 33-11. It has a maximum value $B_{M,\text{max}}$ corresponding to complete alignment of the atomic dipoles in the iron. Thus we can plot, as in Fig. 33-9, the fractional degree of alignment ($= B_M/B_{M,\text{max}}$) as a function of B_0. For this specimen a value of 96.5% saturation is achieved at $B_0 = 0.13$ weber/meter^2 ($= 1300$ gauss; this point is about 16 ft to the right of the origin in the figure); increasing B_0 to 1.0 weber/meter^2 ($= 10,000$ gauss; about 120 ft to the right in Fig. 33-9) increases the fractional saturation only to 97.7%.

B in Eq. 33-11 can be found by changing the current in the main winding of the Rowland ring of Fig. 33-10 from zero to a constant value i during a short interval Δt. This causes an induced current to appear momentarily in secondary coil S and, during interval Δt, a charge Δq passes through this coil. A ballistic galvanometer can be used to measure Δq and then B can be calculated in terms of Δq and fixed circuit parameters; see Problem 33-11.

Magnetization curves for ferromagnetic materials do not retrace themselves as we increase and then decrease the toroid current. Figure 33-11 shows the following operations with a Rowland ring: (1) starting with the iron unmagnetized (point a), increase the toroid current until B_0 ($= \mu_0 n i$) has the value correspond-

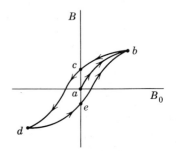

FIGURE 33-11   A magnetization curve (ab) for a specimen of iron and an associated hysteresis loop ($ebcde$).

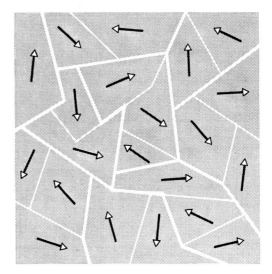

FIGURE 33–12 The separate magnetic domains in an unmagnetized polycrystalline ferromagnetic sample are oriented to produce little external effect. Each domain, however, is made up of completely aligned atomic dipoles, as suggested by the arrows. The heavy boundaries define the crystals that make up the solid; the light boundaries define the domains within the crystals.

ing to point b; (2) reduce the current in the toroid winding back to zero (point c); (3) reverse the toroid current and increase it in magnitude until point d is reached; (4) reduce the current to zero again (point e); (5) reverse the current once more until point b is reached again. The lack of retraceability shown in Fig. 33–11 is called *hysteresis*. Note that at points c and e the iron core is magnetized, even though there is no current in the toroid windings; this is the familiar phenomenon of permanent magnetism.

The magnetization curve for paramagnetism (Fig. 33–5) is explained in terms of the mutually opposing tendencies of alignment with the external field and of randomization because of the temperature motions. In ferromagnetism, however, we have assumed that adjacent atomic dipoles are locked in rigid parallelism. Why, then, does the magnetic moment of the specimen not reach its saturation value for very low— even zero—values of B_0? The modern interpretation is to assume the existence within the specimen of *domains*, that is, of local regions within which there is essentially perfect alignment. The domains themselves, however, as Fig. 33–12 suggests, are not parallel at moderately low values of **B_0**.

As we magnetize a piece of iron by placing it in an external magnetic field, two effects take place. One is a growth in size of the domains that are favorably oriented at the expense of those that are not. Second, the direction of orientation of the dipoles within a domain may swing around as a unit, becoming closer to the field direction. Hysteresis comes about because the domain boundaries do not move completely back to their original positions when the external field **B_0** is removed.

QUESTIONS

1. Two iron bars are identical in appearance. One is a magnet and one is not. How can you tell them apart? You are not permitted to suspend either bar as a compass needle or to use any other apparatus.

2. How could you reverse the magnetism of a compass needle?

3. Two iron bars always attract, no matter the combination in which their ends are brought near each other. Can you conclude that one of the bars must be unmagnetized?

4. If we sprinkle iron filings on a particular bar magnet, they cling both to the ends and to the middle. Sketch roughly the lines of **B,** both outside and inside the magnet.

5. The earth is a huge magnetic diople. (*a*) Is the magnetic pole in the Northern Hemisphere a north or a south magnetic pole? (*b*) In the Northern Hemisphere do the magnetic lines of force associated with the earth's magnetic field point toward the earth's surface or away from it?

6. Cosmic rays are charged particles that strike our atmosphere from some external source. We find that more low-energy cosmic rays reach the earth at the north and south magnetic poles than at the (magnetic) equator. Why is this so?

7. How might the magnetic dipole moment of the earth be measured?

8. Give three reasons for believing that the flux Φ_B of the earth's magnetic field is greater through the boundaries of Alaska than through those of Texas.

9. The neutron, which has no charge, has a magnetic dipole moment. Is this possible on the basis of classical electromagnetism, or does this evidence alone indicate that classical electromagnetism has broken down?

10. Is the magnetization at saturation for a paramagnetic substance very much different from that for a saturated ferromagnetic substance of about the same size?

11. Explain why a magnet attracts an unmagnetized iron object such as a nail.

12. Does any net force or torque act on (*a*) an unmagnetized iron bar or (*b*) a permanent bar magnet when placed in a uniform magnetic field?

13. A nail is placed at rest on a smooth table top near a strong magnet. It is released and attracted to the magnet. What is the source of the kinetic energy it has just before it strikes the magnet?

14. The magnetization induced in a diamagnetic sphere by an external magnetic field does not vary with temperature, in sharp contrast to the situation in paramagnetism. Is this understandable in terms of the description that we have given of the origin of diamagnetism?

15. Compare the magnetization curves for a paramagnetic substance (Fig. 33–5) and for a ferromagnetic substance (Fig. 33–9). What would a similar curve for a diamagnetic substance look like? Do you think that it would show saturation effects in strong applied fields (say 10 weber/meter²)?

PROBLEMS

1(1). The earth has a magnetic dipole moment of 6.4×10^{21} amp-m². (*a*) What current would have to be set up in a single turn of wire going around the earth at its magnetic equator if we wished to set up such a dipole? (*b*) Could such an arrangement be used to cancel out the earth's magnetism at points in space well above the earth's surface? (*c*) On the earth's surface?
Answer: (*a*) 5.0×10^7 amp. (*b*) Yes. (*c*) No.

2(1). The dipole moment of a small current loop is 2.0×10^{-4} amp-m². What is the magnetic field on the axis of the dipole 8.0 cm away from the loop?

3(1). Calculate (*a*) the electric field and (*b*) the magnetic field at a point 1.0×10^{-10} meter (roughly, one atomic diameter) away from a proton, measured along its axis of spin. The magnetic moment of the proton is 1.4×10^{-26} amp-m².
Answer: (*a*) 1.4×10^{11} volt/meter. (*b*) 2.8×10^{-3} weber/meter².

4(1). A total charge q is distributed uniformly on a dielectric ring of radius r. If the ring is rotated about an axis perpendicular to its plane and through its center at an angular speed ω, find the magnitude and direction of its resulting magnetic moment.

5(1). Show, by sketching the magnetic field of a magnetic dipole, that (*a*) if the dipole moments of two nearby dipoles are parallel they will tend to remain that way and (*b*) if they are anti-parallel they will tend to remain that way. In each case consider the torques on the second dipole in the field of the first.

6(1). A simple bar magnet is suspended by a string as shown in Fig. 33–13. If a uniform magnetic field **B** directed parallel to the ceiling is then established, show the resulting orientation of string and magnet.

7(1). Show that a spinning positive charge will have a spin magnetic moment that points in the same direction as its spin angular momentum.

8(1). An electron has a spin angular momentum L of 0.53×10^{-34} joule-sec and a magnetic moment μ of 9.3×10^{-24} amp-m². Compare μ/L and e/m for the electron.

9(3). At what temperature will the average thermal energy of a paramagnetic gas be equal to the magnetic energy in a field of 5000 gauss if the dipole moments of the atoms are about 10^{-23} amp-m²? *Answer:* 0.48° K.

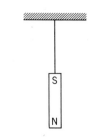

FIGURE 33-13   Problem 6(1).

10(3). (*a*) What is the magnetic moment due to the orbital motion of an electron in an atom when the orbital angular momentum is 1.1×10^{-34} joule-sec.). (*b*) The spin magnetic moment of an electron is 9.3×10^{-24} amp-meter². What is the difference in the magnetic potential energy U between the states in which the magnetic moment is aligned with and aligned in the opposite direction to an external magnetic field of 1.2 weber/meter²? (*c*) What absolute temperature would be required so that the energy difference in (*b*) would equal the mean thermal energy $\frac{1}{2}kT$?

11(4). An electron travels in a circular orbit about a fixed positive point charge in the presence of a uniform magnetic field **B** directed normal to the plane of its motion. The electric force has precisely N times the magnitude of the magnetic force on the electron. (*a*) Determine the two possible angular speeds of the electron's motion. (*b*) Evaluate these speeds numerically if $B = 4.27 \times 10^3$ gauss and $N = 100$.

Answer: (*a*) $(N \pm 1)\dfrac{eB}{m}$. (*b*) 7.43×10^{12} radians/sec; 7.57×10^{12} radians/sec.

12(5). The dipole moment associated with an atom of iron in an iron bar is 1.8×10^{-23} amp-m². Assume that all the atoms in the bar, which is 5.0 cm long and has a cross-sectional area of 1.0 cm², have their dipole moments aligned. (*a*) What is the dipole moment of the bar? (*b*) What torque must be exerted to hold this magnet at right angles to an external field of 15,000 gauss? Take the density of iron to be 7.8 gm/cm³.

13(5). *Dipole-dipole interaction.* The exchange coupling mentioned in Section 33-5 as being responsible for ferromagnetism is *not* the mutual magnetic interaction energy between two elementary magnetic dipoles. To show this (*a*) compute B a distance a (= 1.0×10^{-10} meter, roughly one atomic diameter) away from a dipole of moment μ (= 1.8×10^{-23} amp-m²); (*b*) compute the energy (= $2\mu B$) required to turn a second similar dipole end for end in this field. What do you conclude about the strength of this dipole-dipole interaction? Compare with the results of Example 3. (Note: for the same distance, the field in the median plane of a dipole is only half as large as on the axis; see Eq. 33-2.)

Answer: (*a*) 1.8 webers/meter². (*b*) 6.5×10^{-23} joules.

14(5). Show that B in Eq. 33-11 is given by

$$B = \frac{R\,\Delta q}{NA}.$$

in which R is the resistance of the secondary coil S in Fig. 33-10, N is its number of turns, A is the cross-sectional area of the toroid, and Δq is defined in the text. See Problem 10(3) in Chapter 31.

15(5). A Rowland ring is formed of ferromagnetic material. It is circular in cross section, with an inner radius of 5.0 cm and an outer radius of 6.0 cm and is wound with 400 turns of wire. (*a*) What current must be set up in the windings to attain $B_0 = 2.0 \times 10^{-4}$ weber/meter² in Fig. 33-9? (*b*) A secondary coil wound around the toroid has 50 turns and has a resistance of 8.0 ohms. If, for this value of B_0, we have $B_M = 800B_0$, how much charge moves through the secondary coil when the current in the toroid windings is turned on? (See result of previous problem.)

Answer: (*a*) 0.14 amp. (*b*) 7.9×10^{-5} coul.

CHAPTER 34

Electromagnetic Oscillations

34–1 LC Oscillations

The LC system of Fig. 34–1 resembles a mass-spring system (see Fig. 7–4) in that, among other things, each system has a characteristic frequency of oscillation. To see this, we assume that initially the capacitor C in Fig. 34–1a carries a charge q_m and the current i in the inductor is zero. At this instant the energy stored in the capacitor is given by Eq. 26–16, or

$$U_E = \frac{1}{2}\frac{q_m{}^2}{C}.\tag{34–1}$$

The energy stored in the inductor, given by

$$U_B = \frac{1}{2}Li^2,\tag{34-2}$$

is zero because the current is zero. The capacitor now starts to discharge through the inductor, positive charge carriers moving counterclockwise, as shown in Fig. 34–1b. This means that a current i, given by dq/dt and pointing down in the inductor, is established.

As q decreases, the energy stored in the electric field in the capacitor also decreases. This energy is transferred to the magnetic field that appears around the inductor because of the current i that is building up there. Thus the electric field decreases, the magnetic field builds up, and energy is transferred from the former to the latter.

At a time corresponding to Fig. 34–1c, all the charge on the capacitor will have disappeared. The electric field in the capacitor will be zero, the energy stored there having been transferred entirely to the magnetic field of the inductor. According to

625

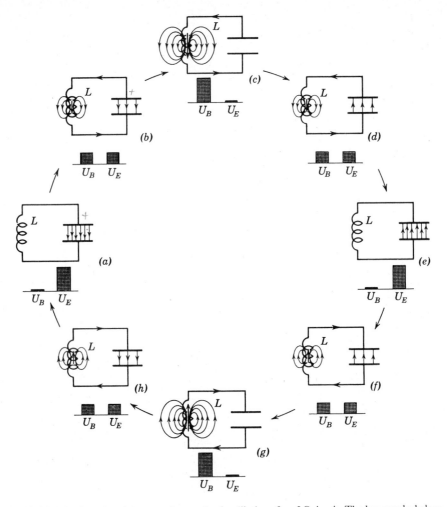

FIGURE 34–1 Showing eight stages in a cycle of oscillation of an *LC* circuit. The bar graphs below each figure show the stored magnetic and electric potential energy. The vertical arrows on the inductor axis show the current. Compare this figure in detail with Fig. 7–4, to which it exactly corresponds.

P. 118

Eq. 34–2, there must then be a current—and indeed one of maximum value—in the inductor. Note that even though q equals zero the current (which is dq/dt) is not zero at this time.

The large current in the inductor in Fig. 34–1c continues to transport positive charge from the top plate of the capacitor to the bottom plate, as shown in Fig. 34–1d; energy now flows from the inductor back to the capacitor as the electric field builds up again. Eventually, the energy will have been transferred completely back to the capacitor, as in Fig. 34–1e. The situation of Fig. 34–1e is like the initial situation, except that the capacitor is charged in the opposite direction.

The capacitor will start to discharge again, the current now being clockwise, as in Fig. 34–1f. Reasoning as before, we see that the circuit eventually returns to its initial situation and that the process continues at a definite frequency ν (measured, say, in cycles/sec) to which corresponds a definite angular frequency ω ($= 2\pi\nu$ and measured, say, in radians/sec). Once started, such *LC* oscillations (in the ideal case described, in which the circuit contains no resistance) continue indefinitely, energy being shuttled back and forth between the electric field in the capacitor and the magnetic field in the inductor. Any configuration in Fig. 34–1 can be set up as an initial condition. The oscillations will then continue from that point, proceeding clockwise around the figure. Compare these oscillations carefully with those of the mass-spring system described in Fig. 7–4.

To measure the charge q as a function of time, we can measure the variable potential difference $V_C(t)$ that exists across capacitor C. The relation

$$V_C = \left(\frac{1}{C}\right) q$$

shows that V_C is proportional to q. To measure the current we can insert a small resistance R in the circuit and measure the potential difference across it. This is proportional to i through the relation

$$V_R = (R)i.$$

We assume here that R is so small that its effect on the behavior of the circuit is negligible. Both q and i, or more correctly V_C and V_R, which are proportional to them, can be displayed on a cathode-ray oscilloscope. This instrument can plot automatically on its screen graphs proportional to $q(t)$ and $i(t)$, as in Fig. 34–2.

▮ **Example 1.** A 1.0-μf capacitor is charged to 50 volts. The charging battery is then disconnected and a 10-mh coil is connected across the capacitor, so that LC oscillations occur. What is the maximum current in the coil? Assume that the circuit contains no resistance.

The maximum stored energy in the capacitor must equal the maximum stored energy in the inductor, from the conservation-of-energy principle. This leads, from Eqs. 34–1 and 34–2, to

$$\frac{1}{2}\frac{q_m{}^2}{C} = \frac{1}{2}Li_m{}^2,$$

where i_m is the maximum current and q_m is the maximum charge. Note that the maximum current and the maximum charge do not occur at the same time but one-fourth of a cycle apart; see Figs. 34–1 and 34–2. Solving for i_m and substituting CV_0 for q_m gives

$$i_m = V_0\sqrt{\frac{C}{L}} = (50 \text{ volts})\sqrt{\frac{1.0 \times 10^{-6} \text{ farad}}{10 \times 10^{-3} \text{ henry}}} = 0.50 \text{ amp.} \qquad ▮$$

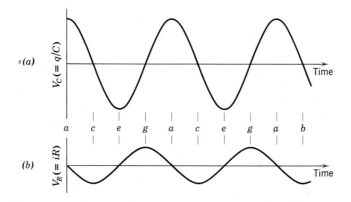

FIGURE 34–2   A drawing of an oscilloscope screen showing potential differences proportional to (a) the charge and (b) the current, in the circuit of Fig. 34–1, as a function of time. The letters indicate corresponding phases of oscillation in that figure. Note that because $i = dq/dt$ the lower curve is proportional to the derivative of the upper.

In an actual *LC* circuit the oscillations will not continue indefinitely because there is always some resistance present that will drain away energy by the Joule effect. The oscillations, once started, will die away, as in Fig. 34–3.

It is possible to have sustained electromagnetic oscillations if arrangements are made to supply, automatically and periodically (once a cycle, say), enough energy from an outside source to compensate for that lost to internal energy. We are reminded of a clock escapement, which is a device for feeding energy from a spring or a falling weight into an oscillating pendulum, thus compensating for frictional losses that would otherwise cause the oscillations to die away. Oscillators whose frequency v may be varied between certain limits are commercially available as packaged units over a wide range of frequencies, extending from low audio-frequencies (lower than 10 cycles/sec) to microwave frequencies (higher than 10^{10} cycles/sec).

34–2 Analogy to Simple Harmonic Motion

Figure 7–4 shows that in a mass-spring system performing simple harmonic motion, as in an oscillating *LC* circuit, two kinds of energy occur. One is potential energy of the compressed or extended spring; the other is kinetic energy of the moving mass. These are given by the familiar formulas in the first column of Table 34–1. The table suggests that a capacitor is in some formal way like a spring and an inductor is like a mass and that certain electromagnetic quantities "correspond" to certain mechanical ones, namely,

q corresponds to x,
i corresponds to v,
C corresponds to $1/k$,
L corresponds to m.

Comparison of Fig. 34–1, which shows the oscillations of the *LC* circuit, with Fig. 7–4, which shows the oscillations in a mass-spring system, indicates how close the correspondence is. Note how v and i correspond in the two figures; also x and q. Note, too, how in each case the energy alternates between two forms, magnetic and electric for the *LC* system, and kinetic and potential for the mass-spring system.

In Section 13–3 we saw that the natural angular frequency of oscillation of an undamped mass-spring system is

$$\omega = 2\pi v = \sqrt{\frac{k}{m}}.$$

p. 230

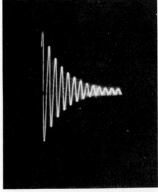

FIGURE 34–3 A photograph of an oscilloscope trace showing how the oscillations in an *LRC* circuit die away because energy is transferred to the resistor as internal energy by the Joule effect. The figure is a plot of the potential difference across the resistor as a function of time.

Table 34–1 SOME ENERGY FORMULAS

Mechanical		Electromagnetic	
spring	$U_P = \dfrac{1}{2} kx^2$	capacitor	$U_E = \dfrac{1}{2} \dfrac{q^2}{C}$
mass	$U_K = \dfrac{1}{2} mv^2$	inductor	$U_B = \dfrac{1}{2} Li^2$
	$v = dx/dt$		$i = dq/dt$

The method of correspondences suggests that to find the natural frequency for the LC circuit k should be replaced by $1/C$ and m by L, obtaining

$$\omega = 2\pi\nu = \sqrt{\frac{1}{LC}}. \tag{34–3}$$

This formula is indeed correct, as we show in the next section.

34–3 Electromagnetic Oscillations—Quantitative

We now derive an expression for the frequency of oscillation of an LC circuit, using the conservation of energy principle. The total energy U at any instant in an oscillating LC circuit is given by

$$U = U_B + U_E = \frac{1}{2} Li^2 + \frac{1}{2} \frac{q^2}{C},$$

which expresses the fact that at any arbitrary time the energy is stored partly in the magnetic field in the inductor and partly in the electric field in the capacitor. If we assume the circuit resistance to be zero, there is no energy transfer to internal energy and U remains constant with time, even though i and q vary. In more formal language, dU/dt must be zero. This leads to

$$\frac{dU}{dt} = \frac{d}{dt}\left(\frac{1}{2} Li^2 + \frac{1}{2} \frac{q^2}{C}\right) = Li\frac{di}{dt} + \frac{q}{C}\frac{dq}{dt} = 0. \tag{34–4}$$

Now, q and i are not independent variables, being related by

$$i = \frac{dq}{dt}.$$

Differentiating yields

$$\frac{di}{dt} = \frac{d^2q}{dt^2}.$$

Substituting these two expressions into Eq. 34–4 leads to

$$L\frac{d^2q}{dt^2} + \frac{q}{C} = 0. \tag{34–5}$$

This is the differential equation that describes the oscillations of a (resistanceless) LC circuit. We could also have obtained Eq. 34–5 directly from the loop theorem (which is completely equivalent to the conservation of energy approach; see Section 28–2) by adding the potentials $-L\,di/dt\ (= -L\,d^2q/dt^2)$ and $-q/C$ as one traverses the circuit.

To solve Eq. 34–5 we note first that it is mathematically of exactly the same form as Eq. 13–6,

$$m \frac{d^2x}{dt^2} + kx = 0, \qquad (13\text{–}6)$$

which is the differential equation for the mass-spring oscillations. Fundamentally, it is by comparing these two equations that the correspondences on p. 628 arise.

The solution of Eq. 13–6 proved to be

$$x = A \cos (\omega t + \phi), \qquad (13\text{–}8)$$

where A ($= x_m$) is the amplitude of the motion and ϕ is an arbitrary phase constant. Since q corresponds to x, we can write the solution of Eq. 34–5 as

$$q = q_m \cos (\omega t + \phi), \qquad (34\text{–}6)$$

where ω is the still unknown angular frequency of the electromagnetic oscillations.

We can test whether Eq. 34–6 is indeed a solution of Eq. 34–5 by substituting it and its second derivative in that equation. To find the second derivative, we write

$$\frac{dq}{dt} = i = -\omega q_m \sin (\omega t + \phi) \qquad (34\text{–}7)$$

and

$$\frac{d^2q}{dt^2} = -\omega^2 q_m \cos (\omega t + \phi).$$

Substituting q and d^2q/dt^2 into Eq. 34–5 yields

$$-L\omega^2 q_m \cos (\omega t + \phi) + \frac{1}{C} q_m \cos (\omega t + \phi) = 0.$$

Canceling $q_m \cos (\omega t + \phi)$ and rearranging leads to

$$\omega = \sqrt{\frac{1}{LC}}.$$

Thus, if ω is given the constant value $1/\sqrt{LC}$, Eq. 34–6 is indeed a solution of Eq. 34–5. This expression for ω agrees with Eq. 34–3, which was arrived at by the method of correspondences.

The phase constant ϕ in Eq. 34–6 is determined by the conditions that prevail at $t = 0$. If the initial condition is as represented by Fig. 34–1a, then we put $\phi = 0$ in order that Eq. 34–6 may predict $q = q_m$ at $t = 0$. What initial condition in Fig. 34–1 is implied if we select $\phi = 90°$?

Example 2. An oscillating LC circuit has inductance $L = 10$ mh and capacitance $C = 1.0$ μf.
(a) What is the frequency of oscillation?

The frequency is obtained from Eq. 34–3,

$$\omega = \sqrt{\frac{1}{LC}} = \sqrt{\frac{1}{(10 \times 10^{-3} \text{ henry})(1.0 \times 10^{-6} \text{ f})}} = 1.0 \times 10^4 \text{ rad/sec}$$

and

$$\nu = \frac{\omega}{2\pi} = 1.6 \times 10^3 \text{ cycle/sec}.$$

(b) If the maximum voltage across the capacitor is 100 volts, calculate the maximum current in the coil.

From Eq. 34–7

$$i = \frac{dq}{dt} = -\omega q_m \sin(\omega t + \phi)$$

so

$$i_{max} = \omega q_m.$$

The maximum charge can be found from the maximum voltage V_m

$$q_m = CV_m$$

so the current i_m is

$$i_m = \omega CV_m = (1.0 \times 10^4 \text{ rad/sec})(1.0 \times 10^{-6} \text{ farad})(100 \text{ volts}) = 1.0 \text{ amp.}$$

Example 3. (*a*) In the oscillating LC circuit of Example 2 what value of charge, expressed in terms of the maximum charge, is present on the capacitor when the energy is shared equally between the electric and the magnetic field?

The stored energy and the maximum stored energy in the capacitor are, respectively,

$$U_E = \frac{q^2}{2C} \quad \text{and} \quad U_{E,m} = \frac{q_m{}^2}{2C}.$$

Substituting $U_E = \frac{1}{2}U_{E,m}$ yields

$$\frac{q^2}{2C} = \frac{1}{2}\frac{q_m{}^2}{2C} \quad \text{or} \quad q = \frac{1}{\sqrt{2}}q_m.$$

(*b*) How much time is required for this condition to arise, assuming the capacitor to be fully charged initially?

We write, assuming $\phi = 0$ in Eq. 34–6,

$$q = q_m \cos \omega t = \frac{1}{\sqrt{2}}q_m,$$

which leads to

$$\omega t = \cos^{-1}\frac{1}{\sqrt{2}} = \frac{\pi}{4} \quad \text{or} \quad t = \frac{\pi}{4\omega} = \frac{1}{8}T,$$

where T is the period. Using the angular frequency ω found from Example 2, we obtain

$$t = \frac{\pi}{4\omega} = \frac{\pi}{(4)(1.0 \times 10^4 \text{ radians/sec})} = 7.9 \times 10^{-5} \text{ sec.}$$

The stored electric energy in the LC circuit, using Eq. 34–6, is

$$U_E = \frac{1}{2}\frac{q^2}{C} = \frac{q_m{}^2}{2C}\cos^2(\omega t + \phi), \tag{34–8}$$

and the magnetic energy, using Eq. 34–7, is

$$U_B = \frac{1}{2}Li^2 = \frac{1}{2}L\omega^2 q_m{}^2 \sin^2(\omega t + \phi).$$

Substituting the expression for ω (Eq. 34–3) into this last equation yields

$$U_B = \frac{q_m{}^2}{2C}\sin^2(\omega t + \phi). \tag{34–9}$$

Figure 34–4 shows plots of $U_E(t)$ and $U_B(t)$ for the case of $\phi = 0$. Note that (*a*) the maximum values of U_E and U_B are the same ($= q_m{}^2/2C$); (*b*) at any instant the sum of U_E and U_B is a constant ($= q_m{}^2/2C$); and (*c*) when U_E has its maximum value, U_B is zero and conversely. This analysis supports the qualitative analysis of Section

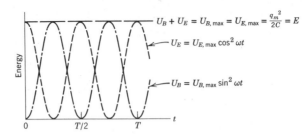

FIGURE 34–4   The stored magnetic and electric energy in the circuit of Fig. 34–1. Note that their sum is a constant.

34–1. Compare this discussion with that given in Section 13–4 for the energy transfers in a mass-spring system.

34–4   Induced Magnetic Fields

Now that we have shown how to generate a time-varying electric field we complete our description of the basic equations of electromagnetism by introducing a new concept: *a changing electric field produces a magnetic field.* This concept is the symmetrical counterpart of Faraday's law of induction. We will develop it by a symmetry argument and will let the agreement with experiment speak for itself. This comparison with experiment, which is worked out largely in Chapter 35, forms one of the chief experimental bases of electromagnetic theory. Its central achievement was the demonstration that the experimentally measured speed c of visible light in free space could be related to purely electromagnetic quantities by

$$c = \frac{1}{\sqrt{\mu_0 \epsilon_0}}. \tag{34–10}$$

This demonstration not only revealed optics as a branch of electromagnetism but led directly to the concept of the electromagnetic spectrum, which in turn resulted in the discovery of radio waves.

Figure 34–5*a* shows a uniform electric field **E** filling a cylindrical region of space.

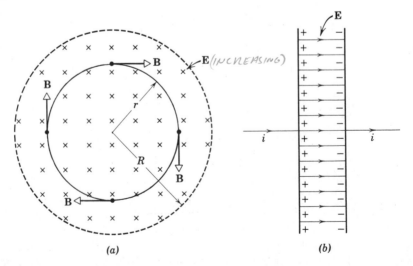

FIGURE 34–5   (*a*) Showing the induced magnetic fields **B** at four points, produced by a changing electric field **E**. The electric field is increasing in magnitude. Compare Fig. 31–9. (*b*) Such a changing electric field may be produced by charging a parallel-plate capacitor as shown.

It might be produced by a circular parallel-plate capacitor, as suggested in Fig. 34–5*b*. We assume that E is increasing at a steady rate dE/dt, which means that charge must be supplied to the capacitor plates at a steady rate; to supply this charge requires a steady current i into the positive plate and an equal steady current i out of the negative plate. Note that this condition does not hold in Fig. 34–1.

If a sufficiently delicate experiment could be performed, it would be found that a magnetic field is set up by this changing electric field. Figure 34–5*a* shows **B** for four arbitrary points. Figure 34–5 suggests a beautiful example of the symmetry of nature. A changing magnetic field induces an electric field (Faraday's law); now we see that a changing electric field induces a magnetic field.

To describe this new effect quantitatively, we are guided by analogy with Faraday's law of induction,

$$\oint \mathbf{E} \cdot d\mathbf{l} = -\frac{d\Phi_B}{dt}, \qquad (34\text{–}11)$$

OPPOSING

which asserts that an electric field (left term) is produced by a changing magnetic field (right term). For the symmetrical counterpart we might write

$$\oint \mathbf{B} \cdot d\mathbf{l} = \mu_0\epsilon_0 \frac{d\Phi_E}{dt}. \qquad (34\text{–}12)$$

NOT OPPOSING

Our system of units requires that we insert the constants ϵ_0 and μ_0. In some unit systems they would not appear. Equation 34–12 asserts that a magnetic field (left term) can be produced by a changing electric field (right term). Compare carefully Fig. 31–9, which illustrates the production of an electric field by a changing magnetic field, with Fig. 34–5*a*. In each case the appropriate flux Φ_B or Φ_E is increasing. However, experiment shows that the lines of **E** in Fig. 31–9 are counterclockwise, whereas those of **B** in Fig. 34–5*a* are clockwise. This difference requires that the minus sign of Eq. 34–11 be omitted from Eq. 34–12.

In Section 30–1 we saw that a magnetic field can also be set up by a current in a wire. We described this quantitatively by Ampère's law:

$$\oint \mathbf{B} \cdot d\mathbf{l} = \mu_0 i,$$

in which i is the conduction current passing through the loop around which the line integral is taken. Thus there are at least two ways of setting up a magnetic field: (*a*) by a changing electric field and (*b*) by a current. In general, both possibilities must be allowed for, or

$$\oint \mathbf{B} \cdot d\mathbf{l} = \mu_0\epsilon_0 \frac{d\Phi_E}{dt} + \mu_0 i. \qquad (34\text{–}13)$$

Maxwell is responsible for this important generalization of Ampère's law.

In Chapter 30 we assumed that no changing electric fields were present so that the term $d\Phi_E/dt$ in Eq. 34–13 was zero. In the discussion just given we assumed that there were no conduction currents in the space containing the electric field. Thus the term i in Eq. 34–13 was zero. We see now that each of these situations is a special case.

Example 4.. A parallel-plate capacitor with circular plates is being charged as in Fig. 34–5. (*a*) Derive an expression for the induced magnetic field for $r \lessgtr R$.

Equation 34–12 is

$$\oint \mathbf{B} \cdot d\mathbf{l} = \mu_0 \epsilon_0 \frac{d\Phi_E}{dt}$$

We pick an imaginary path of integration having radius $r < R$ as shown in Fig. 34–5. The circular path shown is chosen because of the symmetry of the capacitor plates and the electric field which allows an easy evaluation of $\oint \mathbf{B} \cdot d\mathbf{l}$. Since $\mathbf{B} \cdot d\mathbf{l} = B\,dl$ at all points on the path and also since B is constant

$$\oint \mathbf{B} \cdot d\mathbf{l} = (B)(2\pi r).$$

Now, if we can find the time rate of change of the electric flux on the surface bounded by the imaginary loop we can, using Eq. 34–12, obtain the magnetic field at a point on the loop. We assume the electric field is constant over the surface. The flux is given by

$$\Phi_E = \int \mathbf{E} \cdot d\mathbf{S} = \int E\, dS = (E)(\pi r^2)$$

We obtain, therefore,

$$(B)(2\pi r) = \mu_0 \epsilon_0 \frac{d}{dt}[(E)(\pi r^2)] = \mu_0 \epsilon_0 \pi r^2 \frac{dE}{dt}.$$

Solving for B yields
$$B = \tfrac{1}{2}\mu_0 \epsilon_0 r \frac{dE}{dt} \qquad (r \lesssim R).$$

(*b*) Find B at $r = R$ for $dE/dt = 1.0 \times 10^{12}$ volts/m-sec and for $R = 5.0$ cm.

$$B = \tfrac{1}{2}\mu_0 \epsilon_0 R \frac{dE}{dt}$$

$$= (\tfrac{1}{2})(4\pi \times 10^{-7}\ \text{weber/amp-m})(8.9 \times 10^{-12}\ \text{coul}^2/\text{nt-m}^2)$$

$$(5.0 \times 10^{-2}\ \text{meter})(1.0 \times 10^{12}\ \text{volts/m-sec})$$

$$= 2.8 \times 10^{-7}\ \text{weber/meter}^2 = 0.0028\ \text{gauss}.$$

This shows that the induced magnetic fields in this example are so small that they can scarcely be measured with simple apparatus, in sharp contrast to induced electric fields (Faraday's law), which can be demonstrated easily. This experimental difference is in part due to the fact that induced emfs can easily be multiplied by using a coil of many turns. No technique of comparable efficiency exists for magnetic fields. In experiments involving oscillations at very high frequencies dE/dt above can be very large, resulting in significantly larger values of the induced magnetic field.

34–5   Displacement Current

Equation 34–13 shows that the term $\epsilon_0\, d\Phi_E/dt$ has the dimensions of a current. Even though no motion of charge is involved, there are advantages in giving this term the name *displacement current*. Thus we can say that a magnetic field can be set up either by a conduction current i or by a displacement current i_d $(= \epsilon_0\, d\Phi_E/dt)$, and Eq. 34–13 can be rewritten as

$$\oint \mathbf{B} \cdot d\mathbf{l} = \mu_0(i_d + i). \tag{34–14}$$

The concept of displacement current permits us to retain the notion that current is continuous, a principle established for steady conduction currents in Section 27–1. In Fig. 34–5*b*, for example, a current i enters the positive plate and leaves the negative plate. The *conduction* current is *not* continuous across the capacitor gap because no charge is transported across this gap. However, the displacement current i_d in the gap will prove to be exactly i, thus retaining the concept of the continuity of current.

To calculate the displacement current, recall (see Eq. 26–5) that Φ_E in the gap is given by

$$\Phi_E = \frac{q}{\epsilon_0}.$$

But (see Eq. 34–13 and 34–14)

$$i_d = \epsilon_0 \frac{d\Phi_E}{dt} = \left(\epsilon_0\right)\left(\frac{dq/dt}{\epsilon_0}\right) = dq/dt = i.$$

Therefore, the displacement current in the gap is identical with the conduction current in i the lead wires.

Example 5. What is the displacement current for the capacitor of Example 4? From the definition of displacement current,

$$i_d = \epsilon_0 \frac{d\Phi_E}{dt} = \epsilon_0 \frac{d}{dt}[(E)(\pi R^2)] = \epsilon_0 \pi R^2 \frac{dE}{dt}$$

$$= (8.9 \times 10^{-12}\ \text{coul}^2/\text{nt-m}^2)(\pi)(5.0 \times 10^{-2}\ \text{meter})^2(1.0 \times 10^{12}\ \text{volts/m-sec})$$

$$= 0.070\ \text{amp.}$$

Even though this displacement current is reasonably large, it produces only a small magnetic field (see Example 4) because it is spread out over a large area.

34–6 Maxwell's Equations

Equation 34–13 completes our presentation of the basic equations of electromagnetism, called *Maxwell's equations*. They are summarized in Table 34–2. All equations of physics that serve, as these do, to correlate experiments in a vast area and to predict new results have a certain beauty about them and can be appreciated, by those who understand them, on an aesthetic level. This is true for Newton's laws of motion, for the laws of thermodynamics, for the theory of relativity, and for the theories of quantum physics. As for Maxwell's equations, the German physicist Ludwig Boltzmann (quoting a line from Goethe) wrote "Was it a god who wrote these lines. . . ." In more recent times J. R. Pierce,* in a book chapter entitled "Maxwell's Wonderful Equations" says: "To anyone who is motivated by anything beyond the most narrowly practical, it is worthwhile to understand Maxwell's equations simply for the good of his soul." The scope of these equations is remarkable, including as it does the fundamental operating principles of all large-scale electromagnetic devices such as motors, cyclotrons, computers, television, and microwave radar.

---

* *Electrons, Waves and Messages,* Hanover House, 1956. This book is recommended as collateral reading in electromagnetism.

Table 34-2 THE BASIC EQUATIONS OF ELECTROMAGNETISM (MAXWELL'S EQUATIONS)*

Name	Equation	Describes	Crucial Experiment	Text Reference
Gauss's law for electricity	$\epsilon_0 \oint \mathbf{E} \cdot d\mathbf{S} = q$	Charge and the electric field	1. Like charges repel and unlike charges attract, as the inverse square of their separation. 1'. A charge on an insulated conductor moves to its outer surface.	Chapter 24
Gauss's law for magnetism	$\oint \mathbf{B} \cdot d\mathbf{S} = 0$	The magnetic field	2. It is impossible to create an isolated magnetic pole.	Section 33–2
Ampère's law (as extended by Maxwell)	$\oint \mathbf{B} \cdot d\mathbf{l}$ $= \mu_0 \left(\epsilon_0 \dfrac{d\Phi_E}{dt} + i \right)$	The magnetic effect of a changing electric field or of a current	3. The speed of light can be calculated from purely electromagnetic measurements. 3'. A current in a wire sets up a magnetic field near the wire.	Section 35–3 Chapter 30
Faraday's law of induction	$\oint \mathbf{E} \cdot d\mathbf{l} = - \dfrac{d\Phi_B}{dt}$	The electric effect of a changing magnetic field	4. A bar magnet, thrust through a closed loop of wire, will set up a current in the loop.	Chapter 31

* Written on the assumption that no dielectric or magnetic material is present.

NOTE - DOT PRODUCTS BTWN $\vec{B}$, $\vec{E}$, $d\vec{S}$, $d\vec{l}$

636

QUESTIONS

1. Why doesn't the LC circuit of Fig. 34–1 simply stop oscillating when the capacitor has been completely discharged?

2. How might you start an LC circuit into oscillation with its initial condition being represented by Fig. 34–1c? Devise a switching scheme to bring this about.

3. In an oscillating LC circuit, assumed resistanceless, what determines (a) the frequency and (b) the amplitude of the oscillations?

4. Tabulate as many mechanical or electric systems as you can think of that possess a natural frequency, along with the formula for that frequency if given in the text.

5. What constructional difficulties would you encounter if you tried to build an LC circuit of the type shown in Fig. 34–1 to oscillate (a) at 0.01 or (b) at 10^{10} cycles/sec?

6. Why is Faraday's law of induction more familiar than its symmetrical counterpart, Eq. 34–12?

7. Find a rule for relating **B** in Fig. 34.5 from the direction of the displacement current.

8. Devise "hand rules" (see Fig. 2–12b) for predicting the directions of **E** in Fig. 31–9 and of **B** in Fig. 34–5.

9. Why is the quantity $\epsilon_0 \, d\Phi_E/dt$ referred to as a (displacement) *current?*

10. In Fig. 34–1c a displacement current is needed to maintain continuity of current in the capacitor. How can one exist, since there is no charge on the capacitor?

11. Discuss the symmetries that appear between (a) the first two and (b) the second two of Maxwell's equations.

PROBLEMS

1(1). Find the capacitance of an LC circuit if the maximum charge on the capacitor is 1.0×10^{-6} coul and the total energy is 1.4×10^{-4} joules.
Answer: 3.6×10^{-9} farad.

2(2). You are given a 10-mh inductor and two capacitors, of 5.0- and 2.0-μf capacitance. Can you find four resonant frequencies that can be obtained by connecting these elements in various ways? Are there more than four such frequencies?

3(2). Given a 1.0-mh inductor, how would you make it oscillate at 1.0×10^6 cycles/sec?
Answer: Connect a $25\mu\mu$f capacitor across it and use it as the resonant element in an oscillator.

4(2). A 10-kg mass oscillates on a spring that, when extended 2.0 cm from equilibrium, has a restoring force of 5.0 newtons. (a) Find the capacitance of the analogous LC system with $L = 1.0 \times 10^{-3}$ henry. (b) Would it be a simple matter to construct the analogous circuit?

5(2). An inductor is connected across a capacitor whose capacitance can be varied by turning a knob. We wish to make the frequency of the LC oscillations vary linearly with the angle of rotation of the knob, going from 2×10^5 cycles/sec to 4×10^5 cycles/sec as the knob turns through 180°. If $L = 1.0$mh, plot C as a function of angle for the 180° rotation.
Answer: 0°, 45°, 90°, 135°, and 180° correspond respectively to 6.4, 4.1, 2.8, 2.1, and 1.6×10^{-10} f.

6(2). A variable capacitor with a range from 10 to 365 pf is used with a coil to tune the input to an A.M. radio. (a) What ratio of maximum to minimum frequencies may be tuned with such a capacitor? (b) If this capacitor is to tune from 0.54×10^6 to 1.60×10^6 cycles/sec, the ratio computed in (a) is too large. By adding a capacitor in parallel to the variable capacitor this range may be adjusted. How large should this capacitor be and what inductance should be chosen in order to tune the desired range of frequencies?

7(3). An LC circuit has an inductance $L = 3.0$ mh and a capacitance $C = 10 \ \mu$f. (a) Calculate the angular frequency ω of oscillation. (b) Find the period T of the oscillation. (c) At time $t = 0$ the capacitor is charged to 200 μcoul, and the current is zero. Plot the charge on the capacitor as a function of time.

Answer: (*a*) 5.8×10^3 cycles/sec. (*b*) 1.1×10^{-3} sec.

8(3). Find the energy stored in a 1.0-μf capacitor that has been charged to 300 volts.

9(3). A 1.5-mh inductor in an LC circuit stores a maximum energy of 1.0×10^{-5} joules. What is the peak current?

Answer: 0.12 amp.

10(3). In an oscillating LC circuit $L = 1.0 \times 10^{-3}$ henry, $C = 4.0 \times 10^{-6}$ farad and the maximum charge on C is 3.0×10^{-6} coul. Find the maximum current.

11(3). An oscillating LC circuit consisting of an 0.0010-μf capacitor and a 3.0-mh coil carries a peak voltage of 3.0 volts. (*a*) What is the maximum charge on the capacitor? (*b*) What is the peak current through the circuit? (*c*) What is the maximum energy stored in the magnetic field of the coil?

Answer: (*a*) 3.0×10^{-9} coul. (*b*) 1.7×10^{-3} amp. (*c*) 4.5×10^{-9} joules.

12(3). What average power must be supplied to increase the current through a 4.0-henry inductor from 0 to 20 amp in 15×10^{-3} sec?

13(3). Derive the differential equation for an LC circuit (Eq. 34–5) using the loop theorem.

14(3). How long will it take for an uncharged 4.0 pf capacitor in an LC circuit to charge if its final voltage is 1.0×10^{-3} volts and the maximum current is 5.0×10^{-2} amp?

15(3). In an oscillating LC circuit, (*a*) in terms of the maximum charge on the capacitor, what value of charge is present when the energy is shared equally between the electric and the magnetic fields? (*b*) What fraction of a period must elapse following the time the capacitor is fully charged for this condition to arise?

Answer: (*a*) $q = q_m/\sqrt{2}$. (*b*) $t = T/8$.

16(3). An oscillating LC circuit is designed to operate at a peak current i (30 ma). The inductance L (0.042h) is fixed and the frequency is varied by changing C. (*a*) If the capacitor has a maximum peak voltage V_m (50 volts) can the circuit safely operate at a frequency ν of 1.0×10^6 cycles/sec? (*b*) What is the maximum safe operating frequency? (*c*) What is the minimum capacitance?

17(3). Initially the 900-μf capacitor is charged to 100 volts, and the 100-μf capacitor is uncharged in Fig. 34–6. (*a*) Describe in detail how one may charge the 100-μf capacitor to 300 volts using S_1 and S_2 appropriately. (*b*) Describe in detail the mass + spring mechanical analogy of this problem.

Answer: Let T_2 be the period of the inductor and 900 μf capacitor and T_1 the period of inductor and 100 μf capacitor. Then (*a*) close S_2, wait $T_2/4$; quickly close S_1 then open S_2; wait $T_1/4$ and then open S_1.

18(4). Calculate the value of the speed of light c from Eq. 34–10.

19(4). If the capacitor of Example 4 has a radius $R = 3.0$ cm and a separation of 0.50 cm, what is the peak value of B at $r = R$ if the capacitor is connected to a 60-cycle 100-volt line?

Answer: 1.3×10^{-12} webers/meter^2.

20(5). Show that $\epsilon_0 \dfrac{d\Phi_E}{dt}$ has the dimensions of current.

21(5). In Example 4 how does the displacement current through a concentric circular loop of radius R vary with r? Consider both $r < R$ and $r > R$.

Answer: (*a*) $i_d = \epsilon_0 \pi r^2 \dfrac{dE}{dt}$ $(r \leqslant R)$. (*b*) $i_d = \epsilon_0 \pi R^2 \dfrac{dE}{dt}$ $(r \geqslant R)$.

22(5). In Example 5 show that the displacement current density j_d is given, for $r < R$, by

$$j_d = \epsilon_0 \frac{dE}{dt}.$$

23(5). Prove that the displacement current in a parallel-plate capacitor can be written as

$$i_d = C \frac{dV}{dt}.$$

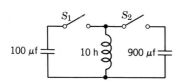

FIGURE 34–6 Problem 17(3).

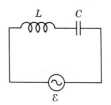

FIGURE 34–7 Problem 27(5).

24(5). A 20-pf capacitor has a peak instantaneous voltage change of 1.4×10^{-3} volts/sec. Find the corresponding displacement current.

25(5). You are given a 1.0-μf capacitor. How would you establish an (instantaneous) displacement current of 1.0 amp in the space between its plates?
Answer: Cause the potential difference to change at 10^6 volts/sec.

26(5). In 1929 Cauwenberghe succeeded in directly measuring (for the first time) the displacement current between the plates of a parallel-plate capacitor. He used a huge pair of circular plates, $C = 1.0 \times 10^{-10}$ farads, upon which he applied 174,000 peak volts at 50 cycles/sec. What maximum displacement current was present between the plates?

27(5). Figure 34–7 shows a parallel plate capacitor with plate area A (30 cm²) with a coil of inductance L (5.0 mh). An alternating current generator completes the circuit. Let the electric field between the capacitor plates be $E(t) = E \sin \omega t$ in which $E = 1.5 \times 10^3$ volts/cm and $\omega = 380$ radians/sec. ($\cong 61$ cycles/sec). (*a*) What is the displacement current i_d as a function of time? (*b*) What is the magnetic energy U_B stored in the coil, as a function of time?
Answer: (*a*) $(1.5 \times 10^{-6}) \cos (\omega t)$ amps. (*b*) $(5.7 \times 10^{-15}) \cos^2 (\omega t)$ joules.

28(6). Collect and tabulate expressions for the following four quantities, considering both $r < R$ and $r > R$. Copy down the derivations side by side and study them as interesting applications of Maxwell's equations to problems having cylindrical symmetry.

(*a*) $B(r)$ for a current i in a long wire of radius R (see Section 30–2).

(*b*) $E(r)$ for a long uniform cylinder of charge of radius R (see Section 24–6).

(*c*) $B(r)$ for a parallel-plate capacitor, with circular plates of radius R, in which E is changing at a constant rate (see Section 34–4).

(*d*) $E(r)$ for a cylindrical region of radius R in which a uniform magnetic field B is changing at a constant rate (see Section 31–5).

29(6). Using the definitions of flux, volume charge density ρ, and current density **j**, write the four Maxwell equations in such a manner that all the fluxes, currents, and charges appear as volume or surface integrals.

CHAPTER 35

Electromagnetic Waves

35–1 Introduction

We have learned how to calculate **E** and **B** at an arbitrary point in space P for any given distribution of static charges and currents. Our tools are Gauss's law and Ampère's law; see Table 34–2. We display them here in the equivalent Coulomb and Biot-Savart forms, using vector notation and considering only differential charge and current elements. Thus:

$$d\mathbf{E} = \frac{1}{4\pi\epsilon_0} \frac{dq\,\mathbf{r}}{r^3} \qquad \text{(see 23–4)}$$

and

$$d\mathbf{B} = \frac{\mu_0 i}{4\pi} \frac{d\mathbf{l} \times \mathbf{r}}{r^3} \qquad \text{(30–9)}$$

in which **r** is the displacement vector from the charge or current element to the point P at which we wish to calculate $d\mathbf{E}$ or $d\mathbf{B}$. These equations show that the electric and magnetic fields fall off as $1/r^2$ for differential charge and current elements respectively. The fields fall off as $1/r^3$ for electric and magnetic dipoles respectively (see Table 30–1) and as higher inverse powers of r for more complicated distributions.

What if the charges and currents in the source are not static but vary with time in position or magnitude? We might think that we should continue to use the static source procedures, calculating $\mathbf{E}(\mathbf{r},t)$ and $\mathbf{B}(\mathbf{r},t)$ corresponding to the source configuration at each time t. This doesn't work, however, for two related reasons: (1) When charges or currents change, the fields **E** and **B** at point P learn about the changes by a field disturbance that moves out from the source with a speed c, which turns out to be the speed of light. Thus the values of **E** and **B** at P at time t depend on the source configuration at the earlier time $t - (r/c)$. This retardation effect becomes greater the farther P is from the source. (2) Because the fields at P are changing with time other fields are generated, described by Faraday's law (loosely: *a changing magnetic field generates an electric field;* see Section 31–5) and the Maxwell extension of Ampère's law (loosely: *a changing electric field generates a magnetic field;* see Section 34–4).

Experiment shows that, *whenever charges are accelerated,* an electromagnetic wave radiates from the source with a speed c in free space. This wave can transfer energy and momentum to gross objects placed in its path.

We can use Maxwell's equations to calculate the fields $\mathbf{E(r,}t)$ and $\mathbf{B(r,}t)$ generated at point P by, say, an oscillating electric dipole. We will not write down these results except to say that: (1) $\mathbf{E(r,}t)$ contains three terms, proportional to $1/r$, $1/r^2$, and $1/r^3$, respectively, and (2) $\mathbf{B(r,}t)$ contains two terms proportional to $1/r$ and $1/r^2$, respectively. As we move away from the dipole the terms proportional to $1/r^2$ and $1/r^3$ rapidly become negligible with respect to terms proportional to $1/r$. We describe the field at large distances as the *radiation field.* Close to the dipole we cannot neglect the $1/r^2$ and $1/r^3$ terms, and we describe the field here as the *near field.* We are interested only in the radiation field, in which $\mathbf{E}(r)$ and $\mathbf{B}(r)$ each fall off with distance as $1/r$.

In Section 35–4 we will see that the rate of energy transport in electromagnetism, measured perhaps in watts/meter^2, is given in magnitude and direction by $(1/\mu_0)(\mathbf{E \times B})$; see Eq. 35–12. In the radiation field, where both E and B vary as $1/r$, this quantity drops off in magnitude as $1/r^2$; the direction of flow proves to be radially outward. This is just the inverse square relationship that we expect for, say, radiation from a radio or a radar antenna.

In the near field, however, the expression for the rate of energy transport contains terms (assuming an oscillating electric dipole) in $1/r^3$, $1/r^4$, and $1/r^5$; the direction of energy transport associated with these terms is not everywhere radially outward. We say again that our present concern is only with the radiation field.

35–2 Radiation Sources

Here we look more closely into the relationship between a traveling electromagnetic wave and its source. We will deal only with large sources, such as radio and radar antennas. We do this because, in such cases, both classical mechanics and classical electromagnetism (Maxwell's equations) give results that agree with experiment. In many parts of the electromagnetic spectrum (visible light, x rays, gamma rays, etc.) the sources are of atomic or nuclear size. Although Maxwell's equations describe the radiated wave in these cases we must use quantum physics (see Chapters 39 and 40) to describe the mechanism by which the wave is generated by the source.

In Chapter 34 we saw that sinusoidally oscillating charges were produced in LC circuits. If a conducting coil is brought near the inductor in the circuit, a current is induced in the coil. As shown in Fig. 35–1 this energy may then be carried in a *transmission line* to an antenna. The antenna shown is called an *electric dipole antenna.* Electromagnetic waves radiate from the antenna. Of course, this would deplete the energy in the LC circuit if we provided no means for restoring it.

The potential difference between the two conductors in the antenna alternates sinusoidally, causing charges to accelerate back and forth. The effect is that of an electric dipole whose dipole moment $\mathbf{p}$ varies with time.

Figure 35–2 shows an electromagnetic wave generated by an oscillating electric dipole. The figure is a figure of revolution about the dipole axis and the wave moves out in any direction from the dipole with speed c. The field shown is the radiation field, the near field being omitted.

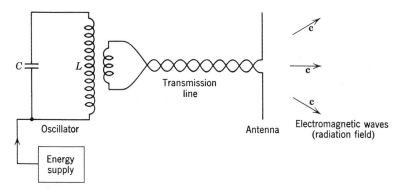

FIGURE 35–1

Figure 35–3 shows "patches" of the wave front as they would appear at different times to a distant observer at a point such as *P* in Fig. 35–2. The wave is moving directly out of the plane of Fig. 35–3 at speed *c*. The figure, which shows one complete cycle of oscillation, suggests how **E** and **B** change periodically with time as the wave sweeps through *P*.

An important characteristic of the radiation is that the electric field **E** is perpendicular to the magnetic field **B;** these fields are each perpendicular to the direction of the wave velocity **c**. The vector **c** has the direction of the cross product **E** ✕ **B** (see Eq. 35–12) as will be discussed later in Sections 35–3 and 35–4.

35–3 Traveling Waves and Maxwell's Equations

In this section we will analyze the wave that passes an observer at point *P* in Fig. 35–3. We will show that this wave is consistent with Maxwell's equations and prove that these equations predict that such an electromagnetic wave travels with a certain speed *c* which turns out to be the speed of light.

We assume, for simplicity, that *P* is very far from the source so that the wavefronts moving past *P* are planar (*r* varying little over a wavelength, allowing us to neglect

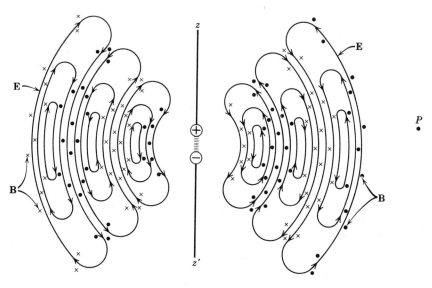

FIGURE 35–2

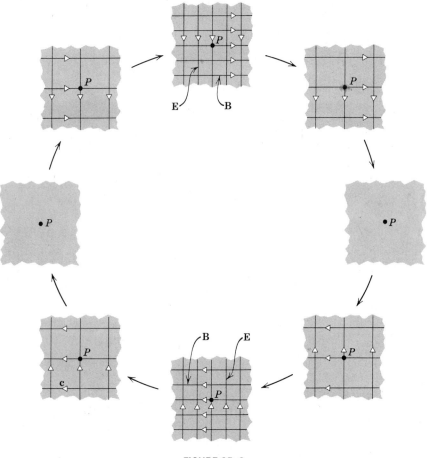

FIGURE 35–3

the $1/r$ dependence of the fields). The observer at P will say that a plane wave (see Section 16–2) is rushing past him.

Figure 35–4 shows a "snapshot" of a plane wave traveling in the x-direction. The lines of **E** are parallel to the y-axis and those of **B** are parallel to the z-axis. The values of **B** and **E** for this wave depend only on x and t (not on y or z). We postulate that they are given in magnitude by

$$B = B_m \sin (kx - \omega t) \tag{35–1}$$

and
$$E = E_m \sin (kx - \omega t). \tag{35–2}$$

Figure 35–5 shows two sections through the three-dimensional diagram of Fig. 35–4. In Fig. 35–5a the plane of the page is the xy-plane and in Fig. 35–5b it is the xz-plane. Note that **E** and **B** are in phase, that is, at any point through which the wave is moving they reach their maximum values at the same time.

The shaded rectangle of dimensions dx and h in Fig. 35–5a is fixed at a particular point P on the x-axis. As the wave passes over it, the magnetic flux Φ_B through the rectangle will change, which will give rise to induced electric fields around the rectangle, according to Faraday's law of induction. These induced electric fields are, in fact, simply the electric component of the traveling electromagnetic wave.

Let us apply Lenz's law. The flux Φ_B for the shaded rectangle of Fig. 35–5a is de-

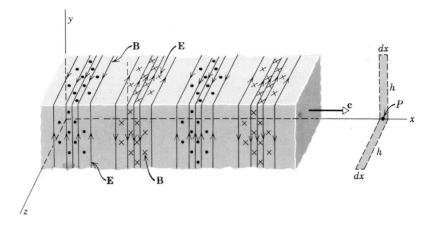

FIGURE 35–4 A plane electromagnetic wave traveling to the right at speed c. Lines of E are parallel to the y-axis; those of B are parallel to the z-axis. The shaded rectangles on the right refer to Fig. 35–5. The observer P of Figs. 35–2 and 35–3 is shown fixed in the coordinate system.

creasing with time because the wave is moving through the rectangle to the right and a region of weaker magnetic fields is moving into the rectangle. The induced electric field will act to oppose this change, which means that if we imagine the boundary of the rectangle to be a conducting loop a *counterclockwise* induced current would appear in it. This current would produce a field of **B** that, within the rectangle, would point out of the page, thus opposing the decrease in Φ_B. There is, of course, no conducting loop, but the net induced electric field **E** does indeed act counterclockwise around the rectangle because $E + dE$, the magnitude of **E** at the right edge of the rectangle is greater than E, the magnitude of **E** at the left edge. Thus the electric field configuration is entirely consistent with the concept that it is induced by the changing mag-

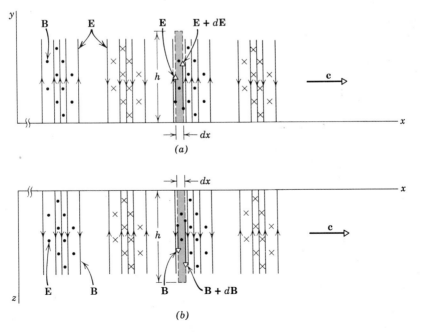

FIGURE 35–5 The wave of Fig. 35–4 viewed (*a*) in the xy-plane and (*b*) in the xz-plane.

netic field. The situation is similar to that of Fig. 31–9, which shows a counterclock-wise configuration of **E** opposing an *increase* of Φ_B.

For a more detailed analysis let us apply Faraday's law of induction, or

$$\oint \mathbf{E} \cdot d\mathbf{l} = -\frac{d\Phi_B}{dt}, \tag{35-3}$$

going counterclockwise around the shaded rectangle of Fig. 35–5a. There is no con-tribution to the integral from the top or bottom of the rectangle because **E** and $d\mathbf{l}$ are at right angles here. The integral then becomes

$$\oint \mathbf{E} \cdot d\mathbf{l} = [(E + dE)(h)] - [(E)(h)] = dE\, h.$$

The flux Φ_B for the rectangle is

$$\Phi_B = (B)(dx\, h),$$

where B is the magnitude of **B** at the rectangular strip and $dx\, h$ is the area of the strip. Differentiating gives

$$\frac{d\Phi_B}{dt} = h\, dx\, \frac{dB}{dt}.$$

From Eq. 35–3 we have then

$$dE\, h = -h\, dx\, \frac{dB}{dt},$$

or

$$\frac{dE}{dx} = -\frac{dB}{dt}. \tag{35-4}$$

Actually, both B and E are functions of x and t. In evaluating dE/dx, we assume that t is constant because Fig. 35–5a is an "instantaneous snapshot." Also, in evalu-ating dB/dt we assume that x is constant since what we want is the time rate of change of B at a particular place, the point P in Fig. 35–5a. The derivatives under these circumstances are called *partial derivatives*, and a special notation is used for them. In this notation Eq. 35–4 becomes

$$\frac{\partial E}{\partial x} = -\frac{\partial B}{\partial t}. \tag{35-5}$$

The minus sign in this equation is appropriate and necessary, for, although E is increasing with x at the site of the shaded rectangle in Fig. 35–5a, B is decreasing with t. Since $E(x,t)$ and $B(x,t)$ are known (see Eqs. 35–1 and 35–2), Eq. 35–5 reduces to

$$kE_m \cos(kx - \omega t) = \omega B_m \cos(kx - \omega t),$$

or

$$\frac{\omega}{k} = \frac{2\pi\nu}{2\pi/\lambda} = \nu\lambda = c = \frac{E_m}{B_m}. \tag{35-6a}$$

Thus the speed of the wave c is the ratio of the amplitudes of the electric and the magnetic components of the wave. From Eqs. 35–1 and 35–2 we see that the ratio of amplitudes is the same as the ratio of instantaneous values, or

$$E = cB. \tag{35-6b}$$

This result will be useful in Section 35–4.

We now turn to Fig. 35–5*b*, in which the flux Φ_E for the shaded rectangle is decreasing with time as the wave moves through it. According to Maxwell's third equation with $i = 0$, because there are no conduction currents in a traveling electromagnetic wave,

$$\oint \mathbf{B} \cdot d\mathbf{l} = \mu_0 \epsilon_0 \frac{d\Phi_E}{dt}, \tag{35–7}$$

this changing flux will induce a magnetic field at points around the periphery of the rectangle. This induced magnetic field is simply the magnetic component of the electromagnetic wave. Thus the electric and the magnetic components of the wave are intimately connected, each depending on the time rate of change of the other. Although there are no conduction currents in the traveling wave we can associate $\oint \mathbf{B} \cdot d\mathbf{l}$ for the shaded rectangle of Fig. 35–5*b* with a displacement current. It points out of the plane of the figure within the rectangle.

The integral in Eq. 35–7, evaluated by proceeding counterclockwise around the shaded rectangle of Fig. 35–5*b*, is

$$\oint \mathbf{B} \cdot d\mathbf{l} = [-(B + dB)(h)] + [(B)(h)] = -h\,dB,$$

where B is the magnitude of **B** at the left edge of the strip and $B + dB$ is its magnitude at the right edge.

The flux Φ_E through the rectangle of Fig. 35–5*b* is

$$\Phi_E = (E)(h\,dx).$$

Differentiating gives $$\frac{d\Phi_E}{dt} = h\,dx\,\frac{dE}{dt}.$$

Equation 35–7 can thus be written

$$-h\,dB = \mu_0 \epsilon_0 \left(h\,dx\,\frac{dE}{dt} \right)$$

or, substituting partial derivatives,

$$-\frac{\partial B}{\partial x} = \mu_0 \epsilon_0 \frac{\partial E}{\partial t}. \tag{35–8}$$

Again, the minus sign in this equation is necessary, for, although B is increasing with x at point P in the shaded rectangle in Fig. 35–5*b*, E is decreasing with t.

Combining this equation with Eqs. 35–1 and 35–2 yields

$$-kB_m \cos (kx - \omega t) = -\mu_0 \epsilon_0 \omega E_m \cos (kx - \omega t),$$

or (see Eq. 35–6*a*) $$\frac{E_m}{B_m} = \frac{k}{\mu_0 \epsilon_0 \omega} = \frac{1}{\mu_0 \epsilon_0 c}. \tag{35–9}$$

Eliminating E_m/B_m between Eqs. 35–6*a* and 35–9 yields

$$c = \frac{1}{\sqrt{\mu_0 \epsilon_0}}. \tag{35–10}$$

Substituting numerical values yields

$$c = \frac{1}{\sqrt{(4\pi \times 10^{-7} \text{ weber/amp-m})(8.9 \times 10^{-12} \text{ coul}^2/\text{nt-m}^2)}}$$

$$= 3.0 \times 10^8 \text{ meters/sec}, \tag{35-11}$$

which is the speed of light in free space! This emergence of the speed of light from purely electromagnetic considerations is the crowning achievement of Maxwell's electromagnetic theory. Maxwell made this prediction before radio waves were known and before it was realized that light was electromagnetic in nature. His prediction led to the concept of the electromagnetic spectrum, which we discuss later, and to the discovery of radio waves by Heinrich Hertz in 1890. It made it possible to view optics as a branch of electromagnetism and to derive its fundamental laws from Maxwell's equations.

A conclusion as fundamental as Eq. 35–10 must be tested rigorously. Of the three quantities in that equation, one, μ_0, has an *assigned* value, namely, $4\pi \times 10^{-7}$ weber/amp-m. The speed of light c is one of the most precisely measured physical constants, having the 1969 accepted value of 2.9979250×10^8 meters/sec. The remaining quantity, ϵ_0, can be found by making measurements on an accurately constructed parallel-plate capacitor, as described in Section 26–2. The best measured value, by Rosa and Dorsey of the National Bureau of Standards (U.S.A.) in 1906, is 8.84025×10^{-12} coul²/nt-m² which yields 2.99781×10^8 meters/sec for the right side of Eq. 35–10. To this accuracy (that is, 0.004%) Eq. 35–10 is verified. Our confidence in electromagnetic theory, bolstered by numerous successful predictions and agreements with experiment, is now such that we reverse the above procedure and calculate the presently accepted value of ϵ_0 from the measured speed of light, using Eq. 35–10.

35–4 Energy and the Poynting Vector

An electromagnetic wave can deliver energy to a body on which it falls. As we show below, the rate of energy flow per unit area in an electromagnetic wave can be described by a vector **S**, called the *Poynting vector* after John Henry Poynting (1852–1914), who first pointed out its properties. We define **S** from

$$\mathbf{S} = \frac{1}{\mu_0} \mathbf{E} \times \mathbf{B}. \tag{35-12}$$

In the mks system **S** is expressed in watts/meter²; the direction of **S** gives the direction in which the energy moves. The vectors **E** and **B** refer to their instantaneous values at the point in question. If we apply Eq. 35–12 to the traveling plane electromagnetic wave of Fig. 35–4, it is clear that **E** × **B**, hence **S**, point in the direction of propagation.

Figure 35–6 shows a traveling plane wave, along with a thin "box" of thickness dx and area A. The box, a mathematical construction, is fixed with respect to the axes while the wave moves through it. At any instant the energy stored in the box, from Eqs. 26–18 and 32–19, is

$$dU = dU_E + dU_B = (u_E + u_B)(A\,dx)$$

$$= \left(\tfrac{1}{2}\epsilon_0 E^2 + \frac{1}{2\mu_0} B^2\right) A\,dx, \tag{35-13}$$

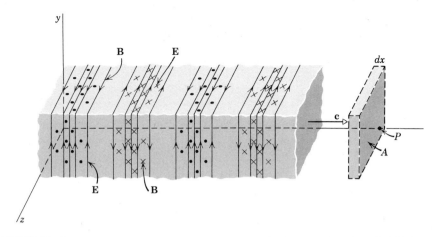

FIGURE 35-6   A plane wave is traveling to the right at speed c through an imaginary box. By knowing the average values of **E** and **B** at the position of the box the energy transport can be found.

where $A\,dx$ is the volume of the box and E and B are the instantaneous values of the field vectors in the box.

Using Eq. 35–6b ($E = cB$) to eliminate one of the E's in the first term in Eq. 35–13 and one of the B's in the second term leads to

$$dU = \left[\tfrac{1}{2}\epsilon_0 E(cB) + \frac{1}{2\mu_0} B\left(\frac{E}{c}\right)\right] A\,dx$$

$$= \frac{(\mu_0\epsilon_0 c^2 + 1)(EBA\,dx)}{2\mu_0 c}.$$

From Eq. 35–10, however, $\mu_0\epsilon_0 c^2 = 1$, so that

$$dU = \frac{EBA\,dx}{\mu_0 c}.$$

This energy dU will pass through the right face of the box in a time dt equal to dx/c. Thus the energy per unit area per unit time, which is S, is given by

$$S = \frac{dU}{dt\,A} = \frac{EBA\,dx}{\mu_0 c(dx/c)A} = \frac{1}{\mu_0} EB.$$

This is exactly the prediction of the more general relation Eq. 35–12 for a traveling plane wave.

This relation refers to values of S, E, and B at any instant of time. We are usually more interested in the average value of S, taken over one or more cycles of the wave. An observer making intensity measurements on a wave moving past him would measure this average value $\bar{S}$. We can show (see Example 1) that $\bar{S}$ is related to the maximum values of E and B by

$$\bar{S} = \frac{1}{2\mu_0} E_m B_m.$$

Example 1. An observer is 1.0 meter from a point light source whose power output P_0 is 1.0×10^3 watts. Calculate the magnitudes of the electric and the magnetic fields. Assume that the source is monochromatic, that it radiates uniformly in all directions, and that at distant points it behaves like the traveling plane wave of Fig. 35–6.

The power that passes through a sphere of radius r is $(\bar{S})(4\pi r^2)$, where $\bar{S}$ is the average value of the Poynting vector at the surface of the sphere. This power must equal P_0, or

$$P_0 = \bar{S}4\pi r^2.$$

From the definition of **S** (Eq. 35–12), we have

$$\bar{S} = \left(\frac{1}{\mu_0}\,\overline{EB}\right).$$

Using the relation $E = cB$ (Eq. 35–6b) to eliminate B leads to

$$\bar{S} = \frac{1}{\mu_0 c}\,\overline{E^2}.$$

The average value of E^2 over one cycle is $\frac{1}{2}E_m{}^2$, since E varies sinusoidally (see Eq. 35–2). This leads to

$$P_0 = \left(\frac{E_m{}^2}{2\mu_0 c}\right)(4\pi r^2),$$

or

$$E_m = \frac{1}{r}\sqrt{\frac{P_0\mu_0 c}{2\pi}}.$$

For $P_0 = 1.0 \times 10^3$ watts and $r = 1.0$ meter this yields

$$E_m = \frac{1}{(1.0\text{m})}\sqrt{\frac{(1.0 \times 10^3 \text{ watts})(4\pi \times 10^{-7} \text{ weber/amp-m})(3.0 \times 10^8 \text{ meters/sec})}{2\pi}}$$

$$= 240 \text{ volts/meter}.$$

The relationship $E_m = cB_m$ (Eq. 35–6a) leads to

$$B_m = \frac{E_m}{c} = \frac{240 \text{ volts/meter}}{3.0 \times 10^8 \text{ meters/sec}} = 8.0 \times 10^{-7} \text{ weber/meter}^2.$$

Note that E_m is appreciable as judged by ordinary laboratory standards but that B_m ($= 0.0080$ gauss) is quite small.

35–5 Momentum

The fact that energy is carried by electromagnetic waves, as described by the Poynting vector **S,** is confirmed by everyday experience. Hands being warmed over a campfire, or the energy received by the earth from the sun are common examples.

Less familiar is the fact that electromagnetic waves may also transport linear momentum. It is possible to exert a pressure (a *radiation pressure**) on an object by shining a light on it. Such forces must be small in relation to forces of our daily experience because we do not ordinarily notice them. The first measurement of radiation pressure was made in 1901–1903 by Nichols and Hull in this country and by Lebedev in Russia, about thirty years after the existence of such effects had been predicted theoretically by Maxwell.

Let a parallel beam of radiation, light for example, fall on an object for a time t and be entirely absorbed by the object. If energy U is absorbed during this time, the magnitude of the momentum p delivered to the object is given, according to Maxwell's prediction, by

$$p = \frac{U}{c} \qquad \text{(total absorption)}, \qquad (35\text{–}14a)$$

* See "Radiation Pressure" by G. E. Henry, in *Scientific American,* June 1957.

where c is the speed of light. The direction of **p** is the direction of the incident beam. If the energy U is entirely reflected, the magnitude of the momentum delivered will be twice that given above, or

$$p = \frac{2U}{c} \qquad \text{(total reflection)}. \qquad (35\text{–}14b)$$

In the same way, twice as much momentum is delivered to a heavy object when a light but perfectly elastic ball is bounced from it as when it is struck by a perfectly inelastic ball of the same mass and speed. If the energy U is partly reflected and partly absorbed, the delivered momentum will lie between U/c and $2U/c$.

Example 2. A parallel beam of light with an energy flux S of 10 watts/cm^2 falls for 1.0 hr on a perfectly reflecting plane mirror of 1.0-cm^2 area. (*a*) What momentum is delivered to the mirror in this time and (*b*) what force acts on the mirror?

(*a*) The energy that falls on the mirror is

$$U = (10 \text{ watts/cm}^2)(1.0 \text{ cm}^2)(3600 \text{ sec}) = 3.6 \times 10^4 \text{ joules}.$$

The momentum delivered after 1.0 hr's illumination is

$$p = \frac{2U}{c} = \frac{(2)(3.6 \times 10^4 \text{ joules})}{3.0 \times 10^8 \text{ meters/sec}} = 2.4 \times 10^{-4} \text{ kg-m/sec}.$$

(*b*) From Newton's second law, the average force on the mirror is equal to the average rate at which momentum is delivered to the mirror, or

$$F = \frac{p}{t} = \frac{2.4 \times 10^{-4} \text{ kg-m/sec}}{3600 \text{ sec}} = 6.7 \times 10^{-8} \text{ nt},$$

a very small force. Would the answers be different if the radiation were in the form of radiowaves?

Nichols and Hull, in 1903, measured radiation pressures and verified Eqs. 35–14, using a torsion balance technique. They allowed light to fall on mirror M as in Fig. 35–7; the radiation pressure caused the balance arm to turn through a measured angle θ, twisting the torsion fiber F. Assuming a suitable calibration for their torsion fiber, the experimenters could arrive at a numerical value for this pressure. Nichols and Hull measured the intensity of their light beam by allowing it to fall on a blackened metal disk of known absorptivity and by measuring the temperature rise of this disk. In a particular run these experimenters measured a radiation pressure of 7.01×10^{-6} nt/meter^2; for their light beam, the value predicted, using Eqs. 35–14, was 7.05×10^{-6} nt/meter^2, in excellent agreement. Assuming a mirror area of 1 cm^2, this represents a force on the mirror of only 7×10^{-10} nt, about 100 times smaller than the force calculated in Example 2.

35–6 Polarization

We have seen that electromagnetic radiation is predicted by Maxwell's theory to be a transverse

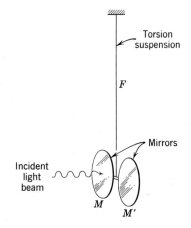

FIGURE 35–7

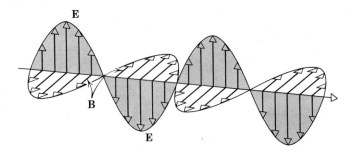

FIGURE 35–8 An instantaneous "snapshot" of a plane-polarized wave showing the vectors **E** and **B** along a particular ray. The wave is moving to the right with speed *c*. The plane containing the vibrating **E**-vector and the direction of propagation is a *plane of vibration*. This figure is simply another representation of Fig. 35–4.

wave, the directions of the vibrating electric and magnetic vectors being at right angles to the direction of propagation instead of parallel to it as in a longitudinal wave. The transverse waves of Figs. 35–8 and 35–4 have the additional characteristic that they are *plane-polarized*. This means that the vibration of the **E**-vectors are parallel to each other for all points in the wave. At any such point the vibrating **E**-vector and the direction of propagation form a plane, called the *plane of vibration;* in a plane-polarized wave all such planes are parallel.

Electromagnetic waves in the radio and microwave range readily exhibit plane-polarization. Such a wave, generated by the surging of charge up and down in the dipole that forms the transmitting antenna of Fig. 35–9, has (at large distances from the dipole and at right angles to it) an electric field vector parallel to the dipole axis. When this plane-polarized wave falls on a second dipole connected to a microwave detector, the alternating electric component of the wave will cause electrons to surge back and forth in the receiving antenna, producing a reading on the detector. If we turn the receiving antenna through 90° about the direction of propagation, the detector reading drops to zero. In this orientation the electric field vector is not able to cause charge to move along the dipole axis because it points at right angles to this axis. We can reproduce the experiment suggested by Fig. 35–9 by turning the receiving antenna of a television set (assumed an electric dipole type) through 90° about an axis that points toward the transmitting station.

Common sources of visible light differ from radio and microwave sources in that the elementary radiators, that is, the atoms and molecules, act independently. The

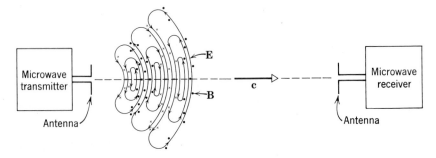

FIGURE 35–9 The vectors **E** in the transmitted wave are parallel to the axis of the receiving antenna so that the wave will be detected. If we rotate the receiving antenna through 90° about the direction of propagation, no signal will be detected.

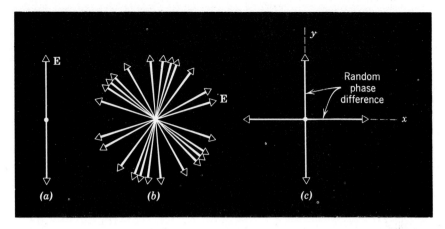

FIGURE 35–10   (*a*) A plane-polarized transverse wave moving toward you, showing only the electric vector. (*b*) An unpolarized transverse wave viewed as a random superposition of many plane-polarized wavetrains. (*c*) A second, completely equivalent, description of an unpolarized transverse wave; here we view the unpolarized wave as two plane-polarized waves with a random phase difference. The orientation of the *x*- and *y*-axes about the propagation direction is completely arbitrary.

light propagated in a given direction consists of independent wavetrains whose planes of vibration are randomly oriented about the direction of propagation, as in Fig. 35–10*b*. Such light, though still transverse, is unpolarized. The random orientation of the planes of vibration conceals the true transverse nature of the waves. To study this transverse nature, we must find a way to unsort the different planes of vibration.

Figure 35–11 shows unpolarized light falling on a sheet of commercial polarizing material called *Polaroid*. There exists in the sheet a certain characteristic polarizing direction, shown by the parallel lines. The sheet will transmit only those wave-train components whose electric vectors vibrate parallel to this direction and will absorb those that vibrate at right angles to this direction. The emerging light will be plane-polarized. This polarizing direction is established during the manufacturing process by embedding certain long-chain molecules in a flexible plastic sheet and then stretching the sheet so that the molecules are aligned parallel to each other. Polarizing sheets 2 ft wide and 100 ft long may be produced.

In Fig. 35–12 the polarizing sheet or *polarizer* lies in the plane of the page and the direction of propagation is into the page. The arrow **E** shows the plane of vibration of a randomly selected wavetrain falling on the sheet. Two vector components, E_x (of magnitude $E \sin \theta$) and $\mathbf{E}_y$ (of magnitude $E \cos \theta$), can replace **E,** one parallel to the polarizing direction and one at right angles to it. Only $\mathbf{E}_y$ will be transmitted; $\mathbf{E}_x$ is absorbed within the sheet.

Let us place a second polarizing sheet P_2 (usually called, when so used, an *analyzer*)

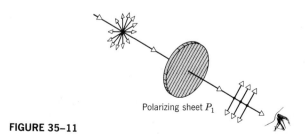

Polarizing sheet P_1

FIGURE 35–11

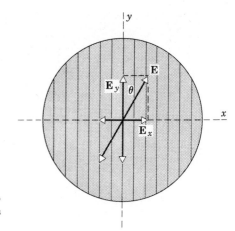

FIGURE 35-12 A wavetrain **E** is equivalent to two component wavetrains **E**$_y$ and **E**$_x$. Only the former is transmitted by the polarizer.

as in Fig. 35–13. If we rotate P_2 about the direction of propagation, there are two positions, 180° apart, at which the transmitted light intensity is almost zero; these are the positions in which the polarizing directions of P_1 and P_2 are at right angles.

If the amplitude of the plane-polarized light falling on P_2 is E_m, the amplitude of the light that emerges is $E_m \cos \theta$, where θ is the angle between the polarizing directions of P_1 and P_2. Recalling that the intensity of the light beam is proportional to the square of the amplitude, we see that the transmitted intensity I varies with θ according to

$$I = I_m \cos^2 \theta, \tag{35-15}$$

in which I_m is the maximum value of the transmitted intensity. It occurs when the polarizing directions of P_1 and P_2 are parallel, that is, when $\theta = 0$ or 180°. Figure 35–14a, in which two overlapping polarizing sheets are in the parallel position ($\theta = 0$ or 180° in Eq. 35–15) shows that the light transmitted through the region of overlap has its maximum value. In Fig. 35–14b one or the other of the sheets has been rotated through 90° so that θ in Eq. 35–15 has the value 90 or 270°; the light transmitted through the region of overlap is now a minimum.

Equation 35–15, called the law of Malus, was discovered by Étienne Louis Malus (1775–1812) experimentally in 1809, using polarizing techniques other than those so far described.

Example 3. Two polarizing sheets have their polarizing directions parallel so that the intensity I_m of the transmitted light is a maximum. Through what angle must either sheet be turned if the intensity is to drop by one-half?

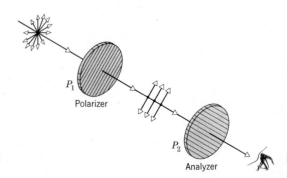

FIGURE 35-13   Unpolarized light is not transmitted by crossed polarizing sheets.

(a)   (b)

FIGURE 35–14 Two square sheets of Polaroid are laid over a book. In (a) the axes of polarization of the two sheets are parallel and light passes through both sheets. In (b) one sheet has been rotated 90° and no light passes through. The book is opened to an illustration of the Luxembourg Palace in Paris. Malus discovered the phenomenon of polarization by reflection while looking at sunlight reflected off the palace windows through a calcite crystal.

From Eq. 35–15, since $I = \frac{1}{2}I_m$, we have

$$\tfrac{1}{2}I_m = I_m \cos^2 \theta$$

or
$$\theta = \cos^{-1} \pm \frac{1}{\sqrt{2}} = \pm 45°, \pm 135°.$$

The same effect is obtained no matter which sheet is rotated or in which direction.

Historically, polarization studies were made to investigate the nature of light. Today we reverse the procedure and deduce something about the nature of an object from the polarization state of the light emitted by or scattered from that object. We can tell, from studies of the polarization of light reflected from them, that the grains of cosmic dust present in our galaxy have been oriented in the weak galactic magnetic field ($\sim 2 \times 10^{-4}$ gauss) so that their long dimension is parallel to this field. Polarization studies have shown that Saturn's rings consist of ice crystals. We can find the size and shape of virus particles by the polarization of ultraviolet light scattered from them. We can learn a lot about the structure of atoms and nuclei from polarization studies of their emitted radiations, in all parts of the electromagnetic spectrum. Thus we have a useful research technique for structures ranging in size from a galaxy ($\sim 10^{+20}$ meters) to a nucleus ($\sim 10^{-14}$ meter).

35–7 The Electromagnetic Spectrum

Figure 35–15 shows the electromagnetic spectrum. All these waves are electromagnetic in nature and have the same speed c in free space. They differ in wavelength, the sources that give rise to them, and the instruments used to make measurements with them. The electromagnetic spectrum has no definite upper or lower limit. The labeled regions in Fig. 35–15 represent frequency intervals within which a common body of experimental technique, such as common sources and common detectors, exists. Each region overlaps others. For example, we can produce radiation of wavelength 10^{-3} meter either by microwave techniques (microwave oscillators) or by infrared techniques (incandescent sources).

"Light" is defined here as radiation that can affect the eye. Figure 35–16, which shows the relative eye sensitivity of an assumed *standard observer* to radiations of various wavelengths, shows that the center of the visible region is about 5.55×10^{-7} meter. Light of this wavelength produces the sensation of yellow-green.

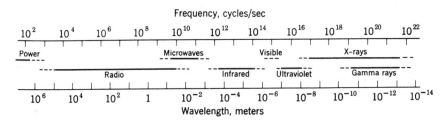

FIGURE 35-15 The electromagnetic spectrum. Note that the wavelength and frequency scales are logarithmic.

In optics we often use the micron (abbr. μ), the millimicron (abbr. mμ), and the Angstrom (abbr. A) as units of wavelength. They are defined from

$$1 \; \mu = 10^{-6} \; \text{meter}$$

$$1 \; \text{m}\mu = 10^{-9} \; \text{meter}$$

$$1 \; \text{A} = 10^{-10} \; \text{meter}.$$

Thus the center of the visible region can be expressed as 0.555 μ, 555 mμ, or 5550 A.

The limits of the visible spectrum are not well defined because the eye sensitivity curve approaches the axis asymptotically at both long and short wavelengths. If we take the limits, arbitrarily, as the wavelengths at which the eye sensitivity has dropped to 1% of its maximum value, these limits are about 4300 A and 6900 A, less than a factor of two in wavelength. The eye can detect radiation beyond these limits if it is intense enough. In many experiments in physics we can use photographic plates or light-sensitive electronic detectors in place of the human eye.

35–8 The Speed of Light

The speed of electromagnetic radiation c (usually called simply the speed of light) is probably the most investigated constant in all of science. According to electromagnetic theory its value, when measured in a vacuum, should not depend upon wavelength. Table 35–1 shows that this is true, at least for microwaves, radiowaves, and light.

Galileo was probably the first person to try measuring the speed of light. In his chief work, *Two New Sciences,* published in the Netherlands in 1638, is a conversation

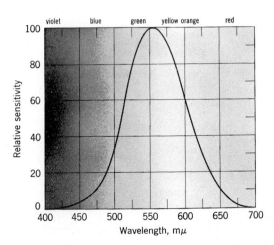

FIGURE 35–16

Table 35–1   THE SPEED OF ELECTROMAGNETIC RADIATION IN FREE SPACE   (SOME SELECTED MEASUREMENTS)

Date	Experimenter	Country	Method	Speed (km/sec)	Uncertainty (km/sec)
1600(?)	Galileo	Italy	Lanterns and shutters	"If not instantaneous, it is extraordinarily rapid"	
1675	Roemer	France	Astronomical	200,000	
1729	Bradley	England	Astronomical	304,000	
1849	Fizeau	France	Toothed wheel	313,300	
1862	Foucault	France	Rotating mirror	298,000	500
1876	Cornu	France	Toothed wheel	299,990	200
1880	Michelson	U.S.A.	Rotating mirror	299,910	50
1883	Newcomb	England	Rotating mirror	299,860	30
1883	Michelson	U.S.A.	Rotating mirror	299,853	60
1906	Rosa and Dorsey	U.S.A.	Electromagnetic theory	299,781	10
1923	Mercier	France	Standing waves on wires	299,782	15
1926	Michelson	U.S.A.	Rotating mirror	299,796	4
1928	Karolus and Mittelstaedt	Germany	Kerr cell	299,778	10
1932	Michelson, Pease, and Pearson	U.S.A.	Rotating mirror	299,774	11
1940	Huettel	Germany	Kerr cell	299,768	10
1941	Anderson	U.S.A.	Kerr cell	299,776	14
1950	Bergstrand	Sweden	Geodimeter	299,792.7	0.25
1950	Essen	England	Microwave cavity	299,792.5	3
1950	Bol and Hansen	U.S.A.	Microwave cavity	299,789.3	0.4
1951	Aslakson	U.S.A.	Shoran radar	299,794.2	1.9
1952	Rank, Ruth, and Ven der Sluis	U.S.A.	Molecular spectra	299,776	7
1952	Froome	England	Microwave interferometer	299,792.6	0.7
1954	Florman	U.S.A.	Microwave interferometer	299,795.1	1.9
1957	Bergstrand	Sweden	Geodimeter	299,792.85	0.16
1958	Froome	England	Microwave interferometer	299,792.50	0.10
1965	Kolibayev	U.S.S.R.	Geodimeter	299,792.6	0.06
1967	Grosse	West Germany	Geodimeter	299,792.5	0.05
1967	Simkin, Lukin, Sikora, and Strelenskii	U.S.S.R.	Microwave interferometer	299,792.56	0.11

among three fictitious persons called Salviati, Simplicio, and Sagredo (who evidentally is meant to be Galileo himself). Here is part of what they say about the speed of light.

Simplicio: Everyday experience shows that the propagation of light is instantaneous; for when we see a piece of artillery fired, at a great distance, the flash reaches our eyes without lapse of time; but the sound reaches the ear only after a noticeable interval.

Sagredo: Well, Simplicio, the only thing I am able to infer from this familiar bit of experience is that sound, in reaching our ear, travels more slowly than light; it does not inform me whether the coming of the light is instantaneous or whether, although extremely rapid, it still occupies time. . . .

Sagredo then describes a possible method for measuring the speed of light. He and an assistant stand facing each other some distance apart, at night. Each carries a lantern which can be covered or uncovered at will. Galileo started the experiment by uncovering his lantern. When the light reached the assistant he uncovered his own lantern, whose light was then seen by Galileo. Galileo tried to measure the time between the instant at which he uncovered his own lantern and the instant at which the light from his assistant's lantern reached him. For a 1-mile separation we now know that the round trip travel time would be only 10^{-5} sec. This is much less than human reaction times, so the method fails.

To measure such a large velocity directly, we must either deal with a small time interval or use a long base line. This suggests that astronomy, which involves great distances, might provide an experimental value for the speed of light.

In 1675 Ole Roemer, a Danish astronomer working in Paris, made some observations of the moons of Jupiter (see Problem 35) from which a speed of light of 2×10^8 meters/sec may be deduced. About fifty years later James Bradley, an English astronomer, made some observations of an entirely different kind, obtaining a value of 3.0×10^8 meters/sec. Actually, the speed of light is now so well known from terrestrial measurements that we use it to measure distances in the solar system. For example, laser pulses reflected from "mirrors" on the moon and microwave pulses reflected from Venus today provide the distances to these bodies.

In 1849 Hippolyte Louis Fizeau (1819–1896), a French physicist, first measured the speed of light by a nonastronomical method, obtaining a value of 3.13×10^8 meters/sec. Figure 35–17 shows Fizeau's apparatus. Let us first ignore the toothed wheel. Light from source S is made to converge by lens L_1, is reflected from mirror M_1, and forms in space at F an image of the source. Mirror M_1 is a so-called "half-silvered mirror"; its reflecting coating is so thin that only half the light that falls on it is reflected, the other half being transmitted.

Light from the image at F enters lens L_2 and emerges as a parallel beam; after passing through lens L_3 it is reflected back along its original direction by mirror M_2. In Fizeau's experiment the distance l between M_2 and F was 8630 meters or 5.36 miles. When the light strikes mirror M_1 again, some will be transmitted, entering the eye of the observer through lens L_4.

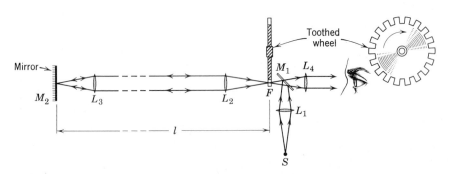

FIGURE 35–17   Fizeau's apparatus for measuring the speed of light.

The observer will see an image of the source formed by light that has traveled a distance $2l$ between the wheel and mirror M_2 and back again. To time the light beam a marker of some sort must be put on it. This is done by "chopping" it with a rapidly rotating toothed wheel. Suppose that during the round-trip travel time of $2l/c$ the wheel has turned just enough so that, when the light from a given "burst" returns to the wheel, point F is covered by a tooth. The light will hit the face of the tooth that is toward M_2 and will not reach the observer's eye.

If the speed of the wheel is exactly right, the observer will not see any of the bursts because each will be screened by a tooth. The observer measures c by increasing the angular speed ω of the wheel from zero until the image of source S disappears. Let θ be the angular distance from the center of a gap to the center of a tooth. The time needed for the wheel to rotate a distance θ is the round-trip travel time $2l/c$. In equation form,

$$\frac{\theta}{\omega} = \frac{2l}{c} \quad \text{or} \quad c = \frac{2\omega l}{\theta}. \tag{35–16}$$

Example 4. The wheel used by Fizeau had 720 teeth. What is the smallest angular speed at which the image of the source will vanish?

The angle θ is $1/1440$ rev; solving Eq. 35–16 for ω gives

$$\omega = \frac{c\theta}{2l} = \frac{(3.00 \times 10^8 \text{ meters/sec})(1/1440 \text{ rev})}{(2)(8630 \text{ meters})} = 12.1 \text{ rev/sec.}$$

The French physicist Foucault (1819–1868) greatly improved Fizeau's method by substituting a rotating mirror for the toothed wheel. The American physicist Albert A. Michelson (1852–1931) conducted an extensive series of measurements of c, extending over a fifty-year period, using this technique.

We must view the speed of light within the larger framework of the speed of electromagnetic radiation in general. It is a significant experimental confirmation of Maxwell's theory of electromagnetism that the speed in free space of waves in all parts of the electromagnetic spectrum has the same value c. Table 35–1 shows some selected measurements that have been made of the speed of electromagnetic radiation since Galileo's day. It stands as a monument to man's persistence and ingenuity. Note in the last column how the uncertainty in the measurement has improved through the years. Note also the international character of the effort and the variety of methods.

The task of arriving at a single "best" value of c from the many listed in the table is difficult, for it involves a careful study of each reported measurement and a selection from among them, based on the reported uncertainty, the selector's judgment of the probable presence or absence of hidden error, and related measurements of other physical constants. By careful analysis of such measurements the "best" value, as of 1969, is

$$c = 2.99792458 \times 10^8 \text{ meters/sec.}$$

The uncertainty of measurement is 0.00000004×10^8 meters/sec.

35–9 Moving Sources and Observers

When we say that the speed of sound in dry air at $0°C$ is 331.7 meters/sec, we imply a reference frame fixed with respect to the air mass. When we say that the speed of electromagnetic radiation in free space is 2.9979250×10^8 meters/sec, what reference frame are we talking about? It cannot be the medium through which the light wave travels because, in contrast to sound, no medium is required.

Physicists of the nineteenth century, influenced as they then were by a false analogy between light waves and sound waves or other purely mechanical disturbances did not accept the idea of a wave requiring no medium. They postulated the existence of an *ether,* which was a tenuous substance that filled all space and served as a medium of transmission for light. The ether was required to have a vanishingly small density to account for the fact that it could not be observed by any known means in an evacuated space.

The ether concept, although it proved useful for many years, did not survive the test of experiment. In particular, careful attempts to measure the speed of the earth through the ether always gave the result of zero. Physicists were not willing to believe that the earth was permanently at rest in the ether and that all other bodies in the universe were in motion through it.

Einstein in 1905 resolved the difficulty of understanding the propagation of light by making a bold postulate: If a number of observers are moving (at uniform velocity) with respect to each other and to a source of electromagnetic radiation such as light and if each observer measures the speed of the radiation emerging from the source, *they will all measure the same value.* This is the fundamental assumption of Einstein's theory of relativity. It does away with the need for an ether by asserting that the speed of light is the same in *all* reference frames; none is singled out as fundamental. The theory of relativity, derived from this postulate, has been tested many times, and agreement with the predictions of theory has always emerged. These agreements, extending over half a century, lend strong support to Einstein's basic postulate about light propagation.

Figure 35–18 focuses specifically on the fundamental problem of light propagation. A source of light emits a light pulse whose speed is measured by two observers. The first, S', is at rest with respect to the source and the second, S, is moving with speed u in the negative x-direction thus seeing the source move with the same speed in the positive x-direction. The speed of light, v', measured by S' would be equal to c. Question: What speed, v, would observer S measure? Einstein's hypothesis asserts that *each* observer would measure the same speed, c, or that

$$v = v' = c.$$

This hypothesis contradicts the classical law of addition of velocities (see Section 4–5) which asserts that

$$v = v' + u. \tag{35–17}$$

This law, which is so familiar that it seems (incorrectly) to be intuitively true, is in fact based on observations of gross moving objects in the world about us. Even the fastest

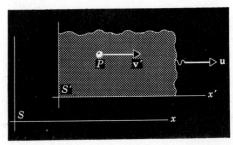

FIGURE 35–18   Observers S and S', who are in relative motion, each observe a pulse P of electromagnetic radiation (light, say). The pulse is emitted from a source, not shown, that is at rest in the S' frame of reference.

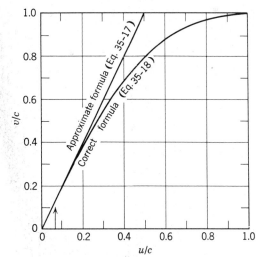

FIGURE 35–19 The speed of a particle P, as seen by observer S in Fig. 35–18, for the special case of $v' = u$. All speeds are expressed as a ratio to c, the speed of light. The vertical arrow corresponds to 5×10^7 miles/hr, about 2500 times the speed of a typical artificial earth satellite.

of these—an earth satellite, say—is moving at a speed that is quite small compared to that of light. The body of experimental evidence that underlies Eq. 35–17 thus represents a severely restricted area of experience, namely, experiences in which $v' \ll c$ and $u \ll c$. If we assume that Eq. 35–17 holds for all particles regardless of speed, we are making a gross extrapolation. Einstein's theory of relativity predicts that this extrapolation is indeed not true and that Eq. 35–17 is a limiting case of a more general relationship that holds for light pulses and for material particles, whatever their speed, or

$$v = \frac{v' + u}{1 + v'u/c^2} .$$ (35–18)

Equation 35–18 is quite indistinguishable from Eq. 35–17 at low speeds, that is, when $v' \ll c$ and $u \ll c$.

If we apply Eq. 35–18 to the case in which the moving object is a light pulse, and if we put $v' = c$, we obtain

$$v = \frac{c + u}{1 + cu/c^2} = c.$$

This is consistent, as it must be, with the fundamental assumption on which the derivation of Eq. 35–18 is based; it shows that *both* observers measure the same speed c for light. Equation 35–17 predicts (incorrectly) that the speed measured by S will be $c + u$. Figure 35–19 shows that we cannot tell the difference between the (correct) Eq. 35–18 and the (approximate) Eq. 35–17 at speeds that are small compared to the speed of light.

Example 5. Two electrons are ejected in opposite directions from radioactive atoms in a sample of radioactive material. Let each electron have a speed, as measured by a laboratory observer, of $0.60c$ (this corresponds to a kinetic energy of 130 KeV). What is the speed of one electron as seen from the other?

Equation 35–17 gives

$$v = v' + u = 0.60c + 0.60c = 1.2c.$$

Equation 35–18 gives

$$v = \frac{v' + u}{1 + v'u/c^2} = \frac{0.60c + 0.60c}{1 + (0.60c)^2/c^2} = 0.88c.$$

This example shows that for speeds that are comparable to c, Eqs. 35–17 and 35–18 yield quite different results. A wealth of indirect experimental evidence points to the latter result as being correct.

35–10   Doppler Effect

We have seen that the same speed is measured for electromagnetic radiation no matter what the relative speeds of the light source and the observer are. The measured frequency and wavelength will change, however, but always in such a way that their product, which is the speed of light, remains constant. Such frequency shifts are called *Doppler shifts*, after Johann Doppler (1803–1853), who first predicted them. In what follows we will focus on visible light although our conclusions hold for all parts of the electromagnetic spectrum.

In Section 17–6 we showed that if a source of sound is moving away from an observer at a speed u, the frequency heard by the observer (see Eq. 17–10, which has been rearranged with u substituted for v_s) is

$$\nu' = \nu\,\frac{1}{1 + u/v}\,. \qquad \begin{cases} 1.\ \text{sound wave} \\ 2.\ \text{observer fixed in medium} \\ 3.\ \text{source receding from observer} \end{cases} \qquad (35\text{–}19)$$

In this equation ν is the frequency heard when the source is at rest and v is the speed of sound.

If the source is at rest in the transmitting medium but the observer is moving away from the source at speed u, the observed frequency (see Eq. 17–9b, in which u has been substituted for v_0) is

$$\nu' = \nu\left(1 - \frac{u}{v}\right). \qquad \begin{cases} 1.\ \text{sound wave} \\ 2.\ \text{source fixed in medium} \\ 3.\ \text{observer receding from source} \end{cases} \qquad (35\text{–}20)$$

Even if the relative separation speeds u of the source and the observer are the same, the frequencies predicted by Eqs. 35–19 and 35–20 are different. This is not surprising, because a sound source moving through a medium in which the observer is at rest is simply not the same thing as an observer moving through that medium with the source at rest, as comparison of Figs. 17–8 and 17–9 shows.

We might be tempted to apply Eqs. 35–19 and 35–20 to light, substituting c, the speed of light, for v, the speed of sound. For light, as contrasted with sound, however, it has proved impossible to identify a medium of transmission relative to which the source and the observer are moving. This means that "source receding from observer" and "observer receding from source" are precisely the same and must exhibit exactly the same Doppler frequency. As applied to light, either Eq. 35–19 or Eq. 35–20 or both must be incorrect. The Doppler frequency predicted by the theory of relativity is, in fact,

$$\nu' = \nu\,\frac{1 - u/c}{\sqrt{1 - (u/c)^2}}\,. \qquad \begin{cases} 1.\ \text{light source} \\ 2.\ \text{source and observer separating} \end{cases} \qquad (35\text{–}21)$$

In all three of the foregoing equations we obtain the appropriate relations for the source and the observer approaching each other if we replace u by $-u$.

Equations 35–19, 35–20, and 35–21 are not so different as they seem if the ratio u/c is small enough. Let us expand Eqs. 35–19, 35–20 and 35–21 by the binomial theorem. The equations, each describing the situation when the source and observer are separating, then become, substituting c for v,

$$v' = v\left[1 - \frac{u}{c} + \left(\frac{u}{c}\right)^2 + \cdots\right], \qquad \left\{\begin{matrix}\text{classical theory; observer}\\ \text{fixed in medium}\end{matrix}\right\} \quad (35\text{–}19a)$$

$$v' = v\left[1 - \frac{u}{c}\right], \qquad \left\{\begin{matrix}\text{classical theory; source}\\ \text{fixed in medium}\end{matrix}\right\} \quad (35\text{–}20a)$$

$$\text{and} \quad v' = v\left[1 - \frac{u}{c} + \frac{1}{2}\left(\frac{u}{c}\right)^2 + \cdots\right]. \quad \{\text{relativistic theory}\} \quad (35\text{–}21a)$$

The ratio u/c for all available monochromatic light sources, even those of atomic dimensions, is small. This means that successive terms in these equations become small rapidly and, depending on the accuracy required, only a limited number of terms need be retained.

Under nearly all circumstances the differences between these three equations are not important. Nevertheless, it is important to test Eq. 35–21a experimentally and thus, in part, the theory of relativity. To do this one must obtain sources having speeds close enough to the speed of light so that at least the term $(u/c)^2$ be measurable since that is where the equations begin to differ.

H. E. Ives and G. R. Stilwell carried out such a precision experiment in 1938. They sent a beam of hydrogen atoms, generated in a gas discharge, down a tube at speed u, as in Fig. 35–20a. In effect, they observed light emitted by these atoms in a direction opposite to **u** (atom 1), using a mirror, and also in the same direction as **u** (atom 2). With a precision spectrograph they observed a particular spectral line having a frequency v, obtaining also lines having the shifted frequencies v_1 and v_2. The rays, including those from atoms at rest, exposed a photographic plate at different positions depending upon their frequencies as shown in Fig. 35–20b.

From Eq. 35–19a we can find the frequency shift of light from atoms 1 and 2, divided by the original frequency, assuming light moves in a classical medium fixed relative to the observer. Source 1 is moving away from the observer. The mirror is

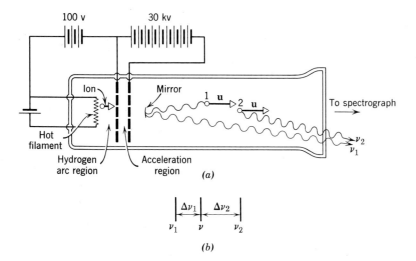

(a)

(b)

FIGURE 35–20   The Ives-Stilwell experiment.

Table 35–2   THE IVES-STILWELL EXPERIMENT*

$\dfrac{\Delta\nu_1 + \Delta\nu_2}{\nu}, 10^{-5}$	Speed of moving atoms ($=u$), 10^6 meters/sec			
	0.865	1.01	1.15	1.33
Theoretical value according to classical theory (observer fixed in medium, Eq. 35–19a)	1.67	2.26	2.90	3.94
Theoretical value according to the theory of relativity (Eq. 35–21a)	0.835	1.13	1.45	1.97
Experimental value	0.762	1.1	1.42	1.9

* The table shows only part of the data taken by Ives and Stilwell.

just used to bring the light to the detector. Source 2 is moving toward the detector, which requires us to introduce $-u$ for the speed in Eq. 35–19a.

$$\frac{\Delta\nu_1}{\nu} = \frac{\nu_1 - \nu}{\nu} = -\frac{u}{c} + \left(\frac{u}{c}\right)^2 + \cdots$$

$$\frac{\Delta\nu_2}{\nu} = \frac{\nu_2 - \nu}{\nu} = \frac{u}{c} + \left(\frac{u}{c}\right)^2 + \cdots .$$

If these two shifts, one positive and one negative, are added and higher-order terms are neglected, we get an expression dependent only upon the velocity squared term, as we desired.

$$\frac{\Delta\nu_1 + \Delta\nu_2}{\nu} = 2\left(\frac{u}{c}\right)^2. \quad \text{(classical theory, fixed observer)} \quad (35\text{–}22)$$

Since the relativistic Doppler equation differs in this term we would expect a different expression assuming relativity theory. Indeed,

$$\frac{\Delta\nu_1 + \Delta\nu_2}{\nu} = \left(\frac{u}{c}\right)^2. \quad \text{(relativistic theory)} \quad (35\text{–}23)$$

The results of the measurements of the frequency shifts by Ives and Stilwell are shown in Table 35–2. Comparison with the theoretical predictions shows agreement only with the theory of relativity.

The Doppler effect for light finds many applications in astronomy, where it is used to determine the speeds at which luminous heavenly bodies are moving toward us or receding from us. Such Doppler shifts measure only the radial or line-of-sight components of the relative velocity. All galaxies** for which such measurements have been made (Fig. 35–21) appear to be receding from us, the recession velocity being greater for the more distant galaxies; these observations are the basis of the expanding-universe concept.

■ **Example 6.** Certain characteristic wavelengths in the light from a galaxy in the constellation Virgo are observed to be increased in wavelength, as compared with terrestrial sources, by about 0.4%. What is the radial speed of this galaxy with respect to the earth? Is it approaching or receding?

** See "The Red-Shift" by Allen R. Sandage, in *Scientific American*, September 1956.

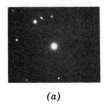
(a)

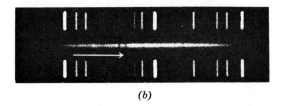
(b)

FIGURE 35–21   (a) The central spot is a nebula in the constellation Corona Borealis; it is 130,000,000 light years distant. (b) The central streak shows the distribution in wavelength of the light emitted from this nebula. The two vertical dark bands show the presence of calcium. The horizontal arrow shows that these calcium lines occur at longer wavelengths than those for terrestrial light sources containing calcium, the length of the arrow representing the wavelength shift. Measurement of this shift shows that the galaxy is receding from us at 13,400 miles/sec. The lines above and below the central streak represent light from a terrestrial source, used to establish a wavelength scale. (Courtesy Mount Wilson and Mount Palomar Observatories.)

If λ is the wavelength for a terrestrial source, then

$$\lambda' = 1.004\lambda.$$

Since we must have $\lambda'\nu' = \lambda\nu = c$, we can write this as

$$\nu' = 0.996\nu.$$

This frequency shift is so small that, in calculating the source velocity, it makes no practical difference whether we use Eq. 35–19, 35–20, or 35–21. Using Eq. 35–20 we obtain

$$\nu' = 0.996\nu = \nu\left(1 - \frac{u}{c}\right).$$

Solving yields $u/c = 0.0040$, or $u = (0.0040)(3.0 \times 10^8 \text{ meters/sec}) = 1.2 \times 10^6$ meters/sec or 2.7×10^6 miles/hr. The galaxy is receding; had u turned out to be negative, the galaxy would have been moving toward us.

QUESTIONS

1. Figure 35–22 shows a magnetic dipole activated by an oscillating LC circuit; see Fig. 35–1. Discuss the nature of the traveling wave at a distant point P.

2. Explain why the term $\epsilon_0\, d\Phi_E/dt$ is needed in Ampère's equation to understand the propagation of electromagnetic waves.

3. In the equation $c = 1/\sqrt{\mu_0\epsilon_0}$ (Eq. 35–10), how can c always have the same value if μ_0 is arbitrarily assigned and ϵ_0 is measured?

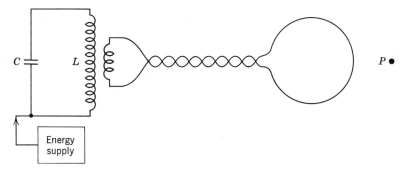

FIGURE 35–22

4. Is it conceivable that electromagnetic theory might some day be able to predict the value of c (3×10^8 meters/sec), not in terms of μ_0 and ϵ_0, but directly and numerically without recourse to any measurements?

5. How can an object absorb light energy without absorbing momentum?

6. A searchlight sends out a parallel beam of light. Does the searchlight experience any force associated with the emission of light?

7. Name two historic experiments, in addition to the radiation pressure measurements of Nichols and Hull, in which a torsion balance was used. Both are described in this book.

8. Show that for complete absorption of a parallel beam of light the radiation pressure on the absorbing object is given by $p = S/c$, where S is the magnitude of the Poynting vector and c is the speed of light in free space.

9. Why do sunglasses made of polarizing materials have a marked advantage over those that simply depend on absorption effects?

10. Unpolarized light falls on two polarizing sheets so oriented that no light is transmitted. If a third polarizing sheet is placed between them, can light be transmitted?

11. Find a way to identify the polarizing direction of a sheet of Polaroid.

12. When observing a clear sky through a polarizing sheet, you find that the intensity varies by a factor of two on rotating the sheet. This does not happen when you look at a cloud through the sheet. Why?

13. How might an eye-sensitivity curve like that of Fig. 35–16 be measured?

14. Why are danger signals in red, when the eye is most sensitive to yellow-green?

15. Comment on this definition of the limits of the spectrum of visible light given by a physiologist: "The limits of the visible spectrum occur when the eye is no better adapted than any other organ of the body to serve as a detector."

16. How could Galileo test experimentally that reaction times were an overwhelming source of error in his attempt to measure the speed of light, described in Sec. 35–8?

17. It has been suggested that the speed of light may change slightly in value as time goes on. Can you find any evidence for this in Table 35–1?

18. A friend tells you that Einstein's postulate (that the speed of light is not affected by the uniform motion of the source or the observer) must be discarded because it violates "common sense." How would you answer him?

19. In a vacuum, does the speed of light depend on (a) the wavelength, (b) the frequency, (c) the intensity, (d) the speed of the source, or (e) the speed of the observer?

20. Can a galaxy be so distant that its recession speed equals c? If so, how can we see the galaxy? Will its light ever reach us?

PROBLEMS

1(1). (a) At a distance of 80 miles from a radio transmitter, how much later would you observe a wave emitted from the antenna? (b) If the radiation were a radio wave emitted from the sun? (c) If it were emitted from a stellar radio source 380 light years distant?
Answer: (a) 4.3×10^{-4} sec. (b) 8.3 min. (c) 380 years.

2(1). If E and B did not drop off as $1/r$ in the radiation field, regardless of the nature of the source, what problem would this make for us?

3(3). Prove that for any point in an electromagnetic wave such as that of Fig. 35–6 the density of energy stored in the electric field equals that stored in the magnetic field.

4(3). A cavity magnetron consists of a metallic parallel-plate capacitor molded onto a metallic inductor, forming a resonant circuit. The capacitor has an area A (0.40 cm²), the spacing between its plates is t (0.20 cm). What must be the inductance to generate electromagnetic waves of wavelength λ (3.0 cm)?

5(3). Show that (a) through (d) below satisfy Eqs. 35–5 and 35–8. In each of these, A is a constant. (a) $E = Ac(x - ct)$, $B = A(x - ct)$. (b) $E = Ac(x + ct)^{15}$, $B = -A(x + ct)^{15}$. (c) $E = Ace^{(x-ct)}$, $B =$

$Ae^{(x-ct)}$. (d) $E = Ac \ln (x + ct)$, $B = -A \ln (x + ct)$. (e) Generalize these examples to show that $E = Acf(x - ct)$, $B = Af(x - ct)$ is a solution where f is any function of $(x - ct)$. What is the corresponding situation for functions of $(x + ct)$?

6(3). How does the displacement current vary with space and time in a traveling plane electromagnetic wave?

7(4). Show that the directions of the fields **E** and **B** in Fig. 35–2 are consistent with the direction of propagation of the radiation.

8(4). Show that the directions of the fields **E** and **B** in the various "patches" in Fig. 35–3 are consistent with the direction of propagation of the radiation.

9(4). An electromagnetic wave is traveling in the negative y-direction. At a particular position and time, the electric field is along the positive z-axis and has a magnitude of 100 volts/meter. What are the direction and magnitude of the magnetic field?
Answer: 3.3×10^{-7} webers/meter^2, in the negative x-direction.

10(4). A plane radio wave has $E_m \cong 10^{-4}$ volt/meter. Calculate (a) B_m and (b) the intensity of the wave, as measured by $\bar{\mathbf{S}}$.

11(4). Sunlight strikes the earth, outside its atmosphere, with an intensity of 2.0 cal/cm^2-min. Calculate E_m and B_m for sunlight, assuming it to be a wave like that of Fig. 35–3.
Answer: $E_m = 1000$ volts/meter; $B_m = 3.4 \times 10^{-6}$ webers/meter^2.

12(4). An airplane flying at a distance of 10 km from a radio transmitter receives a signal of power 10 microwatts/meter^2. Calculate (a) the average electric field at the airplane due to this signal; (b) the average magnetic field at the airplane; (c) the total power radiated by the transmitter, assuming the transmitter to radiate isotropically and the earth to be a perfect absorber.

13(4). If the maximum electric field at a distance of 10 m from a point source of light is 1.0×10^{-6} volt/meter, find (a) the maximum value of the magnetic field and (b) the average value of the Poynting vector at the same distance. (c) What is the power output of the source?
Answer: (a) 3.3×10^{-15} webers/meter^2. (b) 1.3×10^{-15} watts/meter^2. (c) 1.7×10^{-12} watts.

14(4). If the maximum magnetic field at a distance of 1.0 km from a point source of electromagnetic radiation is 3.0×10^{-16} webers/m^2 find (a) the maximum value of the electric field, (b) the intensity of the radiation and (c) the total power output from the source.

15(4). The intensity of direct solar radiation that was unabsorbed by the atmosphere on a lazy summer day is 100 watt/meter^2. How close would you have to stand to a 1.0-kilowatt electric heater to feel the same intensity? Assume that the heater is 100% efficient and radiates equally in all directions.
Answer: 0.89 meters.

16(4). A cube of edge a has its edges parallel to the x-, y-, and z-axes of a rectangular coordinate system. A uniform electric field **E** is parallel to the y-axis and a uniform magnetic field **B** is parallel to the x-axis. Calculate (a) the rate at which, according to the Poynting vector point of view, energy may be said to pass through each face of the cube and (b) the net rate at which the energy stored in the cube may be said to change.

17(4). A #10 copper wire (diameter, 0.10 in.; resistance per 1000 ft, 1.0 ohm) carries a current of 25 amp. Calculate (a) **E** (b) **B** and (c) **S** for a point on the surface of the wire.
Answer: (a) 8.2×10^{-2} volt/meter, parallel to current. (b) 3.9×10^{-3} weber/meter^2, tangent to surface and perpendicular to current. (c) 260 watts/meter^2, radially inward.

18(4). Consider a wire, of length l having current i and resistance R. Show that the Poynting vector on the surface of the wire indicates an energy flow into this region of i^2R.

19(4). Figure 35–23 shows a long resistanceless transmission line, delivering power from a battery to a resistive load. A steady current i exists as shown. (a) Sketch qualitatively the electric and magnetic fields around the line, and (b) show that, according to the Poynting vector point of view,

FIGURE 35–23 Problem 19(4).

energy travels from the battery to the resistor through the space around the line and not through the line itself. (Hint: Each conductor in the line is an equipotential surface, since the line has been assumed to have no resistance.)

20(4). Consider the possibility of standing waves:

$$E = E_m(\sin \omega t)(\sin kx)$$
$$B = B_m(\cos \omega t)(\cos kx)$$

(*a*) Show that these satisfy Eqs. 35–5 and 35–8 if E_m is suitably related to B_m and ω suitably related to k. What are these relationships? (*b*) Find the (instantaneous) Poynting vector. (*c*) Show that the time average power flow across any area is zero. (*d*) Describe the flow of energy in this problem.

21(4). A 15-watt fluorescent light converts 80% of the electrical energy input to light. Estimate the magnitude of the Poynting vector 10 cm from the surface of a tube 45 cm long and 3.0 cm in diameter. What assumptions do you make? Remember that this is not a point light source!
Answer: About 35 watts/meter², depending on the assumptions.

22(5). A plane electromagnetic wave propagating through a vacuum is described by the following electric and magnetic fields: $E_x = E_0 \sin (kz - \omega t)$, $B_y = \dfrac{E_0}{c} \sin (kz - \omega t)$, $E_y = E_z = B_x = B_z = 0$. (*a*) What is the direction of propagation of the wave? (*b*) Find the average power per unit area. (*c*) What is the rate at which momentum is delivered to a perfectly absorbing surface of area A that is normal to the direction of propagation?

23(5). A plane electromagnetic wave, with wavelength 3.0 meters, travels in free space in the $+x$-direction with its electric vector **E**, of amplitude 300 volts/meter, directed along the *y*-axis. (*a*) What is the frequency ν of the wave? (*b*) What is the direction and amplitude of the **B** field associated with the wave? (*c*) If $E = E_m \sin (kx - \omega t)$, what are the values of k and ω for this wave? (*d*) What is the time-averaged rate of energy flow per unit area associated with this wave? (*e*) If the wave fell upon a perfectly absorbing sheet of area 2.0 meter², at what rate would momentum be delivered to the sheet and what is the radiation pressure exerted on the sheet?
Answer: (*a*) 1.0×10^8 cycles/sec. (*b*) 1.0×10^{-6} weber/meter² along the *z*-axis. (*c*) 2.1 meter⁻¹; 6.3×10^8 radians/sec. (*d*) 120 watts/meter². (*e*) 8.0×10^{-7} nt; 4.0×10^{-7} nt/meter².

24(5). Show that $\epsilon_0 \mathbf{E} \times \mathbf{B}$ has the dimensions of momentum/volume. (The vector $\epsilon_0 \mathbf{E} \times \mathbf{B}$ may be used to compute the momentum stored in the fields in the same manner that the scalar $\frac{1}{2}\epsilon_0 \mathbf{E}^2 + \dfrac{1}{2\mu_0} \mathbf{B}^2$ may be used to compute the energy stored in the fields.)

25(5). Show that $\sqrt{\dfrac{\epsilon_0}{\mu_0}} \mathbf{E} \times \mathbf{B}$ has the dimensions of momentum/area-time, whereas $\dfrac{1}{\mu_0} \mathbf{E} \times \mathbf{B}$ has the dimensions of energy/area-time. $\left(\text{The vector } \sqrt{\dfrac{\epsilon_0}{\mu_0}} \mathbf{E} \times \mathbf{B} \text{ may be used for computing momentum flow in the same manner that } \mathbf{S} = \dfrac{1}{\mu_0} \mathbf{E} \times \mathbf{B} \text{ is used to compute energy flow.}\right)$

26(5). What is the radiation pressure 1.0 meter away from a 500-watt light bulb? Assume that the surface on which the pressure is exerted faces the bulb and is perfectly absorbing and that the bulb radiates uniformly in all directions.

27(5). Radiation from the sun striking the earth has an intensity of 1400 watts/meter². (*a*) Assuming that the earth behaves like a flat disk at right angles to the sun's rays and that all the incident energy is absorbed, calculate the force on the earth due to radiation pressure. (*b*) Compare it with the force due to the sun's gravitational attraction.
Answer: (*a*) 6.0×10^8 nt. (*b*) 3.6×10^{22} nt.

28(5). A small spaceship whose mass, with occupant, is 1.5×10^3 kg is drifting in outer space, where no gravitational field exists. If it shines a searchlight, which radiates 10^4 watts into space, what speed would the ship attain in one day because of the reaction force associated with the momentum carried away by the light beam?

29(5). Radiation of power P is incident on an object that absorbs a fraction f of it and reflects the rest. What is the radiation force?
Answer: $P(2 - f)/c$.

30(5). It may be possible that a spaceship can propel itself in the solar system by radiation pressure, using a large sail made of aluminum foil. How large must the sail be if the radiation force is to be equal in magnitude to the sun's gravitational attraction? Assume that the mass of the ship + sail is 1500 kg, that the sail is perfectly reflecting, and that the sail is oriented at right angles to the sun's rays. The sun's mass is 2.0×10^{30} kg and its power output is 4.0×10^{26} watts.

31(5). A particle in the solar system is under the combined influence of the sun's gravitational attraction and the radiation force due to the sun's rays. Assume that the particle is a sphere of density 1.0 gm/cm³ and that all of the incident light is absorbed. (*a*) Show that all particles with radius less than some critical radius, R_0, will be blown out of the solar system. (*b*) Calculate R_0. (*c*) Does R_0 depend on the distance from the earth to the sun? (See the appendices for the necessary constants.) *Answer:* (*b*) About 6×10^{-7} meter. (*c*) No.

32(5). Prove, for a plane wave at normal incidence on a plane surface, that the radiation pressure on the surface is equal to the energy density in the beam outside the surface. This relation holds no matter what fraction of the incident energy is reflected.

33(5). Prove, for a stream of bullets striking a plane surface at right angles, that the "pressure" is twice the kinetic energy density in the stream above the surface; assume that the bullets are completely absorbed by the surface. Contrast this with the behavior of light [Problem 32(5)].

34(6). The magnetic field equations for an electromagnetic wave in free space are $B_x = B \sin(ky + \omega t)$, $B_y = B_z = 0$. (*a*) What is the direction of propagation? (*b*) Write the electric field equations. (*c*) Is the wave polarized? If so, in what direction?

35(6). Unpolarized light falls on two polarizing sheets placed one on top of the other. What must be the angle between the characteristic directions of the sheets if the intensity of the transmitted light is (*a*) one-third the maximum intensity of the transmitted beam or (*b*) one-third the intensity of the incident beam? Assume that the polarizing sheet is ideal, that is, that it reduces the intensity of unpolarized light by exactly 50%.
Answer: (*a*) ±55°. (*b*) ±35°.

36(6). An unpolarized beam of light is incident on a group of four polarizing sheets which are lined up so that the characteristic direction of each is rotated by 30° clockwise with respect to the preceding sheet. What fraction of the incident intensity is transmitted?

37(6). A beam of light is a mixture of plane polarized light and randomly polarized light. When it is sent through a Polaroid sheet, we find that the transmitted intensity can be varied by a factor of five depending on the orientation of the Polaroid. Find the relative intensities of these two components of the incident beam.
Answer: 67% polarized, 33% unpolarized.

38(6). A beam of plane polarized light strikes two polarizing sheets. The characteristic direction of the second is 90° with respect to the incident light. The characteristic direction of the first is at angle θ with respect to the initial beam. Find angle θ for a transmitted beam intensity that is 1/10 the incident beam intensity.

39(6). It is desired to rotate the plane of polarization of a beam of plane polarized light by 90°. (*a*) How might this be done using only Polaroid sheets? (*b*) How many sheets are required in order that the total intensity loss is less than about 40%?
Answer: (*a*) Allow light to pass through a series of two or more Polaroid sheets so that each sheet rotates the axis by less than 90°, but the total rotation adds up to 90°. (*b*) 5 sheets.

40(7). (*a*) At what wavelengths does the eye sensitivity have half its maximum value? (*b*) What are the frequency and the period of the light for which the eye is most sensitive?

41(8). Suppose that light is timed over a 1-mile base line and its speed is measured to the accuracy quoted on p. 657. How large an error in the length of the base line could be tolerated, assuming other sources of error to be negligible?
Answer: 0.021 in.

42(8). Roemer's method for measuring the speed of light consisted in observing the apparent times of revolution of one of the moons of Jupiter. The true period of revolution is 42.5 hr. (*a*) Taking into account the finite speed of light, how would you expect the apparent time of revolution to alter as the earth moves in its orbit from point *x* to point *y* in Fig. 35–24? (*b*) What observations would be needed to compute the speed of light? Neglect the motion of Jupiter in its orbit. Figure 35–24 is not drawn to scale.

43(10). For what value of u/c does Eq. 35–20 differ from Eq. 35–21 by 1%?
Answer: 0.14.

44(10). A galaxy is receding from us at a velocity $v = 0.50c$. Find the observed wavelength of a hydrogen spectral line whose proper wavelength is $\lambda = 6560$ A.

45(10). A rocketship is receding from the earth at a speed of $0.20c$. A light in the rocketship appears blue to passengers on the ship. What color would it appear to be to an observer on the earth? See Fig. 35–16.
Answer: Yellow-orange.

46(10). The period of rotation of the sun at its equator is 24.7 days; its radius is 7.0×10^8 meters. What Doppler wavelength shifts are expected for characteristic wavelengths in the vicinity of 5500 A emitted from the edge of the sun's disk?

47(10). The "red shift" of radiation from a distant nebula consists of the light (H_γ), known to have a wavelength of 4340×10^{-8} cm when observed in the laboratory, appearing to have a wavelength of 6562×10^{-8} cm. (*a*) What is the speed of the nebula in the line of sight relative to the earth? (*b*) Is it approaching or receding?
Answer: (*a*) 1.2×10^8 meters/sec. (*b*) Receding.

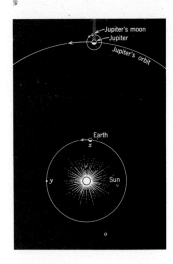

FIGURE 35–24 Problem 42(8).

48(10). Could you go through a red light fast enough to have it appear green? Would you rather get a ticket for speeding or for running the red light? Take $\lambda = 6200 \times 10^{-8}$ cm for red light, $\lambda = 5400 \times 10^{-8}$ cm for green light, and $c = 3 \times 10^{10}$ cm/sec as the speed of light.

49(10). The difference in wavelength between an incident microwave beam and one reflected from an approaching or receding car is used to determine automobile speeds on the highway. (*a*) Show that if v is the speed of the car and ν the frequency of the incident beam, the change of frequency is approximately $2v\nu/c$, where c is the speed of the electromagnetic radiation. (*b*) For microwaves of frequency 2450 megacycles/sec, what is the change of frequency per mile/hr of speed?
Answer: (*b*) 7.3 (cycles/sec)/(mile/hr).

50(10). Microwaves, which travel with the speed of light, are reflected from a distant airplane approaching the wave source. It is found that when the reflected waves are beat against the waves radiating from the source the beat frequency is 990 cycles/sec. If the microwaves are 0.10 meter in wavelength, what is the approach speed of the airplane?

51(10). Show that, for slow speeds, the Doppler shift can be written in the approximate form

$$\frac{\Delta\lambda}{\lambda} = \frac{u}{c},$$

where $\Delta\lambda$ is the change in wavelength.

52(10). In the experiment of Ives and Stilwell the speed u of the hydrogen atoms in a particular run was 8.61×10^5 meters/sec. Calculate $(\Delta\nu_1 + \Delta\nu_2)/\nu$, on the assumptions that (*a*) Eq. 35–21a is correct and (*b*) that Eq. 35–19a is correct. Compare your results with those given in Table 35–2 for this speed. Retain the first three terms only in these equations.

53(10). An earth satellite, transmitting on a frequency of 40×10^6 cycles/sec (exactly), passes directly over a radio receiving station at an altitude of 250 miles and at a speed of 18,000 miles/hr. Plot the change in frequency, attributable to the Doppler effect, as a function of time, counting $t = 0$ as the instant the satellite is over the station. (Hint: The speed u in the Doppler formula is not the actual speed of the satellite but its component in the direction of the station. Use the nonrelativistic formula (Eq. 35–19a) and neglect the curvature of the earth and of the satellite orbit.)

CHAPTER 36

Geometrical Optics

36–1 Geometrical Optics

In Chapter 35 we saw that light is an electromagnetic wave. It is a characteristic of waves of all kinds that, under most conditions, they do not travel in straight lines. We can hear sound waves around a corner. Ocean waves meeting an obstacle bend around it and meet on the other side. In a ripple tank, water waves that meet an opening in a barrier flare out as they pass through; see Fig. 37–2.

In certain circumstances, however, waves do travel in straight lines to a high degree of accuracy. We need only to agree not to place in the path of the wave any obstacle or aperature (mirror, lens, slit, baffle, etc.) unless its dimensions are much larger than the wavelength. Also, we must agree not to look too closely into what goes on at the edges of the obstacle or aperature. This special case, which we call *geometrical optics*, is the subject of this chapter. We will represent the straight lines in which light is traveling by *rays* which are lines perpendicular to the wave fronts. We can, in fact, ignore the wave nature of light completely (see, however, Section 36–3) and concentrate only on the rays.

In later chapters we will remove these restrictions on the dimensions of obstacles and deal with the more general case of *wave optics*.

36–2 Reflection and Refraction—Plane Waves and Plane Surfaces

In Fig. 36–1a a plane light wave falls on a plane water surface. The light beam is both reflected from the surface and bent (that is, *refracted*) as it enters the water. The incident beam is represented in Fig. 36–1b by a line, the *incident ray,* parallel to the direction of propagation. The incident wave is assumed in Fig. 36–1b to be a plane wave, the wavefronts being normal to the incident ray. The reflected and refracted waves

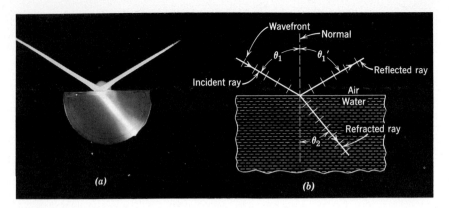

FIGURE 36–1 (*a*) A photograph showing reflection and refraction of plane waves at a (plane) air-water interface. (*b*) A representation using rays.

are also represented by rays. The angles of *incidence* (θ_1), of *reflection* (θ_1'), and of *refraction* (θ_2) are measured between the normal to the surface and the appropriate ray, as shown in the figure.

The laws governing reflection and refraction can easily be found from experiment:

1. The reflected and the refracted rays lie in the plane formed by the incident ray and the normal to the surface at the point of incidence, that is, the plane of Fig. 36–1*b*.

2. For reflection:

$$\theta_1' = \theta_1. \tag{36–1}$$

3. For refraction:

$$\frac{\sin \theta_1}{\sin \theta_2} = n_{21}, \tag{36–2}$$

where n_{21} is a constant called the *index of refraction* of medium 2 with respect to medium 1. Table 36–1 shows the indices of refraction for some common substances with respect to a vacuum for a particular wavelength.

The index of refraction of one medium with respect to another generally varies with wavelength, as Fig. 36–2 shows. Because of this fact refraction, unlike reflection, can be used to analyze light into its component wavelengths.

Table 36–1 SOME INDICES OF REFRACTION* (FOR $\lambda = 5890$ A—SODIUM YELLOW LIGHT)

Medium	Index of Refraction	Medium	Index of Refraction
Air	1.0003	Glass, zinc crown	1.52
Carbon disulfide	1.63	Polyethylene	1.52
Diamond	2.42	Quartz, fused	1.46
Glass, heaviest flint	1.89	Sapphire	1.77
Glass, light barium flint	1.58	Sodium chloride	1.53
		Water	1.33

* Measured with respect to a vacuum. The index with respect to air (except, of course, the index of air itself) will be negligibly different in most cases.

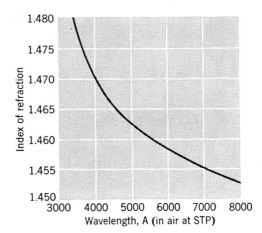

FIGURE 36–2   The index of refraction of fused quartz at 18° C, with respect to a vacuum.

The law of reflection was known to Euclid. The law of refraction was discovered by Willebrod Snell (1591–1626) and deduced from an early corpuscular theory of light by René Descartes (1596–1650); see also Problem 16.

The laws of reflection and refraction can be derived from Maxwell's equations. They hold for all regions of the electromagnetic spectrum.

Example 1. An incident ray in air falls on the plane surface of a block of quartz and makes an angle of 30° with the normal. This wave contains two wavelengths, 4000 and 5000 A. The indices of refraction for quartz with respect to air (n_{qa}) at these wavelengths are 1.4702 and 1.4624, respectively; see Fig. 36–2 for the index with respect to a vacuum. What is the angle between the two refracted rays?

From Eq. 36–2 we have, for the 4000-A ray,

$$\sin \theta_1 = n_{qa} \sin \theta_2,$$

or

$$\sin 30° = (1.4702) \sin \theta_2,$$

which leads to

$$\theta_2 = 19.88°.$$

For the 5000-A ray we have

$$\sin 30° = (1.4624) \sin \theta_2',$$

or

$$\theta_2' = 19.99°.$$

The angle $\Delta\theta$ between the rays is 0.11°, the shorter wavelength component being bent through the larger angle, that is, having the smaller angle of refraction.

Example 2. An incident ray falls on one face of a glass prism in air as in Fig. 36–3. The angle θ is so chosen that the emerging ray also makes an angle θ with the normal to the other face. Derive an expression for the index of refraction of the prism material with respect to air.

Note that $\angle abc = \alpha$, the two angles having their sides mutually perpendicular. Therefore

$$\alpha = \tfrac{1}{2}\phi. \qquad (36\text{–}3)$$

The *deviation angle* ψ is the sum of the two opposite interior angles in triangle *aed*, or

$$\psi = 2(\theta - \alpha).$$

Substituting $\tfrac{1}{2}\phi$ for α and solving for θ yields

$$\theta = \tfrac{1}{2}(\psi + \phi). \qquad (36\text{–}4)$$

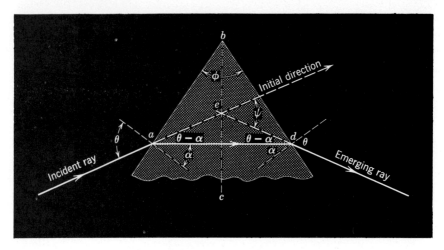

FIGURE 36-3 Example 3.

At point a, θ is the angle of incidence and α the angle of refraction. The law of refraction (see Eq. 36-2) is

$$\sin \theta = n_{ga} \sin \alpha,$$

in which n_{ga} is the index of refraction of the glass with respect to air.

From Eqs. 36-3 and 36-4 this yields

$$\sin \frac{\psi + \phi}{2} = n_{ga} \sin \frac{\phi}{2}$$

or

$$n_{ga} = \frac{\sin \frac{1}{2}(\psi + \phi)}{\sin (\phi/2)},$$

which is the desired relation. This equation holds only for θ so chosen that the light ray passes symmetrically through the prism.

36-3 Huygens' Principle

The Dutch physicist Christian Huygens (1629–1695) in 1678 proposed a wave theory for light, which, among other things, accounted for the laws of reflection and refraction. While much less comprehensive than Maxwell's theory, it was simpler mathematically and was useful for many years. It remains useful today for pedagogic and certain other practical purposes.

Huygens' theory simply assumes that light is a wave rather than, say, a stream of particles. It says nothing about the nature of the wave and, in particular—since Maxwell's theory of electromagnetism appeared only after the lapse of a century—gives no hint of the electromagnetic character of light. Huygens did not know whether light was a transverse wave or a longitudinal one; nor did he know the wavelengths of visible light.

The theory is based on a geometrical construction, called *Huygens' principle*, that allows us to tell where a given wavefront will be at any time in the future if we know its present position; it is: *All points on a wavefront can be considered as point sources for the production of spherical secondary wavelets. After a time t the new position of the wavefront will be the surface of tangency to these secondary wavelets.*

We illustrate this by a trivial example: Given a wavefront, ab in Fig. 36–4, in a plane wave in free space, where will the wavefront be a time t later? Following

Huygens' principle, we let several points on this plane (see dots) serve as centers for secondary spherical wavelets. In a time t the radius of these spherical waves is ct, where c is the speed of light in free space. The plane of tangency to these spheres at time t is represented by de. As we expect, it is parallel to plane ab and a perpendicular distance ct from it. Thus plane wavefronts are propagated as planes and with speed c. Note that the Huygens method involves a three-dimensional construction and that Fig. 36–4 is the intersection of this construction with the plane of the page.

36–4   The Law of Refraction

We shall now use Huygens' principle to derive the law of refraction, Eq. 36–2. In Problem 6 you are asked to use it to derive the law of reflection.

Figure 36–5 shows four stages in the refraction of three wavefronts in a plane wave falling on a plane interface between air (medium 1) and glass (medium 2). We choose the wavefronts in the incident beam to be separated, arbitrarily, by λ_1, the wavelength in medium 1. Let the speed of light in air be v_1 and that in glass be v_2. We assume that $v_2 < v_1$, which happens to be true.

The wavefronts in Fig. 36–5a are related in the same way as those in Fig. 36–4 which show the Huygens construction. The angle of incidence is θ_1. The time ($= \lambda_1/v_1$) for a Huygens wavelet to expand from point e to include point c will equal the time ($= \lambda_2/v_2$) for a wavelet in the glass to expand at the reduced speed v_2 from h to include e'. By equating these times we many obtain wavelength λ_2,

$$\lambda_2 = \lambda_1 \frac{v_2}{v_1} \tag{36–5}$$

which, you can see, is less than that in air.

The refracted wavefront must be tangent to an arc of radius λ_2 centered on h. Since c lies on the new wavefront the tangent must pass through this point also. Note that θ_2, the angle between the refracted wavefront and the air–glass interface, is the same

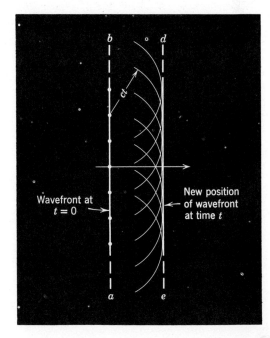

FIGURE 36–4 The propagation of a plane wave in free space is described by the Huygens construction. Note that the ray (horizontal arrow) representing the wave is perpendicular to the wavefronts.

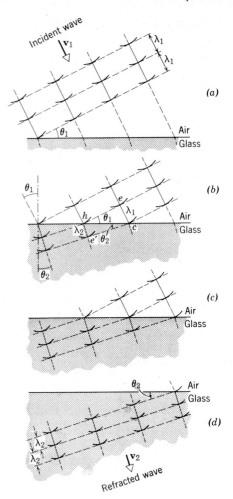

FIGURE 36–5 The refraction of a plane wave as described by the Huygens construction; the reflected wave is omitted for simplicity. Note the change in wavelength on refraction.

as the angle of refraction (that between the refracted ray and the normal to this interface).

For the right triangles *hce* and *hce'* we may write

$$\sin \theta_1 = \frac{\lambda_1}{hc} \qquad \text{(for } hce\text{)}$$

and

$$\sin \theta_2 = \frac{\lambda_2}{hc} \qquad \text{(for } hce'\text{)}.$$

Dividing and using Eq. 36–5 yields

$$\frac{\sin \theta_1}{\sin \theta_2} = \frac{\lambda_1}{\lambda_2} = \frac{v_1}{v_2} \tag{36–6}$$

Since v_1 and v_2 are constants, Eq. 36–6 is just the law of refraction (Eq. 36–2) provided we put

$$n_{21} = \frac{v_1}{v_2}. \tag{36–7}$$

We now define the index of refraction n of a medium with respect to a vacuum, where v equals c. From Eq. 36–7 we have $n = c/v$. From Eq. 36–5 we may then express the wavelength in the medium as

$$\lambda_n = \frac{\lambda}{n} \tag{36–8}$$

where λ is the wavelength in vacuum.

Multiplying Eq. 36–7 by c/c gives the relative index of refraction between any two media as $n_{21} = n_2/n_1$, and the law of refraction becomes

$$n_1 \sin \theta_1 = n_2 \sin \theta_2. \tag{36–9}$$

Referring to Fig. 36–5b, we saw how two rays moving different distances, λ_1 in medium 1 and λ_2 in medium 2, nevertheless remained in phase; in other words the same number of wavelengths (in this case one) were contained in these distances. From Eq. 36–8 we can see that $n_1\lambda_1 = n_2\lambda_2 = \lambda$. This suggests that we define a quantity nl to be the *optical path length* of radiation traveling distance l in a medium of index of refraction n. This quantity, in effect, measures the number of wavelengths contained in distance l and is a useful quantity wherever phase differences between rays traveling in different media must be considered. The number of wavelengths contained in distance l is l/λ_n which, using Eq. 36–8 equals nl/λ. Thus, since λ is the fixed free-space wavelength, a condition that would insure that the same number of wavelengths are contained within distances l_1 and l_2 in two different media is

$$n_1 l_1 = n_2 l_2. \tag{36–10}$$

Equality of the optical path lengths thus implies no relative change of phase.

36–5 Total Internal Reflection

Let rays in an optically dense medium (glass, say) fall on a plane surface on the other side of which is a less optically dense medium (air, say); see Fig. 36–6. As the angle of incidence θ is increased, we reach a situation (see ray e) at which the refracted ray points along the surface, the angle of refraction being 90°. For angles of incidence larger than this *critical angle* θ_c there is no refracted ray, and we speak of *total internal reflection*.

The critical angle is found by putting $\theta_2 = 90°$ in the law of refraction (see

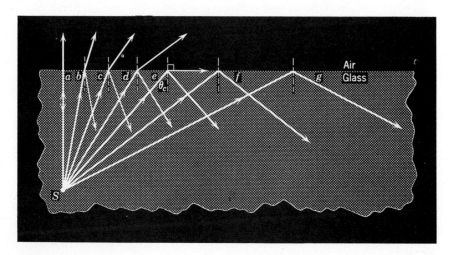

FIGURE 36–6 Showing the total internal reflection of light from a source S; the critical angle is θ_c.

Eq. 36–9):

$$n_1 \sin \theta_c = n_2 \sin 90°,$$

or
$$\sin \theta_c = \frac{n_2}{n_1}. \tag{36–11}$$

For glass and air, $\sin \theta_c = (1.00/1.50) = 0.667$, which yields $\theta_c = 41.8°$. Total internal reflection can not occur when light originates in the medium of lower index of refraction.

▨   **Example 3.** Figure 36–7 shows a triangular prism of glass, a ray incident normal to one face being totally reflected. If θ_1 is $45°$, what can you say about the index of refraction n of the glass?

The angle θ_1 must be equal to or greater than the critical angle θ_c where θ_c is given by Eq. 36–11:

$$\sin \theta_c = \frac{n_2}{n_1} = \frac{1}{n},$$

in which, for all practical purposes, the index of refraction of air ($= n_2$) is set equal to unity. Suppose that the index of refraction of the glass is such that total internal reflection just occurs, that is, that $\theta_c = 45°$. This would mean

$$n = \frac{1}{\sin 45°} = 1.41.$$

Thus the index of refraction of the glass must be equal to or larger than 1.41. If it were less, total internal reflection would not occur.   ▨

36–6   Brewster's Law

The laws of reflection and refraction give us information about the direction of reflected or refracted rays. They don't say anything about the intensities of these rays, except for the fact that the refracted beam intensity is zero for light striking at an angle equal to or greater than the critical angle. The complete intensity relationships can be derived from Maxwell's equations.* The derivation is somewhat difficult; however, from it you will find that the reflection coefficient (the ratio of the reflected intensity to the incident intensity) not only depends upon the angle of incidence but also upon the direction of polarization of the incident beam. Similarly for the refracted beam.

Malus discovered in 1809 that light can be partially or completely polarized by reflection. Anyone who has watched the sun's reflection in water, while wearing a pair of sunglasses made of polarizing sheet, has probably noticed the effect. You need only to tilt your head from side to side, thus rotating the polarizing sheets, to see that the intensity of the reflected sunlight passes through a minimum.

Figure 36–8 shows an unpolarized beam falling on a glass surface. The **E**-vector for each wavetrain in the beam can be resolved into two components, one perpendicular to the plane of incidence (the plane of Fig. 36–8), represented by dots, and one lying in this plane, represented by arrows. We shall call

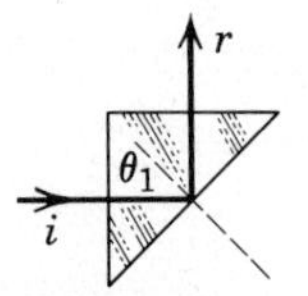

* See Chapter 11 "Plane Waves—The Influence of the Medium" in *Intermediate Electromagnetic Theory* by W. M. Schwarz, John Wiley and Sons, New York, 1964.

FIGURE 36–7   Example 3.

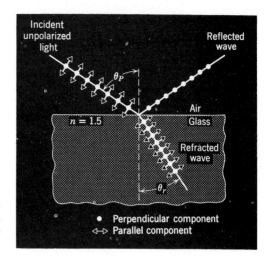

FIGURE 36–8 For a particular angle of incidence θ_p, the reflected light is completely polarized, as shown. The transmitted light is partially polarized.

them the *perpendicular* and *parallel components,* respectively. For completely unpolarized incident light, these two components are of equal amplitude.

Experimentally, for glass or other dielectric materials, there is a particular angle of incidence, called the *polarizing angle* θ_p, at which the reflection coefficient for the parallel component is zero. This means that the beam reflected from the glass is plane-polarized, with its plane of vibration at right angles to the plane of incidence. This polarization of the reflected beam can easily be verified by analyzing it with a polarizing sheet. The refracted beam at this incident angle still has both components, although they are not of equal amplitude.

At the polarizing angle we find experimentally that the reflected and the refracted beams are at right angles, or (Fig. 36–8)

$$\theta_p + \theta_r = 90°.$$

From the law of refraction,

$$n_1 \sin \theta_p = n_2 \sin \theta_r.$$

Combining these equations leads to

$$n_1 \sin \theta_p = n_2 \sin (90° - \theta_p) = n_2 \cos \theta_p$$

or
$$\tan \theta_p = \frac{n_2}{n_1}, \tag{36–12}$$

where the incident ray is in medium one and the refracted ray in medium two. We can write this as

$$\tan \theta_p = n, \tag{36–13}$$

where $n \; (= n_2/n_1)$ is the index of refraction of medium two with respect to medium one.* Equation 36–13 is known as *Brewster's law* after Sir David Brewster (1781–1868), who deduced it empirically in 1812. It is possible to prove this law rigorously from Maxwell's equations.

---

* We will often drop the subscript on n_{21} if it is clear in which medium the incident ray travels.

Example 4. We wish to use a plate of glass ($n = 1.50$) as a polarizer. What is the polarizing angle? What is the angle of refraction?

From Eq. 36–13,

$$\theta_p = \tan^{-1} 1.50 = 56.3°.$$

The angle of refraction follows from Snell's law:

$$(1) \sin \theta_p = n \sin \theta_r$$

or

$$\sin \theta_r = \frac{\sin 56.3°}{1.50} = 0.555 \qquad \theta_r = 33.7° (= 90° - \theta_p).$$

36–7   Spherical Waves—Plane Mirror

In Section 36–3 we considered a special case of the reflection and refraction of light waves; that of plane waves falling upon plane surfaces. We will now proceed in stages to more complex situations culminating in a discussion of refraction in thin lenses. The next topic is spherical waves falling upon a plane mirror.

Figure 36–9 shows a point source of light O, the *object,* placed at a distance o in front of a plane mirror. The light falls on the mirror as a spherical wave represented in the figure by rays emanating from O. At the point in Fig. 36–9 at which each ray from O strikes the mirror we construct a reflected ray. If we extend the reflected rays backward, they intersect in a point I which is the same distance behind the mirror that the object O is in front of it; I is called the *image* of O.

Images may be *real* or *virtual.* In a real image light actually passes through the image point; in a virtual image the light behaves as though it diverges from the image point even though it does not pass through this point; see Fig. 36–9. Images in plane mirrors are always virtual. We know from daily experience how "real" such a virtual image appears to be and how definite is its location in the space behind the mirror, even though this space may, in fact, be occupied by a brick wall.

Figure 36–10 shows two rays from Fig. 36–9. One strikes the mirror at v, along a

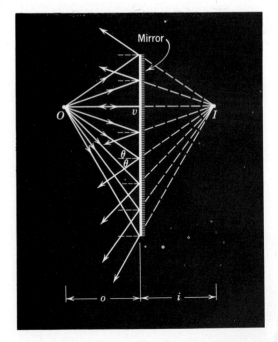

FIGURE 36–9   A point object O forms a virtual image I in a plane mirror. The rays appear to emanate from I, but actually light energy does not pass through this point.

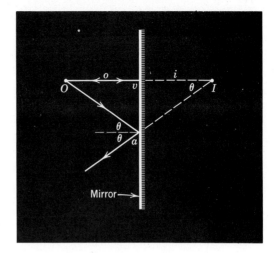

FIGURE 36–10   Two rays from Fig. 36–9; ray *Oa* makes an arbitrary angle θ with the normal.

perpendicular line. The other strikes it at an arbitrary point *a*, making an angle of incidence θ with the normal at that point. Elementary geometry shows that the right triangles *aOva* and *aIva* are congruent and thus

$$o = -i, \tag{36–14}$$

in which the minus sign is arbitrarily introduced to show that *I* and *O* are on opposite sides of the mirror. Equation 36–14 does not involve θ, which means that all rays striking the mirror pass through *I* when extended backward, as we have claimed above. Beyond assuming that the mirror is truly plane and that the conditions for geometrical optics hold, we have made no approximations in deriving Eq. 36–14. A point object produces a point image in a plane mirror, with $o = -i$, no matter how large the angle θ in Fig. 36–10.

Because of the finite diameter of the pupil of the eye, only rays that lie fairly close together can enter the eye after reflection at a mirror. For the eye position shown in Fig. 36–11 only a small patch of the mirror near point *a* is used in forming the image. If we move our eye to another location, a different patch of the mirror will be used;

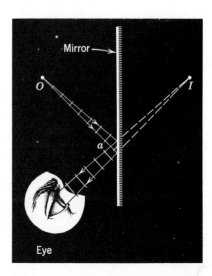

FIGURE 36–11   A pencil of rays from *O* enters the eye after reflection at the mirror. Only a small portion of the mirror near *a* is effective. The small arcs represent portions of spherical wavefronts. The eye "thinks" that the point light source is at *I*.

the location of the virtual image *I* will remain unchanged, however, as long as the object remains fixed.

If the object is an extended source such as the head of a person, a virtual image is also formed. From Eq. 36–14, every point of the source has an image point that lies an equal distance directly behind the plane of the mirror. Thus the image reproduces the object point by point.

Images in plane mirrors differ from objects in that left and right are interchanged. The image of a printed page is different from the page itself. Similarly, if a top is made to spin clockwise, the image, viewed in a vertical mirror, will seem to spin counterclockwise. Figure 36–12 shows an image of a left hand, constructed by using point-by-point application of Eq. 36–14; the image has the symmetry of a right hand.

Example 5. How tall must a vertical mirror be if a person 6.0 ft high is to be able to see his entire length? Assume that his eyes are 4.0 in. below the top of his head.

Figure 36–13 shows the paths followed by two of the many light rays leaving the top of the man's head and the tips of his toes. These rays, chosen so that they will enter the eye *e* after reflection, strike the vertical mirror at points *a* and *b*, respectively. The mirror need occupy only the region between these two points. Calculation shows that *b* is 2 ft, 10 in. and *a* is 5 ft, 10 in. above the floor. The length of the mirror is thus 3.0 ft, or half the height of the person. Note that this height is independent of the distance between the person and the mirror. Mirrors that extend below point *b* show reflections of the floor between the person and the mirror.

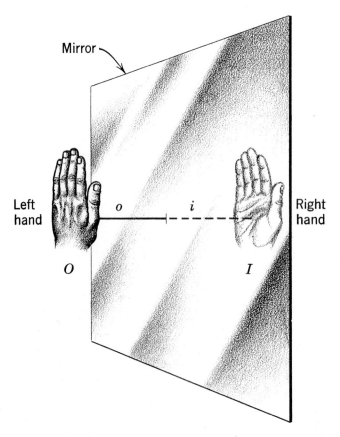

FIGURE 36–12   A plane mirror reverses right and left.

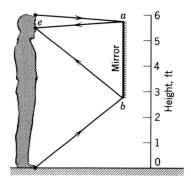

FIGURE 36–13   Example 5. A person can view his full-length image in a mirror that is only half his height.

36–8 Spherical Waves—Spherical Mirror

In Fig. 36–14 a spherical light wave from a point object O falls on a concave spherical mirror whose radius of curvature is r. (We will always judge whether a spherical mirror is concave ["caved in"] or convex by sighting along an incident ray.) A line through O and the center of curvature C makes a convenient reference axis.

A ray from O that makes an arbitrary angle α with this axis intersects the axis at I after reflection from the mirror at a. A ray that leaves O along the axis will be reflected back along itself at v and will also pass through I. Thus, for these two rays at least, I is the image of O; it is a real image because light actually passes through I. Let us find the location of I.

A useful theorem is that the exterior angle of a triangle is equal to the sum of the two opposite interior angles. Applying this to triangles $OaCO$ and $OaIO$ in Fig. 36–14 yields

$$\beta = \alpha + \theta$$

and

$$\gamma = \alpha + 2\theta.$$

Eliminating θ between these equations leads to

$$\alpha + \gamma = 2\beta. \tag{36-15a}$$

In radian measure we can write angles α, β, and γ as

$$\alpha \cong \frac{av}{vO} = \frac{av}{o}$$

$$\beta = \frac{av}{vC} = \frac{av}{r} \tag{36-15b}$$

$$\gamma \cong \frac{av}{vI} = \frac{av}{i}.$$

Note that only the equation for β is exact, because the center of curvature of arc av is at C and not at O or I. However, the equations for α and for γ are approximately correct if these angles are small enough. In all that follows we assume that the rays diverging from the object make only a small angle α with the axis of the mirror. We did not find it necessary to make such an assumption for plane mirrors. Substituting these equations into Eq. 36–15a and canceling av yields

$$\frac{1}{o} + \frac{1}{i} = \frac{2}{r}, \tag{36-16}$$

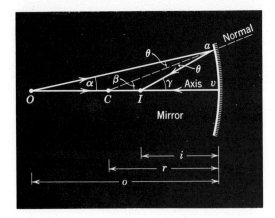

FIGURE 36–14 Two rays from O converge after reflection in a spherical concave mirror, forming a real image at I. In this case, o, i, and r are all positive in Eq. 36–15.

in which o is the object distance and i is the image distance. Both these distances are measured from the *vertex* of the mirror, which is the point v at which the axis intercepts the mirror.

Significantly, Eq. 36–16 does not contain α (or β, γ, or θ), so that it holds for all rays that strike the mirror provided that they make small enough angles with the axis. In an actual case we can insure this by putting a small enough circular diaphragm in front of the mirror, centered about the vertex v; this will impose a certain maximum value of α.

As we let α in Fig. 36–14 become larger, it will become less true that a point object will form a point image; the image will become extended and fuzzy.

Although we derived Eq. 36–16 for the special case in which the object is located beyond the center of curvature, it is generally true, no matter where the object is located. It is also true for convex mirrors, as in Fig. 36–15.

In applying Eq. 36–16 to such general cases we must follow a consistent convention of signs for o, i, and r. We start by fixing our minds on the side of the mirror from which incident light comes. Because mirrors are opaque, the light, after reflection, must remain on this side, and if an image is formed here it will be a real image. Therefore, we call the side of the mirror from which the light comes the *R-side* (for real image). We call the back of the mirror the *V-side* (for virtual image), because images formed on this side of the mirror must be virtual, no light energy being present on this side.

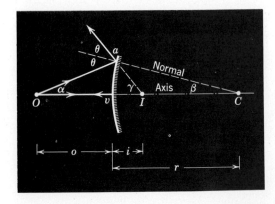

FIGURE 36–15 Two rays from O diverge after reflection in a spherical convex mirror, forming a virtual image at I, the point from which they appear to originate. In this case, o is positive (diverging light), but i and r are negative in Eq. 36–15. Compare Fig. 36–14.

Next, you will find it helpful to start from Fig. 36–14 and, with the aid of a few rough sketches, see what happens as you change the mirror slowly but continuously from concave to convex. This means that you must increase the radius of curvature r steadily, point c moving toward the left. Convince yourself that, if $r > 0$ in Fig. 36–14, the image becomes virtual and I flops over to the V-side.

As the mirror becomes plane, $r \to \infty$, the image remaining virtual as in Fig. 36–9. As soon as the mirror becomes convex, C flops over to the V-side (see Fig. 36–15) and, as you make the mirror more and more convex, C moves closer toward the mirror.

The following sign convention holds throughout this range:

> The image distance i is positive if the image (real) lies on the R-side of the mirror; i is negative if the image (virtual) lies on the V-side. The radius of curvature r is positive if the center of curvature of the mirror lies on the R-side; r is negative if the center of curvature lies on the V-side.

Throughout this book we will always deal with light diverging from a real object as in Figs. 36–14, 15. It is possible, however, to let converging light fall on a mirror; (reverse the directions of all rays in Fig. 36–15, for example.) In such cases the object is called virtual and we take o as negative in Eq. 36–16.

Example 6. A convex mirror has a radius of curvature of 20 cm. If a point source is placed 14 cm away from the mirror, as in Fig. 36–15, where is the image?

A rough graphical construction, applying the law of reflection at a in the figure, shows that the image will be on the V-side of the mirror and thus will be virtual. We may verify this from Eq. 36–16, noting that r is negative here because the center of curvature of the mirror is on its V-side, as it is for all convex mirrors. We have

$$\frac{1}{o} + \frac{1}{i} = \frac{2}{r}$$

or

$$\frac{1}{+14 \text{ cm}} + \frac{1}{i} = \frac{2}{-20 \text{ cm}},$$

which yields $i = -5.8$ cm, in agreement with the graphical prediction. The negative sign for i reminds us that the image is on the V-side of the mirror and thus is virtual.

When parallel light falls on a mirror (Fig. 36–16), the image point (real or virtual) is called the *focal point* F of the mirror. The focal length f is the distance between F and the vertex. If we put $o \to \infty$ in Eq. 36–16, thus insuring parallel incident light, we have

$$i = \tfrac{1}{2}r = f.$$

Equation 36–16 can then be rewritten

$$\frac{1}{o} + \frac{1}{i} = \frac{1}{f}, \tag{36–17}$$

where f, like r, is taken as positive for mirrors whose centers of curvature are on the R-side (that is, for concave; see Fig. 36–16*a*) and negative for those whose centers of curvature are on the V-side (that is, for convex mirrors; see Fig. 36–16c). Figure 36–16b shows an incident plane wave that makes a small angle α with the mirror axis. The rays are focused at a point in the *focal plane* of the mirror. This is a plane at right angles to the mirror axis at the focal point.

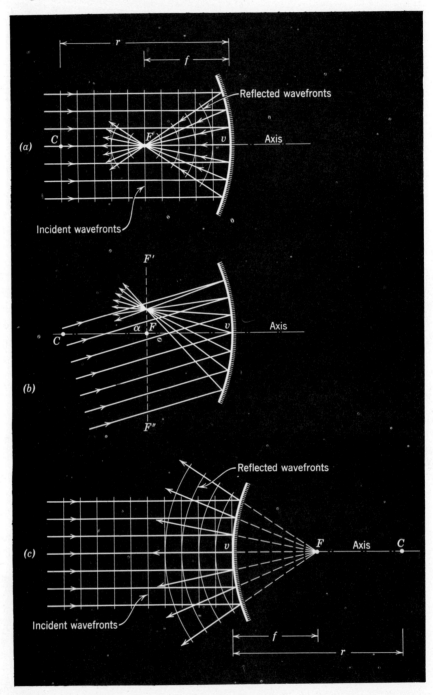

FIGURE 36–16   (a) The focal point for a concave spherical mirror, showing both the rays and the wavefronts. *F* and *C* lie on the R-side, the focal point is real, and the focal length *f* of the mirror is positive (as is *r*). (b) The same, except that the incident light makes an angle α with the mirror axis; the rays are focused at a point in the focal plane *F'F''*. (c) Same as (a) except that the mirror is convex; *F* and *C* lie on the V-side of the mirror. The focal point is virtual and the focal length *f* is negative (as is *r*).

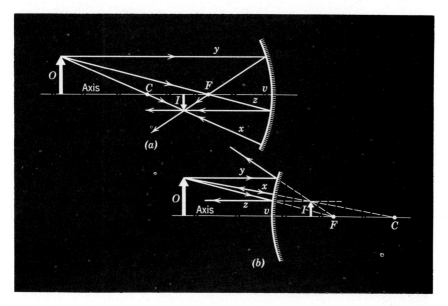

FIGURE 36–17   The image of an extended object in (*a*) a concave mirror and (*b*) a convex mirror is located graphically. Any two of the three special rays shown are sufficient. In (*a*) the image is *inverted;* in (*b*) it is *erect.*

We now consider objects that are not points. Figure 36–17 shows a luminous arrow in front of (*a*) a concave mirror and (*b*) a convex mirror. We choose to draw the mirror axis through the foot of the luminous arrow and, of course, through the center of curvature. We can find the image of any off-axis point, such as the tip of the luminous arrow, graphically by using the following facts:

1. A ray that strikes the mirror after passing (either directly or upon being extended) through the center of curvature *C* returns along itself (ray *x* in Fig. 36–17). Such rays strike the mirror at right angles.

2. A ray that strikes the mirror parallel to its axis passes (or will pass when extended) through the focal point (ray *y*).

3. A ray that strikes the mirror after passing (either directly or upon being extended) through the focal point emerges parallel to the axis (ray *z*).

Figure 36–18 shows a ray (*dve*) that originates on the tip of the object arrow of Fig. 36–17*a*, is reflected from the mirror at point *v*, and passes through the tip of the image arrow. The law of reflection demands that this ray make equal angles θ with the

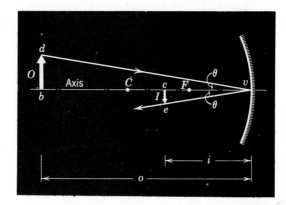

FIGURE 36–18   A particular ray for the arrangement of Fig. 36–17, used to show that the lateral magnification *m* is given by $-i/o$.

mirror axis as shown. For the two similar right triangles in the figure we can write

$$\frac{ce}{bd} = \frac{vc}{vb}.$$

The quantity on the left (apart from a question of sign) is the *lateral magnification m* of the mirror. Since we want to represent an inverted image by a negative magnification, we arbitrarily define m for this case as $-(ce/bd)$. Since $vc = i$ and $vb = o$, we have at once

$$m = -\frac{i}{o}. \tag{36-18}$$

This equation gives the magnification for spherical and plane mirrors under all circumstances. For a plane mirror, $o = -i$ and the predicted magnification is $+1$ which, in agreement with experience, indicates an erect image the same size as the object.

36–9 Spherical Waves—Spherical Refracting Surface

Figure 36–19 shows a point source O near a convex spherical refracting surface of radius of curvature r. The surface separates two media whose indices of refraction differ, that of the medium in which the incident light falls on the surface being n_1 and that on the other side of the surface being n_2.

From O we draw a line through the center of curvature C of the refracting surface, thus establishing a convenient axis which intercepts the surface at vertex v. From O we also draw a ray that makes a small but arbitrary angle α with the axis and strikes the refracting surface at a, being refracted according to

$$n_1 \sin \theta_1 = n_2 \sin \theta_2.$$

The refracted ray intersects the axis at I. A ray from O that travels along the axis will not be bent on entering the surface and will also pass through I. Thus, for these two rays at least, I is the image of O.

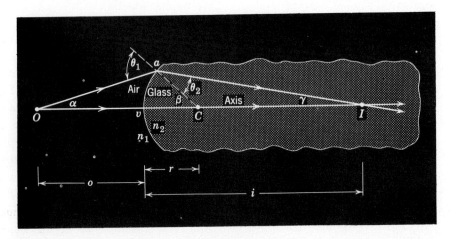

FIGURE 36–19 Two rays from O converge after refraction at a spherical surface, forming a real image at I. In this case o is positive (diverging light), in Eq. 36–24; i and r are also positive.

As in the derivation of the mirror equation, we use the theorem that the exterior angle of a triangle is equal to the sum of the two opposite interior angles. Applying this to triangles *COaC* and *ICaI* yields

$$\theta_1 = \alpha + \beta \tag{36–19}$$

and
$$\beta = \theta_2 + \gamma. \tag{36–20}$$

As α is made small, angles β, γ, θ_1, and θ_2 in Fig. 36–19 also become small. We at once assume that α, hence all these angles, are arbitrarily small. We also made this assumption for spherical mirrors. Replacing the sines of the angles by the angles themselves—since the angles are required to be small—permits us to write the law of refraction as

$$n_1\theta_1 \cong n_2\theta_2. \tag{36–21}$$

Combining Eqs. 36–20 and 36–21 leads to

$$\beta = \frac{n_1}{n_2}\theta_1 + \gamma.$$

Eliminating θ_1 between this equation and Eq. 36–19 leads, after rearrangement, to

$$n_1\alpha + n_2\gamma = (n_2 - n_1)\beta. \tag{36–22}$$

In radian measure the angles α, β, and γ in Fig. 36–19 are

$$\alpha \cong \frac{av}{o}$$

$$\beta = \frac{av}{r} \tag{36–23}$$

$$\gamma \cong \frac{av}{i}.$$

Only the second of these equations is exact. The other two are approximate because *I* and *O* are not the centers of circles of which *av* is an arc. However, for α small enough we can make the inaccuracies in Eqs. 36–23 as small as we wish.

Substituting Eqs. 36–23 into Eq. 36–22 leads readily to

$$\frac{n_1}{o} + \frac{n_2}{i} = \frac{n_2 - n_1}{r}. \tag{36–24}$$

This equation holds whenever light is refracted at a spherical surface, assuming only that the incident rays make a small enough angle α with the axis. In particular, Eq. 36–24 must hold whether the refracting surface is convex (as in Fig. 36–19), plane (which means $r \to \infty$), or concave (as in Fig. 36–21). It also holds whether $n_2 > n_1$ (as in Figs. 36–19, 21) or $n_2 < n_1$ (as in Fig. 36–22).

Before stating the sign convention for *o*, *i*, and *r* in Eq. 36–24, let us fix our attention on the side of the refracting surface from which the incident light falls on the surface. In contrast to mirrors, the light passes *through* a refracting surface to the other side, and if an image is formed on the far side, which we call the R-side, it must be a real image. The side from which the incident light comes is called the V-side because images formed here must be virtual. Figure 36–20 suggests this important distinction between reflection and refraction.

We now simply state that, for all the conditions under which Eq. 36–24 holds, the sign convention is just the same as that for spherical mirrors. If we make the appropriate changes in nomenclature, we have (see indented paragraph on page 683):

> The image distance i is positive because the image (real) lies on the R-side of the refracting surface; i is negative if the image (virtual) lies on the V-side. The radius of curvature r is positive if the center of curvature of the refracting surface lies on the R-side; r is negative if the center of curvature lies on the V-side.

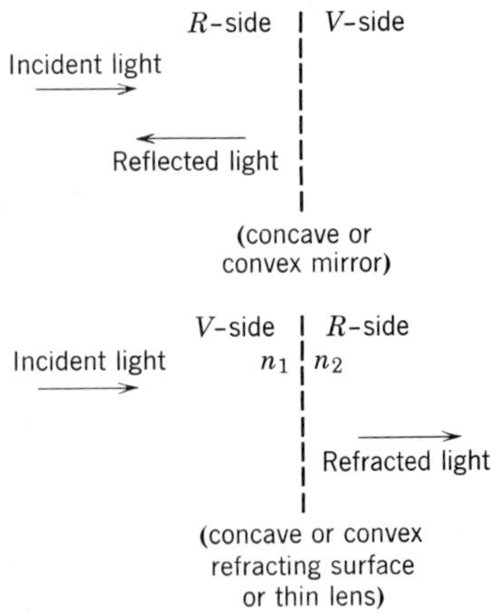

We can use the same sign convention as for mirrors because we have recognized the difference between mirrors and refracting surfaces in defining the concepts of R-side and V-side.

FIGURE 36–20 Real images are formed on the same side as the incident light for mirrors but on the opposite side for refracting surfaces and thin lenses. This is so because the incident light is reflected by mirrors but is transmitted by refracting surfaces and lenses.

In this book, again as for mirrors, we consider only the case of light falling on a refracting surface after *diverging* from a real object; in this case we take o as positive. For converging incident light (reverse all rays in Fig. 36–21 and interchange n_1 and n_2), the object is virtual and o is negative.

    Example 7. Locate the image for the geometry shown in Fig. 36–19, assuming the radius of curvature to be 10 cm, n_2 to be 2.0, and n_1 to be 1.0. Let the object be 20 cm to the left of v.

From Eq. 36–24,

$$\frac{n_1}{o} + \frac{n_2}{i} = \frac{n_2 - n_1}{r},$$

we have

$$\frac{1.0}{+20 \text{ cm}} + \frac{2.0}{i} = \frac{2.0 - 1.0}{+10 \text{ cm}}.$$

Note that r is positive because the center of curvature of the surface lies on the R-side. This relation yields $i = +40$ cm in agreement with the graphical construction. The light energy actually passes through I so that the image is real, as indicated by the positive sign for i.

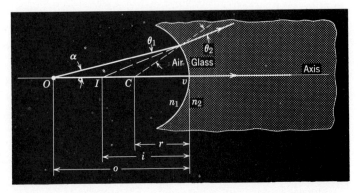

FIGURE 36–21 Two rays from O diverge after refraction at a spherical surface, forming a virtual image at I. In this case o is positive (diverging light), but i and r are negative in Eq. 36–24.

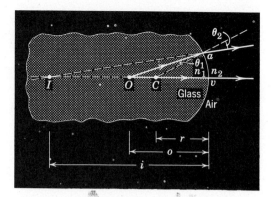

FIGURE 36–22 Two rays from O appear to originate from I (virtual image) after refraction at a spherical surface. In this case o is positive (diverging light) but i and r are negative in Eq. 36–4.

Example 8. An object is immersed in a medium with $n_1 = 2.0$, being 15 cm from the spherical surface whose radius of curvature is -10 cm, as in Fig. 36–22; r is negative because C lies on the V-side. Locate the image.

Figure 36–22 shows a ray traced through the surface by applying the law of refraction at point a. A second ray from O along the axis emerges undeflected at v. The image I is found by extending these two rays backward; it is virtual.

From Eq. 36–24,

$$\frac{n_1}{o} + \frac{n_2}{i} = \frac{n_2 - n_1}{r},$$

we have

$$\frac{2.0}{+15 \text{ cm}} + \frac{1.0}{i} = \frac{1.0 - 2.0}{-10 \text{ cm}},$$

which yields $i = -30$ cm, in agreement with Fig. 36–22 and with the sign conventions. Remember that n_1 always refers to the medium on the side of the surface from which the light comes.

36–10 Thin Lenses

In most refraction situations there is more than one refracting surface. This is true even for a spectacle lens, the light passing from air into glass and then from glass into air. In microscopes, telescopes, cameras, etc., there are often many more than two surfaces.

Figure 36–23a shows a thick glass "lens" of length l whose surfaces are ground to radii r' and r''. A point object O' is placed near the left surface as shown. A ray leaving O' along the axis is not deflected on entering or leaving the lens because it falls on each surface along a normal.

A second ray leaving O', at an arbitrary angle α with the axis, strikes the surface at point a', is refracted, and strikes the second surface at point a''. The ray is again refracted and crosses the axis at I'', which, being the intersection of two rays from O'', is the image of point O', formed after refraction at two surfaces.

Figure 36–23b shows the first surface, which forms a virtual image of O' at I'. To locate I', we use Eq. 36–24,

$$\frac{n_1}{o} + \frac{n_2}{i} = \frac{n_2 - n_1}{r}.$$

Putting $n_1 = 1.0$ and $n_2 = n$ and bearing in mind that the image distance is negative (that is, $i = -i'$ in Fig. 36–23b), we obtain

$$\frac{1}{o'} - \frac{n}{i'} = \frac{n-1}{r'}. \tag{36–25}$$

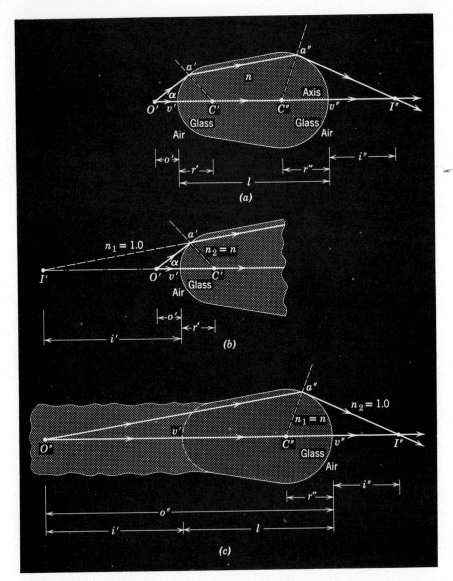

FIGURE 36-23 (*a*) Two rays from O' intersect at I'' (real image) after refraction at two spherical surfaces. (*b*) The first surface and (*c*) the second surface shown separately. The quantities α and n have been exaggerated for clarity.

In this equation i' will be a positive number because we have arbitrarily introduced the minus sign appropriate to a virtual image.

Figure 36–23*c* shows the second surface. Unless an observer at point a'' were aware of the existence of the first surface, he would think that the light striking that point originated at point I' in Fig. 36–23*b* and that the region to the left of the surface was filled with glass. Thus the (virtual) image I' formed by the first surface serves as a real object O'' for the second surface. The distance of this object from the second surface is

$$o'' = i' + l. \tag{36–26}$$

In applying Eq. 36–24 to the second surface, we insert $n_1 = n$ and $n_2 = 1$ because the object behaves as if it were imbedded in glass. If we use Eq. 36–26, Eq. 36–24 becomes

$$\frac{n}{i' + l} + \frac{1}{i''} = \frac{1 - n}{r''}.$$ (36–27)

Let us now assume that the thickness l of the "lens" in Fig. 36–23 is so small that we can neglect it in comparison with other linear quantities in this figure (such as o', i', o'', i'', r', and r''). In all that follows we make this *thin-lens approximation*. Putting $l = 0$ in Eq. 36–27 leads to

$$\frac{n}{i'} + \frac{1}{i''} = -\frac{n - 1}{r''}.$$ (36–28)

Adding Eqs. 36–25 and 36–28 leads to

$$\frac{1}{o'} + \frac{1}{i''} = (n - 1)\left(\frac{1}{r'} - \frac{1}{r''}\right).$$

Finally, calling the original object distance simply o and the final image distance simply i leads to

$$\frac{1}{o} + \frac{1}{i} = (n - 1)\left(\frac{1}{r'} - \frac{1}{r''}\right).$$ (36–29)

This equation holds only for rays that make small angles with the axis and only if the lens is so thin that it essentially makes no difference from which surface of the lens we measure the quantities o and i. In Eq. 36–29, r' refers to the first surface struck by the light as it traverses the lens and r'' to the second surface.

The sign conventions for Eq. 36–29 are the same as those for mirrors and for single refracting surfaces. Because the lens is assumed to be thin, we refer to the R-side and the V-side of the lens itself (see Fig. 36–20) rather than those of its separate surfaces. The sign conventions then are the following:

> The image distance i is positive if the image (real) lies on the R-side of the lens; i is negative if the image (virtual) lies on the V-side of the lens. The radii of curvature r' and r'' are positive if their respective centers of curvature lie on the R-side of the lens; they are negative if they lie on the V-side. See Fig. 36–24.

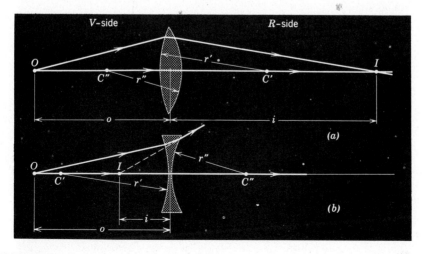

FIGURE 36–24   In (a), o is positive (diverging light) as are i and r'; r'' is negative. In (b), o is positive (diverging light) as is r''; r' and i are negative.

As for mirrors and single refracting surfaces, we restrict ourselves to diverging light from real objects, in which case o is positive.

Figure 36–25a and c shows parallel light from a distant object falling on a thin lens. The image location is called the *second focal point* F_2 of the lens. The distance from F_2 to the lens is called the *focal length f*. The *first focal point* for a thin lens (F_1 in figure) is the object position for which the image is at infinity. For thin lenses the first and second focal points are on opposite sides of the lens and are equidistant from it.

The focal length can be computed from Eq. 36–29 by inserting $o \to \infty$ and $i = f$. This yields

$$\frac{1}{f} = (n - 1) \left(\frac{1}{r'} - \frac{1}{r''} \right). \tag{36–30}$$

This relation is called the *lens maker's equation* because it allows us to compute the focal length of a lens in terms of the radii of curvature and the index of refraction of the material. Combining Eqs. 36–29 and 36–30 allows us to write the thin-lens equa-

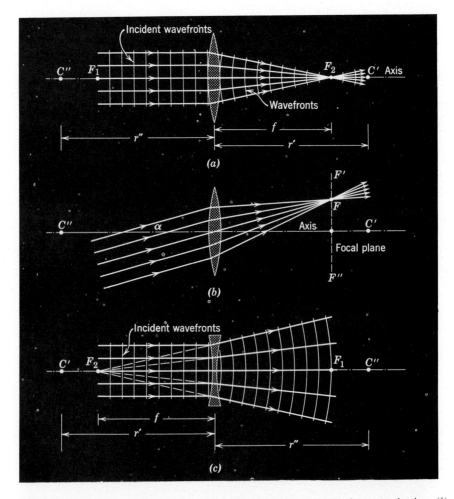

FIGURE 36–25   (*a*) Parallel light passes through the second focal point F_2 of a converging lens. (*b*) The incident light makes an angle α with the lens axis, the rays being focused in the focal plane $F'F''$. (*c*) Parallel light, passing through a diverging lens, seems to originate at the second focal point F_2. C' and C'' are centers of curvature for the lens surfaces; F_1 is the first focal point. A lens shaped like that in (*a*) is called *converging;* one shaped like that in (*b*) is called *diverging;* however, see Question 29.

tion as

$$\frac{1}{o} + \frac{1}{i} = \frac{1}{f}. \tag{36-31}$$

Figure 36–25b shows parallel incident rays that make a small angle α with the lens axis; they are brought to a focus in the *focal plane F'F''*, as shown. This is a plane normal to the lens axis at the focal point.

   Example 9. The lenses of Fig. 36–25 have radii of curvature of magnitude 40 cm and are made of glass with $n = 1.65$. Compute their focal lengths.

   Since C' lies on the R-side of the lens in Fig. 36–25a, r' is positive ($= +40$ cm). Since C'' lies on the V-side, r'' is negative ($= -40$ cm). Substituting in Eq. 36–30 yields

$$\frac{1}{f} = (n - 1)\left(\frac{1}{r'} - \frac{1}{r''}\right) = (1.65 - 1)\left(\frac{1}{+40 \text{ cm}} - \frac{1}{-40 \text{ cm}}\right),$$

or

$$f = +31 \text{ cm}.$$

A positive focal length indicates that, in agreement with Fig. 36–25a, the focal point F_2 is on the R-side of the lens and parallel incident light converges after refraction to form a real image.

   In Fig. 36–25c, C' lies on the V-side of the lens so that r' is negative ($= -40$ cm). Since r'' is positive ($= +40$ cm), Eq. 36–29 yields

$$f = -31 \text{ cm}.$$

A negative focal length indicates that, in agreement with Fig. 36–25c, the focal point F_2 is on the V-side of the lens and incident light diverges after refraction to form a virtual image.

   The location of the image of an extended object such as an illuminated arrow (Fig. 36–26) can be found graphically by using the following three facts:

   1. A ray parallel to the axis and falling on the lens passes, either directly or when extended, through the second focal point (ray x in Fig. 36–26).

   2. A ray falling on a lens after passing, either directly or when extended, through the first focal point will emerge from the lens parallel to the axis (ray y).

   3. A ray falling on the lens at its center will pass through undeflected. There is no deflection because the lens, near its center, behaves like a thin piece of glass with parallel sides. The direction of the light rays is not changed and the sideways displacement can be neglected because the lens thickness has been assumed to be negligible (ray z; see also Problem 17).

   Figure 36–27, which represents part of Fig. 36–26a, shows a ray passing from the tip of the object through the center of curvature to the tip of the image. For the similar triangles abc and dec we may write

$$\frac{de}{ab} = \frac{dc}{ac}.$$

The right side of this equation is i/o and the left side is $-m$, where m is the lateral magnification. The minus sign is required because we wish m to be negative for an inverted image. This yields

$$m = -\frac{i}{o}, \tag{36-32}$$

which holds for all types of thin lenses and for all object distances.

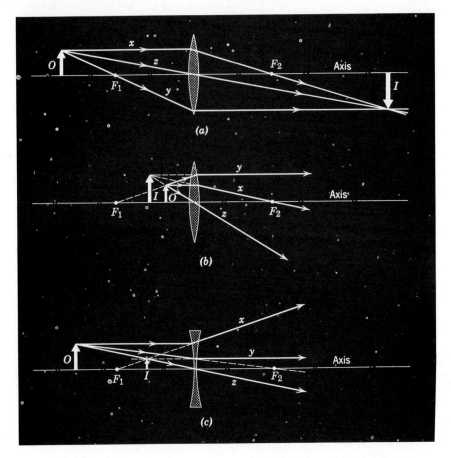

FIGURE 36–26 Showing the graphical location of images for three thin lenses.

Example 10. A converging thin lens has a focal length of $+24$ cm. An object is placed 9.0 cm from the lens as in Fig. 36–26*b*; describe the image.

From Eq. 36–31,

$$\frac{1}{o} + \frac{1}{i} = \frac{1}{f},$$

we have

$$\frac{1}{+9.0 \text{ cm}} + \frac{1}{i} = \frac{1}{+24 \text{ cm}},$$

which yields $i = -14.4$ cm, in agreement with the figure. The minus sign means that the image is on the V-side of the lens and is thus virtual.

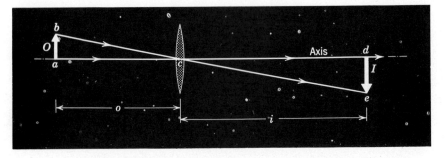

FIGURE 36–27 Two rays for the situation of Fig. 36–26*a*.

The lateral magnification is given by

$$m = -\frac{i}{o} = -\frac{-14.4 \text{ cm}}{+9.0 \text{ cm}} = +1.6,$$

again in agreement with the figure. The plus signifies an erect image.

In real-life optical devices, such as cameras or microscopes, we want the image to be a replica of the object, scaled up or down. Using the methods of this chapter, we can design a camera lens, say, only approximately because, among other things: (1) The object is not usually a plane but we want to record the image on a plane photographic film. (2) A point object produces a point image only approximately. (3) The refracting properties of the lens material vary with wavelength, as Fig. 36–2 shows. To design a high-quality camera lens using the methods of geometrical optics calls for extensive ray-tracing calculations, using an electronic computer. Such lenses contain several components, with different optical properties; the surfaces are not spherical.

Even if we apply geometrical optics exactly (which we did not do here) we must remember that geometrical optics is itself an approximation because it assumes that light travels in straight lines, which it does not. To understand the ultimate performance of optical systems, we must take the wave nature of light into account.

QUESTIONS

1. Discuss the propagation of spherical and of cylindrical waves, using Huygens' principle.

2. Does Huygens' principle apply to sound waves in air?

3. If Huygens' principle predicts the laws of reflection and refraction, why is it necessary or desirable to view light as an electromagnetic wave, with all its attendant complexity?

4. Would you expect sound waves to obey the laws of reflection and of refraction obeyed by light waves?

5. A street light, viewed by reflection across a body of water in which there are ripples, appears very elongated. Explain.

6. The light beam in Fig. 36–1a is broadened on entering the water. Explain.

7. By what per cent does the speed of blue light in fused quartz differ from that of red light?

8. Can (a) reflection phenomena or (b) refraction phenomena be used to determine the wavelength of light?

9. Why does a diamond "sparkle" more than a glass imitation cut to the same shape?

10. Is it plausible that the wavelength of light should change in passing from air into glass but that its frequency should not? Explain.

11. How can one determine the indices of refraction of the media in Table 36–1 relative to water, given the data in that table?

12. You are given a cube of glass. How can you find the speed of light (from a sodium light source) in this cube?

13. Describe and explain what a fish sees as he looks in various directions above his "horizon."

14. Design a periscope, taking advantage of total internal reflection. What are the advantages compared with silvered mirrors?

15. Can polarization by reflection occur if the light is incident on the interface from the side with the higher index of refraction (glass to air, for example)?

16. If unpolarized light is incident on an interface at the polarizing angle the reflected beam is plane polarized. This being the case, how can the refracted beam contain both components?

17. If a mirror reverses right and left, why doesn't it reverse up and down?

18. Is it possible to photograph a virtual image?

19. What approximations were made in deriving the mirror equation (Eq. 36–16):

$$\frac{1}{o} + \frac{1}{i} = \frac{2}{r}?$$

20. Can you think of a simple test or observation to prove that the law of reflection is the same for all wavelengths, under conditions in which geometrical optics prevails?

21. Under what conditions will a spherical mirror, which may be concave or convex, form (*a*) a real image, (*b*) an inverted image, and (*c*) an image smaller than the object?

22. An unsymmetrical thin lens forms an image of a point object on its axis. Is the image location changed if the lens is reversed?

23. Why has a lens two focal points and a mirror only one?

24. Under what conditions will a thin lens, which may be converging or diverging, form (*a*) a real image, (*b*) an inverted image, and (*c*) an image smaller than the object?

25. A skin diver wants to use an air-filled plastic bag as a converging lens for underwater use. Sketch a suitable cross section for the bag.

26. What approximations were made in deriving the thin lens equation (Eq. 36–31):

$$\frac{1}{o} + \frac{1}{i} = \frac{1}{f}?$$

27. Under what conditions will a thin lens have a lateral magnification (*a*) of -1.0 and (*b*) of $+1.0$?

28. How does the focal length of a glass lens for blue light compare with that for red light, assuming the lens is (*a*) diverging and (*b*) converging?

29. Does the focal length of a lens depend on the medium in which the lens is immersed? Is it possible for a given lens to act as a converging lens in one medium and a diverging lens in another medium?

30. Are the following statements true for a glass lens in air? (*a*) A lens that is thicker at the center than at the edges is a converging lens for parallel light. (*b*) A lens that is thicker at the edges than at the center is a diverging lens for parallel light. Explain and illustrate, using wavefronts.

31. Under what conditions would the lateral magnification, ($m = -i/o$) for lenses and mirrors become infinite? Is there any practical significance to such a condition?

32. Light rays are reversible. Discuss the situation in terms of objects and images if all rays in Figs. 36–14, 36–17, 36–19, 36–21, 36–24, and 36–26 are reversed in direction.

PROBLEMS

1(2). Prove that if a mirror is rotated through an angle α, the reflected beam is rotated through an angle 2α. Is this result reasonable for $\alpha = 45°$?

2(2). A 60°-prism is made of fused quartz. A ray of light falls on one face, making an angle of 45° with the normal. Trace the ray through the prism graphically with some care, showing the paths traversed by rays representing (*a*) blue light, (*b*) yellow-green light, and (*c*) red light. See Figs. 35–16 and 36–2.

3(2). Ptolemy, who lived at Alexandria toward the end of the first century A.D., gave the following measured values for the angle of incidence θ_1 and the angle of refraction θ_2 for a light beam passing from air to water:

θ_1	θ_2	θ_1	θ_2
10°	7°45′	50°	35°0′
20°	15°30′	60°	40°30′
30°	22°30′	70°	45°30′
40°	29°0′	80°	50°0′

Are these data consistent with the law of refraction; if so, what index of refraction results? These data are interesting as the oldest recorded physical measurements.

Answer: Even though Ptolemy did not know the law of refraction (Eq. 36–9) his experimental data agree very well with the presently accepted value for the index of refraction for a water-air interface ($n = 1.33$).

4(2). A pole extends 2.0 meters above the bottom of a swimming pool and 0.50 meters above the water. Sunlight is incident at 45°. What is the length of the shadow of the pole on the bottom of the pool?

5(2). In Fig. 36–3 show by graphical ray tracing, using a protractor, that if θ for the incident ray is either increased or decreased, the deviation angle ψ is increased. The symmetrical situation shown in this figure is called the *condition of minimum deviation.*

6(2). Prove that a ray of light incident on the surface of a sheet of plate glass of thickness t emerges from the opposite face parallel to its initial direction but displaced sideways, as in Fig. 36–28. Show that, for small angles of incidence θ, this displacement is given by

$$x = t\theta \, \frac{n-1}{n}$$

where n is the index of refraction and θ is measured in radians.

7(3). One end of a stick is dragged through water at a speed v which is greater than the speed u of water waves. Applying Huygens' construction to the water waves, show that the half-angle α of the conical wavefront that is set up is given by

$$\sin \alpha = u/v.$$

This is familiar as the bow wave of a ship or the shock wave caused by an object moving through air with a speed exceeding that of sound, as in Fig. 17–10.

8(4). What is the speed in fused quartz of light of wavelength 5500 A? See Fig. 36–2.

9(4). The speed of yellow sodium light in a certain liquid is measured to be 1.9×10^8 meters/sec. What is the index of refraction of this liquid, with respect to air, for sodium light?
Answer: 1.6.

10(4). The wavelength of yellow sodium light in air is 5893 A. (*a*) What is its frequency? (*b*) What is its wavelength in glass whose index of refraction is 1.52? (*c*) From the results of (*a*) and (*b*) find its speed in this glass.

11(4). When an electron moves through a medium at a speed exceeding the speed of light in that medium, it radiates electromagnetic energy (the Cerenkov effect, see Section 17–6). What minimum speed must an electron have in a liquid of refractive index 1.54 in order to radiate?
Answer: 1.95×10^8 meters/sec.

12(4). Light of free space wavelength 6000 A travels 1.6×10^{-4} cm in a medium of index of refraction 1.5. Find (*a*) the optical path length, (*b*) the wavelength in the medium, and (*c*) the phase difference after moving that distance, with respect to light traveling the same distance in free space.

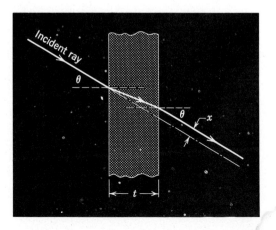

FIGURE 36–28   Problem 6(2).

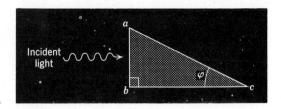

FIGURE 36-29 Problem 14(5).

13(4). Using Huygens' principle, derive the law of reflection.

14(5). A ray of light is incident normally on the face *ab* of a glass prism ($n = 1.52$), as shown in Fig. 36–29. (*a*) Assuming that the prism is immersed in air, find the largest value for the angle φ so that the ray is totally reflected at face *ac*. (*b*) Find φ if the prism is immersed in water.

15(5). A point source is 80 cm below the surface of a body of water. Find the diameter of the largest circle at the surface through which light can emerge from the water.
Answer: 180 cm.

16(5). A light ray falls on a square glass slab as in Fig. 36–30. What must the index of refraction of the glass be if total internal reflection occurs at the vertical face?

17(5). A glass cube has a small spot at its center. (*a*) What parts of the cube face must be covered to prevent the spot from being seen, no matter what the direction of viewing? (*b*) What fraction of the cube surface must be so covered? Assume a cube edge of 1.0 cm and an index of refraction of 1.50. (Neglect the subsequent behavior of an internally reflected ray.)
Answer: (*a*) Cover the center of each face with a circle of radius 0.45 cm. (*b*) 64%.

18(5). A plane wave of white light traveling in fused quartz strikes a plane surface of the quartz, making an angle of incidence θ. Is it possible for the internally reflected beam to appear (*a*) bluish or (*b*) reddish? Roughly what value of θ must be used? (Hint: White light will appear bluish if wavelengths corresponding to red are removed from the spectrum.)

19(5). A glass prism with an apex angle of 60° has $n = 1.60$. (*a*) What is the smallest angle of incidence for which a ray can enter one face of the prism and emerge from the other? (*b*) What angle of incidence would be required for the ray to pass through the prism symmetrically, as in Fig. 36–3?
Answer: (*a*) 36°. (*b*) 53°.

20(5). A given monochromatic light ray, initially in air, strikes the 90° prism at *P* (see Fig. 36–31) and is refracted there and at *Q* to such an extent that it just grazes the right-hand prism surface after it emerges into air at *Q*. (*a*) Determine the index of refraction, relative to air, of the prism for

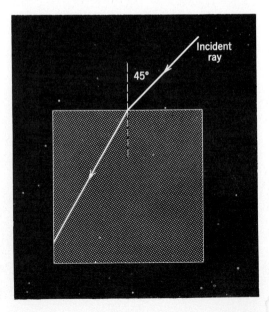

FIGURE 36-30 Problem 16(5).

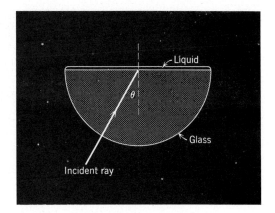

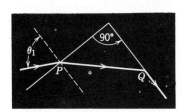

FIGURE 36-31   Problem 20(5).

FIGURE 36-32   Problem 21(5).

this wavelength in terms of the angle of incidence θ_1, which gives rise to this situation. (b) Give a numerical upper bound for the index of refraction of the prism. (c) Show, by a ray diagram, what happens if the angle of incidence at P is slightly greater than θ_1; is slightly less than θ_1.

21(5). A drop of liquid may be placed on a semicircular slab of glass as in Fig. 36–32. Show how to determine the index of refraction of the liquid by observing total internal reflection. The index of refraction of the glass is unknown and must also be determined. Is the range of indices of refraction that can be measured in this way restricted in any sense?

22(6). (a) At what angle of incidence will the light reflected from water be completely polarized? (b) Does this angle depend on the wavelength of the light?

23(6). Calculate the range of polarizing angles for white light incident on fused quartz. Assume that the wavelength limits are 4000 and 7000 A and use the dispersion curve of Fig. 36–2. *Answer:* 55°30′ to 55°46′.

24(7). A small object is 10 cm in front of a plane mirror. If you stand behind the object, 30 cm from the mirror, and look at its image, for what distance must you focus your eyes?

25(7). A small object O is placed one-third of the way between two parallel plane mirrors as in Fig. 36–33. Trace appropriate bundles of rays for viewing the four images that lie closest to the object.

26(7). A point object is 10 cm away from a mirror while the eye of an observer (pupil diameter 5.0 mm) is 20 cm away. Assuming both the eye and the point to be on the same line perpendicular to the surface, find the area of the mirror used in observing the reflection of the point.

27(7). Suppose you wished to photograph an object seen through a mirror. If the object is five feet to your right and one foot closer to the plane of the mirror than you, for what distance must you focus the lens of your camera? *Answer:* $\sqrt{(2D-1)^2 + 5^2}$, in ft, where D is the distance from the camera to the mirror.

28(7). Figure 36–34 shows an incident ray i striking a plane mirror MM′ at angle of incidence θ. Find the angle between i and r′. The two mirrors are at right angles.

29(7). Two plane mirrors make an angle of 90° with each other. What is the largest number of images of an object placed between them that can be seen by a properly placed eye? The object need not lie on the mirror bisector. *Answer:* 3.

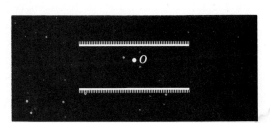

FIGURE 36-33   Problem 25(7).

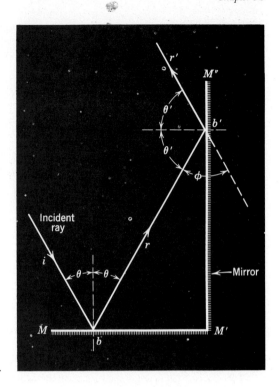

FIGURE 36–34   Problem 28(7).

30(7). Solve Problem 29(7) if the angle between the mirrors is (*a*) 45°, (*b*) 60°, (*c*) 120°, the object always being placed on the bisector of the mirrors.

31(7). Two perpendicular mirrors form the sides of a vessel filled with water, as shown in Fig. 36–35. A light ray is incident from above, normal to the water surface. (*a*) Show that the emerging ray is parallel to the incident ray. Assume that there are two reflections at the mirror surfaces. (*b*) Repeat the analysis for the case of oblique incidence, the ray lying in the plane of the figure. (*c*) Using three mirrors, state and prove the three-dimensional analog to this problem.

32(7). How many images of himself can an observer see in a room whose ceiling and two adjacent walls are mirrors? Explain.

33(8). Fill in this table, each column of which refers to a spherical mirror. Check your results by graphical analysis. Distances are in centimeters; if a number has no plus or minus sign in front of it, find the correct sign.

	a	b	c	d	e	f	g	h
Type	concave						convex	
f	20		+20			20		
r					−40		40	
i					−10		4.0	
o	+10	+10	+30	+60				+24
m		+1.0		−0.50		+0.10		0.50
Real image?		no						
Erect image?								no

Answer: Alternate vertical columns: (*a*) +, +40, −20, +2, no, yes. (*c*) concave, +40, +60, −2, yes, no. (*e*) convex, −20, +20, +0.5, no, yes. (*g*) −20, −, −, +5, +0.80, no, yes.

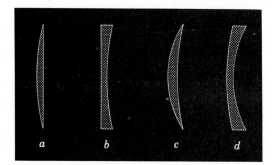

FIGURE 36-35 Problem 31(7).   **FIGURE 36-36**   Problem 40(10).

34(8). A short linear object of length l lies on the axis of a spherical mirror, a distance o from the mirror. (a) Show that its image will have a length l' where

$$l' = l \left(\frac{f}{o - f} \right)^2.$$

(b) Show that the *longitudinal magnification* m' $(= l'/l)$ is equal to m^2 where m is the lateral magnification discussed in Section 36–8. (c) Is there any condition such that, neglecting all mirror defects, the image of a small cube would also be a cube?

35(9). A penny lies on the bottom of a swimming pool 10 ft. deep. What is its apparent depth as viewed from above the water? The index of refraction of water is 1.33.
Answer: 7.5 ft.

36(9). A layer of water ($n = 1.33$) 2.0 cm thick floats on carbon tetrachloride ($n = 1.46$) 4.0 cm thick. How far below the water surface, viewed at normal incidence, does the bottom of the tank seem to be?

37(9). Fill out the following table, each column of which refers to a spherical surface separating two media with different indices of refraction. Distances are measured in centimeters.

	a	b	c	d	e	f	g	h
n_1	1.0	1.0	1.0	1.0	1.5	1.5	1.5	1.5
n_2	1.5	1.5	1.5		1.0	1.0	1.0	
o	+10	+10		+20	+10		+70	+100
i		−13	+600	−20	−6.0	−7.5		+600
r	+30		+30	−20		−30	+30	−30
Real image?								

Draw a figure for each situation and construct the appropriate rays graphically. Assume a point object.
Answer: Alternate vertical columns: (a) −18, no. (c) +71, yes. (e) +30, no. (g) −26, no.

38(9). Define and locate the first and second focal points for a single spherical refracting surface such as that of Fig. 36–19.

39(9). A parallel incident beam falls on a solid glass sphere at normal incidence. Locate the image in terms of the index of refraction n and the sphere radius r.

Answer: Assuming that the light is incident from the left, $i = -\dfrac{r(n-2)}{2(n-1)}$, to the right of the right edge of the sphere.

40(10). Using the lens maker's equation (Eq. 36–30), decide which of the thin lenses in Fig. 36–36 is converging and which diverging for parallel incident light.

41(10). A lens is made of glass having an index of refraction of 1.5. One side of the lens is flat and the other convex with a radius of curvature of 20 cm. (*a*) Find the focal length of the lens. (*b*) If an object is placed 40 cm to the left of the lens, where will the image be located and what will be the magnification?

Answer: (*a*) 40 cm. (*b*) Image to the right at $+\infty$; the nominal magnification is ∞.

42(10). A double-convex lens is to be made of glass with an index of refraction of 1.5. One surface is to have twice the radius of curvature of the other and the focal length is to be 6.0 cm. What are the radii?

43(10). Fill in this table, each column of which refers to a thin lens, to the extent possible. Check your results by graphical analysis. Distances are in centimeters; if a number (except in row *n*) has no plus sign or minus sign in front of it, find the correct sign.

	a	*b*	*c*	*d*	*e*	*f*	*g*	*h*	*i*
Type	converging								
f	10	+10	10	10					
r'					+30	−30	−30		
r''					−30	+30	−60		
i									
o	+20	+5.0	+5.0	+5.0	+10	+10	+10	+10	+10
n					1.5	1.5	1.5		
m			>1	<1				0.50	0.50
Real image?									yes
Erect image?								yes	

Draw a figure for each situation and construct the appropriate rays graphically. Assume a finite object.

Answer: Alternate vertical columns (an *x* means the quantity cannot be determined from the given data. (*a*) +, *x*, *x*, +20, *x*, −1, yes, no. (*c*) converging, +, *x*, *x*, −10, *x*, no, yes. (*e*) converging, +30, −15, +1.5, no, yes. (*g*) diverging, −120, −9.2, +0.92, no, yes. (*i*) converging +3.3, *x*, *x*, +5, *x*, −, no.

44(10). An object is placed at a center of curvature of a double-concave lens, both of whose radii of curvature are the same. (*a*) According to the convention used in this book, what are the signs of the two radii of curvature? (*b*) Find the location of the image in terms of the radius of curvature *r* and the index of refraction, *n*, of the glass. (*c*) Describe the nature of the image. (*d*) Verify your result with a ray diagram.

45(10). An object is 20 cm to the left of a lens with a focal length of +10 cm. A second lens of focal length +12.5 cm is 30 cm to the right of the first lens. (*a*) Using the image formed by the first lens as the object for the second, find the location and relative size of the final image. (*b*) Verify your conclusions by drawing the lens system to scale and constructing a ray diagram. (*c*) Describe the final image.

Answer: (*a*) Coincides in location with original object and is enlarged 5.0 times. (*c*) Virtual and inverted.

46(10). A converging lens with a focal length of +20 cm is located 10 cm to the left of a diverging lens having a focal length of −15 cm. If an object is located 40 cm to the left of the first lens, locate and describe completely the image formed.

47(10). Show that the distance between an object and its real image formed by a thin converging lens is always greater than or equal to four times the focal length of the lens.

48(10). A luminous object and a screen are a fixed distance *D* apart. (*a*) Show that a converging

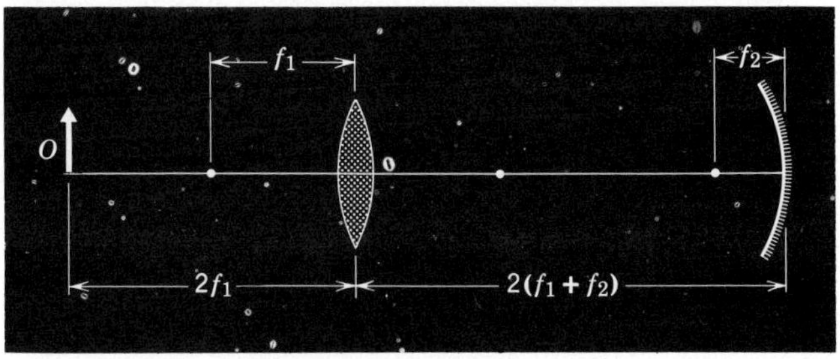

FIGURE 36–37   Problem 50(10).

lens of focal length f will form a real image on the screen for two positions that are separated by

$$d = \sqrt{D(D - 4f)}.$$

(b) Show that the ratio of the two image sizes for these two positions is

$$\left(\frac{D - d}{D + d}\right)^2.$$

49(10). An object is placed 1.0 meter in front of a converging lens, of focal length 0.50 meters, which is 2.0 meters in front of a plane mirror. (a) Where is the final image, measured from the lens, that would be seen by an eye looking toward the mirror through the lens? (b) Is the final image real or virtual? (c) Is the final image erect or inverted? (d) What is the lateral magnification? *Answer:* (a) 0.60 meters on the side of the lens away from the mirror. (b) Real. (c) Erect. (d) 0.20.

50(10). An erect object is placed a distance in front of a converging lens equal to twice the focal length f_1 of the lens. On the other side of the lens is a converging mirror of focal length f_2 separated from the lens by a distance $2(f_1 + f_2)$. (a) Find the location, nature and relative size of the final image. (b) Draw the appropriate ray diagram. See Fig. 36–37.

51(10). Two thin lenses of focal length f_1 and f_2 are in contact. Show that they are equivalent to a single thin lens with a focal length given by

$$f = \frac{f_1 f_2}{f_1 + f_2}.$$

52(10). The formula $\dfrac{1}{o} + \dfrac{1}{i} = \dfrac{1}{f}$ is called the Gaussian form of the thin-lens formula. Another form of this formula, the Newtonian form, is obtained by considering the distance x from the object to the first focal point and the distance x' from the second focal point to the image. Show that

$$xx' = f^2.$$

CHAPTER 37

Interference

37–1 Wave Optics

In Chapter 36 we made a special restriction, namely, that we would place no obstacle in the path of the incident light unless its dimensions were much greater than the wavelength of the light. Under these conditions we saw that light travels in straight lines and can be represented by rays. We called this special case geometrical optics.

In this chapter we remove this restriction and consider the general case of wave optics. In particular, we will deal with slits and baffles whose dimensions are the same order as the wavelength of the incident light. Note that wave optics holds under all circumstances but geometrical optics is a special case.

Figure 37–1a shows schematically a plane wave of wavelength λ falling on a slit of width $a = 5\lambda$. We find that the light flares out into the geometrical shadow of the slit; in other words, the conditions for geometrical optics are not met. Figures 37–1b ($a = 3\lambda$) and 37–1c ($a = \lambda$) show that *diffraction* becomes more pronounced as $a/\lambda \to 0$ and that attempts to isolate a single ray from the incident plane wave are futile.

Figure 37–2 shows water waves in a shallow ripple tank, produced by tapping the water surface periodically and automatically with the edge of a flat stick. We see that the plane wave so generated flares out by diffraction when it encounters a gap in a barrier placed across it. Diffraction is characteristic of waves of all types. We can hear around corners, for example, because of the diffraction of sound waves.

The diffraction of waves at a slit (or at an obstacle such as a wire) is expected from Huygens' principle. Consider the portion of the wavefront that arrives at the position of the slit in Fig. 37–1. Every point on it can be viewed as the site of an expanding spherical Huygens' wavelet. The "bending" of light into the region of the geometrical

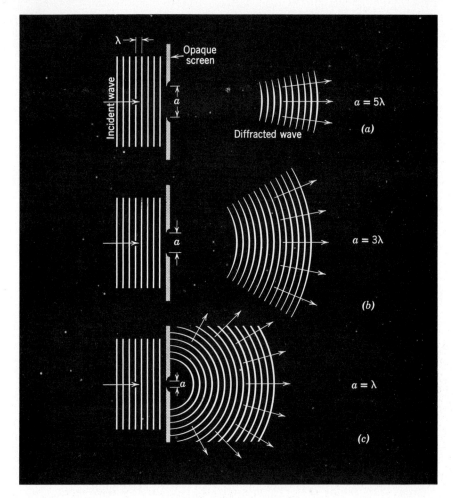

FIGURE 37–1   An attempt to isolate a ray by reducing the slit width *a* fails because of diffraction, which becomes more pronounced as a/λ approaches zero.

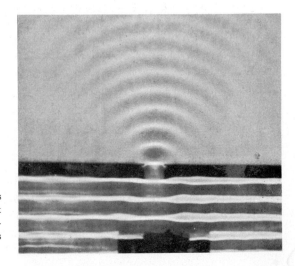

FIGURE 37–2   Diffraction of water waves at a slit in a ripple tank. Note that the slit width is about the same size as the wavelength. (Courtesy of Educational Services Incorporated.)

shadow is associated with the blocking off of Huygens' wavelets from those parts of the incident wavefront that lie behind the slit edges.

37–2 Young's Experiment

Before discussing diffraction we must say something about the interference of waves. In Chapter 16 we discussed interference, leading to an understanding of standing waves, beats, and other phenomena connected with sound. We emphasize again that such effects are not restricted to any particular wave type; they hold for all kinds of waves.

In Section 16–7 we saw that if two waves of the same frequency travel in approximately the same direction and have a phase difference that remains constant with time, they may combine so that their energy is not distributed uniformly in space but is a maximum at certain points and a minimum (possibly even zero) at others. The demonstration of such interference effects for light by Thomas Young in 1801 first established the wave theory of light on a firm experimental basis. Young was able to deduce the wavelength of light from his experiments, the first measurement of this important quantity.

Young allowed sunlight to fall on a pinhole S_0 punched in a screen A. As represented in Fig. 37–3, the emerging light spread out by diffraction and fell on pinholes S_1 and S_2 punched into screen B. Again diffraction occurred and two overlapping spherical waves expanded into the space to the right of screen B.

Interference is not limited to light waves but is a characteristic of all wave phenomena. Figure 37–4, for example, shows the interference pattern of water waves in a

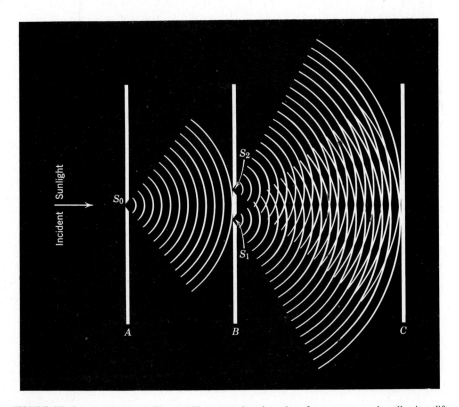

FIGURE 37–3 Showing how Thomas Young produced an interference pattern by allowing diffracted waves from pinholes S_1 and S_2 to overlap on screen C.

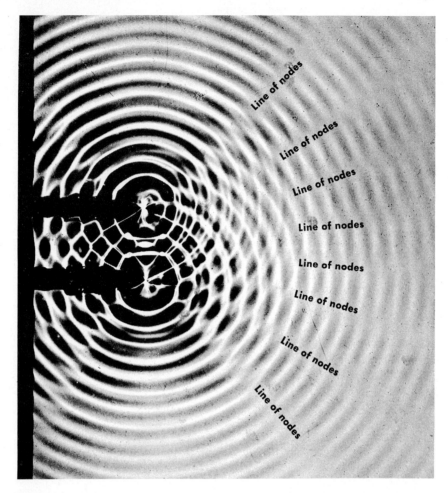

FIGURE 37–4   The interference of water waves in a ripple tank. There is destructive interference along the lines marked "Line of nodes" and constructive interference between these lines. (Courtesy Physical Science Study Committee.)

ripple tank. The waves are generated by two vibrators that tap the water surface in synchronism, producing two expanding spherical waves.

Let us analyze Young's experiment quantitatively, assuming that the incident light consists of a single wavelength only. In Fig. 37–5, P is an arbitrary point on the screen, a distance r_1 and r_2 from the narrow slits S_1 and S_2, respectively. Let us draw a line from S_2 to b in such a way that the lines PS_2 and Pb are equal. If d, the slit spacing, is much smaller than the distance D between the two screens (the ratio d/D in the figure has been exaggerated for clarity), S_2b is then almost perpendicular to both r_1 and r_2. This means that angle S_1S_2b is almost equal to angle PaO, both angles being marked θ in the figure. This is equivalent to saying that the lines r_1 and r_2 may be taken as parallel.

We often put a lens behind the two slits, as in Fig. 37–6, the screen C being in the focal plane of the lens. Under these conditions light focused at P must have struck the lens parallel to the line Px, drawn from P through the center of the (thin) lens. Under these conditions rays r_1 and r_2 are strictly parallel even though the requirement $D \gg d$ may not be met. The lens L may in practice be the lens and cornea of the eye, screen C being the retina.

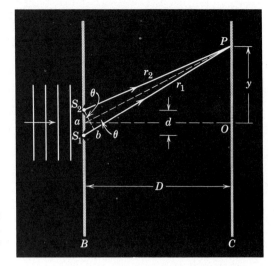

FIGURE 37–5   Rays from S_1 and S_2 combine at P. The light falling on screen B has been taken as parallel. Actually, $D \gg d$, the figure being distorted for clarity.

The two rays arriving at P in Figs. 37–5 or 37–6 from S_1 and S_2 are in phase at the source slits, both being derived from the same wavefront in the incident plane wave. Because the rays contain different numbers of waves, they arrive at P with a phase difference. The number of wavelengths contained in S_1b, which is the path difference, determines the nature of the interference at P. If a lens is used as in Fig. 37–6, it may seem that a phase difference should develop between the rays beyond the plane represented by S_2b, the path lengths between this plane and P being clearly different. However, because of the presence of additional material of higher index of refraction in the path of the ray traveling the shortest distance, the optical path lengths are identical (see Section 36–4). Two rays with the same optical path lengths contain the same number of wavelengths, so that no phase difference will result because of the light passing through the lens.

To have a maximum at P, S_1b ($= d \sin \theta$) must contain an integral number of wavelengths, or

$$S_1b = m\lambda \qquad m = 0, 1, 2, \ldots,$$

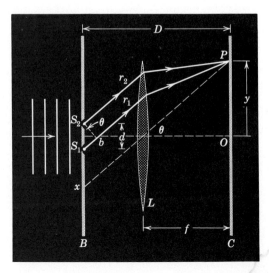

FIGURE 37–6   A lens is normally used to produce interference fringes; compare with Fig. 37–5. The figure is again distorted for clarity in that $f \gg d$ in practice.

which we can write as

$$d \sin \theta = m\lambda \qquad m = 0, 1, 2, \ldots \text{(maxima)}. \qquad (37\text{-}1)$$

Note that each maximum above O in Figs. 37–5 and 37–6 has a symmetrically located maximum below O. There is a central maximum, described by $m = 0$.

For a minimum at P, S_1b ($= d \sin \theta$) must contain a half-integral number of wavelengths, or

$$d \sin \theta = (m + \tfrac{1}{2})\lambda \qquad m = 0, 1, 2, \ldots \text{(minima)}. \qquad (37\text{-}2)$$

Example 1. What is the linear distance on screen C in Fig. 37–5 between adjacent maxima? The wavelength λ is 5460A; the slit separation d is 0.10 mm, and the slit-screen separation D is 20 cm.

If θ is small enough, we can use the approximation

$$\sin \theta \cong \tan \theta \cong \theta.$$

From Fig. 37–5 we see that $\qquad\qquad \tan \theta = \dfrac{y}{D}.$

Substituting this into Eq. 37–1 for $\sin \theta$ leads to

$$y = m\frac{\lambda D}{d} \qquad m = 0, 1, 2, \ldots \text{(maxima)}.$$

The positions of any two adjacent maxima are given by

$$y_m = m\frac{\lambda D}{d}$$

and

$$y_{m+1} = (m + 1)\frac{\lambda D}{d}.$$

Their separation Δy is found by subtracting:

$$\Delta y = y_{m+1} - y_m = \frac{\lambda D}{d}$$

$$= \frac{(5.5 \times 10^{-7} \text{ meter})(20 \times 10^{-2} \text{ meter})}{0.10 \times 10^{-3} \text{ meter}} = 1.1 \text{ mm}.$$

As long as θ in Figs. 37–5 and 37–6 is small, the separation of the interference fringes is independent of m; that is, the fringes are evenly spaced. If the incident light contains more than one wavelength the separate interference patterns, which will have different fringe spacings, will be superimposed.

37–3 Coherence

Analysis of the derivation of Eqs. 37–1 and 37–2 shows that a fundamental requirement for the existence of well-defined interferences fringes on screen C in Fig. 37–3 is that the light waves that travel from S_1 and S_2 to any point P on this screen must have a sharply defined phase difference ϕ that remains constant with time. If this condition is satisfied, a stable, well-defined fringe pattern will appear. At certain points P, ϕ will be given, independent of time, by $n\pi$ where $n = 1, 3, 5, \ldots$ so that the resultant intensity will be strictly zero and will remain so throughout the time of observation. At other points ϕ will be given by $n\pi$ where $n = 0, 2, 4 \ldots$ and the resultant intensity will be a maximum. When a constant phase difference exists the two beams emerging from slit S_1 and S_2 are said to be completely *coherent*.

Let the source in Fig. 37–3 be removed and let slits S_1 and S_2 be replaced by two completely independent light sources, such as two fine incandescent wires placed side by side in a glass envelope. No interference fringes will appear on screen C but only a relatively uniform illumination. We can understand this if we make the reasonable assumption that for completely independent light sources the phase difference between the two beams arriving at P will vary with time in a random way. At a certain instant conditions may be right for cancellation and a short time later (perhaps 10^{-8} sec) they may be right for re-enforcement. This same random phase behavior holds for all points on screen C with the result that this screen is uniformly illuminated. The intensity at any point is equal to the sum of the intensities that each source S_1 and S_2 produces separately at that point. Under these conditions the two beams emerging from S_1 and S_2 are said to be completely *incoherent*.

Note that for completely coherent light beams one (1) combines the amplitudes vectorially, taking the (constant) phase difference properly into account, and then (2) squares this resultant amplitude to obtain a quantity proportional to the resultant intensity. For completely incoherent light beams, on the other hand, one (1) squares the individual amplitudes to obtain quantities proportional to the individual intensities and then (2) adds the individual intensities to obtain the resultant intensity. This procedure is in agreement with the experimental fact that for completely independent light sources the resultant intensity at every point is always greater than the intensity produced at that point by either light source acting alone.

It remains to investigate under what experimental conditions coherent or incoherent beams may be produced and to give an explanation for coherence in terms of the mode of production of the radiation. Consider first a beam of radiowaves emerging from an antenna connected by a transmission line to an LC oscillator as in Fig. 35–1. The oscillations are periodic with time and produce, in the radiation field, a periodic variation of **E** and **B** with time. The radiated wave at large enough distances from the antenna is well represented by Fig. 35–4. Note that (1) the wave has essentially infinite extent in time, including both future time ($t > 0$, say) and past times ($t < 0$); see Fig. 37–7a. At any point, as the wave passes by, the wave disturbance (i.e., **E** or **B**) varies with time in a perfectly periodic way. (2) The wavefronts at points far removed from the antenna are, for a small enough area of observation (see Fig. 35–3), parallel planes at right angles to the propagation direction. At any instant of time the wave disturbance varies with distance along the propagation direction in a perfectly periodic way.

Two beams generated from a single traveling wave like that of Fig. 35–4 will be

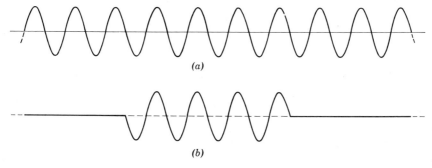

(a)

(b)

FIGURE 37–7   (*a*) A section of an infinite wave and (*b*) a wavetrain.

completely coherent. One way to generate two such beams is to put an opaque screen containing two slits in the path of the beam. The waves emerging from the slits will always have a constant phase difference at any point in the region in which they overlap and interference fringes will be produced.

If we turn from radio sources to common sources of visible light, such as incandescent wires or an electric discharge passing through a gas, we become aware of a fundamental difference. In both of these sources the fundamental light emission processes occur in individual atoms and these atoms do not act together in a cooperative (i.e., coherent) way. The act of light emission by a single atom takes, in a typical case, about 10^{-8} sec and the emitted light is properly described as a *wavetrain* (Fig. 37–7*b*) rather than as a wave (Fig. 37–7*a*). For emission times such as these the wavetrains are a few meters long.

Interference effects from ordinary light sources may be produced by putting a narrow slit (S_0 in Fig. 37–3) directly in front of the source. This insures that the wavetrains that strike slits S_1 and S_2 in screen B in this figure originate from the same small region of the source. The diffracted beams emerging from S_1 and S_2 thus represent the same population of wavetrains and are coherent with respect to each other. If the phase of the light emitted from S_0 changes, this change is transmitted simultaneously to S_1 and S_2. Thus, at any point on screen C, a constant phase difference is maintained between the beams from these two slits and a stationary interference pattern occurs.

The lack of coherence of the light from ordinary sources such as glowing wires is due to the fact that the emitting atoms do not act cooperatively (i.e., coherently). Since 1960 it has proved possible to construct sources of visible light in which the atoms do act cooperatively and in which the emitted light is highly coherent. Such devices are called *lasers* (a coined word meaning *l*ight *a*mplification through *s*timulated *e*mission of *r*adiation). Their light output is extremely monochromatic, intense, and highly collimated. These methods permit, for the first time, a degree of control of visible light approaching that possible for radio and for microwaves. The practical applications of lasers, which include the amplification of weak light signals, the use of light beams as highly efficient carriers of information from point to point (see Problem 17), and the production of high temperatures by intense local heating, remain to be fully exploited.

37–4 Intensity of Interfering Waves

Let us assume that the electric field components of the two waves in Fig. 37–5 vary with time at point P as

$$E_1 = E_0 \sin \omega t \tag{37–3}$$

and
$$E_2 = E_0 \sin (\omega t + \phi) \tag{37–4}$$

where $\omega \ (= 2\pi\nu)$ is the angular frequency of the waves and ϕ is the phase difference between them. Note that ϕ depends on the location of point P, which, in turn, for a fixed geometrical arrangement, is described by the angle θ in Figs. 37–5 and 37–6. We assume that the slits are so narrow that the diffracted light from each slit illuminates the central portion of the screen uniformly. This means that near the center of the screen E_0 is independent of the position of P, that is, of the value of θ.

The resultant amplitude at point P is given by

$$E = E_1 + E_2. \tag{37–5}$$

This is a scalar and not a vector equation because $\mathbf{E}_1$ and $\mathbf{E}_2$ are essentially parallel at point P. We will combine E_1 and E_2 using an analytical, semigraphical method. The method will be especially useful later, when we want to combine a large number of wave disturbances with differing phases.

A sinusoidal wave disturbance E_1 (see Eq. 37–3) can be represented graphically, by a rotating vector. In Fig. 37–8*a* a vector of magnitude E_0 is allowed to rotate about the origin in a counterclockwise direction with an angular frequency ω. Following electrical engineering practice we call such a rotating vector a *phasor*. The alternating wave disturbance E_1 (Eq. 37–3) is represented by the projection of this phasor on the vertical axis.

A second wave disturbance E_2, which has the same amplitude E_0 but a phase difference ϕ with respect to E_1 (see Eq. 37–4) can be represented graphically (Fig. 37–8*b*) as the projection on the vertical axis of a second phasor of magnitude E_0 which makes a fixed angle ϕ with the first phasor. As this figure shows, the sum E of E_1 and E_2, which is the instantaneous amplitude of the resultant wave, is the sum of the projections of the two phasors on the vertical axis. This is revealed more clearly if the phasors are redrawn, as in Fig. 37–8*c*, placing the foot of one arrow at the head of the other, maintaining the proper phase difference, and letting the whole assembly rotate counterclockwise about the origin.

In Fig. 37–8*c* E can also be regarded as the projection on the vertical axis of a phasor of length E_θ, which is the vector sum of the two phasors of magnitude E_0 and makes a phase angle β with respect to the phasor that generates E_1. Note that the (algebraic) sum of the projections of the two phasors is equal to the projection of the (vector) sum of the two phasors.

From the theorem that an exterior angle (ϕ) is equal to the sum of the two opposite

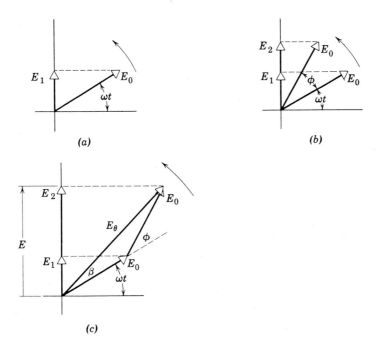

FIGURE 37–8   (*a*) A wave disturbance E_1 is represented by a rotating vector or *phasor*. (*b*) Two wave disturbances E_1 and E_2 with a phase difference ϕ between them are so represented. These two phasors can represent the two wave disturbances in the double-slit problem; see Eqs. 37–3 and 37–4. (*c*) Another way of drawing (*b*).

interior angles $(\beta + \beta)$ we see from Fig. 37–8c that $\beta = \frac{1}{2}\phi$. Thus we have

$$E = E_\theta \sin (\omega t + \beta), \qquad (37\text{–}6a)$$

where

$$\beta = \frac{1}{2}\phi \qquad (37\text{–}6b)$$

and

$$E_\theta = 2E_0 \cos \beta = E_m \cos \beta. \qquad (37\text{–}6c)$$

In most problems in optics we are concerned only with the amplitude E_θ of the resultant wave disturbance and not with its time variation. This is because the eye and other common measuring instruments respond to the resultant intensity of the light (that is, to the square of the amplitude) and cannot detect the rapid time variations that characterize visible light. For sodium light ($\lambda = 5890$ A), the frequency ν ($= \omega/2\pi$) is 5.1×10^{14} cycles/sec. Usually, then, we need not consider the rotation of the phasors but can confine our attention to finding the magnitude of the resultant phasor.

The amplitude E_θ of the resultant wave disturbance, which determines the intensity of the interference fringes in Young's experiment, will turn out to depend strongly on the value of θ, that is, on the location of point P in Figs. 37–5 and 37–6.

In Section 16–5 we showed that the intensity of a wave I, measured perhaps in watts/meter^2, is proportional to the square of its amplitude. For the resultant wave then, ignoring the proportionality constant,

$$I_\theta \propto E_\theta{}^2. \qquad (37\text{–}7)$$

This relationship seems reasonable if we recall (Eq. 26–18) that the energy density in an electric field is proportional to the square of the electric field strength. This is true for rapidly varying electric fields, such as those in a light wave, as well as for static fields.

The ratio of the intensities of two light waves is the ratio of the squares of the amplitudes of their electric fields. If I_θ is the intensity of the resultant wave at P and I_0 is the intensity that a single wave acting alone would produce, then

$$\frac{I_\theta}{I_0} = \left(\frac{E_\theta}{E_0}\right)^2. \qquad (37\text{–}8)$$

Combining with Eq. 37–6c leads to

$$I_\theta = 4I_0 \cos^2 \beta = I_m \cos^2 \beta. \qquad (37\text{–}9)$$

Note that the intensity of the resultant wave at any point P varies from zero (for a point at which $\cos \beta = 0$) to I_m, which is four times the intensity I_0 of each individual wave (for a point at which $\cos \beta = \pm 1$). Let us compute I_θ as a function of the angle θ in Figs. 37–5 or 37–6, that is, of the position of P on screen C in those figures.

The phase difference ϕ in Eq. 37–4 is associated with the path difference $S_1 b$ in Fig. 37–5 or 37–6. If $S_1 b$ is $\frac{1}{2}\lambda$, ϕ will be π; if $S_1 b$ is λ, ϕ will be 2π, etc. This suggests that

$$\frac{\text{phase difference}}{2\pi} = \frac{\text{path difference}}{\lambda},$$

$$\phi = \frac{2\pi}{\lambda} (d \sin \theta),$$

or, finally, from Eq. 37–6b,

$$\beta = \tfrac{1}{2}\phi = \frac{\pi d}{\lambda} \sin \theta. \tag{37–10}$$

This expression for β can be substituted into Eq. 37–9 for I_θ, yielding the latter quantity as a function of θ. For convenience we collect here the expressions for the amplitude and the intensity in double-slit interference.

[Eq. 37–6c]　　　　$E_\theta = E_m \cos \beta$　　　　interference　　(37–11a)

[Eq. 37–9]　　　　$I_\theta = I_m \cos^2 \beta$　　　　from narrow　　(37–11b)

　　　　　　　　　　　　　　　　　　　　slits (that is,

[Eq. 37–10]　　$\beta \,(= \tfrac{1}{2}\phi) = \dfrac{\pi d}{\lambda} \sin \theta$　　$a \ll \lambda)$　　(37–11c)

To find the positions of the intensity maxima, we put

$$\beta = m\pi \qquad m = 0, 1, 2, \ldots$$

in Eq. 37–11b. From Eq. 37–11c this reduces to

$$d \sin \theta = m\lambda \qquad m = 0, 1, 2, \ldots \text{ (maxima)},$$

which is the equation derived in Section 37–2 (Eq. 37–1). To find the intensity minima we write

$$\frac{\pi d \sin \theta}{\lambda} = (m + \tfrac{1}{2})\pi \qquad m = 0, 1, 2, \ldots \text{ (minima)},$$

which reduces to the previously derived Eq. 37–2.

Figure 37–9 shows the intensity pattern for double-slit interference. The horizontal solid line is I_0; this describes the (uniform) intensity pattern on the screen if one of the slits is covered up. If the two sources were incoherent the intensity would be uniform over the screen and would be $2I_0$; see the horizontal dashed line in Fig. 37–9. For coherent sources we expect the energy to be merely redistributed over the screen, since energy is neither created nor destroyed by the interference process. Thus the average intensity in the interference pattern should be $2I_0$, as for incoherent sources. This follows at once if, in Eq. 37–11b, we substitute one-half for the cosine-squared term and if we recall (see Eq. 37–9) that $I_m = 4I_0$. We have seen several times that

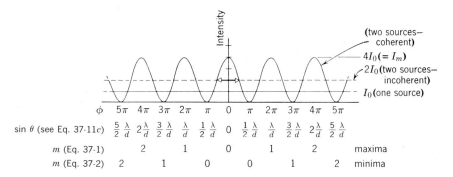

FIGURE 37–9　The intensity pattern for double-slit interference. The heavy arrow in the central peak represents the half-width of the peak. This figure is constructed on the assumption that the two interfering waves each illuminate the central portion of the screen uniformly, that is, I_0 is independent of position as shown.

the average value of the square of a sine or a cosine term over one or more half-cycles is one-half.

In a more general case we might want to find the resultant of a number (>2) of sinusoidally varying wave disturbances. The general procedure is the following:

1. Construct a series of phasors representing the functions to be added. Draw them end to end, maintaining the proper phase relationships between adjacent phasors.

2. Construct the vector sum of this array. Its length gives the amplitude of the resultant. The angle between it and the first phasor is the phase of the resultant with respect to this first phasor. The projection of this phasor on the vertical axis gives the time variation of the resultant wave disturbance.

Example 2. Find graphically the resultant $E(t)$ of the following wave disturbances:

$$E_1 = 10 \sin \omega t$$
$$E_2 = 10 \sin (\omega t + 15°)$$
$$E_3 = 10 \sin (\omega t + 30°)$$
$$E_4 = 10 \sin (\omega t + 45°).$$

Figure 37–10 in which E_0 equals 10, shows the assembly of four phasors that represents these functions. Their vector sum, by graphical measurement, has an amplitude E_R of 38 and a phase ϕ_0 with respect to E_1 of 23°. In other words

$$E(t) = E_1 + E_2 + E_3 + E_4 = 38 \sin (\omega t + 23°).$$

Check this result by trigonometric calculation.

37–5 Interference from Thin Films

The colors of soap bubbles, oil slicks, and other thin films are the result of interference. Figure 37–11 shows interference effects in a thin vertical film of soapy water illuminated by monochromatic light.

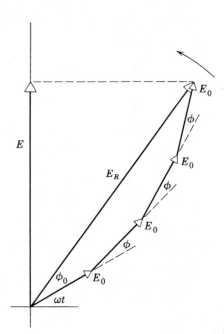

FIGURE 37–10 Example 2. Four wave disturbances are added graphically, using the method of phasors.

FIGURE 37–11 A soapy water film on a wire loop, viewed by reflected light. The black segment at the top is not a tear. It arises because the film, by drainage, is so thin here that there is destructive interference between the light reflected from its front surface and that reflected from its back surface. These two waves differ in phase by 180°.

Figure 37–12 shows a film of uniform thickness d and index of refraction n, the eye being focused on spot a. The film is illuminated by a broad source of monochromatic light S. There exists on this source a point P such that two rays, identified by the single and double arrows, respectively, can leave P and enter the eye as shown, after passing through point a. These two rays follow different paths in going from P to the eye, one being reflected from the upper surface of the film, the other from the lower surface. Whether point a appears bright or dark depends on the nature of the interference between the two waves that diverge from a. These waves are coherent because they both originate from the same point P on the light source.

If the eye looks at another part of the film, say a', the light that enters the eye must

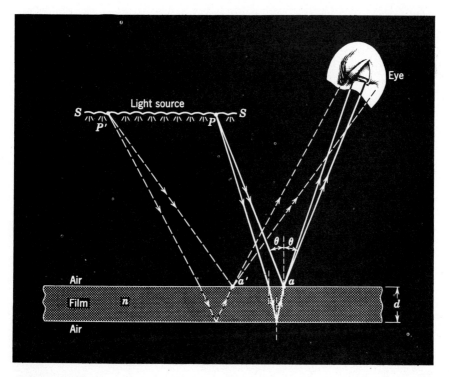

FIGURE 37–12 Interference by reflection from a thin film, assuming an extended source S.

originate from a different point P' of the extended source, as suggested by the dashed lines in Fig. 37–12.

For near-normal incidence ($\theta \cong 0$ in Fig. 37–12) the geometrical path difference for the two rays from P will be close to $2d$. We might expect the resultant wave reflected from the film near a to be an interference maximum if the distance $2d$ is an integral number of wavelengths. This statement must be modified for two reasons.

First, the wavelength must refer to the wavelength of the light in the film λ_n and not to its wavelength in air λ; that is, we are concerned with optical path lengths rather than geometrical path lengths. The wavelengths λ and λ_n (see Eq. 36–8) are related by

$$\lambda_n = \lambda/n. \tag{37–12}$$

To bring out the second point, let us assume that the film is so thin that $2d$ is very much less than a wavelength. The phase difference between the two waves would be close to zero on our assumption, and we would expect such a film to appear bright on reflection. However, it appears dark. This is clear from Fig. 37–11, in which the action of gravity produces a wedge-shaped film, extremely thin at its top edge. As drainage continues, the dark area increases in size. To explain this and many similar phenomena, we assume that one or the other of the two rays of Fig. 37–12 suffers an abrupt phase change of π ($= 180°$) associated either with reflection at the air-film interface or transmission through it. As it turns out, the ray reflected from the upper surface suffers this phase change. The other ray is not changed abruptly in phase, either on transmission through the upper surface or on reflection at the lower surface.

In Section 16–8 we discussed phase changes on reflection for transverse waves in strings. To extend these ideas, consider the composite string of Fig. 37–13, which consists of two parts with different masses per unit length, stretched to a given tension. If a pulse moves to the right in Fig. 37–13a, approaching the junction, there will be a reflected and a transmitted pulse, the reflected pulse being in phase with the incident pulse. In Fig. 37–13b the situation is reversed, the incident pulse now being in the less massive string. In this case the reflected pulse will differ in phase from the incident pulse by π ($= 180°$). In each case the transmitted pulse will be in phase with the incident pulse.

Figure 37–13a suggests a light wave in glass, say, approaching a surface beyond which there is a less optically dense medium (one of lower index of refraction) such as air. Figure 37–13b suggests a light wave in air approaching glass. To sum up the optical situation, when reflection occurs from an interface beyond which the medium has a lower index of refraction, the reflected wave undergoes no phase change; when the medium beyond the interface has a higher index, there is a phase change of π. The transmitted wave does not experience a change of phase in either case.

We are now able to take into account both factors that determine the nature of the interference, namely, differences in optical path length and phase changes on reflection. For the two

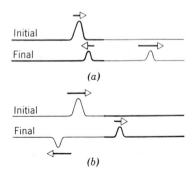

FIGURE 37–13   Phase changes on reflection at a junction between two stretched composite strings. (*a*) Incident pulse in heavy string and (*b*) incident pulse in light string.

rays of Fig. 37–12 to combine to give a maximum intensity, assuming normal incidence, we must have

$$2d = (m + \tfrac{1}{2})\lambda_n \qquad m = 0, 1, 2, \dots.$$

The term $\tfrac{1}{2}\lambda_n$ is introduced because of the phase change on reflection, a phase change of 180° being equivalent to half a wavelength. Substituting λ/n for λ_n yields finally

$$2dn = (m + \tfrac{1}{2})\lambda \qquad m = 0, 1, 2, \dots \text{(maxima)}. \tag{37–13}$$

The condition for a minimum intensity is

$$2dn = m\lambda \qquad m = 0, 1, 2, \dots \text{(minima)}. \tag{37–14}$$

These equations hold when the index of refraction of the film is either greater or less than the indices of the media on each side of the film. Only in these cases will there be a relative phase change of 180° for reflections at the two surfaces. A water film in air and an air film in the space between two glass plates provide examples of cases to which Eqs. 37–13 and 37–14 apply. Example 4 provides a case in which they do not apply.

If the film thickness is not uniform, as in Fig. 37–11, where the film is wedge-shaped, constructive interference will occur in certain parts of the film and destructive interference will occur in others. Lines of maximum and of minimum intensity will appear—these are the interference fringes. They are called *fringes of constant thickness*, each fringe being the locus of points for which the film thickness d is a constant. If the film is illuminated with white light rather than monochromatic light, the light reflected from various parts of the film will be modified by the various constructive or destructive interferences that occur. This accounts for the brilliant colors of soap bubbles and oil slicks.

Example 3. A water film ($n = 1.33$) in air is 3200 A thick. If it is illuminated with white light at normal incidence, what color will it appear to be in reflected light?

By solving Eq. 37–13 for λ,

$$\lambda = \frac{2dn}{m + \tfrac{1}{2}} = \frac{(2)(3200 \text{ A})(1.33)}{m + \tfrac{1}{2}} = \frac{8500 \text{ A}}{m + \tfrac{1}{2}} \qquad \text{(maxima)}.$$

From Eq. 37–14 the minima are given by

$$\lambda = \frac{8500 \text{ A}}{m} \qquad \text{(minima)}.$$

Maxima and minima occur for the following wavelengths:

m	0 (max)	1 (min)	1 (max)	2 (min)	2 (max)
λ, A	17000	8500	5700	4250	3400

Only the maximum corresponding to $m = 1$ lies in the visible region (see Fig. 35–16); light of this wavelength appears yellow-green. If white light is used to illuminate the film, the yellow-green component will be enhanced when viewed by reflection.

Example 4. *Nonreflecting glass.* Lenses are often coated with thin films of transparent substances like MgF_2 ($n = 1.38$) in order to reduce the reflection from the glass surface, using interference.

How thick a coating is needed to produce a minimum reflection at the center of the visible spectrum (5500 A)?

We assume that the light strikes the lens at near-normal incidence (θ is exaggerated for clarity in Fig. 37–14), and we seek destructive interference between rays r and r_1. Equation 37–14 does not apply because in this case a phase change of 180° is associated with each ray, for at both the upper and lower surfaces of the MgF_2 film the reflection is from a medium of greater index of refraction.

There is no net change in phase produced by the two reflections, which means that the optical path difference for destructive interference is $(m + \frac{1}{2})\lambda$ (compare Eq. 37–13), leading to

$$2dn = (m + \tfrac{1}{2})\lambda \qquad m = 0, 1, 2, \dots \text{(minima)}.$$

Solving for d and putting $m = 0$ yields

$$d = \frac{(m + \tfrac{1}{2})\lambda}{2n} = \frac{\lambda}{4n} = \frac{5500 \text{ A}}{(4)(1.38)} = 1000 \text{ A}.$$

37–6 Michelson's Interferometer

An interferometer is a device that can be used to measure lengths or changes in length with great accuracy by means of interference fringes. We describe the form originally built by Michelson in 1881.

Consider light that leaves point P on extended source S (Fig. 37–15) and falls on half-silvered mirror M. This mirror has a silver coating just thick enough to transmit half the incident light and to reflect half; in the figure we have assumed that this mirror, for convenience, possesses negligible thickness. At M the light divides into two waves. One proceeds by transmission toward mirror M_1; the other proceeds by reflection toward M_2. The waves are reflected at each of these mirrors and are sent back along their directions of incidence, each wave eventually entering the eye E. Since the waves are coherent, being derived from the same point on the source, they will interfere.

If the mirrors M_1 and M_2 are exactly perpendicular to each other, the effect is that of light from an extended source S falling on a uniformly thick slab of air, between glass, whose thickness is equal to $d_2 - d_1$. Interference fringes appear, caused by small changes in the angle of incidence of the light from different points on the extended source as it strikes the equivalent air film. For thick films a path difference of

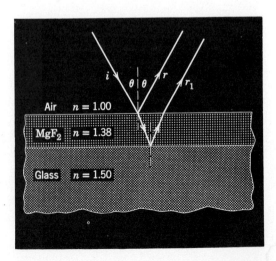

FIGURE 37–14 Example 4. Unwanted reflections from glass can be reduced by coating the glass with a thin transparent film.

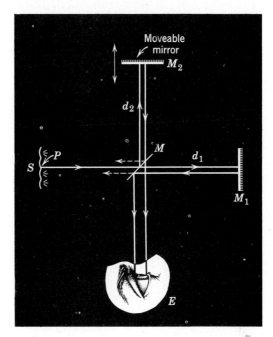

FIGURE 37–15 Michelson's interferometer, showing the path of a particular ray originating at point P of an extended source S.

one wavelength can be brought about by a very small change in the angle of incidence.

If M_2 is moved backward or forward, the effect is to change the thickness of the equivalent air film. Suppose that the center of the (circular) fringe pattern appears bright and that M_2 is moved just enough to cause the first bright circular fringe to move to the center of the pattern. The path of the light beam striking M_2 has been changed by one wavelength. This means (because the light passes twice through the equivalent air film) that the mirror must have moved one-half a wavelength.

The interferometer is used to measure changes in length by counting the number of interference fringes that pass the field of view as mirror M_2 is moved. Length measurements made in this way can be accurate if large numbers of fringes are counted.

Michelson measured the length of the standard meter, kept in Paris, in terms of the wavelength of certain monochromatic red light emitted from a light source containing cadmium. He showed that the standard meter was equivalent to 1,553,163.5 wavelengths of the red cadmium light.

Physicists have long speculated on the advantages of discarding the standard meter bar as the basic standard of length and of defining the meter in terms of the wavelength of some carefully chosen monochromatic radiation. This would make the primary length standard readily available in laboratories all over the world. It would improve the accuracy of length measurements, since one would no longer need to compare an unknown object with a standard object (the meter bar), using interferometer techniques, but could measure the unknown object directly and in an absolute sense, using these techniques. There is the additional advantage that if the standard meter bar were destroyed it could never be replaced, whereas light sources and interferometers will (presumably) always be available.

In 1961 such an atomic standard of length was adopted by international agreement. Quoting from an article* describing the event:

* *Scientific American*, p. 75, December 1960.

The wavelength of the orange-red light of krypton-86 has replaced the platinum iridium bar as the world standard of length. Formerly the wavelength of this light was defined as a function of the length of the meter bar. Now the meter is defined as a multiple (1,650,763.73) of the wavelength of the light.

The light from krypton-86 was used in preference to that from cadmium or other sources because it produces sharper interference fringes in the interferometer over the long optical paths sometimes used in length measurement.

QUESTIONS

1. Is Young's experiment an interference experiment or a diffraction experiment, or both?

2. Do interference effects occur for sound waves? Recall that sound is a longitudinal wave and that light is a transverse wave.

3. In Young's double-slit interference experiment, using a monochromatic laboratory light source, why is screen A in Fig. 37–3 necessary? What would happen if one gradually enlarged the hole in this screen?

4. Describe the pattern of light intensity on screen C in Fig. 37–5 if one slit is covered with a red filter and the other with a blue filter, the incident light being white.

5. Is coherence important in reflection and refraction?

6. Define carefully, and distinguish between, the angles θ and ϕ that appear in Eq. 37–10.

7. If one slit in Fig. 37–5 is covered, what change occurs in the intensity of light in the center of the screen?

8. What changes occur in the pattern of interference fringes if the apparatus of Fig. 37–5 is placed under water?

9. What are the requirements for a maximum intensity when viewing a thin film by transmitted light?

10. Why must the film of Fig. 37–12 be "thin" for us to see an interference pattern of the type described?

11. Why do coated lenses (see Example 4) look purple by reflected light?

12. A person wets his eyeglasses to clean them. As the water evaporates he notices that for a short time the glasses become markedly more nonreflecting. Explain.

13. A lens is coated to reduce reflection, as in Example 4. What happens to the energy that had previously been reflected? Is it absorbed by the coating?

14. If interference between light waves of different frequencies is possible, one should observe light beats, just as one obtains sound beats from two sources of sound with slightly different frequencies. Discuss how one might experimentally look for this possibility.

PROBLEMS

1(2). In a double-slit arrangement the slits are separated by a distance equal to 100 times the wavelength of the light passing through the slits. (a) What is the angular separation between the first and second maxima? (b) What is the linear distance between the first and second maxima if the screen is at a distance of 50 cm from the slits?
Answer: (a) 0.57°. (b) 5.0 mm.

2(2). Design a double-slit arrangement that will produce interference fringes 1.0° apart on a distant screen. Assume sodium light ($\lambda = 5890$ A).

3(2). Refering to Fig. 37–6, in a double-slit experiment, $\lambda = 5460$ A, $d = 0.10$ mm, and $D = 20$ cm. What is the linear distance between the 5th maximum and 7th minimum?

Answer: 2.7 mm.

4(2). In Young's experiment, how are Eqs. 37–1 and 37–2 changed if the light emitted from the two slits has a phase difference of 180°?

5(2). In Young's interference experiment in a large ripple tank (see Fig. 37–4) the coherent vibrating sources are placed 12.0 cm apart. The distance between maxima 2.00 meters away is 18.0 cm. If the speed of ripples is 25.0 cm/sec, find the frequency of the vibrators.

Answer: 23.1 cycles/sec.

6(2). If the distance between the first and tenth minima of a double-slit pattern is 2.4 mm and the slits are separated by 0.15 mm with the screen 50 cm from the slits, what is the wavelength of the light used?

7(2). A double-slit arrangement produces interference fringes for sodium light ($\lambda = 5890$ A) that are 0.20° apart. For what wavelength would the angular separation be 10% greater?

Answer: 6500 A.

8(2). Sodium light ($\lambda = 5890$ A) falls on a double slit of separation $d = 0.20$ mm. A thin lens ($f = +1.0$ meter) is placed near the slit as in Fig. 37–6. What is the linear fringe separation on a screen placed in the focal plane of the lens?

9(2). In a double-slit arrangement the distance between slits is 5.0 mm and the slits are 1.0 meter from the screen. Two interference patterns can be seen on the screen, one due to light of 4800 A and the other 6000 Å. What is the separation on the screen between the third-order interference fringes of the two different patterns?

Answer: 0.072 mm.

10(2). A double-slit arrangement produces interference fringes for sodium light ($\lambda = 5890$ A) that are 0.20° apart. What is the angular fringe separation if the entire arrangement is immersed in water?

11(2). In the front of a lecture hall, a coherent beam of monochromatic light from a helium-neon laser ($\lambda = 6400$ A) illuminates a double slit. From there it travels a distance d (20 meters) to a mirror at the back of the hall, and returns the same distance to a screen. (*a*) In order that the distance between interference maxima be s (10 cm) what should be the distance between the two slits? (*b*) State briefly what you will see if the lecturer slips a thin sheet of cellophane over one of the slits. The optical path length through the cellophane is 2.5 wavelengths longer than the equivalent air path length.

Answer: (*a*) 0.26 mm. (*b*) In place of the central maximum you get a minimum.

12(2). A thin flake of mica ($n = 1.6$) is used to cover one slit of a double-slit arrangement. The central point on the screen is occupied by what used to be the seventh bright fringe. If $\lambda = 5500$ A, what is the thickness of the mica?

13(2). One slit of a double-slit arrangement is covered by a thin glass plate of refractive index 1.4 and the other by a thin glass plate of refractive index 1.7. The point on the screen where the central maximum fell before the glass plates were inserted is now occupied by what had been the

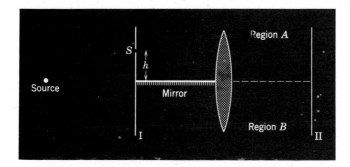

FIGURE 37–16   Problem 14(2).

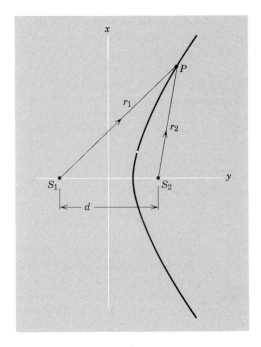

FIGURE 37-17 Problem 15(2).

fifth bright fringe before. Assume that $\lambda = 4800$ A and that the plates have the same thickness t. Find t.

Answer: 8.0×10^{-6} meters.

14(2). In Fig. 37–16, light of wavelength λ falls on S, a very narrow slit in an otherwise opaque screen I. A plane mirror, whose surface includes the axis of the lens shown, is located a distance h below S. Screen II is at the focal plane of the lens. (*a*) Find the condition for maxima and minima brightness of fringes on screen II in terms of the usual angle θ, the wavelength λ, and the distance h. (*b*) Do fringes appear only in region A (above the axis of the lens), only in region B (below the axis of the lens), or in both regions A and B? (Hint: Consider the image of S formed by the mirror.)

15(2). Two point sources in Fig. 37–17 emit coherent waves. Show that curves, such as that given, over which the phase difference for rays r_1 and r_2 is a constant are hyperbolas. Extend the analysis to three dimensions. (Hint: A constant phase difference implies a constant difference in length between r_1 and r_2.)

16(2). Sodium light ($\lambda = 5890$ A) falls on a double slit of separation $d = 2.0$ mm. D in Fig. 37–5 is 4.0 cm. What per cent error do we make in locating the tenth bright fringe because we assume that $D \gg d$?

17(2). As shown in Fig. 37–18 A and B are two identical radiators of waves that are in phase and of the same wavelength λ. The radiators are separated by distance 3.0 λ. Find the largest distance from A, along the line Ax, for which destructive interference occurs. Express this in terms of λ. *Answer:* 8.8 λ.

18(2). Two coherent radio point sources separated by 2.0 meters are radiating in phase with $\lambda = 0.50$ meter. A detector moved in a closed path around the two sources in a plane containing them will show how many maxima?

19(4). S_1 and S_2 in Fig. 37–19 are effective point sources of radiation, excited by the same os-

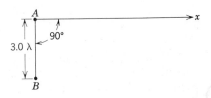

FIGURE 37-18 Problem 17(2).

FIGURE 37-19   Problem 19(4).

cillator. They are coherent and in phase with each other. Placed 4.0 meters apart, they emit equal amounts of power in the form of 1.0-meter wavelength electromagnetic waves. (a) Find the positions of the first (that is, the nearest), the second, and the third maxima of the received signal, as the detector is moved out along Ox. (b) Is the intensity at the nearest minimum equal to zero? *Answer:* (a) 1.2 meters; 3.0 meters; 7.5 meters. (b) No.

20(4). If a radiation source A leads source B in phase by 90° and the distance r_A to a detector is greater than the distance r_B by 100 meters, what is the phase difference at the detector? Both sources have a wavelength of 400 meters.

21(4). Light of wavelength 6000 A is incident normally on a system of two parallel narrow slits separated by 0.60 mm. Sketch the intensity in the pattern observed on a distant screen as a function of angle θ, as in Fig. 37-5, for the range of values $0 \leqslant \theta \leqslant 0.0040$ radians.

22(4). Show that the half-width $\Delta\theta$ of the double-slit interference fringes (see arrow in Fig. 37-9) is given by

$$\Delta\theta = \frac{\lambda}{2d}$$

if θ is small enough so that $\sin \theta \cong \theta$.

23(4). Prove Eq. 37-11a, and hence Eq. 37-11b, by trigonometry, without using phasors.

24(4). One of the slits of a double-slit system is wider than the other, so that the amplitude of the light reaching the central part of the screen from one slit, acting alone, is twice that from the other slit, acting alone. Derive an expression for I_θ in terms of θ, corresponding to Eqs. 37-11b and c.

25(4). Find the sum of the following quantities (a) graphically, by the vector method and (b) by trigonometry:

$$y_1 = 10 \sin \omega t$$
$$y_2 = 8.0 \sin (\omega t + 30°).$$

Answer: $y = 17 \sin (\omega t + 13°)$.

26(4). Add the following quantities graphically, using the vector method:

$$y_1 = 10 \sin \omega t$$
$$y_2 = 15 \sin (\omega t + 30°)$$
$$y_3 = 5.0 \sin (\omega t - 45°).$$

27(5). A soap film on a wire loop held in air appears black at its thinnest portion when viewed by reflected light. On the other hand, a thin oil film floating on water appears bright at its thinnest portion when similarly viewed from the air above. Explain.

28(5). An oil drop ($n = 1.2$) floats on a water ($n = 1.3$) surface and is observed from above by reflected light (see Fig. 37-20). (a) Will the outer (thinnest) regions of the drop correspond to a bright or a dark region? (b) Approximately how thick is the oil film where one observes the third blue region from the outside of the drop? (c) Why do the colors gradually disappear as the oil thickness becomes larger?

29(5). A tanker has dumped a large quantity of kerosene ($n = 1.2$) into the Gulf of Mexico, creating a large slick on top of the water ($n = 1.3$). (a) If you are looking straight down from an airplane onto a region of the slick where its thickness is 4600 A, for which wavelength(s) of visible

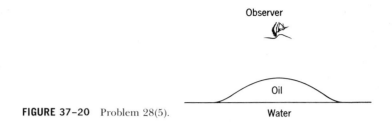

FIGURE 37-20 Problem 28(5).

light is the reflection the greatest? (b) If you are swimming directly under this same region of the slick, for which wavelengths of visible light is the transmitted intensity the strongest?

Answer: (a) 5520 A. (b) 7360 A.

30(5). From a medium of index of refraction n_1, monochromatic light of wavelength λ falls normally on a thin film of uniform thickness and index of refraction n_2. The transmitted light travels in a medium of index of refraction n_3. Find expressions for the minimum film thickness (in terms of λ and the indices of refraction) for the following cases: (a) $n_1 < n_2 > n_3$—minimum reflected light; (b) $n_1 < n_2 > n_3$—maximum transmitted light; (c) $n_1 < n_2 < n_3$—minimum reflected light; (d) $n_1 < n_2 < n_3$—maximum transmitted light; and (e) $n_1 < n_2 < n_3$—maximum reflected light.

31(5). A thin film 4.0×10^{-5} cm thick is illuminated by white light normal to its surface. Its index of refraction is 1.5. What wavelengths within the visible spectrum will be intensified in the reflected beam?

Answer: 4800 A (blue).

32(5). We wish to coat a flat piece of glass ($n = 1.50$) with a transparent material ($n = 1.25$) so that light of wavelength 6000 A (in vacuum) incident normally is not reflected. What thicknesses of the material will bring this about?

33(5). A sheet of glass having an index of refraction of 1.40 is to be coated with a film of material having a refractive index of 1.55 such that green light (wavelength = 5250 A) is preferentially transmitted. (a) What is the minimum thickness of the film that will achieve the result? (b) Will any other parts of the visible spectrum also be preferentially transmitted? (c) Will the transmission of any colors be sharply reduced?

Answer: (a) 1690 A. (b) No. (c) Blue-violet will be sharply reduced.

34(5). A thin film of acetone (refractive index 1.25) is floated on a thick glass plate (refractive index 1.50). Plane light waves of variable wavelength are incident normal to the film. When one views the reflected wave it is noted that complete destructive interference occurs at 6000 A and constructive interference at 7000 A. Calculate the thickness of the acetone film.

35(5). White light reflected at perpendicular incidence from a soap film has, in the visible spectrum, an interference maximum at 6000 A and a minimum at 4500 A, with no minimum in between. If $n = 1.33$ for the film, what is the film thickness, assumed uniform?

Answer: 3380 A.

36(5). A plane wave of monochromatic light falls normally on a uniformly thin film of oil which covers a glass plate. The wavelength of the source can be varied continuously. Complete destructive interference of the reflected light is observed for wavelengths of 5000 and 7000 A and for no other

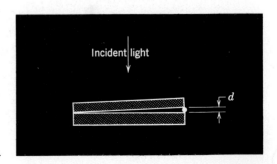

FIGURE 37-21 Problems 37(5) and 38(5).

wavelengths in between. If the index of refraction of the oil is 1.30 and that of the glass is 1.50, find the thickness of the oil film.

37(5). In Fig. 37–21 white light is incident from above. (*a*) Observed from above, why does the region near the edge, where the two glass plates touch, appear black? (*b*) For what part of the visible spectrum does destructive interference next occur? (*c*) What color does an observer see where this destructive interference occurs?
Answer: (*b*) Violet. (*c*) Red.

38(5). A broad source of light ($\lambda = 6800$ A) illuminates normally two glass plates 12 cm long that touch at one end and are separated by a wire 0.048 mm in diameter at the other (Fig. 37–21). How many bright fringes appear over the 12-cm distance?

39(5). Light of wavelength 6300 A is incident normally on a thin wedge-shaped film of index of refraction 1.5. There are ten bright and nine dark fringes over the length of film. By how much does the film thickness change over this length?
Answer: 1.9×10^{-6} meters.

40(5). A plane monochromatic light wave in air falls at normal incidence on a thin film of oil which covers a glass plate. The wavelength of the source may be varied continuously. Complete destructive interference in the reflected beam is observed for wavelengths of 5000 and 7000 A and for no other wavelength in between. The index of refraction of glass is 1.50. Show that the index of refraction of the oil must be less than 1.50.

41(5). In Example 4 assume that there is zero reflection for light of wavelength 5500 A at normal incidence. Calculate the factor by which the reflection is diminished by the coating at 4500 and at 6500 A. Assume that the amplitude of a wave reflected from an air-glass interface is the sum of (the magnitudes of) the amplitudes of the waves reflected from the air-MgF_2 and the MgF_2-glass interfaces.
Answer: The intensity is diminished by 88% at 4500 A and by 94% at 6500 A.

42(6). If mirror M_2 in Michelson's interferometer is moved through 0.233 mm, 792 fringes are counted. What is the wavelength of the light?

43(6). A thin film with $n = 1.40$ for light of wavelength 5890 A is placed in one arm of a Michelson interferometer. If a shift of 7.00 fringes occurs, what is the film thickness?
Answer: 51,500 A.

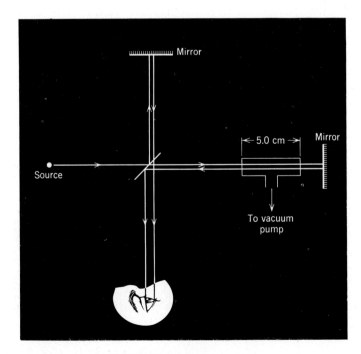

FIGURE 37–22   Problem 44(6).

44(6). An air-tight chamber 5.0 cm long with glass windows is placed in one arm of a Michelson interferometer as indicated in Fig. 37–22. Light of $\lambda = 5000$ A is used. The air is slowly evacuated from the chamber using a vacuum pump. While the air is being removed, 60 fringes are observed to pass through the view. From these data, find the index of refraction of air at atmospheric pressure.

45(6). Write an expression for the intensity observed in Michelson's interferometer (Fig. 37–15) as a function of the position of the moveable mirror. Measure the position of the mirror from the point at which $d_1 = d_2$.

Answer: $I = I_m \cos^2 \left(\dfrac{2\pi x}{\lambda} \right)$.

46(6). A Michelson interferometer is used with a sodium discharge tube as a light source. The yellow sodium light consists of two wavelengths, 5890 and 5896 A. It is observed that the interference pattern disappears and reappears periodically as one moves mirror M_2 in Fig. 37–15. (*a*) Explain this effect. (*b*) Calculate the change in path difference between two successive reappearances of the interference pattern.

CHAPTER 38

Diffraction, Gratings, and Spectra

38–1 Diffraction

If we allow light, assumed monochromatic for illustration, to fall on a single slit whose width a is of the order of magnitude of the wavelength λ, the light will flare out as it passes through the slit. It will form patterns like those of Fig. 38–1, in which we see not only a flaring out but the appearance of secondary maxima; we ignored these secondary maxima in Fig. 37–1. We call this phenomenon *diffraction*. Because the basic assumptions of geometrical optics with regard to the dimensions of the slit are not satisfied (see Section 36–1), we must look to the full power of wave optics for an explanation.

We can see the diffraction of light by looking through a crack between two fingers at a distant light source such as a tubular neon sign or by looking at a distant street light through a cloth umbrella. However, most sources of light have an extended area so that a diffraction pattern produced by one point of the source will overlap that produced by another. Also, common sources of light are not monochromatic. The

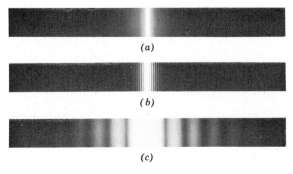

(a)

(b)

(c)

FIGURE 38–1 (*a*) The intensity of monochromatic light diffracted from a single slit of width $a \cong 60{,}000$ A and falling on a screen 50 cm beyond. (*b*) The slit width is reduced by a factor of two. (*c*) The slit width is further reduced by an additional factor of seven.

patterns for the various wavelengths overlap and again the effect is less apparent. We have already assumed the diffraction of light in our study of double-slit interference when we assumed the flaring out of light passing through slits S_1 and S_2 in Figs. 37–5 and 37–6.

Diffraction was discovered by Francesco Maria Grimaldi (1618–1663), and the phenomenon was known both to Huygens (1629–1695) and to Newton (1642–1727). Newton did not see in it any justification for a wave theory for light. Huygens, although he believed in a wave theory, did not believe in diffraction! He imagined his secondary wavelets to be effective only at the point of tangency to their common envelope, thus denying the possibility of diffraction.

Figure 38–2 shows a diffraction situation. Surface A is a wavefront that falls on B, which is an opaque screen containing an aperture of arbitrary shape; C is a diffusing screen that receives the light that passes through this aperture. The pattern of light intensity on C can be calculated by subdividing the wavefront into elementary areas $d\mathbf{S}$, each of which becomes a source of an expanding Huygens wavelet. The light intensity at an arbitrary point P is found by superimposing the wave disturbances (that is, the $\mathbf{E}$-vectors) caused by the wavelets reaching P from all these elementary radiators. Instead of a screen with a hole in it we could equally well use an opaque barrier such as a wire or a coin; the general treatment would be the same.

The wave disturbances reaching P differ in amplitude and in phase because (*a*) the elementary radiators are at varying distances from P, (*b*) the light leaves the radiators at various angles to the normal to the wavefront, and (*c*) some radiators are blocked by screen B; others are not. Diffraction calculations—simple in principle—may become difficult in practice. The calculation must be repeated for every point on screen C at which we wish to know the light intensity. We followed exactly this program in calculating the double-slit intensity pattern in Section 37–4. The calculation there was simple because we assumed only two radiators, the two narrow slits.

Figure 38–3*a* shows the general case of *Fresnel diffraction,* in which the light source and/or the screen on which the diffraction pattern is displayed are a finite distance from the diffracting aperture; the wavefronts that fall on the diffracting aperture in this case and that leave it to illuminate any point P of the diffusing screen are not planes; the corresponding rays are not parallel.

A simplification results if source S and screen C are moved to a large distance from

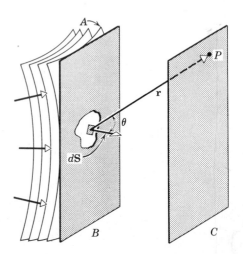

FIGURE 38–2 Light is diffracted at the aperture in screen B and illuminates screen C. The intensity at P is found by dividing the wavefront at B into elementary radiators $d\mathbf{S}$ and combining their effects at P.

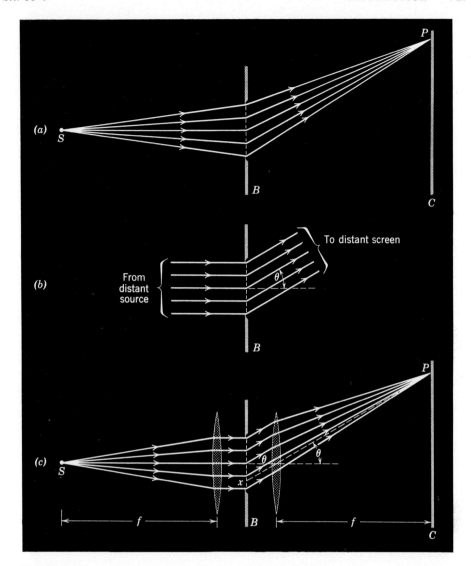

FIGURE 38–3 (*a*) Fresnel diffraction. (*b*) Source *S* and screen *C* are moved to a large distance, resulting in Fraunhofer diffraction. (*c*) Fraunhofer diffraction conditions produced by lenses, leaving source *S* and screen *C* in their original positions.

the diffraction aperture, as in Fig. 38–3*b*. This limiting case is called *Fraunhofer diffraction*. The wavefronts arriving at the diffracting aperture from the distant source *S* are then planes, and the rays associated with these wavefronts are parallel to each other. Fraunhofer conditions can be established in the laboratory by using two converging lenses, as in Fig. 38–3*c*. The first of these converts the diverging wave from the source into a plane wave. The second lens causes plane waves leaving the diffracting aperture to converge to point *P*. All rays that illuminate *P* will leave the diffracting aperture parallel to the dashed line *Px* drawn from *P* through the center of this second (thin) lens. We assumed Fraunhofer conditions for Young's double-slit experiment in Section 37–2 (see Fig. 37–6).

Although Fraunhofer diffraction is a limiting case of the more general Fresnel diffraction, it is an important limiting case and is easier to handle mathematically. This book deals only with Fraunhofer diffraction.

38–2 Single Slit

Figure 38–4 shows a plane wave falling at normal incidence on a long narrow slit of width a. Let us focus our attention on the central point P_0 of screen C. A set of horizontal, parallel rays (not shown in the figure) emerging from the slit will be focussed at P_0. These rays all have the same optical path lengths, as we saw in Section 36–4. Since they are in phase at the plane of the slit, they will still be in phase at P_0, and the central point of the diffraction pattern that appears on screen C has a maximum intensity.

We now consider another point on the screen. Light rays which reach P_1 in Fig. 38–4 leave the slit at an angle θ as shown. Ray r_1 originates at the top of the slit and ray r_2 at its center. If θ is chosen so that the distance bb' in the figure is one-half a wavelength, r_1 and r_2 will be out of phase and will produce no effect at P_1. In fact, every ray from the upper half of the slit will be canceled by a ray from the lower half, originating at a point $a/2$ below the first ray. The point P_1, the first minimum of the diffraction pattern, will have zero intensity (compare Fig. 38–1c).

The condition shown in Fig. 38–4 is

$$\frac{a}{2} \sin \theta = \frac{\lambda}{2},$$

or
$$a \sin \theta = \lambda. \tag{38–1}$$

Equation 38–1 shows that the central maximum becomes wider as the slit becomes narrower, that is, θ increases as a decreases. Figure 38–1 bears this out. For $a = \lambda$ we have $\theta = 90°$, which implies that the central maximum fills the entire forward hemisphere.

In Fig. 38–5 the slit is divided into four equal zones, with a ray shown leaving the

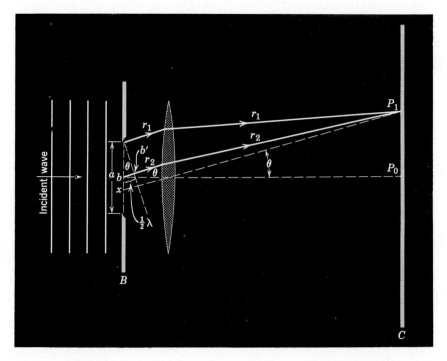

FIGURE 38–4 Conditions at the first minimum of the diffraction pattern.

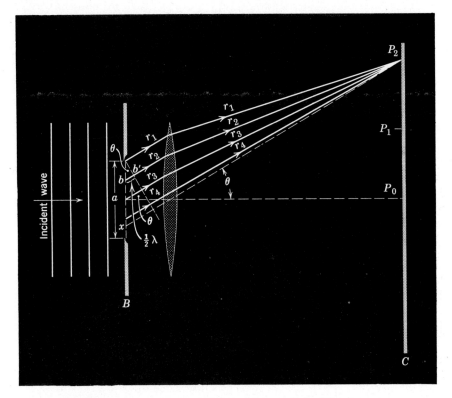

FIGURE 38-5 Conditions at the second minimum of the diffraction pattern.

top of each zone. Let θ be chosen so that the distance bb' is one-half a wavelength. Rays r_1 and r_2 will then cancel at P_2. Rays r_3 and r_4 will also be half a wavelength out of phase and will also cancel. Consider four other rays, emerging from the slit a given distance below the four rays above. The two rays below r_1 and r_2 will cancel uniquely, as will the two rays below r_3 and r_4. We can proceed across the entire slit and conclude again that no light reaches P_2; we have located a second point of zero intensity.

The condition described (see Fig. 38–5) requires that

$$\frac{a}{4} \sin \theta = \frac{\lambda}{2},$$

or

$$a \sin \theta = 2\lambda.$$

By extension, the general formula for the minima in the diffraction pattern on screen C is

$$a \sin \theta = m\lambda \qquad m = 1, 2, 3, \ldots \text{(minima)}. \qquad (38\text{-}2)$$

There is a maximum approximately halfway between each adjacent pair of minima.

■ **Example 1.** A slit of width a is illuminated by white light. For what value of a will the first minimum for red light ($\lambda = 6500$ A) fall at $\theta = 30°$?

At the first minimum we put $m = 1$ in Eq. 38–2. Doing so and solving for a yields

$$a = \frac{m\lambda}{\sin \theta} = \frac{(1)(6500 \text{ A})}{\sin 30°} = 13{,}000 \text{ A}.$$

Note that the slit width must be twice the wavelength in this case.

Example 2. In Example 1 what is the wavelength λ' of the light whose first diffraction maximum (not counting the central maximum) falls at $\theta = 30°$, thus coinciding with the first minimum for red light?

This maximum is about halfway between the first and second minima. It can be found without too much error by putting $m = 1.5$ in Eq. 38–2, or

$$a \sin \theta \cong 1.5\lambda'.$$

From Example 1, however, $a \sin \theta = \lambda.$

Dividing gives $$\lambda' = \frac{\lambda}{1.5} = \frac{6500 \text{ A}}{1.5} = 4300 \text{ A}.$$

Light of this wavelength is violet. The second maximum for light of wavelength 4300 A will always coincide with the first minimum for light of wavelength 6500 A, no matter what the slit width. If the slit is relatively narrow, the angle θ at which this overlap occurs will be relatively large.

38–3   Diffraction from a Single Slit—Qualitative

Figure 38–6 shows a slit of width a divided into N parallel strips of width Δx. Each strip acts as a radiator of Huygens' wavelets and produces a characteristic wave disturbance at point P, whose position on the screen, for a particular arrangement of apparatus, can be described by the angle θ.

If the strips are narrow enough ($\Delta x \ll \lambda$) all points on a given strip have essentially

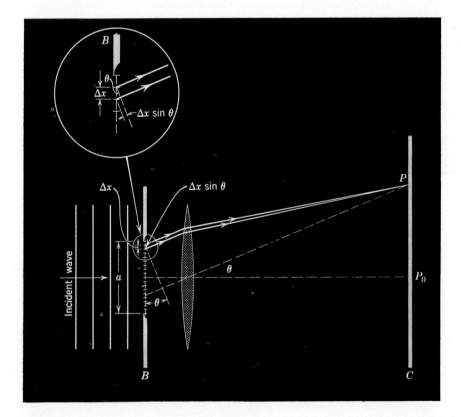

FIGURE 38–6   A slit of width a is divided into N strips of width Δx. The insert shows conditions at the second strip more clearly. In the differential limit the slit is divided into an infinite number of strips (that is, $N \to \infty$) of differential width dx. For clarity in this and the following figure, we take $N = 18$.

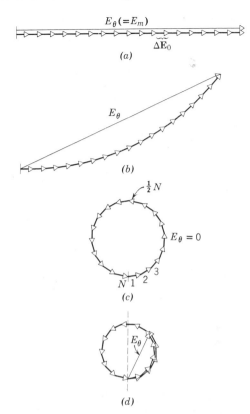

FIGURE 38–7 Conditions at (*a*) the central maximum for Fig. 38–6, (*b*) a direction slightly removed from the central maximum, (*c*) the first minimum, and (*d*) the first maximum beyond the central maximum for single-slit diffraction. This figure corresponds to $N = 18$ in Fig. 38–6.

the same optical path length to P, and therefore all the light from the strip will have the same phase when it arrives at P. The amplitudes ΔE_0 of the wave disturbances at P from the various strips may be taken as equal if θ in Fig. 38–6 is not too large.*

The wave disturbances from adjacent strips have a constant phase difference $\Delta\phi$ between them at P given by

$$\frac{\text{phase difference}}{2\pi} = \frac{\text{path difference}}{\lambda},$$

or
$$\Delta\phi = \left(\frac{2\pi}{\lambda}\right)(\Delta x \sin\theta), \tag{38–3}$$

where $\Delta x \sin\theta$ is, as the figure insert shows, the path difference for rays originating at the top edges of adjacent strips. Thus, at P, N vectors with the same amplitude ΔE_0, the same frequency, and the same phase difference $\Delta\phi$ between adjacent members combine to produce a resultant disturbance. We ask, for various values of $\Delta\phi$ [that is, for various points P on the screen, corresponding to various values of θ (see Eq. 38–3)], what is the amplitude E_θ of the resultant wave disturbance? We find the answer by representing the individual wave disturbances ΔE_0 by phasors and calculating the resultant phasor amplitude, as described in Section 37–4.

At the center of the diffraction pattern θ equals zero, and the phase shift between adjacent strips (see Eq. 38–3) is also zero. As Fig. 38–7*a* shows, the phasor arrows in this case are laid end to end and the amplitude of the resultant has its maximum value E_m. This corresponds to the center of the central maximum.

---

* The socalled *obliquity effect* will be important in some of what follows (Fig. 38–9*a*, Fig. 38–16, Problem 38–7*a*, etc) but we do not take it into account.

As we move to a value of θ other than zero, $\Delta\phi$ assumes a definite nonzero value (again see Eq. 38–3), and the array of arrows is now as shown in Fig. 38–7b. The resultant amplitude E_θ is less than before. Note that the length of the "arc" of small arrows is the same for both figures and indeed for all figures of this series. As θ increases further, a situation is reached (Fig. 38–7c) in which the chain of arrows curls around through 360°, the tip of the last arrow touching the foot of the first arrow. This corresponds to $E_\theta = 0$, that is, to the first minimum. For this condition the ray from the top of the slit (1 in Fig. 38–7c) is 180° out of phase with the ray from the center of the slit ($\frac{1}{2}N$ in Fig. 38–7c). These phase relations are consistent with Fig. 38–4, which also represents the first minimum.

As θ increases further, the phase shift continues to increase, and the chain of arrows coils around through an angular distance greater than 360°, as in Fig. 38–7d, which corresponds to the first maximum beyond the central maximum. This maximum is much smaller than the central maximum. In making this comparison, recall that the arrows marked E_θ in Fig. 38–7 correspond to the amplitudes of the wave disturbance and not to the intensities. The amplitudes must be squared to obtain the corresponding relative intensities (see Eq. 37–7).

38–4   Diffraction from a Single Slit—Quantitative

The "arc" of small arrows in Fig. 38–8 shows the phasors representing, in amplitude and phase, the wave disturbances that reach an arbitrary point P on the screen of Fig. 38–6, corresponding to a particular angle θ. The resultant amplitude at P is E_θ. If we divide the slit of Fig. 38–6 into infinitesimal strips of width dx, the arc of arrows in Fig. 38–8 approaches the arc of a circle, its radius R being indicated in that figure. The length of the arc is E_m, the amplitude at the center of the diffraction pattern, for at the center of the pattern the wave disturbances are all in phase and this "arc" becomes a straight line as in Fig. 38–7a.

The angle ϕ in the lower part of Fig. 38–8 is revealed as the difference in phase between the infinitesimal vectors at the left and right ends of the arc E_m. This means that ϕ is the phase difference between rays from the top and the bottom of the slit of Fig. 38–6. From geometry we see that ϕ is also the angle between the two radii

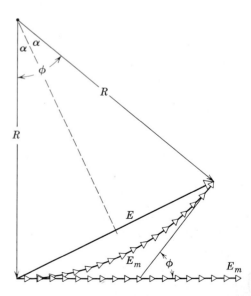

FIGURE 38–8   A construction used to calculate the intensity in single-slit diffraction. The situation corresponds to that of Fig. 38–7b.

marked R in Fig. 38–8. From this figure we can write

$$E_\theta = 2R \sin \frac{\phi}{2}.$$

In radian measure ϕ, from the figure, is

$$\phi = \frac{E_m}{R}.$$

Combining yields
$$E_\theta = \frac{E_m}{\phi/2} \sin \frac{\phi}{2},$$

or
$$E_\theta = E_m \frac{\sin \alpha}{\alpha}, \tag{38–4}$$

in which
$$\alpha = \frac{\phi}{2}. \tag{38–5}$$

From Fig. 38–6, recalling that ϕ is the phase difference between rays from the top and the bottom of the slit and that the path difference for these rays is $a \sin \theta$, we have

$$\frac{\text{phase difference}}{2\pi} = \frac{\text{path difference}}{\lambda},$$

or
$$\phi = \left(\frac{2\pi}{\lambda}\right)(a \sin \theta).$$

Combining with Eq. 38–5 yields

$$\alpha = \frac{\phi}{2} = \frac{\pi a}{\lambda} \sin \theta. \tag{38–6}$$

Equation 38–4, taken together with the definition of Eq. 38–6, gives the amplitude of the wave disturbance for a single-slit diffraction pattern at any angle θ. The intensity I_θ for the pattern is proportional to the square of the amplitude, or

$$I_\theta = I_m \left(\frac{\sin \alpha}{\alpha}\right)^2. \tag{38–7}$$

For convenience we display together, and renumber, the formulas for the amplitude and the intensity in single-slit diffraction.

[Eq. 38–4] $$E_\theta = E_m \frac{\sin \alpha}{\alpha}$$ $$\tag{38–8a}$$

[Eq. 38–7] $$I_\theta = I_m \left(\frac{\sin \alpha}{\alpha}\right)^2 \quad \begin{array}{l} \text{single-slit} \\ \text{diffraction} \end{array}$$ $$\tag{38–8b}$$

[Eq. 38–6] $$\alpha \, (= \tfrac{1}{2}\phi) = \frac{\pi a}{\lambda} \sin \theta$$ $$\tag{38–8c}$$

Figure 38–9 shows plots of I_θ for several values of the ratio a/λ. Note that the pattern becomes narrower as a/λ is increased; compare this figure with Figs. 37–1 and 38–1.

Minima occur in Eq. 38–8b when

$$\alpha = m\pi \qquad m = 1, 2, 3, \ldots. \tag{38–9}$$

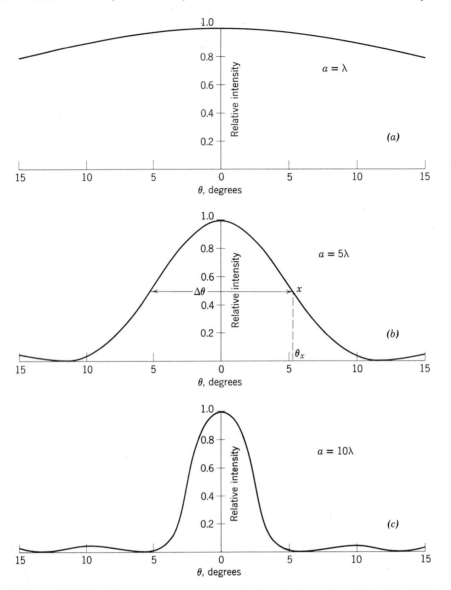

FIGURE 38–9   The relative intensity in single-slit diffraction for three values of the ratio a/λ. The arrow in (*b*) shows the half-width $\Delta\theta$ of the central maximum.

Combining with Eq. 38–8*c* leads to

$$a \sin \theta = m\lambda \qquad m = 1, 2, 3, \ldots \text{(minima)},$$

which is the result derived in a preceding section (Eq. 38–2). In that section, however, we derived only this result, obtaining no quantitative information about the intensity of the diffraction pattern at places in which it was not zero. Here (Eqs. 38–8) we have complete intensity information.

■   **Example 3.** *Intensities of the secondary diffraction maxima.* Calculate, approximately, the relative intensities of the secondary maxima in the single-slit Fraunhofer diffraction pattern.

The secondary maxima lie approximately halfway between the minima and are found (compare Eq. 38–9) from

$$\alpha \cong (m + \tfrac{1}{2})\pi \qquad m = 1, 2, 3, \ldots .$$

Substituting into Eq. 38–8*b* yields

$$I_\theta = I_m \left[\frac{\sin (m + \tfrac{1}{2})\pi}{(m + \tfrac{1}{2})\pi} \right]^2,$$

which reduces to

$$\frac{I_\theta}{I_m} = \frac{1}{(m + \tfrac{1}{2})^2 \pi^2}.$$

This yields, for $m = 1, 2, 3, \ldots$, $I_\theta/I_m = 0.045, 0.016, 0.0083$, etc. The successive maxima decrease rapidly in intensity. Figure 38–1 has been deliberately overexposed to reveal the secondary maxima.

38–5 Diffraction from a Circular Aperture

Here we consider diffraction by a circular aperture of diameter d, the aperture constituting the boundary of a circular lens.

Our previous treatment of lenses was based on geometrical optics, diffraction being specifically assumed not to occur. A rigorous analysis would be based from the beginning on wave optics, since geometrical optics is always an approximation, although often a good one. Diffraction phenomena would emerge in a natural way from such a wave-optical analysis.

Figure 38–10 shows the image of a distant point source of light (a star, for instance) formed on a photographic film placed in the focal plane of a converging lens. It is not a point, as the (approximate) geometrical optics treatment suggests, but a circular disk surrounded by several progressively fainter secondary rings. Comparison with Fig. 38–1*c* leaves little doubt that we are dealing with a diffraction phenomenon in which, however, the aperture is a circle rather than a long narrow slit. The ratio d/λ, where d is the diameter of the lens (or of a circular aperture placed in front of the lens), determines the scale of the diffraction pattern, just as the ratio a/λ does for a slit.

Analysis shows that the first minimum for the diffraction pattern of a circular aperture of diameter d, assuming Fraunhofer conditions, is given by

$$\sin \theta = 1.22 \frac{\lambda}{d}. \qquad (38\text{–}10)$$

This is to be compared with Eq. 38–1, or

$$\sin \theta = \frac{\lambda}{a},$$

which locates the first minimum for a long narrow slit of width a. The factor 1.22 emerges from the mathematical analysis when we integrate over the elementary radiators into which the circular aperture may be divided.

The fact that lens images are diffraction patterns is important when we wish to distinguish two distant point objects whose angular separation is small. Figure 38–11 shows the visual appearances and the corresponding intensity

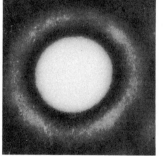

FIGURE 38–10   The image of a star formed by a converging lens is a diffraction pattern. Note the central maximum, sometimes called the Airy disk (after George Airy, who first solved the problem of diffraction at a circular aperture in 1835), and the circular secondary maximum. Other secondary maxima occur at larger radii but are too faint to be seen in this figure.

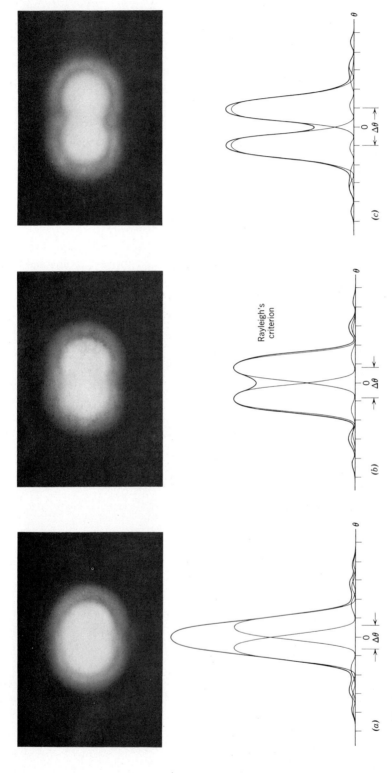

FIGURE 38–11 The images of two distant point objects are formed by a converging lens whose diameter ($= 10$ cm) is 200,000 times the effective wavelength ($= 5000$ Å). Sketches of the images as they appear in the focal plane of the lens are shown with the corresponding intensity plots below them. (*a*) The angular separation of the objects (see vertical ticks) is so small that the images are not resolved. (*b*) The objects are farther apart and the images meet Rayleigh's criterion for resolution. (*c*) The objects are still farther apart and the images are well resolved. The Rayleigh criterion is useful only if the two sources are of comparable brightness.

736

patterns for two distant point objects with small angular separations. In (*a*) the objects are not resolved; that is, they cannot be distinguished from a single point object. In (*b*) they are barely resolved and in (*c*) they are fully resolved.

In Fig. 38–11*b* the angular separation of the two point sources is such that the maximum of the diffraction pattern of one source falls on the first minimum of the diffraction pattern of the other. This is called *Rayleigh's criterion*. This criterion, though useful, is arbitrary; other criteria for deciding when two objects are resolved are sometimes used. From Eq. 38–10, two objects that are barely resolvable by Rayleigh's criterion must have an angular separation θ_R of

$$\theta_R = \sin^{-1} \frac{1.22\lambda}{d}.$$

Since the angles involved are small, we can replace $\sin \theta_R$ by θ_R, or

$$\theta_R = 1.22 \frac{\lambda}{d}. \tag{38–11}$$

If the angular separation θ between the objects is greater than θ_R, we can resolve the two objects; if it is significantly less, we cannot. The objects must be of comparable brightness.

Example 4. A converging lens 3.0 cm in diameter has a focal length f of 20 cm. (*a*) What angular separation must two distant point objects have to satisfy Rayleigh's criterion? Assume that $\lambda = 5500$ A.

From Eq. 38–11,

$$\theta_R = 1.22 \frac{\lambda}{d} = \frac{(1.22)(5.5 \times 10^{-7} \text{ meter})}{3.0 \times 10^{-2} \text{ meter}} = 2.2 \times 10^{-5} \text{ radian}.$$

(*b*) How far apart are the centers of the diffraction patterns in the focal plane of the lens? The linear separation is

$$x = f\theta = (20 \text{ cm})(2.2 \times 10^{-5} \text{ radian}) = 44{,}000 \text{ A}.$$

This is 8.0 wavelengths of the light employed.

When one wishes to use a lens to resolve objects of small angular separation, it is desirable to make the central disk of the diffraction pattern as small as possible. This can be done (see Eq. 38–11) by increasing the lens diameter or by using a shorter wavelength. One reason for constructing large telescopes is to produce sharper images so that celestial objects can be examined in finer detail. The images are also brighter, not only because the energy is concentrated into a smaller diffraction disk but because the larger lens collects more light.

To reduce diffraction effects in microscopes we often use ultraviolet light, which, because of its shorter wavelength, permits finer detail to be examined than would be possible for the same microscope operated with visible light. We shall see in Chapter 40 that beams of electrons can behave like waves under some circumstances. In the electron microscope such beams may have an effective wavelength of 0.04 A, of the order of 10^5 times shorter than visible light ($\lambda \cong 5000$ A). This permits the detailed examination of tiny objects like viruses. If a virus were examined with an optical microscope, its structure would be hopelessly concealed by diffraction.

38–6   Diffraction from a Double Slit

In Young's double-slit experiment (Section 37–2) we asume that the slits are arbitrarily narrow (that is, $a \ll \lambda$), which means that the central part of the diffusing screen was uniformly illuminated by the diffracted waves from each slit, as Fig. 38–9a suggests. When such waves interfere, they produce fringes of uniform intensity, as in Fig. 37–9. This idealized situation cannot occur with actual slits because the condition $a \ll \lambda$ cannot usually be met. Waves from the two actual slits combining at different points of the screen will have intensities that are not uniform but are governed by the diffraction pattern of a single slit. The effect of relaxing the assumption that $a \ll \lambda$ in Young's experiment is to leave the fringes relatively unchanged in location but to alter their intensities. The net result is a combination of interference and diffraction.

The interference pattern for infinitesimally narrow slits is given by Eq. 37–11b and c or, with a small change in nomenclature,

$$I_{\theta,\text{int}} = I_{m,\text{int}} \cos^2 \beta, \tag{38–12}$$

where

$$\beta = \frac{\pi d}{\lambda} \sin \theta, \tag{38–13}$$

d being the distance between the centers of the slits.

The intensity for the diffracted wave from either slit is given by Eqs. 38–8b and c, or, with a small change in nomenclature,

$$I_{\theta,\text{dif}} = I_{m,\text{dif}} \left(\frac{\sin \alpha}{\alpha} \right)^2, \tag{38–14}$$

where

$$\alpha = \frac{\pi a}{\lambda} \sin \theta. \tag{38–15}$$

The combined effect is found by regarding $I_{m,\text{int}}$ in Eq. 38–12 as a variable amplitude, given in fact by $I_{\theta,\text{dif}}$ of Eq. 38–14. This assumption, for the combined pattern, leads to

$$I_\theta = I_m \, (\cos \beta)^2 \left(\frac{\sin \alpha}{\alpha} \right)^2, \tag{38–16}$$

in which we have dropped all subscripts referring separately to interference and diffraction.

Let us express this result in words. At any point on the screen the available light intensity from each slit, considered separately, is given by the diffraction pattern of that slit (Eq. 38–14). The diffraction patterns for the two slits, again considered separately, coincide because parallel rays in Fraunhofer diffraction are focused at the same spot (see Fig. 38–5). Because the two diffracted waves are coherent, they will interfere.

The effect of interference is to redistribute the available energy over the screen, producing a set of fringes. In Section 37–2, where we assumed $a \ll \lambda$, the available energy was virtually the same at all points on the screen so that the interference fringes had virtually the same intensites (see Fig. 37–9). If we relax the assumption $a \ll \lambda$, the available energy is not uniform over the screen but is controlled by the diffraction pattern of a slit of width a. In this case the interference fringes will have intensities that are determined by the intensity of the diffraction pattern at the location of a particular fringe. Equation 38–16 is the mathematical expression of this argument.

Figure 38–12 is a plot of Eq. 38–16 for $d = 50\lambda$ and for $a = 5\lambda$. The interference factor (Fig. 38–12a), the diffraction factor (Fig. 35–12b), and their product (Fig. 38–12c) are clearly shown.

If we put $a = 0$ in Eq. 38–16, then (see Eq. 38–15) $\alpha = 0$ and $\displaystyle\lim_{\alpha\to0} \sin\alpha/\alpha \cong \alpha/\alpha = 1$.

Thus this equation reduces, as it must, to the intensity equation for a pair of vanishingly narrow slits (Eq. 38–12). If we put $d = 0$ in Eq. 38–16, the two slits coalesce into a single slit of width a, as Fig. 38–15 shows; $d = 0$ implies $\beta = 0$ (see Eq. 38–13) and $\cos^2\beta = 1$. Thus Eq. 38–16 reduces, as it must, to the diffraction equation for a single slit (Eq. 38–14).

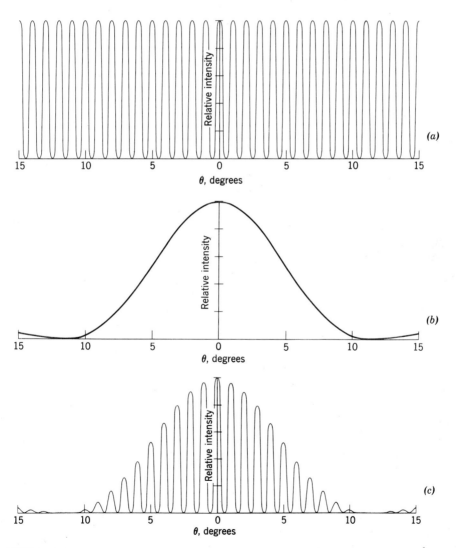

FIGURE 38–12   (*a*) The "interference factor" (due to double slit interference) and (*b*) the "diffraction factor" in Eq. 38–16 (due to diffraction from a single slit) and (*c*) their product; (compare Figs. 37–9 and 38–9(*b*). The figure is drawn for $a = 5\lambda$.

Figure 38–13 shows a double-slit interference photograph. The uniformly spaced interference fringes and their intensity modulation by the diffraction pattern of a single slit is clear. If one slit is covered up, as in Fig. 38–13*b*, the interference fringes disappear and we see the diffraction pattern of a single slit.

Example 5. In double-slit Fraunhofer diffraction what is the fringe spacing on a screen 50 cm away from the slits if they are illuminated with blue light (λ = 4800 A), if d = 0.10 mm, and if the slit width a = 0.020 mm? What is the linear distance from the central maximum to the first minimum of the fringe envelope?

The intensity pattern is given by Eq. 38–16, the fringe spacing being determined by the interference factor $\cos^2 \beta$. From Example 1, Chapter 37, we have

$$\Delta y = \frac{\lambda D}{d},$$

where D is the distance of the screen from the slits. Substituting yields

$$\Delta y = \frac{(480 \times 10^{-9} \text{ meter})(50 \times 10^{-2} \text{ meter})}{0.10 \times 10^{-3} \text{ meter}} = 2.4 \times 10^{-3} \text{ meter} = 2.4 \text{ mm}.$$

The distance to the first minimum of the envelope is determined by the diffraction factor $(\sin \alpha/\alpha)^2$ in Eq. 38–16. The first minimum in this factor occurs for $\alpha = \pi$.

From Eq. 38–15,

$$\sin \theta = \frac{\alpha \lambda}{\pi a} = \frac{\lambda}{a} = \frac{480 \times 10^{-9} \text{ meter}}{0.020 \times 10^{-3} \text{ meter}} = 0.024.$$

This is so small that we can assume that $\theta \cong \sin \theta \cong \tan \theta$, or

$$y = D \tan \theta \cong D \sin \theta = (50 \text{ cm})(0.024) = 1.2 \text{ cm}.$$

There are about ten fringes in the central peak of the fringe envelope.

Example 6. What requirements must be met for the central maximum of the envelope of the double-slit Fraunhofer pattern to contain exactly eleven fringes?

The required condition will be met if the sixth minimum of the interference factor ($\cos^2 \beta$) in Eq. 38–16 coincides with the first minimum of the diffraction factor $(\sin \alpha/\alpha)^2$.

The sixth minimum of the interference factor occurs when

$$\beta = \tfrac{11}{2} \pi$$

in Eq. 38–12.

The first minimum in the diffraction term occurs for

$$\alpha = \pi.$$

Dividing (see Eqs. 38–13 and 38–15) yields

$$\frac{\beta}{\alpha} = \frac{d}{a} = \frac{11}{2}.$$

This condition depends only on the slit geometry and not on the wavelength. For long waves the pattern will be broader than for short waves, but there will always be eleven fringes in the central peak of the envelope.

The double-slit pattern as illustrated in Fig. 38–12*c* combines interference and diffraction in an intimate way. Both are superposition effects

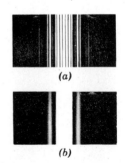

(a)

(b)

FIGURE 38–13 (*a*) Interference fringes for a double-slit system in which the slit width is not negligible in comparison to the wavelength. The fringes are modulated in intensity by the diffraction pattern of a single slit. (*b*) If one of the slits is covered up, the interference fringes disappear and we see the single slit diffraction pattern. (Courtesy G. H. Carragan, Rensselaer Polytechnic Institute.)

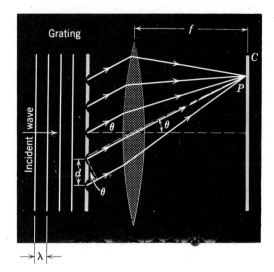

FIGURE 38–14 An idealized diffraction grating containing five slits. The slit width a is shown for convenience to be considerably smaller than λ, although this condition is not realized in practice. The figure is distorted in that f is much greater than d in practice.

and depend on adding wave disturbances at a given point, taking phase differences properly into account. If the waves to be combined originate from a finite (and usually small) number of elementary coherent radiators, as in Young's double-slit experiment, we call the effect *interference*. If the waves to be combined originate by subdividing a wave into infinitesimal coherent radiators, as in our treatment of a single slit (Fig. 38–6), we call the effect *diffraction*. This distinction between interference and diffraction is convenient and useful. However, it should not cause us to lose sight of the fact that both are superposition effects and that often both are present simultaneously, as in Young's experiment.

38–7 Multiple Slits

A logical extension of Young's double-slit interference experiment is to increase the number of slits from two to a much larger number N. An arrangement like that of Fig. 38–14, usually involving many more slits, is called a *diffraction grating*. As for a double slit, the intensity pattern that results when monochromatic light of wavelength λ falls on a grating consists of a series of interference fringes. The angular separation of these fringes are determined by the ratio λ/d, where d is the spacing between the centers of adjacent slits. The relative intensities of these fringes are determined by the diffraction pattern of a single grating slit, which depends on the ratio λ/a, where a is the slit width.

Figure 38–15, which compares the intensity patterns for $N = 2$ and $N = 5$, shows clearly that the "interference" fringes are modulated in intensity by a "diffraction" envelope, as in Fig. 38–13. Figure 38–16 shows the results of a theoretical calculation of the intensity patterns for a few fringes near the centers of the patterns of Fig. 38–15.

FIGURE 38–15 Intensity patterns for "gratings" with (a) $N = 2$ and (b) $N = 5$ for the same value of d and λ. Note how the intensities of the fringes are modulated by a diffraction envelope as in Fig. 38–12; thus the assumption $a \ll \lambda$ is not realized in these actual "gratings." For $N = 5$ three very faint secondary maxima, not visible in this photograph, appear between each pair of adjacent primary maxima.

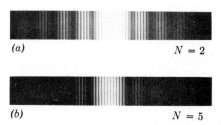

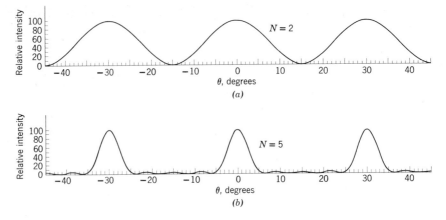

FIGURE 38-16   Calculated intensity patterns for (*a*) a two-slit and (*b*) a five-slit grating for the same value of *d* and λ. This figure shows the sharpening of the principal maxima and the appearance of faint secondary maxima for *N* > 2. The figure assumes slits with *a* ≪ λ so that the principal maxima are of uniform intensity.

These two figures show that increasing *N* (*a*) does not change the spacing between the (principal) interference fringe maxima, provided *d* and λ remain unchanged, (*b*) sharpens the (principal) maxima, and (*c*) introduces small secondary maxima between the principal maxima. Three such secondaries are present (but not readily visible) between each pair of adjacent principal maxima in Fig. 38–15*b*. These secondary maxima can easily be explained but, because they are not important in actual gratings, we will ignore them in what follows.

A principal maximum in Fig. 38–14 will occur when the path difference between rays from adjacent slits (= *d* sin θ) is given by

$$d \sin \theta = m\lambda \qquad m = 0, 1, 2, \ldots \qquad \text{(principal maxima)}, \qquad (38\text{--}17)$$

where *m* is called the *order number*. This equation is identical with Eq. 37–1, which locates the intensity maxima for a double slit. The locations of the (principal) maxima are thus determined only by the ratio λ/*d* and are independent of *N*. As for the double slit, the ratio *a*/λ determines the relative intensities of the principal maxima but does not alter their locations appreciably.

We can understand the sharpening of the principal maxima as *N* is increased by a graphical argument, using phasors. Figures 38–17*a* and *b* show conditions at any of the principal maxima for a two-slit and a nine-slit grating. The small arrows represent the amplitudes of the wave disturbances arriving at the screen at the position of each principal maximum. For simplicity we consider the central principal maximum only, for which *m* = 0, and thus θ = 0, in Eq. 38–17.

Consider the angle $\Delta\theta_0$ corresponding to the position of zero intensity that lies on either side of the central principal maximum. Figures 38–17*c* and *d* show the phasors at this point. The phase difference between waves from adjacent slits, which is zero at the central principal maximum, must increase by an amount Δφ chosen so that the array of phasors just closes on itself, yielding zero resultant intensity. For *N* = 2, Δφ = 2π/2 (= 180°); for *N* = 9, Δφ = 2π/9 (= 40°). In the general case it is given by

$$\Delta\phi = \frac{2\pi}{N}.$$

This increase in phase difference for adjacent waves corresponds to an increase in the path difference Δl given by

$$\frac{\text{phase difference}}{2\pi} = \frac{\text{path difference}}{\lambda},$$

or

$$\Delta l = \left(\frac{\lambda}{2\pi}\right)\Delta\phi = \left(\frac{\lambda}{2\pi}\right)\left(\frac{2\pi}{N}\right) = \frac{\lambda}{N}.$$

From Fig. 38–14, however, the path difference Δl at the first minimum is also given by $d \sin \Delta\theta_0$, so that we can write

$$d \sin \Delta\theta_0 = \frac{\lambda}{N},$$

or

$$\sin \Delta\theta_0 = \frac{\lambda}{Nd}.$$

Since $N \gg 1$ for actual gratings, $\sin \Delta\theta_0$ will ordinarily be quite small (that is, the lines will be sharp), and we may replace it by $\Delta\theta_0$ to good approximation, or

$$\Delta\theta_0 = \frac{\lambda}{Nd} \qquad \text{(central principal maximum).} \qquad (38\text{–}18)$$

This equation shows specifically that if we increase N for a given λ and d, $\Delta\theta_0$ will decrease, which means that the central principal maximum becomes sharper.

We state without proof, and for later use, that for principal maxima other than the central one (that is, for $m \neq 0$) the angular distance between the position θ_m of the principal maximum of order m and the minimum that lies on either side is given by

$$\Delta\theta_m = \frac{\lambda}{Nd \cos \theta_m} \qquad \text{(any principal maximum).} \qquad (38\text{–}19)$$

For the central principal maximum we have $m = 0$, $\theta_m = 0$, and $\Delta\theta_m = \Delta\theta_0$, so that Eq. 38–19 reduces, as it must, to Eq. 38–18.

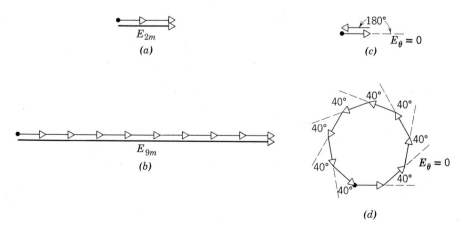

FIGURE 38–17 Drawings (*a*) and (*b*) show conditions at the central principal maximum for a two-slit and a nine-slit grating, respectively. Drawings (*c*) and (*d*) show conditions at the minimum of zero intensity that lies on either side of this central principal maximum. In going from (*a*) to (*c*) the phase shift between waves from adjacent slits changes by 180° ($\Delta\phi = 2\pi/2$); in going from (*b*) to (*d*) it changes by 40° ($\Delta\phi = 2\pi/9$).

38–8   Diffraction Gratings

The grating spacing *d* for a typical grating that contains 12,000 "slits" distributed over a 1-in. width is 2.54 cm/12,000, or 21,000 A. Gratings are often used to measure wavelengths and to study the structure and intensity of spectrum lines. Few devices have contributed more to our knowledge of physics.

Gratings are made by ruling equally spaced parallel grooves on a glass or a metal plate, using a diamond cutting point whose motion is automatically controlled by an elaborate ruling engine. Gratings ruled on metal are called *reflection gratings* because the interference effects are viewed in reflected rather than in transmitted light. Once such a master grating has been prepared, replicas can be formed by pouring a collodion solution on the grating, allowing it to harden, and stripping it off. The stripped collodion, fastened to a flat piece of glass or other backing, forms a good grating.

Figure 38–18 shows a simple grating spectroscope, used for viewing the spectrum of a light source, assumed to emit a number of discrete wavelengths, or *spectrum lines*. The light from source *S* is focused by lens L_1 on a slit S_1 placed in the focal plane of lens L_2. The parallel light emerging from collimator *C* falls on grating *G*. Parallel rays associated with a particular interference maximum occurring at angle θ fall on lens L_3, being brought to a focus in plane *F-F′*. The image formed in this plane is examined, using a magnifying lens arrangement *E*, called an eyepiece. A symmetrical interference pattern is formed on the other side of the central position, as shown by the dotted lines. The entire spectrum can be viewed by rotating telescope *T* through various angles. Instruments used for scientific research or in industry are more complex than the simple arrangement of Fig. 38–19. They invariably employ photographic or photoelectric recording and are called *spectrographs*. Figure 39–12 shows a small portion of the spectrum of iron, produced by examining the light produced in an arc struck between iron electrodes, using a research type spectrograph with photographic recording. Each line in the figure represents a different wavelength that is emitted from the source.

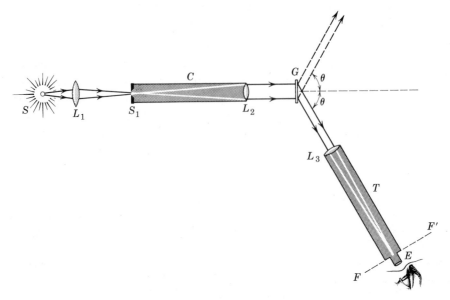

FIGURE 38–18   A simple type of grating spectroscope used to analyze the wavelengths of the light emitted by source *S*.

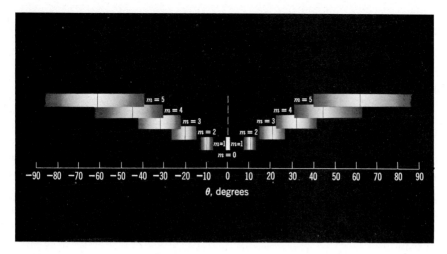

FIGURE 38-19 The spectrum of white light as viewed in a grating instrument like that of Fig. 45-7. The different orders, identified by the order number m, are shown separated vertically for clarity. As actually viewed, they would not be so displaced. The central line in each order corresponds to $\lambda = 5500$ A.

Grating instruments can be used to make absolute measurements of wavelength, since the grating spacing d in Eq. 38-17 can be measured accurately with a microscope. Several spectra are normally produced in such instruments, corresponding to $m = \pm1, \pm2$, etc., in Eq. 38-17 (see Fig. 38-19). This may cause some confusion if the spectra overlap. Further, this multiplicity of spectra reduces the recorded intensity of any given spectrum line because the available energy is divided among a number of spectra. However, by controlling the shape of the grating rulings, a large fraction of the energy can be concentrated in a particular order; this is called *blazing*.

Light can also be analyzed into its component wavelengths if the grating in Fig. 38-18 is replaced by a prism. In a *prism spectrograph* each wavelength in the incident beam is deflected through a definite angle θ, determined by the index of refraction of the prism material for that wavelength. Curves such as Fig. 36-2, which gives the index of refraction of fused quartz as a function of wavelength, show that the shorter the wavelength, the larger the angle of deflection θ. Such curves vary from substance to substance and must be found by measurement. Prism instruments are not adequate for accurate absolute measurements of wavelength because the index of refraction of the prism material at the wavelength in question is usually not known precisely enough. Both prism and grating instruments make accurate comparisons of wavelength, using a suitable comparison spectrum such as that shown in Fig. 39-12, in which careful absolute determinations have been made of the wavelengths of the spectrum lines.

Example 7. A diffraction grating has 10^4 rulings uniformly spaced over 1.00 in. It is illuminated at normal incidence by yellow light from a sodium vapor lamp. This light contains two closely spaced lines (the well-known sodium doublet) of wavelengths 5890.0 and 5895.9 A. (*a*) At what angle will the first-order maximum occur for the first of these wavelengths?

The grating spacing d is 10^{-4} in., or 25,400 A. The first-order maximum corresponds to $m = 1$ in Eq. 38-17. We thus have

$$\theta = \sin^{-1}\frac{m\lambda}{d} = \sin^{-1}\frac{(1)(5890\text{ A})}{25{,}400\text{ A}} = \sin^{-1} 0.232 = 13.3°.$$

(b) What is the angular separation between the first-order maxima for these lines?

The straightforward way to find this separation is to repeat this calculation for $\lambda = 5895.9$ A and to subtract the two angles. A difficulty, which can best be appreciated by carrying out the calculation, is that we must carry a large number of significant figures to obtain a meaningful value for the difference between the angles. To calculate the difference in angular positions directly, let us write down Eq. 38–17, solved for $\sin \theta$, and differentiate it, treating θ and λ as variables:

$$\sin \theta = \frac{m\lambda}{d}$$

$$\cos \theta \, d\theta = \frac{m}{d} \, d\lambda.$$

If the wavelengths are close enough together, as in this case, $d\lambda$ can be replaced by $\Delta\lambda$, the actual wavelength difference; $d\theta$ then becomes $\Delta\theta$, the quantity we seek. This gives

$$\Delta\theta = \frac{m \, \Delta\lambda}{d \cos \theta} = \frac{(1)(5.9 \text{ A})}{(25,400 \text{ A})(\cos 13.3°)} = 2.4 \times 10^{-4} \text{ radian} = 0.014°.$$

Note that although the wavelengths involve five significant figures our calculation, done this way, involves only two or three, with consequent reduction in numerical manipulation.

The quantity $d\theta/d\lambda$, called the *dispersion D* of a grating, is a measure of the angular separation produced between two incident monochromatic waves whose wavelengths differ by a small wavelength interval. From this example we see that

$$D = \frac{d\theta}{d\lambda} = \frac{m}{d \cos \theta}. \tag{38–20}$$

38–9 Resolving Power of a Grating

To distinguish light waves whose wavelengths are close together, the principal maxima of these wavelengths formed by the grating should be as narrow as possible. Expressed otherwise, the grating should have a high *resolving power R*, defined from

$$R = \frac{\lambda}{\Delta\lambda}. \tag{38–21}$$

Here λ is the mean wavelength of two spectrum lines that can barely be recognized as separate and $\Delta\lambda$ is the wavelength difference between them. The smaller $\Delta\lambda$ is, the closer the lines can be and still be resolved; hence the greater the resolving power R of the grating. It is to achieve a high resolving power that gratings with many rulings are constructed.

The resolving power of a grating is usually determined by the same consideration (that is, the Rayleigh criterion) that we used in Section 38–5 to determine the resolving power of a lens. If two principal maxima are to be barely resolved, they must, according to this criterion, have an angular separation $\Delta\theta$ such that the maximum of one line coincides with the first minimum of the other; see Fig. 38–11. If we apply this criterion, we will show that

$$R = Nm, \tag{38–22}$$

where N is the total number of rulings in the grating and m is the order. As expected, the resolving power is zero for the central principal maximum ($m = 0$), all wavelengths being undeflected in this order.

Let us derive Eq. 38–22. The angular separation between two principal maxima whose wavelengths differ by $\Delta\lambda$ is found from Eq. 38–20, which we recast as

$$\Delta\theta = \frac{m\,\Delta\lambda}{d\cos\theta}.$$

(38–20)

The Rayleigh criterion (Section 38–5) requires that this be equal to the angular separation between a principal maximum and its adjacent minimum. This is given from Eq. 38–19, dropping the subscript m in $\cos\theta_m$, as

$$\Delta\theta_m = \frac{\lambda}{Nd\cos\theta}.$$

(38–19)

Equating Eqs. 38–20 and 38–19 leads to

$$R\,(=\lambda/\Delta\lambda) = Nm,$$

which is the desired relation.

Example 8. In Example 7 how many rulings must a grating have if it is barely to resolve the sodium doublet in the third order?

From Eq. 38–21 the required resolving power is

$$R = \frac{\lambda}{\Delta\lambda} = \frac{5890\text{ A}}{(5895.9 - 5890.0)\text{A}} = 1000.$$

From Eq. 38–22 the number of rulings needed is

$$N = \frac{R}{m} = \frac{1000}{3} = 330.$$

This is a modest requirement.

The resolving power of a grating must not be confused with its dispersion. Table 38–1 shows the characteristics of three gratings, each illuminated with light of $\lambda = 5890$ A, the diffracted light being viewed in the first order ($m = 1$ in Eq. 38–17). Verify that the values of D and R given in the table can be calculated from Eqs. 38–20 and 38–22, respectively.

For the conditions of use noted in Table 38–1, gratings A and B have the same dispersion and A and C have the same resolving power. Figure 38–20 shows the intensity patterns that would be produced by these gratings for two incident waves of wavelengths λ_1 and λ_2, in the vicinity of $\lambda = 5890$ A. Grating B, which has high resolving power, has narrow intensity maxima and is inherently capable of distinguishing lines that are much closer together in wavelength than those of Fig. 38–21. Grating C, which has high dispersion, produces twice the angular separation between rays λ_1 and λ_2 that grating B does.

Table 38–1 SOME CHARACTERISTICS OF THREE GRATINGS   ($\lambda = 5890$ A, $m = 1$)

Grating	N	d (A)	θ	R	D (10^{-3} degrees/A)
A	10,000	25,400	13.3°	10,000	2.32
B	20,000	25,400	13.3°	20,000	2.32
C	10,000	13,700	25.5°	10,000	4.64

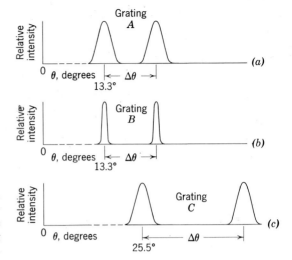

FIGURE 38–20 The intensity patterns for light of wavelengths λ_1 and λ_2 near 5890 A, incident on the gratings of Table 38–1. Grating B has the highest resolving power and grating C the highest dispersion.

Example 9. A grating has 8000 lines uniformly spaced over 1.00 in. and is illuminated by light from a mercury vapor discharge. (*a*) What is the expected dispersion, in the third order, in the vicinity of the intense green line ($\lambda = 5460$ A)? Noting that $d = 31,700$ A (2.54 cm/8000), we have, from Eq. 38–17,

$$\theta = \sin^{-1}\frac{m\lambda}{d} = \sin^{-1}\frac{(3)(5460 \text{ A})}{31,700 \text{ A}} = \sin^{-1}0.517 = 31.1°.$$

From Eq. 38–20 we have

$$D = \frac{m}{d\cos\theta} = \frac{3}{(31,700 \text{ A})(\cos 31.1°)} = 1.1 \times 10^{-4} \text{ radian/A} = 6.3 \times 10^{-3} \text{ deg/A}.$$

(*b*) What is the expected resolving power of this grating in the fifth order? Equation 38–22 gives

$$R = Nm = (8000)(5) = 40,000.$$

Thus near $\lambda = 5460$ A a wavelength difference $\Delta\lambda$ given by Eq. 38–21, or

$$\Delta\lambda = \frac{\lambda}{R} = \frac{5460 \text{ A}}{40,000} = 0.14 \text{ A},$$

can be distinguished.

38–10 X-ray Diffraction

X rays and gamma rays are electromagnetic radiation whose wavelengths are on the order of or smaller than 1 A. The wavelength region extends from 1 A to milliangstroms and less, apparently with no lower limit: compare this with 5500 A for the center of the visible spectrum. Figure 38–21 shows how x rays are produced when electrons from a heated filament F are accelerated by a potential difference V and strike a metal target T.

For such small wavelengths a standard optical diffraction grating, as normally employed, cannot be used to discriminate between different wavelengths. For $\lambda = 1.0$ A and $d = 30,000$ A, for example, Eq. 38–17 shows that the first-order maximum occurs at

$$\theta = \sin^{-1}\frac{m\lambda}{d} = \sin^{-1}\frac{(1)(1.0 \text{ A})}{3.0 \times 10^4 \text{ A}} = \sin^{-1}0.33 \times 10^{-4} = 0.0020°$$

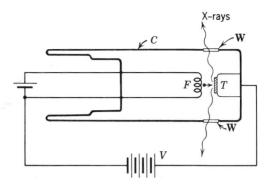

FIGURE 38–21 X-rays are generated when electrons from heated filament *F*, accelerated by potential difference *V*, are brought to rest on striking metallic target *T*. *W* is a "window" —transparent to x rays—in the evacuated metal container *C*.

This is too close to the central maximum to be practical. A grating with $d \cong \lambda$ is desirable, but, since x-ray wavelengths are about equal to atomic diameters, such gratings cannot be constructed mechanically.

In 1912 it occurred to Max von Laue that a crystalline solid, consisting as it does of a regular array of atoms, might form a natural three-dimensional "diffraction grating" for x rays. The idea is that in a crystal, such as sodium chloride (NaCl), there is a basic unit of atoms (called the unit cell) which repeats itself throughout the array. In NaCl four sodium ions and four chlorine ions are associated with each unit cell. Figure 38–22 represents a section through a crystal of NaCl and identifies this basic unit. Since this crystal has a cubic structure, the unit cell is itself a cube measuring a_0 on the side.

Diffraction of electromagnetic radiation is accomplished by the electrons surrounding the atoms in the array. Each unit cell acts as a diffraction center just as a ruled groove acts as a two-dimensional diffraction line in a grating. The crystal, then, forms a three-dimensional array of diffraction centers. Just as in the two-dimensional case, the intensity of any point outside the array is determined by the phase difference and intensity of radiation diffracted from each center to the point in question.

Bragg's law predicts the conditions under which diffracted x-ray beams from a crystal are possible. In deriving it, we ignore the structure of the unit cell, which is related only to the intensities of these beams. The dashed sloping lines in Fig. 38–23a represent the intersection with the plane of the figure of an arbitrary set of planes passing through the elementary diffracting centers. The perpendicular distance

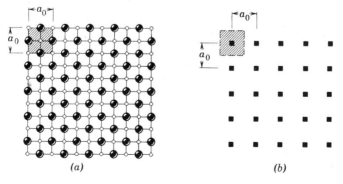

(a) (b)

FIGURE 38–22 (a) A section through a crystal of sodium chloride, showing the sodium and chlorine ions. (b) The corresponding unit cells in this section, each cell being represented by a small black square.

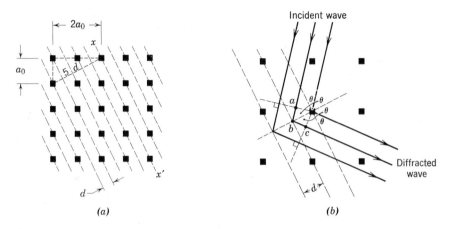

FIGURE 38–23   (*a*) A section through the NaCl unit cell lattice of Fig. 38–22*b*. The dashed sloping lines represent an arbitrary family of planes, with interplanar spacing *d*. (*b*) An incident wave falls on the entire family of planes shown in (*a*). A strong diffracted wave is formed.

between adjacent planes is *d*. Many other such families of planes, with different interplanar spacings, can be defined.

Figure 38–23*b* shows an incident wave striking the family of planes. For a single plane, mirror-like "reflection" occurs for any incident angle as with diffraction gratings. To have a constructive interference in the beam diffracted from the entire family of planes in the direction θ, the rays from the separate planes must reinforce each other. In x-ray diffraction it is customary to specify the direction of a wave by giving the angle between the ray and the plane (the *glancing angle*) rather than the angle between the ray and the normal. This means that the path difference for rays from adjacent planes (*abc* in Fig. 38–23*b*) must be an integral number of wavelengths or

$$2d \sin \theta = m\lambda \qquad m = 1, 2, 3, \ldots . \qquad (38\text{--}23)$$

This relation is called *Bragg's law* after W. L. Bragg who first derived it. The quantity *d* in this equation (the interplanar spacing) is the perpendicular distance between the planes. For the planes of Fig. 38–23*a* you can show that *d* is related to the unit cell dimension a_0 by

$$d = \frac{a_0}{\sqrt{5}}. \qquad (38\text{--}24)$$

If an incident monochromatic x-ray beam falls at an arbitrary angle θ on a particular set of atomic planes, a diffracted beam will not result because Eq. 38–23 will not, in general, be satisfied. However if the incident x rays are continuous in wavelength, diffracted beams will result when wavelengths given by

$$\lambda = \frac{2d \sin \theta}{m} \qquad m = 1, 2, 3, \ldots$$

are present in the incident beam (see Eq. 38–23).

   Example 10. At what angles must an x-ray beam with $\lambda = 1.10$ A fall on the family of planes represented in Fig. 38–23*b* if a diffracted beam is to exist? Assume the material to be sodium chloride ($a_0 = 5.63$ A).

   The interplanar spacing *d* for these planes is given by Eq. 38–24 or

$$d = \frac{a_0}{\sqrt{5}} = \frac{5.63 \text{ A}}{2.24} = 2.52 \text{ A}.$$

Equation 38–23 gives

$$\sin \theta = \frac{m\lambda}{2d} = \frac{(m)(1.10 \text{ A})}{(2)(2.52 \text{ A})} = 0.218m.$$

Diffracted beams are possible at $\theta = 12.6°$ ($m = 1$), $\theta = 25.9°$ ($m = 2$), $\theta = 40.9°$ ($m = 3$), and $\theta = 60.7°$ ($m = 4$). Higher-order beams cannot exist because they require $\sin \theta$ to exceed unity. Actually, the odd-order beams ($m = 1, 3$) prove to have zero intensity because the unit cell in cubic crystals such as NaCl has diffracting properties such that the intensity of the light scattered in these orders is zero.

X-ray diffraction is a powerful tool for studying both x-ray spectra and the arrangements of atoms in crystals. To study spectra a particular set of crystal planes, having a known spacing d, is chosen. Diffraction from these planes disperses different wavelengths into different angles. A detector, then, which can discriminate one angle from another can be used to determine the wavelength of radiation reaching it. On the other hand, using a monochromatic x-ray beam one can study the crystal itself, determining not only the spacings of various crystal planes but even the structure of the unit cell itself.

QUESTIONS

1. Why is the diffraction of sound waves more evident in daily experience than that of light waves?

2. Why do radio waves diffract around buildings, although light waves do not?

3. A loud-speaker horn has a rectangular aperture 4 ft high and 1 ft wide. Will the pattern of sound intensity be broader in the horizontal plane or in the vertical?

4. A radar antenna is designed to give accurate measurements of the height of an aircraft but only reasonably good measurements of its direction in a horizontal plane. Must the height-to-width ratio of the radar reflector be less than, equal to, or greater than unity?

5. A person holds a single narrow vertical slit in front of the pupil of his eye and looks at a distant light source in the form of a long heated filament. Is the diffraction pattern that he sees a Fresnel or a Fraunhofer pattern?

6. In a single-slit Fraunhofer diffraction, what is the effect of increasing (a) the wavelength and (b) the slit width?

7. Sunlight falls on a single slit of width 10^4 A. Describe qualitatively what the resulting diffraction pattern looks like.

8. In Fig. 38–5 rays r_1 and r_3 are in phase; so are r_2 and r_4. Why isn't there a maximum intensity at P_2 rather than a minimum?

9. Distinguish clearly between θ, α, and ϕ in Eq. 38–8.

10. If we were to redo our analysis of the properties of lenses in Section 36–10 by the methods of geometrical optics but without restricting our considerations to rays making small angles with the axis, and to "thin" lenses, would diffraction phenomena, such as that of Fig. 38–10, emerge from the analysis?

11. Distinguish between interference and diffraction in Young's double-slit experiment.

12. In what way are interference and diffraction similar? In what way are they different?

13. In double-slit interference patterns such as that of Fig. 38–13a we said that the interference fringes were modulated in intensity by the diffraction pattern of a single slit. Could we reverse this statement and say that the diffraction pattern of a single slit is intensity-modulated by the interference fringes? Discuss.

14. Discuss this statement: "A diffraction grating can just as well be called an interference grating."

15. For the simple spectroscope of Fig. 38–18, show (*a*) that θ increases with λ for a grating and (*b*) that θ decreases with λ for a prism.

16. You are given a photograph of a spectrum on which the angular positions and the wavelengths of the spectrum lines are marked. (*a*) How can you tell whether the spectrum was taken with a prism or a grating instrument? (*b*) What information could you gather about either the prism or the grating from studying such a spectrum?

17. Assume that the limits of the visible spectrum are 4300 and 6800 A. Is it possible to design a grating, assuming that the incident light falls normally on it, such that the first-order spectrum barely overlaps the second-order spectrum?

18. (*a*) Why does a diffraction grating have closely spaced rulings? (*b*) Why does it have a large number of rulings?

19. The relation $R = Nm$ suggests that the resolving power of a given grating can be made as large as desired by choosing an arbitrarily high order of diffraction. Discuss.

20. Show that at a given wavelength and a given angle of diffraction the resolving power of a grating depends only on its width $W\,(= Nd)$.

21. According to Eq. 38–19 the principal maxima become wider (that is, $\Delta\theta_m$ increases) the higher the order m (that is, the larger θ_m becomes). According to Eq. 38–22 the resolving power becomes greater the higher the order m. Explain this apparent paradox.

22. For a given family of planes in a crystal, can the wavelength of incident x rays be (*a*) too large or (*b*) too small to form a diffracted beam?

23. If a parallel beam of x rays of wavelength λ is allowed to fall on a randomly oriented crystal of any material, generally no intense diffracted beams will occur. Such beams appear if (*a*) the x-ray beam consists of a continuous distribution of wavelengths rather than a single wavelength or (*b*) the specimen is not a single crystal but a finely divided powder. Explain.

24. How would you measure (*a*) the dispersion D and (*b*) the resolving power R for either a prism or a grating spectrograph?

PROBLEMS

1(2). A plane wave ($\lambda = 5900$ A) falls on a slit with $a = 0.400$ mm. A converging lens ($f = +70.0$ cm) is placed behind the slit and focuses the light on a screen. What is the linear distance on the screen from the center of the pattern to (*a*) the first minimum and (*b*) the second minimum? *Answer:* (*a*) 1.03 mm. (*b*) 2.06 mm.

2(2). In a single-slit diffraction pattern the distance between the first minimum on the right and the first minimum on the left is 5.2 mm. The screen on which the pattern is displayed is 80 cm from the slit and the wavelength is 5460 A. Calculate the slit width.

3(2). If the distance between the first and fifth minima of a single slit pattern is 0.35 mm with the screen 40 cm away from the slit and using light having a wavelength of 5500 A, what is the width of the slit? *Answer:* 2.5 mm.

4(2). A single slit is illuminated by light whose wavelengths are λ_a and λ_b, so chosen that the first diffraction minimum of λ_a coincides with the second minimum of λ_b. (*a*) What relationship exists between the two wavelengths? (*b*) Do any other minima in the two patterns coincide?

5(4). In Fig. 38–7*d*, why is E_θ, which represents the first maximum beyond the central maximum, not vertical?

6(4). Show that the values of α at which intensity maxima for single-slit diffraction occur can be found exactly by differentiating Eq. 38–8b with respect to α and equating to zero, obtaining the condition

$$\tan \alpha = \alpha.$$

(b) Find the values of α satisfying this relation by plotting graphically the curve $y = \tan \alpha$ and the straight line $y = \alpha$ and finding their intersections. (c) Find the first few (nonintegral) values of m corresponding to successive maxima in the single-slit pattern. (d) Do the secondary maxima lie exactly halfway between minima?

7(4). The half-width of the central diffraction maximum is defined as the angle between the two points in the pattern where the intensity is one-half that at the center of the pattern. (See Fig. 38–9b.) (a) Show that the intensity drops to one-half of the maximum value when $\sin^2 \alpha = \alpha^2/2$. (b) Verify that $\alpha = 1.34$ radians (about 80°) is a solution to the transcendental equation of part (a). (c) Show that the half-width $\Delta\theta = 2 \sin^{-1}\left[0.443\,\dfrac{\lambda}{a}\right]$. (d) Calculate the half-width of the central maximum for a slit whose width is 1, 5, and 10 wavelengths.

8(5). (a) A circular diaphragm 0.60 meter in diameter oscillates at a frequency of 25,000 cycles/sec in an underwater source of sound. Far from the source the sound intensity is distributed as a Fraunhofer diffraction pattern for a circular hole whose diameter equals that of the diaphragm. Take the speed of sound in water to be 1450 meters/sec and find the angle between the normal to the diaphragm and the direction of the first minimum. (b) Repeat for a source having an (audible) frequency of 1000 cycles/sec.

9(5). How closely may two small objects be located if they are to be resolved when viewed through the telescope of a transit having a 3.0-cm objective lens if the transit is 400 yd from the objects? Take the wavelength to be 5500 A.
Answer: 8.2 mm.

10(5). The two headlights of an approaching automobile are 4.0 ft apart. At what maximum distance will the eye resolve them? Assume a pupil diameter of 5.0 mm and $\lambda = 5500$ A. Assume also that diffraction effects and not retinal structure limit the resolution.

11(5). The wall of a large room is covered with acoustic tile in which small holes are drilled 5.0 mm from center to center. How far can a person be from such a tile and still distinguish the individual holes, assuming ideal conditions? Assume the diameter of the pupil to be 5.0 mm and λ to be 5500 A.
Answer: 37 meters.

12(5). Find the separation of two points on the moon's surface that can just be resolved by the 200-in. telescope at Mount Palomar, assuming that this distance is determined by diffraction effects. The distance from the earth to the moon is 240,000 miles. Assume $\lambda = 5500$ A.

13(5). An astronaut in a satellite claims he can just barely resolve two point sources on the earth, 100 miles below him. What is their separation, assuming ideal conditions? Take $\lambda = 5500$ A, and the pupil diameter to be 5.0 mm.
Answer: 71 ft.

14(5). Under ideal conditions, estimate the linear separation of two objects on the planet Mars that can just be resolved by an observer on earth (a) using the naked eye, (b) using the 200-in. Mt. Palomar telescope. Use the following data: distance to Mars = 50 million miles, diameter of pupil = 5.0 mm; wavelength of light = 5500 A.

15(5). (a) How small is the angular separation of two stars if their images are barely resolved by the Thaw refracting telescope at the Allegheny Observatory in Pittsburgh? The lens diameter is 30 in. and its focal length is 46 ft. Assume $\lambda = 5000$ A. (b) Find the distance between these barely resolved stars if each of them is exactly 10 light years distant from the earth. (c) For the image of a single star in this telescope, find the diameter of the first dark ring in the diffraction pattern, as measured on a photographic plate placed at the focal plane. Assume that the star image structure is associated entirely with diffraction at the lens aperture and not with (small) lens "errors."
Answer: (a) 0.17″. (b) 7.6×10^7 km. (c) 2.2×10^{-5} meters.

16(5). It can be shown that, except for $\theta = 0$, a circular obstacle produces the same diffraction pattern as a circular hole of the same diameter. Furthermore, if there are many such obstacles located randomly, then the interference effects vanish leaving only the diffraction associated with a single obstacle. (*a*) Explain why one sees a "ring" around the moon on a foggy day. (*b*) Calculate the size of the water droplet in the air if the bright ring around the moon appears to have a diameter 1.5 times that of the moon.

17(6). For $d = 2a$ in Fig. 38–24, how many interference fringes lie in the central diffraction envelope?
Answer: 3.

18(6). Suppose that, as in Example 6, the envelope of the central peak contains 11 fringes. How many fringes lie between the first and second minima of the envelope?

19(6). (*a*) Design a double-slit system in which the fourth fringe, not counting the central maximum, is missing. (*b*) What other fringes, if any, are also missing?
Answer: (*a*) Must have $d = 4a$. (*b*) Every fourth fringe is missing.

20(6). (*a*) How many complete fringes appear between the first minima of the fringe envelope to either side of the central maximum for a double slit pattern if $\lambda = 5500$ A, $d = 0.15$ mm and $a = 0.030$ mm? (*b*) What is the ratio of the intensity of the third fringe to the side of the center to that of the central fringe?

21(6). If we put $d = a$ in Fig. 38–24, the two slits coalesce into a single slit of width $2a$. Show that Eq. 38–16 reduces to the diffraction pattern for such a slit.

22(7). Show that in a grating with alternately transparent and opaque strips of equal width all the even orders (except $m = 0$) are absent.

23(7). The central intensity maximum formed by a grating (along with its subsidiary secondary maxima) can be viewed as the diffraction pattern of a single "slit" whose width is that of the entire grating. Treating the grating as a single wide slit and assuming that $m = 0$, derive Eq. 38–18.

24(7). A diffraction grating is made up of slits of width 3000 A with a 9000 A separation between centers. The grating is illuminated by monochromatic plane waves, $\lambda = 6000$ A, the angle of incidence being zero. (*a*) How many diffracted lines are there? (*b*) What is the angular width of the spectral lines observed, if the grating has 1000 slits? *Angular width* is defined to be the angle between the two positions of zero intensity on either side of the maximum.

25(7). Assume that light is incident on a grating at an angle ψ as shown in Fig. 38–25. Show that the condition for a diffraction maximum is

$$d(\sin \psi + \sin \theta) = m\lambda \qquad m = 0, 1, 2, \ldots$$

Only the special case $\psi = 0$ has been treated in this chapter (compare Eq. 38–17).

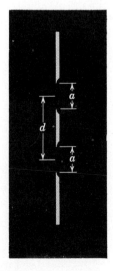

FIGURE 38–24   Problems 17(6) and 21(6).

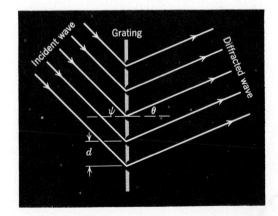

FIGURE 38–25   Problem 25(7).

26(7). Derive this expression for the intensity pattern for a three-slit "grating":

$$I_\theta = \tfrac{1}{9}I_m(1 + 4 \cos \phi + 4 \cos^2 \phi),$$

where

$$\phi = \frac{2\pi d \sin \theta}{\lambda}.$$

Assume that $a \ll \lambda$ and be guided by the derivation of the corresponding double-slit formula (Eq. 37–9).

27(7). (a) Using the result of Problem 26(7), show that the half-width of the fringes for a three-slit diffraction pattern, assuming θ small enough so that $\sin \theta \cong \theta$, is

$$\Delta\theta \cong \frac{\lambda}{3.2d}.$$

(b) Compare this with the expression derived for the two-slit pattern in Problem 22(4), Chapter 37. (c) Do these results support the conclusion that for a fixed slit spacing the interference maxima become sharper as the number of slits is increased?

28(7). A three-slit grating has separation d between adjacent slits. If the middle slit is covered up, will the half-width of the intensity maxima become broader or narrower? See Problem 27(7) above and also Problem 22(4), Chapter 37.

29(7). Derive Eq. 38–19, that is, the expression for $\Delta\theta_m$, the angular distance between a principal maximum of order m and either adjacent minimum.

30(8). A diffraction grating 2.0 cm wide has 6000 rulings. At what angles will maximum-intensity beams occur if the incident radiation has a wavelength of 5890 A?

31(8). With light from a gaseous discharge tube incident normally on a grating with a distance 1.73×10^{-4} cm between adjacent slit centers, a green line appears with sharp maxima at measured transmission angles $\theta = \pm17.6°$, $37.3°$, $-37.1°$, $65.2°$, and $-65.0°$. Compute the wavelength of the green line that best fits the data.
Answer: 5230 A.

32(8). A diffraction grating has 5000 rulings/in., and a strong diffracted beam of visible light is noted at $\theta = 30°$. (a) What are the possible wavelengths of the incident light? (b) What colors are they (see Fig. 35–16)?

33(8). A grating has 8000 rulings/in. For what wavelengths in the visible spectrum can fifth-order diffraction be observed?
Answer: All wavelengths shorter than 6300 A.

34(8). Given a grating with 4000 lines/cm, how many orders of the entire visible spectrum (4000–7000 A) can be produced?

35(8). Light of wavelength 6000 A is incident normally on a diffraction grating. Two adjacent principal maxima occur at $\sin \theta = 0.20$ and $\sin \theta = 0.30$, respectively. The fourth order is a missing order. (a) What is the separation between adjacent slits? (b) What is the smallest possible individual slit width? (c) Name all orders actually appearing on the screen with the values chosen in (a) and (b).
Answer: (a) 60,000 A. (b) 15,000 A. (c) $m = 0$, 1, 2, 3, 5, 6, 7, 9; the tenth order is at $\theta = 90°$.

36(8). Assume that the limits of the visible spectrum are arbitrarily chosen as 4300 and 6800 A. Design a grating that will spread the first-order spectrum through an angular range of 20°.

37(8). Light containing a mixture of two wavelengths, 5000 A and 6000 A, is incident normally on a diffraction grating. It is desired (1) that the first and second principal maxima for each wavelength appear at $\theta \leq 30°$, (2) that the dispersion be as high as possible, and (3) that the third order for 6000 A be a missing order. (a) What is the separation between adjacent slits? (b) What is the smallest possible individual slit width? (c) Name all orders for 6000 A that actually appear on the screen with the values chosen in (a) and (b).
Answer: (a) 24,000 A. (b) 8000 A. (c) $m = 0$, 1, 2 ($m = 4$ at $\theta = 90°$).

38(8). Show that the dispersion of a grating can be written as

$$D = \frac{\tan \theta}{\lambda}.$$

39(8). A grating has 8000 rulings/in. and is illuminated at normal incidence by white light. A spectrum is formed on a screen 30 cm from the grating. If a 1.0-cm square hole is cut in the screen, its inner edge being 5.0 cm from the central maximum, what range of wavelengths passes through the hole?
Answer: 5200 to 6200 A.

40(8). Two spectral lines have wavelengths λ and $\lambda + \Delta\lambda$, respectively, where $\Delta\lambda \ll \lambda$. Show that their angular separation $\Delta\theta$ in a grating spectrometer is given approximately by $\Delta\theta = \Delta\lambda/\sqrt{(d/m)^2 - \lambda^2}$, where d is the separation of adjacent slit centers and m is the order at which lines are observed. Notice that the angular separation is greater in the higher orders.

41(8). A transmission grating with $d = 1.50 \times 10^{-4}$ cm is illuminated at various angles of incidence by light of wavelength 6000 A. [See Problem 25(7).] Plot as a function of angle of incidence (0 to 90°) the angular deviation of the first-order diffracted beam from the incident direction.

42(8). An optical grating with a spacing $d = 15,000$ A is used to analyze soft x-rays of wavelength $\lambda = 5.0$ A. The angle of incidence θ is $90° - \gamma$, where γ is a small angle. The first order maximum is found at an angle $\theta = 90° - 2\beta$. Find the value of β.

43(9). A grating has 6000 rulings/cm and is 6.0 cm wide. (*a*) What is the smallest wavelength interval that can be resolved in the third order at $\lambda = 5000$ A? (*b*) How many higher orders can be seen? Assume normal incidence of light on the grating throughout.
Answer: (*a*) 4.6×10^{-2} A. (*b*) None.

44(9). A grating has 40,000 rulings spread over 3.0 in. (*a*) What is its expected dispersion D for sodium light ($\lambda = 5890$ A) in the first three orders? (*b*) What is its resolving power in these orders?

45(9). A source containing a mixture of hydrogen and deuterium atoms emits a red doublet at $\lambda = 6563$ A whose separation is 1.8 A. Find the minimum number of lines needed in a diffraction grating that can resolve these lines in the first order.
Answer: 3650.

46(9). In a particular grating the sodium doublet (see Example 7) is viewed in third order at 10° to the normal and is barely resolved. Find (*a*) the grating spacing and (*b*) the total width of the rulings.

47(9). A diffraction grating has a resolving power $R = \lambda/\Delta\lambda = Nm$. (*a*) Show that the corresponding frequency range, $\Delta\nu$, that can just be resolved is given by $\Delta\nu = c/Nm\lambda$. (*b*) From Fig. 38–14, show that the "times of flight" of the two extreme rays differ by an amount $\Delta t = Nd \sin \theta/c$. (*c*) Show that $(\Delta\nu)(\Delta t) = 1$, this relation being independent of the various grating parameters.

48(10). Monochromatic high energy x-rays are incident on a crystal. If first order reflection is observed at Bragg angle 3.4°, at what angle would second order reflection be expected?

49(10). Monochromatic x-rays are incident on a NaCl crystal, whose lattice spacing is 3.00 A. When the beam is rotated 60° from the normal, first-order Bragg reflection is observed. What is the wavelength of the x-rays?
Answer: 3.0 A.

50(10). In comparing the wavelengths of two monochromatic x-ray lines, it is noted that line A gives a first-order reflection maximum at a glancing angle of 30° to the smooth face of a crystal.

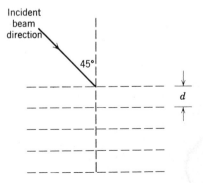

FIGURE 38-26 Problems 51(10) and 52(10).

Line B, known to have a wavelength of 0.97 angstroms, gives a third-order reflection maximum at an angle of 60° from the same face of the same crystal. Find the wavelength of line A.

51(10). Monochromatic x-rays ($\lambda = 1.25$ A) fall on a crystal of sodium chloride, making an angle of 45° with the reference line shown in Fig. 38–26. The planes shown are those of Fig. 38–23a, for which $d = 2.52$ A. Through what angles must the *crystal* be turned to give a diffracted beam associated with the planes shown? Assume that the crystal is turned about an axis that is perpendicular to the plane of the page. Ignore the possibility that some of these beams may be of zero intensity.

Answer: 30.6°, 15.3°, (clockwise); 3.1°, 37.8°, (counterclockwise).

52(10). Assume that the incident x-ray beam in Fig. 38–26 is not monochromatic but contains wavelengths in a band from 0.95 to 1.30 A. Will diffracted beams, associated with the planes shown, occur? Assume $d = 2.75$ A.

53(10). Prove that it is impossible to determine both wavelength of radiation and spacing of Bragg reflecting planes in a crystal by measuring the angles for Bragg reflection in several orders.

CHAPTER 39

Light and Quantum Physics

39–1 Sources of Light

We have studied many properties of light, including its propagation, reflection, refraction, diffraction, and interference. This chapter deals in part with the production of light and with the way that such studies led, in 1900, to the birth of quantum physics.

The most common light sources are heated solids and gases through which an electric discharge is passing. The tungsten filament of an incandescent lamp and the familiar neon sign are examples in each category. By analyzing the light from a source with a spectrometer, we can learn how strongly it radiates at various wavelengths. Figure 39–1, which is typical of spectra for heated solids, shows the results of such measurements for a heated tungsten ribbon at $2000°$ K.

The ordinate $\mathcal{R}_\lambda$ in Fig. 39–1 is called the *spectral radiancy*, defined so that the quantity $\mathcal{R}_\lambda \, d\lambda$ is the rate at which energy is radiated per unit area of surface for wavelengths lying in the interval λ to $\lambda + d\lambda$. Typical units for $\mathcal{R}_\lambda$ are watts/cm²-μ; the corresponding units of $\mathcal{R}_\lambda \, d\lambda$ are watts/cm². In measuring $\mathcal{R}_\lambda$, all radiation emerging into the forward hemisphere is included.

Sometimes we wish to discuss the total radiated energy without regard to its wavelength. An appropriate quantity here is the *radiancy* $\mathcal{R}$, defined as the rate per unit surface area at which energy is radiated into the forward hemisphere, appropriate units being watts/cm². We can find it by integrating the radiation present in all wavelength intervals:

$$\mathcal{R} = \int_0^\infty \mathcal{R}_\lambda \, d\lambda. \tag{39–1}$$

The radiancy $\mathcal{R}$ can be interpreted as the area under the plot of $\mathcal{R}_\lambda$ against λ. In Fig. 39–1 this area—and thus $\mathcal{R}$—is 23.5 watts/cm².

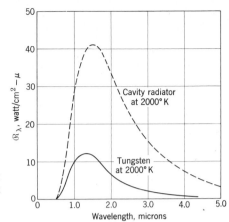

FIGURE 39-1 The spectral radiancy of tungsten at 2000° K. The dashed curve refers to a cavity radiator at the same temperature. One micron $(1\,\mu) = 10^{-6}$ meter $= 10^4$ A.

For every material there exists a family of spectral radiancy curves like that of Fig. 39-1, one curve for every temperature. A quantitative understanding in terms of a basic theory presents serious difficulties. Fortunately, it is possible to work with an idealized heated solid, called a *cavity radiator*. Its light-emitting properties prove to be independent of any particular material and to vary in a simple way with temperature. In much the same way it proved convenient earlier to deal with an ideal gas rather than to analyze the properties of the infinite variety of real gases. The cavity radiator is the ideal solid as far as its light-emitting characteristics are concerned. We shall describe in the next two sections how the theoretical study of cavity radiation in 1900 by Max Planck (1858–1947) laid the foundations of modern quantum physics.

39-2 Cavity Radiators

Let us construct a cavity in each of three metal blocks through the walls of which a small hole is drilled. Let the blocks be made of any suitable materials; for example, tungsten, tantalum, and molybdenum. Let each block be raised to the same uniform temperature, say 2000° K. Finally, let us observe the blocks by their emitted light in a dark room. Measurements of $\mathfrak{R}$ and $\mathfrak{R}_\lambda$ show the following:

1. The radiation from the cavity interior is always more intense than the radiation from the outside wall. Comparison of the two curves in Fig. 39–1 makes this clear for tungsten. For the three materials given, at 2000° K the ratio of the radiancy for the outside surface to that for the cavity is 0.259 (tungsten), 0.212 (molybdenum), and 0.232 (tantalum).

2. At a given temperature the radiancy of the hole is identical for all three radiators, in spite of the fact that the radiancies of the outer surfaces are different. At 2000° K the cavity radiancy (that is, the hole radiance) is 90.0 watts/cm².

3. In contrast to the radiancy of the outer surfaces, the cavity radiancy $\mathfrak{R}_c$ varies with temperature in a simple way, namely as

$$\mathfrak{R}_c = \sigma T^4, \tag{39-2}$$

where σ is a universal constant (the Stefan-Boltzmann constant) whose measured value is 5.67×10^{-8} watt/(meter²)(°K⁴). The radiancy of the outer surfaces varies with temperature in a more complicated way and is different for different materials.

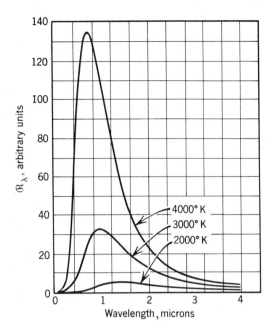

FIGURE 39–2 The spectral radiancy for cavity radiation at three different temperatures.

It is often written as

$$\Re = e\Re_c = e\sigma T^4, \tag{39–3}$$

where e, the *emissivity*, depends on the material and the temperature.

4. $\Re_\lambda$ for the cavity radiation varies with temperature in the way shown in Fig. 39–2. These curves depend only on the temperature and are quite independent of the material and of the shape and size of the cavity.

Figure 39–3 shows an actual cavity, consisting of a hollow thin-walled cylinder of tungsten heated by sending an electric current through it. The cylinder is mounted in an evacuated glass bulb, and a tiny hole is drilled through the cylinder wall. You can see from the photograph that the radiance of the cavity interior is greater than that of the cavity walls.

We can deduce many of the facts just given about cavity radiation from Fig. 39–4, which shows two cavities made of different materials, of arbitrary shapes, and with

FIGURE 39–3 Photograph of an incandescent tungsten tube with a small hole drilled in its wall. The radiation emerging from the hole is cavity radiation.

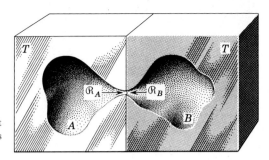

FIGURE 39–4   Two radiant cavities initially at the same temperature are placed together as shown.

the same wall temperature T. Radiation, described by $\mathfrak{R}_A$, goes from cavity A to cavity B and radiation described by $\mathfrak{R}_B$ moves in the opposite direction. If these two rates of energy transfer are not equal, one end of the composite block will start to heat up and the other end will start to cool down, which is a violation of the second law of thermodynamics. (Why?) Thus we must have

$$\mathfrak{R}_A = \mathfrak{R}_B = \mathfrak{R}_c, \tag{39–4}$$

where $\mathfrak{R}_c$ describes the total radiation for all cavities.

Not only the total radiation but also the distribution of radiant energy with wavelength must be the same for each cavity in Fig. 39–4. We can show this by placing a filter between the two cavity openings, so chosen that it permits only a selected narrow band of wavelengths to pass. Applying the same argument, we can show that we must have

$$\mathfrak{R}_{\lambda A} = \mathfrak{R}_{\lambda B} = \mathfrak{R}_{\lambda c}, \tag{39–5}$$

where $\mathfrak{R}_{\lambda c}$ is a spectral radiancy characteristic of all cavities.

39–3   Planck's Radiation Formula

A theoretical explanation for the cavity radiation was the outstanding unsolved problem in physics during the years before the turn of the present century. A number of capable physicists advanced theories based on classical physics, which, however, had only limited success. Figure 39–5, for example, shows the theory of Wien; the fit to the experimental points is reasonably good, but definitely not within the experimental

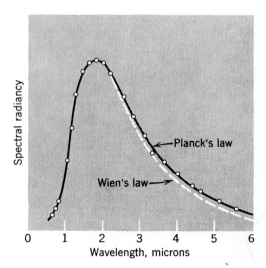

FIGURE 39–5   The circles show the experimental spectral radiance data of Coblentz for cavity radiation. The theoretical formulas of Wien and Planck are also shown, Planck's providing an excellent fit to the data.

error of the data. Wien's formula is

$$\mathcal{R}_\lambda = \frac{c_1}{\lambda^5} \frac{1}{e^{c_2/\lambda T}},$$

where c_1 and c_2 are constants that must be determined empirically by fitting the theoretical formula to the experimental data.

In 1900 Max Planck pointed out that if Wien's formula was modified in a simple way it would fit the data precisely. Planck's formula, announced to the Berlin Physical Society on October 19, 1900, was

$$\mathcal{R}_\lambda = \frac{c_1}{\lambda^5} \frac{1}{e^{c_2/\lambda T} - 1}. \tag{39–6}$$

This formula, though interesting and important, was still empirical at that stage and did not constitute a theory.

Planck sought such a theory in terms of a detailed model of the atomic processes taking place at the cavity walls. He assumed that the atoms that make up these walls behave like tiny electromagnetic oscillators, each with a characteristic frequency of oscillation. The oscillators emit electromagnetic energy into the cavity and absorb electromagnetic energy from it. Thus it should be possible to deduce the characteristics of the cavity radiation from those of the oscillators with which it is in equilibrium.

Planck was led to make two radical assumptions about the atomic oscillators. As eventually formulated, these assumptions are the following:

1. An oscillator cannot have any energy but only energies given by

$$E = nh\nu, \tag{39–7}$$

where ν is the oscillator frequency, h is a constant (now called *Planck's constant*), and n is a number (now called a *quantum number*) that can take on only integral values. Equation 39–7 asserts that the oscillator energy is *quantized*. Later developments show that the correct formula for a harmonic oscillator is $E = (n + \frac{1}{2})h\nu$. This change makes no difference to Planck's conclusions, however.

2. The oscillators do not radiate energy continuously, but only in "jumps," or *quanta*. These quanta of energy are emitted when an oscillator changes from one to another of its quantized energy states. Thus, if n changes by one unit, Eq. 39–7 shows that an amount of energy given by

$$\Delta E = \Delta n h\nu = h\nu \tag{39–8}$$

is radiated. As long as an oscillator remains in one of its quantized states (or *stationary states* as they are called), it neither emits nor absorbs energy.

These assumptions were radical ones and, indeed, Planck himself resisted accepting them wholeheartedly for many years. In his words, "My futile attempts to fit the elementary quantum of action [that is, the quantity h] somehow into the classical theory continued for a number of years, and they cost me a great deal of effort."

Consider the application of Planck's hypotheses to a large-scale oscillator such as a mass-spring system or an LC circuit. It would be a stoutly defended common belief that oscillations in such systems could take place with any value of total energy and not with only certain discrete values. In the decay of such oscillations (by friction in the mass-spring system or by resistance and radiation in the LC circuit), it would seem that the mechanical or electromagnetic energy would decrease in a perfectly continuous way and not by "jumps." There is no basis in everyday experience, however, to dismiss Planck's assumptions as violations of "common sense," for Planck's constant proves to have a very small value; its 1969 "best" value is:

$$h = 6.626196 \times 10^{-34} \text{ joule-sec.}$$

The following example makes this clear.

Example 1. A mass-spring system has a mass $m = 1.0$ kg and a spring constant $k = 20$ nt/meter and is oscillating with an amplitude of 1.0 cm. (*a*) If its energy is quantized according to Eq. 39–7, what is the quantum number n?

From Eq. 13–8 the frequency is

$$\nu = \frac{1}{2\pi} \sqrt{\frac{k}{m}} = \frac{1}{2\pi} \sqrt{\frac{20 \text{ nt/meter}}{1.0 \text{ kg}}} = 0.71 \text{ cycles/sec.}$$

From Eq. 7–8 the mechanical energy is

$$E = \tfrac{1}{2}kx_{\text{max}}^2 = \tfrac{1}{2}(20 \text{ nt/meter})(1.0 \times 10^{-2} \text{ meter})^2 = 1.0 \times 10^{-3} \text{ joule.}$$

From Eq. 39–7 the quantum number is

$$n = \frac{E}{h\nu} = \frac{1.0 \times 10^{-3} \text{ joule}}{(6.6 \times 10^{-34} \text{ joule-sec})(0.71 \text{ cycles/sec})} = 2.1 \times 10^{30}.$$

(*b*) If n changes by unity, what fractional change in energy occurs?

The fractional change in energy is given by dividing Eq. 39–8 by Eq. 39–7, or

$$\frac{\Delta E}{E} = \frac{h\nu}{nh\nu} = \frac{1}{n} = \sim 10^{-30}.$$

Thus for large-scale oscillators the quantum numbers are enormous and the quantized nature of the energy of the oscillations will not be apparent. Similarly, in large-scale experiments we are not aware of the discrete nature of mass and the quantized nature of charge, that is, of the existence of atoms and electrons.

On the basis of his two assumptions, Planck was able to derive his radiation law (Eq. 39–6) entirely from theory. His theoretical expressions for the hitherto empirical constants c_1 and c_2 were

$$c_1 = 2\pi c^2 h \quad \text{and} \quad c_2 = \frac{hc}{k},$$

where k is Boltzmann's constant (see Section 20–5) and c is the speed of light. By inserting the experimental values for c_1 and c_2, Planck was able to derive the values of both h and k. Planck described his theory to the Berlin Physical Society on December 14, 1900. Quantum physics dates from that day. Planck's ideas soon received re-enforcement from Einstein, who, in 1905, applied the concepts of energy quantization to a new area of physics, the photoelectric effect.

Before discussing this effect, it is important to realize that although Planck had quantized the energies of the oscillators in the cavity walls he still treated the radiation within the cavity as an electromagnetic wave. Einstein's analysis of the photoelectric effect first pointed out the inadequacy of the wave picture of light in certain situations.

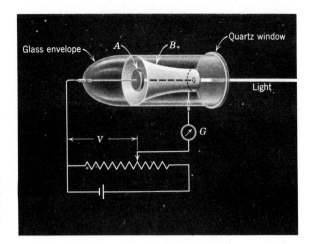

FIGURE 39–6 An apparatus used to study the photoelectric effect. V can be varied continuously and can also be reversed in sign by a switching arrangement not shown.

39–4 Photoelectric Effect

Figure 39–6 shows an apparatus used to study the photoelectric effect. Monochromatic light, falling on metal plate A, will liberate *photoelectrons,* which can be detected as a current if they are attracted to metal cup B by means of a potential difference V applied between A and B. Galvanometer G serves to measure this *photoelectric current.*

Figure 39–7 (curve a) is a plot of the photoelectric current in an apparatus like that of Fig. 39–6, as a function of the potential difference V. If V is made large enough, the photoelectric current reaches a certain limiting value at which all photoelectrons ejected from plate A are collected by cup B.

If we reverse V in sign, the photoelectric current does not immediately drop to zero, which proves that the electrons are emitted from A with a finite velocity. Some will reach cup B in spite of the fact that the electric field opposes their motion. However, if this reversed potential difference is made large enough, a value V_0 (the *stopping potential*) is reached at which the photoelectric current does drop to zero. This potential difference V_0, multiplied by the electron charge, measures the kinetic energy K_{max} of the fastest ejected photoelectron. In other words,

$$K_{max} = eV_0. \tag{39–9}$$

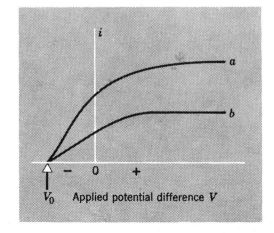

FIGURE 39–7 Some data taken with the apparatus of Fig. 39–6. The applied potential difference V is called positive when the cup B in Fig. 39–6 is positive with respect to the photoelectric surface A. In curve b the incident light intensity has been reduced to one-half that of curve a.

Here K_{max} turns out to be independent of the intensity of the light as shown by curve *b* in Fig. 39–7, in which the light intensity has been reduced to one-half.

Figure 39–8 shows the stopping potential V_0 as a function of the frequency of the incident light for sodium. Note that there is a definite cutoff frequency ν_0, below which no photoelectric effect occurs. These data were taken by R. A. Millikan (1868–1953).

Three major features of the photoelectric effect cannot be explained in terms of the wave theory of light:

1. Wave theory suggests that the kinetic energy of the photoelectrons should increase as the light beam is made more intense. However, Fig. 39–7 shows that K_{max} ($= eV_0$) is independent of the light intensity; this has been tested over a range of intensities of 10^7.

2. According to the wave theory, the photoelectric effect should occur for any frequency of the light, provided only that the light is intense enough. However, Fig. 39–8 shows that there exists, for each surface, a characteristic *cutoff frequency* ν_0. For frequencies less than this, the photoelectric effect disappears, no matter how intense the illumination.

3. If the energy of the photoelectrons is "soaked up" from the incident wave by the metal plate, it is not likely that the "effective target area" for an electron in the metal is much more than a few atomic diameters. Thus, if the light is feeble enough, there should be a measurable time lag (see Example 2) between the impinging of the light on the surface and the ejection of the photoelectron. During this interval the electron should be "soaking up" energy from the beam until it had accumulated enough energy to escape. However, no detectable time lag has ever been measured. This disagreement is particularly striking when the photoelectric substance is a gas; under these circumstances the energy of the emitted photoelectron must certainly be "soaked out of the beam" by a single atom.

▋ **Example 2.** A metal plate is placed 5 meters from a monochromatic light source whose power output is 10^{-3} watt. Consider that a given ejected photoelectron may collect its energy from a circular area of the plate as large as ten atomic diameters (10^{-9} meter) in radius. The energy required to remove an electron through the metal surface is about 5.0 eV. Assuming light to be a

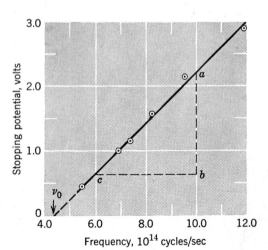

FIGURE 39–8 A plot of Millikan's measurements of the stopping potential at various frequencies for sodium. The cutoff frequency ν_0 is 4.39×10^{14} cycles/sec.

wave, how long would it take for such a "target" to soak up this much energy from such a light source?

The target area is $\pi\,(10^{-9}$ meter$)^2$ or 3×10^{-18} meter^2; the area of a 5-meter sphere centered on the light source is $4\pi\,(5$ meters$)^2 \cong 300$ meters^2. Thus, if the light source radiates uniformly in all directions, the rate P at which energy falls on the target is given by

$$P = (10^{-3}\ \text{watt}) \left(\frac{3 \times 10^{-18}\ \text{meter}^2}{300\ \text{meters}^2} \right) = 10^{-23}\ \text{joule/sec}.$$

Assuming that all this power is absorbed, we may calculate the time required from

$$t = \left(\frac{5\ \text{eV}}{10^{-23}\ \text{joule/sec}} \right)\left(\frac{1.6 \times 10^{-19}\ \text{joule}}{1\ \text{eV}} \right) \cong 20\ \text{hr}.$$

However, no detectable time lag has been measured.

39–5   Einstein's Photon Theory

Einstein succeeded in explaining the photoelectric effect by making a remarkable assumption, namely, that the energy in a light beam travels through space in concentrated bundles, called (by others at a later time) *photons*. The energy E of a single photon (see Eq. 39–8) is given by

$$E = h\nu. \tag{39–10}$$

Recall that Planck believed that light, although emitted from its source discontinuously, travels through space as an electromagnetic wave. Einstein's hypothesis suggests that light traveling through space behaves not like a wave at all but like a particle. Millikan, whose experiments verified Einstein's ideas in every detail, spoke of Einstein's "bold, not to say reckless, hypothesis."

Applying the photon concept to the photoelectric effect, Einstein's concept suggests

$$h\nu = E_0 + K_{\max} \tag{39–11}$$

where $h\nu$ is the energy of the photon. Equation 39–11 says that a photon carries an energy $h\nu$ into the surface. Part of this energy (E_0) is used in causing the electron to pass through the metal surface. The excess energy $(h\nu - E_0)$ is given to the electron in the form of kinetic energy; if the electron does not lose energy by internal collisions as it escapes from the metal, it will exhibit it all as kinetic energy after it emerges. Thus $K_{\max}$ represents the maximum kinetic energy that the photoelectron can have outside the surface; in nearly all cases it will have less energy than this because of internal losses.

Consider how Einstein's photon hypothesis meets the three objections raised against the wave-theory interpretation of the photoelectric effect. As for objection 1 (the lack of dependence of $K_{\max}$ on the intensity of illumination), there is complete agreement of the photon theory with experiment. If we double the light intensity we double the number of photons and thus double the photoelectric current; we do not change the energy $(= h\nu)$ of the individual photons or the nature of the individual photoelectric processes described by Eq. 39–11.

Objection 2 (the existence of a cutoff frequency) follows from Eq. 39–11. If $K_{\max}$ equals zero, we have

$$h\nu_0 = E_0,$$

which asserts that the photon has just enough energy to eject the photoelectrons and none extra to appear as kinetic energy. This quantity E_0 is called the *work function* of the substance. If ν is reduced below ν_0, the individual photons, no matter how many of them there are (that is, no matter how intense the illumination), will not have enough energy to eject photoelectrons.

Objection 3 (the absence of a time lag) follows from the photon theory because the required energy is supplied in a concentrated bundle. It is not spread uniformly over a large area, as in the wave theory.

Although the photon hypothesis certainly fits the facts of photoelectricity, it seems to be in direct conflict with the wave theory of light which, as we have seen in earlier chapters, has been verified in many experiments. Our modern view of the nature of light is that it has a dual character, behaving like a wave under some circumstances and like a particle, or photon, under others. We discuss the wave-particle duality at length in Chapter 40. Meanwhile, let us continue our studies of the firm experimental foundation on which the photon concept rests.

Let us rewrite Einstein's photoelectric equation (Eq. 39–11) by substituting eV_0 for K_{max} (see Eq. 39–9). This yields, after rearrangement,

$$V_0 = \frac{h}{e}\nu - \frac{E_0}{e}. \tag{39–12}$$

Thus Einstein's theory predicts a linear relationship between V_0 and ν, in complete agreement with experiment; see Fig. 39–8. The slope of the experimental curve in this figure should be h/e, or

$$\frac{h}{e} = \frac{ab}{bc} = \frac{2.20 \text{ volt} - 0.65 \text{ volt}}{(10 \times 10^{14} - 6.0 \times 10^{14}) \text{ cycles/sec}} = 3.9 \times 10^{-15} \text{ volt-sec.}$$

We can find h by multiplying this ratio by the electron charge e,

$$h = (3.9 \times 10^{-15} \text{ volt-sec})(1.6 \times 10^{-19} \text{ coul}) = 6.2 \times 10^{-34} \text{ joule-sec.}$$

From a more careful analysis of these and other data, including data taken with lithium surfaces, Millikan found the value $h = 6.57 \times 10^{-34}$ joule-sec, with an accuracy of about 0.5%. This agreement with the value of h derived from Planck's radiation formula is a striking confirmation of Einstein's photon concept.

Example 3. Deduce the work function for sodium from Fig. 39–8.

The intersection of the straight line in Fig. 39–8 with the horizontal axis is the cutoff frequency ν_0. Substituting these values yields

$$E_0 = h\nu_0 = (6.63 \times 10^{-34} \text{ joule-sec})(4.39 \times 10^{14} \text{ cycles/sec})$$

$$= 2.92 \times 10^{-19} \text{ joule} = 1.82 \text{ eV.}$$

39–6   The Compton Effect

Compelling confirmation of the concept of the photon as a concentrated bundle of energy was provided in 1923 by A. H. Compton (1892–1962). Compton allowed a beam of x rays of sharply defined wavelength λ to fall on a graphite block, as in Fig. 39–9, and he measured, for various angles of scattering, the intensity of the scattered x rays as a function of their wavelength. Figure 39–10 shows his experimental results. We see that although the incident beam consists essentially of a single wavelength λ

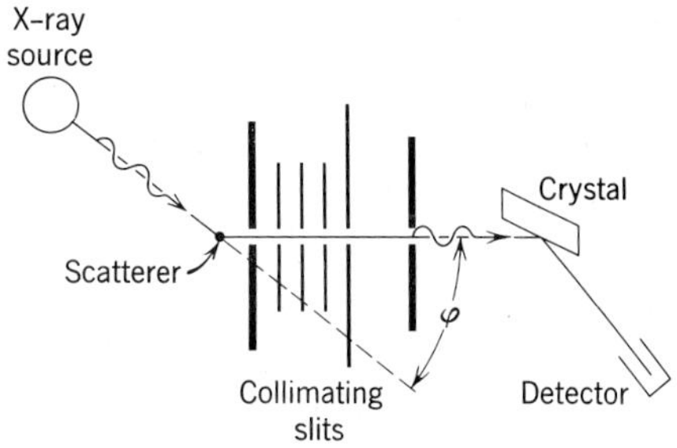

X-ray
source

FIGURE 39–9 Compton's experimental arrangement. Monochromatic x rays of wavelength λ fall on a graphite scatterer. The distribution of intensity with wavelength is measured for x rays scattered at any selected angle φ. The scattered wavelengths are measured by observing Bragg reflections from a crystal; see Eq. 38–23. Their intensities are measured by a detector, such as an ionization chamber.

Crystal

Scatterer

φ

Collimating
slits

Detector

the scattered x rays have intensity peaks at two wavelengths; one of them is the same as the incident wavelength, the other, λ', being larger by an amount $\Delta\lambda$. This so-called *Compton shift* $\Delta\lambda$ varies with the angle at which the scattered x rays are observed.

We cannot understand the presence of a scattered wave of wavelength λ' if we regard the incident x rays as an electromagnetic wave like that of Fig. 35–4. On this picture the incident wave of frequency ν causes electrons in the scattering block to oscillate at that same frequency. These oscillating electrons, like charges surging back and forth in a small radio transmitting antenna, radiate electromagnetic waves that again have this same frequency ν. Thus, on the wave picture the scattered wave should have the same frequency ν and the same wavelength λ as the incident wave.

Compton explained his experimental results by postulating that the incoming x-ray beam was not a wave but an assembly of photons of energy $E \,(= h\nu)$ and that these photons experienced billard-ball-like collisions with the free electrons in the scattering block. The "recoil" photons emerging from the block constitute, on this view, the scattered radiation. Since the incident photon transfers some of its energy to the electron with which it collides, the scattered photon must have a lower energy E'; it must therefore have a lower frequency $\nu' \,(= E'/h)$, which implies a larger wavelength $\lambda' \,(= c/\nu')$. This point of view accounts, at least qualitatively, for the wavelength shift $\Delta\lambda$. Notice how different this particle model of x-ray scattering is from that based on the wave picture. Now let us analyze a single photon-electron collision quantitatively.

Figure 39–11 shows a collision between a photon and an electron, the electron assumed to be initially at rest and essentially free, that is, not bound to the atoms of the scatterer. Let us apply the law of conservation of energy to this collision. Since the recoil electrons may have a speed v that is comparable with that of light, we must use the relativistic expression for the ki-

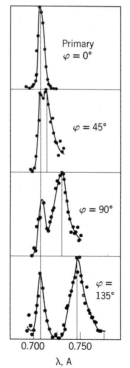

Primary
φ = 0°

φ = 45°

φ = 90°

φ =
135°

0.700   0.750

λ, A

FIGURE 39–10 Compton's experimental results. The solid vertical line on the left corresponds to the wavelength λ, that on the right to λ'. Results are shown for four different angles of scattering φ. Note that the Compton shift $\Delta\lambda$ for φ = 90° is $h/m_0 c = 0.242$ A.

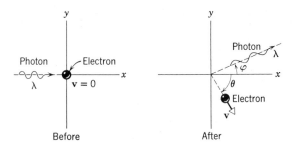

FIGURE 39-11 A photon of wavelength λ is incident on an electron at rest. On collision, the photon is scattered at an angle φ with increased wavelength λ', while the electron moves off with speed v in direction θ.

netic energy of the electron. From Eqs. 39–10 and 7–19 we may write

$$h\nu = h\nu' + (m - m_0)c^2,$$

in which the second term on the right is the relativistic expression for the kinetic energy of the recoiling electron, m being the relativistic mass and m_0 the rest mass of that particle. Substituting c/λ for ν (and c/λ' for ν') and using Eq. 7–18 to eliminate the relativistic mass m leads us to

$$\frac{hc}{\lambda} = \frac{hc}{\lambda'} + m_0 c^2 \left(\frac{1}{\sqrt{1 - (v/c)^2}} - 1 \right). \tag{39-13}$$

Now let us apply the (vector) law of conservation of linear momentum to the collision of Fig. 39–11. We first need an expression for the momentum of a photon. In Section 35–5 we saw that if an object completely absorbs an energy U from a parallel light beam that falls on it the light beam, according to the wave theory of light, will simultaneously transfer to the object a linear momentum given by U/c. In the photon picture we imagine this momentum to be carried along by the individual photons, each photon transporting linear momentum in amount $p = h\nu/c$, where $h\nu$ is the photon energy. Thus, if we substitute λ for c/ν, we can write

$$p = \frac{E}{c} = \frac{h\nu}{c} = \frac{h}{\lambda}. \tag{39-14}$$

This conclusion, that the momentum of a photon is given by h/λ, may also be deduced from the theory of relativity.

For the electron, the relativistic expression for the linear momentum is given by Eq. 8–11, or

$$\mathbf{p}_e = \frac{m_0 \mathbf{v}}{\sqrt{1 - (v/c)^2}}.$$

We can then write for the conservation of the x-component of linear momentum

$$\frac{h}{\lambda} = \frac{h}{\lambda'} \cos \varphi + \frac{m_0 v}{\sqrt{1 - (v/c)^2}} \cos \theta \tag{39-15}$$

and for the y-component

$$0 = \frac{h}{\lambda'} \sin \varphi - \frac{m_0 v}{\sqrt{1 - (v/c)^2}} \sin \theta. \tag{39-16}$$

Our immediate aim is to find $\Delta\lambda$ ($= \lambda' - \lambda$), the wavelength shift of the scattered photons, so that we may compare it with the experimental results of Fig. 39–10.

Compton's experiment did not involve observations of the recoil electron in the scattering block. Of the five collision variables (λ, λ', v, φ, and θ) that appear in the three equations (39–13, 39–15, and 39–16) we may eliminate two. We chose to eliminate v and θ, which deal only with the electron, thereby reducing the three equations to a single relation among the variables.

Carrying out the necessary algebraic steps [(see Problem 30(6)] leads to this simple result:

$$\Delta\lambda \, (= \lambda' - \lambda) = \frac{h}{m_0 c} (1 - \cos\varphi). \qquad (39\text{–}17)$$

Thus the Compton shift $\Delta\lambda$ depends only on the scattering angle φ and *not* on the initial wavelength λ. Equation 39–17 predicts within experimental error the experimentally observed Compton shifts of Fig. 39–10. Note from the equation that $\Delta\lambda$ varies from zero (for $\varphi = 0$, corresponding to a "grazing" collision in Fig. 39–11, the incident photon being scarcely deflected) to $2h/m_0 c$ (for $\varphi = 180°$, corresponding to a "head-on" collision, the incident photon being reversed in direction).

It remains to explain the presence of the peak in Fig. 39–10 for which the wavelength does not change on scattering. We can understand this peak as resulting from a collision between a photon and electrons bound in an ionic core in the scattering block. During photon collisions the bound electrons behave like the free electrons that we considered in Fig. 39–11, with the exception that their effective mass is much greater. This is because the ionic core as a whole recoils during the collision. The effective mass M for a carbon scatterer is approximately the mass of a carbon nucleus. Since this nucleus contains 6 protons and 6 neutrons, we have approximately that $M = 12 \times 1840 m_0 = 22{,}000 m_0$. If we replace m_0 by M in Eq. 39–17, we see that the Compton shift for collisions with tightly bound electrons is immeasurably small.

As in the cavity radiation problem (see Eq. 39–7) and the photoelectric effect (see Eq. 39–11), Planck's constant h is centrally involved in the Compton effect. The quantity h is the central constant of quantum physics. In a universe in which $h = 0$ there would be no quantum physics and classical physics would be valid in the subatomic domain. In particular, as Eq. 39–17 shows, there would be no Compton effect (that is, $\Delta\lambda = 0$) in such a universe.

Example 4. X rays with $\lambda = 1.00$ A are scattered from a carbon block. The scattered radiation is viewed at $90°$ to the incident beam. (*a*) What is the Compton shift $\Delta\lambda$?

Putting $\varphi = 90°$ in Eq. 39–17, we have, for the Compton shift,

$$\Delta\lambda = \frac{h}{m_0 c} (1 - \cos\varphi)$$

$$= \frac{6.63 \times 10^{-34} \text{ joule-sec}}{(9.11 \times 10^{-31} \text{ kg})(3.00 \times 10^8 \text{ meters/sec})} (1 - \cos 90°)$$

$$= 2.43 \times 10^{-12} \text{ meter} = 0.0243 \text{ A}.$$

(*b*) What kinetic energy is imparted to the recoiling electron?

If we put K for the kinetic energy of the electron, we can write Eq. 39–13 as

$$\frac{hc}{\lambda} = \frac{hc}{\lambda'} + K.$$

Since $\lambda' = \lambda + \Delta\lambda$, we obtain

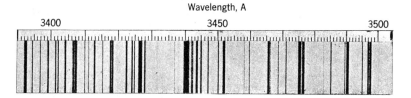

FIGURE 39–12   A small portion of the spectrum of iron, in the region 3400 to 3500 A.

$$\frac{hc}{\lambda} = \frac{hc}{\lambda + \Delta\lambda} + K,$$

which reduces to

$$K = \frac{hc\,\Delta\lambda}{\lambda(\lambda + \Delta\lambda)}$$

$$= \frac{(6.63 \times 10^{-34}\ \text{joule-sec})(3.00 \times 10^8\ \text{meters/sec})(2.43 \times 10^{-12}\ \text{meter})}{(1.00 \times 10^{-10}\ \text{meter})(1.00 + 0.024) \times 10^{-10}\ \text{meter}}$$

$$= 4.73 \times 10^{-17}\ \text{joule} = 295\ \text{eV}.$$

Show that the initial photon energy E in this case ($= h\nu = hc/\lambda$) is 12,400 eV so that the photon lost about 2.3% of its energy in this collision. A photon whose energy was ten times as large ($= 124,000$ eV) can be shown to lose 23% of its energy in a similar collision. This follows from the fact that $\Delta\lambda$ does not depend on the initial wavelength. Hence more energetic x rays, which have smaller wavelengths, will experience a larger per cent increase in wavelength and thus a larger per cent loss in energy.

39–7   Line Spectra

We have seen how Planck successfully explained the nature of the radiation from heated solid objects of which the cavity radiator formed the prototype. Such radiations form *continuous spectra* and are contrasted with *line spectra* such as that of Fig. 39–12, which shows the radiation emitted from iron ions and atoms in an electric arc struck between iron electrodes. We shall see that Planck's quantization ideas, suitably extended, lead to an understanding of line spectra also. The prototype for the study of line spectra is that of atomic hydrogen; being the simplest atom it has the simplest spectrum.

Line spectra are common in all parts of the electromagnetic spectrum. Figure 39–13 shows a spectrum of the γ rays ($\lambda \cong 10^{-12}$ meter) emitted from a particular radioactive nucleus, an isotope of mercury. Figure 39–14 shows a spectrum of x rays ($\lambda \cong 10^{-10}$ meter) emitted from a molybdenum target when struck by a 35-KeV electron beam. The sharp emission lines are superimposed on a continuous background.

Figure 39–15 shows a spectrum associated with the molecule HCl. It occurs in the infrared, with $\lambda \cong 10^{-6}$ meter. This is an absorption spectrum rather than an emission spectrum, as in Fig. 39–12. Experiment shows that isolated atoms and molecules absorb radiation, as well as emit it, at discrete wavelengths.

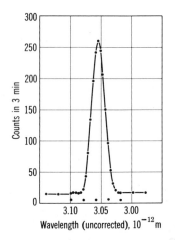

FIGURE 39–13   A wavelength plot for a gamma ray emitted by the nucleus Hg^{198}. (From data by DuMond and co-workers.)

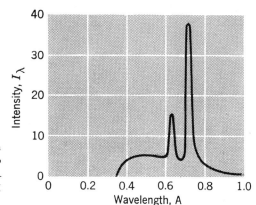

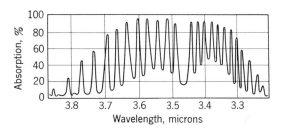

FIGURE 39-14 X rays from a molybdenum target struck by 35-KeV electrons. Note the two sharp lines rising above a broad continuous base. The wavelength of the most intense line is 7.1×10^{-11} meter or 0.71 A. (From data by Ulrey.)

FIGURE 39-15 An absorption spectrum of the HCl molecule near $\lambda = 3.5 \times 10^{-6}$ meter $= 3.5\ \mu$. (From data by E. S. Imes.)

Figure 39–16 shows a portion of the absorption spectrum of ammonia (NH_3) in the microwave region ($\lambda \cong 10^{-2}$ meter). Finally, Fig. 39–17 shows how radiation in the radio-frequency region ($\lambda \cong 43$ meters) is absorbed by hydrogen molecules placed in a magnetic field.

39-8 Atomic Models—The Bohr Hydrogen Atom

The attempts of physicists to explain observable phenomena in terms of theoretical models of the physical world which can be given mathematical expression is nowhere better illustrated than in the development of models of the atom. In this case, key evidence leading finally to the wave-mechanical atom was the line spectrum of hydrogen.

In 1815 Prout (1785–1850) suggested that the elements were made up of hydrogen, using as evidence the fact that the atomic weights of many elements are nearly integer multiples of that of hydrogen. With the discovery of the electron by J. J. Thomson (1856–1940) in 1897 the level of sophistication increased greatly. Thomson proposed the "plum pudding" model where the positive charge of the atom was thought to be

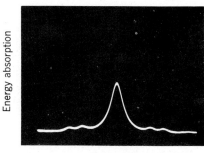

FIGURE 39-16 An oscilloscope trace showing one strong line and four weak lines in the absorption spectrum of ammonia at microwave frequencies.

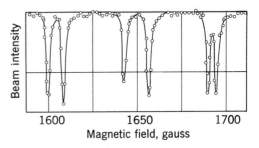

FIGURE 39-17 A portion of the absorption spectrum of the protons in molecular hydrogen at $\lambda \cong 43$ meters. In this technique the frequency is left fixed and the sample is placed in a magnetic field, which is varied to scan the spectrum. (From data by Kellogg, Rabi, and Zacharias.)

spread out through the whole atom (a sphere of radius about 10^{-10} meter) with the electrons located here and there like plums in the pudding. Then in 1911 Ernest Rutherford (1871–1937) showed the inconsistency between the α-particle scattering experiments of Geiger and Marsden and Thomson's atom and proposed the nuclear model of the atom which is the basis of present theories. Here the positive charge is confined to the nucleus, a very small sphere of radius about 10^{-14} meter. The electrons circulate about the nucleus in a volume of the same order of magnitude as Thomson's sphere.

Investigation of the hydrogen spectrum led Niels Bohr (1885–1962) to postulate that the circular orbits of the electrons were quantized, that is, that their angular momentum could have only integral multiples of a basic value. We shall present this *Bohr atom* in some detail here. The Bohr atom, while deficient in several details, illustrates the ideas of quantization within the simpler mathematical framework of classical physics. Before proceeding, however, we should point out that the wave-mechanical atom subsequently replaced the Bohr atom, as we will point out in Sections 40–3 and 40–4. Furthermore, models of the nucleus, while retaining the basic assumptions of Rutherford, have been highly refined and now assume the presence of subnuclear particles which themselves move within and make up the nucleus.

We might wish to associate the frequency of an emitted spectrum line, such as from hydrogen (Fig. 39–18), with the frequency of an electron revolving in an orbit inside the atom. Classical electromagnetism predicts that charges will radiate energy when they are accelerated. In this way electromagnetic waves are emitted from a radio transmitting antenna in which electrons are caused to surge back and forth. This radiation represents a loss of energy for the moving electrons which, in a radio antenna, is compensated for by supplying energy from an oscillator. In an isolated

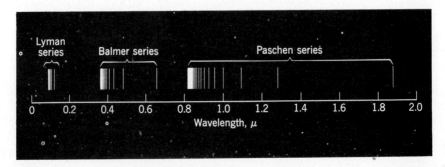

FIGURE 39-18 The spectrum of hydrogen. It consists of a number of series of lines, three of which are shown. Within each series the spectrum lines follow a regular pattern, approaching a so-called *series limit* at the short-wave end of the series.

atom, however, no energy is supplied from external sources. We would expect the frequency of the electron and thus that of the emitted radiation to change continuously as the energy drains away. This prediction of classical theory cannot be reconciled with the existence of sharp spectrum lines. Thus classical physics cannot explain the hydrogen, or any other, spectrum.

Bohr circumvented this difficulty by assuming that, like Planck's oscillators, the hydrogen atom exists in certain stationary states in which it does not radiate. Radiation occurs only when the atom makes a transition from one state, with energy E_k, to a state with lower energy E_j. In equation form

$$hv = E_k - E_j, \qquad (39\text{--}18)$$

where hv is the quantum of energy carried away by the photon that is emitted from the atom during the transition.

In order to learn the allowed frequencies predicted by Eq. 39–18, we need to know the energies of the various stationary states in which a hydrogen atom can exist. This calculation was first carried out by Bohr on the basis of a specific model for the hydrogen atom put forward by him. Bohr's model was highly successful for hydrogen and had a tremendous influence on the further development of the subject; it is now regarded as an important preliminary stage in the development of a more complete theory of quantum physics.

Let us assume that the electron in the hydrogen atom moves in a circular orbit of radius r centered on its nucleus. We assume that the nucleus, which is a single proton, is so massive that the center of mass of the system is essentially at the position of the proton. Let us calculate the energy E of such an atom.

Writing Newton's second law for the motion of the electron, we have (using Coulomb's law)

$$F = ma,$$

or
$$\frac{e^2}{4\pi\epsilon_0 r^2} = m\frac{v^2}{r}.$$

This allows us to calculate the kinetic energy of the electron, which is

$$K = \tfrac{1}{2}mv^2 = \frac{e^2}{8\pi\epsilon_0 r}. \qquad (39\text{--}19)$$

The potential energy U of the proton-electron system is given by

$$U = V(-e) = -\frac{e^2}{4\pi\epsilon_0 r}, \qquad (39\text{--}20)$$

where $V\ (= e/4\pi\epsilon_0 r)$ is the potential of the proton at the radius of the electron.

The total energy E of the system is

$$E = K + U = -\frac{e^2}{8\pi\epsilon_0 r}. \qquad (39\text{--}21)$$

Since the orbit radius can apparently take on any value, so can the energy E. The problem of quantizing E reduces to that of quantizing r.

Every property of the orbit is fixed if the radius is given. Equations 39–19, 39–20, and 39–21 show this specifically for the energies K, U, and E. From Eq. 39–19 we can show that the linear speed v for the electron is also given in terms of r by

$$v = \sqrt{\frac{e^2}{4\pi\epsilon_0 mr}} \,. \tag{39–22}$$

The rotational frequency ν_0 follows at once from

$$\nu_0 = \frac{v}{2\pi r} = \sqrt{\frac{e^2}{16\pi^3\epsilon_0 mr^3}} \,. \tag{39–23}$$

The linear momentum p follows from Eq. 39–22:

$$p = mv = \sqrt{\frac{me^2}{4\pi\epsilon_0 r}} \,. \tag{39–24}$$

The angular momentum L is given by

$$L = pr = \sqrt{\frac{me^2 r}{4\pi\epsilon_0}} \,. \tag{39–25}$$

Thus if r is known, the orbit parameters K, U, E, v, ν_0, p, and L are also known. If any one of these quantities is quantized, all of them must be.

At this stage Bohr had no rules to guide him and so made (after some indirect reasoning which we do not reproduce) a bold hypothesis, namely, that the necessary quantization of the orbit parameters shows up most simply when applied to the angular momentum and that, specifically, L can take on only values given by

$$L = n\frac{h}{2\pi} \,, \qquad n = 1, 2, 3, \ldots . \tag{39–26}$$

Planck's constant appears again in a fundamental way; the integer n is a *quantum number*.

Combining Eqs. 39–25 and 39–26 leads to

$$r = n^2 \frac{h^2\epsilon_0}{\pi me^2}, \qquad n = 1, 2, 3, \ldots , \tag{39–27}$$

which tells how r is quantized. Substituting Eq. 39–27 into Eq. 39–21 produces

$$E = -\frac{me^4}{8\epsilon_0^2 h^2}\frac{1}{n^2}, \qquad n = 1, 2, 3, \ldots , \tag{39–28}$$

which gives directly the energy values of the allowed stationary states.

Figure 39–19 shows the energies of the stationary states and their associated quantum numbers. Equation 39–27 shows that the orbit radius increases as n^2. The upper level in Fig. 39–19, marked $n = \infty$, corresponds to a state in which the electron is completely removed from the atom (that is, $E = 0$ and $r = \infty$). Figure 39–19 also shows some of the quantum jumps that take place between the different stationary states.

Combining Eqs. 39–18 and 39–28 allows us to write a completely theoretical formula for the frequencies of the lines in the hydrogen spectrum. It is

$$\nu = \frac{me^4}{8\epsilon_0^2 h^3}\left(\frac{1}{j^2} - \frac{1}{k^2}\right) \tag{39–29}$$

in which j and k are integers describing, respectively, the lower and the upper sta-

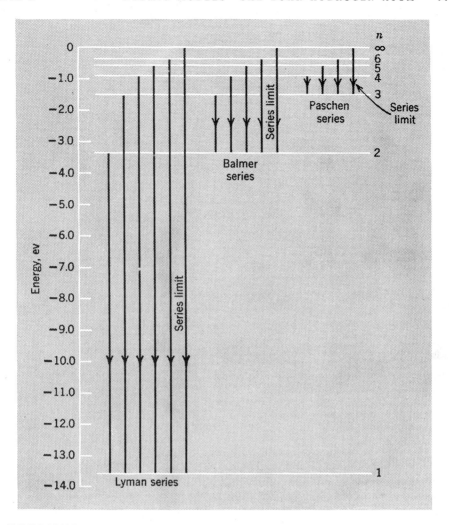

FIGURE 39–19 An energy level diagram for hydrogen showing the quantum number n for each level and some of the transitions that appear in the spectrum. An infinite number of levels is crowded in between the levels marked $n = 6$ and $n = \infty$. Compare this figure carefully with Fig. 39–18.

tionary states. The corresponding wavelengths can easily be found from $\lambda = c/\nu$. Table 39–1 shows some wavelengths so calculated; compare it carefully with Figs. 39–18 and 39–19.

Example 5. Calculate the binding energy of the hydrogen atom (the energy binding the electron to the nucleus) from Eq. 39–28.

The binding energy is numerically equal to the energy of the lowest state in Fig. 39–19. The largest negative value of E in Eq. 39–28 is found for $n = 1$. This yields

$$E = -\frac{me^4}{8\epsilon_0^2 h^2}$$

$$= -\frac{(9.11 \times 10^{-31}\text{ kg})(1.60 \times 10^{-19}\text{ coul})^4}{(8)(8.85 \times 10^{-12}\text{ coul}^2/\text{nt-m}^2)^2(6.63 \times 10^{-34}\text{ joule-sec})^2}$$

$$= -2.17 \times 10^{-18}\text{ joule} = -13.6\text{ eV},$$

which agrees with the experimentally observed energy for hydrogen.

Table 39–1   THE HYDROGEN SPECTRUM   (SOME SELECTED LINES)

Name of series	Quantum Number		Wavelength (A)
	j (lower state)	k (upper state)	
Lyman	1	2	1216
	1	3	1026
	1	4	970
	1	5	949
	1	6	940
	1	∞	912
Balmer	2	3	6563
	2	4	4861
	2	5	4341
	2	6	4102
	2	7	3970
	2	∞	3650
Paschen	3	4	18751
	3	5	12818
	3	6	10938
	3	7	10050
	3	8	9546
	3	∞	8220

39–9   The Correspondence Principle

Although all theories in physics have limitations, they usually do not break down abruptly but in a continuous way, yielding results that agree less and less well with experiment. Thus the predictions of Newtonian mechanics become less and less accurate as the speed is made to approach that of light. A similar relationship must exist between quantum physics and classical physics; it remains to find the circumstances under which the latter theory is revealed as a special case of the former.

The radius of the lowest energy state in hydrogen (the so-called *ground state*) is found by putting $n = 1$ in Eq. 39–27; it turns out to be 5.3×10^{-11} meter. If $n = 10,000$, however, the radius is $(10,000)^2$ times as large or 5.3 mm. This "atom" is so large that we suspect that its behavior should be accurately described by classical physics. Let us test this by computing the frequency of the emitted light on the basis of both classical and quantum assumptions. These calculations should differ at small quantum numbers but should agree at large quantum numbers. The fact that quantum physics reduces to classical physics at large quantum numbers is called the *correspondence principle*. This principle, credited to Niels Bohr, was very useful during the years in which quantum physics was being developed. Bohr, in fact, based his theory of the hydrogen atom on correspondence principle arguments.

Classically, the frequency of the light emitted from an atom is equal to ν_0, its frequency of revolution in its orbit. This can be expressed in terms of a quantum number n by combining Eqs. 39–23 and 39–27 to obtain

$$\nu_0 = \frac{me^4}{8\epsilon_0{}^2 h^3} \frac{2}{n^3}. \tag{39–30}$$

Quantum physics predicts that the frequency ν of the emitted light is given by Eq. 39–29. Considering a transition between an orbit with quantum number $k = n$

and one with $j = n - 1$ leads to

$$\nu = \frac{me^4}{8\epsilon_0^2 h^3} \left[\frac{1}{(n-1)^2} - \frac{1}{n^2} \right]$$

$$= \frac{me^4}{8\epsilon_0^2 h^3} \left[\frac{2n-1}{(n-1)^2 n^2} \right]. \qquad (39\text{--}31)$$

As $n \to \infty$ the expression in the square brackets above approaches $2/n^3$ so that $\nu \to \nu_0$ as $n \to \infty$. Table 39–2 illustrates this example of the correspondence principle.

Table 39–2 THE CORRESPONDENCE PRINCIPLE AS APPLIED TO THE HYDROGEN ATOM

Quantum number (n)	Frequency of revolution in orbit (Eq. 39–30) (cycles/sec)	Frequency of transition to next lowest state (Eq. 39–31) (cycles/sec)	Difference (%)
2	8.20 $\times 10^{14}$	24.6 $\times 10^{14}$	67
5	5.26 $\times 10^{13}$	7.38 $\times 10^{13}$	29
10	6.57 $\times 10^{12}$	7.72 $\times 10^{12}$	14
50	5.25 $\times 10^{10}$	5.42 $\times 10^{10}$	3
100	6.578 $\times 10^9$	6.677 $\times 10^9$	1.5
2,000	8.2224 $\times 10^5$	8.2305 $\times 10^5$	0.10
25,000	4.2099 $\times 10^2$	4.2102 $\times 10^2$	0.007

QUESTIONS

1. "Pockets" formed by the coals in a coal fire seem brighter than the coals themselves. Is the temperature in such pockets appreciably higher than the surface temperature of an exposed glowing coal?

2. The relation $\mathcal{R} = \sigma T^4$ (Eq. 39–2) is exact for true cavities and holds for all temperatures. Why don't we use this relation as the basis of a definition of temperature at, say, $100°$ C?

3. Do all incandescent solids obey the fourth-power law of temperature, as Eq. 39–3 seems to suggest?

4. A cavity radiator is sometimes called a *black body*. Why?

5. It is stated that if we look into a cavity whose walls are maintained at a constant temperature no details of the interior are visible. Does this seem reasonable?

6. How can a photon energy be given by $E = h\nu$ (Eq. 39–10) when the very presence of the frequency ν in the formula implies that light is a wave?

7. In the photoelectric effect, why does the existence of a cutoff frequency speak in favor of the photon theory and against the wave theory?

8. Why are photoelectric measurements so sensitive to the nature of the photoelectric surface?

9. Does Einstein's theory of photoelectricity, in which light is postulated to be a photon, invalidate Young's interference experiment?

10. List and discuss carefully the assumptions made by Planck in connection with the cavity radiation problem, by Einstein in connection with the photoelectric effect, and by Bohr in connection with the hydrogen atom problem.

11. In Bohr's theory for the hydrogen atom orbits, what is the implication of the fact that the potential energy is negative and is greater in magnitude than the kinetic energy?

12. Can a hydrogen atom absorb a photon whose energy exceeds its binding energy (13.6 ev)?

13. Discuss Example 1 in terms of the correspondence principle.

14. According to classical mechanics, an electron moving in an orbit should be able to do so with any angular momentum whatever. According to Bohr's theory of the hydrogen atom, however, the angular momentum is quantized according to $L = nh/2\pi$. Reconcile these two statements, using the correspondence principle.

PROBLEMS

1(2). An oven with an inside temperature T (227° C) is in a room having a temperature T_r (27° C). There is a small opening of area 5.0 cm² in one side of the oven. How much net power is transferred from the oven to the room? Hint: Consider both room and oven as cavities.
Answer: 1.5 watts.

2(2). (*a*) Assuming the surface temperature of the sun to be 5700° K, use the Stefan-Boltzmann law (Eq. 39–2) to determine the rest mass lost per second to radiation by the sun. Take the sun's diameter to be 1.4×10^9 meter. (*b*) What fraction of the sun's rest mass is lost each year from electromagnetic radiation? Take the sun's rest mass to be 2.0×10^{30} kg.

3(2). What is the power radiated from a nichrome wire 1.0 meter long and having a diameter of 0.060 in. at a temperature of 800° C if the emissivity of nichrome is 0.92?
Answer: 330 watts.

4(2). The power rating of a light bulb tells how much electrical power is supplied. (*a*) An incandescent bulb of power rating P (100 watt) has a tungsten filament of diameter d (0.40 mm) and length l (30 cm) (emissivity of tungsten = 0.26). At what temperature does the filament operate? (*b*) A fluorescent tube rated at 40 watts gives as much visible light as a 100-watt incandescent bulb. Explain why.

5(2). Figure 39–1 compares the spectral radiancy of tungsten at 2000° K with that of a cavity radiator. If the radiancy $\mathscr{R}$—the area under the curve for tungsten—is 23.3 watts/cm², verify that the emissivity of tungsten at 2000° K is 0.259.

6(2). The average rate of solar radiation incident per unit area on the earth is 355 watts/meter². (*a*) Explain the difference between this number and the solar constant (the solar energy falling per unit time at normal incidence on unit area of the earth's surface) whose value is 1340 watts/meter². (*b*) Consider the earth to behave like a cavity radiating energy into space at this same rate. What surface temperature would the earth have under these circumstances?

7(3). Show that Wien's law is a special case of Planck's law (Eq. 39–6) for short wavelengths or low temperatures.

8(3). The wavelength, λ_{max}, at which the spectral radiancy has its maximum value per unit wavelength for a particular temperature T is given by the Wien displacement law, $\lambda_{max}T =$ constant. At what wavelength does a cavity radiator at 6000° K radiate most per unit wavelength? The experimentally determined value of Wien's constant is 2.898×10^{-3} meter °K.

9(3). A cavity whose walls are held at 4000° K has a circular aperture 5.0 mm in diameter. At what rate does energy escape through this hole?
Answer: 290 watts.

10(4). In Example 2 suppose that the "target" is a single gas atom of 1.0 A radius and that the intensity of the light source is reduced to 1.0×10^{-5} watt. If the binding energy of the most loosely bound electron in the atom is 2.0 eV, what time lag for the photoelectric effect is expected on the basis of the wave theory of light?

11(5). Show that the energy E of a photon (in eV) is related to the wavelength λ (in A) by

$$E = 1.24 \times 10^4/\lambda.$$

12(5). Solar radiation falls on the earth at a rate of 2.0 cal/cm²-min. How many photons/cm²-min is this, assuming an average wavelength of 5500 A?

13(5). A 100-watt sodium vapor lamp radiates uniformly in all directions. (*a*) At what distance from the lamp will the average density of photons be 10/cm³? (*b*) What is the average density of photons 2.0 meters from the lamp? Assume the light to be monochromatic, with $\lambda = 5890$ A. *Answer:* (*a*) 89 meters. (*b*) 2.0×10^4 photons/cm³.

14(5). An atom absorbs a photon having a wavelength of 3750 A and immediately emits another photon having a wavelength of 5800 A. How much energy was absorbed by the atom in this process? Ease the computation by using the result of Problem 11(5).

15(5). What are (*a*) the frequency, (*b*) the wavelength, and (*c*) the momentum of a photon whose energy equals the rest energy of an electron?
Answer: (*a*) 1.2×10^{20} cycles/sec. (*b*) 0.025 A. (*c*) 2.7×10^{-22} joule-sec/meter.

16(5). The energy required to remove an electron from sodium is 2.3 eV. Does sodium show a photoelectric effect for orange light, with $\lambda = 6800$ A?

17(5). You wish to pick a substance for a photocell operable with visible light. Which of the following will do (work function in parenthesis): tantalum (4.2 eV); tungsten (4.5 eV); aluminum (4.2 eV); barium (2.5 eV); lithium (2.3 eV)?
Answer: Barium and lithium.

18(5). Incident photons strike a sodium surface having a work function E_0 (2.2 eV) causing photoelectron emission. When a stopping potential V_0 (5.0 volts) is imposed there is no photocurrent. What is the wavelength of the incident photons?

19(5). Find the maximum kinetic energy of photoelectrons if the work function of the material is 2.0×10^{-19} joule and the frequency of the radiation is 3.0×10^{15} cycles/sec.
Answer: 1.8×10^{-18} joule.

20(5). (*a*) If the work function for a metal is 1.8 eV, what would be the stopping potential for light having a wavelength of 4000 A? (*b*) What would be the maximum velocity of the emitted photoelectrons at the metal's surface?

21(5). Light of a wavelength 2000 A falls on an aluminum surface. In aluminum 4.2 eV are required to remove an electron. What is the kinetic energy of (*a*) the fastest and (*b*) the slowest emitted photoelectrons? (*c*) What is the stopping potential (*d*) What is the cutoff wavelength for aluminum?
Answer: (*a*) 2.0 eV. (*b*) Zero. (*c*) 2.0 volts. (*d*) 3000 A.

22(5). The work function for a clean lithium surface is 2.3 eV. Make a rough plot of the stopping potential V_0 versus the frequency of the incident light for such a surface.

23(6). Show, by analyzing a collision between a photon and a free electron (using relativistic mechanics), that it is impossible for a photon to give all of its energy to the free electron. In other words, the photoelectric effect cannot occur for completely free electrons; the electrons must be bound in a solid or in an atom.

24(6). Photons of wavelength 0.024 A are incident on free electrons. (*a*) Find the wavelength of a photon which is scattered 30° from the incident direction. (*b*) Do the same if the scattering angle is 120°.

25(6). What is the maximum kinetic energy of the Compton-scattered electrons from a sheet of copper struck by a monochromatic photon beam in which the incident photons each have a wavelength of 1.4×10^{-2} A? (See Example 4).
Answer: 0.68 MeV.

26(6). An x-ray photon of wavelength λ (0.10 A) strikes an electron head on ($\varphi = 180°$). Determine (*a*) the change in wavelength of the photon (*b*) the change in energy of the photon, and (*c*) the final kinetic energy of the electron.

27(6). A 2.0-A photon falling on a carbon block is scattered by a Compton collision and its frequency is shifted by 0.010%. (*a*) Through what angle is the photon scattered? (*b*) How much energy does the electron which scattered the photon gain? [Note that for any wave motion, $\Delta\nu = (c/\lambda^2)\,\Delta\lambda$.]
Answer: (*a*) 7.4°. (*b*) 0.62 eV.

28(6). Calculate the percent change in photon energy for a Compton collision with φ in Fig. 39–11 equal to 90° for radiation in (a) the microwave range, with $\lambda = 3.0$ cm, (b) the visible range, with $\lambda = 5000$ A, (c) the x-ray range, with $\lambda = 1.0$ A, and (d) the gamma-ray range, the energy of the gamma-ray photons being 1.0 MeV. What are your conclusions about the importance of the Compton effect in these various regions of the electromagnetic spectrum, judged solely by the criterion of energy loss in a single Compton encounter?

29(6). A photon "hits" an electron "head-on" and recoils backward directly along the line of incidence. If the electron moves off at a speed βc, where $\beta \ll 1$ ($= 10^{-3}$, for example), show that the ratio of the final electron kinetic energy to the initial photon energy is just β. (Hint: Set up the problem as a nonrelativistic Compton "collision".)

30(6). Carry out the necessary algebra to eliminate v and θ from Eqs. 39–13, 39–15, and 39–16 to obtain the Compton shift relation (Eq. 39–17).

31(7). A spectral emission line, important in radioastronomy, has a wavelength of 21 cm. To what photon energy does this correspond?
Answer: 5.9×10^{-6} eV.

32(8). A line in the x-ray spectrum of gold consists of photons all having nearly the same wavelength 0.185 A. If the energy in these photons comes from the atoms jumping from one specific energy level, -13.7 KeV, to another lower one, what is the second energy?

33(8). Light of wavelength 4863 A is emitted by a hydrogen atom. (a) What transition of the hydrogen atom is responsible for this radiation? (b) To what series does this radiation belong?
Answer: (a) $n = 4$ to $n = 2$. (b) The Balmer series.

34(8). Show on an energy-level diagram for hydrogen the quantum numbers corresponding to a transition in which the wavelength of the emitted photon is 1216 A.

35(8). What are the energy, momentum, and wavelength of the photon that is emitted when a hydrogen atom undergoes a transition from the state $n = 3$ to $n = 1$?
Answer: 12 eV; 6.5×10^{-27} kg-meter/sec; 1030 A.

36(8). (a) Using Bohr's formula for the frequencies of the lines in the hydrogen spectrum, calculate the three longest wavelengths in the Balmer series. (b) Between what wavelength limits does the Balmer series lie?

37(8). In the ground state of the hydrogen atom, according to Bohr's theory, what are (a) the quantum number, (b) the orbit radius, (c) the angular momentum, (d) the linear momentum, (e) the angular velocity, (f) the linear speed, (g) the force on the electron, (h) the acceleration of the electron, (i) the kinetic energy, (j) the potential energy, and (k) the total energy?
Answer: (a) 1. (b) 5.3×10^{-11} meter. (c) 1.1×10^{-34} joule-sec. (d) 2.0×10^{-24} kg-meter/sec. (e) 4.1×10^{16} radians/sec. (f) 2.2×10^{6} meters/sec. (g) 8.2×10^{-8} nt. (h) 9.0×10^{22} meters/sec². (i) $+13.6$ eV. (j) -27.2 eV. (k) -13.6 eV.

38(8). How do the quantities (b) to (k) in Problem 37(8) vary with the quantum number?

39(8). How much energy is required to remove an electron from a hydrogen atom in a state with $n = 8$?
Answer: 0.21 eV.

40(8). A hydrogen atom is excited from a state with $n = 1$ to one with $n = 4$. (a) Calculate the energy that must be absorbed by the atom. (b) Calculate and display on an energy-level diagram the different photon energies that may be emitted if the atom returns to its $n = 1$ state. (c) Calculate the recoil speed of the hydrogen atom, assumed initially at rest, if it makes the transition from $n = 4$ to $n = 1$ in a single quantum jump.

41(8). A hydrogen atom in a state having a *binding energy* (this is the energy required to remove an electron) of 0.85 eV makes a transition to a state with an *excitation energy* (this is the difference in energy between the state and the ground state) of 10.2 eV. (a) Find the energy of the emitted photon. (b) Show this transition on an energy-level diagram for hydrogen, labeling the appropriate quantum numbers.
Answer: (a) 2.6 eV. (b) $n = 4$ to $n = 2$.

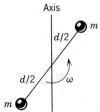

FIGURE 39-20   Problem 49(8).

42(8). A neutron, with kinetic energy of 6.0 eV, collides with a resting hydrogen atom in its ground state. Show that this collision must be elastic (that is, kinetic energy must be conserved).

43(8). From the energy level diagram for hydrogen, explain the observation that the frequency of the second Lyman-series line is the sum of the frequencies of the first Lyman-series line and the first Balmer-series line. This is an example of the empirically discovered *Ritz Combination Principle*. Use the diagram to find some other valid combinations.

44(8). Using Bohr's theory, calculate the energy required to remove the electron from the ground state of singly ionized helium, that is, helium with one electron removed.

45(8). Use Bohr's theory to compare the spectrum of singly ionized helium with the spectrum of hydrogen.
Answer: $\lambda_{He} = \frac{1}{4}\lambda_H$. for corresponding spectrum lines.

46(8). *Muonic atoms.* Apply Bohr's theory to a muonic atom, which consists of a nucleus of charge Ze with a negative muon (an elementary particle with a charge of $-e$ and a mass m that is 207 times as large as the electron mass) circulating about it. Calculate (*a*) the radius of the first Bohr orbit, (*b*) the ionization energy, and (*c*) the wavelength of the most energetic photon that can be emitted. Assume that the muon is circulating about a hydrogen nucleus ($Z = 1$).

47(8). *Positronium.* Apply Bohr's theory to the positronium atom. This consists of a positive and a negative electron revolving around their center of mass, which lies halfway between them. (*a*) What relationship exists between this spectrum and the hydrogen spectrum? (*b*) What is the radius of the ground state orbit? (Hint: It will be necessary to analyze this problem from first principles because this "atom" has no nucleus; both particles revolve about a point halfway between them.)
Answer: (*a*) Corresponding positronium wavelengths are longer by a factor of 2. (*b*) Radius to center of mass is equal to the corresponding radius for hydrogen.

48(8). Perhaps an atom could be formed by an electron and a neutron binding together by gravitational forces. Calculate the ground state radius of an electron in such an atom by using a Bohr-type model in which the Coulomb attractive electrical force is replaced by the attractive gravitational force.

49(8). A diatomic gas molecule consists of two atoms of mass m separated by a fixed distance d rotating about an axis as indicated in Fig. 39–20. Assuming that its angular momentum is quantized as in the Bohr atom, determine (*a*) the possible angular velocities, and (*b*) the possible quantized rotational energies. (*c*) Show on an energy diagram.
Answer: (*a*) $nh/\pi md^2$. (*b*) $n^2h^2/4\pi^2md^2$.

50(9). (*a*) If the angular momentum of the earth due to its motion around the sun were quantized according to Bohr's relation $L = nh/2\pi$, what would the quantum number be? (*b*) Could such quantization be detected if it existed?

51(9). In Table 39–2 show that the quantity in the last column is given by

$$\frac{100(\nu - \nu_0)}{\nu} \cong \frac{150}{n}$$

for large quantum numbers.

52(9). If an electron is rotating in an orbit at frequency ν_0, classical electromagnetism predicts that it will radiate energy not only at this frequency but also at $2\nu_0$, $3\nu_0$, $4\nu_0$, etc. Show that this is also predicted by Bohr's theory of the hydrogen atom in the limiting case of large quantum numbers.

CHAPTER 40

Waves and Particles

40–1 Matter Waves

In 1924 Louis de Broglie reasoned that (*a*) nature is strikingly symmetrical in many ways; (*b*) our observable universe is composed entirely of light and matter; (*c*) if light has a dual, wave-particle nature, perhaps matter has also. Since matter was then regarded as being composed of particles, de Broglie's reasoning suggested that one should search for a wave-like behavior for matter.

De Broglie's suggestion might not have received serious attention had he not predicted what the expected wavelength of the so-called matter waves would be. We recall that about 1680 Huygens put forward a wave theory of light that did not receive general acceptance, in part because Huygens was not able to state what the wavelength of the light was. When Thomas Young rectified this defect in 1800, the wave theory of light started on its way to acceptance.

De Broglie assumed that the wavelength of the predicted matter waves was given by the same relationship that held for light namely, Eq. 39–14, or

$$\lambda = \frac{h}{p}, \tag{40–1}$$

which connects the wavelength of a light wave with the momentum of the associated photons. The dual nature of light shows up strikingly in this equation and also in Eq. 39–10 ($E = h\nu$). Each equation contains within its structure both a wave concept (ν and λ) and a particle concept (E and p) De Broglie predicted that the wavelength of matter waves would also be given by Eq. 40–1, where p would now be the momentum of the particle of matter.

Example 1. What wavelength is predicted by Eq. 40–1 for a beam of electrons whose kinetic energy is 100 eV?

The velocity of the electrons is found from $K = \frac{1}{2}mv^2$, or

$$v = \sqrt{\frac{2K}{m}} = \sqrt{\frac{(2)(100 \text{ eV})(1.6 \times 10^{-19} \text{ joule/eV})}{9.1 \times 10^{-31} \text{ kg}}}$$

$$= 5.9 \times 10^6 \text{ meters/sec.}$$

The momentum follows from

$$p = mv = (9.1 \times 10^{-31} \text{ kg})(5.9 \times 10^6 \text{ meters/sec}) = 5.4 \times 10^{-24} \text{ kg-m/sec.}$$

The wavelength (called the *de Broglie wavelength*) is found from Eq. 40–1 or

$$\lambda = \frac{h}{p} = \frac{6.6 \times 10^{-34} \text{ joule-sec}}{5.4 \times 10^{-24} \text{ kg-m/sec}} = 1.2 \text{ A.}$$

This is the same order of magnitude as the size of an atom or the spacing between adjacent planes of atoms in a solid.

In 1926 Elsasser pointed out that the wave nature of matter might be tested in the same way that the wave nature of x rays was first tested, namely, by allowing a beam of electrons of the appropriate energy to fall on a crystalline solid. The atoms of the crystal serve as a three-dimensional array of diffracting centers for the electron "wave"; we should look for strong diffracted peaks in certain characteristic directions, just as for x-ray diffraction.

This idea was tested by C. J. Davisson and L. H. Germer in this country and by G. P. Thomson in Scotland. Figure 40–1 shows the apparatus of Davisson and Germer. Electrons from a heated filament are accelerated by a variable potential difference V and emerge from the electron gun G with kinetic energy eV. This electron beam is allowed to fall at normal incidence on a single crystal of nickel at C. Detector D is set at a particular angle φ and readings of the intensity of the "reflected" beam are taken at various values of the accelerating potential V. Figure 40–2 shows that a strong beam occurs at $\varphi = 50°$ for $V = 54$ volts.

All such strong "reflected" beams can be accounted for by assuming that the electron beam has a wavelength, given by $\lambda = h/p$, and that "Bragg reflections" occur from certain families of atomic planes precisely as described for x rays in Section 38–10.

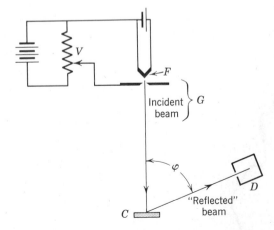

FIGURE 40–1 The apparatus of Davisson and Germer. Electrons from filament F are accelerated by a variable potential difference V. After "reflection" from crystal C they are collected by detector D.

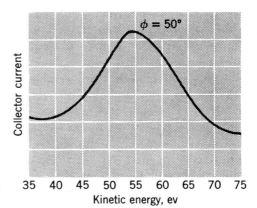

FIGURE 40–2 The collector current in detector *D* in Fig. 40–1 as a function of the kinetic energy of the incident electrons, showing a diffraction maximum. The angle φ in Fig. 40–1 is adjusted to 50°. If an appreciably smaller or larger value is used, the diffraction maximum disappears.

Figure 40–3 shows such a Bragg reflection, obeying the Bragg relationship

$$m\lambda = 2d \sin\theta, \qquad m = 1, 2, 3, \ldots . \tag{40–2}$$

For the conditions of Fig. 40–3 the effective interplanar spacing *d* can be shown by x-ray analysis to be 0.91 A. Since φ equals 50°, it follows that θ equals $90° - \frac{1}{2} \times 50°$ or 65°. The wavelength to be calculated from Eq. 40–2, if we assume $m = 1$, is

$$\lambda = 2d \sin\theta = 2(0.91 \text{ A})(\sin 65°) = 1.6 \text{ A}.$$

The wavelength calculated from the de Broglie relationship $\lambda = h/p$ is, for 54–eV electrons (see Example 1), 1.64 A. This excellent agreement, combined with much similar evidence, is a convincing argument for believing that electrons are wave-like in some circumstances.

Not only electrons but all other particles, charged or uncharged, show wave-like characteristics. Beams of slow neutrons from nuclear reactors are routinely used to investigate the atomic structure of solids. Figure 40–4 shows a "neutron diffraction pattern" for finely powdered lead.

The evidence for the existence of matter waves with wavelengths given by Eq. 40–1 is strong indeed. Nevertheless, the evidence that matter is composed of particles remains equally strong; see Fig. 9–7. Thus, for matter as for light, we must face up to the existence of a dual character; matter behaves in some circumstances like a particle and in others like a wave.

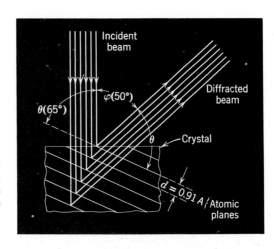

FIGURE 40–3 The strong diffracted beam at $\varphi = 50°$ and $V = 54$ volts arises from wavelike "reflection" from the family of atomic planes shown, for $d = 0.91$ A. The Bragg angle θ is 65°. For simplicity refraction of the diffracted wave as it leaves the surface is ignored.

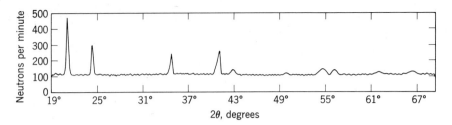

FIGURE 40–4 A diffraction pattern for powdered lead, using a monochromatic neutron beam from a nuclear reactor. The peaks represent "Bragg reflections" from the various atomic planes, θ being the corresponding "Bragg angle."

40–2   Atomic Structure and Standing Waves

The motion of electrons in beams is not bounded or limited in the beam direction. We can make an analogy to a sound wave in a long gas-filled tube, a wave traveling down a long string, or an electromagnetic wave in a long transmission line. All four cases can be described by appropriate traveling waves and, significantly, waves of any wavelength (within a certain range) can be propagated.

Let these last three waves be bounded by imposing physical restrictions. For the sound wave this corresponds to inserting end walls on a section of the long gas-filled pipe, thus forming an acoustic resonant cavity. For the waves in the string it corresponds to removing a finite section of string and clamping it at each end, as a violin string (Section 16–9).

Two important changes occur: (*a*) the motions are now represented by standing rather than traveling waves and (*b*) only certain wavelengths (or frequencies) can now exist. This *quantization* of the wavelength is a direct result of bounding or limiting the wave. We expect that if electrons are limited in their motions by being localized in an atom that (*a*) the electron motion can be represented by a standing matter wave, and (*b*) the electron motion will become quantized, that is, its energy can take on only certain discrete values.

De Broglie was able to derive the Bohr quantization condition for angular momentum by applying proper boundary conditions to matter waves in the hydrogen atom. Figure 40–5 suggests an instantaneous "snapshot" of a standing matter wave associated with an orbit of radius r. The de Broglie wavelength ($\lambda = h/p$) has been chosen so that the orbit of radius r contains an integral number n of the matter waves, or

$$\frac{2\pi r}{\lambda} = \frac{2\pi r}{(h/p)} = n, \qquad n = 1, 2, 3, \ldots.$$

This leads at once to

$$L = pr = n\frac{h}{2\pi}, \qquad n = 1, 2, 3, \ldots,$$

which is the Bohr quantization condition for L.

40–3   Wave Mechanics

The idea that the stationary states in atoms correspond to standing matter waves was taken up

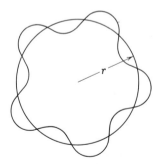

FIGURE 40–5 Showing how an electron wave can be adjusted in wavelength to fit an integral number of times around the circumference of a Bohr orbit of radius r. This concept, like the Bohr orbit concept, is now regarded as oversimplified.

by Erwin Schrödinger in 1926 and used by him as the foundation of *wave mechanics,* one of several equivalent formulations of quantum physics.

An important quantity in wave mechanics is the *wave function* Ψ, which measures the "wave disturbance" of matter waves. For waves on strings the "wave disturbance" may be measured by a transverse displacement y; for sound waves it may be measured by a pressure variation p; for electromagnetic waves it may be measured by the electric field vector **E.**

We make the physical meaning of the wave disturbance Ψ clear in Section 40–4. Meanwhile, let us study the wave function $\Psi(x, t)$ for a simple, one-dimensional problem, that of the possible motions a particle of mass m confined between rigid walls of separation l as in Fig. 40–6b. The wave function can be obtained by analogy with a known mechanical problem, that of the natural modes of vibration of a string of length l, clamped at each end as in Fig. 40–6a.

In the vibrating string the boundary conditions require that nodes exist at each end. This means that the wavelength λ must be chosen so that

$$l = n\frac{\lambda}{2}, \qquad n = 1, 2, 3, \ldots,$$

or that the wavelength λ is "quantized" by the requirement that

$$\lambda = \frac{2l}{n}, \qquad n = 1, 2, 3, \ldots. \qquad (40\text{–}3)$$

The wave disturbance for the string is represented by a standing wave whose equation was shown in Section 16–8 to be

$$y = 2y_m \sin kx \cos \omega t,$$

where $\omega \; (= 2\pi\nu)$ is the angular frequency of the wave and $k \; (= 2\pi/\lambda)$ is the wave number. Since λ is quantized, k must be also, or

$$k = \frac{2\pi}{\lambda} = \frac{n\pi}{l}, \qquad n = 1, 2, 3, \ldots.$$

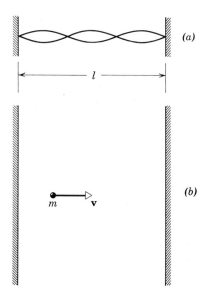

FIGURE 40–6   (a) A stretched string of length l clamped between rigid supports. (b) A particle of mass m and velocity v confined to move betweeen rigid walls a distance l apart.

which leads to

$$y = \left[2y_m \sin \frac{n\pi x}{l} \right] \cos \omega t, \qquad n = 1, 2, 3, \ldots . \qquad (40\text{--}4)$$

Inspection of Eq. 40–4 shows that no matter what value of n is selected nodes exist at $x = 0$ and at $x = l$, as required by the boundary conditions. Figure 40–7 shows plots of the quantity in the square brackets in this equation (the amplitude of the standing wave) for the modes of vibration of the string corresponding to $n = 1$, 2, and 3.

Consider now the particle confined between rigid walls. Since the walls are assumed to be perfectly rigid, the particle cannot penetrate them so that Ψ, which represents the particle motion in some way not yet clearly specified (see Section 40–4), must vanish at $x = 0$ and at $x = l$. The allowed wavelengths of the matter waves must be given by Eq. 40–3, or

$$\lambda = \frac{2l}{n}.$$

Replacing λ by h/p (see Eq. 40–1) leads to

$$p = \frac{nh}{2l}, \qquad (40\text{--}5)$$

which shows that the linear momentum of the particle is quantized. The momentum p ($= mv$) is related to the energy E (which is entirely kinetic and is equal to $\frac{1}{2}mv^2$) by

$$p = \sqrt{2mE}. \qquad (40\text{--}6)$$

Combining Eqs. 40–5 and 40–6 leads to the quantization condition for E, or

$$E = n^2 \frac{h^2}{8ml^2}, \qquad n = 1, 2, 3, \ldots . \qquad (40\text{--}7)$$

The particle cannot have any energy, as we would expect classically, but only energies given by Eq. 40–7.

The matter wave is described, in strict analogy with Eq. 40–4, by

$$\Psi = \left[\Psi_m \sin \frac{n\pi x}{l} \right] \cos \omega t, \qquad n = 1, 2, 3, \ldots .$$

$$(40\text{--}8)$$

Figure 40–7 can serve equally well to show how the amplitude of the standing matter waves for the states of motion corresponding to $n = 1$, 2, and 3 varies throughout the box. We see clearly in this problem how the act of localizing or bounding a particle leads to energy quantization.

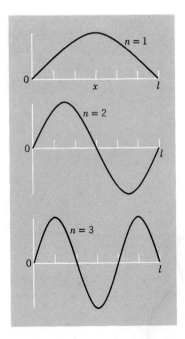

FIGURE 40–7

Example 2. Consider an electron ($m = 9.1 \times 10^{-31}$ kg) confined by electrical forces to move between two rigid "walls" separated by 1.0×10^{-9} meter, which is about five atomic diameters. Find the quantized energy values for the three lowest stationary states.

From Eq. 40–7, for $n = 1$, we have

$$E = n^2 \frac{h^2}{8ml^2} = (1)^2 \frac{(6.6 \times 10^{-34} \text{ joule-sec})^2}{(8)(9.1 \times 10^{-31} \text{ kg})(1.0 \times 10^{-9} \text{ meter})^2}$$

$$= 6.0 \times 10^{-20} \text{ joule} = 0.38 \text{ eV}.$$

The energies for the next two states ($n = 2$ and $n = 3$) are $2^2 \times 0.38 \text{ eV} = 1.5 \text{ eV}$ and $3^2 \times 0.38 \text{ eV} = 3.4 \text{ eV}$.

Example 3. Consider a grain of dust ($m = 1.0 \ \mu\text{gm} = 1.0 \times 10^{-9}$ kg) confined to move between two rigid walls separated by 0.10 mm. Its speed is only 1.0×10^{-6} meter/sec, so that it requires 100 sec to cross the gap. What quantum number describes this motion?

The energy is

$$E (= K) = \tfrac{1}{2}mv^2 = \tfrac{1}{2}(1.0 \times 10^{-9} \text{ kg})(1.0 \times 10^{-6} \text{ meter/sec})^2$$

$$= 5 \times 10^{-22} \text{ joule}.$$

Solving Eq. 40–7 for n yields

$$n = \sqrt{8mE} \ \frac{l}{h} = \sqrt{(8)(10^{-9} \text{ kg})(5.0 \times 10^{-22} \text{ joule})} \left(\frac{10^{-4} \text{ meter}}{6.6 \times 10^{-34} \text{ joule-sec}} \right)$$

$$= 3 \times 10^{14}.$$

Even in these extreme conditions the quantized nature of the motion would never be apparent; we cannot distinguish experimentally between $n = 3 \times 10^{14}$ and $n = 3 \times 10^{14} + 1$. Classical physics, which fails completely for the problem of Example 2, works extremely well for this problem.

40–4 Meaning of Ψ

Max Born first suggested that the quantity Ψ^2 at any particular point is a measure of the probability that the particle will be near that point. More exactly, if we construct a volume element dV at that point, the probability that the particle will be found in the volume element at a given instant is $\Psi^2 \, dV$. This interpretation of Ψ provides a statistical connection between the wave and the associated particle; it tells us where the particle is likely to be, not where it is.

For the particle confined between rigid walls the probability that the particle will lie between two planes that are distance x and $x + dx$ from one wall (see Fig. 40–8) is given by

$$\Psi^2 \, dx = \Psi_m{}^2 \sin^2 \frac{n\pi}{l} x \cos^2 \omega t \, dx.$$

Since we are more interested in the particle's spatial location than in its time behavior, we average Ψ^2 over one cycle of the motion. This is equivalent to replacing $\cos^2 \omega t$ by its average value for one cycle, namely one-half, or

$$\overline{\Psi^2} = \tfrac{1}{2}\Psi_m{}^2 \sin^2 \frac{n\pi x}{l} \qquad n = 1, 2, 3, \ldots \. \qquad (40\text{–}9)$$

Figure 40–8 shows $\overline{\Psi^2}$ for the three stationary states corresponding to $n = 1, 2$, and 3. Note that for $n = 1$ the particle is more likely to be near the center than the ends. This is in sharp contradiction to the results of classical physics, according to which

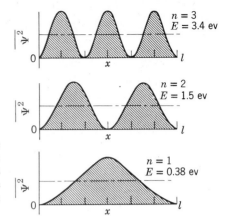

FIGURE 40-8 The "probability functions" for three states of motion of the particle of Fig. 40-6*b*, along with the corresponding quantized energies for the conditions of Example 2. The horizontal lines show the predictions of classical mechanics, in which the probability function is constant for all positions of the particle.

the particle has the same probability of being located anywhere between the walls, as shown by the horizontal line in Fig. 40–8.

The problem of a particle confined between rigid walls has little real application in physics. We would prefer to illustrate the wave mechanics of Schrödinger by applying it to a real situation, such as the hydrogen atom. Only mathematical complexity prevents us from doing this. We state without proof that when we solve this problem by wave mechanics the motion of the electron in the ground state of the atom, defined by putting $n = 1$ in Eq. 39–28, is described by the following wave function,

$$\Psi = \sqrt{\frac{2}{\pi a^3}}\, e^{-r/a} \cos \omega t, \tag{40–10}$$

where

$$a = \frac{h^2 \epsilon_0}{\pi m e^2}.$$

Putting $n = 1$ in Eq. 39–27 shows that a is the radius of the ground-state orbit in Bohr's theory. This special interpretation has little meaning in wave mechanics; a is taken here merely as a convenient unit of length when dealing with atomic problems, having the value 0.529 A.

Example 4. Consider two hypothetical spherical shells centered on the nucleus of a hydrogen atom with radii r and $r + dr$. What is the probability $P(r)$ that the electron will lie between these shells, as a function of r?

This probability is $\Psi^2\, dV$, where dV is the volume between the shells, or $4\pi r^2\, dr$. Thus

$$\Psi^2\, dV = \left(\sqrt{\frac{2}{\pi a^3}}\, e^{-r/a} \cos \omega t \right)^2 (4\pi r^2\, dr) = P(r)\, dr.$$

Averaging over the time (that is, replacing $\cos^2 \omega t$ by $\tfrac{1}{2}$) yields for the probability $\overline{P(r)}$

$$\overline{P(r)} = \frac{4r^2}{a^3}\, e^{-2r/a}.$$

Figure 40–9 shows a plot of this function. Note that the most probable location for the electron corresponds to the first Bohr radius. Thus in wave mechanics we do not say that the electron in the $n = 1$ state in hydrogen goes around the nucleus in a circular orbit of 0.529 A radius but only that the electron is more likely to be found at this distance from the nucleus than at any other distance, either larger or smaller.

40–5   The Uncertainty Principle

Only those quantities that can be measured have any real meaning in physics. If we could focus a "super" microscope on an electron in an atom and see it moving around in an orbit, we would declare that such orbits have meaning. However, we shall show that it is fundamentally impossible to make such an observation—even with the most ideal instruments that could conceivably be constructed. Therefore, we declare that such orbits have no physical meaning.

We observe the moon traveling around the earth by means of the sunlight that it reflects in our direction. Now light transfers linear momentum to an object from which it is reflected. In principle, this reflected light would disturb the course of the moon in its orbit, although a little thought shows that this disturbing effect is negligible in this case.

For electrons the situation is quite different. Here, too, we can hope to "see" the electron only if we reflect light, or another particle, from it. In this case the recoil that the electron experiences when the light (photon) bounces from it completely alters the electron's motion in a way that cannot be avoided or even corrected for.

It is not surprising that the probability curve of Fig. 40–9 is the most detailed information that we can hope to obtain, by measurement, about the distribution of negative charge in the hydrogen atom. If orbits such as those envisaged by Bohr existed, they would be broken up completely in our attempts to verify their existence. Under these circumstances, we prefer to say that it is the probability function, and not the orbits, that represents physical reality.

The fact that we can't describe the motions of electrons in a classical way finds expression in the *uncertainty principle,* enunciated by Werner Heisenberg in 1927. To formulate this principle, consider a beam of monoenergetic electrons of speed v_0 moving from left to right in Fig. 40–10. Let us set ourselves the task of measuring the position of a particular electron in the vertical (y) direction and also its velocity component v_y in this direction. If we succeed in carrying out these measurements with unlimited accuracy, we can then claim to have established the position and motion of the electron (or one component of it at least) with precision. However, we shall see that it is impossible to make these two measurements simultaneously with unlimited accuracy.

To measure y we block the beam with an absorbing screen A in which we put a slit of width Δy. If an electron gets through the slit, its vertical position must be known to this accuracy. By making the slit narrower, we can improve the accuracy of this vertical position measurement as much as we wish.

Since the electron is a wave, it will undergo diffraction at the slit, and a photographic plate placed at B in Fig. 40–10 will reveal a typical diffraction pattern. The

FIGURE 40-9   The probability function for the ground state of the hydrogen atom, as calculated from wave mechanics. The separation between the nucleus and the electron is r; a is the radius of the first Bohr orbit (0.529×10^{-10} meter), used here merely as a convenient unit of distance.

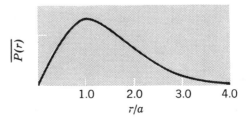

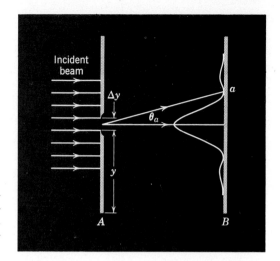

FIGURE 40-10 An incident beam of electrons is diffracted at the slit in screen A, forming a typical diffraction pattern on screen B. If the slit is made narrower, the pattern becomes wider.

existence of this diffraction pattern means that there is an uncertainty in the values of v_y possessed by the electrons emerging from the slit. Let v_{ya} be the value of v_y that corresponds to an electron landing at the first minimum on the screen, marked by point a and described by a characteristic angle θ_a. We take v_{ya} as a rough measure of the uncertainty Δv_y in v_y for electrons emerging from the slit.

The first minimum in the diffraction pattern is given by Eq. 38–2, or

$$\sin \theta_a = \frac{\lambda}{\Delta y}.$$

If we assume that θ_a is small enough, we can write this equation as

$$\theta_a \cong \frac{\lambda}{\Delta y}. \tag{40–11}$$

To reach point a, v_{ya} $(= \Delta v_y)$ must be such that

$$\theta_a \cong \frac{\Delta v_y}{v_0}. \tag{40–12}$$

Combining Eqs. 40–11 and 40–12 leads to

$$\frac{\Delta v_y}{v_0} = \frac{\lambda}{\Delta y},$$

which we rewrite as $\Delta v_y \, \Delta y \cong \lambda v_0.$ (40–13)

Now λ, the wavelength of the electron beam, is given by h/p or h/mv_0; substituting this into Eq. 40–13 yields

$$\Delta v_y \, \Delta y \cong \frac{hv_0}{mv_0}.$$

We rewrite this as $\Delta p_y \, \Delta y \cong h.$ (40–14)

In Eq. 40–14 Δp_y $(= m \, \Delta v_y)$ is the uncertainty in our knowledge of the vertical momentum of the electrons; Δy is the uncertainty in our knowledge of their vertical position. The equation tells us that, since the product of these uncertainties is a constant, we cannot measure p_y and y simultaneously with unlimited accuracy.

If we want to improve our measurement of y (that is, if we want to reduce Δy), we use a finer slit. However (see Eq. 40–11), this will produce a wider diffraction pattern. A wider pattern means that our knowledge of the vertical momentum component of the electron has deteriorated, or, in other words, Δp_y has increased; this is exactly what Eq. 40–14 predicts.

The limits on measurement imposed by Eq. 40–14 have nothing to do with the crudity of our measuring instruments. We are permitted to postulate the existence of the finest conceivable measuring equipment. Equation 40–14 represents a fundamental limitation, imposed by nature.

Equation 40–14 is a derivation, for a special case, of a general principle known as the *uncertainty principle*. As applied to position and momentum measurements, it asserts that

$$\Delta p_x \, \Delta x \gtrsim h$$

$$\Delta p_y \, \Delta y \gtrsim h$$

$$\Delta p_z \, \Delta z \gtrsim h. \tag{40–15}$$

Thus no component of the motion of an electron, free or bound, can be described with unlimited precision.

Planck's constant h probably appears nowhere that has more deep-seated significance than in Eq. 40–15. If this product had been zero instead of h, the classical ideas about particles and orbits would be correct; it would then be possible to measure both momentum and position with unlimited precision. The fact that h appears means that the classical ideas are wrong; the magnitude of h tells us under what circumstances these classical ideas must be replaced by quantum ideas. Gamow* has speculated, in an interesting and readable fantasy, what our world would be like if the constant h were much larger than it is, so that nonclassical ideas would be apparent to our sense perceptions.

Example 5. An electron has a speed of 300 meters/sec, accurate to 0.010%. With what fundamental accuracy can we locate the position of this electron?

The electron momentum is

$$p = mv = (9.1 \times 10^{-31} \text{ kg})(300 \text{ meters/sec}) = 2.7 \times 10^{-28} \text{ kg-m/sec}.$$

The uncertainty in momentum is given to be 0.010% of this, or

$$\Delta p = (0.00010)(2.7 \times 10^{-28} \text{ kg-m/sec}) = 2.7 \times 10^{-32} \text{ kg-m/sec}.$$

The minimum uncertainty in position, from Eq. 40–15, is

$$\Delta x = \frac{h}{\Delta p} = \frac{6.6 \times 10^{-34} \text{ joule-sec}}{2.7 \times 10^{-32} \text{ kg-m/sec}}$$

$$= 2.4 \text{ cm}.$$

If the electron momentum has really been determined by measurement to have the accuracy stated, there is no hope whatever that its position can be known to any better accuracy than that stated, namely about 1 in. The concept of the electron as a tiny dot is not very valid under these circumstances.

---

* *Mr. Tompkins in Wonderland,* Macmillan, New York, 1940.

Example 6. A bullet has a speed of 300 meters/sec, accurate to 0.010%. With what fundamental accuracy can we locate its position? Its mass is 50 gm (= 0.050 kg).

This example is the same as Example 5 in every respect save the mass of the particle involved. The momentum is

$$p = mv = (0.05 \text{ kg})(300 \text{ meters/sec}) = 15 \text{ kg-m/sec}$$

and

$$\Delta p = (0.00010)(15 \text{ kg-m/sec}) = 1.5 \times 10^{-3} \text{ kg-m/sec.}$$

Equation 40–15 yields

$$\Delta x = \frac{6.6 \times 10^{-34} \text{ joule-sec}}{1.5 \times 10^{-3} \text{ kg-m/sec}} = 4.4 \times 10^{-13} \text{ meter.}$$

This is so far beyond the possibility of measurement (a nucleus is only about 10^{-15} meter in diameter) that we can assert that for heavy objects like bullets the uncertainty principle sets no limits whatever on our measuring procedures. Once again the correspondence principle shows us how quantum physics reduces to classical physics under the appropriate circumstances. ▪

The uncertainty relation shows us why it is possible for both light and matter to have a dual, wave-particle, nature. It is because these two views, so obviously opposite to each other, can never be brought face to face in the same experimental situation. If we dream up an experiment that forces the electron to reveal its wave character strongly, its particle character will always be inherently fuzzy. If we change the experiment to bring out the particle character more strongly, the wave character necessarily becomes fuzzy. Matter and light are like coins that can be made to display either face at will but not both simultaneously. Niels Bohr first pointed out in his principle of complementarity how the ideas of wave and of particle complement rather than contradict each other.

QUESTIONS

1. How can the wavelength of an electron be given by $\lambda = h/p$ when the very presence of the momentum p in this formula implies that the electron is a particle?

2. How could Davisson and Germer be sure that the "54-volt" peak of Fig. 40–2 was a first-order diffraction peak, that is, that $m = 1$ in Eq. 40–2?

3. In a repetition of Thomson's experiment for measuring e/m for the electron (see Section 29–8), a beam of 10^4-eV electrons is collimated by passage through a slit of width 0.50 mm. Why is the beam-like character of the emerging electrons not destroyed by diffraction of the electron wave at this slit?

4. Why is the wave nature of matter not more apparent to our daily observations?

5. Apply the correspondence principle to the problem of a particle confined between rigid walls, showing that those features which seem "strange" (that is, the quantization of energy and the nonuniformity of the probability functions of Fig. 40–8) become undetectable experimentally at large quantum numbers.

6. In the $n = 1$ mode, for a particle confined between rigid walls, what is the probability that the particle will be found in a small volume element at the surface of either wall?

7. A standing wave can be viewed as the superposition of two traveling waves. Can you apply this to the problem of a particle confined between rigid walls, giving an interpretation in terms of the motion of the electron?

8. What is the physical significance of the wave function Ψ?

9. Why does the concept of Bohr orbits violate the uncertainty principle?

10. How can the predictions of wave mechanics be so exact if the only information we have about the positions of the electrons is statistical?

11. Make up some numerical examples to show the difficulty of getting the uncertainty principle to reveal itself during experiments with an object whose mass is about 1 gm.

12. Figure 40–8 shows that for $n = 3$ the probability function for a particle confined between rigid walls is zero at two points between the walls. How can the particle ever move across these positions? (Hint: Consider the implications of the uncertainty principle.)

13. The uncertainty principle can be stated in terms of angular quantities (compare Eq. 40–15) as

$$\Delta L \, \Delta \phi \gtrsim h$$

where ΔL is the uncertainty in the angular momentum and $\Delta \phi$ the uncertainty in the angular position. For electrons in atoms the angular momentum has definite quantized values, with no uncertainty whatever. What can we conclude about the uncertainty in the angular position and about the validity of the orbit concept?

PROBLEMS

1(1). A bullet of mass 40 gm travels at 1000 meters/sec. (*a*) What wavelength can we associate with it? (*b*) Why does the wave nature of the bullet not reveal itself through diffraction effects? *Answer:* (*a*) 1.7×10^{-35} meter.

2(1). If the de Broglie wavelength of a proton is 1.0×10^{-13} meters, (*a*) what is the speed of the proton, and (*b*) through what electric potential would the proton have to be accelerated to acquire this speed?

3(1). Sodium ions are accelerated through a potential difference of 300 volts. (*a*) What is the momentum acquired by the ions? (*b*) What is their de Broglie wavelength? *Answer:* (*a*) 1.9×10^{-21} kg meter/sec. (*b*) 3.5×10^{-13} meters.

4(1). What wavelength do we associate with a beam of neutrons whose energy is 0.025 eV?

5(1). An electron and a photon each have a wavelength of 2.0 A. What are their (*a*) momenta and (*b*) energies? *Answer:* (*a*) 3.3×10^{-24} kg-meter/sec for each. (*b*) 38 eV for the electron and 6200 eV for the photon.

6(1). The 50-GeV electron accelerator at Stanford (1-GeV $= 10^9$ eV) provides an electron beam of small wavelength, suitable for probing the fine details of nuclear structure by scattering experiments. What will this wavelength be and how does it compare to the size of an average nucleus? (Hint: At these energies it is necessary to use the extreme relativistic relationship between momentum and energy, namely $p = E/c$. This is the same relationship used for light (Section 35–5) and is justified whenever the kinetic energy of a particle is much greater than its rest energy $m_0 c^2$, as in this case.)

7(1). The highest achievable resolving power of a microscope is limited only by the wavelength used; that is, the smallest detail that can be separated is about equal to the wavelength. Suppose one wishes to "see" inside an atom. Assuming the atom to have a diameter of 1.0 A, this means that we wish to resolve detail of separation about 0.10 A. (*a*) If an electron microscope is used, what minimum energy of electrons is needed? (*b*) If a light microscope is used, what minimum energy of photons is needed? (*c*) Which microscope seems more practical for this purpose? *Answer:* (*a*) 1.5×10^4 eV. (*b*) 12×10^4 eV.

8(1). What accelerating voltage would be required for electrons in an electron microscope to obtain the same ultimate resolving power as that which could be obtained from a gamma-ray microscope using 0.20 MeV-gamma rays?

9(1). In a repetition of Thomson's experiment for measuring e/m for the electron, a beam of 10^4

eV electrons is collimated by passage through a slit of width 0.50 mm. Why is the beamlike character of the emergent electrons not destroyed by diffraction of the electron wave at this slit?
Answer: The deBroglie wavelength (0.12 A) is much smaller than the slit width.

10(1). In the experiment of Davisson and Germer (*a*) at what angles would the second- and third-order diffracted beams corresponding to the strong maximum of Fig. 40–2 occur, provided they are present, and (*b*) at what angle would the first-order diffracted beam occur if the accelerating potential were changed from 54 to 60 volts? The experimenter is free to rotate the crystal.

11(1). A neutron crystal spectrometer utilizes crystal planes of spacing d (0.7323 A) in a beryllium crystal. What must be the Bragg angle θ so that only neutrons of energy K (4.0 eV) are reflected. Consider only first-order reflections.
Answer: 5.6°.

12(1). Make a plot of de Broglie wavelength against kinetic energy for (*a*) electrons and (*b*) protons. Restrict the range of energy values to those in which classical mechanics applies reasonably well. A convenient criterion is that the maximum kinetic energy on each plot be only about 5% of the rest energy m_0c^2 for the particular particle.

13(1). What is the wavelength of a hydrogen atom moving with a velocity corresponding to the mean kinetic energy for thermal equilibrium at 20° C?
Answer: 1.5 A.

14(2). According to the correspondence principle, as $n \rightarrow \infty$ we expect classical results in the Bohr atom. Hence, the de Broglie wavelength associated with the electron (a quantum result) should get smaller compared to the radius of the orbit as n increases. Indeed, we expect that $\lambda/r \rightarrow 0$ as $n \rightarrow \infty$. Is this the case?

15(3). (*a*) What is the separation in energy between the lowest two energy levels for a container 20 cm on a side containing argon atoms? (*b*) How does this compare with the thermal energy of the argon atoms at 300° K? (*c*) At what temperature does the thermal energy equal the spacing between these two energy levels?
Answer: (*a*) 3.9×10^{-22} eV. (*b*) The thermal energy is about 10^{20} times as great. (*c*). 3.0×10^{-18} °K.

16(4). If an electron moves from a state represented by $n = 3$ in Fig. 40–8 to one represented by $n = 2$ in that figure, emitting electromagnetic radiation in the process, what are (*a*) the energy of the emitted single photon and (*b*) the corresponding wavelength?

17(4). A particle is confined between rigid walls separated by a distance l. What is the probability that it will be found within a distance $l/3$ of one wall (*a*) for $n = 1$, (*b*) for $n = 2$, (*c*) for $n = 3$, and (*d*) under the assumptions of classical physics?
Answer: (*a*) 0.20. (*b*) 0.40. (*c*) 0.33. (*d*) 0.33.

18(4). In the ground state of the hydrogen atom show that the probability P_r that the electron lies within a sphere of radius r is given by

$$P_r = 1 - e^{-2r/a}\left(\frac{2r^2}{a^2} + \frac{2r}{a} + 1\right).$$

Does this yield expected values for (*a*) $r = 0$ and (*b*) $r = \infty$? (*c*) State clearly the difference in meaning between this expression and that given in Section 40–4.

19(4). In the ground state of the hydrogen atom, what is the probability that the electron will lie within a sphere whose radius is that of the first Bohr orbit?
Answer: 0.32

20(5). A microscope using photons is employed to locate an electron in an atom to within a distance of 0.10 A. What is the uncertainty in the momentum of the electron located in this way?

21(5). The uncertainty in the position of an electron is given as about 0.50 A, which is the radius of the first Bohr orbit in hydrogen. What is the uncertainty in the linear momentum of the electron?
Answer: 1.3×10^{-23} kg-meter/sec.

22(5). Show that if the uncertainty in the location of a particle is equal to its de Broglie wavelength the uncertainty in its velocity is equal to its velocity.

23(5). In Example 2, the electron's energy was determined exactly by the size of the box. How do you reconcile this with the fact that the uncertainty in the location of the electron cannot exceed 1.0×10^{-9} meter if the uncertainty principle is to be obeyed?
Answer: The kinetic energy and the momentum are not directly connected by the uncertainty principle.

APPENDIX A

*Physical Standards and Constants**

The definition of primary standards is by agreement of the International Conference of Weights and Measures in October, 1964 in Paris. Measured and derived values of the fundamental physical constants summarize hundreds of physical measurements made over the years by scientists in all parts of the world. They have been subjected to exhaustive statistical analysis and, with their accompanying error limits (which are given as standard deviations), represent the best values as of 1969. For most problems in this book three significant figures will do, and the "computational" (rounded) values may be used. The data are those given by Taylor, Parker and Langenberg in their recent adjustment of the fundamental constants as reported in the *Reviews of Modern Physics,* July, 1969.

* See "A Pilgrim's Progress in Search of the Fundamental Constants," by J. W. M. Du Mond, *Physics Today*, October 1965.

DEFINITIONS OF STANDARDS

Standard	Abbreviation	Definition
Meter	m	1,650,763.73 wavelengths in vacuo of the unperturbed transition $2p_{10}$—$5d_5$ in Kr^{86}
Kilogram	kg	mass of the international kilogram at Sèvres, France
Second	sec	9,192,631,770 vibrations of the unperturbed hyperfine transition 4,0—3,0 of the fundamental state $^2S_{1/2}$ in Cs^{133}
Kelvin	K	defined in the thermodynamic scale by assigning 273.16 K to the triple point of water and 0 K to the absolute zero of temperature
Atomic mass unit	amu	$\frac{1}{12}$ the mass of an atom of C^{12}
Mole	mole	amount of substance containing the same number of atoms as 12 gm (exactly) of pure C^{12}
Standard acceleration of free fall	g	9.80665 m/sec^2
Normal atmospheric pressure	atm	101,325 nt/m^2
Thermochemical calorie	cal	4.1840 joules
Liter	li	0.001 m^3 (exactly)
Inch	in.	0.0254 m (exactly)
Pound (mass)	lb	0.453,592,37 kg

FUNDAMENTAL AND DERIVED CONSTANTS

Name	Symbol	Computational value	Best (1969) experimental value*
Speed of light	c	3.00×10^8 m/sec	2.99792458(4)
Permeability constant	μ_0	1.26×10^{-6} henry/m	$4\pi \times 10^{-7}$ exactly
Permitivity constant	ϵ_0	8.85×10^{-12} farad/m	8.8541853(59)
Elementary charge	e	1.60×10^{-19} coul	1.6021917(70)
Avogadro's number	N_0	6.02×10^{23}/mole	6.022169(40)
Electron rest mass	m_e	9.11×10^{-31} kg	9.109558(54)
Proton rest mass	m_p	1.67×10^{-27} kg	1.672614(11)
Neutron rest mass	m_n	1.67×10^{-27} kg	1.674920(11)
Planck's constant	h	6.63×10^{-34} joule sec	6.626196(50)
Electron charge/mass ratio	e/m_e	1.76×10^{11} coul/kg	1.7588028(54)
Quantum/charge ratio	h/e	4.14×10^{-15} joule sec/coul	4.135708(14)
Electron Compton wavelength	λ_e	2.43×10^{-12} m	2.4263096(74)
Rydberg constant	R_∞	1.10×10^7/m	1.09737312(11)
Bohr radius	a_0	5.29×10^{-11} m	5.2917715(81)
Bohr magneton	μ_B	9.27×10^{-24} joule/tesla	9.274096(65)
Nuclear magneton	μ_N	5.05×10^{-27} joule/tesla	5.050951(50)
Proton magnetic moment	μ_p	1.41×10^{-26} joule/tesla	1.4106203(90)
Universal gas constant	R	8.31 joule/°K mole	8.31434(35)
Standard volume of ideal gas	—	2.24×10^{-2} m³/mole	2.24136(30)
Boltzmann's constant	k	1.38×10^{-23} joule/°K	1.380622(59)
Stefan-Boltzmann constant	σ	5.67×10^{-8} watt/m² °K⁴	5.66961(96)
Gravitational constant	G	6.67×10^{-11} nt m²/kg²	6.6732(31)

* Same units and power of ten as the computational value. The numbers in parentheses are the standard deviation uncertainties in the last digits of the quoted value.

Some Terrestrial Data

Standard atmosphere	1.013×10^5 nt/meter^2
	14.70 lb/in^2
	760.0 mm-Hg
Density of dry air at 0° C and 760 mm-Hg	1.293 kg/meter^3
Speed of sound in dry air at 0° C and 760 mm-Hg	331.45 meters/sec
	1087 ft/sec
	741.5 miles/hr
Acceleration of gravity (standard value)*	9.80665 meters/sec^2
	32.1740 ft/sec^2
Solar constant**	1340 watts/m^2
	1.92 cal/cm^2-min
Mean total solar radiation	3.92×10^{26} watts
Equatorial radius of earth	6.378×10^6 meters
	3963 miles
Polar radius of earth	6.357×10^6 meters
	3950 miles
Volume of earth	1.087×10^{21} meter^3
	3.838×10^{22} ft^3
Radius of sphere having same volume	6.371×10^6 meters
	3959 miles
Mean density of earth	5522 kg/meter^3
Mass of earth	5.983×10^{24} kg
Mean orbital speed of earth	29,770 meters/sec
	18.50 miles/sec
Earth's magnetic field at Washington, D. C.	5.7×10^{-5} weber/m^2
	0.57 gauss
Earth's magnetic dipole moment	6.4×10^{21} amp-m^2

* This value, used for barometer corrections, legal weights, etc., was adopted by the International Committee on Weights and Measures in 1901. It approximates 45° latitude at sea level; see Appendix A.

** The solar constant is the solar energy falling per unit time at normal incidence on unit area of the earth's surface.

APPENDIX C

*The Solar System**

Planet →	Mercury	Venus	Earth	Mars	Jupiter	Saturn	Uranus	Neptune	Pluto
Mean diameter km	5,140	12,620	12,756	6,860	143,600	120,600	53,400	49,700	12,700?
Volume (earth volumes)	0.0655	0.967	1.00	0.156	1,428	843	73.6	59.3	0.10
Mass (earth masses)	0.0549	0.807	1.00	0.106	314.5	94.1	14.4	16.7	1.0?
Mean density gm/cm^3	5.61	5.16	5.52	3.95	1.34	0.69	1.36	1.30	?
Surface gravity (earth ratio)	0.40	0.90	1.00	0.40	2.70	1.20	1.00	1.00	?
Velocity of escape, km/sec	3.5	10.4	11.19	5.03	59.7	35.4	21.6	22.8	11?
Length of day (earth days)	58.6^d	30^d?	1^d	1^{d}37^{m}23^s	9^{h}55^m	10^{h}38^m	10.7^h	15.8^h	?
Period of revolution, days	87.97	224.70	365.26	686.98	4,332.59	10,759.20	30,685.93	60,187.64	90,885
Maximum surface temperature, °K	700	700	350	320	153	138	110?	90?	80?
Distance from Sun, 10^6 km	58	108	149	228	778	1426	2869	4495	5900

*The data was taken from *Handbook of Chemistry and Physics*, Chemical Rubber Publishing Co., 50th Edition, 1969.
The Sun 329,390 earth masses, mean density 1.42, mean diameter 1,391,000 km, surface gravity 28 (earth ratio).
The Moon 0.01228 earth masses, mean density 3.36, mean diameter 3,476 km, surface gravity 0.17 (earth ratio), distance from earth 38 × 10^4 km.

Periodic Table of the Elements

Atomic weights are expressed in *atomic mass units* (amu), one atom of C^{12} being defined to have a mass of (exactly) 12 amu. For most unstable elements the mass number of the most stable or best known isotope is given in brackets. Values are as of 1966 with a few adjustments in the Actinide series elements.

Period	Series	I	II	III	IV	V	VI	VII	VIII			0
1	1	1 H 1.00797										2 He 4.0026
2	2	3 Li 6.939	4 Be 9.0122	5 B 10.811	6 C 12.01115	7 N 14.0067	8 O 15.9994	9 F 18.9984				10 Ne 20.183
3	3	11 Na 22.9898	12 Mg 24.312	13 Al 26.9815	14 Si 28.086	15 P 30.9738	16 S 32.064	17 Cl 35.453				18 A 39.948
4	4	19 K 39.102	20 Ca 40.08	21 Sc 44.956	22 Ti 47.90	23 V 50.942	24 Cr 51.996	25 Mn 54.9380	26 Fe 55.847	27 Co 58.9332	28 Ni 58.71	
4	5	29 Cu 63.54	30 Zn 65.37	31 Ga 69.72	32 Ge 72.59	33 As 74.9216	34 Se 78.96	35 Br 79.909				36 Kr 83.80
5	6	37 Rb 85.47	38 Sr 87.62	39 Y 88.905	40 Zr 91.22	41 Nb 92.906	42 Mo 95.94	43 Tc [99]	44 Ru 101.07	45 Rh 102.905	46 Pd 106.4	
5	7	47 Ag 107.870	48 Cd 112.40	49 In 114.82	50 Sn 118.69	51 Sb 121.75	52 Te 127.60	53 I 126.9044				54 Xe 131.30
6	8	55 Cs 132.905	56 Ba 137.34	57–71 Lanthanide series*	72 Hf 178.49	73 Ta 180.948	74 W 183.85	75 Re 186.2	76 Os 190.2	77 Ir 192.2	78 Pt 195.09	
6	9	79 Au 196.967	80 Hg 200.59	81 Tl 204.37	82 Pb 207.19	83 Bi 208.980	84 Po [210]	85 At [210]				86 Rn [222]
7	10	87 Fr [223]	88 Ra [226]	89 Actinide series**	104							

*Lanthanide series:

57 La 138.91	58 Ce 140.12	59 Pr 140.907	60 Nd 144.24	61 Pm [145]	62 Sm 150.35	63 Eu 151.96	64 Gd 157.25	65 Tb 158.924	66 Dy 162.50	67 Ho 164.930	68 Er 167.26	69 Tm 168.934	70 Yb 173.04	71 Lu 174.97

**Actinide series:

89 Ac [227]	90 Th 232.038	91 Pa [231]	92 U 238.04	93 Np [237]	94 Pu [244]	95 Am [243]	96 Cm [247]	97 Bk [247]	98 Cf [251]	99 Es [254]	100 Fm [257]	101 Md [256]	102 [254]	103 Lw [257]

Conversion Factors

Conversion factors may be read off directly from the tables. For example, 1 degree = 2.778×10^{-3} revolutions, so $16.7° = 16.7 \times 2.778 \times 10^{-3}$ rev. The mks quantities are capitalized. The prefix "ab" refers to electromagnetic units (emu); "stat" refers to electrostatic units (esu). Adapted in part from G. Shortley and D. Williams, *Elements of Physics*, Prentice-Hall, Englewood Cliffs, New Jersey, 1965.

PLANE ANGLE

	°	′	″	RADIAN	rev
1 degree =	1	60	3600	1.745×10^{-2}	2.778×10^{-3}
1 minute =	1.667×10^{-2}	1	60	2.909×10^{-4}	4.630×10^{-5}
1 second =	2.778×10^{-4}	1.667×10^{-2}	1	4.848×10^{-6}	7.716×10^{-7}
1 RADIAN =	57.30	3438	2.063×10^{5}	1	0.1592
1 revolution =	360	2.16×10^{4}	1.296×10^{6}	6.283	1

SOLID ANGLE

1 sphere = 4π steradians = 12.57 steradians

LENGTH

	cm	METER	km	in.	ft	mile
1 centimeter =	1	10^{-2}	10^{-5}	0.3937	3.281×10^{-2}	6.214×10^{-6}
1 METER =	100	1	10^{-3}	39.37	3.281	6.214×10^{-4}
1 kilometer =	10^{5}	1000	1	3.937×10^{4}	3281	0.6214
1 inch =	2.540	2.540×10^{-2}	2.540×10^{-5}	1	8.333×10^{-2}	1.578×10^{-5}
1 foot =	30.48	0.3048	3.048×10^{-4}	12	1	1.894×10^{-4}
1 mile =	1.609×10^{5}	1609	1.609	6.336×10^{4}	5280	1

1 angstrom = 10^{-10} meter
1 x unit = 10^{-13} meter
1 nautical mile = 1852 meters
    = 1.151 miles = 6076 ft

1 light-year = 9.4600×10^{12} km
1 parsec = 3.084×10^{13} km
1 fathom = 6 ft

1 yard = 3 ft
1 rod = 16.5 ft
1 mil = 10^{-3} in.

AREA

	METER^2	cm^2	ft^2	in.^{2}	circ mil
1 SQUARE METER =	1	10^4	10.76	1550	1.974 × 10^9
1 square centimeter =	10^{-4}	1	1.076 × 10^{-3}	0.1550	1.974 × 10^5
1 square foot =	9.290 × 10^{-2}	929.0	1	144	1.833 × 10^8
1 square inch =	6.452 × 10^{-4}	6.452	6.944 × 10^{-3}	1	1.273 × 10^6
1 circular mil =	5.067 × 10^{-10}	5.067 × 10^{-6}	5.454 × 10^{-9}	7.854 × 10^{-7}	1

1 square mile = 2.788 × 10^8 ft^2 = 640 acres   1 acre = 43,600 ft^2
1 barn = 10^{-28} meter^2

VOLUME

	METER^3	cm^3	li	ft^3	in.^{3}
1 CUBIC METER =	1	10^6	1000	35.31	6.102 × 10^4
1 cubic centimeter =	10^{-6}	1	1.000 × 10^{-3}	3.531 × 10^{-5}	6.102 × 10^{-2}
1 liter =	1.000 × 10^{-3}	1000	1	3.531 × 10^{-2}	61.02
1 cubic foot =	2.832 × 10^{-2}	2.832 × 10^4	28.32	1	1728
1 cubic inch =	1.639 × 10^{-5}	16.39	1.639 × 10^{-2}	5.787 × 10^{-4}	1

1 U. S. fluid gallon = 4 U. S. fluid quarts = 8 U. S. pints = 128 U. S. fluid ounces = 231 in.^{3}
1 British imperial gallon = 277.4 in.^{3}
1 liter = the volume of 1 kg of water at its maximum density = 1000.028 cm^3

MASS

Quantities in the shaded areas are not mass units but are often used as such. When we write, for example, 1 kg "=" 2.205 lb this means that a kilogram is a *mass* that *weighs* 2.205 pounds under standard condition of gravity (g = 9.80665 meters/sec^2).

	gm	KG	slug	amu	oz	lb	ton
1 gram =	1	0.001	6.852 × 10^{-5}	6.024 × 10^{23}	3.527 × 10^{-2}	2.205 × 10^{-3}	1.102 × 10^{-6}
1 KILOGRAM =	1000	1	6.852 × 10^{-2}	6.024 × 10^{26}	35.27	2.205	1.102 × 10^{-3}
1 slug =	1.459 × 10^4	14.59	1	8.789 × 10^{27}	514.8	32.17	1.609 × 10^{-2}
1 amu =	1.660 × 10^{-24}	1.660 × 10^{-27}	1.137 × 10^{-28}	1	5.855 × 10^{-26}	3.660 × 10^{-27}	1.829 × 10^{-30}
1 ounce =	28.35	2.835 × 10^{-2}	1.943 × 10^{-3}	1.708 × 10^{25}	1	6.250 × 10^{-2}	3.125 × 10^{-5}
1 pound =	453.6	0.4536	3.108 × 10^{-2}	2.732 × 10^{26}	16	1	0.0005
1 ton =	9.072 × 10^5	907.2	62.16	5.465 × 10^{29}	3.2 × 10^4	2000	1

DENSITY

Quantities in the shaded areas are weight densities and, as such, are dimensionally different from mass densities. See note for mass table.

	slug/ft^3	KG/METER^3	gm/cm^3	lb/ft^3	lb/in.^{3}
1 slug per ft^3 =	1	515.4	0.5154	32.17	1.862×10^{-2}
1 KILOGRAM per METER^3 =	1.940×10^{-3}	1	0.001	6.243×10^{-2}	3.613×10^{-5}
1 gram per cm^3 =	1.940	1000	1	62.43	3.613×10^{-2}
1 pound per ft^3 =	3.108×10^{-2}	16.02	1.602×10^{-2}	1	5.787×10^{-4}
1 pound per in.^{3} =	53.71	2.768×10^4	27.68	1728	1

TIME

	yr	day	hr	min	SEC
1 year =	1	365.2	8.766×10^3	5.259×10^5	3.156×10^7
1 day =	2.738×10^{-3}	1	24	1440	8.640×10^4
1 hour =	1.141×10^{-4}	4.167×10^{-2}	1	60	3600
1 minute =	1.901×10^{-6}	6.944×10^{-4}	1.667×10^{-2}	1	60
1 SECOND =	3.169×10^{-8}	1.157×10^{-5}	2.778×10^{-4}	1.667×10^{-2}	1

SPEED

	ft/sec	km/hr	METER/SEC	mile/hr	cm/sec	knot
1 foot per second =	1	1.097	0.3048	0.6818	30.48	0.5925
1 kilometer per hour =	0.9113	1	0.2778	0.6214	27.78	0.5400
1 METER per SECOND =	3.281	3.6	1	2.237	100	1.944
1 mile per hour =	1.467	1.609	0.4470	1	44.70	0.8689
1 centimeter per second =	3.281×10^{-2}	3.6×10^{-2}	0.01	2.237×10^{-2}	1	1.944×10^{-2}
1 knot =	1.688	1.852	0.5144	1.151	51.44	1

1 knot = 1 nautical mile/hr   1 mile/min = 88.00 ft/sec = 60.00 miles/hr

FORCE

Quantities in the shaded areas are not force units but are often used as such, especially in chemistry. For instance, if we write 1 gram-force "=" 980.7 dynes, we mean that a gram-*mass* experiences a *force* of 980.7 dynes under standard conditions of gravity ($g = 9.80665$ meters/sec^2).

	.dyne	NT	lb	pdl	gf	kgf
1 dyne =	1	10^{-5}	2.248×10^{-6}	7.233×10^{-5}	1.020×10^{-3}	1.020×10^{-6}
1 NEWTON =	10^5	1	0.2248	7.233	102.0	0.1020
1 pound =	4.448×10^5	4.448	1	32.17	453.6	0.4536
1 poundal =	1.383×10^4	0.1383	3.108×10^{-2}	1	14.10	1.410×10^{-2}
1 gram-force =	980.7	9.807×10^{-3}	2.205×10^{-3}	7.093×10^{-2}	1	0.001
1 kilogram-force =	9.807×10^5	9.807	2.205	70.93	1000	1

$$1 \text{ kgf} = 9.807 \text{ nt} \qquad 1 \text{ lb} = 32.17 \text{ pdl}$$

PRESSURE

	atm	dyne/cm^2	inch of water	cm-Hg	NT/ METER^2	lb/in.^{2}	lb/ft^2
1 atmosphere =	1	1.013×10^6	406.8	76 *760 Torr*	1.013×10^5	14.70	2116
1 dyne per cm^2 =	9.869×10^{-7}	1	4.015×10^{-4}	7.501×10^{-5}	0.1	1.450×10^{-5}	2.089×10^{-3}
1 inch of water* at 4° C =	2.458×10^{-3}	2491	1	0.1868	249.1	3.613×10^{-2}	5.202
1 centimeter of mercury* at 0° C =	1.316×10^{-2}	1.333×10^4	5.353	1	1333	0.1934	27.85
1 NEWTON per METER^2 =	9.869×10^{-6}	10	4.015×10^{-3}	7.501×10^{-4}	1	1.450×10^{-4}	2.089×10^{-2}
1 pound per in.^{2} =	6.805×10^{-2}	6.895×10^4	27.68	5.171	6.895×10^3	1	144
1 pound per ft^2 =	4.725×10^{-4}	478.8	0.1922	3.591×10^{-2}	47.88	6.944×10^{-3}	1

* Where the acceleration of gravity has the standard value 9.80665 meters/sec^2.

$$1 \text{ bar} = 10^6 \text{ dyne/cm}^2 \qquad 1 \text{ millibar} = 10^3 \text{ dyne/cm}^2$$

1 mm Hg = 1 TORR

ENERGY, WORK, HEAT

Quantities in the shaded areas are not properly energy units but are included for convenience. They arise from the relativistic mass-energy equivalence formula $E = mc^2$ and represent the energy released if a kilogram or atomic mass unit (amu) is completely converted to energy.

	Btu	erg	ft-lb	hp-hr	JOULE	cal	kw-hr	eV	MeV	kg	amu
1 British thermal unit =	1	1.055×10^{10}	777.9	3.929×10^{-4}	1055	252.0	2.930×10^{-4}	6.585×10^{21}	6.585×10^{15}	1.174×10^{-14}	7.074×10^{12}
1 erg =	9.481×10^{-11}	1	7.376×10^{-8}	3.725×10^{-14}	10^{-7}	2.389×10^{-8}	2.778×10^{-14}	6.242×10^{11}	6.242×10^{5}	1.113×10^{-24}	670.5
1 foot-pound =	1.285×10^{-3}	1.356×10^{7}	1	5.051×10^{-7}	1.356	0.3239	3.766×10^{-7}	8.464×10^{18}	8.464×10^{12}	1.509×10^{-17}	9.092×10^{9}
1 horsepower-hour =	2545	2.685×10^{13}	1.980×10^{6}	1	2.685×10^{6}	6.414×10^{5}	0.7457	1.676×10^{25}	1.676×10^{19}	2.988×10^{-11}	1.800×10^{16}
1 JOULE =	9.481×10^{-4}	10^{7}	0.7376	3.725×10^{-7}	1	0.2389	2.778×10^{-7}	6.242×10^{18}	6.242×10^{12}	1.113×10^{-17}	6.705×10^{9}
1 calorie =	3.968×10^{-3}	4.186×10^{7}	3.087	1.559×10^{-6}	4.186	1	1.163×10^{-6}	2.613×10^{19}	2.613×10^{13}	4.659×10^{-17}	2.807×10^{10}
1 kilowatt-hour =	3413	3.6×10^{13}	2.655×10^{6}	1.341	3.6×10^{6}	8.601×10^{5}	1	2.247×10^{25}	2.247×10^{19}	4.007×10^{-11}	2.414×10^{16}
1 electron volt =	1.519×10^{-22}	1.602×10^{-12}	1.182×10^{-19}	5.967×10^{-26}	1.602×10^{-19}	3.827×10^{-20}	4.450×10^{-26}	1	10^{-6}	1.783×10^{-36}	1.074×10^{-9}
1 million electron volts =	1.519×10^{-16}	1.602×10^{-6}	1.182×10^{-13}	5.967×10^{-20}	1.602×10^{-13}	3.827×10^{-14}	4.450×10^{-20}	10^{6}	1	1.783×10^{-30}	1.074×10^{-3}
1 kilogram =	8.521×10^{13}	8.987×10^{23}	6.629×10^{16}	3.348×10^{10}	8.987×10^{16}	2.147×10^{16}	2.497×10^{10}	5.610×10^{35}	5.610×10^{29}	1	6.025×10^{26}
1 atomic mass unit =	1.415×10^{-13}	1.492×10^{-3}	1.100×10^{-10}	5.558×10^{-17}	1.492×10^{-10}	3.564×10^{-11}	4.145×10^{-17}	9.31×10^{8}	931.0	1.660×10^{-27}	1

POWER

	Btu/hr	ft-lb/sec	hp	cal/sec	kw	WATT
1 British thermal unit per hour =	1	0.2161	3.929×10^{-4}	7.000×10^{-2}	2.930×10^{-4}	0.2930
1 foot-pound per second =	4.628	1	1.818×10^{-3}	0.3239	1.356×10^{-3}	1.356
1 horsepower =	2545	550	1	178.2	0.7457	745.7
1 calorie per second =	14.29	3.087	5.613×10^{-3}	1	4.186×10^{-3}	4.186
1 kilowatt =	3413	737.6	1.341	238.9	1	1000
1 WATT =	3.413	0.7376	1.341×10^{-3}	0.2389	0.001	1

CHARGE

	abcoul	amp-hr	COUL	statcoul
1 abcoulomb =	1	2.778×10^{-3}	10	2.998×10^{10}
1 ampere-hour =	360	1	3600	1.079×10^{13}
1 COULOMB =	0.1	2.778×10^{-4}	1	2.998×10^{9}
1 statcoulomb =	3.336×10^{-11}	9.266×10^{-14}	3.336×10^{-10}	1

1 electronic charge = 1.602×10^{-19} coulomb

CURRENT

	abamp	AMP	statamp
1 abampere =	1	10	2.998×10^{10}
1 AMPERE =	0.1	1	2.998×10^{9}
1 statampere =	3.336×10^{-11}	3.336×10^{-10}	1

POTENTIAL, ELECTROMOTIVE FORCE

	abvolt	VOLT	statvolt
1 abvolt =	1	10^{-8}	3.336×10^{-11}
1 VOLT =	10^{8}	1	3.336×10^{-3}
1 statvolt =	2.998×10^{10}	299.8	1

RESISTANCE

	abohm	OHM	statohm
1 abohm =	1	10^{-9}	1.113×10^{-21}
1 OHM =	10^{9}	1	1.113×10^{-12}
1 statohm =	8.987×10^{20}	8.987×10^{11}	1

CAPACITANCE

	abf	FARAD	μf	statf
1 abfarad =	1	10^{9}	10^{15}	8.987×10^{20}
1 FARAD =	10^{-9}	1	10^{6}	8.987×10^{11}
1 microfarad =	10^{-15}	10^{-6}	1	8.987×10^{5}
1 statfarad =	1.113×10^{-21}	1.113×10^{-12}	1.113×10^{-6}	1

INDUCTANCE

	abhenry	HENRY	μh	mh	stathenry
1 abhenry =	1	10^{-9}	0.001	10^{-6}	1.113×10^{-21}
1 HENRY =	10^9	1	10^6	1000	1.113×10^{-12}
1 microhenry =	1000	10^{-6}	1	0.001	1.113×10^{-18}
1 millihenry =	10^6	0.001	1000	1	1.113×10^{-15}
1 stathenry =	8.987×10^{20}	8.987×10^{11}	8.987×10^{17}	8.987×10^{14}	1

MAGNETIC FLUX

	maxwell	WEBER
1 maxwell =	1	10^{-8}
1 WEBER =	10^8	1

MAGNETIC FIELD (**B**)

	gauss	TESLA	milligauss
1 gauss =	1	10^{-4}	1000
1 TESLA =	10^4	1	10^7
1 milligauss =	0.001	10^{-7}	1

1 tesla = 1 weber/meter^2

Mathematical Symbols and the Greek Alphabet

Mathematical signs and symbols

$=$ equals
$\cong$ equals approximately
$\neq$ is not equal to
$\equiv$ is identical to; is defined as
$>$ is greater than ($\gg$ is much greater than)
$<$ is less than ($\ll$ is much less than)
$\geq$ is greater than or equal to
$\leq$ is less than or equal to
$\pm$ plus or minus (for example, $\sqrt{4} = \pm 2$)
$\propto$ is proportional to (for example, Hooke's law: $F \propto x$, or $F = -kx$)
Σ the sum of
$\bar{x}$ the average value of x

The Greek alphabet

Alpha	A	α	Nu	N	ν
Beta	B	β	Xi	Ξ	ξ
Gamma	Γ	γ	Omicron	O	o
Delta	Δ	δ	Pi	Π	π
Epsilon	E	ϵ	Rho	P	ρ
Zeta	Z	ζ	Sigma	Σ	σ
Eta	H	η	Tau	T	τ
Theta	Θ	θ	Upsilon	Υ	υ
Iota	I	ι	Phi	Φ	ϕ, φ
Kappa	K	κ	Chi	X	χ
Lambda	Λ	λ	Psi	Ψ	ψ
Mu	M	μ	Omega	Ω	ω

APPENDIX G

Mathematical Formulas

Quadratic formula

If $ax^2 + bx + c = 0$, then $x = \dfrac{-b \pm \sqrt{b^2 - 4ac}}{2a}$.

Trigonometric functions of angle θ

$$\sin \theta = \frac{y}{r} \qquad \cos \theta = \frac{x}{r}$$

$$\tan \theta = \frac{y}{x} \qquad \cot \theta = \frac{x}{y}$$

$$\sec \theta = \frac{r}{x} \qquad \csc \theta = \frac{r}{y}$$

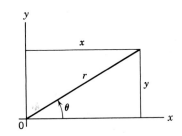

FIGURE APP. G

Pythagorean theorem

$$x^2 + y^2 = r^2$$

Trigonometric identities

$$\sin^2 \theta + \cos^2 \theta = 1 \qquad \sec^2 \theta - \tan^2 \theta = 1 \qquad \csc^2 \theta - \cot^2 \theta = 1$$

$$\sin (\alpha \pm \beta) = \sin \alpha \cos \beta \pm \cos \alpha \sin \beta$$

$$\cos (\alpha \pm \beta) = \cos \alpha \cos \beta \mp \sin \alpha \sin \beta$$

$$\tan (\alpha \pm \beta) = \frac{\tan \alpha \pm \tan \beta}{1 \mp \tan \alpha \tan \beta}$$

$$\sin 2\theta = 2 \sin \theta \cos \theta$$

$$\cos 2\theta = \cos^2 \theta - \sin^2 \theta = 2 \cos^2 \theta - 1 = 1 - 2 \sin^2 \theta$$

$$\sin \theta = \frac{e^{i\theta} - e^{-i\theta}}{2i} \qquad \cos \theta = \frac{e^{i\theta} + e^{-i\theta}}{2}$$

$$e^{\pm i\theta} = \cos \theta \pm i \sin \theta$$

Taylor's series

$$f(x_0 + x) = f(x_0) + \frac{\partial f}{\partial x}\bigg|_{x=x_0} x + \frac{\partial^2 f}{\partial x^2}\bigg|_{x=x_0} \frac{x^2}{2!} + \frac{\partial^3 f}{\partial x^3}\bigg|_{x=x_0} \frac{x^3}{3!} + \cdots$$

Binomial expansion

$$(x + y)^n = x^n + \frac{n}{1!} x^{n-1}y + \frac{n(n-1)}{2!} x^{n-2}y^2 + \cdots \qquad (x^2 > y^2)$$

Derivatives and indefinite integrals

The letters u and v stand for any functions of x, and a and m are constants. To each of the integrals should be added an arbitrary constant of integration. *A Short Table of Integrals* by Pierce and Foster (Ginn and Co.) gives a more extensive tabulation.

1. $\dfrac{dx}{dx} = 1$

2. $\dfrac{d}{dx}(au) = a\dfrac{du}{dx}$

3. $\dfrac{d}{dx}(u + v) = \dfrac{du}{dx} + \dfrac{dv}{dx}$

4. $\dfrac{d}{dx}x^m = mx^{m-1}$

5. $\dfrac{d}{dx}\ln x = \dfrac{1}{x}$

6. $\dfrac{d}{dx}(uv) = u\dfrac{dv}{dx} + v\dfrac{du}{dx}$

7. $\dfrac{d}{dx}e^x = e^x$

8. $\dfrac{d}{dx}\arctan x = \dfrac{1}{1 + x^2}$

9. $\dfrac{d}{dx}\arcsin x = \dfrac{1}{\sqrt{1 - x^2}}$

10. $\dfrac{d}{dx}\text{arcsec } x = \dfrac{1}{x\sqrt{x^2 - 1}}$

11. $\dfrac{d}{dx}\cos x = -\sin x$

12. $\dfrac{d}{dx}\sin x = \cos x$

13. $\dfrac{d}{dx}\tan x = \sec^2 x$

14. $\dfrac{d}{dx}\cot x = -\csc^2 x$

15. $\dfrac{d}{dx}\sec x = \tan x \sec x$

16. $\dfrac{d}{dx}\csc x = -\cot x \csc x$

1. $\int dx = x$

2. $\int au\,dx = a\int u\,dx$

3. $\int (u + v)\,dx = \int u\,dx + \int v\,dx$

4. $\int x^m\,dx = \dfrac{x^{m+1}}{m + 1}$ $(m \neq -1)$

5. $\int \dfrac{dx}{x} = \ln |x|$

6. $\int u\dfrac{dv}{dx}\,dx = uv - \int v\dfrac{du}{dx}\,dx$

7. $\int e^x\,dx = e^x$

8. $\int \dfrac{dx}{1 + x^2} = \arctan x$

9. $\int \dfrac{dx}{\sqrt{1 - x^2}} = \arcsin x$

10. $\int \dfrac{dx}{x\sqrt{x^2 - 1}} = \text{arcsec } x$

11. $\int \sin x\,dx = -\cos x$

12. $\int \cos x\,dx = \sin x$

13. $\int \tan x\,dx = \ln |\sec x|$

14. $\int \cot x\,dx = \ln |\sin x|$

15. $\int \sec x\,dx = \ln |\sec x + \tan x|$

16. $\int \csc x\,dx = \ln |\csc x - \cot x|$

Vector relationships

Let $\mathbf{i}, \mathbf{j}, \mathbf{k}$ be unit vectors in the x-, y-, z-directions. Then

$$\mathbf{i} \cdot \mathbf{i} = \mathbf{j} \cdot \mathbf{j} = \mathbf{k} \cdot \mathbf{k} = 1, \qquad \mathbf{i} \cdot \mathbf{j} = \mathbf{j} \cdot \mathbf{k} = \mathbf{k} \cdot \mathbf{i} = 0,$$

$$\mathbf{i} \times \mathbf{i} = \mathbf{j} \times \mathbf{j} = \mathbf{k} \times \mathbf{k} = 0,$$

$$\mathbf{i} \times \mathbf{j} = \mathbf{k}, \qquad \mathbf{j} \times \mathbf{k} = \mathbf{i}, \qquad \mathbf{k} \times \mathbf{i} = \mathbf{j}.$$

Any vector $\mathbf{a}$ with components a_x, a_y, a_z along the x-, y-, z-axes can be written

$$\mathbf{a} = a_x\mathbf{i} + a_y\mathbf{j} + a_z\mathbf{k}.$$

Let $\mathbf{a}, \mathbf{b}, \mathbf{c}$ be arbitrary vectors with magnitudes a, b, c. Then

$$\mathbf{a} \times (\mathbf{b} + \mathbf{c}) = \mathbf{a} \times \mathbf{b} + \mathbf{a} \times \mathbf{c}$$

$$(s\mathbf{a}) \times \mathbf{b} = \mathbf{a} \times (s\mathbf{b}) = s(\mathbf{a} \times \mathbf{b}) \qquad (s \text{ a scalar}).$$

Let θ be the smaller of the two angles between **a** and **b.** Then

$$\mathbf{a} \cdot \mathbf{b} = \mathbf{b} \cdot \mathbf{a} = a_x b_x + a_y b_y + a_z b_z = ab \cos \theta$$

$$\mathbf{a} \times \mathbf{b} = -\mathbf{b} \times \mathbf{a} = \begin{vmatrix} \mathbf{i} & \mathbf{j} & \mathbf{k} \\ a_x & a_y & a_z \\ b_x & b_y & b_z \end{vmatrix} = (a_y b_z - b_y a_z)\mathbf{i} + (a_z b_x - b_z a_x)\mathbf{j} + (a_x b_y - b_x a_y)\mathbf{k}$$

$$|\mathbf{a} \times \mathbf{b}| = ab \sin \theta$$

$$\mathbf{a} \cdot (\mathbf{b} \times \mathbf{c}) = \mathbf{b} \cdot (\mathbf{c} \times \mathbf{a}) = \mathbf{c} \cdot (\mathbf{a} \times \mathbf{b})$$

$$\mathbf{a} \times (\mathbf{b} \times \mathbf{c}) = (\mathbf{a} \cdot \mathbf{c})\mathbf{b} - (\mathbf{a} \cdot \mathbf{b})\mathbf{c}$$

Values of Trigonometric Functions

TRIGONOMETRIC FUNCTIONS

Radians	Degrees	Sine	Cosine	Tangent	Cotangent		
.0000	0	.0000	1.0000	.0000	∞	90	1.5708
.0175	1	.0175	.9998	.0175	57.29	89	1.5533
.0349	2	.0349	.9994	.0349	28.64	88	1.5359
.0524	3	.0523	.9986	.0524	19.08	87	1.5184
.0698	4	.0698	.9976	.0699	14.30	86	1.5010
.0873	5	.0872	.9962	.0875	11.430	85	1.4835
.1047	6	.1045	.9945	.1051	9.514	84	1.4661
.1222	7	.1219	.9925	.1228	8.144	83	1.4486
.1396	8	.1392	.9903	.1405	7.115	82	1.4312
.1571	9	.1564	.9877	.1584	6.314	81	1.4137
.1745	10	.1736	.9848	.1763	5.671	80	1.3963
.1920	11	.1908	.9816	.1944	5.145	79	1.3788
.2094	12	.2079	.9781	.2126	4.705	78	1.3614
.2269	13	.2250	.9744	.2309	4.332	77	1.3439
.2443	14	.2419	.9703	.2493	4.011	76	1.3265
.2618	15	.2588	.9659	.2679	3.732	75	1.3090
.2793	16	.2756	.9613	.2867	3.487	74	1.2915
.2967	17	.2924	.9563	.3057	3.271	73	1.2741
.3142	18	.3090	.9511	.3249	3.078	72	1.2566
.3316	19	.3256	.9455	.3443	2.904	71	1.2392
		Cosine	Sine	Cotangent	Tangent	Degrees	Radians

TRIGONOMETRIC FUNCTIONS (*Continued*)

Radians	Degrees	Sine	Cosine	Tangent	Cotangent		
.3491	20	.3420	.9397	.3640	2.748	70	1.2217
.3665	21	.3584	.9336	.3839	2.605	69	1.2043
.3840	22	.3746	.9272	.4040	2.475	68	1.1868
.4014	23	.3907	.9205	.4245	2.356	67	1.1694
.4189	24	.4067	.9135	.4452	2.246	66	1.1519
.4363	25	.4226	.9063	.4663	2.144	65	1.1345
.4538	26	.4384	.8988	.4877	2.050	64	1.1170
.4712	27	.4540	.8910	.5095	1.963	63	1.0996
.4887	28	.4695	.8829	.5317	1.881	62	1.0821
.5061	29	.4848	.8746	.5543	1.804	61	1.0647
.5236	30	.5000	.8660	.5774	1.732	60	1.0472
.5411	31	.5150	.8572	.6009	1.664	59	1.0297
.5585	32	.5299	.8480	.6249	1.600	58	1.0123
.5760	33	.5446	.8387	.6494	1.540	57	0.9948
.5934	34	.5592	.8290	.6745	1.483	56	0.9774
.6109	35	.5736	.8192	.7002	1.428	55	0.9599
.6283	36	.5878	.8090	.7265	1.376	54	0.9425
.6458	37	.6018	.7986	.7536	1.327	53	0.9250
.6632	38	.6157	.7880	.7813	1.280	52	0.9076
.6807	39	.6293	.7771	.8098	1.235	51	0.8901
.6981	40	.6428	.7660	.8391	1.192	50	0.8727
.7156	41	.6561	.7547	.8693	1.150	49	0.8552
.7330	42	.6691	.7431	.9004	1.111	48	0.8378
.7505	43	.6820	.7314	.9325	1.072	47	0.8203
.7679	44	.6947	.7193	.9657	1.036	46	0.8029
.7854	45	.7071	.7071	1.0000	1.000	45	0.7854
		Cosine	Sine	Cotangent	Tangent	Degrees	Radians

Nobel Prize Winners in Physics

1901	Wilhelm Konrad Röntgen	1845–1923	Discovery of x rays.
1902	Hendrik Antoon Lorentz	1853–1928	Influence of magnetism on the
	Pieter Zeeman	1865–1943	phenomena of atomic radiation.
1903	Henri Becquerel	1852–1908	Discovery of natural
	Pierre Curie	1850–1906	radioactivity and of the
	Marie Curie	1867–1934	radioactive elements radium and polonium.
1904	John William Strutt	1842–1919	Discovery of argon.
1905	Philipp Lenard	1862–1947	Research in cathode rays.
1906	Joseph John Thomson	1856–1940	Conduction of electricity through gases.
1907	Albert A. Michelson	1852–1931	Spectroscopic and meterological investigations.
1908	Gabriel Lippmann	1845–1921	Photographic reproduction of colors.
1909	Guglielmo Marconi	1874–1937	Development of wireless
	Karl Ferdinand Braun	1850–1918	telegraphy.
1910	Johannes Diderik van der Waals	1837–1923	Equations of state of gases and fluids.
1911	Wilhelm Wien	1864–1928	Laws of heat radiation.
1912	Nils Gustaf Dalén	1869–1937	Automatic coastal lighting.
1913	Heike Kamerlingh-Onnes	1853–1926	Properties of matter at low temperatures; production of liquid helium.
1914	Max von Laue	1879–1960	Diffraction of x rays in crystals.
1915	William Henry Bragg	1862–1942	Study of crystal structure by
	William Lawrence Bragg	1890–1971	means of x rays.
1917	Charles Glover Barkla	1877–1944	Discovery of the characteristic x rays of elements.
1918	Max Planck	1858–1947	Discovery of the elemental quantum.
1919	Johannes Stark	1874–1957	Discovery of the Doppler effect in radiations from moving atoms and the splitting of spectral lines in the electric field.
1920	Charles Edouard Guillaume	1861–1938	Discovery of the anomalies of nickel-steel alloys.
1921	Albert Einstein	1879–1955	Discovery of the law of the photoelectric effect.

1922	Niels Bohr	1885–1962	Study of structure and radiation of atoms.
1923	Robert Andrews Millikan	1868–1953	Work on elementary electric charge and the photoelectric effect.
1924	Karl Manne Siegbahn	1886–1954	Discoveries in the area of x-ray spectra.
1925	James Franck	1882–1964	Laws governing collisions
	Gustav Hertz	1887–	between electrons and atoms.
1926	Jean Perrin	1870–1942	Discovery of the equilibrium of sedimentation.
1927	Arthur H. Compton	1892–1962	Discovery of the scattering of x rays by charged particles.
	Charles T. R. Wilson	1869–1959	Invention of the cloud chamber, a device to make visible the paths of charged particles.
1928	Owen Willans Richardson	1879–1959	Discovery of the law known by his name (the dependency of the emission of electrons on temperature).
1929	Louis-Victor de Broglie	1892–	Wave nature of electrons.
1930	Chandrasekhara Raman	1888–1970	Work on the scattering of light and discovery of the effect known by his name.
1932	Werner Heisenberg	1901–	Creation of quantum mechanics.
1933	Paul Adrien Maurice Dirac	1902–	Discovery of new fertile forms
	Erwin Schrödinger	1887–1961	of the atomic theory.
1935	James Chadwick	1891–	Discovery of the neutron.
1936	Victor Hess	1883–1973	Discovery of cosmic radiation.
	Carl David Anderson	1905–	Discovery of the positron.
1937	Clinton Joseph Davisson	1881–1958	Discovery of diffraction of
	George P. Thomson	1892–	electrons by crystals.
1938	Enrico Fermi	1901–1954	Artificial radioactive elements from neutron irradiation.
1939	E. O. Lawrence	1901–1958	Invention of the cyclotron.
1943	Otto Stern	1888–1969	Work with molecular beams and magnetic moment of proton.
1944	Isidor Isaac Rabi	1898–	Nuclear magnetic resonance.
1945	Wolfgang Pauli	1900–1958	Discovery of quantum exclusion principle.
1946	Percy Williams Bridgman	1882–1961	High-pressure physics.
1947	Edward Appleton	1892–1965	Upper atmosphere physics and discovery of Appleton layer.
1948	Patrick Maynard Stuart Blackett	1897–	Discoveries in cosmic radiation and nuclear physics.
1949	Hideki Yukawa	1907–	Prediction of existence of pion.
1950	Cecil Frank Powell	1903–1969	Photographic method of studying nuclear processes; discoveries about mesons.
1951	John Douglas Cockcroft	1897–1967	Transmutation of atomic nuclei
	Ernest Thomas Sinton Walton	1903–	by artificially accelerated atomic particles.
1952	Felix Bloch	1905–	Nuclear magnetic studies.
	Edward Mills Purcell	1912–	
1953	Frits Zernike	1888–1966	Invention of the phase contrast microscopy.

1954	Max Born	1882–1970	Work in quantum mechanics and the statistical interpretation of the wave function.
	Walther Bothe	1891–1957	Analysis of cosmic radiation using the coincidence method.
1955	Willis E. Lamb, Jr.	1913–	Fine structure of hydrogen.
	Polykarp Kusch	1911–	Magnetic moment of electron.
1956	John Bardeen	1908–	Invention and development
	Walter H. Brattain	1902–	of transistor.
	William B. Shockley	1910–	
1957	Chen Ning Yang	1922–	Non-conservation of parity and
	Tsung Dao Lee	1926–	work in elementary particle theory.
1958	Pavel A. Cerenkov	1904–	Discovery and interpretation
	Ilya M. Frank	1908–	of Cerenkov effect of
	Igor E. Tamm	1895–1971	radiation by fast charged particles in matter.
1959	Owen Chamberlain	1920–	Discovery of the antiproton.
	Emilio Gino Segrè	1905–	
1960	Donald A. Glaser	1926–	Invention of bubble chamber.
1961	Robert L. Hofstadter	1915–	Electromagnetic structure of nucleons from high-energy electron scattering.
	Rudolf L. Mössbauer	1929–	Discovery of recoilless resonance absorption of gamma rays in nuclei.
1962	Lev D. Landau	1908–1968	Theory of condensed matter; phenomena of superfluidity and superconductivity.
1963	Eugene B. Wigner	1902–	Contributions to theoretical atomic and nuclear physics.
	Maria Goeppert-Mayer	1906–1972	Shell model theory and magic
	J. H. D. Jensen	1907–	numbers for the atomic nucleus.
1964	C. H. Townes	1915–	Invention of the maser and
	Nikolai Basov	1922–	theory of coherent atomic
	Aleksandr Prokhorov	1916–	radiation.
1965	Richard Feynman	1918–	Development of quantum
	Julian Schwinger	1918–	electrodynamics.
	Shin-Ichiro Tomonaga	1906–	
1966	Alfred Kastler	1902–	Optical pumping studies.
1967	Hans A. Bethe	1906–	Energy production in stars.
1968	Luis W. Alvarez	1911–	Contributions to elementary particle physics.
1969	Murray Gell-Mann	1929–	Contributions to the theory of elementary particles.
1970	Hannes Alfvén	1908–	Space and plasma physics,
	Louis Néel	1904–	Magnetic studies.
1971	Denis Gabor	1900–	Holography
1972	John Bardeen	1908–	Theory of superconductivity
	Leon N. Cooper	1930–	
	J. Robert Schrieffer	1931–	
1973	Leo Esaki	1925–	Tunneling phenomena in solids
	Ivar Giaever	1929–	
	Brian D. Josephson	1940–	

Index

Aberrations, for lenses, 695
  for mirrors, 686
Absolute zero, 348
Absorption spectrum, 770, 771
Accelerating charges, 640
Acceleration, 43, 60, 31
  angular, 175
  average, 31
  center of mass, 139, 211
  centripetal, 49, 83
  constant, 32, 39, 44
  in free fall, 36
  of gravity, 69, 258, 260
  instantaneous, 31
  radial, 49
  and relative velocity, 51
  in uniform circular motion, 49
  variable, 32
Acceleration of gravity, 258
Acoustic cavity, 748
Action-at-a-distance, 261, 453, 466
Action force, 66
Addition of velocities, 658
Adhesion, surface, 80
Adiabatic process, 368
  entropy changes in, 416
Aerodynamic lift, 291
Airy, Sir George, 1112, 735
Alternating current generator, 592
Ampère, A. M., 552, 563
Ampère, 507
Ampère's law, 557, 560, 606, 633, 639
Amplitude, 226, 304, 327
  of simple harmonic motion, 226, 230
Analyzer, 651
Angstrom, 654
Angular acceleration, 175
Angular frequency, 230
Angular momentum, 186
  conservation of, 183, 199
  of electrons, 774, 784
Angular speed, 175
Angular velocity, 175

Anisotropic, 1155
Annihilation, 173, 660
Antenna, electric dipole, 640
Antinode, 312
Archimedes' principle, 281, 291
Astronomical unit, 9
Atmosphere, 283
  of moon, 395
Atomic clock, 6
Atomic mass unit, 165
Atwood's machine, 76
Audible range, 323
Austern, N. 60
Avogadro's law, 398
Avogadro's number, 383

B, see Magnetic field
Ballistic galvanometer, 620
Banked curves, 85
Barometer, 203
Beats, 332
Bernoulli, Daniel, 289
Bernoulli's equation, 287
  applications of, 289, 290
    Venturi meter, 448
Betatron, 587
Binding energy, 128, 775
Biot-Savart law, 568
Bohr, N., 772, 776, 792
Bohr atom, 772
Bohr model, 77, 129, 864, 55, 93
Boltzmann, L., 376, 365
Boltzmann's constant, 762
Born, M., 787
Boundary conditions, 784
Boyle, Robert, 376
Bradley, J., 656
Bragg, W. L. 750
Bragg's law, 750
Brewster, D., 677
Brewster's law, 676
British engineering system, 68
British thermal unit, 358, 363

818   INDEX

Brush, S. G., 375
Bubble chamber, 24, 540
Bulk modulus, 326
Buoyancy, 232
Bureau of Standards, 250

Calcite,
Calorie, 358, 363
Capacitance, 489, 492
  of cylindrical capacitor, 493
  and dielectric, 494
  and inductance, 598
  of parallel-plate capacitor, 492
  of spherical capacitor, 502
Capacitative time constant, 532, 600
Capacitors, as circuit elements, 530
  and dielectrics, 489
  force on plates of, 505
  in parellel, 503
  in series, 494
Carnot, Sadi, 358, 403, 409
Carnot cycle, 403, 411
Carnot's theorem, 409
Carragan, G. H., 740
Cavendish, H., 250, 455
Cavendish balance, 251
Cavity radiator, 758
Celsius degree, 351
Celsius scale, 348
Center of gravity, 211
Center of mass, 135, 211
  motion of, 139
Centripetal acceleration, 49, 83
Centripetal force, 83
Cerenkov radiation, 337
Cesium clock, 6
Cgs system, 68
Cgs units, 8, 68
Charge, in circular motion, 546
  conservation of, 429
  in electric field, 440
  line of, 457
  and matter, 427
  quantization of, 427
  ring of, 439
  two kinds of, 422
Chizanowski, P., 250
Circuits, 521
  charge conservation in, 529
  current in, 523
  energy conservation in, 523, 529
  potential differences in, 525
  RC, 530
Circular aperture, diffraction at, 735
Circular motion, 48, 82
Circulating charges, 546
Classical mechanics, 79, 122, 59
Clausius, R., 572, 589, 628, 386, 408
Clausius statement of second law, 416
Clock, cesium, 6
  quartz crystal, 6
Cloud chamber, 163
Coaxial cable, 616

Coblentz, 760
Coherence, 708
Colding, L. A., 358
Collisions, 153
  in an ideal gas, 378
  inelastic, 157
  momentum conserved in, 155
  in one dimension, 157
  in two dimensions, 161
Complementary, 792
Compton, A. H., 766
Compton effect, 766
Compton shift, 767
Concave mirror, 681
Conduction electrons, 514
  drift speed of, 515
Conductivity, 510
  electrical, 513
Conductivity, thermal, 361, 514
  table, 361
Conductor, 423, 453
  insulated, 453
Conservation, of angular momentum, 183, 199
  of charge, 429
  of energy, 109, 125
    in circuits, 523, 529
  of linear momentum, 143
  of mechanical energy, 124
Conservative force, 109
  in one dimension, 116
  and path independence, 112
  in two dimensions, 121
Constants of the motion, 270
Constants, table of physical, 795
Continuity, equation of, 286
Continuous spectra, 770
Conversion factors, 800
Convex mirror, 681
Corona discharge, 480
Correspondence principle, 776
Coulomb, C. A., 423
Coulomb's law, 70, 773
Critical angle, 675
Cross product, 18
Cross section, 164
Curie, P., 616
Curie temperature, 619
Curie's law, 616
Current, 507
  displacement, 634
  induced, 578
  magnetic force on a, 541
  sense of, 509
Current balance, 561, 564
Current density, 507
Current element, 569
Current loop, torque on, 542
Cutoff frequency, in photoelectric effect, 764
Cycle, Carnot, 403
Cyclotron, 548
  resonance in, 549
Cyclotron frequency, 547
Cylindrical capacitor, 493

Davisson, C. J., 782
de Broglie, L., 781, 784
de Broglie wavelength, 782
Dees, 548
Degree Celsius, 351
Densities, some measured, 279
Density, 277
  weight, 280
Derivative, 27
Descartes, R., 671
Deuteron, 548
Diamagnetism, 617
Dielectric constant, 495
  table, 495
Dielectric strength, 495
Dielectrics, 423, 496
  and capacitors, 494
  and Gauss's law, 498
Diffraction, 703, 725
  at a circular aperture, 735
  Fraunhofer, 727
  and Huygens' principle, 703
  and interference, 738, 740
Diffraction, Fresnel, 726
  of matter waves, 790
  of neutrons, 783
  from a single slit, 728, 730, 732, 733
  of water waves, 704
  X-ray, 748
Diffraction grating, 741, 744
Dimensions, 35
Dipole antenna, 640
Dipole-dipole interaction, 623
Dipole, electric, *see* Electric dipole
  magnetic, *see* Magnetic dipole
Dipole moment, magnetic, 571
Dispersion, 746, 747
Displacement current, 634, 638
Displacement vector, 26, 43
Domains, 621
Doppler, C. J., 334, 660
Doppler effect, 334, 660
  for light, 660
  for sound, 660
Dot product, 18
Double slit, 738
Drift speed of electrons, 508, 515
Drumhead, vibrations of, 332
Du Mond, J., 770, 795
Duality, of light, 766
  of matter, 781
  wave-particle, 783
Dynamic lift, 291
Dynamics, fluid, 284
  of particles, 59
  rotational, 183
Dyne, 68

E, *see* Electric field
Earth, weight of, 252
Eddy currents, 589
Edgerton, H. E., 154
Efficiency of an engine, 406, 409

Einstein, A., 658, 762, 765, 766
Elastic limit, 227
Electric dipole, alternating, 475, 650
  in dielectrics, 496
  energy of, 444
  field due to, 445, 446
  induced, 475, 496
  potential due to, 474
  properties of, 571
  torque on, 544
Electric dipole, water molecule as, 474
Electric dipole antenna, 640
Electric field, 465, 468
  for a conductor, 459, 507
  definition of, 539
  for dipole, 438
  energy of, 499, 607
  flux of, 449
  induction of, 584
  for line of charge, 457, 561
  of point charge, 437
  and potential, 478
  of ring of charge, 439
  for sheet of charge, 458
  of spherical charge, 456
Electrical-mechanical analogy, 629
Electrical potential, *see* Potential, electric
Electromagnetic force, 521
Electromagnetic induction, 577
Electromagnetic oscillations, 625
  energy in, 631
  quantitative, 629
Electromagnetic radiation, 639
Electromagnetic shielding, 589
Electromagnetic spectrum, 653
Electromagnetic waves, 639, 641, 646
Electromotive force
  gravitational analog of, 522
  induced, 578, 597
Electron, 527
  angular momentum of, 612
  e/m for, 551
  moment of, 613
Electron cloud, 428
Electron microscope, 737
Electron volt, 482
Electrostatic force, 423
Electrostatic generator, 481, 482
Elementary particles, 24
Elsasser, W., 784
Emissivity, 759
Energy, 95
  binding, 128
  in capacitor, 499
Energy in circuits, 516
  conservation of, 109, 125
  in electric field, 499, 500, 522
  in electromagnetic cavity, 646
  equipartition of, 386
  gravitational potential, 265, 268
  and heat, 357
  internal, 366
  in LC oscillations, 625, 629

in magnetic field, 522, 603
mechanical, 114, 123
of oscillations, 225
potential, 113
   see also Potential energy
and Poynting vector, 646
quantization of, 761
of simple harmonic motion, 232
Energy level diagram, 775
Engine, efficiency of, 409
heat, 403, 409
Entropy, 401, 413
and irreversible processes, 413
and reversible processes, 411
and the second law, 415
Environment, 59, 343
Eötvös, 254
Equation of state of ideal gas, 377
Equilibrium, examples of, 213
neutral, 121
of rigid-bodies, 209
stable, 121
thermal, 344
unstable, 121
Equipartition of energy, 386
Equipotential surface, 467
Equivalence, Eratosthenes, 259
principle of, 253
Erect image, 686
Erg, 97
Escape velocity, 267
Essen, L., 4, 7
Ether, 658
Euclid, 671
Euler, Leonard, 284
Evaporation, 395
Exchange coupling, 619
Expanding universe, 662
Expansion, 350
coefficient of linear, 351, 352
coefficient of volume, 353
Eye, sensitivity of, 654

Fahrenheit scale, 348
Farad, 491
Faraday, M., 421, 436, 491, 494, 562, 577, 573
Faraday's law, 577, 640
counterpart of, 632
in traveling waves, 642
Ferromagnetism, 615, 619
Field, near, 640
radiation, 640
Field concept, 261
Field, gravitational, 261
First law of thermodynamics, 365, 366
Fizeau, H. L., 4, 656
Flow, compressible, 285
incompressible, 285
irrotational, 285
nonsteady, 285
nonviscous, 285
rotational, 285

steady, 285
streamline, 286
tube of, 286
viscous, 285
Fluid dynamics, 284
Fluid mechanics, conservation of
momentum in, 277
Fluid statics, 278
Flux, of electric field, 449
of magnetic field, 538
Flux linkages, 597
Focal length, 692
Focal plane, 684
Focal point, 683, 692
Force, action, 66
centripetal, 83
conservative, 109, 116, 121
definition of, 62
Force, dynamic procedure for measuring, 72
frictional, 78
gravitational, 247
laws of, 60, 69
nonconservative, 109, 123
and potential energy, 115
reaction, 66
static measurement of, 72
Force constant, 101, 226
Foucault, J. B. L., 657
Fourier, J., 310
Fourth-power law, 758
Fps units, 8
Franklin, B., 423
Fraunhofer diffraction, 727
Free-body diagram, 73
Free charge, 498
Free-electron model, 423
Free expansion, 369, 414
Free fall, 36
equations of, 37
Freitag, E. H., 78
Frequency, 224
angular, 230
Fresnel diffraction, 726
Friction, 70, 78
kinetic, 79
as nonconservative force, 111
static, 79
Fringes of constant thickness, 717
Fundamental frequency, 329

g, 36, 433
Galileo, 45, 54, 61, 248, 654
Galvanometer, 543
ballistic, 620
Gamma rays, 429
Gamow, G., 791
Gas constant, universal, 377
Gas, ideal, 376
Gas thermometer, 345
Gauge pressure, 283
Gauss, 539
Gaussian surface, 452
Gaussian units, 8

Gauss's law, 449, 452, 568, 639
Gauss's law, applications of, 455
  and Coulomb's law, 452
  and dielectrics, 498
  for fluids, 703
  for gravitation, 461
  for magnetism, 614
Generator, alternating current, 592
Geometrical optics, 669
Germer, L. H., 782
Gibbs, J. Willard, 376
Glancing angle, 750
Geoid, 259
Goudsmit, S. A., 556
Gradient, 122
Gram, 68
Grating, 744
  diffraction, 741
  dispersion of a, 747
  resolving power of, 746
Gravitation, 247
  constant of, 70, 250
  law of, 249
Gravitational field, 261, 433, 435
Gravitational force, 247
Gravitational mass, 253
Gravitational potential energy, 265, 268
Gravity, acceleration due to, 69, 258, 260
  center of,
  screen for, 388, 250
  of a sphere, 255
Grimaldi, F. M., 726
Group speed, 303

Hahn, O., 128
Hall, E. H., 545
Hall effect, 423, 508, 545
Halley, E., 248
Harmonic motion, 223
  combinations of, 241
Harmonic oscillator, 118
Harmonics, 329
  organ pipe, 331
HCl spectrum, 771
Heat, 124, 357
  conduction of, 360, 414
Heat, mechanical equivalent of, 362
  path dependence of, 365
  of vaporization, 368
  and work, 363
Heat capacity, 359
Heat engine, 405, 407
Heaviside, O., 421
Heisenberg, W., 789
Helmholtz coils, 576
Henry, G. E., 648
Henry, J., 577, 598
Henry, W. E., 617
Hertz, H., 224, 422, 646
Heyl, P. R., 250
Hooke, Robert, 227
Hooke's law, 101, 110, 227, 237
Horsepower, 105

Hull, G. F., 648, 649
Huygens, C., 672, 726
Huygens' principle, 672, 673
Hydrogen atom, 771
Hydrogen spectrum, 772, 776
Hysteresis, 621

Iceberg, 282
Ideal gas, adiabatic process in, 386
  diatomic, 388
  equation of state of, 377
  macroscopic definition of, 376
  microscopic definition of, 378
  monatomic, 387
  polyatomic, 388
  specific heats of, 383, 386
Ideal gas temperature scale, 347
Image, 678
  erect, 686
  inverted, 686
  real, 678
  virtual, 678
Image distance, 682
Imes, E. S., 771
Impulse, 155
Incident ray, 669
Index of refraction, 670, 671, 674
  principal, 1157
  table, 670
Induced magnetic fields, 632
Induced surface charge, 496, 498
Inductance, 597
  calculation of, 598
  energy storage in, 605
  of solenoid, 599
  of toroid, 599
Induction, electromagnetic, 577, 581
  Faraday's law of, 578, 584
Inelastic collision, 157
Inertia, moment of, 190
  rotational, 190
Inertial frame, 62
Inertial mass, 253
Insulated conductor, 480
Insulator, 423
Integral, 100
Intensity, 308
Interference, 310, 332
  and diffraction, 738, 740
  double-slit, 713
  intensity of, 710
  in thin films, 714
Interferometer, 4, 718
Internal energy, 366
International Bureau of Weights and Measures, 62
Interplanar spacing, 250
Invar, 352
Irreversible process, 401
  entropy change in, 413
Isobaric process, 367
Isotope separation, 383
Isotopes

Isotropic, 352
Ives, H. F., 661

Joule, J. P., 358, 362
Joule, 97
Joule experiment, 371
Joule effect, 517, 603
Joule's law, 517
Joyce, J., 674
Junction theorem, 529

Kellogg, 772
Kelvin, Lord, 408, 409, 481; see also
    Thomson, W.
Kelvin-Planck statement of second law, 416
Kepler, J., 262
Kilocalorie, 358
Kilogram, standard, 62, 68
Kilowatt-hour, 105
Kinematics, 24
Kinetic energy, 103, 116
Kinetic theory of gases, 375
Kinetic theory of pressure, 379
Kinetic theory of temperature, 382
King-Hele, 260
Krypton-86 meter, 5
Kundt's method, 340

Langenberg, 795
Laser, 656, 710
Lattice, 508, 514
Law, see under specific type, Newton's,
    Lenz's, Thermodynamics, etc.
Laws of force, 60
Lawrence, E. O., 548
Lawton, 455
LC oscillations, 625
    analogy of to SHM, 628
Lebedev, 648
Length, atomic standard of, 4
    standard of, 3
Lengths, some measured, 3
Lens, 689
    equation for, 691
    extended object for, 693
    sign conventions for, 691
Lens maker's equation, 692
Lenz, H. F., 579
Lenz's law, 579, 586, 598
    and radiation, 642
Lift, dynamic, 291
    static, 291
Light, 639
    and the electromagnetic spectrum, 653
    energy of, 648
    momentum of, 648
    moving source of, 657
    polarization of, 649
    pressure of, 648
    propagation of, 639
    and quantum physics, 757
    sensitivity of eye to, 654
    sources of, 757

speed of, 654, 655
    wave theory of,
Light year, 9
Limiting process, 28, 31
Lindsay, R. B., 126
Line integral, 469
Line spectra, 770
Linear momentum, conservation of, 143
    of a particle, 141
    of a system 142
Linear motion with constant acceleration,
    32, 34
Lines of force, 435
Lines of induction, 562
Lissajous figures, 246
Locke, John, 344
Longitudinal wave, 300
Loop theorem, 524
Lorentz, H. A., 421, 540, 550
Lorentz relation, 540
LR circuit, 600
LRC oscillations, 628

Macroscopic description, 343
Magnesia, 537
Magnetic damping, 509
Magnetic dipole, 544, 611
    B for, 571
    of bar magnet, 612
    of electron, 613
    energy of, 545
    properties of, 571
    torque on, 544
Magnetic dipole moment, 544, 571, 613
Magnetic domain, 621
Magnetic field, 433, 537
    for antiparallel wires, 574
    for a circular loop, 570
    for current sheet, 573
    definition of, 538
    for a dipole, 571
    energy and, 603
    energy density and, 605
    for Helmholtz coils, 576
    induced, 632
    for a long wire, 861, 561
    for parallel wires, 563
    for rectangular loop, 576
    for solenoid, 565
    for square loop, 575
    time-varying, 584
    for toroid, 568
    for a traveling wave, 641
Magnetic force, on a current, 541
    on a moving charge, 538
Magnetic pole, 611, 614
Magnetism, and matter, 611
    permanent, 621
Magnetite, 537
Magnetization, 616, 620
Magnetization curve, 619
Magnification, lateral, for lenses, 693
    for mirrors, 686

Malus, E. L., 652, 676
Marconi, 422
Mass, 60, 63
  center of, 135
  inertial, 253
  relativistic, 142, 171
  rest, 126, 429
  variable, 146
  and weight, 70
Mass-energy conversion, 429
Mass-energy equivalence, 166
Mass spectrometer, 556
Masses, some measured, 65
Matter, magnetism of, 611
Matter waves, 781
Maxwell, J. C., 387, 394, 421, 552, 657
Maxwell speed, distribution, 394
Maxwell's equations, 421
  and radiation, 640
  summary of, 635
  table of, 636
  and waves, 641
McNish, A. G., 7
Mean free path, 391, 515
Measurement, 1
Mechanical energy, 114, 123
Mechanical equivalent of heat, 362
Mechanics, 24
  classical, 59
  fluid, 277
Meitner, L., 128
Meter, defined in wavelengths, 719
  standard, 3
Michelson, A. A., 657, 718, 719
Microfarad, 491
Microscope, 737
Microscopic description, 343
Microwaves, 650, 656
Millikan, R. A., 447, 764, 765, 766
Mirror formula, 683
Mirrors, converging, 683
  diverging, 684
Mks system, 68
Mks units, 8
Molar heat capacity, 384
Molecules, diameter of, 391
Moment of inertia, 190
  of common solids, 192
Moment of mass, 138
Momentum, 155
  angular, 199
  conservation of, 155
  linear, 135, 141, 142
Moon, atmosphere of, 395
  laser signals from, 656
Motion, in one dimension, 24, 28, 32
  in a plane, 43
  of projectiles, 45
  rectilinear, 36
  rotational, 173
Motion, of translation, 26
  uniform circular, 48
Multiloop circuits, 528

Multiples and supmultiples-metric, 8
Multipoles, expansion in, 722

Near field, 640
Neutron, 427
Neutron diffraction, 783
Newton, Isaac, 36, 80, 81, 82, 112, 191, 60,
    61, 62, 382, 385, 460, 589, 41, 247, 248,
    726
Newton, 63, 68
Newton's first law, 61
Newton's law of cooling, 355
Newton's law of gravitation, 70
Newton's laws of motion, applications of, 72
Newton's second law, 63
Newton's third law, 65
Nichols, 648, 649
Nobel prize winners, 813
Node, 313
Nonconservative force, 109, 123
Nonreflecting glass, 717
Nuclear force, 429
Nuclear reaction, 165
Nucleus, 427
  radius of, 427

Object distance, 682
Oersted, H. C., 421, 537, 557
Ohm, 510
Ohm's law, 512, 516
Oil drop experiment, 447
Oliver, Jack, 323
Onnes, Kamerlingh, 511
Operational viewpoint, 1
Optical maser, 710
Optical path, 675
Optical reversibility, 1089
Optics, 669
  geometrical, 669
  wave, 669, 703
Order number, 742
Organ pipe, harmonics of, 331
Oscillations, 223, 625
  electromagnetic, 625, 628
  frequency of, 224
  LC, 625, 628
  mechanical, 628
  period of, 229
  simple harmonic, 225
Overtones, 329

Panofsky, W., 418
Parallel axis theorem, 191
Parallel conductors, forces between, 563
Parallel-plate capacitor, 492
  with dielectric, 494
  energy storage in, 500
Paramagnetism, 615
Parker, 795
Parsec, 9
Partial derivative, 644
Particle dynamics, 59
Particle kinematics, 24

Particles and matter, 781
Particles and waves, 781
Pascal, B., 281
Pascal's principle, 281, 380
Path independence, 112
Pendulum, ballistic, 160
  conical, 83
  simple, 122, 236
  torsional, 237
Period, 224, 304
Periodic motion, 223
Periodic table of the elements, 799
Permeability constant, 559
  and speed of light, 645
Permittivity constant, 426, 560
  and speed of light, 645
Phase angle, 630
Phase change, 316
Phase change on reflection, 716
Phase constant, 230
Phase of simple harmonic motion, 230
Phase speed, 303
Phasor, 711, 742
Phillips, M., 418
Photoelectric effect, 763
  cutoff frequency, 764
Photon, 765
Physical constants, table of, 795
Piano, waveform of, 330
Pierce, J. R., 635
Piczoelectric effect, 323
Pitot tube, 290
Planck, M., 408, 758, 760, 762, 765, 770, 773
Planck's constant, 761, 766, 769, 774
Planck's radiation formula, 760
Plane mirror, 669
  spherical waves on, 678
Plane, motion in a, 43
Plane of vibration, 650
Plane-polarization, 650
Plane waves, 301
  reflection, 669
  refraction, 669
Planets, 262
  energy of, 269
Plimpton, 455
Polar materials, 496
Polarization, 475, 496, 649
  by reflection, 676
Polarizer, 651
Polarizing angle, 677
Polaroid, 651
Pole, magnetic, 611, 614
Potential, for charged disk, 473
  for a dipole, 474
  electric, 465
  and electric field strength, 468, 470
  and induction, 587, 589
  inside conductor, 480
  for point charges, 470, 472
Potential difference, 465
  in circuits, 525
  path independence of, 466

Potential energy, 113
  electric, 475
    in electric circuits, 517
    of electric dipole, 444
    of magnetic dipole, 545
Potential energy, and work, 118
Pound, 68
Power, 105
  in wave motion, 308
Poynting, J. H., 250, 646
Poynting vector, 648
Prefixes, metric, 8
Pressure, 277
  in fluid at rest, 278
  gauge, 283
  isobaric, 367
  kinetic theory of, 379
Pressure amplitude, 328
Principia, 61, 141, 248
Principle of equivalence, 253
Prism, 672, 676
Process, adiabatic, 368
  irreversible, 401
  reversible, 401
  thermodynamic, 363
Projectile, motion, 45
Proton, 427
Prout, 771
Pulse, 300

Quantities, physical, 1
Quantization, in Bohr model, 774
  of charge, 427
  of energy, 761
  of momentum, 786
  of wavelength, 784, 786
Quantum number, 761, 774
Quantum physics, 757
Quantum statistics, 376
Quantum theory, 390
Quartz crystal clock, 6

Rabi, I. I., 772
Rabinowicz, Ernest, 78
Radiancy, 757
Radiation, electromagnetic, 639
  sources of, 640
Radiation field, 640
Radiation pressure, 648
Radioactive decay, 145, 154
Ray, 301, 669
Rayleigh's criterion, 736, 737
RC circuits, 530
Reaction force, 66
Reactions, nuclear, 165
Real image, 678
Red-shift, 662
Reference frame, 2, 657
  absolute, 2
  earth as, 413
  equivalence of, 254
Reference frames, and kinetic energy, 104
  and relative velocity, 51
  and work, 105

Reflected ray, 670
Reflection, 669
  phase change on, 716
  at plane surface, 678
  of plane wave, 669
  sign conventions for, 682
  at spherical surface, 681
  of spherical wave, 678
  total internal, 675
  of wave, 315
Refracted ray, 670
Refraction, 669
  at plane surface, 670
  of plane wave, 669
  sign conventions for, 687
  at spherical surface, 686
Refrigerator, 406, 408
  coefficient of performance, 419
Retardation of electromagnetic effects, 639
Rekveld, J., 484
Relative velocity and acceleration, 51
Relativistic mass, 142, 171, 550
Relativity, 658
  and cyclotron, 550
Relativity theory, 2, 126
  general, 253
Resistance, 507, 510
Resistance thermometer, 354
Resistivity, 510
  atomic view of, 514
  coefficient of, 511
  table, 511
  temperature variation of, 511
Resistors, in parallel, 530
  in series, 524
Resolving power, of a grating, 746
  of a lens, 735
Resonance, 316
  in cyclotron, 549
Rest mass, 126, 429
Reversible processes, 401
  entropy for, 411
Right-hand rule, 559
Rigid body, 173
  equilibrium of, 206
Ripple tank, coherent waves in, 709
  diffraction in, 703
  interference in, 706
Rocket, 147
  thrust on, 292
Roemer, O., 656
Root-mean-square speed, 380
Rotational dynamics, 182
Rotational inertia, 190
  of common solids, 192
Rotational motion, 173
Rotor, 84
Rowland ring, 619
Rumford, Count, 357; see also
        Thompson, B.
Russel, B., 248
Rutherford, E., 772
Rutherford, Lord, 772

Rutherford atom, 772

Sandage, A. R., 662
Satellite, 51, 260, 262
  energy of, 269
Scalar product, 18, 23, 97
Scalars, 11
Schrödinger, E., 785
Second law of motion, 63
Second law of thermodynamics, 401
Self-field, 441
Self-induction, 597
Semiconductor, 423
Shielding, electromagnetic, 589
SHM, see Simple harmonic motion
Shock wave, 337
Sign conventions, for mirrors, 683
  for spherical surfaces, 686
  for thin lenses, 691
Simple harmonic motion, 225, 228
  amplitude of, 226, 230
  angular, 237
  applications of, 236
  energy of, 232
  equation of motion of, 227, 228
  frequency of, 230
  and LC oscillations, 628
  period of, 229
  phase of, 230
  phase constant of, 230
  uniform circular motion and, 238
Simple harmonic oscillator, 225
Single slit, diffraction at, 728, 730, 732, 733
Slug, 69
Snell, W., 671
Soap bubbles, 714
Solar system, data, 798
Solenoid, 565
Sound, sources of, 329
  speed of, 326, 340
Sound waves, 323
Specific heat, 358
  at constant pressure, 359
  at constant volume, 359, 385
  of hydrogen, 389
  of an ideal gas, 383
Specific heats, ratio of, 326, 385, 388
Spectral radiancy, 757
Spectrograph, 744
Spectrum, 671, 744
  electromagnetic, 653
  line, 770
Speed, angular, 175
  average, 394
  molecular, 393
  most probable, 394
  root-mean-square, 380, 394
  of sound, 340
  of waves, 305
Speed of light, 654, 657
  and electromagnetism, 632, 646
Speeds, Maxwellian distribution, see
        Maxwellian distribution

Spherical wave, 301
Spin, 612
Standard meter, 719
Standard of time, 4
Standards, 1
Standing waves, 313, 315, 317, 789
Static friction, 79
Statics, 65
Statics, fluid, 277
Stationary state, 761
Statistical mechanics, 343, 375
Stefan-Boltzmann constant, 758
Stilwell, G. R., 661
Stopping potential, 763
Streamlines, 286
Superconductivity, 511
Superposition principle, 309
Surface adhesion, 80
Surface charge density, 480
Surface, gaussian, 452
Surface integral, 451
System, 343
System of mechanical units, 68

Taylor, 795
Telescope, 737
Temperature, 343, 345, 348
    absolute zero of,
    Celsius scale of, 348
    Fahrenheit scale of, 348
    fixed points of, 350
    International Practical scale of, 349
Temperature and kinetic theory, 382
Temperature expansion, 350
Temperature gradient, 360
Terrestrial data, 797
Thermal conductivity, 360
Thermal equilibrium, 344
Thermistor, 512
Thermodynamic process, 363
Thermodynamics, 343, 363, 375
    first law of, 365, 407
    second law of, 401, 407, 411, 415, 416
    zeroth law of, 344, 411
Thermometer, constant volume gas, 345
    resistance, 354
Thin lens, 689
Third law of motion, 65
Thompson, B., 357
Thomson, G. P., 782
Thomson, J. J., 550, 771
Thomson, W., 1, 409
Thomson atom, 771
Thrust, 147
Time, standard of, 4
    universal, 6
Time constant, capacitative, 532, 600
    inductive, 602
Time intervals, some measured, 5
Time-of-flight spectrometer, 556
Tolman, R. C., 344
Toroid, 568
    inductance of, 599

Torque, 183
    on current loop, 542
    on electric dipole, 544
    on magnetic dipole, 545
Torricelli, E., 283
Torricelli's law, 297
Torsion balance, 424, 649
Total internal reflection, 675
Translational motion, 26
Transmission line, 640
Transverse wave, 300, 650
Trigonometric functions, values of, 811
Triple point, 346, 349
Tube of flow, 286
Turning point, 120

Ulrey, 771
Uncertainty principle, 789, 791
Uniform circular motion, 48, 82, 238
Unit cell, 749
Unit vectors, 14
Unit cell, 749
Units, 1, 7, 35
    British engineering, 68
Universal gas constant, 377
Universal time, 6

Vacuum tube, resistance of, 512
Van de Graaff, R. J., 481
Vector field, 433
Vector product, 19, 23
Vectors, 11
    addition of, 12, 13
    components of, 13
    cross product, 18
    displacement, 26
    dot product, 18
    multiplication of, 17
    resolution of, 13
    scalar product, 18
    subtraction of, 13
    unit, 14
    vector product, 19
Velocity, 43
    addition of, 658
    angular, 299
    average, 26
    escape, 267
    instantaneous, 27
    relative, 51
    variable, 28
Venturi meter, 289
Venturi tube, 297
Venus, radar signals from, 656
Vertex, 682
Vibrating systems, 329
Vibration, plane of, 650
Vibration of string, 305
Violin, waveform of, 330
Virtual image, 678
Volt, 465
von Guericke, Otto, 295
von Helmholtz, H., 358

von Laue, M., 749
von Mayer, J., 358

Water, density of, 353
  dipole moment of, 474
  specific heat of, 358
Water molecule, 475
Waterston, J. J., 375
Watt, J., 105
Watt, 105, 517
Wave, intensity of, 308
  plane, 301
  spherical, 301
  standing, 315, 317
  traveling, 302
Wave disturbances, addition of, 309
Wave function, 785
  meaning of, 787
Wave mechanics, 784, 785
Wave number, 304
Wave optics, 669, 703
Wavefront, 300
Wavelength, 304
Waves, audible, 323
  complex, 309
  elastic, 299
  electromagnetic, 639
  infrasonic, 323
  interference of, 310
  longitudinal, 300, 324, 327
  and Maxwell's equations, 641
  mechanical, 299

power in, 308
  sound, 323
  standing, 313
  transverse, 300
  ultrasonic, 323
Waves and particles, 781
Wavetrain, 710
Weber, 539
Weight and mass, 70
Weight density, 280
Wien, W.,
Wien's formula, 760
Work, 95
  as an area on a p-V diagram, 364
  by a constant force, 96
  and heat, 363
  path dependence of, 365
  as a scalar product, 97
  by a variable force, 99, 101
Work-energy theorem, 102, 104, 105
WWV, 7

X-rays, Compton effect for, 766
  diffraction of, 748

Young, T., 705
Young's experiment, 705, 738
Young's modulus, 326

Zacharias, J. K., 772
Zeeman, P., 550
Zeroth law of thermodynamics, 344

Some Useful Numbers

$\sqrt{2} = 1.414$ $\sqrt{3} = 1.732$ $\sqrt{10} = 3.162$ $\pi = 3.142$

$\pi^2 = 9.870$ $\sqrt{\pi} = 1.772$ $\log \pi = 0.4971$ $4\pi = 12.57$

$e = 2.718$ $1/e = 0.3679$ $\log e = 0.4343$ $\ln 2 = 0.6932$

$\sin 30° = \cos 60° = 0.5000$ $\cot 30° = \tan 60° = 1.7321$

$\cos 30° = \sin 60° = 0.8660$ $\sin 45° = \cos 45° = 0.7071$

$\tan 30° = \cot 60° = 0.5774$ $\tan 45° = \cot 45° = 1.0000$

Change of Base

$\log x = \ln x / \ln 10 = 0.4343 \ln x$

$\ln x = \log x / \log e = 2.303 \log x$